Fifth Edition

Clinical Exercise Physiology

Exercise Management for Chronic Diseases and Special Populations

Jonathan K. Ehrman, PhD

Henry Ford Hospital

Paul M. Gordon, PhD, MPH

Baylor University

Paul S. Visich, PhD, MPH

University of New England

Steven J. Keteyian, PhD

Henry Ford Hospital

Editors

HUMAN KINETICS

D1285694

Library of Congress Cataloging-in-Publication Data

Names: Ehrman, Jonathan K., 1962- editor. | Gordon, Paul M., 1960- editor.
| Visich, Paul S., 1955- editor. | Keteyian, Steven J., editor.
Title: Clinical exercise physiology : exercise management for chronic diseases
and special populations / Jonathan K. Ehrman, Paul M. Gordon, Paul S. Visich,
Steven J. Keteyian, editors.
Other titles: Clinical exercise physiology (Ehrman)
Description: Fifth edition. | Champaign, IL : Human Kinetics, [2023] |
Includes bibliographical references and index.
Identifiers: LCCN 2021048958 (print) | LCCN 2021048959 (ebook) | ISBN
9781718200449 (paperback) | ISBN 9781718200456 (epub) | ISBN
9781718200463 (pdf)
Subjects: MESH: Exercise Therapy | Exercise--physiology | Exercise Movement
Techniques
Classification: LCC RM725 (print) | LCC RM725 (ebook) | NLM WB 541 | DDC
615.8/2--dc23/eng/20211101
LC record available at https://lccn.loc.gov/2021048958
LC ebook record available at https://lccn.loc.gov/2021048959

ISBN: 978-1-7182-0044-9 (print)

Acquisitions Editor: Amy N. Tocco; **Developmental Editor:** Judy S. Park; **Managing Editor:** Anna Lan Seaman; **Copyeditor:** Patricia L. MacDonald; **Indexer:** Dan Connolly; **Permissions Manager:** Dalene Reeder; **Senior Graphic Designer:** Joe Buck; **Cover Designer:** Keri Evans; **Cover Design Specialist:** Susan Rothermel Allen; **Photograph (cover):** traffic-analyzer/DigitalVision/Getty Images; **Photographs (interior):** © Human Kinetics, unless otherwise noted; **Photo Asset Manager:** Laura Fitch; **Photo Production Manager:** Jason Allen; **Senior Art Manager:** Kelly Hendren; **Illustrations:** © Human Kinetics, unless otherwise noted; **Printer:** Walsworth

Printed in the United States of America

10 9 8 7 6 5 4 3 2 1

The paper in this book was manufactured using responsible forestry methods.

Human Kinetics
1607 N. Market Street
Champaign, IL 61820
USA

United States and International
Website: **US.HumanKinetics.com**
Email: info@hkusa.com
Phone: 1-800-747-4457

Canada
Website: **Canada.HumanKinetics.com**
Email: info@hkcanada.com

E8177

Tell us what you think!
Human Kinetics would love to hear what we
can do to improve the customer experience.
Use this QR code to take our brief survey.

Contents

Preface

The profession of clinical exercise physiology continues to evolve. Likewise, the role of the clinical exercise physiologist in health care continues to grow. Most cardiac rehabilitation programs employ at least one clinical exercise physiologist, because it is recognized that they are uniquely trained to appreciate normal and abnormal cardiorespiratory responses to exercise as well as prescribe exercise in patients across a variety of clinical conditions. And although it is still true that the day-to-day duties of many clinical exercise physiologists primarily involve patients with cardiovascular disease, clinical exercise physiologists continue to expand their role, working with other patient populations, including those with cancer, musculoskeletal disorders, and metabolic diseases such as chronic kidney disease and diabetes. Cardiac rehabilitation programs regularly enroll patients with peripheral artery disease because insurance companies are now reimbursing for supervised exercise therapy. The growing body of research in clinical exercise physiology is being incorporated into evidence-based guidelines for the treatment of patients with a variety of diseases, many of which are presented in this text. And the key words *clinical exercise physiologist* and *clinical exercise physiology* continue to grow exponentially in PubMed searches.

Professional organizations in the United States, Australia, and Canada provide support, certification, and continuing education opportunities for clinical exercise physiologists. In the United States, the Clinical Exercise Physiology Association (CEPA) is the official professional organization. The CEPA was established in 2008 to serve practitioners in the field through advocacy and education. Importantly, the CEPA has published its official journal, the *Journal of Clinical Exercise Physiology (JCEP)*, since 2012. *JCEP* is also the official journal of Exercise & Sports Science Australia (ESSA), the leading clinical exercise physiology association serving Australia and New Zealand, as well as a member benefit for those in the Canadian Society for Exercise Physiology (CSEP). This exemplifies the continued advancement of the clinical exercise physiology profession around the world. Each of these professional organizations offers continuing education and credits, which are vitally important for maintaining clinical exercise physiology–related certifications. Organizations such as ESSA, CSEP, the American College of Sports Medicine (ACSM), and the American Council on Exercise (ACE) offer clinical certifications that identify individuals as having the required competency as defined by each organization. And the number of clinical exercise physiology programs recognized by the Commission on Accreditation of Allied Health Education Programs (CAAHEP) continues to grow.

The first edition of *Clinical Exercise Physiology* (2003), which was the first clinical exercise physiology text of its kind, was the cornerstone textbook for the field, quickly becoming a primary textbook for both upper-level undergraduate students and graduate students preparing to work as clinical exercise physiologists. The two original purposes for developing this book remain today in the fifth edition: (1) to provide a contemporary review of a variety of chronic diseases and conditions for the clinical exercise physiologist in training and (2) to provide a comprehensive resource for people working in the field. Another use of this text is to serve as a resource for those preparing for the ACSM, ESSA, and CSEP clinical examinations. In the past the ACSM offered two clinical certifications but has since combined these into a single certification titled Clinical Exercise Physiologist (ACSM-CEP). This change was a natural part of the evolution of clinical exercise physiology as a profession.

This fifth edition of *Clinical Exercise Physiology* is fully revised. The initial section of the book presents the foundational chapters, including an excellent review of the history of clinical exercise physiology, a description of the essentials of the physical examination, and a review of the general properties of drugs and pharmacotherapy. A new foundational chapter provides an introduction to clinical exercise programs offered to the specific populations covered in the remaining chapters of the book. These chapters are the core of the book and are specific to 28 conditions and populations of patients. The chapters are organized into seven parts: endocrine conditions, cardiovascular diseases, respiratory diseases, immunologic disorders, orthopedic and musculoskeletal conditions, neuromuscular disorders, and special populations.

Because the format was popular with previous editions, it has been retained for each of the chapters in parts II through VIII of this fifth edition of *Clinical Exercise Physiology*. Each of the chapters in these sections begins with an introduction to the specific disease that includes the definition and scope of the condition and a discussion of the relevant pathophysiology. This is followed by a look at the medical and clinical considerations, including signs and symptoms, diagnosis, exercise testing, and evidence-based treatment. Each chapter concludes with an overview of the exercise prescription for the disorder being discussed, with special emphasis placed on any disease-specific issues that might alter the exercise prescription.

Most chapters also contain practical application boxes. In each of the disease-specific chapters, two of these practical application boxes focus on the exercise prescription and on information to consider when interacting with the patient. A third practical application box reviews the relevant literature and discusses the physiological adaptations to exercise training, including potential mechanisms by which exercise can influence primary and secondary disease prevention.

Each clinical chapter (parts II through VIII) has two companion case studies that can be accessed via the web resource. Each of these cases focuses on an actual patient, progressing from initial presentation and diagnosis to therapy and exercise treatment. Each case study contains several questions that can be used for facilitating group discussion in the classroom or for the individual learner to consider when preparing for a clinical exercise physiology certification examination.

To keep abreast of trends and new research in the field, the chapters on peripheral artery disease, cancer, spinal cord injury, multiple sclerosis, and depression have undergone major revision. Additionally, all of the remaining chapters have undergone a thorough review to ensure the material is consistent with current science and evidence-based practice guidelines.

Few, if any, upper-level undergraduate courses or graduate-level clinical exercise physiology programs currently provide students with the breadth of information required to sit for the ACSM clinical examination. Those who plan to study for this or any similar certification examination should understand that no single text provides in-depth coverage of all the clinical populations that benefit from physical activity and exercise. But this text may be as close as one can come.

Besides serving as a textbook for students, *Clinical Exercise Physiology* is an excellent resource guide for exercise professionals to have in their office. Its consistent organization, case studies, discussion questions, up-to-date references, and feature boxes provide the information required for effective study. In fact, the content was developed based on the knowledge and skills assessed in the ACSM clinical exercise physiologist examination (and the previous registered clinical exercise physiologist examination).

REFERENCES

As a benefit to the reader, we've moved the chapter references online for this edition. You'll find nearly a hundred pages of references covering each of the 35 chapters, all readily accessible. Visit http://courses.humankinetics.com/references/ehrman5E/index.html or scan the QR code using your mobile device or tablet.

INSTRUCTOR RESOURCES

Resources for instructors consist of a test package, chapter quizzes, presentation package, and image bank—all updated and revised for this edition. These instructor ancillaries are available online at **http://hkpropel.humankinetics.com**. Answers to the case study questions in the student web resource are also included here.

- **Test package.** Created with Respondus 2.0, the test package includes over 1,000 true-false, multiple-choice, fill-in-the-blank, multiple-response, and essay questions. With Respondus, instructors can create versions of their own tests by selecting from the question pool, select their own test forms and save them for later editing or printing, and export the tests into a word-processing program. Additionally, instructors can access the test package in RTF format or in a format that can be imported into learning management systems.

- **Chapter quizzes.** Created in the same platform as the test package, the chapter quizzes present 10 to 15 questions per chapter, narrowed down to the key ideas and concepts.

- **Presentation package.** The presentation package includes approximately 950 slides of text, artwork, and tables from the book that instructors can use for class discussion and presentation. The slides in the presentation package can be used directly within PowerPoint or printed to make transparencies or handouts for distribution to students. Instructors can easily add, modify, and rearrange the order of the slides.

- **Image bank.** The image bank provides almost all the figures, photos, and tables from the print book, organized by chapter. Instructors can use these free-standing images to create their own presentations, handouts, or other class materials.

STUDENT RESOURCES

Students will be able to access HK*Propel* to read case studies and view discussion questions. Common questions and their answers are also provided for students. Case studies are presented in a SOAP note format, so students can practice evaluating Subjective and Objective data, Assessments, and Plans.

Acknowledgments

Over the past 20 years, the field of clinical exercise physiology has grown exponentially. I am proud to have worked with Paul, Paul, and Steven on this text as I believe we have contributed to this growth. I thank them and Human Kinetics for their knowledge, dedication, and desire to put this fifth edition together. I also thank all the many contributors for this fifth edition and each of the previous four editions. Finally, I dedicate this book to my growing family: Janel, Joshua, Melissa, Jacob, Laura, Jared, and Johanna. I love you all, and thank you always for your support.

—*Jonathan K. Ehrman*

We are excited to present to you the fifth edition of this text, which brings together a renowned group of international scholars to impart state-of-the-art knowledge of their respective subspecialties. Many thanks to them for their tireless efforts as well as to the reviewers and staff whose behind-the-scenes efforts often remain unnoticed but truly ensure an exceptional product. Also, to my children, Joshua and Emily, Mark and Natalie, and Liam: Thank you all for believing in me and for your timely comic relief along the way. I love you all. Finally, in loving memory of my beautiful late wife and soulmate, Ina, who still inspires us onward.

—*Paul M. Gordon*

As the field of exercise science continues to show how the benefits of exercise in preventing and rehabilitating individuals with chronic disease, it is exciting to see how this textbook continues to evolve as one of the most ideal references for the clinical exercise physiology field. This textbook would not be possible without the hard work of the authors, who are experts in their field and who I am very thankful. As a professor for the past 39 years, I am excited that this comprehensive book in clinical exercise physiology can be offered to colleagues and, most importantly, students in the field of exercise physiology who are interested in improving the health of others. I continue to thank my wife, Diane, my two sons, Matt and Tim—they have always been supportive of my endeavors and provided encouragement. Lastly, I thank my father, Frank, and mother, Mary, in loving memory, who challenged me to always put forth my best effort in whatever I did.

—*Paul S. Visich*

My sincere thanks to the staff at Human Kinetics for again helping us publish a greatly revised edition of our textbook. The role of regular exercise in the assessment and management of patients with a clinically manifest disorder continues to develop, and the fifth edition of our book does much to help those entering and working in the field keep pace with the expanding clinical exercise science. To that end, my thanks to the many contributing authors for their quality writing. My appreciation to Henry Ford Health System for representing an employer that fosters both clinical research and innovative approaches to patient care, and thank you to all past and current staff of the Cardiac Rehabilitation/Preventive Cardiology Unit for their tireless efforts to provide outstanding patient care. As always, many blessings to my wife, Lynette, and our children, Courtland, Jacob, Aram (and Carly), and Stephanie (and Jake, Ian, Ani, and Nia). Finally, and forever, in loving memory of Albert Z. and Virginia Keteyian.

—*Steven J. Keteyian*

PART I

Introduction to Clinical Exercise Physiology

Although the day-to-day duties of people working in our field most often involve patients with coronary heart disease or heart failure, clinical exercise physiologists now, more than ever before, contribute to the evidence-based care provided to patients with cancer, musculoskeletal disorders, peripheral artery disease, and metabolic diseases such as chronic kidney disease and diabetes. As a result, the chapters in this first section of *Clinical Exercise Physiology* review not only the rich history and expanding scope of our profession but also the foundational knowledge that we use for safe and effective exercise testing and exercise prescription. The chapters in this section also address key prerequisite areas that practicing clinical exercise physiologists must demonstrate proficiency in if they are to help care for patients across a broad range of chronic diseases. These areas include behavioral approaches to the maintenance of regular exercise training, essential principles in pharmacology, and general interview and examination skills. A thorough reading of these chapters will provide the necessary background for the important work that lies ahead.

Chapter 1 is a general introduction to the field of clinical exercise physiology. With each passing year, clinical exercise physiology as a profession deepens its roots into the delivery of evidence-based health care. Chapter 1 reviews the profession's history, current practices, and future directions for clinical exercise physiology, particularly in relation to disease prevention and management. The chapter also reviews the various professional organizations and certifications that are continuing to influence the profession.

Chapter 2 focuses on behavioral approaches. Although most of the material in this textbook (and in the classroom) pertains to the knowledge and physical skills needed to write safe exercise prescriptions, interpret exercise responses, and lead or supervise exercise, the fact remains that we are in the business of changing human behavior. This challenge differs little from the one that confronts the practicing physician or nurse. They too play an important role and should be relied on to help educate and motivate patients to take an active part in improving their own health. To that end, this chapter reviews and applies the various behavioral approaches known to help patients adopt long-term habits aimed at improving their health. The information presented here can be applied across demographic categories and disease conditions.

Chapter 3 focuses on pharmacotherapy. Those who want to understand the underlying pathophysiology of a particular condition should also take the time to understand how and why a particular medication is being used to help treat it. Adopting this approach will help clinical exercise physiologists better understand the underlying disease process and will also improve their ability to develop a proper exercise prescription and conduct safe exercise evaluations and training sessions. Learning essential drug properties, along with the key factors that influence drug compliance, will go a long way toward preparing a person to work effectively in the field.

Chapter 4 discusses general interview and examination skills in order to better prepare the clinical exercise physiologist to determine whether a patient can safely exercise. Rarely is a chronic disorder stagnant. Instead, diseases are dynamic. As a result, the clinical exercise physiologist must regularly inquire about or recognize evidence of possible disease progression. This approach applies regardless of whether a patient is being considered for surgery or about to undergo a graded exercise stress test. Safety remains paramount. Although this chapter will help the clinical exercise physiologist learn what questions to ask, what signs to look for, and how to interpret the evaluations performed by others, there is no substitute for taking additional time to practice the skills of observation, interviewing, and examining patients.

Chapter 5 looks at graded exercise testing. Like the other chapters in part I, this chapter reviews and provides the foundational information needed to read through the subsequent chapters of the book. Fortunately, graded exercise testing is a topic that most students have likely had prior coursework in, freeing them up to begin to think about conducting an exercise test in a manner that integrates their growing knowledge of exercise physiology with the unique clinical issues germane to the chronic condition of the patient.

Chapter 6 reviews the primary tenets (specificity and overload) associated with prescribing exercise. This chapter addresses the overload principle by reviewing the proper and safe application of intensity, duration, frequency, and progression for various types of exercise, providing the background knowledge needed to help care for the clinical conditions that are covered in subsequent chapters.

The final chapter in part I is new to this edition of *Clinical Exercise Physiology* and pertains to clinical exercise programming. Chapter 7 provides a summary of how and why regular exercise therapy is delivered to select patient groups in a systematic and organized manner. The five programs addressed in this chapter are cardiac rehabilitation, pulmonary rehabilitation, supervised exercise therapy for patients with peripheral artery disease, cancer or oncology rehabilitation, and renal rehabilitation. For each condition, the clinical contributions associated with exercise testing and exercise training have progressed to the point that these programs now provide structured, patient-centered core components, such as supervised exercise, risk factor and disease management education, outcomes measurement, and behavioral support. The overview of these programs is continued in more detail within the chapters that correspond to the specific disease.

CHAPTER 1

The Profession of Clinical Exercise Physiology

Jonathan K. Ehrman, PhD

Paul M. Gordon, PhD, MPH

Paul S. Visich, PhD, MPH

Steven J. Keteyian, PhD

In the United States and throughout the world, clinical exercise physiology has experienced continual professional growth and change. This chapter provides a brief introduction to relevant issues and, where appropriate, makes comparisons to past experiences.

THE PAST, PRESENT, AND FUTURE OF CLINICAL EXERCISE PHYSIOLOGY

Clinical exercise physiology is a subspecialty of exercise physiology (15). The Clinical Exercise Physiology Association is an affiliate society of the American College of Sports Medicine (ACSM) and has a purpose "to advance the scientific and practical application of clinical exercise physiology." The CEPA states the following on its public page:

> Clinical Exercise Physiology is a relatively new career field that has only been around since the late 1960s. It is found within the healthcare realm, primarily in medical settings, where exercise is used to help clients manage or reduce their risk of chronic disease. It is an exciting and growing profession that has recently been organized on the national level, recognized by the federal government and defined as an occupation. (14)

The discipline of clinical exercise physiology investigates and addresses

- the relationship of exercise and its impact on chronic disease assessment and management,

- the mechanisms and adaptations by which exercise influences the disease process, and

- the role and importance of exercise testing and training in the evaluation, diagnosis, prevention, and treatment of a variety of diseases and conditions.

One purpose of clinical exercise physiology is to develop and implement exercise training programs for those who might benefit. Therefore, an individual who practices clinical exercise physiology (i.e., the clinical exercise physiologist [CEP] in the United States and Canada, or the accredited exercise physiologist [AEP] in Australia and New Zealand) must be knowledgeable about a broad range of exercise responses, including those that occur both within a disease category (e.g., cardiovascular or respiratory diseases), across different chronic diseases (e.g., cardiovascular, respiratory, oncologic, metabolic), and within or between various organ systems (e.g., cardiac, skeletal, muscle). And although the response of the various organ systems is a primary focus, important behavioral, psychosocial, and spiritual issues often play a role in the recovery and maintenance processes. A competent clinical exercise physiologist must have a wide range of knowledge and skills in the following areas of study:

- Anatomy (gross, including cadaver laboratory experience, and microscopic)
- Physiology (organ systems, cellular, molecular, and exercise)
- Chemistry (organic and biochemistry)

Acknowledgment: We thank Francis Neric (ACSM), Antia Hobson-Powell (ESSA), Cedric Bryant (ACE), and Kirstin Lane (CSEP) for a content review of their respective organizations within this chapter.

- Kinesiology (movement, biomechanics, and applied physiology)
- Psychology (counseling and health coaching)

In addition, clinical exercise physiologists must complete an undergraduate or graduate degree in a related major. In the United States, these requirements vary across colleges and universities because the curriculum is not yet standardized. This differs from Australia, where universities preparing students to become accredited exercise physiologists are required to provide a standardized curriculum. To sit for specific credentialing exams (reviewed later in this chapter), students are required to perform a clinical internship that includes a variety of clinical experiences (e.g., diagnostic and functional exercise testing; exercise assessment, prescription, leadership, and supervision; counseling; body composition and anthropometry assessment; and education). And ideally, those desiring to work in the field of clinical exercise physiology will pass a certification examination that is recognized according to standards set by an independent accrediting agency (i.e., outside of the organization offering the certification). See practical application 1.1 for more information.

The Past

A comprehensive review of the history of exercise physiology can be found elsewhere (31). With respect to clinical exercise physiology, the formal use of exercise in the assessment and treatment of chronic disease has existed since at least the early 1970s. Prior to this, in the late 1930s, Dr. Sid Robinson and his colleagues at Harvard and Indiana University studied the effects of the aging process on exercise performance. Outside of the United States, other countries have also contributed greatly to the knowledge base of exercise physiology, notably the Scandinavian and other European countries in the 1950s and 1960s.

The development of the modern-day clinical exercise physiologist dates back to the 1960s, around the time when the term *aerobics* was popularized by Dr. Kenneth Cooper—a time when regular exercise was beginning to be considered important for maintaining optimal health (16) and prolonged bed rest was found to be associated with marked loss of exercise tolerance (28). In addition, pioneers such as Herman Hellerstein were beginning to understand that prolonged sitting was likely detrimental to people with heart disease (21). These findings gave way to the development of inpatient cardiac rehabilitation programs and the subsequent rapid proliferation of outpatient programs throughout the 1970s and into the 2000s. It was also during this era that clinical exercise physiologists, including Karl Wasserman (9) and Jack Wilmore (36), developed breath-by-breath gas analysis.

The Present

Since 1996 when *Physical Activity and Health: A Report of the Surgeon General* was published, a growing number of milestone reports for the field of clinical exercise physiology have been issued (32). The Surgeon General's guidelines used an evidence-based approach and identified numerous chronic diseases and conditions whose risk increases among people who do not participate in regular exercise or physical activity. In 2007, the American College of Sports Medicine (ACSM) and American Heart Association (AHA) jointly published physical activity guidelines for both healthy and older adults (19, 24). And in 2008 the U.S. Department of Health and Human Services published the first edition of the Physical Activity Guidelines for Americans (33). The second edition of these guidelines was published in 2018 (27). In addition to general guidelines, there are specific recommendations for population subgroups, including children, adults, and older adults as well as those with comorbid conditions such as arthritis, diabetes, hypertension, cancer, and physical disabilities. Each guideline suggests a recommended physical activity dose for reducing the risk or improving the status of a number of chronic health diseases and conditions (see the sidebar *Health Benefits Associated With Regular Physical Activity*). Also of note is the addition of an ACSM-certified CEP (Katrina Piercy) as co-executive secretary for the development of these guidelines. Individuals with CEP backgrounds assisted with content development and review of the other previously presented guidelines and reports.

It is accepted that regular physical activity and higher physical fitness help prevent disability and improve outcomes for many diseases and conditions, including cardiovascular, skeletal muscle, metabolic, and pulmonary diseases. For these and many other conditions, exercise training is frequently part of a comprehensive, interdisciplinary treatment plan, often delivered in a supervised setting (see chapter 7). Additionally, some population groups are at increased risk for developing a chronic disease or disability because of physical inactivity; these include women, children, and people of selected races and ethnicities. This risk also increases in all persons as a result of normal aging. The tenets of clinical exercise physiology continue to be at the forefront of advances in clinical care and public policy directed at improving health and lowering future disease risk (i.e., primary and secondary disease prevention) through regular exercise training and increased physical activity.

Published practice guidelines, position statements, and ongoing research have driven cardiac and pulmonary rehabilitation programs to expand into multifaceted exercise training and behavioral management programs administered by multidisciplinary teams that include, among other allied health professionals, clinical exercise physiologists. Additionally, bariatric surgery and weight management, exercise oncology, diabetes management, renal rehabilitation, and other clinical exercise programs and services have developed because of the research and clinical contributions to the field of clinical exercise physiology.

The Future

The profession of clinical exercise physiology has matured appreciably in recent years as evidenced by the growth of vibrant and

Practical Application 1.1

SCOPE OF PRACTICE

The following are definitions of the (clinical) exercise physiologist or specialist from several organizations. Careful reading of these definitions reveals that there continues to be no clear worldwide consensus regarding the title of the person who works with patients in an exercise or rehabilitative setting (e.g., *clinical exercise physiologist* in North America and *accredited exercise physiologist* in Australia).

U.S. Department of Labor, Bureau of Labor Statistics

The following is the description of an exercise physiologist according to the U.S. government (34). This classification was added on March 11, 2010, and was last updated March 31, 2021.

29-1128 Exercise Physiologists. Exercise physiologists assess, plan, or implement fitness programs that include exercise or physical activities such as those designed to improve cardiorespiratory function, body composition, muscular strength, muscular endurance, or flexibility.

Examples of specific job title include kinesiotherapists, clinical exercise physiologists, and applied exercise physiologists. In a previous iteration of this page, under the "duties" section of the job description was the following, which better describes those who consider themselves a clinical exercise physiologist:

- Analyze a patient's medical history to assess their risk during exercise and to determine the best possible exercise and fitness regimen for the patient
- Perform fitness and stress tests with medical equipment and analyze the resulting patient data
- Measure blood pressure, oxygen usage, heart rhythm, and other key patient health indicators
- Develop exercise programs to improve patients' health

Some physiologists work closely with primary care physicians, who may prescribe exercise [regimens] for their patients and refer them to exercise physiologists. The physiologists then work with patients to develop individualized treatment plans that will help the patients meet their health and fitness goals. Exercise physiologists should not be confused with fitness trainers and instructors (including personal trainers) or athletic trainers.

The U.S. Bureau of Labor Statistics provides information about the median annual wage overall and for a few specific work environments (34). The median annual wage

was $54,020 in 2020. Those working for the government made more ($77,300) than those working in hospitals or physician's offices (~$53,000)

Clinical Exercise Physiology Association (CEPA)

The following is from the CEPA website (14):

What is a clinical exercise physiologist (CEP)? A CEP is a certified healthcare professional that utilizes scientific rationale to design, implement and supervise exercise programming for those with chronic diseases, conditions and/or physical shortcomings. They also assess the results of outcomes related to exercise services provided to those individuals.

The CEPA describes CEP services as follows (14):

Clinical Exercise Physiology services focus on the improvement of physical capabilities for the purpose of:

1. chronic disease management;
2. reducing risks for early development or recurrence of chronic diseases;
3. creating lifestyle habits that promote enhancement of health;
4. facilitating the elimination of barriers to habitual lifestyle changes through goal-setting and prioritizing;
5. improving the ease of daily living activities; and
6. increasing the likelihood of long-term physical, social and economic independence.

CEPs work in varied clinical settings, including hospitals, outpatient clinics, physician offices, and university or hospital-based research facilities. Although able to independently work directly with patients, much of the time a CEP is working with a team of health professionals such as physicians, dietitians, psychologists or behavioral coaches, and nurses. The CEP provides a unique skill set to this team, including knowledge of acute exercise response and chronic exercise adaptation as well as the ability to safely prescribe exercise for stable patients with a clinically manifest disease.

A CEP holds a minimum of a master's degree in exercise physiology, exercise or movement science, or kinesiology *and* is either licensed under state law (Louisiana only in the United States) or holds a professional certification from a national organization that is functionally equivalent to the ACSM's Certified Clinical Exercise Physiologist credential. It is important to note that the CEPA is closely aligned with the ACSM as an affiliate society.

(continued)

American College of Sports Medicine (ACSM)

In 2014 the ACSM changed the previously titled Health Fitness Specialist to Exercise Physiologist, and the Clinical Exercise Specialist was renamed Clinical Exercise Physiologist. It is not yet understood whether these changes have helped clarify either the public or academic perception of these certifications.

ACSM Certified Exercise Physiologists® are advanced health fitness (non-clinical) professionals with a minimum of a bachelor's degree in exercise science qualified to pursue a career in university, corporate, commercial, hospital, and community settings. Beyond training, ACSM-EP not only conduct complete physical fitness assessments—they also interpret the results in order to prescribe appropriate, personalized exercise programs. (7)

From 2005 to 2019, the ACSM developed and maintained two clinical exercise physiologist certifications accredited by the National Commission for Certifying Agencies (NCCA): the Clinical Exercise Physiologist (CEP) and the Registered Clinical Exercise Physiologist (RCEP). Each credential could be earned as a stand-alone designation, or a professional could earn both. The ACSM's Committee on Certification and Registry Boards (CCRB) identified persistent market confusion from clinical exercise professionals, educators, and employers as to the differences between the two credentials and the unique meaning of each. The ACSM evaluated each program to determine the best approach for recognizing knowledge and skills in this area.

In 2017 the CCRB conducted a feasibility study that determined minimal differences in the expected knowledge, skills, and abilities as reported by those certified as a CEP compared with those certified as an RCEP. The ACSM determined that these professional designations did not represent substantially different expectations in knowledge and skills based on the similarity between the responses of the two survey groups. As a result, in 2019 the ACSM combined the CEP and RCEP programs into one certification—Certified Clinical Exercise Physiologist (ACSM-CEP).

ACSM Certified Clinical Exercise Physiologists are allied health care professionals with a minimum of a bachelor's degree in exercise science and practical experience, including 1,200 hours of hands-on clinical experience, or a master's degree in clinical exercise physiology and 600 hours of hands-on clinical experience.

While working primarily with individuals with clinically manifest disease, ACSM Certified Clinical Exercise Physiologists build better health outcomes through a combination of assessing risk, managing exercise implementation, and helping individuals attain positive health outcomes while recovering and rehabilitating from disease, injury, or other limiting factors (6). ACSM-CEPs prescribe exercise, use basic health behavior interventions, and promote physical activity for individuals with chronic diseases or conditions, such as cardiovascular, pulmonary, metabolic, orthopedic, musculoskeletal, neuromuscular, neoplastic, immunologic, and hematologic conditions. The ACSM-CEP works with populations ranging from children to older adults, providing primary and secondary prevention strategies designed to improve or maintain fitness and health.

American Council on Exercise (ACE)

In August 2015, ACE announced the launch of a new clinical certification termed ACE Medical Exercise Specialist. This new accreditation is "an evolution of the ACE Advanced Health and Fitness Specialist certification" (4). According to ACE, the ACE Medical Exercise Specialist designation is appropriate for advanced health and fitness professionals who provide in-depth preventive and postrehabilitative fitness programming for people recovering from or at risk for cardiovascular, pulmonary, musculoskeletal, and metabolic diseases and disorders, as well as apparently healthy clients in special population groups, including youth, older adults, and prenatal and postpartum women.

Canadian Society for Exercise Physiology (CSEP)

The essence of the CSEP clinical certification has remained unchanged since 2007. In 2019 members voted to amend the title of the CSEP Certified Exercise Physiologist to the CSEP Clinical Exercise Physiologist to better reflect their knowledge and skills. A CSEP Clinical Exercise Physiologist (CSEP-CEP) performs assessments, prescribes conditioning exercise, and provides exercise supervision, counseling, and healthy lifestyle education. The CEP serves apparently healthy individuals and populations with medical conditions, functional limitations, or disabilities associated with musculoskeletal, cardiopulmonary, metabolic, neuromuscular, and aging conditions (13). The CSEP specifies duties that the CEP is sanctioned or not sanctioned to perform, as follows.

A CSEP-CEP is sanctioned by CSEP to:

1. Conduct pre-participation screening, administer various health and fitness assessments, prescribe and supervise exercise programs, and provide counseling and healthy lifestyle education to general populations, high performance athletes, and those with chronic health conditions, functional limitations, or disabilities across the lifespan.

2. Conduct preparticipation screening using evidence-informed tools that will support recommendations for individual client-tailored physical activity and exercise programs within the CEP's individual knowledge competency.

3. Accept referrals from licensed health care professionals who are trained and licensed to diagnose and treat acute and chronic medical conditions. The CEP may clear clients with one chronic health condition to work with a CSEP Certified Personal Trainer (CPT). Further medical clearance should be sought if the health condition is unstable.

4. Use evidence-informed behavior change models to facilitate physical activity and exercise participation and lifestyle modification.

5. Interpret the results of comprehensive fitness assessment protocols to determine health, physical function, work or sport performance.

6. Monitor the influence of commonly used medications on the response to submaximal and maximal exercise during assessments or training sessions.

7. Use the outcomes from objective health and fitness assessments to design and implement safe and effective physical activity and exercise prescriptions for both healthy and unhealthy populations

8. Make general, evidence-informed, dietary recommendations if within the CEP's individual knowledge competency, recognizing when to refer to a registered dietician for more specialized counselling.

9. Measure and monitor heart rate, electric activity of the heart (using ECG) and blood pressure (by auscultation unless hearing impaired) before, during, and after exercise and post-exercise. These measures can be used to identify, but not diagnose, irregularities during rest, sub-maximal, and maximal exercise.

10. Conduct group physical activity or exercise sessions with appropriately screened participants. Sessions can include any one or combination of aerobic, resistance, balance, or flexibility exercises within the CEP's individual knowledge competency. Additional training and certification must be sought to ensure that the CEP has expert competence for different exercise modalities where necessary.

A CSEP-CEP is not sanctioned by CSEP to:

1. Diagnose pathology based on any assessment or observation.

Exercise & Sports Science Australia (ESSA)

Exercise & Sports Science Australia is a professional organization, founded in 1991, committed to establishing, promoting, and defending the career paths of tertiary trained exercise and sports science practitioners. Part of its mission is to accredit exercise professionals. One of these accreditations is the Accredited Exercise Physiologist (AEP), a designation for professionals who use exercise to manage chronic conditions, disability, and injuries (29).

Professional practice is influenced by many factors, including the context in which practice occurs, individual client needs, and the practice environment as well as local, government, and industry policies. Following is the foundational scope of practice that ESSA accepts as reasonable for AEPs entering the profession:

- Screening, assessing, and applying clinical reasoning to ensure the safety and appropriateness of exercise and physical activity interventions, which includes conducting tests of physiological measures

- Assessing movement capacity in people of all ages and levels of health, well-being, or fitness

- Developing safe, effective individualized exercise interventions

- Providing health education, advice, and support to enhance health and well-being, including nutrition advice in line with national nutrition guidelines and information on relevant prescribed medications

- Providing exercise intervention and education for those at risk of developing a chronic condition or injury

- Providing clinical exercise prescription for those with existing chronic and complex medical conditions and injuries

- Providing exercise-based rehabilitation and advice for patients in the acute or subacute stage of injury, after surgical intervention, or during recovery to restore functional capacity and well-being

These tasks may occur at any level of primary, secondary, or tertiary health care and may include employment or volunteer work at an individual, community, or population health level through various employers or industries.

Although some may consider the differences in these definitions for the clinical exercise professional to be simply a matter of semantics, such differences often lead to confusion among the public about which title or type of clinical exercise professional they should look for when referred to or considering participation in a clinical exercise program. Additionally, there may be confusion among other health care professionals about the job titles and the duties of those who hold them. For instance, in many institutions, someone who performs the technical duties in a noninvasive cardiology laboratory is typically titled a cardiovascular technician (an occupation defined by the U.S. Department of Labor). But since the duties performed by a cardiovascular technician can overlap with those of a clinical exercise physiologist, it is not uncommon for cardiovascular technicians to be hired into exercise physiologist types of positions or for exercise physiologists to be hired into cardiovascular technician types of positions.

Health Benefits Associated With Regular Physical Activity

Children and Adolescents

- Improved bone health (ages 3 through 17 years)
- Improved weight status (ages 3 through 17 years)
- Improved cardiorespiratory and muscular fitness (ages 6 through 17 years)
- Improved cardiometabolic health (ages 6 through 17 years)
- Improved cognition (ages 6 to 13 years)*
- Reduced risk of depression (ages 6 to 13 years)

Adults and Older Adults

- Lower risk of all-cause mortality
- Lower risk of cardiovascular disease mortality
- Lower risk of cardiovascular disease (including heart disease and stroke)
- Lower risk of hypertension
- Lower risk of type 2 diabetes
- Lower risk of adverse blood lipid profile
- Lower risk of cancers of the bladder, breast, colon, endometrium, esophagus, kidney, lung, and stomach
- Improved cognition

- Reduced risk of dementia (including Alzheimer's disease)
- Improved quality of life
- Reduced anxiety
- Reduced risk of depression
- Improved sleep
- Improved immune function contributing to lower risk for contracting the flu or SARS-CoV-19
- Slowed or reduced weight gain
- Weight loss, particularly when combined with reduced calorie intake
- Prevention of weight regain following initial weight loss
- Improved bone health
- Improved physical function
- Lower risk of falls (older adults)
- Lower risk of fall-related injuries (older adults)

Note: The advisory committee rated the evidence of health benefits of physical activity as strong, moderate, limited, or grade not assignable. Only outcomes with strong or moderate evidence of effect are included in this list.

Reprinted from U.S. Department of Health and Human Services, *Physical Activity Guidelines for Americans,* 2nd ed. (Washington, DC: Department of Health and Human Services, 2018).

---

effective national and regional member and accreditation organizations. Professional organizations have achieved an extensive reach across the health industry and the broader community by developing and actively curating practice standards, cultivating accreditation systems to support the professions they represent, and ensuring that research and professional development are the beacon for driving evidence-based practice.

International Collaboration

Many national organizations have explicitly referenced in their strategic plans an increased focus on the international landscape. Although memoranda of understanding (MOUs), bilateral agreements, and informal collaborations between exercise physiology member organizations have historically served international agendas well, these formulations have limitations. For some organizations, the end game has become establishing a formalized global alliance through which a wider and richer influence, across the communities they serve and for the memberships they represent, can be achieved. Strategic initiatives—including global connections for dissemination of information and fostering of research, common professional standards and practice guidelines, and

influence on health policy of clinical exercise physiology—have a common purpose across organizations that could be placed under a broader international coalition. Discussions and development of such an initiative is currently in the early stages.

Labor Outlook

The Bureau of Labor Statistics in the United States projects careers in the field exercise physiology to expand by roughly 10% between 2018 and 2028. This expansion is greater than the average for all other occupations. This demand may be related to hospitals' increased use of exercise therapy for primary and secondary preventive care, particularly for patients with cardiovascular or respiratory disease. However, as covered in chapters 8 through 35, it is becoming increasingly clear that the benefits of exercise are nearly universal for multiple diseases and conditions. There is even evidence that exercise may be important in mitigating the risk and severity of some communicable diseases (e.g., SARS-CoV-19, or COVID-19).

To date, most U.S.-based clinical exercise physiologists work in the cardiac rehabilitation setting (22). A 2016 paper reported that 59% of CEPs primarily work in cardiac rehabilitation (table

1.1). Other places of employment are also listed in table 1.1. More recently opportunities have expanded to include other diseases and conditions, either within a cardiac rehabilitation program or in independent disease management programs in which exercise plays an important role. In Australia, areas of employment for AEPs include workplace health and rehabilitation as well as residential aged care facilities (senior living centers). The conditions and diseases that CEPs report working with often include cardiovascular disease (62%), obesity and diabetes (62%), pulmonary diseases (48%), neuromuscular diseases (26%), and cancer and HIV (13%) (22), but they may also include kidney problems and mental health issues.

The Million Hearts initiative by the Centers for Disease Control and Prevention (CDC) and the Centers for Medicare and Medicaid Services (CMS) in the United States may represent an opportunity for expansion of the CEP profession. A stated goal is an increase cardiac rehabilitation participation from the current level of 20% to 70% by 2022 (1). This would more than triple the current cardiac rehabilitation census and is estimated to save approximately 25,000 lives and prevent roughly 180,000 hospitalizations annually. Since CEPs are often the primary staff in cardiac rehabilitation programs, the potential need for more trained CEPs is evident based on initiatives such as this.

Additionally, in Australia, there is ongoing planning to embed their AEPs into the clinical health care system. As stated in their strategic plan, a major goal by 2020 was to include AEPs as part of standard care for all people with complex and chronic medical conditions or injuries. As part of this strategy, they desire to have Australians understand and value the benefits of exercise as prescribed by an AEP. These types of regional initiatives are very important for advancing the CEP profession.

Opportunities also exist to work with those who are mentally or physically disabled. The National Center on Health, Physical Activity and Disability (NCHPAD) has taken a leading role in this promising focus, partnering with the ACSM to develop and offer the Inclusive Fitness Trainer certification (25):

> An ACSM/NCHPAD Certified Inclusive Fitness Trainer (CIFT) is a fitness professional who assesses, develops and implements individualized exercise programming for healthy or medically cleared persons with a physical, sensory or cognitive disability. ACSM/NCHPAD CIFTs make exercise accessible for all—empowering individuals to achieve fitness goals that previously seemed beyond their own limitations.

In addition to leading and demonstrating safe, adapted exercise techniques as well as necessary precautions, ACSM/NCHPAD Certified Inclusive Fitness Trainers (CIFT) also possess a working knowledge of current American Disability Act (ADA) policy specific to recreation facilities, and standards for accessible facility design. ACSM/NCHPAD CIFTs commonly work in community and public health settings (e.g., YMCA or Departments of Parks and Recreation) to improve access to exercise for all, as well as gyms, health clubs and university or corporate-based fitness centers.

Table 1.1 Areas of Practice for Clinical Exercise Physiologists

Primary work setting[a]	Percentage (2016; n = 1,271)[b]	Percentage (2022, n = 796)[c]
Cardiac rehabilitation	43.0	59.0
Commercial fitness or community center	11.5	1.0
College or university	10.0	9.0
Cardiovascular or exercise stress testing	8.0	10.0
Hospital wellness	7.0	1.0
Corporate (fitness, wellness)	6.0	<1.0
Government (fitness, research, wellness)	4.5	<1.0
Hospital or medical facility (nonrehabilitation, nonwellness)	3.0	3.0
Physical therapy	2.0	<1.0
Weight loss or bariatric clinic	1.5	2.0
Pulmonary rehabilitation	1.5	2.0
K-12 schools	1.0	<1.0
Other	1.0	10.0
Exercise oncology	NA	1.0

NA = non-applicable.

[a,b]Reprinted by permission from D.J. Kerrigan et al., "CEPA Clinical Exercise Physiologist Practice Survey: 2016 Update," *Journal of Clinical Exercise Physiology* 6, no. 1 (2017): 9-16.

[c]Reprinted by permission from T.A. Hargens et al. "CEPA Clinical Exercise Physiology Practice Survey," *Journal of Clinical Exercise Physiology* 11, no. 1 (2022): 2-11.

Throughout the world, clinical exercise professionals continue to gain employment in clinical research trials at both the sponsor and site investigation locations, the latter of which may be increasing due to the increased emphasis on exercise-related research that includes exercise or variables gathered during exercise as an outcome (26). Similarly, noninvasive exercise testing laboratories, including those specializing in cardiopulmonary exercise testing, also often regard clinical exercise professionals as an asset because their academic training is well suited for this type of employment. Finally, the importance of physical activity and health and an increase in both professional and public awareness of the skills of clinical exercise physiologists have led to increasing employment in nonclinical settings. These include personal training, corporate fitness programs, medically affiliated fitness centers, schools and communities, professional and amateur sport consulting, and weight management programs.

A 2016 survey conducted by the Clinical Exercise Physiology Association (CEPA) ($N = 1,271$) reported information about clinical exercise physiologists in the United States (22). Among these, 770 (61%) held at least one clinical certification (ACSM clinical exercise physiologist = 561, ACSM-registered clinical exercise physiologist = 293, American Association of Cardiovascular and Pulmonary Rehabilitation's Certified Cardiac Rehabilitation Professional = 67).

The profession of clinical exercise physiology continues to build a unique body of knowledge that includes developing and implementing exercise prescriptions for both primary and secondary prevention services. This information has been published through the years in a growing number of biomedical journals that contribute to the clinical exercise physiology body of knowledge (see Selected Biomedical Journals), including many high-impact medical journals. in the sidebar Selected Biomedical Journals lists other journals that regularly publish manuscripts written either by or for the clinical exercise physiologist. The CEPA publishes a journal, titled the *Journal of Clinical Exercise Physiology (JCEP)*, that is focused on review articles and original research specific to the field of clinical exercise physiology. *JCEP* is also the official journal of ESSA, demonstrating the interest in CEP-specific research and information in another part of the world. This body of knowledge has led to the emergence of evidence-based recommendations by preeminent CEPs and journals representing specific diseases, some of which have been recently updated and provided earlier in this chapter. These pronouncements and statements are represented throughout the chapters of this text.

In addition, the clinical exercise physiologist is uniquely trained to identify individual lifestyle-related issues that promote poor health, as well as design and implement behavior-based treatment plans aimed at modifying lifestyle habits. Clinical exercise professionals can enhance their counseling and coaching skills specific to health coaching by studying for and taking a number of health coach certification assessments that cater to the exercise professional. In addition to the certifications presented in depth earlier, the Certified Diabetes Educator offered by the American Diabetes Association and the Interdisciplinary Obesity and Weight Management Board Certification offered by the Academy of Nutrition and Dietetics are two examples of health coaching certifications available to qualified CEPs. Thus, the profession of clinical exercise physiology and the clinical exercise physiologist are filling a void in health care and are becoming increasingly important, particularly as the average age of the population rapidly increases. In the United States, for instance, baby boomers (those born between 1946 and 1964) have begun to retire in large numbers, and many of them have one or more chronic diseases or conditions that would benefit from the expertise of a CEP. These individuals now have the time (and, in many cases, the money) to afford care for these conditions. And this situation is certainly not unique to the United States; it is occurring in many other developed countries as well.

In summary, the involvement of clinical exercise physiologists in a variety of clinical settings has grown dramatically over the last 40+ yr. Historically, most exercise physiologists were engaged in human performance–related research or academic instruction. Many now provide professional advice and services in clinical, preventive, and general fitness programs located in health clubs, corporate facilities, and hospital-based complexes. Clinical exercise physiologists provide important services, including fitness assessments or screenings, exercise testing and outcome assessments, exercise prescriptions or recommendations, exercise leadership, and exercise supervision.

Over the past 25 yr the licensure of clinical exercise physiologists has been considered by several U.S. states, including Louisiana, Maryland, Massachusetts, Minnesota, Montana, Oklahoma, California, Texas, Utah, Wisconsin, and Kentucky. However, to date, only Louisiana has opted to offer licensure to clinical exercise physiologists. Other efforts have not succeeded for a variety of reasons. The Clinical Exercise Physiology Association (CEPA) occasionally provides updates on the progress and status for many of these states on its website. Practical application 1.2 presents information about licensure of the exercise physiologist in the United States.

Health Coaching Training and Certifications Ideal for the Clinical Exercise Physiologist

These organizations offer health coaching training and certifications ideal for the CEP.

- American Council on Exercise (ACE): www.ace fitness.org/fitness-certifications/health-coach-certification
- Duke Integrative Medicine: www.dukeintegrative medicine.org/integrative-health-coach-training
- National Society of Health Coaches (NSHC): www.nshcoa.com
- Wellcoaches School of Coaching: www.well coachesschool.com

Selected Biomedical Journals

- *ACSM's Health and Fitness Journal*
- *American Journal of Cardiology*
- *American Journal of Clinical Nutrition*
- *American Journal of Physiology*
- *American Journal of Sports Medicine*
- *Annals of Internal Medicine*
- *Applied Physiology, Nutrition, and Metabolism*
- *Archives of Internal Medicine*
- *British Journal of Sports Medicine*
- *Canadian Journal of Applied Physiology*
- *Circulation*
- *Clinical Journal of Sport Medicine*
- *Diabetes*
- *Diabetes Care*
- *European Journal of Applied Physiology*
- *European Journal of Sport Science*
- *Exercise and Sport Sciences Reviews*
- *International Journal of Obesity*
- *Journal of Aging and Physical Activity*
- *Journal of Applied Physiology*
- *Journal of Cardiopulmonary Rehabilitation and Prevention*
- *Journal of Clinical Exercise Physiology*
- *Journal of Obesity*
- *Journal of Sport Science and Medicine*
- *Journal of the American Medical Association (JAMA)*
- *Mayo Clinic Proceedings*
- *Medicine and Science in Sports and Exercise*
- *New England Journal of Medicine (NEJM)*
- *Pediatric Exercise Science*
- *Pediatric Obesity*
- *The Physician and Sportsmedicine*
- *Translational Journal of the American College of Sports Medicine*
- *Research Quarterly for Exercise and Sport*
- *Sports Medicine*

PROFESSIONAL ORGANIZATIONS AND CERTIFICATIONS THROUGHOUT THE WORLD

Literally hundreds of exercise- and fitness-related certifications are available, but only a few exist for the purpose of certifying exercise professionals to work with people who have chronic diseases. The primary professional organizations in North America that provide clinically oriented exercise physiology certifications are the American Council on Exercise (ACE [3]), the American College of Sports Medicine (ACSM [5]), and the Canadian Society for Exercise Physiology (CSEP [12]). In addition, Exercise & Sports Science Australia (ESSA [17]) accredits exercise physiologists in Australia who have the "freedom to act" as independent health practitioners via their recognition as an allied health profession (29). In the United Kingdom, professional members of the British Association of Sport and Exercise Sciences (BASES [11]) can apply for certified exercise practitioner status, which is a professional certification recognizing exercise professionals with a sport and exercise science degree. It establishes the credibility of practitioners who can demonstrate the necessary practical vocational knowledge and skills to work effectively in a clinical exercise environment underpinned by an academic qualification to degree level. At the time of this writing (i.e., 2021), key stakeholders, including BASES, have submitted a proposal to develop clinical exercise physiology accreditation.

To date there do not appear to be any organized efforts for certification or accreditation of clinical exercise physiologists in mainland Europe, Asia, Africa, or South America. However, ACE, ACSM, and CSEP provide workshops and certifications throughout the world. Tables 1.2a and 1.2b outline aspects of each of these organizations' certifications, which are also reviewed in the next several paragraphs. Of note are several other professional organizations that provide clinically related certifications for which the clinical exercise physiologist may be well suited. These include the American Association of Cardiovascular and Pulmonary Rehabilitation's Certified Cardiac Rehabilitation Professional (CCRP), the Medical Fitness Association's Medical Fitness Facility Director credential, and the National Association of Sports Medicine's Certified Personal Trainer (CPT). Information about these certifications can be found on the respective professional organizations' websites.

Clinical Exercise Physiology Association

Established in 2008 as an affiliate society of the American College of Sports Medicine, the Clinical Exercise Physiology Association (CEPA [14], www.cepa-acsm.org) is an autonomous professional member organization with the sole purpose of advancing "the scientific and practical application of clinical exercise physiology for the betterment of health, fitness, and quality of life among patients at high risk or living with a chronic disease." Although no clinical

THE CHALLENGE OF CEP LICENSURE IN THE UNITED STATES

The profession of clinical exercise physiology is largely unregulated in the United States. The process toward licensure is a long and tedious effort. With its beginnings during Socratic debates at the annual ACSM meetings in the 1990s, the move toward licensure has been slow to sometimes nonexistent. To date, only the state of Louisiana has passed and enacted (in 1995) a licensure process for the clinical exercise physiologist. Some have been critical of the Louisiana enactment, suggesting that it too narrowly limits the scope of practice of the CEP. Others believe it is sufficient and should serve as the model of future legislative attempts. Additionally, other professional groups, such as physical therapists (PTs), have worked within their professional organizations to challenge attempts to license CEPs. Some have argued that the PT professional scope of practice sufficiently covers the suggested scope of practice of the CEP and that licensing CEPs would add to an "already too large" allied health licensing process. Moreover, some state that CEPs are not sufficiently trained in the neuromuscular and orthopedic areas of exercise programming.

Evidence justifying licensure must first be gathered and published and must then be appreciated by reasonable segments of both the public and state legislators. Since 1988 a publication titled *The Exercise, Sports and Sports Medicine Standards and Malpractice Reporter* has been a resource for reporting on standards, case law, new case filings, and legislation affecting the fitness, sport, and sports medicine industries in the United States. Unfortunately, this publication ended in 2016, and no other publication has filled this gap. The final issue reported the following trends related to legal standards, state and federal law, and personal injury suits:

In the United States from 1988 to 2002 the primary legal themes were

- informed consent, waivers, and releases;
- exercise testing and supervision;
- health risk screening prior to exercise;
- group class leadership and related facility issues;
- standards and guidelines from professional organizations (e.g., ACSM, AHA, AACVPR); and
- risk management and unauthorized practice of medicine.

From 2003 to 2016, additional primary legal themes included

- exercise services in health clubs and worksite programs and
- negligence claims related to personal trainers.

A point–counterpoint presentation of the pros and cons of licensure was published in the *Journal of Clinical Exercise Physiology* (10, 18). Although opposition to any attempt at licensure is expected and very likely will occur, one must appreciate the fact that CEPs deliver relevant services, often in a cost-effective manner. Additionally, CEPs are trained to provide unique services (e.g., prescription of aerobic, resistance, and flexibility exercise in patients with known diseases; evaluation of program effectiveness) outside the scope of other health care providers.

In large part, efforts toward licensure will likely focus on whether such legislation is necessary to protect the public from harm due to services provided by CEPs. As noted, (in the list of trends of legal standards, state and federal law, and personal injury suits) very little suggests harm specifically by CEPs. While this suggests that services provided by CEPs are safe, it in turn is an argument against the need for licensure. This good record is possibly related to the close medical supervision and evidence-based exercise interventions used by practicing CEPs (35). But one might also argue that the diligence that has operated to date to keep patient safety at a high level does not suggest that any loosening of the standards for CEPs should occur. In fact, because of the aging of the world's population, along with the increasing incidence of chronic disease risk factors and clinically manifest chronic disease in younger people, some might suggest a need for tightening legal standards via the licensure process.

This lack of applicable state licensing requirements for clinical exercise physiologists has made it difficult for those who attempt to assess and define the practice roles for such professionals in the delivery of related services to consumers. This confusion extends to the legal system, where questions arise relative to a variety of concerns related to service delivery. Important legal questions include the following:

- Given the absence or lack of licensure, what services may clinical exercise physiologists lawfully provide?
- What practices performed by clinical exercise physiologists may be prohibited as a matter of law because of state statutes regarding unauthorized practice of medicine?
- What practices performed by clinical exercise physiologists may be prohibited because they are in the scope of practice of other licensed health care professionals (e.g., physical therapy, nursing, dietetics) and thus prohibited by current law?
- What potential liabilities may clinical exercise physiologists face when their delivery of service results in harm, injury, or death attributable to alleged negligence or malpractice?
- What recognition may be given to clinical exercise physiologists and their opinions in a variety of legal settings (such as evaluating disability or working capacity in matters involving insurance, personal injury, or workers' compensation)? Can they serve as reputable expert witnesses?

Table 1.2a Comparison of Qualifications for Certified Exercise Physiologists

Certification title	Certifying organization	Web	Country	Degree	Exam	Total practical experience[c]
Accredited Exercise Physiologist (AEP)[a]	Exercise & Sports Science Australia (ESSA)	www.essa.org.au	Australia	ESSA-accredited exercise physiology program; 4 yr university or master's degree	No	500 h
Clinical Exercise Physiologist (ACSM-CEP)	American College of Sports Medicine (ACSM)	www.acsm.org/get-stay-certified	United States	Minimum: master's degree in exercise physiology or bachelor's degree in exercise science or equivalent	Yes (written only)	Master's prepared 600 h and bachelor's prepared 1,200 h
Clinical Exercise Physiologist (CSEP-CEP)	Canadian Society for Exercise Physiology (CSEP)	www.csep.ca	Canada	Minimum: 4 yr university degree in exercise science or a related field	Yes (written and practical)	100 h
ACE Certified Medical Exercise Specialist (ACE CMES)	American Council on Exercise (ACE)	www.acefitness.org	United States	Minimum: bachelor's degree in exercise science or a related field at the time of exam registration	Yes	500 h
Medical Fitness Facility Director[b]	Medical Fitness Association (MFA)	www.medicalfitness.org	United States	Not stated but Fitness Facility Director credential is a prerequisite	Yes	Not stated
Certified Exercise Practitioner	British Association of Sport and Exercise Sciences (BASES)	www.bases.org.uk	United Kingdom	Bachelor's or postgraduate study in sport or exercise science that is either BASES endorsed or includes studies in physiology, psychology, and movement analysis	No	500 h

[a]New criteria effective January 1, 2016.

[b]New certification announced November 16, 2016.

[c]See table 1.2b for a breakdown of practical experience requirements.

NUCAP = National Universities Curriculum Accreditation Program.

exercise certifications are offered through Clinical Exercise Physiology Association (CEPA) per se, the focus of this organization is to serve the profession of clinical exercise physiology and practicing clinical exercise physiologists through advocacy, education, and career development. In general, the CEPA supports efforts toward licensure within the United States.

In 2019 the ACSM and the CEPA codeveloped the Clinical Exercise Physiology Registry. The purpose of the registry is to verify the certification of practicing CEPs, recognize the professional experience of CEPs, distinguish areas of specialization, and identify the location of CEPs by city and state. The registry will increase awareness of the value that clinical exercise professionals provide to health care teams and increase exposure to professional opportunities.

American Council on Exercise

The American Council on Exercise (ACE [3], www.acefitness.org) continues its long history of certifying exercise professionals and health coaches, predominantly in the area of personal training, with a mission of significantly reducing the impact of inactivity-related diseases by 2035. ACE certifies thousands of health and fitness professionals and health coaches annually. With the foreseeable growth in the aging population and the rise

Table 1.2b Practical Experience Requirements by Type

Certification title	EXPERIENCE REQUIREMENT TYPES							
	Healthy populations	Cardio-pulmonary	Metabolic	Musculo-skeletal	Neuro-muscular	Aging	Immunological/ hematological	Other
Accredited Exercise Physiologist (AEP)[a]	140 h (compulsory)	200 h (compulsory)	100 h (compulsory) across other pathologies including neuromuscular, aged facilities, and immunologic/hematologic, among others					Compulsory 60 h in any other activities relevant to the AEP Professional Standards
Clinical Exercise Physiologist (ACSM-CEP)	Not specified[a]	MS: 180 h (recommended)[a] BS: 360 h (recommended)[a]	MS: 150 h (recommended)[a] BS: 300 h (recommended)[a]	MS: 40 h (recommended)[a] BS: 80 h (recommended)[a]	MS: 40 h (recommended)[a] BS: 80 h (recommended)[a]	Not specified[a]	Not specified[a]	CPR for the professional rescuer or basic life support certification
Clinical Exercise Physiologist (CSEP-CEP)	Not specified	25 h (recommended)	25 h (recommended)	25 h (recommended)	25 h (recommended)	Not specified	Not specified	CPR level C certification and course in emergency first aid
ACE Certified Medical Exercise Specialist (ACE CMES)	Experience designing and implementing exercise programs for apparently healthy or high-risk individuals, as documented by a qualified professional							Be at least 18 yr old and hold a current CPR-AED certification with a live skills check
Medical Fitness Facility Director[b]	Not stated							Not stated
Certified Exercise Practitioner	Not stated							Must demonstrate practical vocational knowledge and skills by holding an advanced instructors qualification: Skills Active Level 4 Qualification ACSM-EP certification

[a]See http://certification.acsm.org for more information about experience requirements for bachelor's-prepared candidates.

MS = master's degree; BS = bachelor's degree; CPR = cardiopulmonary resuscitation; AED = automated external defibrillator.

in the prevalence of chronic diseases, particularly obesity, ACE has four professional certifications for exercise professionals and health coaches. Besides the ACE Personal Trainer certification, other available certifications include Health Coach, Group Fitness Instructor, and Certified Medical Exercise Specialist.

In 2015 ACE announced a name change for their clinical certification, from Advanced Health and Fitness Specialist to ACE Medical Exercise Specialist. Of the four ACE certifications, this is most closely aligned with the clinical certifications offered by other credentialing bodies. This certification tests individual competencies to work with chronic diseased populations, including patients with cardiovascular, pulmonary, metabolic, and musculoskeletal disorders. Specifically, ACE states this is a credential aimed at postrehabilitative clients. To sit for the computer-based examination, one must be at least 18 yr old, have proof of current cardiopulmonary resuscitation (CPR) and automated external defibrillator (AED) certification, and have a bachelor's degree in exercise science or a related field. In addition, 500 h of practical experience designing and implementing exercise programs for apparently healthy or high-risk people is required and must be documented by a qualified professional. The ACE website states that currently there are more than 90,000 ACE-certified professionals, holding over 100,000 ACE certifications. Along with ACSM, ACE is a member of the Coalition for the Registration of Exercise Professionals (CREP), and all ACE (and ACSM) certified exercise professionals are listed on the United States Registry of Exercise Professionals (www.USREPS.org). ACE has international appeal and collaborates with several international groups dedicated to health and fitness, including the International Confederation of Registers for Exercise Professionals (ICREPs), which includes USREPS, the recognized registry from the United States; Europe Active; and the International Health, Racquet and Sportsclub Association (IHRSA).

American College of Sports Medicine

The American College of Sports Medicine (ACSM [5], www.acsm .org) has been internationally acknowledged as a leader in exercise and sports medicine for almost 70 yr. Besides clinical exercise physiologists, the ACSM comprises professionals with expertise in a variety of other sports medicine fields, including physicians, general exercise physiologists and those from academia, physical therapists, athletic trainers, dietitians, nurses, and other allied health professionals.

Since 1975, the ACSM has been recognized throughout the industry for its certification program and now boasts more than 20,000 certified professionals worldwide. The ACSM's Clinical Exercise Physiologist certification is designed for the professional who prescribes exercise, uses basic health behavior interventions, and promotes physical activity for individuals with chronic diseases or conditions, such as cardiovascular, pulmonary, metabolic, orthopedic, musculoskeletal, neuromuscular, neoplastic, immunologic, and hematologic diseases.

The ACSM also offers specialty certifications for the Cancer Exercise Trainer (CET) and the Certified Inclusive Fitness Trainer (CIFT). Each of these have aspects aligned with the focus of the clinical exercise physiologist.

The ACSM has an emphasis on international outreach not only for membership but also for its professional certifications. The organization also maintains an international subcommittee on the Certification and Registry Board, which handles all issues related to certification.

Canadian Society for Exercise Physiology

The vision of the Canadian Society for Exercise Physiology (CSEP [12], www.csep.ca) is to be "the recognized authority in exercise science and prescription, integrating research into best practice." The CSEP is considered the gold standard for health and fitness professionals who are dedicated to getting Canadian citizens physically active. The CSEP directs a part of its effort toward providing a certification for those who are exercise physiologists.

The CSEP Clinical Exercise Physiologist (CSEP-CEP) title was developed and approved as a certification in 2007. According to the CSEP website, the CSEP-CEP "performs assessments, prescribes conditioning exercise, as well as exercise supervision, counseling and healthy lifestyle education in apparently healthy individuals and/or populations with medical conditions, functional limitations or disabilities associated with musculoskeletal, cardiopulmonary, metabolic, neuromuscular, and ageing conditions." Certification requires a university degree in physical activity, exercise science, kinesiology, human kinetics, or a related field and passing written and practical examinations. The CSEP offers regular workshops to assist candidates in preparing for the national examination process, although attendance is not a requirement to sit for the examination.

Exercise & Sports Science Australia

The website for Exercise & Sports Science Australia (ESSA, formerly the Australian Association for Exercise and Sports Science [17], www.essa.org.au) states that it is a professional organization committed to establishing, promoting, and defending the career paths of exercise and sports science practitioners. Its vision is to enhance the health and performance of all Australians.

ESSA does not offer a certification examination for its clinical exercise professionals; instead, it accredits graduates from accredited university programs, akin to all other allied health professionals in Australia. The Accredited Exercise Physiologist (AEP) designation fits the needs of individuals who are working in a clinical setting. In Australia this accreditation is required for people who seek to become a Medicare (Australia), Department of Veterans Affairs, National Disability Insurance Scheme, State Workers Compensation Scheme, and private health insurance provider. ESSA approves courses and other studies that are delivered to maintain continuing professional development credits. ESSA also approves published journal articles, scientific conference presentations, and grant funding submissions.

PROFESSIONALIZATION OF CLINICAL EXERCISE PHYSIOLOGY

As you read through chapters 8 through 35 you will note that the biomedical literature continues to expand our knowledge of how physical activity plays a direct role in preventing the development of many chronic diseases, as well as an important role in inhibiting disease progression. Thus, physical activity and exercise are emerging as necessary treatments in comprehensive primary and secondary care. In addition, there are several emerging areas of research, including inactivity physiology (i.e., health risk associated with sitting) and its relationship to metabolic and cardiovascular health risk (8), the relationship of exercise and physical activity to cognitive health and progression of neurodegenerative disease (30), and the importance of exercise in the prevention and treatment of cancer (23) and peripheral artery disease (20). Given these and other findings of the importance of exercise and physical activity, in combination with the increasing number of people 65 yr of age and older in the world population, employment opportunities for the clinical exercise physiologist continue to expand. An important task of the clinical exercise physiologist is to review, explain, and provide practical advice and solutions regarding the vast amounts of exercise-related research and recommendations to the public and patients. The field is increasingly moving away from a one-size-fits-all exercise prescription model toward recommendations that are specific to a given population or individual with a specific health condition or chronic disease. The complexity of today's health information, along with the oft-conflicting findings of published studies, can make it difficult for the general public and patients to comprehend. The clinical exercise physiologist should be prepared to be an intermediary between knowledge and understanding of the exercise prescription.

In April 2004 the Committee on Accreditation for the Exercise Sciences (CoAES), under the auspices of the Commission on Accreditation of Allied Health Education Programs (CAAHEP), established guidelines and standards for postsecondary academic institutions for personal fitness trainer, exercise physiology, and exercise science academic programs at both the bachelor's and master's degree levels. To date, 14 academic institutions have attained the CoAES exercise physiology program accreditation. One of these is in Australia, and the remaining 13 are in the United States. Exercise physiology, as it relates to this specific accreditation, is described by CAAHEP as follows:

Exercise Physiology is a discipline that includes clinical exercise physiology and applied exercise physiology. Applied Exercise Physiologists manage programs to assess, design, and implement individual and group exercise and fitness programs for apparently healthy individuals and individuals with controlled disease. Clinical Exercise Physiologists work under the direction of a physician in the application of physical activity and behavioral interventions in clinical situations where they have been scientifically proven to provide therapeutic or functional benefit.

As of January 2022, 72 academic institutions have achieved CoAES exercise science program accreditation. All are from the United States except for the Shanghai University of Sport (Shanghai, China). CAAHEP describes exercise science as it relates to accreditation as follows:

Graduates of Exercise Science programs are trained to assess, design, and implement individual and group exercise and fitness programs for individuals who are apparently healthy and those with controlled disease. They are skilled in evaluating health behaviors and risk factors, conducting fitness assessments, writing appropriate exercise prescriptions, and motivating individuals to modify negative health habits and maintain positive lifestyle behaviors for health promotion. The Exercise Science professional has demonstrated competence as a leader of health and fitness programs in the university, corporate, commercial or community settings in which their clients participate in health promotion and fitness-related activities.

It is anticipated that over the next several years, more academic programs throughout the United States and potentially elsewhere will receive accreditation at the undergraduate (exercise science) and graduate (exercise physiology) levels. No other independent third-party organization currently accredits academic exercise science or exercise physiology programs in North America.

In Australia the Course Accreditation Program is administered by ESSA, and this is the only pathway to accreditation as an AEP. For individuals who were educated in other countries, ESSA assesses the credentials of clinical exercise physiologists by audit of their coursework prior to their taking the theory and practical examinations required for the AEP credential.

Students interested in becoming a clinical exercise physiologist should seek the guidance of professionals in the field to select a program that also provides a well-rounded set of academic, practical, laboratory, and research experiences aimed at best preparing them to successfully work in the field. Enrolling in a program accredited by CoAES (U.S.) or ESSA (Australia) would help ensure the quality of education with respect to those program offerings. Currently, although most clinical exercise physiologists in the United States are employed in cardiac rehabilitation, more programs designed to help care for patients with other chronic diseases and disabilities are being launched each year. These other programs target patients with chronic kidney disease, cancer, chronic fatigue, arthritis, peripheral artery disease, and metabolic syndrome, among others. Because of legislation in Australia allowing accredited CEPs (i.e., AEPs) to provide regulated health services, their employment location is much broader.

Individuals interested in cardiac rehabilitation should become familiar with the American Association of Cardiovascular and Pulmonary Rehabilitation (AACVPR). The AACVPR publishes a directory of programs in the United States and in other countries, which is an ideal resource for identifying potential employers or clinical internship sites. The AACVPR also offers a career hotline to members and nonmembers that lists open positions and provides a site for posting resumes (2). In addition, posted positions can be accessed through the ACSM website (5). The website www.exercisejobs.com lists positions nationwide and provides the opportunity to post a resume online. In Australia, the website www.seek.au provides a listing of potential places of employment for the exercise physiologist.

Another area within the clinical exercise profession involves programs that use exercise in primary prevention. Many hospitals and corporations recognize that regular exercise training can reduce future medical expenses and increase productivity, and many of these programs are implemented in a medical fitness center setting. In fact, most medically-based fitness facilities now exclusively hire people with exercise-related degrees as a means to ensure safety for an increasingly diverse clientele that includes both healthy people and individuals with clinically manifest disease. To learn more about this exercise physiology specialty, visit the website of the Medical Fitness Association (MFA) at www.medicalfitness.org. Of note, the CEPA and the MFA have joined in a mutual operating agreement (MOU) with the goal of further advancing the CEP profession and further promoting established standards and best practices for associated programs, personnel, and facilities.

Also important in the area of primary prevention is the Exercise Is Medicine (EIM) initiative, launched as a joint effort by the ACSM and the American Medical Association. The EIM has become a global entity with recognition in 37 countries worldwide. According to the EIM website (www.exerciseismedicine.org),

> EIM encourages physicians and other health care providers to include physical activity when designing treatment plans and to refer patients to evidence-based exercise programs and qualified exercise professionals. EIM is committed to the belief that physical activity promotes optimal health and is integral in the prevention and treatment of many medical conditions.

The EIM has developed a credential, in conjunction with the ACSM, ACE, and MFA, aimed at exercise professionals. The goal is to accredit individuals who will work closely with the health care community. This three-level credential is based on the ability to

- navigate and apply the EIM solution in health care systems,
- understand the changing health care landscape and the potential opportunity for EIM and credentialed exercise professionals,
- develop physical activity or exercise programs for individuals with various health conditions, and
- develop the skills needed to support sustained behavior change.

Those seeking employment or attempting to maintain professional certification would also be wise to join one or more professional organizations. Besides the national-level organizations previously mentioned, many clinical exercise physiologists in North America also join a regional chapter of the ACSM or a state or provincial affiliate of the AACVPR, CEPA, or CSEP. These organizations hold local educational programs that provide professionals an avenue through which to gain continuing education credits and an opportunity to network with those who make decisions to hire clinical exercise physiologists.

Several other professional organizations support, publish, and advocate for exercise programming and exercise-related research or information that focuses on specific diseases or disorders. These include the American Heart Association (AHA), the American College of Cardiology (ACC), the American Diabetes Association (ADA), the American Academy of Orthopedic Surgeons, the American Cancer Society, the Multiple Sclerosis Society, and the Arthritis Foundation. The National Center on Physical Activity and Disability (NCPAD) rounds out this listing of such organizations with its focus on various disabilities and conditions. Table 1.3 lists the websites for these and other organizations, most of which serve as excellent information resources. Finally, clinical exercise physiologists should stay abreast of current research in general exercise physiology, as well as research specific to their disease area of interest. The body of scientific information specific to clinical exercise physiology continues to grow rapidly. Regular journal reading is an excellent way to keep current in the field. (See the sidebar Selected Biomedical Journals in this chapter.)

A person practicing as a clinical exercise physiologist should consider using several safeguards. Since only Louisiana has a licensing process for clinical exercise physiologists, almost anyone in the other 49 states can claim to be a clinical exercise physiologist. Licensure may help solve this issue, but, as previously mentioned, this is a slow, state-by-state process and may not be the ultimate answer. This is still under debate within the profession. A major issue related to licensure and professionalization is the risk that practicing professionals assume when working with clients, as discussed in practical application 1.2. Professionals must protect themselves against potential litigation related to their practice. Practical application 1.3 describes an interesting case that highlights both the potential risk and the important role that a clinical exercise physiologist can play in litigation. Clinical exercise physiologists should consider malpractice insurance as a safeguard in the event of such incidents. In a hospital setting the institution often covers malpractice insurance; but if working in fitness, small clinical, ambulatory care, or personal training settings, exercise professionals should inquire about and obtain, if necessary, their own insurance coverage. Organizations such as the ACSM offer discounted group rates for malpractice insurance for the exercise professional.

Limiting legal risk is important to the practicing exercise physiologist. Within the confines of the civil justice system, under which personal injury and wrongful death cases are determined, clinical exercise physiologists could be held accountable to a variety of

Table 1.3   Selected Chronic Disease and Disability Organizations and Institutes

Category	Organization or institute	Internet site
Endocrinology and metabolic disorders	American Diabetes Association Diabetes Canada European Association for the Study of Obesity Kidney Health Australia National Institute of Diabetes and Digestive and Kidney Diseases National Kidney Foundation North American Association for the Study of Obesity Obesity Society	www.diabetes.org www.diabetes.ca www.easo.org www.kidney.org.au www.niddk.nih.gov www.kidney.org www.naaso.org www.obesity.org
Cardiovascular diseases	American Association of Cardiovascular and Pulmonary Rehabilitation American College of Cardiology American Heart Association Canadian Association of Cardiovascular Prevention and Rehabilitation Canadian Cardiovascular Society Cardiac Society of Australia and New Zealand European Society of Cardiology Heart and Stroke Foundation of Canada Heart Failure Society of America Hypertension Canada National Heart, Lung, and Blood Institute Vascular Disease Foundation	www.aacvpr.org www.americanheart.org www.acc.org www.cacpr.ca www.ccs.ca www.csanz.edu.au www.escardio.org www.heartandstroke.ca www.hfsa.org www.hypertension.ca www.nhlbi.nih.gov www.vascularcures.org
Respiratory diseases	American Lung Association Canadian Lung Association European Respiratory Society Forum of International Respiratory Societies National Heart, Lung, and Blood Institute	www.lungusa.org www.lung.ca www.ersnet.org www.firsnet.org www.nhlbi.nih.gov
Oncology and immune diseases	American Cancer Society British Society for Immunology Canadian Cancer Society National Cancer Institute	www.cancer.org www.immunology.org www.cancer.ca www.nci.nih.gov
Bone and joint diseases and disorders	American Academy of Orthopedic Surgeons American College of Rheumatology and Association of Rheumatology Health Professionals Arthritis Foundation International Osteoporosis Foundation National Institute of Arthritis and Musculoskeletal and Skin Diseases Osteoporosis Canada Spondylitis Association of America	www.aaos.org www.rheumatology.org www.arthritis.org www.osteofound.org www.niams.nih.gov www.osteoporosis.ca www.spondylitis.org
Neuromuscular disorders	International Collaboration on Repair Discoveries The International Spinal Cord Society MS Society of Canada National Institute of Neurological Disorders and Stroke National Multiple Sclerosis Society National Spinal Cord Injury Association	www.icord.org www.iscos.org.uk www.spinalcord.org www.mssociety.ca www.nmss.org www.ninds.nih.gov
Special populations	Active Aging Canada American Geriatrics Society National Institute on Aging National Institute of Child Health and Human Development National Women's Health Information Center	www.activeagingcanada.ca www.americangeriatrics.org www.nia.nih.gov www.nichd.nih.gov www.4woman.gov
Disabilities	Canadian Disability Participation Project National Center on Physical Activity and Disability	www.cdpp.ca www.ncpad.org

Practical Application 1.3

LEGAL CASE

To appreciate what can be at stake for health professionals who become involved in courtroom litigation, consider this case of an exercise physiologist working in the clinical setting. She wrote of her litigation experiences in the first person. Excerpts follow.

"The hospital where I work is a tertiary care facility. . . . The staff members in Cardiac Rehabilitation are required to have a master's degree in exercise physiology, American Heart Association basic cardiac life support certification, and American College of Sports Medicine certification as an exercise specialist. . . ."

"In May of 1991, Cardiac Rehabilitation was consulted to see a 78 yr old woman 4 d post–aortic valve surgery. Her medical history included severe aortic stenosis, coronary artery disease, left ventricular dysfunction, carotid artery disease, atrial fibrillation, hypertension, and noninsulin dependent diabetes. . . . She was in atrial fibrillation but had stable hemodynamics, and was ambulating without problems in her room."

"The patient was first seen by Cardiac Rehabilitation 7 d postsurgery. Chart review . . . revealed no contraindications to activity. . . . Initial assessment revealed normal supine, sitting, standing, and ambulating hemodynamics. She walked independently in her room, and approximately 100 ft (30 m) in 5 min . . . using the handrails for support. . . . General instructions for independent walking later that day were given. . . ."

"On postop day 8, there was documentation in the chart regarding the patient's need for an assist device while ambulating. . . . The patient ambulated approximately 200 ft (60 m) with a quad cane. . . . She appeared stable. . . ."

"On postop day 9, I was informed by the patient's nurse, as well as by documentation in the chart, that the patient had fallen in her room that morning, apparently hitting her head on the floor. . . . A computed tomography scan of the head was negative."

"On postop day 10, I observed the patient sitting with her daughter in the hall. . . . Her mental status had noticeably changed. She did not recognize her daughter, where she was, or why she was in the hospital. . . . No further formal exercise was performed. . . . The patient was transferred to intensive care and later died of a massive brain hemorrhage."

"Two years after the incident, a hospital lawyer contacted me and informed me that the family had brought suit against the hospital contending that the patient fell because she was not steady enough to walk independently. . . . I was called to testify and explain my participation in her care. . . ."

"I met with the hospital's defense lawyer on two separate occasions. The first meeting, the lawyer inquired about my education, certification, years of employment, and what specifically my job entailed. . . . Two weeks before trial, the hospital's defense lawyer again questioned me. . . . He also played the role of the plaintiff's lawyer and reworded similar questions in a slightly different manner. The defense lawyer encouraged me to maintain consistency in my answers, and remain calm and assured. . . . To prove our case, we had to establish, by a preponderance of the evidence (i.e., more likely than not) that the patient was indeed ambulating independently prior to her fall. My documentation was crucial. . . ."

"Answering the plaintiffs' lawyer's questions during cross-examination was like a mental chess game. . . . My answers were crucial, as was my composure and the belief that all my training and certification *did* qualify me as a professional health care provider. . . . Although the only hard evidence I had to work with in the courtroom was my chart notes, the judge did allow me to supplement them with oral testimony. . . ."

"Whether you document in the patient's chart or on a separate summary sheet that is filed in your office, your documentation is a permanent record that provides powerful legal evidence. . . ."

"Later I was told that my testimony helped the hospital prove the patient's fall was not a result of negligence. The court ruled in favor of our hospital."

professional standards and guidelines. These standards and guidelines, sometimes referred to as practice guidelines, can apply to some practices carried out by clinical exercise physiologists and others who provide services within health and fitness facilities. A variety of professional organizations have developed and published guidelines or standards dealing with the provision of service by fitness professionals, including clinical exercise physiologists. Professionals use standards and guidelines to identify probable and evidence-based benchmarks of expected service delivery owed to patients and clients, which helps ensure appropriate and uniform delivery of service.

In the course of litigation, these standards are used by expert witnesses to establish the standard of care to which providers will be held accountable in the event of patient injury or death. In years past, expert witnesses often used their subjective and personal opinions to establish the standard of care and then provided an opinion as to whether that standard was violated in the actual delivery of service. Today, most professional organizations use an evidence-based approach to establish standards of care for the following purposes:

- To achieve a consensus of standards of care among professionals that practitioners should aspire to meet in their delivery of services in accordance with known and established benchmarks of expected behavior
- To reduce cases of negligence and findings of negligence or malpractice
- To minimize the significance of individually and subjectively expressed opinions in court proceedings
- To reduce potential inconsistent verdicts and judgments that arose from the expression of individual opinions

Practice guidelines are readily available when one wishes to compare the actual delivery of service to what the profession as a whole has determined to be established benchmarks. Consequently, clinical exercise physiologists should review relevant published practice guidelines and then consider in what situations the standards might act as a shield to protect against negligent actions. Others may engage in the same process to determine whether such standards should be used as a sword to attack the care provided to clients in particular cases. Clinical exercise physiologists can use the following strategies to limit exposure and risk.

- Involve physicians in decision making. In programs that include patients with known disease, and in which diagnosis, prescription, or treatment is the probable or actual reason for performing the procedures, a physician should be involved in a meaningful way. This involvement must be meaningful and real if it is to be legally effective. Examples include referring patients, receiving and acting on results provided by rehabilitation staff, and discussing findings with the rehabilitation or therapy staff.

- Involve physicians in initial testing. In programs that have patients who demonstrate a high risk of suffering an adverse event or injury during exercise, the physician should be within immediate proximity to the participant during initial testing. The physician need not be watching the patient or the electrocardiogram monitor, but they should be controlling the staff during the procedure. Thereafter, the physician's proximity of supervision in further exercise testing should be dictated by the medical interpretation of each patient's initial test result.
- Evaluate and screen participants before recommending or prescribing activity, especially in rehabilitative or preventive settings.
- Secure a medically mandated consent document when involved in procedures such as a graded exercise test.
- Develop and assist individuals with implementing a safe and evidence-based exercise prescription that is aimed at a patient's goals and addresses disease-related disability.
- Recognize and refer people who have conditions that need evaluation before commencing or continuing with various activity programs.
- Provide feedback to referring professionals relative to the progress of participants in activity.
- Provide appropriate, timely, and effective emergency care, as needed.

CONCLUSION

Clinical exercise physiology is an allied health profession recognized by other practicing clinicians. It is strongly encouraged that all individuals who consider themselves clinical exercise physiologists become involved in relevant local or national organizations dedicated to promoting the profession; acquire a relevant certification; and regularly update their knowledge and skill level by reading contemporary relevant literature and attending local, regional, national, and international education events when possible. Given the growing evidence for and acceptance of the role of physical activity and exercise in the prevention and treatment of chronic disease, the timing is ideal for permanently solidifying the role of the clinical exercise physiology profession in the health care of the population. Several developments since the previous edition of this text in 2018 have moved the profession toward this goal (e.g., changes in accreditation names, development of a registry, a vision of an international alliance). A primary purpose of this text is to provide aspiring and practicing clinical exercise physiologists throughout the world with up-to-date and practical information to prepare for, or to maintain, certification.

 ## Online Materials

Visit HK*Propel* for a link to the references and to access a quiz to help you review key concepts and test your understanding of the material covered in this chapter.

CHAPTER 2

Promoting a Physically Active Lifestyle

Anna G. Beaudry, BS

Danielle A. Young, PsyD

Annie T. Ginty, PhD

The clinical exercise physiologist can use a number of behavioral strategies in assessing and counseling individual patients or clients about their physical activity behavior change. The behavioral strategies discussed in this chapter are intended to be used in the context of supportive social and physical environments. An unsupportive environment is considered a barrier to regular physical activity participation; consequently, if this barrier is not altered, change is unlikely to occur. Thus, one of the clinician's goals is to identify environmental barriers with the client, include steps on how to overcome these barriers, and build supportive social and physical environments as part of the counseling strategy. Although no guarantees can be made, the literature suggests that if clinicians take a behavior-based approach to physical activity counseling, within the context of a supportive environment, they may indeed experience greater success in getting their clients moving. Therefore, this chapter also presents information about the role of social and contextual settings in promoting health- and fitness-related levels of physical activity. The most important task of the clinical exercise physiologist is to guide a client into a lifelong pattern of regular, safe, and effective physical activity.

BENEFITS OF PHYSICAL ACTIVITY

Exercise physiologists understand the physiological basis for activity as well as the impact of pathology on human performance. However, to be effective counselors, clinical exercise physiologists also need to understand human behavior in the context of the individual client's social and physical milieu. This chapter seeks to underscore some of the important theories and models of behavior that are important adjuncts for clinicians seeking to help people make positive changes in their physical activity behavior.

The historical literature has established evidence that persons who engage in regular physical activity have an increased physical working capacity (64), decreased body fat (46), increased lean body tissue (46), increased bone density (48), and lower rates of coronary heart disease (CHD) (48), diabetes mellitus, hypertension (48), and cancer (48). Increased physical activity is also associated with greater longevity (48, 64), as well as favorable brain and cognitive outcomes (21, 58). Regular physical activity and exercise can also assist persons in improving mood and motivational climate (45), enhancing their quality of life, improving their capacity for work and recreation, and altering their rate of decline in functional status (56). An international review of the health benefits of physical activity reinforces these findings (34). Lengthy periods of sitting can also negatively influence health, even if a person engages in daily physical activity (16). Frequent breaks to stand or move throughout the day seem to offset these challenges in sedentary work environments (50).

When promoting planned exercise and physical activity, heath practitioners must pay attention to specifically designed outcomes, notably the health and fitness outcomes of a well-designed exercise prescription. Finally, a number of physiological, anatomical, and behavioral characteristics should also be considered to ensure a safe, effective, and enjoyable exercise experience for the participant.

Acknowledgment: Much of the writing in this chapter was adapted from earlier editions of *Clinical Exercise Physiology*. Thus, we wish to gratefully acknowledge the previous efforts of and thank Gregory W. Heath, DHSc, MPH and Josh M. Johann, MS.

Health Benefits

Physical activity has been defined as any bodily movement produced by skeletal muscles that results in caloric expenditure (11). Because caloric expenditure uses energy and because energy use enhances weight loss or weight maintenance, caloric expenditure is important in the prevention and management of obesity, CHD, and diabetes mellitus. Healthy People 2030 Physical Activity Objective PA-02 (61) highlights the need for every adult to engage in moderate aerobic physical activity for at least 150 min/wk or vigorous aerobic physical activity for a minimum of 75 min/wk or an equivalent combination. Current research suggests that engaging regularly in moderate aerobic physical activity for at least 150 min/wk or vigorous physical activity for at least 75 min/wk will help ensure that the calories expended confer specific health benefits. For example, daily physical activity equivalent to a sustained walk for 30 min/d for 7 d would result in an energy expenditure of about 1,050 kcal/wk. Epidemiologic studies suggest that a weekly expenditure of 1,000 kcal could have significant individual and public health benefits for CHD prevention, especially among those who are initially inactive. In 2007, the American College of Sports Medicine and the American Heart Association concluded that the scientific evidence clearly demonstrates that regular, moderate-intensity physical activity provides substantial health benefits (24). In addition, following an extensive review of the physiological, epidemiological, and clinical evidence, the 2018 Physical Activity Guidelines Advisory Committee formulated this guideline (62):

> For substantial health benefits, adults should do at least 150 minutes (2 hours and 30 minutes) to 300 minutes (5 hours) a week of moderate-intensity, or 75 minutes (1 hour and 15 minutes) to 150 minutes (2 hours and 30 minutes) a week of vigorous-intensity aerobic physical activity, or an equivalent combination of moderate- and vigorous-intensity aerobic activity. Preferably, aerobic activity should be spread throughout the week.

This guideline emphasizes the benefits of moderate and vigorous aerobic physical activity. Intermittent activity has been shown to confer substantial benefits. Therefore, the recommended minutes of activity can be accumulated in shorter bouts spaced throughout the day. Although the accumulation of 150 min of moderate-intensity or 75 min of vigorous-intensity aerobic physical activity per week has been shown to confer important health benefits, these guidelines are not intended to represent the optimal amount of physical activity for health but instead a minimum standard or base on which to build to obtain more specific outcomes related to physical activity and exercise. Specifically, selected fitness-related outcomes may be a desired result for the physical activity participant, who may seek the additional benefits of improved cardiorespiratory fitness, muscle endurance, muscle strength, flexibility, and body composition. Indeed, Healthy People 2030 affirms these guidelines. Objective PA-03 states that adults need at least 300 min/wk of moderate-intensity aerobic activity or 150 min/wk of vigorous-intensity aerobic activity to gain more extensive health benefits. In regard to muscular strength, objective PA-04 states that adults must engage in muscle-strengthening activities at least 2 d/wk to attain important health benefits. Table 2.1 presents the health benefits of regular physical activity for the brain (62).

Fitness Benefits

Regular vigorous physical activity helps achieve and maintain higher levels of cardiorespiratory fitness than moderate physical activity. There are five components of health-related fitness: cardiorespiratory fitness, muscle strength, muscle endurance, flexibility, and enhanced body composition.

Cardiorespiratory fitness, or aerobic capacity, refers to the body's ability to perform high-intensity activity for a prolonged period without undue physical stress or fatigue. Having higher levels of cardiorespiratory fitness enables people to carry out their daily occupational tasks and leisure pursuits more easily and with greater efficiency. Vigorous physical activities such as the following help achieve and maintain cardiorespiratory fitness and can also contribute substantially to caloric expenditure:

- Very brisk walking
- Jogging, running
- Lap swimming
- Cycling
- Fast dancing
- Skating
- Rope jumping
- Soccer
- Basketball
- Volleyball

These activities may also provide additional protection against CHD over moderate forms of physical activity. People can achieve higher levels of cardiorespiratory fitness by increasing the frequency, duration, or intensity of an activity beyond the minimum recommendation of 20 min per occasion, on three occasions per week, at more than 45% of aerobic capacity (3). Recent research suggests high-intensity interval training may also improve fitness within a shorter duration of exercise (15, 67).

Muscular strength and endurance are the ability of skeletal muscles to perform work that is hard or prolonged or both (3). Regular use of skeletal muscles helps improve and maintain strength and endurance, which greatly affects the ability to perform the tasks of daily living without undue physical stress and fatigue. Examples of tasks of daily living include home maintenance and household activities such as sweeping, gardening, and raking. Engaging in regular physical activity such as weight training or the regular lifting and carrying of heavy objects appears to maintain essential muscle strength and endurance for the efficient and

Table 2.1   Benefits of Physical Activity for Brain Health

Outcome	Population	Benefit	Acute	Habitual
Cognition	Children ages 6 to 13 yr	Improved cognition (performance on academic achievement tests, executive function, processing speed, memory)	✓	✓
	Adults	Reduced risk of dementia (including Alzheimer's disease)		✓
	Adults older than age 50 yr	Improved cognition (executive function, attention, memory, crystallized intelligence, processing speed)		✓
Quality of life	Adults	Improved quality of life		✓
Depressed mood and depression	Children ages 6 to 17 yr and adults	Reduced depressed mood and risk of depression		✓
Anxiety	Adults	Reduced short-term feelings of anxiety (state anxiety)	✓	
	Adults	Reduced long-term feelings and signs of anxiety (trait anxiety) for people with and without anxiety disorders		✓
Sleep	Adults	Improved sleep outcomes (increased sleep efficiency, sleep quality, deep sleep; reduced daytime sleepiness, frequency of use of medication to aid sleep)		✓
	Adults	Improved sleep outcomes that increase with duration of acute episode of physical activity	✓	

Reprinted by U.S. Department of Health and Human Services, *Physical Activity Guidelines for Americans,* 2nd ed. (Washington, DC: U.S. Department of Health and Human Services, 2018), 40.

effective completion of most activities of daily living throughout the life cycle (3). The prevalence of such physical activity behavior is still quite low, with recent prevalence estimates indicating that only 19.6% of the adult population engages in strength training at least twice per week (60).

Musculoskeletal flexibility refers to the range of motion in a joint or sequence of joints. Joint movement throughout the full range of motion helps improve and maintain flexibility (3). Those with greater total body flexibility may have a lower risk of back injury (48). Older adults with better joint flexibility may be able to drive an automobile more safely (3, 66). Engaging regularly in stretching exercises and a variety of physical activities that require one to stoop, bend, crouch, and reach may help maintain a level of flexibility that is compatible with quality activities of daily living (3).

Excess body weight occurs when too few calories are expended and too many consumed for individual metabolic requirements (46). The maintenance of an acceptable ratio of fat to lean body weight is another desired component of health-related fitness. The results of weight loss programs focused on dietary restrictions alone have not been encouraging. Physical activity burns calories, increases the proportion of lean to fat body mass, and raises the metabolic rate (69). Therefore, a combination of caloric control and increased physical activity is important for attaining a healthy body weight. The 2020-2025 United States Dietary Guidelines state (64):

To attain the most health benefits from physical activity, adults need at least 150 to 300 minutes of moderate-intensity aerobic activity, like brisk walking or fast dancing, each week. Adults also need muscle-strengthening activity, like lifting weights or doing push-ups, at least 2 days each week.

PARTICIPATION IN REGULAR PHYSICAL ACTIVITY

In designing any exercise prescription, the professional needs to consider physiological, behavioral, psychosocial, and environmental (physical and social) variables that are related to participation in physical activity (9). Two commonly identified determinants of physical activity participation are self-efficacy and social support.

Self-efficacy, a construct from social cognitive theory, is most characterized by the person's confidence to exercise under a number of circumstances and appears to be positively associated with greater participation in physical activity. **Social support** from family and friends has consistently been shown to be associated with greater levels of physical activity participation. Incorporating some mechanism of social support within the exercise prescription appears to be an important strategy for enhancing compliance with a physical activity plan (31). Common barriers to participation in physical activity are time constraints and injury. The professional can take these barriers into account by encouraging participants to include physical activity as part of their lifestyle, thus not only engaging in planned exercise but also incorporating transportation, occupational, and household physical activity into their daily routine.

Participants can also be counseled to help prevent injury. People are more likely to adhere to a program of low- to moderate-intensity physical activities than one comprising high-intensity activities during the early phases of an exercise program. Moreover, moderate activity is less likely to cause injury or undue discomfort (48, 63).

A number of physical and social **environmental factors** can also affect physical activity behavior (53). Family and friends can be role models, can provide encouragement, or can be companions during physical activity. The physical environment often presents important barriers to participation in physical activity, including inclement weather, unsafe neighborhoods, and a lack of bicycle trails and walking paths away from vehicular traffic (55). Sedentary behaviors such as excessive television viewing or computer use may also deter persons from being physically active (54, 55).

Risk Assessment

An exercise prescription may be fulfilled in at least three different ways:

1. On a program-based level that consists primarily of supervised exercise training (32)
2. Through exercise counseling and exercise prescription followed by a self-monitored exercise program (30)
3. Through community-based exercise programming that is self-directed and self-monitored (70)

Within supervised exercise programs and programs offering exercise counseling and prescription, participants should complete a brief medical history and risk factor questionnaire and a preprogram evaluation (3). More information on the medical history and risk factor questionnaire and preprogram evaluation is presented in chapter 4.

In a self-directed community-based program, medical clearance is left to the judgment of the individual participant. An active physical activity promotion campaign in the community seeks to educate the population regarding precautions and recommendations for moderate and vigorous physical activity (17, 18). These messages should provide information that participants must know before beginning a regular program of moderate to vigorous physical activity. This information should encompass the following:

- Awareness of preexisting medical problems (e.g., CHD, arthritis, osteoporosis, or diabetes mellitus)
- Consultation before starting a program, with a physician or other appropriate health professional, if any of the previously mentioned problems are suspected
- Appropriate mode of activity and tips on different types of activities

- Principles of training intensity and general guidelines as to rating of perceived exertion and training heart rate
- Progression of activity and principles of starting slowly and gradually increasing activity time and intensity
- Principles of monitoring symptoms of excessive fatigue
- Making exercise fun and enjoyable

Theories and Models of Physical Activity Promotion

Historically, the most common approach to exercise prescription taken by health professionals has been direct information. In the past, the counseling sequence often consisted of the following:

1. Exercise assessment, usually cardiorespiratory fitness measures
2. Formulation of the exercise prescription
3. Counseling the patient regarding mode (usually large-muscle activity), frequency (three to five sessions per week), duration (20-30 min per session), and intensity (assigned target heart rate based on the exercise assessment) (33)
4. Review of the exercise prescription by the health professional and participant
5. Follow-up visits (reassessments and revising of the exercise prescription)
6. Phone contact

Most of the research evaluating this traditional approach to exercise prescription has not been too favorable with respect to long-term compliance and benefits (48). That is, most people who begin an exercise program drop out during the first 6 mo. Why has the traditional information-sharing approach been used? Because it's easiest for the clinician, requires less time, and is prescriptive. However, it is not interactive with the client. More recently, contemporary theories and models of human behavior have been examined and developed for use in exercise counseling and interventions (1, 2, 5, 6, 7, 40, 41, 49, 51). Referred to as cognitive–behavioral techniques, these are the most salient theories and models for promoting the initiation of and adherence to physical activity. These approaches vary in their applicability to physical activity promotion. Some models and theories were designed primarily as guides to understanding behavior, not as guides for designing intervention protocols. Others were specifically constructed with a view toward developing cognitive–behavioral techniques for physical activity behavior initiation and maintenance. Consequently, the clinical exercise physiologist may find that most of the theories summarized in table 2.2 will deepen their understanding of physical activity behavior change. Nevertheless, other theories have evolved sufficiently to provide specific intervention techniques to assist in behavior change.

Table 2.2 Summary of Theories and Models Used in Physical Activity Promotion

Theory or model	Level	Key concepts
Health belief model (51)	Individual	Perceived susceptibility Perceived severity Perceived benefits Perceived barriers Cues to action Self-efficacy
Relapse prevention (40, 41)	Individual	Skills training Cognitive reframing Lifestyle rebalancing
Theory of planned behavior (1, 2)	Individual	Attitude toward behavior Outcome expectations Value of outcome expectations Subjective norm Beliefs of others Motive to comply with others Perceived behavioral control
Social cognitive theory (6, 7)	Interpersonal	Reciprocal determinism Behavioral capability Self-efficacy Outcome expectations Observational learning Reinforcement
Social support (6)	Interpersonal	Instrumental support Informational support Emotional support Appraisal support
Ecological perspective (35)	Environmental	Multiple levels of influence: • Intrapersonal • Interpersonal • Institutional • Community • Public policy
Transtheoretical model (10, 38, 39, 49)	Individual	Precontemplation Contemplation Preparation Action Maintenance

Adapted from K. Glanz and B.K. Rimer, *Theory-at-a-Glance: A Guide for Health Promotion Practice* (U.S. Department of Health and Human Services, 1995); K. Glanz, The Application of Behavior Change Theory in the Worksite Setting in *ACSM's Worksite Health Handbook: A Guide to Building Healthy and Productive Companies,* edited by N.P, Pronk (Champaign, IL: Human Kinetics, 2009), 224-230.

The Patient-Centered Assessment & Counseling for Exercise & Nutrition (PACE and PACE+) materials were developed for use by the primary care provider in the clinical setting targeting apparently clinically healthy adults (38). The materials have been evaluated for both acceptability and effectiveness in a number of clinical settings (10, 38, 39). A more recent application of the principles of physical activity assessment and counseling used in PACE has been the development of the Exercise Is Medicine (EIM) protocols (14, 23, 27, 37). This model has been summarized in practical application 2.1, the so-called EIM Solution, demonstrating the interface between the health care provider and the exercise physiology or fitness professional in assessing, referring, and counseling an individual for physical activity. The materials incorporate many of the principles from a number of theoretical constructs reviewed in table 2.2. Wankel and colleagues (65) demonstrated the effectiveness of cognitive–behavioral techniques for enhancing physical activity promotion in showing that using increased social support and decisional strategies improved adherence to exercise classes among participants. McAuley and coworkers (42) emphasized strategies to increase self-efficacy and thereby increase physical

Practical Application 2.1

THE EXERCISE IS MEDICINE MODEL

Telling patients what to do doesn't work, especially over the long term. An effective behavioral model helps facilitate long-term changes by telling patients how to change. The Exercise Is Medicine Solution is a comprehensive approach to physical activity assessment, counseling, and referral materials based on the previous work of PACE+ and other behavior-based approaches to physical activity promotion. The protocol draws heavily on the stages of change model, which suggests that individuals change their habits in stages. Taking into account each person's readiness to make changes, EIM provides tailored recommendations for patients in each stage. The EIM materials have sought to incorporate the recommendations from the Surgeon General's report *Physical Activity and Health* (63) as well as the *Physical Activity Guidelines* (48).

activity levels among adult participants in a community-based physical activity promotion program. These successful strategies included social modeling and social persuasion to improve compliance and exercise adherence. Promoting physical activity through home-based strategies holds much promise and might prove to be cost effective (32). Through tailored mail and telephone interventions, significant levels of social support and reinforcement have been shown to enhance participants' self-efficacy in complying with exercise prescription, thus significantly improving levels of physical activity (17, 18).

Most recently, behavioral approaches to physical activity promotion, such as EIM and PACE+, have been incorporated into the concept of mobile health (mHealth) and use of the smartphone platform in conjunction with electronic medical records. These approaches have proven feasible and may usher in a new era for physical activity promotion, health surveillance, and community-based networks (29, 36, 43, 47).

Other health and fitness professionals can play a supportive role for the clinical exercise physiologist in improving the overall health and increased physical activity levels among individuals outside of the clinical setting. Supervised physical activity and wellness coaching professionals can further enhance participant adherence to physical activity through motivational counseling and by providing education on health behavior, self-efficacy, and goal-setting skills, which often results in improved health and fitness outcomes (13). Hence, with a health and physical activity support team, participants are more likely to engage and adhere in health-promoting levels of physical activity. In addition to improving cardiorespiratory fitness, supervised exercise training programs often demonstrate favorable changes in blood lipid levels and blood pressure, resulting in decreased risk of coronary heart disease morbidity and mortality as well as all-cause mortality by 4.4%, 4.2%, and 1.8%, respectively (25). Further, supervised exercise interventions facilitate greater improvements in BMI, body composition, cardiorespiratory endurance, muscular strength, and flexibility (12, 22, 52). Wellness coaching, an emerging health and fitness profession, may also positively influence health and fitness more broadly. This patient-centered process includes elements of goal setting, encouraging education and self-discovery, and holding individuals accountable (68). Clark and colleagues (13) found wellness coaching to provide a range of health benefits, including increased levels of physical activity, increased confidence to engage in other health behaviors, and improved health-related quality of life and overall health. Therefore, health and wellness coaching professionals can also play a role in enhancing overall participant health and fitness and in lowering risk of adverse health outcomes.

Finally, **lifestyle-based physical activity** promotion increases the levels of moderate physical activity among adults. Lifestyle-based physical activity focuses on home- or community-based participation in forms of activity that include much of a person's daily routine (e.g., transport, home repair and maintenance, yard maintenance) (19). This approach evolved from the idea that physical activity health benefits may accrue from an accumulation of physical activity minutes over the course of the day (63). Because lack of time is a common barrier to regular physical activity, some researchers recommend promoting lifestyle changes whereby people can enjoy physical activity throughout the day as part of their lifestyle. Taking the stairs at work, taking a walk during lunch, and walking or biking for transportation are all effective forms of lifestyle physical activity. Assessing the common barriers (table 2.3) to physical activity among participants can be helpful for developing individual awareness and targeting strategies to overcome the barriers.

Ecological Perspective

A criticism of most theories and models of behavior change is that they emphasize individual behavior change and pay little attention to sociocultural and physical environmental influences on behavior (8). Recently, interest has developed in ecological approaches to increasing participation in physical activity (53). These approaches place the creation of supportive environments on par with the development of personal skills and the reorientation of health services. Creation of supportive physical environments

Table 2.3 Barriers to Being Active Quiz: What Keeps You From Being More Active?

Instructions: Listed here are reasons people give to indicate why they do not get as much physical activity as they should. Please read each statement and circle the number that represents how likely you are to say it.

How likely are you to say?	Very likely	Somewhat likely	Somewhat unlikely	Very unlikely
1. My day is so busy now that I just don't think I can make the time to include physical activity in my regular schedule.	3	2	1	0
2. None of my family members or friends like to do anything active, so I don't have a chance to exercise.	3	2	1	0
3. I'm just too tired after work to get any exercise.	3	2	1	0
4. I've been thinking about getting more exercise, but I just can't seem to get started.	3	2	1	0
5. I'm getting older so exercise can be risky.	3	2	1	0
6. I don't get enough exercise because I have never learned the skills for any sport.	3	2	1	0
7. I don't have access to jogging trails, swimming pools, bike paths, and so forth.	3	2	1	0
8. Physical activity takes too much time away from other commitments—work, family, and so on.	3	2	1	0
9. I'm embarrassed about how I will look when I exercise with others.	3	2	1	0
10. I don't get enough sleep as it is. I just couldn't get up early or stay up late to get some exercise.	3	2	1	0
11. It's easier for me to find excuses not to exercise than to go out to do something.	3	2	1	0
12. I know of too many people who have hurt themselves by overdoing it with exercise.	3	2	1	0
13. I really can't see learning a new sport at my age.	3	2	1	0
14. It's just too expensive. You have to take a class or join a club or buy the right equipment.	3	2	1	0
15. My free times during the day are too short to include exercise.	3	2	1	0
16. My usual social activities with family or friends do not include physical activity.	3	2	1	0
17. I'm too tired during the week, and I need the weekend to catch up on my rest.	3	2	1	0
18. I want to get more exercise, but I just can't seem to make myself stick to anything.	3	2	1	0
19. I'm afraid I might injure myself or have a heart attack.	3	2	1	0
20. I'm not good enough at any physical activity to make it fun.	3	2	1	0
21. If we had exercise facilities and showers at work, then I would be more likely to exercise.	3	2	1	0

Follow these instructions to score yourself:

- Enter the circled numbers in the spaces provided, putting the number for statement 1 in space 1, for statement 2 in space 2, and so on.
- Add the three scores on each line. Your barriers to physical activity fall into one or more of seven categories: lack of time, social influences, lack of energy, lack of willpower, fear of injury, lack of skill, and lack of resources. A score of 5 or above in any category shows that the barrier is an important one for you to overcome.

1 =	8 =	15 =	Row sum =	Lack of time
2 =	9 =	16 =	Row sum =	Social influence
3 =	10 =	17 =	Row sum =	Lack of energy
4 =	11 =	18 =	Row sum =	Lack of willpower
5 =	12 =	19 =	Row sum =	Fear of injury
6 =	13 =	20 =	Row sum =	Lack of skill
7 =	14 =	21 =	Row sum =	Lack of resources

From Center for Disease Control and Prevention. www.cdc.gov/diabetes/ndep/pdfs/8-road-to-health-barriers-quiz-508.pdf

is as important as intrapersonal factors when behavior change is the defined outcome. Stokols (59) illustrated this concept of a health-promoting environment by describing how physical activity could be promoted through establishing environmental supports such as bike paths, parks, and incentives to encourage walking or bicycling to work. An underlying theme of ecological perspectives is that the most effective interventions occur on multiple levels. Interventions that simultaneously influence multiple levels and multiple settings (e.g., schools, worksites) may be expected to lead to greater and longer-lasting changes and maintenance of existing health-promoting habits.

In addition, investigators have recently demonstrated that behavioral interventions primarily work by means of mediating variables of intrapersonal and environmental factors (53). Mediating variables are those that facilitate and shape behaviors—we all have a set of intrapersonal factors (e.g., personality type, motivation, genetic predispositions) and environmental factors (e.g., social networks like family, cultural influences, and the built and physical environments). However, few researchers have attempted to delineate the role of these mediating factors in facilitating health behavior change. Sallis and colleagues (53) have identified how difficult it is to assess the effectiveness of environmental and policy interventions because of the relatively few evaluation

studies available. However, based on the experience of the New South Wales (Australia) Physical Activity Task Force, a model of the steps necessary to implement these interventions has been proposed (44). Figure 2.1 presents an adaptation of this model as prepared by Sallis and colleagues (53) and outlines the necessary interaction between advocacy, coordination, or planning agencies; policies; and environments to make such interventions a reality. Another pragmatic model that appears to have relevance for the promotion of physical activity, identified in the Physical Activity Guidelines Advisory Committee report (48), specifies five levels of determinants for health behavior:

1. Intrapersonal factors, including psychological and biological variables, as well as developmental history
2. Interpersonal processes and primary social groups, including family, friends, and coworkers
3. Institutional factors, including organizations such as companies, schools, health agencies, or health care facilities
4. Community factors, which include relationships among organizations, institutions, and social networks in a defined area
5. Public policy, which consists of laws and policies at the local, state, national, and supranational levels

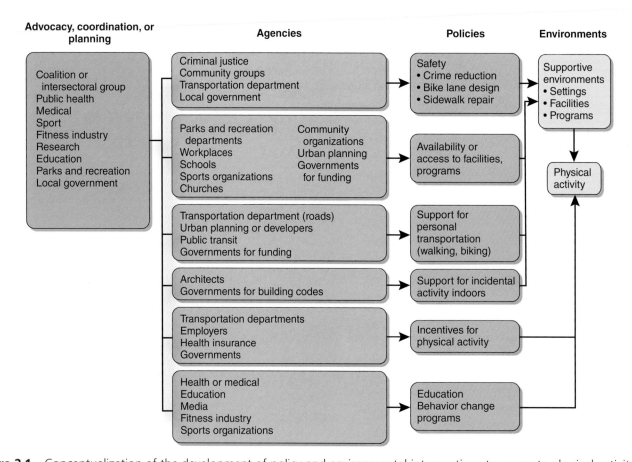

Figure 2.1 Conceptualization of the development of policy and environmental interventions to promote physical activity.

Reprinted from *American Journal of Preventive Medicine*, Vol. 15, J.F. Sallis et al., "Environmental and Policy Interventions to Promote Physical Activity," pgs. 379-397, Copyright 1998, with permission from Elsevier.

Important in implementing this concept of behavioral determinants is realizing the key role of behavioral settings, which are the physical and social situations in which behavior occurs. Simply stated, human behavior such as physical activity is shaped by its surroundings—if you're in a supportive social environment with access to space and facilities, you are more likely to be active. It is important to acknowledge the determinant role of selected behavioral settings: Some are designed to encourage healthy behavior (e.g., sports fields, gymnasiums, health clubs, and bicycle paths), whereas others encourage unhealthy (or less healthy) behaviors (e.g., fast food restaurants, vending machines with high-fat and high-sugar foods, movie theaters). Structures such as fields, gymnasiums, and community centers are part of each of our living environments; people who disregard them are less likely to be active. For physical activity providers, a potentially important adjunct in assessing and prescribing physical activity interventions for participants is understanding the physical and social contexts in which their patients live. This information can be obtained from various sources and can be at the level of the individual or at the more general community level. When individual physical activity behavior information is coupled with sociodemographic, physical, and social context information, physical activity interventions can be further tailored to maximize the participant's physical activity behavior change and maintenance plan. Exercise physiologists cannot alter the client's physical environment; however, they should be able to address environmental barriers and provide insights into how to overcome these barriers. In the long run, we all should be a part of changing our environments for the better.

An example of such tailoring for physical activity promotion that alters physical activity behaviors is the work of Linenger and colleagues (35), an effort to increase physical activity levels among naval personnel through a multifactorial environmental and policy approach to physical activity promotion. These investigators compared an enhanced base with a control base. The enhancements involved increasing the number of bike trails on base, acquiring new exercise equipment for the local gym, opening a women's fitness center, instituting activity clubs, and providing released time for physical activity and exercise (35). The changes were positive for those living on the enhanced base—that is, they increased their physical activity levels.

Another example, this time emphasizing an incentive-based approach to promoting physical activity, is the work of Epstein and Wing (20). Although this work was undertaken quite some time ago, the lessons are very relevant in today's inactive culture. In this study, contracts and the use of a lottery (a popular enterprise today!) were used to boost exercise attendance with the consequence of increasing participants' overall physical activity levels. Compared with results for a "usual care" comparison group, adherence and activity levels were significantly improved and sustained (44). However, some researchers have urged caution in using an incentive-based approach, believing that over the long term, participants never internalize the health behavior—meaning they are likely to stray back to sedentary habits once the incentive is removed or loses its appeal. Nevertheless, incentives have been proven to be effective in the short term. Additional community-based environmental efforts to influence physical activity behavior have included placing signs in public settings to increase use of stairs and walkways (57). These latter studies are examples of single intervention efforts that can be carried out in concert with systematic exercise prescription efforts among individuals. Thus, the increase in stair usage as a result of a promotional campaign can help individuals meet their prescribed energy expenditure requirements. Table 2.4 outlines some suggested solutions to some of the common barriers to becoming more physically active, although solutions can vary from client to client.

Useful resources for the clinical exercise physiologist are the very recent evidence-based recommendations for physical activity promotion in communities (www.thecommunityguide.org). The evidence base for these recommendations provides insights into how exercise practitioners can integrate their clinical efforts to assess and counsel patients while promoting supportive and reinforcing environments. The recommendations are summarized with respect to informational, behavioral–social, and environmental approaches to promoting physical activity (table 2.5) (4, 26, 30). A more recent review further establishes these physical activity intervention domains and expands the number of evidence-based approaches to promoting physical activity (28).

CONCLUSION

Physical activity has emerged as a key factor in the prevention and management of chronic conditions. Although the role of exercise in health promotion has been appreciated and applied for decades, findings regarding the mode, frequency, duration, and intensity of physical activity have modified exercise prescription practices. Included in these modifications has been the delineation between health and fitness outcomes relative to the physical activity prescription. Most importantly, new approaches to physical activity prescription and promotion that emphasize a behavioral approach with documented improvements in compliance have now become available to health professionals. Behavioral science has contributed greatly to the understanding of health behaviors such as physical activity. Behavioral theories and models of health behavior have been reexamined in light of physical activity and exercise. Although more research is needed to further develop successful, well-defined applications that are easily adaptable for intervention purposes, behavioral principles and guidelines have evolved that help the health professional understand health behavior change and guide people into lifelong patterns of increased physical activity and improved exercise compliance.

New frontiers in the application of exercise prescription to specific populations, as well as efforts to define the specific dose

Table 2.4 Tips on Overcoming Potential Barriers to Regular Physical Activity

Barriers	Suggestions for overcoming physical activity barriers
Lack of time	Identify available time slots. Monitor your daily activities for 1 wk. Identify at least three 30 min time slots you could use for physical activity. Add physical activity to your daily routine. For example, walk or ride your bike to work or shopping, organize school activities around physical activity, walk the dog, exercise while you watch TV, park farther from your destination. Make time for physical activity. For example, walk, jog, or swim during your lunch hour, or take fitness breaks instead of coffee breaks. Select activities that require minimal time, such as walking, jogging, or stair climbing.
Social influence	Explain your interest in physical activity to friends and family. Ask them to support your efforts. Invite friends and family members to exercise with you. Plan social activities that involve exercise. Develop new friendships with physically active people. Join a group, such as the YMCA or a hiking club.
Lack of energy	Schedule physical activity for times in the day or week when you feel energetic. Convince yourself that if you give it a chance, physical activity will increase your energy level; then try it.
Lack of motivation	Plan ahead. Make physical activity a regular part of your daily or weekly schedule and write it on your calendar. Invite a friend to exercise with you on a regular basis. Then both of you write it on your calendars. Join an exercise group or class.
Fear of injury	Learn how to warm up and cool down to prevent injury. Learn how to exercise appropriately considering your age, fitness level, skill level, and health status. Choose activities that involve minimum risk.
Lack of skill	Select activities that require no new skills, such as walking, climbing stairs, or jogging. Exercise with friends who are at the same skill level as you are. Find a friend who is willing to teach you some new skills. Take a class to develop new skills.
Lack of resources	Select activities that require minimal facilities or equipment, such as walking, jogging, jumping rope, or calisthenics. Identify inexpensive, convenient resources available in your community (community education programs, park and recreation programs, worksite programs, and so on).
Weather conditions	Develop a set of regular activities that are always available regardless of weather (e.g., indoor cycling, aerobic dance, indoor swimming, calisthenics, stair climbing, rope skipping, mall walking, dancing, gymnasium games). Look at outdoor activities that depend on weather conditions (e.g., cross-country skiing, outdoor swimming, outdoor tennis) as bonuses—extra activities possible when weather and circumstances permit.
Travel	Put a jump rope in your suitcase. Walk the halls and climb the stairs in hotels. Stay in places with swimming pools or exercise facilities. Join the YMCA or YWCA (ask about reciprocal membership agreements). Visit the local shopping mall and walk for 30 min or more. Take a portable audio player and listen to your favorite upbeat music as you exercise.
Family obligations	Trade babysitting time with a friend, neighbor, or family member who also has small children. Exercise with the kids—go for a walk together, play tag or other running games, or get aerobic dance or exercise music for kids (several are on the market) and exercise together. You can spend time together and still get your exercise. Hire a babysitter and look at the cost as a worthwhile investment in your physical and mental health. Jump rope, do calisthenics, ride a stationary bicycle, or use other home gymnasium equipment while the kids are busy playing or sleeping. Try to exercise when the kids are not around (e.g., during school hours or their nap time). Encourage exercise facilities to provide child care services.
Retirement years	Look at your retirement as an opportunity to become more active instead of less. Spend more time gardening, walking the dog, and playing with your grandchildren. Children with short legs and grandparents with slower gaits are often great walking partners. Learn a new skill that you've always been interested in, such as ballroom dancing, square dancing, or swimming. Now that you have the time, make regular physical activity a part of every day. Go for a walk every morning or every evening before dinner. Treat yourself to an exercise bicycle, and ride every day while reading a favorite book or magazine.

Reprinted from Center for Disease Control and Prevention, *Promoting Physical Activity: A Guide for Community Action,* 2nd ed., edited by D.R. Brown, G.W. Heath, and S. Levin Martin (Champaign, IL: Human Kinetics, 2010).

Table 2.5   Summary of Recommended Physical Activity Interventions—Guide to Community Preventive Services

Intervention	Task force finding
BEHAVIORAL AND SOCIAL APPROACHES	
College-based physical education and health education	Insufficient evidence February 2001
Family-based interventions	Recommended October 2016
Enhanced school-based physical education	Recommended December 2013
Individually adapted health behavior change programs	Recommended February 2001
Interventions including activity monitors for adults who are overweight or obese	Recommended August 2017
Social support interventions in community settings	Recommended February 2001
CAMPAIGNS AND INFORMATIONAL APPROACHES	
Community-wide campaigns	Recommended February 2001
Classroom-based health education focused on providing information	Insufficient evidence October 2000
Stand-alone mass media campaigns	Insufficient evidence March 2010
ENVIRONMENTAL AND POLICY APPROACHES	
Built environment approaches combining transportation system interventions with land use and environmental design	Recommended December 2016
Creation of or enhanced access to places for physical activity combined with informational outreach activities	Recommended May 2001
Point-of-decision prompts to encourage use of stairs	Recommended June 2005

Reprinted from Center for Disease Control and Prevention, *Guide to Community Preventive Services.* http://www.thecommunityguide.org/pa/pa.pdf

(frequency, intensity, duration) of physical activity for specific health and fitness outcomes, as well as new platforms for launching physical activity interventions, such as smartphone applications and electronic medical records, are now being explored. As new information becomes available, it must be introduced to the participant via the most effective behavioral paradigms, such as the models discussed in this chapter. Moreover, positive changes in the participant's physical and social environments must occur to enhance compliance with exercise prescriptions. In turn, increased levels of physical activity among all people will improve health and function.

Online Materials

Visit HK*Propel* to access a link to the references, the case studies with discussion questions, and a quiz to help you review key concepts and test your understanding of the material covered in this chapter.

CHAPTER 3

General Principles of Pharmacology

Steven J. Keteyian, PhD

The material in this chapter assumes that many clinical exercise physiologists currently practicing in the field or students interested in becoming a clinical exercise physiologist did not complete a semester-long course on drug therapeutics or pharmacology and exercise. Instead, most people in the field likely received their pharmacology training through several lectures given as part of a broader course on exercise testing or prescription, with additional knowledge gathered during a clinical internship experience, self-study, or on-the-job training.

Obviously, clinical exercise physiologists are not involved in the prescription of medications, but we do work in environments where many, if not all, of the patients we help care for are taking prescribed drugs. That said, a robust presentation on the topic of exercise and medications is beyond the scope of this book; therefore, this chapter instead focuses on essential principles and information pertaining to drug therapy in general. The disease-specific chapters in this textbook provide information about which drugs are used to treat patients with specific clinical conditions, as well as how these agents might interact with exercise testing or training responses. In fact, most of the chapters include a summary table devoted to the pharmacology pertinent to the disease under discussion.

GENERAL PROPERTIES OF DRUGS

Drugs have two names: generic and brand. The generic name identifies the drug no matter who makes or manufactures it. The brand name refers to the name given by the company that manufactures the drug. For example, *carvedilol* is the generic name for a drug that is manufactured by a company and sold under the brand name *Coreg*. When used in a living organism, a drug is a chemical compound that yields a biologic response. By themselves, drugs do not confer new functions on tissues or an organ; instead, they only modify existing functions. This means that a drug attenuates, accentuates, or replaces a response or function (14). For example, a β-adrenergic blocking agent (i.e., a so-called β-blocker) *attenuates* the magnitude of the increase in heart rate during exercise. Conversely, glipizide is used to treat patients with type 2 diabetes because it *accentuates* insulin release from secretory granules by binding to the plasma membrane of beta cells in the pancreas. Finally, a 20 yr old patient with insulin-dependent diabetes is prescribed insulin to *replace* what is no longer produced by their own pancreas.

Another general principle about the effect of drugs is that none exert just a single effect. Instead, almost all drugs result in more than one response; some of these might be well tolerated by a patient whereas others might represent an undesirable response (i.e., side effect).

ROUTES OF ADMINISTRATION

Before discussing the effects of drugs and how they are "moved" through the body, it is important to address the various pathways through which a drug can enter the body. Deciding which route of administration works best for each patient is necessary to ensure that the drug is suitable, well tolerated, effective, and safe. Keep in mind that drugs are usually given for either local or systemic effects. Any significant loss of drug from an intended local site would be considered a negative for locally administered agents, whereas absorption into circulation is a must for agents administered to provide a more generalized or systemic effect.

With respect to drug administration, the two main routes are **enteral** and **parenteral** (see also table 3.1). Certain agents can also be effectively and safely delivered as an inhaled gas or fine mist through the pulmonary system or topically through the skin, eyes, ears, or mucosal membranes of the nasal passages.

Table 3.1 Routes of Drug Administration

Route of administration	Example	Advantages
Oral (PO)	Ibuprofen	Easiest way to administer Reversible
Sublingual (SL) or buccal	Testosterone	Rapid absorption Avoids first-pass metabolism
Transdermal	Nicotine	Easy to administer Avoids first-pass metabolism
Rectal	Acetaminophen	Alternative route when GI system is upset (e.g., vomiting)
Intravenous (IV)	Morphine	Rapid absorption More precise control of drug levels Avoids first-pass metabolism
Intramuscular (IM)	Influenza vaccine	More precise control of smaller amounts of drugs Avoids first-pass metabolism
Subcutaneous (SC)	Insulin	Allows slower release of drug Avoids first-pass metabolism

Enteral

Although rectal absorption is an effective method for drug delivery in certain instances (e.g., patient vomits when given an oral agent), the majority (up to 80%) of all medications prescribed today are delivered enterally through oral ingestion; such a route is safe, economical, and convenient. Once placed in the mouth, drugs can be absorbed through the oral mucosa, in the stomach (gastric absorption), or in the small intestine. Although the mouth has a good blood supply and is lined with epithelial cells that make up the oral mucosa, few agents can be sufficiently disintegrated and dissolved to the extent that the desired systemic response is achieved. One commonly prescribed agent that is rapidly broken down and absorbed through the oral mucosa is sublingual nitroglycerin. When it is placed under the tongue and the patient refrains from swallowing the saliva that contains the dissolved drug, this agent can act within minutes on the angina pectoris that patients with coronary artery disease often experience. Although the stomach has the potential to be an excellent site for the absorption of orally ingested drugs, the small intestine is by far more responsible for this task. The small intestine has the large surface area (i.e., villi), the vascular blood supply, and the pH to facilitate absorption.

Parenteral

Parenteral means delivered by injection. Parenteral routes are many, but the three predominant routes are subcutaneous, intravenous, and intramuscular. Obviously, with the intravenous route, 100% of the drug is delivered into circulation, allowing for a higher concentration of the drug to be the most quickly delivered to target tissues. A few other parenteral routes are intrathecal (directly into the cerebrospinal fluid), epidural, and intra-articular (directly into a joint).

PHASES OF DRUG EFFECT

To produce the desired effect, the molecules in an administered drug must move from the point at which they enter the body to the target tissue. The magnitude of the effect it then renders is heavily influenced by the concentration of these molecules at the target site. For all of this to occur, we usually look at the administered drug passing through three distinct phases, the pharmaceutical phase, the **pharmacokinetic phase**, and the **pharmacodynamic phase** (5, 10, 14).

Pharmaceutical Phase

The pharmaceutical phase pertains to how a drug progresses from the state in which it was administered (e.g., solid), through the stages in which it is disintegrated and then dissolved in solution (i.e., dissolution). Obviously, the pharmaceutical phase applies only to drugs taken orally, because drugs taken subcutaneously, intramuscularly, or intravenously are already in solution. A drug taken orally in liquid form (vs. solid form) is more quickly dissolved in gastric fluid and ready for absorption through the small intestine of the gastrointestinal (GI) tract. Some common factors can accentuate or interfere with the dissolution process for certain drugs that are taken in solid (e.g., pill) form; these include the enteric coating of a tablet, crushing of the tablet, and the presence of food in the GI tract.

Pharmacokinetic Phase

The pharmacokinetic phase pertains to the effect of the body on the drug. The movement of the drug's molecules through the body comprises four discrete subphases: absorption, distribution, metabolism, and excretion (figure 3.1). The absorption of dissolved drugs consumed orally from the GI tract to the blood occurs via either passive absorption (i.e., diffusion; no energy required), active absorption (i.e., involves a carrier such as a protein; requires energy), or pinocytosis. The latter refers to cells engulfing a drug to move it across a cell membrane. Factors that affect absorption include the pH of the drug, local blood flow to the GI tract (which can be negatively influenced during exercises that involve the skeletal muscles of the limbs), hunger, and food content in the GI tract. Certainly, there are drugs given transdermally or intramuscularly, or via eye drops, nasal sprays, and respiratory inhalants, that are also absorbed but without the involvement of the GI system. Because of the number of surrounding blood vessels, drugs administered intramuscularly are taken up more readily than those given subcutaneously.

Bioavailability refers to the percentage of a drug that makes it all the way into the bloodstream. The bioavailability of drugs given intravenously is obviously 100%, whereas in agents taken orally the bioavailability is almost always less than 100% because it is first transported from the GI tract and then to the liver via portal circulation (so-called first-pass metabolism). For example, when swallowed orally, 90% of the drug nitroglycerin is metabolized in the liver before it even enters systemic circulation. Conversely, agents administered parenterally (e.g., subcutaneously, intramuscularly, intravenously) avoid first-pass metabolism; therefore, more of the drug is available to the tissues during its first time through systemic circulation.

The process of moving a now-absorbed drug to the target tissues is influenced by the concentration of the drug in the body, the flow of blood into the target tissue, and the percentage of the drug bound to protein in the plasma. Most drugs are bound to protein (usually albumin) at varying levels (high, moderately high, moderate, low) and while in this bound state are inactive with respect to causing a pharmacologic response. The portion of the drug not bound to protein, or active and able to exert its effect, is replenished from the bound portion as the concentration of unbound drug decreases in circulation.

The biotransformation, or metabolism, of a drug occurs mainly in the liver, within which most drugs are both rendered inactive by the enzymes in this organ and readied for excretion. In some instances, the enzymatic actions that occur on a drug in the liver yield an active metabolite that accentuates the pharmacologic response.

With respect to metabolism, the **half-life (t-1/2)** of a drug refers to the time it takes for one half of the drug concentration to be eliminated. For example, 500 mg of drug A is absorbed, and it has a known half-life of 3 h. That means it takes 3 h to eliminate 250 mg, another 3 h to eliminate another 125 mg, and so on. Knowledge of a drug's half-life helps us compute how long it will take a drug to reach a stable concentration in the blood. Drugs with a half-life less than 8 h or more than 24 h are considered to have short or long half-lives, respectively.

The fourth and final subcomponent within the pharmacokinetic phase pertains to elimination, the main avenue of which is through the kidneys as urine; however, other routes such as bile, feces, saliva, and sweat contribute to drug elimination as well. All drugs that are free in plasma (not bound to protein) are filtered through the kidneys. Factors that can affect the ability of the kidneys to filter (and therefore eliminate) a drug include any disease that decreases glomerular filtration rate, any disorder that hampers renal tubular secretion, or an overall change in blood flow to the kidneys themselves.

It is important to also consider how exercise might alter the body's actions on a drug. Specifically, exercise can influence the pharmacokinetics of many drugs, especially when performed for a long time. Drugs commonly affected include insulin, those with a narrow therapeutic range (e.g., digoxin, theophylline, and warfarin), and those administered via a transdermal delivery system (12).

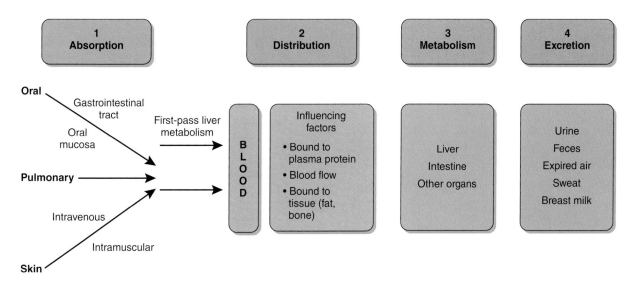

Figure 3.1   The four subphases of pharmacokinetics (the effect of the body on the drug).

Exercise and Drug Absorption

Blood redistribution during exercise is important because the proportion of blood flow to the metabolically more active skeletal muscles shifts from approximately 20% of resting cardiac output to 70% or more, depending on the intensity of exercise (12). This can result in a shunting of blood away from important drug absorption sites in the gastrointestinal system. Exercise intensities at 70% peak oxygen uptake ($\dot{V}O_2$peak) or greater result in decreased gastric emptying; thus, the absorption of oral drugs may be slowed. Also, since the majority of oral drugs are absorbed through the small intestines, decreases in gut transit time (due to blood redistribution during exercise) may, in theory, be a source of exercise–drug interactions, although evidence for this is still lacking.

In addition to redistributing cardiac output (blood flow) to the metabolically more active skeletal muscles, exercise also results in blood flow being redistributed to the skin, to assist with maintenance of body temperature. This is especially important for individuals with transdermal medications, such as the nicotine patch, because increased blood flow to the skin allows for quicker drug absorption. In fact, several studies have shown that exercise increases the amount of drug available in plasma when administered by transdermal methods by as much as twofold when compared with rest (12). This increased transdermal absorption associated with exercise has been shown to increase drug-related side effects (e.g., nausea and palpitations) for those using a nicotine patch (12).

Another common example of how exercise can interact with drug administration, leading to potential adverse reactions, is seen with insulin. Insulin is administered subcutaneously, often in the thigh or abdomen. Several studies have shown that injecting insulin into the thigh prior to exercise on a stationary cycle increased the absorption of insulin when compared with subcutaneous administration at an abdominal site (12). Thus, along with the ability of exercise to translocate GLUT 4 receptors independent of insulin, it can also increase exogenous insulin levels in the body through an increased rate of absorption. Interestingly, in a study that investigated the long-acting insulin glargine, which is known to exert its effects for over 20 h in the body, exercise was shown not to affect insulin absorption (21).

Exercise and Drug Metabolism and Distribution

Since the liver is the main organ responsible for the metabolism of most drugs, reductions in hepatic blood flow during exercise might slow down drug metabolism and elimination, resulting in greater concentrations in the blood (20). Van Baak et al. showed that during a bout of exhaustive exercise at 70% of $\dot{V}O_2$peak, concentrations of the β-blocker propranolol temporarily increased before returning to similar levels of a resting control group (27). Because propranolol has a high hepatic extraction rate, it is possible that the increased concentration was the result of reduced hepatic blood flow. Conversely, regular exercise training may have the opposite effect on drug metabolism, since some studies have found a greater hepatic blood flow and increased liver enzymes associated with habitual exercise (27).

One well-known exercise–drug interaction involves the medication warfarin. Warfarin (common brand name Coumadin) is an anticoagulant used to prevent the formation of a thrombosis. Warfarin has a strong affinity for the plasma protein albumin, and while bonded to albumin, it is essentially inactive. Exercise has been shown to increase albumin synthesis by 51% for up to 22 h after exercise (15). Thus, more plasma albumin is available to bind to warfarin, consequently possibly decreasing its effectiveness and increasing the amount of drug that must be given. In fact, studies have confirmed an association between increased levels of physical activity and higher required doses of warfarin (12, 24).

Exercise and Drug Elimination

The primary method of drug elimination is renal excretion. Therefore, it stands to reason that any effect exercise has on the kidneys will in all likelihood affect drug elimination. Once again, this points to the effects of exercise on blood redistribution, since acute exercise has been shown to reduce renal blood flow by up to 50% (12). One common medication that has been studied during exercise is atenolol (12, 13). Atenolol is a β-blocker that is mostly unchanged in the body and is highly reliant on renal elimination. During exercise, renal elimination of atenolol has been shown to decrease, leading to subsequent increases of plasma concentrations (13, 25). However, these effects seem to be transient and return to baseline shortly after an exercise bout. Because of this temporary effect on the kidneys, any impact exercise may have on atenolol and other drugs that are dependent on renal elimination may pertain only to exercise bouts of longer duration.

Pharmacodynamic Phase

The third phase pertains to the effect of drug molecules on the body, known as the science of pharmacodynamics. Such effects are usually categorized as primary or secondary, with the former referring to the planned or therapeutic effect and the latter to either a desired or an unwanted effect (side effect). Again using β-blockers as an example, in patients sustaining a myocardial infarction, these agents are *primarily* prescribed because they decrease one's future risk for death. Coincidently, these same agents also lessen the frequency and severity of migraine headaches, which could easily be a welcomed *secondary effect* for the patient with heart disease who suffers from migraine headaches.

The association between the amount or dose of a drug and the body's response is referred to as the *dose–response relationship*. Although no two patients yield the exact same response to the same dose of a drug, because of a variety of factors discussed in more detail later in this chapter, in general, less of a drug yields a lesser response and more of a drug (up to a point) yields a greater response. The amount of any drug that is associated with its greatest response (i.e., beyond which no further increase in drug yields an observed increase in response) is called the drug's maximum drug effect or maximal efficacy. For example, two agents might be used to treat the pruritis (itching) and the disrupted skin associated with eczema. The first agent might be an over-the-counter product with a maximal drug effect that helps relieve the itching but has little effect on the disrupted skin. The maximal drug effect of the second medication not only relieves itching but also heals the broken or disrupted skin.

MECHANISM OF ACTION

The means by which a drug produces an alteration in function at the target cells is referred to as its *mechanism of action*, with most drugs either (a) working through a protein receptor found on the cell membrane, (b) influencing the effect of an enzyme, or (c) involving some other nonspecific interaction. The drug–protein receptor mechanism is based on the premise that certain drugs have a selective high affinity for an active or particular portion of the cell. This portion of the cell is referred to as a receptor and once stimulated produces a biologic effect. Simply, the receptor represents the "keyhole" that a unique key (i.e., the drug) can fit into and yield a specific response (i.e., open the door). Those keys that best fit the keyhole yield the greatest response, whereas any other keys that may also fit the keyhole, but just not quite as well as the best-fitting key, yield a lesser response.

With respect to the drug–receptor interaction, a drug that combines with the receptor and leads to physiologic changes or responses is referred to as an **agonist**. Conversely, an agent or drug that interferes with or counteracts the actions of an agonist is called an **antagonist**. A *competitive antagonist* is an agent that inhibits or interferes with an agonist by binding to the exact same receptor the agonist is targeting, thus preventing its action. For example, β-blockers slow heart rate and lessen inotropicity by blocking the β-1 receptors found on cardiac cells—cells that when normally stimulated by the agonist norepinephrine lead to increases in both heart rate (positive chronotropic effect) and force of contraction (positive inotropic effect) (18).

It is also important to know that many agonists and antagonists lack specific effects, in that the response produced by the receptor they influence varies based on the location of the receptor in the body. For example, in the salivary glands, stimulation of the cholinergic receptors with the agonist acetylcholine leads to an increase in saliva production, which is the correct and needed response for someone preparing to chew food. However, stimulation of another cholinergic receptor located elsewhere in the body, again by acetylcholine, results in different responses such as pupillary constriction in the eye or a slowing of heart rate.

As already mentioned, a second type of mechanism of action involves influencing the interaction between an enzyme and a drug. Since enzymes work as catalysts that control biochemical reactions, inhibition of the enzymes via their binding with certain drugs can inhibit subsequent catalyzed chemical reactions. For example, among patients with an elevated blood concentration

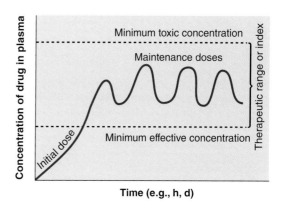

Figure 3.2   The relationship between the regular administration of a drug over time and the maintenance of a plasma concentration that yields the desired therapeutic effect.

of low-density lipoprotein (LDL) cholesterol, a class of drugs commonly referred to as statins are used because they inhibit the normal action of a rate-controlling enzyme in the liver. This enzyme normally facilitates the production of cholesterol by the liver, so the inhibition of this process leads to reductions in both total cholesterol and LDL cholesterol levels in the blood.

Before leaving our discussion about the pharmacodynamic phase of drugs, there is one more term you should become familiar with: therapeutic index (TI). The **therapeutic index** helps describe the relative safety of a drug, representing the ratio between toxic dose$_{50}$ (TD$_{50}$) and effective dose$_{50}$ (ED$_{50}$), where TD$_{50}$ equals the dose of a drug that is toxic in 50% of humans tested and ED$_{50}$ is the dose of a drug needed to be therapeutically effective in 50% of a like population of humans. The closer the TI for a drug is to 1, the greater the danger of toxicity; the higher the TI is above 1, the wider the margin of safety and the less likely the agent will result in harmful or toxic effects. For the drugs with a higher TI, plasma drug levels do not need to be monitored as frequently. Figure 3.2 shows how the initial dose and subsequent doses maintain drug concentration within the therapeutic range.

PHARMACOTHERAPY

The science of prescribing or individualizing drug therapy based on patient-specific characteristics is referred to as *drug therapeutics* or *pharmacotherapy*. Just as with developing a safe exercise prescription, the prescription of a medication by a licensed professional involves choosing the correct mode of administration (e.g., oral, subcutaneous), intensity (i.e., dose), frequency of administration per day or per week, and duration or time (e.g., 3 d, 10 d, permanently). To accomplish this, "orders" are written to communicate the prescribed therapy plan to the patient, the person (e.g., pharmacist) filling the prescription, and other health care professionals. Table 3.2 shows a brief list of common terms and abbreviations used in health care. Since clinical exercise physiologists work in

Table 3.2 Common Terms and Abbreviations Pertinent to Drug Measurements, Delivery, and Times of Administration

DRUG MEASUREMENTS AND DRUG FORMS		ROUTES OF DRUG ADMINISTRATION		TIMES OF ADMINISTRATION	
Abbreviation	Meaning	Abbreviation	Meaning	Abbreviation	Meaning
cap	capsule	A.D., ad	right ear	AC, ac	before meals
dr	dram	A.S., as	left ear	ad lib	as desired
elix.	elixir	A.U., au	both ears	BID, b.i.d.	twice daily
g, gm, G, GM	gram	ID	intradermal	c̄	with
gr	grain	IM	intramuscular	hs	hour of sleep, at bedtime
gtt(s)	drops	IV	intravenous	NPO	nothing by mouth
kg	kilogram	IVPB	intravenous piggyback	PC, pc	after meals
l, L	liter	KVO	keep vein open	PRN, p.r.n.	whenever necessary, as needed
m²	square meter	L	left	Q, q	every
mcg, μg	microgram	NGT	nasogastric tube	Qam, qAM	every morning
mEq	milliequivalent	O.D., od	right eye	Qh, qh	every hour
mg	milligram	O.S., os	left eye	q2h	every 2 hours
mL, ml	milliliter	O.U., ou	both eyes	q4h	every 4 hours
oz	ounce	PO, po, os	by mouth	q6h	every 6 hours
pt	pint	R	right	q8h	every 8 hours
qt	quart	SC, subc, sc, SQ, subQ	subcutaneous	s̄	without
SR	sustained release	SL, sl, subl	sublingual	stat	immediately
supp	suppository	TKO	to keep open	TID, t.i.d.	three times a day
susp	suspension				
T.O.	telephone order				
T, tbsp	tablespoon				
t, tsp	teaspoon				
V.O.	verbal order				

settings where many, if not all, of the patients are under the care of a physician and taking prescribed medications, it is important that you familiarize yourself with this language in order to effectively perform your patient care duties.

The remainder of this chapter provides other important medication-related information that the clinical exercise physiologist needs to be familiar with. Although exercise physiologists play no role in determining what medications are prescribed, how often they are to be taken, or in what form they are to be taken, we may be called on to administer a medication, such as sublingual nitroglycerin, if the agent is part of a program-related standing order approved by a physician. Additionally, we often care for patients over many visits and, as a result, are in a unique position to help ensure that patients are taking their medication as prescribed (see practical application 3.1) and to watch for factors that might influence an agent's efficacy.

Factors That Modify Drug Response or Activity

As one might assume, a host of factors can influence an individual's response to a given drug. Some of these factors are easily recognized, well understood, and predictable, whereas others are less obvious. Here are several of the more common influential factors or characteristics.

Drug–Exercise Pharmacodynamics

When discussing the relationship between drugs and exercise, it is important to consider the pharmacodynamic effects of drugs (i.e., how a drug affects the exercise response). Specifically, while some drugs are taken because they are purported to enhance the exercise response or performance, other drug–exercise interactions can impair exercise performance or blunt the training adaptations. A commonly discussed example of a possible drug–exercise interaction involves β-adrenergic blocking agents and their effect of attenuating heart rate and contractility. As a result, exercise capacity, as measured by peak $\dot{V}O_2$ (where $\dot{V}O_2$ = stroke volume × heart rate × [arterio-mixed venous oxygen difference]), may be affected (26). Interestingly, individuals taking a β-blocker often show a compensatory augmentation of stroke volume during exercise, and while peak $\dot{V}O_2$ may still be the same or slightly reduced (compared with no β-blocker), this increase in stroke volume negates some of the negative chronotropic effects of the drug (26).

Additionally, specific drug side effects can be exacerbated by exercise because of parallel physiological pathways leading to similar outcomes. Examples include antihypertensive agents and several of the drugs used to lower blood glucose levels in patients with diabetes. Exercise can concomitantly lower blood pressure and blood glucose during and after exercise, which can cause adverse responses.

Practical Application 3.1

CLIENT–CLINICIAN INTERACTION

The use of medicine to treat illness is effective and highly complex. Concerns about the polypharmacy experienced by many patients today, complex physician–staff workflows, and lack of patient knowledge about their medications (4) all contribute to increasing the potential risk of harm to patients. To help combat such potential harm, a formal process referred to as *medical reconciliation* compares the medications a patient states they are taking against those in the patient's record or medication orders (4).

Medical reconciliation should be carried out at each encounter between a patient and a health care provider and is solely intended to avoid errors (e.g., omissions, interactions) or conflicts and prevent unwanted medical problems. To that end, clinical exercise physiologists who help care for a patient in any setting (ambulatory clinic, physician's office, cardiac rehabilitation program, medical fitness center) are in an excellent position to advance medication optimization through this process of reconciliation, with the goal of ensuring that all medications are being continued, discontinued, or modified as prescribed (6).

The reason medical reconciliation is so important is that in the United States alone, medication errors and adverse drug events represent a major source of preventable illness and adverse events. Adverse drug events account for nearly 700,000 emergency department visits and 100,000 hospitalizations annually (19). Many of these are due to omitting a medication, taking a drug at the wrong dose or frequency, or taking the wrong drug. The process of medical reconciliation is effective in reducing emergency room visits and adverse drug events.

Although reconciling with patients the medications they are taking against what has been prescribed might seem tedious, it helps to reduce errors and improve documentation (7, 22, 23). To begin the medical reconciliation process, you need a current medication list that should include all drugs being taken at home, regardless of whether they are prescribed or over-the-counter medications (including herbal and nutritional supplements). Also, gather information about dose, route, and frequency; time of last dose; who prescribed or recommended the agent (self, pharmacist, or physician); and how the patient is complying. Don't hesitate to ask patients to bring in all the medications they are currently taking to help ensure that their list reflects actual actions.

There is no perfect list, but given this information, clinicians can go on to the next step in the reconciliation process, which is comparing the information they are provided by the patient against documents in the patient's medical records. Source documents to seek out to obtain the correct information on prescriptions might include hospital discharge summaries, nursing notes, pharmacy records, and notes from a patient's last clinic encounter with a physician or physician extender.

In this process, take the initiative to look beyond the obvious. For example, it may be that a physician has switched a patient's blood pressure–lowering medication because another drug is less expensive, but the patient forgot or was confused and did not stop taking the initial drug that is to be replaced. Any "red flag" concerns like this should be identified for the patient and discussed with the doctor, then discussed with the patient to correct any errors.

- *Age:* Younger children and older adults often demonstrate a greater response to many medications than do postpubescent youth through adults aged 65 yr. As a result, the doses may have to be modified. As well, older adults may be taking multiple agents, and various drug interactions can affect a drug's efficacy or lead to unwanted side effects.

- *Body mass:* Adjusting the dosage of a drug to match a patient's body mass or body surface area helps ensure that the concentration of drug that reaches specific body tissues is not higher or lower than desired.

- *Sex:* Women are generally smaller than men, and the dose may have to be adjusted for this difference in body mass. Other sex-specific reasons that might lead a physician to adjust the dose in women include the states of pregnancy and lactation and the generally greater amount of essential body fat in women.

- *Other factors:* Other factors that can also affect a drug's actions include time of administration, severity of coexisting diseases (e.g., diseases of the liver or kidney), genetics, patient mood or mental state (e.g., placebo effect), and surrounding environment (e.g., ambient temperature, altitude). Concerning time of day, this can be tied to the absence or presence of food, which must be considered if the absorption of a drug in the GI tract is or is not hampered by the presence of food.

Compliance

There are over 10,000 prescription medications available today. Nearly 60% of adults take at least one daily medication (29), and approximately 30% take five or more (19). What happens when a patient stops taking their medication too soon, does not go to the pharmacy and get their prescription filled because they can't afford the co-payment, is taking the wrong medication, or does not take the medications as prescribed? Are there factors or characteristics we can recognize beforehand that would allow us to identify a patient at risk for such noncompliance? These are important questions that not only can affect the health of the patient but also are associated with great public health and economic burdens.

The obvious main concern is that noncompliance may not allow a patient to get better or will worsen a patient's disease state. This can translate to increased morbidity, treatment failure, more frequent physician or emergency room visits, and even premature death. Noncompliance contributes to almost $300 billion of avoidable health care costs annually, comprising up to 10% of the total health care costs in the United States (29). Noncompliance can be as high as 50% for some medications (3, 28). The economic impact of noncompliance also includes increased absenteeism from work, lost productivity at work, and lost revenues to pharmacies and pharmaceutical manufacturers.

Factors that contribute to or help predict noncompliance in the future include prior compliance behaviors, one's ability to integrate the complexity of a drug regimen into daily life, underlying health beliefs, health literacy, and social support. Patients at very high risk for noncompliance include those who are asymptomatic (e.g., hypertension), are suffering from a chronic versus an acute condition, are cognitively impaired (e.g., dementia), are taking frequent doses per day (vs. once a day), are taking multiple medications, and are experiencing drug-related side effects (2, 9, 29).

Health literacy is an important issue that deserves special discussion and pertains to how well an individual is able to obtain, process, and understand basic health information and do so in a manner that allows them to make appropriate health decisions and comply with health-related directions (e.g., restrict physical activity, take medications) (8, 17). Health literacy is related to one's age, physical disabilities, psychological or emotional state, culture, and ability to read, write, and solve problems. Individuals who have limited health literacy are more likely to have chronic diseases, miss appointments, and suffer poorer overall care. As a clinician you can help these patients if you keep technical phrases and information simple by using "living room" language, use multiple teaching modes (e.g., pictures and demonstrations), and assess comprehension by having patients "teach back" what they learned.

Strategies shown to improve compliance include effective counseling and communication between patients and clinicians and between patients and pharmacists, special labels and written materials, self-monitoring, and follow-up (16). The RIM (recognize, identify, manage) technique is a structured tool to facilitate patient compliance.

R = *Recognize*: Use both subjective (i.e., ask the patient) and objective (pill counts) evidence to document whether compliance is, in fact, an issue.

I = *Identify*: Determine the causes of noncompliance using supportive, probing questions and empathetic statements. Responses pertaining to noncompliance may range from side effects and lack of financial resources to doubts about a drug's effectiveness.

M = *Manage*: Use counseling, contracting, and multiple options for teaching (devices, telephones, calendars, daily pill boxes) and communicating to establish effective short-term and intermediate-term plans aimed at improving compliance. Monitor, follow up, and adjust strategies as indicated.

Almost all medications taken orally by patients require some period of time before they exert their desired action in the body. Because of this delay, medications that might influence the exercise response may need to be timed such that they appropriately affect or do not affect the exercise response. For example, in patients with heart disease or heart failure, β-adrenergic blocking agents such as carvedilol should be taken at least 2.5 h before exercise training or an exercise test that is performed for the purpose of developing or revising an exercise prescription. This will help ensure that the medication is sufficiently absorbed and exerting its heart rate–controlling effect. Additionally, some medications are best taken before sleep because any associated mild side effects are better tolerated when the person is less active, inactive, or asleep. Table 3.3 provides a list of common medications that may affect exercise capacity or heart rate or blood pressure response to exercise (1).

Table 3.3 Common Medications and a Summary of Their Effect on Heart Rate and Blood Pressure During Exercise and on Exercise Capacity

Organ system, drug class, or disease state	Heart rate response	Blood pressure response	Exercise capacity	Comments
ANTIARRHYTHMIC AGENTS				
Amiodarone (Cordarone) Sotalol (Betapace)	↓	↔	↔	
β-ADRENERGIC BLOCKING AGENTS				
Atenolol (Tenormin) Metoprolol (Lopressor SR, Toprol XL) Carvedilol (Coreg)	↓↓	↓	↑ in patients with angina ↓ or ↔ in patients without angina	Carvedilol has both α- and β-adrenergic blocking properties
BRONCHODILATORS				
Albuterol (Proventil, Ventolin) Ipratropium (Atrovent)	↑ ↔	↑ ↔	↔	
CALCIUM CHANNEL BLOCKERS				
Diltiazem (Cardizem CD) Verapamil (Calan, Isoptin) Amlodipine (Norvasc) Nifedipine (Adalat, Plendil, Procardia XL)	↓ ↔	↓	↑ in patients with angina ↔ in patients without angina	Negative chronotropic effect more pronounced with diltiazem and verapamil
CARDIAC GLYCOSIDES				
Digoxin (Lanoxin)	↓ ↔	↔	Possible ↑ in patients with atrial fibrillation or heart failure	Negative chronotropic effect more likely in patients with atrial fibrillation or heart failure
Nicotine	↑ ↔	↑	Possible ↓ in patients with angina	
NITRATES				
Isosorbide mononitrate (ISMO, Imdur, Monoket) Isosorbide dinitrate (Isordil)	↑ ↔	↔	↑ in patients with angina ↔ in patients without angina	
PSYCHOTROPICS				
Fluoxetine (Prozac) Sertraline (Zoloft) Paroxetine (Paxil) Citalopram (Celexa) Venlafaxine (Effexor) Bupropion (Wellbutrin)	↑ ↔	↓ ↔	↔	
THYROID AGENTS				
Levothyroxine (Synthroid)	↑ ↔	↑ ↔	↔ but may lead to a ↓ if angina is worsened	
ANGIOTENSIN II RECEPTOR BLOCKERS				
Candesartan (Atacand) Irbesartan (Avapro) Losartan (Cozaar) Valsartan (Diovan)	↔	↔ ↓	↑ in patients with heart failure due to diastolic dysfunction ↑ ↔ in patients with heart failure due to systolic dysfunction	
VASODILATORS				
Angiotensin-converting enzyme inhibitors Captopril (Capoten) Enalapril (Vasotec) Lisinopril (Prinivil, Zestril) β-Adrenergic blocking agents Cardura (Doxazosin) Flomax (Tamsulosin) Minipress (Prazosin) Terazosin (Hytrin)	↔	↔ ↓	↑ ↔ in patients with chronic heart failure	

↑ = increase; ↓ = decrease; ↔ = no change.

Finally, some patients will approach you and ask what they should do if they have missed a dose of their medication: They are unsure whether they should take the missed dose late or simply wait and take the next scheduled dose as planned. Ideally patients should have been informed about how to handle this scenario beforehand, but that does not always occur. One option is for patients to check the literature that came with the prescription, which often includes instructions about what to do if a dose is missed. Another option is to recommend that patients speak with a pharmacist (or their doctor). If neither of these options is available, a general recommendation is for people to take the medication as soon as they realize they missed a dose. However, if the amount of time remaining before the next scheduled dose is less than 50% of the usual time between doses, the patient should not take the missed dose and should instead wait until the next scheduled time. What patients must *not* do is take a double dose for the one they missed; they must not take the dose they missed and the next scheduled dose at the same time.

CONCLUSION

The widespread use of pharmacological agents in our society requires the clinical exercise physiologist to be familiar with many of the essential elements associated with how drugs are administered and handled in the body, as well as factors that affect a drug's action and patient compliance. Since they are likely to encounter patients who are taking pharmacological agents, exercise physiologists should do the extra work needed to learn about agents they are unfamiliar with and assist whenever possible with ensuring that patients comply with their prescribed medication regimens.

Online Materials

Visit HK*Propel* to access a link to the references, the case studies with discussion questions, and a quiz to help you review key concepts and test your understanding of the material covered in this chapter.

CHAPTER 4

General Interview and Examination Skills

Lizbeth R. Brice, MD

Clinical exercise physiologists work in settings that require them to assess patients with various health problems. This chapter focuses on helping clinical exercise physiologists better understand the elements of the clinical evaluation conducted by a physician or physician extender, as well as the measurements they may need to obtain to determine whether the patient can exercise safely.

The clinical evaluation of any patient usually involves several steps. First, a general interview is conducted to obtain historical and current information. A physical assessment or examination follows, the extent of which may vary based on who is conducting the examination and the nature of the patient's symptoms or illness. After these are completed, a brief numerical list is generated to summarize the assessment, relative to both prior and current findings and diagnoses. Finally, a numerical plan is generated to indicate the one, two, or three key actions that are to be taken in the care of the patient. This chapter describes in detail the general interview and physical examination components of the clinical evaluation.

GENERAL INTERVIEW

The general interview is a key step in establishing the patient database, which is the working body of knowledge the patient and the clinical exercise physiologist will share throughout the course of treatment. This database is primarily built from information obtained from the patient's hospital or clinical records. But as the clinical exercise physiologist, you will also need to interview the patient to obtain information that is missing, as well as update data to address any changes in the patient's clinical status since the last clinic visit of record. With experience you will learn which informa-

tion is incomplete or necessitates questioning of the patient. A list of the relevant components, some of which you will need to enter into the patient file or database, is shown in the sidebar Essentials of the Clinical Evaluation for the Newly Referred Patient.

Essentials of the Clinical Evaluation for the Newly Referred Patient

- General interview
  - Reasons for referral
  - Demographics (age, sex, ethnicity)
  - History of present illness (HPI)
  - Current medications
  - Allergies
  - Past medical history
  - Family history
  - Social history
  - Review of systems
- Physical examination
- Laboratory data and diagnostic tests
- Assessment
- Plan

Adapted by permission from P.A. McCullough, *Clinical Exercise Physiology* 1, no. 1 (1999): 33-41.

Acknowledgment: Much of the writing in this chapter was adapted from earlier editions of *Clinical Exercise Physiology.* Thus, we wish to gratefully acknowledge the previous efforts of and thank Peter A. McCullough, MD, Steven J. Keteyian, PhD, Quinn R. Pack, MD, and Hayden Riley, MS.

Reason for Referral

The reason for referral for exercise therapy is generally self-explanatory and may include one or more of the following: improve exercise tolerance, improve muscle strength, increase range of motion, or provide a relevant intervention and behavioral strategies to reduce future risk. But the clinical exercise physiologist may need to reconcile the physician's reason for referral and the patient's understanding of the need for therapy. Differences can exist between the two and, if unaddressed, can result in a failure to create a good working relationship. For example, it is not uncommon that patients have a different or limited understanding of the medical condition(s) for which they were referred for exercise training. Additionally some patients may falsely believe they were "cured" by their medical procedure and surmise that they do not need to engage in exercise or rehabilitation or make lifestyle adjustments. For example, when patients are referred for cardiac rehabilitation, they must understand that coronary artery disease is generally a progressive disorder that is strongly influenced by lifestyle habits and medications. Thus, the clinical exercise physiologist plays an important role in promoting long-term compliance to physical activity, hypertension and diabetes management, smoking cessation, proper nutrition, and medical compliance—all key components of secondary prevention.

Demographics

Patient demographics such as age, sex, and ethnicity are the basic building blocks of clinical knowledge. A great deal of medical literature describes the relationship between these types of demographic information and various health conditions. Age is a nonmodifiable independent risk factor associated with increased incidence of many cardiopulmonary disorders, including myocardial infarction (MI), stroke, and chronic obstructive pulmonary disease (19). According to the American Heart Association, a myocardial infarction occurs every 40 s in the United States. The average age of first MI is 65.6 yr for males and 72.0 yr for females (2). Age is also the most important factor in the development of osteoarthritis, with 68% of those over the age of 65 having some clinical or radiographic evidence of the disorder. Because these and other age-related diseases can influence a patient's ability to exercise, age becomes a key factor to consider when developing an exercise prescription.

Sex also relates to outcomes such as behavioral compliance or disease management in patients with chronic diseases. For example, rheumatoid arthritis (RA) is more common in women than men (3:1 ratio) and seems to have an earlier onset in women. Although the onset of cardiovascular disease is, in general, 10 yr later in women than in men, the morbidity and mortality after revascularization procedures (i.e., coronary bypass or coronary stent) are higher in women (13). Finally, exercise capacity, as measured by peak oxygen uptake, decreases progressively in men and women from the third through the eighth decade. Ades and colleagues (1) showed that the rate of decline in men with age ($-0.242\,mL \cdot kg^{-1} \cdot min^{-1}$ per yr) is greater than the rate of decline for women ($-0.116\,mL \cdot kg^{-1} \cdot min^{-1}$ per yr). Of note, less research is available describing the effects of exercise as an intervention for many chronic diseases in women, especially those with cardiopulmonary disease (5). These and other sex-based differences remain an area of intense investigation as clinical scientists strive to determine which biological or socioeconomic factors account for the poorer outcomes sometimes observed in women. When an exercise prescription is developed for women, unique compliance- and disease-related barriers and confounders need to be solved to improve exercise-related outcomes.

A great deal of information is available about differences in health status between ethnic groups. Most of these differences are attributable to socioeconomic status and access to care, but a few differences related to ethnicity are worth mentioning. For example, obesity, hypertension, renal insufficiency, and left ventricular hypertrophy are all more common in African American patients with cardiovascular disease than in their age- and sex-matched Caucasian counterparts. The prevalence of diabetes is higher among American Indians and Alaska Natives (15.1%), non-Hispanic African Americans (12.7%), and people of Hispanic ethnicity (12.1%) than among non-Hispanic Caucasians (7.4%) and Asians (8.0%) (6). The clinical exercise physiologist should consider these issues when developing, implementing, and evaluating an exercise treatment plan as well as deciding on which risk factor(s) to address first.

History of Present Illness

The purpose of this element is to record and convey the primary information that led to the patient's referral to a clinical exercise physiologist. The history begins with the chief complaint, which is typically communicated as one sentence that sums up the patient's comments (see practical application 4.1). The body of the history of present illness is a paragraph or two that summarizes the manifestations of the illness as they pertain to pain, mobility, nervous system dysfunction, or alterations in various other organ

---

OPQRST and A

This is a useful mnemonic to describe the characteristics of any symptom:

O = Onset

P = Provocation and palliation

Q = Quality

R = Region and radiation

S = Severity

T = Timing

A = Associated signs and symptoms

Practical Application 4.1

THE ART OF THE MEDICAL INTERVIEW

The importance of taking a thorough history cannot be overstated. In many instances, the history alone is sufficient to determine the diagnosis and often guides subsequent testing and treatment (12). In an age of advanced diagnostic testing, *listening to and carefully observing* the patient is usually the most important clinical tool for making a diagnosis and communicating your findings. In general, you should interrupt patients only occasionally and only as necessary. In fact, studies show that physicians tend to interrupt patients within 25 s of the start of an interview, and these interruptions lead to patient dissatisfaction and a failure to address all their concerns (14). By asking questions and thoughtfully listening, you will better understand your patients and be better equipped to help them.

Also, the ideal interview occurs in a quiet, secure environment where distractions are limited and you can focus primarily on the patient. Be sure to use language that is understandable and appropriate for the patient, and try to avoid medical jargon. Interview questions should proceed from the general to the specific, such that you start with open-ended questions and proceed to more specific, closed-ended questions. This might mean starting with questions like "What brings you in for exercise therapy today?" and ending with questions like "On a scale of 1 to 10, how severe is your back pain?" If a patient's response is confusing, ask for clarification or try summarizing your understanding of the answer. It is often helpful to say, "Let me see if I understand you correctly . . ." or "OK, so your pain started yesterday and it is worse with exercise. Is that right?" By starting with open-ended questions, moving to direct questions, and then clarifying your patient's answers, you can move from topic to topic through the entire interview and be confident that you are effectively communicating with your patient.

In addition to verbal communication, you should be aware of nonverbal clues from the patient such as eye contact, body posture, facial expression, and mood. Such clues often suggest important additional features of the history or physical examination that should be directly solicited from the patient and then actively addressed. Furthermore, recognizing and empathizing with a patient's emotions about their medical condition can go a long way toward creating a therapeutic relationship of trust and understanding (see practical application 4.2).

---

system functions (e.g., circulatory, pulmonary, musculoskeletal, skin, and gastrointestinal). Important elements of the illness are reviewed such as the date of onset, chronicity of symptoms, types of symptoms, exacerbating or alleviating factors, major interventions, and current disease status. Traditionally, this paragraph describes events in the patient's own words. A practical approach for the clinical exercise physiologist is to incorporate reported symptoms with information from the patient's medical record.

It is important to fully describe each reported symptom by asking about the characteristics of that symptom. Ask about the symptom onset, provocation, palliation, quality, region, radiation, severity, timing, and associated signs and symptoms. An easy way to remember this is the mnemonic OPQRST&A (see the sidebar OPQRST and A). For example, knee pain could be described as follows: started 3 years ago (O); worse at the end of the day and better with ibuprofen (P); sharp, knife-like (Q); right knee, front of knee, radiation to feet (R); 8 out of 10 pain during walking but 4 out of 10 pain at rest (S); worse after recent fall 3 days ago (T); and swollen, hot to the touch, with a "popping sound" with motion (A).

For patients with cardiovascular disease, the features of chest pain should be characterized (table 4.1). Such a description can help in the future application of diagnostic testing when it comes to assessing pretest probability of underlying obstructive coronary artery disease (9). Standard classifications should be used, if possible, such as the Canadian Cardiovascular Society functional class for angina or the New York Heart Association functional class for patients with heart failure (table 4.2) (4). For patients with pain attributable to muscular, orthopedic, or abdominal problems, the important elements of the illness such as chronicity, type of symptoms, and exacerbating or alleviating factors need to be addressed.

Medications and Allergies

A current medication list is an essential part of the clinical evaluation, especially for the practicing clinical exercise physiologist, because certain medications can alter physiological responses during exercise. Often, asking about current medications is an excellent segue into obtaining relevant past medical history. Compare and reconcile (see chapter 3) the medications that patients state they are taking against what they should be taking given their medical condition and what is stated in their official medical record. For example, expected or typical medical therapy for patients with chronic heart failure usually includes a β-adrenergic blocking agent (i.e., β-blocker) and a vasodilator such as an angiotensin-converting enzyme inhibitor. If during your first evaluation of a new patient with chronic heart failure you learn

Table 4.1 Features of Chest Pain

Type of discomfort	Quality	Radiation	Exacerbating and alleviating factors
Typical*	Heaviness, pressure, squeezing, generalized left to midchest	To neck, jaw, back, left arm, less commonly the right arm	Worsened with exertion or relieved with rest or nitroglycerin
Atypical	Sharp, stabbing, pricking, tingling	None	None clearly present; can happen any time
Noncardiac	Discomfort clearly attributable to another cause	Not applicable	Not applicable

*Note the difference between typical stable and typical unstable angina. The difference between these two types of angina pertains to no change in (or stable pattern for) intensity, duration, or frequency of pain in the past 60 d, as well as no change in precipitating factors.

Reprinted by permission from P.A. McCullough, *Clinical Exercise Physiology* 1, no.1 (1999): 33-41.

Practical Application 4.2

CLIENT–CLINICIAN INTERACTION

During any interview or physical examination, clinical exercise physiologists should remember two important tenets:

1. All interactions are confidential, and any information obtained should remain private and protected. A breach of patient confidentiality represents serious misconduct and in some cases could result in termination of your job.

2. When examining patients, maintaining their modesty is always very important; never become casual about this. Remember that some patients have modesty standards that are very different from those you yourself might hold.

Table 4.2 Classification of Cardiovascular Disability

Class I	Class II	Class III	Class IV
NEW YORK HEART ASSOCIATION FUNCTIONAL CLASSIFICATION			
Patients with cardiac disease but no marked physical limitations (e.g., undue fatigue, palpitation, dyspnea, or anginal pain)	Patients with cardiac disease resulting in slight limitation (fatigue, dyspnea, angina) of physical activity	Patients with cardiac disease with marked limitation of physical activity	Patients with cardiac disease who are unable to carry out physical activity without discomfort
	Comfortable at rest	Symptoms such as fatigue, palpitations, and dyspnea occur with less than ordinary physical activity	Symptoms may even be present at rest
			Physical activity worsens symptoms
CANADIAN CARDIOVASCULAR SOCIETY FUNCTIONAL CLASSIFICATION			
Routine daily activity does not cause angina	Slight limitation of ordinary physical activities such as climbing stairs rapidly or exertion after meals, in cold or windy conditions, under emotional stress, or within a few hours after awakening	Marked limitation of routine daily activities such as walking one or two blocks on level ground or climbing more than one flight of stairs in normal conditions	Inability to carry out physical activity without discomfort
Angina occurs with rapid pace or prolonged exertion	No symptoms when walking more than two blocks on level ground or climbing more than one flight of ordinary stairs at a normal pace and in normal conditions		Angina may be present at rest

Adapted by permission from American Alliance of Health, Physical Education, Recreation and Dance, *Physical Education for Lifelong Fitness: The Physical Best Teacher's Guide* (Champaign, IL: Human Kinetics, 2004), 127.

that the patient is not taking one of these agents, you should ask if their doctor has prescribed it in the past. In doing so, you may learn whether the patient has been found to be intolerant to an agent, is nonadherent to a prescribed medication, or has another medical condition that prevents them from taking the expected medication. If no clear reason for nonadherence is identified, you may discover a potentially correctable gap in the quality of the patient's medical care.

When describing the current medication regimen, be sure to include dose, administration route (e.g., oral, inhaled), frequency, and time taken during the day. The latter may be especially important if a medication affects heart rate response at rest or during exercise, because you must allow sufficient time between when the medication is taken and when the patient begins to exercise. Specifically, most β-blockers and antihypertensive medications require up to 2.5 h to be absorbed and exert their maximal therapeutic effects.

A medical allergy history, with a comment on the type of reaction, is also a necessary part of your patient database. If a patient is unaware of any drug allergies, note this as "no known drug allergies (nkda)" in your database. Note as well medicines the patient does not tolerate (e.g., "Patient is intolerant to nitrates, which cause severe headache").

Medical History

This section should contain a concise, relevant list of past medical problems with attention to dates. Be sure your list is complete because orthopedic, muscular, neurological, gastrointestinal, immunological, respiratory, and cardiovascular problems all have the potential to influence the exercise response and the type, progression, duration, and intensity of exercise. For example, for patients with coronary artery disease, the record must include the severity of coronary lesions, types of conduits (e.g., mammary artery versus saphenous vein graft) used if coronary bypass was performed, target vessels (e.g., left anterior descending artery), and the most recent assessment of left ventricular function (i.e., ejection fraction). For patients with cerebral and peripheral vascular disease, the same degree of detail is needed with respect to the arterial beds treated.

Among patients with intrinsic lung disease, attempt to clarify asthma versus chronic obstructive pulmonary disease attributable to cigarette smoking. Such information may help explain why certain medications are used when the patient is symptomatic and why others are part of a patient's long-term, chronic medical regimen. Investigate other organ systems as well. For example, if a diagnostic exercise test is ordered for a patient with intermittent claudication, consider using a dual-action stationary cycle to elicit a higher heart rate response (versus what might be achieved if a treadmill is used). Additionally, knowing about any current or prior musculoskeletal pain is important when you are developing an exercise prescription.

Family History

This element should be restricted to known relevant heritable disorders in first-degree family members (parents, siblings, and offspring). Relevant heritable disorders include certain cancers (e.g., breast), adult-onset diabetes mellitus, familial hypercholesterolemia, sudden death, and premature coronary artery disease defined as new-onset disease before the age of 55 in men or 65 in women. While assessing family history, you may wish to discuss with the patient that first-degree family members should be screened for pertinent risk factors and possible early disease detection, if indicated.

Social History

The social history section brings together information about the patient's lifestyle and living patterns. It is a key component of the interview for the clinical exercise physiologist and deserves extra time and attention. Always inquire about tobacco use. For patients who currently use tobacco (traditional cigarettes, electronic cigarettes, smokeless tobacco), consider strongly encouraging them to quit (if the patient states they are open to engaging in such a conversation). Suggest over-the-counter smoking cessation aids, refer them for specialized treatment, and follow up later to offer additional support. Ask routinely about alcohol, marijuana, cocaine, and other substance use (including inappropriate opioid use). Ask about marital or significant partner status, occupation, transportation, housing, and routine and leisure activities. Because long-term compliance to healthy behaviors is influenced, in part, by conflicts with transportation, work hours, childcare, and family responsibilities, inquire about and discuss these issues. Conclude by making reasonable suggestions to improve long-term compliance.

Inquire about prior physical activity and exercise habits. Be specific and try to quantify the amount, intensity, and type of activities patients engage in, as well as barriers to regular exercise. Attempt to estimate your patient's aerobic capacity (table 4.3) (17). Inquire about preferred forms of exercise, what patients like and dislike about exercise, and exercise facilities they have access to. Build on each patient's prior successes. If there have been problems in the past, help patients understand why they had problems so they will be more successful in the future when trying to establish daily exercise habits. Try to identify any possible roadblocks to improving their exercise habits, such as an occupational constraints job, lack of access to equipment, safety concerns, and family responsibilities. Work with your patients to help them identify and overcome such obstacles.

Ask about nutrition patterns. Find out what kinds of food patients eat on a typical day. When they cook do they follow a specific nutrition plan? Do they eat fast food? Do they eat breakfast? How many servings of fruit and vegetables do they eat each day? How many desserts do they have each day? Do they try to avoid salt? Have they ever been on a diet, and if so, was it success-

Table 4.3   Estimating Aerobic Capacity by History

Can your patient do the following without symptoms?	Estimated metabolic equivalents of task (METs)
Get dressed without stopping	2 METs
Do general housework such as making the bed, washing the car, hanging laundry	3 METs
Push a lawnmower or grocery cart	3-4 METs
General ballroom dancing	5 METs
Carry a 15 lb (7 kg) load of laundry up two flights of stairs	6 METs
Moderate jogging	7 METs
Running, basketball, swimming	9 METs or more

ful? Finding out this kind of information is essential for correctly encouraging and planning proper nutrition.

Ask about sleep and snoring. Do patients get the recommended 8 h of sleep per night? Do they awake refreshed? Are they tired and fatigued throughout the day? Do they have regular morning headaches? Has anyone ever observed them stop breathing while asleep? Do they have nocturnal choking or gasping arousals? These types of questions may help you identify patients with insufficient sleep or perhaps even **obstructive sleep apnea**. Obstructive sleep apnea is an age- and obesity-related condition that occurs when the upper airway collapses while a patient is sleeping. This prevents the patient from breathing properly while asleep and results in chronic nocturnal hypoxia, inappropriate sleepiness, and substantial fatigue. Increasingly, abnormal sleep patterns are being identified as significant contributors to heart disease, atrial fibrillation, and stroke (15, 16, 18).

In summary, the social history is a rich source of information about a patient's risk factors. Assisting patients with modifying lifestyle behaviors is within the training and expertise of clinical exercise physiologists. You are a professional who is able to help others identify and overcome common and unique barriers that hamper improvement in long-term health and well-being.

Review of Systems

Before concluding the interview, you should take a few moments to try to identify any symptoms the patient has not previously mentioned. The goal is to make sure you have not missed any important organ systems that might affect the exercise prescription. For an exercise physiologist, this might include asking about any difficulties with hearing, vision, weight changes, limb pain, limb numbness, and changes in mood.

PHYSICAL EXAMINATION

Physicians and physician extenders are taught to take a complete head-to-toe approach to the physical examination. For every part of the body examined, an orderly process of inspection, palpation, and, if applicable, **auscultation** and percussion is followed. The

clinical exercise physiologist, however, can take a more focused approach and concentrate on abnormal findings, based on patient complaints or symptoms, as well as information from prior examinations performed by others.

Specifically, you must develop the skills needed to determine whether, on any given day, it is safe to allow exercise by a patient who presents with signs or symptoms that may or may not be related to a current illness. For example, consider the patient with a history of dilated cardiomyopathy who complains of worsening shortness of breath, weight gain, lower extremity edema, and having to sleep on two or three pillows the previous one to two nights in order to breathe more comfortably. These complaints should raise a red flag for you, because it may indicate that the patient is experiencing pulmonary edema attributable to heart failure. Besides asking other questions, you or another clinician should physically assess body weight and peripheral edema, and perhaps lung sounds, before allowing this person to exercise. A telephone conversation with the patient's doctor concerning your findings, if meaningful, might also be warranted.

At no time should your examination be represented as one performed in lieu of the evaluation conducted by a professional licensed to do so. Still, you are responsible for ensuring patient safety. Therefore, the information and data you gather are important and should be communicated to the referring physician and become part of the patient's permanent medical record. Identifying the relevant aspects of an earlier examination performed by another provides a reference point for comparison, particularly if complications occur. Important red flags that should be identified and evaluated by the clinical exercise physiologist and, if needed, a physician, are shown in the sidebar Red Flag Indicators of a Change in Clinical Status. See also chapter 5, which discusses contraindications for exercise testing.

General State

The initial general survey is an important marker of overall illness or wellness. Take a careful look at the patient. Are they comfortable, well nourished, and well groomed? Or do they look ill, fatigued, or worn

Red Flag Indicators of a Change in Clinical Status

The following are red flag indicators that, if detected by the clinical exercise physiologist, should be discussed with a physician before a patient is allowed to exercise:

- New-onset or definite change in pattern of shortness of breath or chest pain
- Complaint of recent syncope (loss of consciousness) or near syncope
- Neurologic symptoms suggestive of transient ischemic attack (vision or speech disturbance)
- Recent fall
- Lower leg pain and bluish skin discoloration at rest (also called critical leg ischemia)
- Severe headache
- Pain in a bone area (i.e., in patients with history of cancer)
- Unexplained resting tachycardia (>100 b · min^{-1}) or bradycardia (<40 b · min^{-1})
- Systolic blood pressure >200 mmHg or <86 mmHg; diastolic blood pressure >110 mmHg
- Pulmonary rales or active wheezing

down? Are they cheerful, engaged, and motivated? Or do they look distressed, anxious, frail, or malnourished? In time, this so-called "eyeball" test will become a valuable tool in your physical examination skills. It will alert you to a change in a patient's clinical status. In fact, you may sometimes find that your intuition tells you something is wrong even though the rest of the examination appears normal. Although this skill can take years to develop and mature, if you get a sense that something might be amiss, take the time to investigate.

Blood Pressure, Heart Rate, and Respiratory Rate

Determinations of blood pressure, heart rate, and respiratory rate are the foundation of any physical evaluation and are sometimes made before and after an exercise session. These vital signs should be reassessed in the event of any clinical change or symptom such as chest pain or dizziness. You should become an expert at taking an accurate blood pressure reading. Ideally, patients should have rested for 5 min and should be seated, with arm and back supported, bare skin exposed, and arm at the level of the heart. Additionally, they should have an empty bladder and not have used caffeine or tobacco products within 30 min.

Blood pressure is categorized into several stages: normal is <120/<80 mmHg; elevated is 120-129/<80 mmHg; **hypertension** stage 1 is 130-139/80-89 mmHg, and hypertension stage 2 is ≥140 or ≥90 mmHg. Hypertension is an important risk factor for many diseases, especially cardiovascular disease, and is discussed fully in chapter 10. **Hypotension** (i.e., blood pressure <90/60 mmHg) is uncommon and may signify an important clinical change. When hypotension is accompanied by symptoms of light-headedness or dizziness it often requires immediate medical attention. The one possible exception to this rule is patients with chronic heart failure, who are treated with blood pressure–lowering medications to the point that a blood pressure of 90/60 is not uncommon and often well tolerated and without symptoms.

Tachycardia (heart rate >100 b · min^{-1}) after 15 min of sitting is almost always abnormal and can be found in many medical conditions, including very low left ventricular systolic function, severely impaired pulmonary function, hyperthyroidism, anemia, volume depletion, pregnancy, infection, or fever, as well as a medication side effect. **Bradycardia** (heart rate <60 b · min^{-1}) can be either abnormal or normal, depending on the patient's condition and medication profile. For example, β-blockers are generally known to reduce the resting heart rate by 10 to 15 b · min^{-1}. However, a heart rate below 40 b · min^{-1} can represent an underlying problem and requires further evaluation even if the patient is taking a higher-dose of a β-blocker.

Tachypnea is a respiratory rate greater than 20 breaths/min. **Bradypnea** is a respiratory rate less than 8 breaths/min. Labored, tachypneic breathing is worrisome when it occurs at rest or with minimal exertion. Measurement of the oxygen saturation using pulse oximetry is important in the case of abnormal breathing rates or bluish skin discoloration. **Hypoxia** is a blood oxygen saturation below 95%. In patients with chronic obstructive pulmonary disease (COPD), an oxygen saturation greater than 88% is acceptable. The presence of tachypnea, bradypnea, or hypoxia usually requires medical attention.

Body Fatness

Height and weight should be measured in each patient to obtain a sense of body fatness. The simplest measure of body fatness is the **body mass index (BMI)**. The BMI is mass in kilograms (weight) divided by the height in meters squared (kg · m^{-2}). A BMI less than 18.5 is underweight, between 18.5 and 24.9 is normal, between 25 and 29.9 is overweight, and 30 or more is obese. You should record in your database the BMI of every patient you are evaluating for the first time. Practically, you accomplish this by weighing the patient in kilograms and then asking for self-reported height in feet and inches. An easy-to-use conversion nomogram, such as the one in figure 4.1, should be accessible for the quick calculation of BMI. Obesity (BMI ≥30) is a major comorbidity that doubles a patient's risk for heart disease, is a risk factor for diabetes and sleep apnea, and worsens bone and joint problems.

Body fatness can be further assessed by measuring the patient's waist circumference with a measuring tape. In general, a waist circumference greater than 40 in. (100 cm) in a male or more than 35 in. (90 cm) in a female is indicative of central adiposity. Central (android) obesity, and more particularly visceral adiposity (fat accumulation in the intra-abdominal cavity), is an important risk factor for heart disease, hypertension, diabetes, and dyslipidemia. Although gynecoid obesity (around the legs) is a significant comorbidity, it does not carry the same disease risk as does android obesity (7). Waist circumference is the best measure of android and visceral obesity and should be used routinely.

Pulmonary System

The thorax should be inspected and deformities of the chest wall and thoracic spine noted. Common abnormalities to look for include kyphosis (i.e., curvature of the upper spine resulting in rounding of the back and appearing as slouching posture), barrel chest (i.e., large torso suggesting upper body strength, or emphysema among patients who smoke cigarettes), and pectus excavatum (i.e., so-called sunken or hollowed chest due to a congenital deformity of the ribs and sternum in the anterior chest wall). Thoracic surgical incisions and implantable pacemaker or defibrillator sites should be inspected and palpated. With the patient sitting, the lungs can be auscultated with the diaphragm of the stethoscope in both anterior and posterior positions, and breath sounds are typically characterized as normal, decreased, absent, coarse, wheezing, or crackling (i.e., rales). Decreased or absent breath sounds often prompt **percussion** of the chest wall for dullness. Dullness signifies a pleural effusion, which is an abnormal collection of fluid in the pleural space that does not readily transmit sound. This finding on examination should generally prompt withholding exercise and notifying the patient's physician. Coarse breath sounds can signify pulmonary congestion or chronic bronchitis. Crackles or rales

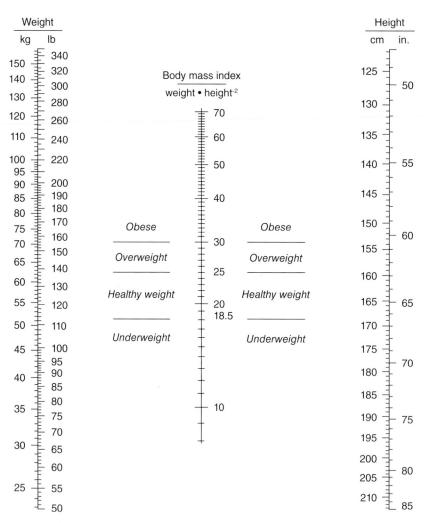

Figure 4.1 Body mass index (BMI) conversion chart that is used with weight and height. To use this nomogram, place a straightedge between the body weight (without clothes) in kilograms or pounds located on the left-hand line and the height (without shoes) in centimeters or inches located on the right-hand line. The point where the straightedge intersects the middle line indicates the BMI.

From Centers for Disease Control (6).

can be caused by atelectasis (inadequate alveolar expansion after thoracic surgery), pulmonary edema attributable to congestive heart failure, or intrinsic lung disease such as pulmonary fibrosis.

Cardiovascular System

With the patient supine, the cardiac examination starts with inspection of the anterior chest wall. A cardiac pulse that can be visualized or easily palpated on the chest wall is often abnormal and represents a left ventricular hyperdynamic state. Be sure to observe the patient's breathing and respiratory rate, and characterize this as normal, rapid, slow, labored, or shallow. Report any rapid, shallow, or labored breathing to a physician.

An essential component in the examination of the heart is auscultation. An introduction to the assessment of heart sounds is reviewed in practical application 4.3. Practice listening to heart and lung sounds in apparently healthy people and patients you help care for who are known to have cardiac and pulmonary disorders as described in their medical records.

The cardiovascular physical examination should also include some evaluation of the peripheral vascular circulation. Extremities that are well perfused with blood are warm and dry. Poorly perfused extremities are often cold and clammy. Measuring and grading the characteristics of the arterial pulse in a region assesses adequacy of blood flow in arteries. Arterial pulses are graded (11) as follows: 3 = bounding, 2 = normal, 1 = reduced or diminished, and 0 = absent or nonpalpable. Using a stethoscope, one can also listen for **bruits**, which are high-velocity swooshing sounds created as blood becomes turbulent when it flows past a narrowing artery or through a tortuous artery. Volume should be assessed, and bruits should be characterized as soft or loud. Bruits detected in the carotid arteries that were not previously mentioned in the medical record should be brought to the attention of a physician, because they may indicate severe carotid atherosclerosis. Bruits in the common femoral arteries are suggestive of peripheral arterial disease but by themselves do not call for immediate physician evaluation.

Peripheral **edema** (e.g., swelling of the lower legs, ankles, or feet) can be a cardinal sign of congestive chronic heart failure. Because of elevated left ventricular end-diastolic pressure and consequently the backward cascade of increased pressure to the left atrium, pulmonary veins, pulmonary capillaries, pulmonary artery, and right-sided cardiac chambers, increased hydrostatic forces shift extracellular fluid from within the blood vessels into the tissue (interstitial) spaces of the extremities and abdomen (ascites). Edema in the lower limbs is graded on a 1 to 3 scale, with 1 being mild, 2 being moderate, and 3 being severe. Additionally, "pitting edema" can be present, which is easily identified by pressing a thumb into an edematous area (e.g., distal anterior tibia) and observing if an indentation remains, as well as the duration of time that it remains. A patient with 3+ pitting edema of the lower legs and ankles obviously has a great deal of fluid that has left the vascular compartment and moved into the surrounding tissue.

Not all edema, however, results from congestive heart failure. Minor edema can be a side effect of many medications such as channel calcium blockers (e.g., nifedipine). In addition, chronic venous incompetence associated with prior vascular surgery, obesity, or lymphatic obstruction can cause edema in the setting of normal cardiac hemodynamics and heart function. A practical point for the clinical exercise physiologist caring for patients with cardiovascular disease is that an increase in edema or body mass (>1.5 kg) over a 2 or 3 d period is often the first sign of volume overload in a patient with congestive heart failure and warrants a call to a physician.

Musculoskeletal System

Approximately one person in seven suffers from some sort of musculoskeletal disorder. The history of present illness or past medical history should note the major areas of discomfort and self-reported limitation of motion. In addition, prior major orthopedic surgeries, such as a hip or knee joint replacement, should be noted.

The approach to the musculoskeletal physical examination should be grounded in observation. For example, observe the patient as they get up from a chair and walk into a rehabilitation area, get onto an examination table, or handle personal belongings. Observe gait and characterize it as normal (narrow based, steady, deliberate), antalgic (limping because of pain), slow, hemiplegic (attributable to weakness or paralysis), shuffling (parkinsonian), wide based (cerebellar ataxia or loss of position information), foot drop, or slapping (sensory ataxia or loss of position information). An antalgic gait is a limp, which reflects unilateral pain and compensation for that pain. A slow gait is often a tip-off for back disease, hip arthritis, or underlying neurological problems. Hemiplegic, shuffling, wide-based, foot-drop, and slapping gaits all represent compensation for underlying neurologic disease such as a spinal cord injury, cerebellar dysfunction (e.g., attributable to alcohol), or midbrain dysfunction (e.g., Parkinson's disease). These neurological gaits are all unsteady and leave the patient prone to falling. For safety reasons, the clinical exercise physiologist must pay special attention to gait and modify the exercise prescription as necessary.

The core of the musculoskeletal physical examination is an assessment of range of motion of the movable joints. Important terminology for describing limb motion is given in table 4.4. As needed, palpation of the major joints (elbows, wrists, hips, knees, and ankles) can be performed to note thickening of the joint capsules, swelling or effusion, and tenderness of ligaments or tendons. Redness, warmth, swelling, and fever are all signs of active inflammation and require evaluation by a physician before you proceed with exercise testing or therapy. If these signs are found in conjunction with a prosthetic joint, such as a total knee replacement, they may indicate an infection and require immediate contact with the patient's surgeon.

Practical Application 4.3

AUSCULTATION OF THE HUMAN HEART

Auscultation of, or listening to the sounds made by, the heart is but one part of a comprehensive cardiac examination. Begin a habit of auscultating the heart in a systematic fashion that is the same for every patient you evaluate. Establishing a uniform approach will more quickly familiarize you with normal heart sounds and help you identify abnormal sounds. To aid your concentration, try to auscultate the heart in a quiet room. Approach the patient from the right side and do all that you can to minimize anxiety. Attempt to warm your stethoscope before using it, and communicate with the patient as you progress through this part of the physical examination. Remember that maintaining patient modesty is always a priority.

Auscultation is usually done with the patient lying on their back, such as before an exercise test; however, auscultation can be performed with the patient in the sitting position. As mentioned earlier, when you auscultate the heart for cardiac sounds, you should do so systematically. Begin by placing the diaphragm of the stethoscope firmly on the chest wall in the lower-left parasternal region. First, characterize the rhythm as regular, occasionally irregular, or regularly irregular, with the latter usually attributable to atrial fibrillation. The diaphragm of the stethoscope is best used to hear high-pitched sounds, whereas the bell portion is used for low-pitched sounds.

Next, move the stethoscope to the point on the chest where the first heart sound (S_1, the sounds of mitral and tricuspid valves closing) is best characterized. For most people, this is found at the apex of the heart at the left midclavicular line and near the fourth and fifth intercostal spaces (figure 4.2). You will hear two sounds. The first sound (S_1) is the louder and more distinct sound of the two. Then, move your stethoscope upward and to the right side of the sternal border at the second intercostal space. This location is generally the best place to characterize the second heart sound (S_2, the sounds of aortic and pulmonic valves closing) because it is louder and more pronounced here.

Soft heart sounds can occur with low cardiac output states, obesity, and significant pulmonary disease (e.g., diseased or hyperinflated lung tissue between the chest wall and the heart). Loud heart tones can occur in thin people and in hyperdynamic states such as pregnancy. The second heart sound normally splits with inspiration as right ventricular ejection is delayed with the increased volume it receives from the augmented venous return of inspiration. This delays or splits the pulmonic component of S_2 (some-

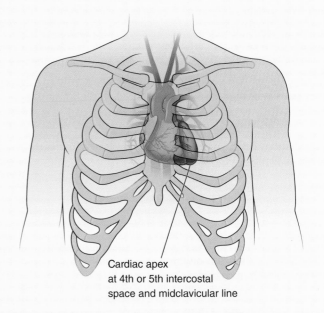

Cardiac apex at 4th or 5th intercostal space and midclavicular line

Figure 4.2   Surface topography of the heart.

times referred to as P_2) from the aortic component of S_2.

With practice and as you begin to care for patients with various cardiac problems, you will become exposed to and appreciate third (S_3) and fourth (S_4) heart sounds. S_3 and S_4 are low-pitched sounds that are heard during diastole and best appreciated with use of the bell portion of the stethoscope. The presence of either of these two heart sounds is most often associated with a heart problem and, if not previously noted in the patient's medical record, should be brought to the attention of a physician. S_3 is best heard at the apex and occurs right after S_2. S_4 is also well heard at the apex and occurs just before S_1. An S_3 commonly indicates severe left ventricular systolic impairment with volume overload and dilation. An S_4 commonly indicates chronic stiffness or poor compliance of the left ventricle, usually attributable to long-term hypertension. As you can appreciate, learning to identify heart sounds requires a great deal of practice.

The clinical exercise physiologist should be able to appreciate systole and diastole and, with advanced training, listen for murmurs in the mitral, tricuspid, pulmonic, and aortic areas (figure 4.3). Murmurs are characterized by the timing in the cardiac cycle (systolic, diastolic, or both), location where best heard, radiation, duration (short or

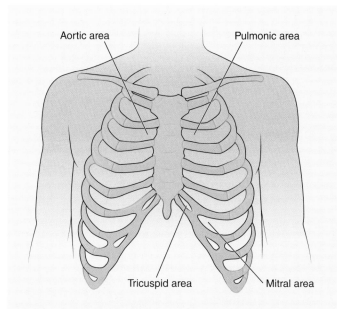

Figure 4.3   Auscultatory areas of the heart.

Reprinted by permission from P.A. McCullough *Clinical Exercise Physiology* 1, no. 1 (1999): 33-41.

long), intensity, pitch (low, high), quality (musical, rumbling, blowing), and change with respiration. Systolic murmurs are more common and are often characterized as ejection type (e.g., diamond shaped or holosystolic). Diastolic murmurs are distinctly less common and are always abnormal.

The most important murmur exercise physiologists should be able to identify is the sound of aortic stenosis. Fortunately, this murmur is not usually difficult to detect. It is usually described as a loud, harsh, crescendo–decrescendo, late-peaking systolic murmur, best heard in the right upper sternal border, with radiation to the carotid arteries, and is associated with a delayed carotid pulse and an absent or greatly diminished S_2. If you hear this murmur, confirm with a physician before performing exercise stress testing or prescribing exercise training, because symptomatic severe aortic stenosis is a contraindication to exercise.

Exercise physiologists working in the clinical exercise testing setting would be well served to acquire the basic skills needed to auscultate the heart. For further information and to hear audio clips of various heart sounds and murmurs go to www.blaufuss.org.

Table 4.4   Definitions for Selected and Common Joint Movements

Motion	Definition
Flexion	Motion in sagittal plane that decreases the angle at a joint (away from zero position)
Extension	Motion in sagittal plane that increases the angle at a joint (toward zero position)
Dorsiflexion	Movement decreases the 90° angle between the dorsum (superior surface) of the foot and the leg (toes are brought closer to the shin)
Plantar (or palmar) flexion	Movement increases the 90° angle between the surface of the foot and the leg (e.g., standing on tiptoes)
Adduction	Movement in the frontal plane that involves moving the body part toward, and perhaps beyond, the central line or midline
Abduction	Movement in the frontal plane that involves moving the body part away from the central line or midline
Plantar inversion/ supination	Movement of turning the sole (plantar surface) of the foot inward
Plantar eversion/pronation	Movement of turning the sole (plantar surface) of the foot outward
Pronation	Movement or rotation of the palm over or to face downward
Supination	Movement or rotation of the palm to face upward
Internal (medial) rotation	Movement of the hip or shoulder joints in the transverse plane; limb surface rotates inward (toward the body) or downward
External (lateral) rotation	Movement of the hip or shoulder joints in the transverse plane; limb surface rotates outward (away from the body) or upward

In addition to joint health, muscle strength can be examined and graded on a scale of 0 to 5, with 0 indicating flaccid paralysis and 5 indicating sufficient power to overcome the resistance of the examiner. Muscle stiffness and soreness (not related to exercise) should be noted because they are often signs of a chronic underlying inflammatory condition that requires medical evaluation.

Functional Fitness and Balance

A vital component of the physical examination is an assessment of the patient's **functional fitness**, which is distinct from the patient's exercise capacity as determined by an exercise stress test. Functional fitness tests, such as the 6 min walk test, Short Physical Performance Battery, sit-to-stand test, and Timed Up and Go (TUG), are often utilized to assess physical function, mobility, risk for falling, and muscular strength. In older individuals, or those suffering from a condition that impairs neuromotor function, balance testing also becomes essential. The Berg Balance Scale is a 14-item evaluation that quantitatively assesses a person's balance proficiency and potential risk for falls (3). Other balance assessments include the TUG test, the Functional Reach Test, and the Activities-Specific Balance Confidence Scale. These tests are useful for assessing balance in a timely and effective manner.

As an exercise physiologist, you should be able to not only identify neuromotor and functional deficiencies but help correct them as well. Exercise prescriptions for individuals with compromised balance should include balance-specific exercises such as single-leg standing, walking heel to toe, side leg raises, weight shifts, tai chi, and even walking. When done successfully, you will help reduce the risk of falls in your patients and improve their quality of life.

Low Back Pain

Because low back pain (see chapter 26) is one of the most common physical complaints in human medicine, it is worth mentioning that the etiology of this problem can range from a mild muscle strain to a life-threatening ruptured abdominal aortic aneurysm. The clinical exercise physiologist must have a rational approach to this problem and tailor aspects of care based on etiology, severity, and prognosis. In the young, stable patient with no symptomatic evidence of neurological compromise (e.g., radiating pain, numbness), a physician evaluation is likely not necessary before exercise testing or therapy. In the geriatric patient (i.e., >65 yr old), however, new-onset low back pain that has not been evaluated by a physician deserves referral to the primary care physician before exercise testing or training because it could be something serious, such as a spontaneous compression fracture attributable to osteoporosis.

Arthritis

A final musculoskeletal condition worth mentioning, and one that is especially common among the elderly, is arthritis (see chapter 24). The two most common types of arthritis are **rheumatoid arthritis (RA)** and **osteoarthritis (OA)**. Rheumatoid arthritis is a chronic inflammatory condition manifested by functional disability and early morning stiffness (>1 h) followed by improvement through the rest of the day. RA has a predilection for the small joints in the hand, especially the metacarpophalangeal joints (knuckles). RA results in pain, inflammation, thickening of the joint capsule, lateral deviation of the fingers, and significant disability. It can involve large joints and the spine, especially the cervical vertebrae.

Patients with RA can generally exercise safely unless neck pain or lancinating pains in the shoulders or arms are reported. These symptoms are indicative of cervical spine involvement and require a physician's attention. In some cases, bracing or surgery is needed to prevent cervical spine subluxation and paralysis.

Osteoarthritis is the most common skeletal disorder in adults. It can result from longstanding wear and tear on large and small joints including shoulders, elbows, wrists, hands, hips, knees, and back. The correlation between the pathological severity of OA and patient symptoms is poor. Patients with OA may show signs of synovitis and secondary muscle spasm.

Exercise rehabilitation may improve the symptoms of OA in one location (e.g., the back) only to worsen pain in another location (e.g., hips and knees). The clinical exercise physiologist should take a pragmatic approach and work in a coordinated fashion with physical and occupational therapy colleagues to find activities that minimize joint loading at involved sites.

Nervous System

Like the examination of the musculoskeletal system, a neurological examination performed by the clinical exercise physiologist mainly involves observation. In general, the clinician should make a comment in the patient file regarding level of understanding, orientation, and cognition. Muscle tremors, obvious disabilities of speech and balance, swallowing difficulties, and disabilities of the eyes, ears, mouth, and face should be cross-referenced for confirmation with the patient's medical record.

The practicing clinical exercise physiologist will almost never perform a detailed neurological examination, so this chapter does not describe the procedure for such an examination. Nevertheless, a notation regarding gross **hemiparesis** or complete paralysis of a limb should be made. As mentioned previously, neurological gaits should be identified as well. For obvious reasons, patients with any of these problems may require an appropriately modified exercise prescription.

Although the clinical exercise physiologist may work with the spinal cord–injured patient (see chapter 27) or patients with multiple sclerosis (see chapter 28), the most common neurological problem that most exercise physiologists encounter is with patients who have suffered a previous cerebrovascular event, or stroke. Stroke is defined as an ischemic insult to the brain resulting in neurological deficits that last for more than 24 h (see chapter 30). The etiology of stroke includes local arterial thrombosis, cardiac and carotid thromboembolism, and intracranial hemorrhage. Risk factors for stroke include the following:

- History of stroke
- Atrial fibrillation
- Left ventricular dysfunction
- Aneurysms
- Carotid artery stenosis
- Uncontrolled hypertension

If the clinical exercise physiologist determines that any of these risk factors are new, they should notify the physician before the patient starts an exercise regimen. Common abnormalities after a stroke are a loss in or diminished limb function, drooping of one side of the face, and garbled speech (dysarthria).

Skin

A full examination of the skin is not appropriate for the clinical exercise physiologist, but examination of the hands, arms, legs, feet, and surgical incisions for coronary bypass or pacemaker implantation should be routine. If the skin is hot, warm, swollen, or draining fluids, this should be reported and may represent a new infection.

For patients with diabetes, examination of the skin should include inspecting the feet for any blisters, cuts, scrapes, ulcers, and discoloration. Patients with diabetes may have a condition called **peripheral neuropathy**, in which the nerves of the legs are damaged and stop functioning properly, with the result that people can no longer feel their toes and feet. They lose the ability to sense pain. Consequently, diabetic patients sometimes develop wounds and infections in their skin that they are completely unaware of. This makes it important to examine the feet of all diabetic patients and discuss appropriate socks and footwear.

When examining the skin of the arms and hands, pay careful attention to discoloration or **clubbing**, which may be seen in patients with chronic lung disease. Clubbing is a condition where a buildup of tissue in the fingers causes the fingernails to become rounded, like an upside-down spoon; it is often seen with diseases (e.g., pulmonary diseases) that result in less oxygen in the blood. Blueness of the hands, lips, and nose suggests hypoxia. These findings should be evaluated by a physician.

Significant bruising of the skin may be the result of medications such as aspirin, clopidogrel, or warfarin or could result from frequent falls. Pale skin may signify **anemia**, in which a patient has a low blood count. Anemia is defined as a hemoglobin less than $12 \text{ g} \cdot \text{dL}^{-1}$ in women and less than $13 \text{ g} \cdot \text{dL}^{-1}$ in men and may result from recent operations, procedures, or bleeding. In addition, some patients have anemia due to prolonged chronic illness or chronic kidney disease. In general, anemia requires further physician evaluation.

Laboratory Data and Diagnostic Tests

This section of the clinical evaluation summarizes the relevant testing that has previously been performed. In general, test results or reports from a patient's resting electrocardiogram, chest X-ray, echocardiogram, pulmonary functions, cardiac catheterization, and operations (e.g., coronary artery bypass grafting) are relevant and should be included. Consultation results from physical therapy, orthopedics, or physical medicine and rehabilitation should be reviewed. Furthermore, laboratory data such as hemoglobin, electrolytes, creatinine, cholesterol, and blood glucose levels should be noted. Detailed discussion of these tests is beyond the scope of this chapter, but they may be relevant and can affect the exercise prescription.

The most important diagnostic test to incorporate into your evaluation is the most recent stress test. What was the level of exercise attained? What was the peak heart rate? Were there any symptoms? Was there any evidence of ischemia or arrhythmia, and if so, at what heart rate did these abnormalities first appear? These kinds of results help you build an appropriate exercise prescription and should be included in this section of the general evaluation. Exercise testing and prescription are discussed fully in chapters 5 and 6, respectively.

An alternative approach to documenting the clinical evaluation (i.e., systematic interview and physical examination) is referred to as SOAP charting, where S = subjective, O = objective, A = assessment, and P = plan. In this approach, S includes all of the elements of the interview (reason for referral, demographics, history of present illness, medications, allergies, and past medical, family, and social histories), and O encompasses the examination and laboratory data.

Research Focus

Patient History Perspective

Although the tools and techniques of the patient history and physical examination are well established and have been practiced for hundreds of years, there is still research being done on the usefulness of the history and physical examination in the era of modern medicine. Most of this research has to do with the clinical reliability and predictive power of various physical examination findings and how those are best incorporated into routine clinical practice. Several references will be particularly valuable to the clinical exercise physiologist interested in learning more about the utility of the history and physical examination (8, 10, 11, 15).

CONCLUSION

This chapter provides a platform for the clinical evaluation of the new patient by the clinical exercise physiologist. We have given special emphasis to the day-to-day interview and examination skills that a practicing clinical exercise physiologist may need in order to help decide whether a patient can exercise safely on any given day. Also, we have emphasized that the information gathered through a clinical evaluation, when viewed in conjunction with existing medical record information and physical examination findings reported by others, provides a point of reference should complications occur during exercise treatment.

Online Materials

Visit HK*Propel* to access a link to the references, the case studies with discussion questions, and a quiz to help you review key concepts and test your understanding of the material covered in this chapter.

CHAPTER 5

Graded Exercise Testing

Steven J. Keteyian, PhD

The material presented in this chapter represents the minimum foundational knowledge the clinical exercise physiologist needs to acquire to be sufficiently prepared to help conduct a graded exercise test (GXT, or stress test) in apparently healthy people and those with clinically manifest disease. The reader is expected to use this material and adapt it based on the population-specific exercise testing material provided in each of the chapters that follow.

The earliest form of a GXT was used in approximately 1846, when Edward Smith began to evaluate the responses of different physiological parameters (heart rate, respiratory rate) during exertion (9). In 1929 Master and Oppenheimer used exercise to evaluate a patient's cardiac capacity (27); however, they did not yet recognize the value of incorporating an electrocardiogram (ECG) during or following exercise in the detection of **myocardial ischemia** (i.e., reduced or inadequate blood flow to the heart). In fact, Master's original two-step test, so called because patients repeatedly walked up and over a box-like device with two steps that stood 18 in. (45 cm) high, counted the number of ascents completed during 90 s as a measure of exercise capacity. It was not until several years later that Master obtained an ECG before and immediately after the test to screen for myocardial ischemia. The Master two-step remained the clinical standard for assessing exercise-induced ischemia until the early 1970s.

Graded exercise testing has evolved greatly over the past 50 yr. In addition to serving as a tool to detect the presence of ischemic heart disease using various exercise-based modalities (e.g., treadmill, stationary bike), such testing yields data that are now also used to assess prognosis. Additionally, in combination with various cardiac imaging modalities, some in conjunction with exercise and others using drugs to induce myocardial stress, these tests have an increased ability to detect coronary artery disease (CAD). All of this notwithstanding, even with the addition of **radionuclide agents** or the **stress echocardiogram** for imaging, the standard GXT with a 12-lead ECG remains an acceptable choice to evaluate myocardial ischemia in most individuals who have a normal resting ECG and the ability to physically exert themselves (42).

INDICATIONS

Before a GXT is performed it is important to understand which patients the GXT is useful for and the reason for completing the test. Further, although the risks for experiencing a clinical event during or immediately after a GXT are rare, this does not mean there is no risk. Knowing this information will help ensure that the benefits and data derived from the test outweigh any potential risks (4, 13, 14, 19, 30). If a question arises concerning the safety or rationale for completing the GXT, the supervising or referring physician should be contacted for clarity.

The American College of Cardiology, the American Heart Association, and other organizations provide guidelines for exercise testing in various clinical situations (7, 14, 29). In general, a GXT is ordered by the physician for two main reasons: first, to evaluate chest discomfort or other cardiac-like symptoms as a means to assist in the *diagnosis* of CAD, and second to identify a patient's future risk or prognosis. In addition, there are several other conditions in which a GXT can assist the physician, several of which are discussed a little later in this section.

As discussed in detail later in this chapter, the diagnostic value of determining the presence of CAD is greatest when one is evaluating individuals with an intermediate probability of CAD, which is based on the person's age and sex, the presence of cardiac-related symptoms (typical or atypical), and the prevalence of CAD in other persons of the same age and sex. To determine a patient's *prognosis* related to CAD or help stratify the person's future risk

Acknowledgment: Much of the writing in this chapter was adapted from earlier editions of *Clinical Exercise Physiology.* Thus, we wish to gratefully acknowledge the previous efforts of Micah Zuhl, PhD.

for experiencing a clinical event, a GXT can again prove helpful. Specifically, in addition to angina and the presence of ST-segment depression evident on the ECG during exercise or in recovery, other factors that influence prognosis include the magnitude of ST depression (1 mm vs. 3 mm, where more ST depression represents greater risk), the number of ECG leads showing significant ST-segment depression, time when ST depression started during and resolved after exercise, and estimated **functional capacity** (FC) as measured by exercise duration or **metabolic equivalents of task (METs)**, where 1 MET = a resting seated oxygen uptake ($\dot{V}O_2$) of ~3 to 4 mL of O_2 per kilogram of body weight per minute. Even after potentially confounding factors known to be related to survival, such as age and comorbidities (e.g., hypertension, diabetes) are accounted for, functional capacity remains a strong independent predictor of survival (figure 5.1) (2, 3).

For example, a patient who complains of anginal pain and demonstrates 2.5 mm of ST-segment depression in four ECG leads at a workload that approximates just 4 METs has a poorer prognosis than the patient who is symptom-free and completes 10 METs of work with no ECG evidence of ST-segment depression. Such information is used to stratify a patient's risk. In people found to be at higher risk (i.e., abnormal GXT response at low workloads, <5 METs), further medical testing and intervention may be required. But if an abnormal ECG response does not occur until the person reaches 10 METs, the prognosis is much more favorable and additional medical or surgical intervention may not be needed. Gathering together all this information allows for use of the Duke nomogram to estimate the average annual mortality rate and 5 yr survival (26). The take-home point is that much other information (FC, blood pressure, symptoms) can be gathered from a GXT, beyond simply the presence of ST depression—information that can be used to help guide a patient's future care (additional testing vs. medical management vs. surgical intervention).

The association between FC and prognosis can be strengthened if the GXT is performed in conjunction with direct measurement of respiratory gas exchange (via a cardiopulmonary exercise test, or CPET, the methods and indications for which are described later in this chapter). A CPET allows for the indirect measurement of peak $\dot{V}O_2$ and other gas exchange parameters during exercise (6, 29).

A regular GXT can also be performed for reasons other than diagnosis or prognosis. These include quantification of change in FC due to an exercise training program or a medical intervention; assessment of exercise-induced arrhythmias and pacemaker or heart rate (HR) response to exercise; assessment of blood pressure response to exercise; and determination of FC for the purpose of guiding return to work, eligibility for disability insurance, or preoperative clearance.

CONTRAINDICATIONS

Although the risk of death or a major complication during a GXT is very small, such a test is not ordered when the risks of safely completing the test outweigh the value of the potentially derived information (4, 13, 14). Current guidelines separate absolute versus relative contraindications. Absolute contraindications are conditions in which the risk for potential serious medical consequences is unsatisfactorily high. Some common examples of absolute contraindications include certain abnormalities on the resting ECG, unstable angina, decompensated heart failure, symptomatic severe aortic stenosis, acute myocarditis or pericarditis, and acute infections.

Relative contraindications suggest the presence of a potential medical issue and indicate that concerns regarding the worsening

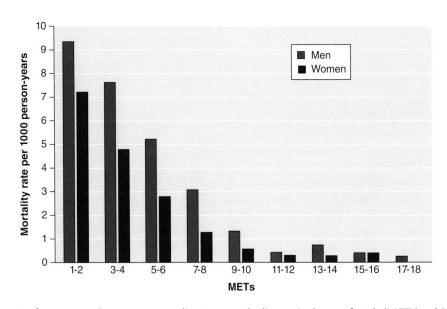

Figure 5.1 Mortality rates for men and women according to metabolic equivalents of task (METs) achieved.

Data adapted from Al-Mallah et al. (2).

of the patient's condition should be considered before the GXT is conducted. Some common examples of relative contraindications include left main coronary artery disease, arterial hypertension at rest (systolic blood pressure >200 mmHg or diastolic blood pressure >110 mmHg), tachycardic or marked bradycardic rhythm at rest, uncontrolled metabolic disease (e.g., diabetes, hyperthyroidism), and chronic infectious disease (e.g., mononucleosis).

In addition to contraindications for GXT, there are conditions in which GXT is not recommended for diagnostic purposes because abnormalities in the resting ECG (e.g., left bundle branch block, ventricular pacemaker) render the exercise ECG insensitive to accurately identify the presence of CAD (i.e., ST-T changes). As discussed later in this chapter, a GXT with nuclear perfusion or echocardiographic imaging is recommended for people who have these abnormalities in their resting ECG. Note that if the primary reason for ordering the GXT is something other than evaluation of exertional myocardial ischemia (e.g., evaluation of FC), then the GXT with ECG alone (without imaging) remains appropriate.

PROCEDURES FOR PREPARING, CONDUCTING, AND INTERPRETING A GRADED EXERCISE TEST

With an introduction to graded exercise testing and a review of who should and should not undergo such testing behind us, we can turn our attention to the procedural and interpretive elements associated with such testing. In general, these elements can be organized or grouped into seven topics: pretest considerations; appearance and quantification of symptoms; assessment and interpretation of HR response during exercise and recovery; assessment and interpretation of blood pressure (BP) response during exercise and recovery; ECG findings; assessment of FC; and interpretation of findings and generation of final summary report.

Pretest Considerations

Preparing a person to undergo any exercise test (i.e., regular or with imaging or with measurement of expired gases) involves several common pretest considerations. These are personnel, informed consent, general interview and examination, assessing pretest likelihood for CHD, pretest instructions and preparation, and selection of exercise protocol and testing modality.

Personnel

Since the 1980s, GXTs have been performed by many types of health care professionals (clinical exercise physiologists, physicians trained in internal medicine or family medicine, physician assistants, nurses, and physical therapists). The use of health care professionals other than a physician results from a better understanding of the risks associated with a GXT, cost-containment initiatives, time constraints on physicians, and more sophisticated ECG analysis (computerized exercise ST-segment interpretation).

The American Heart Association guidelines for exercise testing laboratories state that the health care professionals just mentioned, when appropriately trained and possessing specific performance skills (e.g., a certification through the American College of Sports Medicine), can safely supervise clinical GXTs (30).

Since diagnostic and most other forms of GXT usually necessitate that both healthy people and patients with a clinically manifest disease put forth a near-maximal or maximal effort, the competency, knowledge, skills, and ability of physician and nonphysician professionals to safely supervise and conduct all aspects of the test are essential (29, 30, 35). The incidence of death or a major event requiring hospitalization during a GXT is rare (19, 29), approximately 1 to 10 events per 10,000 tests; and the rate for such events is the same regardless of whether the tests are supervised by physicians or nonphysicians. Interestingly, with respect to the preliminary interpretation of the ECG and other test findings made by nonphysicians, the level of agreement with a cardiologist appears to be quite good (29, 30). Key needed skills for test supervision include

- knowledge of absolute and relative contraindications for testing;
- knowledge of specificity, sensitivity, and predictive value of positive and negative tests;
- understanding of the causes of false-positive and false-negative tests;
- demonstrated ability to select the correct test protocol and modality;
- knowledge of normal and abnormal cardiorespiratory and metabolic responses in normal, healthy individuals and in patients with a clinical condition;
- ability to assess and respond appropriately to important clinical signs and symptoms;
- knowledge of and ability to demonstrate proper test termination skills using established indications;
- knowledge of and ability to correctly respond to normal and abnormal ECG findings at rest and during exercise; and
- appropriate skills for the measurement of BP at rest and during exercise.

Informed Consent

Obtaining adequate informed consent from individuals before exercise testing is an important ethical and legal step. The form you ask the patient to read and sign should lay the foundation for a "meeting of the minds" between the testing staff and the patient relative to the reason for the test, an understanding of test procedures, and safety issues or potential risks. Enough information must be exchanged that the patient knows, understands, and verbalizes *back to you* (a) the test purpose, (b) the test procedures, and (c) the risks. In addition to asking the patient to read the informed consent form, it is appropriate for you to orally summarize the content of the form. The patient should be given several defined opportunities to ask questions. Since most consent forms contain information explaining the risks associated with the test, the likelihood of their occurrence, and the fact that emergency equipment is available should it be required, you should also explain that appropriately trained staff are present in the rare event that an emergency occurs. If the client agrees to participate, they must sign the form; clients under 18 years of age must have a parent or guardian sign the consent form as well.

General Interview and Examination

In hospital-based or outpatient ambulatory care settings, the clinical exercise physiologist will be involved with supervising or assisting with GXTs that were ordered by a licensed physician or physician extender (e.g., nurse practitioner, physician assistant). This means that sometime in the usually not-too-distant past, perhaps yesterday or within the past month or two, a health care provider familiar with the indications and contraindications for GXT considered the patient's clinical status and ordered the test.

Determining Pretest Likelihood for CHD

The general interview, combined with a review of the patient's medical history, will assist the clinical exercise physiologist in determining the pretest probability that the patient's test will be positive for (i.e., indicates) CHD. Methods for evaluating the pretest likelihood of a patient's exhibiting exercise-induced ischemia due to CHD has historically been based on the patient's age, sex, and type of angina (see table 5.1), as well as the prevalence of CHD in people of the same age and sex as the person undergoing the test. The following pretest categories are recognized (42):

- Very low probability = <5%
- Low probability = <10%
- Intermediate probability = 10% to 90%
- High probability = >90%

Using table 5.1, you can see that a 55 yr old male patient who complains of stable typical angina with exertion is classified as high probability, which means a >90% pretest probability of having CHD. In this case, a GXT is used not so much for the purpose of diagnosing the presence of CHD (which is already known to be very likely given the patient's classic symptoms and the high prevalence of CHD in the general population of men his age), but more for the purpose of quantifying the extent of his disease (e.g., moderate to severe myocardial ischemia). Conversely, a 35 yr old male patient, also with typical angina, is classified as having an intermediate (10%-90%) pretest likelihood of CHD, mostly because the prevalence of CHD in 35 yr old men is generally lower overall (certainly lower than in 55 yr old men). A GXT is ordered for the 35 yr old man because the test is helpful for either diagnosing or ruling out CHD in patients who have an intermediate pretest probability of disease. Finally, patients with a very low or low pretest probability most likely do not have CHD; therefore, exercise testing is not typically warranted. The diagnostic value of GXT for uncovering suspected CHD is most helpful in patients with an intermediate pretest probability.

Table 5.1 Pretest Probability of Having Coronary Heart Disease

Age, yr	Nonanginal chest pain	Atypical angina	Typical angina
MEN			
35	Low	Intermediate	Intermediate
45	Intermediate	Intermediate	High
55	Intermediate	Intermediate	High
65	Intermediate	Intermediate	High
WOMEN			
35	Very low	Very low	Intermediate
45	Very low	Low	Intermediate
55	Low	Intermediate	Intermediate
65	Intermediate	Intermediate	High

Adapted from Pryor et al. (34); Fihn et al. (12).

However, it is still the responsibility of the clinical exercise physiologist supervising the test to be sure there have been no changes in the patient's clinical status since they were was last seen by their doctor; any such change might preclude taking the test on the scheduled day for safety reasons.

The elements of the interview and examination performed by a physician are detailed in chapter 4. Fortunately, for the purpose of deciding whether it is safe to conduct the test, the questions that must be asked and any examination needed can be much more focused. The test supervisor should have already reviewed the patient's medical record for past medical and surgical history; this knowledge will help determine whether any new contraindications to the GXT have arisen since the patient was last seen by their physician. It will also aid in choosing an appropriate testing protocol and will help identify areas that may require more attention during the test (e.g., history of hypertension, previous knee injury). A review of the medical record should pay attention to height and weight; cardiovascular risk factors; any recent signs or symptoms of active cardiovascular disease; previous cardiovascular disease; other chronic diseases that may influence the outcome (e.g., osteoarthritis); information about current medications (including dose and frequency), drug allergies, recent hospitalizations or illnesses; and exercise and work history. In addition to collecting information about the patient's medical history and status through both the review of the patient's medical record and your interview of the patient, you may need to perform a brief physical examination, one that focuses on just peripheral edema (and possibly heart and lung sounds).

Obviously, any red flags (e.g., marked recent weight gain or worsening edema over the past 3 d, worsening or unstable symptoms, recent fall or injury) detected by the clinical exercise physiologist supervising the test should be brought to the attention of the physician overseeing testing that day or the referring physician. A decision can then be made about proceeding with the test, having the patient first be seen by a physician, or rescheduling the test.

Patient Pretest Instruction and Preparation

The patient must receive clear instructions on how to prepare for the GXT along with an appropriate explanation of the test. A lack of proper instruction can delay the test and increase patient anxiety. Instructions for patient preparation should include information on comfortable clothing; comfortable and supportive footwear (no sandals, high heels, or open-back shoes); prior food consumption (i.e., limit food for at least one hour prior, with water allowed ad libitum); avoidance of alcohol, cigarettes, caffeine, and over-the-counter medicines that could influence the GXT results; and whether the patient should take prescribed drugs as usual. A prescribed medication may need to be discontinued because certain agents can limit the ability to observe ischemic responses during exercise (e.g., β-adrenergic blocking agents attenuate HR and BP, which could lessen the occurrence of ST-segment changes associated with ischemia). Note that if a physician wants the patient to discontinue medications before GXT, the patient should receive instructions on how to do so because the process of weaning a person off different medications can vary.

The clarity of the ECG tracing during exercise is of utmost importance, especially for diagnostic purposes. Therefore, properly preparing the patient for an ECG at rest is mandatory. For males, hair should be removed from all electrode positions with a disposable razor or battery-operated shaver. Detection of the heart's electrical conduction activity on the skin surface is critical for achieving a clear ECG recording, and shaving away hair allows the electrodes to lie flat on the skin. Removal of body oils and lotions with an alcohol wipe and skin abrasion is also required for optimal conductance. Silver chloride electrodes are ideal because they offer the lowest offset voltage and are the most dependable for minimizing **motion artifact** (7, 29). After electrodes are in place and cables are attached, each electrode site should be assessed for insufficient skin preparation: Tap lightly on the electrode and observe the ECG; if the ECG concomitantly shows a great deal of artifact, check the electrode and lead wire for potential interference and replace if needed. Artifact due to muscle contraction cannot be easily reduced unless upper body skeletal muscle activity is decreased.

A 12-lead ECG and BP should be recorded in both the supine and standing positions; with respect to the latter, at least 2 min should elapse after the patient stands upright and measurements are again taken. Evaluating both conditions allows the test supervisor to assess the effects of body position on hemodynamics and the ECG. Also, the resting ECG should be compared, if available, against a resting ECG taken on a prior day. If any significant differences are found between the two ECGs that contraindicate starting the GXT (see the Contraindications section), the person supervising the test should contact the supervising or referring physician with the updated information so a decision can be made about proceeding with the test.

Graded Exercise Testing Protocols and Testing Modalities

The exercise testing protocol refers to the manner or schedule in which external work is incrementally applied during the test. Exercise test modality refers to the type of equipment (e.g., treadmill, stationary bike, arm ergometer) used to apply the work.

Numerous GXT protocols are available to systematically impose incremental work; a few common examples are shown in table 5.2. The supervisor of the GXT must select an exercise protocol that enables the client to achieve maximal effort without premature fatigue, hopefully allowing the patient to exercise for a minimum of 7 min and preferably no more than 12 min. This "window for exercise" provides enough time for the physiologic adaptations during exercise to occur and reduces the likelihood that the patient will end the GXT because of skeletal muscle fatigue. To help with the selection of the appropriate exercise protocol, before the test the patient should be asked about current exercise and activity habits (e.g., is the patient able to walk three

Table 5.2   Commonly Used Treadmill and Bicycle Protocols

Protocol	Stage	Time (min)	Speed in mph (kph)	Grade (%)	Estimated $\dot{V}O_2$ (mL · kg^{-1} · min^{-1})	METs
Modified-Bruce or Standard Bruce	1	3	1.7 (2.7)	0.0	8.1	2.3
	2	3	1.7 (2.7)	5.0	12.2	3.5
	3	3	1.7 (2.7)	10.0	16.3	4.6
	4	3	2.5 (4.0)	12.0	24.7	7.0
	5	3	3.4 (5.5)	14.0	35.6	10.2
	6	3	4.2 (6.8)	16.0	47.2[b]	13.5[b]
	7	3	5.0 (8.0)	18.0	52.0	14.9
	8	3	5.5 (8.9)	20.0	59.5	17.0
Naughton	1	2	1.0 (1.6)	0.0	8.9	2.5
	2	2	2.0 (3.2)	0.0	14.2	4.1
	3	2	2.0 (3.2)	3.5	15.9	4.5
	4	2	2.0 (3.2)	7.0	17.6	5.0
	5	2	2.0 (3.2)	10.5	19.3	5.5
	6	2	2.0 (3.2)	14.0	21.0	6.0
	7	2	2.0 (3.2)	17.5	22.7	6.5
Standard Balke	1	2	3.0 (4.8)	2.5	15.2	4.3
	2	2	3.0 (4.8)	5.0	18.8	5.4
	3	2	3.0 (4.8)	7.5	22.4	6.4
	4	2	3.0 (4.8)	10.0	26.0	7.4
	5	2	3.0 (4.8)	12.5	29.6	8.5
	6	2	3.0 (4.8)	15.0	33.2	9.5
	7	2	3.0 (4.8)	17.5	36.9	10.5

Protocol	Stage	Time (min)	RPM	kgm/min; watts	Estimated $\dot{V}O_2$ (mL · kg^{-1} · min^{-1})	METs
Stationary bike	1	2 or 3	50.0	150; 25	10.9	3.1
	2	2 or 3	50.0	300; 50	14.7	4.2
	3	2 or 3	50.0	450; 75	18.6	5.3
	4	2 or 3	50.0	600; 100	22.4	6.4
	5	2 or 3	50.0	750; 125	26.3	7.5
	6	2 or 3	50.0	900; 150	30.1	8.6

Note: When trying to determine the appropriate protocol, consider the client's age, current functional capacity, activity level, and medical history and disease status.

[a]The conventional, or standard, Bruce protocol begins at stage 3 (1.7 mph [2.7 kph] and a 10% grade); modified versions start at either stage 1 (1.7 mph [2.7 kph] at 0% grade) or stage 2 (1.7 mph [2.7 kph] and 5% grade).

[b]$\dot{V}O_2$ and METs are calculated for this stage of the Bruce protocol while the client is walking. If the client is running at stage 6, the $\dot{V}O_2$ would be 42.4 mL · kg^{-1} · min^{-1} and 12.1 METs.

METs = metabolic equivalents of task.

blocks or climb a flight of stairs?). Ideally, clinical testing laboratories have several testing protocols because patients present with a variety of functional capacities. For any repeat testing performed on the same patient over time, the supervisor should try to use the same exercise protocol as before. Such consistency between tests makes comparisons of results possible.

For diagnostic purposes, the standard Bruce treadmill protocol (i.e., starting at 1.7 mph [2.7 kph] and a 10% grade) is the most common because of its longstanding history of use in clinical laboratories, frequent citation in the medical literature, and normative data for many populations. In people who are not frail, do not have an extremely low FC (e.g., no difficulty walking a flight of stairs), and are free from orthopedic problems that are exacerbated by walking, the Bruce protocol is acceptable for diagnostic purposes. That said, the Bruce protocol does have some limitations; these include relatively large increments in workload (2 to 3 METs) per stage; 3 min stages that for some patients are too fast for walking and too slow for running; and a large vertical component, which can lead to premature leg fatigue. When trying to determine whether the Bruce protocol is appropriate, the GXT supervisor should ask clients whether they can comfortably walk a flight of stairs without stopping. If no, another treadmill protocol may need to be considered, one that employs smaller MET increments between stages (e.g., modified Bruce or Naughton protocols). These protocols are commonly used with the elderly and people who are very much deconditioned because of a chronic medical problem such as stage C heart failure or end-stage renal disease.

Ramping protocols are also an excellent option for a GXT performed using a treadmill or bicycle ergometer. The advantage of ramping tests is that there are no large incremental changes in work rate. Equipment manufacturers often include a ramping option that allows conversion of a standardized protocol (e.g., Bruce) to a ramping format, in which treadmill speed and grade are increased (i.e., ramped up) in small time increments (i.e., every 6 to 15 s). When the Bruce protocol was compared using the standard 3 min stages to a ramping version that increased work rate every 15 s, similar hemodynamic changes were observed. However, the ramping test produced a significantly greater duration in time and a higher peak MET level. In addition, subjects perceived the ramping protocol to be easier than the standard Bruce protocol (41).

With respect to deciding which mode or type of equipment to use for the GXT, in the United States if a treadmill cannot be used, the bicycle ergometer is the usual second choice. But a comparison between the treadmill and the bicycle ergometer demonstrated that people achieve a 5% to 20% lower FC on the bicycle ergometer, which is due to leg strength and conditioning level (4). Bicycle ergometer testing is appropriate for diagnostic reasons and to assess FC in frail patients, patients with **claudication** due to peripheral arterial disease, and patients who have difficulty weight bearing or have an abnormal gait that can affect their ability to perform the test. In such cases, a bicycle ramping protocol is often used that increases work rate between 15 and 20 W/min. Alternatively, a work rate increase of 25 W every 2 or 3

min stage is also common if a nonramping protocol is used (table 5.2). With respect to pedal rate, or cadence, 50 to 60 rev/min is suggested for standard bicycle ergometers. In electronically braked bicycles, cadence is not a major concern because resistance automatically changes, based on cadence, to maintain a specific work rate. Among heavier and physically active clients able to complete a higher workload to reach maximum effort, increasing work rate by 300 kgm/min or 50 W every 2 or 3 min may be necessary. In these patients, after they achieve 75% of predicted maximal HR, the increases in workload can be reduced to 150 kgm/min or 25 W per stage so that the ending work rate and peak FC can be more precisely determined.

In years past, arm ergometry testing was used for diagnostic purposes in clients with severe orthopedic problems, peripheral vascular disease, lower extremity amputation, and neurological conditions that inhibited their ability to exercise on a treadmill or bicycle ergometer. In a clinical setting, however, the arm ergometer is now routinely replaced by pharmacological stress testing, which means drugs are administered as a method to increase myocardial stress (in place of exercise). Clients for whom an arm ergometry GXT may still be beneficial include those with symptoms of myocardial ischemia during dynamic upper body activity and those who have suffered a myocardial infarction and plan to return to an occupation that requires upper body activity. Because of a smaller muscular mass in the arms versus the legs, a typical arm protocol involves 10 to 15 W increments for every 2 or 3 min stage at a cranking rate, or cadence, of 50 to 60 rev/min.

After the testing staff prepares the patient, chooses the appropriate protocol and modality, and reviews the resting ECG for any abnormalities that may hinder the evaluation, the GXT can start after the patient receives instruction on how to use the mode identified for testing (treadmill, bicycle ergometer). In patients unfamiliar or uncomfortable with walking on a treadmill, the test supervisor should take the needed time to make sure they are comfortable doing so with little or no handrail support for assistance. Clients who are not comfortable may stop prematurely simply because of apprehension. In addition to the supervisor of the GXT (physician or allied health professional) being present, a technician is needed to monitor the patient's BP and potential signs and symptoms of exercise intolerance. The test supervisor is responsible for discontinuing the test at the appropriate time.

Appearance and Quantification of Symptoms and General Monitoring

Symptoms such as angina or claudication should be noted and quantified as to when they first occur, periodically throughout the test, and at peak exercise. The time when symptoms resolve in recovery should also be noted. Heart rate, RPE, and BP should be recorded just before the end of each stage.

Before the GXT begins, patients must understand that they are responsible for communicating the onset of any discomfort. As the test progresses from one stage to the next, testing staff must observe clients and communicate regularly, which includes asking patients how they are feeling and if they are having any discomfort. Testing staff should record any such discomforts, including the absence of any stated discomfort, because doing so implies that the question was asked. During the GXT, the ECG must be continually monitored. A 12-lead ECG should be reviewed at the end of each stage, because it is possible that ECG changes can take place in any of the lead combinations.

If gas exchange is analyzed during the GXT, communication with the patient becomes more difficult because of the use of a mouthpiece or mask. Establishing specific hand signals or using handheld posters is helpful so that important symptoms (e.g., chest pain, **dyspnea**) along with RPE can be accurately determined. Patients can be apprehensive during their first experience with the measurement of gas exchange because breathing through a mouthpiece is uncommon and the ability to speak is inhibited.

The use of RPE with exercise testing is common and provides a monitor of how hard patients perceive the overall work and helps guide how much longer they will be able to continue. A patient's RPE is generally recorded during the last 15 s of each stage. Two RPE scales are commonly used: the 6- to 20-point category scale and the 0- to 10-point category–ratio scale. To improve the accuracy of the RPE scales, the following instructions should be given to the patient:

During the exercise test we want you to pay close attention and rate how you feel *overall*. Your response should rate your total amount of exertion and fatigue, combining all sensations and feelings of physical stress, effort, and fatigue. Don't focus on or concern yourself with any one factor such as leg pain, leg fatigue, or shortness of breath, but try to rate your total, inner feeling of overall exertion.

Other scales that are helpful in evaluating a patient's specific symptoms include those for angina, dyspnea, and claudication (table 5.3), the latter of which is beneficial when evaluating patients with documented or suspected peripheral arterial disease. The angina scale evaluates the intensity of chest discomfort and helps determine whether the test should be terminated based on standard criteria (see Indications for Termination of Graded Exercise Test). Two other important data points to document are the MET level and **rate–pressure product (RPP)** that correspond to the onset of angina. The progression of angina can be evaluated with the angina scale, which can be beneficial when a comparison is made against previous test results (i.e., whether a given intensity of angina occurs at a lower or higher workload). The dyspnea scale, which rates the intensity and progression of dyspnea, is commonly used when testing patients who complain of increased shortness of breath.

Test Termination

As patients approach their maximum effort, they may become anxious and want to stop the test, so the supervisor and technician must be prepared to quickly record peak values (ECG, BP, RPE). A submaximal GXT results when the patient achieves a certain predetermined MET level or the test is terminated because the patient achieved a percentage of maximal HR (i.e., 85% of predicted maximum HR). Symptom-limited or maximal GXTs should be used for diagnostic purposes and the assessment of FC. A symptom-limited test refers to a test that is terminated because

Table 5.3 Angina, Dyspnea, and Peripheral Vascular Disease Scales

	ANGINA SCALE
1	Light, barely noticeable
2	Moderate, bothersome
3	Moderately severe, very uncomfortable
4	Most severe or intense pain ever experienced
	DYSPNEA SCALE
1	Mild, noticeable to patient but not observer
2	Mild, some difficulty, noticeable to observer
3	Moderate difficulty but patient can continue
4	Severe difficulty, patient cannot continue
	PERIPHERAL VASCULAR DISEASE SCALE FOR ASSESSMENT OF INTERMITTENT CLAUDICATION
1	Definite discomfort or pain but only of initial or modest levels (established but minimal)
2	Moderate discomfort or pain from which the patient's attention can be diverted by a number of common stimuli (e.g., conversation, interesting TV show)
3	Intense pain from which the patient's attention cannot be diverted except by catastrophic events (e.g., fire, explosion)
4	Excruciating and unbearable pain

of the onset of symptoms that put the client at increased risk for a medical problem. A maximal test terminates when an individual reaches the maximal level of exertion, without being limited by any abnormal signs or symptoms. When a client gives a maximal effort, the test is normally terminated because of voluntary fatigue. Several criteria are used to determine whether a person has reached maximal effort:

1. A plateau in $\dot{V}O_2$ (<150 mL · min^{-1} or 2 mL · kg^{-1} · min^{-1} increase) despite increasing workload

2. A respiratory exchange ratio value greater than 1.1

3. Venous blood lactate exceeding 8 to 10 mM

4. Rating of perceived exertion (RPE) >17 (6-20 scale)

5. Achieved 100% of age-predicted maximal heart rate and/or plateau in heart rate despite increasing workload

6. Rate pressure product (RPP) ≥24,000 mmHg · min^{-1}

The first two criteria require a metabolic cart and the third requires a lactate analyzer, both of which are not available in most clinical exercise testing laboratories. Therefore, the other criteria are listed because they are commonly used, in conjunction with voluntary fatigue, to evaluate if the patient reached a maximal cardiovascular effort. It is important that patients challenge themselves during the test so results can be interpreted accurately. As long as it is safe, the clinical exercise physiologist may need to encourage patients to continue the test to achieve their maximal effort.

The clinical exercise physiologist supervising the assessment must possess a commanding knowledge of the normal and abnormal physiological responses that occur during a GXT, as well as be aware of and understand potential complications that may occur and know when to terminate the test. Reasons for test termination are presented as absolute or relative criteria (see the sidebar Indications for Termination of Graded Exercise Test). Apart from a patient's request to stop the test or technical difficulties that arise during testing, absolute indications for stopping comprise high-risk criteria that have the potential to result in serious complications if the test is continued. Therefore, the test should be terminated immediately, and if a physician is in the immediate area but not present in the room during testing, they should be contacted and informed of the outcome as soon as possible. With respect to relative indications for stopping, generally the test is not stopped unless other abnormal signs or symptoms occur simultaneously. The reason for performing the GXT may also influence relative reasons for test termination. For example, when evaluating a patient for suspected CAD, if the test supervisor observes left bundle branch block with exertion, this precludes interpretation of ST changes and is a reason for test termination.

Each exercise testing laboratory must have policies and procedures that specify absolute and relative indications for test termination. Having these policies and procedures eliminates any confusion about why a test was terminated and helps protect the facility from potential negligence if complications arise. If a question ever arises about whether a test should be stopped based on what is observed, the tester should always err on the conservative side and consider the safety of the patient as the highest priority (i.e., the test can always be repeated if necessary).

Resting, Exercise, and Recovery ECG Abnormalities

This section discusses analysis of the 12-lead ECG, which is based on recognizing normal and abnormal findings at rest, during exercise, and in recovery. The material is intended as a review of GXT-specific issues. When assessing the resting ECG before the GXT, the clinician should adjudicate their interpretation against the patient's medical history to determine whether any discrepancies are present (e.g., the patient states no previous myocardial infarction, but the current ECG shows significant Q waves throughout the anterior leads). Also important is a comparison of the patient's current resting 12-lead ECG against a previous ECG, especially when an abnormality is detected (e.g., the patient's current resting ECG rhythm shows atrial fibrillation; is this arrhythmia new or old?). When a clinically significant difference is detected on the resting ECG and it appears to be a new finding, the supervising or referring physician should be informed before the test is undertaken.

The most common reason to complete a GXT with a 12-lead ECG is to assess for potential myocardial ischemia due to CAD. However, in certain patients, abnormalities in their ECG at rest preclude use of the ECG during exercise to accurately detect or determine exercise-induced myocardial ischemia. In these patients, an exercise- or pharmacology-induced stress test is completed in conjunction with cardiac imaging by echocardiography or radionuclide testing. The following are common abnormalities in the resting ECG that limit the sensitivity of the exercise ECG to detect ischemia:

- Left bundle branch block
- Right bundle branch block (ST changes in anterior leads, V1-V3, cannot be interpreted, but remaining leads are interpretable)
- Preexcitation syndrome (Wolff-Parkinson-White [WPW])
- Nonspecific ST-T wave changes with >1 mm depression
- Abnormalities due to digoxin therapy or left ventricular hypertrophy
- Electronically paced ventricular rhythm

During the GXT, one person should continuously observe the ECG monitor to identify the onset and nature of any ECG change. The evaluation of the ST segment is of great importance because of its ability to suggest the onset of ischemia. In addition, the onset of arrhythmias should be identified, especially those that are related to indications for test termination (see the sidebar Indications for Termination of Graded Exercise Test). Although the onset and progression of ST depression are subtle in most cases, arrhythmia can occur suddenly and can be brief, intermittent, or sustained.

All 12 ECG leads should be monitored; however, V5 is the lead that is most likely to demonstrate ST-segment depression, whereas the inferior leads (II, III, and aVF) are associated with a relatively higher incidence of false-positive findings (i.e., ST depression occurs, but it is not truly related to ischemia). V5 is the most diagnostic lead because when true ischemia occurs with exertion, it is most commonly observed in this lead. When the test supervisor recognizes ST-segment changes, they should note the exercise time and the work rate (i.e., METs) at onset, the morphology, and the magnitude, as well as document any corresponding symptoms (e.g., chest pain, shortness of breath, dizziness).

ST-Segment Depression

Of the potential ST-segment changes, ST-segment depression is the most frequent response during exercise and is suggestive of subendocardial ischemia. One or more millimeters of horizontal or downsloping ST-segment depression that occurs 0.08 s (i.e., 80 ms) past the J point is recognized as a positive test for myocardial ischemia (figure 5.2). When ST-segment changes of this type occur along with typical angina symptoms, the likelihood of CAD is extremely high. In addition, the earlier the onset, the greater the amount of ST depression, the more leads with ST depression, and the more time it takes for the ST depression to resolve in recovery, the more likely it is that significant CAD is present. In some cases ST-segment depression is observed only in recovery, yet it should be treated as an abnormal response. Additionally, J-point depression with an upsloping ST segment that is more than 1.5 mm depressed at 0.08 s past the J point is also suggestive of exercise-induced ischemia.

ST-Segment Elevation

ST-segment or J-point elevation observed on a resting ECG is often attributable to early repolarization and is not necessarily abnormal in healthy people, but this should be documented on the GXT report. With exertion, this type of ST elevation normally returns to the isoelectric line. New ST-segment elevation with exertion (assuming the resting ECG is normal) is a somewhat rare finding and may suggest transmural ischemia or a coronary artery spasm. This type of ST-segment elevation is an absolute reason for stopping the test (see the sidebar Indications for Termination of Graded Exercise Test). When Q waves are present on the resting ECG from a previous infarction, ST elevation with exertion may simply reflect a left ventricular wall motion abnormality. ST-segment elevation can localize the ischemic area and the arteries involved, whereas this is not always the case with ST-segment depression (31, 40).

T-Wave Changes

In healthy individuals, T-wave amplitude initially decreases gradually with the onset of exercise. Later, at maximal exercise, T-wave amplitude increases. In the past, T-wave inversion with exertion was thought to reflect an ischemic response. But it is now believed that flattening or inversion of T waves may not be associated with ischemia. T-wave inversion is common in the presence of left ventricular hypertrophy. It remains undecided whether inverted T waves present on the ECG at rest that then normalize with exertion reflect myocardial ischemia. Normalization of T waves is also present during ischemic responses associated with coronary spasms; however, this finding has the greatest significance under

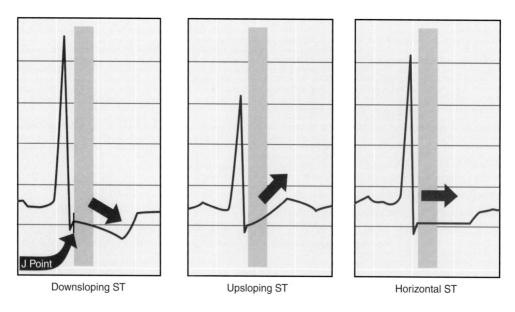

| Downsloping ST | Upsloping ST | Horizontal ST |

Figure 5.2 Location of J point and examples of the three types of ST-segment depression (downsloping, upsloping, and horizontal). See text for criteria for determining a positive test for myocardial ischemia based on the type of ST-segment depression.

resting conditions. Overall, T-wave changes with exertion are not specific to exercise-induced ischemia but should be correlated with ST-segment changes and other signs and symptoms. Note that T-wave changes frequently follow ST changes and can be difficult to isolate with exertion.

Arrhythmia

In relation to ST changes with exertion, arrhythmias are of equal clinical importance and potentially more life threatening, based on the suddenness in which arrhythmias may appear. Although a GXT is most commonly used to diagnose potential CAD, this test can also be used to evaluate symptoms (e.g., near-syncope) attributable to arrhythmia. In addition, a GXT may be used to evaluate the effectiveness of medical therapy in controlling an arrhythmia. The supervisor of the GXT must have a strong knowledge of arrhythmia detection and be able to respond appropriately and quickly. When arrhythmias appear during the GXT, the onset of the arrhythmia and any signs or symptoms associated with the arrhythmia should be documented, along with any other ECG changes (e.g., ST depression). The health care provider supervising the GXT should be knowledgeable about and able to recognize three major types of ECG rhythm or conduction abnormalities during exercise:

1. Supraventricular arrhythmias that compromise cardiac function (e.g., atrial flutter, atrial fibrillation)

2. Ventricular arrhythmias that have the potential to progress to a life-threatening arrhythmia

3. The onset of high-grade conduction abnormalities

Assessment of Functional Capacity

As mentioned earlier in this chapter, FC as measured by estimated METs, exercise duration, or peak $\dot{V}O_2$ is inversely associated with prognosis in several patient groups (1, 2, 3, 18, 22, 26). $\dot{V}O_2$ is determined by indirect open-circuit spirometry, accomplished using a gas collection apparatus (e.g., a mouthpiece) and a metabolic measuring system (figure 5.3). Although manual and semi-automated gas collection systems and analyzers are still in use today, most clinical exercise testing labs use fully automated real-time systems (i.e., a metabolic cart). These systems can be almost desktop in

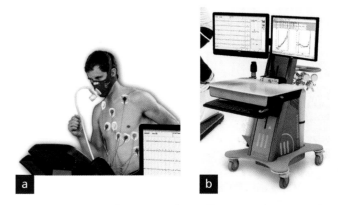

Figure 5.3   Configuration for collection of ECG and gas exchange data during *(a)* a CPET using *(b)* a metabolic cart.

Photos courtesy of MGC Diagnostics Corporation.

Indications for Termination of Graded Exercise Test

Absolute Indications

1. Decrease in systolic BP of >10 mmHg below baseline BP despite an increase in workload and when accompanied by other evidence of ischemia

2. Moderate to severe angina (2 or more on a 4-point scale)

3. Increasing nervous system symptoms (e.g., ataxia, dizziness, or near-syncope)

4. Signs of poor perfusion (cyanosis or pallor)

5. Technical difficulties monitoring the ECG or systolic BP

6. Subject's desire to stop

7. Sustained ventricular tachycardia or other arrhythmia that interferes with hemodynamic stability

8. ST elevation (>1.0 mm) in leads without diagnostic Q waves (other than V1, aVL, or aVR)

Relative Indications

1. Decrease in systolic BP of >10 mmHg below baseline BP despite an increase in workload and in the absence of other evidence of ischemia

1. ST or QRS changes such as excessive ST depression (>2 mm horizontal or downsloping ST-segment depression) or marked axis shift

2. Arrhythmias other than sustained ventricular tachycardia, including multifocal premature ventricular contractions (PVCs), triplets of PVCs, supraventricular tachycardia, heart block, or bradyarrhythmias

3. Fatigue, shortness of breath, wheezing, leg cramps, or claudication

4. Development of bundle branch block or intraventricular conduction delay that cannot be distinguished from ventricular tachycardia

5. Increasing chest pain

6. A systolic BP of >250 mmHg or a diastolic BP of >115 mmHg

Adapted by permission from G.F. Fletcher et al., "Exercise Standards for Testing and Training: A Scientific Statement From the American Heart Association," *Circulation* 128, no. 8 (2013): 873-934.

size; they collect input variables such as barometric pressure and temperature, as well as measure and analyze breath-by-breath exhaled air for ventilation and the concentrations of O_2 and CO_2. The information provided is monitored continuously and can be viewed in real time during a test, with results reported in any desired time frame. This additional testing equipment transforms a regular GXT into a CPET.

A CPET incorporates most of the methods, procedures, and absolute and relative contraindications associated with starting or stopping a regular GXT. Additionally, the gas exchange data that are gathered allow for the accurate calculation of minute ventilation ($\dot{V}_E$), $\dot{V}CO_2$, and $\dot{V}O_2$ as well as other variables of clinical importance, such as breathing reserve, ventilatory-derived anaerobic threshold, breathing frequency, oxygen pulse ($\dot{V}O_2$ divided by heart rate), ventilatory equivalent for O_2, slope of the change in $\dot{V}_E$ relative to change in carbon dioxide ($\dot{V}CO_2$) production (i.e., $\dot{V}_E$–$\dot{V}CO_2$ slope), slope of the change in work rate relative to change in $\dot{V}O_2$, oxygen uptake efficiency slope, and partial pressure of end-tidal CO_2 (6, 15, 16). Finally, the conduct and discontinuation of a test can now be guided based on physiologic effort or cellular metabolism, as determined by respiratory exchange ratio (RER; the ratio of $\dot{V}CO_2$ to $\dot{V}O_2$), instead of some percentage of predicted maximal heart rate. This approach is especially helpful when testing a patient taking a medication that influences heart rate response to exercise, such as a β-adrenergic blocking agent. Using RER to evaluate level of effort to help guide (not determine) when a patient might need to stop exercising is usually associated with an achieved RER value that preferably reaches 1.10 or greater.

Several statements from the American Heart Association on cardiopulmonary gas exchange methods provide a complete review of this topic (6, 15, 16). Besides providing a more accurate measurement of exercise capacity, a CPET is often used for risk stratification and for determining prognosis and need for mechani-cal support or cardiac transplantation in patients with heart failure (6, 20, 25; see also chapter 16). Additionally, CPET is increasingly being used to determine whether the cause of a patient's unexplained dyspnea has a cardiac or pulmonary etiology.

The assessment of FC during testing can also help the clinician guide decision making relative to return to work after an illness or injury. For example, for the patient recovering from a myocardial infarction that occurred 4 wk earlier who now asks to return to an occupation that requires periodic bouts of exertion equivalent to 6.5 METs (e.g., moderately paced shoveling of dirt), FC measured during a GXT can help determine if the patient is able to achieve a peak MET level that is at least 20% to 25% more than the task he wishes to resume—in this example, almost 8 METs. If such a peak work rate cannot yet be achieved, return to work with restrictions or further time in rehabilitation may be indicated.

Finally, assessment of FC during a GXT can help quantify, for clinical or research purposes, the impact of either an exercise training program or a drug or surgical device intervention on FC, be it change in exercise duration, peak $\dot{V}O_2$, or work rate (e.g., watts). Because of the inconvenience and the slight increase in expense of directly measuring a patient's FC using a metabolic cart for respiratory gas exchange, some clinicians and researchers use equations to predict peak $\dot{V}O_2$. The standard American College of Sports Medicine (ACSM) metabolic equations (4) are commonly used; but they rely on achieving a steady state, which does not occur at maximal effort, and therefore can lead to an overestimation of peak FC and potential misclassification of a patient's risk or inaccurate assessment of the response to a therapy. Also, in the absence of a metabolic cart, some clinics will estimate FC (i.e., METs) based on the complete or partial completion of an exercise stage. For example, a patient must complete 1.5 or 2 min of a 3 min stage to be awarded the estimated METs for that stage.

Practical Application 5.1

CLIENT–CLINICIAN INTERACTION

The 6 to 8 min recovery period that follows every GXT provides an excellent opportunity for the testing staff and patients to discuss strategies for modifying existing relevant risk factors, with the goal of decreasing future risk for a lifestyle-related health problem. Begin by simply asking patients if they would mind talking about exercise habits, weight management practices, or smoking cessation. If they agree, you'll have a few minutes to learn about their goals and any steps they are currently taking or have tried in the past, and then bring in your expertise relative to effective behavior methods that may be of help. Despite all the important work that clinical exercise physiologists perform with respect to supervising or conducting exercise tests, prescribing or leading exercise, and measuring changes in fitness, in the end, a big portion of our time is spent counseling others about how to make favorable changes to unhealthy behaviors.

Finally, any major clinical concerns or findings that arose during the test need to be mentioned to patients during recovery. Without going into specific details or interpreting the results, staff should remind patients to follow up on their results with the clinician who ordered the test.

Interpretation of Findings and Generation of Final Summary Report

The recovery period, which follows the exercise portion of the GXT, allows the clinical exercise physiologist to monitor a patient's physiologic and clinical responses after exercise, as well as provides an opportunity to briefly speak with the patient about one or two important recommendations aimed at pertinent lifestyle-related risk factors (see practical application 5.1). Once the GXT is done, the clinical exercise physiologist who helped supervise the test often helps prepare a final summary report that is later read and signed by a physician. The key elements to be included and succinctly discussed and interpreted in this report are (a) angina status, (b) ECG findings (i.e., ST-segment changes) pertaining to ischemia, (c) ECG findings pertaining to arrhythmia, (d) functional capacity, (e) HR response, and (f) BP response. When applicable, each of these elements can be interpreted relative to diagnosis, prognosis, or future risk for developing a disorder.

The statement pertaining to angina status should specify whether or not angina was present and, if so, when it began (at what HR or MET level). Also, include whether angina was the reason for stopping the test, what intensity level it reached (see angina scale in table 5.3), and what intervention was required during recovery to relieve the discomfort (rest, medications). The statement addressing ECG evidence of exercise-induced myocardial ischemia should be based on ST-segment information gathered at rest, during exercise, and in recovery. If the ECG at rest was normal, then simply state, based on the presence or absence of ST depression during exercise, that the test either does or does not meet the criteria to be considered positive for exercise-induced myocardial ischemia.

Any discussion regarding whether an exercise test is positive or negative for exercise-induced ischemia should consider the diagnostic value of such a test, which introduces the concepts of **sensitivity** and **specificity**. Generically, these terms apply to all diagnostic tests and refer to how effectively a test identifies patients with a certain disease (sensitivity) and those free of a certain disease (specificity). Specific to ECG stress testing, sensitivity refers to how well the test identifies patients with CHD. It is computed as the percentage of patients with actual CHD who exhibit ST-segment depression (true positive, TP) during exercise testing divided by TP plus the number of patients with CHD who do not exhibit ST-segment depression during exercise testing (false negative, FN). In summary, sensitivity is computed as follows:

$$\text{Sensitivity (expressed as a \%)} = TP / (TP + FN) \times 100$$

For ECG stress testing, specificity refers to how well the test identifies patients free of CHD; it is the number of patients actually free of CHD who show no ST-segment abnormalities during testing (true negative, TN) divided by TN plus the number of patients free of disease who do exhibit ST-segment abnormalities (false positive, FP). Specificity is computed as follows:

$$\text{Specificity (expressed as a \%)} = TN / (TN + FP) \times 100$$

Information concerning FC, stated as estimated METs or measured peak $\dot{V}O_2$, should be provided in the final report both as an absolute number and in qualified terms (e.g., above average, poor, superior) relative to data for other people of similar age and gender. The ACSM provides tables for making such comparisons (4).

Quantify in the final report whether achieved peak HR was normal (e.g., exceeded 80% of age predicted) or consistent with a chronotropic incompetence response (less than 80% of age predicted) (table 5.4). The prognostic value of a reduced peak HR response is as great as that of an exercise-induced myocardial perfusion deficit. In a similar fashion, the failure of the HR to decrease promptly after exercise also provides independent information related to prognosis (21). The failure of the HR to decrease by at least 12 beats during the first minute or 22 beats by the end of the second minute of recovery is independently associated with an increased risk for mortality over the next 3 to 5 yr (21, 22, 39). However, HR recovery cannot discriminate those with coronary disease and instead should be used to supplement other predictors of mortality, such as the Duke nomogram (26).

Table 5.4 Abnormal Hemodynamic Responses to Peak Exercise and in Recovery From Exercise

Time point	Heart rate	Blood pressure
Peak exercise	Achieves ≤80% of APMHR Achieves ≤62% of APMHR if patient is using a β-adrenergic blocking agent	SBP • Men: SBP >210 mmHg • Women: SBP >190 mmHg DBP • DBP >10 mmHg from rest *or* • DBP >90 mmHg overall
Exercise recovery	Compared with peak exercise, decreased <12 b · min⁻¹ at min 1 of active recovery Compared with peak exercise, decreased <22 bpm at min 2 of active recovery	SBP • SBP at 3 min after exercise should be <90% of SBP at peak exercise

APMHR = age-predicted maximal heart rate; SBP = systolic blood pressure; DBP = diastolic blood pressure.

Adapted from Brubaker and Kitzman (8); Dlin et al. (10); Lauer et al. (23).

Systolic BP normally increases in a negatively accelerated manner during incremental exercise. The magnitude of the increase approximates 8 to 10 mmHg per 1 MET of work. An absolute peak systolic pressure of >250 mmHg is considered an indication to stop the test. In nonathletic populations, an increase in systolic to >210 mmHg in men or >190 mmHg in women (table 5.4), or a relative increase of >140 mmHg above resting levels, is considered a hypertensive response and may be predictive of future hypertension at rest (4, 24, 37). Patients with limitations of cardiac output can show an inappropriately small increase in systolic BP during the exercise test. Note, a decrease of systolic BP to below the resting value or by >10 mmHg after a preliminary increase, particularly in the presence of evidence of ischemia (angina, ST changes), is abnormal (32). There is typically no change or a slight decrease in diastolic BP during an exercise test.

After exercise, systolic BP normally decreases promptly (5). Several investigators have demonstrated that a delay in the recovery of systolic BP is related to both ischemic abnormalities and a poor prognosis (5, 28). As a general principle, the systolic BP at 3 min after exercise should be <90% of the systolic BP at peak exercise. Also, if peak exercise BP cannot be measured accurately, the systolic BP at 3 min of recovery should be less than the systolic BP measured at 1 min of recovery.

GRADED EXERCISE TESTING WITH DIAGNOSTIC IMAGING

Graded exercise combined with an imaging modality is an appropriate method for assessing myocardial ischemia and function. The two most utilized tests with exercise are echocardiography and myocardial radionuclide (perfusion) imaging.

Exercise Echocardiography

An exercise echocardiography test, performed using ultrasound technology, assesses ischemic heart disease by evaluating left ventricular (LV) wall motion during systole (i.e., contraction). Ischemia slows the onset of contraction and relaxation, along with reducing the velocity of the contraction. Segments of the LV are evaluated as normal, decreased contraction (**hypokinetic**), the absence of contraction (**akinetic**), or abnormal movement during contraction (**dyskinetic**). Ischemia is diagnosed if an echocardiogram performed immediately after the exercise test shows any of these three wall motion abnormalities. The test begins with a baseline, or preexercise, echocardiogram. The patient then completes a symptom-limited graded exercise test (either treadmill or bike), following all previously discussed guidelines. On termination of exercise, the patient is rapidly moved from the treadmill to a left lateral decubitus position (lying on left side of body) on a nearby bed, and an echocardiogram is again performed in order to compare the before and after images.

The hemodynamic response, functional capacity, patient-reported symptoms, and ECG are combined with the echocardio-gram findings to provide a thorough evaluation of any exercise-induced myocardial ischemia. As a clinical exercise physiologist, you may be responsible for preparing the patient, along with helping conduct the exercise test. However, a trained echocardiographer will perform the echocardiograms. Recent advances in stress echocardiography include the administration (intravenous injection) of contrast agents to improve the accuracy of the interpretation of the LV-segments (30). Stress echocardiography is indicated for patients with an intermediate probability of CAD who have an uninterpretable ECG, patients being considered for revascularization, and patients who have previously undergone revascularization therapy, as well as for evaluating hemodynamics in patients with aortic stenosis and low cardiac output (11, 42).

Exercise Nuclear Myocardial Perfusion

Myocardial perfusion imaging (MPI) combined with exercise is another important test for evaluating patients with known and unknown coronary artery disease. This test involves single-photon emission computed tomography (SPECT) and requires the injection of radionuclide tracers such as technetium (Tc)-99m or thallous (thallium) chloride-201, which are two agents that are taken up by the perfused myocardium. Any myocardial tissue that is experiencing ischemia has inadequate blood flow, and this shows up as reduced tracer uptake on the SPECT images. A baseline measurement of myocardial perfusion (i.e., tracer uptake) is performed by injecting the radionuclide agent under resting conditions, and then an image is taken using a nuclear imaging camera. The patient then performs a graded exercise test, and another dose of the tracer is injected roughly 1 min before termination of exercise. Additional images are typically taken 10 to 20 min after the completion of exercise to allow for adequate distribution of the tracers. A perfusion defect (i.e., absent or reduced tracer uptake) seen in the postexercise images that was not present in the images taken before exercise indicates myocardial ischemia. A perfusion defect at rest that remains during exercise is a marker of a previous myocardial infarction. Similar to the stress echocardiogram, the clinical exercise physiologist conducts the GXT portion of the study, then all hemodynamic, symptom, and ECG data are combined with the imaging results to support or reject a diagnosis of CHD. In addition, it is important that the patient be able to achieve a minimal exercise hemodynamic level (e.g., >85% of age predicted maximal heart rate) to ensure the accuracy of the test.

Nuclear perfusion testing is indicated for patients who have an intermediate probability of CHD (see table 5.1) and abnormalities in the resting ECG that limit the ability of the exercise ECG to detect myocardial ischemia. It is also a valuable tool for assessing the severity of stable CHD among candidates for revascularization therapy and for evaluating ischemia after revascularization (17, 42).

CONCLUSION

The GXT is often the first diagnostic tool used to assess the presence of significant CHD, with or without nuclear perfusion or echocardiography imaging. In past years, a cardiologist normally supervised the GXT. Today, other health care providers such as clinical exercise physiologists are supervising these tests; the reasons include a better understanding as to which patients are at increased risk for a complication during testing, improved ECG technology, cost-containment initiatives within health care organizations, greater constraints on the cardiologists' time, and improved knowledge and training of other health care providers. Data from the test can be used not only to help diagnose the presence of CHD but also to determine prognosis and provide needed information for the development of a safe exercise prescription.

Online Materials

Visit HK*Propel* to access a link to the references, the case studies with discussion questions, and a quiz to help you review key concepts and test your understanding of the material covered in this chapter.

CHAPTER 6

Exercise Prescription

Steven J. Keteyian, PhD

As the writings and evidence conveyed in this textbook indicate, all physical activity—whether daily (e.g., occupational, household) or purposeful (i.e., exercise training undertaken to improve physical fitness)—has a favorable effect on a variety of health disorders. Regular exercise can lessen disease-related symptoms, improve functional capacity, and improve disease-related outcomes (50, 51). To that end, this chapter provides the minimal knowledge a clinical exercise physiologist needs in order to develop and implement a safe and effective exercise prescription, using methods and principles that are relatively similar, whether prescribing exercise for a high-performance athlete or a patient with a clinically manifest disease. Chapters 8 through 31 adapt these principles in a manner that accommodates the unique pathophysiology and clinical considerations associated with each disease condition, and chapters 32 through 35 address special populations.

First it is important to summarize the updated physical activity recommendations ("prescription") for all Americans (51). Specifically, strong evidence demonstrates that both single and repeated bouts of physical activity or exercise promote improvement in intermediate risk factors (e.g., blood pressure, insulin sensitivity), mood, sleep, executive function, quality of life, and health (e.g., longevity, fall risk). Although increasing one's weekly energy expenditure by almost any amount provides benefit (see the Progress Overload section), the target range suggested is 150 to 300 min of moderate-intensity physical activity per week, a level that approximately half the U.S. adult population does not currently attain. In fact, 30% of the population reports doing no moderate to vigorous activity (51).

Despite the use of the word *prescription*, the development of an exercise prescription does not necessarily require approval by a physician. In some situations, however, it may, especially when the prescription is developed for a patient with a clinically manifest disease. For example, a physician's approval and signature may be required to secure reimbursement from Medicare for a patient participating in a cardiac rehabilitation program or if the clinical exercise professional feels approval is warranted to limit personal liability. Individuals practicing clinical exercise physiology who are charged with developing an exercise prescription often find that doing so is both an art and a science; they must possess the requisite knowledge and skills needed to put into action a prescription that is safe and practical (see practical application 6.1).

The primary purpose of the exercise prescription is to provide a valid and specific guide to help individuals achieve optimal health and physical fitness. The exercise prescription should be specific to the nature of the clinical population. Several authoritative groups have published statements on prescribing exercise in particular populations, including healthy people and those with coronary artery disease, osteoporosis, hypertension, and diabetes, as well as older people (2, 3, 4, 5, 45, 51). The exercise prescription can also address the five health-related components of physical fitness (4):

1. *Cardiorespiratory endurance (aerobic fitness):* ability of the cardiorespiratory system to transport oxygen to active skeletal muscles during prolonged submaximal exercise and the ability of the skeletal muscles to utilize oxygen through aerobic metabolic pathways.

2. *Skeletal muscle strength:* peak ability to produce force. Force may be developed by isometric, dynamic, or isokinetic contraction.

3. *Skeletal muscle endurance:* ability to produce a submaximal force for an extended period.

4. *Flexibility:* the ability of a joint to move through its full, capable range of motion.

5. *Body composition:* the relative percentage of fat and nonfat masses that compose total body weight. Chapter 9 provides details regarding body composition assessment and the principles for tailoring the exercise prescription toward weight loss and reducing body fat.

Each of these components of physical fitness is related to at least one aspect of health, and each component is positively influenced by exercise training, such as reducing the risk of experiencing a primary or secondary chronic disease (5, 6, 7, 40, 50, 51). The general benefits of regular exercise training can be summarized as follows (4, 51):

- Improved cardiorespiratory and musculoskeletal fitness
- Improved metabolic, endocrine, and immune function
- Reduced all-cause mortality
- Reduced risk of cardiovascular disease
- Reduced risk of certain cancers (colon, breast)
- Reduced risk of osteoporosis and osteoarthritis
- Reduced risk of non-insulin-dependent diabetes mellitus
- Improved glucose metabolism
- Improved mood
- Reduced risk of obesity
- Overall improved health-related quality of life
- Reduced risk of falling
- Improved sleep patterns
- Improved health behavior

Three key principles or tenets must be considered in the development of an exercise prescription:

1. Specificity of training
2. Progressive overload
3. Reversibility

The clinical exercise physiologist must also consider several aspects of a person's psychosocial condition, including factors that may be relevant to beginning and adhering to an exercise training program (see chapter 2).

EXERCISE TRAINING SEQUENCE

A comprehensive training program should include flexibility, resistance, and cardiorespiratory (aerobic) exercises. The order of the exercise training routine can be important for both safety and effectiveness; however, scientific data on this topic are lacking. Before engaging in aerobic or resistance training, patients should perform a 4 to 5 min bout of slow full-body activity such as walking, cycling, or swimming. This will proportionately increase bioenergetic activity within, and circulation to, the soon-to-be active skeletal muscles, as well as increase body temperature overall and within the muscle and muscle–tendon unit. Then, after aerobic or resistance training, flexibility exercises can be performed to improve range of motion (2, 31). In a clinical population, if aerobic training and resistance training take place on the same day, the best approach is to first perform the activity that is the primary focus of that day's training.

Practical Application 6.1

THE ART OF EXERCISE PRESCRIPTION

The art of prescribing exercise involves the successful integration of exercise science with behavioral techniques in a manner that results in long-term program compliance and attainment of the individual's goals (2). Unlike the disciplines of chemistry or physics, physiology and behavioral sciences are sometimes less exact. We cannot always precisely predict physiological or psychological responses because numerous factors and confounders can influence the outcome. These include age, physical and environmental conditions, sex, previous experiences, genetics, and nutrition. When developing an exercise prescription, you should follow the basic guidelines provided in this chapter. In doing so, you can help elicit the desired response both during a single exercise training session and over the course of an multiple repeated bouts of training. Keep in mind, however, that not all people respond as expected, especially those with a chronic disease. For example, people with coronary artery disease may require modification of exercise intensity because they experience myocardial ischemia above a certain heart rate (HR). Additionally, those currently undergoing treatment for cancer often fatigue easily and may better tolerate repeated training sessions of shorter duration. There are several reasons for altering an exercise prescription in selected individuals:

- Variance in objective (physiological) and subjective (perceptual) responses to an exercise training bout
- Variance in the amount and rate of exercise training responses
- Differences in goals between individuals
- Variance in behavioral changes relative to the exercise prescription

Each of these reasons should be considered for both the initial development and subsequent review of the exercise prescription. A modified exercise prescription should not be considered adequate unless it is evaluated for effectiveness over time. As a rule, a person's exercise prescription should be reevaluated weekly until its parameters appear to be safe as well as adequate for improving health-related behaviors and selected physiological indexes.

GOAL SETTING

A comprehensive exercise prescription should consider the goals that are specific for each person. Common goals include the following:

- Improving appearance
- Improving quality of life
- Managing weight
- Preparing for competition
- Improving general health to reduce risk for primary or secondary occurrence of disease
- Reducing the burden of a chronic disease or condition (early fatigue, rehospitalization, depression, loss of personal control, economic hardship)

People with specific diseases often have goals that relate directly to reversing or reducing the progression of their disease and its side effects or the side effects of the therapies used to treat the disorder. A clinical exercise physiologist must have a comprehensive understanding of how to alter the general exercise prescription so as to provide the patient with the best chance of success in achieving a desired goal. Also, the exercise physiologist should help assess whether the stated goals are realistic and discuss them with patients when they are not.

PRINCIPLES OF EXERCISE PRESCRIPTION

To gain the optimal benefits of exercise training, regardless of the area of emphasis (i.e., cardiorespiratory, strength, muscular endurance, body composition, range of motion), several principles must be followed; these are discussed next.

Specificity of Training

Long-term changes or adaptations in physiological function occur in response to a chronic or repeated series of stimuli. The principle of specificity of exercise training states that these physiologic changes and adaptations are specific to the cardiorespiratory, neurologic, and muscular responses that are called upon to perform the exercise activity. Specifically, the unique neuromuscular firing patterns and the cardiorespiratory responses that are engaged to perform an activity are the same ones that undergo the greatest degree of adaptation. For example, if a 1600 m college runner wants to do all she can to improve her race performance, she should spend the majority of her training time running middle and longer distances; this type of training stress and combines (a) the cardiorespiratory processes involved in the transport and use of oxygen with (b) the firing of the neurons used during running at a higher velocity. In contrast, an offensive lineman in football should spend very little of his practice time engaged in distance running and instead engage in explosive activities and blocking skill techniques.

A classic example of specificity of training would be observed if we measured a person's peak oxygen uptake ($\dot{V}O_2$) on the treadmill one day and then again on a stationary bike the next day. We know that for this person, $\dot{V}O_2$peak will be 5% to 15% higher on a treadmill versus a cycle ergometer (26, 36, 52, 53). The majority of this difference is related to both the weight independence associated with sitting on a bike and the smaller total muscle mass used during cycling (vs. running). However, if we repeated

---

Questions to Ask a Person When Developing an Exercise Prescription

Specificity
- What are your specific goals when performing exercise (health, fitness, performance)?
- Do you want to exercise more?
- Do you want to be able to do more activities of daily living?
- Do you want to perform something that you currently cannot? If yes, describe.

Mode
- What types of exercise or activity do you like the best?
- Do you already have any exercise equipment in your home?
- What types of exercise do you like the least?

Frequency
- Do you know how many days per week of exercise or physical activity are required for you to reach your goals?
- How many days during a week do you have 30 to 60 min for exercise?

Intensity
- Do your goals include optimal improvement of your fitness level, or are your goals primarily related to your health?
- Do you have any musculoskeletal problems that would limit how hard or often you are able to exercise?

Time
- What is the best time of the day for you to exercise?
- Can you awaken earlier on some days or take 30 to 40 min at lunchtime for exercise?

this same experiment using highly trained competitive cyclists, we might observe that they achieve a strikingly similar peak $\dot{V}O_2$ regardless of whether tested on the bike or the treadmill. This observation among competitive cyclists is attributable to the fact that they spend virtually all of their training time cycling and as a result, their physiologic adaptations are specific in response to such. Some modest crossover adaptations likely do occur from one mode to another, partly the result of a combination of improvements in both central cardiac function (e.g., increased stroke volume and cardiac output) and engaging the skeletal muscles used in the alternative exercise mode (e.g., training the leg muscles by cycling and then using many of those same leg muscles when running), which is partly the basis for the concept of **cross-training**.

The sidebar Questions to Ask a Person When Developing an Exercise Prescription will help ensure that exercise professionals develop an exercise prescription specific to each person being assessed.

Progressive Overload

The progressive overload principle refers to the relationship between the magnitude of the dose or exercise stimulus and the benefits gained. In general, an increasing amount or volume of exercise is associated with commensurately greater gains in health and fitness; however, there appears to be a level of exercise beyond which benefits plateau or possibly even diminish. For instance, in the Harvard Alumni Health Study, an apparent dose–response relationship was observed between all-cause death and the number of kilocalories expended each week (40), such that for between 500 and 3,500 kcal expended each week, the risk for mortality fell as

the kilocalories expended increased (figure 6.1). Note the somewhat steep decrease in deaths between the energy expenditures of 500 and 1,500 kcal/wk, suggesting that even small to modest increases in weekly energy expenditure provides meaningful health benefits (51).

Overload refers to an increase in total work above and beyond that normally performed on a day-to-day basis. For example, when a person performs walking as part of an exercise training regimen to improve fitness, the pace and duration for one or both of these parameters should be above that typically experienced on a daily basis. Progressive overload is the gradual increase in the amount of work performed in response to the continual adaptation of the body to the work.

Applying this principle to a walking program would require progressively walking more often, farther, or at a faster pace. Overload is often expressed using the FITT principle, in which the FITT-VP acronym stands for **F**requency, **I**ntensity, **T**ime (i.e., duration), **T**ype, **V**olume, and **P**rogression of exercise.

Before we discuss frequency, intensity, and duration separately, it is important to understand how these three parameters can be combined as a means to quantify the total volume (the V in the FITT-VP acronym) of exercise an individual is engaging in. Although kilocalories per week can be used, as shown in figure 6.1, other common units for expressing total exercise volume are MET-min/wk and MET-h/wk, where MET is defined as metabolic equivalents of task and 1 MET approximates a resting $\dot{V}O_2$ of 3.5 mL · kg⁻¹ · min⁻¹. For example, a 50 yr old person who square-dances at a moderate pace (~4 METs) for 40 min three times a week engages in 480 MET-min/wk (4 × 40 × 3 = 480) or 8 MET-h/wk (480 MET-min/wk divided by 60 min/h = 8 MET-h/wk). Although it is not easy to

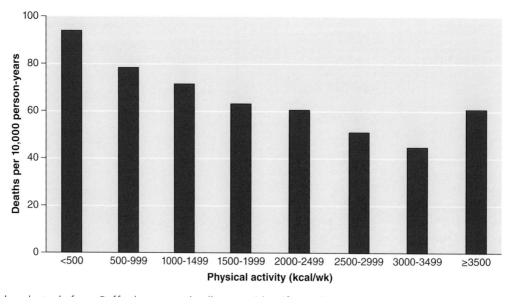

Figure 6.1 Landmark study from Paffenbarger and colleagues identifies a dose–response relationship between all-cause death and weekly energy expenditure (kcal/wk) among Harvard alumni.

Adapted from M.L. Foss and S.J. Keteyian, *Fox's Physiological Basis for Exercise and Sport*, 6th Ed. (Pittsburgh, PA: William C. Brown, 1998). By permission of authors.

use MET-min or MET-h/wk to prescribe exercise to individuals or patients, this unit of measure does come in handy when comparing estimates of exercise volume across research studies or conveying broad recommendations for public health. Current public health recommendations are that all adults engage in at least 8 MET-h/wk of exercise (4, 5, 51).

Concerning exercise progression (the *P* in the FITT-VP acronym), especially for the patient who has not regularly exercised before, frequency and duration of activity are typically increased first to desired levels. Then, as tolerated, intensity is increased. With respect to cardiorespiratory fitness, Gormley and colleagues (23) showed that when the volume of exercise is held constant, training at greater exercise intensities is associated with greater gains in fitness.

Frequency

Frequency is the number of times an exercise routine or physical activity is performed (per week or per day).

Intensity

The intensity of exercise or physical activity refers to either the objectively measured work or the subjectively determined level of effort performed by an individual. Typical objective measures of work that are important to the clinical exercise professional include HR, oxygen uptake ($\dot{V}O_2$ or METs), caloric expenditure (kilocalories [kcal] or joules [J]), mass or weight lifted (kilograms, pounds), and power output (kilograms per minute [$kg \cdot min^{-1}$] or watts [W]). The anaerobic, lactate, or ventilatory thresholds may also be used to determine exercise intensity, but they are often impractical during exercise training in the clinical setting. The subjective level of effort can be evaluated through (a) a verbal statement from the person performing exercise (e.g., "I'm tired" or "This is easy"), (b) the so-called talk test (i.e., fastest pace possible while still able to carry on a conversation), or (c) a standardized scale (e.g., Borg rating of perceived exertion). Concerning the Borg scale, the patient must be taught the proper use of this assessment tool to obtain accurate indications of perceived effort (38).

Duration

Duration (or *time*, the first *T* in the FITT acronym) refers to the amount of time that is spent performing exercise or physical activity. During aerobic-type exercise training, the duration (e.g., 30 min or more) is typically accumulated without interruption (i.e., continuously) or with very short rest periods. However, the 1996 Surgeon General's report on physical activity and health included an important recommendation for exercise and better health (50): All children and adults should *accumulate* a minimum of 30 min of moderate physical activity on most and preferably all days of the week. That means 30 min of exercise can be accomplished in a single continuous session (e.g., one 30 min bout of Nordic skiing) or accumulated during the day (e.g., three 10 min bouts

of walking, or one 15 min bout of stationary cycling plus one 15 min bout of stationary rowing). Compared with the continuous method, the discontinuous (i.e., accumulative) model appears to also improve cardiorespiratory fitness and blood pressure, decrease frailty risk, and lower risk for all-cause mortality, with its effect on body composition, obesity, blood lipids, blood glucose, and mental health inconclusive (30, 39). The discontinuous approach may be more appealing for currently inactive people trying to become more active or for people with very low fitness levels.

Reversibility

Typically, with an 8 to 12 wk period of training, most untrained people can expect a 10% to 30% improvement in $\dot{V}O_2$peak and work capacity, and a 25% to 30% increase in muscle strength (4, 5, 19); less fit people generally achieve gains at a faster rate and to a greater relative degree than more fit people. Alterations in other physiological variables, such as body weight and blood pressure, may take a variable amount of time. Maintaining these improvements requires a minimal volume of training, and the principle of reversibility describes the loss of these acquired gains due to inactivity.

The reversal of fitness due to inactivity or increased sedentary lifestyle habits is often called deconditioning or detraining and is in addition to any loss of fitness with aging. Concerning cardiorespiratory fitness and detraining, Saltin and colleagues (43) reported on the effects of bed rest over a 3 wk period in five subjects. Mean $\dot{V}O_2$peak was decreased 28% at wk 3, and all subjects had reductions in peak cardiac output caused by reduced stroke volumes. This deleterious trend was reversed when exercise training was again implemented (so-called retraining). In another study of seven subjects who exercised for 10 to 12 mo and then engaged in 3 mo of detraining, $\dot{V}O_2$peak decreased by 7% between day 0 and day 12 of detraining and by another 7% between day 21 and day 56 (13). The initial 7% reduction in $\dot{V}O_2$peak was the result of a near-equal 9% reduction in stroke volume, with the subsequent reduction mostly due to a decline in arteriovenous oxygen difference.

The complete loss of training-related skeletal muscle strength and endurance adaptations occurs after several weeks to months of inactivity. A reduction in resistance training volume without complete cessation of training does help with maintenance of much of the gained resistance training effects (24). Concerning rate of loss of flexibility or range of motion (ROM) with detraining or inactivity, many factors are involved including injury, specific individual physiology, degree of overall inactivity, and posture. The reintroduction of a flexibility training routine (i.e., retraining) typically results in rapid regain of ROM.

CARDIORESPIRATORY ENDURANCE

Exercise training to improve cardiorespiratory, or aerobic, endurance requires that individuals perform activities that use large muscle groups and are continuous and repetitive or rhythmic (4). Satisfactory modes of exercise that involve the large muscle groups include walking, running, cycling, skating, stair stepping, rope skipping, and group aerobics (e.g., dance, step, cycling, water). Exercises using strictly the arms are more limited but include upper body crank ergometry, dual-action stationary cycling using only the arms, and wheelchair ambulation. Several popular modes of exercise use both the arms and legs: rowing, swimming, and exercise on some types of stationary equipment (e.g., dual-action [arm and leg] cycles, stationary rowers, cross-country skiing, elliptical trainers, seated dual-action steppers).

Specificity of Training

The general benefits gained from aerobic exercise training appear to be independent of any specific type or mode of training. For instance, the Harvard Alumni Health Study reported that men who were physically active and had a high weekly caloric expenditure, regardless of mode, had a lower incidence of all-cause mortality than those who were less active (see figure 6.1) (40). A subsequent analysis of the Harvard alumni database showed that among those already active or exercising, all-cause mortality rate is lower in those who perform higher-intensity exercises compared to those who perform less vigorous activity (35). Taken together, these reports suggest that the specific type of physical activity is less important for general mortality benefits than the amount and intensity of the activity.

Progressive Overload

To derive benefits from cardiorespiratory training, people must follow the principle of progressive overload, which involves the appropriate application of frequency, intensity, and duration (or time).

Frequency

For the development of cardiorespiratory fitness, a frequency of at least 5 d/wk of moderate-intensity training, at least 3 d/wk of vigorous-intensity training, or a combination of 3 to 5 d/wk is recommended (2, 4, 45, 51). This frequency of regular exercise is sufficient to induce improvements in both cardiorespiratory fitness and health. Exercise performed just one or two times per week at a moderate to vigorous intensity may lead to some improvements in health and fitness, especially if each bout is performed for a long period of time (e.g., 60 min).

Relative to improving cardiorespiratory fitness, research suggests that exercising more than 3 or 4 d/wk may not be an efficient use of time for the nonathlete with little time to spare (29, 41). In fact, in a study that held total exercise volume constant, no difference in $\dot{V}O_2peak$ was reported for those who exercised 3 versus 5 d/wk (44). That said, exercising 5 or more d/wk may play a positive role by increasing total caloric expenditure, optimizing health improvements, and reducing all-cause mortality rates (51).

Intensity and Duration

Intensity and duration of training are often interdependent with respect to the overall cardiorespiratory training load and adaptations. Generally, the higher the intensity of an exercise training bout, the shorter the duration, and vice versa. The selection of the intensity and duration for a given patient should consider several factors, including their current cardiorespiratory conditioning level of the individual; the existence of underlying chronic diseases such as cancer, CAD, chronic heart failure, or obesity; the risk of an adverse event; and the individual's goals.

To achieve an adequate training response, most people must exercise at an intensity between 50% and 85% of their oxygen uptake reserve ($\dot{V}O_2R$, computed as $\dot{V}O_2peak - \dot{V}O_2rest$) in order for the cardiorespiratory system to be sufficiently stimulated to adapt and for aerobic capacity to increase (2, 32, 45, 51). In patients with a clinically manifest disorder and suffering from marked deconditioning (e.g., those with heart failure or currently undergoing treatment for cancer), improvements in aerobic fitness may be observed with training intensities as low as 40% of $\dot{V}O_2R$ (2). Generally, the lower a person's initial fitness level, the lower the required intensity level to produce initial adaptations. The upper-level intensity for continuous training for currently active, healthy individuals who want to improve $\dot{V}O_2peak$ can be set as high as 90% of $\dot{V}O_2R$ (4). This upper level might be regarded as the threshold between optimal gains in fitness and increased risk of orthopedic injury or adverse cardiovascular event. Training at too high an intensity may be difficult for patients with a chronic disease. Practical application 6.2 provides additional information about determining appropriate training HR, and practical application 6.3 highlights a growing area of clinical practice and research interest that pertains to the use of higher-intensity interval training in patients with a chronic disease.

Several authoritative groups (2, 4, 45, 51) recommend a minimum of 30 min of moderate exercise per exercise session (≥150 min/wk), or 15 to 20 min per session if a more vigorous exercise regimen (≥75 min/wk) is used. These recommended exercise durations can be accumulated in one continuous exercise session or in bouts of 10 min or more throughout the day. And like training at an exercise intensity less than 40% of $\dot{V}O_2R$, in patients suffering from marked deconditioning, starting with shorter exercise durations of just 5 to 10 min/d will likely be sufficient to initially improve fitness and health.

Figure 6.2 summarizes much of the discussion concerning frequency, duration, and intensity of training aimed at improving cardiorespiratory fitness. Here we can observe that among very fit people (e.g., athletes) that train seven or more times per week, more exercise is associated with continued improvement, but the magnitude of the gains begin to diminish (see the upper flat portion of figure 6.2). For these individuals, even small gains in fitness are appreciated and may be sufficient to improve on their previous best performance.

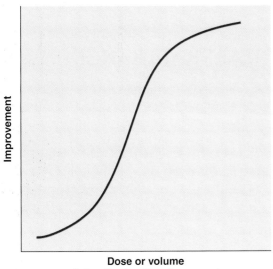

Dose or volume
(intensity, duration, frequency)

Figure 6.2   Relationships between exercise volume or dose (intensity, duration, or frequency) and improvement (e.g., $\dot{V}O_2$, resting HR). Note that both the initial and terminal portions of the curve are relatively flat, suggesting that very low and high volumes of exercise are associated with relatively mild improvements.

Reprinted by permission from J. Ehrman, D. Kerrigan and S. Keteyian, *Advanced Exercise Physiology: Essential Concepts and Applications* (Champaign, IL: Human Kinetics, 2015), 137.

Conversely, although other people or patients with a chronic disease who are poorly fit can experience an improvement in fitness from training only 2 d/wk, in general, a low dose of exercise is not associated with much improvement (see the lower flat portion of figure 6.2). Therefore, most professional organizations have established three sessions per week as the minimum number of bouts needed. For general purposes, a 3 d/wk training frequency may be sufficient if intensity of effort is vigorous, whereas a 5 d/wk regimen may be needed for a moderate-intensity training regimen.

Another common method used to safely prescribe exercise, especially in patients with a chronic disease, is rating of perceived exertion (RPE). Generally, for aerobic-type activities, an RPE between 12 and 16 on the 6 to 20 scale is used to guide exercise intensity (4, 45). Further, the RPE during exercise testing that is associated with a subsequent prescribed training HR range can help the clinician determine whether the patient can achieve and sustain that level of intensity during training. Be aware, however, that differences may exist when mode of testing differs from mode of training (e.g., the patient was tested on a cycle ergometer but walks for exercise training).

SKELETAL MUSCLE STRENGTH AND ENDURANCE

Resistance training improves muscular strength and power and reduces levels of muscular fatigue. The definition of *muscular*

strength is the maximum ability of a muscle to develop force or tension. The definition of *muscular power* is the maximum ability to apply a force or tension per a given unit of time. See the sidebar Benefits of Resistance Training.

Traditional resistance training typically includes both eccentric (lengthening) and concentric (shortening) phases of movement across a set of repetitions (15), with an emphasis on improving the concentric portion. In general, the focus of a resistance training program should be on the primary muscle groups. Proper lift technique is important to reduce the risk of injury and increase the effectiveness of an exercise. General recommendations include the following:

- Lift throughout the range of motion unless otherwise specified.
- Breathe out (exhale) during the lifting phase, and breathe in (inhale) during the recovery phase.
- Do not arch the back.
- Do not recover the weight passively by allowing weights to "crash down" before beginning the next lift (i.e., always control the recovery phase of the lift).
- Train with a partner.

In certain clinical populations, it may be prudent for the clinical exercise physiologist to follow these recommendations:

- Initially monitor blood pressure before and after a resistance training session and periodically during a session.
- Try to involve the same exercise professional who assisted with a patient's initial orientation and evaluation with regular reevaluations of lifting technique.
- Regularly assess for signs and symptoms of exercise intolerance during resistance training (e.g., excessive fatigue, decreased ability to maintain similar or increasing workloads, light-headedness).

Resistance exercises should be sequenced so that large muscle groups are worked first and smaller groups thereafter (24). If smaller muscle groups are trained first (i.e., those associated with fine movement), the large muscle groups may become fatigued earlier when they are used. In circuit resistance training, the individual completes a series of planned exercises and then repeats the first exercise to begin another circuit. Such a program is a valid alternative (to conventional resistance training) to increase muscle strength (11) and may improve adherence in middle-aged and older adults (10).

Resistance training, like other types of training, can be general or specialized (e.g., for athletes) to result in the specific adaptations one seeks. For patients with a chronic disease, a standard resistance training program is more than sufficient to improve muscle strength and endurance (4, 45). In fact, circuit programs that incorporate both aerobic and resistance training are a popular way of adding a cardiorespiratory stimulus to a resistance training program (21, 37). However, although impressive strength gains

DETERMINING THE APPROPRIATE HEART RATE RANGE

It is impractical and nearly impossible for individuals to guide their exercise intensity using $\dot{V}O_2$. But because HR has a near-linear relationship with $\dot{V}O_2$ at levels between 50% and 90% of $\dot{V}O_2$peak in both healthy people and most patients with a chronic disease, HR becomes an excellent surrogate for $\dot{V}O_2$ and can be used to guide exercise intensity. Several methods exist to determine an exercise training HR range. The following outlines how to develop an exercise prescription based on HR.

When the true maximal HR (HRmax) is unknown, in apparently healthy people it can be approximated using either of these two equations:

Equation 1: 220 − age = estimated HRmax

Equation 2: 208 − 0.7 (age) = estimated HRmax (49)

Example: To predict HRmax in a 32 yr old person:

Using equation 1: 220 − 32 = 188

Using equation 2: 208 − (0.7 × 32) = 186

A disadvantage of estimating HRmax is that the standard deviation (SD) can be as high as ±10 to 12 b · min⁻¹, and thus actual HRmax could differ greatly for two people of the same age. Another disadvantage is the inaccuracy of equations 1 and 2 in people taking a β-adrenergic blocking agent. Equation 3 is for use in patients with stable CAD taking a β-blocker (8).

Equation 3: 164 − 0.7 (age) = estimated HRmax

Given the drawbacks associated with estimating HRmax, using the HRmax from a graded exercise stress test is always preferred as long as the patient

- attained a true peak effort on the exercise stress evaluation (e.g., did not stop because of intermittent claudication, arrhythmias, poor motivation, severe dyspnea) and

- took any prescribed chronotropic-effecting medication (e.g., β-blocker) before the exercise stress test and no change has occurred in the type or dose of this medication since the test was completed.

Once a maximal HR has been estimated using equation 1, 2, or 3 or measured during a GXT, two methods can be used to determine a client's or patient's training HR range (THRR): the HR reserve (HRR) method and the %HRmax method (see table).

HRR Method (aka Karvonen Method)

Equation 4: Step I. HRmax − HRrest = HRR

Step II. (HRR × desired $\dot{V}O_2$ percentages) + HRrest = THRR

Example: HRmax (estimated or measured maximal) = 170, HRrest = 68, and desired percentages for training are 50% and 80%.

Step I. 170 − 68 = 102 (HRR)

Step II. 102 × .5 + 68 = 119 (lower end)

102 × .8 + 68 = 150 (upper end)

Note: $\dot{V}O_2$ reserve is defined as the difference between resting $\dot{V}O_2$ and peak exercise $\dot{V}O_2$ ($\dot{V}O_2$max − $\dot{V}O_2$rest). The %HRR method has a closer 1:1 relationship with %$\dot{V}O_2$ reserve than with % peak $\dot{V}O_2$ (9, 47, 48). This means that a 50% HRR ≈ 50% $\dot{V}O_2$reserve and 75% HRR ≈ 75% $\dot{V}O_2$ reserve. The range used for intensity with the HRR method should be 50% to 80% of peak $\dot{V}O_2$ reserve for most patient populations (2, 4, 45).

%HRmax Method

Equation 5: HRmax × desired HR percentages = THRR

Example: HRmax (estimated or measured) = 170, and desired HR percentage range is 60% to 90%.

170 × .6 = 102 (lower end)

170 × .9 = 153 (upper end)

Note: Because the relationship between $\dot{V}O_2$ and HR is not perfect, % peak exercise HR does not match up to the exact same %$\dot{V}O_2$peak. At lower intensity levels, % peak HR is approximately 10 percentage points higher than the %$\dot{V}O_2$peak (i.e., 60% of peak HR is equivalent to about 50% of $\dot{V}O_2$peak). This difference in percentages is reduced at higher intensity levels (i.e., 90% of peak HR is roughly equivalent to 85% of $\dot{V}O_2$peak).

	MAXIMAL HEART RATE (B · MIN⁻¹)	RESTING HEART RATE (B · MIN⁻¹)	SELECTED EXERCISE INTENSITY (%)		COMPUTED EXERCISE TRAINING HEART RATE (B · MIN⁻¹)	
			Lower limit	Upper limit	Lower limit	Upper limit
Heart rate reserve method	150	55	50%	80%	103	131
% maximal heart rate method	150	Not applicable	50%	80%	75	120

Practical Application 6.3

CLINICAL FOCUS: HIGHER-INTENSITY INTERVAL TRAINING

Several clinical care and research settings now use intermittent higher-intensity aerobic interval training (in addition to continuous moderate-intensity exercise) to improve or further improve exercise capacity and other selected cardiometabolic parameters (12). This method of training involves a series of repeated bouts of higher-intensity work intervals (i.e., training up to 95% of HR reserve vs. up to 80% of HR reserve with continuous exercise) interspersed with periods of mild- or moderate-intensity relief. Higher-intensity interval training results in the completion of more total work during a single training session and typically a greater improvement in cardiorespiratory fitness.

Common among competitive athletes, interval training is now being increasingly applied to patients with a chronic disease. For example, among patients with stable chronic heart failure, Wisloff and colleagues (54) showed a 6 mL · kg⁻¹ · min⁻¹ (+46%) increase in peak $\dot{V}O_2$ after 12 wk of intermittent higher-intensity aerobic interval training. In patients with coronary heart disease, Keteyian and colleagues (33) studied the effects of 10 wk of higher-intensity interval training in the standard cardiac rehabilitation setting and showed a 3.6 mL · kg⁻¹ · min⁻¹ increase in peak $\dot{V}O_2$ among subjects randomized to the interval training group compared with a 1.7 mL · kg⁻¹ · min⁻¹ increase in patients who underwent standard cardiac rehabilitation using moderate continuous training. Although higher-intensity interval training demonstrates improvement with relatively shorter training regimens, more work is needed to evaluate its safety, its ability to maintain training adaptations over longer periods (i.e., months, years), and the incremental effect (if any) it has on lowering a patient's future risk for a clinical event (16).

have been reported, improvement in cardiorespiratory fitness may be mild to moderate at best (22, 28).

Specificity of Training

Resistance training for improvement in specific muscles should follow a lifting routine that closely mimics the activity or muscular movements in which gains in muscular fitness are desired. Because the components of skeletal muscle fitness are related, any type of resistance training program will provide some benefits in each area of muscular fitness (i.e., strength, power, endurance). The resistance training program for general health should emphasize dynamic exercises involving concentric (shortening) and eccentric (lengthening) muscle actions that recruit multiple muscle groups (multijoint) and target the major muscle groups of the chest, shoulders, back, hips, legs, trunk, and arms. Single-joint

exercises involving the abdominal muscles, lumbar extensors, calf muscles, hamstrings, quadriceps, biceps, and triceps should also be included.

Progressive Overload

Resistance training programs that sufficiently load the eccentric phase of movement are a potent stimulus for enhancing muscle mechanical function and the muscle–tendon unit. Eccentric training has application for athletic performances that require speed, power, and strength (15). However, for the client or patient seeking general or overall muscular fitness, the ACSM and the American Association of Cardiovascular and Pulmonary Rehabilitation (AACVPR) recommend traditional resistance training, which focuses on improving concentric strength. With respect to training repetitions, the ACSM advises 8 to 12 repetitions per set, while the

Benefits of Resistance Training

- Improved muscular strength and power
- Improved muscular endurance
- Modest improvements in cardiorespiratory fitness
- Reduced effort for activities of daily living as well as leisure and vocational activities
- Improved flexibility
- Reduced skeletal muscle fatigue
- Elevated density and improved integrity of skeletal muscle connective tissue
- Improved bone mineral density and content
- Reduced risk of falling
- Improved body composition
- Possible reduction of blood pressure
- Improved glucose tolerance

AACVPR recommends 10 to 15 (4, 45). In general, the greater the overload, the greater the improvement. Excessive and prolonged overload, however, can lead to worsening performance or increased risk for skeletal muscle injury.

Frequency

Most studies report optimal gains when patients engage in resistance training 1 to 3 d/wk (14, 20, 25), which supports the 2 or 3 d/wk recommendation from both the ACSM and AACVPR when performing a general or circuit resistance training program (2, 4, 45). There is little evidence that substantial additional gains are realized from performing resistance exercise on more than 3 d/wk.

Intensity and Duration

The intensity of resistance training is also important in determining the load or overload placed on the skeletal muscle system to produce adaptation. For improvement in strength and endurance, the resistance should be at 40% to 60% of a person's one-repetition maximum (1RM), with progression up to 70% or 80% of 1RM for leg exercises in selected patients. Alternatively, lift intensity can be set at an RPE of 11 to 13. Another approach is to prescribe 8RM to 12RM, such that the person is at or near maximal exertion at the end of these repetitions (although this level of resistance may be too strenuous for some clinical conditions). The RM can be determined using either a direct or an indirect method (11, 34, 46).

The total training load placed on the skeletal muscle system is a combination of the number of repetitions performed per set and the number of sets per resistance exercise, with the recommendation for the latter being between one and three sets per exercise (4, 45). Several well-designed studies indicate little benefit from performing resistance training for more than one set per resistance exercise (18). If more than one set is performed per exercise, it may be prudent to keep the between-set period to a minimum to reduce the total exercise time. Generally, a 2 min rest is sufficient between sets.

FLEXIBILITY TRAINING

Flexibility is the ability to move a joint through its full ROM. Proper flexibility is associated with good postural stability and balance, especially when exercises aimed at improving flexibility are performed in conjunction with a resistance training program. Despite what one might expect, there is no consistent link between regular flexibility exercise and reduction of incidence of musculotendinous injuries or prevention of low back pain (4).

Several devices can be used to assess ROM. These include a goniometer, which is a protractor-type device; the Leighton flexometer, which is strapped to a limb and reveals the ROM in degrees as the limb moves around its joint; and the sit-and-reach box, which assesses the ability to forward flex the torso while in a seated position and measures lower back, hamstring, and calf flexibility. An excellent review of the methodology of the sit-and-reach test is provided by Adams (1). In patients with certain diseases, flexibility may decrease as the course of the disease progresses (e.g., multiple sclerosis, osteoporosis, obesity). These and other populations benefit from regular ROM assessment and an exercise training program designed to enhance flexibility. A well-rounded flexibility program focuses on the neck, shoulders, upper trunk, lower trunk and back, hips, knees, and ankles.

The following are brief descriptions of the three primary modes of flexibility training, usually performed after 4 to 10 min of slower full-body activity such as walking or cycling to increase temperature within the skeletal muscle and muscle–tendon unit.

1. *Static:* A stretch of the muscles surrounding a joint that is held without movement for a period of time (e.g., 10-30 s) and may be repeated several times. Within 3 to 10 wk, static stretching yields improvements in joint ROM of 5° to 20°.

2. *Ballistic:* A method of rapidly moving ("bouncing") a muscle to stretch and relax quickly for several repetitions, often used for sports that involve ballistic movements such as basketball. This method uses the momentum of the body segment to produce the stretch.

3. *Proprioceptive neuromuscular facilitation (PNF):* A method whereby a muscle is isometrically contracted, relaxed, and subsequently stretched. The theory is that the contraction activates the muscle spindle receptors, or Golgi tendon organs, which results in a reflex relaxation (i.e., inhibition of contraction) of either the agonist or the antagonist muscle (27). There are two types of PNF stretching:

   - Contract-relax occurs when a muscle is contracted at 20% to 75% of maximum for 3 to 6 s and then relaxed and passively stretched. Enhanced relaxation is theoretically produced through the muscle spindle reflex.

- Contract-relax with agonist contraction begins in the same manner as contract-relax, but during the static stretch the opposing muscle is contracted. This action theoretically induces more relaxation in the stretched muscle through a reflex of the Golgi tendon organs.

Static and ballistic types of stretching are simple to perform and require only basic instruction. Proprioceptive neuromuscular facilitation stretching is somewhat complex, requires a partner, and may require close supervision by a clinical exercise professional. Static stretching is typically believed to be the safest method for enhancing the ROM of a joint. Both ballistic and PNF stretching may increase the risk for experiencing delayed-onset muscle soreness and muscle fiber injury. Generally, PNF is the most effective of the three methods of stretching for improving joint ROM (17, 42).

Specificity of Training

The flexibility of a joint or muscle–tendon unit depends on the joint structure, the surrounding muscles and tendons, and the use of that joint for activities. Improved flexibility and ROM of a joint are developed through a flexibility training program that is specific to that joint. A joint used during an activity, especially if it requires a good ROM, typically demonstrates good flexibility.

Progressive Overload

As ROM increases, people should enhance or increase the stretch to a comfortable level. This practice will produce optimal increases in ROM.

Frequency

An effective stretching routine should be performed a minimum of 2 or 3 d/wk (4). As stated previously, however, daily stretching is advised for optimal improvement in ROM.

Intensity and Duration

Static and PNF stretches should be held for 10 to 30 s. For PNF, this should follow a 6 s contraction period. Each stretch should be performed for three to five repetitions and to a point of only mild discomfort or a feeling of stretch.

CONCLUSION

Any type of exercise training routine, whether it is cardiorespiratory conditioning, resistance training, or ROM training, should follow the FITT principle to ensure an optimal rate of improvement and safety during training. When an exercise physiologist is working with specific clinical populations, modifications of these general principles may be necessary to accommodate any unique characteristics or concerns associated with the clinical condition. In fact, subsequent chapters in this textbook provide information pertinent to adapting the general exercise training principles discussed here to specific patient populations.

Online Materials

Visit HK*Propel* to access a link to the references, the case studies with discussion questions, and a quiz to help you review key concepts and test your understanding of the material covered in this chapter.

CHAPTER 7

Clinical Exercise Programming

Jonathan K. Ehrman, PhD

A purpose of this textbook is to describe how regular exercise training can help treat many of the most common chronic diseases and conditions. Although it would be ideal for individuals with health issues to understand how to safely perform physical activity and exercise for optimal benefit, and be self-motivated to do so on their own, this is not usually the case. Instead, more than 80% of adults do not meet the U.S. guidelines for physical activity (19). About half of adults over age 18 have at least one chronic disease or condition for which performing regular exercise would be beneficial (19). Unfortunately, many chronically ill people do not exercise regularly. For instance, less than 30% of eligible Medicare-aged individuals who qualify for cardiac rehabilitation actually participate (1). Reasons given for not participating in regular physical activity include lack of understanding about the benefits of exercise, fear, socioeconomic issues, lack of availability, and lack of perceived need (11).

Supervised organized programs are offered for several of the clinical conditions presented within this text. These include well-developed programs for those with cardiac and pulmonary diseases as well as emerging programming for those with peripheral arterial, oncologic, and renal disease. Additionally, many cardiac rehabilitation programs offer a maintenance option (traditionally known as phase 3 cardiac rehab) that is often made available to people with other chronic diseases. These may include obesity, diabetes, human immunodeficiency virus (HIV), and skeletal or muscle disorders (osteoporosis, arthritis, low back pain, sarcopenia) that may improve or stabilize with regular exercise, as well as individuals with neuromuscular disorders or stroke who have previously participated in physical rehabilitation and are considered recovered.

Exercise programs for patients with a chronic disease often take place in a clinical setting (e.g., hospital, outpatient clinic) and are staffed by allied health professionals such as **clinical exercise physiologists**, **accredited exercise physiologists**, **exercise physiologists**, physical therapists, and nurses. Some program costs are covered by a health insurance benefit (cardiac, pulmonary,

and peripheral artery disease). This is also true in other countries with national health care programs including Canada, the United Kingdom and Australia. And some of these programs are emerging as key variables for assessing hospital discharge measures. For instance, in the United States, the National Committee for Quality Assurance (NCQA) has developed Healthcare Effectiveness Data and Information Set (HEDIS) measures specifically for cardiac rehabilitation attendance. Cardiac and pulmonary rehabilitation programs have an excellent research base that over time has demonstrated improvements in functional capacity, disease or condition progression, morbidity, and/or mortality. Other programs such as oncology rehab are more recent and have not yet demonstrated adequate benefits to ensure insurance coverage based on improved long-term outcomes.

Despite these differences, there is an ever-growing interest in using structured exercise to improve many chronic diseases and conditions. This chapter briefly focuses on several of these programs and in doing so highlights the effectiveness, general processes, and policies of each.

CARDIAC REHABILITATION

Cardiac rehabilitation (CR) is the longest established structured exercise program among those used to treat patients with a chronic disease, with origins dating back to the late 1960s. CR is part of the continuum of care for patients with cardiovascular disease and is typically offered in three phases (table 7.1). Participation in CR is considered a class Ia recommendation in those who have had a myocardial infarction or have been revascularized, which is the highest recommendation possible based on existing research showing a consistent benefit in both functional capacity and outcomes including future **myocardial infarction** and death (23). CR is also recommended for many other cardiac conditions. Although much is known about the benefits of CR and the referral rate is high, the enrollment rates are low, at about 25% in the United States (24) and 10% to 30% in Australia (6). There are strong efforts in the United

States to address this low participation rate, with the expectation that a 70% CR participation rate would save up to 25,000 lives and prevent 180,000 hospitalizations annually (1). CR programs in the U.S. can become certified by the American Association of Cardiovascular and Pulmonary Rehabilitation (AACVPR), which ensures a minimum standard for participating programs.

The following is a brief presentation of the effectiveness of CR, as well as its implementation and policies.

Program Effectiveness

The benefits of CR are well documented and include improved functional capacity and disease-related behaviors, reduced subsequent heart-related events (e.g., heart attack), and decreased risk of death. Improvements of 10% to 30% are common for cardiorespiratory fitness, and studies have demonstrated an approximate 30% to 40% reduction in risk of serious clinical events for each 1 unit increase in peak METs (2, 8, 15). Traditional exercise training involves continuous moderate-intensity activity. Recommendations now include higher-intensity interval training at 85% to 90% of heart rate reserve to elicit a greater improvement in peak exercise capacity and submaximal exercise endurance (14). A more recent focus provides home-based CR using telehealth platforms, including voice and video, for single patients or groups of patients connected through their medical records (13). A goal of the telehealth option is to improve enrollment and adherence to CR. It has not been established whether the benefits of telehealth-delivered CR are comparable to those of center- or clinic-based CR, although there is evidence that heart rate–based training intensities are similar between programs (13). Improvements in functional capacity have also been noted in a small, randomized study (5).

Program Structure and Processes

A typical early outpatient CR (phase 2) program offers structured exercise and disease-specific and risk factor education on 2 or 3 d/wk. Maintenance CR (phase 3) is often programmed on the same or alternate days. Programs with adequate staffing and facilities may offer programming on all weekdays. Aerobic machines (e.g., treadmill, cycle ergometer, seated stepper, elliptical, arm ergometer) are the typical exercise modalities used in phase 2 CR. Most patients wear a single-lead ECG system for some of their CR sessions, viewed via telemetry by the CR staff, primarily clinical exercise physiologists and nurses (figure 7.1a). Some patients with an indication (e.g., high risk; chest pain or other indicators of exertional ischemia) may have a stress test before the program, but many can participate without one. The test provides information about peak heart rate, which can be used to develop the exercise prescription. When stress test data are not available, exercise intensity is often guided using a HR-based approach set at between 20 and 30 b · min^{-1} above standing resting HR (or the rating of perceived exertion scale is used).

Most programs include several bouts of exercise (10-20 min each) between a low-intensity warm-up and cool-down period. Patients are evaluated daily, and as exercise becomes easier for the patient over time, the intensity is increased in small, consistent amounts to ensure improvement. Some programs incorporate resistance training, range of motion exercise, and group classes (e.g., chair aerobics, walking club), but these more often occur in the phase 3 setting (figure 7.1b). It is typical to measure weight, heart rate, blood pressure, blood glucose in those with diabetes, and symptoms before and after an exercise session, which helps assess both exercise safety and progress.

Program Policies

The following conditions and procedures are covered by most insurances in the United States: myocardial infarction in past 12 mo, stable angina, coronary revascularization (angioplasty with or without stent and bypass surgery), stable chronic heart failure, heart valve repair or replacement, and heart or heart–lung transplant. Patients are typically referred after a cardiac event requiring hospitalization (figure 7.2). Per insurance requirements, an

Table 7.1 Cardiac Rehabilitation Review

Phase	Setting	Steps
1	Inpatient	1. Discussion with physician 2. Visit from CR staff, which may include clinical exercise physiologist, nurse, or physical therapist 3. Referral to phase 2 4. Ambulation and education in preparation for discharge
2 (early outpatient)	Outpatient, in hospital or clinic	1. Intake assessment 2. Individualized treatment plan (ITP) developed and implemented 3. Exercise 2 or 3 times/wk 4. ECG telemetry monitoring as needed 5. Education about risk factor modification and other secondary prevention
3 (maintenance)	Outpatient, in hospital, clinic or fitness center	1. Continuation of risk factor modification and prevention 2. Encouragement and reinforcement of independence 3. ECG telemetry not used

Figure 7.1   Examples of cardiac rehabilitation exercise classes for *(a)* early outpatient, phase 2, and *(b)* maintenance, phase 3.

Photos courtesy of Henry Ford Cardiac Rehabilitation Program.

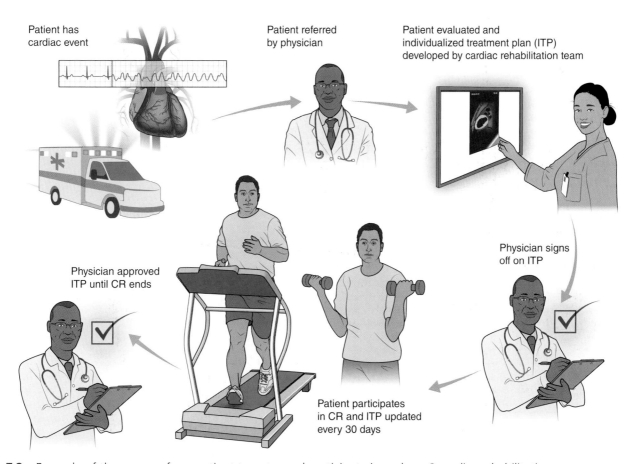

Patient has cardiac event

Patient referred by physician

Patient evaluated and individualized treatment plan (ITP) developed by cardiac rehabilitation team

Physician signs off on ITP

Physician approved ITP until CR ends

Patient participates in CR and ITP updated every 30 days

Figure 7.2   Example of the process for a patient to enter and participate in a phase 2 cardiac rehabilitation program.

individualized treatment plan (ITP) is developed and updated regularly (figure 7.3). The ITP must be signed monthly by either the patient's physician or the CR medical director and is typically part of a patient's medical record.

Decisions are made daily regarding exercise participation. If any safety-related issues arise, the CR staff may defer exercise and communicate with the referring physician about the safety concern. Each program develops its own situation-specific decision processes for issues that may put patients at risk of complication during exercise. These policies are typically reviewed and updated annually.

Individualized Treatment Plan: Exercise Intervention for Hypertension

Name: _____   Physician: _____   ICD: _____

Initial Assessment   Date: _____   Initials: _____

Exercise history prior to event

❏ No regular exercise

❏ Irregular exercise: 1 or 2 times/wk, 20+ min for <6 mo prior to event

❏ Irregular exercise: 1 or 2 times/wk, 20+ min for >6 mo prior to event

❏ Regular exercise: 3+ times/wk, 20+ min for <6 mo prior to event

❏ Regular exercise: 3+ times/wk, 20+ min for >6 mo prior to event

Target goals

- ≥2 MET increase by completion of CR
- Individual exercise prescription ExRx every 30 d
- BP <130/80 mmHg
- Aerobic exercise ≥150 min/wk

Exercise and BP education

❏ Rating of Perceived Exertion (RPE) scale

❏ Equipment orientation and safety

❏ Warm-up and cool-down

❏ Sodium and hypertension

❏ Signs or Symptoms (S or S) explained

❏ Discussed home exercise

❏ Home exercise equipment

❏ Discussed physical activity

❏ Is using pedometer (Y/N)

❏ Daily step goal

❏ Discussed exercise barriers

❏ Barrier 1

❏ Barrier 2

❏ Discussed resistance training

❏ Started resistance training

Date: _____   Initials: _____

Exercise prescription in cardiac rehab

Frequency: _____

Intensity: recent stress test Heart Rate Reserve _____ - _____; rest % 20-30; RPE

Time (duration): _____

Mode: ❏ Treadmill ❏ NuStep ❏ Recumbent Bike ❏ Airdyne ❏ Rower ❏ Elliptical ❏ Arm Ergometer ❏ Bike

Progression: _____

S or S with exercise: _____

Limitations: _____

Initial MET level: _____

Figure 7.3   Example of an individualized treatment plan (ITP).

30-day exercise intervention

Date: _____   Initials: _____   S or S with exercise: _____

At Training Heart Rate (THR): Y N   Change in MET from initial: _____

At MET goal: Y N   Change in ExRx from initial? Y N

Home exercise

Type: _____   Frequency: _____   Duration: _____

Resistance training: Y N

60-day exercise intervention

Date: _____   Initials: _____   S or S with exercise: _____

At THR: Y N   Change in MET from initial: _____

At MET goal: Y N   Change in ExRx from initial? Y N

Home exercise

Type: _____   Frequency: _____   Duration: _____

Resistance training: Y N

90-day exercise intervention

Date: _____   Initials: _____   S or S with exercise: _____

At Training Heart Rate (THR): Y N   Change in MET from initial: _____

At MET goal: Y N   Change in ExRx from initial? Y N

Home exercise

Type: _____   Frequency: _____   Duration: _____

Resistance training: Y N

Notes (include initial goals): _____

Figure 7.3 *(continued)*

PULMONARY REHABILITATION

Pulmonary rehabilitation (PR) is a comprehensive multidisciplinary program considered a core part of the overall management plan for patients with chronic lung disease who are functionally limited. Among patients with lung disease, reductions in physical function are often related to limitations in breathing reserve from either obstructive or restrictive processes; patients can lose skeletal muscle mass, causing reductions in strength, power, and endurance. This can lead to a spiraling effect of increasing sedentary living, which can further accelerate the decline in physical function (27). Similar to many individuals with other chronic diseases, patients referred to PR have higher rates of depression and anxiety and a poorer quality of life than the general population. And similar to patients in CR programs, PR patients tend to have a high level of social isolation. These issues increase the risk for hospitalization and death (9).

Program Effectiveness

Although mortality benefits have not been observed, PR programs offer a number of functional and psychological benefits. These include a reduction in the discomfort of dyspnea at a given workload and throughout daily life, improvements in functional capacity, reductions in self-reported depression and anxiety, and reductions in health care utilization and costs (4). Upon entering a PR program, patients tend to be more physically debilitated when compared to patients entering CR. Evidence about the benefits of PR are not as strong for benefits in patients with interstitial lung disease, but improvements in this patient group have been observed (7).

Program Structure and Processes

PR programs are usually in the outpatient setting and often share space with CR programs, either concurrently or during nonconflicting hours or days of the week. Programs typically run 3 d/

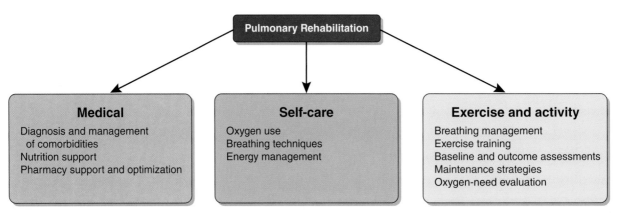

Figure 7.4   Components of a pulmonary rehabilitation program.

wk, and daily sessions tend to be longer than in CR. A patient can spend 1.5 to 4 h at each PR session. The longer duration is related to additional lung-specific exercise training and self-care taught to patients in a PR program (figure 7.4).

On the first day of PR, patients may be assessed for medical history, fitness (6 min walk test or gas exchange exercise test), and goal setting. Daily, each patient performs exercise training, attends an education session, and receives psychosocial support. Exercise training can include aerobic activity (moderate continuous to higher-intensity interval); muscular strength and endurance exercises, with a focus on the respiratory muscles; and balance and flexibility exercises. A common goal is for pulse oximetry (SpO_2) values to remain >88% during exercise. If oxygen saturation drops below 88%, supplemental oxygen can be administered or the delivery of oxygen increased (20). The education sessions cover a wide range of topics (e.g., nutrition, breathing techniques, energy conservation, medication review, home and ambulatory O_2 use) over the course of up to 36 visits.

Program Policies

To be eligible, patients must have moderate to very severe chronic obstructive pulmonary disease (COPD). This is defined as global obstructive lung disease (GOLD) classification II to IV (see chapter 19 for information about the GOLD classification) (10). Patients must be referred by their treating physicians. Programs are most often staffed by respiratory therapists but may also include clinical exercise physiologists, nurses, and physical therapists. Similar to CR programs, PR programs can become AACVPR certified to demonstrate a minimal level of program competency.

SUPERVISED EXERCISE THERAPY FOR PATIENTS WITH PERIPHERAL ARTERY DISEASE

Exercise training is a very effective type of treatment for symptomatic **peripheral artery disease (PAD or peripheral vascular disease)**. The primary symptom of PAD is leg pain from ischemia, known as claudication. Exercise training is most commonly used when a patient complains of significant lifestyle limitations, which are primarily pain and fatigue with walking during normal daily activities. Ideally, **supervised exercise therapy (SET)** is implemented before any attempts of revascularization, but this is not usually the case. In fact, of those who are candidates, <2% participate in SET therapy.

Insurance coverage for SET was approved in 2017 in the United States, so compared with cardiac and pulmonary rehabilitation, it is a relatively newly covered therapy. Although insurance coverage allows for SET in either a clinic (e.g., private physician's office) or an outpatient setting, the most common setting is within a CR program. Depending on the CR program, SET may be incorporated into a phase 2 class or a maintenance (phase 3) setting, or it can be a separate class open only to those with symptomatic PAD. Chapter 17 provides in-depth information about PAD.

Program Effectiveness

Extensive data indicate a 100% to 300% improvements in functional capacity following 3 to 6 mo of exercise training, as measured by total distance or time walked to either onset of symptoms or intolerance of symptoms(26). This is comparable to a 30 to 35 m improvement in 6 min walk distance. These gains in functional capacity are durable and similar to improvements after lower-extremity angioplasty and stenting (18).

Program Structure and Processes

Most patients can begin SET without a previous exercise test. However, a test may be performed if there is a known underlying cardiac condition or one is suspected. Tests can also help when developing an exercise prescription. Walking is the desired exercise mode for optimal improvement (figure 7.5), at a pace that will elicit a moderate to moderately severe level of claudication (e.g., 2 or 3 on a 0- to 4-point scale of pain severity) in 5 to 10 min (27). At this point, the individual should rest until the pain dissipates and then

Goals for Exercise Training Dictate Modality of Exercise

Walking: Gold Standard

Maximal improvement in pain-free walking distance (PFWD) and maximal walking distance (MWD)

Non-Weight-Bearing Exercise: Optional

Primarily improves cardiorespiratory function, with some effect on PFWD and MWD

Resistance Training: Recommended

Improves muscle strength, with some effect on PFWD and MWD

Claudication Pain Scale

    0 = no pain (resting or early exercise effort)

    1 = mild pain (first feeling of any pain in legs)

    2 = moderate pain (pain level at which exercise training should cease)

    3 = moderately severe pain (nearly maximal pain; may require patient to stop)

    4 = severe/unbearable pain (most severe pain experienced; requires patient to stop)

Common Walking Protocol for SET PAD

The patient walks to bring on moderate claudication, ideally within 5 to 10 min.

- Upon reaching moderate-intensity (2 or 3) pain, the patient rests (sitting or standing).
- The patient resumes walking when the pain has completely subsided.
- This process is repeated for a total time (walking + resting) of 30 to 45 min of walking within 60 min.

Figure 7.5   Exercise during SET PAD participation.

repeat, with a goal of walking 30 to 45 min within a 60 min period. Patients should be regularly assessed for improvement; speed or grade or both should be increased when patients are able to walk more than 10 min without significant pain. For those who cannot tolerate walking when beginning SET, or who have not improved upon monthly evaluation, a nonwalking mode of exercise (e.g., cycle ergometer, seated stepper, arm ergometer) can allow for functional improvement until walking can be implemented (26, 30). Since patients often lose muscle mass and this affects walking and mobility, resistance training is often incorporated into the exercise program.

Program Policies

Insurances typically cover up to 36 visits over a 12 wk period. Note that this is different from CR, which allows 36 wk for 36 visits. For this reason, most programs offer SET on 3 d/wk. In the United States, there is a 72-visit lifetime cap on SET coverage, and because of co-payments of up to 20% per session, it may be financially beneficial for patients to transition to self-pay CR programs (i.e., maintenance CR, or phase 3), which often cost less than $100 per month. Per insurance coverage requirements in the United States, referrals for SET should be submitted only after the patient has seen a physician and received treatment for symptomatic PAD, including education about disease management.

CANCER REHABILITATION

Programs for patients diagnosed with cancer may begin during the treatment phase but are more often implemented after treatment is complete. The rationale for beginning exercise training during the treatment phase of cancer care relates to the mitigation of side effects associated with chemotherapy and radiation treatment. However, because of a lack of understanding about the potential benefits of exercise among both treating physicians and patients, there is an overall reluctance to recommend exercise training during treatment. Other barriers to participation include financial constraints due to self-pay, transportation issues and distance to the exercise location, and ongoing cancer and cancer treatment side effects. The rationale for beginning exercise training after treatment completion includes an assumption that patients will tolerate the exercise better, and at that point the exercise is less likely to interfere with clinical care effectiveness. Data also show the importance of improving and maintaining overall fitness for a better quality of life and to potentially reduce disease recurrence during the cancer survivor phase.

Despite the benefits of cancer rehabilitation, it is estimated that less than 5% of cancer patients in the United States are referred to an exercise program (22). The overall benefits of exercise in those with various types of cancers can be found in chapter 22. The changes in functional ability during the phases of cancer diagnosis, treatment, and recovery are provided in figure 7.6.

Program Effectiveness

The primary complaints after cancer treatment are most likely related to fatigue and weakness, likely due to increased sedentary lifestyle habits, loss of muscle mass, or the development of anemia. Some patients experience cancer- or treatment-related pain as a result of surgery or possibly metastases, among other potential reasons. Wonders reported a reduction in emergency room visits, hospital admissions at 30 d after treatment, and length of hospital stay in those who exercised during treatment versus those who remained sedentary (28). The study also reported reductions in health care costs associated with office visits, emergency room visits, and hospital readmissions in the exercise group.

Program Structure and Processes

There are generally two types of structured exercise programs for patients with cancer or in recovery. Although few in number, some stand-alone cancer rehabilitation programs are available. A more common setting is use of the maintenance (phase 3) program of cardiac rehabilitation, deemed ideal because of the staffing expertise and existing facilities and equipment. Since cancer rehabilitation is not yet insurance reimbursable, all patients must self-pay for services. The CR maintenance program setting is likely to be less expensive than a stand-alone program. Lower-risk patients or those with a previous exercise history may do well with general exercise counseling and performing their exercise either at home or in another nonclinical location. However, facility-based exercise settings offer the benefits of both improved physical functioning and increased long-term adherence, which may be related to the socialization that occurs in the group setting of a CR program (12, 25).

The structure of the clinical setting for those with cancer does not differ much from a CR program setting. There is a focus on both aerobic and resistance exercise. Some patients may require specific training for deficits acquired through disease progression (weight loss, balance issues, loss of strength, loss of endurance) or treatment (e.g., reduced shoulder range of motion or **lymphedema** after breast cancer surgery). Exercise in groups or classes can make it easier for facilitators to handle larger numbers of patients and also increases the socialization aspect of a program.

Program Policies

Although the qualifications of exercise professionals vary, many programs require staff to have a minimum of a bachelor's degree and certification as a clinical exercise professional (see chapter 1). There are also several cancer-specific certifications from both professional exercise associations and university settings (17). Program referral should be made by a patient's treating or primary care physician. Facilitators can either develop a medical screening process to determine safety to begin exercise or rely on information provided by the referring physician. Since some patients may be in an immunocompromised state, each exercise setting should implement processes (e.g., cleaning equipment) to lower patient risk of acquiring a communicable disease.

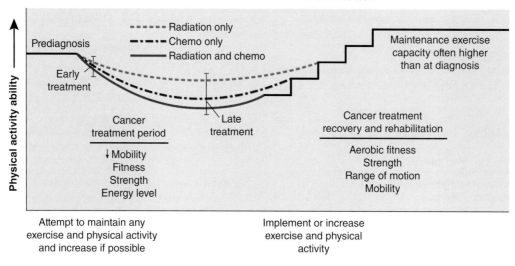

Cancer Treatment and Rehabilitation

Figure 7.6 Changes in functional ability along the cancer continuum.

RENAL REHABILITATION

As patients move through the spectrum of kidney disease, they experience a progressive decline in physical function and physical activity level; the lowest point of this continuum is **end-stage renal disease (ESRD)** (29). Beginning and maintaining exercise training during the earlier stages of kidney disease when physical activity levels, cardiorespiratory fitness, and muscle mass begin to decrease can attenuate functional losses (29). Importantly, maintaining muscle quality is associated with renal protection (29). Similar to options for cancer rehabilitation, patients with ESRD can exercise during dialysis treatment (intradialytic; figure 7.7) or more typically on a nontreatment day. Exercise is recommended for ESRD patients to improve physical functioning, particularly muscle strength and mobility, as these individuals tend to be sedentary and frail and have poor functional abilities (3). Although it makes sense that those who have had a renal transplant would do well by performing structured exercise, this area has not been well researched. Chapter 13 provides information about renal disease and its treatment.

Program Effectiveness

Patients participating in a renal rehabilitation program may see an increase in urea clearance and reductions in hospitalizations and length of stay. Those with stage 3 and 4 chronic kidney disease (predialysis) demonstrated improvements in 6 min walk distance (+19%), sit-to-stand ability (+29%), and overall quality of life

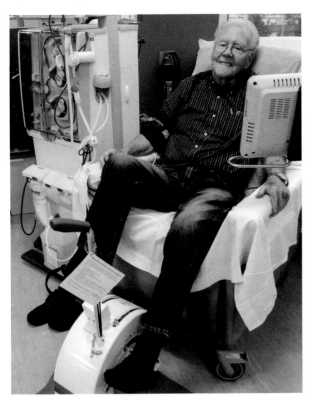

Figure 7.7 An intradialytic exercise program.

Photo courtesy of South Calgary.

scores from two exercise sessions per week (progressive resistance and aerobic-type training conducted in the presence of a clinical exercise physiologist or physical therapist) over 12 wk (21). These patients would likely do well in a CR maintenance program (phase 3) setting. Physical performance (6 min walk distance, sit-to-stand time) and quality of life (cognitive function, social interaction) may improve in patients on dialysis who perform a home walking program on nondialysis days, as directed by the dialysis staff (16).

Program Structure and Processes

Patients not requiring dialysis can typically participate in structured programs in a CR setting (figure 7.8). They do well with group exercise and have similar limitations as the typical cardiac patient with respect to mobility, endurance, and strength and therefore benefit from CR programming. Patients who advance to ESRD can also participate in these programs on nondialysis days. They are often considerably less functional than patients not yet requiring dialysis, and in a CR setting they can resemble patients with moderately advanced heart failure in terms of physical abilities. These patients will have dialysis ports that must be cared for with respect to cleanliness and ensuring they are not damaged by the exercise equipment or a particular exercise. Patients requiring dialysis who have the option to exercise at their dialysis appointment are typically offered a leg cycle to pedal. Upper body strengthening exercises using elastic bands and ankle weights are another option (3).

Program Policies

Unlike cardiac rehabilitation, it is not typical for hospital discharge to have an automatic referral option for renal rehabilitation. This is even the case in hospitals with established programs. Renal rehabilitation is not covered by insurance. If a patient is referred to an outpatient exercise program is usually means referral to a maintenance CR program, self-pay is required. Most programs charge nominal fees in the range of $50 to $75 per month for 3 d/wk attendance. There is usually no charge for exercise training offered in the dialysis setting.

GENERAL EXERCISE PROGRAMS FOR PATIENTS WITH CHRONIC DISEASE

As mentioned, CR programs have traditionally developed and maintained other exercise offerings for their patients who complete an early outpatient (i.e., phase 2) exercise program. These other classes are known as phase 3, or maintenance, programs. From their inception in the 1970s through much of the 1990s, maintenance programs primarily targeted those with cardiac conditions or at-risk patients attempting to avoid complications of cardiac disease. More recently, when space and time are available, these programs have expanded the enrollment criteria to include other patients with various chronic diseases and conditions.

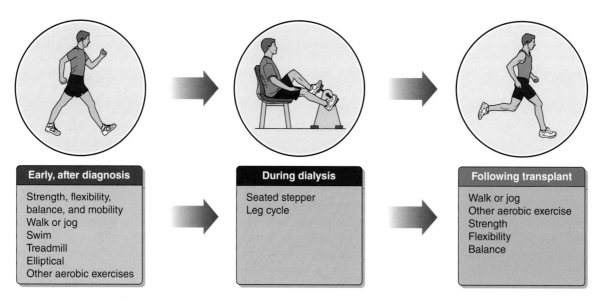

Early, after diagnosis	During dialysis	Following transplant
Strength, flexibility, balance, and mobility Walk or jog Swim Treadmill Elliptical Other aerobic exercises	Seated stepper Leg cycle	Walk or jog Other aerobic exercise Strength Flexibility Balance

Figure 7.8   Exercise recommendations throughout the continuum of care for patients with renal disease, including during dialysis and following transplant.

CONCLUSION

Supervised exercise training, when performed properly, can be beneficial for most individuals with a chronic disease or condition. Programs for cardiac and pulmonary patients have a rich research base indicating improvements in function and quality of life and, in some cases, reductions in disease progression, morbidity, and mortality. Quality of life improvements may be related to increased functionality, but in many older individuals they may also be associated with the social aspects of these programs. More clinicians are enrolling their patients in supervised programs for cardiac, pulmonary, and peripheral artery disease, which are reimbursed by health insurance payers in the United States and some other countries. It is incumbent upon clinical exercise physiology professionals to foster relationships with clinicians and develop programs for a wide variety of persons. There is recent interest in providing supervised exercise training for other chronic diseases including cancer, obesity, and neuromuscular, metabolic, and skeletal muscle issues. To be successful, research must continue to demonstrate improvements in physical and mental health, disease progression, and disease management from these programs. Ideally these efforts should be led by clinical exercise physiologists.

Online Materials

Visit HK*Propel* for a link to the references and to access a quiz to help you review key concepts and test your understanding of the material covered in this chapter.

PART II

Diseases of the Endocrine System and Metabolic Disorders

The dysfunctions of any metabolic system in the human body is often complex and can result in detrimental health outcomes. Some of these conditions, such as diabetes, high blood pressure, and renal failure, result in their own specific health hazards. But these and the other disorders addressed in part II (metabolic syndrome, obesity, and dyslipidemia) also increase the underlying risk for many other chronic diseases, including cardiovascular disease and cancer.

Diabetes has risen to epidemic proportions in the United States, with about 1 in every 11 individuals afflicted. And this affliction is increasingly being observed throughout the world. Behaviors such as poor eating habits and lack of exercise or physical activity influence the risk and development of type 2 diabetes. Chapter 8 presents information about the rise in the prevalence of diabetes and the role that the clinical exercise physiologist can play in the prevention and treatment of the condition. As we continue to better understand the role of exercise in the prevention and treatment of diabetes, this chapter is very important for any practicing clinical exercise physiologist.

Paralleling and itself contributing to the rise in the incidence of diabetes is the increase in obesity—that is, having a body mass index (BMI) of 30 kg · m^{-2} or greater. The link between obesity and diabetes is very strong. Obesity is also strongly linked to an increased risk of heart disease, cancer, arthritis, disability, hypertension, and many other chronic diseases. As shown in chapter 9, weight loss strategies can be effective, and clinical exercise physiologists must play an active role in guiding and implementing these plans. The clinical exercise physiologist will be called upon more and more to develop and implement sustainable exercise programs, particularly for those patients actively involved in a weight management program.

Although hypertension awareness, diagnosis, and treatment have been enhanced tremendously since the 1960s, hypertension remains a leading contributor to as many as 10% of all deaths in the United States. New guidelines recommend taking an aggressive approach to identifying and treating hypertension. Exercise training can undoubtedly enhance both its prevention and treatment. Chapter 10 provides specific exercise recommendations for those with hypertension and those at risk for hypertensive disease.

Although the extent of the role that dyslipidemia plays in the development of atherosclerosis continues to be debated, the association is irrefutable. Therefore, therapies such as statins and nutrition counseling are traditionally used in the care of patients with abnormal blood lipid values. Exercise can play a vital role in this treatment regimen with respect to controlling weight and raising HDL levels. Chapter 11 presents this information in detail.

Chapter 12 delves into how conditions such as prediabetes, obesity, hypertension, and dyslipidemia can cluster to comprise a condition that is termed *metabolic syndrome*. Although specific definitions of metabolic syndrome are still subject to debate, it is clear that when these conditions are combined they generate greatly increased risks for cardiovascular and other diseases.

In patients with renal disease, kidney function deteriorates by stages, with many people unknowingly already in the early stages of the disease; chapter 13 thoroughly discusses those patients with chronic kidney disease (CKD). This includes the nearly 600,000 patients in the United States suffering from end-stage kidney disease. Patients with CKD experience fatigue and exercise intolerance, and those with end-stage disease are often treated with hemodialysis. Exercise is an important treatment modality for the comorbid conditions associated with CKD as well as for the exercise intolerance and depression that often accompany the condition. Additionally, because hemodialysis requires a 4 h block of time several times per week, exercise is increasingly being incorporated into the hemodialysis setting.

CHAPTER 8

Diabetes

Sheri R. Colberg, PhD

Exercise has long been recognized as an important component of diabetes management and is a central component in the prevention or delay of type 2 diabetes (32, 102). However, many adults without diabetes are physically inactive (108), but likely even more of those with diabetes are sedentary because of age, physical limitations, health comorbidities, fear of hypoglycemia, depression, and other reasons (61, 65). Exercise professionals must have the knowledge to assist people with diabetes and those at risk for diabetes in adopting and maintaining more physically active lifestyles, both safely and effectively.

DEFINITION

Diabetes mellitus (diabetes) is a group of metabolic diseases characterized by an inability to produce sufficient insulin or use it properly, resulting in **hyperglycemia**, or elevations in blood glucose (9). Insulin, a hormone produced by the beta cells of the **pancreas**, is needed by the skeletal muscles, adipose tissue, and liver to absorb and use glucose, the primary simple sugar (single sugar molecules) found in the blood (blood sugar) that is essential for proper functioning of the brain and nerves. The elevations in blood glucose associated with inadequately managed diabetes increase the risk for acute and chronic health issues, including cardiovascular (macrovascular) disease, microvascular diseases like **retinopathy** and **nephropathy**, and nerve damage (both autonomic and peripheral neuropathy) (14, 15).

SCOPE

More than 34 million people in the United States have diabetes (as of 2018). Worldwide, the total is approaching 500 million. In the United States, almost one quarter of these cases are undiagnosed, in large part because symptoms of the most common form of dia-

betes, type 2, may develop gradually, and years can pass without discernible symptoms (22). The percentage of undiagnosed cases worldwide may be even higher than that in the United States because of inadequate testing efforts in some parts of the globe. Another 88 million Americans are estimated to have **prediabetes**, an insulin-resistant state that puts them at high risk of developing type 2 diabetes. Even asymptomatic individuals with diabetes and prediabetes are at increased risk for developing long-term health complications (16, 21, 69). Although diabetes is a global problem, the Centers for Disease Control and Prevention considers it to be at epidemic proportions in the United States. The reasons for the epidemic are likely threefold:

1. An increasingly sedentary lifestyle and poor eating practices, resulting in a rise in overweight and obesity

2. An increase in high-risk ethnic populations in the United States

3. Aging of the adult population

Diabetes is currently the seventh leading cause of death in the United States (22). African Americans, Hispanics, American Indians, Alaskan Natives, Native Hawaiians, other Pacific Islanders, and some Asians have higher rates of diabetes than the approximate 7.5% rate in the non-Hispanic white population. The death rate of people with diabetes is twice that of people without diabetes of the same age. As serious as these mortality statistics are, they underestimate the effect of diabetes. Given that many people with this disease actually die from its related complications, diabetes is underreported as the cause or underlying cause of death. In one large study, diabetes was reported as listed on 39% of death certificates and as the underlying cause of death for only 10% of decedents with diabetes (67); other studies have associated diabetes with an increased risk of death from several diseases, including

Acknowledgment: Much of the writing in this chapter was adapted from earlier editions of *Clinical Exercise Physiology,* for which we gratefully acknowledge Ann Albright, PhD.

common complications like renal (kidney) disease, heart disease, and stroke (116). In addition, the economic effect of diabetes is staggering, and a large portion of the economic burden is attributable to long-term complications and hospitalizations.

PATHOPHYSIOLOGY

The various types of diabetes affect the options for treatment, although all share the risk for developing associated complications. Type 1 and type 2 diabetes are heterogeneous diseases in presentation, and their progression varies from person to person. While classification assists in determining the best treatment options, diabetes cannot always be clearly classified by type at the time of diagnosis. This section reviews the types of diabetes and associated complications.

Diabetes Categories

The American Diabetes Association classifies diabetes into four general categories (9), as listed in the sidebar Classification of Diabetes Mellitus.

Type 1 Diabetes

Type 1 diabetes comprises two subgroups: immune-mediated and **idiopathic**. Type 1 immune-mediated diabetes was formerly known as juvenile-onset or insulin-dependent diabetes and accounts for approximately 5% to 10% of those with the disease (9). This form of diabetes is considered an **autoimmune** disease in which the immune system attacks the body's own beta cells, resulting in absolute deficiency of insulin. Three distinct stages of type 1 diabetes can be identified. Multiple autoantibodies are detectable in the blood during the first stage before blood glucose levels rise; the second stage results in blood glucose levels in a prediabetes range; and the third stage is characterized by overt hyperglycemia and classic symptoms of diabetes. By this final stage, insulin must be supplied by regular injections or an insulin pump. Type 1 immune-mediated diabetes more frequently occurs in childhood and adolescence but can occur at any age. Symptoms appear to develop quickly in children and adolescents, who may present with ketoacidosis, and more slowly in adults (9). Type 1 idiopathic diabetes has no known etiologies and is present in only a small number of people, most of African or Asian ancestry. This form of type 1 diabetes is strongly inherited and lacks evidence of beta cell autoimmunity. The requirements for insulin therapy in individuals with type 1 idiopathic diabetes are sporadic (9).

Type 2 Diabetes

Type 2 diabetes was formerly called adult-onset or non-insulin-dependent diabetes. This is the most common form of the disease and affects approximately 90% to 95% of all those with diabetes (9). The onset of type 2 diabetes usually occurs after age 40, although it is seen at increasing frequency in adolescents (9). Most patients with type 2 diabetes are overweight or have obesity. Excess weight

---

Classification of Diabetes Mellitus

Type 1 Diabetes
Caused by autoimmune pancreatic beta cell destruction, usually leading to absolute insulin deficiency; most often diagnosed at a young age (i.e., children and teens) but can occur during young to middle adulthood

Type 2 Diabetes
Caused by a progressive loss of adequate beta cell insulin secretion, frequently on the background of insulin resistance; most often diagnosed after age 45 but can occur in younger individuals

Gestational Diabetes Mellitus
Diabetes diagnosed in the second or third trimester of pregnancy that was not clearly overt diabetes prior to gestation

Specific Types of Diabetes Due to Other Causes
- Monogenic diabetes syndromes (such as neonatal diabetes and maturity-onset diabetes of the young)
- Diseases of the exocrine pancreas (such as cystic fibrosis and pancreatitis)
- Drug- or chemical-induced diabetes (such as with glucocorticoid use, in the treatment of HIV/AIDS, or after organ transplantation)

Note: People with any type of diabetes may require insulin treatment at some stage of their disease. Such use of insulin does not, of itself, classify the type (e.g., insulin dependent).

Based on American Diabetes Association (9).

---

independently causes some degree of insulin resistance. Patients who do not have obesity or are not overweight by traditional weight criteria may have an increased percentage of body fat distributed predominantly in the abdominal region.

The pathophysiology of type 2 diabetes is complex and multifactorial. Insulin resistance of the peripheral tissues and defective insulin secretion are common features. With insulin resistance, the body cannot effectively use insulin in the muscles or liver even though sufficient insulin is being produced early in the course of the disease (9). This form encompasses individuals who have relative (rather than absolute) insulin deficiency and have peripheral insulin resistance. At least initially, and often throughout their lifetimes, these individuals may not need insulin treatment to survive. The treatment options include lifestyle management, including **medical nutrition therapy (MNT)** and physical activity; if medication is needed to reach glycemic targets, oral agents, insulin, or other injectable diabetes medications are prescribed (10, 11, 13). Bariatric surgery may also be added to the treatment plan for those who are obese (body mass index [BMI] ≥35 kg·m⁻²), especially if the diabetes or comorbidities are unmanageable with lifestyle and medication (12). Ketoacidosis rarely occurs. Table 8.1 provides an overview of the types of diabetes.

A genetic influence is present for type 2 diabetes. The risk for type 2 diabetes among offspring with a single parent with type 2 diabetes is 3.5-fold higher, and the risk for those with two such parents is 6-fold higher compared with offspring of people who don't have the disease (68). In addition, obesity contributes significantly to insulin resistance, and most people (80%) with type 2 diabetes are overweight or obese at disease onset (55). An abdominal distribution of body fat (e.g., belt size >40 in. [>102 cm] in men and >35 in. [>89 cm] in women) is associated with type 2 diabetes (55, 79). The risk of developing type 2 diabetes also increases with age, lack of physical activity, history of gestational diabetes, and presence of hypertension or dyslipidemia (9).

Gestational Diabetes

Since many women who become pregnant may have prior undiagnosed diabetes, **gestational diabetes** is defined as "diabetes diagnosed in the second or third trimester of pregnancy that is not clearly either type 1 or type 2 diabetes" (9). It is usually diagnosed with oral glucose tolerance testing, performed routinely at 24 to 28 wk of pregnancy. Risk factors for developing this form of diabetes include family history of gestational diabetes, previous delivery of a large birth weight (>9 lb [4 kg]) baby, and obesity. Although glucose tolerance usually returns to normal after delivery, women who have had gestational diabetes have a greatly increased risk of conversion to type 2 diabetes over time. They are recommended to have lifelong screening for the development of diabetes or prediabetes at least every 3 yr (9). Structured moderate physical exercise training during pregnancy decreases the risk of gestational diabetes, diminishes maternal weight gain, and is safe for the mother and the neonate (89).

Other Specific Types

The final category of diabetes, termed *other specific types*, accounts for only a small proportion of all diagnosed cases of diabetes (9). In these cases, certain diseases, injuries, infections, medications, or genetic syndromes cause the diabetes. This type may or may not require insulin treatment.

Complications of Diabetes

Health complications associated with diabetes are categorized as acute and chronic. This section reviews those complications.

Acute Complications

The acute complications of diabetes are hyperglycemia (high blood glucose) and **hypoglycemia** (low blood glucose). Each of these acute complications must be quickly identified to ensure proper treatment and reduce the risk of serious consequences (11).

Hyperglycemia   The manifestations of hyperglycemia are as follows:

1. Blood glucose levels poorly managed
2. Diabetic ketoacidosis
3. Hyperosmolar nonketotic syndrome

Blood glucose levels may be considered poorly managed when they are frequently above the recommended **glycemic goals** (see following discussion and table 8.2). High blood glucose levels cause the kidneys to excrete glucose and water, which causes increased

Table 8.1   Types of Diabetes and Stages of Glycemic Management

Stages of blood glucose management	TYPES OF DIABETES			
	Type 1	**Type 2**	**Gestational**	**Others**[b]
Normal glucose regulation	✓[a]	✓	✓	✓
Impaired glucose tolerance	✓	✓	✓	✓
Diabetes • No insulin needed • Insulin needed for glucose management • Insulin needed for survival	✓[a] ✓ ✓	✓ ✓	✓ ✓	✓ ✓

[a]Even after presenting in diabetic ketoacidosis, these patients can briefly return to normoglycemia without requiring continuous therapy (i.e., "honeymoon" remission).

[b]In rare instances, patients in these categories (e.g., Vacor toxicity, type 1 diabetes presenting in pregnancy) may require insulin for survival.

Based on American Diabetes Association (11).

Criteria for the Diagnosis of Diabetes

The World Health Organization, the Centers for Disease Control and Prevention, and the American Diabetes Association indicate one or more of the following test results as markers for a diagnosis of diabetes.

Test	Prediabetes	Diabetes
Fasting plasma glucose	Impaired fasting glucose (IFG): • ADA and CDC: 5.6-6.9 mmol · L⁻¹ (100-125 mg · dL⁻¹) • WHO: 6.1-6.9 mmol · L⁻¹ (110-125 mg · dL⁻¹)	≥7.0 mmol · L⁻¹ (126 mg · dL⁻¹)
Two-hour plasma glucose	Impaired glucose tolerance (IGT): 7.7-11.0 mmol · L⁻¹ (140-199 mg · dL⁻¹)	≥11.1 mmol · L⁻¹ (200 mg · dL⁻¹)
A1c	5.7-6.4% (39-47 mmol · mol⁻¹)	≥6.5% (48 mmol · mol⁻¹)

Based on American Diabetes Association (9).

urine production and dehydration. Symptoms of high blood glucose levels and dehydration may include

- headache,
- blurred vision,
- increased thirst,
- weakness, and
- fatigue.

The best treatment for anyone with frequently elevated blood glucose levels with ineffective management includes drinking plenty of non-carbohydrate-containing beverages, regular self-monitoring of blood glucose, and, when instructed by a health care professional, an increase in diabetes medications. Frequent high blood glucose levels damage target organs or tissues over time, which increases the risk of chronic complications (11).

Diabetic ketoacidosis occurs in people whose blood glucose levels are elevated and whose effective **circulating insulin** is very low or absent. This result is much more likely to occur in those with type 1 diabetes (9). **Ketones** form because without insulin, the body cannot use glucose effectively, and alterations in the pathways of typical fat metabolism occur to provide necessary energy. A byproduct of fat metabolism in the absence of adequate carbohydrate is ketone body formation by the liver, causing an increased risk of coma and death. Ketone levels in the blood are approximately 0.1 mmol · L⁻¹ in a person without diabetes and can be as high as 25 mmol · L⁻¹ in someone with diabetic ketoacidosis. When **individuals without diabetes** follow a low-carbohydrate diet that typically raise ketone body production, ketone levels do not become this elevated or result in ketoacidosis because insulin levels are still sufficient (25). Excessive levels of ketone body formation related to insulin insufficiency can be evaluated with a urine dipstick test or a fingerstick blood test. Other symptoms of ketoacidosis include abdominal pain, nausea, vomiting, rapid or deep breathing, and sweet- or fruity-smelling breath. Exercise is contraindicated in anyone experiencing diabetic ketoacidosis because blood glucose can rise further with any physical activity under these metabolic conditions (86, 87).

Hyperglycemic hyperosmolar nonketotic syndrome occurs in individuals with type 2 diabetes when hyperglycemia is profound and prolonged. This circumstance is most likely to happen during periods of illness or stress, in the elderly, or in people who are undiagnosed (2, 11). This syndrome results in severe dehydration attributable to rising blood glucose levels, causing excessive urina-

Table 8.2   Summary of Glycemic Recommendations for Nonpregnant Adults With Diabetes

Glycemic management		Key concepts in setting glycemic goals:
A1c	<7.0%	More or less stringent glycemic goals may be appropriate for individual patients. Goals should be individualized based on duration of diabetes, age and life expectancy, comorbid conditions, known cardiovascular disease or advanced microvascular complications, hypoglycemia unawareness, and individual patient considerations. Postprandial glucose may be targeted if A1c goals are not met despite reaching preprandial glucose goals. Postprandial glucose measurements should be made 1-2 h after the beginning of the meal, generally the peak level in patients with diabetes.
Preprandial capillary plasma glucose	80-130 mg · dL⁻¹ (4.4-7.2 mmol · L⁻¹)	
Peak postprandial capillary plasma glucose	<180 mg · dL⁻¹ (<10.0 mmol · L⁻¹)	

Based on American Diabetes Association (11).

tion. Extreme dehydration eventually leads to decreased mentation and possible coma. Exercise is also contraindicated in anyone in a hyperglycemic hyperosmolar metabolic status without ketosis.

Hypoglycemia   Hypoglycemia is a potential side effect of diabetes treatment and usually occurs when blood glucose levels drop below 65 mg · dL^{-1} (3.6 mmol · L^{-1}). Hypoglycemia may occur in the presence of the following factors, in isolation or combination:

- Too much insulin or use of selected antidiabetic oral agents
- Too little carbohydrate (or other food) intake
- Missed meals
- Excessive or poorly planned exercise

Hypoglycemia can occur either during exercise or hours to days later. Late-onset postexercise hypoglycemia generally occurs within 2 to 48 h following moderate- to high-intensity exercise that lasts longer than 30 min. This kind of hypoglycemia results from increased insulin sensitivity, ongoing glucose use, and physiological replacement of glycogen stores through gluconeogenesis (46). Individuals should be instructed to monitor blood glucose before and periodically after exercise to assess glucose response. This approach is also recommended in clinical exercise programs, such as cardiac rehabilitation, especially in anyone new to exercise. Recommendations for pre- and postexercise blood glucose assessment are provided later in this chapter.

The two categories of symptoms of hypoglycemia are adrenergic and neuroglycopenic. As blood glucose decreases, glucose-raising hormones (i.e., **glucagon**, epinephrine, growth hormone, and cortisol) are released to help increase circulating blood glucose levels (4). Adrenergic symptoms such as shakiness, weakness, sweating, nervousness, anxiety, tingling of the mouth and fingers, and hunger result from epinephrine release. As blood glucose delivery to the brain decreases, neuroglycopenic symptoms like headache, visual disturbances, mental dullness, confusion, extreme fatigue, amnesia, seizures, or coma may occur.

Some people with diabetes lose their ability to sense hypoglycemic symptoms (termed *hypoglycemia unawareness*), which is essentially onset of **neuroglycopenic** symptoms before the appearance of adrenergic warning symptoms related to glucoregulatory hormones (96). Instituting tight management of blood glucose may lower the threshold so that symptoms do not occur until blood glucose drops quite low. Intensity of management may need to be slightly reduced to alleviate hypoglycemia unawareness. In contrast, individuals whose blood glucose has been inadequately managed may have symptoms associated with low blood glucose at levels much higher than 70 mg · dL^{-1}.

In association with prior exercise or prior hypoglycemia, some individuals with type 1 diabetes may experience a related condition called *hypoglycemia-associated autonomic failure*, or HAAF. In this case, their lack of adrenergic response leading to hypoglycemia unawareness may be related to having a hypoglycemic episode in the previous 24 to 48 h, especially one that was severe or long lasting, or having exercised during that time (39). Both prior exercise and prior hypoglycemia may blunt the normal adrenergic response to the next bout of exercise or hypoglycemia and set individuals up for potentially dangerous hypoglycemic episodes (90, 91).

Treatment of hypoglycemia consists of testing blood glucose to confirm hypoglycemia and then consuming approximately 15 to 20 g of carbohydrate (e.g., glucose, sucrose, or lactose) that contains minimal or no fat (11). Commercial products (glucose tablets, gels, or liquids) allow the consumption of precise amounts of carbohydrate, and glucose increases blood glucose levels more rapidly than other simple sugars or carbohydrate sources. Other treatment options include 1 C (240 mL) of nonfat milk, 1/2 C (120 mL) of orange juice, 1/2 can (180 mL) of regular soda, six or seven Life Savers, 2 tbsp (30 mL) of raisins, or 1 tbsp (15 mL) of sugar, honey, or corn syrup. Individuals should wait about 15 or 20 min to allow the symptoms to resolve and then recheck blood glucose levels to determine if additional carbohydrate or other food sources are necessary. If the individual becomes unconscious because of hypoglycemia, glucagon should be administered. Mini-doses of glucagon via subcutaneous injection with a pen and via nasal sprays are now available as delivery options, potentially for prevention of hypoglycemia during exercise and for mild to moderate hypoglycemia treatment (85, 95). Individuals who received full doses of glucagon frequently experienced nausea and vomiting after regaining consciousness, but using mini-doses for less severe hypoglycemia prevents the majority of those symptoms. If glucagon is not available and an individual is unconscious or unable to self-treat hypoglycemia, 911 should be called immediately.

Given that exaggerated fluctuations in blood glucose can cause health problems both short term (hypoglycemia, hyperglycemia) and long term (e.g., heart disease, neuropathy, diabetic kidney disease, and more), an international consensus panel has recommended that most individuals with type 1 or type 2 diabetes aim to spend at least 17 h a day (i.e., more than 70% of their time) in a blood glucose range of 70 to 180 mg · dL^{-1} (3.9-10.0 mmol · L^{-1}); during pregnancy, the goal is 63 to 140 mg · dL^{-1} (3.5-7.8 mmol · L^{-1}) (17). This so-called *time in range* is typically measured via continuous glucose monitor (CGM) or flash monitor readings in individuals who are wearing those devices. Everyone else can simply strive to keep their self-monitored fingerstick glucose values in those ranges. The specific targets for time spent out of range were set accordingly, as shown in the table 8.3.

Individuals should try to limit how much time they spend out of that range, as well as extreme glucose fluctuations. The recommendation is to be below 70 mg · L^{-1} (3.9 mmol · L^{-1}) less than 1 h/day, or under 4% of the time, and below 54 mg · dL^{-1} (3.0 mmol · L^{-1})—the cut point for potentially serious hypoglycemia—less than 15 min/d (1%). Time spent at or above 180 mg · dL^{-1} (10.0 mmol · L^{-1}) should be less than 6 h daily (25%), and above 250 mg · dL^{-1} (13.9 mmol · L^{-1}), which is considered potentially serious hyperglycemia, less than 1 h and 15 min (5%).

Table 8.3   Recommended Daily Time in Range Goals for Users of Continuous or Flash Glucose Monitoring Devices

Condition	Time in range goals	Glucose target
Optimal (for most youth and adults)	70% (almost 17 h)	70-180 mg · dL^{-1} (3.9-10.0 mmol · L^{-1})
Pregnancy	70% (almost 17 h)	63-140 mg · dL^{-1} (3.5-7.8 mmol · L^{-1})
Older or sicker adults	50% (12 h)	70-180 mg · dL^{-1} (3.9-10.0 mmol · L^{-1})
Hypoglycemia	<4% (1 h) <1% for older or sicker	<70 mg · dL^{-1} (<3.9 mmol · L^{-1})
Severe hypoglycemia	<1% (15 min)	<54 mg · dL^{-1} (<3.0 mmol · L^{-1})
Hyperglycemia	<25% (6 h)	≥180 mg · dL^{-1} (≥10.0 mmol · L^{-1})
Severe hyperglycemia	<5% (1 h, 15 min)	>250 mg · dL^{-1} (>13.9 mmol · L^{-1})

Based on Battelino et al. (17).

Chronic Complications

Diabetes is the leading cause of adult-onset blindness, nontraumatic lower limb amputation, and end-stage renal failure (22). In addition, those with diabetes are at two to four times the normal risk of heart disease and stroke. Chronic exposure to hyperglycemia is considered of primary importance in the development of chronic complications, along with hypertension and hyperlipidemia. In determining appropriate exercise programming for people with diabetes, it is helpful to divide the potential chronic complications into three categories:

1. Macrovascular (large vessel or atherosclerotic) disease, which includes coronary artery disease with or without angina, myocardial infarction, cerebrovascular accident, and peripheral arterial disease

2. Microvascular (small vessel) disease, which includes proliferative diabetic retinopathy (eye disease) and nephropathy (diabetic kidney disease)

3. Neuropathy that involves either the peripheral or the autonomic nervous system

Intensive blood glucose management leading to glycemic levels closer to normal has long been proven to lower the risk of developing microvascular complications in anyone with type 1 or 2 diabetes (31, 104). Evidence from the ACCORD trial, however, does not support intensive glycemic management (as indicated by a hemoglobin A1c value <6%) to reduce the rate of cardiovascular events in middle-aged and older people with type 2 diabetes and either established cardiovascular disease or additional cardiovascular risk factors (2). In that trial, the mortality rate was higher in the intensive therapy group. For adults with type 1 diabetes, glycemic management is a predictor of macrovascular disease,

independent of diabetes duration, with lower A1c levels affording more protection against disease; A1c may be less of a risk factor in people with long-term type 1 diabetes (50 yr or more) (3). The recent new focus on time in range (using continuous and flash glucose monitoring devices) may prevent certain health problems related to frequent glycemic fluctuations, even when A1c values are in acceptable ranges.

The clinical exercise physiologist who is involved in the exercise training of people with diabetes must obtain information about the presence and stage of complications. The clinician should then use this information when developing an exercise prescription and behavior modification plan designed to help those with diabetes reduce their risk of developing or amplifying the complications of the disease. Cardiac rehabilitation programs may be suitable for individuals, including those at high risk, who wish to incorporate an exercise program into their lifestyle. These programs often provide pre- and postexercise blood glucose monitoring for safety and teaching purposes. Note that isolated diabetes is not a reimbursable diagnosis for cardiac rehabilitation, but many programs have low-cost options for patients without insurance coverage or reimbursement for specific diagnoses.

Macrovascular Disease   Diabetes is a major risk factor for **macrovascular disease** (14). The vessels to the heart, brain, and lower extremities can be affected. Blockage of the blood vessels in the legs results in peripheral artery disease, which can cause intermittent claudication, and exercise intolerance. Reduction and management of vascular risk factors are especially important in those with diabetes. The methods used for this purpose are similar to those used for coronary heart disease. The chapter on myocardial infarction (chapter 14) in this text reviews the vascular risk factor management methods in detail. The symptoms of peripheral arterial disease can be improved with exercise training, as reviewed in chapter 17.

Microvascular Disease

Microvascular disease causes retinopathy and nephropathy, which result in abnormal function and damage to the small vessels of the eyes and kidneys, respectively. The ultimate result of retinopathy can be blindness, whereas end-stage renal failure is the most serious complication of diabetic kidney disease (see chapter 13). Prevention or appropriate management requires periodic dilated eye examinations and kidney function tests, along with optimal blood glucose and blood pressure management. The clinical exercise physiologist must give careful attention to the stage of complications when prescribing exercise for those with microvascular involvement; this topic is discussed in detail in the diabetes-specific exercise prescription section of this chapter.

Peripheral and Autonomic Neuropathy

Peripheral neuropathy typically affects the legs before the hands. Individuals may initially experience sensory symptoms (paresthesia, burning sensations, and hyperesthesia) and loss of tendon reflexes (29). As the complication progresses, the feet become insensate, putting them at high risk for foot trauma that can go undetected. Muscle weakness and atrophy can also occur. Foot deformities can result, causing areas to receive increased pressure from shoe wear or foot strike, placing them at risk for injury. The large number of lower limb amputations from diabetes results from loss of sensation that places the individual at risk for injury and from diminished circulation attributable to peripheral artery disease. This circumstance impairs healing and can lead to severe reductions in blood flow, potential gangrene, and amputation.

People with diabetes of any type must be given instruction on how to examine their feet and practice good foot care (29). Foot care is especially important when someone with peripheral neuropathy begins an exercise program because increased walking and cycle pedaling increases the risk of foot injury. Having peripheral neuropathy may also change gait and ability to balance, increasing the risk for falling, but balance training may help lower fall risk (70).

Diabetic autonomic neuropathy may occur in any system of the body (e.g., cardiovascular, respiratory, neuroendocrine, gastrointestinal). Many of these systems are integral to the ability to perform exercise. **Cardiovascular autonomic neuropathy** is manifested by a high resting heart rate, an attenuated exercise heart rate response and variability, an abnormal blood pressure, and redistribution of blood flow response during exercise (106). This combination can severely limit exercise capacity and physical functioning.

CLINICAL CONSIDERATIONS

In the clinical setting, laboratory tests and examinations are used to diagnose diabetes or to facilitate ongoing monitoring. The following sections review these tests.

Signs and Symptoms

The classic signs and symptoms of diabetes include excessive thirst (**polydipsia**), frequent urination (**polyuria**), unexplained weight loss, infections and cuts that are slow to heal, blurry vision, and fatigue. Many who develop type 1 diabetes (especially during childhood or adolescence when onset is typically more rapid) have some or all of these symptoms, but those with type 2 diabetes may remain asymptomatic. In the United States, over 20% of those with diabetes do not know they have the disease (22).

History and Physical Examination

Individuals with diabetes having an annual physical examination should be evaluated for potential indicators of complications. These may include elevated resting pulse rate, loss of sensation or reflexes especially in the lower extremities, foot sores or ulcers that heal poorly, excessive bruising, and retinal vascular abnormalities. Exercise testing may be appropriate before beginning an exercise program (see the Exercise Testing section) (88). See chapter 4 for additional information about the physical examination.

Although a clinical exercise physiologist (or other exercise professional) cannot give medical clearance to someone who desires to start exercise training, they can review the medical history of anyone with diabetes and request that the person seek medical clearance from a health care provider prior to starting training (88). In making a determination regarding the necessity of such clearance, the clinical exercise professional should consider the following:

- Exercise participation and training status
- Body weight and body mass index
- Resting blood pressure
- Laboratory values for hemoglobin A1c, plasma glucose, lipids, and **proteinuria**
- Presence or absence of acute and chronic complications
- If chronic complications exist, the severity of those complications
- Other non-diabetes-related health issues

Of more immediate concern, prior to each exercise training session, the clinical exercise professional may inquire about the following to help the individual prevent acute complications like hypoglycemia or hyperglycemia:

- Starting blood glucose level (self-monitored)
- Timing, amount, and type of most recent food intake
- Medication use and timing

Exercise Testing

Most people with diabetes can benefit from participating in regular exercise. Participation in exercise is not without risk, however, and each individual should be assessed for safety. Priority must be given to minimizing the potential adverse effects of exercise through appropriate screening, program design, monitoring, and education (8). Exercise testing may be viewed as a barrier or as unnecessary for some individuals. Discretion must be given to an individual's health care provider when determining the need for exercise testing.

For participation in low-intensity exercise, health care professionals should use clinical judgment in deciding whether to recommend preexercise testing (26, 27). Conducting exercise testing before starting participation in most low- or moderate-intensity activities (<60% heart rate reserve or $\dot{V}O_2$ reserve) is considered unnecessary (26, 27). For exercise more vigorous than brisk walking (>60% heart rate reserve or $\dot{V}O_2$ reserve) or exceeding the demands of everyday living, sedentary people and individuals 40 yr or older with diabetes will likely benefit from being assessed for conditions that might be associated with cardiovascular disease, that contraindicate certain activities, or that predispose to injuries. The assessment should include a medical evaluation and screening for blood glucose management, physical limitations, medications, and macrovascular and microvascular complications. It may also include a graded exercise test. In general, electrocardiogram (ECG) stress testing may be indicated for individuals meeting one or more of these criteria (26):

1. Age >40 yr, with or without CVD risk factors other than diabetes

2. Age >30 yr and any of the following:
   - Type 1 or type 2 diabetes of >10 yr
   - Hypertension
   - Cigarette smoking
   - Dyslipidemia
   - Proliferative or preproliferative retinopathy
   - Nephropathy including microalbuminuria

3. Any of the following, regardless of age:
   - Known or suspected coronary artery disease, cerebrovascular disease, or peripheral artery disease
   - Autonomic neuropathy
   - Advanced nephropathy with renal failure

There is no evidence to determine if stress testing is necessary or useful before participation in anaerobic or resistance training. Coronary ischemia is less likely to occur during resistance compared with aerobic training at the same heart rate response (38, 41). Contraindications and general procedures for exercise testing are listed in chapter 5. The clinical exercise professional must be prepared to assist health care providers in the decision-making process to determine the need for exercise testing. Table 8.4 summarizes exercise testing specifics. Practical application 8.1 provides important information about the client–clinician interaction.

Table 8.4　Exercise Testing Review

Test type	Mode	Protocol specifics	Clinical measures	Clinical implications	Special considerations
Cardiovascular	Treadmill Ergometer (leg or arm)	Low level for many (≤2 METs per stage or 20 W/min increases in work rate)	Peak $\dot{V}O_2$ or estimated METs Heart rate and blood pressure responses 12-lead ECG	Watch for ischemia and arrhythmias because these are often undiagnosed and patients are at high risk for heart disease.	Chest pain due to myocardial ischemia may not be perceived in those with neuropathy (and also may blunt peak HR achieved). Patients with peripheral vascular disease or peripheral neuropathy may need to be assessed using the cycle ergometer mode. Consider testing blood glucose before exercise test to reduce the risk of hypoglycemia or hyperglycemia.
Resistance	Machine weights Isokinetic dynamometer	1RM or indirect 1RM method	Strength and power	Perform to allow for safe initiation of resistance training and to compare for change after implementing a training routine.	1RM may not be recommended in those with severe disease and those who are sedentary. Those with severe or unstable proliferative retinopathy most typically should not perform heavy resistance training.
Range of motion	Sit-and-reach test Goniometry		Range of motion in major muscle groups	Perform to demonstrate deficiencies and to compare for change after implementing a training routine.	Patients should not hold their breath at any time; any exercise may result in excessive blood pressure response.

W = watts; HR = heart rate; RM = repetition maximum.

Data from T.G. Lohman, *Advances in Body Composition Assessment* (Champaign, IL: Human Kinetics, 1992), 94.

Practical Application 8.1

CLIENT–CLINICIAN INTERACTIONS AND MOTIVATION FOR BEHAVIOR CHANGE

The interaction between the client and the clinician at the time of exercise evaluation is important, especially during ongoing exercise training visits. Living with diabetes poses many challenges and creates fears for individuals and their families. The exercise professional must be aware of the psychosocial components of living with a chronic disease and be able to apply strategies to help the person maintain participation in exercise.

The clinician should treat individuals as much more than their diagnoses. Be cautious about referring to anyone as "a diabetic" (used as a noun) because this terminology labels the individual by the disease; using "a person with diabetes" is more appropriate. Remember that the person usually needs to apply a great deal of effort and discipline to live with diabetes. Acknowledge that diabetes is challenging, and listen to the person's particular challenges. In general, do not use terms like *noncompliance* when discussing an exercise program. Inherent in the definition of noncompliance is the concept that a person is not following rules or regulations enforced by someone else. This concept is incongruent with self-management and empowerment, which consider the individual the key member of

the health care team. The health care professional should not make decisions for clients. Instead, the clinical exercise professional should equip them with information so they can make their own decisions. Looking for opportunities to engage an individual appropriately is important. Some may not understand or have yet made the connection between having diabetes and effective management of their blood glucose with the risk of developing a poor health outcome.

Having a defined strategy is helpful for exercise maintenance. Ask individuals to consider the following: How easily can they engage in their activity of choice where they live? How suitable is the activity in terms of their physical attributes and lifestyle? Have individuals identify exercise benefits they find personally motivating. Avoid physical activity goals that are too vague, ambitious, or distant; instead, have them set goals that are specific, measurable, attainable, realistic, and time-bound (i.e., SMART). Establish a routine to help exercise become more habitual. Have individuals identify any social support systems they may have. Provide positive feedback. Help them troubleshoot how to overcome specific obstacles they have to being active related to their diabetes (such as fear of hypoglycemia) or other concerns.

Treatment

Diabetes can be managed with a program of regular physical activity, medical nutrition therapy, self-monitoring of blood glucose, diabetes self-management education, and, when needed, medication (always needed in type 1) or significant weight loss from bariatric surgery or a complete meal replacement diet. When medication or bariatric surgery is used, it should be added to lifestyle improvements, not replace them. The individual and the health care team must work together to develop a program to achieve treatment goals.

Few diseases require the same level of ongoing daily personal involvement as diabetes. Individuals should receive information and training on disease management. Other members of the health care team may include a person's primary care physician or an endocrinologist, a nurse practitioner, a physician assistant, a diabetes care and education specialist, a registered dietitian, a clinical exercise physiologist, a behavioral or psychosocial counselor, and a pharmacist. In many instances, these health care professionals work together in a diabetes education program (and these programs may be connected to clinical exercise and weight management programs as added resources for the individual).

The individual must understand and be involved in developing appropriate treatment goals, which take into consideration their desires, abilities, willingness, cultural background, and comorbidities. Suggested glycemic targets from the American Diabetes Association and other organizations are provided earlier in the chapter in tables 8.2 and 8.3.

Medical nutrition therapy, often the most challenging aspect of therapy, is essential to the management of diabetes. A consensus report from the American Diabetes Association on this topic recommends nutrition counseling that works toward improving or maintaining blood glucose targets, weight management goals, and cardiovascular risk factors within individualized treatment goals for all adults with diabetes and prediabetes (37). These guidelines promote individually developed dietary plans based on metabolic, nutrition, and lifestyle requirements in place of a calculated caloric prescription because a single diet cannot adequately treat all types of diabetes or individuals equally.

Although macronutrient (i.e., protein, fat, carbohydrate) types and quantities should be considered in the development of a nutrition plan for the person with diabetes, many food choices and eating patterns can help people achieve health goals and increase quality of life. The total amount of carbohydrate consumed and the type must always be considered because all carbohydrates can raise blood glucose (although some have a more immediate effect), but fat and protein intake may also affect blood glucose and overall health (18, 114). Nutritional value must also be considered. A higher-fiber, less refined diet that is plant based is considered healthier for anyone with diabetes (76). In general, it is recommended that people with diabetes eat more nonstarchy vegetables, minimize their intake of added sugars and refined grains, and choose whole foods over highly processed ones. For most, focusing on at least reducing overall carbohydrate intake improves glycemia and works in a variety of eating patterns.

ADA Consensus Recommendations for Macronutrients (2019) (37)

- Evidence suggests there is no ideal percentage of calories from carbohydrate, protein, and fat for every person with or at risk for diabetes; therefore, macronutrient distribution should be based on individualized assessment of current eating patterns, preferences, and metabolic goals.

- When counseling people with diabetes, a key strategy for achieving glycemic targets is to assess current dietary intake and then provide individualized guidance on self-monitoring carbohydrate intake to optimize meal timing and food choices and to guide medication and physical activity recommendations.

- People with diabetes and those at risk for diabetes should consume at least the amount of dietary fiber recommended for the general public; increasing fiber intake, preferably through food (vegetables, pulses [beans, peas, and lentils], fruits, and whole intact grains) or through dietary fiber supplements (pill or soluble), may modestly lower A1c.

Monitoring of Blood Glucose

Self-monitoring of blood glucose is also an important part of managing diabetes. No standard frequency for self-monitoring has been established, but it should be performed frequently enough to help the individual meet treatment goals. Increased frequency of testing is often required when people begin an exercise program to assess blood glucose before and after exercise and to allow safe exercise participation. Individuals must be given guidance about how to use the information to make exercise, food, and medication adjustments. Those who require glucose-lowering medication must understand how their medications work with food and exercise to ensure the greatest success and safety. Clinical exercise physiologists must understand diabetes medications so they can safely prescribe exercise and provide guidance on exercise training to people with diabetes. Refer to table 8.5 for specific information on glucose-lowering medications and to chapter 3 for information on other pharmacology-related issues. In addition, consult the chapters that address blood pressure and lipid medications, since these considerations are often part of diabetes management.

Monitoring blood glucose is vital for the long-term maintenance of glycemic control and is especially important around all physical activity. Such monitoring may also provide positive feedback regarding the regulation or progression of the exercise prescription, which may result in greater subsequent long-term adherence to exercise. In addition to traditional fingerstick blood glucose self-monitoring, individuals with both type 1 and type 2 diabetes may be able to use continuous or flash glucose monitoring devices to get more frequent feedback about their glycemic levels and responses (7, 17). These newer devices are invasive, meaning a small sampling port must be placed into the skin for a period of days to weeks that allows the device to regularly track interstitial glucose levels, typically every 5 min. Such systems allow users to see dynamic data for glucose values, trends, rate of change, and time in range and can alert them to glucose levels that are too low or too high.

When people with diabetes begin an exercise program, it is helpful to closely monitor their glycemic levels. Monitoring and recording their levels before and after exercise (regardless of the method) may do the following:

- Allow for early detection and prevention of hypoglycemia or hyperglycemia

- Help determine appropriate preexercise levels to lower risk of hypoglycemia or hyperglycemia

- Identify those who can benefit from monitoring during and after exercise

- Provide information for modifying the exercise prescription

- Allow for better adjustment of diabetes regimens to manage all activities

- Motivate patients to remain more active to better manage their diabetes

Many factors can affect blood glucose responses to physical activity, particularly in insulin users, as shown in figure 8.1 (94, 101, 118). Timing of exercise is particularly important for glycemic balance, especially in individuals taking insulin, because their circulating levels during and after activity greatly influence their need for food intake. Consider the following for anyone using basal or bolus insulin:

- Changing insulin timing, reducing insulin doses, and increasing carbohydrate intake are effective strategies to prevent hypoglycemia and hyperglycemia during and after exercise.

- Early morning exercise may result in elevations in blood glucose levels instead of the usual decrease with moderate activity.

- Food intake must be considered with respect to the timing of exercise, particularly for anyone who takes insulin.

Table 8.5 Pharmacology

ORAL GLUCOSE-LOWERING MEDICATIONS			
Medication class and name	**Primary effects**	**Exercise effects**	**Important considerations**
Sulfonylureas (1st generation—only generics still available) • Tolbutamide • Tolazamide	Increase insulin production in the pancreas	Risk of hypoglycemia	Use extreme caution with elderly adults or anyone with hepatic or renal dysfunction. Not recommended without an established history of use.
Sulfonylureas (2nd generation) • Glyburide (Micronase, DiaBeta, Glynase) • Glipizide (Glucotrol, Glucotrol XL) • Glimepiride (Amaryl)	Increase insulin production in the pancreas Have an insulin-sensitizing effect	Risk of hypoglycemia (but lower than for 1st generation sulfonylureas)	Second-generation sulfonylureas provide more predictable results with fewer side effects and more convenient dosing. Clearance may be diminished in patients with hepatic or renal impairment; glipizide is preferred with renal impairment. Doses >15 mg should be divided. Glimepiride can be used with insulin.
Meglitinides • Repaglinide (Prandin) • Nateglinide (Starlix)	Increase insulin release from pancreas	Risk of hypoglycemia during exercise that follows meals for which these are taken	*Repaglinide:* Use with caution in patients with hepatic or renal impairment. Patients should be instructed to take medication no more than 30 min before a meal. If meals are skipped or added, the medication should be skipped or added as well. *Nateglinide:* Use with caution with moderate to severe hepatic disease. Has only 2 h duration of action. If meals are skipped or added, the medication should be skipped or added as well.
Biguanides Metformin (Glumetza, Glucophage, Glucophage XR, Riomet)	Primarily decrease hepatic glucose production May improve insulin resistance in muscles	Very low risk of hypoglycemia (possibly after prolonged or strenuous exercise)	Because of increased risk of lactic acidosis, should not use if frequent alcohol use, liver or kidney disease, or CHF is suspected. Contraindicated if serum creatinine is >1.5 mg · dL^{-1} in men or >1.4 mg · dL^{-1} in women. Do not use if creatinine clearance is abnormal.
Thiazolidinediones • Pioglitazone (Actos) • Rosiglitazone (Avandia) • Rosiglitazone and metformin (Avandamet) • Rosiglitazone and glimepiride (Avandaryl)	Decrease insulin resistance, increasing glucose uptake, fat redistribution Minor decrease in hepatic glucose output Preserve beta cell function Decrease vascular inflammation		Should not be used in patients with CHF or hepatic disease. Can cause mild to moderate edema.
α-glucosidase inhibitors • Acarbose (Precose) • Miglitol (Glyset)	Slow absorption of starch, disaccharides, and polysaccharides from GI tract	Gas and bloating; sometimes diarrhea	Should not be used if GI disorders are present. Avoid if serum creatinine is >2.0 mg · dL^{-1}.

(continued)

Table 8.5   *(continued)*

ORAL GLUCOSE-LOWERING MEDICATIONS			
Medication class and name	**Primary effects**	**Exercise effects**	**Important considerations**
DPP-4 inhibitors • Sitagliptin (Januvia) • Saxagliptin (Onglyza) • Vildagliptin (Galvus) • Alogliptin (Nesina) • Linagliptin (Tradjenta)	Inhibit the DPP-4 enzyme that degrades GLP-1 and GIP, resulting in two- to threefold increased levels of these incretins Increase insulin secretion in presence of elevated plasma glucose Reduce postmeal glucagon secretion		Observe patients carefully for signs and symptoms of pancreatitis. Rare reports of hypersensitivity reactions. Not for use in type 1 diabetes. Assessment of renal function is recommended before initiation and periodically thereafter.
SGLT2 inhibitors • Canagliflozin (Invokana) • Canagliflozin and metformin (Invokamet) • Dapagliflozin (Farxiga) • Dapagliflozin and metformin extended-release (Xigduo XR) • Empagliflozin (Jardiance) • Empagliflozin and linagliptin (Glyxambi) • Ertugliflozin (Steglatro)	Lower blood glucose levels by causing the kidneys to remove sugar from the body through the urine		Carry FDA warning of possible ketoacidosis and serious urinary tract infections. Other possible side effects include dehydration, kidney problems, increased cholesterol in the blood, and yeast infections. Approved only for type 2 diabetes; off-label use in type 1 diabetes.
INJECTED (NON-INSULIN) MEDICATIONS			
GLP-1 receptor agonists • Exenatide (Byetta) • Exenatide XR (Bydureon) • Dulaglutide (Trulicity) • Liraglutide (Victoza) • Semaglutide (Ozempic/Rybelsus—oral version) • Lixisenatide (Adlyxin, Lyxumia)	Decrease postmeal glucagon production Delay gastric emptying Increase satiety, leading to decreased caloric intake	Possible delayed treatment of hypoglycemia when taken near initiation of exercise	Not for use in patients with type 1 diabetes, severe renal disease, or severe GI disease. Consider lowering dose of sulfonylurea to avoid hypoglycemia when starting. May reduce the rate of absorption of oral medication.
Amylin analogue Pramlintide (Symlin)	Decrease postmeal glucagon production Delay gastric emptying Increase satiety, leading to decreased caloric intake Degree of response dependent on plasma glucose levels	Possible delayed treatment of hypoglycemia when taken near initiation of exercise	Indicated for insulin-treated type 2 diabetes or for type 1 diabetes. Contraindicated in patients with hypoglycemia unawareness, gastroparesis, or poor adherence. Should never be mixed with insulin and should be injected separately. Reduce insulin dose by 50% when starting. Requires patient testing of blood sugars before and after meals, frequent physician follow-up, and thorough understanding of how to adjust doses of insulin and pramlintide. May reduce the rate of absorption of orally administered medication. Medications requiring threshold concentrations should be taken 1 h before injection.

INSULIN, RAPID ACTING			
Medication class and name	**Primary effects**	**Exercise effects**	**Important considerations**
Lispro (Humalog) Lispro biosimilar (Admelog) Lispro new formulation (Lyumjev)	Onset <15 min Peak 1-2 h Effective duration 2-4 h	Risk of hypoglycemia	Should be taken just before or just after eating.
Aspart (Novolog) Fast-acting aspart (Fiasp)	Onset <15 min Peak 1-3 h Effective duration 3-5 h	Risk of hypoglycemia	Should be taken just before or just after eating.
Glulisine (Apidra)	Onset 0.5-1 h Peak 1-2 h Effective duration 3-5 h	Risk of hypoglycemia	Should be taken just before or just after eating.
Inhaled insulin (Afrezza)	Onset 12-15 min Peaks by 30 min Effective duration 3 h	Risk of hypoglycemia	Should be taken just before or just after eating. Available only in 4- or 8-unit increments.
INSULIN, SHORT ACTING			
Regular (Novolin R, Humulin R)	Onset 0.5-1 h Peak 2-4 h Effective duration 3-5 h	Risk of hypoglycemia	Best if taken 30 min before a meal.
INSULIN, INTERMEDIATE ACTING			
NPH (Novolin N, Humulin N)	Onset 2-4 h Peak 4-10 h Effective duration 10-16 h	Risk of hypoglycemia	Bedtime dosing minimizes nocturnal hypoglycemia.
INSULIN, LONG ACTING AND ULTRA LONG ACTING			
Glargine (Lantus, Toujeo) Glargine biosimilar (Basaglar/Abasaglar)	Onset 1-2 h Peak none (or minimal) Duration 24 h	Risk of hypoglycemia	Cannot be mixed in same syringe with other insulins; do not use same syringe used with other insulins. Small doses usually do not last 24 h.
Detemir (Levemir)	Onset 1-3 h Peak 50% in 3-4 h, lasting up to 14 h Effective duration 5.7-23.2 h	Risk of hypoglycemia	Cannot be mixed in same syringe with other insulins; do not use same syringe used with other insulins Administer twice daily.
Degludec (Tresiba)	Onset 30-90 min No peak Minimal effective duration over 42 h	Risk of hypoglycemia	Cannot be mixed in same syringe with other insulins; do not use same syringe used with other insulins. Once daily injections can be given at any time of day. Recommended time between dose increases is 3 or 4 d.

Combination oral medications and combination insulins, such as insulin degludec/insulin aspart injection (Ryzodeg 70/30), are available.

CHF = congestive heart failure; GI = gastrointestinal.

Adapted from Diabetes Medications Supplement. National Diabetes Education Program.

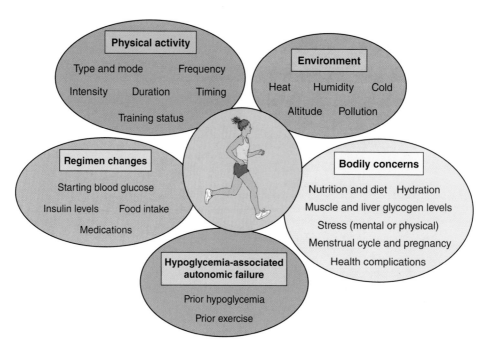

Figure 8.1 Blood glucose responses to physical activity are influenced by a large number of factors, especially in insulin users, who must precisely manage circulating insulin levels before, during, and after exercise.

Reprinted by permission from S.R. Colberg, *Athlete's Guide to Diabetes* (Champaign, IL: Human Kinetics, 2020), 22.

EXERCISE PRESCRIPTION

Exercise is a vital component of diabetes management and is considered a method of treatment for type 2 diabetes because it can positively affect how well insulin works in the body. Although exercise alone is not considered a method of treating type 1 diabetes because of the absolute requirement for exogenous insulin, it is still an important part of a healthy lifestyle for everyone.

Special Exercise Considerations

When developing an exercise prescription for persons with diabetes, the clinical exercise professional should consider the topic of fitness versus the health benefits of exercise. Methods to enhance maximal oxygen uptake are often extrapolated to the exercise prescription for disease prevention and management. Changes in health status, however, do not necessarily parallel increases in maximal oxygen uptake. In fact, evidence strongly suggests that regular participation in light- to moderate-intensity exercise may help prevent diseases such as coronary artery disease, hypertension, and type 2 diabetes, but it does not necessarily have an optimal effect on maximal oxygen uptake or overall fitness levels (32). Therefore, when frequency and duration are sufficient, exercise performed at an intensity below the threshold for an increase in maximal oxygen uptake can be beneficial to health in persons with diabetes.

Exercise must be prescribed with careful consideration given to risks and benefits. The consequences of physical inactivity combined with the complications of diabetes are likely to lead to more disability than the complications alone. Exercises that can be readily maintained at a constant intensity may be preferred for individuals with health complications who require more precise control of intensity, whereas higher-intensity intervals may lead to greater gains in overall fitness in those who are able to undertake such training safely (62).

Macrovascular Disease

Macrovascular disease is a complication that often affects people with diabetes. The primary macrovascular diseases are coronary artery disease and peripheral artery disease. Chapters 14 and 17 review specifics regarding preexercise evaluation and exercise prescription for coronary and peripheral artery disease. These approaches should be incorporated for individuals with diabetes and macrovascular disease.

Peripheral Neuropathy

The major consideration in individuals with peripheral neuropathy is the loss of protective sensation in the feet and lower legs that can lead to musculoskeletal injury and infection. Those without acute foot wounds or injuries can undertake moderate weight-bearing exercise, but anyone with a foot injury, open sore, or unhealed foot ulcer should be restricted to non-weight-bearing activities (e.g., chair exercises, arm exercises) (66). Proper footwear and daily examination of the feet are especially important. The clinical exercise professional should reinforce instruction given to the

Practical Application 8.2

MANAGEMENT OF BLOOD GLUCOSE BEFORE AND AFTER EXERCISE

Preexercise Hypoglycemia or Euglycemia

In most cases, blood glucose levels should be monitored before an exercise session to determine whether the person can safely begin exercising, especially someone using insulin or selected glucose-lowering oral agents that cause the pancreas to release insulin (such as sulfonylureas). Consideration must be given to how long and intense the exercise session will be, as well as where and when it will take place. Some individuals will need to take in supplemental carbohydrate to prevent hypoglycemia (25, 27, 86).

If diabetes is managed by diet or oral glucose-lowering medications with little to no risk of hypoglycemia (which is most of them), the majority of individuals will not need to consume supplemental carbohydrate for exercise lasting less than 60 min. Supplemental carbohydrate is mainly required during longer efforts, during exercise when circulating insulin levels are higher, or when starting out with blood glucose levels on the low side or even up to 180 mg $\cdot$ dL^{-1} (10.0 mmol $\cdot$ L^{-1}). If preexercise blood glucose levels are less than 100 mg $\cdot$ dL^{-1} (5.6 mmol $\cdot$ L^{-1}), the need for carbohydrate will vary widely with exercise intensity and duration, with vigorous exercise often conversely causing a rise in blood glucose if performed when fasted or when insulin levels are low. If blood glucose levels are moderately elevated prior to exercise (up to 180 mg $\cdot$ dL^{-1}, or 10.0 mmol $\cdot$ L^{-1}), an activity that would normally require supplemental carbohydrate intake may instead require less or none at all.

Active individuals without diabetes benefit from consuming carbohydrate during longer-duration and higher-intensity workouts, and most can use up to 70 to 80 g per hour during such activities (98). In people with diabetes, supplemental carbohydrate or other macronutrient intake needs can be similar but also largely determined by circulating insulin levels at the time of the activity. The best advice is to supplement with carbohydrate on a trial-and-error basis and customize intake for each individual and each situation (86). More detailed suggested intake of carbohydrate in insulin users can be found in a consensus statement from 2017 in The Lancet (87). Keep in mind that for anyone trying to lose weight through physical activity, medication adjustments may be preferable to increased food and calorie intake to prevent exercise-associated hypoglycemia.

Preexercise Hyperglycemia

If the preexercise blood glucose is greater than 250 mg $\cdot$ dL^{-1} (13.9 mmol $\cdot$ L^{-1}), urine or blood should be checked for ketones. If ketones are present (moderate to high), exercise should be postponed until glucose management is improved (86). An individual with a blood glucose greater than 300 mg $\cdot$ dL^{-1} (16.7 mmol $\cdot$ L^{-1}) without ketones is safe to exercise, but be sure they are adequately hydrated and generally feeling well. These blood glucose values are guidelines, and actions should be verified with the individual's physician. Individuals who use medication as part of diabetes treatment should be assessed to determine whether the timing and dosage of medication will allow exercise to have a positive effect on blood glucose. For example, a person who uses insulin and has a blood glucose of 270 mg $\cdot$ dL^{-1} (15.0 mmol $\cdot$ L^{-1}), has no urinary or blood ketones, and took mealtime insulin within 30 min of starting the activity will see a reduction in blood glucose from the additive glucose-lowering effect of both insulin and muscle contractions. If this individual has only basal levels of circulating insulin (meaning no mealtime insulin in the previous 2-3 h), more insulin will likely be needed to help reduce blood glucose before exercise and keep it from rising. Those with type 2 diabetes who are appropriately managed by diet and exercise alone usually experience a reduction in blood glucose with low to moderate exercise. Timing of exercise after meals can help many individuals with type 2 diabetes reduce **postprandial** hyperglycemia. Blood glucose should be monitored after an exercise session to determine the individual's response to exercise.

Postexercise Hypoglycemia

Individuals may experience hypoglycemia (<65 mg $\cdot$ dL^{-1}, or 3.6 mmol $\cdot$ L^{-1}) after exercise due to replacement of muscle glycogen, which largely comes from blood glucose (6). Periodic monitoring of blood glucose is necessary in the hours following exercise to determine whether blood glucose is decreasing. More frequent monitoring is especially important when undertaking new activities or training regimens. Later onset of hypoglycemia can be prevented with decreases in insulin doses (if taken) or increased intake of a variety of macronutrients. Overnight hypoglycemia may be avoided if individuals eat a bedtime snack, often one containing some protein or fat (111, 112). If hypoglycemia occurs, it should be treated as previously presented.

Postexercise Hyperglycemia

In poorly managed diabetes or when insulin doses are missed for any reason, low circulating insulin levels may result in a greater release of counterregulatory hormones (e.g., glucagon, epinephrine, cortisol, and growth hormone)

(continued)

Practical Application 8.2 *(continued)*

with exercise, especially when vigorous. This circumstance causes glucose production by the liver, enhanced free fatty acid release by adipose tissue, and reduced muscle uptake of glucose. The result, which is more likely in type 1 than type 2 diabetes, can be an increased blood glucose level during and after exercise. High-intensity exercise (done continuously or as intervals) can also result in hyperglycemia (51, 103). In this case, the intensity and duration of exercise should be reduced as needed. In insulin users, a small dose of correction insulin at mealtime insulin given postexercise may be needed to lower blood glucose levels.

---

individual on self-examination of the feet, stress the importance of seeking early treatment for injuries, and encourage annual foot examinations by a qualified health care professional (e.g., podiatrist).

Autonomic Neuropathy

Autonomic neuropathy, or central nerve damage, can affect anyone with long-standing diabetes. Such damage can affect the heart, digestive system, and more, often leading to inconsistent digestion of solid foods (gastroparesis), dehydration, impaired thermoregulation, hypotension and dizziness with changed positions, and even sudden death. If individuals have central nervous system damage that makes them dizzy when standing up, they will also be more likely to become dehydrated during exercise without realizing it (28). Normally, thirst centers in the brain are not activated until 1% to 2% of body water has already been lost, and autonomic neuropathy can make people even less likely to experience thirst. For most, an active cool-down reduces the possibility of a postexercise hypotensive response. Individuals with impaired thermoregulation should stay hydrated and not exercise in hot or cold environments (28).

Central nerve damage that affects heart function is known as cardiac autonomic neuropathy (CAN), which is manifested by abnormal heart rate, abnormal blood pressure, and redistribution of blood flow. Individuals with CAN have a higher resting heart rate and lower maximal exercise heart rate than those without this condition (106). Thus, estimating peak heart rate may lead to an overestimation of the training heart rate range if heart rate–based methods are used (see chapter 6 for general exercise prescription methods). In those with CAN, exercise intensity may be more accurately prescribed using the heart rate reserve method with maximal heart rate directly measured, but rating of perceived exertion (RPE) can also be used (28). Early warning signs of ischemia may be absent in these individuals, and the risk of exercise hypotension and sudden death increases (106). Moderate-intensity aerobic training can improve autonomic function in those with and without CAN (29). These individuals should have an exercise stress test and physician approval before starting an exercise program.

Retinopathy

Although many mild or background cases of diabetic proliferative retinopathy exist, only individuals with active proliferative or severe nonproliferative diabetic retinopathy need to be carefully screened before beginning an exercise program. Activities that increase intraocular pressure (e.g., high-intensity aerobic and resistance training with large increases in systolic blood pressure), head-down activities, and jumping and jarring activities are also not advised with severe nonproliferative or active proliferative disease (27). The clinician should advise people to never exercise during an active retinal hemorrhage (i.e., bleeding from breakage of abnormal, weak vessels that have grown in the back of the eye, leading to blood in the vitreous fluid of the eye that can cloud vision). Once proliferative retinopathy has stabilized following treatments and the risk of hemorrhages is low, most physical activities are safe to undertake or resume.

Diabetic Kidney Disease

Elevated blood pressure is related to the onset and progression of diabetic kidney disease, which can progress over time from small amounts of protein in the urine (microalbuminuria) to overt nephropathy and end-stage renal disease. Placing limits on low- to moderate-intensity activity is not necessary, but strenuous exercises should likely be discouraged in those with more severe disease because of the elevation in blood pressure and general fatigue (27). Individuals on renal dialysis or anyone who has received a kidney transplant can also benefit from exercise (53), and in most cases mild or moderate activities can be safely undertaken during dialysis treatments (60). See chapter 13 for details about renal failure.

Exercise Recommendations

Endurance, resistance, balance, and range of motion exercise training are all appropriate modes for most people with diabetes. Individuals trying to lose weight (especially those with type 2 diabetes) should expend a minimum cumulative total of 2,000 kcal/wk in aerobic activity and participate in a well-rounded resistance training program (12). Individual interests, goals of therapy, type of diabetes, medication use (if applicable), and presence and severity of complications must be carefully evaluated in developing the exercise prescription. The following exercise prescription recommendations are guidelines, and individual circumstances must always determine the specific prescription. Table 8.6 presents a summary of the general exercise prescription recommendations.

Cardiorespiratory Exercise

The health value of cardiorespiratory, or aerobic, exercise for persons with diabetes is strong. Given the high risk for developing atherosclerotic disease in those with diabetes, the ameliorating effects of cardiorespiratory exercise may help reduce this risk. In most individuals, such exercise should be performed daily, if possible.

Mode   Personal interests and the desired goals of the exercise program should drive the type of physical activity selected. Caloric expenditure is often a key goal for those with diabetes. Walking is the most commonly performed mode of activity in adults with type 2 diabetes as it is a convenient, low-impact activity that can be used safely and effectively to maximize caloric expenditure. Non-weight-bearing modes should be used if necessary, such as when

Table 8.6   Exercise Prescription Review

Training method	Frequency	Intensity	Time (Duration)	Type (Mode)	Progression	Important considerations
Cardiorespiratory	3-7 d/wk No more than 2 consecutive days between bouts of activity	40%-59% of $\dot{V}O_2$ reserve (moderate), RPE 11-13; or 60%-89% of $\dot{V}O_2$ reserve (vigorous), RPE 14-16	Minimum of 150-300 min/wk of moderate activity or 75-150 min/wk of vigorous activity	Walking, cycling, swimming and other aquatic activities, dancing	Rate of progression depends on many factors including baseline fitness, age, weight, health status, and individual goals Gradual progression of both intensity and volume is recommended	Avoid or take special precautions for exercise undertaken during insulin peak time. Be aware of any signs and symptoms for vascular and neurological complications, including silent ischemia. Warm-up and cool-down are important. Promote patient education. Assess for proper footwear and inspect feet daily. Avoid extreme environmental temperatures. Avoid exercise when blood glucose management is poor. Adequate hydration should be maintained. Instruct patient on blood glucose monitoring and on following guidelines to prevent hyper- and hypoglycemic events.
Resistance	2 or 3 d/wk but never on consecutive days	Moderate at 50%-69% of 1RM or vigorous at 70%-85% of 1RM	10-15 reps per set, 1-3 sets per type of specific exercise	Free weights, machines, elastic bands, or body weight as resistance	As tolerated, increase resistance first, followed by a greater number of sets, and then increased training frequency	
Range of motion and balance training	2 or 3 d/wk or more Usually done along with other types of training when muscles and joints are warmed up Lower body and core resistance exercises may double as balance training	To the point of tightness or slight discomfort	10-30 s per exercise of each muscle group; 2-4 reps of each; balance exercises can be practiced daily or as often as possible	8-10 exercises involving the major muscle groups	As tolerated, may increase range of stretch as long as patient does not complain of pain (acute or chronic) Balance training should be done carefully to minimize the risk of inadvertent falls	

Note: All important considerations listed in the chapter apply to all these training methods.

RPE = rating of perceived exertion; 1RM = one-repetition maximum; PNF = proprioceptive neuromuscular facilitation.

sores are noted on the feet. For a given level of energy expenditure, the health-related benefits of exercise appear to be independent of the mode in those with diabetes.

Intensity Programs of moderate intensity are preferable for most people with diabetes because the cardiovascular risk and chance for musculoskeletal injury are lower and the likelihood of maintaining the exercise program is greater. Exercise intensity appears to predict improvements in overall blood glucose management to a greater extent than exercise volume (20). With attention to safety, it may be beneficial for those already exercising at a moderate intensity to consider adding some vigorous physical activity or at least including some faster intervals interspersed into less intense training (27). Exercise should generally be prescribed at an intensity of 40% to 59% of $\dot{V}O_2$ reserve, or a rating of perceived exertion of 11 to 13, to be considered moderate (8). Vigorous exercise is 60% to 89% of $\dot{V}O_2$ reserve, or a rating of perceived exertion of 14 to 16 for most individuals. Chapter 6 provides specifics for determining and calculating proper exercise intensity.

Frequency The frequency of exercise should be 3 to 7 d/wk. Exercise duration, intensity, weight loss goals, and personal interests determine the specific frequency. Additionally, the blood glucose improvements with exercise in those with diabetes last only 2 to 72 h (19), suggesting that activity should be done minimally on 3 nonconsecutive days each week, with no more than 2 consecutive days between bouts.

Those who take insulin and have difficulty balancing caloric needs with insulin dosage may prefer to exercise daily for consistency. This schedule may result in less daily adjustment of insulin dosage and caloric intake than a schedule in which exercise is performed every other day or sporadically, and it will reduce the likelihood of a hypoglycemic or hyperglycemic response. In addition, individuals who are trying to lose weight will be more successful in their weight loss efforts by maximizing caloric expenditure (52), and continuing to participate in regular physical activity will help prevent weight regain (100).

Duration and Rate of Progression Exercise duration for those with diabetes should be 150 to 300 min/wk or more of moderate activity (ideally spread throughout the week, such as 30 min on 5 d/wk) or 75 to 150 min of vigorous activity (27, 82). Bouts of exercise should ideally be a minimum of 10 min, but some benefits are still gained from shorter bouts (82). Gradual progression of both intensity and volume is recommended to reduce the risk of injury.

Timing Exercise should be performed at the time of day that is most convenient for the participant. Because of the risk of hypoglycemia, those taking insulin should give consideration to the time of day they perform exercise and should avoid activity when insulin is peaking (unless compensating with additional food intake). Muscle contractions promote peripheral glucose

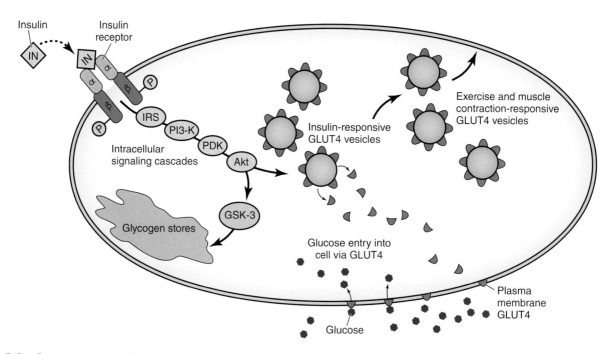

Figure 8.2 Separate but additive mechanisms stimulate blood glucose uptake via GLUT 4 into active muscles and increase hypoglycemia risk, especially when circulating insulin levels are higher.

Reprinted by permission from R.K. Dishman, G.W. Health, M.D. Schmidt, and I-M. Lee, *Physical Activity Epidemiology,* 3rd ed. (Champaign, IL: Human Kinetics, 2022), 268.

uptake via a contraction-induced mechanism involving calcium release and translocation of GLUT 4 glucose transport proteins, as shown in figure 8.2 (80). Given that insulin itself can bind to a cellular receptor and promote the uptake of glucose from blood into muscles (both at rest and during exercise), and this insulin-associated uptake involves a separate pool of GLUT 4 proteins (92), the risk of hypoglycemia greatly increases during activity. The independent effects of exercise and insulin on blood glucose levels are known to be additive.

The replacement of muscle glycogen after exercise also increases hypoglycemia risk (47). This happens because blood glucose continues to be taken up into muscle cells after exercise to restore glycogen, and this uptake requires very little insulin in the first couple of hours following the activity (80). Exercising late in the evening for those on insulin and some oral medications may be more difficult to manage because of the risk of hypoglycemia overnight during sleep.

Resistance Exercise

Resistance training should be recommended for almost everyone because training can improve cardiovascular function, insulin sensitivity, strength, and body composition in people with diabetes (44, 99), and persons with type 1 diabetes may even experience more stable blood glucose levels during resistance training compared with aerobic activities (40). All individuals should be screened for contraindications before they begin resistance training. Proper instruction and monitoring are also needed. A recommended resistance training program consists of 5 to 10 exercises involving major muscle groups performed with one to three sets of 8 to 15 repetitions to near fatigue. Resistance training exercises should be done 2 or 3 d/wk on nonconsecutive days (27, 82). Progression of intensity, frequency, and duration should occur slowly: increases in resistance should be made first and only once the target number of repetitions per set can be exceeded, followed by a greater number of sets and then increased training frequency (27). Modifications such as lowering the intensity of lifting and eliminating the amount of sustained gripping or isometric contractions should be considered to ensure safety given that people with diabetes are more prone to overuse and other tendon, ligament, and joint injuries. Chapter 6 provides specific information about resistance training that should be considered for these individuals.

Range of Motion and Balance Exercises

Range of motion exercises, or flexibility training, can be included as part of an exercise program but should not be substituted for aerobic or resistance exercise. Flexibility exercises combined with resistance training increase range of motion in people with type 2 diabetes (45). At least one study reported that home workouts consisting of range of motion, stretching, and strengthening exercises could improve motion around foot joints (frequently limited in adults with diabetes) and plantar pressure distribution during walking to prevent or lessen potential foot complications (23). Stretching is best done when muscles and joints have been warmed up, such as after an initial warm-up or a bout of physical activity. Both static and dynamic stretching can improve joint flexibility and range of motion.

In addition to flexibility training, it is recommended that all individuals over the age of 40 engage regularly in balance training, which is known to reduce the risk of falls in people with diabetes and even with peripheral neuropathy (70). Balance training can be as simple as standing on one leg at a time, but it can also include many exercises, some using balance training equipment. Most core and lower body resistance exercises also lead to improved balance and reduced fall risk, as can the practice of yoga, tai chi, qigong, and other activities.

EXERCISE TRAINING

When beginning a regular exercise training program, those with diabetes can anticipate significant and meaningful improvements. Exercise training considerations for diabetes include the following:

- Exercise training promotes blood glucose uptake by the skeletal muscles. People who use insulin or selected oral glucose-lowering medications may run the risk of hypoglycemia.
- The heart rate response of those with longstanding diabetes may be impaired. They may have a higher than typical resting heart rate and a lower peak heart rate on an exercise test.
- Do not allow anyone to exercise with a blood glucose >250 mg · dL^{-1} (13.9 mmol · L^{-1}) if moderate or higher levels of blood or urinary ketones are present. If it is determined safe for a person to exercise with a blood glucose >300 mg · dL^{-1} (16.7 mmol · L^{-1}) (i.e., no ketones are present), be sure the individual is hydrated and feels well (27, 86).

Cardiorespiratory and Resistance Exercise

Benefits for persons with diabetes are seen with both acute and chronic cardiorespiratory and resistance exercise training. Acute bouts of exercise typically improve blood glucose, although the actual response can be influenced by preexercise blood glucose levels, the duration and intensity of exercise, and other factors. For people with type 2 diabetes, a reduction in blood glucose levels that is sustained into the postexercise period following mild to moderate exercise is typical, with the reduction attributed to an attenuation of hepatic glucose production along with a normal increase of muscle glucose use (93, 105). The effect of acute exercise on blood glucose levels in those with type 1 diabetes and in lean individuals with type 2 diabetes is more variable and unpredictable, but glycemic benefits are still possible with concomitant dietary management (119). A rise in blood glucose with exercise can be seen in people who are extremely insulin deficient (usually type 1) and with short-term, high-intensity exercise (64, 71).

Most of the benefits of exercise for those with diabetes of any type come from regular, long-term activity. These benefits can include improvements in metabolic health (glucose management and insulin resistance), hypertension, lipids, body composition and weight loss or maintenance, and psychological well-being (49, 119). Both the frequency of aerobic training and the volume of resistance training appear to be important in lowering overall blood glucose levels in type 2 diabetes (105). Table 8.7 provides a brief review of the effects of exercise training for individuals with diabetes.

Like acute exercise, exercise training can improve blood glucose. Exercise training (both aerobic and resistance) improves glucose management as measured by hemoglobin A1c or glucose tolerance, primarily in those with type 2 diabetes (36, 113). In a meta-analysis, regular exercise was demonstrated to have a significant benefit on insulin sensitivity in adults with type 2 diabetes that may persist beyond 72 h after the last exercise session in some cases (109).

Following exercise training, insulin-mediated glucose disposal is improved. Insulin sensitivity of both skeletal muscle and adipose tissue can improve with or without a change in body composition (35, 63). Exercise may improve insulin sensitivity through several mechanisms, including changes in body composition, muscle mass, fat oxidation, capillary density, and GLUT 4 glucose transporters in muscle (24, 43, 48, 73, 107, 109). The effect of exercise on insulin action is lost within a few days, again emphasizing the importance of consistent exercise participation.

Hypertension affects approximately 74% of adults with diabetes (57). Myriad studies have shown that exercise training can lower blood pressure in adults with diabetes. A meta-analysis and systematic review suggest that while both aerobic and resistance exercise appear effective in reducing systolic and diastolic blood pressure, subgroup analyses indicate that aerobic activity may have a greater effect (81). Reduction in blood pressure occurs more commonly in systolic levels, whereas other research has shown no change in blood pressure with training (77, 83).

Information about the effect of exercise on lipids in diabetes shows mixed results and may vary with the activity and other lifestyle changes (such as diet). In general, supervised aerobic exercise leads to more significant lowering of total cholesterol, triglycerides, and low-density lipoprotein cholesterol than no exercise. Supervised resistance training provides more benefit than inactivity in improving total cholesterol (77). Lipid profiles may benefit more from a combination of exercise training and weight loss. The Look AHEAD (Action for Health in Diabetes) trial found that lifestyle participants with type 2 diabetes had greater decreases in triglycerides and high-density lipoprotein cholesterol than controls, although both groups experienced decreases in low-density lipoprotein cholesterol (63). For adults with type 1 diabetes, exercise training may reduce low-density lipoproteins (75) and total cholesterol levels (115). Resistance and aerobic exercise are generally considered to have a similar impact on cardiovascular risk markers and are equally safe for adults with diabetes (117).

Table 8.7 Exercise Training Review

Cardiorespiratory endurance	Skeletal muscle strength	Skeletal muscle endurance	Flexibility and balance	Body composition
Associated with a lower risk of all-cause mortality and cardiovascular mortality. Prevention or delay of type 2 diabetes. Improves insulin action and fat oxidation and storage in muscle, and can improve blood glucose management. Exercise intensity affects blood glucose levels to a greater extent than exercise volume. May result in a small reduction in LDL cholesterol. May result in a small reduction in systolic blood pressure, but reductions in diastolic blood pressure are less common.	Improves insulin action and fat oxidation and storage in muscle and can improve blood glucose management, primarily through increases in muscle mass; prevention of loss of muscle due to aging, disuse, and diabetes; and improved muscle quality (less marbling with fat).	Improves insulin action and fat oxidation and storage in muscle and can improve blood glucose management, although its effects last only 2-72 h (mostly related to muscle glycogen repletion following activities).	No effect on diabetes management. Flexibility training combined with resistance training can increase range of motion. Flexibility training has not been shown to necessarily reduce risk of injury. Balance training can reduce the risk of falling.	Helps produce and maintain weight loss and, consequently, blood glucose management. Combined weight loss and exercise may be more effective with respect to lipids than aerobic training alone.

Weight loss is often a therapeutic goal for those with type 2 diabetes because most are overweight or obese, as are many adults with type 1 diabetes. Moderate weight loss improves glucose management and decreases insulin resistance in adults with type 2 diabetes (19). Medical nutrition therapy and exercise combined are more effective than either alone in achieving moderate weight loss (5). Visceral fat (i.e., deep abdominal) decreases peripheral insulin sensitivity. This body fat is a significant source of free fatty acids and may be preferentially oxidized over glucose, contributing to hyperglycemia. Exercise results in preferential mobilization of visceral body fat, likely contributing to the metabolic improvements even in the absence of significant weight loss (54, 74). Exercise is one of the strongest predictors of long-term weight loss maintenance (58). This is extremely important because weight lost is often regained. The amount of exercise required to maintain significant weight loss is probably much larger than the amount needed for improved blood glucose management and cardiovascular health (34, 50).

Epidemiological evidence has long supported the role of exercise in the prevention or delay of type 2 diabetes. Some of the early data in support of exercise in the prevention of type 2 diabetes come from a 6 yr clinical trial in which subjects with impaired glucose tolerance were randomized into one of four groups: exercise only, diet only, diet plus exercise, or control. The exercise group was encouraged to increase daily physical activity to a level that was comparable to a brisk 20 min walk. The incidence of diabetes in the exercise intervention groups was significantly lower (78). A randomized multicenter clinical trial of type 2 diabetes prevention (the Diabetes Prevention Program, or DPP) in those with impaired glucose tolerance at numerous sites around the United States showed a 58% reduction in the incidence of type 2 diabetes with lifestyle intervention that had goals of 7% weight loss and 150 min of physical activity per week (32, 33). A 10 yr follow-up showed that the lifestyle group continued to have a 34% lower risk of developing type 2 diabetes (59). DPP lifestyle modification programs have achieved clinically meaningful weight and cardio-metabolic health improvements (72). Given its success, the DPP program has been implemented in various communities around the United States (primarily in YMCAs and hospitals) with a goal to prevent type 2 diabetes in the at-risk adult population (1). The Centers for Medicare and Medicaid Services (CMS) has deemed these programs eligible for reimbursement. The clinical exercise physiologist may be an appropriate team member to deliver some version of the DPP, even if done through online apps and other social networks like Prevent, an online social network–based translation (97).

Psychological Benefits

Psychological benefits of regular exercise are realized in people without and with diabetes, including reduced stress and anxiety, improvements in mild to moderate depression, and increased self-esteem (30, 110). However, those who exercise to prevent disease have significantly improved psychological well-being, while it deteriorated in those who exercise for management of diseases (e.g., cardiovascular disease, end-stage renal disease, and cancer) (42). These findings suggest that benefits may vary and that people with fewer existing health concerns benefit the most. Mechanisms for the impact of exercise include psychological factors such as increased self-efficacy and changes in self-concept, as well as physiological factors like increased norepinephrine transmission and endorphins (30).

Research Focus

Exercise and Glycemic Management

Most clinical trials (such as Look AHEAD) have focused on the impact of an active lifestyle on cardiovascular risk factors in adults with type 2 diabetes, but individuals with type 1 diabetes also have a significant risk of suffering from cardiovascular diseases. At least one recent multicenter clinical trial has focused on adults with type 1 diabetes, with important findings. In this study, 18,028 adults (ages 18-79 yr) from Germany and Austria with type 1 diabetes were stratified according to their self-reported frequency of physical activity (inactive; one or two times per week; more than two times per week). Among participants, 63% reported being physically inactive.

Greater self-reported activity was associated with a lower body mass index, better glycemic management (A1c levels), fewer diabetes-related comorbidities (specifically retinopathy and microalbuminuria), and a lower cardiovascular risk related to contributing factors like hypertension, elevated blood lipid levels, and overweight and obesity, all without an increase in the risk of severe hypoglycemic episodes. Such encouraging results suggest that regular participation in physical activity should be promoted in individuals with type 1 diabetes to reduce their cardiovascular risk and improve their glycemic management and overall health.

Based on B. Bohn et al. "Impact of Physical Activity on Glycemic Control and Prevalence of Cardiovascular Fisk Factors in Adults With Type 1 Diabetes: A Cross-Sectional Multicenter Study of 18,028 Patients," *Diabetes Care* 38, no. 8 (2015): 1536-1543.

Practical Application 8.3

AMPUTATION

Diabetes is the leading cause of nontraumatic lower limb amputations. These types of amputations occur because of impaired sensation from nerve damage and reduced circulation in the extremities, resulting in wounds that are not able to heal properly. Walking capacity and performance decrease with progression of foot complications (56), and the energy cost of walking increases markedly for someone with an amputation and a prosthetic limb. In addition, prolonged walking may cause trauma and ulceration of the stump. Various upper body exercises, including chair exercises, weights, and arm ergometry, or non-weight-bearing activities such as swimming or other aquatic exercises, may be better choices. The clinical exercise physiologist should closely monitor individuals with amputations, who should have regular visits with a health care professional.

Clinical Exercise Bottom Line

- Exercise testing or training promotes blood glucose uptake into active muscles through a contraction-induced mechanism. Anyone using exogenous insulin may experience hypoglycemia (due to greater glucose uptake via dual mechanisms) or hyperglycemia (due to exercise intensity) either during or after activities. Check blood glucose levels before and after exercise.

- Anyone with a blood glucose >250 mg · dL^{-1} (13.9 mmol · L^{-1}) should not exercise if blood or urine ketones are at moderate or higher levels, as the person is likely in a state of insulin deficiency. Use caution implementing exercise when blood glucose is >300 mg · dL^{-1} (16.7 mmol · L^{-1}) without ketones, and allow that person to exercise only if they are feeling well and properly hydrated.

- Both resting and exercise heart rates may be impaired in individuals with cardiac autonomic neuropathy, which can result in a higher than typical resting heart rate (usually 100 b · min^{-1} or higher) and a blunted peak heart rate during an exercise test. Because of this, developing a heart rate–based exercise prescription without an exercise test may result in an inappropriately high heart rate range.

- Peripheral neuropathies and poor circulation may increase the risk of foot injury and cause slow healing. Have patients inspect their feet regularly for redness or areas of trauma, and avoid weight-bearing activities with an unhealed ulcer on the plantar foot surface.

CONCLUSION

Living with a chronic disease like diabetes of any type poses special issues, and its management requires ongoing dedication. In addition, the individual must cope with diabetes-related health complications if they develop. Exercise training of various types should be an essential component of the treatment plan for all individuals with diabetes or prediabetes. Although certain precautions may be necessary to participate, physical activity can usually improve blood glucose management, lipid levels, blood pressure, and body weight; lower mental stress, anxiety, and depression; and potentially reduce the physical and emotional burden of this chronic metabolic disease.

Online Materials

Visit HK*Propel* to access a link to the references, the case studies with discussion questions, and a quiz to help you review key concepts and test your understanding of the material covered in this chapter.

CHAPTER 9

Obesity

David C. Murdy, MD

Dennis J. Kerrigan, PhD

Jonathan K. Ehrman, PhD

The World Health Organization reports that the prevalence of obesity has tripled worldwide since 1975. In 2016 it was estimated that upwards of 2.5 billion adults were either overweight or obese and that being overweight or obese leads to the death of more people than does being underweight (211). The prevalence of overweight and obesity in the United States is at an all-time high, increasing from 30.5% in 2000 to 42.4% in 2018, with only Hawaii, Colorado, and Washington, D.C., at lower than 25% prevalence (see figure 9.1). One in two U.S. adults is predicted to have obesity by the year 2030 (197).

The U.S. National Center for Chronic Disease Prevention and Health Promotion, part of the Centers for Disease Control and Prevention (CDC), developed an At a Glance report in 2010 titled "Obesity: Halting the Epidemic by Making Health Easier" (37). In this document the U.S. government considers obesity an epidemic that is associated with—if not the principal cause of—much disease and disability in the United States, and even some forms of discrimination. An example of the latter is that many see obesity as a social or moral issue and stigmatize its management. This can add to the barriers related to understanding and treating this mounting epidemic. Additionally, there are and likely will continue to be disparities among demographic groups with respect to the incidence and effects of obesity (197).

Short-term interventions have limited effectiveness, and long-term success is rare without ongoing medical efforts or weight loss surgery (136). Clinicians should approach obesity as a chronic illness and manage its comorbidities, as well as work to help individuals lose weight. Many of the comorbidities can be controlled or eliminated with sufficient weight loss (17). But few medical treatments have such far-ranging positive health effects as assisting an overweight or obese patient achieve and sustain medically significant weight loss.

Exercise, diet modifications, and lifestyle changes are the foundation of a behavioral-based intervention for the management of obesity (figure 9.2). These three elements of therapeutic lifestyle change are critical first steps in the prevention and treatment of many medical conditions but are commonly overlooked because of the complexity of their practical application and the assumption of only short-lived success. Exercise combined with calorie reduction, lifestyle change, and, in some cases, weight loss medication and surgery is best provided by medical clinicians who work in a focused, integrative team environment with a long-term-horizon approach. The clinical exercise physiologist must be a key part of this group because of their skill in exercise assessment, planning, and implementation. Using a behavioral approach, it is possible to reduce the individual burden of obesity in a cost- and care-efficient manner, and its comorbidities can be controlled or possibly eliminated (101). Considering that the level of obesity in the United States and around the world will not decrease significantly in the short term, the continued use and growth of clinical exercise physiologists in this type of programming are promising.

DEFINITION

While there are methods and standards by which a person can be categorized as obese, the simplest definition from *Stedman's Medical Dictionary* is "an excess of subcutaneous fat in proportion to lean body mass." More specifically, overweight and obesity can be defined as weight that exceeds the threshold of a criterion standard or reference value (104). Criteria have been set using several methods including body mass index, waist circumference, and body composition (body fat percentage). Weight ranges based on sex and height are also available but typically used for health insurance purposes rather than for clinical determination of excessive weight.

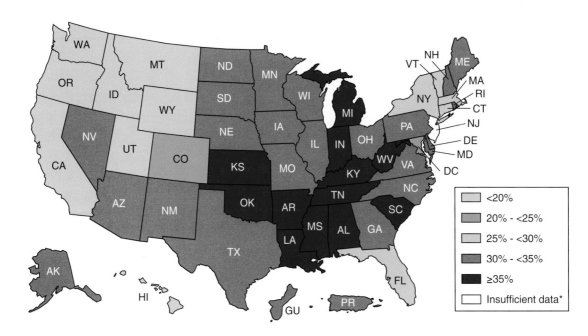

Figure 9.1   Prevalence of self-reported obesity among U.S. adults by state and territory.

From Center for Disease Control and Prevention, BRFSS (2018).

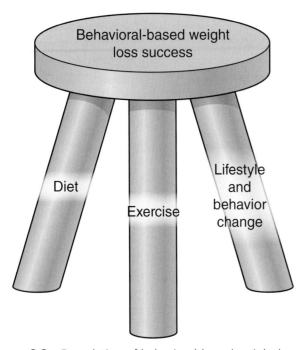

Figure 9.2   Foundation of behavioral-based weight loss.

The World Health Organization (WHO) defines **overweight** and **obesity** as "abnormal or excessive fat accumulation that presents a risk to health" (212).

The most common current approach in the clinical setting for determining overweight and obesity is body mass index (BMI), the measurement recommended by the U.S. Preventive Services Task Force (USPSTF) (185), the U.S. CDC (36), and the WHO (212). BMI indicates overweight for height but does not discriminate between fat mass and lean tissue, as demonstrated in figure 9.3.

The BMI does, however, significantly correlate with total body fat, with one report of a moderately strong correlation coefficient of 0.65 (148). Despite this moderately strong correlation between BMI and body fat percentage, the diagnostic accuracy of a BMI ≥ 30 kg · m^{-2} to diagnose obesity is limited (low sensitivity) but does well to identify those who are not obese (high specificity) (156). However, since BMI in general does not overestimate obesity prevalence (156), it should be considered an acceptable general screening method for obesity in the clinical setting. Body mass index is calculated as weight in kilograms divided by height in meters squared. It can also be determined using pounds and inches. The following are the BMI formulas:

$$BMI = \frac{Weight\ (kg)}{(Height\ [m])^2}$$

or

$$BMI = \frac{Weight\ (lb) \times 703}{(Height\ [in.])^2}$$

Reference values used to define overweight and obesity for BMI are derived from population data and may be specific to certain characteristics, including sex. Distributions of these data are determined, and criteria are then based on threshold values associated with increased risk for morbidity and mortality. The classification of overweight and obesity by BMI for U.S. adults is based on the 1998 *Clinical Guidelines on the Identification, Evaluation, and Treatment of Overweight and Obesity in Adults*, with concurrence by the WHO (212, 213) and is shown in table 9.1 (135). Prior to 1987, the U.S. government used BMI values of 27.8 (males) and 27.3

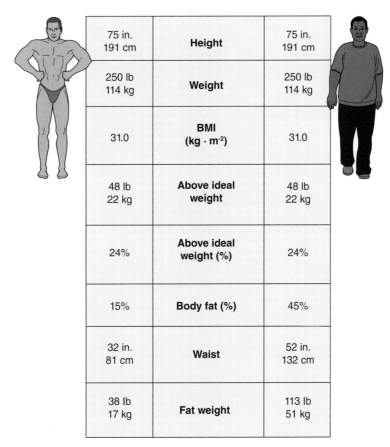

75 in. 191 cm	**Height**	75 in. 191 cm
250 lb 114 kg	**Weight**	250 lb 114 kg
31.0	**BMI** **(kg · m⁻²)**	31.0
48 lb 22 kg	**Above ideal weight**	48 lb 22 kg
24%	**Above ideal weight (%)**	24%
15%	**Body fat (%)**	45%
32 in. 81 cm	**Waist**	52 in. 132 cm
38 lb 17 kg	**Fat weight**	113 lb 51 kg

Figure 9.3   The two men in this example are the same height and weight and thus have the same BMI and body weight relative to ideal weight. But it is clear from their physical appearance that the person on the left has a lower body fat percentage than the man on the right. The values provided for body fat percentage, waist circumference, and total fat and nonfat weight clarify this difference.

(females) kg · m⁻² to define overweight in the population (132, 133). Threshold determinations may vary internationally by race. For instance, in Japan an overweight BMI ranges from 23.0 to 24.9 kg · m⁻², and obesity is any BMI ≥25.0 kg · m⁻² (163, 210). The reason is the higher incidence of diseases associated with weight and body fat occurring at lower BMI values in Asian populations (221). This has also been observed in the Chinese population (218, 219). In the United States, Australia, Canada, and Europe, *overweight* is defined as a BMI ranging from 25.0 to 29.9, and obesity is ≥30 (181). Importantly, the increased mortality risk with each unit over 25.0 kg · m⁻² is similar (29%-39%) among those in Europe, the United States, East Asia, Australia, and New Zealand (75).

Despite its ease of use, there may be important inaccuracies with respect to determining if a given BMI is unhealthy or if a person with a BMI categorized as overweight (or obese) has an elevated body fat percentage. For instance, some studies have reported that a BMI range of 25.0 to 29.9 may be no different from a BMI <25.0 or even superior with respect to all-cause death (60, 100, 141). Others report that all-cause mortality is lowest for the BMI range of 20.0 to 24.9 (19, 29, 51, 100). In general, the relationship between BMI and body fat percentage is good, as shown in figure 9.4. Most of the 8,550 individuals assessed were determined to be

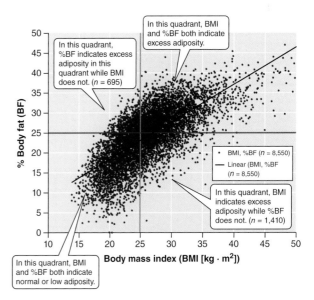

Figure 9.4   Relationship between BMI and percent body fat (% BF) in 8,550 individuals.

Reprinted by permission from A. Romero-Corral et al., "Accuracy of Body Mass Index in Diagnosing Obesity in the Adult General Population," *International Journal of Obesity* 32, no. 6 (2005): 959–966.

Table 9.1 Body Weight–Related Classifications

Category	BMI	% over ideal weight[a]	% fat[b]	Disease risk relative to normal weight and waist circumference in inches		Children and adolescents (ages 2-19 yr), BMI for age % range
				Men ≤40 in. (102 cm) Women ≤35 in. (88 cm)	Men ≥40 in. (102 cm) Women ≥35 in. (88 cm)	
Underweight	<18.5	NA	<20			<5th percentile
Normal	18.5-24	0%-10%	20-25			5th to <85th percentile
Overweight	25-29	10%-20%	26-31	Increased	High	85th to 95th percentile
Mildly obese (class I)	30-34	20%-40%	32-37	High	Very high	≥95th percentile
Moderately obese (class II)	35-39	40%-100%	38-45	Very high	Very high	
Morbidly obese (class III)	≥40	>100%	>45	Extremely high	Extremely high	

[a]Weight: percentage over standard height–weight tables.

[b]Fat (%): calculated body fat expressed as percentage of total weight. Ranges provided suggest men are at the lower end of the range and women at the upper end.

BMI = body mass index (kg · m^{-2}); NA = non-applicable.

Adapted from National Institutes of Health and U.S. Departments of Health and Human Services and Agriculture; U.S. Department of Health and Human Services and U.S. Department of Agriculture, *2015–2020 Dietary Guidelines for Americans*, 8th ed. (2015). http://health.gov/dietaryguidelines/2015/guidelines/.).

at a normal or high adiposity by both body fat analysis and BMI.

Many methods of assessing body composition (i.e., lean and nonlean tissue) are available (e.g., skinfolds, bioelectrical impedance, water or air displacement, and dual-energy X-ray absorptiometry [DXA]). But using these techniques requires various levels of technical skill, patient cooperation, time, cost, and space, all of which may preclude their use. It is for this reason that BMI, despite some criticism, remains the most common method for categorization of those who are overweight or obese (93).

Although BMI is recommended to evaluate obesity, patients may request that their healthy weight range for their height be determined. *Dietary Guidelines for Americans, Ninth Edition,* recommends children and adults attempt to achieve and maintain a healthy body weight (181). Additionally, the guidelines recommend a lower waist circumference and body fat to reduce the risk of chronic diseases associated with overweight and obesity. The current guidelines use the previously mentioned BMI categories to categorize for normal weight, underweight, overweight, and obesity; however, this is only in the context of weight management during pregnancy.

SCOPE

Data for the United States and worldwide were provided in the chapter introduction. Other countries and regions such as Australia (67% of population overweight or obese, a 4% increase from 2015 to 2018 [11]), Europe (22% of males and 25% of females obese [215]), and India (upwards of 36% obese [3]) have high levels or increasing rates of obesity. With the increase in overweight and obesity throughout the world, the problem has been considered a pandemic (18). Using 2014 data, the WHO predicts that 2.5 billion people worldwide are at least overweight; of those, 600 million are classified as obese. Also ~39% of the world's adults are overweight and ~13% are obese, with the United States leading the way in percentage of obesity among its population (figure 9.5).

Figure 9.6 demonstrates the amount of overweight, obesity, and extreme obesity in men and women in the United States since 1960 (133). Of note is the rate of extreme (grade 3) obesity (i.e., BMI ≥40 kg · m^{-2}), which has risen in both sexes from 1999 to 2014, with women (8.9%) being almost twice as likely as men (4.9%) to have grade 3 obesity (91). Extreme obesity appears to be increasing in younger people and is greater among Black women than Black men, as well as among Black people than in the non-Hispanic white population. In the United States, Black women have the highest rates of extreme obesity (~15% in 2006) in terms of ethnic and gender makeup (91).

Although obesity often begins in childhood and early adolescence (14%-21% of children in the United States are obese, 25% of Australian children are overweight or obese, and one in three 11-yr-olds in Europe is overweight or obese), approximately 70% of all obesity begins in adulthood. There is a strong association between childhood obesity and the risk of adulthood obesity; this association is affected by the severity of childhood obesity, the age of onset, and whether the parents are also obese (28, 82, 133, 138, 164, 165, 204). An increasing prevalence of childhood obesity appears to be a worldwide issue. The WHO reports that from 1990

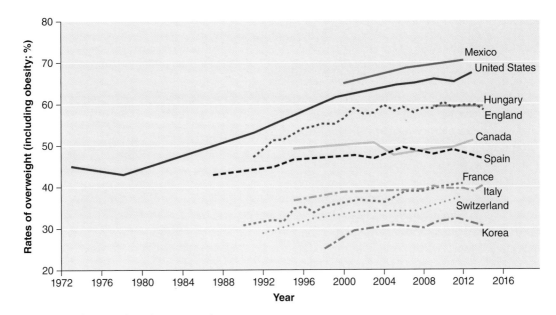

Figure 9.5 Obesity rate changes in select countries.

Reprinted by permission from OECD, *Obesity Update 2017.* https://www.oecd.org/els/health-systems/Obesity-Update-2017.pdf.

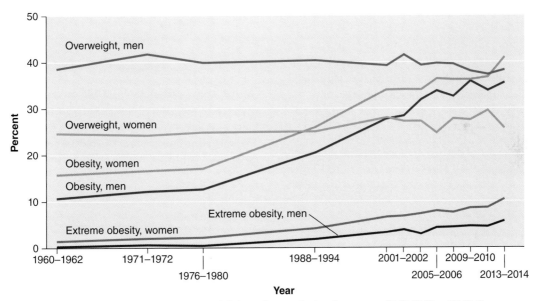

NOTES: Age-adjusted by the direct method to the year 2000 U.S. Census Bureau estimates using age groups 20–39, 40–59, and 60–74. Overweight is body mass index (BMI) of 25 kg/m2 or greater but less than 30 kg/m2; obesity is BMI greater than or equal to 30; and extreme obesity is BMI greater than or equal to 40. Pregnant females were excluded from the analysis.
SOURCES: NCHS, National Health Examination Survey and National Health and Nutrition Examination Surveys.

Figure 9.6 Trends in adult overweight, obesity, and extreme obesity among men and women aged 20 to 74 in the United States from 1960-62 through 2013-14.

Reprinted from C.D. Fryar, M.D. Carroll, and C.L. Ogden, "Prevalence of Overweight, Obesity, and Extreme Obesity Among Adults Aged 20 and Over: United States, 1960–1962 Through 2013–2014," *National Health Examination Survey and National Health and Nutrition Examination Surveys.*

to 2016 the number of obese infants and young children (up to age 5) increased globally from 32 to 41 million (211, 212). In Africa during the same time, the number of overweight and obese children increase by 125% (from 4 to 9 million). Although the WHO predicted this trend to continue, some suggest there has been no significant changes in the prevalence of obesity in the United States for youth or adults between 2003 and 2012 (138). Both Black and Hispanic youth in the United States have a higher rate of obesity and severe obesity (137). Weight gain tends to be more rapid in early adulthood for Black and white men and for Black women and can be as much as 22 to 35 lb (10 to 16 kg) over a 10 yr period (1.0 to 1.6 kg, annually) (111). For white women, a similar weight gain occurs but later in early adulthood. This weight gain is likely the result of a transition in adulthood to an environment that deemphasizes physical activity and allows for easy excessive food intake. Despite these data there is some evidence that the rate of increase in obesity prevalence may be slowing in the United States and Australia (56, 126).

Weight that is 20% above ideal weight (or ~BMI = 30 kg · m^{-2}) carries increased health risk for obesity-related comorbidities (figure 9.7). The patterns of fat distribution also affect risk (122). The enzyme lipoprotein lipase, which regulates the storage of fats as triglyceride, is more active in abdominal obesity and therefore increases fat storage. Consistently it has been noted that abdominal (aka central or android) obesity is more closely related to elevated cardiometabolic risk (risk of heart disease, diabetes, hypertension, dyslipidemia) than is gluteofemoral (aka gynoid or lower body) obesity (87, 196) and this risk may have genetic mechanisms (116). Figure 9.8 provides fat distribution patterns in white, Black, and Asian populations.

Central obesity can also be grossly assessed by measuring the waist-to-hip ratio (W/H). Increased health risk is present in women when the W/H ratio exceeds 0.8 and in men when the W/H ratio exceeds 1.0 (167). An additional assessment of risk is simply the waist measurement. If it exceeds 88 cm (35 in.) for a female and 102 cm (40 in.) for a male it is related to an increased risk of type 2 diabetes, hypertension, and cardiovascular disease (57). Accurate waist and hip measurements are important for proper risk assessment. The following describes these measurements performed with a Gulick-type spring-loaded tape measure (154):

- Waist: With the subject standing, arms at the sides, feet together, and abdomen relaxed, a horizontal measure is taken at the narrowest part of the torso (above the umbilicus and below the xiphoid process). The National Obesity Task Force

(NOTF) suggests obtaining a horizontal measure directly above the iliac crest as a method to enhance standardization. Unfortunately, current formulas are not predicated on the NOTF's suggested weight measurement site.

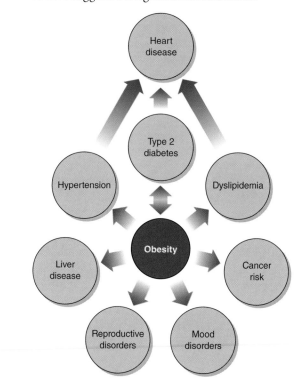

Figure 9.7   Obesity health risks.

Adapted from I. Kyrou et al., "Clinical Problems Caused by Obesity." [Updated 2018 Jan 11]. In: K.R. Feingold KR, Anawalt B, Boyce A, et al., editors. Endotext [Internet]. South Dartmouth (MA): MDText.com, Inc.; 2000. https://www.ncbi.nlm.nih.gov/books/NBK278973/. Distributed under the terms of the Creative Commons Attribution (CC-BY-NC-ND) license.

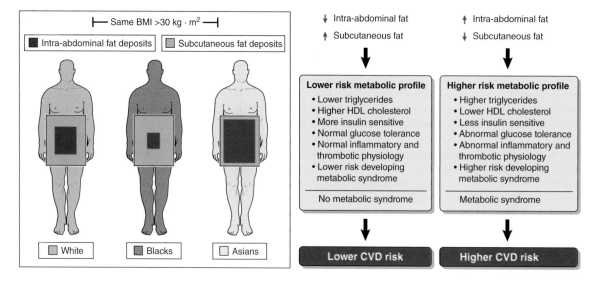

Figure 9.8   Fat pattern distribution in select races.

Reprinted by permission from International Chair on Cardiometabolic Risk, myhealthywaist.org. http://www.myhealthywaist.org/index.php?id=11&no_cache=1&tx_stdoccenter_pi1[mod_type]=5&tx_stdoccenter_pi1[uid]=35

- Hips: With the subject standing, legs slightly apart (10 cm), a horizontal measure is taken at the maximal circumference of the hips (the proximal thighs), just below the gluteal fold.

A more important factor in fat distribution is distinguishing abdominal visceral fat (i.e., intra-abdominal) from subcutaneous fat. Visceral fat lies deep within the body cavities, surrounding the internal organs (i.e., liver, pancreas, intestines), and in the case of the liver can be stored within this organ and lead to nonalcoholic fatty liver disease (NAFLD). It is believed that NAFLD is currently the fastest growing reason for liver dysfunction (88). Visceral fat accumulation is associated with a higher cardiometabolic risk than subcutaneous fat because of its metabolic characteristics, which include insulin resistance and glucose intolerance (122). Visceral fat is best measured by magnetic resonance imaging (MRI), ultrasound, or computed tomography (CT), which are expensive and unavailable to most clinical exercise practitioners.

A more practical measurement of visceral fat uses sagittal diameters (145). This technique requires the patient to lie on their back with bent knees and feet on the surface. The arms should be in a relaxed position. The clinician obtains the sagittal diameter by measuring the distance from the examination table or floor to a horizontal level placed over the abdomen at the level of the iliac crest (see figure 9.9). The measure should be taken during a normal exhale. This promising technique is currently the best practical predictor of visceral fat. Measures exceeding 30 cm (12 in.) are related to an increased cardiovascular disease risk and to insulin resistance (145).

The estimated annual health care cost for obesity-related illness in 2013 was $342.2 billion, or about 28% of all health care expenditures in the United States annually (20). This is a rise from 21% of all health care spending in 2005 (34). The effect on businesses who experience obesity-related absenteeism (but not overweight absenteeism) is estimated at $8.65 billion annually (6, 35). Andreyeva and colleagues suggested that each increment of 5 in BMI (e.g., 30 to 35, 35 to 40) results in a 25% to 100% increase in the lifetime health care expenditures for an individual (7). Per capita medical spending in the United States is as much as $3,508 (U.S. dollars) higher each year for an obese person compared with a person of normal weight (33). In Australia, per capita spending

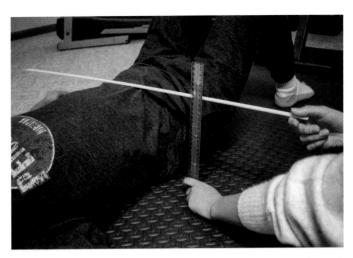

Figure 9.9 Sagittal diameter measurement to assess visceral fat.

is approximately $830 (AUS dollars) more for an obese individual versus someone at normal weight (44). Colagiuri also reported an increase in government subsidies (i.e., welfare) of $1,305 in the obese versus normal population (44). As noted in table 9.2, Medicaid and Medicare pay up to half of obesity-related medical care expenditures in the United States, which exemplifies the growing burden of overweight and obesity on federal and state spending (59). Indirect costs associated with morbidity- and mortality-related costs are difficult to capture. It is predicted that a 5% reduction in weight of a person with a BMI of 40 kg · m^{-2} reduces healthcare expenditures by $2,137, and this number drops to $528 and $69 for those with a starting BMI of 35 and 30 kg · m^{-2}, respectively (33).

Obesity reduces life expectancy throughout the world, and this effect is most powerful in those who develop obesity earlier in life (22, 66). Flegal and colleagues performed a meta-analysis and reported that compared with a BMI <25 kg · m^{-2}, grades 2 and 3 obesity are related to excess premature death, and overweight is associated with a lower all-cause mortality (62). Flegal previously estimated 112,000 excess deaths per year because of obesity and reported that as obesity levels became more severe (class I to II to III), the mortality rate increased (61). Conversely, several epide-

Table 9.2 Estimated U.S. Medical Spending, in Billions of Dollars, Attributable to Overweight and Obesity

Insurance category	OVERWEIGHT AND OBESITY	
	MEPS	NHEA
Private	$49.4	$74.6
Medicaid	$8.1	$27.6
Medicare	$19.7	$34.3
Total	$85.7	$146.6

MEPS = Medical Expenditure Panel Survey; NHEA = National Health Expenditure Accounts.

From 2006 Medical Expenditure Panel Survey (MEPS): 2006 National Health Expenditure Accounts (NHEA).

miological studies have found a protective effect on mortality for overweight individuals (termed the "obesity paradox") when other conditions are present such as, but not limited to, heart failure and revascularization per percutaneous intervention (47, 83, 90).

It is vital to understand that there are very important limitations to this obesity paradox, in which there appears to be a mortality benefit for being overweight. These include the cross-sectional nature of the observations, limited data on change in weight over the follow-up period, and assessment of variables that may affect weight including pulmonary disease, human immunodeficiency virus (HIV) infection, alcohol and drug use, smoking, and cancer (1). In fact, recent data have suggested a reverse causation (i.e., a cause and effect that are in the opposite direction of the common presumption) due to smoking (172). Others have challenged the paradoxical findings as an anomaly of data collection and analysis bias (173, 152). Additionally, the obesity paradox may, in part, be explained by cardiorespiratory fitness levels in overweight and obese individuals. Higher fitness values based on estimated metabolic equivalents (METs) during an exercise test are associated with a lower all-cause mortality rate (123, 124). Findings from the renowned Harvard Alumni Health Study reported a doubling of heart disease risk in those who were obese in early childhood (78). But, importantly, this risk was reduced when weight loss occurred in midlife or later. This study is one of the few to provide these data longitudinally, with prospectively measured BMI in both early and later life.

Americans may spend as much as $66 billion each year in efforts to control their weight. These attempts include participation in commercial and clinical weight loss programs and purchases of food and nonfood supplements. A health care market research report published in 2017 by Markets and Markets (www.markets andmarkets.com) estimated that the global market for weight loss was approximately US$672 billion in 2015. The U.S. estimate for 2022 is $246 billion. In recent years several sets of professional guidelines have been written that have focused on obesity, including *Dietary Guidelines for Americans* (181); *Physical Activity and Health: A Report of the Surgeon General* (184); NIH's "Clinical Guidelines on the Identification, Evaluation, and Treatment of Overweight and Obesity in Adults" (135); the American College of Sports Medicine position stand (55); the American Heart Association (150) and American Dietetic Association (now known as the Academy of Nutrition and Dietetics) scientific statements (161a); "Healthy People 2020: National Health Promotion and Disease Prevention Objectives" (183); *Clinical Practice Guidelines for the Management of Overweight and Obesity in Adults, Adolescents and Children in Australia-2013* (134); and the European Association for the Study of Obesity (EASO) (216).

PATHOPHYSIOLOGY

Obesity results from longstanding positive energy balance. The simplicity of this restatement of the first law of thermodynamics can blind us to the complex physiology of food intake and calorie expenditures and tempt us to tell patients only to "eat less and run more." Such an approach is overly simplistic and ineffective. Positive energy balance has a myriad of contributors in current societies, ranging from increasing availability of lower-cost foods that some refer to as toxic to decreasing physical activity at work and in leisure. Average daily calorie intake has increased by over 200 kcal over the last several decades (67) as food costs have fallen dramatically and as more calories are consumed outside the home. At the same time, physical activity has fallen because of advances in equipment and ergonomics in the workplace.

Beyond daily behavioral and dietary influences, research has identified an ever-increasing number of genetic and physiologic factors pointing to a large array of neurological and peripheral endocrine messengers that influence food intake and nutrient utilization and that regulate weight in a way that often frustrates patients' weight loss efforts (87). Advanced neuroimaging research has recently shown similar patterns of brain activation between addictive-like eating and substance abuse (73).

A revolution in the understanding of obesity began with the identification of **leptin**, particularly when it was first produced in a laboratory in the 1990s (27). Leptin is secreted by fat cells and was found to regulate body weight in mice. Administration of leptin to mice genetically altered to be leptin deficient reduced their extreme obesity and poor growth, resulting in marked reduction in food intake and weight, as well as return of normal growth and metabolic function. This research proved that a molecular defect could be the basis of some forms of obesity and could provide a new approach for its treatment (15). Humans produce leptin from their fat cells in proportion to their weight and particularly their girth. Humans are not leptin deficient, however, and exogenous leptin administration has limited benefit for weight reduction except in the rarest of cases of human obesity because of a genetic absence of this fat cell hormone. Nevertheless, leptin research changed the perception of many clinicians and researchers. Instead of shunning obesity or considering it an untreatable moral failing, they now see it as a complex behavioral and neuroendocrine disorder that may be unlocked with additional study (26).

Genetic causes of obesity are a feature of rare disorders such as Prader-Willi syndrome (31) but are thought to be a factor in at least half of human obesity based on studies of twins raised separately or children who were adopted (174). Genetics plays a significant part in explaining responses to overfeeding or weight loss in metabolic lab settings, which points to variation in inheritable control of food intake, fat storage, and energy expenditure (25). Patients do not conveniently follow equations of calorie intake or energy utilization.

Body fat distribution is also genetically determined and gender specific, and predictable changes throughout life may confound patients' efforts at weight control. Increasing age reduces growth hormone and gonadal hormone secretion, which may predispose to greater visceral fat storage with its links to metabolic and cardiovascular abnormalities that in turn are linked to hypertension, coronary artery disease, and diabetes mellitus (24).

Energy expenditure has complex genetic and environmental variation that can promote weight gain in some and frustrate weight loss in others. The **set-point theory** of weight regulation can be seen in metabolic lab studies of overfeeding compared with calorie restriction. Weight loss decreased total and resting energy expenditure, which slowed further weight loss. Weight gain through overfeeding was associated with an increase in energy expenditure, which slowed further weight gain. These studies show that after weight loss, individuals require 15% fewer calories to maintain their lower weight. This tendency causes weight loss patients to return to higher weights unless they restrict calories over the long term or expend calories through greater physical activity (108, 160). Research has found defects in energy homeostasis that contribute to increased obesity caused by high circulating triglycerides, with ensuing inflammatory markers of cellular damage in the hypothalamus of experimental animals fed high-fat diets in excess of energy balance requirements (177).

Variation in energy regulation is seen in genetic studies of the Pima Indians, who are at significantly increased risk of obesity and its comorbid conditions (176). Efforts to augment energy expenditure through medications have been complicated by adverse effects of increased blood pressure and heart rate. This area will remain a target for ongoing pharmaceutical research because resting metabolic activity accounts for about 70% of energy expenditure. The metabolic cost of food digestion accounts for about 10% of energy expenditure. The remainder is physical activity, which is influenced by sedentary pursuits, such as television and computer viewing, compared with intentional exercise or dedicated increases in lifestyle activities.

Of the millions of calories eaten or expended over a lifetime, a mere fraction of 1% can result in clinically significant obesity. The neurological and peripheral control system for this process is biased toward preserving weight in times of famine and receives signals from fat cells and the gastrointestinal tract. An interesting observation in patients following gastric bypass weight loss surgery is the dramatic reduction in **ghrelin**, a potent appetite-increasing gut hormone, which may explain, in part, the success of this surgical procedure (45). Ghrelin, produced during stomach distention, triggers the central nervous system appetite stimulant **neuropeptide Y (NPY)**, growth hormone, and norepinephrine. Neuropeptide Y and norepinephrine predominantly stimulate carbohydrate intake. Weight loss can increase serum ghrelin levels, which may explain part of the process that makes sustaining weight loss difficult.

Cholecystokinin, serotonin, and peptide YY are among a group of central nervous system and gut peptides involved in satiety and reduced food intake (109). Serotonin has been targeted in the treatment of depression with selective serotonin reuptake inhibitors such as fluoxetine, but these medications have had limited effect in weight loss treatment. Nutrient selection during weight loss treatment may be a factor because appetite-reducing peptide YY is increased during high-protein diets. This may partially explain the appetite-reducing effect of high-protein diets. Table 9.3 lists the monoamines and peptides that affect appetite. The carbohydrate–insulin model of obesity has highlighted how increased consumption of processed, high-glycemic-load carbohydrate produces hormonal changes that promote fat deposition, exacerbate hunger, and lower energy expenditure. This has implications not only for the cause of the current obesity epidemic but also for alternatives to the conventional focus on dietary fat and calorie restriction in reversing obesity (119, 153).

The pathophysiology of obesity has been known for millennia. The degree of obesity predicts the morbidity and mortality for a long list of common afflictions in humans (29). Obesity and extreme obesity are associated with an increased rate of death from all causes, particularly from cardiovascular disease. In the United States, estimates of annual excess deaths due to obesity range from 112,000 to 365,000 (61). Obesity and sedentary lifestyle is considered the second leading cause of preventable death in the United States and may overtake tobacco abuse within the next decade. During the last 45 yr, other cardiovascular risk factors have decreased: High total cholesterol levels ($\geq$240 mg · dL^{-1}) decreased from 34% to 17%; high blood pressure (BP $\geq$140/$\geq$90 mmHg) decreased from 31% to 15%; and smoking dropped from 39% to 26%. Weight-related cardiometabolic risks have increased at an accelerated rate; diagnosed diabetes mellitus rose from 1.8% to 6.0%, and metabolic syndrome increased from 27% to 56% for men (79). The effects of obesity are so dramatic that projections of life expectancy, which have risen with declining cardiovascular risks over the past 50+ yr, are predicted to decline as the full effect of increasing obesity is felt in society (140). For obese individuals, life expectancy decreased by about 7 yr compared with normal-weight individuals in the Framingham Study (144). In 2016 life expectancy fell slightly for the first time, reflecting the effects of obesity-related chronic illness on society (214).

Table 9.3 Monoamines and Peptides That Affect Appetite

Stimulatory	Inhibitory
Norepinephrine	Leptin
Neuropeptide Y	Cholecystokinin
Opioids	Serotonin
Melanin-concentrating hormone	Corticotropin-releasing hormone (CRH) or urocortin

Type 2 diabetes mellitus is strongly associated with obesity; up to 80% of cases are weight related. Compared with people with a BMI of 22, those with a BMI of 35 have a 61-fold increased risk of developing diabetes (43). The Diabetes Prevention Project (DPP) showed that even a modest 7% weight loss reduced the progression of prediabetes to diabetes by 58% over 4 yr, exceeding medication treatment (i.e., metformin) by 25% (103). Obesity, particularly abdominal (or visceral) obesity, is also among the current diagnostic criteria for **metabolic syndrome**, as explained in chapter 12 (80).

Intensive lifestyle interventions have been studied in the multicenter Look AHEAD trial sponsored by the NIH (114). This decade-long study evaluated the benefit of modest weight loss through low-calorie diets, meal replacement programs, and weight loss medications coupled with thorough behavioral education, support, and exercise over 5 yr of initial treatment and 6 yr of follow-up in over 5,000 individuals. The 1 yr results of the Look AHEAD trial showed an average 8.6% weight loss in the group treated with intensive lifestyle interventions (ILI) compared with a 0.7% loss in the diabetes support and education group (DSE). While the Look AHEAD trial failed to show reduced cardiovascular mortality at 10 yr, some reanalysis suggests the intervention didn't achieve sufficient early weight loss in comparison with other methods (bariatric surgery) to produce the desired outcome (115). Importantly, in Look AHEAD trial participants who lost more than 10% of their initial weight in the first year, there was a 20% risk reduction in death from cardiovascular causes, nonfatal myocardial infarction, nonfatal stroke, and hospitalization for angina (113).

Hypertension rates often increase in obese people and may be present in up to half of those seeking weight loss treatments. In the Framingham Heart Study, excess body weight explained over 25% of the incidence of hypertension (207). A 10 kg weight loss can reduce blood pressure by 10 mmHg, which is comparable to the effect required to achieve Food and Drug Administration (FDA) approval for a new antihypertensive drug.

Cardiovascular morbidity and mortality are greater in obese individuals, and the role of adipose tissue, particularly visceral fat mass, is becoming more clearly established as a key coronary disease risk factor. Women in the Nurses' Health Study showed a fourfold increased risk of death from cardiovascular disease if their BMI was over 32 compared with those with a BMI under 19 (205). The INTERHEART study showed that central obesity, particularly elevated waist circumference, explained 20% of the risk of first myocardial infarction (217). Obesity may also increase cardiovascular disease through elevated low-density lipoprotein (LDL) cholesterol levels and reduced high-density lipoprotein (HDL) cholesterol levels. Heart failure, particularly associated with preserved ejection fraction (PEF), is increased twofold in obese people. This consequence is often largely reversible with significant weight loss (105). Atrial fibrillation and flutter are increased about 50% in obese individuals. Stroke risk is also doubled for obese women, based on the findings of the Nurses' Health Study. In addition, venous thromboembolic disease, including deep venous thrombosis and pulmonary embolus, is increased in obese individuals, particularly those who are sedentary (195).

Respiratory illness, with adverse effects on exercise capacity, is also common in obese people (151). Obesity is the greatest predictor of obstructive sleep apnea, a syndrome of interrupted sleep with snoring and apneic periods that lowers oxygen saturation levels to those associated with potentially lethal ventricular arrhythmias. Dyspnea and asthma are increased with obesity through adverse effects on respiratory mechanics and through gastroesophageal reflux disease (GERD)-associated bronchospasm. Obesity is also a risk factor for severe illness and mortality from H1N1 influenza and SARS-CoV-2, which can cause COVID-19 (54, 153).

Obesity is associated with increased gastroesophageal disease and hepatobiliary illness. GERD is increased through reflux of stomach acid into the esophagus because of increased intra-abdominal pressure, which may cause esophagitis and lead to esophageal cancer. Gallstones, cholecystitis, and biliary dyskinesia (biliary colic without cholelithiasis) are increased in obese individuals because of increased biliary excretion of cholesterol (171). Active weight loss may increase the risk of cholelithiasis and cholecystitis because cholesterol is removed when reducing fat stores and secreted; it may then crystallize in the gallbladder. Weight loss diets with daily modest amounts of fat may empty the gallbladder and reduce this risk. Nonalcoholic fatty liver disease is linearly related to obesity and can lead to cirrhosis and liver failure but is often reversible with weight loss (8).

Osteoarthritis, in particular weight-bearing joint disease, is increased with increasing obesity. Elevated weight multiplies the effects on the hips, knees, and ankles and increases the risk of foot pain and plantar fasciitis, which may further increase weight through secondary reductions in physical activity. Significant weight loss is associated with a 50% reduction in arthritic joint pain and may postpone or obviate the need for joint replacement in some individuals.

Depression and eating disorders, particularly binge-eating disorder, are increased in obese women, especially those seeking professional help with weight loss. Evaluation and management of these issues are important to eliminate them as barriers to weight loss. Psychosocial function is decreased by obesity, and outright prejudice is common toward obese people, particularly women, who may be significantly less likely to marry or to complete their education and thus are likely to face increased poverty and lower annual incomes (77).

CLINICAL CONSIDERATIONS

Although medically significant obesity exists in most patients who seek medical care, only a minority present requesting medical help with weight reduction. Increasing the awareness of those not aware of the relationship between their weight and medical problems is an effective way of helping them become determined to address their weight. Compassion and understanding coupled with flexible and practical weight loss recommendations are essential to building rapport with obese patients. Most individuals seeking professional help with weight loss have a BMI of 38 or more, and invariably they have attempted to lose weight several times in the past. Proper assessment of the barriers to and benefits of weight loss should be the initial step in the clinical evaluation of the obese patient. Comprehensive assessment of obesity can lead to appropriate treatment and effective long-term control of obesity and its related comorbidities, as shown in the sidebar Health Consequences of Obesity.

Signs and Symptoms

Although obesity would seem the most obvious medical condition, its key determinants, weight and height, are often omitted in medical records or ignored in patient problem lists. Ideally, patients should be measured at each visit in lightweight clothing without shoes. Comparable measurements would ideally be at the same time of day (intraday variation of up to 2% is common) with empty bladders. Appropriate scales that can measure extreme weights (up to 600 lb [270 kg]) are essential, or wheelchair scales should be used, because many patients who weigh more than 600 lb cannot stand in order to be weighed. A wall-mounted stadiometer can be used for heights, although patients typically report accurate height (but not weight). Waist circumference should be measured at the umbilicus while the patient is standing, because measurement at that location is the best predictor of central, or visceral, obesity, and its significance is not related to a patient's height. Patients should not suspend their breathing during waist measurements, and the measuring tape should not be overtightened.

Body fat measurements can be helpful for individuals who are outliers on the BMI chart (e.g., individuals with greater than average muscle mass or those suspected of cachexia), as well as for determining the composition of weight loss. However, in general they add little to therapeutic management of weight loss and can have significant methodological errors and associated costs. Two common methods of body composition analysis are air displacement plethysmography (commercially known as the Bod Pod) and dual-energy X-ray absorptiometry (DXA). Although both methods rely on certain assumptions (e.g., an assumed density of bone for the Bod Pod), user error with these devices is generally low, contributing to both a high reliability and overall validity. The U.S. Surgeon General recommends that BMI be added to blood pressure and pulse as a routinely recorded vital sign because further evaluation and management of obesity and its related comorbidities depend on it (182).

History and Physical Examination

Most clinicians and patients are aware that a blood pressure of ≥150/90 mmHg or a cholesterol of ≥240 mg · dL^{-1} is associated with increased cardiovascular risk, and immediate aggressive therapy would be started. Few are aware, however, that a BMI of 33 conveys the same risk of cardiovascular death (26). Additionally, only 25% of physician visits with patients who are obese address issues of weight reduction (170).

Therefore, many overweight and obese patients who begin to work with a clinical exercise physiologist have not been approached about their weight by a medical professional. The clinical exercise physiologist should address this issue by simply asking the patient about the subject. One can assess the patient's readiness to address their weight with the simple question "Have you been trying to lose weight?" Potential responses are the following:

- No, and I do not intend to in the next 6 mo. (precontemplation)
- No, but I intend to in the next 6 mo. (contemplation)
- No, but I intend to in the next 30 d. (preparation)
- Yes, but for less than 6 mo. (action)
- Yes, for more than 6 mo. (maintenance)

The patient's response should be used to assess their readiness to lose weight and develop a tailored plan. For instance, patients who are not intending to lose weight (precontemplators) might simply be educated about the health risks of being overweight and the benefits of losing weight. Those who are preparing to lose weight

---

Health Consequences of Obesity

Greatly increased risk (relative risk >3)

- Diabetes
- Hypertension
- Dyslipidemia
- Breathlessness
- Sleep apnea
- Gallbladder disease

Moderately increased risk (relative risk about 2-3)

- Coronary heart disease or heart failure
- Osteoarthritis of the knees
- Hyperuricemia and gout
- Complications of pregnancy

Increased risk (relative risk about 1-2)

- Cancer
- Impaired fertility or polycystic ovary syndrome
- Low back pain
- Increased risk of anesthesia-related complications
- Fetal defects arising from maternal obesity

Based on Haslam, Sattar, and Lean (86).

soon might be best served by being directed into a clinically based weight loss program.

The evaluation of the obese person, whether they are seeking medical help for weight control or for routine care of unrelated medical concerns, should include relevant factors of history and a physical exam as well as laboratory testing that can better characterize the person's risks from obesity, its often silent comorbid conditions, and barriers to effective treatment. The medical approach to obesity should consider, in compassionate cooperation with the patient, the risks of obesity, appropriate treatment, and the most effective form of treatment.

Medical causes of obesity, including illnesses, medications, and lifestyle changes known to increase weight, must be identified. Examples include diabetes mellitus, hypothyroidism, polycystic ovary syndrome, Cushing's syndrome (although rare), many medications (some neuroleptics, psychotropics, anticonvulsants, antidiabetics, β-blockers, ACE inhibitors, and hormone therapies), and behavioral changes such as smoking cessation, job changes, injuries, sleep habits, and life stressors. History of obesity should be obtained whenever possible, including the age of onset of obesity, weight gain since adolescence, changes in weight distribution, peak and lowest maintained adult weight, any history of weight loss (intentional or unintentional), and methods used to lose weight. Assessment of readiness for weight loss, social support, weight control skills, and history of any past or present eating disorders (binge eating, bulimia nervosa, or, rarely, anorexia nervosa) should be considered. Family history of obesity predisposes toward weight problems, and social history can identify key issues such as divorce, job changes, and upcoming weddings or class reunions that can affect weight loss efforts. An early family history of cardiovascular disease may motivate overweight patients to control their cardiovascular risk factors. A thorough review of systems can help identify medical conditions such as joint pain or asthma that may be overlooked barriers to weight control.

The clinical exercise physiologist should identify current intentional exercise or opportunities for increased lifestyle activities, evaluate exercise barriers and patient-specific benefits to help in the assessment of any contraindications to moderate or vigorous exercise, and develop a comprehensive exercise plan. Exercise testing may be appropriate for the assessment of cardiovascular disease or to explain dyspnea on exertion, which is common in obese patients.

Diagnostic Testing

Laboratory testing should screen for diabetes (fasting blood sugar), elevated cholesterol and triglycerides as well as LDL and HDL cholesterol levels (lipid profile), and clinical or subclinical hypothyroidism (thyroid-stimulating hormone [TSH] and free T4). Comprehensive chemistry profiles can assess for nonalcoholic fatty liver and renal disease. Electrocardiograms are rarely necessary except to evaluate specific cardiovascular problems such as elevated blood pressure or palpitations.

Exercise Testing

There are no specific obesity-related guidelines that provide recommendations for exercise testing. The ACSM states that routine exercise testing in the overweight and obese population is often not necessary, particularly for those beginning a low- to moderate-intensity exercise program (154). It has been suggested that pharmacological stress testing using dipyridamole with PET imaging is preferred for evaluating obese individuals for myocardial ischemia (147). When an exercise test is performed in those who are overweight or obese, the purpose is typically to assess for the presence of coronary artery disease (64). This is because these individuals are at risk for poor outcomes associated with cardiovascular disease risk (63). Exercise testing may also be performed to determine functional capacity for risk assessment and to develop an exercise prescription based on heart rate (125). Several studies demonstrate that obese and morbidly obese patients can exercise to maximal exertion on a variety of treadmill protocols (64, 68, 125).

Although walking is the preferred mode of exercise for testing, it is not always practical in those who are obese. Patients in this group, especially those with BMI values greater than $40 \, \text{kg} \cdot \text{m}^{-2}$, often have concomitant gait abnormalities, joint-specific pain during weight-bearing exercise, and other musculoskeletal or orthopedic conditions that require a non-weight-bearing mode such as leg or arm ergometry. Seated devices such as upper body ergometers, stationary cycles, or recumbent stepping machines offer excellent alternatives that allow patients to achieve maximal exercise effort in a non-weight-bearing mode. It is important to make sure the weight-bearing capacity of the equipment is sufficient for very heavy individuals for maximum safety. Despite the suggested use of non-weight-bearing modes, McCullough and colleagues reported that in a group of 43 consecutive patients referred for bariatric surgery (mean BMI = $48 \pm 5 \, \text{kg} \cdot \text{m}^{-2}$), only one could not perform a walking protocol (125). Because of the potential for a low peak exercise capacity, a low-level protocol with small increments (e.g., 0.5-1.0 MET) may be preferred, although if increments are too small there may be a risk of skeletal muscle fatigue, related to moving and supporting excessive weight, prior to achieving cardiorespiratory maximum.

If exercise testing can be performed it should generally follow the normal routine (see chapter 5). Prediction equations for METs from the work rate achieved on an exercise device are typically inaccurate in people who are obese. Assessment of cardiopulmonary gas exchange provides an accurate measurement of exercise ability. Gallagher reported a peak oxygen consumption (peak $\dot{V}O_2$) of $17.8 \pm 3.6 \, \text{mL} \cdot \text{kg}^{-1} \cdot \text{min}^{-1}$ (equivalent to 5.1 METs) in a morbidly obese group of patients who achieved peak respiratory exchange ratio (RER) values greater than 1.10 (72). No complications related to exercise testing were reported in this cohort, suggesting that exercise testing is safe in this extremely obese population. See the sidebar Exercise Testing: Important Considerations for a review of exercise testing methods.

Exercise Testing: Important Considerations

Cardiorespiratory

Testing is not generally required for those beginning a comprehensive weight management program, particularly those performing low- to moderate-intensity physical activity; consider weight limitations of certain equipment, particularly for people weighing more than 300 lb (136 kg); non-weight-bearing modes may minimize joint-related reasons for stopping exercise; low-level protocols are commonly used because of deconditioning and anticipated low fitness level (154).

Strength

Standard assessments using machine-based equipment can be performed; in those who are significantly deconditioned, consider a 2RM to 10RM assessment.

Range of Motion

The sit-and-reach test and specific joint goniometry can be used to set a baseline and to assess for improvements; most standard assessments can be performed, but body habitus and the individual's ability to perform simple movements and to even sit on the floor to perform the sit-and-reach test may be limited; consider testing as an individual loses weight because results will likely improve.

There are no specific recommendations for assessing muscular strength, muscular endurance, and range of motion. A normal sequence of testing can generally be followed. Some equipment may be difficult for some patients to use because of excessive body habitus and limited flexibility and movement. However, because many overweight and obese individuals have significant early exercise training improvements, retesting strength and particularly range of motion might be informative.

Treatment

Although all weight loss treatment should include specific diet, behavioral, and exercise prescriptions, and in some cases pharmacotherapy or weight loss surgery, the specifics of treatment must be matched to the patient's circumstances, including BMI and related considerations. Patients at lower BMI and without comorbid conditions should be offered less intensive treatment than patients with extreme obesity (BMI ≥40) who have significant obesity-related comorbidities such as diabetes, hypertension, and obstructive sleep apnea. Table 9.4 illustrates the strategy used for weight management as discussed in this chapter.

Treatment goals need to consider both the medical benefits of modest (10%) weight loss and the patient's expectations. The NIH has recommended a 10% weight loss within 4 to 6 mo and weight loss maintenance as an initial weight loss goal because this amount is associated with several health-related benefits (see the sidebar Health Benefits of a 10% Weight Loss). Improvements in obesity-related functional limits and medical comorbidities should be identified. Patients, on the other hand, commonly want to lose 35% of their weight to attain their dream weight, seek to lose 25%

Table 9.4 A Systematic Approach to Management Based on BMI and Other Risk Factors

BMI	Suggested weight loss	Deciding factors for treatment level
<18.5	None Consider weight gain	Keep at same or move to less intensive strategy if: • No cardiovascular risk factors • Lower body obesity or overweight (i.e., gynoid) • No previous weight loss attempts • <25 lb to lose or goal of less than 5% to 10% weight loss
18.5-26.9	Exercise Diet modification Counseling	
27-29.9	Weight loss program Behavioral health services Self-help materials or program	
30-34.9	Weight loss program Pharmacotherapy Meal replacement	Consider move to more intensive strategy if: • Any cardiovascular risk factors present • Metabolic syndrome present • Abdominal obesity or overweight • Previous weight loss attempt failure at current level • >25 lb to lose or goal of more than 10% weight loss
35-39.9	Very low-energy diet Residential programs Pharmacotherapy	
≥40	Very low-energy diet Suggest surgery	

Adapted from World Health Organization (213).

Health Benefits of a 10% Weight Loss

Blood Pressure
Decline of about 10 mmHg in systolic and diastolic blood pressure in patients with hypertension (equivalent to that with most BP medications)

Diabetes
Decline of up to 50% in fasting glucose for newly diagnosed patients

Prediabetes
>30% decline in fasting or 2 h postglucose insulin level
>30% increase in insulin sensitivity
40% to 60% decline in the incidence of diabetes

Lipids
10% decline in total cholesterol
15% decline in LDL cholesterol
30% decline in triglycerides
8% increase in HDL cholesterol

Mortality
>20% decline in all-cause mortality
>30% decline in deaths related to diabetes
>40% decline in deaths related to obesity

Data from Haslam, Sattar, and Lean (86).

to reach a satisfactory weight, and would consider a weight loss of 17% disappointing (70). Few commonly prescribed weight loss programs achieve average weight losses that match patients' expectations. Physicians' expectations are not much lower; they consider a weight loss level of 13% disappointing. Patient expectations have been remarkably resistant to change, which may contribute to treatment dissatisfaction, treatment discontinuation, and treatment recidivism.

Diet Therapy

To lower weight, energy balance must be negative (see figure 9.10). Calorie reduction is the essential first step. For normal adults, ~10 kcal/lb (22 kcal/kg) is required daily to maintain weight. A bell-shaped curve describes a variation in average energy expenditure of about 20% (figure 9.11). Hence, for example, some individuals will require 12 kcal/lb (26 kcal/kg) and others only 8 kcal/lb (18 kcal/kg) per day to maintain weight. The lowest calorie level for weight maintenance is about 1,200 kcal daily, even for those at bed rest. The minimum calorie intake to ensure adequate intake of essential micro- and macronutrients is ~500 kcal/d, commonly provided under medical supervision in very low-calorie diets for weight loss. Exercise and activity levels affect maintenance calorie levels by 25% or more, depending on the degree of physical activity, and physical activity needs to be considered in determining maintenance calorie levels.

Hypocaloric diets for weight loss typically set intake at 500 to 750 kcal less than predicted maintenance requirements. It is often estimated that a deficit of 3,500 kcal is needed to lose 1 lb (7,700 kcal to lose 1 kg), and such diets can average about a 1 lb (0.45 kg) weight loss per week. However, this may overestimate weight loss, and dynamic estimates might better predict weight loss in obese individuals (178). If losses are slow during a weight loss effort, more

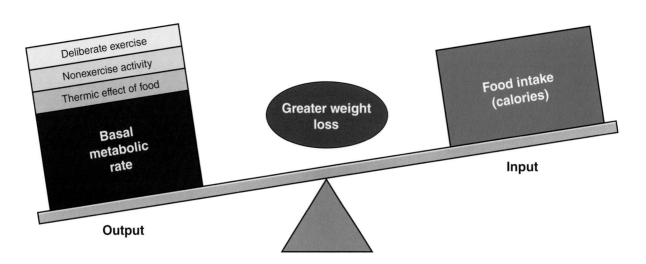

Figure 9.10 Energy balance shifted to increased caloric expenditure versus caloric consumption as required to lose weight.

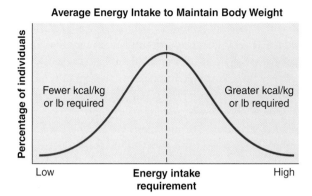

Average Energy Intake to Maintain Body Weight

Percentage of individuals

Fewer kcal/kg
or lb required

Greater kcal/kg
or lb required

Low   **Energy intake
requirement**   High

Figure 9.11 Energy intake requirements to maintain body weight.

aggressive diet therapy should be considered. Many popular variations in hypocaloric diet composition have been developed. Some of these are popular because of the restrictions used and promoted as a means of weight loss (e.g., Paleo, low carb, low fat, Zone, belief based, vegetarian, plant based). Because of the high calorie content of fat (9 kcal per gram) compared with carbohydrate and protein (4 kcal per gram) and the heart health benefits for cholesterol lowering, most national guidelines recommend low-fat diets.

For some, higher-protein and lower-carbohydrate diets are favored because of greater weight losses and improved satiety; higher-protein diets tend to promote satiety, and lower carbohydrate levels can promote greater fat utilization (ketosis), which can boost the rate of weight loss (10, 53). Overall reviews of popular commercial diet programs have shown about 5% weight loss at 1 yr and a 3% weight loss for standard hypocaloric diets at 2 yr (180). Although published data are supportive of higher-protein and lower-carbohydrate diets, no specific low-calorie diet composition is clearly superior to any other, and all fall short of patient and provider expectations. There is evidence that an unhealthy low-carbohydrate is associated with a higher all-cause mortality rate than a healthy low-carbohydrate diet (161b).

Intermittent fasting has made a resurgence within the longevity research community, showing benefits for health, aging, and disease (49). Periodic energy restriction through alternate-day fasting, daily time-restricted feeding of 500 to 750 calories, or other forms increases serum ketone bodies on those days. The switch from glucose and stored glycogen to fatty acids and ketone bodies improves energy transfer from stored fat and signals many proteins and molecules that may enhance health and aging. These benefits extend to improved cardiovascular risk factors including glucose regulation, abdominal fat loss, decreased blood pressure, decreased heart rate, and efficacy of endurance training. Weight loss with intermittent fasting regimens are modest, consistent with other standard calorie-deficit diets (2.5%-4+%), but they do show greater improvement in insulin resistance initially and heart rate variability by enhancing parasympathetic tone (117, 220).

Recent research has focused on more structured lower-calorie diets using meal replacement supplements. These diets are combined with exercise and lifestyle efforts to increase weight loss and achieve greater long-term results (71). Better portion control is achieved in these programs, which often use higher protein content to promote satiety and reduce snacking. Most use 1,000 kcal diets initially and combine weight loss medications to optimize results.

Because most patients who present for medical help to lose weight have BMIs of 38 or more, even structured meal replacement programs may not be sufficient. When rapid and substantial weight loss is critical, medically supervised very low-calorie diets (<800 kcal/d) can be used. These diets routinely consist of medically engineered powdered supplements rich in protein. Weight losses can begin at 1 lb (0.45 kg) per day but average 3 to 5 lb (1.4 to 2.3 kg) weekly for a typical 16 wk period (131). First developed in the 1920s as a means to treat seizures, these diets resurfaced in the 1960s in surgical nutrition research centers. Popularized in the 1970s, such "last-chance" diets fell from favor when excess protein losses led to deaths from ventricular arrhythmias (199). These types of plans can also be safely used by those with liver disease (nonalcoholic steatohepatitis, or NASH) and chronic kidney disease (199) and have been shown to be beneficial in those with type 2 diabetes (76). Reformulated in the 1980s with better-absorbed nutrients, these diets, often called liquid protein diets, are an important option for select individuals. Secondary to carbohydrate depletion leading to substantial lipolysis, ketone bodies are generated as described previously for intermittent fasting. These low levels of serum ketones tend to suppress appetite, which can be beneficial for maintaining the very low caloric intake. In studies comparing meal replacement diets with low-calorie diets, outcomes were similar at 4 mo to 1 yr (76, 192). Among completers, weight loss ranged from 10% to 35%, averaging over 15% at 2 and 4 yr (5, 101, 110, 121). Ketogenic diets that provide weight loss but are considered heart-unhealthy (e.g., Atkins) are generally not recommended in a clinical setting.

Practical application 9.1 presents a study that assessed weight loss effects on lean mass and aerobic fitness.

Behavioral Therapy

Providing any level of diet advice without behavior change is typically futile. Most medically significant weight loss efforts require frequent contact and support in order for people to adopt the healthier weight behaviors necessary for weight loss and maintenance of weight loss. Regular accountability, problem solving, and skill building are necessary over a 20 wk period to establish long-term success. Such behavioral change can be supported by individual or group therapy and augmented by phone, mobile or wearable device, and Internet follow-up. Weight loss efforts typically move from precontemplation to contemplation to preparation and then to action phases. Maintenance efforts must follow action steps in weight loss; otherwise, relapse is common. Motivation and realistic goal setting must be supported in a compassionate environment focused on measurable progress (99b, 128).

Record keeping and review predicts success because most people make better choices when they are made aware of the significance of those choices. Stimulus control helps patients identify stress and emotional eating cues and make other choices. Unhealthy eating behaviors like eating while driving or eating in front of a television or computer screen should be discouraged. Increased intentional exercise or lifestyle activity should be planned and monitored. Cognitive restructuring is used to detect black-and-white thinking and to help patients avoid an all-or-nothing pattern. Addressing emotional issues such as depression and shame with supportive therapy is essential, and referral for significant mental health or eating disorders may be necessary. Nutrition education and planning for maintenance, including relapse, can help reduce recidivism and the need for retreatment (50).

Exercise Therapy

Certainly aerobic-based exercise and physical activity are important in order for people to avoid becoming overweight or obese (107). But for the treatment of overweight and obesity, aerobic exercise alone has not shown long-term weight loss success (71, 175, 208). However, there is some evidence that exercise alone can result in significant weight loss in the short term (2, 157). Exercise in conjunction with diet therapy, however, is effective in accelerating weight loss, possibly by up to 11.5 lb (5.3 kg) (84, 175, 209). There is no evidence that resistance training alone results in clinically significant weight loss, likely because of its inability to produce a negative energy balance (46, 69, 99a, 175). However, resistance training can improve body composition (159). The National Weight Control Registry (NWCR) reported that over 90% of successful subjects (i.e., who lost 30 lb and kept the weight off for at least 1 yr) combined regular exercise with diet therapy (102). And the Look AHEAD study reported that 4 yr after an intensive weight loss intervention, there was a graded weight loss maintenance directly related to the amount of exercise energy expenditure (191). This is important because more than 80% of those who lose weight are unable to maintain the weight loss (208).

Evidence from the NWCR suggests that regular exercise of 60 to 90 min on most days of the week, expending 2,500 to 2,800 kcal/wk, may be required to maintain large amounts of weight loss for the long term (i.e., 5 yr or longer) (102). The Look AHEAD researchers suggest that at least a 2,000 kcal/wk expenditure is required to maintain a 10% weight loss, with about 1,500 kcal/wk needed to maintain between 5% and 10% loss (191). Regardless of the timing of exercise and its effectiveness in causing or maintaining weight loss, all overweight and obese patients will likely demonstrate improvements in cardiorespiratory function and physical fitness as a result. Additionally, exercise can improve self-esteem, which may improve adherence to weight loss–based treatments. The general exercise prescription for overweight and obese persons is reviewed later in this chapter.

Pharmacotherapy

Medications have been used to reduce weight for over 100 yr. The first medication advertised for weight loss was thyroid extract. Now known to cause more muscle loss than fat losses, thyroid replacement medication, necessary for hypothyroidism, is no longer used for weight loss. Weight loss drugs have faced significant prejudice. Many people expect long-term benefits from short-term treatment (something not expected from antihypertensives, for example) and are reluctant to use them at all. Although not a cure, some weight loss medications are an important tool for achieving and maintaining medically significant weight loss. Weight loss drugs are appropriately recommended for individuals with a BMI ≥30 or with a BMI ≥27 if they have obesity-related comorbidities (see table 9.5).

Practical Application 9.1

DIABETES REMISSION WITH SUBSTANTIAL WEIGHT LOSS

A resurgence of interest in complete meal replacement programs followed publication of the DiRECT Trial findings in 2018 (106). At 12 mo almost half of participants in this open-label, cluster-randomized trial achieved remission to a nondiabetic state and were taken off antidiabetic drugs. No previous trial had studied weight loss as an intervention in the same way as for a new pharmaceutical, with remission of type 2 diabetes as its targeted end point. This 600- to 700-calorie complete meal replacement diet normalized liver insulin resistance and fat content in 7 d and first-phase insulin response and pancreas fat content in 8 wk. At 12 mo 24% had a sustained weight loss of 15 kg or more compared with none in the control group. Diabetes remission at 12 mo was achieved in 46% in the intervention group compared with 4% in the control group. This trial was extended and showed durability of diabetes remission for more than a third of people with type 2 diabetes, linked to the extent of sustained weight loss. A subsequent paper will analyze the effects of exercise and physical activity on the ability to sustain weight loss and diabetes remission (106).

Phentermine (Adipex-P, Ionamin) is the most commonly prescribed weight loss drug in the United States. Approved in 1958 and currently recommended for short-term use, it acts as an appetite suppressant but can cause dry mouth, palpitations, and anxiety. It was part of the famous phen–fen combination therapy effort pioneered in the mid-1990s by a study published by Weintraub until the fenfluramines (Pondimin and Redux) were withdrawn from the market after fears of cardiac valve abnormalities were discovered. Cleared by the FDA, phentermine remains the least expensive common weight loss medication.

Orlistat (Xenical, Alli), approved in 1999, is an intestinal lipase inhibitor that causes a 30% malabsorption of dietary fat. It is not absorbed into the bloodstream, and patients who eat high-fat meals may have oily diarrhea. Patients who follow a typical fat diet (30% fat) do not have significant gastrointestinal symptoms from fat malabsorption. Orlistat produces an energy deficit of about 180 kcal daily (48). It has been approved by the FDA for over-the-counter sale.

One FDA-approved oral therapy for chronic weight loss and weight loss maintenance has been on the market for approximately 5 yr. Qsymia (phentermine and topiramate ER) is capable of between 12% and 14% weight loss. Belviq (lorcaserin hydrochloride) is capable of 4% weight loss at 1 yr in completers; it is sometimes combined off-label with phentermine. However, the FDA removed Belviq from the market in 2020 after a postmarketing analysis showed an increased risk of cancer in lorcaserin users. Contrave (bupropion and naltrexone) and an injectable diabetes agent used at higher doses, Saxenda (liraglutide), are the newest weight loss medications available by prescription; they achieve modest weight loss in completers (~4%-5% of baseline weight) in trials of up to 1 yr but are currently in limited use because of cost. Additional combination medications are pending FDA approval, and some central-acting agents plus those targeting gut hormones are on the horizon (169).

Surgical Therapy

Surgical procedures to restrict the stomach or cause malabsorption of food, or both, are considered the current gold standard for weight loss and weight loss maintenance (13). Frustration and failure with less invasive medical therapies lead over 250,000 people to have one of these types of procedures each year. Surgery for weight loss has been performed for over 50 yr, and newer laparoscopic techniques can cause patients to lose a third of their weight (>50% of their excess weight) within 18 mo. Such weight loss results in marked reduction in obesity-related medical conditions and improved life expectancy in these seriously obese patients (39). Research has confirmed low short-term procedure-related mortality and shows evidence of potential reduction in longer-term all-cause mortality as well as in cancer- and cardiac-related mortality (30).

Surgery is the best treatment for long-term weight loss. Importantly, patients undergoing bariatric surgery must commit to a lifelong program of restricted diet, lifestyle changes, vitamin supplementation, and follow-up testing to ensure safety. Weight regain after 2 yr can cancel out some of the initial benefits such that, in the Swedish Obesity Study, longer-term weight loss at 10 yr averaged between 15% and 24% of initial weight depending on the procedure chosen (166).

Surgery is typically restricted to those with a BMI ≥40 or those with a BMI ≥35 if they have obesity-related comorbid conditions. Many insurers require a 6 to 12 mo trial of comprehensive medical therapy before surgery, and many surgeons require that patients attempt to lose 5% to 10% of initial weight before surgery to reduce complications, aid exposure intraoperatively, and allow a laparoscopic approach (168). In very rare and selective cases, an adolescent may be allowed to have bariatric surgery.

The initial intestinal bypass procedures (jejunoileal bypass) of the 1950s were abandoned because of excessive nutritional deficiencies and were replaced by gastric banding procedures (gastric stapling) and vertical banded gastroplasty from the 1960s through the early 1990s (130). Roux-en-Y gastric bypass, combining the restrictive effect of gastric stapling with the malabsorptive effect of intestinal bypass, became the gold standard in the mid-1990s (120). Now done laparoscopically, this procedure is still recommended for those with BMIs ≥50. A newer approach, the gastric sleeve procedure, achieves intermediate results between Roux-en-Y gastric bypass and the largely abandoned laparoscopic adjustable gastric banding procedure. Intragastric balloon systems and vagal blockade electrical devices are also available options (9).

Table 9.5   Pharmacology

Medication name	Primary effects	Exercise effects	Important considerations
Phentermine	Appetite suppression	None	May increase HR and BP acutely
Orlistat	Intestinal lipase inhibitor	None	Causes flatulence and oily stools; may result in intestinal discomfort
Qsymia	Appetite suppression	None	May increase HR and BP acutely; average weight loss 12%-14%
Belviq	Satiety enhancer	None	Average weight loss 3%-4%

HR = heart rate; BP = blood pressure.

Comprehensive Long-Term Therapy

The CARDIA study reported an average weight gain of 0.5 to 0.8 kg/yr in adults. Black men and women gained ~4.5 and 7 kg more during a 25 yr period (1986-2011) than their white counterparts, respectively (56). The imperative for health providers interested in obesity, and for those who face the many comorbid conditions directly related to obesity, is to match patients to treatment options that will achieve meaningful medical and personal benefits and not merely tell patients to "eat less and exercise more." This level of treatment also requires combinations of diet, exercise, behavioral and pharmacologic therapy, and in many cases weight loss surgery for those who fail with less invasive approaches.

The maintenance of weight loss also requires diligent follow-up to offset the metabolic penalty of the reduced obesity (108). The Study to Prevent Regain (STOP) looked at face-to-face and Internet options for long-term weight loss management and found that both have a role in preventing the regain of weight so commonly seen with termination of treatment (208). This is important in today's culture of online capabilities and attempting to reduce patient burden and overall costs. The chronic disease model applies to obesity as much as it does to the comorbidities of obesity such as diabetes mellitus. Comprehensive approaches that depend on long-term lifestyle training can tame obesity and reduce its effect on mortality in this century.

EXERCISE PRESCRIPTION

When developing an exercise prescription for the prevention and management of obesity, clinical exercise physiologists must initially consider the aim of the prescription. There are four common scenarios:

1. Weight gain prevention
2. Weight loss using exercise alone
3. Weight loss with combined exercise and caloric reduction
4. Weight loss maintenance

Each of these situations will result in somewhat different individual recommendations. This section focuses on the recommendations for exercise to produce weight loss. Another factor not covered, but that could be considered when treating a patient, is a desired change in body composition. There are few professional society statements regarding the use of physical activity and exercise for the treatment and prevention of obesity. The recommendations in table 9.6 come from statements from the American College of Sports Medicine (154), the American Dietetic Association (161a), the European Association for the Study of Obesity (216), and the National Health and Medical Research Council of Australia (134). Unfortunately, none of them cover any of the aforementioned four common scenarios.

The following general recommendations can be made for the four scenarios:

- *Exercise for weight gain prevention:* Prevention of weight gain needs to begin as early as possible. An American Heart Association scientific statement focuses on reducing sedentary behaviors (e.g., screen time, sitting, lack of recreational pursuits) in children as young as preadolescence to begin preventing the accumulation of adiposity (12). In adults there is evidence of a dose–response of exercise and the prevention of weight gain that is most pronounced when an individual performs moderate-to vigorous- intensity exercise for more than 150 min/wk (96). The ACSM position stand on weight loss lists the scientific evidence for physical activity and exercise in preventing weight gain at its highest level, meaning there are substantial randomized trials with a consistent finding of weight gain prevention (55). The required amount of exercise to prevent weight gain is in the range of 150 to 200 min/wk (~1,200-2,000 kcal/wk), performed at a moderate intensity. This is a general recommendation; no evidence supports specific needs for individuals based on race, ethnicity, or socioeconomic status (96). The exercise effect does not vary between sexes.

- *Exercise alone for weight loss:* High amounts of total caloric expenditure, likely in the range of 1,500 to 3,000 kcal/wk (225-420 min/wk), are required to lose weight by exercise without a caloric intake reduction (55, 85). There is a dose–response relationship between physical activity and the magnitude of weight loss achieved. The ACSM position stand on weight loss lists the scientific evidence supporting physical activity for weight management as a level B, meaning there is some evidence for support, but strong, randomized trials are lacking and suggest that although weight loss can be achieved, the amount of loss is likely to be only modest, in the range of 1.5 to 2.0 kg (55, 162, 179). Hansen and colleagues, who developed a more contemporary (2018 vs. 2009) consensus statement for the Expert Working Group on exercise prescription, rate the scientific evidence as level A (85). Of course a key to weight gain prevention throughout life is to make this level of exercise a permanent lifestyle change.

- *Exercise and caloric reduction for weight loss:* Along with caloric reduction, aerobic exercise increases the weekly rate of weight loss by approximately 2 lb (1 kg) over caloric reduction alone (42, 162). The ACSM position stand on weight loss lists scientific evidence supporting exercise plus caloric restriction at its highest level (substantial randomized trials with consistent findings) (55). Minimally, this requires 150 min/wk but will vary based on the amount of caloric reduction. There is evidence that if exercise intensity is increased, additional weight loss will occur (162). Exercise will have less of an additive weight loss effect when caloric restriction is severe (i.e., >700 kcal/d deficit needed to meet resting metabolic rate).

- *Exercise for weight loss maintenance:* Regular exercise of 60 to 90 min/d on most days of the week is generally recommended for long-term weight loss maintenance. The ACSM position stand on weight loss lists the scientific evidence supporting this recommendation as a level B (some evidence but a lack of strong, randomized trials) because "there are no correctly designed, adequately pow-

Table 9.6 Exercise Recommendations for Weight Loss and Maintenance From Selected Position Stands

	ACSM (55)	ADA (161a)	European (215, 216)	Australian (134)
AEROBIC				
Modes	Not specified but uses walking data examples as only mode mentioned	Not specified but suggests using a pedometer and step goal	Only example is brisk walking for exercise but also suggests increasing physical activity and suggests walking, cycling, and stairs	Stairs, walking, swimming, biking, jogging, joining a gym
Intensity	Moderate to moderately vigorous	Moderate to vigorous	Moderate (brisk walking)	Moderate to vigorous depending on duration
Frequency	Daily	30-60 min on most days of week for health and weight control; 60-90 min/d for weight gain prevention	Not specified	Not specified
Duration	150-200 min/wk to prevent gain; 150 to 225-420 min/wk for weight loss; 200-300 min/wk for long-term control	150 min/wk for health benefits and 300+ min/wk for weight loss and long-term control	At least 150 min/wk	300 min/wk moderate intensity or 150 min/wk vigorous intensity
RESISTANCE				
	Does not support resistance training for weight loss; suggests only limited evidence for lean mass maintenance during weight loss	Not mentioned	States that resistance training can increase lean mass in middle-aged overweight and obese individuals and should be performed 3 times per week but also states that aerobic exercise is the only mode associated with reducing fat and body mass	Only mention is that initial weight gain is associated with muscle-strengthening exercise secondary to increases in skeletal muscle size

ered, energy balance studies to provide the needed evidence"(55). Exercise at the level recommended for health in adults (i.e., 150 min/wk) is not enough to prevent weight regain; volumes exceeding 200 min/wk are probably needed (175).

The ACSM provides the most comprehensive recommendations for exercise to control body weight, recommending an exercise program focused on physical activities and intentional exercise for 60 to 90 min/wk for the loss and maintenance of weight (55, 154). This is beyond the general recommendation of expending 1,000 to 2,000 kcal/wk (30 min on most days) for general health benefits. Exercising for 250 to 300 min/wk (or ~60 min/d) is equivalent to about a 2,000 kcal energy expenditure per week. But this figure is less than the 2,500 to 2,800 kcal/wk expenditure recommended by the National Weight Control Registry mentioned previously in this chapter (208). Therefore, the following exercise prescription recommendations are based on a weekly caloric expenditure of 2,000 to 2,800 kcal. This goal range is appropriate

for all obese individuals, although some class II patients and most class III patients will need to progress gradually to these higher levels of daily energy expenditure. A recommendation is to aim for 30 min/d initially and to progress as tolerated. Physical activity counseling provided by a clinical exercise physiologist will help people develop realistic goals, establish appropriate exercise progression schedules, and gain control of their exercise programs (143). See table 9.7 for a summary of the exercise prescription for overweight and obese individuals.

Cardiorespiratory Exercise

Initially, exercise and physical activity should focus on cardiorespiratory (i.e., aerobic) modes. The primary reason for this approach is that aerobic exercise is the only type linked to reductions in body weight. Additionally, to achieve a target of 200 to 300+ min/wk (2,000-3,000+ kcal/wk), exercise must be predominantly aerobic. Although resistance training does provide added benefits, particu-

Table 9.7   Exercise Prescription Review

Training method	Frequency	Intensity	Time (Duration)	Type (Mode)	Progression	Important considerations
Cardiorespiratory	≥5 d/wk	Initially moderate increasing to vigorous as measured by heart rate reserve method	Depending on goals: • ≥150 min/wk for weight loss with diet • Up to 300 min/wk for loss and maintenance	Prolonged rhythmic activity using large muscles (walking, cycling, etc.)	As tolerated, increase duration initially followed by intensity	Walking is generally best but some people may need a non-weight-bearing mode. Many have not exercised regularly in years.
Resistance	2 or 3 d/wk	60%-70% of 1RM or at a moderate intensity	2-4 sets ranging from 8-12 reps	Machines are best; free weights and devices (medicine balls, bands) can also be used, especially if an individual is too large for a machine	As tolerated when soreness is resolved and 12 reps is considered an easy effort by the individual	Never exercise the same muscle groups on consecutive days; this will not promote weight loss.
Range of motion	Daily	Stretch only to the point of tightness or minor discomfort	Static stretches of 10-30 s for 2-4 reps Dynamic for up to 10 reps	Static, dynamic, or proprioceptive neuromuscular facilitation	As range of motion allows, attempt to increase range of stretches	For morbidly obese individuals, modifications may need to be made to avoid stretches requiring extreme ranges of motion or being on the floor.

1RM = one-repetition maximum; reps = repetitions.

larly with improving body composition, the caloric expenditure of resistance training is less than that of aerobic exercise for several reasons: Resistance training is performed discontinuously, a single training session incorporates less exercise time than an aerobic session does, and resistance training is commonly performed only 2 or 3 d/wk because it should not be done on consecutive days. For those with a very low functional capacity, consider adding resistance training only after the individual has been performing a regular aerobic program for a minimum of 1 mo and is approaching the weekly duration goal.

Exercise mode selection is important for enhancing adherence and reducing injury risk. People with preexisting musculoskeletal conditions (e.g., arthritis, low back pain) may be limited to certain modes of cardiorespiratory exercise. However, these limitations, particularly those related to the lower back or joints, may improve as weight is lost. The clinical exercise physiologist should assess any painful conditions and make recommendations to attempt to avoid this type of pain and any exacerbation of a condition. When necessary, they should advise on general treatment (rest, ice, compression, elevation) and when to seek medical care.

In general, aerobic exercise and physical activity should be categorized as either weight bearing or non–weight bearing. When possible, walking is the best form of exercise for several reasons. It has few disadvantages; all patients have experience with the activity and a goal to remain functional and independent. Walking can easily range from low to moderate or vigorous intensity with a low risk of injury. It is available to most patients and does not require special facilities. Neighborhoods, parks, walking trails, shopping malls, fitness centers, and the like offer walking opportunities, often allowing an individual to meet others with similar goals. A minimum amount of attention is necessary, so socializing is easy and convenient. And for those who may be extremely limited, walking in their home such as in a hallway may be an option. This would allow them to do multiple shorter bouts throughout a day until they are able to do more outside of the house.

If a patient wishes to walk on a treadmill, care should be taken to assess the weight limits of the treadmill. Many are rated to handle only up to 350 lb (160 kg). Issues relate to the ability of the walking board and motor to handle a large amount of weight. If a motor seems to bog down when an individual is walking on it, then the weight of the person is likely too much. Treadmills specially designed for 500 to 750 lb (227-341 kg) are available. Individuals new to using a treadmill should be taught how to control the speed and grade because this is a common source of difficulty, with inadvertent increases beyond the individual's capability. Elliptical equipment may be used in place of a treadmill, but again weight limitation must be determined to prevent not only equipment breakage but also potential injury. In general, jogging should be avoided, especially for those beginning an exercise routine and for those with no previous jogging history or who have a preexisting musculoskeletal issue. Some obesity class I patients may be appropriate candidates for jogging.

Non-weight-bearing exercise options include stationary cycling, recumbent cycling, seated stepping, upper body ergometry, seated aerobics, and water activities. The equipment for this type of exercise is also subject to weight limitations, and some is available for use by those in obesity classes II and III. These activities are useful at any time but are particularly advantageous for those with joint injury or pain. The clinical exercise physiologist should adapt these modes as needed (e.g., by providing larger seats and stable exercise machines). People who are obese can have difficulty getting on or off these types of equipment or difficulty moving through the range of motion required by a given piece of equipment. For some individuals, seated aerobics may be an excellent option to reduce the typical orthopedic limitations that some people experience, including back, hip, knee, and ankle pain, and they can be performed in the comfort of a person's home. For extremely obese individuals (i.e., obesity class II and III), it is important to use a chair that is rated to handle a very heavy body weight because the force on the chair from the movement and body weight can be quite large.

Water provides an alternative to walking or aerobic dance activities performed on land. The buoyancy of water takes much of the body weight off the joints. Additionally, patients who experience heat intolerance with other activities are often more comfortable performing water-based exercise. Most patients are not efficient swimmers, so swimming laps should generally be avoided. An experienced exercise leader can make a workout session fun and effective. For example, the resistance of the water can be used creatively to increase intensity. Many patients do not consider water activities because of the effort necessary to get into and out of the pool and because of their concern about their appearance in a bathing suit. The clinical exercise physiologist should help each individual overcome these issues by using zero-entry pools and locations where the public does not have a direct view of the entire facility.

Frequency

Behavioral changes in activity must be consistent and long lasting if the patient is to lose weight and maintain weight loss over the long term. Some individuals cannot initially perform daily exercise, but many will be able to achieve this goal. However, daily exercise and physical activity at the recommended levels of duration and intensity should be the goal for everyone. Daily exercise is most likely required to achieve and sustain long-term, significant weight loss. A key factor is to initially minimize the duration and intensity

to prevent excessive fatigue or muscle soreness (i.e., DOMS) that may sabotage the patient's willingness to return to exercise. Altering the exercise mode between sessions may also help reduce the risk of injury or allow any mode-specific pain to subside.

Intensity

The intensity of exercise must be adjusted so the patient can endure up to 1 h of activity each day. For those who have never exercised previously, a moderate intensity in the range of 50% to 60% of peak $\dot{V}O_2$ (50%-60% of heart rate reserve) is typically low enough for sustained exercise. However, an intensity closer to 40% may be required by some individuals, particularly those who have not exercised recently. As an individual progresses, a goal of 60% to 80% of heart rate reserve (i.e., vigorous intensity) is adequate. People without significant comorbid conditions can often perform at these intensities in either a supervised or a nonsupervised setting. Some individuals who are obese are hypertensive and may be taking a β-blocker. This possibility must be considered if intensity is prescribed using heart rate. Typical rating of perceived exertion values of 11 to 15 (6-20 scale) may be substituted when assessing heart rate is not convenient. Interval training is typically not part of an initial exercise routine, but some individuals who establish and tolerate a regular exercise routine and who are successful with weight loss may enjoy this form of training.

Duration

For those with little or no recent exercise history, beginning with 20 to 30 min each day is appropriate. Breaking this exercise time into two or three sessions per day of shorter duration (~10 min) may be required for extremely deconditioned people. The focus should be on duration and frequency of exercise before increasing intensity, with a progression of approximately 5 min every 1 to 2 wk until the person can perform at least 60 min of exercise. This progression scheme is intended to increase compliance to the duration of each session as well as to daily exercise because it may reduce the risk of extreme fatigue or soreness. An accumulation of time over several sessions in a day (i.e., three or four 10 min bouts) is as beneficial as one continuous work bout with respect to total caloric expenditure. Besides performing this intentional exercise duration, all obese people should be continually encouraged to maximize daily physical activity by considering all options. For instance, they could park at the far end of the parking lot when visiting a store or get off one or two stops early when taking public transportation. The duration of daily physical activity should typically not be restricted unless the individual appears to be suffering from effects of excessive activity.

As exercise progresses and an individual is able to better tolerate an exercise routine, higher-intensity activities should be encouraged. Individuals should be encouraged to increase the duration from 20 consecutive min/d to 60 min/d or more on every day of the week so they are expending >2,000 kcal/wk. Ideally, an exercise program for obese patients should include both supervised and nonsupervised phases with adaptations in modes, intensity, duration, and frequency to maximize caloric expenditure and prevent soreness and injury. Supervised exercise may improve adherence and progression.

Resistance Exercise

If resistance training is incorporated, careful attention must be given to beginning this type of program. Strength equipment may not be an option for some morbidly obese individuals. In general, the exercise prescription for obese people should include resistance intensity in the range of 60% to 80% of an individual's one-repetition maximum (1RM) performed for 8 to 12 repetitions for two sets each (progressing to as many as four sets), with 2 to 3 min of rest after each bout. This plan will allow the person to perform 6 to 10 exercises in a 20 to 30 min session. Resistance exercises can be performed maximally on 2 or 3 d/wk, ideally without working the same muscle groups on consecutive days. The major muscle groups to involve include the chest, shoulders, upper and lower back, abdomen, hips, and legs. The primary acute benefit of the prescribed resistance program is to improve muscle endurance; the secondary benefits include increased muscle strength and the preservation of lean mass during weight loss.

Practical application 9.2 provides guidance for interactions with those who are overweight or obese.

Range of Motion

Obese patients have a reduced range of motion primarily due to an increased fat mass surrounding joints of the body, in conjunction with a lack of movement and routine stretching. As a result, these patients often respond slowly to changes in body position, which limits their ability to function in daily activities. They also tend to have poor balance. Persons who are obese are at a greater risk of low back pain and joint-related osteoarthritis because of their condition (58, 190, 194). Therefore, range of motion may

---

Practical Application 9.2

CLIENT–CLINICIAN INTERACTION

Many clinicians who might be an important part of the process of helping an individual lose weight may be uncomfortable or even too embarrassed to bring up the topic. This includes physicians, nurses, clinical exercise physiologists, and other allied health staff. They often find the topic difficult to broach, even in a clinical setting (e.g., clinic visit, clinical exercise program) in which a patient might be expecting to hear an assessment of their weight and body habitus and the associated risk. Alternatively, some clinicians do not bring up overweight issues because they feel it is a lost cause; most patients neither heed their advice nor wish to do anything about their weight. Despite these situations, there is ample evidence that clinicians who provide advice (either directly or using a behavior change technique such as motivational interviewing) can have a meaningful impact.

As a first step, even if the patient is specifically there for help with losing weight, the interaction should begin by asking if it is OK to discuss their weight. Receiving this permission is vital for establishing an effective client–clinician interaction. Always be sensitive about the language used. Many of these individuals do not want to hear the terms *obese* or *fat* when discussing their health condition because they find them demeaning, even though *obesity* is a clinical term.

Research has demonstrated that people who are overweight or obese often have lower self-esteem than those of normal weight. They also are less likely to be married or in a relationship and are less likely to be employed; if employed, they tend to make less money than similar-aged individuals, often because they are underemployed. Also, many may have psychological issues that hamper their ability to function from day to day. These might include eating disorders, depression, and anxiety. When attempting to help someone lose weight, it is important to be sensitive to each of these issues. And if there are behavioral issues beyond the scope of a clinical exercise physiologist, a referral should be made to a behavioral health specialist.

Regarding an exercise program, the clinical exercise physiologist must be able to discuss realistic expectations about exercise for weight loss and to design a program that begins at a person's level of readiness and comfort, provides proper progression, and provides an avenue for regular follow-up for program adjustments to enhance adherence. One must also be realistic with the client so she does not have an unrealistic expectation as to the amount of weight loss to be expected from regular exercise. The clinical exercise physiologist must be comfortable with obese clientele, especially when discussing exercise options that may not be appropriate. For instance, some patients may wish to jog, and the clinical exercise physiologist must be prepared to discuss the issues that may confront a morbidly obese individual when jogging. Others may be reluctant to consider using a swimming pool because of issues of appearance or ease of entering and exiting the pool. Depending on the program and counseling sessions, time may not be available to build rapport, so the clinical exercise physiologist must be able to discuss these issues almost immediately with any patient.

improve spontaneously with weight loss (16, 92, 112). Still, to the degree possible, patients should perform a brief flexibility routine focused on the legs, lower back, and arm and chest regions. Normal flexibility routines (see chapter 6) are recommended as tolerated.

EXERCISE TRAINING

Low cardiorespiratory fitness adds to the risk of cardiovascular disease and mortality in overweight and obese people, including those with heart failure (41, 123, 200). In fact, low fitness is similar in risk to diabetes, hypertension, elevated cholesterol, and smoking. Ample evidence indicates that regular aerobic exercise training improves physical functioning, independent of weight loss, in those who are obese or overweight (40). Although their data are not specific to the obese and overweight individual, Blair and colleagues have reported that men who increase their cardiorespiratory fitness level by one quartile have significant reductions in their long-term morbidity and mortality profile (21). Gulati and colleagues report similar benefits for women (81). Barry and colleagues reported on the fitness versus fatness issue in a meta-analysis and found that fitness was an important protector against mortality in both normal weight, overweight, and obese individuals (14). Additionally, the obesity paradox (suggesting that overweight and class I obesity protect against death) has partially been explained by differences in fitness levels (123, 124, 146, 203).

As previously discussed, exercise training is likely most important for the weight maintenance process of weight control (127). Data from the National Weight Control Registry project indicate that regular exercise training expending more than 2,000 kcal/wk is a strong predictor of long-term weight loss maintenance (102). Walking is effective for long-term (2 yr) weight loss maintenance in women (65). However, there is evidence that weight loss maintenance can be achieved for up to 12 mo without regular intentional exercise (52) and that weight regain may be similar between those attempting to maintain their weight loss by diet alone or by diet plus exercise (198).

Cardiorespiratory Exercise

With respect to balancing time required for implementation of weight loss methods, the focus of exercise training in the overweight or obese person should initially be on caloric expenditure, which is best achieved with aerobic-based training (216). Although dietary changes appear to be more effective than structured exercise alone for reducing body weight, limited data suggest that exercise alone can result in weight loss similar to caloric restriction (157). The amount of exercise required (e.g., expenditure of ~700 kcal/d for 3 mo), however, is likely beyond the capability, time allotment, and willingness of most individuals, particularly those with no recent exercise history. However, a study of obese cardiac rehabilitation patients used a high-caloric-expenditure exercise routine, compared with standard cardiac rehabilitation, to assess the effects on weight loss (2). Energy expenditure in the groups was

---

Exercise Training: Important Considerations

Cardiorespiratory Endurance
Although fitness (i.e., peak $\dot{V}O_2$) is generally low in those who are overweight and obese, an improvement of 15% to 30% is expected from a regular exercise routine; some very deconditioned individuals may improve 50% to 100% (158, 188). Moderate- and high-intensity interval training has been shown to increase peak $\dot{V}O_2$ in overweight and obese individuals (186). However, high-intensity interval training (HIIT) may result in increased inflammatory response in those who are obese.

Skeletal Muscle Strength
Resistance training appears to be very effective at maintaining or reducing the rate of loss of lean mass during a weight loss effort, resulting in improved body composition and skeletal muscle strength (94); however, resistance training alone or in combination with caloric restriction does not affect weight loss (32, 188).

Skeletal Muscle Endurance
Muscular endurance is likely to improve similarly to an individual of healthy weight.

Flexibility
Range of motion (ROM) will likely improve with weight loss because of reduced mechanical limitation secondary to body fat reduction (112).

Body Composition
Regular aerobic exercise alone can reduce weight and fat mass, but the amount of exercise required to achieve this effect is very high (~700 kcal/d) (2). Aerobic exercise that does not result in weight loss may reduce intra-abdominal (i.e., visceral) fat (129).

3,000 to 3,500 versus 700 to 800 kcal/wk. The high-expenditure group lost significantly more weight (8.2 ± 4 vs. 3.7 ± 5 kg) over 5 mo of intervention. Importantly, this group of patients with a low initial fitness level (22 mL $\cdot$ kg^{-1} $\cdot$ min^{-1}) and little exercise experience was apparently able to tolerate and adhere to this type of exercise regimen.

Data from a systematic review support a synergistic effect of combined caloric restriction and exercise for weight loss in a summary of 14 randomized controlled studies (89). This finding, however, has not been universally replicated (193). Jakicic and colleagues reported that exercise intensity (moderate or vigorous) and duration (moderate or high) adjusted to expend between 1,000 and 2,000 kcal/wk, combined with calorie intake restriction, did not have an effect on the amount of weight lost over 12 mo in a group of obese women (97). Others have demonstrated that diet combined with increased daily lifestyle physical activity may be as effective as a program of diet and intentional exercise (5).

Because obesity can negatively affect cardiorespiratory function, it is reasonable to believe that weight loss will result in improvement (4). The expectations for cardiorespiratory exercise training with regard to hemodynamic and other physiologic system adaptations are similar to those for people who are not obese or overweight. For instance, Pouwels and colleagues reported on 56 obese patients undergoing bariatric surgery and 24 normal-weight individuals (149). Three months after surgery, significant changes occurred in mean arterial pressure (-18 mmHg, $p = .001$), systolic blood pressure (-17 mmHg, $p = .001$), diastolic blood pressure (-18 mmHg, $p = .001$), stroke volume ($+82$ mL per beat, $p = .03$), and heart rate (-8 bpm, $p = .02$) as compared with the normal-weight group (149). Acute blood pressure declines following exercise, and accumulation of exercise over time, may chronically reduce blood pressure independent of weight loss (142).

An interesting consistent observation is the relationship between exercise and visceral fat reduction in obese individuals, occurring with or without weight loss. It is well known that weight loss by any method will preferentially reduce visceral fat (38, 187). However, there is evidence that aerobic, but not resistance, exercise will reduce visceral fat even if there is no weight loss (95, 98). However, the dose of exercise to produce visceral fat reductions is not well understood. Ohkawara and colleagues suggested that at least 10 MET-h/wk is required (e.g., 4 d/wk × 5 METs intensity × 30 min per session = 600 MET-min/wk = 10 MET-h/wk) (139); and Ismail and colleagues stated that aerobic exercise in amounts below current public health recommendations for overweight and obesity management may be sufficient to reduce visceral fat (95). Weight loss via caloric restriction not only affects fat tissue but also can lead to undesirable weight loss effects on the skeletal muscle, bones, heart, liver, and kidneys (23, 155). These changes can have negative effects on muscle and bone strength, cardiorespiratory capacity (i.e., peak $\dot{V}O_2$), and resting metabolic rate. Regular aerobic exercise has been shown to protect against each of these potential effects (189, 201, 202).

Resistance Exercise

Although resistance training is suggested for those losing weight, only aerobic exercise has substantial and solid evidence supporting the outcome of weight loss or fat mass loss. However, resistance training is recommended for most weight loss programs, particularly for those individuals who desire to lose large amounts, because it may offset potential losses in lean mass. This is despite the paucity of well-designed studies comparing the effects of aerobic, resistance, or combined aerobic and resistance training on fat mass and lean mass changes in obese individuals. The STRRIDE AT/RT study randomized 119 sedentary overweight or obese adults to either resistance training alone, aerobic training alone, or a combined resistance and aerobic training protocol (206). The investigators attempted to design this study so the different training regimens were comparable for the amount of exercise performed and the caloric intake. The study found that resistance training alone was not effective at reducing total body weight or fat mass. However, when combined with aerobic training, a similar

Research Focus

High-Intensity Interval Training and Weight Loss

High-intensity interval training (HIIT) has been studied frequently over the past several decades. Data suggest it can improve cardiorespiratory fitness, as measured by peak $\dot{V}O_2$, more so than continuous training in a variety of clinical populations including those attending cardiac rehabilitation. Less frequently investigated is the effect of HIIT on body weight in those who are overweight or obese. A pilot study performed by Grossman and colleagues evaluated the effects of HIIT versus continuous endurance training for 12 wk in 11 obese (BMI = 32.0 ± 2.53 kg · m^{-2}) women who were postmenopausal (age = 59 ± 5.3 yr). Both groups followed a calorie-restricted meal plan, had their weight recorded weekly, and attended monthly in-person meetings. The rationale for performing HIIT was to provide an exercise benefit similar to that of continuous training in a shorter time per session. The women wore Fitbit sensors to assess adherence to exercise training. Adherence was 100% in the HIIT group and 60% in the endurance exercise group. Interestingly, those in the HIIT group lost 8.7% of their body weight, while the endurance group lost 4.3%. This difference was statistically significant ($p < .05$). Other parameters including loss of fat mass and a reduction in waist circumference occurred for both groups, but the amounts were not different between the groups. However, because this was a pilot study, it was not statistically powered to assess between-group differences. The investigators concluded that this pilot study demonstrated the potential feasibility of HIIT for improving exercise adherence and factors related to weight loss in the obese, postmenopausal population.

Reprinted by permission from J.A. Grossman, D. Arigo, and J.L. Bachman, "Meaningful Weight Loss in Obese Postmenopausal Women: A Pilot Study of High-Intensity Interval Training and Wearable Technology," *Menopause* 25, no. 4 (2018): 465-470.

total weight loss was achieved versus aerobic training alone. But the resistance training provided an additional benefit of a greater percent fat reduction, primarily due to a greater fat mass loss combined with a gain in lean mass, thus having a greater impact on improving body composition versus aerobic training alone. A primary drawback of this type of training is that the combined protocol took twice as long for the same amount of weight loss as the aerobic training alone.

Exercise Training Review

- Prescribed exercise must be specific to the desired outcomes.
- Exercise can be prescribed alone for weight loss (least effective) or as part of a calorie-restricted meal plan (most effective).
- Regular exercise is an excellent predictor of long-term weight loss maintenance and must be implemented for success. People must approach 60+ min/d for best maintenance results. Adding resistance training to aerobic exercise along with caloric restriction will not result in more weight loss but will improve body composition.
- Obesity is a risk factor for cardiovascular disease, diabetes, hypertension, dyslipidemia, and many other chronic diseases. This must be considered in those beginning an exercise program and treated appropriately. Consider aggressive techniques for weight loss in the extremely obese. These include meal replacement (partial or complete) and surgery.

A study that compared aerobic training and aerobic training plus resistance training in patients with type 2 diabetes showed that aerobic training plus resistance training provided no additional weight loss benefit (118). And a more recent study compared caloric restriction alone with caloric restriction plus either aerobic or resistance training on weight loss and anthropometrics in a group of overweight and obese subjects (74). There was no difference in total weight loss, BMI, or body fat percentage between the groups. However, researchers reported a trend toward a preservation of lean mass in the group that performed resistance training. No evidence suggests that resistance training during caloric restriction for weight loss results in additive weight loss (55).

Range of Motion Exercise

Few studies have dealt with weight loss and changes in range of motion. One would anticipate that any mechanical impediment to range of motion would improve with weight loss. However, little is known about potential structural effects (e.g., plasticity) on connective tissue and muscle. In a nonrandomized study, 10 obese adults had lost $27.1 \pm 5.1\%$ of their weight (34 ± 9.4 kg) by 7 mo and an additional $6.5 \pm 4.2\%$ (8.2 ± 6 kg) at 13 mo following bariatric surgery (92). Assessment of range of motion showed significant increases for the hip during the swing phase of walking, for maximal knee flexion, and for ankle joint function toward plantar flexion. This was also associated with a greater swing time during self-paced walking with increased stride length and speed. Similarly, Benetti and colleagues reported improved range of motion (sit and reach) in 16 subjects who underwent bariatric surgery. Interestingly, this improved flexibility did not result in improved static postural balance (16). These findings suggest that weight loss in the morbidly obese (BMI = 43 ± 6.5 kg $\cdot$ m^{-2}) can positively affect range of motion via mechanical plasticity and that this may have positive effects on ambulation but not on balance.

Clinical Exercise Bottom Line

- Overweight and obesity continue to rise worldwide and are associated with the development of chronic disease and disability.
- Many overweight and obese individuals want to lose more than 15% of their current weight.
- Bariatric surgery is the current gold standard for weight loss. However, behavioral weight loss strategies can be effective.
- Tailoring the behavioral weight loss strategy, particularly for the meal plan, is important to meet individual expectations.
- Exercise can be supplemental to a behavioral weight loss effort with a greater rate of weight loss and less loss of lean tissue (i.e., skeletal muscle).

CONCLUSION

Regular exercise and behavioral modifications specific to eating and exercise are the foundations of effective weight management. Clinical exercise physiologists, particularly those who have a solid background in behavioral or lifestyle counseling, should be involved in the primary prevention, treatment phase, and secondary prevention of overweight and obesity, particularly in medically supervised programs. Given the number of overweight individuals in the United States and worldwide, it is very important to continue to develop competent clinical exercise physiologists to work with these patients.

Online Materials

Visit HK*Propel* to access a link to the references, the case studies with discussion questions, and a quiz to help you review key concepts and test your understanding of the material covered in this chapter.

CHAPTER 10

Hypertension

Yin Wu, PhD

Linda S. Pescatello, PhD

Hypertension is the most common, costly, and modifiable cardiovascular disease (CVD) risk factor in the United States and world (1). CVD is the leading global cause of death, accounting for approximately one in three deaths; while high **systolic** blood pressure (SBP) is the top-ranked health risk factor for the global burden of disease, accounting for 14% of the deaths of 8.69 million people in 154 countries (1-3). The American College of Cardiology (ACC) and American Heart Association (AHA) guidelines for the prevention, treatment, and management of hypertension lowered the **blood pressure (BP)** threshold for having hypertension to ≥130 mmHg for SBP *or* ≥80 mmHg for **diastolic** BP (DBP) (4) from the Joint National Committee Seven (JNC7) threshold of ≥140 mmHg for SBP or ≥90 mmHg for DBP (5). Now 46% of adults (116 million) in the United States live with hypertension according to these new BP thresholds. The prevalence of hypertension is highest in Black men (59%) and women (56%) followed by white men (48%), Hispanic men (47%), Asian men (46%), white and Hispanic women (41%), and Asian women (36%) (1, 4). The annual medical costs for a person with hypertension in the United States are approximately $2,000 more compared with a person without hypertension (6). The most recent estimated direct and indirect costs of high BP are $55.9 billion in the United States, while the directs costs of high BP alone are projected to rise to $220.9 billion in 2035 (1). Curbing this public health epidemic is a national and global priority (1-3, 7).

Using the 2017 ACC/AHA BP thresholds, the lifetime risk for developing hypertension is 86% for Black men, 84% for Black women and white men, and 69% for white women (8). Given that baby boomers represent the fastest-growing age segment of the United States population, reducing the proportion of individuals diagnosed with hypertension continues to be a major priority for all leading public health organizations (9). Early diagnosis with accurate and repeated BP measurements following standard professional methodology and procedures (see figure 10.1) is advised (4). Lifestyle modifications such as exercise are regarded as cornerstone approaches to prevent, treat, and control hypertension (4, 9). In patients who cannot achieve their goal BP with lifestyle approaches alone, antihypertensive medication may be initiated, but medication regimens will vary depending on initial BP values, the presences of other medical conditions, age, race, and atherosclerotic CVD (ASCVD) risk as determined from the ASCVD risk calculator (4). An understanding of the unique aspects surrounding diagnosis, treatment, preexercise health participation screening, exercise prescription, and potential exercise–drug interactions in patients with hypertension is imperative to significantly reduce the individual, national, and global burden of hypertension.

DEFINITION

BP classification is based on the average of two or more accurately measured, seated BP readings on each of two or more office visits as demonstrated in figure 10.1. The ACC/AHA classification scheme of BP for adults 18 yr or older is shown in figure 10.2. Hypertension is defined as having a resting SBP of ≥130 mmHg or DBP of ≥80 mmHg, taking antihypertensive medication, being told by a physician or health professional on at least two occasions that one has high BP, or any combination of these criteria (4). The ACC/AHA guidelines also define an additional class of patients, those with SBP ranging from 120 to 139 mmHg and DBP <80 mmHg, as having elevated BP; these patients are at heightened risk of developing hypertension. Last, adults with a SBP <120 mmHg and a DBP <80 mmHg are defined as having normal BP (4).

Hypertension increases the risk for CVD, stroke, heart failure, peripheral artery disease, and chronic kidney disease (CKD).

Acknowledgment: We wish to gratefully acknowledge the efforts of the previous authors of this chapter, Amanda L. Zaleski, PhD, Antonio B. Fernandez, MD, and Beth A. Taylor, PhD, and thank Aashish S. Contractor, MD, MEd, Terri L. Gordon, MPH, and Neil F. Gordon, MD, PhD, MPH.

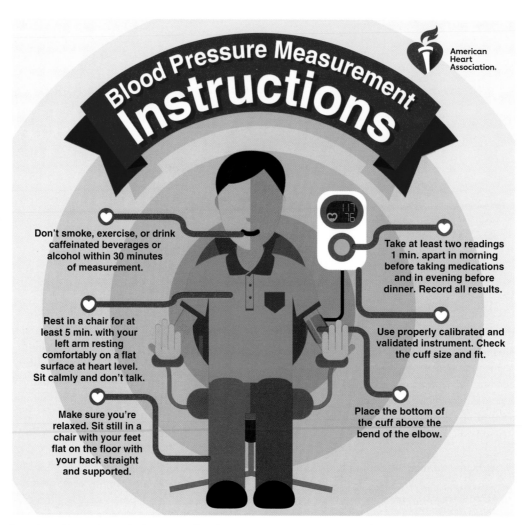

Figure 10.1   The American Heart Association steps for the accurate measurement of blood pressure.

Reprinted with permission https://www.heart.org/-/media/Files/Health-Topics/High-Blood-Pressure/How_to_Measure_Your_Blood_Pressure_Letter_Size.pdf ©2018 American Heart Association, Inc.

The risk for CVD doubles for every increment increase in SBP of 20 mmHg or DBP of 10 mmHg above 115/75 mmHg (5). In approximately 90% of cases, the **etiology** of hypertension is unknown, and it is called essential, **idiopathic**, or primary hypertension. **Systemic** hypertension with a known cause is referred to as secondary or inessential hypertension. Systemic hypertension primarily involves disorders and diseases of the renal, endocrine, or nervous systems, such as Cushing's syndrome and Guillain-Barre syndrome. Other causes of secondary hypertension are obstructive sleep apnea, tumors (e.g., pheochromocytoma), hypothyroidism, and medication side effects.

Although essential and secondary hypertension are the major classifications, several other descriptive terms are used to define various types of the condition. Isolated systolic hypertension is defined as SBP of ≥140 mmHg and DBP <90 mmHg. Patients with white-coat hypertension have normal BP outside of a physician's office (or other clinical setting) but have hypertension in the physician's office. Masked hypertension is defined as having normal BP in the physician's office (or other clinical setting) but having hypertension outside of the physician's office. Sustained hypertension is defined as having hypertension in both the clinic and outside the physician's office. Pulmonary hypertension is characterized by elevated pulmonary arterial pressure accompanied by dyspnea, fatigue, syncope, or substernal chest pain. Resistant hypertension is the failure to achieve goal BP in patients who are adhering to maximum doses of an appropriate three-drug regimen that includes a diuretic or when BP control is achieved with ≥four medications (4). Resistant hypertension may be caused by improper BP assessments, volume overload (i.e., fluid retention, excess sodium intake), other drugs (i.e., side effect of a prescription medication), antihypertensive medication noncompliance, obesity, excessive alcohol intake, or other identifiable causes of hypertension (i.e., secondary hypertension) (4). Finally, malignant hypertension is defined by markedly elevated BP levels (i.e., SBP >200 mmHg or DBP >140 mmHg) due to **papilledema**, a condition of optic nerve swelling that is secondary to elevated intracranial pressure (10).

Normal Blood Pressure

- <120 mmHg (systolic) AND <80 mmHg (diastolic)
- Recommendations: Healthy lifestyle choices and yearly checks.

Elevated Blood Pressure

- 120-129 mmHg (systolic) AND <80 mmHg (diastolic)
- Recommendations: Healthy lifestyle changes; reassess in 3-6 months.

High Blood Pressure/Stage 1

- 130-139 mmHg (systolic) OR 80-89 mmHg (diastolic)
- Recommendations: 10 yr heart disease and stroke risk assessment. If less than 10% risk, lifestyle changes and reassess in 3 to 6 mo. If higher, lifestyle changes and medication with monthly follow-ups until BP controlled.

High Blood Pressure/Stage 2

- ≥140 mmHg (systolic) OR ≥90 mmHg (diastolic)
- Recommendations: Lifestyle changes and two different classes of medicine, with monthly follow-ups until BP is controlled.

Figure 10.2 The American College of Cardiology and American Heart Association adult blood pressure classification scheme and management guidelines. Individual recommendations need to come from your doctor.

Reprinted by permission from P.K. Whelton et al., "2017 ACC/AHA/AAPA/ABC/ACPM/AGS/APhA/ASH/ASPC/NMA/PCNA Guideline for the Prevention, Detection, Evaluation, and Management of High Blood Pressure in Adults: A Report of the American College of Cardiology/American Heart Association Task Force on Clinical Practice Guidelines American Heart Association's journal Hypertension," *Hypertension* 71, no. 6 (2018): e13-3115.

SCOPE

Several definitions of hypertension exist in the literature, resulting in slight variations in hypertension prevalence and control rates (4, 11). Using a clear and consistent definition of hypertension is essential for guiding its diagnosis, treatment, control, and surveillance. According to 2017 ACC/AHA guidelines, hypertension affects 116 million (or 46% of) American adults (1, 4).

In addition, another ~12% of American adults have elevated BP. Among adults with elevated BP, the progression to hypertension is rapid, with about one in four developing the condition within 5 yr (12). The prevalence of hypertension increases substantially with age: 26% among adults 20 to 44 yr, 59% among adults 45 to 64 yr, and 78% among adults ≥65 yr (4). The prevalence is higher in men than women up to 64 yr, but for those ≥65 yr the prevalence is higher in women than men. Black Americans have the highest prevalence of hypertension of all the races in the United States. Compared with Caucasians, African Americans develop hypertension earlier in life and their BP is higher over their lifetimes (4), which translates to a 1.8 times greater rate of fatal stroke and 4.2 times greater rate of end-stage renal disease among Black versus white populations (5).

The awareness, treatment, and control of hypertension have been improving since the 1960s (table 10.1). The National Health and Nutrition Examination Survey (NHANES), spanning from 1999 to 2016, indicates there has been an overall increase in the awareness of hypertension by 11.9%, treatment by 12.5%, and control by 10.5% over these years. Yet these statistics are somewhat misleading considering that hypertension is the most common primary diagnosis in the United States and the leading cause for medication prescriptions among adults >50 yr (13). Of the 46% of adults (116 million) in the United States with hypertension, 19% are recommended to try lifestyle modifications and 81% are recommended to use lifestyle modifications plus antihypertensive medications (14). In those using antihypertensive medications, 71% of cases are not properly controlled; half of these individuals are not taking their recommended medications. The elimination of hypertension could reduce CVD mortality by 30% among men and 38% among women (1). Clearly, there is a critical need to improve the use of lifestyle and pharmacological approaches to prevent, treat, and control hypertension.

PATHOPHYSIOLOGY

The regulation of BP is complicated and multifaceted because it is influenced by renal, hormonal, vascular, and peripheral and central adrenergic systems. According to Ohm's law, mean arterial pressure (MAP) is the product of cardiac output (CO) and total peripheral resistance (TPR): MAP = CO × TPR. Therefore, the pathogenic mechanisms leading to hypertension must increase TPR, increase CO, or increase both. Established hypertension is frequently associated with a normal CO and elevated TPR (15).

Essential hypertension tends to develop gradually, emphasizing the importance of prevention and early diagnosis. The devel-

Table 10.1   Trends in the Awareness, Treatment, and Control of High Blood Pressure in Adults in the United States From 1999 Through 2016

	1999-2004	2005-2010	2011-2016
Awareness	51.8	60.2	63.7
Treatment	40.3	50.7	52.8
Control	14.1	22.4	24.6

Note: Data are based on the National Health and Nutrition Examination Survey; all numbers are percentages.

Adapted from Whelton et al. (4).

opmental stages of essential hypertension are characterized by elevated CO, normal TPR, and enhanced endothelial-dependent dilation. As essential hypertension progresses, the condition becomes characterized by normal CO, elevated TPR, endothelial dysfunction, and left ventricular hypertrophy (16). Factors considered important in the genesis of essential hypertension include family history, genetic predispositions, salt sensitivity, imbalances in the major BP regulatory systems that favor vasoconstriction over vasodilation, aging, and the environment. Other factors that have been implicated in the development of hypertension include obesity, physical inactivity, excessive salt intake and alcohol consumption, and stress. All these lifestyle factors are modifiable and can elicit BP reductions ranging from 2 to 20 mmHg (17).

Although secondary hypertension forms a small percentage of the cases of hypertension, recognizing these types is important because they can often be improved by targeted medical interventions. Most of the secondary forms of hypertension are renal- or **endocrine**-related hypertension. Renal hypertension is usually attributable to a derangement in the renal handling of sodium and fluids, leading to volume expansion or an alteration in renal secretion of vasoactive materials that results in systemic or local changes in arteriolar tone (16). Endocrine-related hypertension is usually attributable to an abnormality of the adrenal glands of the kidney.

Untreated hypertension is the most important risk factor that contributes to and accelerates the pathological processes that lead to premature death from heart disease, stroke, and renal failure (18). In 2011, untreated hypertension was responsible for 54% and 47% of deaths due to stroke and ischemic heart disease, respectively (19). Hypertension damages the **endothelium**, which predisposes the individual to **atherosclerosis** and other **vascular pathologies**. Increased **afterload** on the heart caused by hypertension may lead to left **ventricular hypertrophy** and can ultimately lead to **heart failure (HF)**, or **congestive or chronic heart failure (CHF)**. Hypertension-induced vascular damage can lead to stroke and transient ischemic attacks as well as end-stage renal disease (16).

CLINICAL CONSIDERATIONS

The clinical evaluation of an individual with hypertension should include assessing secondary causes of hypertension, assessing factors that may influence therapy, determining whether target organ damage is present, and identifying other risk factors for CVD. Establishing an accurate pretreatment baseline BP is also vital.

Signs and Symptoms

Hypertension is often referred to as the silent killer because most patients do not have specific symptoms related to their high BP. Although rare, any symptoms associated with or attributed to high BP occur in individuals with abrupt elevations in BP, such as during a **hypertensive crisis** (SBP ≥180 mmHg or DBP ≥120 mmHg), rather than in individuals with high average BP levels (20). Symptoms of a hypertensive crisis can include headache, dizziness, palpitations, **epistaxis**, **hematuria**, and blurring of vision (20). Headache is popularly regarded as a symptom of hypertension, although most studies have not supported an association between headache and BP <179/110 mmHg (21).

History and Physical Examination

A thorough medical history should include questions concerning the individual's risk factors for CVD, as well as symptoms and signs of CVD, heart failure, renal disease, and endocrine disorders. The history should focus on behavioral and lifestyle habits as well as common comorbidities that might predispose an individual to hypertension. Diets high in sodium and saturated fats or low in potassium, calcium, fruits, and vegetables have been identified as contributing factors to the development of elevated BP. Excessive alcohol consumption (consistent use of >2 oz daily), heavy caffeine intake, tobacco use (in any form), or use of illicit drugs (particularly stimulants such as cocaine or amphetamines) have also been identified as contributing factors (17).

Obtaining a detailed medication history is also essential. For example, nonsteroidal anti-inflammatory drugs (NSAIDs), a class of drugs commonly used for a variety of common diseases and health conditions, can increase BP by impairing natriuresis when used regularly in high doses. Oral contraceptives and high doses of decongestants can also raise BP. Some herbal supplements and over-the-counter supplements used for losing weight or boosting energy may contain high levels of caffeine and can elevate BP. Other risk factors for hypertension include African American ethnicity, obesity, high mental stress levels, and a family history of hypertension or CVD (17). Symptoms of episodic headaches, palpitations, diaphoresis, and large variations in BP may indicate pheochromocytoma (i.e., a tumor of the adrenal gland). A history of fatigue and heavy snoring in an individual who is overweight or obese should prompt the consideration of obstructive sleep apnea as a secondary cause of hypertension. A comprehensive medical history will ultimately aid in the treatment of primary hypertension and in the diagnosis and treatment of causes of secondary hypertension.

The physical examination should include evaluation of target organ damage and secondary causes of hypertension. Therefore, it may include a careful fundoscopic examination to evaluate the possibility of hypertensive retinopathy. The thyroid gland should be examined by palpation to determine potential nodules or enlargement, because hypothyroidism has been recognized as a cause of secondary hypertension. Auscultation for carotid, abdominal, or femoral bruits should be performed to rule out vascular disease. An abnormal heart examination may demonstrate a displaced point of the maximal impulse and the presence of a fourth heart sound during auscultation, particularly when there is left ventricular hypertrophy secondary to long-standing uncontrolled hypertension. Last, abdominal striae or a Cushingoid body habitus may suggest an excess of cortisol; these are signs and symptoms indicative of Cushing's syndrome, a potential cause of secondary hypertension.

Diagnostic Testing

A diagnosis of hypertension is based on the average of two or more accurately measured BP readings (see figure 10.1) that have been obtained during each of two or more visits (4). BP is highly variable, and therefore accurate BP assessment following established guidelines is an essential part of the diagnostic evaluation of hypertension (4). Several factors related to the patient, health care provider, and BP-measuring equipment can cause errors in measured BP by as much as ±40 mmHg (22). The conventional method for clinic BP assessment is the auscultatory technique with a trained observer and mercury sphygmomanometer (22). Automated and ambulatory BP monitors can be used to confirm a diagnosis or differentiate among the various forms of hypertension mentioned earlier in this chapter (e.g., masked, white coat). Regardless of the BP-measuring device, frequent calibration of equipment is of the utmost importance for ensuring valid prognostic value of the BP readings (23). Proper training of the observer will limit intertester variability in BP measurements. Ideally, the same patient should be assessed by the same observer and on the same BP-measuring device.

Proper patient positioning and preparation are critical for ensuring BP accuracy. If possible, caffeine, exercise, and smoking should be avoided at least 24 h before the measurement. Prior to the first reading, the patient should be seated quietly for at least 5 min, with the legs uncrossed, the bladder empty, and the back and arm supported, such that the middle of the cuff on the upper arm

is at heart level (figure 10.1). The patient should be instructed to relax as much as possible and not talk during the measurement. An appropriate-sized cuff (i.e., cuff bladder encircling at least 80% of the arm) should be used to ensure accuracy. At the first visit, record BP in both arms. For subsequent readings, use the arm that gives the higher BP reading. Separate repeated BP measurements by 1 to 2 min, taking two or more readings that are averaged and obtained on two or more occasions to determine an individual's resting BP level (4). Consistent discrepancies between arm readings of 10 mmHg or more should raise the possibility of aortic coarctation and prompt measurement of BP in the lower extremities. For research purposes, additional repeated BP measurements may be warranted following the AHA standards, such that BP should be measured three times in each arm, separated by at least 1 min, and averaged (22). Additional readings may need to be taken, if necessary, until three readings agree within 5 mmHg.

A major limitation of the auscultatory method is the subjective nature of the observer's having to hear and appraise the appearance and disappearance of Korotkoff sounds (22). SBP is defined as the point at which the first of two or more Korotkoff sounds is heard (phase 1), and DBP is defined as the point before the disappearance of Korotkoff sounds (phase 5) (22). Although both SBP and DBP are important, in individuals >50 yr, SBP becomes a more important CVD risk factor than DBP (4, 5, 22).

Clinic BP assessment via auscultation is the mainstay method for screening and diagnosing hypertension, but there are several limitations to this method. In addition to observer bias, two other drawbacks of clinic BP assessment are the inability to capture masked hypertension and the tendency for patients to experience white-coat hypertension, resulting in underdiagnosis and overdiagnosis of hypertension, respectively. Ambulatory BP monitoring circumvents the shortcomings of clinic BP assessment by allowing BP to be measured multiple times over a 24 h period and under normal conditions of daily living. A cross-sectional analysis of 14,143 mostly European patients compared clinic-measured to ambulatory-measured BP and found that 23% of patients who had a diagnosis of hypertension by clinic BP did not have hypertension by ambulatory BP (i.e., white-coat hypertension); and 10% of patients were revealed to have masked hypertension as evident by normal clinic BP but hypertension by ambulatory BP (24).

Ambulatory BP monitoring is a superior predictor of CVD morbidity and mortality than BP measured in the clinic. It also provides additional predictive information that is beyond the capabilities of clinic BP, such as nighttime BP levels, which can be used to calculate nocturnal dipper status, an indicator of BP reactivity (25). For these reasons, ambulatory BP monitoring is considered the gold standard of BP assessment, yet it is not universally integrated into patient care (11). Common barriers to ambulatory BP monitoring include patient burden, lack of clinician knowledge, and lack of resources to apply the monitor and interpret the readings. Furthermore, ambulatory BP monitors are costly and not universally covered by health insurance. It should be noted, however, that cost-effectiveness analyses have shown ambulatory BP monitoring to reduce the cost of care for hypertension by 10% (26).

Home BP monitoring with an automated BP device is now recommended as a user-friendly, low-cost, and accessible method to substantiate a diagnosis of hypertension made in the clinic. Depending on the frequency and standardization of BP measurements, home BP monitoring can also differentiate among sustained, white-coat, and masked hypertension. Home BP monitoring is unique in that it can be used to confirm (or refute) an initial diagnosis of hypertension, but it can also be utilized as a long-term adjunct to traditional outpatient care. A randomized controlled trial of 348 patients with hypertension examined BP over the course of 6 mo in patients assigned to usual care in the clinic (i.e., education and follow-up with primary care physician) versus home BP monitoring (27). Patients in the home BP monitoring group were asked to assess their home BP at least three times per week and electronically share their BP values with their physician using the AHA Heart360 web application (www.heart360.org) (27). After 6 mo, mean BP was significantly lower in both groups; however, patients in the home BP monitoring group experienced a greater reduction in SBP (12.4 mmHg) and DBP (5.7 mmHg) and higher rates of BP control (54.1% vs. 35.4%) when compared with usual care (*p* <.01) (27).

Because of the inherent differences in the specificity and sensitivity of various BP devices, the diagnostic threshold to diagnose hypertension differs slightly among clinic, home, and ambulatory BP as shown in table 10.2 (4).

The initial laboratory tests in the assessment of newly diagnosed hypertension should include urinalysis, **hematocrit (Hct)**, blood chemistry (including sodium, potassium, creatinine, and the lipid–lipoprotein profile), thyroid function, a random cortisol level, and serum metanephrines. A baseline electrocardiogram is also recommended (28). These routine tests will help determine the presence of target organ damage and other CVD risk factors as well as guide certain therapeutic decisions. The electrocardiogram is an important test because it may reveal the presence of left ventricular or atrial hypertrophy, arrhythmias, and baseline ST-segment changes. Other tests that can be of value, depending on indication, include **creatinine clearance**, microscopic urinalysis, chest X-ray, echocardiogram, hemoglobin A1c, fasting plasma glucose, serum calcium, phosphate, and uric acid.

Exercise Testing

Although small, the risk for acute exercise-related cardiovascular events is highest among sedentary adults with known or underlying CVD who perform unaccustomed vigorous-intensity exercise (29). Therefore, the goal of exercise preparticipation health screening is to assess and mitigate this risk but not to present unnecessary obstacles or excessive burden to begin or maintain an exercise program on the individual or clinician. Consequently, the American College of Sports Medicine (ACSM)'s exercise preparticipation health screening recommendations emphasize the Surgeon General's message that regular physical activity is important for all individuals, particularly those who stand to benefit the most, such as individuals with hypertension (30).

Recommendations for exercise testing in individuals with hypertension vary depending on their BP level; signs and symptoms or presence of cardiovascular, metabolic, or renal disease; and the health care provider's clinical discretion (29, 31, 32). Clinicians are encouraged to evaluate the need for a medical examination, exercise stress test, or diagnostic imaging using their own clinical judgment and on an individualized basis. For example, a diagnosis of hypertension alone is not an indication for exercise testing; however, determining the BP response to exercise may be valuable in guiding an exercise prescription (33). Based on guidelines from the ACSM and other expert professional groups, the following would appear to be prudent recommendations (29, 33, 34).

Table 10.2 Corresponding SBP/DBP Values for BP Measured in the Clinic, Home, and Under Ambulatory Conditions During the Day and Night and Over 24 H

Clinic	HBPM	Daytime ABPM	Nighttime ABPM	24 h ABPM
120/80	120/80	120/80	100/65	115/75
130/80	130/80	130/80	110/65	125/75
140/90	135/85	135/85	120/70	130/80
160/100	145/90	145/90	140/85	145/90

SBP = systolic blood pressure; DBP = diastolic blood pressure; BP = blood pressure; HBPM = home blood pressure monitoring; ABPM = ambulatory blood pressure monitoring.

Adapted from Whelton et al. (4).

- Individuals with controlled hypertension as defined by JNC 7 (i.e., SBP <140 and DBP <90 mmHg) or prehypertension (i.e., SBP of 120-139 mmHg or DBP of 80-89 mmHg) who are asymptomatic and have no cardiovascular, metabolic, or renal disease do not require medical evaluation or exercise testing prior to beginning an exercise program of any intensity, regardless of their physical activity status. Of note, individuals who have been performing planned, structured physical activity for at least 30 min at moderate intensity on ≥3 d/wk for at least the last 3 mo are considered physically active. Otherwise, they are considered physically inactive (29).

- Physically inactive individuals with controlled hypertension or prehypertension who are asymptomatic but have known cardiovascular, metabolic, or renal disease are recommended to have a medical evaluation and consult with their health care provider to determine if an exercise test with electrocardiogram monitoring is medically indicated before beginning an exercise program of any intensity.

- Physically inactive individuals with controlled hypertension or prehypertension who are symptomatic (i.e., one or more major signs or symptoms suggestive of cardiovascular, renal, or metabolic disease), regardless of disease status, are recommended to have a medical evaluation and consult with their health care provider to determine if an exercise test with electrocardiogram monitoring is medically indicated before beginning an exercise program of any intensity.

- Physically active individuals with controlled hypertension or prehypertension who are asymptomatic but have known cardiovascular, metabolic, or renal disease are recommended to have a medical evaluation and consult with their health care provider to determine if an exercise test with electrocardiogram monitoring is medically indicated for engaging in vigorous-intensity exercise. Until the medical examination takes place, most of these individuals may participate in light- to moderate-intensity exercise programs.

- Physically active individuals with hypertension or prehypertension who are symptomatic (i.e., one or more major signs or symptoms suggestive of cardiovascular, renal, or metabolic disease), regardless of disease status, should discontinue exercise and are recommended to have a medical evaluation and consult with their health care provider to determine if an exercise test with electrocardiogram monitoring is medically indicated prior to resuming exercise.

Special considerations include the following:

- Individuals with stage 1 hypertension as defined by JNC 7 (i.e., SBP ≥140-<160 mmHg or DBP ≥90-<100 mmHg) are recommended to have a medical evaluation and consult with their health care provider regarding adequate BP management and to determine if an exercise test is needed prior to starting or continuing their exercise program.

- Individuals with stage 2 hypertension as defined by JNC 7 (i.e., SBP ≥160 mmHg or DBP ≥100 mmHg) should not engage in any exercise; they are recommended to have a medical evaluation and consult with their health care provider regarding adequate BP management and to determine if an exercise test with electrocardiogram monitoring is medically indicated prior to starting or continuing their exercise program.

Contraindications

The ACC, AHA, and ACSM guidelines on exercise testing indicate that severe arterial hypertension, defined as SBP >200 mmHg or DBP >110 mmHg at rest, is a relative contraindication to exercise testing (34, 35).

Additional Recommendations and Anticipated Responses

Standard exercise testing methods and protocols may be used for individuals with hypertension (35). Before an individual with hypertension undergoes graded exercise testing, a detailed health history and baseline BP in both the supine and standing positions should be obtained. Certain medications, especially β-blockers, may affect BP and heart rate (HR) at rest and during exercise. Thiazide diuretics may cause hypokalemia, electrolyte imbalances, cardiac dysrhythmias, or a false-positive test. See table 10.3 for the commonly used antihypertensive medications. The reader is referred to the ACC/AHA guidelines for more detailed information (4). When exercise testing is conducted for diagnostic purposes, antihypertensive medications may be withheld before testing with physician approval. When exercise testing is performed for the purpose of designing an exercise prescription, patients should take their usual antihypertensive medications.

Abnormal BP Response

BP is the product of CO and TPR. During exercise, TPR normally decreases; however, the increase in CO more than compensates for the decrease in TPR, resulting in an increase in SBP. DBP usually remains the same or may decrease slightly because of decreased TPR. Individuals with hypertension may experience an increase in DBP both during and after exercise and are often unable to reduce TPR to the same extent as individuals with normal BP. Impaired endothelial function in the early stages and a reduced lumen-to-wall thickness as hypertension progresses may be responsible for the increased TPR seen during exercise among individuals with hypertension (5).

Indications for Terminating a Graded Exercise Test

A significant decrease in SBP of >10 mmHg from baseline SBP despite an increase in workload is an indication for terminating an exercise test. An excessive increase in BP, defined as a peak SBP >250 mmHg or DBP >115 mmHg, is also an indication for terminating an exercise test (35).

Table 10.3 Pharmacology of Most Commonly Used Antihypertensive Medications

Medication class	Primary effects and treatment population	Exercise hemodynamic, ECG, and capacity effects	Important considerations
Thiazide-type diuretics	Decrease resting BP; first-line antihypertensive therapy in the general population*	Decrease exercise BP; may increase exercise capacity in patients with CHF	May result in serum potassium depletion and thereby accentuate the risk for exercise-induced arrhythmias; may adversely affect thermoregulatory function
Angiotensin-converting enzyme (ACE) inhibitors and angiotensin II receptor blockers (ARBs)	Decrease resting BP; first-line antihypertensive therapy in the non-Black general population or patients with stable CKD with or without diabetes of all races	Decrease exercise BP; may increase exercise capacity in patients with CHF	ACE: cough, angioedema (most common in the Black population), hyperkalemia, renal impairment (pertinent during volume depletion with exercise) ARB: hyperkalemia, renal impairment, thrombocytopenia
Calcium channel blockers	Decrease resting BP; may decrease resting HR; first-line antihypertensive therapy in the general population	Decrease exercise BP; may decrease exercise HR	May predispose to postexertion hypotension (an adequate cool-down may be especially important); can cause lower extremity edema
β-blockers	Decrease resting BP and HR; typically reserved for post-MI or CHF patients	Decrease exercise BP and HR; decrease ischemia during exercise; may decrease exercise aerobic capacity acutely but increase aerobic exercise capacity with chronic administration particularly in individuals with CHF	May blunt exercise training–induced lowering of triglycerides and increase in HDL cholesterol; may adversely affect thermoregulatory function; may increase predisposition to hypoglycemia in patients taking insulin or insulin secretagogues
Other vasodilators, including nitrates and α-blockers	Decrease resting BP; may increase resting HR (nitrates)	Decrease exercise BP; may increase exercise HR (nitrates); decrease exercise ischemia; may increase exercise capacity in patients with angina and CHF (nitrates)	May predispose to postexertion hypotension (an adequate cool-down may be especially important)

*No diabetes or CKD present.

ECG = electrocardiogram; BP = blood pressure; CHF = congestive heart failure; CKD = chronic kidney disease; HR = heart rate; MI = myocardial infarction; HDL = high-density lipoprotein.

Adapted from Whelton et al. (4).

Predictive Value of the BP Response to Exercise

An exaggerated BP response to a maximal or a submaximal graded exercise test in individuals with normal BP has prognostic value for future hypertension and CVD (36, 38). A normal BP response to endurance exercise is a progressive increase in SBP, typically 8 to 12 mmHg per metabolic equivalent of task (MET), whereas DBP usually falls slightly or remains unchanged. Therefore, the expected peak BP on an exercise test is 180 to 210 mmHg for SBP and 60 to 85 mmHg for DBP. Although the BP thresholds to classify an exaggerated BP response to maximal exercise can vary, SBP/DBP values above these upper thresholds provide guidance for determining whether the BP response to maximal exercise is exaggerated when considering the individual's baseline BP levels. Table 10.4 provides a summary of exercise testing considerations for adults with hypertension.

Treatment

The overarching goal in the treatment, control, and management of hypertension is to reduce the risk of CVD morbidity and mortality. Target BP levels may be achieved through lifestyle modification alone or in combination with pharmacological treatment. The accepted BP goal for most individuals with hypertension of <140/90 mmHg (5, 11, 17, 39-41) has been lowered to <130/80

Table 10.4 Summary of Exercise Testing Considerations for Adults With Hypertension

Test type	Mode	Protocol specifics	Clinical measures	Clinical implications	Special considerations
Cardiorespiratory	Aerobic	Cycle (ramp protocol 17 W/min; staged protocol 25-50 W/3 min stage); treadmill (2-3 METs/3 min stage)	12-lead ECG; HR; SBP and DBP; rate pressure product; RPE; ventilatory expired gas responses	Evaluate physical signs and symptoms,[a] abnormal BP or HR responses, arrhythmias, and myocardial ischemia.	Use standard end points for test termination,[b] including SBP >250 mmHg or DBP >115 mmHg; cardioactive medications such as vasodilators or β-blockers should be taken at usual time relative to exercise session.
Resistance	Free weights or resistance machines	Determine 1RM or MVC; 1RM may be estimated from a higher RM (e.g., 5RM)	1RM; MVC	Evaluate physical signs and symptoms and abnormal BP or HR responses; if ECG is monitored, evaluate for arrhythmias and myocardial ischemia.	Observe for exaggerated pressor response (i.e., SBP >250 mmHg or DBP >115 mmHg); avoid Valsalva maneuver.
Flexibility	As for otherwise healthy persons	As for otherwise healthy persons	As for otherwise healthy persons	As for otherwise healthy persons	As for otherwise healthy persons

[a]Degree of chest pain, burning, discomfort, dyspnea, light-headedness, leg discomfort or pain.

[b]Failure of HR to increase with further increases in intensity; a plateau in $\dot{V}O_2$ with increased workload; respiratory exchange ratio ≥1.10; RPE >17 on the 6-20 scale.

ECG = electrocardiogram; HR = heart rate; SBP = systolic blood pressure; DBP = diastolic blood pressure; RPE = rating of perceived exertion; BP = blood pressure; RM = repetition max; MVC = maximal voluntary contraction.

Based on Fletcher et al. (33); Pescatello et al. (9).

mmHg in the ACC/AHA guidelines. The BP goal may vary based on the presence of other comorbid conditions, overall CVD risk, 10 yr ASCVD risk, age, race, and individual clinician approaches to BP control. The 2017 ACC/AHA guidelines emphasize that published recommendations to achieve a BP goal should not be a substitute for good clinical judgment, and treatment should be individualized to the patient, with the overarching goal of reducing overall CVD risk. Nonetheless, there is a lack of consensus and considerable controversy among professional organizations regarding the threshold for which to treat and control hypertension.

The 2017 ACC/AHA guidelines' departure from the traditional and long-standing treatment recommendations was in part owing to the results of the Systolic Blood Pressure Intervention (SPRINT) trial (42). SPRINT was a randomized trial of patients ($N = 9,361$) ≥50 yr with prehypertension and high CVD risk. The trial assigned patients to a standard treatment group (target SBP of <140 mmHg) or an intensive treatment group (target SBP of <120 mmHg). Among individuals in the intensive treatment group, there was a substantial reduction in CVD-related events by 25% and death by 43% after achieving an overall SBP of 122 mmHg in ~3.3 yr (43). Taken altogether, the findings from SPRINT and other recent reports have challenged the previous BP threshold

for treatment by providing evidence to support a target SBP ~15% lower than the recommended level of <140/90 mmHg. Although preliminary, the landmark SPRINT study has spearheaded an international call to action to reappraise BP treatment thresholds and may continue to inform evidence-based guidelines (39).

Adoption of a healthy lifestyle is essential for the primary prevention, treatment, and control of hypertension. Lifestyle modifications are advocated as a first line of defense to treat and control high BP among individuals with hypertension. When lifestyle modifications are executed properly, it may have a positive effect on additional coexisting CVD risk factors, translating to an even greater reduction in overall CVD risk. The relationship between BP and CVD risk is linear, continuous, and consistent starting at 115/75 mmHg. Because patients with elevated BP are particularly vulnerable to developing future hypertension, early and aggressive lifestyle intervention is critical to prevent or delay the rapid, progressive rise in BP (4). Table 10.5 displays recommended lifestyle modifications to prevent, treat, and control hypertension.

Individuals who are overweight or obese are 2 to 2.5 times more likely to have hypertension than individuals who are normal weight, and it has been estimated that 78% of the risk for developing hypertension in men and 65% of the risk in women is attributed to

Table 10.5 Lifestyle Modifications to Prevent, Treat, and Control Hypertension

Nonpharmacologic intervention	Recommendation
Healthy diet: Use the DASH dietary pattern.	Consume a diet rich in fruits, vegetables, whole grains, and low-fat dairy products, with reduced saturated fat and total fat.
Weight loss: Focus on losing excess weight and body fat.	Ideal body weight is the best goal for most overweight adults. Expect about 1 mmHg reduction in blood pressure for every 1 kg reduction in body weight.
Sodium: Reduce intake of dietary sodium.	Optimal intake is <1,500 mg/d, but aim for at least 1,000 mg/d reduction in most adults.
Potassium: Increase intake of dietary potassium.	Consume 3,500-5,000 mg/d, preferably through a diet rich in potassium.
Physical activity: Add aerobic exercises to weekly routine.	Exercise for 90-150 min/wk at 56%-75% HRR.
Physical activity: Add dynamic resistance training to weekly routine.	Exercise for 90-150 min/wk at 50%-80% 1RM. Perform 6 exercises, 3 sets/exercise, 10 reps/set.
Physical activity: Add isometric resistance training to weekly routine.	Perform handgrip exercises (4 × 2 min, with 1 min of rest between sessions) at 30%-40% maximum voluntary contraction. Do 3 times per week for 8-10 wk.
Stress: Reduce stress with mind–body exercise and mindfulness.	Although more research is needed, mind–body exercise and meditation appear to lower blood pressure.
Alcohol: Reduce consumption of alcohol.	Consume no more than two drinks for men and one drink for women per day.

Note: For overall cardiovascular risk reduction, stop smoking.

DASH = Dietary Approaches to Stop Hypertension; HRR = heart rate reserve; 1RM = one-repetition maximum.

Based on Unger et al. (17), Whelton et al. (4), and Wu and Pescatello (110).

excess body mass (44). Weight loss of as little as 1 kg corresponds to reductions in SBP of 1.2 mmHg and DBP of 1.0 mmHg; these reductions are dose dependent, such that greater BP reductions are experienced with greater weight loss (5, 45, 46). Weight reduction can also enhance the effects of antihypertensive medications and positively affect other CVD risk factors or conditions such as diabetes mellitus and dyslipidemia (47). For these reasons, the ACSM emphasizes the importance of increased caloric expenditure coupled with reduced caloric intake in individuals with hypertension who are overweight or obese to facilitate weight reduction and the resultant reductions in BP (35).

Dietary modifications such as the Dietary Approaches to Stop Hypertension (DASH) diet (i.e., a plant-focused diet rich in fruits, vegetables, nuts, whole grains, low-fat and nonfat dairy, lean meats, fish, poultry, and heart-healthy fats) and sodium restriction can reduce BP by 2 to 14 mmHg (5). High dietary salt causes hypertension in about 30% of those with high BP (48). Individuals with hypertension may be classified as salt sensitive or salt resistant, based on the absolute changes in BP that originate from dietary salt intake (49, 50). African Americans, older adults, and individuals with hypertension or diabetes mellitus are more sensitive to changes in dietary sodium than are others in the general population. The 2017 ACC/AHA guidelines recommend <1,500 mg/d as an optimal goal but suggest aiming for at least a 1,000 mg/d reduction in sodium intake for most adults.

Excessive alcohol consumption is a risk factor for hypertension, and reducing alcohol intake can lower BP by 2 to 4 mmHg (5, 51). It is recommended that men limit their daily intake to no more than 1 oz (30 mL) of ethanol and women and lighter-weight individuals limit their daily intake to no more than 0.5 oz (15 mL) (5). These quantities are equivalent to two drinks per day in men and one drink per day in women. One drink is defined as 12 oz (360 mL) of beer, 5 oz (150 mL) of wine, or 1.5 oz (45 mL) of 80-proof liquor.

Regular cardiorespiratory endurance exercise can reduce BP on average by 5 to 7 mmHg among individuals with hypertension. This rivals the magnitude of the BP reductions obtained from first-line antihypertensive medications and lowers CVD risk by 20% to 30% (5, 52, 53). Exercising as little as 1 d/wk is as effective as pharmacotherapy (or even more so) for reducing all-cause mortality among those with hypertension (54). Naci and colleagues performed a comparative effectiveness network meta-analysis and found exercise to be as effective as common antihypertensive medications in terms of its mortality benefits in the secondary prevention of coronary heart disease, rehabilitation after stroke, treatment of heart failure, and prevention of diabetes mellitus (55). This investigative team performed another comparative effectiveness network meta-analysis on the BP effects of exercise alone and medication alone in the treatment of hypertension (56). They found that exercise alone reduced SBP by ~9 mmHg among adults with hypertension, with no detectable difference from the

SBP reductions resulted from medication alone. These findings have been confirmed in a more recent comparative effectiveness network meta-analysis by Noone et al. (57). The optimal exercise prescription for the primary prevention and treatment of hypertension is discussed in detail later in this chapter.

When lifestyle interventions are not effective in achieving treatment BP goals, the decision to initiate antihypertensive therapy should be guided by the presence of disease (i.e., CVD, diabetes mellitus, or CKD) and 10 yr risk for heart disease and stroke as assessed by the ASCVD risk calculator according to the 2017 ACC/AHA treatment algorithm in figure 10.3 (4).

Specifically, the ACC/AHA endorse the following recommendations for the management of hypertension in adults:

- In adults with SBP of 130 to 139 mmHg or DBP of 80 to 89 mmHg (i.e., stage 1 hypertension), assess the 10 yr risk for ASCVD. If risk is <10%, start with healthy lifestyle recommendations and reassess in 3 to 6 mo.

- If risk is ≥10% or the patient has known clinical CVD, diabetes mellitus, or CKD, recommend lifestyle changes and BP-lowering medication and reassess in 1 mo for effective-

ness of medication therapy. If the BP goal is met after 1 mo, reassess in 3 to 6 mo. If the BP goal is not met after 1 mo, consider different medication or titration. Continue monthly follow-up until control is achieved.

- In adults with systolic SBP ≥140 mmHg or DBP ≥90 mmHg (i.e., stage 2 hypertension), recommend healthy lifestyle changes and BP-lowering medication (two medications of different classes).

- If the BP goal is met after 1 mo, reassess in 3 to 6 mo. If the BP goal is not met after 1 mo, consider different medication or titration. Continue monthly follow-up until control is achieved.

Most individuals who require treatment for hypertension require two or more antihypertensive medications to achieve goal BP. If goal BP is not achieved within 1 mo treatment, it is recommended to increase the dose of the initial drug or add a second class of drug. The clinician should continue to adjust the treatment regimen until goal BP is reached. If a goal BP cannot be achieved with two drugs, a third drug of a different medication class, such

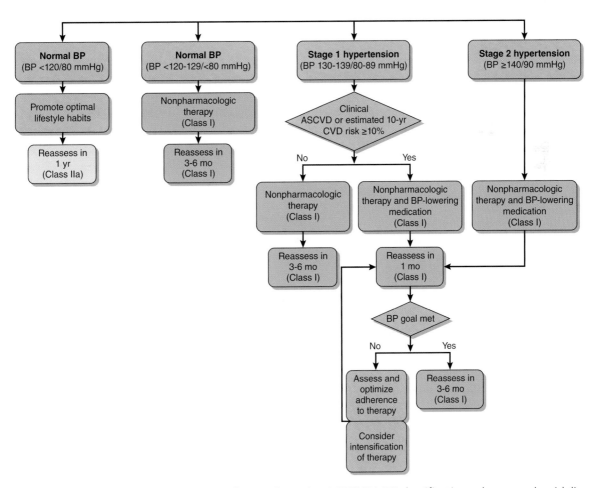

Figure 10.3 Treatment and follow-up recommendations from the ACC/AHA BP classification scheme and guidelines.

as a β-blocker or aldosterone antagonist, should be added. Individuals who cannot achieve a goal BP despite the concurrent use of three antihypertensive agents of different classes, including a diuretic, may require a hypertension specialist because resistant hypertension is almost always multifactorial in etiology. Successful treatment of resistant hypertension requires appropriate lifestyle modifications, diagnosis, and appropriate treatment of secondary causes of hypertension and use of effective multidrug regimens (58). In cases of treatment-resistant hypertension, the importance of drug adherence should be assessed and discussed with patients, because a portion of treatment-resistant hypertension cases may be attributed to suboptimal adherence to antihypertensive drug therapy (59).

After initiation of drug therapy, most patients should return for follow-up and adjustment of medications at approximately monthly intervals until goal BP is reached. Follow-up visits should be at 3 to 6 mo intervals after goal BP is achieved, with more frequent follow-up among patients with stage 2 hypertension or with complicated comorbid conditions. Serum potassium and creatinine should be monitored at least once or twice per year (5). Many antihypertensive medications affect the physiological response to exercise and should be taken into consideration during exercise testing and training (table 10.3). Furthermore, lifestyle modifications such as exercise should be encouraged and continued during antihypertensive therapy (4, 123). Understanding potential interactions between lifestyle interventions and antihypertensive therapy is important to guide clinical care. Clinicians must weigh the benefit of antihypertensive medication alone, and in combination with lifestyle modifications, to tailor an optimal treatment plan.

Practical application 10.1 provides advice on delivering patient-centered care among individuals with hypertension. See practical application 10.2 for discussion of other comorbidities typically observed in patients with hypertension, such as dyslipidemia and metabolic syndrome.

EXERCISE PRESCRIPTION

Participation in regular exercise is a key modifiable determinant of hypertension and is recognized as a cornerstone therapy for the primary prevention, treatment, and control of high BP (1, 4, 9, 17, 41, 63-65). Large-scale prospective studies consistently demonstrate that physical activity and cardiorespiratory fitness are inversely associated with the development of hypertension (66-68). Meta-analyses of randomized controlled intervention trials conclude that regular cardiorespiratory endurance and resistance exercise alone or combined lowers resting BP on average by 5 to 8 mmHg; however, in certain subpopulations the BP reductions are even greater (69-72). For these reasons professional organizations throughout the world universally endorse exercise for the primary prevention and treatment of hypertension (table 10.5) (1, 4, 9, 17, 41, 63, 64). Of note, the recommended **F**requency (how often?), **I**ntensity (how hard?), **T**ime (how long?), **T**ype (what kind?), **V**olume (what is the total amount?) and **P**rogression (how

to advance?) (or *FITT-VP* principle) of exercise prescription vary slightly across the professional guidelines. However, the consensus from these professional recommendations is that adults with elevated BP to established hypertension should participate in moderate-intensity cardiorespiratory endurance and dynamic resistance exercise on most (preferably all) days of the week for 20 to 30 min per session (or more) to total 150 min/wk (or more) of exercise (9, 73, 74).

The strength of the evidence on which these professional recommendations are based is influenced by their methodological quality (30, 73, 75, 76). Methodological limitations of this literature include small sample sizes; admixture of study populations with normal and high BP; not accounting for major confounders to the BP response to exercise, such as timing of the last bout of exercise and detraining effects; and lack of integration of standardized protocols for the assessment of BP and exercise intervention, among others. Because of the law of initial values (77), admixture of populations with normal and high BP underestimates the effectiveness of exercise as antihypertensive lifestyle therapy because the subjects who experience the greatest BP reductions from exercise are those with elevated BP (63, 73, 75, 78). Furthermore, large randomized clinical trials that examine and account for both the acute (immediate, short term, or postexercise hypotension [PEH]) and chronic (long term, or training) BP-lowering effects of exercise among diverse populations are needed before professional organizations can definitively determine the optimal FITT-VP exercise prescription for individuals with hypertension. New and emerging research is highlighted later in this chapter; such studies may expand on the FITT-VP exercise prescription for individuals with hypertension, providing more exercise options.

Presently, the ACSM recommends the following FITT-VP exercise prescription for individuals with hypertension (table 10.6) (63, 79):

- *Frequency:* On most, preferably all, days of the week consisting of two or three sessions (or more) per week of cardiorespiratory endurance, dynamic resistance, neuromotor, and flexibility exercise.

- *Intensity:* Emphasize moderate intensity, which is 40% to <60% of maximum oxygen uptake reserve ($\dot{V}O_2R$), or 12 to 13 on a scale of 6 (no exertion) to 20 (maximal exertion), or an intensity that causes noticeable increases in HR and breathing for cardiorespiratory endurance exercise; 60% to 80% of one-repetition maximum (1RM) for dynamic resistance exercise; and stretching to the point of feeling tightness or slight discomfort for flexibility exercise. An effective intensity for neuromotor exercise has not been determined; however, low to moderate intensity is consistent with the other recommended modalities and seems prudent for this patient population.

- *Time:* For cardiorespiratory endurance exercise, ≥30 min of continuous or accumulated aerobic activity. For dynamic resistance exercise, ≥20 min per session, with rest days interspersed depending on the muscle group being exercised.

Practical Application 10.1

CLIENT–CLINICIAN INTERACTION

How satisfied a patient is with the clinician managing their hypertension influences their adherence to treatment and consequently hypertension control (5). A positive, long-term patient–clinician partnership that empowers the patient to actively participate in the decision-making process is the foundation for successful, enduring hypertension management. Patient education regarding the risks of high BP and the importance of achieving BP goals may be tailored according to the patient's level of understanding and beliefs. Hypertension is the single most powerful risk factor in the development of CVD and the leading cause of death in the United States (1, 4). Suboptimal BP control contributes to 62% and 49% of cerebrovascular disease and ischemic heart disease, respectively (5). Therefore, it is imperative that clinicians be committed to a holistic management approach and provide counsel on lifestyle factors that influence *overall* cardiovascular health such as nutrition, stress, healthy weight, smoking, alcohol, salt intake, and physical activity (5, 60). Lifestyle therapies are an integral part of hypertension management, with regular cardiorespiratory endurance exercise recognized as a "polypill" that mutually supports other lifestyle modifications and positively improves many aspects of overall health. Although exercise-related cardiac complications are rare, the risk is higher in physically inactive individuals who attempt to engage in vigorous activities they are unaccustomed to (29). Therefore, clinicians should educate patients about the importance of proper progression and the warning symptoms and signs of an impending cardiac event.

Drug therapy is often needed to optimize hypertension management and to facilitate CVD risk reduction. Individuals with hypertension are often treated with one or more medications. The clinician should educate the patient about the effect, if any, of specific medications on exercise performance and training (table 10.3). The overarching goal is to adequately control BP without reducing exercise capacity, producing medication-related or other adverse effects, or using banned substances. The clinician should also empha-

size to the patient the importance of taking medications as prescribed and not discontinuing drug therapy without notifying their primary care physician. If the clinician believes the patient may be experiencing medication-related adverse effects, they should refer the patient to their personal physician. Referral to a hypertension specialist or cardiologist is appropriate for any individual with stage 2 hypertension. These patients may require a workup for secondary causes of hypertension, as previously discussed. Referral to a hypertension specialist or cardiologist is also advisable if BP remains difficult to control after 6 mo of standard treatment strategies. For athletes, referral should be considered earlier if there is difficulty with BP control because of limited drug choices for athletes, unacceptable side effects, or other comorbidities. In fact, the ACC/AHA and European Association of Preventive Cardiology recommend that athletes with hypertension who wish to engage in training for competitive sports have a clinical assessment, including BP, before participating (61, 62).

In specialty clinics, patients often present with pseudo-resistance, which is characterized by poorly controlled BP that appears resistant but is due to other factors. Common causes of pseudo-resistance include incorrect BP measurement techniques (e.g., incorrect cuff size), clinical inertia (i.e., insufficient treatment), and inadequate adherence to lifestyle and medication therapies. Clinicians are likely to interact with patients on many occasions throughout the course of a year and therefore will have an opportunity to frequently measure a patient's BP. It is important to document the cuff size used at each encounter. An undersized cuff creates a falsely increased BP and results in unnecessary referrals to physicians for evaluation and consideration of antihypertensive therapy. Self-monitoring of BP with a home BP monitor is encouraged; however, clinicians should be cognizant of frequent disparities between in-office and out-of-office (self-measured) BP. Frequent office visits involving family members or significant others in these encounters enforces accountability and improves adherence in difficult cases (4).

For flexibility exercise, ≤10 min per session. For neuromotor exercise, ≥20 min per session.

- *Type:* For cardiorespiratory endurance exercise, emphasize prolonged rhythmic activities using large muscle groups such as walking, cycling, or swimming. For dynamic resistance exercise, machines, free weights, resistance bands, and functional body weight exercise. For flexibility exercise, static stretching, dynamic stretching, or proprioceptive neu-

romuscular facilitation. For neuromotor exercise, combinations of motor skill, functional body weight, and flexibility exercises such as yoga, Pilates, and tai chi.

- *Volume:* To total 90 to 150 min/wk or more of continuous or accumulated exercise of any duration that emphasizes cardiorespiratory endurance and dynamic resistance exercise alone or added neuromotor and flexibility exercise depending on personal preference.

Table 10.6   ACSM FITT Exercise Prescription Recommendations for Hypertension

	Cardiorespiratory[a]	Resistance	Neuromotor[c]	Flexibility	New ACSM FITT exercise recommendations
Frequency	2 or 3 sessions per week (or more)	2 or 3 sessions per week (or more)	2 or 3 sessions per week (or more)	2 or 3 sessions per week (or more), with daily being most effective	On most, preferably all, days of the week[d]
Intensity	Moderate (i.e., 40%- 59% $\dot{V}O_2R$ or HRR; RPE of 12-13 on a 6-20 scale) to vigorous (i.e., 60%-80% $\dot{V}O_2R$ or HRR; RPE of 14-16 on a 6-20 scale)[b]	Moderate (i.e., 60%-70% 1RM); may progress to 80% 1RM; older adults and novice exercisers: begin with 40%-50% 1RM	Low to moderate	Stretch to the point of feeling tightness or slight discomfort	Low, moderate, or vigorous, with an emphasis on moderate
Time	≥30 min per session of continuous or accumulated exercise of any duration	2-4 sets of 8-12 repetitions of 8-10 resistance exercises for each of the major muscle groups per session to total ≥20 min per session, with rest days interspersed depending on the muscle groups being exercised	20-30 min per session (or more)	Hold static stretch for 10-30 s for 2-4 repetitions of each exercise, targeting the major muscle–tendon units, to total 60 s of total stretching time for each exercise; ≤10 min per session	90 to 150+ min per week of continuous or accumulated exercise of any duration
Type	Prolonged, rhythmic activities using large muscle groups (e.g., walking, cycling, swimming)	Resistance machines, free weights, resistance bands, functional body weight exercises	Exercise involving motor skills or functional body weight and flexibility exercise such as yoga, Pilates, and tai chi	Static stretching, dynamic stretching, or proprioceptive neuromuscular facilitation	An emphasis on aerobic and resistance exercise alone or combined with neuromotor and flexibility exercise depending on personal preference

Note: The 2018 Physical Activity Guidelines Advisory Committee adhered to the JNC 7 blood pressure (BP) classification scheme (5) in its pronouncement (79) because the literature reviewed was based on this blood pressure classification scheme.

[a]May be aerobic and resistance, or aerobic or resistance.

[b]The magnitude of the BP reductions resulting from aerobic exercise is directly proportional to intensity such that the greatest blood pressure reductions occur after vigorous-intensity exercise if the person is willing and able to perform vigorous-intensity exercise.

[c]Neuromotor functional body weight exercise can be substituted for resistance exercise, and depending on the amount of flexibility exercise integrated into a session, neuromotor flexibility exercise can be substituted for flexibility exercise depending on patient preference. The evidence is promising but limited for neuromotor exercise to be recommended alongside aerobic and resistance exercise as a primary exercise modality at this time.

[d]This frequency recommendation is made because of the immediate blood pressure–lowering effects of exercise, termed *postexercise hypotension.*

FITT = *F*requency, *I*ntensity, *T*ime, *T*ype; $\dot{V}O_2R$ = oxygen uptake reserve; HRR = heart rate reserve; RPP = rating of perceived exertion; 1RM = one-repetition maximum; ACSM = American College of Sports Medicine.

Reprinted by permission from L. Pescatello, *What's New in the ACSM Pronouncement on Exercise and Hypertension?* (2019), https://www.acsm.org/blog-detail.

- *Progression:* Progress gradually, avoiding large increases in any of the FITT components. Increase exercise duration over the first 4 to 6 wk, and then increase frequency, intensity, and time (or some combination of these) to achieve the recommended volume of 700 to 2,000 kcal/wk over the next 4 to 8 mo. Progression may be individualized based on tolerance and preference in a conservative manner.

Consideration should be given to the level of BP control, recent changes in antihypertensive drug therapy, medication-related adverse effects, the presence of target organ disease and other comorbidities, and age as well as other risk factors. Adjustments to the exercise prescription should be made accordingly. In general, progression should be gradual, avoiding large increases in any of the FITT components of the exercise prescription, especially intensity.

The following is a list of special considerations for exercise prescription for individuals with hypertension.

- An exaggerated BP response to relatively low exercise intensities and at HR levels <85% of the age-predicted maximum HR is likely to occur in some individuals, even after resting BP is controlled with antihypertensive medication (<130 and <80 mmHg). In some cases, an exercise test may be beneficial to establish the exercise HR corresponding to the exaggerated BP in these individuals. See the previous section on exercise testing.
- Exercise is contraindicated if resting SBP exceeds 200 mmHg or DBP exceeds 100 mmHg.
- When exercising, it is prudent to maintain an SBP <220 mmHg and a DBP <105 mmHg.
- Although vigorous-intensity cardiorespiratory endurance exercise is not necessarily contraindicated in patients with hypertension, moderate-intensity cardiorespiratory endurance exercise is generally recommended to optimize the benefit-to-risk ratio.
- Individuals with hypertension are often overweight or obese. If that is the case, the exercise prescription should focus on increasing caloric expenditure coupled with reducing caloric intake to facilitate weight reduction and minimize weight gain (see chapter 9).
- Inhaling and breath holding while lifting a weight (i.e., Valsalva maneuver) can result in extremely high BP responses, dizziness, and even fainting. Thus, this practice should be avoided during resistance exercise.
- Individuals taking antihypertensive medications should be monitored during and after exercise for potential adverse interactions (table 10.3).
- β-blockers and diuretics may adversely affect thermoregulatory function or increase the predisposition to hypoglycemia in certain individuals, so appropriate precautions should be taken in these situations.

- Individuals taking β-blockers and diuretics should be well informed about signs and symptoms of heat intolerance and hypoglycemia and should be educated on how to make prudent modifications in their exercise routines to prevent adverse events.
- β-blockers may also attenuate the HR response to exercise, while α-blockers, calcium channel blockers, and vasodilators may lead to sudden excessive BP reductions at the conclusion of exercise. Therefore, an adequate cool-down may be especially important for patients taking these medications.
- The BP-lowering effects of exercise are immediate, a physiological response termed *postexercise hypotension (PEH)*. Educate patients about PEH to possibly enhance adherence (see Research Focus).

EXERCISE TRAINING

Exercise training recommendations for individuals with hypertension should take into consideration their medical history, their current BP levels, and the presence of CVD and its risk factors. Comorbid conditions such as diabetes mellitus, coronary heart disease, and heart failure should be adequately controlled before the start of exercise training. The program should include cardiorespiratory endurance, dynamic resistance, flexibility, and neuromotor exercise.

Cardiorespiratory Exercise

This section provides an overview of the mechanisms and effects of *acute* (i.e., immediate, short-term effects, or PEH) and *chronic* (i.e., long-term effects, or training) cardiorespiratory endurance exercise on BP among individuals with hypertension. When appropriate, the ACSM pronouncement on exercise and hypertension (63, 79)—and other emerging research that has the potential to expand on the way exercise may be prescribed to prevent, treat, and control hypertension—will be discussed.

Mechanisms for the BP-Lowering Effects of Exercise

An isolated bout of cardiorespiratory endurance exercise results in immediate reductions in SBP of 5 to 7 mmHg that persist for up to 24 h, or PEH (78). Arterial pressure is determined by CO and TPR; therefore, the reductions in BP that occur after exercise are likely due to reductions in TPR since resting CO is generally not changed in healthy people. The reductions in TPR appear to be the result of a combination of centrally mediated decreases in sympathetic outflow, reduced neurovascular transduction, and local vasodilator mechanisms that all contribute to the drop in TPR that generates PEH (80, 81).

Vascular resistance is mediated by neurohumoral and structural adaptations, such as increased vasodilatory factors (e.g., nitric oxide), decreased vasoconstrictor factors (e.g., norepinephrine), increased vessel diameter, and increased vessel distensibility (15). Therefore, reductions in TPR after acute exercise are likely to be predominantly due to exercise-induced alterations involving the sympathetic nervous and renin–angiotensin systems and endothelial nitric oxide synthase pathways and their influence on vascular, renal, and baroreceptor function (80, 81). Also, hypertension and hyperinsulinemia, along with insulin resistance, abdominal obesity, increased triglycerides, and decreased high-density lipoprotein cholesterol, often cluster together to form the metabolic syndrome (82). Hyperinsulinemia has been postulated to raise BP through renal sodium retention, sympathetic nervous activation, and induction of vascular smooth muscle hypertrophy (83, 84). Even a single bout of exercise has well-established insulin-like effects and markedly increases skeletal muscle glucose transport. More long-term exercise training also results in centrally mediated decreases in sympathetic outflow, reduced neurovascular transduction, enhanced vasodilatory capacity, and improved insulin sensitivity that all lower BP.

Relationship Between the BP Effects of Acute and Chronic Exercise

Physiological responses to acute, or short-term, exercise (i.e., PEH) translate into functional adaptations that occur during and for some time following an isolated exercise session, a phenomenon termed the last bout effect. It has been previously hypothesized that frequent repetition of acute cardiorespiratory endurance exercise sessions produces functional and more permanent structural adaptations, forming the exercise training response. These persistent alterations in structure and function remain for as long as the training regimen is continued and then dissipate quickly, returning to pretraining values (85, 86). Several studies have supported the notion that the magnitude of BP reductions experienced immediately after acute exercise correlates to those experienced after cardiorespiratory endurance exercise training, an observation that suggests the BP benefits attributed to exercise training are largely the result of PEH (73, 78, 87-89).

Liu and colleagues were the first to study whether PEH could be used to predict the BP response to exercise training among middle-aged men (n = 8) and women (n = 9) with elevated BP (88). Participants completed a 30 min acute cardiorespiratory endurance exercise session at moderate intensity (65% maximum oxygen consumption [$\dot{V}O_2$max]) prior to beginning a supervised 8 wk cardiorespiratory endurance exercise training program, performed 4 d/wk for 30 min per session at 65% $\dot{V}O_2$max. After exercise training, SBP/DBP was reduced to similar magnitudes after acute (7/4 mmHg) and chronic (7/5.2 mmHg) exercise, and the BP response to acute exercise was strongly correlated with the BP response to exercise training (88). This finding was subsequently confirmed by Hecksteden, Grutters, and Meyer in a small sample of overweight to obese middle-aged men and women with elevated BP (87). Together, these findings support the long-held notion that PEH may account for a significant portion of the BP reduction attributed to exercise training. Further research in a

larger, more diverse sample of adults with hypertension is needed to substantiate this premise.

In summary, PEH is clinically important for several reasons: It occurs immediately; a person does not need to be physically fit to experience its benefits; because of its correlation with the BP response to exercise training, PEH holds promise as a screening tool to identify individuals with hypertension who respond to more long-term exercise training as antihypertensive therapy; it lowers BP on exercise days; it occurs after cardiorespiratory endurance and dynamic resistance exercise alone and combined (i.e., concurrent); and it can be used as a behavioral self-regulation strategy to increase exercise adherence (78, 90). Since BP is lower on exercise than nonexercise days because of PEH, the ACSM and other professional organizations recommend that individuals with hypertension exercise on most, preferably all, days of the week (63, 79).

BP Effects of Cardiorespiratory Endurance Exercise

The 2018 Physical Activity Guidelines Advisory Committee (PAGAC) was charged to make evidence-based conclusion statements based on the newest, best informed science for prioritized chronic conditions, including hypertension (30); their report was the scientific foundation for the *Physical Activity Guidelines for Americans, Second Edition* (91). The seminal portions of the 2018 PAGAC scientific report on the role of physical activity in the prevention and treatment of hypertension were subsequently published as an ACSM pronouncement on physical activity and hypertension (79). The systematic review methods used to compile the evidence that supported the conclusions made in the ACSM pronouncement are described in detail elsewhere (30, 63, 92), and the protocol is registered at PROSPERO #95748.

The ACSM pronouncement was a meta-review that included systematic reviews, meta-analyses of controlled trials, and pooled analyses meeting the following criteria: involved healthy adults ≥18 yr with hypertension; investigated the relationship between all types and intensities of exercise and BP; had a nonexercise, nondiet control group; reported pre- and postintervention BP; and were published in English. Trials involving diet modifications in addition to exercise, populations with chronic diseases (e.g., cancer, coronary artery disease, HIV/AIDS), or animals were excluded. The evidence was graded as strong, moderate, limited, or not assignable based on applicability, generalizability, risk of bias, study limitations, quantity and consistency of results across studies, and magnitude and precision of effect (30, 63). We now highlight the evidence used to answer the following question among adults with normal BP, prehypertension, and hypertension by the JNC 7 BP criteria, because the literature reviewed was based on this BP classification scheme (5): Does the relationship between physical activity and BP vary based on the modality of exercise?

Cornelissen and Smart (69) examined 59 cardiorespiratory endurance training trials with almost 4,000 participants. Cardiorespiratory endurance exercise training was performed, on average, at moderate to vigorous intensity for 40 min per session on 3 d/wk for 16 wk. SBP/DBP values were reduced −8.3 (95% confidence interval [CI]: −10.7 to −6.0)/−5.2 (95% CI: −6.9 to −3.4), −4.3 (95% CI: −7.7 to −0.9)/−1.7 (95% CI: −2.7 to −0.7), and −0.8 (95% CI: −2.2 to +0.7)/−1.1 (95% CI: −2.2 to −0.1) mmHg among adults with hypertension, prehypertension, and normal BP, respectively. The moderator analyses revealed that men experienced greater BP reductions than women (3-5 mmHg vs. 1 mm Hg, respectively), concluding that sex may influence the BP response to exercise training. In addition, Cornelissen and Smart identified several moderators related to the FIT of the cardiorespiratory endurance exercise intervention, such that exercise training <24 wk reduced BP to a greater extent than training programs ≥24 wk (3-6 mmHg vs. 1-2 mmHg, respectively). They also found the BP reductions were greater in those who trained ≥30 min per session than <30 min and who accumulated a weekly exercise volume of ≤210 min compared with a weekly volume >210 min. Last, exercise intensity appeared to alter the BP response to cardiorespiratory endurance exercise training such that BP reductions were greatest as a result of moderate to vigorous intensity compared with low-intensity cardiorespiratory endurance exercise training (4-5/2-3 mmHg vs. ~0/0 mmHg, respectively).

High-intensity interval training (HIIT) is characterized by brief periods of very high-intensity cardiorespiratory endurance exercise (>90% $\dot{V}O_2$max) separated by recovery periods of lower-intensity exercise or rest (93). Although no meta-analyses were available on the BP-lowering effects of HIIT in patients with hypertension, limited lines of evidence in the primary literature indicate that vigorous-intensity cardiorespiratory endurance exercise may be superior to moderate-intensity cardiorespiratory endurance exercise in lowering BP (94), and the magnitude of reductions after HIIT appear greater among individuals with prehypertension and hypertension versus normal BP (8 vs. 3 mmHg, respectively) (78, 93, 95, 96). HIIT holds promise for some people with hypertension because it allows individuals to perform brief periods of vigorous-intensity exercise that would not be tolerable for longer periods of time. In addition, HIIT can yield an equal amount of work (i.e., energy expenditure) compared with continuous moderate-intensity exercise in a shorter amount of time (78).

In summary, cardiorespiratory endurance exercise training lowers BP by 5 to 8 mmHg among adults with hypertension, 2 to 4 mmHg among adults with prehypertension, and 1 to 2 mmHg among those with normal BP. Exercise intensity appears to be an important moderator of the BP response to cardiorespiratory endurance exercise training in dose–response fashion (69). HIIT shows promise as a viable alternative to the current ACSM exercise prescription recommendations for hypertension; however, further investigation is warranted among individuals with hypertension to more definitively determine the benefit-to-risk ratio of exercising at vigorous intensity among a population that is predisposed to heightened CVD risk (78, 93).

Resistance Exercise

It was previously thought that individuals with hypertension should avoid resistance exercise because of marked elevations in BP following the Valsalva maneuver (97). The Valsalva maneuver is characterized by a strenuous and prolonged expiratory effort when the glottis is closed, resulting in decreased venous return and an increase in peripheral venous pressures, quickly followed by increased venous return to the heart and subsequent elevations in arterial pressure (97). Increases in central BP as high as 480/350 mmHg have been reported among bodybuilders immediately after lifting weights at or above 80% 1RM (98). However, such BP surges are known to return to initial values within 10 s of the last repetition of each set (98). A growing body of literature claims that the BP benefits of dynamic resistance exercise that does not elicit the Valsalva maneuver may be comparable to cardiorespiratory endurance exercise among adults with hypertension.

BP Effects of Dynamic Resistance Exercise

MacDonald and colleagues meta-analyzed 64 controlled studies (71 interventions) involving middle-aged adults ($N = 2,344$) with prehypertension (126/76 mmHg), the majority of whom were white (57%) (72). The dynamic resistance exercise training interventions were of moderate intensity and performed for 32 min per session on 3 d/wk for 14 wk. Overall, the interventions elicited BP reductions of 3/2 mmHg. However, when examined by BP group, greater BP reductions were found among individuals with hypertension (−5.7 [95% CI: −9.0 to 2.7]/−5.2 [95% CI: −8.4 to −1.9]), followed by prehypertension (−3.0 [95% CI: −5.1 to −1.0]/−3.3 [95% CI: −5.3 to −1.4]) and normal BP (0.0 [95% CI: −2.5 to 2.5]/0.9 [95% CI: −2.1 to 2.2]) mmHg. Furthermore, MacDonald et al. found that the antihypertensive effects of dynamic resistance exercise training were moderated by race. Among nonwhite samples with hypertension, SBP/DBP values were reduced by ~14/10 mmHg, and among white samples with hypertension, SBP/DBP values were reduced by ~9/10 mmHg.

In addition, the antihypertensive effects of dynamic resistance exercise were moderated by medication use, such that greater SBP/DBP reductions occurred among samples exercising and not on medications than among samples exercising and on medications (difference ~4/2 mmHg); by number of exercises per session, such that eight or more exercises elicited greater SBP reductions than fewer than eight exercises (difference ~3 mmHg); and by number of days per week, such that ≥3 d/wk resulted in greater DBP reductions than 3 d/wk (difference ~4 mmHg).

These promising findings suggest that the magnitude of the BP reductions resulting from dynamic resistance exercise training are comparable to those achieved from cardiorespiratory endurance exercise training, and for some populations dynamic resistance exercise training may elicit BP reductions even greater than those from cardiorespiratory endurance exercise training (i.e., nonwhite samples with hypertension) (72). The findings by MacDonald and colleagues indicate that dynamic resistance exercise can be a stand-alone exercise modality choice for patients with hypertension (72), which has led to the expansion of the ACSM FITT-VP recommendations for adults with hypertension to include dynamic resistance exercise alongside cardiorespiratory endurance exercise training, providing additional exercise options (table 10.6).

BP Effects of Isometric Resistance Exercise

Carlson and colleagues (99) investigated the BP response among adults with hypertension ($n = 61$) and normal BP ($n = 162$) to ≥4 wk of isometric resistance training at 30% to 50% maximal voluntary contraction (four contractions held for 2 min, with 1 to 3 min rest between contractions). SBP/DBP values were reduced among the adults with hypertension, all of whom were on medication, by −4.3 (95% CI: −6.6 to −2.2)/−5.5 (95% CI: −7.9 to −3.3) mmHg, and by −7.8 (95% CI: −9.2 to −6.4)/−3.1 (95% CI: −3.9 to −2.3) mmHg among adults with normal BP, respectively. Thus, it appears the BP-lowering effects of isometric resistance training may be greater in adults with normal BP than patients with hypertension.

The ACC/AHA guidelines endorse isometric resistance training as an exercise option for adults with hypertension, as do the Exercise and Sport Science Australia position stand (100) and the Canadian 2018 hypertension guidelines (101). Until a large, well-designed randomized controlled trial is done among adults with hypertension comparing isometric exercise training against cardiorespiratory endurance, dynamic exercise training, or both, any conclusions about the use of isometric exercise training in the treatment of hypertension should be made with caution (65). Nonetheless, the BP-lowering effects of isometric resistance training appear to be greater in adults with normal BP than patients with hypertension and thus may prove useful in the prevention of incident hypertension.

BP Effects of Concurrent Exercise

Concurrent exercise training is defined as cardiorespiratory endurance and resistance exercise being performed in proximity to each other (i.e., in a single session or on separate days) (73). Although the ACSM does not provide specific guidelines for concurrent exercise training among adults with hypertension, new and emerging evidence suggests that concurrent exercise training may be as effective as cardiorespiratory endurance and resistance exercise training alone as antihypertensive therapy among individuals with hypertension (63, 70).

Corso and colleagues pooled 68 trials and examined the influence of concurrent exercise training on BP among 4,110 subjects (70). Concurrent exercise training was performed, on average, at moderate intensity (cardiorespiratory endurance = 55% $\dot{V}O_2$max; resistance = 60% 1RM) 3 d/wk for 58 min per session for 20 wk. Overall BP was significantly lowered by 3.2/2.5 mmHg. SBP/DBP

values were reduced −5.3 (95% CI: −6.4 to −4.2)/−5.6 (95% CI: −6.9 to −3.8), −2.9 (95% CI: −3.9 to −1.9)/−3.6 (95% CI: −5.0 to −0.2), and +0.9 (95% CI: 0.2 to 1.6)/−1.5 (95% CI: −2.5 to −0.4) mmHg among adults with hypertension, prehypertension, and normal BP, respectively. Among individuals with hypertension in higher-quality trials that examined BP as the primary outcome, SBP/SBP reductions were as great as 9.2/7.7 mmHg (70). BP reductions of this magnitude are clinically meaningful.

In summary, new and emerging research examining the influence of cardiorespiratory endurance exercise training and dynamic resistance exercise alone or combined (i.e., concurrent exercise training) has resulted in the expansion of the ACSM exercise and hypertension FITT-VP guidelines to include dynamic resistance exercise alongside cardiorespiratory endurance exercise training to prevent, treat, and control hypertension (table 10.6). This new recommendation is based on this conclusion made by the PAGAC in their scientific report (30): Moderate evidence indicates the relationship between resting BP level and the BP response to physical activity does not vary by traditional type (i.e., cardiorespiratory endurance, dynamic resistance, combined) of physical activity among adults with normal BP, prehypertension, and hypertension. PAGAC Grade: Moderate.

PAGAC also made the following two recommendations that are pertinent to the content of this section of this chapter: "Insufficient evidence is available to determine whether the relationship between physical activity and the disease progression indicators of BP and CVD mortality varies by the frequency, intensity, time, or duration of physical activity among adults with hypertension." PAGAC Grade: Grade not assignable. Limited evidence suggests that, among adults with hypertension, the blood pressure response to physical activity varies by resting blood pressure level, with the greatest blood pressure reductions occurring among those adults who have the highest resting blood pressure levels. PAGAC Grade: Limited.

Range of Motion Exercise

The present recommendation for range of motion (i.e., flexibility) training among individuals with hypertension is to engage in flexibility training on 2 or 3 d/wk (or more), which is consistent with the exercise prescription recommendation for healthy adults. Although there is limited evidence to support flexibility exercise training for the purposes of reducing BP, the goal of this training is to reduce the risk of musculoskeletal injury, increase exercise adherence, and improve overall health.

Neuromotor Exercise

At the time of publication of the PAGAC scientific report (30), the PAGAC made the following conclusion: "Moderate evidence indicates the relationship between physical activity and the disease progression indicator of blood pressure does not vary by type of physical activity, with the evidence more robust for traditional types (modes, i.e., cardiorespiratory endurance, dynamic resistance, combined) of physical activity than for other types (tai chi, yoga, and qigong) among adults with hypertension." PAGAC Grade: Moderate.

The four meta-analyses that the PAGAC reviewed reported SBP reductions of 12 to 17 mmHg and DBP reductions of 2 to 11 mmHg (102-105). Although these findings are promising, the PAGAC urged caution in making any conclusions about complementary and alternative types of exercise in the prevention and treatment of hypertension because of the low methodological quality of the studies, the lack of disclosure of important study design considerations, the considerable heterogeneity in this literature, the inability to generalize findings to racial and ethnic groups other than Asian, and the lack of long-term follow-up.

Since the publication of the PAGAC scientific report (30), four moderate- to high-quality meta-analyses have been published (106-109) involving 124 trials with 10,841 participants performing yoga, tai chi, qigong, Baduanjin, Wuqinxi, and YiJinjing. Among adults with hypertension, SBP reductions ranged from 9 to 15 mmHg and DBP reductions ranged from 5 to 7 mmHg. Consistent with traditional forms of exercise, the BP reductions were greatest for adults with hypertension followed by prehypertension and normal BP. Other moderators identified by these four meta-analyses collectively included the following: age, with greater BP reductions among adults <50 yr than ≥50 yr; intervention length, with greater BP reductions among interventions lasting 12 to 24 wk than <12 wk; antihypertensive medication use, with greater BP reductions among exercise combined with medication than medication or exercise alone; breathing techniques, meditation, and relaxation, with yoga interventions that included these features resulting in greater BP reductions than those that did not; and publication language, with trials published in Chinese reporting greater BP reductions than those published in English.

The expansion of the ACSM FITT-VP recommendations for hypertension to include dynamic resistance exercise training alongside cardiorespiratory endurance exercise training delivered a valuable public health message by providing adults with hypertension more evidence-based exercise options. In line with this public health message, newer evidence indicates that neuromotor exercise such as yoga and tai chi are promising candidates to be considered as additional exercise options to prevent and treat hypertension; their antihypertensive effects appear comparable to more traditional forms of exercise (figure 10.4) (110). Therefore, health care and exercise professionals should consider recommending yoga and tai chi to their patients and clients with hypertension, especially for those who prefer neuromotor exercise, or for those who are resistant or unable to engage in cardiorespiratory endurance and resistance exercise.

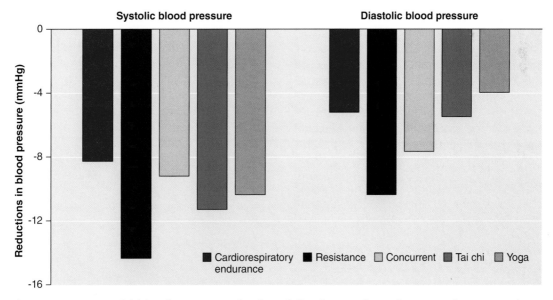

Figure 10.4 The greatest potential blood pressure reductions following cardiorespiratory endurance, resistance, concurrent, tai chi, and yoga exercise training among adults with hypertension. The BP reductions after resistance (72), concurrent (70), and yoga (109) exercise were generated from additive statistical models that capture the combination of study-level moderators that elicit the optimal BP benefits.

Adapted from Cornelissen and Smith (69), MacDonald et al. (72), Corso et al. (70), Wu et al. (107), Wu et al. (109).

Practical Application 10.2

OTHER COMMON HEALTH CONDITIONS EXHIBITED BY INDIVIDUALS WITH HYPERTENSION

Typically, hypertension does not occur in isolation but is accompanied by other chronic diseases and health conditions and CVD risk factors. Indeed, 80% of patients with hypertension have additional CVD risk factors (111), while 25% of those who are overweight and 40% of those who are obese also have hypertension compared with 15% of those who are normal weight (44, 112). In addition, the prevalence rates of hypertension in adults diagnosed with dyslipidemia, metabolic syndrome, and diabetes mellitus are 52%, 62%, and 77%, respectively (82). Excess body weight increases overall CVD risk such that for every one unit increase in body mass index (BMI), the risk of CVD pathologies such as heart failure and stroke increases 5% to 7% (112). Consequently, the frequent presence of cardiovascular and metabolic comorbidities in patients with hypertension influences the prescription of both pharmacological therapy and lifestyle modifications (i.e., diet, weight loss, exercise) as well as the interaction between these two types of therapies.

Lifestyle modifications are effective treatments for mitigating hypertension, with weight loss, the DASH diet, and cardiorespiratory endurance exercise all evoking reductions in SBP and DBP (table 10.6). These individual strategies when combined may augment their BP effects alone, such that a combined weight management approach (e.g., weight loss and cardiorespiratory endurance exercise combined with diet modification) can evoke the greatest reductions in BP (113, 114). Furthermore, adding these intensive lifestyle modifications to pharmacological treatment for hypertension may augment reductions in BP compared with medication alone (102-104, 106). For example, in patients treated with one antihypertensive medication, the addition of cardiorespiratory endurance exercise and a weight loss program that elicited an 11 lb (5 kg) weight loss over 9 wk resulted in an additional 9/5 mmHg decrease in BP compared with those receiving antihypertensive medication without cardiorespiratory endurance exercise and a weight loss program (115).

Similar interactions can be observed in patients with hypertension and dyslipidemia. Hydroxy-methyl-glutaryl coenzyme A reductase inhibitors (i.e., statins) are the most effective cholesterol-lowering drugs; the average reduction in low-density lipoprotein cholesterol (LDL-C) with routine statin monotherapy ranges from 25% to 50% (116). Statins thus reduce cardiac events by 20% to 44% in both coronary artery disease patients and previously healthy subjects (116-118). Statins also have pleiotropic effects unrelated to the magnitude of LDL-C reduction. For people with high BP, these pleiotropic effects include a reduction of SBP up to 8 mmHg in patients with dyslipidemia and normal BP, 6 mmHg in patients without dyslipidemia and with hypertension, and ~14 mmHg in patients with dyslipidemia and hypertension (119). In addition, the combination of exercise training and cholesterol-lowering drugs may be most efficacious for patients with dyslipidemia and hypertension, because reductions in both LDL-C and mortality with statin use are augmented by simultaneously participating in exercise training (120, 121). Moreover, in animal data, combined statin therapy and exercise training evokes greater reductions in BP than statin therapy alone (122).

In summary, patients with hypertension and other comorbidities may exhibit different and possibly greater impacts of pharmacological treatment and prescription of exercise when combined than with either alone (120-122). Thus, the clinician should be aware of these potential beneficial interactions when prescribing pharmacological and lifestyle treatments for hypertension.

Research Focus

Using the Immediate Blood Pressure Benefits of Exercise to Improve Exercise Adherence Among Adults With Hypertension

Hypertension is a unique health condition in that no signs or symptoms are associated with high BP levels. However, the advent of home BP monitoring has enabled BP to be a relatively easy and reliable vital sign to self-assess and track with the proper tools and education. As such, the ACC and AHA recommend that adults with hypertension use a home BP monitor to measure BP at least twice daily. Self-monitoring of BP is associated with greater BP control and lower CVD and all-cause mortality rates compared with usual care. Surprisingly, despite its immediate and observable BP benefits, no study has leveraged PEH as a condition-specific self-monitoring behavioral strategy to improve exercise adherence and BP control among adults with hypertension. Self-monitoring of PEH may foster positive outcome expectations of exercise and thus enhance exercise adherence among adults with hypertension. Zaleski and colleagues compared the efficacy of self-monitoring of exercise (EXERCISE) versus exercise plus PEH (EXERCISE+PEH) for exercise adherence and BP control among adults with hypertension.

Adults with high BP were randomized to EXERCISE ($n = 12$) or EXERCISE+PEH ($n = 12$). Subjects underwent supervised moderate-intensity cardiorespiratory endurance exercise training for 40 to 50 min per session, 3 d/wk for 12 wk, and were encouraged to exercise at home unsupervised for ≥30 min/d, 1 or 2 d/wk. All subjects self-monitored exercise using a calendar recording method. EXERCISE+PEH also self-monitored BP before and after exercise. Adherence was calculated as follows: [(# of exercise sessions performed ÷ # of possible exercise sessions) × 100%]. BP was measured before and after training.

Healthy, middle-aged (52.3±10.8 yr) men ($n = 11$) and women ($n = 13$) with hypertension (136.2±10.7/85.2±8.9 mmHg) completed an overall exercise training adherence rate of 87.9±12.1%. EXERCISE+PEH demonstrated greater adherence to supervised training (94.3±6.6%) than EXERCISE (81.6±13.2%; $p = .007$). In addition, EXERCISE+PEH performed 32.6±22.5 min/wk more unsupervised home exercise than EXERCISE ($p = .004$), resulting in greater overall study exercise adherence (107.3±18.7%) than EXERCISE (82.7±12.2%; $p = .002$). Post- versus pretraining, SBP/DBP values were reduced by −7.4±11.3/−4.9±9.9 mmHg ($p < .025$), with no statistical difference between EXERCISE (−5.2±13.3/−3.6±6.1 mmHg) and EXERCISE+PEH (−9.9±11.3/−6.1±6.9 mmHg; $p > .344$).

This study is the first to demonstrate that using condition-specific BP self-monitoring for PEH is an efficacious behavioral strategy to improve exercise adherence among adults with hypertension. Indeed, Zaleski et al. found that adults with hypertension who self-monitored their BP, daily and before and after exercise, were ~24% *more* adherent to a 12 wk structured cardiorespiratory endurance exercise training program. In addition, those adults using BP self-monitoring maintained 37% more of their supervised exercise training volume at 1 mo follow-up than those who were not using BP self-monitoring. Future research among a larger, more diverse sample is needed to confirm these novel findings and determine whether EXERCISE+PEH translates to better BP control relative to EXERCISE self-monitoring alone.

Based on Zaleski et al. (90).

Clinical Exercise Bottom Line

- BP is lower on exercise than nonexercise days because of postexercise hypotension, so the ACSM and other professional organizations recommend that individuals with hypertension exercise on most, preferably all, days of the week.

- Individuals with hypertension are recommended to emphasize moderate-intensity exercise.

- Individuals with hypertension are recommended to exercise for 20 to 30 min per session to total 90 to 150 min/wk or more of continuous or accumulated exercise of any duration.

- The ACSM now recommends dynamic resistance exercise training alongside cardiorespiratory endurance exercise training as an exercise option. In addition, newer evidence indicates that neuromotor exercises such as yoga and tai chi are promising candidates to be considered as additional exercise options to prevent and treat hypertension.

CONCLUSION

Hypertension is the most common, costly, and modifiable CVD risk factor in the United States and world, affecting 116 million Americans (45%) (1, 4). For these reasons, there is an international call to action for better treatment, control, and management of hypertension to reduce the markedly elevated risk of CVD morbidity and mortality associated with high BP. The traditionally accepted BP goal for proper control for most individuals with hypertension was <140/90 mmHg (5); however, this threshold has been lowered to <130/80 mmHg by the ACC/AHA guidelines (4).

Lifestyle modifications that include exercise are recommended as cornerstone approaches to prevent, treat, and control hypertension. Accordingly, adults with hypertension are encouraged to engage in moderate-intensity cardiorespiratory endurance and resistance exercise training on most days of the week to total 90 to 150 min/wk or more. In patients who cannot achieve goal BP with lifestyle management alone, BP-lowering medication may be initiated. The good news for those with hypertension is that antihypertensive medications, when combined with exercise, can facilitate greater improvements in health outcomes and CVD risk factors than medication or exercise alone. For these reasons, clinicians and exercise professionals should understand the potential adverse interactions that can occur when individuals taking antihypertensive medications perform exercise. An interdisciplinary, patient-centered approach by the clinician, exercise professional, and other health care professionals involved with the treatment of a patient with hypertension will ultimately improve adherence to lifestyle and pharmacological interventions and translate to better BP control and lower overall CVD risk, which is the ultimate goal in the treatment of hypertension.

Online Materials

Visit HK*Propel* to access a link to the references, the case studies with discussion questions, and a quiz to help you review key concepts and test your understanding of the material covered in this chapter.

CHAPTER 11

Hyperlipidemia and Dyslipidemia

Paul G. Davis, PhD

Peter W. Grandjean, PhD

Stephen F. Crouse, PhD

J. Larry Durstine, PhD

Lipids are fats or derivatives of fat that are insoluble in water and must be bound to protein to be transported in blood. **Cholesterol** is a lipid that is an essential component of cell membranes and a number of vital substances, including bile acids, steroid hormones, and vitamin D. High plasma cholesterol concentrations present a major public health problem, given that 38% of U.S. adults have a concentration ≥200 mg/dL and this substantially increases a person's risk for atherosclerotic cardiovascular disease (ASCVD) (200). **Triglycerides** are other important blood lipids; they are major carriers of energy-containing fatty acids. A high triglyceride concentration can increase one's risk of vascular and other diseases. ASCVD risk is also affected by the concentrations of different types of **lipoproteins** (lipid–protein complexes that transport cholesterol and triglycerides in the blood). Lipid and lipoprotein concentrations are majorly affected by genetics, diet, and, to a lesser extent, physical activity. Because of the high prevalence of elevated plasma cholesterol and triglycerides, professionals providing exercise testing and training, as well as patients themselves, should be keenly aware of ASCVD signs and symptoms, as well as the side effects of medications used to treat abnormal lipid concentrations.

DEFINITION

The terms **hypercholesterolemia** and **hypertriglyceridemia** represent unhealthily elevated plasma concentrations of cholesterol and triglycerides, respectively. **Hyperlipidemia** is a collective term that can represent either or both of the above. Although often used interchangeably with hyperlipidemia, the term **dyslipidemia** refers to abnormal lipid concentrations that *may* be high, but it can also refer to *low* levels of cholesterol associated with lipoproteins that are mostly cardioprotective in nature. Similar to dyslipidemia, **dyslipoproteinemia** describes the condition of abnormal lipoprotein levels. The most severe forms of dyslipidemia and dyslipoproteinemia are linked to genetic defects in cholesterol or triglyceride metabolism (26, 50, 172). Secondary dyslipidemia usually results from metabolically dysfunctional conditions that arise from the accumulation of ectopic fat (199), insulin resistance (80), diabetes mellitus (61, 146, 193), hypothyroidism (168), and renal insufficiency and nephrotic kidney disease (85, 90), some of which can be due to a high-fat or hypercaloric diet.

Since lipids are not soluble in body fluids, they must combine with proteins to be transported in blood and other fluids in the body. Lipoproteins are spherical macromolecules composed of an outer shell of hydrophilic apolipoproteins and phospholipids, with the more hydrophobic triglycerides and cholesterol packaged in the core. In addition to providing structural integrity, **apolipoproteins** possess functional roles, such as activating or inhibiting enzymes and serving as ligands for receptors (table 11.1) (94).

Scientists and clinicians use a variety of techniques to classify and study lipoproteins (77). In fact, the measurement technique dictates how lipoproteins are identified and described. The clinician should remember that the lipid and apolipoprotein composi-

Acknowledgment: We wish to gratefully acknowledge and thank Benjamin Gordon, PhD, ACSM-CEP, for his previous work on this chapter.

Table 11.1 Major Human Apolipoproteins

Apolipoprotein	Major function	ASCVD risk factor
A-I	LCAT activator; SR-BI ligand	Inversely related with ASCVD risk
A-II	LCAT inhibitor or activator of heparin releasable hepatic triglyceride hydrolase or both; SR-BI ligand	Not associated with ASCVD risk
B-48	Required for synthesis of chylomicron	Directly associated with ASCVD risk
B-100	LDL receptor–binding ligand	Directly associated with ASCVD risk
(a)	Similar characteristics between apo(a) and plasminogen, thus may have a prothrombotic role by interfering with function of plasminogen; possible acute phase reactant to tissue damage	Directly associated with ASCVD risk
C-I	LCAT activator	Not associated with ASCVD risk
C-II	LPL activator	Not associated with ASCVD risk
C-III	LPL inhibitor, several forms depending on content of sialic acids	Not associated with ASCVD risk
D	Core lipid transfer protein, possibly identical to the cholesteryl ester transfer protein	Not associated with ASCVD risk
E	Remnant receptor–binding ligand, present in excess in the β-VLDL of patients with type III hyperlipoproteinemia and exclusively with HDL cholesterol	Not associated with ASCVD risk

ASCVD = atherosclerotic coronary artery disease; LCAT = lecithin-cholesterol acyltransferase; SR-BI = scavenger receptor class B type I; LDL = low-density lipoprotein; LPL = lipoprotein lipase; VLDL = very low-density lipoprotein; HDL = high-density lipoprotein.

tion constantly changes in the circulation as these macromolecules interact with each other and with the surrounding body tissues. Clinicians generally discuss blood lipids and lipoproteins based on a traditional measurement technique called ultracentrifugation. To be consistent, we'll describe general classes of lipoproteins based on gravitational density determined by ultracentrifugation methodology (table 11.2) (77).

Development of arterial atherosclerotic lesions, characterized first by fatty streaks and later by fibrous plaque, is significantly related to concentrations of total blood cholesterol, **low-density lipoprotein cholesterol (LDLc)**, and triglycerides and inversely with levels of **high-density lipoprotein cholesterol (HDLc)**. As described in the Pathophysiology section, LDL is primarily responsible for the delivery of cholesterol to tissues throughout the body, including atherosclerotic plaque. HDL, on the other hand, has been shown to harvest cholesterol from plaque and deliver it to the liver. **Chylomicrons** and **very low-density lipoproteins (VLDLs)** are the major carriers of triglyceride although, unless triglyceride concentration is unusually high, chylomicrons are present only in trace amounts in fasted plasma.

SCOPE

A wealth of data supports the relationship between dyslipidemias and greater risk for ASCVD and other conditions (e.g., cerebrovascular disease, kidney disease). Arterial atherosclerotic lesions

begin in childhood and increase in prevalence and size with age. As such, relationships between blood lipid concentrations and ASCVD morbidity and mortality have been regularly demonstrated across the life span (20, 149, 171). A meta-analysis of 49 randomized controlled trials of various lipid-lowering medications and supplements demonstrated a 20% reduction of major ASCVD events (myocardial infarction, coronary revascularization, or sudden cardiac death) per 1 mmol/L (39 mg/dL) decrease in LDL-c and a 16% reduction of events per 1 mmol/L (89 mg/dL) decrease in triglycerides. In an earlier meta-analysis, Genser and Marz (72) estimated that LDLc lowering in adults accounted for 75% of the variance in risk reduction for cardiovascular end points. Randomized controlled trials of various methods to lower non-HDLc (i.e., atherogenic cholesterol, LDLc + VLDLc) consistently demonstrate that achieving lower on-treatment levels is associated with lower absolute risk of ASCVD events (21, 140). Although low HDLc concentrations (<1 mmol/L or 39 mg/dL) are associated with increased ASCVD mortality (88), pharmaceutically increasing HDLc does not typically decrease ASCVD event incidence or mortality (111). Furthermore, abnormally high HDLc (≥2.5 mmol/L or 97 mg/dL) is associated with increased all-cause mortality risk (88, 137).

Relationships between blood lipid concentrations and ASCVD are consistent in men and women and in individuals with and without documented ASCVD. In older adults, the relationship between cholesterol and ASCVD becomes nuanced. In men 65

Table 11.2 Characteristics of Plasma Lipids and Lipoproteins

Lipid/lipoprotein	Source	Protein %	Total lipid %	TG	Chol	Phosp	Free chol	Apolipoprotein
Chylomicron	Intestine	1-2	98-99	88	8	3	1	Major: A-IV, B-48, B-100, H Minor: A-I, A-II, C-I, C-II, C-III, E
VLDL	Major: liver Minor: intestine	7-10	90-93	56	20	15	8	Major: B-100, C-III, E, G Minor: A-I, A-II, B-48, C-II, D
IDL	Major: VLDL Minor: chylomicron	11	89	29	26	34	9	Major: B-100 Minor: B-48
LDL	Major: VLDL Minor: chylomicron	21	79	13	28	48	10	Major: B-100 Minor: C-I, C-II, (a)
HDL$_2$	Major: HDL$_3$	33	67	16	43	31	10	Major: A-1, A-II, D, E, F Minor: A-IV, C-I, C-II, C-III
HDL$_3$	Major: liver and intestine Minor: VLDL and chylomicron remnants	57	43	13	46	29	6	Major: A-1, A-II, D, E, F Minor: A-IV, C-I, C-II, C-III
Chol	Liver and diet		100			70-75	25-30	
TG	Diet and liver		100	100				

Header note: COMPOSITION spans Protein %, Total lipid %, and PERCENTAGE OF TOTAL LIPID (TG, Chol, Phosp, Free chol); Apolipoprotein is separate.

TG = triglyceride; chol = cholesterol; phosp = phospholipid; VLDL = very low-density lipoprotein; IDL = intermediate-density lipoprotein; LDL = low-density lipoprotein; HDL = high-density lipoprotein.

to 80 yr, the relative risk for ASCVD incidence increases by 28% and ASCVD mortality by 22% for each 1 mmol/L increase in total cholesterol. After 80 yr of age, the positive association no longer exists. In fact, an inverse relationship between total cholesterol and all-cause mortality is found. In older women, the relationship between total cholesterol and ASCVD incidence and mortality is not as strong as in men (7).

The association between triglyceride concentrations and the incidence of ASCVD is a topic of ongoing debate. Part of this debate is based on the inverse and biologically plausible relationships among triglyceride, HDLc, and non-HDLc concentrations (48, 149). Because of these relationships, these three lipid classes account for much of the same variance when predicting ASCVD. Therefore, some would suggest that triglycerides do not predict ASCVD outcomes independently after HDLc and non-HDLc have been accounted for. This argument is countered by Austin (11), who calculated that the relative risk for incident ASCVD was increased 76% in women and 32% in men for every 1 mmol/L (~89 mg/dL) increase in triglyceride. These risk estimates were reduced to 37% and 14% in women and men, respectively, after

adjusting for HDLc; however, the association between triglyceride and ASCVD incidents remained significant (11, 100). A meta-analysis found that lowering triglyceride concentration lowered the risk of major ASCVD events after controlling for the reduction in LDLc, although the effect was not as large as that of lowering LDLc (138). In addition, nonfasting, or postprandial (i.e., "after meal"), triglyceride concentrations predict ASCVD risk in men and women independent of other established risk factors and at least as well as fasting triglyceride levels (126, 143, 149, 150). Meta-analyses of large population studies report a strong and consistent relationship between nonfasting triglyceride levels and a variety of cardiovascular events, such as myocardial infarction, ischemic stroke, coronary revascularization, and cardiovascular mortality (48, 52, 100, 143, 177). The relationship between nonfasting triglyceride concentrations and ASCVD outcomes remains after adjusting for age, body weight, blood pressure, smoking, HDLc, LDL particle size, and hormone therapy use in women.

According to results from these meta-analyses, triglyceride concentrations appear to provide predictive information beyond that obtained for cholesterol and LDLc, especially in women. On the other hand, results from a meta-analysis that included 68 long-term prospective studies and over 302,000 participants indicate that triglyceride and non-HDLc account for the same variation in predicting future ASCVD outcomes (48). The most recent guidelines identify non-HDLc as the primary target for lipid-lowering interventions because this broad category includes all apo B lipoproteins containing cholesterol associated with ASCVD outcomes (84). Here, the clinical exercise physiologist is reminded to keep current with the available literature and to be mindful that traditional and emerging risk factors will vary in their predictive efficacy among subgroups within our population. Race, gender, genetics, disease states, and physical characteristics, among other factors, will influence the strength of association between blood lipids and lipoproteins, the atherosclerotic process, and ASCVD outcomes (28, 71, 108, 172, 199).

Data from the National Health and Nutrition Examination Survey (NHANES; 2011-2014) indicate that fewer than two-thirds of U.S. adults who were eligible for cholesterol-lowering therapy based on current guidelines actually took medication, and medication use varied considerably among race, gender, and disease state (84). NHANES data from 2015 to 2018 indicate that 38% of American adults 20 yr of age or older—35% of males and 40% of females—had high total cholesterol (≥200 mg/dL or ≥5.2 mmol/L) (200). Among males, high cholesterol prevalence was 39% in non-Hispanic (NH) Asian males, 38% in Hispanic males, 35% in NH white males, and 31% in NH Black males. Among females, the greatest prevalence of high serum cholesterol was found in NH white females (42%), followed by NH Asian females (39%), Hispanic females (37%), and NH Black females (33%). One in eight, or roughly 12%, of American adults has hypercholesterolemia, defined as total cholesterol >240 mg/dL.

LDLc and non-HDLc are the primary targets for lipid-lowering interventions (84, 105, 188). Approximately 69.6 million adults, or 29% of the U.S. adult population, have LDLc >130 mg/dL (200). Elevated LDLc is present in 29% of males and 30% of females, with Hispanic and NH Asian males having slightly higher averages (34% and 32%, respectively) and NH Black, Hispanic, and NH Asian females slightly lower (23%, 24%, and 25%, respectively). Low serum HDLc, generally defined as <40 mg/dL, is present in 41.9 million U.S. adults, or 17% of the population (200). Low HDLc is three times more prevalent in males (27%) than in females (9%), with Hispanic males and females having the highest sex-specific prevalence (32% and 12%, respectively) (84). Elevated triglyceride levels (e.g., serum triglyceride >150 mg/dL) are observed in approximately 22% of the U.S. adult population and tend to be higher than average in Hispanic, NH Asian, and NH white males (200).

The percentage of U.S. adults aged 20 to 74 yr with hypercholesterolemia has decreased from 33% to 12% since 1960. Tracking over the same time, the population-wide average total cholesterol has decreased from 222 mg/dL to 191 mg/dL (188 mg/dL in males, 193 mg/dL in females) (32, 169, 179, 200). This reduction in individuals with hypercholesterolemia represents a Healthy People 2010 objective that was attained, although the Healthy People 2020 target of lowering the nation's average cholesterol concentration to 177.9 mg/dL has not been attained at the time of this writing (200). Nevertheless, these population statistics are promising from a public health perspective, since the American Heart Association (AHA) estimates that the incidence of ASCVD can be reduced by 30% for every 10% decrease in total cholesterol among U.S. adults (169).

Average LDLc values have also declined, from 129 to 112 mg/dL. Much of the reduction in our population-wide total cholesterol and LDLc values has been attributed to dietary changes (30) and the increased use of lipid-lowering drugs (102, 104, 123). Indeed, the use of lipid-lowering medication has increased dramatically over the last few decades, and these drugs have had a greater prescription volume than any other drug class over this century. Statins are recommended by the major cardiology associations both before (primary prevention) and after (secondary prevention) cardiovascular events (73, 84, 97). As such, statins are by far the most frequently prescribed of the lipid-lowering agents, and approximately 75% of individuals taking statins do so for primary prevention (104, 123).

Although the trends for recommended concentrations of total cholesterol and LDLc are promising, continued and more aggressive lipid-lowering interventions appear to be warranted. Of U.S. adults who have a 10 yr predicted ASCVD diagnosis risk of at least 7.5%, only about one-third are taking a statin, which is recommended as the first-line medication for hyperlipidemia. Self-reported statin use is only 66%, 65%, and 46% for individuals with LDLc ≥190 mg/dL, ASCVD, and diabetes, respectively (200). This problem is greater among younger adults compared with those aged 65 yr and older (102). Issues associated with reduced compliance in taking statins include occasional muscle pain and, since hyperlipidemia itself generally does not present symptoms, lack of "feeling better" after starting the medication. Moreover, the average HDLc and triglyceride values have not changed appreciably over the last five decades (32). This result is likely due to the increased incidence of overweight and obesity among our adult population over the same period (32, 62). In fact, Burke and colleagues (28) reported that fasting glucose and triglyceride concentrations were greater (8 to 17 mg/dL and 18 to 55 mg/dL), HDLc was lower (4 to 14 mg/dL), and concentrations of small LDL particles were greater (109 to 173 mmol/L) in obese adults versus individuals of normal weight across ethnicities.

PATHOPHYSIOLOGY

Atherosclerotic disease processes originate in early childhood and continue to develop through the interactions of genes with modifiable and nonmodifiable environmental exposures. These exposures influence a complex network of lipid and lipoprotein pathways that provide transportation for delivery, metabolism, or elimination.

Lipoprotein Metabolic Pathways

Blood lipoproteins provide a system of transportation for the movement of lipids between the intestine, liver, and extrahepatic tissue (figure 11.1). Several reviews deal with the complex interactions that take place within the circulation (45, 71, 74, 132, 183). Important enzymes and transfer proteins facilitate these interactions: **lipoprotein lipase (LPL)**, **hepatic lipase (HL)**, **lecithin–cholesterol acyltransferase (LCAT)**, and **cholesteryl ester transfer protein (CETP)**. Although scientists and clinicians recognize the dynamic exchange of lipids and lipoproteins among the various lipoprotein classes and their subfractions, the transport of cholesterol and triglycerides is generally described in terms of two general processes (26, 45, 50, 198). The LDL receptor pathway consists of a sequence of chemical steps designed for the delivery of lipids to extrahepatic tissue (45). A second pathway, reverse cholesterol transport, involves a sequence of chemical reactions necessary for returning lipids from peripheral tissue to the liver for metabolism and excretion (53, 198). These two intravascular transport processes are described next in further detail.

LDL Receptor Pathway

Dietary, or exogenous, fat is absorbed by the small intestine as fatty acids and free cholesterol. During absorption, these lipids combine with apolipoproteins B-48, A-I, A-II, A-IV, and E and are internalized into the core of the chylomicron and enter the circulatory system via the lymphatic system and the thoracic duct. Blood chylomicrons then react with LPL (apo C-II facilitates this reaction), hydrolyzing the triglyceride core and releasing fatty acids. During this process, some of the phospholipids and apolipoproteins are transferred to nascent HDL particles, while other apolipoproteins of the chylomicron remnants bind to apo E and apo B-48 receptors on the surface of hepatic (liver) cells, allowing these remnants to be removed from circulation and catabolized. A similar endogenous pathway exists for VLDLs synthesized by the liver. In this pathway, LPL breaks down the VLDL triglyceride core, releasing fatty acids that are taken up by extrahepatic tissue. The remnant intermediate-density particles interact with LPL and HL to form LDL, with the remaining VLDL remnants being removed from the blood by hepatic apo E receptors. Low-density lipoprotein particles formed in this process are the primary cholesterol carriers and deliver LDLc to extrahepatic tissue, where LDL receptors located on the cell's surface mediate its uptake. Once the LDLc is recognized by the LDL-apo B-100/apo E receptor, LDL is moved inside the cell and used for cellular metabolic needs. This process also initiates a negative feedback response within the cell, causing a reduction in cellular cholesterol synthesis while promoting cholesterol storage. As these processes occur, cellular LDL receptor synthesis is suppressed and further cellular LDL uptake is halted.

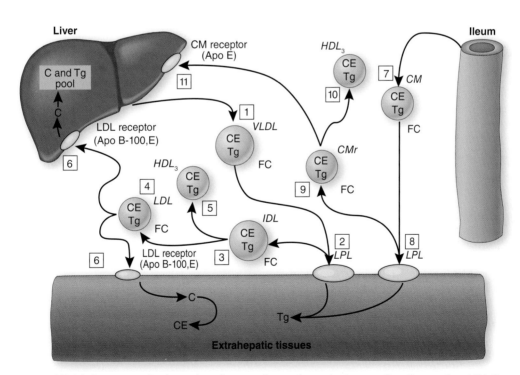

Figure 11.1   Transport of lipids between the intestine, liver, and extrahepatic tissues. See the section LDL Receptor Pathway for details. Apo = apolipoprotein; C = cholesterol; CE = cholesteryl ester; FC = free cholesterol; TG = triglyceride; CM = chylomicron; CMr = chylomicron remnant; HDL = high-density lipoprotein; IDL = intermediate-density lipoprotein; LDL = low-density lipoprotein; VLDL = very low-density lipoprotein; LPL = lipoprotein lipase.

Reverse Cholesterol Transport

Cholesterol is moved to the liver for catabolism by a process termed reverse cholesterol transport (figure 11.2). This elimination of cholesterol from the peripheral tissue involves several processes. The most notable pathway uses nascent HDL particles secreted by the liver and enriched with free cholesterol and phospholipid derived from LPL-mediated chylomicron and VLDL catabolism. In this process, free cholesterol is esterified by the action of LCAT, with apo A-I as a cofactor. The cholesteryl ester is moved into the HDL_3 core, causing a chemical gradient with a constant cholesterol supply for the LCAT reaction. As the HDL_3 core expands, the HDL_3 is transformed into a more lipid-rich HDL_2 particle. While this process is happening, reverse cholesterol transport proceeds in both indirect and direct pathways. In an *indirect* pathway, CETP facilitates an exchange for the newly formed HDL_2 cholesteryl ester with triglyceride obtained from triglyceride-rich lipoprotein (TRL) remnants. The remaining lipid-depleted TRL remnants are transported to the liver for metabolism and removal from circulation. A second set of reactions mediated by hepatic HL removes triglyceride from the HDL_2 particle that previously gained triglyceride by CETP action. Smaller, denser HDL_3 particles return to the circulation once HL-mediated triglyceride removal is complete.

In *direct* pathways, HDLc (especially from cholesteryl ester–rich HDL_2 particles, because of the content and morphology of their apolipoproteins) is withdrawn from circulation through various hepatic receptors. One important receptor in this process is the *scavenger receptor class B type I* (SR-BI). The SR-BI exists on both hepatic and extrahepatic cells (including atherosclerotic macrophages and vascular endothelial cells), has high affinity for HDL-associated apo A-I and apo A-II, and allows bidirectional flux of cholesterol between cells and HDL. As such, SR-BI, which contains a lipophilic tunnel-like structure, allows HDL to both "harvest" cholesterol from atherosclerotic lesions and "unload" it into the liver and steroid hormone–producing tissues. Unlike some of the other hepatic lipoprotein receptors, SR-BI takes cholesterol into the liver while preserving the spherical structure of HDL, allowing it to remain in circulation to participate further in reverse cholesterol transport.

As mentioned in the Scope section of this chapter, although HDL can participate in reverse cholesterol transport, raising HDLc concentration pharmaceutically does not typically decrease ASCVD risk, and this risk may be paradoxically elevated in people with abnormally high HDLc levels. Hence, HDLc concentration does not tell the whole story. For example, optimal reverse cholesterol transport relies on proper structure and function of both the HDL-associated apolipoproteins and their receptors. Malfunction of either could cause more cholesterol to be trapped within HDL, with the unloading of cholesterol from HDL to the liver being disrupted (175). In fact, a rare SR-B1 mutation has been associated with higher HDLc and is more likely to be present in people with a history of ACSVD (217). In addition to its role in cholesterol efflux from atheroma and transport to the liver, HDL may possess antioxidative, anti-inflammatory, and antithrombotic properties and may also contribute to vascular endothelial repair.

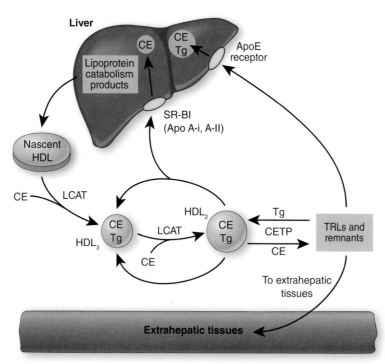

Figure 11.2 Reverse cholesterol transport. See text for details. Apo = apolipoprotein; CE = cholesteryl ester; TG = triglyceride; HDL = high-density lipoprotein; TRL = triglyceride-rich lipoprotein; LCAT = lecithin–cholesterol acyltransferase; CETP = cholesteryl ester transfer protein; SR-BI = scavenger receptor class B type I.

Postprandial Lipemia

Normally, blood triglycerides will not appreciably increase or will quickly and modestly rise after a meal and then steadily decline (121, 122). Generally, the time needed after a meal for blood triglyceride levels to return to fasting levels is 6 to 8 h. Exaggerated or prolonged lipemia is associated with increased ASCVD risk (121, 126). Elevated **postprandial lipemia (PPL)** may initiate harmful events associated with endothelial dysfunction and arterial plaque buildup (47, 153). Possibly the most significant finding is that PPL fosters the formation of the highly atherosclerotic small, dense LDL particles while reducing the concentration of HDLc. Prolonged PPL also promotes inflammation, oxidative stress, and thrombosis (blood clot) formation (204). Genetic factors affect the magnitude of the postprandial responses (156), but postprandial lipid responses are also affected by exercise (142, 158) and nutrient composition (128, 153).

Metabolic Dyslipidemia

The metabolic syndrome (see chapter 12) is a well-recognized but contentious concept in the scientific literature and in clinical practice (70, 166, 187). There are ongoing large-scale efforts to understand the complex pathophysiology that arises from the ectopic fat accumulation, insulin resistance, and dysfunction in glucose and lipid metabolism that result (55, 167, 199). An accompanying complication is hypertriglyceridemic dyslipidemia, which occurs mostly because of overproduction of TRLs. In recent years scientists have focused on the mechanisms regulating the overproduction of these atherogenic apo B–containing lipoproteins, and evidence from animal models and human studies has identified hepatic VLDL overproduction as a critical underlying factor in the development of hypertriglyceridemia and metabolic dyslipidemia (which is characterized by the combination of high triglyceride and low HDLc concentrations) (12, 13). Metabolic dyslipidemia is a common component of metabolic syndrome and was associated with a 48% higher risk of ASCVD events in overweight or obese adults with type 2 diabetes mellitus in the Look AHEAD (Action for Health in Diabetes) study (110).

Elevation of bloodborne fatty acids observed with metabolic dyslipidemia affects the vascular endothelium by reducing nitric oxide production, inducing adhesion characteristics, facilitating oxidative damage and inflammation, and resulting in diminished vascular compliance and reactivity (34, 204). Oversupply of these fatty acids to the liver facilitates triglyceride synthesis, reduces apolipoprotein B-100 degradation, and enhances the production of triglyceride-rich VLDL. Elevated plasma triglyceride concentrations and TRLs can be further exacerbated by impaired LPL activity (145).

The increased actions of HL and CETP, observed with insulin resistance and an overabundant amount of liver fat, potentiate the transfer of triglyceride from VLDL to HDL and result in the formation of small, dense LDL. These smaller, dense LDL are less likely to be taken up by the normal LDL receptor pathway, more readily penetrate the vascular endothelium, and are more susceptible to oxidative damage (12). Once in the subendothelial tissue, the modified LDL induces vascular inflammation and contributes to atherosclerotic plaque accumulation and lesion instability (127). At the same time, the activities of HL and CETP and dysfunctional apolipoprotein metabolism contribute to greater hepatic HDL clearance and a reduction in HDLc and HDL particle numbers (34). Effects on HDL number and composition further impair reverse cholesterol transport and attenuate HDL antioxidant potential.

CLINICAL CONSIDERATIONS

Elevated LDLc proportionally increases one's risk for ASCVD (112). Although national guidelines for treating dyslipidemias have existed since the early 1980s and some improvement has been seen, many individuals at the turn of the century were still not achieving previously recommended goals or residual ASCVD risk remained despite LDLc lowering (161). Accordingly, in 2002, the Adult Treatment Panel III (ATP III) of the National Institutes of Health's National Cholesterol Education Program (161) called for aggressive clinical management of dyslipidemias and emphasized achievement of targeted lipid-lowering goals based on ASCVD risk. These recommendations have been updated by various organizations over the years, based on evolving treatments and knowledge, culminating in consensus guidelines (2018) that tailor treatment based on an individual's lipid and lipoprotein profile combined with their risk for or presence of ASCVD (84). Although proper nutrition and physical activity are recognized as staples of good health and ASCVD prevention, these guidelines pay particular attention to the implementation of pharmaceutical treatment, which is discussed in the History and Physical Examination, Diagnostic Testing, and Treatment sections.

Signs and Symptoms

Given that most physical manifestations of dyslipidemic diseases are absent, these conditions are considered silent diseases. Because signs and symptoms are not physically evident, dyslipidemic diagnosis is based almost solely on the lipid profile itself; thus, the primary means for dyslipidemia diagnosis are beyond the scope of practice for the typical clinical exercise physiologist. Most importantly, each dyslipidemic condition requires laboratory testing to determine the specific abnormal blood lipid or lipoprotein level, allowing for classification and diagnosis. Remember that each type of dyslipidemia carries a different level of atherosclerotic disease risk, making the evaluation of individual lipid and lipoprotein levels crucial.

In addition to rare symptoms caused by dyslipidemia, pharmaceutical treatment of dyslipidemia may cause side effects that, although not highly common, can sometimes be quite serious. These are discussed in the Treatment section.

Hypercholesterolemia

Typical hypercholesterolemia has no outward physical symptoms; however, more severe forms of hypercholesterolemia are accompanied by physical signs. For example, familial hypercholesterolemia, a rare genetic condition of hypercholesterolemia, results in drastically reduced LDL clearance rates. As a result, blood total cholesterol levels can reach 500 mg/dL or more. Such extreme blood cholesterol levels can sometimes cause **xanthomas** (depositions of yellowish cholesterol-rich material) to form in tendons and subcutaneous tissues. Another more common genetic form is familial polygenic hypercholesterolemia, the most common cause of elevated serum cholesterol concentrations. In this case LDLc is mildly to markedly elevated (140-300 mg/dL), with serum triglyceride concentrations generally within the normal range (172). Another prominent genetic form is familial combined hypercholesterolemia, which is typified by elevations of both total cholesterol and triglyceride concentrations (27). All forms of hypercholesterolemia increase the risk for premature ASCVD; however, familial hypercholesterolemia carries the highest risk (75).

Hypertriglyceridemia

In addition to familial combined hypercholesterolemia, other genetic conditions can result in elevated triglycerides. Three primary isoforms of apo E exist—E2, E3, and E4. Apo E2 promotes hypertriglyceridemia and is associated with increased ASCVD risk due to poor binding to the apo E receptor. Because receptor binding is required for triglyceride hydrolysis and removal from blood, both blood chylomicrons and VLDL–triglyceride levels increase (46). Deficient levels of blood apo C-II, a cofactor for LPL activity, are also associated with elevated blood triglyceride levels because of limited clearance of chylomicrons and VLDL (25, 67). Familial LPL deficiency, usually found in childhood, is characterized by severe hypertriglyceridemia with episodes of abdominal pain, recurrent acute pancreatitis, and development of cutaneous xanthomas (14). Blood chylomicron clearance is impaired, causing blood triglycerides to accumulate and the plasma to appear milky. These physical signs usually decrease or resolve with restriction of total dietary fat to 20 g/d or less (67).

Other genetic defects involving single nucleotide polymorphisms and haplotypes directly influence lipoprotein–triglyceride transport in the blood (156). Familial hypertriglyceridemia is caused by a genetic defect and is passed on in an autosomal-dominant fashion. In this condition, VLDLs are overproduced, causing elevated blood VLDL–triglyceride concentrations. Familial hypertriglyceridemia does not usually become noticeable until puberty or early adulthood. Obesity and high blood glucose and insulin concentrations often accompany this genetic form of dyslipidemia (67).

History and Physical Examination

Chapter 4 provides general interview and physical examination procedures. For the most part, the patient with hyperlipidemia or dyslipidemia will not present differently from the average person.

An exception is the rare presence of xanthomas. If xanthomas are suspected in a patient who has not been previously diagnosed with hyperlipidemia or dyslipidemia, this should be brought to the attention of their physician.

In some cases, lipid-lowering medications such as statins and fibric acid, when used individually or in combination, have the potential to cause muscle damage, releasing proteins such as myoglobin into the blood (e.g., rhabdomyolysis) and causing reduced exercise performance. Besides reducing exercise performance, these proteins are harmful to the kidneys. Elevated levels may lead to kidney dysfunction or failure and, in extreme cases, death (2, 57). Although rhabdomyolysis is fairly rare, patients on lipid-lowering medications who are experiencing sudden or increasing muscle pain should curtail exercise training and refrain from undergoing exercise testing until they have been cleared by their physician.

Although hyperlipidemia and dyslipidemia can exist on their own, they are often accompanied by other conditions, such as obesity, diabetes mellitus, or ASCVD. Such conditions should obviously be taken into consideration when planning exercise testing or training. As long as potential comorbidities have been screened appropriately, patients with hyperlipidemia or dyslipidemia can safely complete exercise testing and begin a moderate-intensity exercise program that progresses to vigorous intensity (5).

Diagnostic Testing

Common dyslipidemias are diagnosed by obtaining a blood sample and measuring the blood lipid profile. Dyslipidemias traditionally were classified according to the Frederickson classification system, in which the pattern of lipoproteins was described using electrophoresis or ultracentrifugation (68). Later clinical classification for dyslipidemias was adopted from the NCEP-ATP III guidelines; updates establish priorities and rationale for treating and managing blood lipids for primary and secondary ASCVD prevention (161).

Clinical screening for dyslipidemia includes analysis of the blood lipid profile measured every 5 yr, beginning at age 20. LDLc levels are measured directly or most likely estimated using the Friedewald formula (LDLc = total cholesterol – HDLc – [triglyceride/5]) (69). However, the Friedewald formula is not used when an individual's triglyceride levels exceed 400 mg/dL. The basic lipid profile of total cholesterol, LDLc, HDLc, and triglyceride concentrations may not provide the clinician with enough information to properly diagnose some dyslipidemias. In these cases, the clinician may order more sophisticated analytic procedures to determine lipoprotein particle numbers and sizes or apolipoprotein concentrations (77). Another strategy for determining whether intermediate-risk or selected borderline-risk adults should undergo statin therapy is to obtain a coronary artery calcification score via computerized tomography, with detection of calcification generally indicating pharmaceutical treatment (84).

Some clinicians include the measurement of postprandial lipids in patients who exhibit refractory hypertriglyceridemia (i.e., hypertriglyceridemia that remains after treatment). Assessment of triglyceride and other lipid parameters, such as the number and size of TRLs, apolipoprotein responses, and calculated non-HDLc during the postprandial period, may help the clinician develop a specific therapeutic strategy for ameliorating the dyslipidemia (143, 149-151). Exaggerated triglycerides can prohibit an accurate assessment of the primary lipid target (LDLc), contribute to vascular endothelial dysfunction, and increase ASCVD risk (121, 149). A strong argument can be made that PPL is a serious but treatable condition since most people live in a postprandial state throughout the day and those with PPL may not fully clear blood lipids from a previous meal before eating again (121, 143, 149). Typically, a patient undergoes 8 to 12 h of overnight fasting before consuming a test meal with known amounts of fats, protein, and other macronutrients. Before ingestion of the test meal, blood samples are taken, and after the meal is ingested, blood samples are taken at timed intervals for up to 8 h. The changes in blood triglyceride are plotted and a removal curve is developed. The more quickly blood triglycerides are removed after the fat test meal, the less the ASCVD risk (122, 151).

In 2018, the AHA and the American College of Cardiology (ACC) spearheaded the efforts of 12 medical associations to release the Guideline on the Management of Blood Cholesterol (referred to as "the Guideline" in the remainder of this chapter) (84). To assist with pharmaceutical decision making, the Guideline employs pooled cohort equations to determine 10 yr ASCVD risk. Patients who meet criteria for one of four high-risk groups are treated with statins using dosages supported by evidence from randomized controlled trials (188). The four groups identified in the Guideline are patients with ASCVD, patients with LDLc ≥190 mg/dL, patients between 40 and 75 yr of age with diabetes mellitus and not classified in groups 1 or 2, and patients between 40 and 75 yr of age with a 10 yr ASCVD risk of ≥7.5% and not classified in any of the other groups. LDLc is the primary target for intervention through the use of statins for high-intensity therapy (intended to lower LDLc by ≥50% of the baseline lipid profile) or moderate-intensity therapy (intended to lower LDLc by 30 to 49% of baseline), depending on risk group. Because evidence from randomized clinical trial data does not support specific LDLc targets for reducing ASCVD, monitoring LDLc is recommended to assess adherence to and need for adjustment of treatment regimens, rather than to reach specific LDLc concentrations.

Although LDLc is the primary treatment target, severe hypertriglyceridemia (fasting triglycerides ≥500 mg/dL or ≥5.6 mmol/L) with a 10 yr ASCVD risk of ≥7.5% can also indicate statin therapy. When statin therapy does not perform to the extent expected, or when the optimal statin dose cannot be tolerated because of side effects, additional medications may be prescribed to help manage elevated LDLc or triglyceride. These medications are discussed in the Treatment section (84).

The Guideline focuses on pharmaceutical treatment, but it also emphasizes lifestyle behavior modification to address both the primary and secondary treatment of hyperlipidemia and dyslipidemia. Physical activity, weight loss and weight management, dietary modification, and the consumption of functional foods are described in the Guideline. In addition, proper treatment (both behavioral and pharmaceutical) of secondary conditions (e.g., diabetes mellitus, hypothyroidism) can help control hyperlipidemia and dyslipidemia (84).

Exercise Testing

Secondary forms of dyslipidemia are void of signs and symptoms for other comorbidities (e.g., ASCVD or renal insufficiency), and in these cases exercise testing should follow established protocols used in populations at risk for ASCVD. Exercise testing protocols for individuals at risk for ASCVD have been developed by the AHA and the American College of Sports Medicine (ACSM) (see chapter 5). The possibility of dyslipidemia patients having latent ASCVD is higher than in the healthy population. Considering this possibility, the clinical exercise physiologist should be aware of and understand the contraindications for exercise and exercise testing for this group.

Contraindications for general exercise testing have been identified and described by the ACSM and AHA (64). An in-depth review of each patient's medical history is performed to find possible contraindications and to stratify the patient's level of risk. As stated before, exercise is not contraindicated for individuals with a dyslipidemic blood profile. Rather, attention is given to possible existing comorbidities. When diseases are suspected or present, the protocols for testing a patient are likely to change. The clinical exercise physiologist should pay careful attention to the facility's procedures for determining eligibility for exercise testing, including obtaining a physician's release and having the appropriate personnel on hand for exercise and functional tests. The clinical exercise physiologist should be cognizant and observant of signs and symptoms that may suggest underlying or latent ASCVD while testing patients with dyslipidemia. Special considerations for cardiovascular, musculoskeletal, and flexibility testing of patients with hyperlipidemia and dyslipidemia follow.

Cardiovascular Testing

The purpose of cardiovascular testing in an individual with dyslipidemia is to help diagnose ASCVD, determine the functional capacity of the individual, and help determine an appropriate exercise intensity range. Various appropriate cardiovascular tests exist for the dyslipidemic population, and as previously stated, cardiovascular testing should follow established protocols unless comorbidities are present (see chapter 5). When comorbidities are present, testing protocols should allow for the needs of the individual and their condition(s). Also, when exercise testing and training are being considered for patients with a medical condition, a referral and a medical evaluation performed by a physician are required before testing (64).

If a patient with dyslipidemia is otherwise healthy, the anticipated responses to an exercise test are no different from what is expected in healthy individuals. However, some special circumstances may exist. Prescribed medications can alter anticipated exercise responses. Unexplained muscle pain can indicate rhabdomyolysis, and the patient should be evaluated by a physician before exercise testing. Also, chronically high cholesterol levels can result in the formation of tendon xanthomas. Individuals with this condition may experience biomechanical problems that contribute to reduced exercise performance. Exercise testing is a crucial step in evaluating the health and fitness status of patients with dyslipidemia. This information is extremely valuable for creating an exercise prescription.

Musculoskeletal Testing

The purpose of musculoskeletal testing in a patient with dyslipidemia is to determine musculoskeletal strength, endurance, and performance. Various appropriate muscular strength and endurance tests are available for use in the dyslipidemic client. Most often one-repetition maximum tests are used; however, if comorbidities are present, other tests may be employed (159). Other less hazardous tests include the estimation of one-repetition maximum (see chapter 5). Special attention should always be given to any possible orthopedic issues the patient may have, as well as to potential tendon xanthomas in patients with severe hypercholesterolemia and to statin-induced myalgia (muscle soreness).

Flexibility Testing

The purpose of flexibility testing in a patient with dyslipidemia is to determine the range of motion in all joints and, most practically, to assist in the assessment of functional abilities as they relate to exercise prescription and activities of daily living. Patients with dyslipidemia can perform testing protocols for the healthy population (see chapter 5).

Treatment

The primary goal in the clinical management of dyslipidemias is to reduce the global risk for ASCVD. Attention focuses on reducing the severity and number of traditional ASCVD risk factors. The current goals and recommendations for managing dyslipidemia are described in consensus statements for dyslipidemia, high blood pressure, obesity, diabetes, and physical activity (51, 84, 181, 208). Aggressive treatment, often including the use of lipid-lowering medication, is of primary importance for preventing future ASCVD outcomes in dyslipidemia patients with documented or known disease (8, 63, 85, 97, 99, 195). The AHA advocates teamwork by physicians and allied health professionals (e.g., nutritionists, nurses, clinical exercise physiologists) to help patients normalize their blood lipids and improve their health.

The fundamental interventions for individuals with dyslipidemia are to engage in regularly practiced physical activity, consume a heart-healthy diet, lose weight, and prevent weight regain after weight loss. All characteristics of secondary dyslipidemia are mitigated when these lifestyle behaviors are consistently practiced (63). In addition, efforts should be made to quit smoking, improve stress

management, and practice individually appropriate behavioral techniques for long-term adherence to healthy lifestyle changes (8, 63, 106). A clinician's guide explains how to prescribe exercise for most patients and delineates useful strategies for promoting healthy lifestyle changes (29). Strategies for improving patient adherence to lifestyle behavior changes are also available to the physician and members of the collaborative health care team (59, 63, 106).

Diet

Adopting a healthy diet is crucial for treating dyslipidemia and managing healthy blood lipid concentrations (8, 82, 106, 116, 130, 219). Evidenced-based dietary recommendations from the AHA, the NCEP-ATP III, and the Dietary Approaches to Stop Hypertension (DASH) diet are widely used in clinical practice (36, 106, 125, 161). A Mediterranean-style diet is consistent with the National Lipid Association's recommendations for a dietary fat intake of 25% to 35% of total calories (106). This type of diet, composed of whole grain foods, fruit, vegetables, fish, fiber, and nuts, has demonstrated efficacy in mitigating ASCVD risk and reducing future ASCVD events (179). The benefits of a Mediterranean-style diet partly come from the low content of saturated fats, trans fats, cholesterol, sodium, red meat, simple sugars, and refined grains (19, 63, 195). In addition, the Mediterranean diet provides nutrients such as omega-3 fatty acids, polyunsaturated fatty acids, soluble fiber, and carotenoids, all of which are associated with cardiometabolic health (19). Although red wine is often associated with the Mediterranean diet, the health benefits of moderate alcohol consumption, such as an increase in HDLc and enhanced vascular endothelial function, appear to be unrelated to the type of alcoholic beverage that is consumed (49).

Adopting a Mediterranean diet will help individuals with dyslipidemia balance macronutrient intake and avoid the temptation to try one of the ever-present fad diets. A fad diet may generally be characterized as one that is based on omitting an important food group and promising unsupported benefits. Current fad diets include those that recommend high or low-fat intake, high carbohydrate consumption, or high protein intake. Except for Mediterranean fare, diets with fat intake exceeding 35% are likely to include too much saturated fat, contributing to elevated LDLc, insulin resistance, and weight gain (82, 125). Low-fat, high-carbohydrate diets promote features of atherogenic dyslipidemia, such as elevated triglycerides and lower HDLc. High-protein diets can increase blood phosphorus levels and may lead to hypercalciuria, acidosis, and insulin resistance, especially in individuals with poor renal function (82, 125, 219).

Functional foods contain biologically active substances that impart medicinal or health benefits beyond their basic nutritional components. These foods appear promising in the prevention and treatment of ASCVD (184). What makes functional foods attractive to the clinician and patient is that they do not appear to have major side effects and therefore are safely incorporated into the diet (63). In addition, functional foods can impart health benefits when substituted into or added to the current diet without major alterations in eating habits or changes in nutrient intake. For individuals who are taking lipid-lowering medications, the effects of drug therapy can be augmented when combined with functional foods because these foods exert their effects differently from medications (15, 106). Functional foods that can be effective for blood lipid management include soluble fiber, plant stanols and sterols, psyllium, flaxseed, a variety of nuts, omega-3 fatty acids, garlic, flavonoids found in dark chocolate, and soy protein. Literature reviews on the topic provide a more thorough understanding of the proposed mechanisms of action and lipid-lowering effects (15, 93, 173, 184). The AHA's recommendations for managing dyslipidemias delineate a practical approach to including functional foods in an individualized plan (63).

Dietary habits are hard to break, and the clinician must think beyond itemized recommendations for dietary nutrient composition. Patients should be provided with resources to make incremental and lasting behavioral changes in meal planning, food selection, and food preparation. Contingencies for dietary challenges, such as food choices and portion control when dining out, and habits associated with altered eating patterns need to be addressed (8, 59, 63).

Physical Activity and Exercise

The amount of physical activity recommended for improving and maintaining health in most adults is very attainable. The health benefits of regularly practiced physical activity, such as improved triglyceride and HDLc concentrations, blood pressure, blood glucose control, and cardiovascular function, can be enjoyed regardless of weight change or in the absence of noticeable improvements in fitness (8, 33, 92, 107, 124, 131). The following section on exercise prescription discusses the amount and type of physical activity appropriate for individuals with dyslipidemia.

Weight Loss

Weight loss is a top priority for overweight and obese patients with dyslipidemia. The current recommendations are to achieve a 7% to 10% reduction in body weight over a 6- to 12-month period through caloric restriction and increased physical activity. Most scientists and practitioners agree that this target is best attained through a modest reduction in daily caloric intake (i.e., lowering total calories by 500 to 1,000 kcal) (51, 96). Achieving a modest weight loss of 7% to 10% will positively affect all lipid characteristics (96, 160, 218). Daily physical activity adds to the caloric deficit and imparts health benefits that may not be achieved through hypocaloric diets alone (35, 51, 92, 116, 206). Weight loss achieved through healthy lifestyle behavior can be the most effective means of preventing and treating dyslipidemia (194, 209, 212, 214). Additional health benefits can be realized with greater weight loss and long-term maintenance of the lower body weight (51, 135, 160, 165, 206). Therefore, individuals who respond well to the initial weight loss goal should be encouraged to continue behaviors conducive to further weight loss or maintenance of their new lower body weight.

Pharmacology: Lipid-Altering Medications

Table 11.3 summarizes the major medications prescribed to treat hyperlipidemia and dyslipidemia. The primary targets for lipid-lowering therapy are LDLc and non-HDLc (105, 188). Statin medications are the therapy used to lower LDLc (104). In standard dosages, statins typically reduce LDLc by 18% to 55%, but they vary in their efficacy for lowering triglycerides, increasing HDLc, and modifying other aspects of atherogenic dyslipidemia like small, dense LDL particles (83). In addition to its lipid-lowering effects, statin treatment can reduce serum uric acid levels (9), improve renal function (10, 144), and lower ASCVD events in patients with metabolic syndrome and diabetes mellitus (1, 18,

Practical Application 11.1

EFFECTS OF GRAPEFRUIT

Although grapefruit and grapefruit juice are healthful, providing many important nutriments, they can interact with dozens of medications and have undesirable effects. The chemicals in grapefruit do not interact directly with the medication. Rather these chemicals bind to an intestinal enzyme, blocking its action and making the passage of the medication from the gut to the bloodstream easier. As a result, blood medication levels may rise faster and remain at higher than normal levels; in some cases, the abnormally high levels are dangerous. Regarding statins

and grapefruit use, some statins are affected more than others. For example, blood levels of atorvastatin (Lipitor), simvastatin (Zocor), and lovastatin (Mevacor) are boosted more than fluvastatin (Lescol), pravastatin (Pravachol), or rosuvastatin (Crestor) after grapefruit juice consumption. As a general recommendation, anyone taking a statin should not consume grapefruit or grapefruit products. The patient should contact their dietitian, pharmacist, or physician with any questions concerning grapefruit or grapefruit juice consumption and statin use.

39, 61, 163, 180, 186). Statins are so named because *statin* is the common suffix (e.g., atorva*statin*, rosuva*statin*) of the medication class more formally called 3-hydroxy-3-methylglutaryl coenzyme A (HMG-CoA) reductase inhibitors. These medications interfere with a key enzyme responsible for cholesterol synthesis. As such, statins inhibit cholesterol production by the liver, and because of the resulting lower intracellular cholesterol availability, the number of LDL receptors increases in hepatic and other tissues. Downstream effects are stabilization of atherosclerotic lesions and decreased endothelial dysfunction. The most common side effects reported with statin therapy are elevated liver enzymes, myopathy, myalgia, and dyspepsia (190, 191). Statins should be avoided or discontinued in patients with liver disease or elevated liver enzymes and during pregnancy (58, 123, 155).

When very high-risk ASCVD patients (history of multiple major ASCVD events or a single ASCVD event with multiple high-risk conditions) and patients with primary hypercholesterolemia (LDLc ≥190 mg/dL or ≥4.9 mmol/L) do not achieve desired LDLc concentrations with maximally tolerated statin doses, supplemental lipid-lowering medications may be prescribed (84). The first such drug recommended is ezetimibe. Ezetimibe inhibits cholesterol uptake by the intestine, resulting in a typical reduction in LDLc of 13% to 20% and upregulation of LDL receptors. Compared with statins, ezetimibe's side effects are less common and less severe, but they may include gastrointestinal discomfort and worsening of statin-induced myalgia. A similar class of medications, bile acid sequestrants, may be prescribed instead of or in addition to ezetimibe. Bile acid sequestrants promote cholesterol elimination through the digestive tract by binding bile acids in the intestine. Although LDLc tends to decrease by 15% to 30%, compliance is poor because the medicine is unpalatable to most patients (it is administered in powder form) and it can cause considerable gastrointestinal discomfort. The use of bile acid sequestrants may not be indicated for some patients because these medications interfere with the absorption of several pharmacological agents, such as digoxin, thyroid hormones, and warfarin. They can also increase triglyceride concentration, especially in patients who are already hypertriglyceridemic (58). Otherwise, the potential for harmful side effects is rather low, so bile acid sequestrants may be prescribed to younger individuals facing long-term pharmacologic management of dyslipidemia (83). In addition to elevated triglyceride concentrations, constipation and gastrointestinal disturbances are commonly reported.

If the addition of an intestinal-acting medication to statin therapy does not result in the desired LDLc concentration in high-risk patients, a newer medication class, proprotein convertase subtilisin/kexin type 9 (PCSK9) inhibitors, may be prescribed (84). PCSK9 is an endogenous protease that binds to the LDL receptor and renders it inactive once taken into the hepatocyte via endocytosis. PCSK9 inhibitors, administered through subcutaneous injection once or twice monthly, are monoclonal antibodies that bind to PCSK9, which prevent it from binding the LDL receptor,

allowing the LDL receptor to be recycled back to the hepatocyte surface once LDL has been removed. PCSK9 inhibitors are usually very effective in lowering LDLc (43%-64%) and appear to be fairly safe (upper and lower respiratory tract infections and injection site reactions are the most common side effects). However, at this writing, safety has not been validated past 3 yr. The medications are also quite expensive, with projected lifetime cost exceeding the projected monetary savings from the prevention of cardiac events. Studies in the United States revealed that only 1.5% of patients with primary hypercholesterolemia receive PCSK9 medications, and when prescribed, this therapy was approved for reimbursement in only 61% of Medicare cases and 24% of third-party provider cases (98).

In 2020, the U.S. Food and Drug Administration approved bempedoic acid for the treatment of hyperlipidemia. Like statins, this medication class inhibits hepatic production of cholesterol, but it interrupts the process earlier in the synthetic pathway. Clinical trials have shown bempedoic acid to reduce LDLc by 15% to 25% (16) and up to 38% when combined with ezetimibe (192). Evidence suggests that side effects are relatively low, with the most common being elevated uric acid, nasopharyngitis, and urinary tract infections. Although there are no current guidelines on the use of bempedoic acid, it is anticipated that it might substitute for statin therapy when that is not well tolerated. At this writing, a clinical trial is ongoing to determine whether bempedoic acid can reduce ASCVD events (17).

Elevated fasting and postprandial triglycerides, low HDLc, and small, dense LDL particles can still contribute to residual ASCVD risk after LDLc goals are achieved (22, 83). This atherogenic dyslipidemia is approximated in the clinic through calculation of non-HDLc and measurement of triglyceride concentrations (48, 121). A triglyceride concentration of ≥175 mg/dL (1.9 mmol/L) indicates significant elevation of VLDL in the blood, and a level ≥500 mg/dL (5.6 mmol/L) indicates significant presence of chylomicrons (84). Although statins are typically considered as first-line therapy for hypertriglyceridemia, an additional medication is sometimes needed to bring triglyceride concentration below 500 mg/dL, which is important to prevent the onset of pancreatitis. In addition to pharmaceutical therapy, attention should be paid to secondary factors that can cause blood triglyceride levels to rise, such as obesity, diabetes, hypothyroidism, and medications that might increase triglycerides (e.g., oral estrogens, β-blockers, thiazide diuretics, glucocorticoids, rosiglitazone, bile acid sequestrants). Omega-3 fatty acid supplementation can also reduce triglyceride concentration. Importantly, a meta-analysis showed that each 1 g/d of eicosapentaenoic acid was associated with a 7% reduction in major ASCVD events, while no significant association was shown with docosahexaenoic acid (138).

Fibrates and niacin are both effective for lowering triglycerides, increasing HDLc, and increasing LDL particle size (82, 130, 195, 203). Fibrates reduce fatty acid uptake by the liver, thereby slowing hepatic triglyceride synthesis and VLDL production. They may

Table 11.3 Pharmacology

Medication class and name	Primary effects	Exercise effects	Important considerations
HMG-CoA reductase inhibitors • Atorvastatin (Lipitor) • Pravastatin (Pravachol) • Rosuvastatin (Crestor) • Simvastatin (Zocor) • Others	Blocks cholesterol synthesis; increases tissue LDL receptors	No direct effect	Benefits: convenient dosage schedule; useful in those with multiple risk factors; low drug interaction potential; low level of system toxicity Side effects: increased liver enzymes; myopathy; myalgia; dyspepsia Contraindications: liver disease; known or suspected pregnancy; in some instances, use with erythromycin, antifungal agents, cyclosporine
Cholesterol uptake inhibitor Ezetimibe [Zetia])	Inhibits intestinal cholesterol uptake	No direct effect	Benefits: indicated when LDLc goal is not met with maximally tolerated statin dose; side effects generally less severe than those of statins Side effects: fever; gastrointestinal discomfort; headache; myalgia; upper respiratory tract infections Contraindications: liver disease; known or suspected pregnancy
Bile acid sequestrants • Cholestyramine (Prevalite, Questran) • Colestipol (Colestid)	Binds bile acids in the intestine (facilitates cholesterol excretion)	No direct effect	Benefits: low system toxicity; use permitted in children and during pregnancy Side effects: GI distress; constipation; interferes with absorption of warfarin, digoxin, and thyroxine Contraindications: bile duct obstruction; dysbetalipoproteinemia; triglyceride >200 mg/dL
PCSK9 inhibitors • Alirocumab (Praluent) • Evolocumab (Repatha)	Inhibits LDL receptor catabolism	No direct effect	Benefits: can be prescribed when LDLc goal is not met with maximally tolerated statin dose and ezetimibe or bile acid sequestrant Side effects: upper and lower respiratory tract infections; injection site reaction Contraindication: hypersensitivity to the medication
Bempedoic acid (Nexletol)	Blocks cholesterol synthesis	No direct effect	General: Recently approved by FDA; not yet shown to lower ASCVD events (research is ongoing); available as a combination medication with ezetimibe Benefits: may be prescribed when LDLc goal is not met with maximally tolerated statin dose and/or ezetimibe; converted to active form in liver but not muscle, lowering likelihood of myalgia Side effects: hyperuricemia and gout symptoms in patients with history of gout; tendon damage, especially if over age 60 or if taking corticosteroids or fluroquinolones; upper respiratory tract infection; urinary tract infection
Fibric acid derivatives • Fenofibrate (Fibricor, Triglide, Trilipix, others) • Gemfibrozil (Lopid)	Decreases triglyceride synthesis and VLDL production; increases lipoprotein lipase activity	No direct effect	Benefits: used with well-controlled diet to reduce the number of small, dense atherogenic LDL particles; is generally better tolerated than niacin; also decreases fibrinogen levels Side effects: GI distress; gallstones; myopathy; increased excretion of uric acid Contraindications: renal disease; liver disease
Nicotinic acid	Decreases VLDL production; decreases liver clearance of HDL particles	No direct effect	Benefits: low cost and availability (OTC); favorably improves all atherogenic components of the lipid profile; often used with bile acid binding resins Side effects: flushing; GI distress; hyperglycemia; hyperuricemia Contraindications: severe hypotension; liver disease; diabetes; gout; hyperuricemia; active peptic ulcer

LDL = low-density lipoprotein; LDLc = LDL cholesterol; GI = gastrointestinal; PCSK9 = proprotein convertase subtilisin/kexin type 9; VLDL = very low-density lipoprotein; HDL = high-density lipoprotein.

also lower serum triglyceride and increase HDLc by stimulating the activity of LPL (155, 157). Common side effects of fibrates include constipation and those reported for statins. Fibrates are contraindicated for people with hepatic or renal dysfunction (81). Of the two main fibrates, fenofibrate may be prescribed as an adjunct to statin therapy, but the Guideline recommends against prescribing gemfibrozil with statins because of its higher risk of severe myopathy (84).

Niacin primarily reduces serum triglyceride and LDLc by suppressing hepatic synthesis of VLDL. A long-standing limitation for niacin has been poor tolerance of its side effects (23), although extended-release forms have improved tolerance. Niacin therapy often results in flushing, skin rashes, gastrointestinal problems, and pruritus (58, 157). It is not recommended for individuals with hypotension, liver dysfunction, or peptic ulcers or for patients with diabetes mellitus because it can increase blood glucose concentrations (81). Prescription of niacin has decreased in recent years after two large, randomized trials failed to demonstrate a reduction of ASCVD events or mortality when niacin was added to statin therapy in order to increase HDLc (79, 103). However, a systematic review and meta-analysis suggests that niacin monotherapy may reduce the incidence of acute coronary syndrome, stroke, and revascularization (44). Even so, the Guideline does not mention niacin as a major option in lipid management (84). When encountering patients on niacin therapy, the clinical exercise physiologist should remind the patient to take this medication before going to sleep at night to avoid experiencing the symptoms of flushing and hypotension during the day and, especially, with exercise. The clinical exercise physiologist should make the patient aware of these side effects so they may be distinguished from symptoms of heavy exertion.

Up to 25% of patients taking lipid-lowering medications, particularly statins and fibric acid derivatives, may experience some form of medication intolerance that includes muscle inflammation and muscle damage (2). Rhabdomyolysis, a severe side effect arising from medication-induced muscle damage, is characterized by the presence of myoglobin in the blood. Rhabdomyolysis is also characterized by reduced and dark urine output and a general feeling of weakness. Since kidney damage may result, a patient experiencing these symptoms should immediately contact their physician.

Choosing to follow a Mediterranean-style diet, incorporating functional foods in the diet, exercising regularly, and even modest weight loss are each effective means for improving blood lipid levels. When these strategies are combined, they may work in concert to improve blood lipids. If patients respond conservatively to each of these therapeutic lifestyle changes, they may expect to lower LDLc and triglyceride levels and improve HDLc to the same magnitude that would be expected with lipid-lowering medications (63, 106). For patients who are not on lipid-lowering medication, the lifestyle strategies could be enough to keep them off these medicines. For patients already taking these drugs, lifestyle strategies may add to their medicine's effect. However, this does

not always occur, and further research on combination therapy is clearly needed (3, 158). For all patients, these therapeutic lifestyle changes will provide health benefits above and beyond what can be attained with lipid-lowering medication alone.

EXERCISE PRESCRIPTION

Exercise can have a profound impact either directly or indirectly (i.e., weight changes) on blood lipid and lipoprotein levels. Although various exercise modalities can provide health benefits that may ultimately improve lipid status, the majority of evidence suggests that cardiorespiratory exercise, or endurance-type activity, is preferred, although a yet-defined combination program may be optimal. Nevertheless, the specific doses for optimizing improvements have not been completely defined. Here we review currently recommended programming.

Cardiorespiratory Exercise

The ranges of physical activity volumes recommended in the 2008 U.S. Department of Health and Human Services' Physical Activity Guidelines for Americans (182) and informed by position statements from the ACSM (37, 92, 148) are appropriate for individuals with dyslipidemia (table 11.4). These guidelines suggest ranges of 150 to 300 min of moderate-intensity or 75 to 150 min of vigorous-intensity physical activity per week, or some combination of the two. A dose–response relationship is recognized, with physical activity at the upper ends of these ranges likely to result in more positive health benefits. Indeed, some of the earliest research investigating the effect of exercise training on lipid and lipoprotein concentrations indicated that a threshold of 8 to 10 mi (12.9 to 16 km) of running per week is necessary to significantly change HDLc concentration (211, 213). Because the lipid- and lipoprotein-related benefits of physical activity rely more on the total volume of activity performed than on the intensity of the activity (54, 124), walking an equivalent distance is thought to yield similar benefits. Assuming common walking and running speeds (e.g., 3 and 6 mph [4.8 and 9.7 kph], respectively), this 8 to 10 mi threshold falls within the lower end of the moderate-intensity and vigorous-intensity time ranges mentioned.

In support of a dose–response relationship, Kraus and colleagues (124) reported greater favorable changes in LDL particle size and HDLc and triglyceride concentrations in dyslipidemic men and women who averaged 20 mi (32 km) of running per week versus those who ran or walked an average of 12 mi (19 km) weekly. In addition, only doses of exercise training as high as or higher than that listed in table 11.4 resulted in improved cholesterol efflux from cultured macrophages exposed to plasma or serum from overweight or obese adults, suggesting that higher exercise training volumes may be necessary to improve certain aspects of HDL function (178). Therefore, although fulfilling the minimum recommendations in the Physical Activity Guidelines for Americans is likely beneficial, exercising at or slightly above the upper end of these guidelines may be necessary to optimize

Practical Application 11.2

CLIENT–CLINICIAN INTERACTION

When interacting with patients who have hyperlipidemia and dyslipidemia, the clinical exercise physiologist must clearly outline the dangers and implications of having altered levels of cholesterol. This will help communicate the seriousness of the patient's condition and the importance of ameliorating it. The patient can then be given information about how to improve their condition, including the benefits of high-volume aerobic exercise. The clinical exercise physiologist needs to be simple and direct in highlighting the improvements seen in HDL and triglycerides with exercise training. In addition, emphasis should be placed on the beneficial effects of exercise on overall ASCVD risk. This information is particularly important for patients with multiple risk factors. Regarding aerobic training, the patient needs to understand that the benefits are increased with greater exercise energy expenditure (i.e., approximately 300 min/wk of moderate-intensity physical activity or 150 min/wk of vigorous-intensity physical activity). Daily or near-daily exercise is advocated. Daily physical activity may be split into two or more sessions, if needed, to attain the desired exercise volume (181).

The safest approach to accomplishing this goal is to begin an exercise program with a nonfatiguing amount of moderate-intensity exercise and to gradually build the volume (frequency, duration, intensity, or a combination) over a number of weeks (5). This method is prudent in terms of both safety and program compliance and may also help distinguish normal exercise-induced muscle soreness from myalgia that can occur from statin or fibrate treatment. Patients should be instructed to inform the clinical exercise physiologist whenever their medications are changed or the dosages are increased. In such cases, increases in exercise volume should be withheld for a couple of weeks, and any unusual muscle soreness should be noted. In addition, any sudden or severe muscle pain that cannot be logically explained by a recent increase in physical activity should be brought to the attention of both the patient's physician and clinical exercise physiologist.

Along with aerobic exercise, a calorie-restricting diet can be an effective tool in decreasing total cholesterol, non-HDLc, LDLc, and triglycerides while increasing HDLc. Therefore, the clinical exercise physiologist should ultimately encourage the client to combine high volumes of aerobic exercise of moderate to vigorous intensity with a calorie-restricting diet to result in high weekly caloric deficits. The collaborative team approach, invoking the knowledge and expertise of the clinical exercise physiologist and the clinical nutritionist and the advocacy and guidance of a physician, is strongly recommended.

---

lipid and lipoprotein metabolism. In fact, in some cases, practice of physical activity at the lower end of the recommended levels may result in a stabilization of lipid and lipoprotein levels rather than an actual improvement. It should be noted, however, that a sedentary person's lipid and lipoprotein profile is likely to worsen over time, particularly if they are gaining weight, so such stabilization should be recognized as a positive effect if lipid and lipoprotein levels had previously been getting worse. In addition, as mentioned in the next section on exercise training, regularly practiced physical activity can directly affect the metabolism of some lipids and lipoproteins, while alteration of others may depend on changes in body composition. Therefore, energy expenditure at the upper end of the continuum, combined with prudent nutrition, is the best behavioral prescription for improving the overall blood lipid and lipoprotein profile.

Although progressing toward 300 min/wk of moderate-intensity or 150 min/wk of vigorous-intensity physical activity is recommended, little research is available to suggest the number of days per week the activity should be performed. Studies employing single sessions of exercise typically show that lipid and lipopro-

tein concentrations change the most 1 or 2 d afterward and that they return toward baseline by 3 d. In addition, these short-term benefits likely accumulate from one exercise session to the next (i.e., exercising before lipid and lipoprotein concentrations return to baseline may stimulate further improvement) (6, 91, 202). Thus, a sensible approach might be to allow no more than 1 or 2 d of inactivity between exercise sessions. This routine would require performing leisure-time physical activity on at least 3 d across the week. Because of the amount of time involved, most people require 5 d/wk or more to complete the optimal amount of moderate-intensity physical activity. On the other hand, a person's daily dose of physical activity does not need to be completed all at once. Limited research indicates that breaking physical activity up into three sessions of 10 min or more within a day is at least as effective in altering blood lipid and lipoprotein concentrations as is continuous physical activity (4, 141, 147).

In summary, people should accumulate 150 to 300 min of moderate-intensity or 75 to 150 min of vigorous-intensity dynamic physical activity using large muscle groups, or some combination of the two, throughout each week. To optimize lipid and lipopro-

Table 11.4   Exercise Prescription Review

Training method	Frequency	Intensity	Time (Duration)	Type (Mode)	Progression	Important considerations
Cardiorespiratory	At least every other day	Work toward continuous moderate-intensity activity: 40%-59% HRR or $\dot{V}O_2R$; in absolute terms, 3-6 METs is generally recognized as moderate intensity	At least 30 min/d	Dynamic, large-muscle exercise (walking, jogging, cycling, elliptical training)	Increase duration and frequency. Work toward daily activity beyond moderate intensity; intensity does not appear to be as influential as duration.	More positive benefits are observed as the total volume of activity increases, which can be best reflected by increasing duration. Follow the general prescription guidelines for cardiorespiratory exercise found in chapter 6.
Resistance	2 or 3 sessions per week on nonconsecutive days	8-12 reps for most exercises; for those new to resistance training and older individuals, 8-15 reps for most exercises	1 set of 8-10 exercises involving all major muscle groups	Body weight, free weights, bands, water exercises	Increase repetitions to achieve upper end of range (e.g., 12-15 reps) and then add resistance so that 8 reps can be achieved. Sets may be added, but training time is doubled with this option.	Follow the general prescription guidelines for resistance exercise found in chapter 6
Range of motion	At least 2 sessions per week	Stretching to the point of feeling tightness or slight discomfort	Static stretches: hold for 10-60 s PNF stretches: 3-6 s of light to moderate contraction (20%-75% of MVC) followed by 10-30 s of assisted stretching; work toward 60 s for each exercise Repeat each stretch 2-4 times	Static (active or passive), dynamic, ballistic, and PNF stretches are all effective	Increase frequency to a daily routine.	Also consider neuromotor exercises (e.g., activities such as yoga and tai chi that focus on balance agility, coordination, and gait) to maintain ROM and physical function.

HRR = heart rate reserve; $\dot{V}O_2R$ = oxygen uptake reserve; METs = metabolic equivalents of task; MVC = maximal voluntary contraction; PNF = proprioceptive neuromuscular facilitation.

tein changes, clients are encouraged to progress to the upper end of the exercise dose range. Leisure-time physical activity should be practiced on at least 3 d throughout the week, and 5 d or more are required for most people to meet the dose recommendations. Moderate- to vigorous-intensity physical activity may be accumulated throughout the day and does not necessarily need to be accomplished within a single daily session.

Resistance Exercise

Resistance training alone may have a very limited effect on improving blood lipid and lipoprotein concentrations (54, 114). However, given that resistance exercise has numerous benefits not associated with lipoprotein metabolism, patients with dyslipidemia should still follow the recommendations presented in chapter 6. Focusing on a higher than minimally recommended volume of resistance training is not necessary, unless needed to improve other aspects of health.

Range of Motion Exercise

Although it provides no known benefit to lipid and lipoprotein profiles, range of motion exercise is important to overall fitness and should be practiced as discussed in chapter 6. Unless there are certain comorbidities, no special considerations for dyslipidemia patients exist.

EXERCISE TRAINING

Exercise is a valuable therapeutic treatment for improving blood lipids and reducing ASCVD risk (38, 40, 95, 131, 205). Several studies and decades of research provide accumulating evidence of the benefits of exercise training on blood lipids and lipoproteins. Although many initial studies were either cross-sectional or were conducted in individuals with normal lipid concentrations, more recent approaches have attempted to clarify the expected improvements among dyslipidemia patients.

Cardiorespiratory Exercise

Blood lipid profiles of physically active groups generally reflect a reduced risk for the development of ASCVD compared with those of their inactive counterparts (53, 65, 154). Strong evidence exists for the presence of lower triglyceride and greater HDLc concentrations in physically active individuals. Triglyceride levels are almost always lower in endurance athletes, aerobically trained people, and physically active individuals when compared with sedentary controls. Significant triglyceride differences of up to 50% exist between these groups in over half of all related cross-sectional studies. Blood levels of HDLc are between 9% and 59% higher in those having physically demanding jobs and in individuals engaged in endurance exercise compared with their less active counterparts (53, 54, 205). Only limited evidence suggests that people who are physically active exhibit lower concentrations of total cholesterol and LDLc concentrations than those who are less active.

In longitudinal studies, total cholesterol and LDLc infrequently change with exercise training in either men or women. When these lipid fractions are altered with exercise training, the reductions are minimal or moderate, averaging only 4% to 7% when compared with values in nonexercising control subjects (53, 54, 86). HDLc and triglyceride are more responsive to regular exercise than are total cholesterol and LDLc, based on the frequency of reported changes. Significantly greater HDLc concentrations are reported after exercise training in over half of the reviewed publications, while reductions in triglyceride levels are found in a third of the related literature (53, 54, 205). When HDLc is significantly elevated after exercise training, the increases are similar in men and women, ranging from 4% to 22%. In addition to increasing HDLc, exercise boosts the antioxidant properties of HDL (189, 197). Combined with reductions in triglyceride, the elevated antioxidant potential of HDL is thought to attenuate the inflammation and oxidative stress that contribute to vascular dysfunction and ASCVD (139, 164, 170, 174, 207).

Significant reductions in triglyceride concentrations range from 4% to 37% after aerobic exercise training in males; the magnitude of change is similar in women but is seen less frequently (table 11.5). Resistance training also has positive but more modest effects on blood lipid and lipoprotein concentrations than observed for aerobic exercise (113, 114).

Exercise and Dyslipidemia

A meta-analysis of lipid changes in normo- and hyperlipidemic groups suggested that exercise training may have only limited influence in lowering total cholesterol and LDLc. The effects of exercise training on HDLc and triglyceride also favored more conservative estimates, with HDLc increases averaging 4% and triglyceride decreases averaging 6% to 19% (86).

LDLc is lowered by a modest 5.5 mg/dL (95% CI: −9.9 to −1.2 mg/dL) through regularly practiced walking programs lasting 8 wk or more (119). In men, the pooled average reduction in total cholesterol and increase in HDLc were only 2%, whereas triglyceride was decreased by 9% (113). In both men and women, walking lowered non-HDLc by a pooled mean of 5.6 mg/dL (95% CI: −8.8 to −2.4 mg/dL), which equated to a modest 4% decrease (118). The pooled estimated changes in total cholesterol, LDLc, HDLc, and triglyceride concentrations indicate that individuals with documented ASCVD respond similarly to regular exercise training (115). Solid evidence exists that regular moderate-intensity aerobic exercise can increase LDL and HDL particle sizes in men and women (87, 124). In women, however, the lipid responses appear to be more favorable among those who are at greater risk for heart disease versus their healthy counterparts (134).

Overweight and Obesity

Meta-analytic results of 13 studies and 31 groups of overweight and obese adults were strikingly similar to those reported for most adults regardless of weight status (86, 113, 118). Total cholesterol

Table 11.5 Lipid and Lipoprotein Changes Associated With Exercise

Lipid/Lipoprotein	Single exercise session	Exercise training
Triglyceride	Decrease of 7% to 69% Approximate mean change 20%	Decrease of 4% to 37% Approximate mean change 24%
Cholesterol	No change[a]	No change[b]
LDL cholesterol	No change	No change
Small, dense LDL cholesterol particles	Literature unclear	Can increase LDL particle size, usually associated with triglyceride lowering
Lp(a)	No change	No change
HDL cholesterol	Increase of 4% to 18% Approximate mean change 10%	Increase of 4% to 18% Approximate mean change 8%

[a]No change unless the exercise session is prolonged (see text).

[b]No change if body weight and diet do not change (see text).

decreased by 2% with endurance exercise training of at least 8 wk. Triglycerides were lowered by 11%, and LDLc and HDLc were not significantly affected (117).

More rigorous physical activity programs are required for weight loss, and they result in greater health benefits (51); healthful changes in blood lipids and lipoproteins, particularly total cholesterol and LDLc, are greatest when physical activity is accompanied by weight loss (209, 210, 212). These findings are consistent with an initial meta-analysis of 95 studies examining the question of exercise-induced lipid changes with and without weight loss (196).

Immediate and Transient Effects of a Single Exercise Session

The characteristic antiatherogenic lipid profile of physically active and exercise-trained individuals was recognized long ago as primarily a transient response to the last session of physical activity or exercise and independent of chronic exercise training (152). Early support for this acute response hypothesis came from studies showing that blood triglyceride concentration was reduced for up to 2 d after a single aerobic exercise session, and this beneficial reduction was evident even in men with hyperlipidemia (31, 101, 152). Pioneering work in this field provided evidence that HDLc concentrations were higher and total cholesterol concentrations lower shortly after a single session of exhaustive exercise (56, 120, 190).

Total cholesterol and LDLc responses to a single exercise session are highly variable. Small postexercise reductions in total cholesterol (3%-5%) have been reported for male and female hyper- and normocholesterolemic subjects (24, 41, 42, 78, 109, 162). Lower serum LDLc has been reported in trained men immediately and up to 72 h after completion of intense endurance events (24, 60) and in women after exercise of relatively high intensity and volume (76, 162). In contrast, LDLc remained unaffected in normolipidemic obese women after 1 h of exercise (216), and LDLc concentration may increase or decrease 5% to 8% in hypercholesterolemic men after exercise (41, 42, 78).

Low-density lipoprotein particle size, a measure of the atherogenic potential of LDL, was not altered in either normal or hypercholesterolemic women by a single aerobic exercise session (42, 215). However, an increase in LDL particle size has been shown to follow completion of a marathon in men (129). The transient effects of exercise on LDL may not always benefit health, because circulating LDL may be more susceptible to harmful oxidation after very intense, long-duration exercise such as running a marathon (133).

A single aerobic exercise session of sufficient volume can raise serum HDLc. The postexercise increase in HDLc is strikingly like what is generally attributed to long-term endurance exercise training. However, HDLc levels peak 24 to 48 h after exercise and last up to 72 h before returning to preexercise levels (54). More stable lipid and lipoprotein changes sometimes take several months to achieve (131). Acutely, an exercise energy expenditure threshold of about 350 kcal is enough to elevate HDLc in deconditioned individuals (41), but a caloric expenditure threshold of 800 kcal or more may be needed in well-conditioned individuals (60). The average density of HDL is reportedly reduced for up to 2 d after aerobic exercise (suggesting larger HDL particle size), a finding that provides further evidence that exercise acutely influences lipoprotein metabolism (42).

A single endurance exercise session can lower blood triglyceride concentrations (176, 201), and this exercise effect is observed in apparently healthy normo- and hyperlipidemic men (24, 41, 42, 60, 66, 78). Similar to the HDLc responses to a single exercise session, the exercise effect on serum triglyceride is influenced by the training status of the subjects and volume of exercise performed. Regardless of mode or intensity, the exercise effect is lost after about 48 to 72 h (54). Existing evidence also supports a relationship between preexercise triglyceride concentration and the magnitude of the postexercise change. In other words, people with elevated preexercise serum triglyceride concentrations exhibit the greatest postexercise reductions, while those with relatively low preexercise triglyceride concentrations show only modest or no change after exercise (42, 43, 78, 109). Genetic variations are known to play a role in triglyceride metabolism (108, 156); however, very little data exist on the transient lipid-altering effects of a single exercise session in individuals with genetically determined hypertriglyceridemia.

Postprandial lipemia is lower in the hours after completion of aerobic exercise of sufficient volume (136, 142, 158) but is increased when exercise is withdrawn for several days (89). Together, these data suggest that the beneficial influence of exercise on circulating lipids and lipoproteins is an acute phenomenon that is lost rather quickly after cessation of exercise, even in the most highly trained individuals (95). The message for the clinical exercise physiologist is that exercise must be repeated regularly to maintain the acute benefit.

Resistance Exercise

Resistance exercise appears to have a small beneficial influence on blood lipids that is somewhat similar to that reported for endurance exercise. Meta-analyses on up to 69 studies including up to 2,158 male and female adult participants provide evidence that regular resistance training modestly lowers total cholesterol, LDLc, non-HDLc, and triglyceride while increasing HDLc, as long as the training volume (resistance, repetitions, or sets, or a combination) increases progressively over the course of the exercise program (40, 114). The clinician should interpret these findings with caution and continue to review the literature in this area; relatively fewer calories are expended in resistance versus aerobic activity, and therefore resistance training per se may be less effective than endurance activities for modifying blood lipid levels (185).

Research Focus

Exercise and HDL Function

Exercise is thought to enhance reverse cholesterol transport and increase HDLc. However, evidence provides insight on how aerobic exercise improves antioxidant properties of HDL and the role HDL plays in transporting lipid peroxides (197). The purpose of this investigation was to determine the influence of exercise on endogenous antioxidant potential and lipid oxidation in LDL and HDL fractions.

Twenty-four male endurance runners underwent a progressive treadmill running protocol to exhaustion. Exercise was strenuous (between 20 and 22 min, with the last 6-8 min above the anaerobic threshold). Blood samples were obtained after standardized preexercise procedures and just before exercise, 15 min after exercise, and again at 90 min postexercise. Lipoproteins were separated from serum obtained from the blood samples. Lipids were extracted from the lipoprotein fractions, and measures of LDLc, HDLc, and oxidized lipids were estimated from standardized benchtop methods. Serum was measured for total antioxidant capacity, the HDL-associated antioxidant paraoxonase, and lipoprotein oxidation resistance.

Greater HDLc and lower LDLc concentrations were observed within 90 min of completing the treadmill exercise. Total antioxidant capacity was enhanced along with HDL-associated paraoxonase. LDL lipid peroxides were reduced; however, HDL lipid peroxides increased after exercise. The lipid peroxide changes favored greater HDL peroxide transport.

Exercise acutely increases the concentrations of oxidized HDL lipids while having an opposite effect on oxidized LDL lipid concentrations. The distribution shift is opposite of what occurs in the hours after eating a high-fat meal. Results from the serum analyses were interpreted to suggest that the increase in HDL lipid peroxides was not attributed to a transfer among LDL and HDL fractions. Rather, greater oxidized HDL lipids were thought to arise from enhanced HDL lipid peroxide clearing function, where products of lipid oxidation were scavenged by HDL at sites of formation and transported back to the liver. The reduced LDL lipid peroxide levels were not due to a decrease in LDLc, because the ratio of oxidized LDL lipid to LDLc decreased after exercise. The authors speculate that an increased HDL antioxidant capacity may have protected LDL lipids from oxidizing postexercise; however, this interpretation was not directly tested.

This study offers strong evidence that exercise induces biochemical changes beyond altering lipoprotein lipid concentrations. Potentially harmful lipid peroxides in LDL are diminished, and HDL transport of lipid peroxides and antioxidant function are improved. The study highlights the key role exercise plays in modifying lipoprotein lipids and the inflammation and oxidative stress associated with lipid transport.

Based on Valimaki et al. (197).

Clinical Exercise Bottom Line

- Dyslipidemia patients have an elevated risk of cardiovascular disease. Therefore, the patients themselves and professionals providing exercise testing and training should be keenly aware of ASCVD signs and symptoms.

- A dyslipidemic condition alone does not necessitate a change in exercise testing protocols or intensity for exercise training. Patients with dyslipidemia could have preexisting comorbidities, such as coronary artery disease, obesity, and diabetes, that would significantly change exercise testing or training protocols.

- Up to 25% of patients taking lipid-lowering medications, particularly statins and fibric acid derivatives, may experience some form of medication intolerance that includes muscle inflammation and muscle damage. Rhabdomyolysis, a severe side effect arising from medication-induced muscle damage, is characterized by the presence of myoglobin in the blood, reduced and dark urine output, and a general feeling of weakness. Since kidney damage may result, a patient experiencing these symptoms should immediately contact their physician.

- Flushing, sweating, and nausea are associated with niacin use. The clinical exercise physiologist should remind the patient to take this medication before going to sleep at night to avoid these unwanted side effects during the day and especially with exercise. The patient should be made aware of these side effects so they can distinguish between the effect of the drug and sensations they might experience during exercise.

CONCLUSION

Exercise is essential for management of hyperlipidemia and dyslipidemia, with proven benefits that likely extend to a reduction in mortality and morbidity. Transient improvements in some lipid fractions, such as HDL cholesterol and triglyceride, may occur with just one exercise session, while more stable changes sometimes take several months to achieve. A comprehensive lifestyle treatment approach should be of utmost priority to best attain targeted patient cholesterol goals. As such, in addition to exercise programming, better nutrition choices and body composition improvements are needed along with appropriate medications, when necessary. A multidisciplinary team approach may prove most beneficial for helping patients achieve long-term success in meeting stringent guidelines.

Online Materials

Visit HK*Propel* to access a link to the references, the case studies with discussion questions, and a quiz to help you review key concepts and test your understanding of the material covered in this chapter.

CHAPTER 12

Metabolic Syndrome

James R. Churilla, PhD, MPH, MS

Metabolic syndrome is an aggregation of interrelated **cardiometabolic risk factors** that are present in a given individual more frequently than may be expected by a chance combination. Patients with metabolic syndrome are usually characterized as overweight or obese and have significantly greater prospective risk for developing atherosclerotic cardiovascular disease (ASCVD) (131) and type 2 diabetes (T2D) (202). Although metabolic syndrome has received a great deal of research attention since the beginning of the 21st century, physicians and researchers have documented and tracked the accumulation and clustering of obesity-related cardiometabolic risk factors since the early 20th century. In fact, the seminal syndrome involving a constellation of cardiometabolic risk factors was originally conceived by the Swedish physician Kylin, who in 1923 reported an association between hypertension (HTN), hyperglycemia, and **hyperuricemia** (gout) (113). Moreover, in 1947, Vague documented a robust link between android (i.e., central or abdominal) adiposity and metabolic abnormalities (188); and in 1965 and 1967, Avogaro and colleagues confirmed the co-occurrences of obesity, hyperinsulinemia, hypertriglyceridemia, and HTN (13, 14). Camus (31) and Jahnke and colleagues (97) subsequently reported on similar ensembles of these risk factors, coining the terms *deadly quintet* and *special metabolic syndrome*, respectively. However, it was not until 1977 that the term *metabolic syndrome* was applied by Haller, who suggested that obesity, diabetes mellitus, hyperlipoproteinemia, hyperuricemia, and hepatic steatosis (fatty liver) cluster to pose significant risk for developing or worsening arteriosclerosis (88).

On the basis of those findings, and during his Banting lecture to the American Diabetes Association in 1988 (156), the late Dr. G.M. (Jerry) Reaven proposed that hyperinsulinemia and insulin resistance were the pathophysiologic mechanisms linking obesity and these risk factors, in a condition for which he coined the term *syndrome X*. Since then, metabolic syndrome has also been referred to as *plurimetabolic syndrome* (47), *dysmetabolic syndrome* (54), *the deadly quartet* (105), and *insulin resistance syndrome* (59), among other names. Although at present *metabolic syndrome* is a widely recognized public health term, the clinical diagnosis and classification of this condition have been debated. This is in large part due to historical inconsistencies regarding the operationalization and classification of the condition.

DEFINITION

The American Association of Clinical Endocrinology (AACE) championed the first International Classification of Diseases code (code 277.7) for dysmetabolic syndrome (54). This code was approved in 2001 and added to the ninth revision (ICD-9) of the coding system (158). In 2015, ICD-9 code 277.7 was converted to E88.81 (metabolic syndrome) in the ICD-10. An initial formal definition for metabolic syndrome was proposed by a consultation group for the World Health Organization (WHO) in 1998; however, it was ultimately deemed impractical because of the requirement of direct measures of insulin resistance (9). Since the original definition by the WHO, numerous other medical and public health institutions have proposed alternative criteria to provide a basis for clinical diagnostics and population health surveillance. Specifically, the European Group for the Study of Insulin Resistance (also requiring insulin measurement) (16), the AACE (54), the International Diabetes Federation (IDF) (7), the National Cholesterol Education Program Adult Treatment Panel III (NCEP-ATP III) (184), and the American Heart Association/National Heart, Lung, and Blood Institute (AHA/NHLBI) (82) have each contributed separate definitions. Whereas most of these demonstrated overlap regarding the individual risk factors

Acknowledgment: We wish to gratefully acknowledge the efforts of the previous authors of this chapter, Mark D. Peterson, PhD, Paul M. Gordon, PhD, MPH, and Flor Elisa Morales, MS.

of metabolic syndrome (i.e., dysglycemia, HTN, elevated triglyceride levels, low high-density lipoprotein cholesterol (HDL-C), and central obesity or augmented body mass index (BMI), the conflicting criteria for cut points across definitions led to confusion in the clinical context for identifying patients at risk, as well as inconsistencies in prevalence estimates across population-representative studies (39).

In 2009, a harmonized definition was proposed by the IDF Task Force on Epidemiology and Prevention, AHA, NHLBI, World Health Federation, International Atherosclerosis Society, and International Association for the Study of Obesity (7). The purpose of the harmonized definition was to standardize the criteria for metabolic syndrome by publishing an agreed-upon set of cut points for individual factors (table 12.1). Further, whereas several previous definitions incorporated requisite criteria (e.g., insulin resistance and abdominal obesity [8, 9, 39]), the newest definition recommends that a diagnosis of metabolic syndrome reflect the presence of any three or more abnormal findings. Although this diagnostic scheme is superior to previous recommendations, there is still some disagreement pertaining to the definition and cut points for elevated waist circumference (WC). This is due to the significant variability in abdominal obesity phenotypes between sexes and among races and ethnic groups, as well as the subsequent association with other metabolic risk factors and predictive values for ASCVD and diabetes (7, 8, 91). Therefore, the harmonized definition still defers to the IDF recommendations for cut points

of abdominal obesity to be used as a single potential factor in the diagnosis of metabolic syndrome (table 12.2) (7, 8). Table 12.2 also includes the WC thresholds being recommended in several different populations and ethnic groups and by different organizations.

SCOPE

Before the harmonized definition was proposed, the reported prevalence estimates for metabolic syndrome varied widely depending on the criteria used for diagnosing the condition (39). During the late 1980s and early 1990s, approximately 47 million U.S. adults were affected by metabolic syndrome (60). At the beginning of the 21st century, nearly 65 million Americans possessed a minimum of three of the five clinical criteria for metabolic syndrome (61). According to a study of U.S. adults participating in the 2007-2014 National Health and Nutrition Examination Survey (NHANES), 34.3% of adults and nearly 55% of older adults (>60 yr) satisfied the AHA/NHLBI criteria for metabolic syndrome (176). Overall prevalence of metabolic syndrome decreased from 36.6% in 2007-2008 to 33.8% in 2013-2014 in the United States, but this attenuation was not statistically significant. The relatively stable prevalence estimates of metabolic syndrome during this 8 yr period may be due to significant decreases in elevated triglyceride and blood glucose levels, accompanied by stable HDL-C levels, in U.S. adults during this time. Additionally, global estimates indicate that approximately one in four adults in European and Latin American

Table 12.1 Harmonized Clinical Criteria for Metabolic Syndrome

Measure	Categorical cut points
Elevated waist circumference[a]	Population- and country-specific definitions
Elevated triglycerides (drug treatment for elevated triglycerides is an alternative indicator)[b]	≥150 mg · dL⁻¹ (1.7 mmol · L⁻¹)
Reduced HDL-C (drug treatment for reduced HDL-C is an alternative indicator)[b]	<40 mg · dL⁻¹ (1.0 mmol · L⁻¹) in males <50 mg · dL⁻¹ (1.3 mmol · L⁻¹) in females
High blood pressure (antihypertensive drug treatment in a patient with a history of HTN is an alternative indicator)	Systolic ≥130 mmHg, diastolic ≥85 mmHg, or both[d]
Elevated fasting glucose[c] (drug treatment of elevated glucose is an alternative indicator)	≥100 mg · dL⁻¹

[a]It is recommended that the IDF cut points be used for non-Europeans and either the IDF or AHA/NHLBI cut points be used for people of European origin until more data are available.

[b]The most commonly used drugs for elevated triglycerides and reduced HDL-C are fibrates and nicotinic acid. A patient taking one of these drugs can be presumed to have high triglycerides and low HDL-C. High-dose omega-3 fatty acids presumes high triglycerides.

[c]Most patients with type 2 diabetes mellitus have metabolic syndrome by the proposed criteria.

[d]According to American Heart Association and American Stroke Association classifications, an elevated blood pressure is a systolic blood pressure of 120-129 mmHg and a diastolic blood pressure <80 mmHg; stage 1 high blood pressure is a systolic blood pressure of 130-139 mmHg or a diastolic BP of 80-89 mmHg. Taking BP-lowering medication carries the elevated or high blood pressure diagnosis.

HDL-C = high-density lipoprotein cholesterol.

Reprinted by permission from K.G. Alberti et al. "Harmonizing the Metabolic Syndrome: A Joint Interim Statement of the International Diabetes Federation Task Force on Epidemiology and Prevention; National Heart, Lung, and Blood Institute; American Heart Association; World Heart Federation; International Atherosclerosis Society; and International Association for the Study of Obesity," *Circulation* 120 (2009): 1640-1645.

countries have metabolic syndrome, and the constellation is rising rapidly in Asia (196). Results from the 2015-2016 and 2017-2018 NHANES illustrate that 32% of U.S. adults are overweight, and 42% were classified as obese based on BMI (www.cdc.gov/nchs/data). This contrasts data reported by the WHO in 2016, illustrating global overweight and obesity estimates being 39% and 13%, respectively (204).

Overweight and obesity have been identified as the major driver of metabolic syndrome. Depending on the criteria, they affect up to one-half to three-quarters of all U.S. adults (62, 85, 175). Furthermore, evidence continues to mount in support of the robust association between excess abdominal, visceral, and hepatic adiposity and a broad spectrum of deleterious cardiometabolic outcomes (48), prompting the alarming public health message that obesity leads to more chronic diseases and poorer health-related quality of life than smoking (182). The effort to target the diagnosis, prevention, and treatment of both obesity and metabolic syndrome as separate yet overlapping conditions remains a vital public health agenda.

Previous epidemiological studies have used the definitions set forth by either the WHO, AHA/NHLBI, or IDF guidelines (39, 42). Whereas the 2009 harmonized definition (7) represents a significant departure from the WHO definition, the only modifications to previous NCEP-ATP III (now AHA/NHLBI) guidelines

are a lower cutoff for systolic blood pressure ($\geq$130 mm Hg) and an endorsement of IDF race- and ethnicity-specific cutoffs for abdominal obesity (table 12.2) (i.e., rather than abdominal obesity defined uniformly across all races and ethnicities as $\geq$102 cm for men and $\geq$88 cm for women).

Many reports have suggested similar prevalence estimates of metabolic syndrome between men and women, with a significant disproportion by age and ethnicity. Earlier estimates from the 2011-2012 NHANES indicated metabolic syndrome prevalence in men (32.8%) to be slightly lower than in women (36.6%), increasing with age in both (5). For those aged 20 to 29 yr, the prevalence of metabolic syndrome was 18.3%, while for those aged 60 and older, the estimated prevalence significantly increased to 46.7%, being higher for women at each time point. Additionally, from 2003 to 2012 the highest incidence rates of metabolic syndrome in the United States were detected in Hispanics (35.4%), followed by non-Hispanic white (33.4%) and non-Hispanic Black (32.7%) individuals. Although the specific mechanisms underlying these disparities have yet to be fully elucidated, the differential susceptibility to abdominal obesity and insulin resistance may play a substantial role (2). Based on current estimates, prevalence of obesity in men is lower among non-Hispanic Asians, followed by non-Hispanic Black, non-Hispanic white, and Hispanic populations, while the lower prevalence of obesity in women has also been

Table 12.2 Current Recommended Waist Circumference Thresholds for Abdominal Obesity by Organization

Population	Organization	RECOMMENDED WAIST CIRCUMFERENCE THRESHOLD FOR ABDOMINAL OBESITY	
		Men	Women
Europid	IDF	94 cm	80 cm
Caucasian	WHO	94 cm (increased risk) 102 cm (higher risk)	80 cm (increased risk) 88 cm (higher risk)
United States	AHA/NHLBI (ATP III)	102 cm	88 cm
Canada	Health Canada	102 cm	88 cm
European	European Cardiovascular Societies	102 cm	88 cm
Asian (including Japanese)	IDF	90 cm	80 cm
Asian	WHO	90 cm	80 cm
Japanese	Japanese Obesity Society	85 cm	90 cm
China	Cooperative Task Force	85 cm	80 cm
Middle East, Mediterranean	IDF	94 cm	80 cm
Sub-Saharan Africa	IDF	94 cm	80 cm
Ethnic Central and South American	IDF	$\geq$90 cm	80 cm

Recent AHA/NHLBI guidelines for metabolic syndrome recognize an increased risk for CVD and diabetes at waist circumference thresholds of 94 cm in men and 80 cm in women and identify these as optional cut points for individuals or populations with increased insulin resistance.

Reprinted by permission from K.G. Alberti et al. "Harmonizing the Metabolic Syndrome: A Joint Interim Statement of the International Diabetes Federation Task Force on Epidemiology and Prevention; National Heart, Lung, and Blood Institute; American Heart Association; World Heart Federation; International Atherosclerosis Society; and International Association for the Study of Obesity," *Circulation* 120 (2009): 1640-1645.

observed among non-Hispanic Asians, followed by non-Hispanic white, Hispanic, and non-Hispanic Black populations (www.cdc.gov/nchs/data).

Normal-Weight Obesity

Abdominal obesity (defined as ≥88 cm for women and ≥102 cm for men by the AHA/NHLBI) increases with age even more so than BMI-assessed obesity. Prevalence estimates of abdominal obesity have trended significantly upward, going from 49% in 2001 to 58.4% in 2016, indicating a 19.1% increase (175). As previously mentioned, there is a well-documented and robust association between obesity and metabolic syndrome; however, there are exceptions in which these do not overlap. One exception is the metabolically healthy obese (90, 104) phenotype, characterized by high BMI yet subclinical (i.e., normal) cardiometabolic health. A primary example is athletes with greater muscle mass proportional to adipose tissue. However, research has shown that strength trained athletes with BMI values greater than 32 kg · m^{-2} possess greater levels of adiposity (89). Furthermore, there are also cases of **normal-weight obesity** (125, 164), in which an individual is classified in the desirable range for body mass or BMI but has excess adiposity or risk factors (or both) for metabolic syndrome, ASCVD, or diabetes.

According to the 2007-2010 NHANES data and based on the AHA/NHLBI criteria for metabolic syndrome diagnosis, the age-adjusted prevalence of metabolic syndrome among normal-weight (BMI <25) adults (12.1%) is lower than among overweight (BMI 25.0-29.9) (32.8%) and obese (BMI ≥30) (63.3%) adults (69). Normal-weight obesity is strongly associated with cardiometabolic dysregulation, high prevalence of metabolic syndrome, and increased risk for cardiovascular mortality (164). The vast majority of population-based studies use surrogate indicators of adiposity (e.g., BMI and WC) (163), which can be somewhat problematic since such metrics do not discriminate between adipose and muscle tissue and lack sensitivity to identify nonobese individuals with excessive body fat (137).

Metabolic Syndrome and the Aging Adult

Incidence of normal-weight obesity is of particular concern for older adults with age-related losses of bone mass (i.e., osteopenia or osteoporosis) and muscle mass (i.e., sarcopenia) or frailty, because these conditions often coincide with subclinical chronic inflammation and oxidative stress (36), insulin resistance (180), myosteatosis (i.e., intra- and intermuscular adipose tissue infiltration) (46), and increased overall fat mass (i.e., sarcopenic obesity) (169). Therefore, under these conditions an individual may have a normal BMI and yet still have excessive body fat or increased risk of cardiometabolic decline or both (148). Indeed, the prevalence of metabolic syndrome increases with age and peaks among individuals around age 60 (69).

Pediatric Metabolic Syndrome

Previous studies of pediatric populations have classified metabolic syndrome as presenting three of the following factors: a waist circumference in the 90th percentile for age and sex; either systolic or diastolic blood pressure in the 90th percentile for height, age, and sex; triglyceride concentration ≥110 mg · dL^{-1}; HDL-C concentration ≤40 mg · dL^{-1}; and glucose concentration ≥100 mg · dL^{-1}. Earlier estimates from the NHANES 2001-2006 data indicated that approximately 42% of adolescents aged 12 to 19 yr had one or two metabolic syndrome components, with 8.6% satisfying the criteria for metabolic syndrome (100). More recent estimates adding 4 yr of data from NHANES (2007-2010) illustrated that 47% and 51% of male and female adolescents, respectively, possessed at least one metabolic syndrome component; the estimated prevalence of metabolic syndrome was reported as 10.1%, thus revealing a 17% increase in U.S. adolescents (128). Furthermore, Hispanic adolescents continue to have the greatest prevalence of an augmented WC, followed by non-Hispanic white and non-Hispanic Black individuals. Regardless, the utility of identifying and diagnosing metabolic syndrome in pediatrics has been heavily debated, largely because of the instability of dichotomized diagnoses in this population (181). According to the Princeton School District Study using NCEP-ATP III criteria, over half of the adolescents who had previously been diagnosed with metabolic syndrome lost the diagnosis within 3 yr (76).

Conversely, the occurrence of obesity and insulin resistance during childhood has been demonstrated to track into adulthood (86, 179) and appears to be an antecedent to adult-onset T2D (134) and cardiovascular disease (CVD) (130). Several ASCVD risk factors among children, including obesity, glucose intolerance, and HTN, have been associated with higher risk of premature death (67); and there is evidence to suggest that risks for ASCVD originate during childhood. Specifically, in the Bogalusa Heart Study, Berenson and colleagues (21) observed preclinical coronary atherosclerotic lesions in young autopsies, and the extent of the lesions was robustly associated with BMI, blood pressure, and serum concentrations of total cholesterol, triglyceride, low-density lipoprotein cholesterol (LDL-C), and HDL-C.

Since the risk for cardiometabolic diseases can originate during childhood, screening children to identify emergent risk is crucial for early intervention and public health preventive efforts. However, consistent guidelines for research and clinical practice are still lacking with regard to the structure and norms of CVD risk factors for children and adolescents (182). In 2009 the AHA released a scientific statement about the progress and challenges pertaining to metabolic syndrome in children and adolescents (181); however, there is no consensus or harmonized definition for metabolic syndrome in children and adolescents (121). Barriers to a universally accepted definition include the fact that clinical end points emerge infrequently at early ages, and there is substantial variability in the established normal values across different ages, between the sexes, and across races. Moreover, there are no established cut points for abdominal obesity (thus forcing the use of sex- and age-adjusted BMI cut points), as well as for a normal range of insulin concentration or cutoffs for physiologic insulin resistance during puberty (181).

Although there has also been debate regarding the designation of high BMI (136), a recent expert committee (111) has recommended that children and adolescents aged 2 through 19 yr at or above the 95th percentile of BMI for age be labeled obese and that children between the 85th and 95th percentiles be labeled overweight. The use of current adult criteria to stratify children and adolescent patients with metabolic abnormalities could significantly underestimate cardiometabolic risk. Because of the progressive nature of cardiometabolic disease, dichotomizing individual risk factors may lead to losses of valuable information and possible misclassification (91). Rather than simple binary definitions for classification or diagnosis of metabolic syndrome, the use of aggregate continuous scores (55, 103, 138) has been reported to be potentially superior in pediatric populations (56, 185), particularly for epidemiological studies (200, 201). Using this method, each factor (i.e., BMI, blood pressure or mean arterial pressure, glucose or homeostasis model assessment of insulin resistance, HDL-C and triglyceride levels) is weighted (e.g., through principal component analysis, summing of standardized residuals or z-scores, or a combination of the two), and a higher aggregate score is indicative of diminished metabolic health (see the sidebar Research Focus). Analysis of the validity of a continuous metabolic syndrome score among a nationally representative sample of 1,239 adolescents (2003-2004 and 2005-2006 NHANES) revealed high construct and predictive validity for metabolic syndrome (138).

Economic Burden

Metabolic syndrome has been referred to as "the disease of the new millennium" (64). Although controversy and debate have surrounded the definitions, individual criteria, and diagnostic utility of metabolic syndrome, high prevalence rates combined with the morbidity and mortality associated with this condition imply significant economic consequences (183, 124). There is general consensus that the increasing incidence of metabolic syndrome represents a huge burden on current and future medical costs coincident with obesity, CVD, and diabetes. The most recent analyzed data from three health care delivery systems in the United States indicate 60% greater health care costs ($5,732 vs. $3,581) in adults with metabolic syndrome compared against those without metabolic syndrome (27). Additionally, health care costs for adults with diabetes were 76% greater independent of metabolic syndrome status. Associations between metabolic syndrome, individual criteria of the syndrome, long-term clinical outcomes, and future health care costs need continued investigation because of their potential impact on public health.

Metabolic syndrome increases the risk of additional health complications or conditions (e.g., allergies, asthma, arthritis, back pain, depression), health care costs, pharmacy costs, and costs stemming from short-term disability up to double the amount compared with people without metabolic syndrome (25, 82, 172). Individual health care costs are augmented ~10% to 15% for each additional metabolic syndrome criterion up to three (135). Excess body mass (manifested as an elevated BMI [not the result of heavy strength training] or as android obesity) and insulin resistance are the primary drivers behind metabolic syndrome.

Research Focus
Muscle Strength for the Detection of Cardiometabolic Risk in Adolescents

This study (147) assessed the sex-specific low-strength cut points for the detection of cardiometabolic risk in adolescents. Using a cross-sectional research design, a large cohort ($N = 1,326$) of 6th grade students was assessed for cardiometabolic profiles. A metabolic syndrome score (MetScore) was computed from the following cardiometabolic components: percent body fat, fasting glucose, blood pressure, plasma triglyceride levels, and HDL-C. A high-risk cardiometabolic phenotype was characterized by having greater than the 75th percentile of the MetScore. Conditional inference tree analyses were used to identify sex-specific low normalized strength (grip strength / body mass) thresholds and risk categories. The findings demonstrate that lower strength was strongly associated with increased odds of the high-risk cardiometabolic phenotype; for every 0.05 lower normalized strength, the odds of having high cardiometabolic risk increased by 1.51 and 1.48 in boys and girls, respectively. These findings reveal the cutoff points for high, intermediate, and low cardiometabolic risk categories in both boys and girls. The highest risk was observed at a normalized strength of ≤33% and ≤28% of body mass for boys and girls, respectively; an intermediate risk was present at a strength of >33% and ≤45% of body mass for boys, and >28% and ≤36% of body mass for girls; while the strength values that conferred the lowest cardiometabolic risk was >45% and >36% of body mass for boys and girls, respectively. This study suggests that normalized grip strength may be a valuable clinical screening tool to assess cardiometabolic risk among adolescent boys and girls.

Based on Peterson et al. (147).

Insulin resistance, measured by the homeostatic model assessment of insulin resistance (HOMA-IR), is positively associated with BMI and WC (28). In the United States, annual health care costs are elevated by $3,429 and $9,601 for each U.S. adult with obesity and diabetes, respectively (12, 24). The total annual health care costs in the United States for obesity and diabetes are $342.2 billion (year 2013) and $327 billion (year 2017), respectively. The relationship between BMI and health care costs is nonlinear (follows a J shape), which means that in the range of obesity classes II and III, the increment in health care costs is exponential. This is because morbid obesity increases the risk for diabetes mellitus, CVD, and other comorbidities, thus imposing a large economic burden as BMI increases beyond $30\,kg \cdot m^{-2}$. Data indicate there is a six-fold increase in diabetes risk in patients with class III obesity (BMI $>40\,kg \cdot m^{-2}$) in comparison with normal-weight individuals. The presence of diabetes plus obesity significantly increases health care utilization. The health care cost of a 50 yr old patient with diabetes, for example, is approximately three times higher than for someone of the same age without diabetes (119).

PATHOPHYSIOLOGY

Metabolic syndrome is characterized by a co-occurrence of atherogenic dyslipidemia, HTN, elevated glucose, chronic low-grade inflammation, and prothrombosis (81). In conjunction with several behavioral factors (i.e., sedentary behavior and atherogenic diet), genetic predisposition, and advancing age, the clustering of multiple risk components within metabolic syndrome is widely thought to occur as a result of obesity (more specifically, android obesity) and insulin resistance (81, 152, 181) (figure 12.1). However, although obesity and insulin resistance are considered two hallmarks of chronic health risk, not all obese or insulin-resistant individuals develop metabolic syndrome (141). The specific trajectory of cardiometabolic decline leading to or coinciding with excessive accumulation of adiposity, altered fat partitioning, and diminished insulin sensitivity is a multifactorial, complex issue to disentangle.

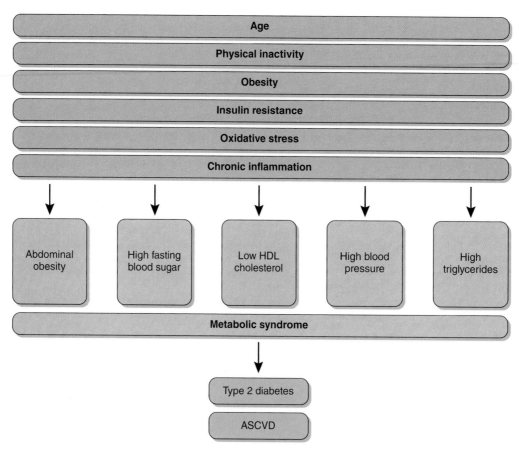

Figure 12.1   Schematic of the components of metabolic syndrome.

Insulin Resistance and Metabolic Syndrome

Gradual decreases in cardiometabolic health start to occur long before an individual reaches obesity or is diagnosed as insulin resistant. For that reason, metabolic syndrome is often referred to as a premorbid condition (177). In a healthy, insulin-sensitive person, glucose stimulates the release of insulin from pancreatic beta cells, which in turn reduces plasma glucose concentration through suppression of hepatic glycogenolysis and gluconeogenesis and simultaneous glucose uptake, utilization, and storage by the liver, muscle, and adipose tissue. Conversely, under conditions of insulin resistance, there is a chronic failure of insulin to maintain glucose homeostasis. The role of insulin resistance and hyperinsulinemia in the development of metabolic syndrome has been controversial because a direct causal link has not yet been identified (103); however, some data suggest that hyperinsulinemia may be more sensitive than BMI or WC in identifying those with metabolic syndrome (39). Part of this confusion has been driven by the consistently oversimplified definition of insulin resistance, which, although useful in the clinical context, does not account for the fact that insulin regulates various other processes in addition to glucose metabolism (87) and cannot discern the origin of the resistance (i.e., defects in insulin signaling vs. those in the insulin receptor [23]). It is postulated that the insulin resistance associated with metabolic syndrome is indeed pathway specific (87) and that multiple forms of molecular insulin resistance could contribute to abnormal glucose homeostasis (22). In the clinical setting, insulin resistance is characterized as the extent to which the liver manifests insulin resistance in proportion to the periphery (i.e., skeletal muscle insulin resistance) (181).

Over time, resistance to insulin and compensatory hyperinsulinemia in metabolic syndrome result in increased lipogenesis, hypertriglyceridemia, HTN, and **steatosis** (22, 87, 181). Although a thorough discussion of insulin action and insulin signaling is beyond the scope of this chapter, research continues to emerge that confirms a pathophysiologic link not only between insulin resistance, metabolic syndrome, and glucose intolerance but also with ASCVD (87). Evidence also links systemic inflammation (142), oxidative stress (195), and endothelial dysfunction in both animal (187) and human models (151) to metabolic syndrome. Thus, if metabolic syndrome is indeed a premorbid condition, as it has been suggested (103), identifying and treating at-risk individuals before the emergence of categorical insulin resistance or hyperglycemia is a vitally important directive.

Obesity and Adiposity Distribution Abnormalities

Obesity is an independent risk factor for insulin resistance, hyperglycemia, dyslipidemia, and HTN. Left untreated, this combination of pathophysiologic factors precipitates increased risk for chronic diseases and premature all-cause mortality (48, 112). Despite the robust association between obesity and poor cardiometabolic health, it is a heterogeneous condition that must be considered in a broader biological and public health context. For example, BMI is suggested to account for only 60% of the variance in insulin resistance among adults (3). Rather, in conjunction with several abnormalities in adipose tissue metabolism, abnormal regional fat distribution and partitioning may actually be the pathophysiologic link between obesity and the numerous hormonal and metabolic derangements that compose metabolic syndrome (66).

Accumulation of fatty acids in nonadipose tissue depots is a dynamic, lipotoxic (170) process that occurs as a result of chronic disequilibrium between energy intake and energy expenditure—and it is robustly associated with skeletal muscle insulin resistance (95, 109, 115). Specifically, visceral adipose tissue (VAT) is broadly recognized to have metabolic, endocrine, and immune system interactions; and increases in VAT precipitate heightened risk for metabolic and cardiovascular disorders. However, during conditions in which fat infiltrates the muscle (i.e., intra- and intermuscular adipose tissue [**IMAT**]), it appears to independently contribute to impaired glucose metabolism and decreased insulin sensitivity (115). Fat infiltration has been identified in aging adults (i.e., muscle attenuation on computed tomography [77] or localized IMAT with magnetic resonance imaging [166]), as well as in certain diseases and morbid conditions such as Duchenne muscular dystrophy (118), T2D (71), spinal cord injury (79, 173), cerebral palsy (146) obesity (80, 106, 178), excessive sedentary behavior (123), and vitamin D insufficiency (73). IMAT is a dynamic tissue with both paracrine and endocrine properties (74) and is suggested to arise from satellite stem cells or distinct fibrocyte–adipocyte progenitor cells, which form adipocytes within skeletal muscle during conditions of metabolic dysregulation and hyperglycemia (4). Moreover, evidence reveals a robust link between IMAT and elevated levels of proinflammatory, adipocyte-derived hormones and cytokines (18, 207), which also lead to insulin resistance (108) and muscle dysfunction (194). An important distinction is when lipids (sequestered in lipid droplets) are stored within muscle fibers as intramyocellular triglycerides (IMTGs). These IMTGs are a vital energy source during skeletal muscle contraction and are often seen in greater levels in endurance athletes. Evidence indicates that unlike IMAT, IMTGs do not cause insulin resistance (191).

Mitochondrial Dysfunction

Concurrent with an increased storage of ectopic adiposity, reductions in mitochondrial size, density, and function (144, 150, 171) have been implicated in the etiology of insulin resistance, metabolic syndrome, and diabetes. More specifically, obese, sedentary, and insulin-resistant individuals have smaller and fewer mitochondria, with impaired function. Diminished mitochondrial density and function may lead to or coincide with decreased or incomplete lipid oxidation and subsequent accumulation of lipid metabolites (diacylglycerol, ceramides, and acyl coenzyme A [CoA]), impaired insulin signaling, metabolic inflexibility, and oxidative stress (35, 129, 140). Moreover, each of these outcomes is also thought to cause further impairment of mitochondrial function and thus provokes a chronic circular cause and consequence of cardiometabolic events. Impaired mitochondrial function may also lead to diminished adenosine triphosphate (ATP) production, energy deficit, and decreased functional capacity. Sedentariness may lead to hypokinetic disease, which is a robust predictor of mitochondrial dysfunction (51); and more important, evidence has long suggested (92) improvements in ATP synthesis and fatty acid oxidation after exercise interventions (96), independent of weight loss.

Proinflammatory and Prothrombotic Characteristics

Adipose tissue is considered a dynamic organ with pleiotropic properties (197). Previous research among obese adults with diabetes has revealed significantly elevated levels of adipocyte-derived hormones and cytokines, which are significant contributors to insulin resistance (108). Specifically, ectopic adiposity is known to play a role in secreting proinflammatory cytokines (e.g., tumor necrosis factor alpha [TNF-α] and interleukin-6 [IL-6]), adipocytokines (e.g., leptin, resistin, and adiponectin), and chemokines (e.g., monocyte chemoattractant protein [MCP-1]). Produced by adipose tissue macrophages, the inflammatory cytokines IL-6 and TNF-α are positively associated with triglycerides and total cholesterol, are inversely associated with HDL-C, are capable of interfering with insulin signaling (174), and can result in cellular oxidative stress (i.e., accumulation of reactive oxygen species [ROS]) (206). Moreover, IL-6 stimulates hepatic production of C-reactive protein (CRP) (17), an acute-phase protein and robust predictor of various features in metabolic syndrome (68). Clinically, high-sensitivity CRP (hs-CRP) has become accepted as a useful biomarker for chronic, low-grade inflammation. Augmented levels of hs-CRP (>3.0-<10.0 mg · L^{-1}) have been shown to be robustly associated with ASCVD (161) and inversely associated with vigorous physical activity (PA) (159).

Markers of thrombosis (i.e., clotting factors) are also known to be increased in metabolic syndrome, thus representing an additional link with ASCVD. Specifically, elevated levels of plasma plasminogen activator inhibitor-1 (PAI-1) (102) and fibrinogen (68) are linked to thrombosis and fibrosis, insulin resistance, and abdominal obesity. However, it is plausible that the proinflammatory and prothrombotic states associated with metabolic syndrome are elevated before the presentation of standard syndrome risk criterion. Adipocyte proliferation and accumulation occur long before an individual meets the criteria for obesity or is dichotomized as at risk for cardiometabolic disease. Therefore, in conjunction with pronounced changes in the hormonal-metabolic milieu (78, 80), increases in adiposity could yield a pathological environment that contributes to attenuated insulin sensitivity even in the absence of a clinical diagnosis of metabolic syndrome. A study among prepubescent children demonstrated that obesity without additional established comorbidities of metabolic syndrome was associated with proinflammatory and prothrombotic states (127). These findings are significant because they support the need for aggressive preventive strategies early in life in those with obesity, even in the absence of comorbidities.

Although various pathophysiologic mechanisms for metabolic syndrome have been identified, debate pertaining to the nature of these associations has raised questions about reverse causality—that is, whether hepatic and cardiac steatosis (29) and chronic inflammation cause mitochondrial dysfunction or insulin resistance (or both) or vice versa (101, 132, 144)—as well as about primary versus secondary targets for intervention. Unraveling these associations has been considered a critical directive for future studies (120) and may be fundamental in optimizing therapeutic interventions to reverse the sequelae caused by—or coinciding with—obesity, insulin resistance, and metabolic syndrome.

CLINICAL CONSIDERATIONS

The state of the literature pertaining to the clinical utility of metabolic syndrome is in constant flux, and thus several points of clarification and caution are warranted. First, the AHA/NHLBI (updated NCEP-ATP III), the IDF, and the new harmonized definitions do not exclude hyperglycemia in the diabetes range as a potential criterion for diagnosis of metabolic syndrome (7, 8, 184). Therefore, most patients with T2D also meet the criteria for metabolic syndrome; however, this overlap between diabetes and metabolic syndrome has not been without controversy (103, 177).

In fact, ever since the original definition was published (9), there have been many opponents to initiatives to impose a clinical definition, as well as to initiatives to derive threshold cut points for the syndromic clustering of the respective risk factors (103, 168). The most frequent argument against the use of a diagnosable metabolic syndrome pertains to the clinical relevance of the condition (26, 50), with data spanning pediatrics (117) to older adults (190). According to the most recent WHO Expert Consultation Group (177), the diagnostic criteria for metabolic syndrome do not add predictive value beyond the sum of the individual risk factors in forecasting future CVD, diabetes, or disease progression. In this 2010 WHO report, various other rationales against the clinical and epidemiological use of metabolic syndrome were presented. Most notably, criticisms from previous reports (103) were reiterated to conclude that although metabolic syndrome may be a useful educational concept, "it has limited practical utility as a diagnostic or management tool" (177: 600 and 604). Specific limitations that were proposed as rationale against widespread clinical adoption of metabolic syndrome include the following (177):

- Lack of a single agreed-upon pathophysiologic mechanism
- Reliance of definitions on dichotomization of the diagnosis and individual risk factors
- The fact that the syndrome describes relative risk as opposed to absolute risk
- Differing predictive value of various risk factor combinations
- Inclusion of individuals with established diabetes and CVD
- The omission of important risk factors for predicting diabetes and CVD

Despite these long-standing criticisms regarding the clinical utility of metabolic syndrome, and since the release of the 2010 WHO report, a large study demonstrated that among nearly one million patients ($N = 951,083$), metabolic syndrome was associated with a twofold increase in cardiovascular outcomes and a 1.5-fold increase in all-cause mortality (131). Moreover, this investigation revealed that ASCVD risk was significantly higher among patients with metabolic syndrome even in the absence of diabetes. This is consistent with two previous meta-analyses (70, 72), and thus these studies collectively represent strong support for metabolic syndrome as a robust predictor of cardiovascular events, even when adjusting for the individual risk components.

In conjunction with the elevated risk for cardiometabolic disease and mortality, several secondary clinical outcomes shown to be associated with metabolic syndrome include nonalcoholic fatty liver disease (NAFLD), cholesterol gallstones, polycystic ovary syndrome, and sleep apnea (81, 82). In addition to general increased risk of cardiovascular and metabolic complications in patients with metabolic syndrome, a wide variety of pharmacological treatments might coincide with secondary comorbidities (table 12.3). Moreover, among older adults, studies suggest that metabolic syndrome is associated with an increased risk of age-related dementia and overall cognitive decline (154, 155, 205). Raffaitin and colleagues (154) demonstrated that among a prospective cohort of 4,323 women and 2,764 men (≥65 yr), the presence of metabolic syndrome at baseline was predictive of global cognitive decline and in particular executive function, independent of previous CVD or depression. This may be due to microstructural white matter abnormalities consistent with cardiometabolic abnormalities, and it could explain the underlying mechanisms linking metabolic syndrome and the hallmark impairments to verbal learning and memory performance in older adults (10).

Although children, adolescents, and adults with metabolic syndrome have significantly elevated risk for chronic diseases and secondary comorbidities, it stands to reason that not everyone diagnosed will eventually develop insulin resistance. However, since metabolic syndrome is known to double the risk for ASCVD and is a strong precursor to the development of diabetes, there are profound long-term clinical implications regarding the diagnosis of the syndrome. Moreover, there is an urgent need to develop preventive and treatment strategies for the syndrome, as well as the underlying causes (84). Identifying individuals with emergent risk for and early-phase diagnosis of metabolic syndrome is an exceedingly important clinical and public health directive, as this represents a critical opportunity to intervene. Healthy lifestyle interventions (192) are able to reverse most of the metabolic risk factors; however, drug therapies or bariatric surgery are often necessary (85).

Signs and Symptoms

Considering the pathophysiology and trajectory of metabolic syndrome, much overlap exists between the symptoms of metabolic syndrome and those associated with diabetes (chapter 8) and obesity (chapter 9). In addition to the actual diagnostic risk components of metabolic syndrome (i.e., elevated glucose, HTN, elevated triglyceride levels, low HDL-C levels, and abdominal obesity), many patients also present with **microalbuminuria**, hyperuricemia, NAFLD, high levels of PAI-1 and fibrinogen (i.e., prothrombotic state), elevated hsCRP (i.e., chronic proinflammatory state), cholesterol gallstones, polycystic ovary syndrome, and disordered sleeping (e.g., sleep apnea). Among older adults with metabolic syndrome, patients may report memory loss, impaired cognition, general confusion, or a constellation of factors.

History and Physical Exam

Many patients diagnosed with metabolic syndrome also have chronic hyperglycemia (i.e., T2D). Thus, it is extremely important to take into consideration specific precautions related to the screening, diagnosis, and physical exam for patients with diabetes (chapter 4). In general, one should identify the core risk components of metabolic syndrome during a routine physical examination. However, since many physicians and paramedical professionals (i.e., physician assistants, nurse practitioners) have not traditionally monitored abdominal obesity (height, weight, and BMI calculations are more commonly measured anthropometric indices), a WC measurement needs to become a fundamental component of the patient history and physical exam. In fact, some evidence suggests that WC is a better predictor of metabolic syndrome and its individual criteria than is BMI, thus illustrating the importance of measuring and tracking abdominal obesity (98). Furthermore, abdominal obesity is a fundamental risk component in the now widely utilized harmonized definition for metabolic syndrome (7); thus, accounting for WC should now be a staple in preventive medicine.

Moreover, with regard to individual parameters, estimates of relative risk for ASCVD incrementally and significantly increase as the number of syndrome components increases (198). Thus, individuals who present with one or two of the components are at significantly lower risk for cardiac events, as well as for mortality from coronary heart disease and CVD, than individuals with three or more risk components (figure 12.2) (122). Estimates of relative risk are also contingent upon the particular combination of components. Specifically, numerous combinations of the syndrome components warrant a diagnosis, and recent evidence reveals that ASCVD and mortality risk is highest when elevated

Table 12.3 Pharmacology

Medication class and name	Primary effects	Exercise effects	Important considerations
Angiotensin II receptor antagonists Telmisartan, irbesartan, losartan, olmesartan, valsartan	Blocks the action of angiotensin, dilate blood vessels, and reduce blood pressure	Reduces blood pressure during exercise	Among individuals with impaired glucose tolerance, this class of drug may decrease the risk of developing type 2 diabetes.
Statins (HMG-CoA reductase inhibitors) Atorvastatin, fluvastatin, lovastatin, simvastatin, pravastatin, rosuvastatin, pitavastatin	Inhibits the enzyme HMG-CoA reductase, which results in decreased hepatic cholesterol Improve lipid profile (LDL-C) and may inhibit the progression of atherosclerosis	No effect	May improve (increase) HDL-C and lower triglycerides. Many statins are metabolized by cytochrome p450 and therefore have higher risks of drug–drug interactions. Patients should avoid eating grapefruit (contains furanocoumarins) because it inhibits the metabolism of statins.
Fibrates (fibric acid sequestrants) Bezafibrate, gemfibrozil, fenofibrate	Activates PPAR-α signaling and leads to decreased hepatic cholesterol secretion and increased β-oxidation Improves LDL-C but also may improve HDL-C and decrease triglycerides	Combined with statins, may increase risk of rhabdomyolysis (i.e., severe muscle damage)	
Bile acid sequestrants Cholestyramine, colesevelam, colestipol	Antilipemic agents	Increases risk of gastrointestinal distress and constipation, which may be exaggerated during exercise	May bind with fat-soluble vitamins (i.e., vitamins A, D, E, and K) in the gut and lead to insufficiency.
Nicotinic acid Niacin	Increases HDL-C	May help decrease blood pressure with exercise	Can worsen glycemic control in patients with metabolic syndrome or diabetes.
Lipase inhibitor Orlistat	Prevents absorption of fats from the diet and thus reduces caloric intake	Increases risk of gastrointestinal distress (e.g., fecal incontinence and loose, oily stools)	Absorption of fat-soluble vitamins and nutrients is inhibited.
ACL inhibitor Bempedoic acid (Nexletol)	Inhibits ACL enzyme in liver, inhibiting cholesterol synthesis and lowering LDL-C; ACL enzyme is higher in cholesterol biosynthesis pathway than HMG-CoA)	No effect	Add-on therapy to statin for genetic hypercholesterolemia or established atherosclerotic cardiovascular disease to further lower LDL-C. May cause hyperuricemia, so monitoring is recommended.
Proprotein convertase subtilisin/kexin type 9 (PCSK9) inhibitor Alirocumab (Praluent), evolocumab (Repatha)	Increases number of LDL-C receptors in liver to remove LDL-C from bloodstream by blocking receptor degradation, which improves LDL-C	Possible injection-site reactions and myalgia	Add-on therapy to statin for genetic hypercholesterolemia or established atherosclerotic cardiovascular disease to further lower LDL-C. Expense and injectable route of administration may limit use. Human monoclonal immunoglobulin may cause severe allergic reactions.
Omega-3 fatty acid supplement Icosapent ethyl-EPA (Vascepa), omega-3 ethyl esters [EPA and DHA] (Lovaza)	Both decreases triglycerides (Icosapent ethyl no effect on LDL-C and HDL-C, while omega-3 ethyl esters may increase LDL-C and HDL-C)	Possible indigestion	Use caution in patients with fish or shellfish allergies. May increase bleeding in patients also taking antithrombotic medications. Vascepa may increase risk of atrial fibrillation.
Antidiabetic (biguanides) Metformin	Antihyperglycemic: Decreases hepatic glucose production and intestinal glucose absorption	GI disturbances common during initiation of therapy	

Medication class and name	Primary effects	Exercise effects	Important considerations
Thiazolidinediones Pioglitazone	Increases insulin sensitivity of liver, muscle, and adipose tissues via activation of PPAR-α receptors to modulate transcription of insulin-responsive genes	No effect	Avoid use in patients with heart failure. Monitor for edema. Other meds in class off market due to heart failure risks.
α-glucosidase inhibitors Acarbose	Slows intestinal carbohydrate digestion and absorption, lowering postprandial glucose levels	Possible GI disturbances	

ACL = adenosine triphosphate-citrate lyase; DHA = docosahexaenoic acid; EPA = eicosapentaenoic acid; GI = gastrointestinal; HMG-CoA = 3-Hydroxy-3-methylglutaryl-coenzyme A; HDL-C = high-density lipoprotein cholesterol; LDL-C = low-density lipoprotein cholesterol; PPAR = peroxisome proliferator-activated receptors.

Acknowledgement to Dr. Kimberly Richardson of the University of North Florida for her contributions to this table.

blood pressure, abdominal obesity, and elevated glucose are simultaneously present (65, 162). A careful review of a patient's medical history will provide insights relevant to starting a comprehensive lifestyle intervention for the patient. In conjunction with standard metabolic syndrome components and medical history, the exercise professional should also take into consideration patient exercise history.

Diagnostic Testing

In addition to the five risk components of metabolic syndrome, and especially in the case of a positive diagnosis, laboratory screening for diabetes should also take place (chapter 4). This is particularly necessary for patients over the age of 45 yr. Individuals may also need to be screened for prediabetes, which is characterized by impaired fasting glucose (fasting plasma glucose levels 100 mg · dL^{-1} [5.6 mmol · L^{-1}] to 125 mg · dL^{-1} [6.9 mmol · L^{-1}] [the WHO demarcation point for IFG is 110 mg · dL^{-1} or 6.1 mmol · L^{-1}]) or impaired glucose tolerance (2 h values in the oral glucose tolerance test of 140 mg · dL^{-1} [7.8 mmol · L^{-1}] to 199 mg · dL^{-1} [11.0 mmol · L^{-1}]) (49). Moreover, depending on age, the number of risk components identified, and the specific combination of those components, additional testing may be warranted to screen for clinical inflammation and thrombosis, **hyperandrogenemia** (to rule out polycystic ovary syndrome), microalbuminuria, hyperuricemia, cholesterol gallstones, and sleep apnea.

Exercise Testing

In accordance with the American College of Sports Medicine (ACSM), patients with metabolic syndrome do not require an exercise test prior to beginning a low- to moderate-intensity exercise program. However, if an exercise test is performed, the general recommendation is to evaluate metabolic risk based on the presence of dyslipidemia, HTN, and hyperglycemia (1). Because patients with metabolic syndrome are typically overweight or obese, exercise testing recommendations specific for those individuals should be followed. Standardized treadmill protocols starting with a low initial workload (2-3 METs) with small increments (0.5-1.0 MET) are recommended and are often well tolerated by those with obesity and metabolic syndrome. However, for individuals with morbid obesity, walking may not be practical because of gait abnormalities. In such instances, testing for aerobic fitness may be performed using a seated recumbent cycle ergometer or upper body ergometer (functional capacity estimates will be lower when using upper body aerobic testing, and this must be taken into consideration when developing the exercise prescription).

Patients diagnosed with metabolic syndrome should also be assessed for muscular fitness (i.e., muscle strength and local muscular endurance), as well as for flexibility and range of motion (ROM). Exercise testing should follow standard procedures as described in chapter 5. All exercise testing should be performed with caution and should be completed for the purpose of designing individualized PA and exercise prescriptions. Because of the risk for HTN in this population, it is important for practitioners to carefully follow the blood pressure protocols during and after exercise testing. Additionally, appropriate cuff size (false elevated readings can occur if cuff size is too small) should be used to measure blood pressure in individuals who are overweight or obese to minimize the potential for inaccurate measurement (1). See table 12.4 for information on exercise testing methods for individuals with metabolic syndrome.

Treatment

Although the primary goals in the management of metabolic syndrome are to reduce the risk for clinical ASCVD and prevent T2D, clinical and public health efforts should lend equivalent support to treating the established risk factors as well as their underlying causes. The individual metabolic syndrome criteria are modifiable, but abdominal obesity is often considered the primary driver of

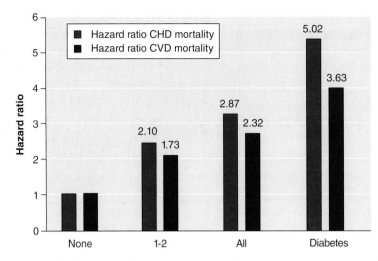

Figure 12.2 Coronary heart disease (CHD) and cardiovascular disease (CVD) mortality increases with number of cardiometabolic risk factors (RF).

Adapted from Malik et al. (122).

the other criteria (50). Thus, treatment strategies should address the contributing risk factors for abdominal obesity, which in turn should positively influence all other criteria. Indeed, interventions designed to promote weight loss (e.g., caloric restriction, increased PA, pharmacological agents, and even surgical procedures when necessary) should be a central feature of the treatment of metabolic syndrome (see practical application 12.1). The NHLBI recommends a minimum weight loss of 10% (43); however, ample evidence also exists to demonstrate significant health benefit after as little as 2% to 3% weight reduction (52, 114). According to Grundy,

a reasonable weight loss goal for the first year of intervention is 10 kg, which people may attain by creating a caloric deficit of about 400 calories per day through reduced calorie intake (e.g., 300 kcal/d) and increased energy expenditure (e.g., 100 kcal/d) (81).

In addition to managing abdominal obesity through improvements in diet quality and participation in PA, patients should quit smoking, where applicable. Smoking is a lifestyle risk factor for ASCVD that should continue to be directly targeted by public health and clinical interventions.

Practical Application 12.1

CLIENT–CLINICIAN INTERACTION

Most patients with metabolic syndrome are obese and lead sedentary lifestyles. As a result, establishing PA or structured exercise interventions can be a challenging endeavor for the exercise physiologist and physician. Since the health decline associated with cardiometabolic conditions is an insidious, often asymptomatic process, many people with metabolic syndrome do not perceive significant deficits in physical function or general well-being and thus are not likely to understand the seriousness of the condition. Therefore, it is vital to educate patients regarding the accumulation and clustering of risk factors, as well as the role of appropriate strategies to help prevent further health declines. At present, weight loss is the inevitable clinical recommendation for treatment of obesity-related cardiometabolic declines; thus PA is generally regarded as

an adjunct to dietary restriction. However, whereas weight loss is indeed effective for ameliorating health risk, very few studies have documented long-term sustainability or efficacy of weight loss; on the contrary, the vast majority of individuals (approximately 80%) who lose significant weight experience complete weight regain (15, 203). This underscores the importance of educating patients about the relative value of PA and exercise as a complement to healthy dietary practices for improving insulin sensitivity and cardiometabolic health. Although the expected weight loss attributed to PA without dietary manipulation is generally small, improvements in cardiovascular risks such as resting blood pressure, lipid profiles, and glucose tolerance may provide the motivation people need to adhere to a physically active lifestyle.

Table 12.4   Exercise Testing Review

Test type	Mode	Protocol specifics	Clinical measures	Clinical implications	Special considerations
Cardiorespiratory	Step test, cycle ergometer, or graded treadmill test	Standard tests (e.g., Harvard step test, YMCA bicycle test, modified Bruce treadmill protocol, or some combination of these) The initial workload should be around 2 or 3 METs with small increments that are equivalent to 0.5-1.0 METs	Oxygen uptake ($\dot{V}O_2$), RER, METs, RPE	$\dot{V}O_2$ is typically diminished among patients with metabolic syndrome. Normative data from healthy populations may be used for each test to provide stratification and percentile rank. Each test may also be used to track longitudinal progress of and adaptation to cardiorespiratory exercise interventions.	Patients with diabetes or HTN or both should follow specific guidelines for each condition to avoid undue risk of cardiovascular events. Obese individuals may not be able to complete the step test or treadmill tests and thus may require use of the cycle ergometer assessment. Careful adherence to blood pressure protocol before and during the exercise session is necessary.
Strength	MVC, 1RM, or multiple RM (e.g., 5RM)	Dynamometer (e.g., Biodex isokinetic dynamometer), selectorized or pneumatic machines (e.g., Nautilus, Keiser), free weights (e.g., chest press, squat), resistance bands, or some combination	Isometric torque (N·m), maximum dynamic force (newtons), or maximal strength (kg or lb)	Lower extremity body mass–adjusted strength (e.g., 1RM leg press / body mass) and muscle quality (i.e., muscle cross-sectional area and strength) are valuable clinical measures that have direct relevance to functional mobility and activities of daily life. Results may be used to prescribe relative intensity for resistance exercise interventions.	Individuals with preexisting HTN should avoid maximal strength (i.e., 1RM) testing. Rather, multiple RM testing (e.g., 5RM-10RM) is a safe and valid alternative technique.
Range of motion (ROM)	Sit-and-reach, overhead squat test	Sit-and-reach or modified (i.e., unilateral) sit-and-reach Overhead squat test provides comprehensive qualitative assessment of functional mobility and hip, low back, knee, ankle, and shoulder ROM and stability	Sit-and-reach (distance in cm) Overhead squat test is a qualitative assessment with the following requirements: Upper torso is parallel with shin or toward vertical; hips are lower than knees; knees are aligned over feet; and bar/dowel or hands are aligned directly over feet, with arms outstretched	Results from the overhead squat test can indicate tightness or weakness in respective musculature through coordinated full ROMs, excessive joint stiffness (e.g., limited ankle dorsiflexion), and strength or ROM asymmetries.	Sit-and-reach may not be possible for individuals with abdominal obesity. Individuals with a history of low back pain or injury may need to refrain from sit-and-reach assessment.

1RM = one-repetition maximum; HTN = hypertension; METs = metabolic equivalents; MVC = maximal voluntary contraction; N·m = newton meters; RER = respiratory exchange ratio; RPE = rating of perceived exertion; YMCA = Young Men's Christian Association.

EXERCISE PRESCRIPTION

Although numerous studies have addressed the value of PA and exercise for patients with metabolic syndrome, in-depth guidelines focusing directly on this topic are not yet available. The following exercise prescription review for metabolic syndrome is representative of the collective recommendations by the ACSM guidelines for weight loss (53) and diabetes (11) as well as their recent book, *ACSM's Guidelines for Exercise Testing and Prescription 2018* (1). Chapter 6 presents additional detail on the fundamentals of exercise prescription. Since a subset of the metabolic syndrome population also has diabetes, specific suggestions pertaining to exercise prescription and contraindications in chapter 8 may be relevant. Table 12.5 identifies exercise prescription recommendations for the management of metabolic syndrome.

Cardiorespiratory Exercise

According to the ACSM recommendation for metabolic syndrome, cardiorespiratory exercise training should be performed at a moderate intensity (40%-59% $\dot{V}O_2R$ or heart rate reserve [HRR]), increasing intensity when appropriate ($\geq$60% $\dot{V}O_2R$ or HRR). For optimal health and fitness improvements, a minimum of 150 min/wk or 30 min/d most days of the week is recommended. However, to reduce body weight, a gradual increment to 250 to 300 min/wk or 50 to 60 min/d on at least 5 d/wk is necessary. This daily PA can be subdivided into multiple daily bouts (any exercise or PA bout length provides health benefits). Longer duration (60-90 min/d on at least 5 d/wk) may be necessary for some individuals (1). Gradual progression in duration and intensity may be effective for chronic weight maintenance, additional weight loss, or further

Table 12.5   Exercise Prescription Review

Training method	Frequency	Intensity	Time (Duration)	Type (Mode)	Progression	Important considerations
Cardiorespiratory	5-7 d/wk	Light (<40% $\dot{V}O_2R$ or HRR) to moderate (40 to <60% $\dot{V}O_2R$ or HRR)	30-60 min (no minimum bout length)	Walking, jogging, swimming, elliptical, stair climbing, hiking, rowing	Short-duration, low-intensity PA may progress to longer-duration or moderate-intensity exercise bouts or both.	For obese individuals, a mix of weight-bearing and non-weight-bearing modalities is better tolerated and may reduce risk of overuse injury or joint pain.
Resistance	2-4 d/wk with a minimum of 48 h between resistance training sessions	Progression from very light (40% 1RM) to light (50% 1RM), moderate (60% 1RM), and vigorous intensity (70% 1RM)	Low volume (2-4 sets per muscle group), repetitions should vary by goal (strength, endurance)	Exercises for major muscle groups: full ROM body weight movements, selectorized or pneumatic machines, free weights, and resistance bands	After familiarization with the movements, progress from low relative intensities and volumes to moderate to vigorous relative intensities and higher volumes.	Full-body program can be split into a combination of upper and lower program to accommodate progression.
Range of motion (ROM)	At least 2 d/wk after a warm-up or workout	Stretching to the point of discomfort	2-4 times per muscle group, holding 10-30 s for each stretch	Dynamic ROM exercises and static stretching	NA	Static stretching should be the preferred method in the initial exercise training period, considering that patients with metabolic syndrome might have low ROM due to sedentary behavior.

1RM = one-repetition maximum; HRR = heart rate reserve; NA = non-applicable; PA = physical activity; $\dot{V}O_2R = \dot{V}O_2$ reserve.

improvement in aerobic fitness capacity beyond the baseline requirements for health. Caution is warranted, however, since progressing sedentary individuals too quickly may exacerbate untoward responses such as musculoskeletal strain and significant soreness. According to the most recent ACSM guidelines, PA for weight loss is dose dependent, such that a minimum of a 150 min/wk optimizes health and fitness benefits but leads to minimal weight loss, while a PA level of >250 min/wk results in long-term weight control (1). Activities such as brisk walking, swimming, and cycling are usually well tolerated by those with metabolic syndrome; however, unless limitations are specified by a primary care physician or cardiologist, individuals may also progress to other modalities such as jogging, running, hiking, rowing, and stair climbing.

Resistance Exercise

Resistance training has been inexplicably understudied, receiving very little attention from the clinical and public health community regarding its relative value for obesity treatment because it is generally believed to be ineffective for weight loss. However, research has begun to shed light on the utility of resistance exercise in stimulating positive cardiorespiratory, endocrine, metabolic, neuromuscular, and morphological adaptations, independent of significant weight loss (199). These adaptations are briefly discussed in the next section on exercise training. With regard to exercise prescription, current minimum recommendations call for resistance exercise training to supplement cardiorespiratory exercise and to be performed on 2 (preferably 3) nonconsecutive days per week, using a single set of 5 to 10 exercises for the whole body, and at a moderate level of intensity that allows 10 to 15 repetitions (1). As is generally accepted for novice trainees, prescription of resistance exercise should include a familiarization period in which very low dosage training (i.e., minimal sets and intensity) takes place 1 or 2 d/wk. Following familiarization, one can expect adults with metabolic syndrome to benefit from gradual increases in dosage to accommodate improvements in muscle strength and hypertrophy. In particular, although the established guidelines provide a basis for maintaining muscular fitness, there is now ample evidence to confirm the viability of progressive resistance exercise for improving strength and muscle mass among adults. Additional suggestions on progression in resistance exercise include gradual increases in intensity from very light (40% of one-repetition maximum [1RM]) to light (50% 1RM), moderate (60% 1RM), and vigorous intensity (70% of 1RM); gradual increases in the number of sets from two sets to as many as four sets per muscle group; gradual decreases in the number of repetitions performed to coincide with progressively heavier loading, going from 10 to 15 repetitions per set to approximately 8 to 12 repetitions per set; and progression and diversity in mode from primarily machine-based resistance exercise to machine plus free-weight resistance exercise (resistance bands and body weight also provide effective stimuli) (1).

Range of Motion Exercise

Range of motion and flexibility training may be included as an adjunct modality to supplement cardiorespiratory and resistance exercise for individuals with metabolic syndrome. Current ACSM recommendations for stretching and ROM exercises suggest that stretching be performed at least 2 d/wk following a warm-up or workout, when muscles are warm. Stretching should be performed to the point of feeling muscular tightness or mild discomfort. Static, dynamic, or proprioceptive neuromuscular facilitation activities are suggested for all major muscles and joints of the body; however, considering that most patients with metabolic syndrome have been sedentary, static stretching may be preferable during the initial phases of exercise participation. For static stretching, the recommendation is that each stretch be repeated two to four times and held for 10 to 30 s to the point of mild discomfort (1).

EXERCISE TRAINING

Increased adiposity (30, 115), insufficient PA (57, 139), and sedentary behavior (63, 126, 157, 167) have all been implicated in the etiology of CVD, insulin resistance, metabolic syndrome, and diabetes. However, with regard to correcting the obesity-related milieu, research is yet to elucidate an optimal PA strategy not only to confer benefits across standard physiological markers of health (e.g., lipid profiles, fasting plasma glucose) but also to reverse various etiologic characteristics of physiological decline. In the clinical context, since weight loss is known to elicit improvements in overall patient health (153, 186), blanket recommendations regarding healthy dietary practices and PA to promote weight reduction have become the inevitable norm. Weight loss is indeed effective for ameliorating health risk, but little research has documented its long-term sustainability or efficacy, and the vast majority of people who lose a significant amount of weight gain it back (15, 203). Further, as compared with interventions with PA, data suggest that purely diet-induced weight loss may have limited utility for establishing a sustainable, insulin-sensitive phenotype (44, 116, 165). This underscores the importance of identifying alternative choices (e.g., resistance training) for improving cardiometabolic health among obese individuals with metabolic syndrome.

Emerging evidence suggests improved cardiometabolic profile and improved insulin sensitivity after cardiorespiratory exercise (133) and resistance exercise (189), independent of weight loss. These changes may in part be attributable to enhanced muscle function and cardiorespiratory fitness, insulin-stimulated glucose disposal, and fatty acid oxidation (93, 149). Evidence has identified an inverse association between adiposity and muscle function even among nonobese individuals (145), underscoring the value of general PA and targeted exercise as primary elements in prescriptive metabolic syndrome treatment. Therefore, comprehensive clinical and public health interventions to treat metabolic syndrome must include simultaneous directives for improved dietary habits and gradual weight loss, increased daily PA, and targeted exercise prescriptions. Table 12.6 provides a review of exercise training and benefits for metabolic syndrome.

Cardiorespiratory Exercise

Physical inactivity has been recognized as a primary contributing factor to progression of obesity (40) and subsequent risk of metabolic syndrome, ASCVD, and diabetes (81, 83). Conversely, when combined with dietary interventions, cardiorespiratory exercise or moderate-intensity PA has been widely regarded as the most acceptable means to lose excess adiposity and reduce cardiometabolic health risk (10, 53). Cardiorespiratory exercise is suggested to be dose dependent, such that greater volumes of PA are associated with greater cardiometabolic health risk reduction (38, 53). It is commonly thought that the health benefits of PA for individuals with metabolic syndrome are directly related to mobilization of free fatty acids and gross decreases in adiposity. Indeed, since extended-duration cardiorespiratory exercise is effective for absolute energy expenditure, recommendations to prioritize this type of exercise are intuitive for the promotion of weight loss. However, research suggests that improvements in metabolic disturbances are possible with exercise interventions independent of overall weight loss or changes in body composition (133). This is an important message for clinical and public health constituents to accept and disseminate to patients, because it underscores the relative value of healthy lifestyles and fitness rather than focusing on body weight or aesthetics.

The positive benefits of exercise on cardiometabolic health in those with metabolic syndrome are multifaceted and can be explained by three phenomena (160). During the acute phase, a single bout of exercise can significantly increase whole-body glucose disposal and thus temporarily attenuate hyperglycemia. Moreover, for several hours after a given bout of exercise, insulin sensitivity is augmented. Lastly, repeated bouts of exercise lead to a chronic adaptive response, characterized by enhanced cardiorespiratory function and global improvements in insulin action (45, 133). Cardiorespiratory exercise is also effective for improving blood pressure and lipid profiles, decreasing visceral adiposity even in the absence of weight loss, enhancing fatty acid oxidation, increasing mitochondrial function and content, and attenuating the proinflammatory state (11, 53, 143).

Resistance Exercise

According to the ACSM position stand, resistance exercise does not promote clinically significant weight loss (evidence category A) (53). However, with regard to the restoration of cardiometabolic health among at-risk, obese individuals, mounting evidence indicates that resistance training may be a viable treatment option, comparable to aerobic exercise. Some evidence suggests that independent of participation in aerobic PA, sufficient resistance training volume is associated with more favorable metabolic risk profiles (41) (figure 12.3). Among adults with unfavorable cardiometabolic risk profiles, several longitudinal studies examining the role of exercise in metabolic disturbance have reported significantly improved insulin sensitivity and glucose tolerance with structured, progressive resistance exercise interventions (32, 58, 110).

Resistance exercise has also been associated with improvements in various ASCVD risk factors in the absence of weight loss, including decreased LDL-C (75, 94), decreased triglycerides (75), reduced blood pressure (107), and increased HDL-C (94). Moreover, several studies have documented the superiority of resistance exercise over traditional aerobic exercise for glycemic control and insulin sensitivity among adults with T2D (34). Chronic resistance training has been traditionally regarded as an appropriate means to augment or preserve skeletal muscle tissue and thus improve 24 h energy expenditure and decrease body fat in the long term (figure 12.4) (53, 58). Combining diet-induced energy restriction and resistance training has been shown to prevent loss in muscle tissue—an effect that

Table 12.6   Exercise Training Review

Cardiorespiratory endurance	Skeletal muscle strength	Skeletal muscle endurance	Flexibility	Body composition
Low-intensity, long-duration physical activity and structured cardiorespiratory exercise are beneficial for energy expenditure, weight loss and management, and improving cardiometabolic risk factors. Moderate-intensity, medium- and short-duration structured cardiorespiratory exercise is beneficial for aerobic fitness ($\dot{V}O_2max$), enhancing body composition, and improving cardiometabolic risk factors.	Resistance exercise for strength is important for preservation of muscle function, lean body mass, and bone mineral density and can improve certain parameters of cardiometabolic health (e.g., HbA1c, insulin sensitivity, glucose tolerance, lipid profiles, and blood pressure).	Resistance exercise for local muscular endurance is beneficial for fatigue resistance during activities of daily living and bouts of physical activity, as well as for low back health and posture.	Flexibility and ROM exercises are vital for musculoskeletal and joint integrity as well as for low back health.	A decrease in relative adiposity (i.e., % body fat) is vital for improved overall cardiometabolic health. An increase in absolute and relative LBM is necessary for preservation of muscular function and basal (resting) metabolic rate.

HbA1c = Glycosylated hemoglobin; LBM = lean body mass; ROM = range of motion; $\dot{V}O_2max$ = maximal oxygen consumption.

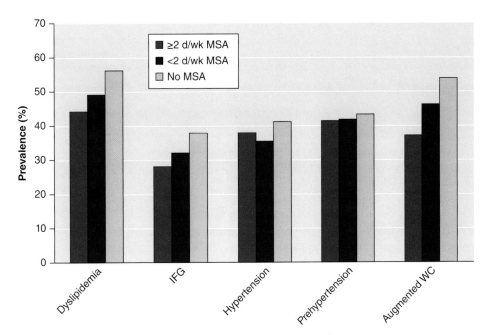

Figure 12.3   Age-adjusted prevalence estimates of metabolic syndrome criteria and prehypertension in U.S. adults aged ≥20/ yr, by level of muscular strengthening activity, National Health and Nutrition Examination Survey 1999-2004. MSA = muscular strength activity; IFG = impaired fasting glucose; WC = waist circumference.

Reprinted by permission from J.R. Churilla et al., "Muscular Strengthening Activity Patterns and Metabolic Health Risk Among U.S. Adults," *Journal of Diabetes* 4 (2012):77-84.

seems to be augmented when sufficient protein is consumed in the diet (53). Despite the well-known independent value of resistance exercise for cardiometabolic health, strong evidence confirms the superiority of combined aerobic and resistance exercise over either individual modality (37, 99).

Among children and adolescents, research evidence suggests that muscular strength is an important component of metabolic fitness and provides protection against insulin resistance (20). Stronger children are 98% less likely to be insulin resistant compared with those who are in the lowest percentiles for strength, even after adjustment for central adiposity, body mass, and maturation. Part of the reason may be that the high-force muscle actions associated with strength capacity are generally attributable to a higher ratio of myosin heavy chain II isoforms (i.e., type II muscle fibers, or fast twitch fibers) over myosin heavy chain I isoforms (i.e., type I muscle fibers, or slow twitch fibers) (193). Animal evidence suggests type IIa muscle fibers have a greater abundance of GLUT4 than type IIb and IIx muscle fibers, suggesting a greater potential for skeletal muscle contraction glucose uptake (33). Increases in strength capacity have been suggested to promote superior glucose metabolism among individuals with impaired baseline glucose tolerance following an exercise intervention (193). Moreover, research has documented the efficacy of isolated high-intensity resistance training (i.e., without dietary intervention or aerobic exercise) to safely reduce adiposity among overweight children (19). ACSM exercise guidelines for children and adolescents recommend ≥60 min of daily exercise of both cardiorespiratory and resistance training combined. Cardiorespiratory activities recommended for children and adolescents should be enjoyable

and developmentally appropriate for their age (e.g., aerobic-based sports, walking, running, swimming, dancing, bicycling, and other recreational options), while resistance training activities can be either unstructured (e.g., playing on playground equipment, climbing trees, tug-of-war) or structured (lifting weights, exercising with resistance bands) (1).

Of particular relevance to the generalizability of structured exercise in obese individuals with metabolic syndrome is the sustainability of such treatments. Some evidence indicates that resistance exercise elicits an earlier adaptive response in muscular function and anthropometric characteristics than traditional aerobic exercise, and as such may be associated with greater adherence rates over the long term. It is conceivable that early positive reinforcement could enhance self-efficacy and physical self-perception (body image), which may substantially influence the success of adhering to a regular PA regimen as part of an overall healthy lifestyle (6).

Range of Motion Exercise

Regular ROM exercises may be particularly relevant to older adults as a way to maintain or improve balance and posture and may help prevent slip-and-fall accidents. Although stretching is commonly recommended as a means to improve joint ROM, very little if any research supports the utility of this modality in the treatment or prevention of cardiometabolic disease. Rather, and in conjunction with full ROM resistance training, stretching exercises may allow patients to preserve physical function and musculoskeletal integrity during activities of daily living.

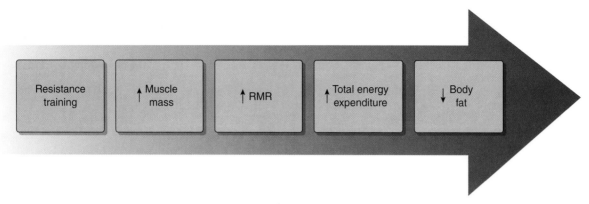

Figure 12.4   Effect of resistance training on body fat. Resistance training builds muscle mass, which is a highly active metabolic organ. An increase in muscle mass raises the resting metabolic rate (RMR), which translates into a daily increase in total energy expenditure (TEE). The increased energy output helps decrease body fat and improve weight control in the long term. The combined increment in muscle mass and decrement in body fat helps improve body composition when following a resistance training program.

Clinical Exercise Bottom Line

- Metabolic syndrome is a collection of interrelated cardiometabolic risk factors that are present in a given individual more frequently than may be expected by a chance combination.

- Individuals with metabolic syndrome are at heightened risk of T2D, atherosclerotic CVD, and early mortality.

- Diagnosing metabolic syndrome requires the presentation of three or more cardiometabolic risk factors, including elevated waist circumference, elevated triglycerides, reduced HDL-C, elevated blood pressure, and elevated fasting glucose.

- Exercise training promotes weight loss, weight management, or both, as well as improvements in cardiometabolic risk profile.

- Obese individuals with metabolic syndrome benefit most from a combination of exercise and healthy dietary intervention.

- Combined aerobic and resistance exercise is superior for improvements in serum lipid profiles, insulin sensitivity, glucose tolerance, body composition, and physical function.

- For individuals who have diabetes, HTN, or both, cardiorespiratory and muscular strengthening exercises, as well as exercise testing protocols, should be modified or closely monitored to avoid undue risk of cardiovascular or other untoward events.

CONCLUSION

The most recent prevalence estimates indicate approximately one in three U.S. adults have metabolic syndrome, representing an enormous financial burden on public health. Although no uniform criteria exist for diagnosing metabolic syndrome among pediatric populations, overweight or obese children and adolescents have a significantly elevated risk for poor cardiometabolic health. Treatment for metabolic syndrome must be multifaceted, and behavioral modification including diet manipulation and PA are fundamental. Indeed, when combined with dietary interventions, cardiorespiratory exercise has been widely regarded as the most acceptable means to induce weight loss and reduce cardiometabolic health risk. However, and as an important point of clarification, ample research exists to demonstrate that improvements in metabolic disturbances are possible with exercise interventions even in the absence of overall weight loss or changes in body composition. Moreover, with regard to the restoration of cardiometabolic health among at-risk obese individuals, mounting evidence indicates that resistance training may also be a viable treatment option, comparable to aerobic exercise. Thus, there is a critical need to inform the appropriate clinical and public health audiences and practitioners in an attempt to promote tailored, evidence-based prevention and treatment efforts specific to individuals with metabolic syndrome.

Online Materials

Visit HK*Propel* to access a link to the references, the case studies with discussion questions, and a quiz to help you review key concepts and test your understanding of the material covered in this chapter.

CHAPTER 13

Chronic Kidney Disease

Samuel Headley, PhD

Kenneth Wilund, PhD

Michael Germain, MD

Chronic kidney disease (CKD) results from structural damage to the kidneys and progressively diminished renal function. Chronic kidney disease is divided into six stages, depending on the extent of kidney damage, the **glomerular filtration rate (GFR)**, and the presence of albumin in the urine (37,42) (table 13.1). After CKD begins, it can progress to end-stage renal disease (ESRD), requiring some form of **renal replacement therapy (RRT)** such as **dialysis** (either hemo- or peritoneal), transplantation, or conservative management if RRT is not done.

DEFINITION

Chronic kidney disease is defined as the presence of kidney damage (usually detected as urinary albumin excretion of ≥30 mg/d, or equivalent) or decreased kidney function (defined as estimated glomerular filtration rate [eGFR] of <60 mL $\cdot$ min^{-2} $\cdot$ 1.73 m^{-2}) for 3 mo or more, irrespective of the cause (37). Many CKD patients do not progress to ESRD but instead die prematurely of cardiovascular disease (32). In recent years, the progression of patients with CKD to ESRD has stabilized (64).

SCOPE

According to current estimates, approximately 15% of the U.S. adult population, or approximately 37 million people, have CKD (11). Data collected by the United States Renal Data Systems indicate that 14.5% of Medicare recipients (individuals ≥65 yr) have CKD (64). Ninety percent of individuals with CKD do not know they have it, and the disease is more common among those ≥65 (11). According to data reported from the Centers for Disease

Control and Prevention, in 2016 approximately 726,000 people were on dialysis or living with a kidney transplant (11, 64). In 2017, the most recent year for which there are reliable estimates, the total Medicare cost to treat both ESRD and predialysis kidney disease was reported as $120 billion (64).

In the United States before 1972, access to dialysis treatment was limited, and selection of patients for treatment was made by committees of medical professionals, clergy, and laypeople. Essentially, these committees decided who would receive the lifesaving therapy of dialysis. In 1972, Congress passed landmark legislation that extended Medicare coverage to patients with ESRD. This legislation hinged on the expectation of successful vocational rehabilitation of these patients (an expectation that has not been realized). Renal replacement therapy is expensive. In 2017, the reported cost of hemodialysis was $91,795 per patient per year; a kidney transplant costs less over time ($35,817/yr) (64).

Although the overall outcomes and well-being of patients with renal failure have significantly improved because of advances in technology and medical therapy, with an accompanying increased potential for rehabilitation, it is generally acknowledged that the rehabilitation of these patients has not been addressed nationally in a sustained, consistent, and integrated fashion (34, 63, 71). Low levels of physical functioning contribute significantly to the low levels of rehabilitation, thus indicating the need for physical rehabilitation as an adjunct to the routine medical therapy in these patients. An exercise intervention is critical since the sedentary behavior that is characteristic of patients with CKD is believed to contribute to the excess morbidity and mortality observed in this population (4, 6, 32)

Acknowledgment: Much of the writing in this chapter was adapted from the first and second editions of *Clinical Exercise Physiology*. As a result, we wish to gratefully acknowledge the previous efforts of and thank Patricia Painter, PhD, and Sahil Bawa, MBBS.

Table 13.1   Staging Criteria for Chronic Kidney Disease

GFR stages	GFR (mL · min^{-2} · 1.73 m^{-2})	Terms
G1	>90	Normal or high
G2	60-89	Mildly decreased
G3a	45-59	Mildly to moderately decreased
G3b	30-44	Moderately to severely decreased
G4	15-29	Severely decreased
G5	<15	Kidney failure (add D if treated by dialysis)
Albumin stages	**AER (mg/d)**	**Terms**
A1	<30	Normal to mildly increased (may be subdivided for risk prediction)
A2	30-300	Moderately increased
A3	>300	Severely increased (may be subdivided into nephrotic and non-nephrotic for differential diagnosis, management strategies, and risk prediction)

GFR = glomerular filtration rate; AER = albumin excretion rate.

PATHOPHYSIOLOGY

Many conditions damage the kidneys including longstanding diabetes mellitus or hypertension, autoimmune diseases (e.g., lupus), glomerulonephritis, pyelonephritis, some inherited diseases (i.e., polycystic kidney disease, Alport's syndrome), and congenital abnormalities. The initial injury to the kidney may result in a variety of clinical manifestations, ranging from asymptomatic hematuria to renal failure requiring dialysis. Many individuals fully recover from acute kidney injury and subsequently suffer little or no sequelae. However, no such recovery occurs once chronic kidney disease develops.

The kidney is able to adapt to damage by increasing the filtration rate in the remaining normal **nephrons**, a process called adaptive hyperfiltration. As a result, the patient with mild renal insufficiency often has a normal or near-normal serum creatinine concentration. Additional homeostatic mechanisms (most frequently occurring within the renal tubules) permit the serum concentrations of sodium, potassium, calcium, and phosphorous and the total body water to also remain within the normal range, particularly among those with mild to moderate renal failure. Adaptive hyperfiltration, although initially beneficial, appears to result in long-term damage to the glomeruli of the remaining nephrons, which is manifest by proteinuria and progression to renal failure. The gradual decline in renal function in patients with CKD is initially asymptomatic. The damaged kidney initially responds with higher filtration and excretion rates per nephron, which masks symptoms until only 10% to 15% of renal function remains.

Progressive renal failure causes loss of both excretory and regulatory functions, which leads to ESRD and results in the uremic syndrome. Clinical manifestations of the uremic state include anorexia, nausea, vomiting, fatigue, pericarditis, peripheral neuropathy, and central nervous system abnormalities (ranging from loss of concentration and lethargy to seizures, coma, and death). No direct correlation exists between the absolute serum levels of blood urea nitrogen (BUN) or creatinine and the development of these symptoms. Please note that creatinine is a marker used to estimate eGFR but is not a uremic toxin. Some patients have relatively low BUN levels (e.g., 60 mg · dL^{-1} [21.4 mmol · L^{-1}] in an older patient) yet are markedly symptomatic. Conversely, others have marked elevations of BUN (e.g., 140 mg · dL^{-1} [50 mmol · L^{-1}]) but remain asymptomatic. Patients present with the previously identified symptoms, as well as peripheral edema, pulmonary edema, or congestive heart failure. To continue life, uremic patients require the institution of RRT using hemodialysis, peritoneal dialysis, or renal transplantation.

The loss of the excretory function of the kidney results in the buildup of toxins in the blood, any of which can negatively affect cellular enzyme activities and inhibit systems such as the sodium–potassium pump, resulting in altered active transport across cell membranes and altered membrane potentials. The loss of the regulatory function of the kidneys results in the inability to regulate extracellular volume and electrolyte concentrations, which adversely affects cardiovascular and cellular functions. Most patients with advanced CKD are volume overloaded; this results in hypertension and often congestive heart failure. Other malfunctions in regulation include impaired generation of ammonia and impaired hydrogen ion excretion, which results in metabolic acidosis. Decreased production of and responsiveness to erythropoietin are the primary reasons for the anemia observed in patients with ESRD (7, 50).

Other hormones that are normally released may be excessively produced or inappropriately regulated in response to renal failure. For example, hyperphosphatemia causes excess parathyroid hormone to be produced, which leads to metabolic bone disease. There is also a decrease in Klotho, an increase in fibroblast growth hormone (FGF)-23, and a decrease in 1,25 vitamin D (7, 50, 61).

Several metabolic abnormalities are associated with uremia, including insulin resistance and hyperglycemia; dialysis is associated with hypertriglyceridemia with a normal (or low) total cholesterol concentration. Some of the treatments associated with dialysis or immunosuppression therapy (following transplant) can contribute to these metabolic abnormalities (7, 50).

CLINICAL CONSIDERATIONS

The management of renal failure is multifaceted. First, the cause of the kidney disease needs to be determined, and either cure or control of the primary kidney abnormality is pursued. If this is not possible, as is the case in CKD, then the clinician should strive to slow or arrest the progression of the decline in kidney function. Treatment is multipronged and typically involves control of blood pressure and diabetes, use of an angiotensin-converting enzyme inhibitor/angiotensin receptor blocker (ACEI/ARB), lowering urine protein, weight loss as needed, smoking cessation, and increased physical activity levels. Dietary adjustments for protein, sodium, and fluid intake play an important role in the initial management of renal failure. If these treatments are not successful, RRT is required. Transplantation is the preferred method, but patients need to be free of other life-threatening illnesses to be considered for transplantation.

In-center **hemodialysis** is the most common therapy for renal failure, although it requires significant time throughout the week at a renal center. The third RRT option, **peritoneal dialysis**, is the method least used in United States, but it is more frequently used in other countries (28). Home dialysis, either as hemodialysis or peritoneal dialysis, is the modality of choice. Randomized controlled trials have demonstrated a better outcome for patients treated with five-plus dialysis treatments a week, completed either in a center or at home (14).

Signs and Symptoms

CKD patients often present with symptoms and signs resulting directly from diminished kidney function. These include edema, hypertension, and decreased urine output. However, many patients have no clinical symptoms. In such patients, kidney disease is detected by laboratory tests that are obtained as part of an evaluation of an unrelated disorder. Deterioration in renal function results in an overall decline in physical well-being. The major laboratory findings in patients with CKD include an increased serum creatinine concentration and increased BUN. Hyperkalemia and metabolic acidosis may be present. Other common laboratory abnormalities include anemia, elevated parathyroid hormone (PTH), elevated serum phosphorus, and reduced serum calcium. The degree to which these abnormalities are present depends on the severity of the renal dysfunction. Signs include anemia, fluid buildup in tissues, loss of bone minerals, and hypertension. Patients complain of fatigue, shortness of breath, loss of appetite,

restlessness, change in urination patterns, and overall malaise. Muscle mass, muscle endurance, and peak oxygen uptake ($\dot{V}O_2$) all decline as the disease progresses.

History and Physical Examination

For the clinical exercise physiologist participating in the care of a patient with CKD, the patient's medical history should be carefully reviewed. Longstanding diabetes and severe hypertension are common causes of CKD. A history of severe peripheral vascular disease and cardio- and cerebrovascular disease may also suggest renovascular issues. Family history is typically explored for familial diseases such as polycystic kidney disease (PKD) or **glomerulonephritis** such as C3 glomerulonephritis or immunoglobulin A (IgA) nephropathy. The medical history usually focuses on renal-related issues, including cause of renal failure, current renal function (if any), and current treatment. The comorbidities should be listed along with the treatments for each.

There is rarely any information available on physical functioning or recommendations for activity; however, for the patient receiving dialysis, additional information may be found in the evaluations by the social worker and dietitian, both of whom often address limitations in physical functioning (in terms of activities of daily living or need for assistance in the home). A regular assessment of quality of life (i.e., kidney disease–specific quality of life) is now required for all dialysis patients.

Finally, the short- and long-term multidisciplinary patient care plans completed for these patients may also prove insightful, offering the clinical exercise physiologist a better understanding of the overall plan for the patient in terms of RRT and social considerations. The clinical exercise physiologist should pay particular attention to any cardiac history and the type and frequency of dialysis treatment, information needed in order to develop the best strategy for exercise that considers the treatment burden experienced by the patient.

Diagnostic Testing and Treatment

Initial testing includes serum creatinine for the estimation of the GFR, reagent strip urinalysis (dipstick) with urine microscopy, and the quantification of urine protein or albumin. A renal ultrasound is completed in all patients with an increased serum creatinine of unclear duration. The results of the urinalysis and ultrasound generally direct the remainder of the diagnostic evaluation. Patients who have a progressive increase in serum creatinine usually undergo a biopsy unless imaging studies show significant evidence of chronicity (such as small kidneys or increased echogenicity).

Treatment of chronic renal failure consists of medical management until the creatinine clearance is less than 10 mL · min^{-1}, at which time RRT may be required. Management is directed at minimizing the consequences of accumulated uremic toxins that are normally excreted by the kidneys. Dietary measures play a primary role in the initial management, with very low-protein diets being prescribed to decrease the symptoms of uremia and possibly to delay the progression of the disease. In addition to protein restriction, dietary sodium and fluid restrictions are critical as well (50), because the fluid-regulating mechanisms of the kidney are deteriorating. Therefore, excess fluid consumed remains in the system, and with progressive deterioration in renal function, ultimately results in peripheral edema, congestive heart failure, and pulmonary congestion.

The decision to begin dialysis is determined by many factors, including cardiovascular status, electrolyte levels (specifically potassium), chronic fluid overload, severe and irreversible oliguria (i.e., urine output less than $0.5 \text{ mL} \cdot \text{kg}^{-1}$ of body weight divided by height) or anuria (i.e., absence of urine output), significant uremic symptoms, abnormal laboratory values (usually creatinine >8-12 $\text{mg} \cdot \text{dL}^{-1}$, blood urea nitrogen >100-120 $\text{mg} \cdot \text{dL}^{-1}$), and creatinine clearance (<10 $\text{mL} \cdot \text{min}^{-1}$). Renal replacement therapy does not correct all signs and symptoms of uremia and often results in other concerns and side effects. Table 13.2 lists the laboratory values for healthy patients versus those undergoing dialysis.

Hemodialysis

Hemodialysis is the most common form of RRT in the United States. Approximately 95% of all patients undergo hemodialysis in a center or at home. In other countries, some patients prefer more home-based treatments such as peritoneal dialysis (discussed later). Hemodialysis is a process of ultrafiltration (fluid removal) and clearance of toxic solutes from the blood. It necessitates vascular access by way of an arteriovenous connection (i.e., fistula) that uses either a prosthetic conduit or native vessels. Two needles are placed in the fistula; one directs blood out of the body to the artificial kidney (dialyzer), and the other returns blood back into the body. The dialyzer has a semipermeable membrane that separates the blood from a dialysis solution, which creates an osmotic and concentration gradient to clear substances from the blood. Factors such as the characteristics of the membrane, transmembrane pressures, blood flow, and dialysate flow rate determine removal of substances from the blood. Manipulation of the blood flow rate, dialysate flow rate, dialysate concentrations, and time of the treatment can be used to remove more or less substances and fluids (50).

The duration of the dialysis treatment is determined by the degree of residual renal function, body size, dietary intake, and clinical status. A typical dialysis prescription is 3 to 4 h, three times per week. Complications of the dialysis treatment include hypotension, cramping, problems with bleeding, and fatigue. Significant fluid shifts can occur between treatments if the patient is not careful with dietary and fluid restrictions. Table 13.3 identifies the complications associated with hemodialysis.

Peritoneal Dialysis

Approximately 10% of dialysis patients in the United States are treated with peritoneal dialysis (66). Other parts of the world tend to have a higher percentage of patients treated with this form of dialysis (e.g., Hong Kong and Thailand) (28). This form of therapy is accomplished via introduction of a dialysis fluid into

Table 13.2   Normal Laboratory Values Compared With Acceptable Values for Dialysis Patients

Laboratory value	Normal range	Target range for dialysis patients[a]
Hemoglobin (g · dL^{-1})	12.0-16.0	Goal 10.0-12.0
Sodium (mEq · L^{-1})	136.0-145.0	135.0-142.0
Potassium (mEq · L^{-1})	3.5-5.3	4.0-6.0
Chloride (mEq · L^{-1})	95.0-110.0	95.0-100.0
HCO$_3$ (mEq · L^{-1})	22.0-26.0	23.0-28.0
Albumin (g · dL^{-1})	3.7-5.2	Goal >4.0
Calcium (mg · dL^{-1})	9.0-10.6	Goal 8.5-9.5
Phosphorus (mg · dL^{-1})	2.5-4.7	Goal <5.5
BUN (mg · dL^{-1})	5.0-25.0	60.0-110.0
Creatinine (mg · dL^{-1})	0.5-1.4	3.0-25.0
pH	7.35-7.45	7.38-7.39
Creatinine clearance (mL · min^{-1})	85.0-150.0	0.0 (or residual of <10)
Glomerular filtration rate (mL · min^{-1})	90.0-125.0	0.0 (or residual of <10)

[a]Assuming well-dialyzed, stable patient.

[b]Hemoglobin levels depend on the level of erythropoietin treatment.

BUN = blood urea nitrogen.

Table 13.3 Complications Associated With Hemodialysis Treatment

Complication	Pathophysiology/Mechanism
Hypotension	Decreased plasma volume with slow refilling Impaired vasoactive or cardiac responses Vasodilation Autonomic dysfunction
Cramping	Contraction of intravascular volume Reduced muscle perfusion
Anaphylactic reactions	Reaction to dialysis membrane (usually at first use)
Pyrogen- or infection-induced fever	Bacterial contamination of water system Systemic infection (often at the access site)
Cardiopulmonary arrest	Dialysis line disconnection Air embolism Aberrant dialysate composition Anaphylactic membrane reaction Electrolyte abnormalities Intrinsic cardiac disease
Itching	Unknown etiology
Restless legs	Unknown etiology
Fatigue	Most prevalent symptom
Myocardial stunning	Myocardial ischemia leading to left ventricular dysfunction

Data from Johansen (34).

Long-Term Complications of Dialysis

- Metabolic abnormalities
  - Metabolic acidosis
  - Hyperlipidemia (type 4)
    - Increased triglycerides
    - Increased very low-density lipoprotein cholesterol
    - Decreased high-density lipoprotein cholesterol
    - Normal total cholesterol
  - Hyperglycemia
  - Other hormonal dysfunction
- Malnutrition
- Cardiovascular disease
  - Hypertension
  - Ischemic heart disease
  - Congestive heart failure
  - Pericarditis

- Renal-related metabolic bone disease (secondary hyperparathyroidism, osteoporosis, osteomalacia, adynamic)
- Peripheral neuropathy
- Amyloidosis
- Severe physical deconditioning
- Frequent hospitalizations
- Continuation of progressive complications in diabetic patients

the peritoneal cavity through a permanent catheter placed in the lower abdominal wall. The peritoneal membranes are effective for ultrafiltration of fluids and clearance of toxic substances in the blood of uremic individuals. The peritoneal barrier is composed of three layers: the peritoneal mesothelium, the interstitium, and the capillary endothelium. According to the three-pore model of solute transport, the capillary endothelium contains three different-sized pores that are size selective in restricting solute transport.

The dialysis fluid is formulated to provide gradients to remove fluid and substances. The fluid is introduced either by a machine (cycler), which cycles fluid in and out over an 8 to 12 h period at night, or manually with 2 to 2.5 L bags that are attached to tubing and emptied by gravity into and out of the peritoneum. The latter process, known as continuous ambulatory peritoneal dialysis, allows the patient to dialyze continuously throughout the day. Continuous ambulatory peritoneal dialysis requires exchange of fluid every 4 h using a sterile technique (50). Table 13.4 lists complications associated with peritoneal dialysis.

Patients choose peritoneal dialysis so they can experience more freedom and less dependency on facility-based care. This method of treatment allows patients to travel and dialyze on their own schedules. Patients who have cardiac instability may also be placed on peritoneal dialysis because this method does not involve the major fluid shifts experienced with hemodialysis.

Complications of peritoneal dialysis are due to increased intra-abdominal pressure resulting from instillation of dialysate into the peritoneal cavity and include problems with the catheter or catheter site (e.g., infection), hernias, low back pain, obesity, gastroesophageal reflux, and delayed gastric emptying. Hypertriglyceridemia is a problem caused by the exposure and absorption of glucose from the dialysate. Patients may absorb as many as 1,200 kcal from the dialysate per day, contributing to the development of obesity and hypertriglyceridemia (36). Although preliminary, among ESRD patients who have diabetes there is evidence of a survival advantage that favors hemodialysis over peritoneal dialysis (73).

Kidney Transplant

Transplantation of kidneys is the preferred treatment for ESRD. In 2016 in the United States, 20,161 transplants were performed (64). The source of the kidneys can be a living relative, an unrelated individual, or a cadaver. Because of the shortage of organs available for transplantation and improvements in immunosuppression medications, living nonrelated transplants are becoming more frequent. Patients considered for transplant are typically healthier and younger than the general dialysis population, although there are no age limits to transplantation. Patients with neoplasia or severe cardiac, cerebrovascular, or pulmonary disease are not considered candidates. Table 13.5 lists long-term complications of transplantation.

Following transplantation, patients are placed on immunosuppression medications, which include combinations of glucocorticosteroids (prednisone), a calcineurin inhibitor (cyclosporine or tacrolimus), a TOR inhibitor (rapamycin or everolimus), a cell cycle inhibitor (azathioprine or mycophenolate), a costimulatory

Table 13.4 Complications Associated With Peritoneal Dialysis Treatment

Complication	Comments
Infections	Possible at exit site or along catheter (tunnel infection; may be the result of a break in sterile procedures during exchange
Peritonitis	Most frequent complication of peritoneal dialysis
Hypotension	Excessive ultrafiltration and sodium removal
Hernia, leaks	Associated with the increased intra-abdominal pressure

Table 13.5 Long-Term Complications Associated With Transplantation

Complication	Comments
Rejection	Can be acute or chronic; in most cases treated with increased immunosuppression dosages.
Cardiovascular disease	Most frequent cause of death after transplant. All known risk factors for cardiovascular disease are often prevalent, and immunosuppression medications may exacerbate the associated risk.
Infections	Immunosuppression may increase infection risk.
Musculoskeletal disorders	Glucocorticoid therapy (prednisone) reduces bone density, increases risk for aseptic necrosis of the hip, and causes muscle protein breakdown. Many centers now use a steroid-free protocol.
Obesity	Very prevalent, often associated with prednisone therapy, but more likely attributable to imbalance between calorie intake and expenditure (i.e., lifestyle issues).

blocker (belatacept), and usually an induction agent (Thymoglobulin or IL-2 inhibitors). Many centers now use a steroid-free protocol. Patients may experience rejection either early (acute) or later (chronic), which is detected by elevation of creatinine. Rejection is treated immediately with increased dosing of immunosuppression (mostly prednisone), with a subsequent tapering back to a maintenance dose. Patients must remain on immunosuppression for the lifetime of the transplanted organ.

In 2015, 1 yr graft survival ranged from 93% to 98% for a deceased or living donor (64). Five-year rates have increased over time, going from 66% in 1999 to 75% in 2011 (64). Causes of graft loss include chronic rejection (25%), cardiovascular deaths (20.3%), infectious deaths (8.7%), acute rejection (10.2%), technical complications (4.7%), and other deaths (10.2%) (table 13.5). Short-term transplant survival has been improved with new immunosuppression medications, leaving the major challenges of long-term survival of graft and patients to be investigated. Loss of kidney function in transplant recipients results in the need to return to dialysis (14).

Complications of kidney transplantation are primarily related to immunosuppression therapy and include infection, hyperlipidemia, hypertension, obesity, steroid-induced diabetes, and osteonecrosis. The incidence of atherosclerotic cardiovascular disease is four to six times higher in kidney transplant recipients than in the general population, and cardiovascular risk factors are prevalent in most patients (8,13,76).

Exercise Testing

The functional capacity (i.e., peak oxygen uptake, $\dot{V}O_2$) of CKD patients before starting dialysis is ~21 ± 5 mL · kg^{-1} · min^{-1}, which represents 50% to 80% of age-matched healthy controls (20, 25, 65). As the disease progresses, functional capacity decreases, such that by the time stage 5 is reached, peak $\dot{V}O_2$ is only 17 to 20 mL · kg^{-1} · min^{-1}, which can be as low as 39% of age-matched health

controls (2, 34, 40, 52, 74). It is important to note that peak $\dot{V}O_2$ is an independent risk factor of mortality in these patients (20). In recent years, researchers have also examined the impact of exercise training on functional capacity and disease progression (see practical application 13.1) in patients with CKD.

The degree to which exercise capacity is limited in patients with CKD is difficult to determine because reduced exercise capacity is almost certainly a multifaceted problem that is influenced by anemia, muscle blood flow, muscle oxidative capacity, myocardial function, autonomic dysfunction, the person's activity levels, and other factors. In ESRD, muscle function may be affected by nutrition status, dialysis adequacy, inactivity, hyperparathyroidism, and other clinical variables (34). The estimation of $\dot{V}O_2$ peak from submaximal responses is not currently recommended in CKD patients since prediction equations have not been validated in this population (40).

Most studies that have measured $\dot{V}O_2$ peak have included only the healthiest patients with CKD; thus, the average CKD patient may have an even lower exercise capacity. One school of opinion holds that information obtained from graded exercise testing in this patient group, particularly patients with stage 5 CKD, is not diagnostically useful since most patients stop exercise because of leg fatigue and do not achieve age-predicted maximal heart rates. Furthermore, many patients have abnormal left ventricular function (15), and some have conditions that make interpretation of the stress electrocardiogram difficult. Thus, exercise stress testing in ESRD patients may not be routinely performed before initiation of exercise training, and requiring that such a test first be completed may needlessly prevent some patients from becoming more physically active (59). However, some experts still recommend the use of exercise testing, particularly with gas exchange, since functional capacity is considered useful for prognostic purposes (36, 40).

Because exercise capacity is so markedly reduced in stage 5 patients, most patients do not train at an intensity level that is

Practical Application 13.1

EXERCISE TRAINING AND THE PROGRESSION OF CKD

Evidence from animal studies shows that exercise training can favorably alter kidney function, but similar evidence is inconsistent in human studies (4,17, 24, 77). A meta-analysis by Barcellos et al. (4) found little evidence of improvements in eGFR from exercise interventions, while a meta-analysis by Zhang et al. (77) showed improvements in eGFR after interventions less than 3 mo in duration, but this effect was not seen in interventions lasting between 3 and 12 mo. These inconsistent findings have been attributed to the small sample sizes used in most of these studies, the relatively

short duration of these studies, and the known variability in the measurement of GFR (77). However, what has consistently emerged from all the exercise studies performed in CKD patients not yet treated with dialysis is that exercise training does not worsen kidney function. Apart from its effects on the kidney, exercise training in CKD improves overall fitness and seems to be of some benefit on resting blood pressure and visceral fat (3), but its effect on serum lipids remains unsure (22, 23,70).

much greater than the energy requirements of their daily activities. Therefore, the risk associated with exercise training is minimal. However, patients with CKD who are not yet undergoing dialysis can exercise at moderate and even higher intensities, in much the same manner as healthy individuals. Caution should be taken if heart rate is used for determining training intensity in patients with stage 5 disease because the heart rate response can be influenced by medication and uremia. If an exercise test is used, an individualized heart rate range can be developed if both the test and the exercise training are completed after the patient has taken their medications as prescribed. Table 13.6 summarizes a few of the important issues associated with various fitness assessments. Table 13.7 lists the medications commonly used to treat patients

with CKD and ways in which these medications might affect the responses observed during exercise. Practical application 13.2 address the client–clinician interaction for patients with CKD.

Cardiorespiratory fitness can be assessed in predialysis CKD patients using cardiopulmonary exercise testing (CPET). The utility of CPET in patients with ESRD may not be as apparent; instead, in such patients it may be better to measure physical performance using standard functional tests such as the short physical performance battery, stair climbing, 6 min walk test, shuttle walk test, sit-to-stand-to-sit test, or gait speed testing. These tests, which have been standardized and used in many studies of elderly people, have been shown to predict outcomes such as hospitalization, discharge to a nursing home, and mortality rate (1, 40, 59).

Table 13.6   Exercise Testing Review

Cardiorespiratory endurance	Skeletal muscle strength	Skeletal muscle endurance	Flexibility	Body composition
Functional tests may be more appropriate for very deconditioned ESRD patients and some predialysis CKD patients. Abnormal cardiovascular responses are common. Cardiopulmonary exercise tests are good for predialysis patients.	Isotonic RM testing is not recommended in ESRD. It can be used in early stage predialysis patients. Isokinetic testing and functional testing are recommended for CKD patients.	Because of the low weights used, this can be similar to testing in healthy individuals.	Standardize the timing of testing in relation to dialysis treatments or in relation to bag changes in peritoneal patients. Can be used normally for predialysis patients.	Specific equations have not been validated in ESRD population. Measurement must be taken in relation to dialysis treatment and may be affected by fluid status. Normal methods can be used in predialysis CKD patients.

Table 13.7   Pharmacology

Medication name and class	Primary effects	Exercise effects	Important considerations
Diuretics (e.g., furosemide)	↓ blood pressure	↓ blood pressure	Hypotensive response postexercise
ACE inhibitor (e.g., lisinopril)	↓ blood pressure	↓ blood pressure	Hypotensive response postexercise
ARB (losartan)	↓ blood pressure	↓ blood pressure	Hypotensive response postexercise
Calcium channel blockers (diltiazem)	↓ blood pressure	↓ blood pressure ↓ heart rate	Hypotensive response postexercise
β-blockers (e.g., metoprolol)	↓ heart rate and blood pressure	↓ heart rate and blood pressure	Can blunt the exercise response
Statins (e.g., pravastatin, simvastatin)	Modify lipids	None	Muscle myalgia
Oral hypoglycemic agents (e.g., glyburide)	↓ blood glucose	↓ blood glucose	Must monitor blood sugars to avoid hypoglycemia
Insulin (e.g., Humalog, Lantus)	↓ blood glucose	↓ blood glucose	Must monitor blood sugars to avoid hypoglycemia
Antiplatelet (e.g., aspirin)	↓ blood clotting	None	Increased tendency to bruise or bleed with any trauma
Phosphate binders (sevelamer)	↓ phosphate	None	None
Erythropoietin (Epogen)	↑ red blood cells	Small ↑ in aerobic capacity	None

↑ = increase; ↓ = decrease; ACE inhibitor = angiotensin-converting enzyme inhibitors; ARB = angiotensin receptor blockers.

Practical Application 13.2

CLIENT–CLINICIAN INTERACTION

Exercise professionals can best serve the needs of patients with CKD if they understand the medical complications associated with the disease, the treatment regimens, and the setting in which they receive treatment. The exercise professional who works with predialysis CKD patients needs to understand that these patients are likely to be deconditioned, diabetic, hypertensive, and obese and, as a result, at increased risk of having cardiovascular disease (4). Since the ideal exercise prescription for CKD patients has not yet been identified, recommendations that are typically used for individuals with the aforementioned comorbidities can be followed.

Concerning patients undergoing hemodialysis, most receive their treatments three times per week for 3 to 4 h, and following these treatments patients tend to feel fatigued for much of the day. They are often transported either by friends and family or by medical transportation services, so they have minimal flexibility with their schedule. Also, the clinical staff working in these clinics are usually quite busy, so the time and opportunity for patient education is diminished. These circumstances may make it challenging for patients to attend exercise class in a supervised program held at another facility. Thus, to optimize adherence, the clinician should consider prescribing supervised exercise classes on those days when the patient is not scheduled for dialysis or implementing an intradialytic exercise program at the dialysis clinic. In addition to performing structured exercise, patients should be encouraged to increase lifestyle physical activity levels, including simple tasks such as shopping, cleaning their homes, or doing yardwork. These activities can be performed without supervision and have been shown to improve physical function and patient quality of life (67).

For a patient's exercise regimen to be successful, regardless of its location, the exercise professional should first interact with and educate the dialysis staff. To do this requires the support of the administration to coordinate the required in-service training. The dialysis staff are close to the patients and can be influential in their participation (or nonparticipation) in exercise. This support is important for both the long-term adherence of the patient and the integration of the exercise personnel in the patient's overall care. Educational and motivational programs at the dialysis clinic not only increase staff and patient awareness of the importance of exercise but also can help change the environment of the clinic from one of illness to the pursuit of optimal health.

After kidney transplant, some patients may be afraid to exert themselves vigorously. This concern may be due to the absence of information and recommendations from health care professionals regarding exercise, general feelings of weakness despite a significant improvement in overall health, fears or concerns on the part of the patient's family, or lack of experience with exercise (because of health concerns before transplant). Exercise is not routinely addressed following kidney transplant—either at the time of transplant or during routine follow-up care in the outpatient clinic. This is despite a growing body of literature documenting that 3 to 12 mo of exercise training in kidney transplant patients can improve cardiorespiratory fitness, muscle strength, and quality of life as well as possibly increase weight loss in obese patients (8, 13, 45); however, hard outcomes such as cardiovascular outcomes and mortality have not been assessed. With this in mind, the exercise professional should attempt to educate the transplant team about the importance of exercise and, as much as possible, become part of the routine care team so that exercise counseling is incorporated into a patient's care plan following transplantation.

Also, a walking–stair-climbing test was validated in hemodialysis patients by Mercer and colleagues (47). However, more research needs to be done on the use of these physical performance tests in patients with ESRD because the minimum difference in scores that is meaningfully linked to important clinical outcomes remains elusive (40). Finally, self-reported physical functioning scales such as those on the SF-36 Health Status Questionnaire are highly predictive of outcomes in dialysis patients, specifically hospitalization and death (16). Exercise training improves the functional scores on these self-reported scales in hemodialysis patients (55).

Exercise capacity is similarly low in peritoneal dialysis patients, who tend to be extremely inactive (5, 28, 43, 58). Following successful renal transplant, exercise capacity increases significantly, to near-inactive normal predicted values (5, 10, 18, 56, 58). This improvement in aerobic capacity after kidney transplantation peaks at 16 mo after transplant, following which there is a progressive decline in fitness that seems to be linked to poorly controlled blood pressures (75). Renal transplant recipients who were active and who participated in the 1996 U.S. Transplant Games had exercise capacities that averaged 115% of normal age-predicted values (57). Along with findings of improved survival, this underscores the need for transplant recipients to be physically active after transplant (75, 76). A meta-analysis confirms that exercise training in transplant recipients increases exercise tolerance and improves quality of life. There is some indication that regular exercise improves arterial stiffness scores, but it does not seem to favorably alter traditional cardiovascular risk factors (13).

EXERCISE PRESCRIPTION

The exercise prescription for patients with CKD should be comprehensive and include aerobic, resistance, balance, and flexibility exercises. Weight management considerations may also be needed for any patients who are overweight or obese, which is common in this population. When prescribing exercise for patients on dialysis, it is important to appreciate that multiple barriers to exercise often exist. These include general feelings of malaise, time requirements of treatment, fear, and adaptation of lifestyles to low levels of functioning. Thus, any prescription should begin with low-volume exercise and gradually progress to prevent discouragement and additional feelings of fatigue or muscle soreness. The research focus sidebar contains a brief review of the literature about exercise training in CKD patients.

Special Exercise Considerations

Many CKD patients live sedentary lives and are not accustomed to being physically active. Because patients with CKD are more likely to die from cardiovascular disease than to progress to ESRD, health care providers are now being encouraged to discuss ways in which CKD patients can incorporate more physical activity into their lives (4).

In general, patients with CKD not on dialysis are likely to be diabetic, hypertensive, and overweight or obese. Therefore, exercise recommendations should be made with these comorbid conditions in mind (31, 65). Patients with ESRD often do not know how much activity is too much, and because they may not feel well and are easily fatigued, they avoid physical activity altogether.

Dialysis patients interact primarily with their dialysis providers; therefore, the exercise professional must take the extra time to learn more about dialysis and kidney transplant to appreciate the burden these patients experience because of their circumstances.

Research Focus

Benefits of Physical Activity, Aerobic Exercise Training, and Resistance Training in CKD

Patients with CKD demonstrate low levels of physical activity and spend as much as 67% of their awake time being sedentary (i.e., <1.5 METs, in a seated or lying posture), which is associated with a greater risk of mortality and a faster rate of disease progression, independent of the amount of time spent engaged in moderate to vigorous physical activity (MVPA) (6, 21). There is evidence to indicate that replacing as little as 2 min/h of sedentary behavior with even light-intensity activity (i.e., 2-2.9 METs) significantly reduces mortality risk in CKD patients (6).

On the basis of available data, CKD patients, across the spectrum of the disease, should be encouraged to adhere to the physical activity guidelines for all Americans in addition to reducing sedentary time. Among patients with CKD, exercise training interventions conducted over 4 to 6 mo are associated with marked improvements in cardiorespiratory fitness (up to 20%), reductions in markers of inflammation, and improvements in muscle strength (4, 25). A meta-analysis indicates that exercise training has a small effect on resting nonambulatory systolic blood pressure values and no effect on 24 hr ambulatory readings (68).

Resistance training programs have led to improvements in muscular strength in all stages of CKD (2, 25, 45). Changes in plasma lipids in response to exercise training seem to be attenuated in patients with CKD. Body composition changes are variable, although there is evidence of significant reductions in visceral fat after aerobic exercise training in both obese predialysis patients (4) and hemodialysis patients (72). The effects of exercise training on vascular health and function in the CKD population have been inconsistent. For example, Kirkman et al. (38) showed that endurance training in CKD patients not on dialysis improves microvascular function but not arterial stiffness (38), and a few small studies have suggested improvements in arterial stiffness from endurance exercise in ESRD (49, 69). However, other studies on both CKD (27, 23) and ESRD (30, 39, 68) suggest no change in arterial stiffness with exercise training across the spectrum of CKD. Taken together, this suggests that the vascular remodeling that occurs in many CKD patients may limit the ability of vessels to respond to exercise interventions in a robust manner.

Other benefits associated with exercise training include a reduction in depression scores and improvements in health-related quality of life indices (4). Improvements in heart rate variability have also been reported, with a shift toward greater vagal tone following exercise training (4, 25). In general, significant benefits from exercise training in CKD patients seem to be stronger in the ESRD population, possibly because most of the exercise training studies to date have been done in this cohort (2, 4, 51). This is surprising given that the number of predialysis CKD patients far exceeds the number of ESRD patients.

This learning experience could entail watching a patient being put on the dialysis machine and visiting with a few patients and the dialysis staff during a patient's treatment. Patient support groups are also a good source of information. Patients often talk freely about their experiences. By listening carefully, the exercise professional will better appreciate the various challenges a patient may encounter when asked to start an exercise program, and as a result, they may be able to devise more effective strategies to motivate patients.

Most exercise professionals practice in a setting that is outside nephrology and depend on the referral of patients with CKD to their exercise facility, or they may counsel patients who participate in home-based, independent exercise. Thus, the clinical exercise physiologist should educate other health care providers about the benefits of exercise for their patient groups and the exercise-related services that exercise professionals can offer. Although the nephrology community is becoming more interested in improving the physical functional capabilities of patients with CKD, most nephrologists and kidney transplant staff are unfamiliar with how to evaluate physical function or prescribe exercise. An exercise professional who appreciates the problems associated with exercise and dialysis or transplant represents a valuable addition to the patient care team.

Exercise Recommendations

Patients with CKD who are not on dialysis have been shown to benefit from exercise training programs using standard exercise prescriptions for the general population, so long as the individuals are screened properly, start at a low to moderate intensity, and progress gradually (1, 4, 25, 26). For CKD patients on dialysis, the timing of exercise in relation to the dialysis treatment should be considered (9, 54). Intradialytic exercise is recommended since it may have unique physiological benefits (e.g., reduced likelihood of cramping, hypotension, and myocardial stunning during dialysis), enhances exercise compliance, and reduces the boredom associated with dialysis treatments (12, 19, 41, 44, 46, 49, 53, 60, 65).

Exercising immediately before dialysis (when patients may have fluid overload) or after dialysis (when fatigue and hypotension are common) is generally not well tolerated. However, responses to exercise are highly variable, so patients should not be discouraged from being physically active if asymptomatic. Exercise should be deferred if the patient is experiencing shortness of breath related to excess fluid status. No specific guidelines regarding the upper limit of fluid weight gain contraindicate exercise, although the guidelines for blood pressure and other CV risk factors established by the American College of Sports Medicine (62) should be followed. Practical application 13.3 discusses the exercise prescription for dialysis patients and kidney transplant recipients.

Intradialytic cycling is the most feasible mode of exercise because it does not interfere with the dialysis treatment. It should be encouraged in most dialysis clinics. Patients can also perform intradialytic resistance exercises with resistance bands, balls, and light weights. Exercise during dialysis is usually best tolerated during the first 1 to 1.5 h of the treatment because after that time the patient has a greater risk of becoming hypotensive even while sitting in the chair, which makes exercise difficult (41). This response is caused by the continuous removal of fluid (including intravascular) throughout the treatment, which decreases cardiac output, stroke volume, and mean arterial pressure at rest (48). Therefore, after 2 h of dialysis, cardiovascular decompensation may contraindicate exercise in some patients (48). Conversely, other research indicates that intradialytic exercise, even during the third hour of dialysis, might not cause hemodynamic instability in many patients (29). As a result, the ideal time for a patient to exercise during dialysis should be determined on an individual basis. This is especially true for patients who dialyze during morning shifts and prefer to sleep when starting their session. One concern with intradialytic exercise is that the volume, intensity, and variety of activities that can be performed are limited. As a result, patients should also be provided guidance on how to implement home exercise or physical activity programs.

Special Considerations for Clients With Chronic Kidney Disease and End-Stage Renal Disease

- CKD patients should be encouraged to reduce the time spent in sedentary behavior by increasing time in light activity (1.5-2.9 METs) daily. ESRD patients are very deconditioned and complain of fatigue, partly due to anemia. Chronic volume overload also contributes to fatigue and may limit some patients' ability to exercise in the hours after a dialysis session.

- Counsel patients on how to build exercise into their schedules whenever possible, including an intradialytic plan when feasible.

- Patients with ESRD are often diabetic and hypertensive.

- Patients with CKD or ESRD are at higher than normal risk of cardiovascular disease. Exercise training should be encouraged to help manage and decrease this risk.

- Renal disease may impose a ceiling on the magnitude of improvement that can occur with exercise training, so both the patient and the exercise professional should not be discouraged if the change in fitness-related variables is less than hoped for over time. The patient's quality of life improves markedly.

EXERCISE TRAINING REVIEW FOR ESRD PATIENTS TREATED WITH DIALYSIS OR RECEIVING A TRANSPLANT

Training method	Frequency	Intensity	Time (Duration)	Type (Mode)	Progression	Important goals
Cardiorespiratory	3-5 d/wk	RPE of 12-15 (on 6- to 20-point scale)	Work up to 30 min of continuous exercise, predialysis patients can work for 20-60 min	Intradialytic cycling, walking, swimming, low-level aerobics, stepping	In ESRD patients, start with intervals of intermittent exercise and gradually increase the work intervals; normal methods can be used in predialysis patients	Improve cardiorespiratory fitness Reduce cardiovascular risk
Resistance	2 or 3 d/wk	10RM	Repetitions: 10-15 per set	For ESRD patients, machine weights, resistance bands, isometric exercises, very low-level hand and ankle weights, body weight resistance	Start with 1 set of 10 reps with 1 or 2 lb (0.5-1 kg) weights; increase gradually	Increase muscular strength Decrease muscle wasting Improve physical function
Range of motion	Daily	To the point of mild discomfort	10-30 s	Predialysis patients can use the same modalities as nondiseased individuals	Start at 10 s and increase to 30 s	Same benefits in predialysis patients as normal Decrease stiffness after dialysis

Special Considerations for Hemodialysis Patients

- Patients have extremely low fitness levels.
- Timing of exercise sessions should be coordinated with dialysis sessions; nondialysis days are often preferred by patients.
- Patients will experience frequent hospitalizations and setbacks.
- Gradual progression is critical.
- Guiding exercise intensity based on heart rate alone is typically invalid—use of rating of perceived exertion (RPE) is recommended.
- Maximal exercise testing may not be feasible in all patients but should be used whenever appropriate.
- Performance-based testing should be used in patients who cannot tolerate graded exercise testing.
- One-repetition maximum testing for strength is not recommended because of secondary hyperparathyroidism-related bone and joint problems.
- Prevalence of orthopedic problems is common.
- Motivating patients is often a challenge. It is important to work with patients to identify activities in which they are willing to engage, instead of mandating a standard or "on-size fits all" exercise prescription.
- Every attempt should be made to educate the dialysis staff about the benefits of exercise so they can help motivate patients to participate.

Special Considerations for Transplant Patients

- Patients are initially weak, so gradual progression is recommended.
- Exercise heart rate responses are normalized after transplant; however, the increase in blood pressure during exercise is often excessive or abnormal.
- Patients may experience a lot of orthopedic and musculoskeletal discomfort with strenuous exercise.
- Weight management often becomes an issue following transplant.
- Patients and their families are often fearful of overexertion; thus, gradual progression should be stressed.
- Prednisone may delay adaptations to resistance training.
- During periods of rejection, both the intensity and duration of exercise should be decreased, not eliminated completely.
- Patients may experience frequent hospitalizations during the first year after the transplant. Because patients are immunosuppressed, every effort must be made to avoid infectious situations (e.g., strict sterilization procedures must be followed for exercise testing and training equipment).

For patients treated with continuous ambulatory peritoneal dialysis, exercise may be best tolerated at a time when the abdomen is drained of fluid (dry). This allows for greater diaphragmatic excursion and less pressure against the catheter during exertion, reducing the risk of hernias or leaks around the catheter site (9, 28, 65). Patients may choose to exercise in the middle of a dialysis exchange—after draining of fluid and before introduction of the new dialysis fluid. This option requires capping off the catheter prior to exercise, a technique that must be discussed with the dialysis staff. Patients can also opt to perform less vigorous activities when they are fluid loaded (i.e., wet) (28). It is important to note that few randomized clinical trials have been performed on these CKD patients, and much is still not definitively known (28). However, increasing their levels of physical activity is highly recommended (28).

Mode

There are no restrictions on the type of activity that can be prescribed for patients across the spectrum of CKD. Range of motion and resistance training exercises are important for all patients because they tend to be both deconditioned and at increased risk for muscle wasting. Also, many ESRD patients have poor musculoskeletal function and experience joint discomfort; therefore, non-weight-bearing cardiorespiratory-type activity may be more easily tolerated.

The vascular access site used for hemodialysis may be in the arm or upper leg. The location should not inhibit physical activity at all, although many patients are initially told not to use the arm with the arteriovenous fistula for 6 to 8 wk after it is implanted to ensure sufficient time for healing after surgery. The only precaution for the fistula is to avoid any activity that would close off the flow of blood (e.g., having weights lying directly over the top of the vessels). Although the patient should be protective of the access site, use of the extremity will increase flow through it and actually help develop muscles around the access site, which makes the placement of needles easier.

Patients with a peritoneal catheter should avoid full sit-ups and activities that involve full flexion at the hip. They can accomplish abdominal strengthening by performing isometric contractions and partial sit-ups, or crunches. Swimming may be a challenge for those with peritoneal catheters because of the possibility of infection. Patients must be advised to cover the catheter with protective tape and to clean around the catheter exit site after swimming. In years past, the recommendation was to avoid swimming in fresh water; however, a report from France suggests swimming in both lakes and swimming pools is allowable as long as patients clean their access site and change the dressing afterward (28).

Although renal transplant recipients are often told not to participate in vigorous activities, the actual primary concern is avoiding any contact sport (e.g., football) that may result in a direct hit to the area of the transplanted kidney. Vigorous noncontact sports and activities are generally well tolerated by transplant recipients who have exercise trained to attain the needed muscle strength and cardiorespiratory endurance. This was demonstrated by Aries Merritt of the United States, who won a gold medal in the 110 m hurdles at the 2012 Summer Olympics in London. Merritt then had a kidney transplant in September 2015 and yet was still able to perform at an elite level, narrowly missing a berth on the team for Rio de Janeiro, finishing fourth at the U.S. Olympic trials.

Frequency

Cardiorespiratory and resistance exercise should be prescribed 3 to 5 d/wk and 2 or 3 d/wk, respectively. Flexibility and balance exercises should be performed whenever CKD patients exercise train; however, because of the stiffness that patients with ESRD experience after prolonged periods of sitting in dialysis chairs, they should be encouraged to stretch daily.

Intensity

Patients with CKD who are not on dialysis can be prescribed aerobic exercise training at a moderate to vigorous intensity (i.e., 50%-80% $\dot{V}O_2$ reserve). However, since many of these patients could have diabetes or take medications that affect heart rate, these issues need to be considered if an exercise professional wants to use a heart rate–based method to guide exercise intensity. In patients with ESRD, exercise intensity should be guided by rating of perceived exertion (RPE), because peak heart rates are lower than age predicted, and exercise heart rates are highly variable because of the fluid shifts and vascular adaptations to fluid loss during the dialysis treatment (48).

Many ESRD patients may initially tolerate only a few minutes of very low-level exercise, which means that any formal warm-up and cool-down intensities are less relevant. These individuals should be encouraged to increase duration gradually at whatever level of effort they can tolerate. After they achieve 20 min of continuous exercise, the warm-up, conditioning, and cool-down phases can be incorporated, using an RPE of 9 to 10 (on a 6- to 20-point scale) for both warm-up and cool-down and an RPE of 12 to 15 for the conditioning phase. For resistance exercise, CKD patients can be prescribed workloads up to 75% of their estimated 1RM (≈10RM load).

Duration and Progression

CKD patients not on dialysis can often start with 15 to 20 min of lower-intensity continuous exercise, but this may need to be modified for some individuals who are deconditioned, for which an interval-type method may be necessary. Extremely deconditioned patients may need to start with flexibility exercises and a resistance program of lower workloads (based on 10RM) and higher repetitions (10-15) before beginning any cardiorespiratory activity. To progress, they should gradually work up to 20 to 30 min of continuous activity at an RPE of 12 to 15.

When a patient begins cycling during dialysis, the initial session is usually limited to 10 min, even if the patient is able to tolerate a longer duration. This precaution ensures the dialysis staff and the patient that cycling does not have any adverse effects on the dialysis treatment. The patient can then increase duration in subsequent sessions according to tolerance.

EXERCISE TRAINING

Among hemodialysis patients (and possibly transplant recipients with diabetes), particularly those who initially have a very low functional capacity and multiple comorbidities, the improvement in fitness due to exercise training may be relatively small. However, since a deterioration in physical function over time is common in patients with CKD, the maintenance or prevention of loss in exercise capacity and physical function over time may be viewed as a favorable outcome.

Most patients experience improvement in muscle strength and often an increase in lean muscle mass. This increase in lean mass may have implications for hemodialysis patients, since the amount of fluid removed during dialysis is based on total body weight. Thus, if body weight (lean mass) increases because of exercise training, the target weight for dialysis may need to be adjusted. Many dialysis patients also experience improvements in blood pressure and glucose control, often requiring a reduction in the medications used to treat these disorders.

For transplant recipients on prednisone, improvements in muscle strength may be slower, and absolute gains may take longer than in healthy individuals; however, these patients can achieve normal levels of muscle strength, partially counteracting the negative effects of prednisone on the muscles (i.e., sarcopenia).

Most patients report significant improvements in their energy level and ability to perform activities of daily living, and they may experience fewer symptoms or problems (e.g., muscle stiffness, cramping, hypotension) when undergoing dialysis. If exercise is performed during the hemodialysis treatment, the clearance of toxins may be improved (41). Finally, overall quality of life is improved, particularly in the physical domain.

Clinical Exercise Bottom Line

- CKD is characterized by kidney damage or decreased kidney function, graded based on urinary albumin excretion rate or the extent of reduction in estimated glomerular filtration rate for 3 mo or more.

- The causes of CKD are many, including diabetes, hypertension, autoimmune disorders (e.g., lupus), and inherited and congenital abnormalities.

- Treating CKD involves managing the treatable underlying causes (e.g., diabetes, hypertension); lifestyle changes, such as stopping cigarette smoking, weight loss as needed, and adjustments in nutrition to manage protein, sodium, and fluid intake; and if needed, renal replacement therapy (dialysis or transplant).

- Poor exercise capacity, 20% to 50% below age-matched healthy people, is a hallmark characteristic of patients with CKD.

- Increasing daily activity, performing regular cardiorespiratory or aerobic-type exercise, and doing resistance training are important treatments to improve exercise tolerance, manage contributing comorbidities, improve quality of life, and possibly improve vascular function.

CONCLUSION

The exercise prescription for patients with CKD depends on where they are in the disease process (i.e., before dialysis vs. ESRD). The prescription must be individualized to the patient's limitations; this includes the type of exercise (cardiorespiratory, flexibility, resistance), frequency of exercise, timing of exercise relative to any dialysis treatment, and duration, intensity, and progression of exercise. The progression should be gradual in those who are extremely debilitated. The starting levels and progressions must be as tolerated, because fluctuations in well-being, clinical status, and overall functional ability frequently change with changes in medical status. Hospitalization or a medical event (e.g., clotting of the fistula or placement of a new fistula) may require that patients with ESRD return to lower levels of exercise, which requires frequent reevaluation of the exercise prescription. The goal is for patients to become more active in general, to decrease time spent being sedentary, and, if possible, to work toward a regular exercise program of 3 to 5 d/wk of cardiorespiratory exercise and 2 or 3 d/wk for resistance training. Exercise training can lead to improvements in exercise capacity in compliant CKD patients, but despite their improvement, CKD patients are likely to have lower fitness values than age- and gender-matched healthy individuals.

Online Materials

Visit *HKPropel* to access a link to the references, the case studies with discussion questions, and a quiz to help you review key concepts and test your understanding of the material covered in this chapter.

PART III

Diseases of the Cardiovascular System

According to the World Health Organization, cardiovascular disease (CVD) is the number one cause of death globally, accounting for 17.9 million deaths annually. Of the global deaths associated with CVD, four out of five are due to heart disease and stroke, and 33% occurred in individuals under 70 yr of age. The American Heart Association reported 2,354 deaths each day associated with CVD in 2017; CVD claims more lives each year in the United States than all forms of cancer and chronic lower respiratory diseases. Of the different forms of CVD, coronary heart disease (CHD, aka coronary artery disease) is the leading cause of death (50%), followed by stroke (17%), high blood pressure (11%), and heart failure (10%). With respect to the cost burden associated with CVD, it is staggering, with direct and indirect costs in the United States estimated at $363.4 billion between 2016 and 2017. Although the number of deaths attributable to CVD has declined (by 9.8%), the risk factors associated with CVD continue to remain extremely high.

The reduction in the mortality due to CVD observed over the last 20 yr is largely because of advances in the technology of assessing and correcting CVD diseases. However, given the societal trends toward overweight and obesity, smoking, physical inactivity, poor eating habits, and elevated cholesterol, we can expect the relatively high rates of CVD to continue. The clinical exercise physiologist should become well familiar with the various common forms of CVD that are addressed in part III, because all these forms of CVD can be positively altered by an appropriate exercise-based rehabilitation program that addresses specific lifestyle interventions.

Chapter 14 centers on acute coronary syndromes, with a primary focus on unstable angina pectoris and acute myocardial infarction (MI). The incidence of acute myocardial infarction continues to be very high in our society; each year, approximately 605,000 individuals suffer their first MI and an additional 200,000 a recurrent MI. The value of secondary prevention has been shown to reduce the risk of mortality. After an MI, patients greatly benefit from positive lifestyle alterations by participating in a cardiac rehabilitation program. The clinical exercise physiologist plays an important role in assessing the patient's cardiovascular limitations through graded exercise testing and developing an appropriate exercise prescription. This chapter provides a clear understanding of the disease process and preventive measures that can be used to prevent further events.

Chapter 15 discusses revascularization of the heart. Revascularization procedures are a common way to address obstructive CHD. Emergency percutaneous coronary intervention is a standard method used to decrease the risk of myocardial damage. Clinical exercise physiologists must be familiar with and prepared to address challenges related to the clinical procedure. Furthermore, they must have a good understanding of potential issues that may arise when performing a graded exercise test and prescribing exercise.

Chapter 16 deals with chronic heart failure. Given improved emergency care for cardiac events and the increasing number of people entering the age group greater than 65 yr, the prevalence of chronic heart failure in our society is increasing. Current estimates suggest that these trends and the public health burden associated with heart failure (HF) will increase even more over the next 10 yr. As a result, the clinical exercise physiologist needs to have a strong knowledge base to work with patients with HF in both the rehabilitation and exercise testing settings.

Chapter 17 explores peripheral artery disease. Peripheral artery disease (PAD) is common in our society, with a global prevalence of 200 to 236 million cases; most alarmingly, there has been a 23% increase from 2000 to 2010 worldwide. Significant PAD can severely limit a person's exercise tolerance, depending on when the subject develops symptoms such as intermittent claudication. Exercise testing can be beneficial to determine a patient's functional capacity prior to the onset of symptoms, and exercise therapy is very beneficial for those with PAD. The clinical exercise physiologist must learn specific strategies to assess disease severity and develop an appropriate exercise program for reducing a patient's symptoms of PAD.

Chapter 18 considers pacemakers and implantable cardioverter defibrillators (ICDs). Numerous advancements have been made in implantable cardiac devices, which are commonly used to regulate heart rate and rhythm, synchronize chambers in patients with heart failure, and defibrillate the heart in patients with life-threatening dysrhythmias. As our population becomes older, the need for pacemakers and ICDs becomes more common. Because of the complexity of these devices, the clinical exercise physiologist needs to be knowledgeable in how they work, how they influence the conduct of a graded exercise test, and what issues may arise under normal exercise training (e.g., what exercise heart rate should be avoided to prevent the risk of premature firing of an ICD).

CHAPTER 14

Acute Coronary Syndromes

Unstable Angina Pectoris and Acute Myocardial Infarction

Ray W. Squires, PhD

The burden of diseases of the cardiovascular system on our society is horrendous. Since 1900, cardiovascular diseases (CVDs), such as coronary heart disease, stroke, heart failure, and hypertension, have been the leading cause of death in the United States every year with the exception of 1918, the year of the great influenza epidemic (1). In 2017, CVDs resulted in 2,354 deaths each day in the United States (859,125 total deaths), more than all forms of cancer and chronic lung disease combined (2). It is the most common cause of death for both men and women (2, 3). Approximately 50% of all cardiovascular deaths are a result of coronary heart disease (CHD) (2). For 2017, CHD caused 365,914 deaths (1,003 deaths each day). Acute coronary syndromes (myocardial infarction, unstable angina pectoris, and some forms of sudden cardiac death), forms of CHD, are most commonly the result of coronary artery atherosclerosis and subsequent thrombosis. Between 2005 and 2014, the annual incidence of myocardial infarction (heart attack) was 605,000 new attacks and 200,000 recurrent attacks. Silent (painless) myocardial infarction will occur in 160,000 persons per year. Every 42 s someone in the United States suffers a myocardial infarction. The average age at the first myocardial infarction is 65 yr for men and 72 yr for women. The number of Americans who underwent percutaneous coronary intervention (stenting primarily) or coronary bypass graft surgery for treatment of CHD in 2014 was 515,000 and 371,000, respectively (2). Approximately 16,300,000 Americans have a history of myocardial infarction, and 9,000,000 experience angina pectoris (chest pain). Estimated total costs for CHD were estimated at $363.4 billion between 2016 and 2017.

This chapter provides information regarding the following topics:

1. The disease process of atherosclerosis and thrombosis
2. Myocardial blood flow and ischemia: angina pectoris
3. Acute coronary syndromes: definition, acute myocardial infarction, clinical assessment, diagnosis, classification by ECG, management strategies, potential complications, right ventricular infarction
4. Factors associated with poor prognosis
5. Stress testing after acute myocardial infarction
6. Exercise training and cardiac rehabilitation for acute coronary syndromes

PATHOPHYSIOLOGY

The pathology of acute coronary syndromes is complex; it most commonly involves the development of atherosclerotic lesions in the walls of the coronary arteries, with subsequent thrombosis formation resulting in an abrupt decrease in vessel blood flow.

Atherosclerosis and Thrombosis

Atherosclerosis is a disease process that may result in blood-flow limiting lesions in the epicardial coronary, carotid, iliac, and femoral arteries, as well as the aorta. Some arteries are resistant to atherosclerosis (brachial, internal thoracic, intramyocardial) for unknown reasons (4a). The processes of atherosclerosis and thrombosis are interrelated, and the term *atherothrombosis* has been adopted by some investigators to emphasize this point (4a). In addition, inflammation plays a central role in the development of atherosclerosis (4b).

The Normal Artery

The channel for the flow of blood within the artery is the *lumen*. The inner, single-cell layer of the artery is the *endothelium*. The endothelium plays a critical role in maintaining vasomotion (the degree of vasoconstriction) and regulating hemostasis (balancing pro- and antithrombotic properties). When intact, the endothelium produces nitric oxide, a vasodilator, and substances such as plasminogen that inhibit thrombosis formation. Various receptors, such as those for low-density lipoprotein and growth factors, are located on the endothelial cells (5). Under normal circumstances, the endothelium protects against the development of atherothrombosis, but when damaged it plays a central role in the development of the disease (5).

Underneath the endothelial basement membrane is the *intima*, consisting of a thin layer of connective tissue with an occasional smooth muscle cell. The lesions of atherosclerosis form in the intima (5).

The *media* contains most of the smooth muscle cells of the arterial wall, in addition to elastic connective tissue, and is located underneath the intima between the internal and external elastic laminae. The smooth muscle cells maintain arterial tone (partial vasoconstriction). Smooth muscle cells have receptors for low-density lipoprotein, insulin, and growth factors. When appropriately stimulated, smooth muscle cells are capable of functioning as synthetic tissue, producing connective tissue (6).

The outermost layer of the arterial wall is the *adventitia*, consisting of connective tissue (collagen, elastin), fibroblasts (cells capable of synthesizing connective tissue), and a few smooth muscle cells. This tissue is highly vascularized (its blood supply is provided by small vessels called the vasa vasorum) and provides the media and intima with oxygen and nutrients (5).

Atherogenesis

Our understanding of the development and progression of atherosclerosis (atherogenesis) is incomplete. However, it is clear that endothelial injury resulting in endothelial dysfunction and a subsequent inflammatory response play critical roles (7). The disease process may begin in childhood and progress for decades before a clinical event occurs. The rate of progression of atherosclerosis may not be consistent over time and impossible to predict.

Under normal conditions, the endothelium may experience periodic minimal amounts of injury. In these situations, the inherent repair processes of the endothelium are adequate to restore normal function. However, chronic, excessive injury to endothelial cells initiating the process of atherogenesis may result from multiple causes, such as the following (6, 7, 8, 9, 10, 11, 12, 13):

- Tobacco smoke and other chemical irritants from tobacco
- Low-density lipoprotein cholesterol (LDL-C)
- Hypertension
- Glycated substances resulting from hyperglycemia and diabetes mellitus
- Plasma homocysteine
- Infectious agents (e.g., *Chlamydia pneumoniae*, herpes viruses)

Endothelial dysfunction may result from these potentially injurious factors, leading to the following abnormalities characteristic of an inflammatory response:

- Increased adhesiveness resulting in platelet deposition and monocyte adhesion
- Increased permeability to lipoproteins and other substances in the blood
- Impaired vasodilation and increased vasospasm

Platelets adhere to the damaged endothelium (platelet aggregation), form small blood clots on the vessel wall (mural thrombi), and release growth factors and vasoconstrictor substances, such as thromboxane A2 (5, 6). These changes indicate a switch in endothelial function favoring a prothrombotic, vasoconstrictive state.

Monocytes, a type of white blood cell, also adhere to the injured endothelium and migrate into the intima. LDL-C enters the arterial wall and undergoes the process of oxidation. Monocytes accumulate LDL-C, augmenting the oxidation process, and become transformed into a distinctly different type of cell, the macrophage (5, 7).

Growth factors expressed by platelets, monocytes, and damaged endothelium result in growth and proliferation (increase in cell numbers and cell size) of certain types of cells (mitogenic effect) as well as the migration of cells into the area of injury (chemotactic effect) (5, 7). In response to the growth factors, smooth muscle cells and fibroblasts (undifferentiated connective tissue cells that can synthesize fibrous tissue) migrate from the media to the intima. Smooth muscle progenitor cells from bone marrow also migrate to the intima (14). Some of these cells, in addition to monocytes, accumulate cholesterol, forming foam cells that may release their cholesterol into the extracellular space, giving rise to fatty streaks, the earliest visually detectable (yellow macroscopic appearance) lesion of atherosclerosis (11, 15). Immune system cells (T lymphocytes) are also present in fatty streaks and are part of the inflammatory state in the arterial wall (16).

With continued migration, proliferation, and growth of tissue, the lesion progresses in complexity and size and becomes a fibromuscular plaque (15). The composition of the plaque now includes a fibrous cap, connective tissue extracellular matrix, lipids, inflammatory cells such as macrophages and T lymphocytes, smooth muscle cells, thrombus, and calcium resulting from smooth muscle cell death. The typical plaque is firm in texture and pale gray in color, and it may contain a yellow cholesterol core.

As the intimal lesions of atherosclerosis progress and thicken the vessel wall, a compensatory outward expansion of the vessel occurs (to a point) and lumen size remains unchanged. This is called arterial remodeling and may be effective in compensating for plaques whose bulk may represent up to 40% of the vessel diameter (17). With continued progression in plaque bulk, the area of the lumen decreases, which may ultimately result in a reduction in blood flow.

The progression of the size and volume of atherosclerotic lesions is highly variable. Some lesions appear relatively stable over many years, other plaques may slowly progress in size, while still other areas of atherosclerosis may enlarge very rapidly (6). The slowly progressing plaques are thought to gradually internalize monocytes and lipids, while rapidly progressing lesions incorporate thrombus into the plaque (4a). Local stressors (e.g., from turbulent blood flow or vasoconstriction) or chemical factors (enzymes such as metalloproteinases that weaken the fibrous cap) within the lesion may result in plaque rupture, erosion, or fissuring of the fibrous cap, exposing the internal contents of the plaque to the blood (6, 18). Various amounts of thrombus form in response to this prothrombotic environment and may be incorporated into the plaque. The scenario of plaque rupture, subsequent thrombus formation, and incorporation into the arterial wall may repeatedly occur, giving a layered appearance to the lesion and resulting in rapid progression in the size of the plaque. These lesions, which include organized thrombus, are called advanced atherosclerotic plaques.

Atherosclerosis affects arteries in an extremely diffuse manner, with occasional discrete, localized areas of more pronounced narrowing of the vessel lumen (19). Selective coronary angiography is the gold standard (best available test) for determination of the severity of coronary lesions. However, based on comparisons of angiographic and autopsy findings, with the exception of complete occlusion of the vessel in question (100% stenosis), the degree of stenosis is greatly underestimated by angiography because of the diffuse nature of the disease process (20). Obstructive coronary lesions (severe enough to reduce blood flow) occur most frequently in the first 4 to 5 cm of the epicardial coronary arteries, although more distal disease may also be seen. Obstructive lesions at the origin (ostial lesions) of the left main and right main coronary arteries may also occur. For reasons not fully understood, women generally lag 5 to 20 yr behind men in the extent and severity of coronary atherosclerosis (21).

Risk Factors for Atherosclerosis

Risk factors are associated with an increased likelihood that atherosclerosis will develop over time. Such factors have been identified on the basis of observational studies evaluating common characteristics of persons with the disease (22, 23a). Ninety percent of myocardial infarctions occur in persons with at least one risk factor (23b). Possible mechanisms of atherogenic effect have been identified for some risk factors. The effects of reducing the severity of some risk factors, especially LDL-C, have been demonstrated to reduce progression of the disease. Predicting whether an individual patient will or will not develop atherosclerosis based on the presence and severity of risk factors is very imprecise, however. Less than half of future cardiovascular events can be predicted using conventional risk factors (24a-d):

- Tobacco use
- Dyslipidemia, especially elevated LDL-C and low levels of high-density lipoprotein cholesterol (HDL-C), but also elevated triglycerides, increased non-HDL-C, and elevated lipoprotein(a)

- Hypertension
- Sedentary lifestyle
- Unhealthy diet: high consumption of foods with a high glycemic index, low consumption of fruits and vegetables, high consumption of red meat, low consumption of fiber, high consumption of trans fats
- Obesity
- Diabetes mellitus
- Metabolic syndrome (a combination of conventional risk factors associated with obesity and insulin resistance)
- Family history of premature coronary disease (male first-degree relatives <55 yr of age, female first-degree relatives <65 yr of age)
- Male sex
- Obstructive sleep apnea
- Chronic kidney disease
- Psychosocial factors, such as depression, anxiety, social isolation, coronary prone personality (type A), lower socioeconomic status, and chronic life stressors
- Mediastinal or chest wall radiation during cancer treatment
- Risk factors unique to women: menopause, polycystic ovary syndrome, gestational hypertension, and gestational diabetes mellitus

Less well-established risk factors, termed emerging risk factors, are being investigated, including the following (24a, 25, 26):

- Elevated plasma homocysteine, an intermediary in the metabolism of the essential amino acid methionine
- Fibrinogen, a protein factor in the blood coagulation cascade
- LDL particle concentration
- High-sensitive C-reactive protein, a marker for systemic inflammation

Myocardial Blood Flow, Metabolism, and Ischemia

Normal contraction and relaxation of cardiac myocytes requires the presence of adequate amounts of adenosine triphosphate (ATP; a high-energy phosphate molecule) in the myocardium. The heart is a highly aerobic organ with an extensive circulatory system and abundant mitochondria (27). Figure 14.1 illustrates the epicardial coronary arteries. The coronary arterial system includes epicardial arteries that bifurcate into intramyocardial and endomyocardial branches. At rest, coronary blood flow averages 60 to 90 mL · min^{-1} · 100 g^{-1} of myocardium and may increase five- to sixfold during exercise (28). Under usual conditions, the heart regenerates ATP aerobically, and myocardial cells are not well adapted to anaerobic energy production. At rest, myocardial oxygen uptake is approximately 8 to 10 mL per 100 g of tissue per minute. During intense exercise, the oxygen requirement may increase by 200% to 300% (29). The myocardium extracts nearly all its oxygen from the capillary blood flow (unlike skeletal muscle), and coronary blood flow must be closely regulated to the needs of the myocardium for oxygen (30). With an increase in myocardial work, oxygen demand increases, and coronary blood flow must also increase to provide the necessary amount of oxygen.

Blood flow through any regional circulation, including the coronary system, is determined by the blood pressure and the vascular resistance (30). During systole, intramyocardial pressure is increased (as is vascular resistance) and the intramural vessels are compressed. Therefore, most coronary blood flow occurs during diastole, when intramyocardial pressure is lower (lower vascular resistance).

Before a decrease in flow can be measured distal to a narrowed, atherosclerotic coronary artery segment, a substantial reduction in vessel luminal diameter must occur. When plaque bulk reduces the luminal cross-sectional area by 75% or more, flow is reduced under resting conditions (a hemodynamically significant lesion) (28). Beyond this amount of critical stenosis, further small decreases in cross-sectional area of the vessel result in large reductions in flow.

A reduction in the lumen diameter may be caused by several factors (31, 32):

- Significant atherosclerotic plaque

- Vasospasm without underlying plaque

- Vasospasm superimposed over a plaque

- Thrombus associated with plaque rupture, erosion, or fissure

Coronary vasospasm may result from several factors, such as endothelial dysfunction, sympathetic nervous system activation (e.g., vasospasm resulting from exposure to very cold ambient temperatures), and bloodborne substances such as epinephrine (33).

Myocardial ischemia results when myocardial blood flow is inadequate to provide the required amounts of oxygen for ATP regeneration (oxygen supply < oxygen demand) (27, 28). Ischemia may result in progressive abnormalities in cardiac function, termed the ischemic cascade (34). The first abnormality is stiffening of the left ventricle, which impairs diastolic filling of the heart (diastolic dysfunction). Second, systolic emptying of the left ventricle becomes impaired (systolic dysfunction). Localized areas of the myocardium develop abnormal contraction patterns such as hypokinesis (reduced systolic contraction). **Left ventricular ejection fraction (LVEF)** may decrease. Third, electrocardiographic abnormalities associated with altered repolarization (ST-segment depression or elevation, T wave inversion) or arrhythmias may occur. Finally, symptoms of angina pectoris may develop.

Angina pectoris is transient referred cardiac pain resulting from myocardial ischemia (35). A minority of patients with substantial amounts of ischemia do not report pain (silent ischemia). The pain of angina may be located in the substernal region, jaw, neck,

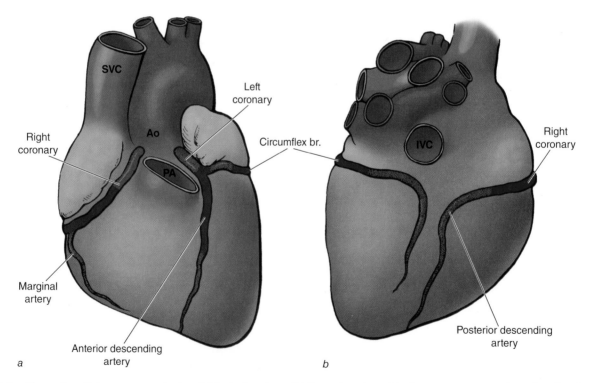

Figure 14.1   The epicardial coronary arteries. (a) Anterior view. (b) Posterior view. Black segments are prime sites for the development of obstructive atherosclerotic plaques. Ao = aorta; IVC = inferior vena cava; SVC = superior vena cava; PA = pulmonary artery.

Reprinted by permission from J.T. Lie, "Pathology of Coronary Artery Disease," in *Cardiology: Fundamentals and Practice,* 2nd ed., edited by E.R. Giuliani et al. (Chicago: Mosby Yearbook, 1991).

or arms, although pain may also occur in the epigastrium and interscapular regions. It is usually described as a feeling of pressure, heaviness, fullness, squeezing, burning, aching, or choking. The pain may vary in intensity and may radiate. The patient may experience dyspnea (anginal equivalent) if the ischemia results in increased left ventricular end-diastolic pressure and increased pulmonary vascular pressure. Typical angina is provoked by exertion, emotions, cold and heat exposure, meals, and sexual intercourse; it is relieved by rest, nitroglycerin, or both. Atypical angina involves similar symptoms but has features that set it apart from typical angina, such as no relationship with exertion. Stable angina is reproducible and predictable in onset, severity, and means of relief. Unstable angina is defined as new onset of typical angina; increasing frequency, intensity, or duration of previously stable angina; or angina that occurs at rest or in the first few days after acute myocardial infarction.

If an episode of ischemia is brief, the contractile abnormalities described previously are quickly reversible. Brief postischemic left ventricular dysfunction is called stunned myocardium (36). Chronic, substantial, nonlethal ischemia may result in prolonged but reversible left ventricular dysfunction called hibernating myocardium. Myocytes remain viable but exhibit depressed contractile function. Elimination of chronic ischemia with revascularization results in a gradual return of normal contractile function, although resolution may require up to a year (37a). Prolonged, severe ischemia results in myocyte necrosis (irreversible damage, myocardial infarction).

Definition of Acute Coronary Syndromes

Unstable angina pectoris, acute myocardial infarction, and instances of sudden cardiac death due to myocardial ischemia comprise the acute coronary syndromes (ACS) (4a). The most common underlying mechanism resulting in these syndromes is atherosclerotic plaque erosion, rupture, or other type of plaque disruption resulting in thrombus formation and possibly vasoconstriction, with subsequent vessel occlusion and acute myocardial ischemia. However, some patients develop an ACS without clinically significant atherosclerosis as a result of vasospasm, microvascular disease, myocarditis, cocaine use, or stress cardiomyopathy (37b). Both unstable angina pectoris and acute myocardial infarction present with symptoms of myocardial ischemia; electrocardiographic signs of ischemia (ST-T wave abnormalities, usually transient in unstable angina) are often present. Myocardial infarction results in biomarker evidence of myocardial necrosis (discussed later), while unstable angina does not result in biomarker evidence of necrosis (37b).

The type of acute coronary syndrome that occurs is related to the duration of vessel occlusion. Unstable angina is probably the result of transient vessel occlusion (<10 min) followed by spontaneous **thrombolysis** (clot dissolution) and vasorelaxation. Vessel occlusion persisting for more than 60 min results in acute myocardial infarction. Ischemia resulting from atherothrombotic vessel occlusion may trigger ventricular tachycardia or ventricular fibrillation and sudden cardiac death (4a).

Why do some plaques rupture and thrombose? Approximately two-thirds of patients with ACS have high-risk or vulnerable atherosclerotic lesions with thin fibrous caps overlying a lipid-rich core with an abundance of macrophages (4a). Inflammation mediated by various cytokines, including proteases and tumor necrosis factor, erodes the plaque from within. The physical forces assisting with plaque disruption include increased blood pressure or heart rate, local vasoconstriction, and nicotine or immune complexes. After plaque rupture, circulating blood platelets come in direct contact with the thrombogenic internal environment of the plaque, resulting in clot formation (38).

Angiographic studies have demonstrated that most of these rupture-prone lesions are less than 50% occlusive before they become disrupted (39). This explains why many patients who experience an ACS do not have warning symptoms. However, angiographically severe coronary atherosclerosis does increase the likelihood of a coronary event by serving as a marker for the presence of extensive disease, including rupture-prone lesions. Autopsy studies have also demonstrated that many patients have disrupted plaques but no history of an ACS. Thus, not all plaque disruption results in clinical events. In approximately one-third of cases, an ACS results from only superficial erosion of a severely stenotic and fibrotic plaque, with clot formation due to a hyperthrombotic state caused by factors such as smoking, hyperglycemia, or elevated LDL-C (4a).

Acute Myocardial Infarction

Acute myocardial infarction is the necrosis (death) of cardiac myocytes resulting from prolonged ischemia caused by complete vessel occlusion (40). The key event in distinguishing reversible from irreversible (infarction) ischemia is disruption of the myocyte membrane (a lethal event). The myocyte cannot recover if membrane disruption occurs and cytoplasmic contents spill into the circulation.

In some patients, a precipitating event, or trigger, for the myocardial infarction may be determined, such as physical exertion, emotional stress, or anger (41, 42). It may also occur in the setting of surgery associated with substantial loss of blood. There is evidence for circadian variation, with slightly more myocardial infarctions occurring in the early morning hours than at other times, suggesting a role for sympathetic nervous system activation as a trigger (40, 42).

CLINICAL CONSIDERATIONS

When a patient presents to a medical facility and there is suspicion of ACS, the following assessment is warranted.

- History of symptoms: Symptoms of myocardial infarction include chest pain or other anginal sensations, gastrointestinal upset, dyspnea, sweating, anxiety, or syncope. It may be painless (silent myocardial infarction) in approximately 25% of cases (43). Pain is often severe, but all intensities of discomfort may be experienced. Elderly patients report more dyspnea, while women are more likely to report atypical symptoms, such as shoulder, middle back, or epigastric pain; fatigue; and general weakness (44).

- Physical examination: Patients with ACS may demonstrate the following findings on physical examination (45):
  - Systolic hypotension
  - Diaphoresis
  - Sinus tachycardia
  - Tachypnea
  - New murmur of mitral regurgitation
  - Third, fourth heart sounds
  - **Pulmonary rales**
- Electrocardiogram: The ECG may show ST-segment elevation or nonspecific ST-T wave abnormalities (see ECG Classification of Myocardial Infarction section).
- Chest radiograph: This is useful for patients with evidence of hemodynamic instability or pulmonary edema.
- Laboratory results: The biomarker cardiac troponin (cTn) is measured twice, 6 to 12 h apart. An ECG is performed on admission and repeated serially, as needed. A blood lipid profile should be obtained within 24 h of symptom onset.

Diagnosis of Acute Myocardial Infarction

The diagnosis of myocardial infarction is based on the presence of elevated cardiac necrosis biomarkers plus at least one additional factor (46):

- Symptoms of ischemia
- ECG evidence of myocardial ischemia (ST-segment elevation or depression, or new left bundle branch block)
- New pathological Q waves on the ECG
- Imaging evidence (usually echocardiography) of infarction

cTn is the preferred biomarker for the detection of cardiomyocyte necrosis. It is highly sensitive and specific for cardiac necrosis. Elevation occurs 2 to 3 h after the onset of infarction and remains elevated for 1 to 2 wk (47). It should be noted that cTn is elevated after percutaneous intervention or cardiac surgery. The prior standard cardiac biomarker was the MB fraction of creatine kinase (CK-MB) (48a). While not as specific and sensitive as cTn, it is still an acceptable diagnostic test if cTn is not available.

After myocardial infarction, myocardial cells do not regenerate, and healing occurs via scar formation. Depending on the extent of infarction, scar formation may take days to weeks for completion.

ECG Classification of Myocardial Infarction

Infarctions are classified as ST-segment elevation (STEMI) or non-ST-segment elevation (NSTEMI) (46). Electrocardiographic criteria for STEMI and NSTEMI are as follows:

- STEMI: ST-segment elevation of at least 1mV in two contiguous leads or new left bundle branch block
- NSTEMI: ST-segment depression or T wave inversion persisting at least 24 h

STEMI infarctions are the result of an occluded epicardial coronary artery with more extensive myocardial damage and a worse prognosis. NSTEMI events have less myocardial damage because of spontaneous thrombolysis (clot dissolution). Figure 14.2 shows the evolution of the electrocardiogram after STEMI, with the formation of a Q wave indicating infarction of all or most of the thickness of the ventricular wall. The ST-segment elevation results from ischemic injury, and inverted T waves are due to ischemia around the outside borders of the infarct. Anatomic localization of myocardial infarctions is possible if Q waves are formed, as shown in the sidebar Criteria for Anatomic Localization of Myocardial Infarction (MI) by Q Wave Appearance.

Criteria for Anatomic Localization of Myocardial Infarction (MI) by Q Wave Appearance

1. Inferior wall MI (usually right coronary artery occlusion): Q wave (>40 ms duration, amplitude >25% of the R wave) in leads II, III, and aVF
2. Anterior wall MI (left anterior descending coronary artery occlusion): Q wave in leads V1 through V3 (anteroseptal), QS pattern in leads V1 through V3 (anteroseptal), Q wave in leads V2 through V4 (anterior), QS pattern in leads V2 through V4 (anterior)
3. Lateral wall MI (usually circumflex coronary artery occlusion): Q wave in leads V4 through V6 or QS pattern in leads V4 through V6
4. Posterior wall MI (usually right coronary artery occlusion): prominent R wave in leads V1 through V2 with positive T waves
5. High lateral wall MI (usually circumflex coronary artery occlusion): Q wave in leads I and aVL or QS pattern in leads I and aVL

Mechanistic Classification of Myocardial Infarction

Infarctions are also classified based on the presumed cause of the myocardial ischemia (46, 48b):

- Type 1: Due to pathology of the wall of the coronary artery, commonly plaque rupture with resultant atherothrombosis; rarely due to spontaneous coronary dissection
- Type 2: Due to imbalance in the oxygen supply and myocardial demand as a result of hyperthyroidism, fever, sepsis, acute heart failure, sustained arrhythmias, myocarditis, respiratory failure, hypertension, coronary artery spasm, coronary embolism, microvascular dysfunction, coronary artery embolus, severe anemia, hypotension

- Type 3: Sudden unexpected cardiac death before availability of cardiac biomarker analysis
- Type 4a: Associated with percutaneous coronary intervention (PCI)
- Type 4b: Associated with stent thrombosis (clot formation within a stent)
- Type 4c: Due to stent restenosis
- Type 5: Associated with coronary artery bypass graft surgery (CABG)

Diagnosis of Unstable Angina

Unstable angina has three presentations: prolonged rest angina, new-onset angina, and accelerated angina (46). Differentiation of unstable angina from acute myocardial infarction is based on biomarker elevation, as discussed previously.

Management of Acute Coronary Syndromes

The following are therapeutic approaches to treating patients (both men and women) with acute coronary syndromes (48a, 49):

Anti-Ischemic Therapy

- Bed or chair rest with continuous ECG monitoring
- Supplemental oxygen when arterial saturations are less than 90% or in the presence of acute heart failure
- Nitroglycerin; sublingual, oral, or intravenous administration
- β-blocker
- Angiotensin-converting enzyme inhibitor (ACE inhibitor) if left ventricular ejection fraction (LVEF) is ≤40% or with anterior wall myocardial infarction
- Angiotensin receptor blocker for patients with ejection fractions ≤40%, or anterior wall myocardial infarction, if intolerant of angiotensin-converting enzyme inhibitors
- Aldosterone antagonists for patients with STEMI and an LVEF of ≤40% who are already receiving ACE inhibitor
- Statin to improve blood lipid profile and reduce the risk of recurrent events
- Intra-aortic balloon pump counterpulsation to maintain diastolic pressure and coronary perfusion

Dual Antiplatelet Therapy

- Aspirin
- Clopidogrel, prasugrel, or ticagrelor

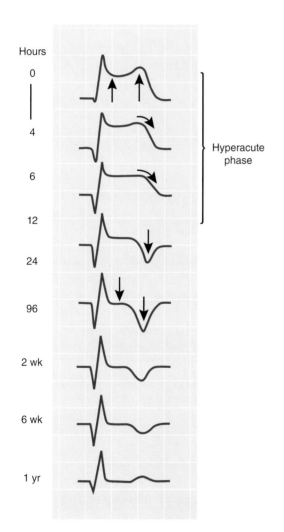

Figure 14.2   The evolution of the electrocardiogram in acute ST-segment elevation myocardial infarction resulting in Q wave formation.

Reprinted by permission from G.T. Gau, "Standard Electrocardiography, Vectorcardiography and Signal-Averaged Electrocardiography," in *Cardiology: Fundamentals and Practice,* 2nd ed. edited by E.R. Giuliani et al. (St. Louis: Mosby Year Book, 1991).

Anticoagulants

- Unfractionated heparin
- Low molecular weight heparin
- Bivalirudin (direct thrombin inhibitor)

Pain Relief (Despite Anti-Ischemic Therapy)

- Intravenous morphine should be avoided, if possible, in the setting of acute myocardial infarction

Reperfusion Therapy

Reperfusion of the **infarct-related artery** may be accomplished via thrombolytic therapy or via revascularization with either PCI or, much less commonly, CABG (for selected patients with either left main or severe three-vessel coronary artery disease) (45, 50). Thrombolytic therapy has not been demonstrated to be beneficial in NSTEMI.

Non-ST-Segment Elevation Myocardial Infarction/Unstable Angina Pectoris

Three management options are available for NSTEMI/unstable angina (45, 46, 48b):

1. Early invasive strategy (within 24 h of presentation to the hospital): Depending on the results of coronary angiography, the patient may undergo PCI (by far the most common occurrence) or CABG if the anatomy is not favorable for PCI. Most patients undergo the early invasive strategy if they are hospitalized at a center capable of PCI.

2. Delayed invasive strategy: Coronary angiography is performed 24 to 48 h after presentation.

3. Conservative treatment: Optimal medical management with imaging stress testing is performed either before discharge or early after discharge. Angiography is indicated if the patient has symptoms or if the stress test is positive for ischemia.

ST-Segment Elevation Myocardial Infarction

The optimal management of patients with STEMI is prompt reperfusion of the infarct-related artery (50). The earlier the reperfusion, the better the outcome for the patient in terms of limiting infarct size, preserving left ventricular function, and improving survival.

Two reperfusion strategies are available: thrombolysis and primary PCI (50). PCI is the best option for the patient if available. Thrombolysis is performed using tissue intravenous plasminogen activators, such as tenecteplase, reteplase, alteplase, or streptokinase, which break down coronary thromboses. It is most effective if given during the first 3 h after symptom onset. These are the benefits of thrombolysis:

1. It is readily available in rural and community hospitals.

2. It doesn't require a cardiologist or a cardiac catheterization laboratory.

3. It can usually be administered within minutes of the patient's reaching the hospital.

However, thrombolytic therapy has substantial limitations (50):

1. Restoration of full coronary blood flow occurs in only 60% to 70% of cases treated within 3 h of symptom onset.

2. Contraindications to thrombolytic agents, such as prior intracranial hemorrhage, prior ischemic stroke, or active bleeding, occur in 30% to 40% of patients.

Patients treated with thrombolysis should be transferred immediately to a PCI-capable facility (facilitated PCI) unless impossible because of weather or other circumstances. PCI is beneficial after both successful or failed thrombolysis. Primary PCI has impressive benefits (50):

1. It improves myocardial salvage (less necrotic myocardium).

2. It leads to less reinfarction and vessel reocclusion.

3. It identifies coronary anatomy more suitable for CABG, such as left main or severe three-vessel coronary disease.

The invasive strategy helps identify causes of ST-segment elevation other than atherosclerosis, such as pericarditis, myocarditis, stress cardiomyopathy (also known as apical ballooning syndrome, or Takotsubo cardiomyopathy), cocaine use, and coronary vasospasm resulting from endothelial dysfunction (48a).

The major limitation of primary PCI is the delay in opening the infarct-related artery once symptoms begin (50). For every 30 min delay, there is an 8% increase in mortality at 1 yr. Delays to reperfusion may be patient or hospital dependent. Patient-dependent delays involve failure to recognize and appreciate the importance of cardiac symptoms and delays in activating the emergency response system (911). Typical patient-related delays are 2 to 3 h. Hospital-dependent delays are related to system and process failures for the prompt delivery of reperfusion therapy (51).

The open-artery hypothesis indicates that even after 12 to 48 h have elapsed from symptom onset, PCI may still limit infarct size and improve survival (50). Another term used in **invasive cardiology** is *no-reflow*, which is angiographically observed slow flow in the infarct-related coronary artery after primary PCI as a result of swollen endothelial cells and subsequent red blood cell plugging of the microcirculation (52b).

Right Ventricular Myocardial Infarction

Most commonly associated with inferior wall myocardial infarction, affecting 30% of these patients, RV infarction is the result of occlusion of the proximal right coronary artery (53). Patients with RV infarction present with hypotension and signs of right heart failure. It is associated with high-degree AV block and increased in-hospital morbidity and mortality.

Factors Associated With Poor Prognosis

The following characteristics (one or more) are associated with a poor prognosis (54a, 54b):

1. Left ventricular ejection fraction ≤35% (heart failure with reduced ejection fraction) or chronic heart failure due primarily to diastolic dysfunction (heart failure with preserved ejection fraction)

2. Extremely poor exercise capacity, <5 METs, assessed with graded exercise testing

3. Evidence of extensive, severe myocardial ischemia during exercise or pharmacologic stress testing

4. Having survived primary (not in the setting of an acute myocardial infarction) sudden cardiac death without treatment with an implantable cardioverter-defibrillator

5. Severe nonrevascularized coronary artery disease (left main, severe proximal three-vessel disease)

6. Complications such as renal failure, stroke

Therapy consists of prompt reperfusion of the right coronary artery and maintenance of adequate RV function. For patients who survive the period of hospitalization, the RV dysfunction resolves and there is no additional long-term increase in mortality compared with patients with inferior wall myocardial infarction without RV involvement.

Medications for Outpatients After Myocardial Infarction

Table 14.1 provides a summary of common medications prescribed for patients after myocardial infarction. Several classes of drugs are associated with improved survival, such as aspirin, β-blockers, statins, and angiotensin-converting enzyme inhibitors or angiotensin receptor blockers and aldosterone antagonists (for patients with depressed left ventricular ejection fraction or anterior wall myocardial infarction) (49).

Stress Testing After Acute Myocardial Infarction

Exercise testing is helpful after myocardial infarction for the following reasons (55):

- To evaluate symptoms and potential myocardial ischemia
- To determine the need for coronary angiography in patients treated initially with a noninvasive strategy

Table 14.1 Common Medications Used in Outpatients After Myocardial Infarction

Medication class and name	Primary effects	Exercise effects	Side effects
Antiplatelet (aspirin, clopidogrel)	Blocks platelet aggregation, improves survival	None	Increases bleeding
β-blocker (metoprolol)	Reduces heart rate and blood pressure; improves survival, antiarrhythmic	Decreases heart rate and blood pressure	Fatigue, hypotension, bradycardia
ACE inhibitor (lisinopril)	Reduces blood pressure, improves survival	Reduces blood pressure	Cough, hypotension
ARB (losartan)	Reduces blood pressure, improves survival	Reduces blood pressure	Hypotension
Aldosterone antagonist (spironolactone)	Decreases blood pressure, diuresis	Decreases blood pressure	Hypotension
Statin (atorvastatin)	Decreases blood cholesterol, improves survival	None	Muscle pain, weakness
Nitrate (isosorbide mononitrate)	Coronary vasodilation	Raises ischemic threshold	Headache, hypotension

ACE inhibitor = angiotensin-converting enzyme inhibitor; ARB = angiotensin receptor blocker.

COMPLICATIONS OF ACUTE MYOCARDIAL INFARCTION

Arrhythmia

Several types of supraventricular arrhythmias are common after myocardial infarction.

- Sinus bradycardia due to excessive vagal tone or ischemia of the sinoatrial node
- Sinus tachycardia related to pain, fear, heart failure, or excessive sympathetic nervous system activation
- Premature atrial contractions—provide no prognostic information
- Atrial fibrillation—observed in up to 20% of patients, usually transient, more frequent in older patients, associated with increased mortality

Ventricular arrhythmias are also common after myocardial infarction.

- Ventricular fibrillation occurs in approximately 5% of hospitalized patients. β-blockers are effective in decreasing the incidence of this arrhythmia in the peri-infarct period.
- Ventricular tachycardia is observed in 10% to 40% of hospitalized patients; it is usually transient and benign in the early postinfarct period.
- Accelerated idioventricular rhythm is also observed in 10% to 40% of hospitalized patients. It is not associated with increased mortality.
- Premature ventricular contractions are very common during the peri-infarct period; there is no clear relationship to the risk of ventricular tachycardia or ventricular fibrillation.
- Asystole or electromechanical dissociation is rare, but it portends an extremely poor prognosis.

Conduction Disturbances

- First-degree AV block occurs in 5% to 10% of hospitalized patients.
- Type I second-degree AV block (Wenckebach) occurs in 10% of hospitalized patients.
- Type II second-degree AV block is rare and usually requires pacemaker implantation.
- Third-degree AV block requires at least temporary pacing; it may resolve spontaneously in inferior wall MI.

Bundle Branch Block (BBB)

BBB occurs in approximately 15% of hospitalized patients; right bundle branch block is more common than left bundle branch block (LBBB). LBBB is associated with an increased risk of third-degree heart block and increased mortality.

Cardiogenic Shock

Cardiogenic shock is the result of inadequate cardiac output, with signs of persistent hypotension (systolic blood pressure <80 mmHg for more than 30 min) and a cardiac index <2.0 $L \cdot min^{-1} \cdot m^{-2}$ (normal is approximately 3.0 $L \cdot min^{-1} \cdot m^{-2}$) in the presence of adequate intravascular volume. It is usually the result of a large myocardial infarction. Treatment options include direct monitoring of pulmonary capillary wedge pressure with a Swan-Ganz catheter to determine the effects of positive inotropic agents such as dobutamine, insertion of an intra-aortic counterpulsation balloon pump to maintain blood pressure, and prompt percutaneous **coronary revascularization**. Mortality is high with this condition.

Infarct Extension and Expansion

Infarct extension is recurrent necrosis occurring 2 to 10 d after myocardial infarction in an area remote from the original infarction. Infarct expansion is thinning and dilatation of the infarcted myocardium without new necrosis. It occurs most commonly with anterior wall myocardial infarctions and may result in aneurysm formation, congestive heart failure, and serious ventricular arrhythmias.

Myocardial Rupture

Rupture of the ventricular free wall presents catastrophically with either sudden death due to electromechanical dissociation (pulseless electrical activity) or cardiac tamponade with cardiogenic shock. Fortunately it is rare and occurs most commonly within 4 d of the myocardial infarction. Left ventricular rupture is much more common than right ventricular rupture. Predisposing factors include advanced age and female sex. The only available treatment is emergency surgery.

Mitral Valve Regurgitation

Mitral valve regurgitation is most often a result of papillary muscle or chordae rupture resulting from the infarction. It may also occur because of left ventricular dilatation due to heart failure. It typically occurs during the first few days after infarction and presents abruptly with hypotension and right ventricular failure. Treatment is prompt surgical intervention, with repair or replacement of the mitral valve.

Pericardial Effusion and Pericarditis

Pericardial effusion is fluid accumulation in the pericardial space. In the setting of acute myocardial infarction, it is associated with pericarditis and occurs in approximately 10% of hospitalized patients. The usual treatment is high-dose aspirin.

Postinfarction Syndrome

Also known as Dressler's syndrome, postinfarction syndrome is pleuritic or pericardial chest pain associated with a friction rub heard on auscultation of the heart. A form of pericarditis, it usually occurs several weeks after myocardial infarction and is treated with high-dose aspirin.

Based on Murphy et al. (52a).

Left Ventricular Mural Thrombus

Fifty percent of patients with anterior wall myocardial infarction will develop a blood clot on the endocardial surface of the left ventricle. Anticoagulation with warfarin for a period of time is the usual treatment.

- To determine the effectiveness of medical therapy
- To assess future risk and prognosis
- To objectively determine exercise capacity (this information is used for exercise prescription, entry into an outpatient cardiac rehabilitation program, and return to work and other activities)

See the Contraindications section in chapter 5 for absolute contraindications to exercise testing (55, 56). Exercise test factors associated with an increased risk of a recurrent cardiac event and poor prognosis include the following (55, 56):

- Inability to exercise
- Substantial exercise-induced myocardial ischemia
- Exercise capacity <5 METs
- Failure of systolic blood pressure to increase at least 10 mmHg

Table 14.2 provides an overview of stress testing after myocardial infarction. The timing of the performance of a postmyocardial infarction graded exercise test varies, depending on the clinical situation and the preferences of the treating cardiologist. For patients treated medically, a predischarge or early postdischarge exercise test may be performed to determine the need for coronary angiography (55, 56). Generally, if testing occurs before 7 d postinfarction, a submaximal protocol is selected, although some studies have evaluated symptom-limited protocols as early as 4 d after infarction. Symptom-limited tests are generally performed 7 or more days after infarction. These tests may be performed at 14 to 21 d or 6 or more weeks after the event, depending on the practice patterns of the individual cardiologists. Exercise testing

after myocardial infarction is safe if the previously mentioned contraindications are not present. Ideally, a symptom-limited graded exercise test should be performed before starting outpatient cardiac rehabilitation, although local practice patterns may preclude this.

Assessment of exercise capacity and the presence and extent of myocardial ischemia may be accomplished using several different techniques (55, 56):

- Standard exercise electrocardiogram, with or without expired air analysis (cardiopulmonary exercise testing)
- Nuclear imaging modalities to measure myocardial perfusion
- Echocardiographic imaging of ventricular systolic function and regional wall motion

In all testing modalities, the clinical interpretation of the test result requires integration of all the available clinical data. These tests are not infallible in detecting the presence or absence of myocardial ischemia. An excellent discussion of test interpretation is provided by Ellestad and colleagues (57).

Standard exercise electrocardiography, with ST-segment depression (or less commonly, ST-segment elevation) of >1 mm at 0.08 s after the J point required for the diagnosis of ischemia, provides a sensitivity of approximately 65% to 70% (55). Other factors, such as the time of onset of ST depression (early in exercise versus near-maximal exertion), the maximal amount of ST-segment change, and the presence of typical angina, increase the accuracy of the assessment of ischemia. Limitations of the exercise ECG are its inability to diagnose ischemia in the setting of digoxin use or an abnormal rest ECG (particularly left bundle

Table 14.2 Stress Testing After Myocardial Infarction

Test type	End point	Advantages
Predischarge exercise test	Submaximal effort	Determines need for coronary angiography in patients not treated initially with an invasive strategy
Standard exercise test	Symptom-limited maximal effort	Inexpensive; diagnoses ischemia; assesses symptoms, exercise capacity, and blood pressure
Cardiopulmonary exercise test	Symptom-limited maximal effort	Provides extensive information regarding the oxygen transport system; diagnoses ischemia; assesses symptoms, exercise capacity, and blood pressure
Nuclear exercise test	Symptom-limited maximal effort	Localizes and quantifies areas of ischemia; provides left ventricular ejection fraction and infarct size
Echocardiographic exercise test	Symptom-limited maximal effort	Localizes and quantifies areas of ischemia; provides left ventricular ejection fraction and infarct size
Pharmacologic stress test; nuclear or echocardiographic imaging	Test type dependent	Localizes and quantifies areas of ischemia; provides left ventricular ejection fraction and infarct size in patients who cannot exercise adequately

branch block), inability to localize the area of the myocardium that is ischemic by ST-segment depression, lower sensitivity than with imaging techniques (sensitivities of 85%+), and the lack of information provided regarding the extent of ischemia. The exercise ECG does provide evidence of the ischemic threshold (the heart rate and systolic blood pressure that corresponds to the first evidence of ischemia), which is valuable in prescribing physical activity for patients.

Direct measurement of oxygen uptake, carbon dioxide production, minute ventilation, and associated variables during graded exercise (cardiopulmonary exercise testing) is particularly useful in determining prognosis in patients with chronic heart failure (58). The technique is also helpful in determining the cause of unexplained dyspnea with exertion. In addition, the direct measurement of maximal oxygen uptake is much more accurate in determining aerobic exercise capacity than is estimating exercise capacity based on the achieved workload.

Myocardial perfusion imaging using radioisotopes (technetium sestamibi or technetium tetrofosmin) is based on the premise that myocardial uptake of these substances is proportional to myocardial blood flow (55, 56). Images are obtained at rest and after exercise with a single-photon emission computed tomography camera system. Reversible defects (better perfusion at rest than with exercise) represent ischemia. Fixed defects (present at rest and with exercise) represent infarct scar or, less commonly, stunned or hibernating myocardium. The images provide quantification of infarct size. In exercise echocardiography, quantitative echo images of LVEF and end-systolic volume, as well as subjective echo images of regional wall motion and thickening, are obtained before and immediately after maximal exercise (55, 56). The nuclear and echocardiographic imaging techniques are capable of localizing areas of ischemia as well as quantifying the extent and severity of ischemia. Imaging techniques do not provide serial information regarding ischemia during graded exercise and thus do not provide information regarding the ischemic threshold.

Exercise for standard exercise electrocardiography (with or without cardiopulmonary measurements) and the imaging techniques is usually performed on a motorized treadmill, although cycle ergometry, arm-only, or combination arm and leg ergometry may be preferable in certain situations (56). Exercise testing protocols and procedures for patients with coronary artery disease are described in several excellent references and will not be reviewed here (56, 57).

Pharmacologic stress using intravenous administration of a coronary vasodilator (adenosine) or a positive chronotropic and inotropic agent (dobutamine) may be used in conjunction with imaging techniques for detection of myocardial ischemia in patients who cannot exercise adequately (55, 56). Obviously, pharmacologic stress testing provides no information regarding exercise capacity and the hemodynamic responses to exercise. Adenosine is the most common coronary vasodilator used in conjunction with nuclear perfusion imaging techniques. An abnormal flow reserve (less flow in a particular region of the myocardium relative to other regions) in the territory supplied by a stenotic coronary artery represents an ischemic response. Dobutamine is a synthetic sympathomimetic that increases both myocardial contractility and heart rate, thus elevating myocardial oxygen requirement in a manner analogous to exercise. It is most commonly used in conjunction with echocardiographic assessment of ventricular function and potential ischemia. If at the maximum dose of dobutamine the heart rate is below 85% of age-predicted level, atropine may be given to further increase the heart rate.

EXERCISE PRESCRIPTION

The prescription of exercise after an acute coronary syndrome should be individualized, based on factors such as the following (54a, 56, 65b):

- Exercise capacity
- Ischemic or anginal threshold or both
- Cognitive and psychological impairments
- Vocational and avocational requirements
- Musculoskeletal limitations
- Obesity
- Prior physical activity history
- Patient goals

Table 14.3 gives an overview of the components of an exercise prescription for a coronary patient.

In an ideal world, patients with a recent acute coronary syndrome would exercise with medical supervision in an outpatient cardiac rehabilitation program for at least several weeks. Unfortunately, some geographical areas do not have sufficient programs to ensure easy accessibility for all patients. Patients may present multiple barriers for attendance at the standard three supervised exercise sessions per week of a typical cardiac rehabilitation program, such as return to full-time employment within a few days of hospital dismissal, difficulty securing reliable transportation to and from the rehabilitation center, and excessive out-of-pocket costs (e.g., coinsurance, parking fees). High-risk and debilitated patients are best served with frequent supervised rehabilitation sessions. Non-high-risk patients may benefit from flexibility in program design. Medicare will reimburse for up to 36 rehabilitation sessions over a 36 wk period, as will most private insurance providers. Patients may attend one to three sessions weekly, with additional independent exercise as prescribed by cardiac rehabilitation staff. Some patients will not be able to attend a formal program and will require an entirely independent program. Recently published performance measures for the referral to and delivery of cardiac rehabilitation and secondary prevention services should help patients obtain appropriate posthospital coronary disease prevention care (74).

After completion of a standard cardiac rehabilitation program of 36 sessions, most patients will need to continue exercise training independently (outside of a formal program). Compliance with independent exercise may be improved with follow-up visits with health care providers. In my own program, patients are encouraged to return for regular follow-up with the cardiac rehabilitation staff on a continuing basis (64). Some cardiac rehabilitation programs

Table 14.3 Exercise Prescription After an Acute Coronary Syndrome

Training method	Frequency	Intensity	Time (Duration)	Type (Mode)	Progression	Important considerations
Cardiorespiratory	3-7 d/wk	RPE 11-16, 40%-80% of exercise capacity	20-60 min	Walking, jogging, cycling, arm–leg ergometer, elliptical, arm ergometer, water-based exercise, stair stepper, rower	Begin with 5-10+ min at RPE 11-13; gradually increase 1-5 min per session and RPE, as tolerated	Use intensity below ischemic threshold. Take medications on schedule before exercise. Use intermittent exercise for very deconditioned patients. Use high-intensity interval training for selected patients.
Resistance	2 or 3 d/wk	RPE 11-14, 30%-80% of 1RM	8-10 exercises, 8-15 slow reps, 1-4 sets	Free weights, weight machines, elastic bands, stability ball (include major muscle groups)	Begin with 1 set of 8 reps at RPE 11-12 for each exercise; gradually increase numbers of repetitions, sets, resistance, and RPE of 13-14, as tolerated	Avoid Valsalva maneuver or straining. Use circuit training for select patients.
Range of motion	Daily	Hold to point of mild discomfort	5-15 min	Static stretching, major muscle groups	Begin with 15 s per static stretch exercise; gradually increase to 30-60 s per exercise, as tolerated	Use full range of motion. Avoid breath holding.

RPE = rating of perceived exertion (Borg scale 6-20); 1RM = one-repetition maximum.

have sufficient capacity to offer patients a maintenance exercise program (usually patient funded) for a period of months up to lifelong participation. Some programs include spouses to aid in retention of patients in maintenance programs.

Frequency

Exercise sessions should occur at least 3 d/wk and ideally on most days of the week (i.e., 4-7d/wk). With the exception of extremely high-risk patients, one or more sessions per week should be performed independently, outside of the supervised environment of a cardiac rehabilitation program. Patients with very limited endurance may perform multiple short-duration (1-10 min) sessions daily.

Intensity

Exercise intensity may be prescribed using one or more of the following:

- Rating of perceived exertion (RPE) of 11 to 16 on a scale of 6 to 20
- 40% to 80% of exercise capacity using the heart rate reserve method, or the percent peak oxygen uptake reserve technique if maximal exercise test data are available
- When exercise test or pharmacologic stress data are not available, upper-limit heart rate of resting heart rate + 20 b · min^{-1} and RPE of 11 to 14, gradually titrating the heart rate to higher levels according to RPE, signs and symptoms, and normal physiologic responses

Exercise intensity should be kept below the ischemic threshold, if one has been demonstrated to exist, unless the responsible physician recommends otherwise in patients with ischemia, even at very low exercise intensities.

Patients should take their medications on schedule for performance of exercise training sessions. Changes in the doses of β-blockers or other drugs that affect the chronotropic response to exercise may occur. In these cases, it is unlikely that a new exercise test will be performed solely for the purpose of determining a new target heart rate. Using RPE and signs and symptoms, as well as determining the new target heart rate at previously performed work intensities, is recommended. It is common for cardiac rehabilitation professionals to diagnose hypotension, either orthostatic, at rest, during exercise, or after exercise, due to excessive medication dosage.

Duration

Patients should perform warm-up and cool-down activities, including static stretching to improve range of motion and low-intensity aerobic exercise, for 5 to 10 min before and after the conditioning phase of exercise. The goal duration for conditioning aerobic exercise is 20 to 60 min per session. Patients may begin with an easily tolerable duration of 5 to 10+ min, with a gradual increase of 1 to 5 min per session or 10% to 20% per week, as tolerated. The progression in duration should be individualized for each patient; some patients may be able to progress much more rapidly than others based on fitness, exercise habits immediately prior to the coronary event, symptoms, and musculoskeletal limitations. Frail and otherwise extremely deconditioned patients may require intermittent exercise, alternating short periods of exercise with rest breaks, although most patients can perform continuous exercise.

Type

Aerobic forms of exercise should include rhythmic large-muscle group activities. Ideally, both upper and lower extremity exercises should be incorporated into the program, such as these:

- Treadmill or track for walking (jogging or inclined walking for more fit patients)
- Upright or recumbent cycle
- Combination upper and lower extremity ergometer
- Elliptical
- Stair stepper
- Arm ergometer
- Rower

Independent exercise often includes walking outdoors, with or without hills, or walking indoors at shopping malls or schools that are open to the public for this purpose.

Resistance Training

The vast majority of patients with treated acute coronary syndromes should perform strength training as a component of their exercise program. Patients with severe ischemia or hemodynamic instability should be adequately stabilized before beginning strengthening exercises. The purposes of strength training for patients include the following (56):

- To increase muscular strength and endurance
- To decrease cardiac demands of muscular work (reduce rate–pressure product [RPP] with lifting and carrying activities of daily life)
- To improve self-confidence
- To maintain independence (enable patient to perform household and personal care duties)
- To prevent diseases such as osteoporosis, type 2 diabetes, and obesity
- To slow the age-related declines in skeletal muscle mass and strength

Although guidelines for strength training in cardiac patients traditionally recommend waiting 5 wk from the date of myocardial infarction before beginning strength training, in my own program patients begin such training within 2 wk of the event. Several thousand patients have followed this practice, and there have been no complications associated with strength training.

Equipment for strength training may include free weights, elastic bands, weight machines with pulley devices, stability balls, and weighted wands. Patient body weight may also be used as resistance for selected exercises. Movements should be performed in a slow, controlled manner while maintaining a regular breathing pattern and avoiding straining and musculoskeletal pain. RPE ratings initially should range from 11 to 14 on the 6 to 20 scale. Initial resistances should allow 12 to 15 repetitions (approximately 30%-60% of 1RM). The resistance may be gradually increased, and most patients may progress to 8 to 12 repetitions with a resistance of 60% to 80% of 1RM, with one to four sets of repetitions. Exercises should be performed for the major muscle groups, usually 8 to 10 different exercises. Frequency should be 2 or 3 sessions per week on nonconsecutive days.

Aerobic High-Intensity Interval Training

Over the past 35 years, there has been a gradual easing of restrictions regarding exercise training for patients with cardiovascular diseases. For example, outpatient exercise training after myocardial infarction now begins within a few days of hospital dismissal (rather than several weeks), with more aggressive aerobic exercise and strength training (forbidden in the past). Patients with chronic heart failure are now encouraged to exercise rather than to rest. The latest exercise taboo to be lifted is high-intensity interval training (HIIT) for cardiac patients. Investigators have documented superior improvements in $\dot{V}O_2$peak, measures of endothelial function, and left ventricular systolic and diastolic function in coronary patients with high-intensity interval training (up to 95% of maximal heart rate for short periods of time interspersed with lower-intensity exercise) compared with conventional continuous aerobic training (66, 67).

In my own program, we have used HIIT in several thousand cardiac patients for the past 12 yr and have not encountered safety issues or patient reluctance to try it. We start with approximately 2 wk of standard continuous aerobic training, increasing the duration to at least 20 min before beginning HIIT. We start with 30 s intervals, three to five times during the exercise session, interspersed with lower-intensity exercise for 2 to 3 min. Over a period of days to weeks, the length of the high-intensity intervals is increased to 60 to 120 s or longer, as tolerated. Interval training occurs 2 or 3 d/wk.

Progression of the Volume of Exercise

An appropriate, gradual progression of the dose of both aerobic and strengthening exercise is essential to optimize potential improvement in cardiorespiratory fitness and skeletal muscle strength while minimizing potential adverse complications (65b).

Balance Exercises

Many participants in cardiac rehabilitation programs are older than 65 yr. With increasing age, the neuromuscular reflexes involved with proprioception and balance become less effective. Falls are a common cause of morbidity and mortality in older individuals. A balance abnormality was found in over half of a consecutive sample of 284 participants in my program (68). Women and patients over 65 yr of age were more likely to exhibit poor balance. As part of the baseline assessment of patients beginning outpatient cardiac rehabilitation, a balance assessment using simple techniques such as the single-leg stand and the tandem gait (walking heel to toe in a straight line) should be performed. For patients with poor balance, specific exercises may be prescribed (56).

Lifestyle Physical Activity

In addition to formal exercise sessions, patients should be encouraged to gradually return to general activities of life, such as walking for transportation, household tasks, shopping, gardening, nonsedentary hobbies, and sport activities (56). Measuring lifestyle activity is relatively easy using a pedometer (steps/day), wrist-worn activity sensor, or various smartphone applications. The more time spent sitting, such as watching television or on the computer, the greater the risk of cardiovascular and all-cause mortality as well as nonfatal cardiovascular events (69, 70). Greater energy expenditure resulting from increased lifestyle physical activity facilitates weight loss.

Exercise Programming for Overweight Coronary Patients

The vast majority of coronary patients who enter outpatient cardiac rehabilitation are either overweight or obese. Standard exercise protocols used in most programs—30 to 45 min of moderate-intensity aerobic exercise three times per week (approximately 800 kcal/wk) for 3 mo—are generally not adequate to result in substantial body fat loss. Investigators used a high-calorie exercise protocol (approximately 3,000 kcal/wk), which required 45 to 60 min of walking more than 4 d/wk (71). Compared with standard cardiac rehabilitation, the high-calorie exercise protocol resulted in twice the weight loss (8.2 kg vs. 3.7 kg).

In my own program, we encourage patients who desire to lose considerable weight to consult with our registered dietitian, gradually increase exercise duration to 60 min five or six d/wk, incorporate high-intensity interval training 2 or 3 d/wk, and perform moderate strength training using free weights and weight machines 2 or 3 d/wk. This high-volume exercise protocol requires very motivated patients.

Safety of Exercise Training for Coronary Patients

Supervised exercise training has been demonstrated to be safe for patients with cardiovascular diseases. Franklin and colleagues reported the average incidence of cardiac arrest, nonfatal myocardial infarction, and death as one for every 117,000, 220,000, and 750,000 patient-hours of participation, respectively (72). A large trial of over 1,000 higher-risk patients with chronic heart failure (more than half with coronary disease) studied the safety of supervised and independent moderate-intensity aerobic exercise training. Over approximately 30 mo, there were 37 events associated with exercise that resulted in hospitalization. In over several million patient-hours of exercise, only five deaths occurred (73). The benefits of exercise training far outweigh the risks for coronary patients without absolute contraindications to exercise.

Underutilization of Cardiac Rehabilitation Services

In spite of the impressive benefits of exercise-based cardiac rehabilitation, most patients do not attend after an acute coronary syndrome. Among Medicare beneficiaries, only 19% of eligible patients participate in outpatient cardiac rehabilitation (83).

Research Focus

Benefits of Cardiac Rehabilitation and Secondary Prevention Programs

The following list summarizes the important benefits available to patients with coronary disease who participate in cardiac rehabilitation (75, 76, 77, 78). As a result of training, $\dot{V}O_2$peak may improve 10% to 20% or more (75). In general, the magnitude of the relative improvement is inversely proportional to the baseline exercise capacity. The rate of improvement is greatest during the first 3 mo of training, but increases in aerobic capacity may continue for 6 mo or more. High-intensity interval training results in greater improvements in $\dot{V}O_2$peak than traditional continuous intensity training (66). Here is a summary of the benefits of cardiac rehabilitation programs:

- Improved aerobic capacity
- Increased submaximal exercise endurance
- Increased muscular strength
- Reduction in symptoms: angina pectoris, dyspnea on exertion, fatigue, claudication
- Vascular regeneration via bone marrow–derived endothelial progenitor cells
- Decreased myocardial ischemia and potential increase in myocardial perfusion
- Improved endothelial function
- Improved left ventricular systolic and diastolic function in chronic heart failure
- Potential retardation of coronary disease progression and actual regression of plaque
- Incorporation of heart-healthy dietary practices
- Improved blood lipid profile
- Improved indices of obesity
- Decreased blood pressure
- Decreased tobacco use
- Improved glucose intolerance and insulin resistance
- Decreased inflammatory markers (e.g., high-sensitive C reactive protein, hsCRP)

- Improved autonomic tone: less sympathetic activity, more parasympathetic activity
- Reduction in ventricular arrhythmias
- Improved blood platelet function and blood rheology
- Improved psychosocial function (less depression, anxiety, somatization, hostility)
- Repeated surveillance of blood pressure, symptoms, arrhythmias, and so on, leading to earlier treatment
- Improved patient compliance with taking cardioprotective medications
- Reduced rehospitalizations
- Reduced health care costs
- Reduced mortality

Strength training results in significant gains in skeletal muscle strength. Symptoms related to coronary disease usually improve with cardiac rehabilitation. Left ventricular function may improve. Improvement in coronary risk factors is observed. Atherogenesis is slowed, endothelial function is improved, and myocardial ischemia may be lessened. Psychosocial function may improve (75, 76, 77, 78). A meta-analysis of several randomized trials demonstrated a 31% reduction in mortality for patients who participated in cardiac rehabilitation (79). For my own program, located at an academic medical center, we have demonstrated a 56%, 45%, and 46% reduction in all-cause mortality for participants with either acute myocardial infarction, PCI, or CABG, respectively, compared with nonparticipants (80, 81a, 81b). In addition, in a large sample of Medicare beneficiaries, a dose–response relationship between the number of cardiac rehabilitation sessions attended and various cardiovascular outcomes was determined (i.e., the more sessions attended, the better the outcomes) (82).

Older adults, nonwhites, patients with comorbidities, patients with low socioeconomic status, the unemployed, single parents, and women are less likely to participate (84, 85, 86). Suggested steps to improve the enrollment rate in cardiac rehabilitation programs include automating the referral process, designating referral and enrollment in cardiac rehabilitation as a quality indicator in cardiovascular care, designing programs that utilize the telephone or Internet for patients who live in areas without programs, determining how best to include underserved populations, and developing a multimedia education program on the benefits of cardiac rehabilitation directed at both patients and health care providers (87). A joint scientific statement from the American Association of Cardiovascular and Pulmonary Rehabilitation, the American Heart Association, and the American College of Cardiology addressed the underutilization of cardiac rehabilitation and provided guidance for the development of home-based cardiac rehabilitation for patients who cannot attend traditional medical center–based programs (88).

EXERCISE TRAINING: INPATIENT CARDIAC REHABILITATION

Lengths of hospital stay after acute myocardial infarction have declined over the past two decades. Currently, patients are hospitalized for no more than 2 or 3 d unless complications arise. The vast majority of patients with unstable angina or acute myocardial infarction are treated invasively with PCI, as noted previously. The detailed, expansive inpatient rehabilitation protocols described by this author in 1987 seem archaic compared with today's practice patterns (59). There is little opportunity for formal exercise training during hospitalization.

To minimize the deleterious effects of bed rest, patients are mobilized as soon as they are stable. Exposure to the normal stress of gravity is emphasized in this stage of rehabilitation to prevent orthostatic intolerance (60). Patients sit, stand, perform active range of motion exercises for the major joints, and walk short distances as soon as possible to prevent further deconditioning. Various allied health professionals may be involved in inpatient cardiac rehabilitation, such as registered nurses, physical therapists, occupational therapists, and exercise physiologists. Patients with neuromuscular diseases or other conditions that limit their ability to ambulate benefit from formal physical therapy treatments. After hospital dismissal, frail elderly and other debilitated patients may be referred to outpatient postacute or transitional care facilities for various time periods before returning home (61a).

A critical aspect of inpatient cardiac rehabilitation is the introduction of the concepts of **secondary coronary prevention** of atherosclerosis to patients and family members (61b). Basic information is provided regarding the importance of cardioprotective medications, avoidance of tobacco, heart-healthy eating patterns, blood pressure and blood lipid goals, exercise (including any temporary restrictions imposed by the coronary event), potential depression after the cardiac event, and return to usual activities. Patients should be referred to an outpatient cardiac rehabilitation program if one is available in their home area. A home exercise prescription providing guidance as to recommended exercise types, intensity, duration, frequency, and progression for the first few weeks after hospital dismissal should be given to the patient. Typically, the prescription recommends walking or stationary cycling at a comfortable intensity, beginning with 10 to 20 min once or twice daily, progressing to 30 to 45 min daily.

EXERCISE TRAINING: EARLY OUTPATIENT CARDIAC REHABILITATION

Outpatient cardiac rehabilitation is recommended for patients with coronary heart disease by the American Heart Association, the American College of Cardiology, and the American Association of Cardiopulmonary Rehabilitation and has been awarded a class I indication (the highest indication; the treatment is effective and should be provided) (61b). This phase of rehabilitation can begin soon after hospital dismissal, often within 1 or 2 wk (56, 61a). During the interval between hospital dismissal and beginning outpatient rehabilitation, patients are encouraged to exercise independently, using their home exercise prescription provided in the hospital. A variety of health care professionals are typically involved in early outpatient cardiac rehabilitation, such as physicians or midlevel providers, exercise physiologists, registered nurses, registered dietitians, physical or occupational therapists, social workers, and psychologists.

The initial assessment, performed by cardiac rehabilitation professionals, may involve the following (56, 61a):

1. Review of the medical records, with emphasis on the recent coronary event

2. Physical examination, performed by a qualified medical professional, usually an MD, advanced practice provider (nurse practitioner, physician assistant), or registered nurse; ideally should include a review of the neuromuscular systems in terms of potential limitations to exercise

3. Graded exercise test, which ideally should be performed to assist with exercise prescription and with risk and prognosis assessment; may not be available in a timely fashion depending on local practice patterns, so a 6 min walk may be performed as a check of submaximal exercise response

Clinical Exercise Bottom Line

- After an acute coronary syndrome, regular exercise training with aerobic, skeletal muscle strengthening, and flexibility components provides multiple clinically important benefits, including improved exercise capacity, reduced symptoms, fewer rehospitalizations, and improved survival.

- Ideally, exercise training should initially be medically supervised in an outpatient cardiac rehabilitation program and should commence within 2 wk after hospital dismissal.

- Medicare and most private insurances will cover up to 36 cardiac rehabilitation sessions.

- Exercise sessions should occur at least 3 d/wk and ideally on most days of the week (i.e., 4-7 d/wk).

- Intensity should range from 11 to 16 on a 6 to 20 RPE scale. If heart rate or aerobic capacity is available, typically use ranges of 40% to 80% of heart rate or oxygen consumption reserve.

- After a 5 to 10 min warm-up, start in the range of 5 to 10 min, with the goal of achieving 20 to 60 min. Intermittent training should be considered in frail and extremely deconditioned patients.

- High-intensity interval training should be considered in stable patients who are willing to work at higher intensities, because greater benefits have been observed.

- Balance training should be incorporated in older (>65 yr) patients.

- Resistance training should be incorporated in stable patients to help reduce skeletal muscle loss with aging.

4. Medication review, with emphasis on compliance, side effects, and optimal cardioprotective medication use

5. Review of standard coronary risk factors, with emphasis on avoidance of tobacco, proper nutrition habits, control of blood pressure and blood lipids, body weight optimization, and stress and depression management

6. Patient-selected and provider-selected goals to work on during the program

7. Anticipated return to work date, if needed

8. Explanation of the short-term and long-term benefits of exercise training, optimal control of coronary risk factors, and the use of cardioprotective medications

9. Anticipated rehabilitation schedule, including number of rehabilitation sessions per week and number of weeks in the program

Cardiac rehabilitation programs are ideal for helping patients achieve the secondary prevention goals of taking appropriate cardioprotective medications; avoiding tobacco; following heart-healthy eating patterns; performing regular exercise and lifestyle physical activity; achieving a desirable body weight; optimally controlling blood lipids, blood pressure, and blood glucose; and maintaining psychosocial health (61a). Typical early outpatient cardiac rehabilitation programs provide up to 36 supervised sessions over 3 or more months. Each patient's clinical status is evaluated periodically (e.g., heart rate, ECG, blood pressure, symptoms, mood, medication compliance and side effects), and ongoing medical surveillance is provided.

Counseling is a major component of patient visits to cardiac rehabilitation. Case management, more recently renamed disease management, involves one or more cardiac rehabilitation staff members taking responsibility for oversight of an individual

patient's secondary prevention program (62, 63). Disease management by cardiac rehabilitation professionals has been demonstrated to be highly effective in promoting patient adherence to secondary prevention measures. My own program includes disease management over 1 or more years of follow-up after completion of the standard 36 cardiac rehabilitation sessions covered by insurance (64).

A technique used by health care providers to help patients make healthy changes in behavior is motivational interviewing (65a). The technique is not difficult to learn and involves developing a partnership with the patient that honors the individual's perspectives and talents. There are four general principles of motivational interviewing:

1. Express empathy.

2. Develop discrepancy: The patient needs to perceive a discrepancy between present behavior and important goals.

3. Roll with resistance: Avoid arguing for change.

4. Support self-efficacy: An important motivator for change is the patient's belief in the possibility of change.

CONCLUSION

Acute coronary syndromes are most commonly the result of atherosclerotic plaque development, with subsequent plaque disruption and thrombus formation leading to myocardial ischemia and potential necrosis. Unstable angina pectoris is the result of transient coronary artery occlusion with spontaneous clot dissolution and no demonstrable myocardial necrosis. Vessel occlusion persisting for more than 1 h results in myocardial necrosis, which is the hallmark of acute myocardial infarction. Myocardial infarctions are categorized, based on electrocardiographic findings, as either ST-segment elevation or non-ST-segment elevation. Preferred treatment is prompt reperfusion of the occluded vessel, and this results in less myocardial damage. Several classes of cardioprotective medications are given to survivors of acute coronary syndromes. In addition to these medications, comprehensive cardiac rehabilitation, including exercise training, forms the basis for secondary prevention of future cardiac events for these patients. Cardiac rehabilitation results in impressive benefits for patients, including reduced mortality.

Online Materials

Visit HK*Propel* to access a link to the references, the case studies with discussion questions, and a quiz to help you review key concepts and test your understanding of the material covered in this chapter.

CHAPTER 15

Revascularization of the Heart

Neil A. Smart, PhD

Heart disease remains the leading cause of death in the United States, according to the Centers for Disease Control and Prevention (CDC) (1). Considering this reality is of utmost importance in understanding the procedures used to potentially alleviate symptoms, reduce risk of significant heart injury, or even save the lives of those afflicted with heart disease. The choice of how to manage revascularization varies with each individual. As surgical and nonsurgical revascularization techniques have evolved, the number of treatment choices has increased, and careful clinical reasoning is required to optimize the management of the revascularized patient. It is also important to know how to use exercise as a rehabilitation therapy, after revascularization, to maintain or improve physical functioning of the heart and body to maintain lower risk of further adverse cardiac events and improve a person's quality of life. A relatively new strategy, remote ischemic conditioning (RIC), can minimize the risk and reduce the impact of myocardial injury in people undertaking planned cardiac surgery.

DEFINITION

The term *revascularization* refers to a surgical procedure to help provide new or additional blood supply to a body part or organ. Several organs, such as the heart, lungs, kidneys, liver, and muscles (in situations such as gangrene), can benefit from this procedure. Typically, diagnostic tests may involve using stress electrocardiography with or without echocardiography, coronary artery angiography or cardiac catheterization, magnetic resonance imaging (MRI), CT scans (coronary artery calcium scan), chest X-ray, or X-ray fluoroscopy to identify the need for revascularization or help in guidance of the procedure.

Although there are multiple parts of the body and organs for which revascularization may be indicated, the focus of this chapter is on procedures concerning the heart.

You will see terms such as *percutaneous transluminal coronary angioplasty* (PTCA) or *percutaneous coronary intervention* (PCI), with or without stenting, and *coronary artery bypass surgery* (CABS), all techniques for establishing proper blood flow back to the heart. Several new methods have been developed to minimize surgical impact and reduce recovery time. These include, an on–off pump (2) and minimally invasive or single thoracic entry (3a).

SCOPE

When a person has coronary artery disease, clinical procedures may be elected to restore myocardial blood flow, with the specific intent of symptom relief and improved morbidity and mortality. The two most commonly used techniques are coronary artery bypass surgery (CABS)—in the form of coronary artery bypass grafting (CABG)—and percutaneous transluminal coronary angioplasty (PTCA), also known as percutaneous coronary intervention (PCI), with or without stenting. A 2018 study estimated that 340,000 CABGs are performed each year in the United States (3b). In 2020 a downward trend was documented, with approximately 202,000 U.S. adults undergoing CABS and 440,000 PTCA. This equates to a decrease from 159 to 82 surgeries per 100,000 U.S. adults and a drop in the annual CABG volume from 366 to 180 per 100,000 U.S. adults (4).

In people with multiple coronary vessel disease, the first question is whether to manage the condition with medication or surgery. If surgery is indicated, the next decision is whether

Acknowledgment: Some of the writing in this chapter was adapted from the third edition of *Clinical Exercise Physiology.* Thus, we wish to gratefully acknowledge the previous efforts of and thank Mark Patterson, PhD.

minimally invasive PTCA or CABS is preferred. This decision has significance because a common side effect of CABS is neurological complication (5).

Compared with PTCA, CABS significantly reduces the rates of death and myocardial infarction but is associated with a higher risk of stroke. Determining the optimal revascularization strategy for these patients is challenging because many clinical factors influence the decision to pursue either CABS or PTCA. However, the current consensus is that CABS should be the preferred revascularization strategy in certain patient populations, such as people with diabetes and extensive multivessel CAD (6).

Research from the United Kingdom shows that referrals to cardiac rehabilitation have increased to almost 50% of eligible patients (7). Today, because of advances in coronary invasive technology, cardiac rehabilitation programs are seeing a growing number of people who have experienced percutaneous interventions, including percutaneous transluminal coronary angioplasty alone and, more frequently, in combination with stent therapy, which involves the placement of a mesh tube along the artery wall to prevent reocclusion (8). A review of both short- and long-term studies reported that in patients with multivessel disease, CABG is associated with better survival and lower rates of major cardiovascular events (specifically myocardial infarction or stroke) and repeat revascularization as compared with PTCA with drug-eluting stents (11).

Cardiac rehabilitation referral at discharge may be less prevalent after PTCA than CABS (9). Even though minor convalescent differences exist among the percutaneous interventional procedures, the standards of practice and expected outcomes for cardiac rehabilitation may be slightly better for CABS (10). A network meta-analysis analyzed surgical revascularization versus aggressive medical management among patients with stable coronary artery disease. This analysis concluded that coronary artery bypass grafting reduces the risk of death, myocardial infarction, and subsequent revascularization compared with medical treatment. All stent-based coronary revascularization technologies reduce the need for revascularization to a variable degree (12).

Some evidence does exist suggesting that regular exercise training may result in superior event-free survival and exercise capacity, compared with PTCA, in people with coronary artery disease (13). Regardless of whether patients with coronary artery disease are revascularized or not, previous work demonstrated that for each 1 MET (metabolic equivalent) improvement in exercise capacity, there appears to be a significant mortality reduction (14). Nevertheless, recent work analyzed mortality after 0.5 to 13.4 yr of follow-up (mean 6.4 yr) and found about a quarter of those undertaking cardiac rehabilitation did not improve their cardiorespiratory fitness (CRF). After adjustment for body mass index, age, gender, left ventricular ejection fraction, and baseline CRF, those who did not improve their CRF had a statistically significant two- to threefold higher mortality when compared with those who did (15).

PATHOPHYSIOLOGY

Coronary artery disease (CAD) involves a buildup of lipids, macrophages, platelets, calcium, and fibrous connective tissue within the coronary arteries. This results in the formation of a plaque that progressively narrows the lumen. This may eventually cause a limitation or obstruction of normal blood flow. Although symptoms or changes on an electrocardiogram (ECG) during a stress test may not occur until a coronary artery has a 75% or greater stenosis, lesions that compromise 50% or more of the lumen in a major coronary artery might be considered clinically significant. However, pooled analyses suggest only about one-third of people who experience a non-ST-segment elevated myocardial infarction (STEMI) have an occluded culprit artery (16). Multiple factors such as location of the lesion, stability of the plaque, symptoms, short- and long-term prognosis, and quality of life may all influence the decision on whether a particular patient is a candidate for revascularization procedures.

Coronary Artery Bypass Surgery

Coronary artery bypass surgery involves revascularization using a venous graft from an arm or leg or an arterial graft—both ends free or from a regional intact native vessel (e.g., internal mammary; **gastroepiploic artery**)—to provide blood flow to the myocardium beyond the site of the occluded or nearly occluded area in a coronary artery. Although CABS has traditionally involved a **sternotomy** and the use of a **heart and lung bypass**, technical advances now permit a growing number of procedures to be performed. Advances in CABS include:

1. Preferential use of the left internal mammary artery for left anterior descending (LAD) artery grafting, as opposed to the saphenous vein. The former has superior 10 yr patency for LAD grafts (17).

2. CABS can be performed using an on-pump method, where a cardioplegic solution is used to arrest the heart and the patient's circulatory system is temporarily maintained via a cardiopulmonary bypass system, or off-pump, also known as beating heart surgery. A meta-analysis has shown that avoidance of on-pump bypass surgery with an off-pump technique reduces the incidence of postoperative atrial fibrillation by 13% to odds ratio 0.87, as well as decreases ventilation time mean difference, ICU stay, and hospital stay (3a).

3. A minisurgery procedure performed through small port incisions using microscopic techniques.

4. Procedures performed with robotic technology.

5. Surgery performed on the beating heart without the use of cardiopulmonary bypass.

Subsequent to these technical advances, postoperative morbidity has significantly decreased. The postsurgical hospital stay for

CABS patients without complications is now less than 5 d. As a result of the evolution of revascularization procedures (particularly percutaneous intervention), the role for CABS has changed; it is now reserved for the following patients:

1. Patients who are post-PTCA or stenting (or both) with restenosis
2. Patients who are no longer candidates for angioplasty but still have target vessels offering preservation of left ventricular systolic function
3. Those with multivessel disease not amenable to angioplasty or stenting
4. Those with technically difficult vessel lesions (e.g., a lesion on the curve of a vessel or in a distal location not readily amenable to angioplasty or stenting)

The number of surgical revascularization procedures has declined, but these procedures still play an essential role in higher-risk occlusive disease. Successful CABS results in a myriad of improved exercise responses; and when combined with medical therapy, CABS may more effectively relieve significant residual exercise-induced symptomatic or silent myocardial ischemia. There is also some evidence that in persons with diabetes, the need for future revascularization procedures is less with CABS than with PTCA or stenting (18, 19). Thus, the symptom relief, improved functional capacity, and improved quality of life may be the most practical and important patient benefits of CABS.

Percutaneous Interventions

Coronary angioplasty (PTCA) is less invasive than CABS. Several techniques have been developed for use in restoring adequate blood flow in diseased coronary arteries. Often the procedure is combined with stent therapy to reduce the likelihood that the artery will reocclude.

Percutaneous Transluminal Coronary Angioplasty

PTCA is a well-established, safe, and effective revascularization procedure for patients with symptoms attributable to CAD. The procedure may use one or more techniques alone or in combination to open the vessel:

1. Balloon dilation is most commonly used in conjunction with stent placement (figure 15.1).
2. Rotational atherectomy, a rotational device used for removing plaque, may be applied to central bulky lesions in a minority of cases.
3. Directional atherectomy and laser may be used to debulk large lesions, but the risk of vessel wall perforation or dissection may be greater. The use of these devices is limited to a few centers in contemporary practice.

The complications of angioplasty are acute vessel closure (rebound vasoconstriction) or chronic restenosis, thrombotic distal **embolism**, myocardial infarction (MI), arrhythmias, dissection of the coronary artery, and bleeding. A 2019 meta-analysis found no evidence to support the beneficial effects of exercise after PTCA to improve heart function or to reduce the incidence of adverse cardiovascular events (20), although fitness and quality of life benefits are likely; these are discussed later in this chapter.

Stent Therapy

Intracoronary stents have been shown to reduce the risk of acute closure and restenosis of coronary arteries after PTCA (21). Stent therapy is frequently used in conjunction with one of the previously described techniques to preserve the patency of the vessel. The first stents were stainless steel mesh tube bridges. The stent is advanced on the end of a balloon catheter, passed across the culprit lesion, and expanded in order to compress the lesion, opening the vessel and serving as a permanent intravascular prosthesis. This is the final treatment following balloon angioplasty or debulking in over 95% of cases. After removal of the balloon catheter, the stent remains permanently in the coronary artery and is eventually covered with endothelium, becoming part of the luminal wall structure. However, a thrombus can form or scar tissue can accumulate within the stent; either of these events can lead to loss of patency.

In 2003, drug-eluting stents have become available (22). The metal scaffolding prevents acute closure and also provides

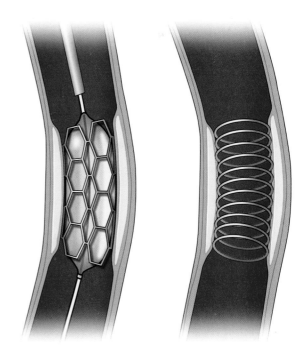

Figure 15.1 Percutaneous transluminal coronary angioplasty balloon catheter and two types of stents: latticed steel (left) and coiled stent (right).

a vehicle for local drug delivery. This innovation plus the use of aspirin and thienopyridines (clopidogrel and prasugrel) has reduced the problem of in-stent restenosis to a great extent (23, 24). Bare-metal stents are susceptible to intra-stent restenosis, which occurs in up to one-third of cases, and stent thrombosis. Although neointimal hyperplasia has certainly been delayed with drug-eluting stents, healthy endothelium will eventually grow over the struts, resulting in stent thrombosis in 0.5% to 1.5% of cases, despite dual antiplatelet therapy. The need for permanent coronary scaffolding and a drug-eluting implant is not justified beyond the first 6 to 12 mo, when the process of intimal hyperplasia and acute or chronic recoil is completed. Other drawbacks of the persistence of metallic stents include interference with the ability of noninvasive techniques such as computed tomography (CT) or magnetic resonance imaging (MRI) to assess the in-stent patency, occlusion or impaired access to ostia of side branches, impairment of physiological vessel tone reactivity, and inability to use the stented segment to anastomose grafts during bypass surgery. To circumvent these issues, absorbable stents were designed that are intended to degrade within the coronary artery, akin to dissolving sutures, while providing the vessel with temporary scaffolding until endothelialization has been established (25).

Many stent placements are same-day procedures or require a one-night hospital stay. The loss in functional capacity following PTCA is less than that following a bypass procedure. Subsequently, PTCA patients begin cardiac rehabilitation and achieve improvements in higher functional capacity, quality of life, and self-efficacy (28, 29).

CLINICAL CONSIDERATIONS

The success rate of a revascularization procedure may be predicted, in part, by the patient's age, existing comorbidities, and severity and location of the lesion.

Coronary Artery Bypass Surgery

Elective CABS improves the likelihood of long-term survival in patients who have the following:

- Significant left main CAD
- Three-vessel disease
- Two-vessel disease with a proximal left anterior descending stenosis
- Two-vessel disease and impaired left ventricular function (32)

In patients experiencing failed angioplasty with persistent pain or hemodynamic instability, acute MI with persistent or recurrent ischemia refractory to medical therapy, **cardiogenic shock**, or failed PTCA with an area of myocardium still at risk, revascularization by CABS offers effective relief of angina pectoris and improves the quality of life (32). The occlusion rate of grafts varies greatly, by 2% to 21% depending on location and length and severity of stenosis, so the decision to bypass or leave a native vessel with intermediate stenosis should cautiously be considered (33).

The CABS patient's postoperative education should include wound care, appropriate management of recurring symptoms, and risk factor modification.

History and Future of PTCA Surgery

The first PTCA surgeries used only a balloon and no stent; this resulted in poor restenosis rates compared with today. Bare metal stents were first used in the early 1980s, and dual antiplatelet medications were required in an attempt to prevent restenosis. The introduction of new antiplatelet drugs, such as prasugrel, reduced endothelial hyperplasia (scarring around the stent, leading to partial or complete restenosis). Administration of anticancer drugs to reduce mitosis around the stent solved the problem of restenosis but introduced a new problem: Stents were designed to become part of the endothelium, and antimitotic medication stopped the endothelialization of the stents, causing late stent thrombosis. A meta-analysis has shown that asking patients to continue antiplatelet medication beyond 3 mo, preferably for 12 mo, reduces this complication (30), although <12 mo of drug therapy is likely to be required if bioabsorbable stents are used.

A 2017 meta-analysis suggests that PTCA with drug-eluting stents is a safe and durable alternative to CABS for the revascularization of unprotected left main coronary artery stenosis in select patients. Long-term follow-up showed mortality, stroke, and MI rates were similar in PTCA with drug-eluting stents versus CABG, but PTCA was associated with higher rates of repeat revascularization (31).

Research Focus

Improving Stent Outcomes

A review (26) and meta-analysis (27) found mixed results when comparing restenosis and thrombosis rates for bioabsorbable-versus drug-eluting-stents, so it seems further work is required to refine stent design and improve outcomes.

Percutaneous Transluminal Coronary Angioplasty

After PTCA, restenosis occurs in 30% to 50% of patients undergoing simple balloon angioplasty and in 10% to 30% of patients who receive an intravascular stent (34). Following the development of drug-eluting stents, restenosis rates have dropped dramatically in many cases, to below 5% after 2 yr (35). Table 15.1 lists several predictors of restenosis (36).

Other very important potential predictors include comorbidities such as diabetes mellitus and whether patients are on optimal medical therapy that includes use of aspirin and thienopyridines such as Plavix (35).

Patients who have had PTCA in the setting of unstable angina should have close surveillance after hospital discharge and should be advised to seek prompt medical attention in the event of a recurrence of the symptoms that were occurring before their PTCA (37).

Stent Therapy

Improved technology currently confers procedural success rates in excess of 95% in most centers. Acute closure and restenosis remain as limitations to short- and long-term success, respectively, although the incidence of both these complications has decreased dramatically in recent years. The incidence of thrombosis and acute closure is in the range of 1% to 2% with use of thienopyridines (clopidogrel or ticlopidine); therefore, chronic **anticoagulation therapy** is no longer required. Restenosis rate ranges are higher with bare metal stents compared with drug-eluting stents (38). These results vary depending on comorbidities and efficacy of aggressive medical management as well as on surgical time, type of surgery, and on- or off-pump surgical methods.

Hybrid Coronary Artery Revascularization

Hybrid coronary artery revascularization (HCR) refers to complete or near-complete revascularization using the combination of single-vessel coronary artery bypass graft (CABG), with the left internal mammary artery (LIMA) placed to the left anterior descending coronary artery (LAD), and percutaneous coronary intervention (PCI) of significant coronary lesions in other vessels. The largest observational study compared 200 patients who underwent HCR with 98 patients who underwent multivessel PCI (39). At 12 mo, there was no difference in the rate of major adverse cardiac and cerebrovascular events. HCR is a reasonable approach to multivessel coronary artery revascularization in selected patients at facilities with significant expertise.

Table 15.2 provides a summary of medications commonly prescribed.

Exercise Testing

The graded exercise test (GXT) is commonly used for continued diagnosis of possible ischemic myocardium, prognostication, and the establishment of functional status for exercise prescription purposes. Although an integral component for exercise prescription, the timing of the GXT is somewhat controversial. Standard administration procedures and contraindications to testing, discussed in chapter 5, should be followed. Practical application 15.1 outlines exercise testing for revascularized patients.

Coronary Artery Bypass Surgery

It is now recommended that all patients undertake a baseline exercise test after successful bypass surgery for the purpose of beginning a supervised and monitored exercise program, but this occurs in only about 30% of referred patients (42), probably because of cost implications. At the very least a submaximal tolerance test should be conducted to determine heart rate and blood pressure, and preferably ECG, responses for the purposes of precision in exercise prescription. The patient's coronary anatomy is known, and unless surgical complications or postsurgical symptoms are present, the chance of detecting unknown ischemia is extremely low. In addition, because of the acute convalescent period, the patient may not be able to give a physiological maximal effort, sacrificing test sensitivity.

A more opportune time for testing the patient is after incisional pain has resolved, blood volumes and hemoglobin concentrations

Table 15.1   Predictors of Restenosis After PTCA and Stent

PTCA	Stent
Degree of residual stenosis after PTCA	Lesion eccentricity
Diameter of the parent vessel	Diameter of the parent vessel
Number of diseased vessels	Type of vessel stented (artery vs. vein)
Degree of reduction of the stenosis	Location of stent in vessel
Presence or type of coronary dissection	Presence of multiple stents
Presence of documented variant angina	Recurrence of unstable angina
Presence of comorbid disease (e.g., diabetes, hyperlipidemia)	Presence of comorbid disease (e.g., diabetes, hyperlipidemia)
Optimal medical therapy and adherence	Optimal medical therapy and adherence

Note: All predictors are positively associated with risk for the revascularized vessel to reocclude.

PTCA = percutaneous transluminal coronary angioplasty.

Table 15.2 Pharmacology

Medication name and class	Primary effects	Exercise effects	Important considerations
β-blockers Atenolol, metoprolol, propranolol, carvedilol*	Used for hypertension, angina, arrhythmias; increases atrioventricular (AV) block to slow ventricular response	Lower heart rate and blood pressure at submaximal and maximal exercise; increase exercise tolerance in patients with angina and heart failure	Traditional heart rate prescriptions are not valid; use of rating of perceived exertion (RPE) may be more desirable
α-blockers	Used for hypertension, chronic heart failure, noncardiac treatment of benign prostatic hyperplasia, Raynaud's syndrome, erectile dysfunction, and anxiety and depression	Possible postexercise hypotension or reflex tachycardia	Side effects include induction of reflex tachycardia, orthostatic hypotension, or heart palpitations via alterations of the QT interval and possible onset of cardiac failure; may reduce serum lipids
Nitrates Nitroglycerin, isosorbide (can come in pills, sprays, and patches)	Used for angina; vasodilator	May decrease heart rate (HR) with exercise; occasionally no effect	Hypotension; may increase exercise tolerance in patients with angina
Calcium channel blockers	Used for angina, hypertension; increases AV block to slow ventricular response	Decreases in or no effect on submaximal and maximal heart rates; decreases in blood pressure at rest and with exercise	Increase exercise tolerance in patients with angina
Digitalis Digoxin	Increases contractility; used primarily in chronic heart failure	Decreases in resting heart rate in people with atrial fibrillation and heart failure; not significantly altered in patients with sinus rhythm	Exercise tolerance improvement in patients with atrial fibrillation and heart failure only; can cause ST depression on ECG leading to false-positive stress test results
Diuretics Hydrochlorothiazide, furosemide, triamterene, spironolactone	Used for edema, chronic heart failure, certain kidney disorders	No effects on resting or exercise heart rates; decrease resting blood pressure but not with exercise	No effects on exercise response except for possibly in patients with heart failure; watch for increased ventricular ectopy due to hypokalemia and hypomagnesemia
ACE inhibitors: captopril, enalapril, lisinopril	Used for hypertension, coronary artery disease, chronic heart failure, diabetes, chronic kidney disease	No effects on exercise heart rate; may lower exercise blood pressure	Some improvement in exercise tolerance in patients with heart failure
Angiotensin II receptor antagonists (ARBs)	Used for hypertension and heart failure, especially in people intolerant of ACE inhibitors	No effects on exercise heart rate; may lower exercise blood pressure	Equivocal evidence that ARBs increase risk of myocardial infarction
Antiarrhythmic agents Procainamide, lidocaine, flecainide	Specific for individual drug but may be used for suppression of arrhythmias such as atrial fibrillation	May increase HR at rest; may decrease blood pressure at rest; typically no effects on exercise HR or blood pressures	Watch for QRS widening with exercise; some agents may cause false-negative stress tests (e.g., quinidine)
Antilipemics and statins Fenofibrate, atorvastatin, lovastatin, simvastatin, ezetimibe	Used for elevated blood cholesterol, triglycerides, and metabolic syndrome	No effects on heart rate and blood pressure at rest or with exercise	Myalgia may result; sometimes difficult to discern whether muscle pain is from exercise or statins
Blood modifiers (anticoagulant or antiplatelet) Clopidogrel, prasugrel, cilostazol, pentoxifylline	Prevent blood clots, heart attack, stroke, and intermittent claudication	No effects on heart rate and blood pressure at rest and with exercise	Cilostazol and pentoxifylline may increase ability to walk in people with claudication from PAD

*Carvedilol has β as well as α-blocking properties.

Practical Application 15.1

EXERCISE TESTING REVIEW FOR SURGERY PATIENTS

For severe regurgitant (leaking) or stenotic (narrowing) valve disease, valve replacement may be essential for symptom relief and improved exercise tolerance (40). Regarding the open-heart surgical process, precautions similar to those for the bypass surgery patient should be followed, with attention to a few special considerations. The introduction of trans-catheter aortic valve implantation (TAVI) has reduced the incidence of open-heart valve surgery. TAVIs are introduced via a femoral artery incision or by minimally invasive surgery. An initial problem with insertion was difficulty placing the TAVI because of movement associated with the beating heart. A pacing wire is used to create ventricular tachycardia; perhaps counterintuitively, increasing heart rate reduces cardia movement and allows precise insertion of the TAVI. Once the TAVI is correctly placed it is deployed. Minimally invasive surgery reduces time spent in the intensive care unit (3a).

Test type	Mode	Protocol specifics	Clinical measures	Clinical implications	Special considerations
Cardiorespiratory	Treadmill Cycle (if treadmill not possible)	Bruce or Ellestad (for younger or more physically fit individuals) Naughton or Balke-Ware (for older, deconditioned, or symptomatic patients) Ramping protocols appear more tolerable for many patients Pharmacologic testing for those unable to exercise	Heart rate and rhythm Blood pressure 12-lead electrocardiogram Symptoms Rating of perceived exertion Nuclear or echocardiographic imaging as prescribed Gas exchange analysis as prescribed	Rhythm disturbances Hemodynamics Myocardial ischemia Ischemic threshold Perception of work difficulty Ischemic myocardium Left ventricular function Dyspnea with exertion	CABS: • Chest and leg wounds (4-12 wk for complete healing) • PTCA and stent • Reocclusion (recurrence of previous symptoms)
Strength	Isometric Isotonic Isokinetic	Peak force RM procedures as described in the text 3RM-10RM dependent on patient conditioning, experience, and clinical condition Peak torque	Maximal strength of the muscle or muscle group tested Blood pressure Heart rate Symptoms	Functional fitness	CABS: • Incisional healing • No Valsalva
Range of motion	Trunk flexion Shoulder flexion, extension, abduction	Sit-and-reach test Goniometer	Posterior leg and lower back flexibility Shoulder flexibility	Functional fitness Maintenance of ability to perform ADLs after sternotomy	Orthopedic complications that may preclude testing

ADLs = activities of daily living; CABS = coronary artery bypass surgery.

Special considerations: Symptom resolution may not occur immediately after surgery. Symptoms may gradually resolve because of heart remodeling after valve replacement (e.g., for aortic stenosis). Avoid strength training for severe stenotic or regurgitant valvular disease. Several meta-analyses have shown isometric exercise may offer an opportunity to manage hypertension in patients who cannot perform, or wish to use an adjunct to, aerobic exercise (41). Common exercise-induced symptoms include shortness of breath, fatigue, and dizziness or light-headedness.

Exercise prescription: Follow procedures similar to those for revascularization surgery.

have normalized, and skeletal muscular strength and endurance have improved from participating in low-level exercise and activity. At least 4 wk postsurgery, the patient will be able to give a near-maximal physiological effort, providing test results with greater diagnostic accuracy for assessing functional capacity, determining return-to-work status, or recommending the resumption of physically vigorous recreational activities. For patients whose surgical revascularization was not successful or who are experiencing symptoms suggestive of ischemia, a clinical exercise test before starting an exercise program is recommended. All testing procedures should follow professional guidelines (43, 44).

Percutaneous Transluminal Coronary Angioplasty

Debate exists regarding the proper timing of stress testing in PTCA patients. One study used exercise testing 6 mo post-PTCA and found 32% of patients presented angina during the exercise test before the procedure and 19% afterward. The exercise test for the detection of restenosis or new lesions presented 61% sensitivity, 63% specificity, 62% accuracy, and 67% and 57% positive and negative predictive values, respectively. In patients without restenosis, the exercise duration after percutaneous coronary intervention was significantly longer (460 ± 154 s vs. 381 ± 145 s). Only the exercise duration permitted identification of patients with and without restenosis or a new lesion (45).

The mechanisms for the apparently abnormal test responses are unclear but are possibly related to the following:

1. Higher levels of **platelet aggregation** during exercise testing
2. An increase in thromboxane A2
3. Platelet activation and hyperreactivity increase during exercise
4. Increased arterial wall stress associated with increased coronary blood flow

A meta-analysis found that some studies reported a reduction in platelet aggregation in the postexercise period, compared with nonexercise controls. However, other studies reported intensified platelet aggregation or platelet activation, while some found no difference. Exercise also seemed to impair the effect of aspirin during or shortly after exercise. The researchers concluded that strenuous short-term exercise increases platelet activation, also implying a reduced effect of aspirin during short-term exercise (46).

On the other hand, exercise testing of patients with PTCA has been accepted standard practice, particularly for those with incomplete revascularization. Guidelines support the use of early postprocedure exercise testing to evaluate the functional status of the PTCA patient (47, 48).

Stent Therapy

Controversy with regard to safety of early testing after stent placement has essentially been laid to rest. Most authorities now accept that performing a stress test after coronary stenting is safe. There has been increasing support for exercise testing before the start of an exercise program (49, 50, 51). The prognostic accuracy of these tests, however, can still be debated, but it improves if peak $\dot{V}O_2$ is measured (52). Exercise testing is especially useful for patients who either cannot participate in a supervised program or choose to exercise independently and need some reassurance and initial guidance. The primary concerns with PTCA and stents are **reocclusion** and **restenosis**. Subsequent restenosis may not be detected immediately after the procedure; however, testing allows assessment of functional capacity and prediction of return to work (53). Additionally, in those undertaking exercise training, an improvement in functional capacity may be indicative of lumen patency retention (54).

Practical application 15.2 provides advice on helping revascularized patients adhere to a long-term exercise program.

Practical Application 15.2

CLIENT–CLINICIAN INTERACTION

Medical advances have come a long way to assist the clinical exercise professional in helping individuals return to exercise and more active lifestyles sooner. Despite all these advances, a pooled data analysis reported that the majority of revascularized patients eventually return to a more sedentary lifestyle (55). The field of behavior change has started to become an increasingly important part of helping these individuals maintain their exercise programs and more active lifestyles.

One of the most important tools you can use to increase exercise adherence is taking the time to listen to your patients. Understand what it is they have gone through. Listen to their fears and concerns when you are testing or designing exercise programs. Understand their physical, emotional, and environmental barriers. If you take the time to consider their sources of support and their resources for activity and exercise, and then remember to reassess their goals, to be flexible, and to follow up in a timely fashion, they will be more likely to succeed.

Ischemic Preconditioning

A number of meta-analyses have noted the benefits of ischemic preconditioning before planned cardiac surgery. Ischemic preconditioning is exposure to four cycles of 5 min of cuff inflation at suprasystolic pressure (often 200 mmHg), with 3 to 5 min of recovery between inflations. Reduced incidence and size of peri- and postoperative myocardial infarcts have been reported, among other acquired benefits. Repeated exposure to ischemic preconditioning has also been investigated for managing hypertension and wound healing (56).

EXERCISE PRESCRIPTION AND TRAINING

The average length of hospital stay for cardiovascular patients has decreased dramatically. Currently, the hospital stay for uncomplicated cases of CABS is usually 2 to 5 d. For PTCA stents, the stay is 1 or 2 d; this procedure is also done on an outpatient basis, with the patient managed in an acute recovery suite and discharged on the same day. Although cardiac rehabilitation begins as soon as possible during hospital admission, the shorter length of hospital stay has changed the inpatient program to basic range of motion exercises and ambulation; and the educational focus is on discharge planning—teaching about medications, home activities, and follow-up appointments. Educational topics previously covered in the inpatient setting are now the responsibility of the outpatient program. Moreover, cardiac rehabilitation professionals must make every effort to enroll patients in an outpatient program. Interval training may offer an alternative to continuous exercise in severely deconditioned patients. Later, in phase II and beyond, high-intensity interval training (HIIT) may offer superior benefits to moderate-intensity continuous exercise (58, 59). A hybrid approach to cardiac rehabilitation may offer a compromise in those cautious about using HIIT (60). Table 15.3 reviews the exercise prescription for the revascularized patient.

Incident sarcopenia during hospital stay is relatively common and is associated with nutrition status and the number of days of bed rest (61), supporting the need for early, phase I exercise intervention during hospital admission in the revascularized cardiac patient (62). Patients who perform typical ward activities and moderate, supervised ambulation do not suffer the magnitude of loss in lean body tissue seen in those who remain inactive. Early standing and low-level activities, including range of motion and slow ambulation, may be all that is required to deter postsurgical lean body mass loss (63).

After hospital discharge, many positive physiological adaptations occur in revascularized patients who participate in a supervised exercise program (64):

- Improved cardiac performance at rest and during exercise
- Improved exercise capacity (aerobic and strength)
- Greater total work performed

- Improved angina-free exercise tolerance (65) and coronary flow reserve (66)
- Improved neurohumoral tone (67)

Patients in such a program gain in several ways:

- They more often achieve full working status.
- They have fewer hospital readmissions.
- They are less likely to smoke at 6 mo following completion of exercise therapy (68).

When we compare the physiological and psychosocial outcomes between CABS, PTCA and stent, and MI patients at the beginning and end of 12 wk of cardiac rehabilitation, some group trends are apparent. CABS patients may begin with lower functional capacities and lower ratings of quality of life and self-efficacy attributable to the surgical recuperative process (69) but they show greater improvement during the program and obtain similar or greater values compared with other cardiac patients at program completion, regardless of age (70) This result may reflect lower rates of **ischemia** than in MI patients, greater confidence in their own ability, and the potential psychological feeling that something was done about their heart disease and that they are "cured." Regardless of age, the CABS patients demonstrate functional improvement but may require a longer training period to obtain the same magnitude of effect (71). The PTCA and stent groups have not suffered the loss in functional capacity because of the more prolonged recuperative process following an MI or bypass surgery and have better clinical status when starting cardiac rehabilitation (72).

Exercise prescription guidelines for **revascularization** patients have been jointly published by the American College of Cardiology and American Heart Association (73).

Special Exercise Considerations

Although revascularized patients are just as knowledgeable about risk factors as post-MI patients, they are less compelled to make changes (74). Post-MI patients are motivated to effect lifestyle change (75). Patients undergoing revascularization may be less motivated to adhere to risk factor behavior change because of a perception that they are less sick or have been cured, which has a negative effect on compliance with risk factor modification (74).

Revascularized patients may encounter, or anticipate, restrictions differently than other cardiac patients do. Most PTCA and stent patients are capable of resuming normal activities of daily living after hospital discharge, but patients frequently perceive considerable restrictions after the procedure with respect to all activities of daily living—leisure activities, sexual activity, and early return to work (76).

Depression is prevalent in patients with coronary heart disease after major cardiac events (CABS and PTCA included). Cardiac rehabilitation does reduce the prevalence and severity of depression. Therefore, cardiac patients should be routinely screened and offered the benefits of comprehensive cardiac rehabilitation,

Table 15.3 Exercise Prescription Review

Training method	Frequency	Intensity	Time (Duration)	Type (Mode)	Progression	Important considerations
Cardio-respiratory	Daily is ideal; at least 4 d/wk or 3 d/wk for HIIT	Asymptomatic: 40%-85% of HRmax, RPE: 11-16 Symptomatic: below ischemic or anginal threshold, RPE: 14-17, or 85%-95% HRmax for HIIT	At least 30 min; may be intermittent (e.g., 3 × 10) or continuous depending on patient tolerance or 4 × 4 min with 3 min recovery for HIIT If time is short, 1 × 4 min has also been shown to improve peak $\dot{V}O_2$	Treadmill, walking, cycling, combined arm and leg exercise, rowing, stair stepper, swimming, elliptical; combination of above or others to ensure adequate utilization of major muscle groups and distribution between upper and lower extremities	If appropriate, start with some ambulation in hospital; first couple of weeks posthospitalization, progress to 5-10 min of very light intensity multiple times a day; after 4 wk posthospitalization, increase intensity to moderate levels and increase time to 15-30 min 1 or 2 times a day; at 6 wk or more posthospitalization, should be working toward 30 min or more at moderate or better intensities or slowly raise workload for HIIT.	Initially, need to be concerned with incisional discomfort in chest, arm, and leg of surgical patient. May need to restrict upper extremity exercises until soreness resolves. Also, those with PTCA may have some groin soreness at the catheter insertion site that may restrict certain physical movements. Avoid HIIT in people who exhibit arrhythmia at near-maximal HR or symptoms of angina or abnormal blood pressure responses with exercise.
Resistance	2 or 3 d/wk	Select a weight such that the last repetitions feel somewhat or moderately hard without inducing significant straining (bearing down and breath holding)	12-15 reps, slowly increasing weight and intensity so that 8-10 reps provide the appropriate response	Elastic bands, hand weights, free weights, multistation machines; equipment selection based on patient progress (a rational progression is to use the equipment in the order listed)	Generally, in the first 4 wk posthospitalization, CABS patients are asked to use little to no resistance and primarily perform ROM exercises and some strengthening exercises that do not produce significant strain on the incision site. PTCA patients can generally start a bit earlier using light weights that can be completed for 12-15 reps without producing a Valsalva effect. After 4 wk posthospitalization, CABS patients can start to increase the amount of weight—can start with 12-15 reps without Valsalva and should not have any clicking or grinding of the sternotomy. PTCA patients should look to increase efforts to moderate levels at this point, and ultimately all revascularization patients should progress to multiple muscle group exercises, using enough resistance to produce muscular fatigue in 10-12 reps. Days per week and sets of each exercise should be individualized according to patient needs and goals.	For surgical patients, the initial upper extremity exercises may be range of motion without resistance—progressing initially with elastic bands or 1-3 lb (0.5-1.5 kg) increments. Slightly higher weight may be employed for muscle groups and movements that do not put sternal healing at risk. Further progression depends on sternal healing and stability. Exercises should be selected that employ muscle groups involved in lifting and carrying.

Training method	Frequency	Intensity	Time (Duration)	Type (Mode)	Progression	Important considerations
Range of motion	Daily	Static stretching	5-10 min	Range of motion and flexibility exercises	In hospital and immediately posthospitalization, CABS patients should be doing daily ROM exercises to the point of mild stretch on the sternum but should not feel any pain. As they continue to heal, the exercises should continue to progress to greater ranges of motion to tolerance. PTCA patients should resume or start a daily stretching routine without restrictions.	Exercises should emphasize major muscle groups, especially lower back and posterior leg muscles.

CABS = coronary artery bypass surgery; HIIT = high-intensity interval training; HR = heart rate; PTCA = percutaneous transluminal coronary angioplasty; RPE = rating of perceived exertion; ROM = range of motion.

including psychosocial support and counselling (77). There has also been some research on preprocedure exercise and counseling (prehabilitation) (78, 79) and extended telephone counseling after the procedure and rehabilitation sessions have ended; these studies have shown promise in reduction in depression surrounding revascularization procedures (80).

Active spouse support during the rehabilitation program may improve participation rates, but in the secondary prevention program, active spouse support may encourage early dropout (81). Also, patients and spouses differ in their views on the causes of CAD and about the responsibility for lifestyle changes and the management of health and stress (82). Therefore, assessing both the patient's and the spouse's educational needs is important. The sidebar Health Care Checklist for PTCA and Stent Patients lists important concerns in the rehabilitation of the post-PTCA patient, taking into account that exercise may be contraindicated for patients who continue to be symptomatic postevent or postprocedure, particularly at relatively low workloads.

Coronary Artery Bypass Surgery

Primary concerns for the CABS patient when entering outpatient cardiac rehabilitation are the state of incisional healing and sternal stability, hypovolemia, and low hemoglobin concentrations. During the initial patient interview, the rehabilitation professional needs to ensure that the surgical wound has no signs of infection, significant draining, or instability. Questions should focus on the following:

- Excessive or unusual soreness and stiffness
- Cracking, grinding, or motion in the sternal region
- Whether the patient is sleeping at night
- How the patient's chest and leg incisions are responding to current activities of daily living since discharge

Also, knowing how patients performed during the inpatient program may help determine how soon they can begin the outpatient program and at what level they can begin exercising. For example, was the patient out of bed, upright, and walking soon after surgery without problems? If not, was the patient's lack of activity attributable to extreme physical discomfort, clinical or orthopedic difficulties, or lack of motivation? Research has shown that chest soreness can reduce walking distance, which can have an impact on functional fitness (62).

Other Conditions

Cardiac disease progresses more rapidly in people with diabetes. Diabetes worsens or accelerates atherosclerosis and can lead to autonomic nervous system (ANS) deficits, which can lead to cardiac arrythmia or chronotropic incompetence. Additionally, people with diabetes are more likely to experience angina or myocardial infarction without chest pain if their ANS is impaired. The lack of pain as a warning sign often carries with it poorer prognosis. Another consideration is the potential uncovering of claudication in peripheral artery disease (PAD) with ambulation after revascularization (83). It is well established that persons with CAD are at higher risk than others for PAD (and vice versa). This may be more of an issue with patients who have been very sedentary and then start a new exercise regimen after the procedure and develop symptoms of claudication.

Health Care Checklist for PTCA and Stent Patients

- Control of hypertension, obesity, and smoking
- Progressive exercise and weight reduction
- Awareness of other cardiac risk factors
- Identification of stressful factors
- Counseling services for weight reduction, stress management, and smoking cessation
- Maintaining close contact between health professionals
- Organizing and maintaining long-term follow-up records
- Reinforcing the noncurative nature of PTCA and stent as cardiac treatment modality
- Encouraging revascularization patients and PTCA and stent patients to take a proactive approach to improve health outcomes

Historically, surgical patients did not begin cardiac rehabilitation for 4 to 6 wk postsurgery or longer and avoided upper extremity exercise for even longer periods. Today, standard practice is for patients to begin the outpatient program prior to, or soon after, discharge, often within a week of surgery. For the uncomplicated revascularized patient, light upper extremity exercises are now prescribed, including range of motion exercises, light hand weights progressing to light resistive machinery, and gradually progressive upper extremity ergometry beginning at zero resistance. Research indicates that women tend not to respond to inpatient cardiac rehabilitation as much as men do (84).

Percutaneous Transluminal Coronary Angioplasty

The primary concern for the PTCA or stent patient is restenosis. At the patient's initial orientation session, questioning should address the presence of signs or symptoms indicative of angina or the person's particular anginal equivalent. Education should include information about symptoms, including anginal equivalents; management of angina (e.g., how to use nitroglycerin, going to the emergency department); precipitating factors (exertion or anxiety related); and care of the catheter insertion site. Patients with PTCAs and stents may begin the outpatient program as soon as they are discharged from the hospital or immediately following the procedure if it has been performed on an outpatient basis (62).

The combination of exercise rehabilitation and education may alleviate the progression of coronary artery stenosis after PTCA by improving endothelial function (85), lowering serum lipids, and improving insulin resistance and glucose intolerance (86).

Traditional aerobic training for 30 to 40 min, four to six times per week for 12 wk, improves treadmill time and myocardial perfusion and may reduce the late (9 mo) restenosis rate after PTCA (54). The use of repeated shorter intervals of vigorous- or high-intensity exercise has been trialed with promising results (87, 88).

As a result of angioplasty with improved techniques of revascularization, more patients with low-risk profiles are being referred to cardiac rehabilitation (i.e., patients with a greater exercise capacity, no evidence of ischemia, normal left ventricular function, and no arrhythmias). Specific examples include patients who are younger, have single-vessel disease, and did not experience an MI before their PTCA. Regarding exercise prescription, these individuals may be treated similarly to apparently healthy individuals with the addition of education concerning the recognition of anginal equivalents, self-monitoring, self-care, and risk factor modification. Optimized medical therapy along with appropriately prescribed exercise training can be an alternative approach to interventional strategies in selected cardiac patients who are asymptomatic (89). When PTCA is the therapy of choice, it should be combined with daily physical exercise and increased physical activity to optimize success (13). Cardiac rehabilitation results in early and sustained improvement in quality of life and is cost-effective (90, 91). In addition, angioplasty patients commonly experience restenosis. Supervised exercise training and education improve recognition of signs and symptoms associated with closure. Most important, angioplasty patients need instruction concerning appropriate exercise training, dietary modifications, medications, and general risk factor reduction to slow or reverse the coronary disease process.

Because the PTCA patient remains on complete bed rest while the sheath is in situ for approximately 18 to 24 h, the immobilization often causes back pain. Appropriate flexibility exercises that enhance range of motion often help to resolve low back pain.

Stent Therapy

Placement of stents uses the same catheter procedure as in the PTCA, so the same considerations exist. But the risk for thrombosis is greater following stent therapy. Consequently, patients are often placed on anticoagulant therapy for preventive purposes. Although no specific contraindications preclude exercise after recent stent placement, proceeding with similar caution is prudent.

Immediate Postsurgical Exercise Advice

CABS patients are very protective of their sternotomy sites and are sometimes reluctant to do range of motion and stretching exercises. Be very persistent in ensuring they are performing exercises to restore normal function to upper body movements as soon as possible to limit future issues with upper body strength and flexibility. Reduced arm swing will reduce stride length and hence walking speed; if not corrected relatively quickly, this could have an adverse effect on weekly exercise.

Equipment and Home Exercise

Access to equipment that can be titrated at minute increments in workload is very important as cardiac patients have low levels of fitness, especially in the immediate postoperative period, so equipment such as stationary cycles may need to be adjusted in relatively small 2 Watt increments in workload to accommodate severely deconditioned patients.

As patients progress they may wish to transition to home exercise. Elastic exercise bands may allow strength to be maintained through resistance exercises.

Medications and Heart Rate

Revascularized patients sometimes feel cured of their disease; the rehabilitation team needs to reinforce proper lifestyle and medical management to ensure continued lower risk of future cardiac events and hospitalizations.

Many revascularized patients are on β-blocker medications, which makes exercise prescription by heart rate difficult if not impossible. It is better to use rating of perceived exertion; spend time helping them "feel" how hard to be exercising when they are on their own, and help them get some practical experience in understanding the sensation of proper exercise intensity. Related to this, about half of all cardiac patients may exhibit an inability to raise heart rate to correspond with exertional requirements; this is called chronotropic incompetence (CI) (99).

Disease Progression

People who need revascularization have ischemic heart disease (IHD), which often leads to worsening symptoms with time. IHD is the leading cause of chronic heart failure (CHF). The etiology of CHF is IHD in >70% of cases, and patients can decompensate into their first experience of heart failure rather suddenly. Exercise can be one cause of decompensation, although exercise should not be avoided by those with IHD and CHF; in fact, it should be encouraged, but not when the person is unstable (angina at rest or uncontrolled heart failure). One of the key signs that a person with IHD is decompensating is sudden weight gain of 4 lb (1.8 kg) or more in less than 48 h. It is therefore advisable to weigh patients before each exercise session and keep a record of this, along with their resting blood pressure and heart rate. If any of these measurements is higher than the normal pattern shows (refer to that person's exercise session recorded data), then it may indicate decompensation. If pops or crackles are noted upon auscultation at the base of the lungs, this may indicate fluid retention on the lungs, and it may be wise to postpone exercise until the person has been assessed by a physician.

Mental Health

Finally, although many revascularized patients struggle with depression, some are on the other end of the spectrum and feel better than they have in years; these patients may be too aggressive in progressing exercise.

Aerobic Exercise

The multiple improvements in patients' tolerance to acute bouts of exercise after revascularization include the following (92):

- Improved myocardial blood flow
- Increased functional capacity
- Improved cycle ergometer or treadmill performance
- Variable improvements in left ventricular function
- Increased maximal heart rate
- Increased rate–pressure product
- Reduced ST-segment depression
- Relief or improved anginal symptoms with exercise
- Improved heart rate recovery
- Reduction in exertional hypotension

The initial exercise prescription is based on information gained from the patient's orientation interview for the outpatient cardiac rehabilitation program. Patients are questioned concerning the presence of signs or symptoms, their activity while in the hospital, and their activity level since their return home from the hospital. Of equal importance are their level and consistency of conditioning before their cardiac event. Depending on how long they were in the hospital and the amount of deconditioning, better-conditioned, higher-functioning patients may be able to return to higher levels of exercise volume and intensity more quickly than most patients. Initially, patients are closely observed and monitored to establish appropriate exercise intensities and durations that are within their tolerance.

A starting program may include treadmill walking (5-10 min), cycle ergometry (5-10 min), combined arm and leg ergometry (5-10 min), and upper body ergometry (5 min). Initial intensities may approximate 2 to 3 METs (multiple of resting oxygen uptake of $3.5 \text{ mL} \cdot \text{kg}^{-1} \cdot \text{min}^{-1}$), but starting MET levels may be a bit higher depending on exercise history and prior conditioning. METs are an absolute, not relative, workload. Many cardiac patients have a maximal exercise capacity of 3 to 4 METs, so a low to moderate percentage (40%-60%) of maximal heart rate or a moderate (11-14 out of 20) rating of perceived exertion may be a good starting point (93). Because many cardiac patients have low functional capacity, especially after cardiac surgery, their daily general living activities probably exceed their maximum cardiorespiratory fitness (peak $\dot{V}O_2$). For this reason, in people who are stable, HIIT may be indicated to quickly improve fitness; over time, frequent exposure to vigorous high-intensity activity has been shown to reduce MI risk (94). A common exercise prescription is four work intervals of 4 min, with 3 min rest or recovery between each work interval. An intensity of 85% to 95% peak $\dot{V}O_2$ is commonly used (89), with a 5 to 10 min warm-up, 5 min cool-down, and recovery bouts all completed at about 55% to 65% peak $\dot{V}O_2$.

The patient's heart rate, blood pressure, rating of perceived exertion, and signs and symptoms are monitored and recorded.

Programs are gradually titrated during the initial sessions to a rating of perceived exertion of 11 to 14 in the absence of any abnormal signs or symptoms.

In general, exercise intensity is progressed by 0.5 to 1.0 MET increments (i.e., 0.5 mph [0.8 kph] or 2.0% grade on the treadmill or 12.5-25.0 Watts on the cycle). The rate of progression is based on the patient's symptoms, signs of overexertion, rating of perceived exertion, indications of any exercise-induced abnormalities, and prudent clinical judgment on the part of the cardiac rehabilitation staff. Patients with greater exercise capacities (PTCA and stent patients with no MI) are started according to their exercise capacities and progressed more rapidly. The selection of exercise modality depends on the person's program objectives. For example, those who are employed in a labor-type occupation or perform many upper extremity activities at home spend a greater portion of their exercise time doing upper extremity exercises. If specific limitations preclude certain exercise modalities, program modifications are made that allow more time on tolerable equipment to obtain the greatest cardiorespiratory and muscular advantage. For patients who have an exercise test, standard recommended procedures for exercise prescription are followed (37).

Resistance Training

Muscular strength and endurance exercise training should be incorporated equally with cardiorespiratory endurance and flexibility exercise training during the early outpatient recovery period. After revascularization, low-risk patients can perform muscular strength and endurance exercise training safely and effectively (95). Depending on the patient's clinical and physical status, successful approaches for upper and lower extremity strength enhancement include 10 to 12 repetitions with a variety of types of equipment that may include elastic bands, Velcro-strapped wrist and ankle weights, hand weights, and various multistation machines. Usual guidelines include maintenance of regular breathing patterns (avoiding the **Valsalva maneuver**), selection of weights so that the last repetition of a set is moderately or somewhat hard, and progression when the perception of difficulty decreases. CABS patients may start range of motion exercises with light weights of 1 to 3 lb (0.5-1.5 kg) within 4 wk of surgery as long as sternal stability is ensured and excessive incisional discomfort is not present. PTCA and stent patients may start resistance training immediately. Exercises should be selected that will strengthen muscle groups used during normal activities of daily living for lifting and carrying and occupational or recreational tasks.

Weights are selected that allow the completion of 12 to 15 repetitions initially; the patient then progresses to higher weights and 10 to 12 repetitions, with the last 3 repetitions feeling moderately hard. Those who cannot securely hold hand weights should use wrist weights with Velcro straps. Typically, exercises are selected that use upper and lower extremity muscle groups involved in routine lifting and carrying and other activities of daily living. Patients are progressed from stretch bands and light hand weights to resistance machines, again using resistances that result in a perception

of difficulty of moderately hard for two or three sets of 10 to 12 repetitions. Figures 15.2 through 15.5 outline possible exercises for a rehabilitative exercise progression for revascularized patients.

Progression of exercise is based on patients with open-heart procedures; patients undergoing PTCA and stenting can perform all exercises listed for the beginning of the program with proper progression based on prior exercise history and clinical judgment of the cardiac rehabilitation team. The list that follows is based partly on information from Adams and colleagues (96).

- *Early-phase rehabilitative exercises (2-4 wk postdischarge):* Traditionally patients at this phase have been told to avoid strength training and do only range of motion or very light, if any, strength training. You may consider having patients start these exercises (see figure 15.2) earlier in their recovery if sternal healing is going well and they are ambulatory.
- *Midphase rehabilitative exercises (4-6 wk postdischarge):* As the patient progresses, these midphase exercises (see figure 15.3) can be added or can replace early-phase exercises that work the same muscle groups.
- *Late-phase rehabilitative exercises (6 wk or more postdischarge):* As the patient progresses, these late-phase exercises (see figure 15.4) can be added or can replace early- and midphase exercises that work the same muscle groups.

The potential benefits of resistance training in the revascularized population include improving quality of life (97), muscular strength, and endurance and possibly attenuating the heart rate and blood pressure response to any given workload (lower workload on the heart). General resistance training guidelines for cardiac rehabilitation are presented in the sidebar Patient's Guide for Resistance Training. Risk stratifying patients to determine eligibility is important.

Patient's Guide for Resistance Training

- Choose an initial weight you can comfortably lift for 12 to 15 repetitions; as you progress over time, increase the weight accordingly so that your muscles will get significantly tired by 10 to 12 repetitions but you are not struggling to complete them.
- Do not hold your breath during the activity. Exhale during the exertion phase, and avoid straining.
- Perform two or three sets of each exercise, and train three times per week.
- Rest 30 to 45 s between sets.
- Increase weight modestly (1-2 lb, or 0.5-1.0 kg) after you can easily perform 12 to 15 repetitions of a given weight.

Figure 15.2 Early-phase rehabilitative exercises: *(a)* seated leg extension, *(b)* standing hamstring curl, *(c)* standing toe raise, *(d)* dumbbell curl (start), and *(e)* triceps push-down.

Photos courtesy of Neil Smart.

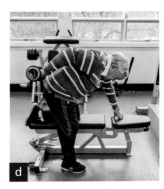

Figure 15.3 Midphase rehabilitative exercises: *(a)* dumbbell bent-over row, *(b)* seated row, *(c)* seated shoulder raise, and *(d)* triceps kickback.

Photos courtesy of Neil Smart.

Figure 15.4 Late-phase rehabilitative exercises: *(a)* lat pull-down, *(b)* seated chest press, *(c)* standing shoulder raise, and *(d)* front shoulder raise.

Photos courtesy of Neil Smart.

Range of Motion

Each exercise session begins with a series of range of motion and flexibility exercises designed to maintain or improve the range of motion around joints and maintain or improve flexibility of major muscle groups (figure 15.5). The exercises begin at the neck with the person standing (or seated in a chair if the person has difficulty standing), progressing downward to the shoulders and trunk and eventually to the lower extremities. The final stretches for the improvement of posterior leg muscles and lower back flexibility may be performed on the floor, or in a chair for those with difficulty getting to the floor.

Patients who underwent an open-heart procedure should perform stretches with caution in the early stages after hospital discharge. Patients should do the stretches daily, and ideally multiple times a day, but should take each stretch only to the point where they may feel some mild tugging on the sternotomy site; the stretch should not be painful.

Table 15.4 outlines exercise prescription for revascularized patients.

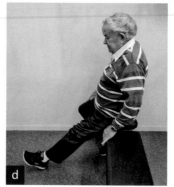

Figure 15.5   Stretching exercises: *(a)* triceps, *(b)* lower back, *(c)* neck stretch, *(d)* seated hamstring stretch, and *(e)* calf stretch.
Photos courtesy of Neil Smart.

Table 15.4   Exercise Training Review

Cardiorespiratory endurance	Skeletal muscle strength	Skeletal muscle endurance	Flexibility and balance	Body composition
Increases $\dot{V}O_2$max and improves RPE, HR, and BP at absolute submaximal levels; reduces possibility of multiple risk factors for further progression of CAD; also very important in overall CAD and all-cause mortality risk	Maintains or increases muscular strength, improves ability to perform activities of daily living, helps in return to work after hospitalization, improves self-esteem; key component in decreasing risk of injuries and fall risk in the elderly	Maintains or increases muscular endurance, improves ability to perform activities of daily living, assists in reduction of fatigue with repetitive activities	Maintains or increases flexibility, help in return to work after hospitalization; very important in proper recovery from open-heart procedures; key component in decreasing risk of injuries and fall risk in the elderly	If loss of weight and body fat is a goal, then exercise programming alone may not be sufficient to reduce body mass. In order to reduce overall body mass >1 h of exercise daily may be required. For weight loss it may also be necessary to reduce caloric intake, and this may be best achieved with the advice of an accredited dietitian.

Research Focus

Earlier Progression and Fewer Restrictions Concerning Strength Training

Guidelines for starting and progressing strength training after revascularization may occasionally be considered overly restrictive, as many tasks of daily living demand lifting or pulling objects that exceed the suggested strength training weight limits. Clinicians may use clinical judgment to start strength training at earlier stages while taking into consideration the patient's prior conditioning and experience with strength training exercises, as well as the knowledge that no absolute contraindications to strength training are present. A position stand suggests early progressive use of strength training even for cardiac patients when exercises are selected individually, with safety and efficacy of resistance exercise in mind (98). This method puts various strength training exercises into categories by their potential to cause harm to the patient (such as damage to the sternum of the CABS patient).

Clinical Exercise Bottom Line

- Programs for cardiac and pulmonary patients have been in existence since the 1970s and have a rich research base indicating improvements in function, in quality of life, and in some cases reductions in disease progression, morbidity, and mortality.

- Supervised exercise programs for cardiac, pulmonary, and peripheral artery disease are reimbursed by insurances.

- Interest has emerged in providing supervised exercise training to a wide variety of persons with other chronic diseases including cancer, obesity, and neuromuscular, metabolic, and skeletal muscle conditions.

- Most supervised programs are staffed by clinical exercise physiologists and nurses and take place in clinical settings.

CONCLUSION

Advances in coronary revascularization procedures and an aging population have led to a greater number of patients presenting for rehabilitation following CABS, PTCA, and stenting. Prehabilitation and other interventions may better prepare patients for surgery and result in better adherence to postsurgery exercise rehabilitation. In addition to exercise programming, risk factor modification is essential for prevention of recurrent events. Furthermore, barring no new symptoms, the GXT, although a good prognostic tool, may better serve its purpose of assessing functional status if it is postponed until later in the rehabilitation program. Provided that no untoward events occur over the course of rehabilitation, CABS and PTCA patients usually outperform their MI counterparts, achieving greater fitness improvements at a faster rate.

Online Materials

Visit HK*Propel* to access a link to the references, the case studies with discussion questions, and a quiz to help you review key concepts and test your understanding of the material covered in this chapter.

CHAPTER 16

Chronic Heart Failure

Steven J. Keteyian, PhD

With advancing age we are subject to a variety of disorders that disable us, impair our quality of life, and contribute to loss of exercise capacity. Perhaps chief among these is heart failure, the final common pathway for a multitude of cardiovascular disorders. Both the incidence and prevalence of heart failure have increased over the past decade, due in part to (a) the aging of our population (increasing percentage of people who reach greater than 65 yr) and (b) improved survival of patients with cardiovascular disease due to advancements in pharmacotherapy, mechanical interventions (e.g., implantable cardioverter-defibrillators), and surgical techniques.

DEFINITION

Heart failure (HF) is a clinical syndrome with signs and symptoms caused by a structural or functional cardiac abnormality, corroborated by elevated natriuretic peptide levels and objective evidence of pulmonary or systemic congestion, or both.

The inability of the left ventricle (LV) to pump blood adequately can be due to a failure of either systolic or diastolic function. Patients with HF due to **systolic dysfunction**, or the inability of cardiac myofibrils to contract or shorten against a load, have a reduced **ejection fraction (EF)** (HFrEF; heart failure due to reduced ejection fraction).

Alternatively, about one-half of patients with HF have normal or near-normal systolic contractile function but still suffer from HF (HFpEF; heart failure with preserved ejection fraction). In these patients, the disorder is not associated with an inability of the heart to contract; instead it is attributable to an abnormal increase in resistance to filling of the LV—referred to as **diastolic dysfunction**. A key characteristic of diastolic dysfunction is a stiff or less compliant LV that is partially unable to relax and expand as blood flows in during diastole.

SCOPE

The public health burden associated with HF is immense. In the United States, nearly 6.5 million people are afflicted with the syndrome, and in people >55 yr of age, approximately 1 million new cases are identified each year. Because of the aging U.S. population and increased survival of patients with cardiovascular disorders, the prevalence of HF will continue to increase over the next 10 yr (84).

Heart failure accounted for more than 800,000 hospitalizations in the United States in 2016 and directly or indirectly contributes to more than 350,000 deaths annually. The 5 yr mortality rate for a person newly diagnosed with HF is ~48%. The worldwide economic burden imposed by HF is enormous at $108 billion annually, with the United States alone accounting for almost $32 billion. Because of the increase in the number of people reaching age 65 in the next 10 yr, the cost burden in the United States is projected to increase to $70 billion by 2030.

PATHOPHYSIOLOGY

Heart failure remains a final common denominator for many cardiovascular disorders, yet the pathophysiology of HFrEF and HFpEF differ in many ways. Note that in patients with HFrEF, some degree of diastolic dysfunction is usually present as well.

Figure 16.1 depicts the complex abnormalities and changes that occur following loss of systolic function. When cardiac cells (myocytes) die because of infarction, chronic alcohol use, longstanding hypertension, viral infections, or unknown factors, diminished LV systolic function results. Among all cases of HFrEF, approximately 60% are caused by **ischemic heart disease** (i.e., coronary atherosclerosis). For this reason, these patients are commonly referred to as having an ischemic versus a **nonischemic cardiomyopathy**, where a nonischemic cardiomyopathy refers to other disease processes that affect the heart muscle (e.g., viral cardiomyopathy or alcoholic cardiomyopathy).

Figure 16.1 also shows the various physiological adaptations and compensatory changes that occur in response to LV systolic dysfunction. Most of the medical therapies used to treat patients with chronic HFrEF are aimed at modifying one or more of these abnormalities. The development of clinical HFpEF involves comorbidities (e.g., diabetes, hypertension) that promote systemic inflammation, leading to endothelial dysfunction, loss of capillarity, and mitochondrial dysfunction in the myocardium as well as other organs. There is a characteristic inability of the left ventricle to relax and fill properly, due greatly to the abnormal handling of calcium ions inside the cardiac cells and intercellular fibrosis.

Key characteristics particular to the pathophysiology of HFrEF, HFpEF, or both include the following:

1. An ejection fraction that is reduced (systolic) or unchanged or slightly increased (diastolic) at rest

2. An increase in LV mass, with end-diastolic and end-systolic volumes that are increased (systolic failure) or decreased (diastolic failure)

3. Edema or fluid retention because of elevation of diastolic filling pressures or activation of the renin–angiotensin–aldosterone system, causing sodium retention

4. More commonly with systolic (vs. diastolic) HF, an imbalance of the autonomic nervous system such that parasympathetic activity is inhibited and sympathetic activity is increased

Additionally, abnormalities of other hormones and chemicals such as an increased release of brain natriuretic peptide (BNP), diminished production of nitrous oxide (endothelium-derived relaxing factor), increased endothelin-1, and increased cytokines (e.g., tumor necrosis factor-alpha) all contribute to adverse cardiac and vascular remodeling or function and changes within and around the skeletal muscles.

Substantial clinical evidence now indicates that many of those factors contribute to the remodeling of the LV, reshaping it from a more elliptical form to a spherical form. This change in shape itself contributes to a further loss in LV systolic function. Currently, several of the treatment strategies used in patients with HFrEF interrupt this process, referred to as reverse remodeling.

In patients with HFpEF, LV end-diastolic and -systolic volumes are generally reduced, and ejection fraction may be normal or increased. Table 16.1 describes normal LV characteristics and those associated with LV systolic and diastolic dysfunction.

CLINICAL CONSIDERATIONS

Before a patient with HF can be cleared for exercise rehabilitation, their past and current medical history should be reviewed and functional status evaluated. Exercise testing can provide important information about the patient's functional status, but signs, symptoms, and medications must also be considered.

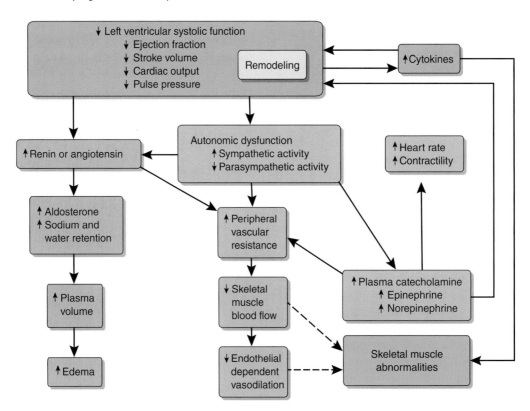

Figure 16.1 Schematic representation of some of the main physiological and pathophysiological adaptations that occur at rest in patients with heart failure attributable to left ventricular systolic dysfunction.

Table 16.1 Comparison of Typical Left Ventricle Characteristics

	End-diastolic volume (mL)	End-systolic volume (mL)	Stroke volume (mL)[a]	Ejection fraction (%)[b]
Normal	120	55	65	55
HFrEF	160	110	50	30
HFpEF	85	35	50	60

[a]Stroke volume = end-diastolic volume – end-systolic volume.

[b]Ejection fraction = stroke volume / end-diastolic volume.

HFrEF = heart failure due to reduced ejection fraction; HFpEF = heart failure with preserved ejection fraction.

Courtesy of Henry Ford Health System.

Signs and Symptoms

Clinically, patients with HF present with several key characteristics or findings, two of which are

1. exercise intolerance, as manifested by fatigue or shortness of breath on exertion; and

2. fluid retention, as evidenced by peripheral edema, recent weight gain, or both.

These signs and symptoms are often associated with complaints of difficulty sleeping flat or awakening suddenly during the night to "catch my breath." Sudden awakening caused by labored breathing is referred to as **paroxysmal nocturnal dyspnea**.

Labored or difficult breathing during exertion is called dyspnea on exertion (DOE), and difficulty breathing while lying supine or flat is referred to as **orthopnea**. Clinically, the severity of orthopnea is rated based on the number of pillows that are needed under a patient's head to prop them up sufficiently to relieve dyspnea. For example, three-pillow orthopnea means a patient needs three pillows under their head and shoulders to breathe comfortably while recumbent.

History and Physical Examination

The signs and symptoms just discussed represent findings that the clinical exercise physiologist may need to evaluate to ensure that on any given day, the patient can safely exercise. For example, a patient's complaint of increased DOE or recent weight gain may or may not be clinically meaningful, but such comments warrant further inquiries by the clinical exercise physiologist to ascertain whether they are associated with other important signs such as new or increased peripheral (e.g., ankle) edema or fluid accumulation in the lungs. As discussed in chapter 4, the severity of ankle edema is typically evaluated on a scale of 1 to 3. Lung sounds, called rales, are associated with pulmonary congestion and are best heard using the diaphragm portion of a stethoscope. Rales appear as a crackling noise during inspiration.

In patients with HFrEF, abnormal heart sounds can also be heard with a stethoscope. A third heart sound (S_3) can often be heard when the bell portion of the stethoscope is lightly placed on the chest wall over the apex of the heart. This S_3 sound occurs soon after S_2 and is most likely attributable to vibrations caused by the inability of the LV wall and chamber to accept incoming blood during the early, rapid stage of diastolic filling. Listening to audio materials is an effective way to learn normal and abnormal breath and heart sounds (see also www.blaufuss.org).

Diagnostic Testing

Although the diagnosis of HF attributable to HFrEF or HFpEF includes the presence of signs and symptoms, it also requires the measurement of characteristics of left ventricular contraction (ejection) and relaxation. Most often, these properties are measured during an **echocardiogram** or a diagnostic cardiac catheterization procedure.

Normally, LV ejection fraction is greater than 55%, which means that slightly more than one-half of the blood present in the LV at the end of diastole is ejected into the systemic circulation during systole. In patients with HFrEF, LV ejection fraction is typically reduced below 45%. Severe LV dysfunction may be associated with an ejection fraction of 40% or lower. The decrease in ejection fraction is qualitatively proportional to the amount of myocardium that is no longer functional. Chronic HFrEF is also usually associated with an enlarged LV.

In patients with HFrEF caused by a recent large acute anteroseptal myocardial infarction due to coronary artery (i.e., ischemic heart) disease, a marked increase in troponin is typically observed in the blood. In the electrocardiograms (ECGs) of these patients, changes such as Q waves are eventually observed in leads V2 through V4. In these patients, a cardiac catheterization with coronary angiography is performed to evaluate the extent of myocardial damage as well as to determine whether other blockages are present and jeopardizing additional myocardial tissue. Ischemic heart disease is the underlying cause for HF in about 60% of cases.

The diagnosis of HFpEF also requires an echocardiogram to evaluate unique myocardial characteristics (e.g., elevated diastolic filling pressures, E/e'; left atrial volume index >34 mL · m^{-2}), the severity of which is typically graded as level I, II, or III diastolic dysfunction (66). Like HFrEF, abnormal echocardiographic findings in patients with HFpEF are often accompanied by the clinical syndrome of congestive HF (i.e., fatigue, dyspnea with exertion, and peripheral and pulmonary edema).

Although much of the diagnosis of HF relies on the use of echocardiography and patient symptoms, some laboratory tests, such as measuring brain natriuretic peptide (BNP) levels in the plasma, may be used to help support the suspected diagnosis of HF. Synthesized in excess and released by the myocardium when the ventricles themselves are stretched by mechanical and pressure overload, a BNP >100 pg · mL^{-1} begins to become a marker for decompensated HF even when other clinical signs are nebulous. Serum BNP can, for example, distinguish dyspnea from HF versus pulmonary disease, a particularly useful tool in patients whose physical examination is nonspecific or in people with poor echocardiographic images.

Despite their utility, echocardiography, BNP, and other assessments of cardiac function do not quantify the full scope of HF as a disease that has both cardiac and systemic effects. However, the assessment of exercise capacity can be very helpful in this regard, measured as peak oxygen uptake ($\dot{V}O_2$) during a cardiopulmonary exercise test (CPET).

The routine evaluation of patients with HF via a CPET involves the analysis of expired gases (i.e., O_2 and CO_2) using indirect spirometry (see chapter 5, figure 5.3), allowing the clinician to quantify peak $\dot{V}O_2$, expressed as either mL · kg^{-1} · min^{-1} or percent of gender- and age-predicted normal (% predicted peak). Ventilatory efficiency (slope of the change in minute ventilation to change in volume of carbon dioxide produced, $\dot{V}_E$-$\dot{V}CO_2$ slope) is measured as well. Among patients with HFrEF treated with guideline-based therapies, these variables are used to assess prognosis by stratifying a patient's risk for mortality over the next 1 to 3 yr. Specifically, a peak $\dot{V}O_2$ <10 to 12 mL · kg^{-1} · min^{-1} (or <50% of age- and gender-predicted peak) or >17 mL · kg^{-1} · min^{-1} (or >70% of predicted peak) is associated with higher (10%) and lower (3%) 1 yr mortality rates, respectively (50, 62, 65, 71) (see figure 16.2). For patients identified as having a higher risk of mortality, advanced procedures known to extend survival may be considered, such as cardiac transplant or implantation of a mechanical left ventricular assist device (LVAD). Although further work is needed, Shafiq and colleagues suggest that peak $\dot{V}O_2$ and percent predicted peak $\dot{V}O_2$ may also have some clinical application relative to determining prognosis in patients with HFpEF (77).

Ventilatory efficiency, as measured by $\dot{V}_E$-$\dot{V}CO_2$ slope during exercise, is also an excellent predictor of future risk of death, with a value >30 indicating moderate increased risk and a value >45 indicating very high risk (5). Clearly, measured $\dot{V}O_2$peak, % predicted peak $\dot{V}O_2$, and $\dot{V}_E$-$\dot{V}CO_2$ are important variables for physicians to consider, helping them determine which patients require continued close monitoring and medical therapy versus more advanced treatment options such as an LVAD or possible cardiac transplant.

Although clinicians often use the New York Heart Association (NYHA) functional class (see chapter 4 and table 16.2) to help

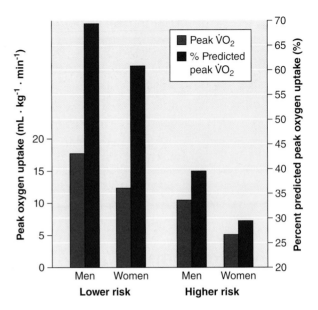

Figure 16.2 Lower (3%) and higher (10%) risk for mortality at 1 yr in men and women based on achieved peak $\dot{V}O_2$ (left axis) and percent predicted peak $\dot{V}O_2$ (right axis)

Data from Keteyian et al. (50).

characterize a patient's clinical status, that system does not fully reflect the breadth of the disorder. Table 16.2 shows the various stages in the development of HF as designated by the American College of Cardiology and American Heart Association (89). This staging system covers all patients with HF, regardless of whether symptoms are present. Note that those patients experiencing symptoms (New York Heart Association classes II-IV) fall within stage C and stage D alone.

Exercise Testing

The use of CPET in patients with HFrEF provides an enormous amount of useful information. As mentioned previously, information is gathered not only on severity of illness and estimated 1 and 3 yr mortality risk but also on response to medications and response to an exercise training program. Information needed to guide exercise training intensity is obtained as well.

Medical Evaluation

The methods for CPET in patients with HF differ little from exercise testing in patients with other types of heart disease. Most exercise tests conducted in patients with heart failure use a steady-state (2-3 min per stage) protocol; however, they typically use a lower workload to start the test and smaller work increments (1-2 METs) during the test. Common protocols are the modified Bruce or Naughton (23); a ramp protocol can also be performed with a stationary cycle that increases exertional work rate 10 to 15 W every minute (64). A ramp protocol generally results in less variable data during submaximal exercise because of the more gradual increments in work rate that it provides. $\dot{V}O_2$peak is approximately

Table 16.2 Stages in the Development of Heart Failure Based on American College of Cardiology/ American Heart Association Guidelines

Stage	Description	Examples	New York Heart Association functional class
A (patient at risk)	High risk for heart failure; no anatomic or functional abnormalities; no signs or symptoms	Hypertension, coronary artery disease, diabetes, alcohol abuse, family history	
B (patient at risk)	Structural abnormalities associated with heart failure but no symptoms	Left ventricular hypertrophy, prior myocardial infarction, asymptomatic valvular disease, low ejection fraction	
C (heart failure present)	Current or prior signs and structural abnormalities	Left ventricular systolic dysfunction with or without dyspnea on exertion or fatigue, reduced exercise tolerance	II or III
D (heart failure present)	Advanced structural heart failure with symptoms at rest despite maximal medical therapy	Frequent hospitalizations, awaiting transplant, intravenous support	III or IV

10% to 15% lower when measured with cycle ergometry than with a treadmill (69).

Because an accurate measure of exercise capacity is needed in these patients, $\dot{V}O_2$ is assessed directly, and the use of exercise and nonexercise prediction equations to estimate $\dot{V}O_2$ is discouraged (23, 67). Exercise capacity is measured using a cardiopulmonary metabolic cart (see chapter 5, figure 5.3); such equipment is usually available in either the cardiac noninvasive or pulmonary function laboratories of many hospitals. In addition to measuring $\dot{V}_E$-$\dot{V}CO_2$, $\dot{V}O_2$peak, and % predicted peak $\dot{V}O_2$, determining heart rate (HR) and $\dot{V}O_2$ at **ventilatory-derived lactate threshold (V-LT)** might also be helpful. An adequate explanation of this variable is provided elsewhere (23), but one approach commonly used to determine V-LT is the V-slope method (6).

Contraindications

Thirty years ago, standard teaching in most clinical settings was that moderate or vigorous physical activity should be avoided or withheld in patients with HF. It was feared that the increased hemodynamic stress that exercise places on an already weakened heart would further worsen heart function. As a result, most guidelines listed HF as a contraindication to exercise testing. Today, however, patients with stable HF routinely undergo symptom-limited maximum cardiopulmonary exercise testing to evaluate cardiorespiratory function, and such testing has been shown to be safe. Specifically, a study involving 2,037 patients with HF who completed 4,411 cardiopulmonary exercise tests revealed no deaths and fewer than 0.5 nonfatal major cardiovascular-related events per 1,000 tests (47).

All other contraindications to exercise testing still apply to patients with HF. Since arrhythmias are common in patients with HF, the person supervising the exercise test should review prior examination and testing reports to determine if arrhythmias have been previously reported in the person being tested. Note that the use of exercise testing to assess myocardial ischemia can be problematic because many patients with HF present with ECG findings at rest that invalidate or reduce the sensitivity of the test (e.g., left bundle branch block, left ventricular hypertrophy, and nonspecific ST-wave changes attributable to digoxin therapy).

Anticipated Exercise Responses

Compared with healthy normal people, patients with HFrEF or HFpEF exhibit some differences in their central and peripheral responses at rest (table 16.1) and during exercise (table 16.3). Resting stroke volume and cardiac output are both lower in patients with HFrEF or HFpEF versus healthy controls (stroke volume: approximately 50 vs. approximately 75 mL · b−1, respectively; cardiac index: <2.5 vs. >2.5 L · min−1 · m−2, respectively). Resting HR may be increased and systolic blood pressure may be reduced, attributable to both the underlying LV dysfunction and, in the case of HFrEF, the common use of afterload-reducing agents such as angiotensin-converting enzyme inhibitors.

Compared with healthy normal persons, at peak exercise patients with HF exhibit a lower power output (30%-40% lower), lower cardiac output (40% lower), lower stroke volume (up to 50% lower, more often observed in HFrEF than HFpEF), lower HR (20% lower), and often a higher level of plasma norepinephrine (an endogenous **catecholamine** released by the adrenal medulla and sympathetic postganglionic fibers) (11, 12, 35, 53, 55, 80). Concerning the finding of a lower peak HR, if it is <62% of age predicted (52) it is referred to as **chronotropic incompetence**, a characteristic that occurs in 20% to 40% of patients with HF and one that is, like $\dot{V}O_2$peak, an independent predictor of mortality (73). For patients not taking a β-adrenergic blocking agent, peak HR often still does not exceed 150 b · min−1.

Exercise capacity, as measured by peak $\dot{V}O_2$, is also decreased in patients with HFrEF or HFpEF, approximately 30% to 35% or more below that of normal persons (55). Depending on severity

Table 16.3 Determinants of Exercise Capacity in Normal People and Patients With HFrEF or HFpEF

	Normal	HFrEF	HFpEF
CENTRAL			
Cardiac output	↑↑↑	↑/↑↑	↑/↑↑
Heart rate	↑↑↑↑	↑↑/↑↑↑	↑↑/↑↑↑
Stroke volume	↑↑	↑	↑/↑↑
Peripheral artery flow-mediated dilation	↑↑↑↑	↑/↑↑	↑↑/↑↑↑
VASCULAR/EXERCISING SK. MUSCLE INTERFACE			
Metabolic mediated microvascular dilation	↑↑↑	↑/↑↑	↑/↑↑
Capillary density and size	N	N/↓	↓
SKELETAL MUSCLE			
Muscle mass	N	↓↓	↓↓
% Type I oxidative fibers	N	↓↓	↓↓
Mitochondrial enzymes/volume density	N	↓↓	↓

HFrEF = heart failure with reduced ejection fraction; HFpEF = heart failure with preserved ejection fraction; N = normal; ↑ = increase; ↓ = decrease.

of illness, $\dot{V}O_2$peak typically ranges from 8 to 21 mL · kg^{-1} · min^{-1}. Although the diminished peak exercise capacity of these patients is certainly partly due to the LV dysfunction, abnormalities other than cardiac function contribute as well (21), such as the inability to sufficiently dilate peripheral vasculature as a means to increase blood flow to the metabolically active muscles (87) and histological and biochemical abnormalities within the skeletal muscle itself (16, 61, 79).

Increasing blood flow to metabolically active skeletal muscles during exercise is part of a complex interplay between blood pressure and peripheral vascular activity (vasoconstriction–vasodilation). Patients with HFrEF have a reduced ability to vasodilate during exercise, which is attributable to both increased plasma norepinephrine (53) and impaired endothelial function. The latter is the result of the lesser or impaired release of local vasoactive chemicals such as endothelium-derived relaxing factors (i.e., nitrous oxide) (14, 58).

Regarding skeletal muscle function, in both HFpEF and HFrEF patients there is clear evidence of a decrease in the percentage of myosin heavy chain type 1 isoforms, diminished oxidative enzymes, and a potential reduction in capillary density. As a result, and in comparison with normal people, the patients rely more on anaerobic pathways to produce energy earlier during exercise.

Finally, participation in traditional aerobic-based exercise and symptom-limited CPET also appears to be safe and well tolerated in patients who have received an LVAD (1, 37, 38, 40, 41). The LVAD can increase cardiac output during exercise, and contemporary models are able to increase flow up to approximately 8 to 10 L · min^{-1} (37).

Treatment

Over the past 30 yr great strides have been made in the medicines and devices used to treat patients with HF. These advances have led to fewer HF-related deaths, fewer symptoms, improved quality of life, and increased exercise tolerance. In patients who are refractory to the standard drug therapies and who demonstrate a deteriorating clinical state consistent with a poor 1 yr survival rate, a mechanical LVAD or cardiac transplant is a possible consideration.

This section summarizes the guidelines for the treatment of patients with HF, with its greatest focus on patients with HFrEF (89). The reason for the greater focus on HFrEF is that, comparatively, very few medications improve symptoms or clinical events in patients with HFpEF. Beyond the medications used to treat the risk factors that are known to worsen HFpEF, such as hypertension and diabetes, only diuretics are used to prevent fluid overload. In fact, the one treatment shown to regularly improve exercise tolerance, lessen symptoms, and improve quality of life in patients with HFpEF is regular exercise training.

Table 16.4 lists many of the current medical therapies used to treat patients with HFrEF, along with any potential influence they might have on the physiologic responses to exercise. Vasodilating agents (e.g., angiotensin-converting enzyme [ACE] inhibitors or angiotensin II receptor blockers [ARB]), β-adrenergic receptor blockers, digoxin, neprilysin inhibitors combined with ARBs, and aldosterone antagonists have all been shown to improve survival and lessen risk for hospitalization. Since congestion and fluid overload represent important complications in many patients with HF, diuretic therapy is commonly used; in some patients, fluid intake

Table 16.4 Pharmacology

Medication class and name	Primary effects	Exercise effects	Important considerations
Diuretics (e.g., furosemide)	↓ volume overload	↓ blood pressure	Attenuate blood pressure response during exercise and may cause possible hypotensive response after exercise
Angiotensin-converting enzyme inhibitors (e.g., lisinopril)	↓ afterload and blood pressure at rest; improved survival	↓ blood pressure	Attenuate blood pressure response during exercise and may cause possible hypotensive response after exercise
Angiotensin receptor blockers (e.g., losartan)	↓ afterload and blood pressure at rest	↓ blood pressure	May cause possible hypotensive response after exercise
Combination drug (Entresto) of angiotensin receptor blocker (valsartan) plus neprilysin inhibitor (sacubitril)	↓ afterload and blood pressure at rest; ¯ sodium and fluid in the body; improved survival and lower rehospitalizations	↓ blood pressure	Attenuates blood pressure response during exercise and may cause possible hypotensive response after exercise
β-adrenergic blockers (e.g., carvedilol)	↓ heart rate and blood pressure at rest; improved survival	↓ heart rate and possibly blood pressure	Improve exercise capacity in patients with angina
Digitalis	Reduces hospitalization in patients with symptomatic heart failure; antiarrhythmic	↓ heart rate at rest and slight decrease, if any, during exercise	Can result in nonspecific ST-T wave changes in resting ECG (so-called dig effect), which can reduce the sensitivity of any ST-segment changes observed during an ECG stress test

↓ = decreased response compared with healthy subjects; ECG = electrocardiogram..

may also be restricted to 2,000 mL or less per day. And the role sodium restriction plays in minimizing fluid congestion cannot be overemphasized. Specifically, sodium restriction (i.e., to between 1,500 and 3,000 mg per day) can help reduce the need for diuretics, to the point that it might actually allow for the discontinuation of diuretic therapy.

For patients with NYHA class II through IV HF due to systolic dysfunction who also have electrocardiographic evidence of dyssynchronous contractions of the left and right ventricles (based on a QRS duration greater than 120 ms), a special type of pacemaker therapy (called cardiac resynchronization therapy) is commonly used. As the name implies, cardiac resynchronization therapy involves implanting pacemaker leads in both the right and left ventricles and pacing the ventricles to reestablish the correct firing pattern of the right ventricle firing milliseconds before the left ventricle. Resynchronization therapy often improves cardiac function and exercise tolerance and lowers future hospitalizations and deaths (35). Additionally, the benefits of implantable defibrillators, either by themselves or combined with cardiac resynchronization pacemakers, are well documented for patients who have dilated hearts with low ejection fractions, particularly as a means to reduce mortality that results from the common occurrence of ventricular arrhythmias.

Despite aggressive attempts to optimize medical therapy, a patient's clinical condition may deteriorate to the point that special medications, an LVAD, or cardiac transplantation is considered.

Continuous-flow LVADs (figure 16.3) represent a standard therapy option, either as a bridge to transplant or as destination therapy for patients who do not qualify for transplant, extending survival to above 70% at 2 yr (31, 74, 75). LVADs provide circulatory support to underperfused organs and partially reverse the pathophysiological sequelae of HF.

In brief, the LVAD is a mechanical, continuous flow pump that is implanted in the chest and attached to the apex of the LV. Blood is drawn from the LV into the device, which then sends the blood into the ascending aorta distal to the aortic valve. A driveline is passed through the skin of the abdomen and connected to a controller and power supply. The design of continuous flow LVADs has evolved dramatically over the past 25 yr, such that current-generation devices are much smaller and rely on magnetic levitation to move impellers within the pump housing. Although there are clinical challenges unique to having an LVAD (e.g., required anticoagulation and increased risk for stroke, bleeding, infection), functional capacity and health-related quality of life are mildly to moderately improved following device implant; in some patients the increases are comparable to the changes observed after heart transplantation (74).

Cardiac transplantation is a surgical procedure that also represents standard therapy for patients with end-stage HF refractory to maximal medical therapy. Practical application 16.1 provides an overview of this procedure.

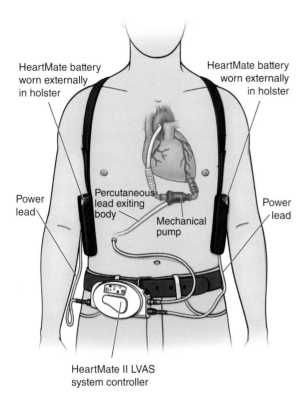

Figure 16.3   HeartMate II left ventricular assist system (LVAS).

Reprinted with permission from Thoratec Corporation.

An important part of the care for any chronic disease remains secondary prevention. For the exercise professional working with patients with HF, this care includes counseling to help manage behavioral habits known to exacerbate their condition. For example, in patients with an ischemic cardiomyopathy, every attempt should be made to help stabilize existing coronary atherosclerosis through aggressive risk factor management as a means to prevent further loss of cardiac cells (myocytes) attributable to reinfarction. Also, as mentioned previously, most patients with HF are asked to follow a low-sodium diet as an important step toward preventing congestion and fluid overload, thus reducing chances of subsequent hospitalizations. The exercise physiologist should support healthy behaviors as well as reinforce compliance with all prescribed medications. See practical application 16.2.

EXERCISE PRESCRIPTION

For patients with HF, continuously monitoring exercise via ECG can be considered (1); however, in the HF-ACTION trial (Heart Failure: A Controlled Trial Investigating Outcomes of Exercise Training), such monitoring was not performed at all participating sites and there were very few exercise-related clinical events. After demonstrating for 1 to 3 wk that they can tolerate supervised exercise three times per week, patients can begin a one or two times per week home exercise program. As patients continue to improve

and demonstrate no complications to exercise, they can transfer to an all-home-based exercise program as needed.

Training for cardiorespiratory endurance is an obvious strategy for patients with HF, but muscular strength, muscular endurance, and flexibility training can also improve functional capacity and foster independence.

Special Considerations

Despite the increased attention given to using moderate exercise training in the treatment plan of patients with HFrEF, few HF-unique guidelines are needed relative to prescribing exercise in these patients. This is especially true if enrolling patients with stable NYHA class II and III symptoms who are receiving standard drug therapy and are appropriately progressed. With respect to patients who received an LVAD or cardiac transplantation, starting regular exercise in these patients within 3 mo after surgery may be associated with residual deconditioning, as well as unique medical and surgical concerns requiring additional attention from the rehabilitation staff.

Perhaps the most important exercise consideration for patients with HF is maintaining compliance over time. In the HF-ACTION trial, after 12 mo only ~40% of patients randomized to the exercise arm of the trial were exercising at or above the prescribed number of minutes per week (69). One reason for the lower compliance in HF patients might be that they often have comorbidities or other illnesses that interrupt or prevent regular program attendance. For example, arrhythmias, pneumonia, fluctuations in edema, adjustments in medications, depression, and hospitalizations can all affect regular participation. Patients should be told to expect interruptions in their exercise therapy caused by both HF- and non-HF-related issues. Then, when they feel better, they should make every attempt to resume regular exercise habits.

Cardiorespiratory Training

This section addresses unique issues pertinent to prescribing exercise in patients with HFrEF. Refer to chapter 6 for a review of the general common premises associated with prescribing exercise.

• *Mode:* Improvements in cardiorespiratory fitness or exercise capacity occur similarly (10, 15, 22, 28, 29, 30, 43, 48, 88), regardless of whether the exercise training was completed using a stationary cycle or walking; these gains transfer fairly well to improvements in routine activities of daily living. Consistent with this, upper body ergometry activities can improve function in the upper limbs as well.

• *Frequency:* After being in an exercise program for several weeks, patients with HF often comment that they notice less DOE and less fatigue during routine daily activities. Consider also educating patients about the detraining effect, which means informing them that many of the benefits they notice from regular exercise will diminish if they stop exercising regularly.

Practical Application 16.1

CARDIAC TRANSPLANTATION

Cardiac transplantation is an effective therapeutic alternative for persons with end-stage HF. Each year, approximately 4,200 transplants are performed worldwide, with a median survival of approximately 14 yr for those patients who survive the first year after transplant (60).

After surgery, many patients with cardiac transplant continue to experience exercise intolerance. In fact, $\dot{V}O_2$peak is typically between 14 and 22 mL · kg^{-1} · min^{-1} among untrained patients. For this and other reasons, these patients are commonly enrolled in a home-based or supervised cardiac rehabilitation program starting as soon as 4 wk after surgery. The increase in $\dot{V}O_2$peak is approximately 2.5 mL · kg^{-1} · min^{-1}, which represents a relative increase of between 10% and 20% (4, 45, 57).

There is an increasing probability of developing an accelerated vasculopathy in the coronary arteries of the donor heart. For this and other reasons, the traditional risk factors for ischemic heart disease such as hypertension, obesity, diabetes, and hyperlipidemia are aggressively treated. Additionally, immune system–mediated rejection of the donor heart remains a concern for heart transplant recipients, and therefore most patients receive a variety of immunosuppressive medications. Common agents include cyclosporine, tacrolimus, sirolimus, mycophenolate mofetil, mycophenolic acid, and prednisone (72).

Patients with cardiac transplant represent a unique physiology in that the donated heart they received is decentralized. This means that except for the parasympathetic postganglionic nerve fibers that are left intact, other cardiac autonomic fibers (i.e., parasympathetic preganglionic and sympathetic pre- and postganglionic) are severed. Partial regeneration of these severed fibers may occur after 1 yr in some patients, but it is likely best to assume that decentralization, for the most part, is permanent.

Because of the decentralized myocardium, the cardiovascular response of cardiac transplant patients to a single bout of acute exercise differs from the response of normally innervated people. At rest, HR may be elevated to between 90 and 100 b · min^{-1} because parasympathetic (i.e., vagal nerve) influence is no longer present on the sinoatrial node. When exercise begins, HR changes little because (a) no parasympathetic input is present to be withdrawn and (b) no sympathetic fibers are present to stimulate the heart directly. During later exercise, HR slowly increases because of an increase in norepinephrine in the blood. At peak exercise, HR is lower, and both cardiac output and stroke volume are approximately 25% lower than in age-matched controls.

Because of the absence of parasympathetic input, the decline of HR in recovery takes longer than normal. This effect is most observable during the first 2 min of recovery, during which HR may stay at or near the value achieved at peak exercise. Systolic blood pressure recovers in a generally normal fashion after exercise (18), which is why this measure can be used to assess adequacy of recovery.

- *Duration:* Most exercise trials involving patients with HF increased training duration up to 30 to 60 min of continuous exercise. On occasion, it may be necessary to start an individual patient with two, three, or four intermittent bouts of exercise that are each 4 to 10 min in duration.

- *Intensity:* After patients are able to comfortably exercise for 25 min at least 3 d/wk, exercise intensity could be progressed (as tolerated) from initially low to moderate levels, and then up to moderate to vigorous work levels. A meta-analysis by Ismail and colleagues that involved patients with HF associated low- to moderate-intensity training with a 5% to 10% increase in peak $\dot{V}O_2$, whereas moderate to vigorous training was associated with a 10% to 15% increase in peak $\dot{V}O_2$ (36). Several single-site trials used intermittent higher-intensity aerobic interval treadmill training (vs. continuous moderate-intensity training) to improve exercise capacity and showed training intensities as high as 90% to 95% of peak HR or HR reserve may yield an even greater increase in $\dot{V}O_2$peak (88). However, a 2017 multicenter study by Ellingsen and colleagues of patients with HFrEF showed that the increase in exercise capacity with higher-intensity interval training was no greater than what was observed with moderate continuous training (19).

In general, for the first few exercise sessions, initially guide intensity using the HR reserve method set at about 60% and then progress to 70% to 80% of heart rate reserve, as tolerated. The exercise professional can also have patients titrate their training intensity using the rating of perceived exertion scale set at 11 to 14. One reason to start these patients at a lower intensity is to allow them an opportunity to adjust to the exercise. Fatigue later in the day is not uncommon, and restricting exercise intensity at first may help minimize this effect.

Concerning patients with an LVAD, Kerrigan and colleagues (40) showed that in patients without an electronic pacemaker, exercise intensity can be prescribed and progressed just like other HF patients using the HR reserve method. However, in patients with an LVAD and a permanent pacemaker, intensity of effort

Practical Application 16.2

CLIENT–CLINICIAN INTERACTION

One of the important responsibilities for an exercise professional working with a patient with HF is to ensure that on any given day, the patient is free of any signs or symptoms that might indicate the need to withhold exercise.

To accomplish this, you must not only persuade the patient to verbalize any problems they might be having but also take the initiative to ask the patient key questions. Following are examples of these types of questions:

- How did you sleep over the weekend? Have you had any more bouts of waking up during the night short of breath? Are you still sleeping using just one pillow, or have you had to increase the number of pillows you use to avoid being short of breath while you sleep?

- How has your body weight been over the past three days, stable or increasing?

- Have you had any increased difficulty breathing?

- Do you ever get that increased swelling in your ankles anymore?

Each question enables you to assess change in HF-related symptoms, and with time and experience, you will develop a sense about when and how to further assess these patients.

Also, some patients with HF may not tolerate the first few days of their exercise rehabilitation well. Therefore, encouragement and guidance on your part are important. Explain that, over time, the patient can expect to be able to perform routine activities of daily living more comfortably (i.e., less fatigue and shortness of breath). For some patients, achieving only this goal may be quite fulfilling. Other patients may wish to improve their exercise capacity to the point that they can resume activities they previously had to avoid. Although such a goal may be realistic, be sure to emphasize that the prudent approach is to advance their training volume in a progressive manner. Trying to do too much too soon may lead to disappointment if their functional improvement does not keep pace with their self-assigned interests.

Finally, when working with patients with HF, emphasize the importance of regular attendance at their exercise program. If they need to miss exercise for personal or medical reasons, let them know you look forward to seeing them back when they feel better. Because you often see the same patient several times each week over several weeks, you are in a unique position to provide ongoing support, motivational counseling, and monitor medical compliance and symptom status over time.

is best titrated using rating of perceived exertion set at 11 to 14. Factors that can influence the overall functional capacity of a patient and rate of progression include initial workloads chosen for rehabilitation, time from LVAD implantation, current activity habits, comorbidities, and age.

Resistance Training

Patients with stable stage B or C (table 16.2) HF who have demonstrated they can tolerate aerobic exercise training, which should be evident within 4 wk, also represent likely candidates for participation in a resistance training program. To date, studies show that mild to moderate resistance training is safe and improves muscle strength 20% to 45% in stable patients with HF. In 2007, Feiereisen and colleagues (22) compared standard moderate-intensity aerobic endurance-type training, strength training, and combined strength and aerobic training before and after 40 exercise sessions and observed, regardless of training type, improvements in peak $\dot{V}O_2$, thigh muscle volume, and knee extensor endurance for all three groups. None of the three regimens appeared to be superior to the other two. Given that loss of both muscle strength and endurance is common in these patients (63, 85), it is appropriate to assume that improvements in these measures through resistance training are helpful.

Because no specific guidelines address resistance training in patients with HF, weight training recommendations in these patients are often drawn from consensus statements or guideline documents (2, 3, 78, 86), which are detailed in chapter 6. The specific exercises should address the individual needs of the patient but will likely include all the major muscle groups. As patients improve, load should be increased 5% to 10%.

For patients with an LVAD, the safety and efficacy of resistance training is an area void of definitive research. Similar to the situation with other patients who have undergone a thoracotomy, rehabilitation staff should wait >12 wk after device implant before beginning patients on a light resistance program. Other considerations include avoiding any exercises or maneuvers (e.g., Valsalva) that may dramatically increase intra-abdominal pressure (e.g., sit-ups, seated leg press) or have the potential for physical trauma (i.e., contact sports); the latter can cause a fracture of the driveline or create damage to the LVAD itself. For upper body strengthening activities, prescribe resistance bands and light hand weights. For the upper legs, include 1/4 wall sits or 1/4 double-leg squats.

Summaries of the exercise prescriptions used for patients with HF and for patients with cardiac transplant are presented in the following section.

Exercise Recommendations for Patients With Heart Failure

Before participation in a regular exercise program, a limited evaluation should be performed to identify acute signs or symptoms that may have developed between when the patient was referred for exercise by their doctor and program participation. Elements to screen for might include an increase in body mass greater than 1.5 kg during the previous 3 to 5 d, complaints of increased difficulty sleeping while lying flat, sudden awakening during the night because of labored breathing, or increased swelling in the ankles or legs. Additionally, a graded exercise test may be helpful to identify any important exercise-induced arrhythmias and develop a target HR range that allows determination of a safe exercise intensity.

Exercise Program

The exercise professional should progressively modulate (as tolerated) type, frequency, duration, and intensity of activity so that the patient attains an exercise regimen of 150 min of moderate-intensity work per week. Table 16.5 provides a summary of the aerobic, resistance training, and range of motion recommendations.

Special Considerations

- Many patients with HF are inactive and possess a low tolerance for activity. Exercise should be progressively increased in an individualized manner for every patient.
- For the first week or so of exercise training, some patients may be a bit tired later in the day. To compensate for this, they could temporarily limit the amount of other home-based activities.

- For patients with an LVAD and without a pacemaker, titrate exercise intensity based on the HR reserve method or rating of perceived exertion. For patients with an LVAD and a pacemaker, guide exercise intensity by rating of perceived exertion alone (40).

Exercise Recommendations for Patients With Cardiac Transplant

Because patients with cardiac transplant present with a decentralized heart, use of a graded exercise test within 12 mo after surgery may not be helpful if the test is being performed for the purpose of developing an exercise training HR range. Such a test, however, will help screen for exercise-induced arrhythmias, quantify exercise tolerance, and serve as a marker to assess exercise outcomes at a future date.

Exercise Program

Like patients with stable HFrEF, exercise should be progressively increased in cardiac transplant recipients so that the patient attains an exercise regimen of 150 min of moderate-intensity work per week.

Special Considerations

- Loss of muscle mass and bone mineral content occurs in patients on long-term corticosteroids such as prednisone. As a result, resistance training programs play an important role in partially restoring and maintaining bone health and muscle strength (7). But high-intensity resistance training should be avoided because of increased risk of fracture in bone that has compromised bone mineral content.

Table 16.5   Exercise Prescription Review for Patients With Heart Failure

Training method	Frequency	Intensity	Time (Duration)	Type (Mode)	Progression
Cardiorespiratory	4 or 5 times per week	60%-75% of heart rate reserve or 11-14 on RPE scale	40 min or more per session; use interval training method as needed	Cycle or treadmill; arm ergometer as needed	10 min or more per session up to 40 min or more per session
Resistance	1 or 2 times per week	Begin with 40% of 1RM for upper body and 50% of 1RM for lower body; progress both over time to 70% of 1RM	1 set of 12-15 reps for each of the involved muscle groups	Fixed machines, free weights, and bands; 6-8 regional exercises	Increase 5%-10% as tolerated
Range of motion	Before and after each aerobic or resistance training workout	—	5 min before and 5-10 min after each workout, with 10-30 s devoted to the major muscle groups and joints	Static stretching	—

RPE = rating of perceived exertion (Borg scale 6-20); 1RM = one-repetition maximum.

- Because the transplanted heart is decentralized from the patient's autonomic nervous system, rating of perceived exertion should be used to guide exercise intensity (targeting 11-14 on 6-20 scale) instead of HR (45).

- Although isolated cases of chest pain or angina associated with accelerated graft atherosclerosis have been reported, decentralization of the myocardium essentially eliminates angina symptoms in most cardiac transplant patients.

- A regular exercise program performed within the first few months after surgery may result in an exercise HR response during training that is equal to or exceeds the peak HR achieved during an exercise test taken before a patient started their training program. This response is not uncommon and further supports using rating of perceived exertion as the method to guide intensity of effort in these patients.

- Marked increases in body fat leading to obesity sometimes occur in cardiac transplant patients. This increase is likely caused by both long-term corticosteroid use and restoration of appetite following illness. Exercise modalities chosen for training should take into account any possible limitation caused by excessive body mass.

- Because of the sternotomy, postoperative range of motion in the thorax and upper limbs may be limited for several weeks.

EXERCISE TRAINING

Improvements with exercise training in patients with HFrEF can be seen not only in clinical and cardiorespiratory measures but also in the skeletal muscle and other organ systems. Figure 16.4 provides a summary of these adaptations. Conversely, less information is available regarding the physiologic effects and safety of exercise training in patients with HFpEF. Thus, the material in this section focuses predominantly on exercise training responses in patients with HFrEF.

Exercise Capacity and Clinical Outcomes

Although exercise is an important component in the treatment of HFrEF today, not until the late 1980s and early 1990s did sufficient research show that patients could safely derive benefit. Since that time, dozens of case-reports and single-site randomized controlled trials showed that exercise training safely results in a 15% to 30% increase in $\dot{V}O_2$peak (10, 29, 39, 42, 43, 46, 49, 82).

Presently, evidence-based guidelines for patients with HFrEF include regular exercise training as a standard recommendation (89). In addition to partially reversing many of the physiologic abnormalities that accompany HF (42), exercise training has a

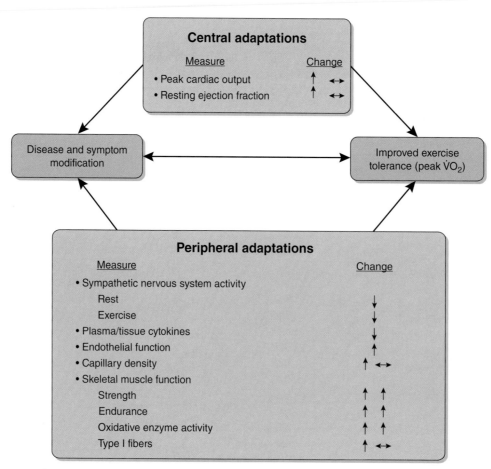

Figure 16.4 Summary of the various exercise training responses in patients with HFrEF.

Research Focus

Exercise Training Trials in Patients With HFpEF: A Good Start, but Much More Clinical Research Is Needed

Among patients with HFpEF, the impact of physical training on exercise capacity, quality of life, and cardiac function has been evaluated in nearly a dozen smaller exercise intervention trials (13, 24, 83). The vast majority of these clinical trials demonstrated improved exercise capacity, as measured by peak $\dot{V}O_2$ (13, 24, 56, 83) and quality of life. Other parameters important to any discussion involving the effects of exercise training in patients with HFpEF are endothelial function and left ventricular diastolic function. In both instances, the few studies addressing these variables have yielded somewhat mixed findings (13, 24, 33, 34, 54).

One area of increasing study, and likely responsible for a good portion of the improved exercise capacity that has been reported, is skeletal muscle plasticity. Like patients with HFrEF, patients with HFpEF suffer abnormalities in skeletal muscle function, which may include size, strength, capillary density, and mitochondrial function (32, 33, 83). In summary, and considering both the relatively small amount of evidence available to date and the absence of any effective medical therapies to improve quality of life and exercise tolerance in HFpEF, exercise training is emerging as an effective means to improve exercise capacity and delay fatigue. Clearly, many avenues continue to exist for future research.

favorable impact on many important clinical outcomes. The HF-ACTION trial compared the effects of exercise training plus usual care with usual care alone and showed an exercise training–related 10% to 15% reduction in the adjusted risk for all-cause mortality or hospitalization and cardiovascular mortality or HF hospitalization (68). Conversely, in 2018, a patient-data meta-analysis showed a nonsignificant 17% reduction in all-cause mortality (81). A secondary analysis of the HF-ACTION trial that involved just those patients assigned to the exercise group showed exercise training at just 3 to 5 MET-h per week was associated with a 40% to 60% reduction in clinical events (51).

Myocardial Function, Skeletal Muscle, and Other Adaptations

Exercise training in patients with HFrEF results in important central and peripheral adaptations as well, including no change or a modest increase in peak cardiac output (mostly due to an exercise training–induced increase in peak HR (17); an improved ability to dilate the small blood vessels that nourish the metabolically active muscles (29, 30); a decrease (at rest and during submaximal exercise) in plasma norepinephrine (10, 43) and other vasoconstrictor agents (8); increased levels of endothelium-derived relaxing factor (27); and flow-mediated improvement in endothelial function (20, 59). Within the skeletal muscle, the volume density of mitochondria and the enzymes involved with aerobic metabolism (e.g., cytochrome c oxidase) is improved (30, 44), and there is improvement in skeletal muscle strength and endurance (9).

Finally, several studies show that exercise training partially normalizes autonomic, immune, and hormonal function in patients with HF (8, 10, 70). Downregulation of the sympathetic nervous system occurs (76), and parasympathetic activity increases. Gielen and associates showed the anti-inflammatory effects associated with exercise training (25, 26).

Clinical Exercise Bottom Line

- HF is a clinical syndrome with signs or symptoms caused by a cardiac abnormality that results in pulmonary or systemic congestion. Ejection fraction may be reduced (HFrEF) or preserved (HFpEF).

- The causes of HF are many, with the most frequent being necrosis, or death, of cardiac myocytes due to ischemic heart disease. Other (nonischemic) causes responsible for loss of cardiac myocytes include viral infection, alcohol abuse, and an unwanted side effect of medications used to treat certain cancers.

- The treatment for HF differs based on the type of disorder (HFrEF vs HFpEF). For patients with HFrEF, vasodilating agents, β-adrenergic blocking agents, and aldosterone antagonists are used to interrupt the cardiac, autonomic, and hormonal abnormalities that develop and contribute to further worsening of the disorder. In contrast, few medical therapies have proved effective in directly treating HFpEF; instead, the factors (e.g., hypertension, diabetes) that worsen the condition are aggressively treated. Diuretics are used with both disorders to control pulmonary and systemic congestion.

- Patients with HF experience marked exercise intolerance, often up to 45% below that of age-matched persons. Regular exercise training is very effective in partially reversing the reduced exercise capacity and improving quality of life in patients with HFrEF or HFpEF. Among patients with HFrEF, exercise training also lessens the risk for mortality and hospitalizations due to HF.

CONCLUSION

Heart failure, one of today's major cardiac-related diagnoses, represents an immense health burden. Current research establishes that among eligible patients with stable HFrEF, regular exercise training improves exercise tolerance and quality of life and moderately reduces the risk for future clinical events.

Online Materials

Visit HK*Propel* to access a link to the references, the case studies with discussion questions, and a quiz to help you review key concepts and test your understanding of the material covered in this chapter.

CHAPTER 17

Peripheral Artery Disease

Ryan J. Mays, PhD, MPH

Ivan P. Casserly, MB, BCh

Judith G. Regensteiner, PhD

The clinical manifestation of peripheral artery disease (PAD) occurs as a result of the development of atherosclerotic plaque in the intimal layer of the arteries of the lower limbs. PAD is associated with an increased risk of adverse cardiovascular and cerebrovascular events and with an increased risk of mortality (2). Walking performance, functional ability, and patient-reported outcomes are impaired in those with PAD (13, 132, 171). This chapter focuses on the diagnosis, assessment, and treatment of PAD.

DEFINITION

PAD refers to the partial or complete blockage of the leg arteries by plaque, which subsequently leads to narrowing of the arteries in the lower extremities. The resulting **stenoses** or **occlusions** (or both) cause decreased blood flow to the muscles of the legs.

SCOPE

The estimated global prevalence of PAD is between 200 and 236 million cases (58, 198). In the United States, prevalence of the disease is estimated to be 8.5 million for people ≥40 yr of age (8). Approximately 35% to 40% of patients with PAD have claudication, which is characterized by pain, cramping, or aching in the calves, thighs, or buttocks, while 1% to 2% have the most severe form of the disease, **critical limb ischemia (CLI)** (156, 212). The majority of patients with PAD are asymptomatic or have atypical symptoms. According to the Trans-Atlantic Inter-Society Consensus II international guidelines, the prevalence of PAD with claudication is approximately 2% to 7% in patients 50 to 70 yr of age (156). Patients with PAD have a high burden of cardiovascular and cerebrovascular events (e.g., myocardial infarction, stroke), with an annual event rate of approximately 5% to 7% (156). Peripheral artery disease constitutes a major burden to national health care

expenditures, regardless of the stage of disease. A seminal review of Medicare costs based on diagnosis and procedure codes revealed that treatment for PAD was estimated to cost $4.37 billion annually, with 88% of the expenditures being for inpatient care and 6.8% of the elderly Medicare population being treated for PAD (101). Previous estimates of 2 yr cumulative hospitalization and treatment costs in the United States ranged from $7,000 to $11,693 per patient, with an overall annual cost in excess of $21 billion (122, 123). More recent evaluations of the financial burden have found annual expenditures for individuals with PAD to be $11,553, compared with $4,219 for patients without the disease (196).

PAD and claudication have overall prevalence rates that are slightly higher in men than in women (18). However, the evidence for this conclusion is equivocal and may be based on the method of diagnosis; a number of studies have depicted similar prevalence rates of PAD between men and women or an even slightly higher presentation of the disease in women (145, 203). To further confound the lack of consensus as to PAD prevalence in men and women, it has been suggested that a much greater presentation of the disease may exist in women ≥85 yr of age compared with men (39.2% vs. 27.8%) (49). Race and ethnicity also play a significant role in the prevalence of PAD (8, 156). For instance, it is suggested that non-Hispanic Blacks have a higher prevalence of PAD than non-Hispanic whites (2, 34). This is evidenced by several studies reporting odds ratios ranging from 1.5 to 3.1 for Black versus white populations (7, 36, 104, 154). Asian population prevalence rates of PAD are generally lower than in other ethnicities (36, 37). This is evidenced by several studies reporting odds ratios ranging from 1.5 to 3.1 for Black versus white populations (7, 36, 104, 154). The prevalence of PAD in American Indian and Alaskan Native populations was initially estimated to be ~6.0%, with similar prevalence rates among tribal groups (54, 112, 124). However, more recent prevalence estimates of the disease in a large general population cohort (over 3 million) indicate that American Indians and Alaskan Natives have higher rates of mild to moderate PAD compared with white populations (7.8% vs. 4.58%) (14).

PAD is often accompanied by multiple comorbidities in addition to cardiac and cerebrovascular disease. For instance, the prevalence of PAD is estimated to be as high as 29% in patients over 50 yr of age with diabetes, which is one of the most common risk factors (smoking being another) (85, 100, 126, 220). Other concurrent health problems such as heart failure (HF) and severe pulmonary disease may prevent sufficient physical activity to produce limb symptoms, so that PAD may be underdiagnosed in these subgroups. Thus, the exercise physiologist must be aware of the patient's comorbid conditions when conducting stress tests and prescribing exercise.

PATHOPHYSIOLOGY

The process of developing clinically significant PAD likely begins with endothelial damage in the arteries of the periphery (93). The following risk factors, which are similar to those for coronary artery and cerebrovascular disease, can lead to increased levels of oxygen-derived free radicals and ultimately endothelial injury (18): smoking, diabetes and insulin resistance, hypertension, hypercholesterolemia, high levels of homocysteine and fibrinogen, chronic kidney disease, and elevated C-reactive protein levels. Reducing these risk factors can aid in normalizing vascular superoxide anion production and improve endothelium-dependent vascular relaxation (157). However, direct links have not been made between controlling these risk factors and improvement in the clinical presentation of PAD.

After damage to the endothelium, a series of further events occurs; these are described elsewhere and are beyond the scope of this chapter (118). It is critical, however, to understand that the events following endothelial damage (e.g., accumulation of macrophages and foam cells within a fibrous cap) result in plaque formation with luminal compromise and subsequent abnormal blood flow in the peripheral arteries. The abnormal blood flow in the lower extremities is predicated on the severity of the stenosis (32, 99). The blockages impair adequate blood flow to meet metabolic demands of active muscles with even low-intensity exercise. For patients with claudication, blood flow is typically adequate at rest; but for patients with CLI, perfusion pressure to the tissues even at rest is impaired because of the critical nature of the disease, thus resulting in rest pain or tissue loss or both (167, 225).

CLINICAL CONSIDERATIONS

A patient presenting with signs and symptoms suggestive of PAD should first be assessed for the presence of PAD, most often via measurement of the **ankle–brachial index (ABI)**. If PAD is present, the patient should be treated as if cardiovascular disease is also present in terms of risk factor modification, since PAD is a heart disease equivalent. Supervised exercise training is a first-line medical therapy if claudication is present; it is also beneficial for those

who are asymptomatic since all patients with PAD are at risk for cardiovascular disease. If symptoms are more severe, noninvasive and invasive imaging studies are typically performed in the event that surgical or endovascular interventions are planned. Thus, evaluating the severity of the patient's clinical condition is essential for planning and conducting appropriate medical treatment as well as for providing an optimal individualized exercise program.

Signs and Symptoms

As mentioned previously, the majority of patients with PAD have atypical symptoms or an asymptomatic presentation of the disease (137). In clinical practice, a vascular specialist uses the term *atypical* to describe symptoms that are felt to be consistent with PAD but do not meet the classic definition of claudication. In essence, the primary components of such a symptom complex would include symptoms that are largely exertional, relieved with rest, and are primarily located in muscle groups of the lower extremity (15).

For patients who have claudication, many terms are often used to describe leg pain symptoms, including cramping, aching, tightening, and fatigue. The calf is the most commonly reported location for leg pain because the gastrocnemius muscle consumes more oxygen during ambulation than the other leg muscles (18). However, because PAD is the result of blood flow restriction to the muscles of the lower limbs, the site of stenoses and occlusions may dictate the symptomatic presentation of patients. Calf claudication is often due to flow-limiting lesions in the femoral and popliteal arteries. Buttock pain is predicated on compromised flow to the internal iliac arteries (i.e., disease in the internal iliac artery itself or disease in the aorta or common iliac artery). Thigh claudication results from compromised flow to the profunda femoral artery (i.e., disease in the profunda artery itself, or disease in the common femoral or iliac arteries). Thus, discussion with the patient regarding symptoms may provide some insight into the anatomic location of PAD underlying the clinical presentation, although specific diagnostic tools should be used to assess disease localization and severity when needed, such as for noninvasive or invasive angiography. Additionally, a number of questionnaires have been developed to identify and detect severity of disease based on symptom presentation. These include the Edinburgh Claudication Questionnaire (116) and San Diego Claudication Questionnaire (35). Similar to using the patient's story regarding symptom presentation, questionnaires alone may lack the sensitivity necessary to detect and diagnose PAD (195) and thus should be used in conjunction with objective diagnostic methods.

CLI is defined as the presence of chronic ischemic rest pain, foot ulcers (nonhealing wounds), or gangrene attributable to objectively proven arterial occlusive disease (212). The burden of atherosclerosis in patients with CLI is such that perfusion to the lower extremity is compromised even at rest. Typically, the disease is present at multiple levels including the tibial vessels in the leg and the small vessels of the foot (183, 197). The diagnosis of CLI carries a grave prognosis, in terms of both the risk for limb loss and overall cardiovascular morbidity and mortality. It is estimated that upon initial presentation of CLI, mortality rates are 25% in the first year, while another 25% of patients will require amputation (156). Mortality rates are estimated to exceed 50% in 5 yr, indicating a poor prognosis for patients with CLI (52).

Patients with ischemic rest pain usually report severe pain in the forefoot that is precipitated by assuming the recumbent position. Hence, many patients with CLI report significant difficulty with sleep since it is interrupted by episodes of severe foot pain, which they relieve by hanging their lower extremity over the edge of the bed. The use of narcotic analgesia is commonly required to provide adequate relief from ischemic rest pain. Tissue loss in patients with CLI is most commonly located in the forefoot (156). Arterial wounds are classically described as being painful and having a pale base with an irregular margin. Superimposed soft tissue infections and underlying bony infections (e.g., osteomyelitis) are complications that may occur, particularly in wounds that are not aggressively managed.

A variety of symptom severity scales have been devised to provide consistency for the grading of PAD in clinical practice and during clinical investigation. The two widely accepted scales are the Fontaine stages and Rutherford categories (56, 184).

The Fontaine Stages (Stage: Symptoms)

- I: Asymptomatic
- IIa: Mild claudication
- IIb: Moderate to severe claudication
- III: Ischemic rest pain
- IV: Ulceration or gangrene

The Rutherford Categories (Grade: Category: Symptoms)

- 0: 0: Asymptomatic
- 0: 1: Mild claudication
- I: 2: Moderate claudication
- I: 3: Severe claudication
- II: 4: Ischemic rest pain
- II: 5: Minor tissue loss
- III: 6: Major tissue loss

History and Physical Examination

The key components of the clinical history and physical examination that should be documented in a patient with PAD are summarized in tables 17.1 and 17.2. The primary element of the clinical history should be the characterization of the patient's leg symptoms and an assessment (both qualitative and quantitative) of the impact of these symptoms on the patient's functional capacity. A careful history should be able to determine with a reasonable degree of certainty whether a patient's symptoms are due to ischemic or nonischemic causes. The remainder of the history centers on the documentation of atherosclerotic risk factors, a survey of conditions that may contribute to functional limitation, and a review of evidence for atherosclerotic involvement of other vascular beds beyond the lower extremities.

Table 17.1 Components of Medical History Important in Assessment of a Patient With Peripheral Artery Disease

Medical history	Comments
Assessment of leg pain site, severity of pain or discomfort, precipitating and relieving factors, effect of leg exertion and rest, temporal pattern of pain	Assess whether leg discomfort represents true ischemia of the leg versus other nonischemic causes of leg pain.
Inquiry about rest pain in foot or tissue loss	Determine if patient has any evidence of CLI.
Perceived impact of leg symptoms on patient's quality of life	This affects treatment decisions.
Patient expectations about ambulation	Expectations play an important component in decision making.
Walking distance, claudication onset distance,[a] peak walking distance[b]	These provide a more objective measure of severity of PAD and degree of functional impairment that the patient can relate to (e.g., how many blocks walked before pain).
Medical comorbidities	Assess whether comorbidities may be contributing limitations to exercise (e.g., HF, COPD, prior stroke). Information on comorbidities has an important role in making treatment decisions.
Atherosclerotic risk factors	These are important targets for medical treatment.
Medications	Make sure the patient is on optimal medical therapy (e.g., antiplatelet, statin).
Social history	Assess lifestyle of patient and need for ambulation to meet patient's needs (e.g., role of walking in current employment).

[a]Distance walked at the initial onset of any claudication pain.

[b]Distance at which claudication becomes so severe that the patient is forced to stop.

CLI = critical limb ischemia; PAD = peripheral artery disease; HF = heart failure; COPD = chronic obstructive pulmonary disease.

Table 17.2 Summary of Primary Components of Physical Exam in Patients With Peripheral Artery Disease

Physical examination	Comments
Signs of limb ischemia	Examples: hair loss, dystrophic nails
Signs of CLI	Examples: dependent rubor, nonhealing wounds, gangrene, associated soft tissue infection
Pulse exam	
Palpation of femoral pulse	Palpable femoral pulse suggests infrainguinal disease. Diminished femoral pulse suggests suprainguinal disease.
Palpation of popliteal pulse	Rule out presence of popliteal aneurysm—increased incidence in patients with PAD.
Palpation of radial pulse	Asymmetry of radial pulses suggests unilateral subclavian disease. Bilateral diminished radial pulse suggests bilateral subclavian disease and that ABI will likely underestimate severity of PAD.
Doppler of DP and PT arteries	Objective assessment of presence of tibial flow with much better inter- and intra-observer variability compared with palpation. Quality of Doppler signal is also important—biphasic signal suggests superior flow compared with monophasic signal.
ABI measurement	Objective assessment of severity of PAD: >1.40 = noncompressible/uninterpretable; 1.0-1.40 = normal; 0.91-0.99 = equivocal; 0.40-0.90 = mild to moderate PAD; <0.40 = severe PAD. Note that ankle cuffs should not be placed over recent bypass grafts (because of risk of graft thrombosis) or over wounds. In the case of wounds, impermeable dressings may be used if indicated.
Pulmonary	Assess for presence of significant pulmonary disease.
Cardiac	Assess for evidence of structural heart disease (e.g., valvular heart disease) or HF.
Abdomen	Palpate for presence of AAA; use auscultation for renal or mesenteric bruits.
Neck	Use auscultation for carotid bruits.
Other	Assess for any other issues that might contribute to functional limitation (e.g., osteoarthritis of knee, peripheral neuropathy).

CLI = critical limb ischemia; PAD = peripheral artery disease; ABI = ankle–brachial index; DP = dorsalis pedis; PT = posterior tibial; HF = heart failure; AAA = abdominal aortic aneurysm.

The physical examination should be centered on the objective assessment of the lower extremity arterial circulation. Palpation of lower extremity pulses (common femoral, popliteal, posterior tibial, and dorsalis pedis) and documentation of the presence and quality of the Doppler signal from the tibial pulses are mandatory. The ABI provides an objective measure of the severity of PAD. Proficiency of ABI assessment among medical practitioners is poor, emphasizing the need for appropriate education of those assigned to perform this critical part of the lower extremity examination (28, 149, 223). In patients with CLI, the ABI has a number of limitations (e.g., greater interobserver variability) (127) and should rarely be used to make clinical decisions in this patient subset. Beyond the assessment of the lower extremity circulation, the physical examination should carefully document any evidence of atherosclerotic involvement of other vascular territories (e.g., abdominal aortic aneurysm) and any other physical manifestations of pathologies that might influence the functional performance of the patient.

As to laboratory testing in patients with PAD, routine testing should include a lipid panel and assessment of glucose control in patients with diabetes. When imaging using iodinated contrast agents is considered, such as computed tomography (CT) **angiography**, a renal profile is mandatory to assess the risk of contrast-induced nephropathy, which is a major cause of acute renal failure in hospital settings (72, 110).

Diagnostic Testing

Diagnostic testing in patients with PAD falls into two broad categories: pressure measurements (i.e., hemodynamic assessment) and imaging studies. Hemodynamic testing provides important functional information about disease severity (181), whereas imaging studies provide anatomic detail about the vascular obstruction. These tests are complementary in allowing optimal decision making in patients with PAD.

Hemodynamic Testing

The ABI is the simplest hemodynamic assessment in a patient with PAD. As such, it is the most widely used screening tool for PAD.

A more in-depth ABI measurement and interpretation review is provided elsewhere (3, 28), but briefly, the ABI measurement is executed as follows:

1. Brachial pressures are acquired through the placement of a blood pressure cuff around the arm and a Doppler probe (8-10 MHz) over the brachial artery. The blood pressure cuff is inflated 20 mmHg above the obliteration of the Doppler signal. The cuff is then slowly deflated (generally at a rate of 2-3 mmHg $\cdot$ s^{-1}), and the pressure at which the Doppler signal returns is recorded for each arm.

2. Using a similar technique, the ankle pressure is recorded from the dorsalis pedis (DP) and posterior tibial (PT) arteries of both lower extremities. It is important to apply an appropriately sized blood pressure cuff (typically a 10 cm cuff) over the lower half of the calf 2 cm above the superior aspect of the medial malleolus. Use of an inappropriately small-sized cuff or placement over the bulk of the calf muscles will result in a falsely elevated ankle pressure. Note the cuff pressure should not be inflated above 300 mmHg; if the cuff is inflated to this high level, it should be promptly deflated to circumvent pain.

In calculation of the ABI, the brachial pressure is used as a surrogate for the intra-aortic pressure, which is a reasonable assumption in the absence of significant subclavian artery disease. Therefore, the higher of the two brachial pressures is used for the ABI calculation of both lower extremities. By convention, the higher of the two ankle pressures (i.e., DP or PT) is used for the ankle pressure component of the calculation. The higher of the two ankle pressures is used in order to increase the specificity of the ABI measurement for making the diagnosis of PAD and to prevent overdiagnosis (3, 89). This may seem counterintuitive for a screening test, particularly because of the potential for higher rates of false negatives, but it remains the standard method for ABI calculation in clinical practice. The standardized interpretation of ABI values is provided in table 17.2 (71).

Additional hemodynamic testing beyond the ABI measurement requires specific equipment and is generally performed in vascular laboratories. These tests include the following:

3. *Great toe pressure:* The great toe pressure is typically acquired in the vascular laboratory using a small pneumatic cuff that is applied to the great toe of each foot. The absolute toe pressure is normally 30 mm lower than the ankle pressure, and a toe–brachial index of ≥0.70 is regarded as normal. In clinical practice, the toe pressure is rarely measured in patients with claudication. It may be used to confirm the diagnosis of PAD in a patient with claudication with a noncompressible vessel and therefore uninterpretable ABI measurement, which has been rated as class I, level of evidence B-NR, by the American College of Cardiology (ACC)/ American Heart Association (AHA) PAD management guidelines (71). In contrast, this measurement is widely used in patients with CLI to confirm the diagnosis (i.e., toe pressure of <50 mmHg). In addition, the absolute toe pressure is helpful in predicting the likelihood of limb salvage and the need for revascularization.

4. *Segmental limb pressures (SLPs):* Segmental limb pressures help provide hemodynamic evidence of the anatomic location of PAD (71). They are acquired via placement of blood pressure cuffs at various locations along the length of the leg (i.e., upper thigh, lower thigh, upper calf, ankle). The SLP at each level is recorded by inflation of the cuff of interest to above systolic pressures and recording the pressure at which the Doppler signal returns to the DP and PT arteries at the ankle. Generally, the thigh pressures are higher than brachial pressures, with a progressive decrease in pressure measured at each subsequent cuff. However, a drop of ≥20 mmHg between adjacent SLPs suggests significant disease in the intervening arterial segment.

5. *Pulse volume recordings (PVRs):* Pulse volume recordings are recorded with the same cuffs used to perform SLPs. For the purpose of recording PVRs, each of the cuffs is attached to a plethysmograph. Each cuff is sequentially inflated up to a predefined reference pressure of ~65 mmHg (70), and the volume change over time at that cuff level is recorded. Since arterial inflow is the only significant volume changing over time in the lower extremity, the waveform generated resembles an arterial waveform. The amplitude of the waveform provides important information about the adequacy of arterial inflow at that level, and comparison with the PVR above and below that level is also helpful in assessing the location of disease in that arterial segment. Table 17.3 provides a summary of presumed anatomic location of disease based on a drop of ≥20 mmHg in pressure between SLPs or change in amplitude of PVRs between adjacent pressure cuffs.

6. *Transcutaneous oxygen pressure (TcPO$_2$) and skin perfusion pressure:* These measurements are performed in a small number of laboratories and provide functional information about a number of key determinants of flow and tissue perfusion, including wound healing and microcirculation (182, 218). As such, they are used primarily in patients with CLI. Clinically, these measurements are especially useful in patients who have had prior digital or transmetatarsal amputation, where the toe pressure measurement is not possible. A cutoff of 30 mmHg for both TcPO$_2$ and skin perfusion pressure is believed to be consistent with the diagnosis of CLI. Figure 17.1 depicts a full hemodynamic evaluation of the lower extremities.

As indicated in figure 17.1, the left ABI measured 0.40 (55/144) and the right ABI measured 0.89 (128/144). This is consistent with the diagnosis of severe PAD in the left lower extremity. The left great toe pressure of 0 mmHg supports the diagnosis of CLI and suggests that wound healing will not occur without revascularization. The PVRs suggest that the location of the disease is located predominantly in the tibial vessels of the leg and small vessels of the foot. The SLPs in the left lower extremity are not helpful in locating the site of disease.

Imaging Studies

Computed tomography and magnetic resonance (MR) angiography are both noninvasive and offer anatomic detail rivaling that obtained from invasive angiography (11, 24, 51, 117, 147, 180). As a result, in contemporary practice, the majority of invasive angiographic studies are performed as a preamble to planned interventional procedures based on the diagnostic findings of noninvasive studies (117).

Computed tomography and MR angiography are performed

Table 17.3   Summary of Presumed Anatomic Location of Disease Using Segmental Limb Pressures and Pulse Volume Recordings

Location of drop in SLP or change in PVR*	Presumed location of diseased arterial segment
Brachial–upper thigh	Aortoiliac, CFA
Upper thigh–lower thigh	SFA
Lower thigh–upper calf	Distal SFA, popliteal artery
Upper calf–ankle	Tibial arteries
Ankle–toe	Small vessels of the foot

*Based on a drop of ≥20 mmHg in pressure between sequential limb pressures or change in amplitude of pulse volume recordings between adjacent pressure cuffs.

SLP = segmental limb pressure; PVR = pulse volume recordings; CFA = common femoral artery; SFA = superficial femoral artery.

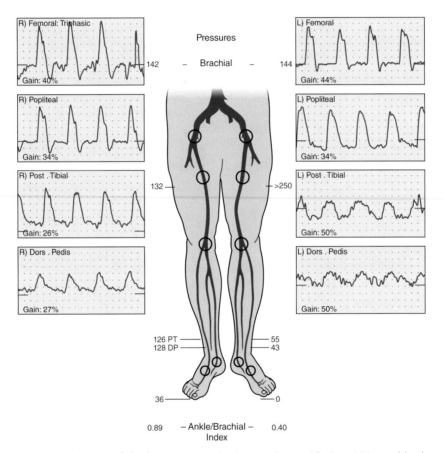

Figure 17.1   Full hemodynamic evaluation of the lower extremity in a patient with CLI. ABI = ankle–brachial index; CLI = critical limb ischemia; PVR = pulse volume recording; SLP = segmental limb pressure.

Courtesy of Ryan J. Mays, Ivan P. Casserly, and Judith G. Regensteiner.

following the administration of iodinated and gadolinium-based contrast agents through a peripheral intravenous cannula. Each of these imaging techniques has specific advantages and disadvantages (table 17.4). Most vascular specialists use CT angiography as the first-line option for noninvasive imaging. Nephrogenic systemic fibrosis is a potential complication of exposure to gadolinium-based contrast agents during MR angiography. Thus, similar thresholds of renal insufficiency as a contraindication to

imaging by MR and CT angiography have been defined, removing the major clinical advantage previously held by MR angiography in this patient population (169, 219).

Duplex ultrasound is unique among imaging studies in that it offers both functional and anatomic detail about the peripheral arteries (1). B-mode imaging provides some basic information regarding vessel size and the presence of plaque. Doppler examination allows an assessment of flow within the arterial segment

Table 17.4 Summary of Advantages and Disadvantages of Noninvasive Imaging Using Computed Tomography or Magnetic Resonance Angiography

Imaging study	Advantages	Disadvantages
CT angiography	Widespread availability Simplicity of imaging protocols Higher spatial resolution Rapid scanning times Large gantry size Ability to visualize calcium Ability to visualize lumen within stents Lower cost	Artifact in presence of severe calcium Exposure to ionizing radiation Use of nephrotoxic contrast
MR angiography	Absence of ionizing radiation No interference from calcium Ability to obtain some information with specialized protocols without use of contrast agents	Higher cost More complicated imaging protocols Contraindicated in patients with ferromagnetic materials (e.g., pacemakers, orthopedic hardware, shrapnel) Small gantry size, limiting studies for patients with claustrophobia or marked obesity Scanning times prolonged, requiring greater degree of patient cooperation Venous contamination in legs Difficulty visualizing stents; artifact from previously placed stents

CT = computed tomography; MR = magnetic resonance.

of interest and provides information regarding the functional significance of a stenosis (figure 17.2). The clues to the presence of functionally significant stenosis are as follows:

1. *Alteration of the normal triphasic Doppler waveform:* The earliest alteration in the waveform is a loss of the transient reverse flow component in systole, resulting in a biphasic waveform. This is followed by loss of the late diastolic forward flow component, resulting in a monophasic waveform. These changes are qualitative and need to be interpreted in the context of the assessment of flow velocities.

2. *Peak systolic velocity (PSV):* A predictable increase in the PSV occurs at the site of an arterial obstruction. A PSV of ≥ 200 cm $\cdot$ s^{-1} is generally regarded as indicating a stenosis of $\geq 50\%$ in the vessel segment (108). However, there is significant variation in the PSV along the length of the lower extremity. Hence a comparison of the PSV at the site of a stenosis with the PSV in the segment of the vessel proximal to the site of stenosis appears to be a more specific measure of the functional significance of a stenosis. A peak velocity ratio of ≥ 2.0 is generally regarded as indicating the presence of a significant stenosis. Another sign of turbulent blood flow indicative of a stenosis and related to high flow velocities is spectral broadening, which is, briefly, the filling of the spectral window seen in the Doppler waveform (70).

Invasive angiography remains an important diagnostic imaging tool for a subset of patients. These include patients who have significant renal insufficiency that precludes the use of a bolus of iodinated contrast or gadolinium-based contrast for CT or MR

angiography, respectively. In such patients, targeted invasive angiography may be used to answer a specific clinical question. In addition, in patients with CLI, invasive angiography remains the gold standard for the assessment of the infrapopliteal and small vessel anatomy of the foot (26, 197). This information is critical in clinical decision making for this patient group. Magnetic resonance angiography suffers from significant issues with venous contamination in the assessment of tibial anatomy, and the resolution of CT angiography is not sufficient to provide this anatomical information in most individuals. Thus, invasive angiography is often used to confirm any clinically relevant findings from noninvasive diagnostic tests that may ultimately lead to potential intervention (27).

Exercise Testing

Exercise testing is useful for determining the degree of functional limitation in patients with PAD. Exercise testing using the treadmill is a valuable tool for the exercise physiologist in designing a walking program for patients with PAD. Assessment of a patient's walking ability is critical in determining the clinical, functional, and quality of life limitations typically associated with the disease. This section summarizes recommended exercise testing modalities and appropriate procedures for assessing the most important health-related fitness components in PAD (i.e., cardiorespiratory fitness, walking performance and functional ability, muscular strength, and endurance).

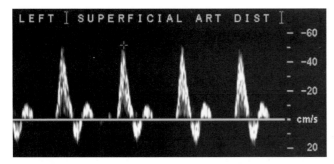 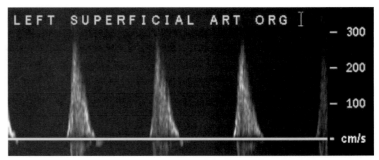

Figure 17.2 *(a)* Doppler ultrasound of normal flow pattern in superficial femoral artery. *(b)* Doppler ultrasound of flow in patient with severe stenosis in superficial femoral artery: high peak systolic velocity and loss of normal triphasic waveform (to monophasic waveform). Also note spectral broadening, indicating turbulent flow.

Photos courtesy of Ryan J. Mays, Ivan P. Casserly, and Judith G. Regensteiner.

Cardiorespiratory and Functional Testing

The hemodynamic assessments outlined earlier are predicated on the detection of PAD based on the presence of a pressure gradient across a diseased segment of the artery at rest. Occasionally, however, moderate arterial obstructions (i.e., causing ~50%-70% stenosis) are not associated with a resting gradient, but the increased flow across the lesion due to the increased cardiac output during exercise may provoke an exercise-induced gradient. It has traditionally been taught that this is most likely associated with moderate lesions in the aortoiliac segment but can be observed with moderate disease in any arterial segment. As a result, an exercise study is crucial in identifying this subset of patients, who typically provide a good clinical history that is consistent with claudication and have a normal resting ABI (202).

Exercise testing is also helpful in evaluating patients with atypical symptoms where a significant drop in perfusion pressures with exercise is associated with the onset of symptoms, supporting the conclusion that PAD is the probable underlying etiology. The presence of PAD is established by measuring an ABI at rest. If the resting ABI is >0.90 but ≤1.40, with the patient describing exertional non-joint-related leg pain, they can walk on a treadmill using a standard protocol (71) or use an alternative method to activate the gastrocnemius muscle (e.g., active pedal plantar flexion (144). The ABI measurement is repeated immediately following the exercise to determine the effects of walking on disease severity. The study is considered abnormal if the ankle pressure decreases >30 mmHg or drops by ≥20% from the baseline measurement and takes more than 3 min to normalize (3). The current PAD management guidelines provide a class IIa recommendation (level of evidence B-NR) for the use of postexercise ABI testing for patients with abnormal resting measurements to assess functional status (71). Evidence suggests that abnormal postexercise ABIs are related to an increase in the incidence of lower limb revascularization (81). Thus, the exercise physiologist may consider this type of test when working with those who have PAD to appropriately stratify their functional limitations and identify the potential and need for procedures.

The most objective modality for testing the patient's walking ability and subsequently determining the effectiveness of any therapy for PAD is a graded treadmill walking test. Exercise stress testing can also be helpful for determining whether heart disease is present (189). This type of test has been well validated for use in patients with claudication (97). Treadmill testing can be used to establish **claudication onset time** (the time point of initial onset of any claudication pain) as well as **peak walking time** (the time point at which claudication pain becomes so severe that the patient is forced to stop). Although claudication onset and peak walking distances (walking distance achieved for each respective category) have been and continue to be used as end points in PAD clinical trials, vascular disease experts have proposed specific nomenclature using time-based outcomes as opposed to distance (91).

Patients with PAD should perform a familiarization session with the treadmill regardless of experience with treadmill walking. During the actual test, the end point is determined by the patient's perception of pain using the Claudication Symptom Rating Scale, which ranges from 1 to 5 (1 = no pain, 2 = onset claudication, 3 = mild, 4 = moderate, and 5 = severe pain) (87). Alternative pain scales have also been proposed (4, 207) to help patients decide when they need to stop. During testing, patients should not use handrail support because claudication distances are falsely elevated and thus do not reflect true exercise performance ability (67). However, because patients may have balance issues in addition to a lack of familiarity with treadmill exercise, they may use handrails for balance only.

Several treadmill protocols have been used in PAD populations to test for leg pain and evaluate other potential limiting factors in response to exercise (68). A typical exercise protocol for patients uses either a graded treadmill test or a constant-load test. The graded exercise test keeps speed constant at 2 mph throughout. Grade begins at 0% and increases 2% every 2 min thereafter (66). The constant-workload or single-stage treadmill test holds speed and grade constant throughout. Gardner and colleagues (66) reported that the severity of PAD is better assessed with a graded protocol because walking performance end points demonstrated better reliability than constant-load tests. However, both types of test have been widely used.

Although optimal, testing using walking exercise is not appropriate for all patients and may be unavailable in some clinic settings. Patients with abdominal aortic aneurysms, uncontrolled hypertension, or other exercise-limiting comorbidities, such as severe HF or chronic obstructive pulmonary disease (COPD), may need aerobic capacity evaluation using a different modality (156). Wolosker and colleagues (221) determined that 16% of patients with PAD were unable to complete a treadmill test because of various limiting factors due to comorbid conditions. This situation, in addition to some patients' lack of familiarity with treadmills, could lead to testing without useful clinical results. Thus, other modalities have been used to evaluate exercise capacity in patients with PAD, including arm and leg ergometry, stair stepping, and active pedal plantar flexion (61, 68, 224, 226). Gardner and colleagues (65) demonstrated similar peak oxygen consumption ($\dot{V}O_2$peak) values between testing protocols when patients with claudication performed level walking, graded walking, or progressive stair climbing. Thus, while graded exercise tests using the treadmill are preferred, other cardiorespiratory fitness assessments using different modalities are available if needed.

In addition to walking impairment, patients with PAD have more overall functional impairment than persons without PAD (137, 140). Therefore, the ability to assess functional impairment is of key importance in assessing the patient. Those with PAD and declining functional performance are at increased risk for mobility loss with increasing age, and women may experience a faster functional decline than men (135, 139). Functional impairment can be subjectively assessed through validated questionnaires and objective measurements such as walking, balance, and other tests of general function. These performance test outcomes are more strongly associated with the volume of physical activity during daily life than are treadmill walking measures and thus may provide complementary information to augment that obtained with a peak aerobic exercise test (69, 134).

In the 6 min walk test, the patient walks a defined course for 6 min, and the distance achieved is recorded. The validity of this test in patients with PAD has been examined extensively in single-site studies (115, 133, 136). However, the 6 min walk test has not been validated in multicenter trials, where standardization within trial sites is critically important to establish appropriate generalizability to diverse sites and real-world locations (e.g., health and fitness settings, other countries). Incremental- and constant-speed shuttle walking tests demonstrate valid and reliable measures of a patient's functional ability (38, 226). Patients typically walk a defined distance (e.g., 10 m) back and forth between two destinations (marked by cones or tape); speed is held constant or a timer signals to increase the speed, depending on the protocol. Finally, the Short Physical Performance Battery has also been used for functional performance testing in PAD (73, 129, 213). It combines several functional tests that provide a comprehensive objective measure of leg function, balance, leg strength, and ability to walk (78). This is important because poor lower limb functioning is strongly related to disability, mortality rates, or both in the aging patient, regardless of disease. Functional performance tests may reflect usual walking and overall limb function during activities of daily life while providing clinicians with objective information for evaluating the progress of exercise training without expensive laboratory equipment.

Muscular Strength and Endurance Testing

Because of the advanced cardiovascular disease in people with PAD, cardiorespiratory fitness and functional testing remain the most important evaluation methods. However, McDermott and colleagues (142) examined the relationship between upper and lower limb strength and ABI of patients with and without PAD. Testing included handgrip, knee extension, and active pedal plantar flexion isometric strength as well as leg power during knee extension exercise. Results indicate that impaired strength may be limited to the lower extremities and that patients with PAD had lower plantar flexion strength and lower knee extension power than patients without PAD (142). Several other studies have evaluated strength and endurance in patients, such as a one-repetition maximum using the leg press (151, 214), high-repetition muscular

Assessment of Patient-Reported Outcomes

The subjective assessment of the health and well-being of patients with PAD can be captured via validated questionnaires. Exercise physiologists should strongly consider using questionnaires when working with these patients because the response to training as well as the appropriate therapeutic strategy is often guided by individual patient perspective. The Medical Outcomes Study Short Form 36 (SF-36) (216) and the Walking Impairment Questionnaire (WIQ) (174) are most commonly used in PAD (128, 209). The 36-item SF-36 is a general tool describing health domains (e.g., physical function, mental health, bodily pain) and is useful for comparing results across patient populations. The WIQ is a disease-specific metric that assesses ability to walk defined distances, increase speeds, and climb stairs. The WIQ in particular has been used extensively in PAD research trials as well as in the clinical setting and is often regarded as the optimal tool for characterizing patients' ambulatory limitations (153). A number of other metrics have been used to evaluate quality of life in PAD (128), including the Peripheral Artery Questionnaire (PAQ) (200), the Intermittent Claudication Questionnaire (31), and the Peripheral Artery Disease Quality of Life (PAD-QoL) questionnaire (208).

endurance testing (164), and isokinetic strength and endurance testing (25, 177). Table 17.5 reviews testing options for patients with PAD to evaluate health-related fitness components.

Treatment

A full discussion of the treatment of PAD is beyond the scope of this chapter. In general, the therapies offered can be broadly divided into two categories: optimal medical treatment and revascularization. Medical treatment focuses on therapies that improve the systemic cardiovascular complications associated with the diagnosis of PAD and agents that specifically may improve claudication symptoms (10, 77, 82). Aggressive modification of atherosclerotic risk factors is mandatory for those with PAD, with the targets for therapy as outlined in table 17.6.

It is generally agreed that all patients with PAD should be treated with an antiplatelet agent, such as aspirin or clopidogrel, as well as other medications that treat risk factors for cardiovascular disease (e.g., statins) (18, 178). Patients with PAD should have their risk factors treated with the same intensity as patients with other cardiovascular diseases. In contrast to the clear importance of drugs for optimally treating cardiovascular risk factors in PAD, pharmacological treatment of claudication has been largely ineffective. Currently, despite numerous studies of proposed agents using many different mechanisms to treat claudication, only two drugs have been approved by the Food and Drug Administration specifically for these patients: pentoxifylline and cilostazol.

Pentoxifylline improves the hemorrhagic profile of patients by improving red cell deformability, lowering fibrinogen levels, and decreasing platelet aggregation (168, 175). However, because

Table 17.5   Exercise Testing Review

Cardiorespiratory endurance	Skeletal muscle strength and endurance	Flexibility	Body composition

1. Graded exercise testing
Primary modality
   TM
Secondary modalities
   CE
   AE
   Active plantar pedal flexion
2. Functional testing
• 6 min walk test
• SPPB
• Shuttle walking tests
	Upper and lower body muscle group testing, using free weights or machines per standard guidelines outlined in chapter 33 for older adults
	1. Isometric testing
	• Knee extension
	• Plantar flexion
	• Handgrip dynamometry
	2. Maximal strength testing
	   1RM or multi-RM (e.g., 5 RM) of upper and lower body muscle groups
	3. Isokinetic testing
		Refer to chapter 33 for standard testing guidelines in older adults.	Refer to chapter 8 for standard testing guidelines.

TM = treadmill; CE = cycle ergometer; AE = arm ergometer; SPPB = Short Physical Performance Battery; 1RM = one-repetition maximum.

Table 17.6   Major Components of Medical Management in Patients With Peripheral Artery Disease

Treatment	Comments
GENERAL MEDICAL TREATMENT	
Antiplatelet agents	Aspirin 75 to 325 mg daily po; clopidogrel 75 mg daily po in patients with contraindication to or intolerance of aspirin
Statins	Statins indicated in patients with diagnosis of PAD independent of LDL cholesterol level; may have favorable impact on leg symptoms independent of lipid-lowering properties
MODIFICATION OF ATHEROSCLEROTIC RISK FACTORS	
Hypertension	Target <140/85; ACE inhibitors should be used as first-line therapy
Hyperlipidemia	Target LDL cholesterol <70 mg · dL^{-1}
Smoking	Smoking cessation strongly recommended
Diabetes	HbA1c <7%
IMPROVING CLAUDICATION DISTANCE	
Cilostazol	Dose 100 mg twice daily po
ACE inhibitors	May have favorable impact on leg symptoms
Exercise	See the Exercise Prescription section

po = orally; PAD = peripheral artery disease; LDL = low-density lipoprotein; ACE = angiotensin-converting enzyme; HbA1c = glycated hemoglobin.

CLIENT–CLINICIAN INTERACTION

Many patients with PAD have been sedentary in their lifetimes and may not have the knowledge to start a safe and effective exercise program. Because patients may fear the symptoms and limitations associated with claudication (59, 76), the exercise physiologist needs to work collaboratively with clients to prescribe a detailed exercise plan and to reassure them that walking will not cause harm. More specifically, neither high-intensity shorter bouts nor low-intensity longer bouts of exercise cause detriment to muscle beyond the possibility of overuse, which is commonly experienced by all populations, especially with aging (62, 155).

The clinician should explain to patients that the arterial and venous systems are separate, because deep vein thrombosis and PAD are distinct problems and present different treatment options and risks. Patients need to understand that routine exercise is one of the most beneficial therapies for PAD, in addition to reducing risk factors associated with their underlying atherosclerosis. Smoking cessation should be strongly encouraged by the exercise physiologist because the relative risk of claudication was demonstrated to be 3.7 in current smokers versus 3.0 in ex-smokers who had discontinued smoking for less than 5 yr (57, 156). Finally, adequate foot care must be stressed to patients with PAD, particularly those who have diabetes. Some patients with claudication may be hesitant to walk on a treadmill for fear of damaging their legs and feet. The risk is that ulceration from trauma, specifically from poorly fitting footwear or improperly cut toenails, may result in a nonhealing wound that could lead to amputation (204).

there is minimal efficacy of this drug in terms of improving walking distance, it is no longer recommended for treating claudication (186). Cilostazol, a phosphodiesterase inhibitor, is currently the only effective approved medication used to improve walking distance in those with claudication in the United States (16, 88). One study in patients with claudication who were randomized to receive cilostazol, pentoxifylline, or placebo for 6 mo showed that cilostazol improved maximal walking distance by 54% compared with a 30% increase in the pentoxifylline group and 34% in the placebo group (40). Additionally, a meta-analysis of six randomized controlled trials indicated that cilostazol improved walking ability and patient-reported outcomes in patients with claudication compared with placebo (176). Many other investigative pharmacological agents have been evaluated in PAD research trials in an attempt to establish an effective treatment for claudication; however, most have not shown significant benefits or are associated with significant side effects and intolerability (table 17.7).

The evolution of endovascular techniques to revascularize lower extremity arterial disease has resulted in a paradigm shift in the treatment of patients with claudication (47, 86, 111). Previously, the primary option for revascularization was surgical bypass, which was associated with significant morbidity and mortality (5, 6, 45, 109, 166). As a result, the risk–benefit ratio of surgical revascularization restricted its use to a small subset of patients with PAD. These included low-risk surgical patients with severe claudication symptoms and patients with CLI. In contrast, **endovascular revascularization** is characterized by very low rates of morbidity and mortality and thus is now considered the current standard of care for patients who are candidates for revascularization (159). With modern technology and techniques, acute technical success rates approach 100% for most procedures (179, 217). As a result, endovascular revascularization may be offered to a much broader spectrum of patients, although exercise training remains the gold standard treatment for improving peak walking time of patients with claudication (152).

The major limitation of endovascular compared with surgical revascularization is lower long-term **patency** rates. In the aortoiliac arterial segments, this difference in long-term patency is modest, whereas in the femoropopliteal segment, the difference is significant. Since restenosis generally results in a recurrence of claudication, the lower patency rates associated with endovascular treatments are a concern in the treatment of patients with claudication and warrant careful consideration before this treatment strategy is offered. In general, a more aggressive approach is adopted in younger, active patients without comorbidities that may contribute to functional limitation and with more favorable disease anatomies (e.g., aortoiliac vs. femoropopliteal, stenosis vs. occlusion, focal vs. diffuse disease, noncalcified vs. calcified plaque). In contrast, since CLI is associated with a significant risk of limb loss, and restenosis is often clinically silent, an aggressive approach to endovascular revascularization is warranted in most patients with CLI.

Table 17.8 summarizes the endovascular and surgical treatment options available for revascularization for disease in the major arterial territories of the lower extremity. Figure 17.3 depicts angiographic images of the lower extremity arterial system of a patient with CLI before and after angioplasty of the peroneal artery.

EXERCISE PRESCRIPTION

Walking exercise is particularly beneficial for helping patients with claudication to improve their walking ability, although all types of exercise are beneficial for general cardiorespiratory health. A walking exercise program causes physiological adaptations leading

Table 17.7 Pharmacology

Medication class and name	Primary effects	Exercise effects	Important considerations
Statins Atorvastatin, lovastatin, simvastatin	May improve plaque stabilization, endothelial function, platelet activity, and inflammation (50)	Atorvastatin and simvastatin found to improve maximum walking distances (148, 150); not all trials successful for improving treadmill walking times in PAD (92)	Side effects may include myalgia with risk of rhabdomyolysis, increased liver enzymes, and polyneuritis (41, 79)
Prostaglandins Beraprost, iloprost, others	Increases cAMP levels, subsequently increasing peripheral vasodilation and platelet inhibition (50)	PAD management guidelines (71) do not recommend its use (class III, evidence level B-NR)	Discontinuation rate 20%, due to adverse effects such as headache, flushing, and gastrointestinal distress (17, 48, 50)
Peripheral vasodilators and cerebral activators Buflomedil	Platelet aggregation and leukocyte adhesion inhibitor; increases tissue tolerance to ischemia (103, 185)	Recent findings conclude lack of evidence for treating claudication (125)	Safety concerns cited as well as publication bias (41, 43, 156)
Phosphodiesterase inhibitors Cilostazol	Vasodilatory effects; inhibits the aggregation of platelets and smooth muscle proliferation (19)	Effective for improving walking and patient-reported outcomes in patients with claudication (16, 161, 176)	Contraindicated in patients with heart failure (30, 160); poor patient compliance, possibly due to side effects such as dyspnea, headache, palpitations, and diarrhea (19, 94)
Amino acids l-carnitine, propionyl-L-carnitine	Improves the oxidation of glucose and enhances oxidative metabolism (19)	Small to moderate improvements in walking ability for most trials (21, 46); propionyl-L-carnitine in combination with monitored home exercise not superior to exercise alone (90)	No major safety concerns for treating claudication reported (90, 120)
Serotonin receptor antagonists Naftidrofuryl, sarpogrelate, others	Reduces erythrocyte and platelet aggregation (156, 201)	Mixed results depending on the agent, but some evidence for improvement in exercise tolerance (42, 119)	Naftidrofuryl is recommended as a treatment option for claudication (grade A) (156); approved in Europe but unavailable in the United States
Hemorrheologic agents Pentoxifylline	Lowers fibrinogen levels, improves RBC deformity, and enhances erythrocyte distensibility (131, 156)	Lack of efficacy for improving walking outcomes (40, 44, 186)	First drug approved by FDA for treating claudication (early 1980s); no longer in general use for treating PAD (class III, evidence level B-R) (71)

PAD = peripheral artery disease; cAMP = cyclic adenosine monophosphate; RBC = red blood cell; FDA = Food and Drug Administration.

to higher $\dot{V}O_2$, increased walking time, and enhanced quality of life (113, 162, 173, 210). PAD patients with claudication are often not able to walk very well when they begin an exercise program, and their exercise prescription is set according to their individual limitations. The exercise physiologist should be aware that exercise before and after peripheral revascularization is highly recommended (71). If patients receive peripheral endovascular therapy, for example, exercise can begin after clearance from the interventionalist, who ensures that the arterial access site has healed and no other complications are present.

Detailed discussions of optimal exercise programs and their implementation in PAD serve as key references for exercise physiologists (209, 210). Briefly, the ideal program for patients should begin with a warm-up and end with a cool-down. Bouts of exercise are intermittent, with the length of time for each bout limited by the onset of moderate to moderately severe claudication (minimum

of 4 on the Claudication Symptom Rating Scale) (87). Each bout is followed by a rest period to allow the leg pain to subside. These exercise and rest sessions are repeated to accumulate 30 to 45 min of exercise during a 60 min session, at least three times a week, for a total duration of 3 to 6 mo, although some programs have been carried out for up to a year.

Initially, the prescription for exercise is to walk (with no incline) at the speed that elicits moderate claudication (209). Patients then sit and rest until the pain abates. For the first few sessions, the goal is for the patient to walk and rest for a total session time of 20 min (with the time from all walking and resting periods added together). Sessions typically become longer with further training until the goal of 60 min has been reached. For patients who have received endovascular therapy or bypass surgery and may have other factors limiting exercise (e.g., orthopedic limitations or muscular fatigue), rating of perceived exertion (RPE) using the

Table 17.8   Endovascular and Surgical Options for the Revascularization of Arterial Segments of the Lower Extremity

Arterial segment	Endovascular	Surgical
Aortoiliac	Stenting[a]	Aortobifemoral bypass
CFA	PTA, atherectomy[b]	CFA endarterectomy
Femoropopliteal	PTA, atherectomy, stenting[c]	Femoropopliteal bypass
Tibial[d]	PTA, atherectomy, stenting[e]	Femorotibial bypass

[a]Typical stent types in use include balloon expandable, self-expanding, covered self-expandable, and covered balloon expandable.

[b]Surgical treatment of common femoral artery disease is preferred over endovascular approach. Stents should be avoided in the common femoral artery.

[c]Typical stent types include self-expanding and covered self-expanding. Stents should be avoided in popliteal artery.

[d]Tibial revascularization typically restricted for treatment of patients with critical limb ischemia.

[e]Typical stent types include bare metal and drug-eluting balloon, expandable, and self-expanding.

CFA = common femoral artery; PTA = percutaneous transluminal angioplasty.

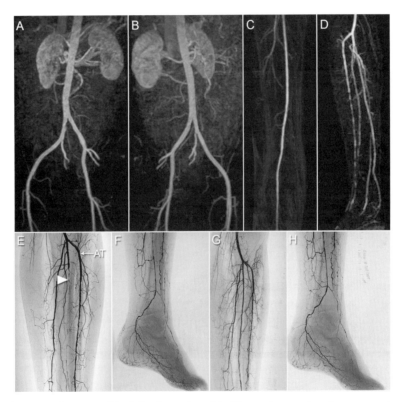

Figure 17.3   Anatomic evaluation of 68 yr old diabetic male with CLI and a nonhealing wound on the great toe of his right foot. *(a-d)* Magnetic resonance angiogram of pelvic region and right lower extremity demonstrating widely patent iliac system, right common femoral artery, superficial femoral artery, and popliteal artery. *(e, f)* Routine angiography of infrapopliteal vessels delineating anatomy of tibial pedal vessels and showing that the peroneal artery (arrowhead) should be the target for revascularization. *(g, h)* Final angiogram of the tibial and pedal vessels following angioplasty of the peroneal artery. AT = anterior tibial.

Photo courtesy of Ryan J. Mays, Ivan P. Casserly, and Judith G. Regensteiner.

Borg 15-category scale or other walking perceived exertion scales (e.g., OMNI Walk/Run Scale) can be used to regulate exercise intensity at a moderate to vigorous level (20, 211), as has been done in the past for exercise training in asymptomatic PAD (133). Regulating exercise intensity based on age-predicted exercise heart rate responses rather than claudication symptoms is not typically recommended for patients with PAD.

Patients should exercise in structured environments, such as supervised hospital-based settings or structured home or community exercise programs (71). Supervised exercise training improves walking performance, physical function, and patient-reported outcomes (113, 162). Additionally, several trials have shown the lasting benefits of walking for patients with PAD, even after completion of the defined supervised training period (146, 152).

However, only recently have supervised programs been reimbursable by the Centers for Medicare and Medicaid Services (105). The decision memo and details for the process of referral and coverage of supervised exercise training is described elsewhere (105, 209). Despite this critically important step to providing access for effective training in PAD, it is important for the exercise physiologist to be aware of the limitations of these programs. Supervised exercise training is often not possible because of barriers, including specific patient inclusion criteria and low levels of physician referrals into programs, patient out-of-pocket expenses, time constraints, and habitual sedentary behavior of patients (84, 209). Other barriers to successful adoption and adherence of supervised exercise training include poor proximity to clinics, the need for patients to navigate large health care facilities, and lack of adequate transportation (170, 188).

Until recently, unsupervised exercise programs (whether the patient was exercising alone or transitioning from supervised hospital-based settings) have primarily involved instructing patients to exercise at home or in the community, with relatively little follow-up (29, 55, 107, 130, 172, 192). As a result, home-based programs have not, for the most part, been successful. Factors resulting in the failure of many of these programs may include a less resource-intensive approach, a lack of motivation in patients, and reliance on patient self-monitoring with no feedback (33, 159). Kakkos and colleagues (107) randomized patients with claudication into supervised and unsupervised exercise training programs and examined resulting claudication onset and peak walking times (baseline, 6 wk, and 6 mo). Patients in the home exercise program were advised to exercise daily by walking, without further advice. The supervised exercise training group improved both claudication onset time and peak walking time, versus no improvement in either outcome for patients in the home training group. Other studies have demonstrated similarly poor results after patients were instructed to exercise at home or in the community with no feedback.

However, a number of structured community-based exercise programs have provided more focused and intensive approaches to improve health outcomes in PAD (63, 64, 75, 129, 141, 205). Briefly, McDermott and colleagues (141) conducted a 6 mo randomized controlled group–based exercise program for patients with PAD at an indoor community facility (e.g., track), led by a trained exercise facilitator. The program applied a group-mediated cognitive behavioral approach. Patients met with other patients once weekly and completed 45 min of walking exercise and 45 min of facilitator-led discussions on various topics (e.g., benefits of walking for PAD, setting goals, monitoring exercise). The control group of patients with PAD received health education sessions each week but did not receive any exercise intervention or feedback from staff regarding physical activity. Significant differences were found for 6 min peak walking distance change scores (primary outcome) between the intervention and control groups (+42.4 vs. –11.1 m [95% CI difference scores, 33.2 to 73.8] $p < 0.001$) at follow-up. Additionally, at 1 yr follow-up where patients were contacted by phone (biweekly for months 7-9 and once per month for months 10-12), those patients in the intervention group were able to maintain their improvements in walking performance

(138). Thus, offering supervised programs in community settings may improve compliance and provide benefits. Barriers to home- and community-based programs do exist and include poor weather or unsuitable walking environments for outdoor exercise, lack of direct patient monitoring by staff, poor self-efficacy of patients, and difficulty managing leg pain (12, 59, 130, 141).

Strength training alone, treadmill walking alone, and combinations of the two have been examined in patients with claudication to determine whether one modality of training was superior (98). Patients with symptomatic PAD showed improvements in peak walking times with treadmill exercise or with weight training for skeletal muscles of the legs (163). The treadmill group showed improvement in $\dot{V}O_2$peak, but no changes were observed in $\dot{V}O_2$peak or claudication onset time for those participating in weight training alone. More recent data have demonstrated improvements in walking and functional outcomes following muscular strength and endurance training in patients with PAD (121, 162, 163). However, although strength training is important, it should only complement the patient's exercise program. Walking is still preferred because of the beneficial effects on the cardiorespiratory system (71).

Other exercise modalities that have been evaluated for use in treating PAD include cycle exercise, arm ergometry, active pedal plantar flexion, pole striding, and total-body recumbent stepping (74, 114, 187, 206, 215). In particular, arm ergometry may be a valid alternative for treating patients with PAD (23, 206) and may have merit for use in patients with CLI. Table 17.9 presents a brief review of the optimal training modalities for improving various health and fitness components of those who have PAD.

EXERCISE TRAINING

In most studies, exercise training has resulted in excellent improvements in walking distances in patients with claudication, as well as a reduction in adverse cardiovascular event risk in a limited number of studies (113). Low-intensity exercise training programs have been shown to be similar to high-intensity exercise rehabilitation for improving initial claudication onset distances and peak walking distances (62). However, achieving moderate exercise levels (guided by pain or general exercise discomfort) is still more often recommended (4, 71, 209). As patients improve their functional status and are able to exercise more, they may experience some adverse cardiac symptoms, at which time they should be reevaluated by their primary care physician, who may order a comprehensive ischemic evaluation that includes an exercise stress test (158).

Both walking distance on the treadmill and health status assessed by questionnaires are improved by supervised exercise training (128). Thus, supervised exercise training is considered the gold standard therapy for patients with PAD and claudication (71, 156). Additionally, supervised exercise training programs improve peak walking time in patients who do not experience classic claudication (133, 143). Overall, exercise training improves the ability to walk for longer periods, and the consistency of these findings suggests that exercise training programs have a clinically

Table 17.9 Exercise Prescription Review

Training method	Frequency	Intensity	Time (Duration)	Type (Mode)	Progression	Important considerations
Cardiorespiratory	At least 3 d/wk	Moderate leg pain in 5-10 min; rest until pain dissipates; repeat	Work up to 30-45 min of actual exercise	Walking exercise (primary)	Increase a few min up to 60 min; increase speed or incline if bouts are >10 min	RPE may be used to guide intensity in patients without leg pain.
Resistance	*	*	*	Machine and free weights	*	*
Range of motion	*	*	*	Stretching	*	*

*Refer to chapter 6 for standard resistance training and flexibility training guidelines.

important impact on functional capacity in those with PAD.

To date, the mechanisms by which exercise training may improve walking in patients with claudication have not been completely elucidated (83). From a physiological perspective, studies have largely shown that exercise rehabilitation for claudication does not increase blood flow (106, 199). Data suggest that the benefits of exercise conditioning for patients with claudication appear more likely attributable to an improvement in calf muscle oxidative metabolism and rate of adenosine triphosphate production rather than to changes in skeletal muscle blood flow (18, 80, 96). Proposed mechanisms for the improvement in walking distance following exercise training include the following:

- Improved biomechanics of walking resulting in decreased metabolic demands (190, 194, 222)

- Increased angiogenesis and collateralization flow (9, 191)
- Reduced blood viscosity (53, 60)
- Increased blood cell filterability and decreased red blood cell aggregation (60)
- Activation of fibrinolysis (165)
- Attenuation of atherosclerosis (95)
- Increased extraction of oxygen and metabolic substrates resulting from improvements in skeletal muscle oxidative metabolism (39, 102, 199)
- Increased pain tolerance (95)
- Improved endothelial function (22, 193)
- Improved carnitine metabolism (96)

Research Focus

Exercise Performance and PAD

Gardner and colleagues (64) conducted a randomized controlled intervention trial assessing exercise performance outcomes of patients with PAD at baseline and following 12 wk of supervised exercise training in a hospital setting ($n = 33$), home-based exercise training ($n = 29$), and no exercise training in a control group ($n = 30$). Patients were included if they had a history of any type of leg pain upon exertion, limitation in ambulation caused by leg pain during a graded treadmill exercise test, and an ABI of ≤0.90 at rest or ≤0.73 after exercise. Eligible patients performed a maximal treadmill walking test to determine peak walking time, claudication onset time, and $\dot{V}O_2$peak. Additionally, functional status and patient-reported outcomes were assessed pre- and postintervention using the WIQ and the physical component of the SF-36. Exercise compliance was monitored in patients randomized to the supervised exercise group and the home-based exercise group using an activity monitor. Patients were also asked to complete diaries or log books outlining the exercise sessions.

In the supervised exercise program, patients walked for 15 min for the first 2 wk and increased duration each week until

they had accomplished 40 min of walking during the final 2 wk of the program. Patients walked at a grade equal to 40% of the final workload from the baseline maximal treadmill test to the point of near-maximal claudication pain; they then stopped to relieve the leg pain and began walking again when pain had attenuated. This process was repeated until the goal time was reached. The exercise program for the home-based group consisted of intermittent walking to near-maximal claudication pain 3 d/wk at a self-selected pace. Walking duration began at 20 min for the first 2 wk and progressively increased 5 min biweekly until a total of 45 min of walking was accomplished. Patients randomized to the control group were encouraged to increase their walking activity but did not receive specific recommendations about an exercise program during the study.

Results indicated differences in pre- and postintervention change scores for claudication onset time and peak walking time for both the supervised exercise group (+165 s and +215 s) and the home-based exercise group (+134 s and +124 s) versus the control group (−16 s and −10 s; $p < 0.05$), with no differences between the two exercise groups. $\dot{V}O_2$peak change

(continued)

Research Focus *(continued)*

scores were significantly lower for the control group (13.7 to 12.8 mL · kg⁻¹ · min⁻¹) compared with both exercise groups ($p < 0.05$); however, there were no differences in pre- and postintervention $\dot{V}O_2$peak within each individual exercise group (supervised: 11.4 to 11.7 mL · kg⁻¹ · min⁻¹; home based: 11.8 to 12.4 mL · kg⁻¹ · min⁻¹). Patients in the supervised and home-based exercise groups demonstrated significant improvements in the WIQ distance (+13% and +10%), speed (+9% and +11%), and stair climbing (+12% and +10%) scores, but improvements were not significantly higher compared with the control group WIQ change scores (distance: +1%; speed: +4%; stair climbing: +3%). Finally, patient-reported outcomes change scores assessed by the physical component of the SF-36 were higher

in the supervised exercise group compared with the control group only (+9% vs. −1%; $p < 0.05$).

Supervised walking exercise is considered the primary choice for physical activity to treat PAD, and the study by Gardner et al. (64) adds to the existing body of literature for this recommendation. However, in contrast to early investigations of home-based exercise, this study also demonstrates that a home-based exercise program may be beneficial for improving the health of patients with PAD. Because compliance was similar for the two exercise groups (supervised: 84.8%; home based: 82.5%), home-based exercise may be a feasible alternative to exercise programs at hospitals and clinics and should be examined in more detail.

Clinical Exercise Bottom Line

- Supervised exercise training has been given a class IA recommendation by American College of Cardiology/American Heart Association PAD treatment guidelines. This indicates that exercise training is a powerful therapeutic option for patients with PAD.
- Claudication is potentially attenuated sooner if the patient sits rather than stands during rest periods. Sitting may impose

less physical demand on patients and allows them to resume exercise sooner.

- Patients with PAD who receive endovascular therapy or surgical bypass may not experience leg pain during physical activity. The use of RPE scales rather than pain scales may be an option to aid in regulating exercise intensity.

CONCLUSION

This chapter provides a general review of PAD, focusing on symptomatic PAD and the benefits of exercise training. Optimal medical therapy coupled with exercise is an important treatment option for patients because of its low risk and high benefits. It is important for the exercise physiologist to be aware of concomitant diseases, as they may directly affect the type of exercise prescribed. Risk factor modification, such as smoking cessation, is critical for reducing cardiovascular morbidity and premature mortality for those with PAD. Exercise training can improve patient-reported outcomes, functional capacity, and

the metabolic risk profile of these patients while also potentially reducing health care costs.

It is clear that supervised exercise training improves patients' health and is also cost-effective. Although insurance reimbursement is now available to offset costs of hospital-based supervised exercise training, these programs are still met with barriers for successful completion. Thus, other options such as home- and community-based exercise programs should be emphasized as a potential treatment option for patients with PAD. More research is needed to improve the effectiveness of these programs, particularly from the standpoint of increasing ease of adoption and long-term adherence.

Online Materials

Visit HK*Propel* to access a link to the references, the case studies with discussion questions, and a quiz to help you review key concepts and test your understanding of the material covered in this chapter.

CHAPTER 18

Cardiac Electrical Pathophysiology

Kerry J. Stewart, EdD

David D. Spragg, MD

Implantable cardiac devices are used for regulating the heart rate, synchronizing the chambers of the heart in patients with heart failure, or defibrillating the heart in the case of life-threatening arrhythmias like ventricular fibrillation and tachycardia. Many patients with an implantable cardiac device can resume their normal daily activities, including regular exercise. A growing number of patients referred to cardiac rehabilitation programs have a cardiac rhythm device—either a pacemaker (PM) or implantable cardioverter-defibrillator (ICD). Cardiac rehabilitation for patients, mostly with heart failure, with either pacers or ICDs affords a unique opportunity to optimize medical treatment, increase exercise capacity, improve clinical condition, and supervise the correct functioning of the device (17). This chapter describes some of the major indications for these devices, guidelines for physical activity, and strategies and precautions for living with a cardiac device to increase physical function and improve quality of life.

DEFINITION

The heart of the average person beats about 100,000 times per day. Each contraction results from an electric impulse that is initiated in the sinoatrial (SA) node, passes through the atrioventricular (AV) node, and then spreads through the ventricles. An artificial pacemaker maintains a normal heart rate when the intrinsic electrical circuitry of the heart fails. The most common indications for **pacemaker** implantation are a slow heart rate because of SA node dysfunction (i.e., a failure of impulse formation) or because of conduction block in the AV node (i.e., a failure of impulse propagation). When either of these conditions results in a heart rate that is slow enough to cause symptoms (symptomatic bradycardia), permanent pacemaker implantation is indicated. Pacing systems have been developed to normalize conduction and resynchronize the ventricles in patients with heart failure and myocardial conduction slowing, usually seen clinically as left bundle branch block (9, 36, 38). This mode of pacing, known as biventricular pacing, uses an additional pacing lead that can restore cardiac synchrony and mechanical activation, leading to improvement in hemodynamics, ventricular remodeling, mitral regurgitation, exercise capacity, and quality of life as well as reduced mortality.

SCOPE

Although some infants require a pacemaker from birth, about 85% of people who need a pacemaker are over the age of 65 yr, with an equal distribution among men and women. About 200,000 cardiac devices are implanted in the United States each year (7), a number that is expected to rise because of the growing number of elderly people in the population. Since the first pacemaker implantation in the 1950s, cardiac pacemaker technology has rapidly advanced (7). Early pacemakers were primarily used for bradycardia. Today, with microcircuitry, improved battery longevity, and advanced programming, (36) pacemakers improve quality of life by optimizing the hemodynamic state at rest and can produce an appropriate heart response to meet the physiological demands of exercise.

Many patients can maintain or even begin to exercise after pacemaker implantation. Thus, clinical exercise physiologists should understand how pacemakers work to emulate normal cardiac rate, conduction, and rhythm in response to physiological and metabolic needs. They should also know about pacemaker programmed settings, how these settings can affect exercise capacity and the **exercise prescription**, and how to determine whether the patient's response to exercise is appropriate. The clinical exercise physiologist should communicate observations of the patient's responses to exercise to the pacemaker physician, who can reprogram the pacemaker to optimal settings.

PATHOPHYSIOLOGY

Rhythm disorders that involve the SA node are classified under the broad term **sick sinus syndrome**. This condition includes the inability to generate a heartbeat or increase the heart rate in response to the body's changing circulation demands. SA node dysfunction can cause fatigue, light-headedness, **exercise intolerance**, and syncope. A potentially more serious condition is failure of conduction through the AV node (heart block). When impulses are blocked at the AV node, ventricular activation is dependent on subsidiary **cardiac activation** arising from tissue below the level of block (typically from low in the AV node or from ventricular tissue). These escape rhythms can be unreliable, and heart block can lead to presyncope, syncope, or even death. Heart block may also cause a loss of *AV synchrony*, a term that refers to the sequence and timing of the atria and ventricles. Normally, the ventricles contract a fraction of a second after they have been filled with blood following an atrial contraction. Asynchrony may not allow the ventricles to fill with enough blood before contracting, thereby decreasing cardiac output.

Depending on the patient's specific condition, the artificial pacemaker may replace SA node signals that are too slow or that are delayed or blocked along the pathway between the upper and lower heart; maintain a normal timing sequence between the upper and lower heart chambers; and ensure that the ventricles contract at a sufficient rate. The guidelines for device-based therapy of cardiac rhythm abnormalities issued in 2019 by the American College of Cardiology, the American Heart Association, and the Heart Rhythm Society provide a review of the scientific literature and recommendations regarding which dysrhythmias may optimally be treated with a pacemaker (20). Table 18.1 shows pacemaker features that are used as therapy for medical conditions (22).

Pacing System

A **pacing system** consists of separate but integrated components that stimulate the heart to contract with precisely timed electrical impulses. The pacemaker (also known as the pulse generator) is a small metal case that contains the circuitry controlling the electrical impulses, along with a battery. Pacemakers use lithium batteries that can last for many years, depending on the extent to which the pacemaker is used. Some patients depend on the pacemaker

Type 2 Diabetes

Although patients with type 2 diabetes often experience chronotropic incompetence during exercise because of cardiovascular autonomic neuropathy, pacemakers are not currently a standard treatment for this condition alone for several reasons (19). First, pacemaker implantation is a surgical procedure, which may be associated with more complications in patients with diabetes. Second, it remains uncertain whether reductions in mortality and morbidity benefits exceed the risks and costs of having a pacemaker and required follow-ups. Third, the etiology of chronotropic incompetence in diabetes, which is primarily hyperglycemia rather than atherosclerosis or ventricular stiffness, is not targeted by a pacemaker; thus, medical and lifestyle treatment of hyperglycemia may be more appealing (19).

For example, exercise training reduces cardiac risk factors; improves glycemic control, body composition, endothelial function, LV diastolic function, and physical fitness; and decreases arterial stiffness and systematic inflammation in patients with diabetes. Because many of these factors increase the risk of chronotropic incompetence, it may be hypothesized that exercise may prevent or treat this condition (19, 40, 41). One study reported that patients with type 2 diabetes were able to increase their peak heart rate during exercise testing after a 7 wk walking program. Improvements in LV ejection fraction, end-diastolic volume, and stroke volume were also reported (24).

Table 18.1   Pacemaker Therapies for Medical Conditions

Pacemaker features	Treatment conditions
Rate responsive for chronotropic incompetence	Feature is used by patients who need to sustain a heart rate that matches their metabolic needs to their daily lifestyle or condition.
Mode switch for managing atrial arrhythmias in patients with bradyarrhythmia	Many patients with sinus node dysfunction and atrioventricular (AV) block experience atrial fibrillation. Mode switch therapy reduces symptoms of atrial fibrillation during dual-chamber pacing.
Rate drop response for neurocardiogenic syncope	Patients with carotid sinus syndrome and vasovagal syncope are treated for their symptoms by preventing their heart rate from falling below a prescribed level.
Ablation and pacing for atrial fibrillation	Patients with drug-refractory atrial fibrillation have been shown to benefit from ablation of the AV junction and implantation of a pacing system to maintain an appropriate heart rate.
Variable AV timing for patients with intermittent or intact AV conduction	Intrinsic AV activation is generally preferred to a ventricular-paced contraction because it provides improved hemodynamics and extended pacemaker longevity.

Based on Medtronics. www.medtronic.com/patients/bradycardia/index.htm.

to provide cardiac rhythm and conduction at all times. In others, the pacemaker acts as a backup that fires as needed when the sinus node fails to produce an appropriate rate or when the conduction system fails to transmit the impulses. The pacing leads are insulated wires connected to a pacemaker. The leads carry the electrical impulse to the heart and carry the response of the heart back to the pacemaker. These leads are very flexible to accommodate both the moving heart and the body. Depending on the type of pacemaker, one, two, or three leads may be present. Each lead has at least one electrode that can deliver energy from the pacemaker to the heart and sense information about the electrical activity of the heart.

Advances in pacemaker technology have allowed for further miniaturization, with ultra-low power circuit designs and even the elimination of pacemaker leads. Such systems have the advantage of few complications, such as systemic infections, dislodgements, and lead fractures, while providing similar performance to traditional pacemakers (30). Cardiac rehabilitation providers should be aware of these new devices, although patient management in cardiac rehabilitation as outlined throughout this chapter is the same as for traditional pacemakers.

Pacemaker Implantation

Implantation is performed by surgeons or cardiologists, usually in an hour or less, and typically requires a single day of postoperative monitoring in the hospital. Increasingly, the implantation is done as an outpatient procedure in appropriate patients. Most patients receive a local anesthetic and remain awake during the surgery. The pulse generator is usually implanted below the collarbone just beneath the skin. The leads are threaded into the heart through a vein located near the collarbone. The tip of each lead is then positioned inside the heart. In rare cases, the pulse generator is positioned in the abdomen and the pacemaker leads are attached to the outside of the heart. After implantation, the pacemaker can be adjusted with an external programming device, using radio frequency signals to the pacemaker via a transmitter placed on the chest.

Temporary External Pacemakers

An external pacemaker pulse generator is a temporary device that is used in emergency and critical care settings, after open heart surgery, or until a permanent pacemaker can be implanted. This device is used outside the body as a temporary substitute for the heart's intrinsic pacing.

Permanent Pacemaker Types

The two basic types of permanent pacemakers are single chamber and dual chamber. Both monitor the heart and send out pacing signals as needed to meet physiological demands.

1. A **single-chamber pacemaker** usually has one lead to carry signals to and from either the right atrium or, more commonly, the right ventricle. This type of pacemaker can be used for a patient whose SA node sends out signals too slowly but whose conduction pathway to the lower heart is intact. A single-chamber pacemaker is also used if there is a slow ventricular rate in the setting of permanent atrial fibrillation. In this case, the tip of the lead is usually placed in the right ventricle.

2. A **dual-chamber pacemaker** has two leads. The tip of one lead is positioned in the right atrium, and the tip of the other lead is located in the right ventricle. This type of pacemaker can monitor and deliver impulses to either or both of these heart chambers. A dual-chamber pacemaker is used when the conduction pathway to the lower chamber is partly or completely blocked. When pacing does occur, the contraction of the atria is followed closely by a contraction in the ventricles, resulting in timing that mimics the heart's natural way of working. Pacemakers are categorized by a standardized coding system developed by the Heart Rhythm Society (formerly the North American Society of Pacing and Electrophysiology) and the British Pacing and Electrophysiology Group.

Figure 18.1 shows the coding system for pacemaker functions. The letters refer to the chamber paced, the chamber sensed, what the pacemaker does when it senses an event, and other programmable features. This example is a ventricular demand pacemaker that is also rate responsive. The first V indicates that the ventricle is paced. The second V indicates that the pacemaker is programmed to sense for an impulse in the ventricle. The I indicates that when the pacemaker senses the patient's native ventricular impulse, the pacemaker is inhibited. The R indicates that the pacemaker is rate responsive or rate adaptive. A sensor in the pacemaker senses physical activity and adjusts the patient's pacing rate according to the level of activity.

Figure 18.2 shows the code for a dual-chamber pacemaker that is also rate responsive. The first D stands for dual and indicates that the pacemaker can pace both the atria and the ventricles. Likewise, the second D indicates that the pacemaker can sense both the atria and ventricles. The third D indicates that the pacemaker can be either inhibited or triggered by the patient's own cardiac activity. The pacemaker watches for atrial activity, and if it detects none, it will pace the atrium. After an appropriate AV time interval, the pacemaker will watch for a ventricular depolarization. If this is sensed, the pacemaker will be inhibited. If no ventricular activity is present, the pacemaker will pace the ventricle. The fourth letter, R, indicates that the pacemaker is also rate responsive.

CLINICAL CONSIDERATIONS

The physiology of exercise for patients with pacemakers is generally the same as for other patients. The difference is in how their physiology interacts with the device. For patients who cannot provide an appropriate cardiac output response to exercise, pacemakers attempt to increase the cardiac output to meet physiological demands. The increase in oxygen uptake from rest to maximal exercise follows this formula: Oxygen uptake is equal to cardiac output multiplied by arteriovenous oxygen difference. From rest

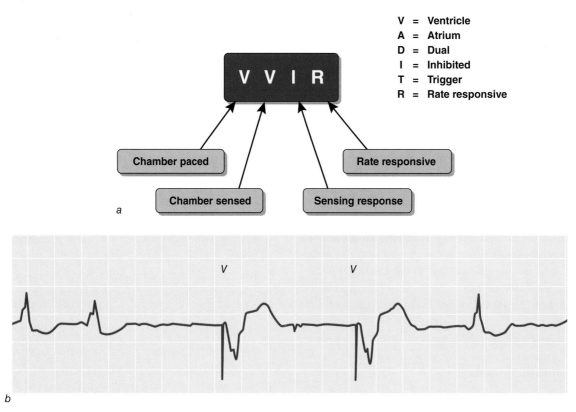

Figure 18.1 *(a)* Coding for a ventricular demand pacemaker that is also rate responsive. *(b)* VVI operation during atrial fibrillation. Ventricular pacing (V) occurs at the programmed lower rate limit of 60 b · min⁻¹ when the intrinsic ventricular activity falls below that level. Intrinsic ventricular activity at a faster rate inhibits ventricular pacing.

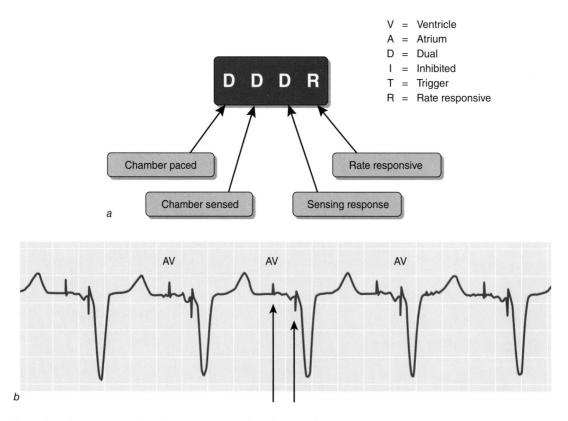

Figure 18.2 *(a)* Coding for a ventricular demand pacemaker that is also rate responsive. *(b)* DDD operation. Atrial pacing (A) occurs at the programmed lower rate limit of 75 b · min⁻¹. Because of complete heart block, the pacemaker tracks the atrial rate to pace the ventricle (V) at the same rate after a programmed AV delay.

to maximal exercise, oxygen uptake can increase 700% to 1,200%, arteriovenous oxygen difference by 200% to 400%, and cardiac output by 200% to 400%. Cardiac output is equal to heart rate multiplied by stroke volume. With exercise, stroke volume can increase by 15% to 20%, whereas heart rate can increase by 200% to 300%. Thus, heart rate is the most important component for increasing cardiac output and is most closely related to metabolic demands. Although AV synchrony contributes to cardiac output, this factor is more important at rest and less important with exercise. However, a 2020 study found that biventricular pacing in symptomatic patients with nonobstructive hypertrophic cardiomyopathy increased their exercise capacity and quality of life via augmentation of diastolic filling during exercise rather than contractile improvement (2).

Physiological Pacing

The term **physiological pacing** refers to the maintenance of the normal sequence and timing of the contractions of the upper and lower chambers of the heart. AV synchrony provides higher cardiac output without increasing myocardial oxygen uptake. Dual-chamber pacemakers attempt to provide this physiologically beneficial function. The pacemaker senses the patient's sinus node and, in complete heart block, sends an impulse to the ventricle following an appropriate AV timing interval. Although the specific change in cardiac output depends on many factors, the optimal AV delay to produce the maximum cardiac output in normal people is about 150 ms from the beginning of atrial depolarization. The efficiency of cardiac work decreases with a shorter or longer AV interval. In normal subjects, the AV interval shortens with increased heart rate. The pacemaker can set the AV interval based on heart rate. A dual-chamber pacer can also initiate an atrial impulse in sick sinus syndrome. AV synchrony augments ventricular filling and cardiac output, improves venous return, and assists in valve closure. The loss of atrial function increases atrial pressure and pulmonary congestion. The benefit of AV synchrony is independent of any measure of left ventricular function. The maintenance of normal AV synchrony allows for improved hemodynamic responses with a more normal increase in cardiac output (16). Because of higher cardiac output at any given level of work with synchronous pacing, the arteriovenous oxygen difference is narrower and the serum lactate is lower. Thus, synchronous pacing results (figure 18.3) in less anaerobic metabolism at the same level of work (18). AV synchrony and stroke volume provide their most important contributions to cardiac output at rest, whereas an increase in heart rate is the predominant factor contributing to cardiac output during exercise (figure 18.4).

Mode Switching and Maximal Tracking Rates in Dual-Chamber Pacemakers

Many patients with sinus node dysfunction and AV block develop atrial arrhythmias, the most common being atrial fibrillation. To prevent the dual-chamber pacemaker from tracking or matching every atrial impulse with a ventricular pacing pulse, **mode switching** controls the ventricular rate. Mode switching temporarily reverts to a nontracking mode so that irregular or excessive atrial activity does not drive the ventricles to an extremely high rate. The mode-switching feature is programmable and, depending on the specific pacemaker model, can be adjusted for optimal performance in any given patient.

A different but related concept is the pacemaker maximal tracking rate. This rate refers to the maximal atrial rate that will trigger ventricular pacing. As the atrial rate begins to exceed the maximal tracking rate, many pacemakers allow individual atrial impulses to be ignored (resulting in single "dropped" ventricular beats). As the atrial rate continues to surpass the maximal tracking rate, the pacemaker reverts to 2:1 atrioventricular conduction, in which every other atrial beat results in ventricular pacing. As the atrial rate slows to below the maximal tracking rate, 1:1 atrioventricular activation resumes.

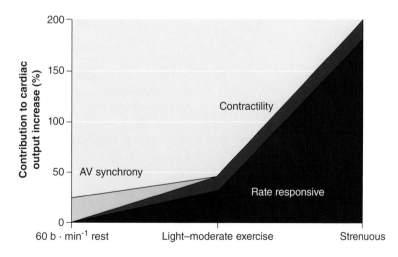

Figure 18.3 The relative contributions of atrioventricular (AV) synchrony, stroke volume (contractility), and heart rate to cardiac output at rest and exercise. Heart rate is the most important contributor to cardiac output during exercise.

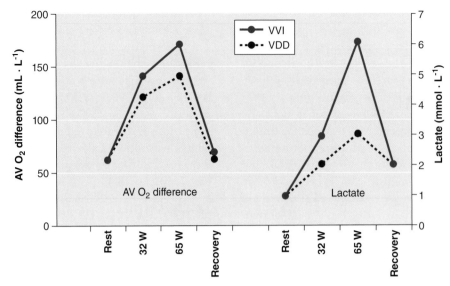

Figure 18.4 Synchronous pacing (VDD) results in less anaerobic metabolism at a given level of work compared with nonsynchronous pacing (VVI).

Rate-Responsive Pacing

The development of **rate-responsive pacing** (also called rate adaptive or rate modulated) has dramatically changed the application of pacing with regard to physical activity. The rate-responsive function is used when the native sinus node cannot increase heart rate to meet metabolic demands. Increasing heart rate in response to exercise is probably the single most important factor for increasing cardiac output and oxygen uptake. A sudden increase in exercise requires the heart rate to adjust quickly to the workload. Rate-responsive pacemakers can sense the body's physical need for increased cardiac output and produce an appropriate cardiac rate in patients with chronotropic incompetence. The highest rate at which the pacemaker will pace the ventricle in response to a sensor-driven rate is known as the maximum sensor rate. When rate-modulated pacing is compared with non-rate-modulated pacing (figure 18.5), exercise capacity is extremely limited without an appropriate increase in heart rate (8).

Several physiological and metabolic changes occur during exercise as the demand for energy increases:

- Movement that produces vibration and acceleration
- Respiration rate and respiratory muscle electromyographic activity that increase with exercise
- Heat that raises body temperature
- Electrocardiographic sensing of an atrial rate to assure ventricular pacing
- Carbon dioxide and lactic acid levels that increase with exercise
- Intracardiac pressures that increase with exercise

Sensors detect these changes and, based on computer algorithms, generate the electrical impulses that are used to pace the heart. The development of optimal sensors and algorithms for rate-modulated pacing systems must meet several requirements:

- The sensor should rapidly detect acceleration and deceleration of physiological changes.
- The response should be proportionate to the exercise workload and metabolic demands.
- The response should be sensitive to both exercise and non-exercise requirements such as posture, anxiety and stress, vagal maneuvers, circadian variations, and fever.
- The response should be specific and not be falsely triggered.

The most common rate-responsive pacemakers detect motion in response to physical activity. Vibration sensors use a piezoelectric crystal located in the pulse generator to detect forces generated during movement. These forces are transmitted to the sensor through connective tissue, fat, and muscle. Acceleration sensors detect body movement in anterior and posterior directions (12). The circuitry is also located in the pulse generator. Because the sensor is not in direct contact with the pacemaker case, no reaction to vibration or pressure occurs.

Producing an appropriate heart rate in response to certain work tasks poses a technological challenge. The simple task of walking up and down stairs can produce different heart rate responses, based on the type of sensor used to sense motion. Compared with a normal heart rate response, an accelerometer produces similar results going up stairs but overestimates the metabolic demand going down stairs. The vibration sensors produce a heart rate that is too low for stepping up and a rate that is too high for stepping down. This result occurs because the vibration of walking down stairs is greater than the vibration of walking up stairs, although the metabolic demand is greater for walking up stairs (5). In contrast to single sensors, dual-sensor rate response provided by

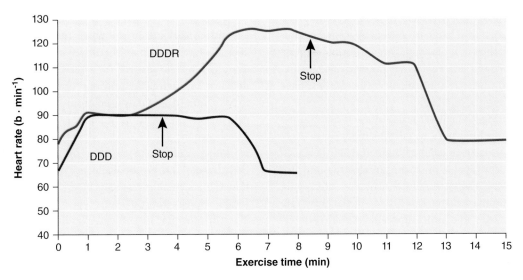

Figure 18.5 A comparison of rate-modulated (DDDR) and non-rate-modulated (DDD) pacing during treadmill exercise. Without an appropriate increase in heart rate, exercise capacity is extremely limited in a patient with chronotropic incompetence.

Adapted from Bodenhamer and Grantham (8).

activity and minute ventilation may help overcome these types of problems (3, 31, 36).

Advances in pacemaker technology may allow the use of combined or blended sensors and advanced algorithms to improve rate performance over a single-sensor system (21, 36). Blended sensors measure the patient's workload through respiration and motion, providing optimal rate response during changing levels of activity. One study found that a blended sensor tracked the intrinsic sinus rate during treadmill testing more accurately than an accelerometer or minute ventilation sensors alone, thus providing a more physiological rate response during physical activity (10). A 2020 study demonstrated a similar benefit of a blended sensor in patients with heart failure and atrial fibrillation (36).

Another feature in some sensors is closed-loop stimulation, which monitors and processes intracardiac impedance signals associated with myocardial contraction dynamics on a beat-to-beat basis. These signals are controlled by the autonomic nervous system and are directly related to other physiological responses including heart rate, blood pressure, and respiration. This approach has showed promise compared with an accelerometer and no rate sensors in subjects requiring ≥80% pacing (1). Overall benefits included better performance on several activities of daily living, a reduced prevalence of orthostatic hypotension, and a brisker chronotropic response closer to a normal heart rate.

Besides their therapeutic function, most implanted devices can automatically collect and store daily physical activity data obtained from an internal accelerometer, providing new opportunities for monitoring a patient's physical activity. These data can be used to develop cost-effective and resource-efficient strategies to proactively prevent patient decompensation and hospitalization (32).

Exercise Testing

Patients with pacemakers capable of rate modulation should undergo **exercise testing** to ensure appropriate rate responses (42). Exercise capacity and quality of life are improved by appropriately programmed rate-responsive pacemakers compared with fixed-rate units (43). These devices can be programmed to match the needs of the patient more closely. The primary pacemaker settings can be adjusted to optimize responses to physical activity:

- Sensitivity of the sensor
- Responsiveness to a physiological change
- Rate at which the cardiac rate changes
- Minimum rate at rest and maximal rate at peak activity

Exercise testing is used to guide the adjustment of these settings to improve exercise capacity and reduce symptoms. Exercise testing helps establish upper rate limits and adjust the sensitivity and responsiveness of the sensor. Exercise testing is also used to determine the anginal threshold, if any. Pacing the heart rate beyond the point at which ischemia would occur would not be prudent. Several approaches to exercise testing can be used. These include informal or formal protocols with or without real-time electrocardiogram (ECG) monitoring and with or without determination of optimal rate-responsive parameters. The patient's health status and lifestyle, the type of pacemaker, and the facilities and experience of personnel also help determine the specific approach to exercise testing. A 2020 study developed an algorithm that incorporated data from cardiorespiratory exercise testing and pacemaker stress echocardiography. After 3 mo of exercise training, patients who had the algorithm incorporated into their pacemakers to optimize the sensor rate had greater improvement in peak oxygen uptake compared with controls, who had conventional pacemaker programming (36). Nevertheless, for many patients, informal exercise testing is a reasonable and less expensive alternative to formal treadmill testing (15). Empiric adjustment of the rate-response parameters is common.

With informal testing, the patient walks at a self-determined casual pace and then at a brisk pace, usually for about 3 min each. The sensor-driven cardiac rate can be obtained from the ECG. Because pacemakers are capable of storing an electronic record of pacemaker activity, a special computer, can also interrogate the pacemaker to examine a histogram display of the heart rate response during the walk. The optimal pacemaker rate is determined empirically. For casual exercise, the target is often 10 to 20 $b \cdot min^{-1}$ above the lower rate limit. For brisk exercise, the target can be 20 to 50 $b \cdot min^{-1}$ above the lower rate limit. This approach to exercise testing is best suited for less active patients who are unlikely to reach their upper rate limit.

By examining a display of the sensed atrial rate as measured by an event counter in the pacemaker, or by measuring the heart rate by ECG and asking the patient about symptoms, the physician makes a clinical judgment about whether the patient is chronotropically competent. If not, the pacemaker will need to be programmed to elicit an appropriate response. Formal exercise testing allows a chronotropic evaluation that seeks to match the pacemaker-augmented response of the chronotropically incompetent patient to the metabolic requirements of the body (37). The 6 min walk can also be used as an alternative test and has good validity for predicting maximal oxygen uptake (28).

Formal exercise testing is typically best for active patients likely to reach the programmed **maximum sensor rate**. Programming the upper rate of rate-adaptive pacing improves exercise performance and exertional symptoms during both low and high exercise workloads compared with a standard nominal value of 120 $b \cdot min^{-1}$ (11).

With formal exercise testing, the protocol selected requires careful consideration. Many of the traditional protocols for exercise testing such as the Bruce and Naughton protocols are designed to test for coronary artery disease. Their usefulness in defining optimal programming for rate-responsive pacemakers may be limited. A widely used protocol for assessing patients with pacemakers is the chronotropic assessment exercise protocol (45, 46) shown in table 18.2. The advantage of this protocol is that the workload gradually increases to mimic the range of activities of daily living. This protocol allows a more complete assessment of how the pacemaker responds at the lower metabolic equivalent (MET) ranges where patients typically spend most of their time. The chronotropic assessment exercise protocol has five stages at a lesser exercise intensity than the Bruce produces in the second stage (47). Because the Bruce protocol increases by 2 to 3 METs during each 3 min stage, assessing the patient's work capacity and the ability of the pacemaker sensor to provide an adequate hemodynamic response would be difficult. Table 18.3 provides a brief review of exercise testing.

EXERCISE PRESCRIPTION AND TRAINING

Dual-chamber pacemakers are in greater use today than in the past. Clinical exercise physiologists need to be familiar with the normal behavior of these devices during exercise. Figure 18.6 shows DDDR operation. The rate at which the sensor-driven heart rate increases follows algorithms that are programmed into the pacemaker. Among the key parameters are the slope of the heart rate increase and decline and the sensitivity of the sensor. With increased physical activity, the pacemaker will follow the sinus rate up to a maximal tracking rate. The activity sensor can be programmed to allow a further increase in the paced rate to the maximum sensor rate in response to physical activity. If the patient

Table 18.2 Chronotropic Assessment Exercise Protocol

Stage	Speed (mph)	Grade (%)	Cumulative time	Metabolic equivalents
1	1.4	2	2	2.0
2	1.5	3	4	2.8
3	2.0	4	6	3.6
4	2.5	5	8	4.6
5	3.0	6	10	5.8
6	3.5	8	12	7.5
7	4.0	10	14	9.6
8	5.0	10	16	12.1
9	6.0	10	18	14.3
10	7.0	10	20	16.5
11	7.0	15	22	19.0

Table 18.3   Exercise Testing Review

Mode	Protocol specifics	Clinical measures	Clinical implications	Special considerations	Cardiovascular
Walking	Informal or formal	Heart rate response, blood pressure response, symptoms, time, METs, or maximal oxygen uptake	Used to guide adjustments of the device to optimize heart rate response to improve exercise capacity and reduce symptoms	Informal or formal testing depending on the patient's lifestyle and health status; formal testing best for active patients likely to reach higher heart rates	Testing focused on conduction system of the heart but can also be used for diagnosis of coronary artery disease; cardiac imaging may be needed because the ECG may not allow for assessment of ischemic changes

continues to exercise, the pacemaker may reach its maximal tracking rate or maximum sensor rate. When this occurs, the pacemaker will not further increase the heart rate. If the patient's native sinus rate continues to increase beyond this point, the pacemaker will switch to an AV block mode because the sinus rate now exceeds the rate at which the pacemaker will permit tracked ventricular pacing. The pacemaker first switches to a Wenckebach-type block to cause a gradual slowing of the ventricles, and 2:1 AV block ensues if the sinus rate continues to rise. This feature protects against nonexercise sinus tachycardia that might otherwise force the pacemaker to produce ventricular tachycardia.

At higher levels of exercise, the metabolic demands will be high, but 2:1 block may occur and slow the ventricular rate. In this situation, the development of 2:1 block is a normal feature of the pacemaker but can cause symptoms because of the sudden drop in heart rate. If this occurs, the patient is exercising too hard or the maximal tracking rate is set too low. If the pacemaker is also rate adaptive, the sensor rate may be too low.

The exercise physiologist should record and communicate episodes of abrupt decreases in heart rate to the patient's pacemaker physician so that programmed settings can be evaluated for possible change. Several of the exercise modalities that are commonly prescribed in cardiac rehabilitation and adult fitness programs may pose a particular challenge in some patients with activity sensors (4). In the example shown in figure 18.7, because most of the increase in work is accounted for by raising the slope rather than speed on a treadmill, little change in the generated forces would be detected by the vibration sensor. Thus, the heart rate determined by the vibration sensor is too slow for the metabolic demand of the increased work. In this case, the accelerometer sensor is better able to provide a heart rate that more closely matches an appropriate rate response.

The clinical exercise physiologist should also be aware of how a vibration sensor responds to outdoor and stationary cycling (figure 18.8). The response of this type of sensor is particularly relevant to cardiac rehabilitation because stationary cycling and seated steppers are prevalent modes of exercise in many programs. This sensor response may explain why some patients complain of unusual fatigue and shortness of breath during stationary cycling but not other types of exercise such as treadmill walking.

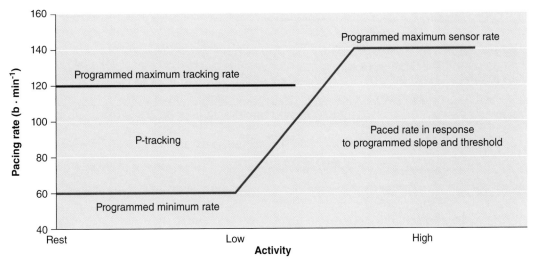

Figure 18.6   DDDR pacing and physical activity. The pacemaker follows the sinus rate to a maximal tracking rate. In response to physical activity, the sensor-driven response can drive the rate to the maximum sensor rate. The paced rate increases in response to programmed slope and threshold settings.

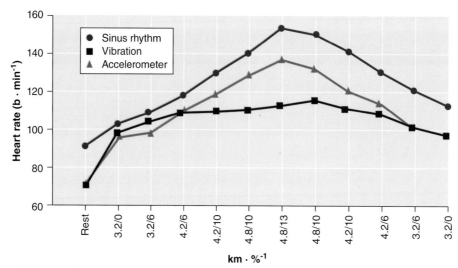

Figure 18.7 Generated forces. The circles show a normal sinus response at rest and with increased treadmill work. Work is increased primarily by raising the slope rather than speed. At rest, sinus rate is about 20 beats above pacemaker rate. This difference is maintained throughout the test. The accelerometer sensor (triangles) is able to produce a heart rate that better matches the workload compared with the vibration sensor (squares).

Adapted from Alt and Matula (4).

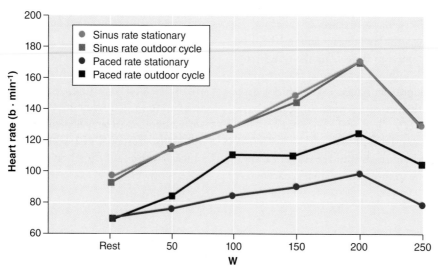

Figure 18.8 Vibration sensor. The upper lines represent a normal sinus rate with outdoor cycling (open squares) and stationary cycling (open circles) at increasing workloads and recovery. The lower lines represent the sensor. The difference of 20 beats at rest is maintained throughout the test during outdoor cycling (filled squares). During stationary cycling (filled circles), the paced cardiac rate is considerably slower than both the sinus rate and the rate during outdoor cycling. This occurs because stationary cycling produces less body motion and vibration.

Adapted from Alt and Matula (4).

Patients with an artificial pacemaker require long-term surveillance by their physicians to ensure optimal adjustment of the programming for their individual needs, to maximize the life expectancy of the pacemaker through adjustment of pacemaker output settings, and to identify and treat complications. Pacemaker follow-up relies on clinical, electrocardiographic, and device assessment. Other tests may include exercise testing, Holter monitoring, and echocardiography. The device assessment requires a specialized programmer to verify pacemaker functions. In some cases, remote interrogation of the pacemaker over telephone lines is done periodically to provide useful information about selected functions of the pacemaker when a more complete test is not deemed necessary. The clinical exercise physiologist can play an important role in the overall evaluation of the patient by providing feedback to the physician about heart rate, blood pressure, and symptomatic responses to exercise. The case study in the web resource illustrates the role of the clinical exercise physiologist in the management of the patient with a pacemaker.

Exercise prescription requires special attention to the type of pacemaker that is implanted. With fixed-rate pacemakers, cardiac output and arterial pressure are increased by stroke volume (1). Target heart rate cannot be used to guide exercise intensity. Instead, the patient follows ratings of perceived exertion (RPEs). It is also important to monitor blood pressure to ensure an appropriate intensity. When the sinus node is normal, it is desirable to have the pacemaker track native sinus activity by pacing the ventricle after an appropriate AV delay (1). In all cases, the target heart rate must be lower than the anginal threshold in a patient with ischemia (6). Tailoring the exercise prescription and modifying the response rate of the pacemaker based on cardiopulmonary stress testing that determines the anaerobic threshold have been shown to provide functional advantages for patients in cardiac rehabilitation (14).

Because rate-responsive pacemakers mediate the heart rate response to exercise, the type of sensor must be taken into account when exercise is prescribed (37). Sensors that detect movement may respond slowly to stationary cycling and increased treadmill slope. Again, the RPE and MET equivalents are extremely useful in establishing the exercise prescription. In some pacemakers, pacing occurs only when needed. In many patients, such as those with normal sinus function with intermittent heart block, the exercise prescription can be written in the same way as for most other patients. Regarding heart rate monitoring, one study found that the use of dry-electrode heart rate monitors that transmit a signal to a monitor, such as those worn on the wrist, had no adverse effect on pacemaker function (18). Modern fitness watches that monitor heart rate at rest and during exercise, and in some cases cardiac rhythm, are unlikely to interfere with an implanted device because they are receiving the electrical impulses from the heart and not transmitting them. Some of these watches also use optical sensors to detect the pulse. The only possible concern is when these devices transmit the signal to a smartphone, although the radio energy is very low. To minimize this risk, the smartphone should not be placed in a shirt pocket.

Exercise Recommendations

In general, patients with pacemakers can derive benefits from an exercise program similar to those gained by other people. In a 2015 report of 2,331 outpatients from Heart Failure: A Controlled Trial Investigating Outcomes of Exercise Training (HF-ACTION) (48), exercise training in patients with heart failure and implanted cardiac devices appeared to be safe and did not lead to increased mortality or hospitalization. The improvements in aerobic capacity and health-related quality of life were generally similar in patients with and without an implanted device. Nevertheless, the aggregate data from HF-ACTION suggested that the beneficial effects of exercise on reductions in clinical events were observed only in patients without an implanted device despite similar improve-

ments in exercise capacity (48). Of note, the interaction of events in patients with and without a device was borderline ($p = 0.058$). Whether this trend was because patients with devices were older, had lower exercise capacity, or had more severe disease requiring a device is uncertain.

The area of greatest consideration in the prescription of exercise is the issue of exercise intensity. Because of the variety of pacemakers, sensors used to detect activity, and mode of exercise prescribed, the appropriate heart rate response can vary considerably. Therefore, when patients with pacemakers start exercising, they should be monitored to make sure the pacemaker is responding appropriately.

The exercise prescription for those who use pacemakers is generally the same as for others, and many of the same cautions apply. The prescription must consider comorbidities such as angina and chronic heart failure, for example. The area of greatest consideration when one is prescribing exercise is the issue of exercise intensity, which is determined by the underlying reason for the pacemaker and the type of pacemaker implanted. For patients with an internal cardioverter-defibrillator, the exercise heart rate should be kept at least 20 beats below the firing threshold. Activities that might result in contact with an implanted device should be avoided.

Frequency

The frequency of activity is based on the goals of the program. If someone is interested in improving their health and is exercising at an intensity less than 60% of maximal aerobic capacity, daily activity is recommended. If the person is able and willing to exercise at a higher intensity (60%-85% of maximal aerobic capacity), activity on 3 to 5 d/wk is recommended.

Intensity

The intensity should be in the recommended range of 40% to 85% of maximal aerobic capacity but is primarily dependent on comorbid conditions (e.g., angina, chronic heart failure). Also, upper limits of the pacemaker (tracking and sensing) can influence the upper limit of exercise intensity. Because heart rate does not increase in a patient with a fixed-rate pacemaker, RPE and MET equivalents need to be used to evaluate exercise intensity. Additionally, blood pressure should be monitored to show appropriate increases with increasing workload. A patient with a fixed-rate pacemaker should not exceed exercise intensity above the point where blood pressure begins to plateau with increasing workload. With dual-chamber and rate-responsive pacemakers, heart rate can be used to determine exercise intensity and should be used along with RPE and METs. Knowledge of maximal tracking or sensing rates determines the upper intensity level. Patients should be monitored closely, and activities should be chosen based on the ability of the pacemaker to adjust heart rate with increasing metabolic demands.

Duration

The duration of activity is similar to that specified in the general recommended guidelines for promoting health and fitness (20-60 min). Duration depends on goals. Ideally, the duration should be adjusted so that the individual achieves an energy expenditure of at least 1,000 kcal/wk.

Mode

Generally all forms of exercise are acceptable in patients with pacemakers, except activities that can cause direct contact with the pacemaker. Therefore, contact sports such as football, soccer, and hockey are generally not recommended. All forms of aerobic exercise are generally acceptable, and these most likely carry the greatest benefit with regard to improving overall health and decreasing risk factors for cardiovascular disease. When a patient with a pacemaker performs any form of aerobic exercise, rate-responsive pacemakers should be evaluated to see that they are increasing the heart rate appropriately relative to the intensity of exercise. Pacemakers that rely on vibration or accelerometer sensors to detect body motion during exercise may not produce an adequate response for activities such as stationary cycling and increased treadmill slope. Unusual shortness of breath and fatigue may indicate a lack of rate-responsive pacing and need to be monitored during different forms of activity. Weight training may also be acceptable, although weights or bars must not come in contact with the pacemaker.

Important Considerations

If a patient goes into second-degree type 1 block (Wenckebach) while exercising, the patient's native sinus rate likely exceeded the pacemaker's maximal tracking or sensor rate. If this occurs, the intensity of exercise should be reduced. If the exercise professional notices a decrease in heart rate well below the patient's tolerable limits, this information should be forwarded to the pacemaker physician so that programmed settings can be evaluated. Practical application 18.1 provides more information about living with a pacemaker.

Automatic Internal Cardioverter-Defibrillators

Subsets of cardiac patients are at high risk for potential lethal ventricular tachycardias. This group includes patients who have survived a previous sudden death event and are at high risk for recurrent cardiac arrest, as well as patients who have never had sudden death but are at high risk for cardiac arrest because of prior myocardial injury. Although medications that stabilize heart rhythm are available, they are not entirely effective and often produce serious side effects. Increasingly, **automatic internal cardioverter-defibrillators (AICDs or ICDs)** are being used to control life-threatening ventricular arrhythmias. An ICD is a battery-driven implanted device, similar to a pacemaker, that is programmed to detect and then stop a life-threatening ventricular arrhythmia by delivering an electrical shock directly to the heart. Some models provide tiered therapy by being able to provide anti-tachycardia pacing, cardioversion, and defibrillation as needed.

ICDs are commonly implanted beneath the skin and muscle of the chest or abdomen, and electrodes that sense the heart rhythm and deliver the shock are inserted into the heart through veins. The site and placement of the electrode wires vary, depending on the patient and model of ICD used. In some cases, electrode patches are sewn to the surface of the heart. Some newer devices are implanted in the midaxillary line and have a lead tunneled under the skin to a parasternal position; electrodes are placed under the skin of the chest near the heart. These devices avoid the need for a transvenous lead directly to the heart and function very similarly to a standard ICD. Many devices today can also serve as both a pacemaker and an ICD and can be programmed to the patient's individual needs. Microchips inside the device record rhythms and shocks to be used to determine optimal therapy.

Special Considerations

In many cases, the failure of the heart's intrinsic pacing and conduction system is associated with comorbid conditions such as myocardial infarction and chronic heart failure. Many patients with artificial pacemakers or ICDs are elderly and have limited exercise capacity. The exercise prescription must consider not only the indications for the type of pacemaker implanted but also the limitations to exercise associated with comorbidities. Besides monitoring the patient for appropriate heart rate responses, the exercise physiologist must pay close attention to signs and symptoms that might occur with increased heart rate such as exercise-induced angina, failure of blood pressure to increase or decrease, and marked shortness of breath.

Exercise Recommendations

The American College of Cardiology/American Heart Association/Heart Rhythm Society 2018 guidelines for device-based therapy of cardiac rhythm abnormalities provide recommendations for ICD therapy (20). These guidelines emphasize the need for the physician to establish limitations on the patient's specific physical activities. The guidelines also refer to policies on driving, advising the patient with an ICD to avoid operating a motor vehicle for a minimum of 3 mo and preferably 6 mo after the last symptomatic arrhythmic event. After appropriate evaluation and observation, many patients with an ICD can participate in exercise programs. In most cases, the guidelines for exercise prescription are similar to those for any other patient with cardiovascular disease and should consider the patient's underlying diagnoses, medications, and symptoms. An exercise stress test is essential for establishing an appropriate exercise prescription. The prescribed target heart rate should be at least 20 beats below the heart rate cutoff point at which the device will shock.

The exercise prescription must also take into account the existence of an **angina threshold** or exercise-induced hypotension,

PATIENT EDUCATION ABOUT LIVING WITH A PACEMAKER

The clinical exercise physiologist is often a primary source of patient education. Pacemaker patients frequently ask about what they should be aware of in their day-to-day lives. Advances in pacemaker technology have resulted in continuing improvements. Pacemakers are smaller and better shielded from external interference and magnetic fields than ever before. Research has shown that with appropriate protocols, magnetic resonance imaging (MRI) can be performed safely in patients with certain pacemakers and internal cardioverter-defibrillator systems (23, 26). Nevertheless, the clinical exercise physiologist should communicate some basic precautions to the patient. This section presents some common questions and issues.

Sports and Recreational Activities

Many patients, after appropriate medical clearance, can travel, drive, bathe, shower, swim, resume sexual activities, return to work, walk, hike, garden, golf, fish, and participate in other similar activities. But contact sports that include jarring, banging, or falling such as football, baseball, and soccer should be avoided. Also, patients should avoid bracing a rifle butt against the implant site if they go hunting.

Work Activities

Most office equipment is unlikely to generate the type of electromagnetic interference that can affect a pacemaker, but equipment with large magnets should not be carried if the magnet is held near the pacemaker. Patients who work with heavy industrial or electrical equipment need to consult their physicians about resuming work because this equipment may produce high levels of electromagnetic interference that could affect pacemaker function.

Home Activities

People with pacemakers can participate in most activities of daily living and can be reassured that most home electrical devices will not interfere with pacemaker operation (25). The evolution of new mobile communication standards such as LTE/4G and other technologies makes it difficult to fully estimate the risk of electromagnetic interference. Thus, their unrestricted use cannot be recommended and some precautions are warranted. Cellular phones should be kept at least 6 in. (15 cm) away from the pacemaker site, and phones transmitting above 3 W should be kept at least 12 in. (30 cm) away. The patient should hold the cell phone to the ear opposite the pacemaker site and should not carry a phone in a pocket or on a belt within 6 in. (15 cm) of the pacemaker. Ordinary cordless, desk, and wall telephones are considered safe. The patient should not lift or move large speakers because their large magnets may interfere with the pacemaker. General household electrical appliances like televisions and blenders, and outdoor tools such as electric hedge clippers, leaf blowers, and lawn mowers, do not usually interfere with pacemakers. But the patient should avoid using tools such as chain saws that require the body to come into close contact with electric spark–generating components. In addition, caution is advised when people are working near the coil, distributor, or spark plug cables of a running engine. The safe approach is to turn off the engine before making any adjustments.

Travel

Most people with pacemakers can travel but should tell airport security personnel that they have a pacemaker or other implanted medical device before going through security systems. Although airport security systems do not affect the pacemaker, the pacemaker's metal case could trigger the metal detection alarm. Home, retail, or library security systems are unlikely to be set off by the pacemaker.

---

because many of these patients have severe coronary artery disease and poor left ventricular function. Furthermore, many patients with ICDs take β-blockers to limit heart rate to control symptoms and to prevent firing of the device. The benefits of pacing are

- alleviation or prevention of symptoms,
- restoration or preservation of cardiovascular function,
- restoration of functional capacity,
- improved quality of life,

- enhanced survival, and
- participation in exercise training programs with many forms of physical activity.

Cardiac Resynchronization Therapy

Cardiac resynchronization therapy (CRT), or biventricular pacing, is an adjunctive therapy for patients with advanced heart failure (29, 33). Many of these patients have left bundle branch block or an intraventricular conduction delay, resulting in left

Practical Application 18.2

CLIENT–CLINICIAN INTERACTION

In a supervised exercise program, the clinician needs to observe whether the client can achieve the desired level of exercise without undue fatigue. Because the increase in heart rate during exercise is the largest contributor to cardiac output, limited exercise capacity may be a result of inappropriate heart rate response. Depending on the type of pacemaker, adjustment of the settings may allow the heart rate to respond more appropriately to the exercise demand. Carefully observing the client's exercise performance, asking the client about fatigue level, and reporting those findings to the client's physician are key responsibilities of the exercise physiologist.

ventricular dyssynchrony and a high mortality rate. The efficacy of CRT is based on the reduction in the conduction delay between the two ventricles. CRT is designed to keep the right and left ventricles pumping together by regulating how the electrical impulses are sent through the leads. This therapy contributes to the optimization of the ejection fraction, decrement in mitral regurgitation, and left ventricular remodeling, thus resulting in symptom improvement and enhanced quality of life. Exercise testing is of particular value in patients with CRT to confirm proper functioning of the device during exercise (42).

Several studies have shown the benefit of CRT in a subgroup of patients with heart failure with conduction delays (12, 16, 22, 34). Improvements have been found in the mean distance walked in 6 min, quality of life, New York Heart Association (NYHA) functional class, peak oxygen uptake, total exercise time, LV function, and LV end-diastolic diameter; there was also a reduction in the number of hospital admissions. Data suggest that improvements in functional capacity with CRT can be maintained long-term (35). In addition, ventricular–arterial coupling, mechanical efficiency, and chronotropic responses are improved after 6 mo of CRT. These findings may explain the improved functional status and exercise tolerance in patients treated with CRT (38). Exercise training after CRT helps improve exercise tolerance, hemodynamic measures, and quality of life (27).

An adverse consequence of long-term right ventricular pacing is adverse remodeling of the left ventricle, which can contribute to or cause left ventricular systolic dysfunction (LVSD) or overt heart failure. One study (13) demonstrated that LSVD in the presence of unavoidable right ventricular pacing responded favorably to CRT when CRT was used at the time of pacemaker generator replacement in terms of cardiac function, exercise capacity, quality of life, and blood markers of cardiac strain. Thus, CRT implanted opportunistically at the time of pacemaker replacement might be cost-effective in preventing or delaying worsening heart failure.

Note that patients with these devices have serious heart disease, so the usual precautions regarding exercise participation should be applied.

Literature Review

Although it has been shown that cardiac resynchronization therapy has beneficial effects on clinical outcomes and cardiac remodeling, little is known about longer-term effects on myocardial function and exercise tolerance. Steendijk and colleagues (39) studied CRT in 22 patients with chronic heart failure. After 6 mo of the device therapy, marked improvements occurred in NYHA class, quality of life scores, 6 min walk distance, left ventricular ejection fraction, and stroke work at rest and with increased heart rates; there was also evidence of reverse remodeling. In another study, CRT benefited aerobic function and ventilation–perfusion mismatching most in those patients with the greatest physiologic impairment (44). These results demonstrate that hemodynamic improvements with CRT were associated with improved functional status and exercise tolerance. In another study, patients with CRT were randomized to 3 mo of exercise training or a control group (27). Exercise training resulted in further improvements in exercise capacity, hemodynamic measures, and quality of life in addition to the improvements seen after CRT. These studies suggest that exercise training allows maximal benefit to be attained after CRT.

Research Focus

Pacemaker Programming Mode and Exercise Capacity

The cardiac pacing mode influences atrioventricular synchronicity and the response of heart rate to physical activity. Patients with a pacemaker implanted for sick sinus syndrome or for AV block would benefit from cardiac rehabilitation to improve their exercise capacity and quality of life. These patients need a comprehensive assessment during exercise testing to ensure their pacemakers are programmed to best meet the demand for cardiac output during exercise.

Caloian and colleagues compared the influence of the most common pacemaker programming modes on the exercise capacity of the subjects (9). Fifty-two patients with either single- or dual-chamber pacemakers set by their personal cardiologists underwent exercise stress testing. Four types of patients were submitted to an exercise stress test based on their pacemaker programming mode: single chamber without rate-responsive function (RRF) activated (19 patients), single chamber with RRF activated (12 patients), dual chamber without RRF activated (9 patients), and dual chamber with RRF activated (12 patients). Patients with dual-chamber devices with RRF had the best exercise capacity (5.6 ± 1.2 METS), while patients with single-chamber devices without RRF had the lowest exercise capacity (4.0 ± 1.6 METS). Patients in groups with single-chamber pacemakers with RRF and dual-chamber pacemakers without RRF had intermediate values (5.1 ± 1.6 METS and 5.3 ± 1.7 METS, respectively).

With adjustment for demographic and cardiac function parameters from a resting echocardiogram, the programmed pacing mode was an independent predictor of exercise capacity in men ($r = 0.51$, $p = 0.01$) but not in women. These results highlight the influence of the pacemaker programming mode on exercise capacity. Dual-chamber pacing is superior to single-chamber pacing (ventricular pacing), and the activation of the rate-responsive function in the single-chamber pacemakers has a similar impact on exercise capacity as does preservation of the AV synchronicity in dual-chamber pacemakers. These data also highlight the need for physicians to consider the programming mode in patients who are interested in increasing their physical activity.

Based on Caloian et al. (9).

Clinical Exercise Bottom Line

- The benefits of exercise training apply to patients with implanted cardiac devices.

- The exercise prescription is generally the same as for others. However, the prescription needs to account for the device settings, the patient's comorbidities such as cardiovascular disease, and medications the patient may be taking.

- Depending on the device, its settings, and patient medications, target heart rate may not be useful for determining exercise intensity. In some patients, rating of perceived exertion and MET equivalents can be used to guide exercise intensity.

- In patients with a limited heart rate response during exercise, blood pressure should be monitored more frequently to make sure it increases in response to increasing exercise workloads.

- If the patient's implanted cardiac device has an activity sensor, the clinical exercise physiologist needs to take that into account when recommending a mode of exercise training.

CONCLUSION

Because of the increased prevalence of pacemakers and ICDs, clinical exercise physiologists and cardiac rehabilitation professionals need to know how these devices function and what their limitations are. The two types of pacemakers are single-chamber units and dual-chamber units. Knowledge of the universal coding system is required to understand appropriate pacemaker function. Initial pacemakers operated at fixed rates and were primarily used for patients who were symptomatic because of bradycardia or high-degree AV block. Because of the inability to increase heart rate with exertion in fixed-rate pacemakers, rate-responsive pacemakers have been developed so that cardiac output can appropriately increase under physical activity. Motion sensors (vibration and accelerometers) are used in rate-responsive pacemakers. Each has advantages and disadvantages. In addition, dual-sensor (activity and ventilation) pacemakers have been developed to enhance normal heart rate response with exertion. In patients with rate-responsive pacemakers, exercise testing should be used to ensure an appropriate increase in heart rate and to allow for adjustment if the unit is not properly functioning. Exercise testing allows optimal programming of the pacemaker to provide maximal hemodynamic benefit and quality of life.

When prescribing exercise training with rate-responsive pacemakers, the clinical exercise physiologist must make sure that heart rate increases appropriately with exertion. In activities that do not involve a great deal of change in body movement (cycling, uphill walking), vibration sensors or accelerometers are not able to detect the real difference in activity level. Therefore, if possible, the physician should determine the type of activity that a person is planning to do before implanting a pacemaker. In addition, when prescribing different modes of activity, the clinical exercise physiologist must consider the type of pacemaker. Ideally, patients with rate-responsive pacemakers should be monitored to determine whether the physiological response is acceptable with exertion (i.e., heart rate, ECG, blood pressure, RPE, and METs). In addition, patients should avoid contact sports that carry a risk of direct contact with the pacemaker. Overall, not much in the way

of limitation applies to prescribing exercise in pacemaker patients, other than making sure an appropriate physiological response (increase in heart rate or blood pressure) occurs with increasing levels of physical exertion.

The use of ICDs is increasing to control life-threatening ventricular arrhythmias. Before patients with ICDs start an exercise program, an exercise test is recommended to determine the safety of exercise and rule out any other underlying diagnoses. When prescribing exercise, the major concern with patients with ICDs is to avoid reaching the threshold heart rate that will cause the device to shock. Training heart rate should stay $20 \text{ b} \cdot \text{min}^{-1}$ below the preset heart rate that produces a shock. Otherwise, no specific limitations govern the prescription of exercise in this select population. The use of CRT to restore the coordinated pumping action of the ventricles is becoming more widespread in patients with chronic heart failure and delayed conduction. This type of device allows patients to be more active and have a better quality of life. Further research is needed to examine the long-term benefit of this therapy and to identify patients who are most likely to benefit from it.

Online Materials

Visit HK*Propel* to access a link to the references, the case studies with discussion questions, and a quiz to help you review key concepts and test your understanding of the material covered in this chapter.

PART IV

Diseases of the Respiratory System

Pulmonary diseases and conditions can play havoc with a person's activities of daily living and ability to be physically active. The clinical exercise physiologist can play a role in functional assessment and the recommendation of exercise training, which can be very challenging with respiratory diseases. Often these patients present in a rehabilitative setting in a pulmonary rehabilitation program. But given the level of association with other diseases (e.g., heart disease, peripheral artery disease, cancer), these patients may also seek exercise therapy in cardiac rehabilitation programs and in general fitness center facilities. This section provides excellent chapters on three of the most common pulmonary conditions.

Chapter 19 examines chronic obstructive pulmonary disease (chronic bronchitis and emphysema), also known as COPD. Currently COPD is the third leading cause of death in the United States and the fourth leading cause worldwide (although it is estimated to be the third leading cause by 2030). Patients with COPD are most likely to seek out or be referred to pulmonary rehabilitation. Most are current or previous smokers, and they are often debilitated to the point of being unable to perform much physical activity without shortness of breath. The clinical exercise physiologist must learn the skills to educate and motivate these patients and implement appropriate exercise programming. This chapter is an excellent resource pertaining to these skills.

Chapter 20 explores asthma. Approximately 300 million people worldwide have asthma, and the condition affects over 25 million people in the United States (8% of adults and 7% of children). Asthma is associated with a combination of genetic, environmental, and lifestyle-related factors. Increasing numbers of people in the United States and the world are being diagnosed with asthma. This trend is likely a result of better recognition and diagnostic testing, although worsening environmental factors that trigger asthma attacks may also be contributing. Exercise-induced bronchospasm, a form of asthma, is also increasing in prevalence. People with asthma can benefit from exercise training and often are able to perform exercise independent of clinical supervision. As this chapter demonstrates, however, the clinical exercise physiologist can be a great asset to these patients in developing and implementing an exercise program to limit potential bouts of asthma and in helping treat attacks associated with exercise.

Chapter 21 deals with cystic fibrosis (CF), which causes inflammation and excess mucus production. Patients with CF have multiple medical issues of the lungs, gastrointestinal tract, sinuses, and sweat glands. The lungs are particularly affected, so a daily process of breaking up and expelling the mucus is a way of life for some people. Exercise is important for these individuals if they wish to maintain independence and functionality, but they must deal with potential respiratory difficulties and impaired thermoregulation. The clinical exercise physiologist can play an important role in helping these people deal with the effects of CF so they can live more active and productive lives.

CHAPTER 19

Chronic Obstructive Pulmonary Disease

Satvir S. Dhillon, MSc, CCRP

Dennis Jensen, PhD

Jordan A. Guenette, PhD

Chronic obstructive pulmonary disease includes the clinical conditions of chronic bronchitis and emphysema. This chapter discusses the underlying pathology of these conditions as well as the procedures and benefits of exercise testing and training for individuals with these conditions.

DEFINITION

Chronic obstructive pulmonary disease (COPD) is defined by the American Thoracic Society (ATS) and the European Respiratory Society (ERS) as a preventable and treatable progressive lung disease, with significant extrapulmonary effects (e.g., cardiovascular disease, osteoporosis, anxiety and depression, metabolic syndrome and diabetes, impaired cognitive function, weight loss, and limb muscle dysfunction), mainly characterized by the presence of persistent expiratory airflow limitation that is progressive and not fully reversible (41). According to the Global Initiative for Chronic Obstructive Lung Disease (GOLD), the major risk factors for the development of COPD include significant exposure to noxious particles or gases from smoking cigarettes or other types of tobacco, marijuana, environmental tobacco smoke, occupational exposures, and outdoor and indoor air pollution (85). In addition, strong evidence implicates several rare genetic syndromes, including α_1-**antitrypsin deficiency** (63), and other host factors, including abnormal lung development (121) and oxidative stress-induced accelerated aging (148).

The mechanisms underlying the chronic airflow limitation in patients with COPD involve a combination of factors contributing to structural abnormalities in the small airways (e.g., obstructive bronchiolitis) and pulmonary parenchyma (e.g., emphysema) (85). Chronic bronchitis is a clinical diagnosis for patients who may have various symptoms including chronic sputum production, mucous or bronchial hypersecretion, or chronic cough with sputum production (117). The ATS arbitrarily defines **chronic bronchitis** as the presence of a chronic productive cough on most days for a period of at least 3 consecutive months in each of 2 successive years (99). The cough is a result of mucus hypersecretion, which in turn results from enlargement of mucus-secreting glands. In contrast to the clinical diagnosis for chronic bronchitis, **emphysema** is a pathological or anatomical diagnosis marked by abnormal permanent enlargement of the respiratory bronchioles and the alveoli, the air spaces distal to the terminal bronchioles, and is accompanied by destruction of the lung parenchyma without obvious fibrosis (206, 207, 231). Most patients with COPD have characteristics of either chronic bronchitis or emphysema or both, and the relative extent of each varies among patients (figures 19.1 and 19.2). Chronic respiratory symptoms and structural evidence of lung disease may be present in current or former smokers without expiratory airflow limitation (184, 253). The World Health Organization's International Classification of Diseases and Related Health Problems (ICD) code used by nosologists to classify COPD is CA22 (255).

Patients with COPD experience acute **exacerbations**, or periods of worsening respiratory symptoms resulting in the need for supplemental therapy (e.g., bronchodilators, glucocorticoids, antibiotics, oxygen therapy, ventilatory support, and mechanical

Acknowledgment: Much of the writing for this chapter was adapted from earlier editions of *Clinical Exercise Physiology*. As a result, we wish to gratefully acknowledge the previous efforts of and thank Michael J. Berry, C. Mark Woodard, Ann M. Swank, and N. Brian Jones.

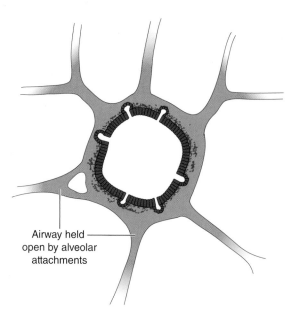

Figure 19.1   A normal airway that has little inflammation or mucus plugging and is being held open by parenchymal lung tissue.

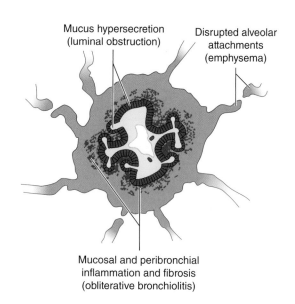

Figure 19.2   An obstructed airway that has significant inflammation and mucus plugging. Also shown is a loss of alveolar attachments, making airway collapse more likely.

ventilation) (85). The pathogenesis of an exacerbation is poorly understood, and it may be difficult to define clinically. These exacerbations can lead to respiratory failure and are a major cause of hospitalizations in the United States. Exacerbations of COPD can also lead to reduced quality of life, increased rate of pulmonary function decline, deterioration of symptoms with prolonged recovery times, and increased mortality (85). As a result, prevention of COPD exacerbations is critical to the clinical management of patients with COPD.

At times, asthma has been subsumed under the rubric of COPD. Asthma is characterized by chronic airway inflammation as well as hyperresponsiveness of the tracheobronchial tree to a variety of stimuli (e.g., dust or allergen exposure) (84). Although people with asthma experience exacerbations or attacks, these exacerbations are interspersed with symptom-free periods when airway narrowing is completely or almost completely reversed (see chapter 20 on asthma). In contrast, most patients with COPD do not exhibit significant reversibility of airway narrowing, and they present with residual symptoms between exacerbations. Although patients with COPD and asthma share similar clinical characteristics, the pathology of the two syndromes differs considerably, suggesting they are different diseases (135). According to GOLD (85), pharmacotherapy for patients experiencing recurrent airflow obstruction with a combination of the usual features associated with asthma and the usual features associated with COPD should primarily be guided by the Global Initiative for Asthma recommendations (84), supplemented with pharmacological and nonpharmacological therapies to manage their COPD.

SCOPE

COPD is the third leading cause of death in the United States (10, 102). According to the World Health Organization (WHO), COPD is the fourth leading cause of death worldwide and is projected to rise to the third leading cause of death by 2030 (254). COPD is a major source of morbidity and mortality in the United States (79). According to Sullivan and colleagues (220), 5.9% of adults in the United States (almost 16 million people) self-reported that they were diagnosed with COPD by a physician or other health professional; however, population-based studies showed that spirometrically defined COPD prevalence was more than twice as high as self-reported COPD prevalence in Canada (224), and over 60% of adults with spirometric evidence of possible COPD or other obstructive lung diseases were undiagnosed in the United States (141). This suggests that a large percentage of the general population is unaware that they have objective evidence of chronic airflow limitation.

In addition to being a major cause of morbidity and disability, COPD is a major health care cost in the United States, estimated to have cost the system $32.1 billion in 2010 and projected to increase to $49 billion in 2020 (80). In 2010, COPD was given as the diagnosis at the time of discharge for 699,000 hospitalizations (79). The average length of stay for COPD hospitalizations has been estimated to be 4.8 d, with an average cost of a single COPD-related hospitalization of $7,500 (249). Also in 2010, 10.3 million physician office visits were attributable to COPD (79). Exacerbations of COPD have a major effect on use of health care resources (175).

PATHOPHYSIOLOGY

The anatomical, physiological, and pathological abnormalities associated with COPD often result in debilitation for the COPD patient. Because of the direct insult that tobacco smoke, the primary risk factor for the development of COPD (41, 118), has on the lungs, airways, and pulmonary vasculature, it has long been thought that COPD primarily affects the respiratory system. Recent research suggests, however, that the disease process itself, or certain aspects of the disease process (e.g., systemic inflammation), adversely affect not only the lungs but also skeletal muscle (138).

Cigarette smoking affects the large airways (**bronchi**), the small airways (**bronchioles**), the pulmonary **parenchyma**, and the pulmonary vasculature. The pathological conditions that develop are a result of the effect that cigarette smoke has on each of these structures. Additionally, the degree of airway reactivity of the individual patient has an effect. Within the large airways, cigarette smoke causes the bronchial mucus glands to become enlarged and the gland ducts to become dilated. Excessive cough and sputum production are a result of these factors and are the characteristic symptoms of chronic bronchitis (230).

These alterations in the large airways have little effect on airflow, or **spirometry**. The airflow obstruction that is characteristic of COPD does not occur until additional damage is incurred by the small airways and the lung parenchyma. The changes in the smaller airways include mucus plugging, inflammation, and an increase in airway smooth muscle tone (figure 19. 1). These changes decrease the cross-sectional area of the airways and can have a profound effect on airflow and the COPD patient's spirometry (230). The spirometry of a COPD patient is characterized by reductions in the **forced expiratory volume in 1 s (FEV$_1$)**, the FEV$_1$/**forced vital capacity (FVC)** ratio, and midexpiratory flows.

Destructive changes occur to the alveolar walls with emphysema. The net effect of these changes is twofold. First, destruction of the alveolar walls results in a loss of the tethering, or supportive, effect that the alveoli have on the smaller airways. This tethering effect helps keep the airways open during expiration. Without this alveolar support, the smaller airways are likely to collapse during expiration, further adding to the airway obstruction (figure 19.2). The second effect of destruction of the alveolar walls is diminished elastic recoil of the lungs, which in turn decreases the force that moves air out of the lungs. The combination of these two effects reduces airflow and increases the amount of work the respiratory muscles must perform to meet the ventilatory demands of the body.

The combined effects of airway obstruction and the reduced expiratory driving force increase the time needed for expiration. If inspiration occurs before the increased expiratory time requirement can be met, then the normal end-expiratory lung volume will not be reached, resulting in increased functional residual capacity and **hyperinflation** of the lungs. Furthermore, the diaphragm will assume a shorter, more flattened position. Because the diaphragm is a skeletal muscle, it operates according to the length–tension relationship (144), whereby the tension developed by skeletal muscle is a function of its resting length. As a muscle is shortened or lengthened beyond its optimal length, the potential for tension development decreases (88). Because the diaphragm is shortened with hyperinflation, it has less force-generating potential (178, 205); that is, it becomes functionally weakened. Some evidence suggests that the diaphragm adapts to these chronic changes by shortening the optimal length of its fibers (202). As such, each fiber would have the potential to generate its maximal force at its new length (75). There is also evidence of mitochondrial adaptations in the diaphragm muscle fibers, resulting in higher oxidative capacity compared with age-matched controls (127). These adaptations may also explain, at least in part, why exercise-induced diaphragmatic fatigue is uncommon in patients with COPD (134). Despite these adaptive changes, evidence still suggests that COPD patients have decreased capacity to generate diaphragmatic pressure.

For individuals with normal lung function, the end-expiratory lung volume decreases from its resting value by approximately 200 to 400 mL with moderate exercise (101, 198). In contrast, most patients with COPD demonstrate an increase in the end-expiratory lung volume in the rest to exercise transition, leading to dynamic hyperinflation of the lungs (217). This dynamic hyperinflation leads to further diaphragm shortening and weakness and may contribute to dyspnea (shortness of breath) and reduced exercise tolerance. The link between hyperinflation and dyspnea is supported by studies showing strong interrelationships between indices of lung hyperinflation (e.g., increase in end-expiratory lung volume from resting values), abnormal constraints on tidal volume expansion, intensity ratings of exertional dyspnea, and exercise tolerance in patients with COPD (167, 168). Because the diaphragm is a skeletal muscle, it has been hypothesized that positive adaptations may result from training of this particular muscle (124, 190) through ventilatory muscle training (see the Exercise Training section).

In addition to the damaging effect on the lungs, COPD contributes to skeletal muscle dysfunction (138), particularly of the lower limbs (e.g., quadriceps). The skeletal muscle dysfunction may contribute to the exercise intolerance seen in COPD patients, as many COPD patients prematurely stop exercise before reaching ventilatory limits because of complaints of intolerable leg fatigue (116). Patients with COPD have diminished peripheral muscle strength (21, 46, 57, 69, 90, 98), as evidenced by a 20% to 30% reduction in quadriceps strength compared with age-matched controls (57, 90, 98), regardless of disease severity (196). The decrease in strength is accompanied by reductions in muscle cross-sectional area (21, 69) and muscle mass (68, 194, 256), both of which are independent predictors of mortality in COPD (142, 193).

A recent statement by the ATS/ERS on skeletal muscle dysfunction in COPD has identified the following abnormalities of the limb muscles in patients with COPD compared with healthy controls (138): a shift in fiber type distribution from type I fibers, the proportion of which has been shown to be inversely associated with COPD disease severity, to type IIx fibers; a reduction in capillary density and quantity per muscle fiber; a reduction in mitochondrial density and function, with reduced oxidative capacity of the locomotor muscles; and increased oxidative stress. Both chronic **hypoxemia** (low oxygen content in the blood) (103) and physical inactivity (248) may contribute to the limb muscle abnormalities in COPD. Chronic steroid use has also been suggested as a contributor to muscle weakness (56, 57). Other possible contributors to the skeletal muscle abnormalities seen in COPD patients include chronic **hypercapnia** (high levels of carbon dioxide in the blood), systemic inflammation, nutritional depletion, and comorbid conditions that may impair skeletal muscle function (5, 138). Practical application 19.1 discusses malnutrition in COPD.

One hallmark of patients with COPD is a reduction in expiratory airflow, which is most prominent during maximal expiratory efforts. This reduction is typically quantified using the results of pulmonary function tests; the FEV_1 is one of the standards used to assess disease severity and monitor disease history. Additionally, the FEV_1 has been shown to be a strong predictor of mortality in patients with COPD (229). Smokers have an increased rate of decline in FEV_1 compared with healthy nonsmokers (118, 223). Kohansal and colleagues (118) reported that smokers had an increased mean rate of decline in FEV_1 by 38.2 mL/yr for males and 23.9 mL/yr for females compared with their healthy nonsmoking counterparts. People who have characteristics compatible with emphysema have a rate of decline in the FEV_1 of 70 mL/yr (31).

In those who quit smoking, the decline of FEV_1 is less pronounced than the decline observed in those who continue to smoke (15). In fact, ex-smokers show rates of decline of FEV_1 similar to those of nonsmokers (15), and those who quit smoking early may derive greater beneficial effects on FEV_1 decline than those who quit later (118); in younger ex-smokers with mild to moderate COPD, the FEV_1 has been shown to initially increase following smoking cessation (87). Although complete cessation of smoking has beneficial effects on the rate of decline of FEV_1, the effect on FEV_1 decline of attempting to quit smoking and relapsing is equivocal. Sherrill and colleagues (199) reported that the rate of decline of the FEV_1 is greater in ex-smokers who resume smoking compared with those who continue to smoke. Murray and colleagues (152) reported that attempts to quit smoking in patients with mild COPD slow the rate of FEV_1 decline compared with those who continue to smoke. In addition, smoking cessation has been shown to decrease the risk of mortality in patients with COPD compared with those who continue to smoke (87).

CLINICAL CONSIDERATIONS

This section discusses signs and symptoms, history and physical examination, diagnostic exercise testing, and treatment specifically related to COPD.

Signs and Symptoms

The acute respiratory illness associated with COPD is characterized by increased cough, purulent sputum production, wheezing, chest tightness, dyspnea, and occasional fever. Airflow limitation may develop after many years of chronic cough and sputum production or may develop without the presence of these characteristic symptoms of COPD (85). In addition, sputum may or may not be produced with chronic coughing in COPD (83). With progression of the disease, the interval between these illnesses

Practical Application 19.1

MALNUTRITION IN COPD

Malnutrition, or nutritional depletion, is a problem for as many as 25% of all COPD patients (119, 194, 252), and the prevalence of malnutrition is higher in those with increased disease severity (194, 252). In addition, 50% of patients hospitalized for treatment of COPD demonstrate protein as well as calorie malnutrition (107). In a review of 90 COPD patients, researchers found that patients who required hospitalization and mechanical ventilation demonstrated the most severe nutritional decrements (76). Whereas the causes of malnutrition in COPD patients have not been clearly defined (250), the results of malnutrition have. Weight loss in COPD patients has been shown to be a predictor of mortality (252). Additionally, prolonged malnutrition results in deleterious changes to the diaphragm muscle such that its ability to generate force is decreased (115, 128). This finding, coupled with the fact that individuals with COPD exhibit dysfunction in other skeletal muscles, suggests the need for nutritional support. Even if dietary intake is adequate, some COPD patients may still have nutritional depletion because of increased caloric demand (30). When individuals with COPD are given sufficient calories in excess of their needs, they gain weight and achieve significant improvement in ventilatory and peripheral muscle strength (251). Given the need for nutritional intervention and the positive outcomes that can result, the clinical exercise physiologist should consult with a nutritionist or dietitian when working with adults with COPD, particularly those who are underweight or sarcopenic.

decreases (12). In the early disease stages, slowed expiration and wheezing are noted during the physical examination. Additionally, breath sounds are decreased, heart sounds may become distant, and coarse crackles may be heard at the base of the lungs (12). In advanced disease stages, slowed inspiration (131) and nutritional depletion are also common problems (194).

History and Physical Exam

The diagnosis of COPD is made based on patient history, a physical examination, and the results of laboratory and radiographic studies (**roentgenogram**). The diagnosis is suspected in patients who have a history of smoking and who present with an acute respiratory illness or respiratory symptoms such as a productive cough or dyspnea on exertion. A smoking history of more than 40 **pack-years** has been reported to be the best predictor of airflow obstruction (182). According to GOLD recommendations (85), a comprehensive medical history of a patient who is suspected to have COPD should investigate the patient's exposure to potential risk factors (e.g., smoking, occupational exposures, and environmental exposures), past medical history, family history of COPD or other chronic lung diseases, pattern of symptom development, exacerbation history or past hospitalizations for respiratory problems, presence of comorbidities, impact of the symptoms on the patient's life, social and family support available, and possibilities for reducing risk factors (e.g., smoking cessation). Although many physical signs of airflow obstruction (e.g., wheezing on auscultation) may be present during a physical examination, their absence should not exclude the diagnosis of COPD, as these signs are typically not present until the disease is well advanced, and findings on physical examinations have poor sensitivity and specificity for detecting airflow obstruction (85).

Diagnostic Testing

Results from pulmonary function tests are necessary for establishing a diagnosis of COPD and for determining the severity of the disease, but they cannot be used to distinguish between chronic bronchitis and emphysema. The FEV_1, FVC, FEV_1/FVC, and single-breath diffusing capacity of the lungs for carbon monoxide are the primary pulmonary function tests recommended to aid in the diagnosis of COPD (11). In patients with COPD, the results from all these tests are less than what would be predicted for a person of the same sex and similar age and stature. Other recommended tests include lung volume measurements and determination of arterial blood gas levels. Lung volume measurements often reveal increases in total lung capacity, functional residual capacity, residual volume, and decreases in inspiratory capacity. Arterial blood gases may reveal hypoxemia in the absence of hypercapnia in the early stages of the disease, with a worsening of hypoxemia and hypercapnia presenting in the later stages of the disease (12).

Because emphysema is defined in anatomic terms, the chest roentgenogram can sometimes be used to differentiate between emphysema and chronic bronchitis. Patients with chronic bronchitis often have a normal chest roentgenogram; the roentgenogram of patients with advanced emphysema may reveal large lung volumes, hyperinflation, a flattened diaphragm, and vascular attenuation (12). Computed tomography has greater sensitivity and specificity than the chest roentgenogram and can be used for both qualitative and quantitative assessment of emphysema. Because the additional information gained from computed tomography rarely alters therapy, it is infrequently used in the routine care of COPD patients. Although computed tomography is not recommended for routine use with COPD patients, it is the best way of recognizing emphysema and probably has a significant role in recognizing localized emphysema that is amenable to surgical treatment (231).

The presence of persistent expiratory airflow limitation requires a postbronchodilator FEV_1/FVC ratio of <0.7 (85). GOLD proposes using the postbronchodilator FEV_1 to diagnose COPD and classify the severity of airflow limitation (i.e., disease severity) in patients with COPD (85). According to GOLD recommendations (85), an FEV_1 ≥80% of predicted indicates grade 1, mild airflow limitation; FEV_1 ≥50% and <80% of predicted indicates grade 2, moderate airflow limitation; FEV_1 ≥30% and <50% of predicted indicates grade 3, severe airflow limitation; and FEV_1 <30% of predicted indicates grade 4, very severe airflow limitation. Recent updates to the GOLD classification system further categorize patients based on their exacerbation history and prior hospitalizations, as well as their self-reported dyspnea assessed using the modified Medical Research Council dyspnea scale (mMRC) (23) or disease-specific quality of life assessed using the COPD Assessment Test (CAT) (114).

According to GOLD recommendations (85), patients are considered a low risk of exacerbation if they had no or one exacerbation within the last 12 mo and had no hospitalizations due to exacerbation within the last 12 mo. These low-risk patients are categorized as group A if they have a CAT score <10 or mMRC grade of 0 to 1 (indicating fewer symptoms) and as group B if they have a CAT score ≥10 or mMRC grade ≥2 (indicating more symptoms). Patients are considered a high risk of exacerbation if they had more than two exacerbations within the last 12 mo or more than one hospitalization due to exacerbation within the last 12 mo. These high-risk patients are categorized as group C if they have a CAT score <10 or mMRC grade of 0 to 1 (indicating fewer symptoms) and as group D if they have a CAT score ≥10 or mMRC grade ≥2 (indicating more symptoms).

Exercise Testing

Exercise testing is an integral component in the evaluation of patients with COPD. In patients with mild or moderate disease, symptoms generally do not present until increased demand is placed on the respiratory system (e.g., with exercise). In fact, recent studies in smokers with normal spirometry and patients with mild COPD (GOLD grade 1) show evidence of small airway dysfunction at rest as well as abnormal respiratory mechanical and gas exchange responses to exercise, which contribute to impaired exercise tolerance and increased dyspnea compared with healthy controls (64, 65, 66, 94, 96, 172). In patients with more severe disease, functional capacity is further reduced to such a level that even simple activities of daily living may impose a challenge to the respiratory system. Most patients with moderate to severe COPD have reduced exercise capacity because of reduced ventilatory capacity in the face of increased ventilatory demand (60, 113). Because exercise places increased demand on the respiratory system, exercise testing provides an objective means of evaluating the functional capacity of an individual COPD patient; that is, exercise testing can be used to identify the factors contributing to dyspnea and exercise intolerance. Additionally, exercise testing is a powerful clinical tool to track the progression of disease, detect exercise hypoxemia, determine the need for supplemental oxygen during exercise training, evaluate the response to treatment, and prescribe exercise (218).

Table 19.1 outlines procedures and guidelines for exercise testing of the patient with COPD (9, 151). Several variables should be measured before the start of the test and monitored continuously throughout the test, at the termination of the test, and during recovery as needed. The minimum monitoring required during the test is detailed in chapter 5 on graded exercise testing.

Blood pressure should be measured with the patient's arm relaxed and the manometer mounted at eye level. Automatic monitors for blood pressure measurement during exercise are available, and their use has been found acceptable (93). Placement for the 12-lead ECG is typically the Mason-Likar. The gold standard for the measurement of arterial oxygen saturation is co-oximetry using arterial blood; the use of a pulse oximeter is acceptable as long as the device has been validated during exercise. At oxygen saturations greater than 90%, these devices have a high degree of reliability; however, as oxygen saturation drops below 90% their reliability decreases (157). Because of the problems with precisely defining the degree of hypoxemia with pulse oximeters, they should be used to determine whether desaturation is occurring and then to correct it with supplemental oxygen. The final variables that should be monitored during the exercise test in patients with COPD are dyspnea and leg fatigue. A number of valid and reliable scales are available for use (4, 9, 136). One particular scale of interest to the exercise specialist is the modified 0 to 10 category-ratio Borg scale (25). This scale has been adapted for use during exercise testing and training with COPD patients (137).

If the equipment is available, gas exchange and ventilatory measurements should be obtained. These measures provide valuable information that can be used to prescribe exercise intensity more accurately, to evaluate the effectiveness of an exercise intervention, and to provide information regarding the extent of the respiratory disease (9, 13). The measurement of inspiratory capacity at rest and during exercise can be used to track changes in dynamic operating lung volumes (e.g., end-expiratory and end-inspiratory lung volumes) (95), which provides valuable information regarding the presence of abnormal constraints on tidal volume expansion as well as the presence and magnitude of dynamic hyperinflation (40, 59, 163). Of special concern with measuring gas exchange and ventilatory parameters for patients with COPD is the use of supplemental oxygen. These patients should be tested on an elevated fraction of inspired oxygen such that it equates with the flow rate established for the use of supplemental oxygen. Most commercially available gas exchange measurement systems have established procedures that allow for use of elevated fractions of inspired oxygen during exercise testing and conversions of oxygen flow rates to inspired oxygen fractions.

Exercise testing has been shown to be extremely safe, even in high-risk populations, with one or fewer deaths per 10,000 tests (204). In patients with coronary heart disease, there is a low risk of cardiovascular events after moderate- and high-intensity exercise (189). To minimize risk to patients, recommended guidelines from the American College of Sports Medicine (9), the American Association of Cardiovascular and Pulmonary Rehabilitation (7, 8), and the American Heart Association (78) should be closely followed. In general, the procedures for testing patients with COPD follow those for testing other at-risk populations. Before conducting the exercise test, the clinician should review information from a medical exam and history to identify contraindications to testing (see chapter 5). Patients with resting oxygen saturation ≤85% while breathing room air must perform the exercise test with supplemental oxygen (13). An additional concern for patients with COPD is accompanying pulmonary hypertension. Some experts advise caution with these individuals because of the risk of serious cardiac arrhythmias or even sudden death during testing (8).

The exercise mode, the test protocol, and the monitoring equipment are all fundamental considerations in exercise testing. The exercise mode should increase total body oxygen demands by requiring the use of a large muscle group. The most common exercise testing modalities for the patient with COPD are the motorized treadmill and cycle ergometer. Field walking tests, such as the 6 min walk test, and incremental and endurance shuttle walk tests are other commonly used modes of exercise in COPD patients in both clinical and research settings (105). One exercise mode that should not be used routinely to test COPD patients is arm ergometry because patients with severe COPD often use the accessory muscles of inspiration for breathing at rest. Any additional burden placed on these muscles could result in significant symptoms and distress for the patient (42).

The testing protocol should start at a work rate the patient can easily accomplish, use increments in the work rate that are progressively difficult, and last a total duration of 8 to 10 min (9) or 5 to 9 min in severe and very severe disease (20). The initial stages should be of an intensity that allows the patient adequate time to warm up and become accustomed to the exercise. The work rate increments should be small and based on characteristics of the patient (e.g., sex, size, severity of disease, and previous level of physical activity). A commonly used approach for identifying the appropriate work rate increments is described by Wasserman and colleagues

Table 19.1 Exercise Testing Recommendations, Procedures, and Guidelines for the COPD Patient

Test type	Mode	Protocol specifics	Clinical measures	Clinical implications	Special considerations
Cardiorespiratory	Treadmill or cycle ergometer (preferred) 6 min walk Incremental and endurance shuttle walk	Duration of 8-10 min (5-9 min in severe and very severe COPD), small incremental increases in workload and slow progression individualized to the patient Treadmill: 1-2 METs per stage Cycle ergometer: unloaded cycling for 3 min followed by ramped protocol 5, 10, 15, or 20 W/min, stage protocol Constant work rate protocols for treadmill or cycle ergometer testing Field walking tests can be either self-paced or externally paced and conducted over a predetermined time or distance	HR, 12-lead ECG (Mason-Likar placement) BP RPE, rating of perceived dyspnea, and leg fatigue using the modified 0-10 category-ratio Borg scale Oxygen saturation (pulse oximetry/ arterial PaO_2) Ventilation measures and gas exchange Blood lactate Distance Duration Power output Inspiratory capacity Walking distance	Serious dysrhythmias, >2 mm ST-segment depression or elevation, ischemic threshold, T-wave inversion with significant ST change SBP >250 mmHg or DBP >115 mmHg Maximum ventilations, $\dot{V}O_2$peak, lactate/ventilatory threshold Note rest stop distance or time, dyspnea index, vitals	Arm ergometer testing results in increased dyspnea, which may limit the intensity and duration of exercise. Clients with COPD often have coexisting CAD. Breathing pattern and inspiratory capacity maneuvers may help identify COPD patients with dynamic hyperinflation. Lactic acidosis may contribute to exercise limitation in some patients. Exercise should be terminated in the event of severe arterial oxyhemoglobin desaturation ≤80%. Exercise testing in mid- to late afternoon is desirable. Constant work rate cycle ergometer testing is useful for measuring improvements in physiological and exercise performance responses following therapeutic interventions. Constant work rate treadmill testing is useful for evaluating the effects of inhaled bronchodilators on exercise endurance. Walk testing may be more suitable in severe COPD.
Muscular strength and endurance	Isokinetic, isotonic, or both Sit-to-stand Stair climb– descent Lifting Push-up Curl-up	Cable tensiometers Handgrip dynamometers Repetition maximum (RM)	Peak force development or maximum voluntary contraction Maximum number of reps without rest Time to 10 reps Absolute 1RM or multiple RM Duration of static contraction before fatigue		Clients may become more dyspneic when lifting objects (teach appropriate breathing strategies for lifting). Specific evaluation and training may be needed. 10RM-15RM within the appropriate training recommendations may be more appropriate in COPD.
Flexibility	Sit-and-reach Goniometry Gait analysis Balance	Hip, hamstring, and lower back flexibility	ROM Furthest distance reached with fingertips on sit-and- reach test		Body mechanics, coordination, and work efficiency are often impaired.

(244), whereby the work rate increments per minute (W/min) equal the patient's estimated peak oxygen consumption in 10 min minus the patient's estimated oxygen consumption at unloaded pedaling divided by 100. The increment of work rate affects the exercise response of patients with COPD (55). For example, the peak work rate achieved for a given level of oxygen consumption is greater when the work rate is increased quickly. Unfortunately, because of severe deconditioning and extreme dyspnea that some patients with COPD experience when performing even mild physical activity, having a test that lasts the minimum recommended duration may not be possible.

Generally speaking, the physiological and perceptual responses of the patient with COPD to an exercise test are consistent across the spectrum of disease severity, although the magnitude of the responses vary as a function of disease severity. Commonly reported exercise responses for patients with COPD compared with age-matched healthy controls are shown in table 19.2. As expected, peak oxygen uptake and peak work rates are abnormally low in patients with moderate or severe COPD (36, 140, 171, 200, 247). Concomitant with the lower peak oxygen consumption and work rate is a lower peak heart rate and a greater heart rate reserve (predicted peak heart rate minus measured peak heart rate). Oxygen pulse, a surrogate measure of cardiac stroke volume, is also low in patients with COPD because they terminate exercise at abnormally low work rates.

In normal and deconditioned individuals and in cardiac patients, the ventilatory reserve (maximal voluntary ventilation minus the peak exercise minute ventilation) is high at peak exercise. In contrast, the patient with COPD has a low ventilatory reserve, and in some cases, peak minute ventilation is equal to or even greater than the maximal voluntary ventilation (244). Additionally, the peak minute ventilation is lower than predicted (36). The ventilatory equivalent for carbon dioxide is elevated across all stages of COPD severity, reflecting greater exercise ventilatory inefficiency compared with healthy controls (153). The partial pressure of oxygen in the arterial blood and the percentage saturation of hemoglobin in the arterial blood are often low at maximal exercise in the patient with moderate or severe COPD (12).

Patients with COPD develop a significant anaerobiosis and demonstrate a lactate threshold, although these occur at relatively low work rates compared with healthy controls (36, 219). In some cases, the lactate threshold may not be detected because patients stop exercise at very low work rates because of intolerable symptoms (e.g., dyspnea or leg fatigue). Because of their ventilatory impairment, patients with moderate and severe COPD do not show a disproportionate increase in minute ventilation with the development of anaerobiosis (219). Thus, the detection of a ventilatory threshold may not be possible in some patients, particularly those with severe to very severe COPD.

Exercise testing is an important tool in assessing the patient with COPD. The test and equipment used must be designed to meet the needs of the patient and the clinician administering the test. Additionally, because of the abnormal responses of the COPD patient, one must exercise care when interpreting the results of

Table 19.2 Exercise Test Responses in Patients With COPD Compared With Normal Healthy Subjects

Parameter	Finding
Peak work rate	Decreased
Peak oxygen consumption	Decreased
Peak heart rate	Decreased
Peak ventilation	Decreased
Heart rate reserve	Increased
Ventilatory reserve	Decreased
Arterial partial pressure of oxygen	Decreased
Arterial oxygen saturation	Decreased
Lactate threshold	Occurred at a lower work rate
Ventilatory threshold	Decreased or absent
Ventilatory equivalent for carbon dioxide	Increased
Leg fatigue	Increased
Dyspnea	Increased
Inspiratory capacity	Decreased
Dynamic hyperinflation	Increased
Breathing frequency	Increased
Tidal volume	Decreased

these tests and also consider the potential impact of age, sex, and body composition as well as the presence of comorbidities, both physiological (e.g., cardiovascular, metabolic, and musculoskeletal disorders) and psychological (e.g., anxiety and depression).

Treatment

After a diagnosis of COPD has been made, a multifaceted approach to the treatment and management of the patient should be adopted. Comprehensive treatment of stable COPD should include smoking cessation, oxygen therapy, ventilatory support, surgical treatments, pharmacological therapy, pneumococcal and influenza vaccination, physical activity, and pulmonary rehabilitation (including exercise training, self-management education, assessment and follow-up, and nutritional support) (85). Because smoking is a major risk factor for the development of COPD, smoking cessation is a major therapy in the treatment of COPD patients. Smoking cessation is one of the few interventions known to improve patient survival (87, 203). Several pharmacological therapies have been shown to increase quit rates in COPD patients, including nicotine replacement products (215, 232), varenicline (228), sustained release bupropion (227, 242), and nortriptyline (242). Behavioral support (e.g., self-help interventions, telephone or face-to-face counseling, brief advice delivered by physician or other health care professional, and group therapy) is also effective in helping people quit smoking (17, 77, 212, 214). Combining pharmacological therapies and behavioral support increases smoking cessation rate compared with usual care or minimal interventions (213).

Another therapy that has been shown to improve survival in stable COPD patients is long-term oxygen therapy. The British Medical Research Council study (145) and the Nocturnal Oxygen Therapy Trial of the National Heart, Lung and Blood Institute (NHLBI) (158) showed that patients with severe resting hypoxemia (partial pressure of oxygen in arterial blood <55 mmHg) who received long-term oxygen therapy (>15 h/d) experienced a significant reduction in mortality rates. Other benefits of long-term oxygen therapy include a reduction in **polycythemia** (126), decreased pulmonary artery pressure (1, 2, 3), and improved neuropsychiatric function (100). Despite these reported benefits of long-term oxygen therapy for COPD, there is no evidence for the effectiveness of this treatment in patients with less severe hypoxemia (51). The NHLBI's Long-term Oxygen Treatment Trial showed that long-term oxygen therapy did not provide a survival advantage in patients with stable COPD and moderate resting or exercise-induced oxygen desaturation (129).

More recently, domiciliary oxygen therapy has been demonstrated to prolong survival for patients with COPD but only for those with resting hypoxia (203). Other investigators are exploring the role of stem cells for the treatment of COPD, and this work is in the preliminary stages (97). Despite the therapeutic potential of stem cell treatment, there is growing concern about unproven stem cell–based treatment options in the United States increasingly being offered to people with COPD and other respiratory diseases

(108). Results from studies on the efficacy of stem cell treatment for patients with COPD have so far been inconclusive (222).

Acute administration of supplemental oxygen has been shown to enhance exercise tolerance in hypoxemic (162) and nonhypoxemic patients (208). Whether patients with COPD undergo an acute bout of exercise with administration of supplemental oxygen (29, 48, 54, 149, 177) or are trained with supplemental oxygen (58, 177, 192, 241, 258), the benefits are significant. The goal of oxygen therapy is to reverse hypoxemia and prevent tissue hypoxia (12). Nonhypoxemic COPD patients have been shown to benefit from supplemental oxygen therapy during training as well (67). These benefits include improved exercise endurance (159, 208) and reduced intensity ratings of perceived dyspnea and leg fatigue (6, 159, 208, 235). Inconsistent results have been reported on the benefits of supplemental oxygen therapy on mortality and maximal exercise capacity (6, 159). If oxygen is to be prescribed for COPD patients, the goal is to maintain the partial pressure of oxygen in arterial blood above 60 mmHg or the percentage saturation above 90. Therefore, the delivery method and the dosage of oxygen, or the flow rate, need to be considered. COPD patients who need supplemental oxygen during exercise often use a liquid oxygen supply. These systems, although more expensive, are lightweight and easily refilled from larger stationary sources. Oxygen concentrators cannot be used during exercise because of their weight and need for an electrical supply.

Pharmacological therapy (table 19.3) in patients with COPD is aimed at inducing bronchodilation, decreasing airway inflammation, managing and preventing acute exacerbations, alleviating symptoms, and improving exercise capacity and quality of life (12, 85, 234). **Bronchodilator** therapy includes the use of β_2-**agonists**, anticholinergic agents, and theophylline. Inhaled bronchodilators have been reported to reduce exertional dyspnea and improve exercise capacity (19, 165). However, inhaler misuse or poor compliance to inhaler therapy is associated with worse treatment outcomes and increased risk of exacerbations (237). Given the association of inhaler misuse with increased utilization of health care services and poor symptom control in patients with COPD (146), training in proper inhaler technique is important to reduce inhaler mishandling (85). The use of β_2-agonists may result in tremors, anxiety, palpitations, and arrhythmias. Because of these problems, careful dosing and monitoring of patients with known cardiovascular disease are necessary (12). After a patient develops persistent symptoms, anticholinergic agents such as ipratropium bromide may be prescribed because their effect is more intense and of longer duration. Additionally, they may have fewer potentially deleterious side effects than β_2-agonists. Combining the different classes of bronchodilators may result in greater therapeutic benefits with fewer side effects than if either drug was used alone (238). For example, a combination of a long-acting β_2-agonist and a long-acting anticholinergic muscarinic antagonist has been associated with significant improvements in pulmonary function measures, quality of life indicators and breathlessness, as well as reductions in exacerbations and need for emergency rescue medications, compared with using the bronchodilator therapies separately in COPD (170, 236). Combining the bronchodilators may allow the drugs to work synergistically to promote smooth muscle relaxation in the airway (38, 39).

Theophylline, one of the methylxanthines taken orally in sustained release form, is a third agent that may be used alone or in combination with β_2-agonists or anticholinergic agents or both to induce bronchodilation. Besides its bronchodilator effects, theophylline increases cardiac output, decreases pulmonary vascular resistance, and may have anti-inflammatory effects (221, 260). Despite the beneficial effects of theophylline, its popularity has declined because of its toxicity (104, 197) and potential for adverse interaction with other drugs (188). Inhaled corticosteroids are prescribed for more severe disease with frequent exacerbations. Oral corticosteroids may be necessary to treat acute exacerbations of COPD (32, 85, 234) and to prevent hospital readmission because of recurrent acute exacerbations (50). Combining inhaled corticosteroids with long-acting bronchodilators may provide a greater benefit compared with using each therapy alone in COPD patients with exacerbations and moderate to very severe disease (85). Phosphodiesterase-4 inhibitors, which inhibit intracellular cyclic adenosine monophosphate breakdown, have been shown to decrease airway inflammation and prevent acute exacerbations of COPD (44, 74, 183, 257). Combination therapy with long-acting bronchodilators is recommended in COPD when using phosphodiesterase-4 inhibitors (85). However, no additional benefits in exercise capacity are elicited from the addition

Table 19.3   Pharmacology

Medication class and name	Primary effects	Exercise effects	Important considerations
β_2-agonist short- and long-acting bronchodilators Salbutamol (albuterol), arformoterol, fenoterol, formoterol, indacaterol, levalbuterol, olodaterol, salmeterol, terbutaline, tulobuterol	Bronchodilation (selectively stimulate β_2 adrenoreceptors to relax airway smooth muscle)	May increase or have no effect on HR, ECG; may increase, decrease, or have no effect on BP; increases exercise capacity by limiting bronchospasm	May increase resting HR and produce cardiac dysrhythmias; may cause palpitations and tremulousness; may decrease potassium levels in the blood
Anticholinergic short- and long-acting bronchodilators Aclidinium bromide, glycopyrronium bromide, ipratropium bromide, oxitropium bromide, tiotropium, umeclidinium	Bronchodilation (selectively inhibit muscarinic receptors to prevent or reduce bronchoconstriction)	May increase or have no effect on HR, ECG; no change in BP	May cause mouth dryness; may increase risk of cardiac events; may lead to acute glaucoma when administered using a face mask
Methylxanthine bronchodilators Aminophylline, theophylline SR	Bronchodilation (nonselectively inhibit the phosphodiesterase enzymes)	May increase or have no effect on HR, ECG; no change in BP	May produce cardiac dysrhythmias; may increase risk of intentional or accidental overdose and seizures; may cause nausea, headaches, heartburn, and insomnia; may increase respiratory drive
Inhaled corticosteroids Beclomethasone, budesonide, fluticasone	Anti-inflammatory	May increase exercise capacity when used in combination with long-acting bronchodilators	May cause oral candidiasis, skin bruises, and hoarseness; increases risk of pneumonia
Oral corticosteroids Methyl prednisolone, prednisone	Anti-inflammatory (systemic)	May decrease muscle strength	May cause fragility, osteoporosis, and skin atrophy; may cause respiratory failure in very severe COPD patients
Thiazide and loop diuretics Furosemide, hydrochlorothiazide	Diuresis (control right heart failure in COPD patients with cor pulmonale)	Decreases BP	May decrease potassium levels in the blood; may cause cardiac dysrhythmias and muscle weakness
Phosphodiesterase-4 inhibitors Roflumilast	Anti-inflammatory	No effect on exercise capacity	Should be combined with one or more long-acting bronchodilators; should not be combined with theophylline; may cause nausea, decrease in appetite, abdominal pain, headaches, diarrhea, and disturbed sleep

of phosphodiesterase-4 inhibitors to long-acting bronchodilators compared with long-acting bronchodilator monotherapy or dual therapy (111, 161).

Pharmacological therapy in stable COPD patients may be guided individually according to symptom burden and future risk of acute exacerbation (85, 164). According to GOLD recommendations (85), strategies for initiating pharmacological therapies in patients with COPD should be individualized based on the patient's GOLD group, and then subsequent escalation or de-escalation strategies should be chosen to predominantly target either persistent dyspnea or exacerbations (table 19.4).

Pulmonary rehabilitation is an evidence-based, multidisciplinary, and comprehensive intervention for patients with chronic respiratory diseases who are symptomatic and often have decreased activities of daily living. Pulmonary rehabilitation is a recognized and globally accepted therapeutic option in COPD disease management (85). The ATS/ERS has defined pulmonary rehabilitation as "a comprehensive intervention based on a thorough patient assessment followed by patient-tailored therapies that include, but are not limited to, exercise training, education, and behavior change, designed to improve the physical and psychological condition of people with chronic respiratory disease and to promote the long-term adherence to health-enhancing behaviors" (211). The various components of each of these services are listed in table 19.5. Practical application 19.2 deals with client–clinician interaction for patients with COPD.

According to GOLD recommendations (85), participation in a comprehensive pulmonary rehabilitation program has various benefits, including improved exercise capacity, health status, dyspnea, survival, and recovery following hospitalization for an exacerbation. As a result of participating in a pulmonary rehabilitation program, patients have also demonstrated improvements in sense of well-being (16, 211), self-efficacy (186, 211), and functional capacity (211). Determining which of the specific components of pulmonary rehabilitation is responsible for these improvements is difficult, because they are all integrally related.

Table 19.4 Recommended Treatment Strategies for Optimal Pharmacological Management by GOLD Group

GOLD group	Initial treatment	Strategy to predominantly target persistent dyspnea	Strategy to predominantly target persistent exacerbations[a]
A	Bronchodilator[b]	Add a second long-acting bronchodilator[c] or Long-acting β_2-agonist plus long-acting muscarinic antagonist plus inhaled corticosteroid[d]	Long-acting β_2-agonist plus long-acting muscarinic antagonist[e] or Long-acting β_2-agonist plus long-acting muscarinic antagonist plus inhaled corticosteroid[f,g,h]
B	Long-acting β_2-agonist (LABA) or Long-acting muscarinic antagonist (LAMA)		
C	Long-acting muscarinic antagonist		
D	Long-acting β_2-agonist plus long-acting muscarinic antagonist or Long-acting muscarinic antagonist or Long-acting β_2-agonist plus inhaled corticosteroid		

[a]If both dyspnea and exacerbations need to be targeted, then the strategy to predominantly target exacerbations is recommended.

[b]Patients can be on either a short- or long-acting bronchodilator.

[c]Escalation strategy for patients on long-acting bronchodilator monotherapy. If there is no effect on symptoms, then de-escalation back to long-acting bronchodilator monotherapy or switching to a different class of bronchodilator monotherapy is recommended.

[d]Escalation strategy for patients on a combination of LABA and ICS. An alternative choice is to switch to a combination of LABA and LAMA if the original indication for ICS was inappropriate or the patient has adverse effects (e.g., pneumonia) or there was a lack of response to ICS.

[e]Escalation strategy for patients on long-acting bronchodilator monotherapy. An alternative choice is to switch to a combination of LABA and ICS (in patients with blood eosinophil counts ≥300 cells · microL^{-1} or patients with blood eosinophil counts ≥100 cells · microL^{-1} and ≥2 moderate exacerbations or at least 1 hospitalization for exacerbation in the previous year).

[f]Escalation strategy for patients on a combination of LABA and LAMA (in patients with blood eosinophil counts ≥100 cells · microL^{-1}). An alternative choice is to add roflumilast or azithromycin (in patients with blood eosinophil counts <100 cells · microL^{-1}).

[g]Escalation strategy for patients on a combination of LABA and ICS. An alternative choice is to switch to a combination of LABA and LAMA if there was a lack of response to ICS or the patient has adverse effects (e.g., pneumonia).

[h]If there are persistent exacerbations in patients on a combination of LABA, LAMA, and ICS, then the addition of roflumilast (in patients with GOLD grade 3 or 4 and chronic bronchitis) or a macrolide antibiotic such as azithromycin (in patients who are former smokers) is the recommended escalation therapy. An alternative choice is de-escalation to a combination of LABA and LAMA if there was a lack of response to ICS or the patient has adverse effects (e.g., pneumonia).

Adapted from Global Initiative for Chronic Obstructive Lung Disease (GOLD) (85).

Table 19.5 Various Components of the Services Offered in Pulmonary Rehabilitation

Patient assessment	Collaborative self-management education	Exercise training	Psychosocial interventions	Patient follow-up
Patient interview	Anatomy and physiology	Mode, duration, frequency, and intensity	Building support systems	Outcome measurements of exercise capacity, symptoms, and health-related quality of life
Medical history	Pathophysiology of chronic lung disease	Upper and lower extremity endurance training	Treatment of depression	Support groups
Physical exam	Description and interpretation of medical tests	Upper and lower extremity strength training	Treatment of anxiety	Maintenance programs
Diagnostic tests (e.g., pulmonary function tests, chest radiograph, electrocardiogram, bone density measurement, and blood chemistry)	Breathing strategies	Respiratory muscle training	Stress management and relaxation techniques	—
Symptom assessment (e.g., dyspnea, fatigue, and pain)	Secretion clearance	Flexibility and posture	Sexuality issues	—
Physical function assessment (e.g., musculoskeletal, exercise, and activities of daily living)	Medications	Orthopedic limitations	Adaptive coping strategies	—
Nutritional assessment	Symptom management	Home exercise program	Adherence to lifestyle modifications	—
Educational assessment	Energy conservation	Emergency procedures	Smoking cessation and relapse prevention	—
Psychosocial assessment	Benefits of exercise and maintaining physical activity	Documenting the evaluation and treatment session	—	—
Multicomponent assessment	Activities of daily living	—	—	—
Goal development	Nutrition	—	—	—
—	Smoking cessation and other irritant avoidance	—	—	—
—	Early recognition and treatment of exacerbations	—	—	—
—	Leisure activities	—	—	—
—	Coping with chronic lung disease	—	—	—
—	End-of-life planning	—	—	—

Adapted from American Association of Cardiovascular and Pulmonary Rehabilitation (7, 8).

Practical Application 19.2

CLIENT–CLINICIAN INTERACTION

Often, the first interaction between the clinical exercise physiologist and the patient with COPD occurs when the patient has been referred for pulmonary rehabilitation. Unfortunately, patients with COPD are often referred to pulmonary rehabilitation only after they have experienced an exacerbation of their disease or when their dyspnea has become so oppressive that they are severely disabled. As a result, these patients are often anxious, scared, frustrated, and depressed. The clinical exercise physiologist must be aware of these problems when working with the COPD patient and be able to present a positive, yet realistic, picture of the benefits of exercise training.

Dyspnea, the primary symptom of COPD, often results in a vicious cycle of fear and anxiety followed by physical inactivity (avoidance) resulting in deconditioning, which results in further dyspnea. Unless this cycle can be broken, patients with COPD are destined to become dependent on others to meet their most basic needs. Patients with COPD need to be made aware that they can learn to live with their dyspnea and that exercise can help reduce the intensity of this symptom and the associated distress. They should be taught strategies that will help them manage their dyspnea on a daily basis, including monitoring the effects of various medications on dyspnea, performing pursed-lip breathing, and avoiding or minimizing factors that can result in dyspnea.

The clinical exercise physiologist must also be aware that patients with COPD often experience exacerbations or periods of worsening symptoms. As a result, these patients may not be able to exercise at their prescribed intensity or may miss exercise sessions completely. The patient should be encouraged to continue exercising, even if at a much lower intensity. If even extremely mild exercise is not possible, the patient should be encouraged to resume exercising after recovering from the exacerbation.

The successful clinical exercise physiologist can effectively interact with each person on a one-to-one basis. Such a professional is sensitive to the particular needs of patients and is able to tailor each program to meet individual needs. As a result of effective clinician–client interaction, patients will be able to take greater control of their disease with less fear and anxiety.

Since the ATS/ERS considers pulmonary rehabilitation an important component within the concept of integrated care, patients with COPD may be referred to a pulmonary rehabilitation program during any stage of their disease progression, regardless of their clinical stability or exacerbation status. However, placing patients on optimal medical therapy prior to initiating exercise training may maximize the effectiveness of the exercise training component of pulmonary rehabilitation (211). GOLD considers pulmonary rehabilitation an essential component of nonpharmacological management of patients with COPD and GOLD group B, C, or D (85).

Although patients with COPD who are referred to pulmonary rehabilitation typically have severe disease, research indicates that pulmonary rehabilitation has positive effects on symptoms, quality of life, and exercise capacity regardless of disease severity (22, 71, 246). Participation in a pulmonary rehabilitation program that includes at least 4 wk of exercise training can result in improvements that are clinically significant for patients with COPD, including increased exercise capacity and health-related quality of life as well as improvements in symptoms, emotional function, and perceptions of how well they can control their disease (143). These improvements are greater with longer programs (120). Given the strong evidence for the benefits of pulmonary

rehabilitation, further randomized controlled trials comparing pulmonary rehabilitation programs with usual care are no longer required in COPD (143).

Despite the reported benefits of pulmonary rehabilitation, evidence for the long-term adherence to physical activity and other health-enhancing behaviors following participation in pulmonary rehabilitation—which is required to maintain benefits over time—has been inconsistent (210). According to GOLD recommendations (85), it is important to offer patients who are completing pulmonary rehabilitation a maintenance program or to support patients in remaining physically active at home. Physical inactivity, an independent predictor of mortality in COPD (243), is a growing area of interest in COPD management and has been reported to be greater in COPD patients compared with age-matched controls, and this inactivity increases with increasing disease severity (245). COPD self-management interventions, which involve repeated communications, or health coaching sessions, between health care professionals and patients to improve self-management skills (62), can help patients engage in the long-term adoption of physically active lifestyles (27, 233). Moreover, health coaching interventions that include a COPD exacerbation action plan have been shown to reduce the risk of hospitalizations and improve health-related quality of life in COPD patients (125).

EXERCISE PRESCRIPTION

Table 19.6 summarizes exercise prescription recommendations for the patient with COPD from American and international evidence-based pulmonary rehabilitation guidelines (81). Because heart rate is not a reliable indicator of exercise tolerance for patients with COPD, intensity is often monitored by rating of perceived dyspnea or exertion (106), with a suggested 4 to 6 rating on the modified 0 to 10 category-ratio Borg scale. At least 3 d/wk of aerobic training is recommended. Other recommendations include ≥2 d/wk of resistance and flexibility training and at least 4 d/wk of inspiratory muscle training, particularly for those with inspiratory muscle weakness (9).

EXERCISE TRAINING

As previously discussed, COPD patients are physically inactive and deconditioned, and they have high symptom burden, impaired exercise tolerance, and skeletal muscle dysfunction. All of these factors contribute to adverse health status in COPD and can partially be remedied by rehabilitative exercise training. The main components of conventional exercise training programs in patients with COPD are endurance (or aerobic) and resistance (or strength) training (81). In general, three exercise training strategies are recommended for use in COPD: lower extremity aerobic training, resistance training (including whole-body and upper extremity resistance modalities), and ventilatory (or inspiratory)

Table 19.6 Exercise Prescription Review

Training method	Frequency	Intensity	Time (Duration)	Type (Mode)	Progression	Important considerations
Cardiorespiratory	3-5 d/wk	Light: 30%-40% peak work Vigorous: 60%-80% peak Alternative: RPE 4-6 on 10-point scale (comfortable pace and endurance) Monitor dyspnea	20-60 min sessions; duration based on COPD severity and may be just a few min long at initial training stages	Large-muscle activities (walking, cycling, swimming, seated aerobics, arm ergometry, water exercises, step exercises)	Emphasize progression of duration more than intensity; 2-3 mo to ensure compliance; titrate to symptoms, dyspnea scale rates, or selected RPE or MET level; increase 5-10 min/wk during first 4-6 wk	Exercise compliance should be considered in the determination of exercise intensity. Shorter intermittent sessions may be necessary initially. Consider interval training for those who cannot tolerate continuous high-intensity exercise.
Resistance	2 or 3 d/wk	Light: 40%-50% 1RM Moderate: 60%-70% 1RM or 100% of 8RM-12RM Low resistance, high reps (fatigue by 8-15 reps)	1-4 sets, 8-10 exercises	Free weights, isokinetic or isotonic machines	Gradual; resistance and reps should be increased as strength increases Monitor RPE, fatigue, and pain	ACSM and ATS/ERS guidelines indicate following resistance and flexibility guidelines for older adults. Respiratory muscle weakness is common in pulmonary patients. Upper body exercise contributes to dyspnea. Inspiratory muscles may require training.
Range of motion	≥2 d/wk	Stretch to the point of feeling tightness	10-60 s static stretch, 2-4 reps for each exercise	Emphasize functional activities	Gradual; ROM should be increased as flexibility increases	Research needed to support this component, but it is commonly included in prescription.

Based on Garvey et al. (81).

muscle training. A brief discussion with review of the supporting literature for these three exercise training strategies follows.

Lower Extremity Aerobic Training

Endurance (or aerobic) training using walking or cycling modalities is the most common exercise intervention in pulmonary rehabilitation programs (37, 139, 166). The goal of endurance training is to improve exercise capacity as well as the function of ambulatory muscles to increase physical activity levels and reduce dyspnea and fatigue (211). Walking exercise (e.g., treadmill- or ground-based) is the preferred training modality for increasing functional exercise capacity and walking endurance because of its importance in activities of daily living. Cycling exercise (e.g., stationary cycle ergometer) has the advantage of eliciting less oxygen desaturation compared with walking exercise (179). Interval training consisting of brief periods of high-intensity exercise interspersed with bouts of low-intensity exercise or rest may be used as an alternative to continuous endurance training in patients limited by comorbidities or in those who find it too difficult to perform longer-duration exercise (181, 240). Both interval and continuous endurance training modes elicit similar effects on exercise capacity, health-related quality of life, and symptoms (18, 259).

Another alternative strategy to conventional lower extremity endurance training is partitioned exercise training using one-leg cycling. One-leg cycling is more effective for improving overall aerobic capacity than conventional cycling (24, 61). Despite the barriers to routine incorporation of partitioned exercise training into pulmonary rehabilitation (e.g., equipment modification concerns, trainer expertise, and safety), one-leg cycling has been shown to be a feasible and effective aerobic training modality in pulmonary rehabilitation (70).

The research focus sidebar reviews the literature about lower extremity exercise and COPD. Cumulative results indicate strong evidence for the use of lower extremity exercise as a therapeutic intervention for patients with COPD.

Resistance Training

Dyspnea and reduced exercise capacity are two of the most common complaints of COPD patients. In addition to the diaphragm muscle, the accessory muscles of inspiration (scalene, sternocleidomastoid, and serratus anterior) are activated during exercise. Even at low work rates, unsupported arm exercise results in greater levels of dyspnea compared with lower extremity exercise in patients with COPD (42). Arm exercise requires the use of the accessory muscles of inspiration, thereby decreasing their participation in ventilation and increasing the work of the diaphragm. This observation may explain, in part, why patients with COPD complain of dyspnea when performing activities of daily living with their upper extremities (225). Thus, strategies aimed at improving the function of the accessory muscles of inspiration, such as resistance training, could benefit COPD patients.

Skeletal muscle dysfunction contributes to the reduced exercise tolerance seen in COPD patients (35, 209). Impaired muscle strength has been found to be a significant contributor to symptom intensity during exercise in these patients (98, 216). Additionally in COPD, quadriceps muscle strength has been shown to be positively correlated with both the 6 min walk distance and maximal oxygen consumption (90, 98). These observations suggest that resistance training may also prove beneficial for the rehabilitation of patients with COPD.

Resistance (or strength) training is more effective than endurance training for increasing muscle strength and mass (45, 46, 133, 169, 209). It can also improve or maintain bone mineral density (155), which is lower than normal in many patients with COPD (91, 92). In addition, resistance training elicits less dyspnea as well as lower levels of oxygen consumption and minute ventilation at peak exercise compared with endurance training (166, 180). As a result, patients with more advanced disease and comorbidities who may be unable to complete continuous or interval endurance training because of symptom limitations (e.g., intolerable dyspnea) may obtain greater benefits from resistance training strategies (139,

Research Focus

Efficacy of Pulmonary Rehabilitation

The American College of Chest Physicians and the American Association of Cardiovascular and Pulmonary Rehabilitation have released evidence-based guidelines for pulmonary rehabilitation (185). This document contains recommendations for pulmonary rehabilitation and reviews the supporting scientific evidence. Lower extremity exercise training received a grade of A, meaning that strong scientific evidence supports lower extremity exercise training in COPD patients. Well-designed and well-conducted controlled (both randomized and nonrandomized) trials with statistically significant results support the use of lower extremity exercise training to improve exercise capacity, as evaluated from time on the treadmill or a timed

walk distance (185). According to a recent discussion at a workshop convened by the National Institutes of Health to investigate the efficacy of pulmonary rehabilitation, limiting the evaluation of interventions to outcomes such as timed walks or physiological measures is a myopic approach that provides incomplete measures of medical outcomes. The conclusion from this group was that the success of therapeutic interventions should be based on a variety of medical outcomes such as health-related quality of life, respiratory symptoms, frequency of exacerbations, activities of daily living, cost–benefit relationships, use of health care resources, and mental, social, and emotional function.

180). Resistance training programs may be designed to specifically improve upper body, lower body, and whole-body (incorporating both upper and lower body muscles) strength in COPD patients. Given these distinct advantages of resistance training over other forms of exercise training, resistance training should be included in a comprehensive exercise rehabilitation program. A reasonable recommendation on exercise dosage for demonstrating improvement in outcomes includes training ≥2 d/wk for one to four sets, with 10 to 15 repetitions per set and loads of 40% to 50% (or 60%-70% for patients who can tolerate a moderate initial intensity) of one-repetition maximum (9).

Upper extremity resistance training has been proposed as a training modality to help reduce dyspnea for patients with COPD. Ventilatory muscle fatigue and dyspnea occur when patients with COPD use their upper extremities to perform activities of daily living. Fatigue results from the additional work the accessory muscles of inspiration must perform in helping to support the arms during such activities (42). Although upper extremity resistance training improves upper limb strength in patients with COPD, the effect of upper limb training on health-related quality of life and dyspnea during activities of daily living is less clear (109). Janaudis-Ferreira and colleagues (110) reported that in patients with COPD who underwent upper limb resistance training, improvements were seen in upper extremity function, exercise capacity, and muscle strength compared with a sham intervention control group; however, no improvements were seen in health-related quality of life, dyspnea during activities of daily living, or symptoms. Preliminary results, however, support the recommendation that upper extremity resistance training be included in a comprehensive rehabilitation program (185).

These preliminary studies do not provide clear recommendations on the specific upper body exercises that would benefit this population or on the resistance or number of repetitions that will provide the optimal benefits. The exercises should probably involve the accessory muscles of inspiration. Examples of exercises targeting upper extremity muscles (e.g., biceps, triceps, deltoids, trapezius, pectorals, and latissimus dorsi) include training with free weights and elastic bands, using weightlifting machines (e.g., chest press and latissimus pull-down exercise), lifting dowels, performing functional tasks, performing suspension training with a pulley system, and using a wall (e.g., pushing off and throwing a ball against a wall) (123). With respect to the amount of resistance used and the number of sets and repetitions to be completed, the American College of Sports Medicine guidelines (9) and the recommendations of Evans (72) and Storer (216) should be followed.

Ventilatory Muscle Training

Ventilatory muscle training is recommended for COPD patients to increase ventilatory muscle strength and endurance. The ultimate goal is to improve exercise capacity, alleviate dyspnea, and improve health-related quality of life. Three strategies have been used to train the ventilatory muscles: voluntary isocapnic **hyperpnea**, **inspiratory resistive loading**, and **inspiratory threshold loading**.

With voluntary isocapnic hyperpnea, the patient is instructed to breathe at the highest sustainable level of minute ventilation for 10 to 15 min. With this technique, the patient is hyperventilating, and therefore a rebreathing circuit must be used to maintain **isocapnia**. The rebreathing circuit is complex and not always portable, and the patient requires constant monitoring to ensure isocapnia with use of these devices. Because of these problems, this type of training has not been used or studied extensively.

During inspiratory resistive loading, the patient breathes through inspiratory orifices of smaller and smaller diameter while attempting to maintain a normal breathing pattern. A potential problem with the use of this device is that the patient may slow breathing frequency in an attempt to decrease the sensation of respiratory effort. Because of this change in the breathing pattern, the load on the inspiratory muscles is reduced such that a training response may not occur.

With inspiratory threshold loading, the patient breathes through a device that permits air to flow through it only after a critical inspiratory pressure has been reached. These devices are small, do not require supervision, and avoid the problems associated with changing breathing patterns during inspiratory resistive loading.

Results from studies on the efficacy of ventilatory muscle training for patients with COPD patients are equivocal. Although ventilatory muscle training improves inspiratory muscle strength and endurance, alleviates dyspnea, and enhances health-related quality of life, no additional benefits in exercise capacity or health-related quality of life are elicited when ventilatory muscle training is added to whole-body exercise training (14, 82, 89, 160). In the evidence-based guidelines for pulmonary rehabilitation (185), ventilatory muscle training received a grade of B, reflecting that the scientific evidence from both observational and controlled clinical trials yielded inconsistent results. Because of this grade, it was recommended that ventilatory muscle training not be considered an essential component of pulmonary rehabilitation. But in patients who have decreased respiratory muscle strength and breathlessness and who remain symptomatic despite optimal therapy, ventilatory muscle training may be considered an adjunctive exercise therapy (47, 173, 211).

Specific recommendations regarding the intensity, frequency, and duration of ventilatory muscle training have not been developed. Most studies reporting improvements in ventilatory muscle function have had patients perform ventilatory muscle training at a minimum of 30% of their maximal inspiratory pressure (82, 89). The duration of the training has been at least 15 min and the frequency at least 3 d/wk. These appear to be the minimal requisites for an exercise prescription if ventilatory muscle strength and endurance are to be improved.

Adjunct Interventions

Other therapeutic modalities, or adjuncts, to conventional exercise training (e.g., cycling and treadmill) in patients with COPD include additional strength training interventions, neuromuscular electrical stimulation (NMES), eccentric exercise, whole-body vibration, noninvasive ventilation (NIV), supplementary oxygen, supplementary helium and oxygen gas mixtures, pharmacotherapy, nutrition, breathing exercises (e.g., ventilatory muscle training, ventilation feedback, and controlled breathing), coaching, osteopathy, harmonica playing, psychotherapy, written disclosure therapy, activity training and lectures, and acupuncture (34). A brief discussion with review of the supporting literature for some of these adjunct therapies follows.

The NMES technique involves eliciting muscular contractions to train selected skeletal muscles without conventional exercise by electrical stimulation of the muscles (211). Although NMES reduces dyspnea and improves muscle strength and exercise capacity in stable patients with severe COPD and reduced baseline exercise capacity (28, 154, 201), it may not be an appropriate alternative intervention for stable patients with COPD who have higher baseline exercise capacity and are able to tolerate conventional endurance and resistance training programs (52, 226). Eccentric cycling is another potentially effective alternative training strategy to conventional concentric cycling exercise in COPD patients with severe airflow limitation (187) or exercise hypoxemia (191). In fact, studies in people with advanced COPD show evidence of increased quadriceps muscle strength in patients participating in eccentric cycling exercise compared with those participating in conventional concentric cycling exercise (26, 132). Eccentric cycling exercise also allows patients with COPD to train at higher intensity levels with lower metabolic and ventilatory demands (156) as well as reduced intensity ratings of perceived dyspnea and leg fatigue (156, 132) compared with concentric cycling exercise. Thus, the reduced symptom burden elicited by eccentric exercise modalities may be especially beneficial for patients with advanced disease whose exercise may be limited by intolerable dyspnea. Additionally, downhill walking (an eccentric exercise) has been shown to elicit greater beneficial effects on functional exercise capacity (33, 150) and health-related quality of life (150) in patients with COPD compared with conventional flat walking.

Whole-body vibration is another method that may improve functional exercise capacity in patients with COPD (86, 261). NIV can also be used during lower limb exercise training to allow patients with severe and very severe COPD to exercise at higher intensity levels to improve exercise capacity and exercise endurance; however, the effect of NIV on functional exercise capacity and health-related quality of life is not clear (147).

Studies show that supplemental helium and oxygen gas mixtures (i.e., heliox) reduce dynamic hyperinflation (43, 122, 174) and respiratory muscle work (130, 239) as well as increase exercise endurance (43, 174) and alleviate exertional symptoms, including dyspnea (174, 239) and leg fatigue (239) in patients with COPD during acute bouts of exercise. In addition, heliox improves oxygen delivery to the muscles of ambulation in patients with COPD (43, 130, 239). Despite these reported benefits of heliox, studies evaluating the effectiveness of supplemental helium and oxygen gas mixtures during exercise training in patients with COPD have shown inconsistent results (73, 112, 195). As a result, the use of helium and oxygen gas mixture supplementation as adjunct interventions to conventional pulmonary rehabilitation is an underexplored strategy requiring further study (85, 211). As discussed previously, pulmonary rehabilitation improves not only physiological symptoms but psychological symptoms as well (176). Incorporation of psychotherapy into conventional pulmonary rehabilitation programs has been shown to improve symptoms of anxiety and depression (53), which are present in up to 40% of patients with COPD (49).

Table 19.7 summarizes the benefits associated with exercise training on fitness components for the patient with COPD. COPD affects a large number of older people. The disease process spans several decades and eventually results in significant morbidity and mortality rates. Research suggests that exercise can be used as an effective therapeutic intervention in these patients. The review presented here supports the notion that participation in an exercise program will decrease dyspnea and increase exercise capacity, addressing two of the most common complaints of COPD patients.

Table 19.7 Benefits Associated With Exercise Training on Fitness Components

Cardiorespiratory endurance	Skeletal muscle strength	Skeletal muscle endurance	Flexibility	Body composition
Cardiorespiratory reconditioning	Improved muscle mass and muscle force	Improved ambulatory muscle endurance to increase activities of daily living	Improved range of motion	Improved body composition
Desensitization to dyspnea and fear of exertion	Better balance	Improved lactate and ventilatory threshold	Reduced loss of mobility	Enhanced body image
Improved ventilatory efficiency	Increased facilitation of activities of daily living	—	—	Reduced risk of comorbidities and mortality associated with obesity

Clinical Exercise Bottom Line

- Patients often experience exacerbations, or flare-ups, of their respiratory symptoms. During these periods of worsening symptoms, patients should be encouraged to exercise at lower intensities, if they are willing and able, or to resume training as soon as they have recovered.

- Because of symptom burden, some patients with COPD may not be able to tolerate exercise testing for adequate durations. Work rate increments should be carefully selected before ini-

tiating exercise testing by factoring in the disease severity of the patient.

- Exercise testing conducted immediately before and after a structured exercise training program can provide detailed insight into possible therapeutic benefits.

CONCLUSION

This chapter presents background information regarding etiology, clinical history, and signs and symptoms of COPD, as well as exercise testing and prescription strategies for the clinical exercise physiologist and other exercise professionals. Exercise testing and training any patient with chronic disease involves individualization of treatment based on all available patient information. Thus, it is crucial that the exercise professional be familiar with each patient's history before developing an exercise program and that the program be individualized for each patient.

Online Materials

Visit HK*Propel* to access a link to the references, the case studies with discussion questions, and a quiz to help you review key concepts and test your understanding of the material covered in this chapter.

CHAPTER 20

Asthma

Louis-Philippe Boulet, MD

Simon L. Bacon, PhD

Andréanne Côté, MD

Asthma is a very common airway disease that affects people worldwide (21, 41, 93). Although most people with asthma have mild disease, it can result in significant disability, reduced quality of life, and health care costs. However, asthma can be controlled in the majority of cases, therefore reducing the consequences of the disease and allowing an active and normal life. Once asthma develops, remission can be observed in about half of children, although this is not frequently observed in adults (18, 50). Many factors, mostly a combination of genetics, environment, and lifestyle, are involved in the development of asthma:

- Having another allergic disease
- Prematurity, low birth weight, or cesarean section
- Duration of breastfeeding or use of antibiotics in early childhood
- Viral infections in childhood
- Low socioeconomic status and stress
- Allergens
- Urban habitat
- Modern lifestyle (Western lifestyle)
- Sensitizing agents at workplace
- Cigarette smoke (particularly maternal smoking)
- Obesity (including obesity of the mother during pregnancy)
- Diet (low levels of antioxidants, high salt intake)

Nonpharmacologic, predominantly behavioral, and pharmacologic strategies can be used to ensure that asthma is well controlled (37, 41, 42). However, many care gaps persist, explaining why about half of asthma patients have insufficient control of the disease.

This chapter discusses the definition, pathophysiology, clinical presentation, and treatment options available for asthma in addition to proper follow-up, with an emphasis on physical exercise.

DEFINITION

The Global Initiative for Asthma (GINA) provides the following definition: Asthma is a heterogeneous disease, usually characterized by chronic airway inflammation. It is defined by the history of respiratory symptoms such as wheeze, shortness of breath, chest tightness, and cough that vary over time and in intensity, together with variable expiratory airflow limitation (41). Asthma is usually associated with hyper-responsiveness of the airway to direct or indirect stimuli and often involves chronic airway inflammation. These changes in airflow limitation can be caused by a variety of factors, including respiratory infections, airborne particular matter, exercise, and weather changes.

In regard to exertion, exercise-induced asthma (EIA) is a transient narrowing of the airways after exercise that is reversible after inhalation of a β_2-agonist bronchodilator (5, 17, 76). Exercise-induced bronchoconstriction (EIB), however, reflects a narrowing of the airways induced by exercise, with or without asthma symptoms.

EIB occurs in many patients who have asthma, particularly when the condition is suboptimally controlled (17, 58). The incidence of EIB was studied in 134 children with asthma and 102 children with atopy but not asthma, and compared to 56 children without asthma or atopy. The incidence among those with asthma and atopy but not asthma was 63% and 41%, respectively (58).

As with airflow limitation, respiratory symptoms are also variable either within a single day or over time and may resolve spon-

Acknowledgment: The writing in this chapter was adapted from the first and second editions of *Clinical Exercise Physiology.* We therefore wish to gratefully acknowledge the previous efforts of and thank Brian W. Carlin, MD.

taneously or in response to treatment. Some patients with asthma may be asymptomatic most of the year while others have seasonal or permanent symptoms. In some instances, mainly when asthma is not optimally controlled, exacerbations can occur, particularly after respiratory infections or relevant allergen exposures (20). These exacerbations may be life threatening, even if asthma is not severe.

SCOPE

Asthma is a worldwide problem affecting more than 300 million individuals (41). Studies show that the prevalence worldwide has been increasing over the last several decades (18, 53, 73). The incidence is higher in some patient populations, with up to 23% of inner-city African Americans being noted to have asthma compared with 5% of Caucasians (73). Asthma can also be found in elite-sport athletes (17, 33, 46).

Despite the availability of good medical therapy, the morbidity and mortality associated with asthma continues to increase, particularly in the African American population. Although the exact reasons for this remain unclear, the variable nature of the pathophysiology of the disease process, the influence of environmental factors, the lack of accessible health care, and a lack of patient adherence to the medical regimen all potentially contribute. In addition to race, the morbidity associated with asthma also depends on geographic locale, with asthma mortality rates in Hispanics being the highest in the northeast United States (49, 59).

PATHOPHYSIOLOGY

Even in mild asthma, airway inflammation is persistent in the vast majority of patients (98). Because asthma is heterogeneous, various phenotypes (specific characteristics of a type of asthma) and endotypes (mainly related to underlying mechanisms) of asthma have been described (13, 103). Phenotypes are recognizable clusters of patients with similar demographic, clinical, pathophysiological, and inflammatory or even molecular characteristics. They may be influenced by associated comorbidities.

There are a variety of asthma phenotypes, including allergic, nonallergic, late onset, asthma with fixed airway obstruction, and asthma with obesity (13, 72, 103). To date, there is no strong relationship between a specific pathological feature and the clinical pattern or response to treatment. In regard to inflammation, asthma has been categorized as eosinophilic, neutrophilic, mixed, and paucigranulocytic. Many of these phenotypes and their effects on clinical outcomes or on treatment need to be further characterized. However, the presence of airway eosinophilia suggests a potential for better corticosteroid response (13).

Phenotypes may change over time (e.g., asthma in a smoker becomes more neutrophilic; obesity-associated asthma can improve after bariatric surgery) or may overlap. Severe asthma is usually differentiated as type 2 (T2) and non-T2 asthma, in relation to the involvement of specific cytokines produced mainly by lymphocytes. The T2 subgroup is believed to promote inflamma-

tion through immunoglobulin E, eosinophils, basophils, and mast cells, involving an increased expression of cytokines such as IL3, IL4, IL5, and IL13 (9, 44). In non-T2 asthma, no inflammatory cells apart from neutrophils are found in sputum, and there are no eosinophils or other T2 markers in airway fluid or tissue.

The inflammatory process is thought to induce structural changes within the airways, although other mechanisms such as airway pressure changes, resulting in stretching of the epithelial cells, may be involved. Airway remodeling includes various processes such as subepithelial fibrosis, hypertrophy and hyperplasia of the airway smooth muscle, angiogenesis, and increased submucosal glands (31, 48, 68, 89). Airway edema and mucus hypersecretion (8) contribute to airway obstruction by increasing resistance to airflow and decreasing expiratory flow rates, often associated with air trapping. These pathophysiologic features then result in the clinical symptoms experienced by a patient who has asthma.

Asthma in the High-Level Athlete

High-level athletes have a higher prevalence of asthma compared with the general population (10, 17). Among athletes practicing winter sports, between 14% and 28% are identified as having asthma (26, 28, 60). The prevalence among those training mainly in an aquatic environment varies from 21% to 31% (60, 78, 87); in athletes practicing an outdoor sport in summer, it is around 15% to 23% (46, 61). These prevalence's are high when compared against those of general population control groups, which is usually less than 10% (41, 53, 67). The reasons remain partially unknown, but it seems related to the high ventilatory volumes during exercise in environments that expose the athletes to pollutants, allergens, and cold, dry air (17, 90).

Similarly, the prevalence of airway hyperresponsiveness (AHR) is also higher among athletes than the general population (60, 61, 90). In athletes performing aerobic sports, training in a winter environment or in a swimming pool, the prevalence of AHR varies from 35% to 80% (16, 61, 78, 90).

Finally, EIB has an increased incidence in athletes, and this phenomenon can also be observed in athletes without evidence of symptomatic asthma (17, 84, 90). The significance of this, however, remains unknown.

Exercise-Induced Asthma

The main mechanism involved in EIA and EIB is the response to dehydration of the airways with an increased osmolarity of the fluid lining the airway (6, 56). This leads to mast cell degranulation, eosinophil activation, and release of cellular mediators of inflammation (e.g., histamine, leukotrienes, and prostaglandins).

Bronchoconstriction then occurs, which results in airflow limitation (see figure 20.1). Although previously considered a main mechanism, airway temperature alterations and postexercise airway rewarming are thought to have only a minor effect on the degree of bronchoconstriction (17, 56, 67).

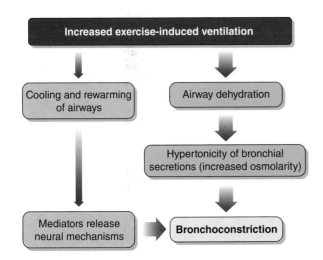

Figure 20.1 Mechanisms of exercise-induced asthma.

CLINICAL CONSIDERATIONS

Establishing a correct diagnosis of asthma is initially based on the clinical features (18, 41). Characteristic respiratory symptoms include wheezing, shortness of breath, chest pain or tightness, and cough. However, many symptoms (e.g., cough) are not specific to asthma. There is usually more than one type of symptom, and they occur variably over time and can vary in intensity. The symptoms are often worse at night or on awakening. They can be triggered by allergens, other sensitizing substances, respiratory irritants, exercise, and cold air and can appear or worsen with viral infections (18, 41). In many instances, the significance of a patient's symptoms may be overlooked by both patients and care providers (2). In addition, overdiagnosis is common—no evidence of asthma was found in about one-third of patients with a diagnosis of asthma in primary care (2).

The physical examination of patients who have asthma is usually normal unless the patient has very poorly controlled asthma or an exacerbation. (Refer to chapter 4 for a complete presentation of interview skills and examination of a patient who wants to get involved in an exercise program.) Most frequently, expiratory wheezing is noted on auscultation but may be heard only on forced expiration. Depending on the degree of airflow obstruction (e.g., in more severe cases of airflow limitation), wheezing may be absent. Crackles and inspiratory wheezing are not common features of asthma. An inspiratory stridor may suggest the presence of glottic (or vocal cord) dysfunction (51). Evaluation of the nose may reveal signs of nasal polyposis or allergic rhinitis (66).

The symptoms associated with other diseases may mimic the symptoms associated with asthma. Allergic rhinitis, sinusitis, and other lower airway diseases and conditions (e.g., foreign body aspiration, vocal cord dysfunction, vascular rings, cystic fibrosis, viral bronchiolitis, chronic obstructive pulmonary disease [COPD], chronic bronchitis, congestive heart failure, bronchiectasis, pul-

monary embolism) can all present with similar clinical findings. To confirm a diagnosis of asthma, further laboratory testing is required (18, 41).

Diagnostic Testing

Lung function testing is required to document the variable expiratory airflow limitation that is characteristic of asthma. This can be confirmed in one of several ways. Spirometry, before and after bronchodilator, is usually performed initially. In addition, bronchoprovocation tests and measurement of peak expiratory flow (PEF) can be used to verify the presence of airway hyperresponsiveness and variable airway obstruction (41). If a patient is suspected of having exercise-induced asthma, the diagnosis can be confirmed by demonstration of a significant bronchodilator response (an increase in forced expiratory volume in 1 s [FEV_1] of more than 10%—some prefer 15%), signifying a variable airway obstruction, or a significant fall in expiratory flows after bronchoprovocation tests (18, 43, 101).

Spirometry and Bronchodilator Response

A forced expiratory maneuver with measurement of FEV_1 and forced vital capacity (FVC) before and after bronchodilator administration should initially be performed if a patient is suspected to have asthma (see figure 20.2). Documentation of a low FEV_1 in the presence of a reduced FEV_1/FVC ratio (<0.75-0.80 in adults; <0.90 in children) along with a positive bronchodilator response—characterized as an increase in FEV_1 of >12% and >200 mL from baseline (adults) or an increase in FEV_1 >12% predicted (children)—helps confirm the diagnosis of asthma. Flow volume loops can help to differentiate other forms of airflow limitation (e.g., vocal cord dysfunction, restrictive disease) (80). In COPD, airway obstruction may show some reversibility, but expiratory flows usually do not normalize.

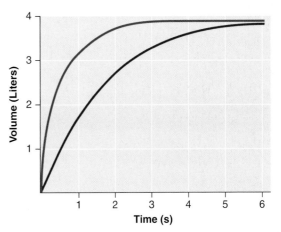

Figure 20.2 Spirometry tracings of a patient with asthma with a significant bronchodilator response (red line).

Peak Expiratory Flow Measurements

Peak expiratory flow measurements that show significant variability (more than 10%) and bronchodilator response (usually more than 15%) suggest the presence of variable airway obstruction. Performed with a peak flow meter, these measures can be helpful in both the diagnosis and management of patients who have asthma. The meters are inexpensive, portable, and ideal for patients to use. However, they are more effort dependent. Peak expiratory flow should be compared against the patient's own previous best measurement and serve as a reference value for monitoring the effects of treatment. Although adherence to regular long-term use is often poor, it is certainly useful in patients with a poor perception of symptoms, providing an objective measure of the degree of airway obstruction.

Bronchoprovocation Tests

In some instances, airflow limitation may not be present at the time of a patient's initial assessment. Airflow limitation may be seen only after bronchial provocation testing or exercise testing. Bronchoprovocation tests (BPT) may be direct, acting directly on airway smooth muscle (e.g., methacholine test), or indirect, acting through the release of mediators (e.g., exercise, eucapnic voluntary hyperpnea [EVH], mannitol) (54). The principle is to detect the intensity of the stimulus required to induce a specific (e.g., 20%) fall in expiratory flows (usually FEV_1). Recommendations base the methacholine challenge result on the delivered dose of methacholine, causing a 20% fall in FEV_1, called the provocative dose, or PD_{20} (21). It is increasingly accepted that PD_{20} provides a more consistent correlation of results than the provocative concentration (PC_{20}) of methacholine (25, 27). A greater than 10% decline in FEV_1 after exercise or EVH indicates the presence of exercise-induced bronchospasm (76, 101). These tests (see table 20.1) are moderately sensitive for a diagnosis of asthma but have limited specificity.

Exercise Testing

The ideal protocol involves a rapid increase in exercise intensity over approximately 2 to 4 min to achieve a high level of ventilation (preferably >21 times the FEV_1). General precautions before starting an exercise test are explained in chapter 5. Most protocols recommend breathing dry air (<10 mg $H_2O \cdot L^{-1}$) with a nose clip in place while running or cycling to achieve this level of ventilation. Once it is attained, exercise should continue for another 4 to 6 min. Ideally, a protocol should be chosen that will elicit a patient's maximal effort at between 8 and 12 min. In most instances, a goal of achieving a heart rate of 85% of maximum predicted value should be achieved. Immediately after exercise, spirometry is performed at 5, 10, 15, and 30 min post-exercise and the values compared with the pre-exercise values. The airway response should be expressed as the percent fall in FEV_1 from the baseline value. A fall of 10% in the FEV_1 from the pre-exercise level supports the diagnosis of EIB (43, 76). It is preferable to use FEV_1 instead of peak flow measures as part of the testing regimen. FEV_1 is more discriminating in the determination of the presence of airflow limitation. EIB severity can be graded based on the percent fall in FEV_1 from the pre-exercise level (mild: at least 10% but <25%; moderate: more than 25% but <50%; severe: 50% or more).

For most patients with asthma, there are few contraindications to exercise testing. Acute bronchospasm, active chest discomfort, and increased shortness of breath above that usually experienced are the primary contraindications to exercise testing. Severe exercise deconditioning or other comorbid conditions (e.g., unstable angina, orthopedic limitations) may limit the ability of a patient to exercise.

From a pathophysiologic point of view, in severe or uncontrolled patients with EIB without prior treatment, exercise increases ventilation and perfusion inequality, physiological dead space, and arterial blood lactate levels (5, 6). From a metabolic response perspective, a blunted sympathoadrenal response to exercise (100), an alteration in potassium homeostasis, and an excessive secretion of growth hormone have all been demonstrated. The role each of these metabolic responses may play in the exercise limitation in some patients with asthma is, however, unknown. Again, given the wide variety of pathophysiological processes in each patient who has asthma, one can expect a wide variety of cardiopulmonary responses to exercise.

Methacholine Challenge Test (MCT)

In the MCT, the patient breathes increasing doses or concentrations of methacholine, with an assessment of airway narrowing

Table 20.1 Bronchoprovocation Tests and Positivity Criteria to Detect Asthma, EIA, and EIB

Test	Positivity criteria
Laboratory or field test exercise	$\geq 10\%$ fall in FEV_1
Eucapnic voluntary hyperpnea (EVH)	$\geq 10\%$ fall in FEV_1
Mannitol or hypertonic saline challenge	$\geq 15\%$ fall in FEV_1
Methacholine challenge	$\geq 20\%$ fall in FEV_1
Patient who does not use inhaled glucocorticoids	$PC_{20} < 4$ mg $\cdot$ mL^{-1}
Patient who uses inhaled glucocorticoids	$PC_{20} < 16$ mg$\cdot$ mL^{-1}

Reprinted by permission from *The New England Journal of Medicine*, "Asthma and Exercise-Induced Bronchoconstriction in Athletes," L.P. Boulet and P.M. O'Byrne, 372, no 7: 641-648, Copyright © 2015 Massachusetts Medical Society. Reprinted with permission from Massachusetts Medical Society.

after each inhalation. This measures direct AHR in a safe way. MCT can be affected by medications and technical factors that influence the delivery of methacholine to the lower airways. The tidal breathing method is commonly used. The interpretation of a methacholine challenge test is based on the PC_{20} or PD_{20} and the pretest probability of disease (21, 23). This test has been well standardized. A PC_{20} of 4 mg · mL^{-1} or less is suggestive of airway hyperresponsiveness, while it is borderline between 4 and 16 mg/mL or 100 to 400 µg. Abnormal responses are further categorized as mild AHR (1-4 mg/mL or 25-100 µg), moderate AHR (0.25-1 mg/mL or 6-25 µg), and marked AHR (<0.25 mg/mL or <6 µg).

Eucapnic Voluntary Hyperpnea

EVH is an alternative to exercise challenges in that it reproduces the main mechanism of EIB, hyperpnea. The patient breathes medical dry air from a reservoir with an admixture of 4.9% CO_2, allowing high ventilation without the adverse consequences of hypocapnia (81, 83, 84), for 6 min at a target ventilation of 85% of MVV, with a minimum ventilation threshold of 60% of MVV (7, 83). Asthma medications should be stopped before the challenge for a given time according to their direction of action because they can inhibit the response (47, 88). When standardized, this test has a high degree of reproducibility and a higher sensitivity than exercise challenges for the detection of induced bronchoconstriction (7, 81, 84). However, some patients may find it difficult to achieve the target ventilation.

Other Tests

Inhalation of hyperosmolar aerosols of dry air or dry powder mannitol have been used. None of these are completely sensitive or specific for EIB, but they may help document airway hyperresponsiveness (54, 78).

Chest roentgenograms (CXRs) are rarely useful. They may show lung hyperinflation in severe or uncontrolled asthma or other conditions mimicking or associated with asthma.

To help identify the inflammatory phenotype involved, forced exhaled nitric oxide (FENO) measured in exhaled breath or induced sputum analysis is a noninvasive test that may determine whether eosinophilic or neutrophilic inflammation is present (41). Allergy skin prick tests may show sensitization to common airborne allergens.

Treatment

The general goals of asthma therapy are to achieve good symptom control, minimize the risk of future exacerbations and fixed airflow obstruction, and minimize the side effects of treatment (37, 41). The patient and family should always be included as part of treatment management. Determining goals of therapy as well as providing an effective management program that addresses the health literacy of the patient (see chapter 3), current resources, and cultural aspects is essential.

Assessing Asthma Control and Severity

Before starting a patient's initial treatment for asthma, the clinical exercise physiologist should record evidence for the diagnosis of asthma and comorbid conditions (e.g., rhinitis, obesity, gastroesophageal reflux disease, obstructive sleep apnea, depression), document symptom control and risk factors, assess lung function (when possible), train the patient to use the inhaler correctly, and check inhaler technique. A follow-up visit should be scheduled.

The assessment of each patient with asthma should include asthma control, including exacerbations, in addition to treatment issues. Overall control can be assessed according to specific criteria based on the patient's symptoms, risk factors for exacerbations, and measures of lung function. Patient understanding of current therapy and adherence to maintenance medication, inhaler technique, and provision of a written action plan for management of exacerbations should be evaluated with every patient–caregiver interaction. A variety of asthma control tools (e.g., Asthma Control Questionnaire [ACQ], Asthma Control Test [ACT], Asthma Control Scoring System [ACSS]) are also available (4, 41, 63).

Pharmacologic and nonpharmacologic/behavioral treatments are the components of a control-based asthma management program. When control-based guidelines are implemented (e.g., controlling symptoms and reducing exacerbation risk), asthma outcomes have been shown to improve (1, 3, 37, 41, 42). A stepwise approach to management is recommended.

Behavioral Interventions

Cessation of smoking and avoidance of environmental exposures (indoor or outdoor allergens or pollution) is of prime importance. Regular physical activity, as discussed later in this chapter, should be promoted. A healthy diet and good weight control are also important parts of the treatment plan. Finally, influenza (and more recently COVID-19) vaccinations should be encouraged (41). Ideally, patients should be referred to an experienced asthma educator.

Preventive interventions include the following:

- *Pre-exercise warm-up:* This leads to a refractory period that can last up to 2 h, reducing the influence of hyperpnea on airways (47, 71). The warm-up can involve various intensities and exercise durations. Often, a warm-up period of 15 min of continuous exercise at 60% of maximal oxygen consumption can significantly decrease post-exercise bronchoconstriction in moderately trained athletes. This type of exercise warm-up may not be effective in all patients, however. Several randomized types of warm-up trials (interval, low-intensity continuous, high-intensity continuous, and combination) have been performed. Statistically significant improvements were noted in the groups using interval and combination therapy; thus, either type of warm-up can precede planned exercise. For those patients with EIB who exercise in cold weather, a device (e.g., mask) that warms and humidifies the air during exercise is recommended (77).

- *Appropriate environment:* Ideally, exercise should not be performed in an environment with unfavorable conditions, such as high exposure to relevant allergens or bad air quality (e.g., during intense pollution episodes) and cold or warm temperatures.

Pharmacologic Measures

The pharmacologic management of a patient with EIB involves various types of drugs, including bronchodilator, anti-inflammatory, or leukotriene modifier therapy (41, 92). Pharmacologic options fall into two categories (41). Controller medications are used for regular maintenance therapy. They reduce airway inflammation, provide bronchodilation, control symptoms, and reduce the risk of future exacerbations. Reliever medications provide as-needed relief of symptoms (e.g., during an exacerbation). They are also recommended for short-term prevention of EIB. However, the main agent to prevent EIB and EIA is a regular daily inhaled corticosteroid (ICS), which helps control asthma and reduce AHR to make airways less responsive to hyperpnea induced by exercise (17). At low to moderate doses, ICS have few adverse effects (manly local, e.g., oropharyngeal candidiasis) but at high doses, they may have some, particularly in children, so the lowest dose required should always be determined (29, 41, 55).

The Global Initiative for Asthma (GINA) has outlined a stepwise approach for reviewing a patient's symptoms and assessing and adjusting treatment (41). GINA provides guidelines for stepping up for uncontrolled symptoms and stepping down for lower-risk patients with controlled symptoms (www.ginasthma.org). GINA's recommendations have significantly changed in recent years, particularly in regard to stage 1 and 2 asthma (mild to very mild asthma).

According to the 2021 GINA report, in the majority of adults and adolescents with asthma, treatment can be started at step 1 or 2 with an as-needed low-dose ICS-Formoterol (formoterol is the first choice; if not available, a low-dose ICS whenever an inhaled short-acting β_2-agonist [SABA] is taken, or an ICS regularly with a SABA as reliever) (41). This preference is for the general population and has yet to be tested in athletes. So, these three options are probably adequate for an athlete and a regular ICS intake has the benefit in those last to reduce progressively responses to exercise (62, 89). Treatment may start at higher steps if, at initial presentation, the patient has troublesome asthma symptoms on most days or is waking from asthma once or more a week.

If the patient has severely uncontrolled asthma at initial asthma presentation, or the initial presentation is during an acute exacerbation, regular controller treatment at step 4 (e.g., medium-dose ICS-LABA [long-acting β_2-agonist]) can be started, although in some severe cases, a short course of oral corticosteroids (OCSs) may also be needed. Stepping-down treatment should be considered after asthma has been well controlled for 3 mo. However, in adults and adolescents, the ICS should not be completely stopped as the risk of recurrence is high.

For severe long-standing asthma, add-on medications are available, such as anticholinergics, anti-IgE, anti-IL5/5R, and in some countries, anti-IL4R agents. However, asthma is rarely severe in athletes, usually being mild; and sometimes moderate.

Ideally, medication should not be needed before each exercise session if asthma is well controlled—otherwise suggesting uncontrolled airway inflammation. However, a SABA or an ICS+ formoterol, which is a long but also fast-acting bronchodilator and should always be used in association with an inhaled corticosteroid in asthma, if the patient is using such medication, can be administered 15 min before planned exercise (17, 62, 43). If the patient's symptoms persist despite use of a bronchodilator before exercise or if the patient requires an inhaled SABA more than twice per month (according to the last GINA report), then daily administration of an inhaled corticosteroid or on-demand low-dose budesonide associated with formoterol in the same inhaler, or as a second choice, a leukotriene modifier, should be considered (41) (see figure 20.3).

Inhaled corticosteroid therapy may take 4 wk (often longer) to have a maximal effect. Regular daily use of β_2-agonists alone or in combination with an inhaled corticosteroid can lead to tolerance to the medication. Tachyphylaxis (tolerance to a drug) means a reduction in response to the therapeutic agent when administration is repeated, which can create a problem. Tachyphylaxis to β_2-agonists occurs in all athletes with asthma, and even nonathletes. However, this problem is mostly relevant for athletes in regard to the daily or almost daily exercise they perform. Frequent use can also lead to an increase in the severity of EIB due to a loss of its protective effect and possibly increased penetration of irritants or allergens into the airways (41). To avoid developing tachyphylaxis, it is essential to minimize the need for SABAs in achieving good asthma control. A tolerance can develop to long-acting bronchodilators, but they may still help control asthma. However, long-acting inhaled β_2-agonists should always be used with a concomitant inhaled corticosteroid (36, 41).

EXERCISE PRESCRIPTION

The vast majority of patients with asthma can perform exercise and should be encouraged to do so regularly. Given the variability of symptoms and airflow limitation among people who have asthma, the response to exercise may vary widely as well (39, 45, 75, 94). Patients with asthma often refrain from performing exercise because of the fear of inducing symptoms; this can promote deconditioning and weight gain, which could interfere with exercise tolerance. The presence of comorbidities (e.g., heart problems, dysfunctional breathing) can also affect exercise performance. General details about exercise prescription are found in chapter 6.

Some patients may be able to exercise at a level associated with what an elite athlete can attain, while others may be taxed simply by walking up a flight of steps or across the room. In addition, an individual patient's exercise ability may vary from time to time depending on the current level of control of the disease. This is

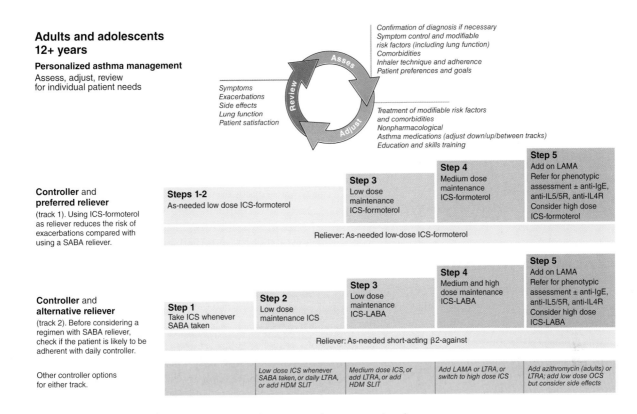

Figure 20.3   Summary of the GINA 2021 stepwise approach to control asthma.

ICS = inhaled corticosteroid; HDL SLIT = house dust mite sublingual immunotherapy; LABA = long-acting β_2 agonist; LAMA = long-acting muscarinic antagonist; LTRA = leukotriene receptor antagonists; OCS = oral corticosteroids; SABA = short-acting β_2 agonist.
Reprinted by permission from *Global Initiative for Asthma Report* (2021), 59.

particularly true during an exacerbation, when exercise ability may be extremely limited (20, 65). The mode, frequency, time, intensity, and progression for health and fitness benefits should be similar to those for a patient who does not have lung disease, taking into account other comorbid conditions that may be present.

For cardiorespiratory fitness, the exercise training should be at least 20 to 30 min in duration for a minimum of 2 d/wk (5). The mode of exercise should take into consideration the patient's interests, past exercise experience, and availability of equipment while acknowledging the effects of the surrounding environment. Exposure to cold air, low humidity, or air pollutants should be mini-

mized. Intermittent exercise or lower-intensity sports performed in warm, humid air are generally better tolerated. There is, however, no consensus on the optimal intensity level at which a patient with asthma should train. The intensity prescription should be based on the clinical and exercise test data in conjunction with the patient's goals. A symptom-limited exercise test with measurement of expired gases is one method to determine the optimal starting level of exercise (5). For patients with more limitations, a target intensity based on perceived dyspnea (such as a Borg scale) could also be considered (14). A sample exercise prescription summary is shown in table 20.2.

Table 20.2   Sample Exercise Prescription Summary

Mode	Frequency	Intensity	Duration	Important considerations
Treadmill (aerobic)	5 d/wk	Just below anaerobic threshold	20-30 min per session	Optimize medication therapy before exercise.
Walking or running (track, sidewalk)	5 d/wk	Just below anaerobic threshold	20-30 min per session	Optimize medication therapy before exercise.
Swimming	5 d/wk	Just below anaerobic threshold	20-30 min per session	Optimize medication therapy before exercise.

If maximum level of exercise has been determined by measurement of oxygen consumption and carbon dioxide production (cardiopulmonary exercise testing), begin exercise prescription at an initial intensity just below the anaerobic threshold. For training purposes, various intensity levels (e.g., just below anaerobic threshold or 50%-85% of heart rate reserve [maximal heart rate minus resting heart rate]) can be used. If such measurements are unavailable, begin exercise at a level the patient is comfortable performing for 5 min. Instruct the patient to continue exercise for 20 to 60 min per session. Have the patient perform sessions three to five times per week and then increase exercise intensity by 5% with each session. When a maximal level of intensity is attained, increase exercise duration by 5%.

EXERCISE TRAINING

A wide variety of physiological outcomes might be expected as a result of exercise training for patients with asthma. No adverse effect has been reported in regard to long-term lung function. However, the benefit of exercise training on static lung function measurements (spirometry), including bronchial hyperresponsiveness, is variable from one study to another (20-22). Exercise avoidance is common in people with asthma but that can be counterproductive because evidence indicates it improves asthma control and quality of life (QOL) (52, 95, 97). However, literature on the benefit of exercise to reduce EIB is still limited. Exercise training plus a weight loss program has been shown to improve daily life physical ability (DLPA), sleep efficiency, depression, and asthma symptoms in adults with obesity and asthma (30, 32, 38, 69). This group also reported that exercise improved asthma symptoms, QOL, exercise capacity, airway responsiveness, EIB, and FEV_1 in people with asthma, suggesting that physical activity should be recommended as a supplementary therapy to medication (34).

Furthermore, several physiological changes have been observed after training. These include increases in maximal oxygen uptake, oxygen pulse, and anaerobic threshold. Significant reductions in blood lactate level, carbon dioxide production, and minute ventilation at maximal exercise have been demonstrated. Subjective responses have also been noted, particularly a reduction in perceived breathlessness at equivalent workloads, following exercise training. This latter response could be attributable to a central nervous system desensitizing effect, a decrease in minute ventilation at submaximal workloads, or an increase in the endorphin levels without a concomitant reduction in ventilatory chemosensitivity.

A variety of cardiopulmonary and metabolic responses to exercise in patients with asthma have been reported, although patients whose asthma is controlled do not seem to have any pathological differences from individuals without asthma.

One of the most confounding variables in patients with asthma who are attempting to exercise is the effect of dyspnea on exercise capability (70). A wide variability between the degree of airway obstruction, exercise tolerance, and the severity of breathlessness has been noted in several studies (19, 79), but this accounts for up to only 63% of the variance in breathlessness that people with asthma noted during progressive incremental exercise. This complexity concerning the development of symptoms and exercise tolerance might be one reason the diagnosis of EIB is obscured. Individual assessment of each patient is thus important when one is trying to determine the degree (and thus subsequent effects) of exercise intolerance.

Exercise Training Schedules

A variety of training schedules have been used for patients with asthma. Various types of exercise, including gym, games, distance running, swimming, cycling, altitude training, and treadmill run-

---

Practical Application 20.1

CLIENT–CLINICIAN INTERACTION

A correct diagnosis of asthma (or exercise-induced bronchospasm) must be made and followed up with appropriate use of medications before exercise to allow the patient to optimize their exercise capabilities. To appropriately determine the level of exercise capability that a patient with asthma might be able to attain, the examiner should review the following features of the clinical assessment:

- Presence or absence of symptoms (e.g., cough, wheezing, shortness of breath) at rest or with exercise. A history of any exercise limitation (attributable to these symptoms) in an otherwise asymptomatic patient should alert the examiner to the possibility of exercise-induced bronchospasm.

- The main intervention to reduce the effects of exercise on airways of patients with asthma is to achieve adequate asthma control with an inhaled corticosteroid, with or without an added β_2-agonist.

- Use of medication before exercise (e.g., inhaled fast-acting β_2-agonist).

- Correct use of the medication and good inhaler technique.

- Correct use of warm-up and cool-down periods during exercise.

ning, have all been shown to improve exercise capability (11). The frequency of exercise training varied from study to study, ranging from once weekly to daily for periods of 20 min up to 2 h per training session. Training periods of 6 to 8 wk were generally used. The intensity of exercise also varied, from a gradual increase in exercise endurance to a short, heavy increase in exercise endurance (22). In one study, 26 adults with mild to moderate asthma (FEV$_1$ 63%) underwent a 10 wk supervised rehabilitation program with emphasis on individualized physical training. Daily exercise (swimming) for 2 wk was followed by twice-weekly exercise. Exercise training intensity was measured by a target heart rate during the first 2 wk and then by perceived exertion as measured by a Borg scale (1-10) during the latter 8 wk. Each participant was encouraged to exercise to a Borg level of 7 or 8. All participants were able to perform high-intensity exercise (80%-90% of their maximum predicted heart rate), and improvements in cardiorespiratory conditioning and walk distance were observed after the program. A decrease in asthma symptoms and a decrease in anxiety were noted after the training period (35).

A Cochrane review evaluated 21 studies (N = 772 patients) (19) in which patients with asthma were randomized to undertake physical training or not. Physical training (at least 20-30 min of exercise 2 or 3 d/wk for a minimum of 4 wk) was well tolerated, with no adverse effects reported. The patients showed an improvement in cardiorespiratory fitness as measured by an increase in maximum oxygen uptake of 4.9 mL · kg^{-1} · min^{-1} (95% confidence interval, 4.0-5.9) and tended to improve maximum expiratory ventilation of 3.1 L · min^{-1} (95% confidence interval, -0.6-6.8). However, no statistically significant effects were observed for FEV$_1$ or FVC. There was some evidence to suggest that physical training may have a positive effect on health-related quality of life, with four of five studies showing a statistically and clinically significant benefit (19). Since that review there have been several other systematic reviews and subsequent randomized controlled trials. In general, they not only supported the findings of improved cardiorespiratory fitness and minimal change in lung function but also demonstrated that exercise training can significantly improve asthma control and asthma-related quality of life (45, 82). These more recent studies have also explored inflammatory changes, with generally little evidence of a benefit of exercise training (45, 85), although one study comparing a weight loss intervention with exercise against a weight loss intervention without exercise found the addition of exercise to improve inflammation, as measured by FeNO and several blood-based markers (38).

Ongoing exercise after the initial training program has been shown to be effective for patients with asthma. Of 58 patients who had previously undergone a 10 wk rehabilitation program, 39 reported continuation of regular exercise. Cardiorespiratory conditioning (as measured by a 12 min walk distance) and lung function values remained unchanged in all patients. There was, however, a significant decrease in the number of emergency department visits over the 3 yr period compared with the year before entry into the rehabilitation program in these 39 patients. A decrease in asthma symptoms was noted only in a subgroup of patients (n = 26) who exercised one or two times per week. Continued exercise after a supervised rehabilitation program is helpful for patients with mild to moderate asthma (34).

Research Focus

Inspiratory Muscle Training and Breathing Exercises

Breathing exercises (85) to strengthen the respiratory muscles have been used by some, but their overall effectiveness is controversial (92, 102). Deep diaphragmatic breathing was used in 67 patients with asthma and significantly decreased the use of medical services and the intensity of asthma symptoms. However, there was no significant change in overall physical activity as measured by an inventory scale (40). Subjects who used a threshold inspiratory muscle training device in a double-blind sham trial showed a significant increase in inspiratory muscle strength (as expressed by the maximum inspiratory pressure measured at residual volume) and respiratory muscle endurance. The training group also had a significant reduction in the number of asthma symptoms, number of hospitalizations, and absence from work or school compared with the sham group (97).

As for adults, most children with asthma can safely perform regular moderate-intensity exercise. In 62 children with mild to moderate asthma (mean age 10.4 ± 2.1 yr) who were randomly assigned to exercise or usual care, the group who underwent exercise had a significant improvement in both exercise capacity and reduction of symptoms (11). Regular exercise should be encouraged in patients who have asthma under good medical control to help improve exercise capacity, improve quality of life, and reduce symptoms (64).

Rehabilitation and Asthma

Comprehensive pulmonary rehabilitation programs are useful in patients with COPD but are not formally required for the vast majority of patients with asthma because they can perform exercise at the same intensity/volume as individuals without asthma. It may, however, be useful in some patients, particularly for severe asthma or when a component of COPD is associated to asthma (asthma–COPD overlap). The initial assessment when considering a patient for entry into a program should include patient interview, medical history, diagnostic testing, symptoms and physical assessment, nutrition evaluation, activities of daily living assessment, educational and psychosocial history, and goal development. Program content should include education about the disease process, triggers of asthma, self-management of the disease (medication use, warning signs and symptoms associated with exacerbations, peak flow monitoring, medication delivery device techniques, importance of exercise warm-up and cool-down), activities of daily living, psychosocial intervention, and dietary intake and nutrition counseling. Follow-up and evaluation of outcomes are of vital importance as part of the rehabilitation process. Questionnaires used to assess the patient's asthma quality of life (measuring variables such as symptoms, emotions, exposure to environmental stimuli, and activity limitation) have been well validated and should be used as part of this assessment process (4). In addition, clinicians can assess cost of medications and equipment, time lost from work or school, and use of health care resources (e.g., emergency department visits, calls to the patient's physician) as part of the follow-up.

World Anti-Doping Regulations for High-Level Athletes

Competing athletes subjected to the WADA regulations should check whenever a medication is prescribed for them (www.wada-ama.org). The table that follows summarizes antidoping regulations for the main asthma medications.

Medication (example of commercial name)	Anti-doping regulations	Maximum allowable dose	Maximum concentration (urine)
Salbutamol (Albuterol)	Allowed with conditions	800 mcg/12 h	>1,000 ng/mL
Terbutaline (Bricanyl)	Prohibited*	NA	NA
Ipratropium (Atrovent)	Allowed	NA	NA
Formoterol (Oxeze)	Allowed with conditions	54 mcg/24 h	>40 ng/mL
Beclomethasone (Qvar)	Allowed	NA	NA
Budesonide (Pulmicort)	Allowed	NA	NA
Ciclesonide (Alvesco)	Allowed	NA	NA
Fluticasone propionate (Flovent)	Allowed	NA	NA
Mometasone (Asmanex)	Allowed	NA	NA
Fluticasone furoate (Arnuity)	Allowed	NA	NA
Prednisone (Prednisone, Deltasone)	Prohibited*	NA	NA
Prednisolone (Medrol, Pediapred)	Prohibited*	NA	NA
Leukotriene receptor antagonists	Allowed	NA	NA
Montelukast (Singulair)	Allowed	NA	NA
Zafirlukast (Accolate)	Allowed	NA	NA
Formoterol + Budesonide (Symbicort)	Allowed with conditions	54 mcg/24 h	>40 ng/mL
Salmeterol + Fluticasone (Advair)	Allowed with conditions	200 mcg/24 h*	NA
Formoterol + Mométasone (Zenhale)	Allowed with conditions	54 mcg/24 h	>40 ng/mL
Fluticasone furoate* + Vilanterol (Breo)	Allowed with conditions	Inhaled vilanterol: maximum 25 mcg/24 h	NA

*These medications are needed to treat a specific medical condition; a therapeutic use exemption (TUE) should be provided in antidoping instances.

Based on World Anti-Doping Agency. www.wada-ama.org/en/what-we-do/adams

NA = non-applicable.

Once asthma is under adequate medical control, improvements in aerobic capacity, muscle strength, and endurance can be maximized through the use of rehabilitation. The exercise prescription should be based on objective measurement of exercise capabilities and individualized for each patient. A variety of training modalities are appropriate, including treadmill, stationary bicycle, walking, and swimming. If weakness of a specific group of muscles is noted, exercises to address that particular muscle group should be offered. As previously mentioned, improvements in physical fitness, asthma symptoms, anxiety, depression, and quality of life were noted after exercise training in patients with asthma (30, 32, 38).

A study compared the effects of an outpatient pulmonary rehabilitation program on exercise tolerance and asthma control in patients with ($n = 53$) and without ($n = 85$) obesity (96). The multidisciplinary comprehensive (exercise and education) program was 12 wk in duration. Each participant performed three 1 h training sessions per week (30 min cardiorespiratory and 30 min strength training) under supervision. Every patient received individualized training, with the intensity and duration of each exercise based on maximal oxygen uptake, anaerobic threshold, and Borg rating. In general, the training intensity was just below the anaerobic threshold for 4 min alternating with maximal training for 1 min. After training, the 6 min walk distance improved in 71% of patients without obesity and 60% of those with obesity, and asthma control also improved in 47% of patients without obesity and 52% of those with obesity (96).

Potential Exercise Risks

There are few risks to exercise in patients who have asthma. However, high-intensity exercise may trigger EIB by increasing minute ventilation (56, 74). Some sports such as cross-country skiing and competitive swimming in chlorinated pools expose individuals to dry, cool air and to chloramines, respectively; setting up the potential for developing airway remodeling and hyperresponsiveness after many years of high-intensity training (15, 48): although hyperresponsiveness seems to be only transient in most athletes (16). Although very rare, sport-related deaths due to asthma have been reported in younger individuals (12). Care must also be afforded for those patients who have underlying comorbid conditions (e.g., cardiovascular or orthopedic limitations), and appropriate adjustments of the training program should be made under such circumstances.

Clinical Exercise Bottom Line

- Although exercise can induce respiratory symptoms in patients with asthma, associated with bronchoconstriction, exercise training can improve general health, quality of life, and asthma control.

- Because many conditions can mimic exercise-induced asthma, its diagnosis should be confirmed by objective airway function tests.

- Avoiding asthma triggers, having good asthma control, warming up before exercise, and occasionally using a preventive drug (e.g., a fast-acting inhaled β_2-agonist) can allow the vast majority of patients with asthma to perform exercise without troublesome asthma symptoms.

CONCLUSION

Asthma represents a complex process involving airway narrowing secondary to airway inflammation and hyperresponsiveness. Environmental risk factors (such as indoor allergens, viral infections) or other triggers (such as exercise, cold air) can initiate an allergic response, resulting in airway inflammation and airway hyperresponsiveness. Although a patient with controlled asthma should have minimal symptoms after exercising, if airflow limitation occurs, the patient may develop symptoms such as chest tightness, wheezing, and shortness of breath. Exercise limitations and decreased levels of fitness are frequently noted in patients with asthma but in many instances are not considered important until sometime after the initial development of symptoms.

The widespread misbelief that patients with asthma cannot and should not exercise caused unnecessary restrictions, particularly in children with asthma. Clinicians must attempt to educate patients and their families about the importance of exercise and how it can be safely performed. Using appropriate pharmacologic and behavioral therapies, aiming to maintain an optimal control of asthma, and avoiding conditions known to precipitate or worsen symptoms (particularly relevant allergens) are important mainstays of the treatment. Many people with asthma have successfully performed at high levels of competition, and most people who have asthma can perform exercise regularly—and even have a successful career in sports—with all the associated health benefits.

Online Materials

Visit HK*Propel* to access a link to the references, the case studies with discussion questions, and a quiz to help you review key concepts and test your understanding of the material covered in this chapter.

CHAPTER 21

Cystic Fibrosis

Kelley Crawford, PT, DPT, MS, CCS

A common genetic disorder, cystic fibrosis results in severe mucus blockage that affects multiple organs and drastically reduces life expectancy. Exercise as part of a therapeutic plan has been proven to offer several physiological and psychological benefits, including enhanced quality of life.

DEFINITION

Cystic fibrosis (CF) is a genetic disorder that affects the respiratory, gastrointestinal, metabolic, and reproductive systems. Excessively viscid secretions cause obstruction of passageways including the pancreatic and bile ducts, intestines, and bronchi. In addition, the sodium and chloride contents of sweat are increased.

SCOPE

Cystic fibrosis is the most common life-shortening genetic disease in the Caucasian population. Currently, more than 30,000 patients in the United States (70,000 worldwide) have CF, and nearly 1,000 new patients are diagnosed each year (21). Cystic fibrosis is inherited as an autosomal recessive disorder that affects approximately 1 in 2,500 to 3,500 live births in the Caucasian population and 1 in 17,000 in the African American population, with a carrier rate of 1 in 25. Individuals with other ethnic or racial backgrounds are also affected but with less frequency. For the first time in the history of CF, there are more individuals age 18 yr and older than those less than 18. At nearly 51% this growth in the adult-based CF population presents unique challenges in the care of these patients and is further emphasized by the increased diagnosis of **adult variant CF** (71). The median survival age continues to improve, measuring in the late 30s (39.3 yr). The estimated total cost to treat CF in the United States continues to be noteworthy at nearly $1.2 billion per year, representing a cost per CF patient of $40,000 per year (22).

PATHOPHYSIOLOGY

The gene for CF is located on chromosome 7 and results in the altered production and function of a protein called the **cystic fibrosis transmembrane conductance regulator (CFTR)**; the protein functions as a chloride channel regulated by cyclic adenosine triphosphate. More than 1,800 unique mutations of the CF gene have been identified; more than 87% of people with CF have at least one delF508 mutation. The primary role of CFTR appears to be as a chloride channel, although other functions for CFTR have been documented; most importantly it regulates reabsorption of sodium and water along the respiratory epithelium. The abnormal CFTR leads to abnormal sodium chloride and water movement across the cell membrane. When this occurs in the lungs, thick and dry mucus ensues, resulting in bronchial airway obstruction, bacterial infection, and inflammation. As this vicious cycle continues, lung tissue is progressively destroyed, leading to eventual respiratory failure (figure 21.1).

Lung disease accounts for more than 95% of the morbidity and mortality associated with CF. Through aggressive intervention and early diagnosis, however, survival has been extended, with adults living well into their 30s and 40s. The average age at time of diagnosis in the United States is approximately 6 mo (22), but now that all states include CF on newborn screening panels, the diagnosis is made much earlier. In patients who are not diagnosed by newborn screening, typically, one or more symptoms lead to diagnosis, including symptoms in the respiratory, gastrointestinal, sinus, and sweat gland systems. The underlying theme in all these systems is the cellular abnormality of ion transport necessary for proper function of epithelial structures.

Respiratory System

At birth, the lungs are normal on a histological basis. As the vicious cycle of infection, inflammation, and impaired mucus clearance

Acknowledgment: Much of the writing in this chapter was adapted from the previous editions of *Clinical Exercise Physiology.* Thus, we wish to gratefully acknowledge the previous efforts of and thank Julie Biller, MD, Lauren Camarada, MD, and Michael J. Danduran.

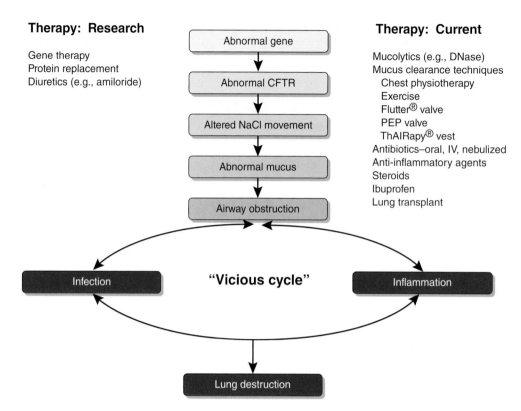

Figure 21.1 The vicious cycle of cystic fibrosis. CFTR = cystic fibrosis transmembrane conductance regulator.

ensues, the lungs become colonized with bacteria. Nearly 50% of all patients are currently infected with *Pseudomonas aeruginosa*, a rate that has shown a recent decline (22). Bacteria such as *Staphylococcus aureus* currently affects 26% of patients, with other bacteria occurring less frequently. Two-thirds of newly diagnosed infants were established through newborn screening (22). When infants are not diagnosed by newborn screening, a young child most commonly presents with failure to thrive or recurrent respiratory tract infections. Most children present with signs of chronic infection including cough, sputum production, crackles, wheeze, fever, and failure to thrive at the time of diagnosis. Many infants or young children with CF have been previously misdiagnosed as having asthma, bronchitis, allergies, pneumonia, or bronchiolitis. Chest radiographs may indicate the presence of acute or chronic changes such as infiltrates, **bronchiectasis** (irreversibly irregular and dilated airways), or hyperlucency.

When pulmonary function is assessed in older children (>5 yr) at the time of diagnosis, early evidence of airway obstruction (forced expiratory flow between 25% and 75% of forced vital capacity [FEF_{25-75}]) or hyperinflation (elevated residual volume and ratio of residual volume to total lung capacity) may exist. Because the disease causes more damage to lung tissue, a reduced forced expiratory volume in 1 s (FEV_1) will be seen on pulmonary function tests. In CF, the lower the FEV_1, the more severe the lung disease (normal: >90%; mild: 70%-89%; moderate: 40%-69%; severe: <40%). Newer multiple-breath washout testing can

be performed in the preschool-aged child and can identify early lung disease as measured by lung clearance index, a measurement of ventilation inhomogeneity (54). There is also technology that allows for infant assessment of pulmonary function, inviting therapeutic interventions to occur early on in patients affected with CF. Exercise tolerance may become significantly compromised as judged against normative values. Ultimately, the progressive loss of lung tissue and airway obstruction lead to respiratory failure. The time course for this progression is variable. Some adults with CF experience little lung damage, and some children experience extensive lung disease. This is thought to be due to modifier genes.

Gastrointestinal and Nutritional Systems

In approximately 85% of individuals with CF, exocrine pancreatic insufficiency is present, resulting in malabsorption of important nutrients including fat, protein, and vitamins. Malabsorption can lead to frequent fatty stools (steatorrhea), malodorous stools, and abdominal pain. The combination of the need for increased caloric intake (attributable to increased resting energy expenditure, cough, and infection) and poor utilization of nutrients with malabsorption often leads to malnutrition or a constant struggle to maintain body weight. Maintaining a desirable body mass index (>22 in adult women, >23 in adult men, ≥50th percentile BMI in children, ≥50th percentile weight for length in infants) has been associated with improved pulmonary outcomes in the CF population (22, 99). In patients with more advanced nutritional issues, a gastrostomy tube placed in the abdomen allows for alternative ways to offer food. Additionally, other organs can be affected, resulting in liver disease, endocrine pancreatic insufficiency (CF-related diabetes mellitus), and gallbladder disease.

Metabolic System

As individuals with CF age, they incur an increasing risk of developing **cystic fibrosis–related diabetes** (approximately 20% of adolescents with CF and almost 50% of those older than 30 yr) (89). Scarring of the pancreas, which produces insulin, often occurs in CF patients. This progressive scarring frequently prevents insulin from entering the bloodstream and can result in diabetes. In patients who develop CF-related diabetes, careful monitoring of blood sugar should occur, especially during times of increased activity. It is also important to be able to recognize symptoms of hypoglycemia and hyperglycemia.

Bone Disease

Bone disease, in particular osteoporosis, is common in individuals with CF. The risk of developing bone disease increases significantly in the second decade of life, with an incidence of more than 20% in individuals aged 18 to 34 yr. A further increase is observed over time, with nearly 40% of patients greater than 35 yr demonstrating decreased bone health (22). This phenomenon is likely related to poor gastrointestinal absorption of bone-building nutritional elements as well as chronic corticosteroid use. Early identification of reduced bone mass may permit early intervention and help prevent the development of osteoporosis (70).

Psychosocial

Symptoms of depression and anxiety affect more than 25% of adults with CF. Although it is not uncommon in individuals with chronic disease, people with CF, as well as their families and caregivers, need to be cognizant of the increased risk of depression, and early treatment and intervention is recommended. Depression becomes even more prominent later in life. Current recommendations are that individuals age 12 and older should receive yearly depression and anxiety screenings during routine clinic visits (82).

Sinuses

The development of pansinusitis and **nasal polyposis** is common in people with CF. This finding may be inconsequential, although some individuals may experience difficulty breathing through the nose (because of nasal congestion) as well as chronic headaches and impaired sleep. Additionally, pansinusitis with associated bacterial colonization may contribute to the extent of lung disease. Some people require aggressive medical intervention (e.g., antibiotics, nasal irrigation, and endoscopic surgery). The incidence of sinus disease is approximately 10% in children >10 yr, increasing incrementally to almost 50% in patients >35 yr (22).

Sweat Glands

All epithelial cells demonstrate the chloride transport defect. This defect in the sweat glands has been turned into a diagnostic test for CF. The basis of the **sweat test** (i.e., pilocarpine iontophoresis analysis) rests on the presence of extremely high salt content in the sweat of individuals with CF. A sweat chloride concentration greater than $60 \, mEq \cdot dL^{-1}$ is highly suggestive of diagnosis of CF.

CLINICAL CONSIDERATIONS

The clinical manifestations of CF are variable, with differing involvement of the pulmonary and gastrointestinal organ systems. Comprehensive evaluation that includes assessment of the signs and symptoms, diagnostic studies, and pulmonary function testing helps determine the severity of disease.

Signs and Symptoms

Cystic fibrosis is usually diagnosed by the presence of classic signs and symptoms (table 21.1). Because of the expansive nature of the disease, many systems are affected. Patient care must often be coordinated by a CF care team that may include pulmonologists, gastroenterologists, nurses, respiratory therapists, physical therapists or exercise clinicians, a social worker, a nutritionist, a psychologist, a genetic counselor, and a pulmonary function technologist. Improved survival and a larger number of new patients diagnosed in adulthood have led many care teams to form adult care practices designed to address adult-specific needs. Nevertheless, respiratory and gastrointestinal support is the mainstay of therapy for patients with CF.

History and Physical Exam

A thorough medical history is necessary to identify potential risk factors that may limit exercise performance in individuals with CF. The history should focus on factors that may be present and can alter the pulmonary–cardiovascular–peripheral systems necessary for effective oxygen delivery and utilization during exercise. The most important consideration before testing a patient with CF is to determine the patient's level of pulmonary disease. Prior pulmonary function data can help predict which patients are likely to experience oxyhemoglobin desaturation with exercise testing. An FEV_1 less than 50% of predicted or a low resting **oxyhemoglobin saturation (SaO_2)** places the person with CF at much greater risk of oxygen desaturation during exercise (32, 46, 59). A history of wheezing, chest tightness, or chest pain during exercise may indicate the presence of exercise-induced bronchoconstriction, which is not uncommon in patients with CF (22). Additional considerations, such as a history of pneumothorax or **hemoptysis** (coughing up blood) at rest or during exercise, should

Table 21.1   Clinical Signs and Symptoms of Cystic Fibrosis

System	Signs and symptoms
Respiratory	Chronic productive cough, pneumonia, wheezing, hyperinflation, exercise intolerance, *Pseudomonas aeruginosa* bronchitis
Gastrointestinal and nutritional	Steatorrhea, failure to thrive, biliary cirrhosis, intestinal obstruction, abdominal pain, bone disease resulting in osteoporosis, vitamin deficiencies
Sinuses	Chronic sinusitis, nasal polyps
Metabolic	CF-related diabetes—pancreatic scarring inhibits insulin secretion
Sweat glands	Salty taste, recurrent dehydration, chronic metabolic acidosis
Other	Depression, infertility, pubertal delay, digital clubbing, family history

be reviewed. A history of nocturnal headaches or cyanosis may suggest advanced lung disease with associated hypercarbia or hypoxemia, respectively. Because exercising at altitude may exaggerate hypoxemia, caution should be taken if the exercise testing or training program will occur at altitude (6).

Few cardiovascular limitations exist for people with CF. However, in advanced lung disease, the incidence of pulmonary hypertension and cor pulmonale is much higher; these conditions require consultation with a cardiologist before testing. Signs and symptoms of right-side heart failure should be sought (e.g., peripheral edema, venous congestion, hypoxemia). Peripheral factors such as **scoliosis, kyphosis,** barrel chest, and tight hamstrings are commonly present in individuals with CF and may reduce mechanical efficiency during exercise.

Other organ systems can be affected by CF and become an issue during acute exercise. Liver disease with associated ascites (abdominal distension) may interfere with respiratory muscle effectiveness, whereas liver-related bleeding disorders may be exacerbated with increased blood pressure during exercise. Cystic fibrosis–related diabetes may result in abnormal blood sugars with exercise and can manifest as clinical symptoms associated with hyper- or hypoglycemia, including excessive fatigue, confusion, or dizziness (106). Increased incidence of bone disease in patients with CF has been noted, with progression beginning in the second decade of life. Aerobic exercise and strength training may assist in retarding the progressive loss of bone mineral density over time (70).

Patients with CF are also at increased risk of dehydration state, because excessive salt loss with physical exertion is commonly seen in this disease (75). Clinical signs of early dehydration should be discussed with those who plan to exercise. These signs include light-headedness, heat intolerance, flushed skin, decreased urine output, concentrated (dark yellow) urine production, nausea, headaches, and muscle cramps. Adequate hydration needs to be stressed for those who will exercise in warm, humid climates. Consumption of 4 oz (120 mL) of fluid every 20 min is a good general rule. For children who cannot readily quantify fluid amounts, eight gulps of fluid equals approximately 4 oz (120 mL) (56).

The extent of malnutrition and body composition (e.g., lean muscle mass) should be noted, because these considerations may alter the mechanical load applied during exercise testing. Finally, before developing a precise prescription, the exercise professional can use validated physical activity questionnaires or diaries to determine how physically active the individual is. Options for assessing activity vary in format (e.g., recall questionnaires, activity diaries). Some tools may not be appropriate for younger individuals, but some are designed specifically for use in the pediatric population (e.g., previous-day physical activity recall) (78). The sensitivity of these tools in this population or in children in general has been questioned, especially when recall is required. In adults, monitoring activity may be less complicated. The section Special Exercise Considerations details other common conditions associated with CF and more specifically how they affect the exercise clinician's decisions.

Diagnostic Testing

In the United States, CF is included in screening panels for newborns, allowing for earlier detection. In utero diagnosis has also become available. A positive sweat test confirms the diagnosis of CF. Genetic mutation analysis can be useful in diagnosing the disease (30). Additional laboratory testing should be performed when the diagnosis of CF is considered:

- Sputum culture (positive for *P. aeruginosa* or other CF bacterial pathogens)
- Chest radiograph
- Sinus computed tomography
- Static and dynamic lung assessment if age appropriate
- Blood sampling for complete cell count
- Liver function
- Nutritional parameters (e.g., total protein, albumin)
- Renal function (e.g., blood urea nitrogen, creatinine)
- Fat-soluble vitamins A, D, E, and K (prothrombin time/international normalized ratio [PT/INR] for vitamin K)
- Glucose

Assessment of static pulmonary function, as defined by the properties of the lung at rest or baseline, is essential in the acute and chronic management of individuals with CF. Simple spirometry, as well as assessment of lung volumes, diffusion capacity, and bronchodilator responsiveness, assists in the detection of an acute **pulmonary exacerbation**. Although some individuals with CF have mild lung disease, most demonstrate varying degrees of airway obstruction with signs of hyperinflation. Additionally, one-third of people with CF will demonstrate signs of reversible airway hyperreactivity when exposed to a bronchodilator. Declines in FEV_1 or indices of smaller-airway function (e.g., FEF_{25-75}, FEF_{50}, FEF_{75}) over time may serve as warning signs of acute or chronic lung deterioration.

In conjunction with static pulmonary function assessment, **dynamic pulmonary function**, as defined by lung function in response to changing physiological state (e.g., work, exercise, physiologic stress), also plays an important role as a diagnostic tool and in monitoring the patient's clinical condition. In fact, a single measure of aerobic exercise tolerance has been strongly correlated with long-term survival in patients with CF (68, 73). Moreover, researchers studied patients who had undergone repeated exercise evaluations over a 5 yr period to look at mortality in years to follow. Individuals who maintained exercise tolerance had greater long-term survival when compared with those who showed a decline in exercise capacity (80). When health care workers assess the lungs under measurable stress (e.g., exercise), ventilatory limitations, as well as impairment in other parameters such as oxygen saturations that depend on dynamic lung function, may become apparent that are not noted when the patient is at rest. Regular assessment of exercise tolerance in patients with CF is an integral component of their medical care. The exercise clinician who treats and assesses individuals with CF should comprehensively understand both dynamic and static lung function measurements and the role exercise testing plays in the management of this population.

Exercise Testing

The importance of performing a complete exercise evaluation before developing an exercise prescription for individuals with CF cannot be overestimated. Guidelines for exercise testing are summarized in table 21.2. This evaluation can help with routine monitoring and management of a patient's clinical status, act as a pretransplant assessment, allow for counseling regarding activity, and finally identify new or recurrent symptoms (43). Despite potential limitations, an effective exercise program should optimize all aspects of fitness including overall well-being, both physical and psychological. The exercise clinician plays an important role in conjunction with the patient, parents, and CF medical team in establishing realistic goals and developing an achievable exercise program to enhance the individual's quality of life.

The best practice assessment of individuals with CF involves standardized treadmill or bicycle protocols to maximum while measuring oxygen consumption at baseline, throughout exercise,

and into recovery. A baseline maximal exercise challenge should be administered with the addition of pulse oximetry and electrocardiogram to ensure appropriate oxygen saturation and cardiac rhythm throughout exercise. If the patient completes the challenge without desaturation (defined as a value <88% on room air), supplemental oxygen is not required. Should an individual desaturate to <88%, the point at which this occurs becomes critical. Because most aerobic exercise prescriptions use submaximal intensity levels (60%-75% of maximal), the exercise clinician should determine whether the patient desaturated before reaching this submaximal level. If so, supplemental oxygen may be desired for subsequent exercise training. If supplemental oxygen is required, a repeat exercise challenge should be administered after an appropriate recovery period to document that the patient remains normoxic during exercise, with the level of required oxygen supplementation recorded.

Special considerations for the person requiring oxygen for exercise are warranted. The choice of activity may need to be modified to allow for the presence of oxygen tanks. Activities using stationary modalities (e.g., treadmills, bicycles) may be more appropriate for this group. Many health clubs can accommodate people with these special needs.

Individuals with advanced lung disease can obtain the beneficial effects of exercise training. Improved gas exchange, ventilation, aerobic tolerance, peripheral muscle adaptations, and sense of well-being have all been documented after exercise training (45, 65). Appropriate exercise prescriptions can be developed for even the most debilitated person with CF. This population has not traditionally had the benefit of interacting with an exercise clinician. One could argue that individuals with severe CF have the most to gain from exercise that would help reestablish functional ability. By understanding the particular needs of this population, the exercise clinician can play a major role in achieving this goal.

Management of CF may require varying types and amounts of medications. Table 21.3 lists the common pharmacologic agents used in patients with CF and includes special considerations with regard to exercise.

Contraindications

Absolute contraindications to exercise testing exist and are consistent with American College of Sports Medicine (ACSM) guidelines. In a study from Germany that surveyed CF centers, the incidence of exercise-related serious adverse events was less than 1% (87). Although the clinical guidelines offered by the ACSM discourage maximal exercise testing in individuals with FEV_1 less than 60% of predicted (109), with appropriate medical direction patients with significant CF-associated lung disease have safely undergone clinical evaluation using a 6 min walk test or alternative submaximal evaluation. Despite the low risk and absence of significant absolute contraindications within the CF population, special considerations need to be observed for individuals with a history of pulmonary hypertension, acute hemoptysis, pneumo-

Table 21.2 Exercise Testing

Test type	Mode	Protocol specifics	Clinical measures	Clinical implications	Special considerations
Cardiorespiratory endurance	Treadmill Bicycle 6 min walk	Treadmill: Bruce, Balke-Ware, Naughton Bicycle: James, Godfrey	HR, BP, ECG, $\dot{V}O_2$, SaO_2	Assessment of endurance, PWC, risk of desaturation, extent of ventilatory limitations	FEV_1 <50% = increased risk of desaturation In severe patients: reduced peak HR, increased $\dot{V}_E$, RR
Muscular strength and endurance	Bicycle 1RM	Bicycle: Wingate anaerobic test 1RM, grip strength Respiratory muscle function	Peak and mean anaerobic power Force production PImax, PEmax	Assessment of muscular power and endurance Inspiratory and respiratory muscle strength	Measures significantly affected by nutritional status Muscular and respiratory strength typically reduced
Flexibility	Stretching Sit-and-reach	Range of motion testing using goniometer Sit-and-reach	Joint range of motion measured in degrees or inches (cm) in the case of classic sit-and-reach	Thoracic kyphosis develops as disease severity progresses	Early detection of inflexibility can lead to stabilization of the abnormality May enhance chest wall mechanics with exercise
Body composition	Height, weight BMI Body composition	Calculation of BMI Triceps skinfold assessment Three-site skinfold assessment	Stature, mass, BMI, % body fat, lean body mass	Desired BMI: • Adults: Males: >23 Females: >22 • Children: >50th percentile • Infants: >50th percentile for length	Significant correlation between nutrition and exercise performance as well as long-term prognosis

HR = heart rate; BP = blood pressure; ECG = electrocardiogram (3-lead or 12-lead based on equipment); $\dot{V}O_2$ = oxygen consumption; SaO_2 = oxygen saturation (measured in percent); PWC = peak work capacity; FEV_1 = forced expiratory volume in 1 s; $\dot{V}_E$ = minute ventilation; RR = respiratory rate; 1RM = one-repetition maximum (to assess strength); PImax = maximal inspiratory pressure; PEmax = maximal expiratory pressure; BMI = body mass index.

thorax, oxygen dependence, a bleeding disorder secondary to liver disease, and severe malnutrition. Care should also be taken when the exercise testing is performed at altitude. Monitoring during testing should include continuous pulse oximetry, electrocardiogram, and vital sign assessment. Supplemental oxygen and a short-acting bronchodilator (e.g., albuterol by inhaler) should be available to all patients with CF during or after testing as needed.

Recommendations

Typical cardiopulmonary responses in CF patients are listed in table 21.4. The precise protocol and testing location depend on several factors, including the indications for testing, the age of the person, and the resources available. Finally, goals for testing should be clearly identified so that individualized exercise programs can be tailored to meet those goals. Such goals may include determining the heart rate at which oxyhemoglobin desaturation occurs, monitoring for improvement in response to medical therapy, or comparison to prior performance.

Cardiorespiratory Exercise

Numerous reproducible exercise protocols exist for testing maximal exercise performance in children and adults: the Godfrey, McMaster, and James protocols for bicycle testing and the Bruce, modified Bruce, and Balke-Ware protocols for treadmill testing (7, 43). Measurement of oxygen consumption and monitoring of pulse oximetry, electrocardiogram, and blood pressure response should be considered in all patients with CF. Despite minimal risk associated with testing, standard practices for emergency management should be followed, including access to a crash cart and supplemental oxygen as well as personnel trained in advanced life support and cardiopulmonary resuscitation.

The comprehensive information provided by a maximal aerobic test allows the exercise clinician to develop an appropriate exercise prescription. Measures of maximal oxygen consumption should be made in all patients with CF. Adjunctive tests including a 12-lead electrocardiogram, assessment of blood gases and immunologic markers, and pre- and posttest spirometry have been performed

Table 21.3   Pharmacology

Medications	Primary effects	Exercise effects	Important considerations
BRONCHODILATORS			
β_2-agonist (albuterol) Anticholinergics (ipratropium, tiotropium)	Relax muscles in the airways and increase airflow into the lungs Prevent bronchospasm, or narrowing in the lungs, in patients with chronic lung disease	Improve airway patency, allowing for increased ventilation during exercise	β_2-agonists may cause tachycardia at rest, pounding or premature heartbeats, feeling jittery or nervous. Phase 3 trials have been completed for tiotropium to evaluate efficacy in CF; FDA approved for use in COPD.
ANTIBIOTICS			
Tobramycin inhalation solution (TOBI) Aztreonam inhalation solution (Cayston) Colistimethate	Treat all severities of infections, including those of the airways and lower respiratory tract Inhaled into the lungs using a nebulizer; used to treat lung infections that are commonly seen in patients with CF	No effects on physiologic responses to exercise; may cause bronchospasm after use	Exercise participation during times of severe infection should be limited. Bronchospasm after use may affect exercise ability. Several other antibiotics are currently in phase 3 trials for patients with CF, including levofloxacin, ciprofloxacin, and amikacin.
AIRWAY HYDRATORS			
Hypertonic saline (HyperSal) Mannitol (Bronchitol)	Counteract the dehydration of airway surfaces associated with defective mucus clearance	Can cause cough or bronchospasm that may be exacerbated with exercise	Encouraging coughing and sputum production during exercise is important as an airway clearance technique.
ANTI-INFLAMMATORIES			
Azithromycin Prednisone Inhaled corticosteroids Combination inhaled corticosteroids	In the patient with CF, used to treat chronic respiratory infections by preventing the release of substances that cause inflammation (many brands are in use)	May improve chronic inflammation of airways, resulting in improved ventilation during exercise	Prednisone is typically not for routine use. Examples of inhaled corticosteroids are LABA, Symbicort, and Advair. Example of a combination corticosteroid is Symbicort, which contains budesonide to reduce inflammation and formoterol to relax the airways.
CFTR-DIRECTED THERAPIES			
Ivacaftor Lumacaftor	Increase functional CFTR protein	May improve cardiopulmonary fitness	Limited research data with regard to exercise outcomes.
MUCOLYTICS			
Dornase alfa (rhDNase) (Pulmozyme)	Break down the excess DNA in the pulmonary secretions of patients with CF Improve lung function by thinning the pulmonary secretions and reducing the risk of respiratory infections	May improve airway obstruction with improvement in sputum mobilization, enhancing tolerance during a bout of exercise	Mucolytics may help maximize the effects of exercise as an airway clearance technique. The use of exercise in combination with other airway clearance techniques has been shown to be effective in mobilizing secretions.
Pancreatic enzymes	Maximize utilization of important nutrients that may go unabsorbed because of exocrine pancreatic insufficiency	No effects on physiologic response to exercise	Lack of patient compliance in taking pancreatic enzymes negatively affects nutritional status and may affect exercise tolerance and lung function.
VITAMIN AND NUTRITIONAL SUPPLEMENTS			
Vitamin D and other fat-soluble vitamins Bisphosphonates	Improve nutrition and bone mineralization	Together with exercise can improve bone mineralization and decrease risk of fractures	Limited research data for bisphosphonates with regard to exercise outcomes.

Table 21.4 Cardiopulmonary Parameter Changes Among Individuals With Cystic Fibrosis

Parameter	Change
STATIC PULMONARY FUNCTION (REST OR BASELINE)[a]	
Spirometry: FEV_1, FEF_{25-75}, tidal volume	Decreased
Lung volumes: RV, FRC, RV/TLC	Increased
Diffusion capacity: DL_{CO}	Decreased
Oxyhemoglobin saturation: SpO_2	Decreased
DYNAMIC CARDIAC AND PULMONARY FUNCTION (IN RESPONSE TO EXERCISE OR STRESS)[b]	
Aerobic capacity: PWC, $\dot{V}O_2max$	Decreased
Breathing response: $\dot{V}_E$, $\dot{V}_E/MVV$, RR	Increased
Gas exchange: $\dot{V}CO_2$, $EtCO_2$	Increased
Blood pressure response	Normal
Heart rate at rest	Increased
Heart rate during peak exercise	Decreased

[a]Static pulmonary function is decreased and declines with advancing lung disease.

[b]Abnormal parameters tend to follow extent of lung disease; aerobic performance is weakly correlated with static lung function parameters.

FEV_1 = forced expiratory volume in 1 s; FEF_{25-75} = forced expiratory flow between 25% and 75% of the forced vital capacity; RV = residual volume; FRC = functional residual capacity; RV/TLC = ratio of residual volume to total lung capacity; DL_{CO} = diffusion capacity of the lung by the carbon monoxide technique; SpO_2 = pulse oximetry; PWC = peak work capacity; $\dot{V}O_2max$ = maximal oxygen consumption; $\dot{V}_E$ = minute ventilation; $\dot{V}_E/MVV$ = ratio of minute ventilation to maximal voluntary ventilation; RR = respiratory rate; $\dot{V}CO_2$ = carbon dioxide production; $EtCO_2$ = end-tidal carbon dioxide.

as a part of clinical and research evaluations. However, this form of testing requires sophisticated exercise equipment and highly trained technical staff and can result in a significant financial cost to the patient.

A submaximal aerobic test may be useful in determining whether exercise desaturation or breathlessness occurs and can help verify the effectiveness of exercise prescriptions established from maximal tests. Submaximal assessment is used infrequently but may be easier to perform for the young child or adult with significant ventilatory limitations. A treadmill or bicycle similar to that used during a maximal test is the ideal equipment for the submaximal test. Traditional submaximal protocols require workloads consistent with 75% of the age-predicted maximal heart rate. Individuals with more advanced CF may experience difficulty with these tests secondary to ventilatory limitations and should be allowed to terminate the exam short of reaching these physiologic criteria if symptoms or marked desaturation occurs (72). Heart rate, blood pressure, and pulse oximetry should be monitored during submaximal testing. The lowered technical demands and financial costs and the ease of repeat testing make the submaximal test an attractive alternative.

Lab-based exercise tests may not always be convenient or available for individuals with CF. Several walking and running tests have been developed in an attempt to mimic real life more accurately and offer simple-to-administer testing protocols. Walk tests for 2, 6, and 12 min have been used for people with CF (37, 39, 88). These protocols allow patients to walk over a set period at their own pace while heart rate and oxygen saturations are monitored. Total distance traveled, the development of exercise breathlessness, and oxygen desaturation are recorded and can be compared over time against prior tests. Outcome variables have been relatively well correlated to standardized maximal tests for individuals with mild to severe CF lung disease. Serial walk tests may provide a simple yet valuable assessment of the usefulness of supplemental oxygen or pulmonary rehab in the patient severely affected with CF. Shuttle tests have also been used for people with CF (13, 14). In one version, the patient walks (or runs) at increasing speeds (set by an audio signal) over a set course until voluntary exhaustion occurs. Heart rate, pulse oximetry, and distance traveled are monitored during the test.

Although both the walk tests and shuttle tests are relatively simple to perform, a limitation is that they are heavily dependent on patient effort. Motivation by the test administrator is therefore essential. A 3 min step test has been developed as a modification of the Master two-step exercise test used in adult cardiac testing (7). The total number of steps can be tabulated along with change in heart rate, oxygen desaturation, and sensation of breathlessness (81). All these noninvasive tests are easy to perform and do not require sophisticated equipment. These field tests have been shown to have value in assessing exercise abilities but should not be considered as a substitute or replacement for traditional maximal testing (43).

Muscular Endurance

Although many tests have been proposed to assess muscular endurance in healthy individuals, relatively few protocols have been used for individuals with CF. The Wingate anaerobic test (WAnT) has been used to assess both short-term mechanical power or strength and leg muscle endurance over a brief, intense period of time. The test consists of a 30 s all-out sprint on a cycle ergometer against fixed resistance. Determination of resistance depends on lean muscle mass, but a standard starting point is 75 g of resistance per kilogram of body weight (48). The test is demanding, but patients with CF have been able to complete it (12, 15). Although sophisticated equipment is available to perform the WAnT, a mechanical cycle ergometer (e.g., Monark or Fleisch) can be adapted for this test.

Alternative protocols for testing muscle endurance in individuals with CF include use of an isokinetic cycle ergometer and cycling at supramaximal levels. Both of these protocols have been used in the research setting for testing children with CF (57, 96) but are not readily accessible outside the academic exercise laboratory. Measurements of respiratory and peripheral muscle fatigue have been used as research tools to assess both respiratory and peripheral muscle function in individuals with CF (53, 58). Many school-based fitness testing measures (push-ups, pull-ups, long jump, and high jump) can be performed by CF patients with mild to moderate disease because they generally do not have restrictions or limitations to any activities. Finally, alternative measures, including balancing, performance accuracy, flexibility, standing vertical jump, and timed exercises, have been used to further quantify fitness ability either as a one-time measure or as an outcome variable (35). These field tests may provide a fun and effective alternative for assessing both muscular strength and muscular endurance in young patients with CF. For adults with CF, a number of functional tests have been found to correlate well with measures of leg muscle strength and power, particularly the stair climb power test (97). Specifically, peak torque and power for the quadriceps muscles, as measured by dynamometer, were found to correlate well with the stair climb power test.

Muscular Strength

Strength of both respiratory muscle and peripheral skeletal muscle groups has been assessed in people with CF. Peak inspiratory pressure determination specifically measures the muscles used for inspiration and consists of the patient's inspiring a breath of air at residual volume against an occluded airway. The greatest inspiratory subatmospheric pressure that can be developed is recorded. Similarly, peak expiratory pressure measures the strength of the abdominal and accessory muscles of breathing and consists of the patient's exhaling forcefully against an occluded airway, usually at total lung capacity. Inspiratory muscle training when corrected for workload can be effective in improving inspiratory muscle function and work capacity (29). These maneuvers are relatively easy to perform. The equipment required is usually part of a standard body plethysmography system. Alternatively, handheld direct-reading manometers or electronic pressure transducers and recorders can be used.

Peripheral skeletal muscle has been shown to respond to training in individuals with CF. Home-based strength training programs, as well as inpatient rehabilitation efforts, have resulted in increased strength and physical work capacity (35, 42, 76). Standard techniques, including use of dynamometers, cable tensiometers, isokinetic muscle testing, and free weights, can be applied to test specific muscle groups. The age of the patient often determines the choice of test. For very young children, the child's own body weight can be used as a resistance tool, according to the testing criteria of the President's Council on Physical Fitness and Sport (e.g., push-ups, pull-ups) (28). Expected muscular endurance and strength responses in patients with CF are listed in table 21.5.

Body Composition

Individuals with CF tend to be lower in both body weight and height than those without CF. From a clinical perspective, monitoring body mass index as well as body composition is an important part of the nutritional assessment. Typical body composition responses in CF relative to age- and gender-matched peers are listed in table 21.5. Clinically, the desirable body mass index for adult males with CF is greater than 23 and for adult females with CF is greater than 22. For children the goal is a BMI at or above the 50th percentile (99). Patients below these values tend to have decreased clinical standing and poor long-term prognosis when compared with those above the desired value (28).

For the exercise clinician, documenting body composition is essential. Many techniques exist for determining fat distribution in healthy adults, but the underlying assumptions of these techniques may be in question for children, for individuals with chronic lung disease, and in conditions associated with electrolyte disturbances. Currently, single-site (triceps) or multiple-site skinfold assessment is the most commonly used technique for monitoring body fat in children with CF. Skinfold calipers (e.g., Harpenden, Lange) are an inexpensive means for determining body composition. Use of pediatric reference equations for a child is mandatory (61). Alternative techniques include dual-energy X-ray absorptiometry (DEXA) and bioelectrical impedance. In light of the aging CF population and the need to assess bone density as well as body composition, DEXA is recommended by the Cystic Fibrosis Foundation in order to screen for osteoporosis and osteopenia. All individuals with CF are advised to have a DEXA scan by the age of 18 yr and to be rescreened every 1 to 5 yr thereafter. Bioelectrical impedance analysis through commercially available systems has been performed. Both skinfold and bioelectrical impedance assessments are easy to perform, inexpensive, and reproducible in the hands of a trained technician. A common practice in some CF specialty centers is to use both techniques as a means of establishing internal reliability of measurements.

Table 21.5 Typical Muscle Endurance and Strength, Body Composition, and Flexibility Levels for Individuals With Cystic Fibrosis

Parameter	Change
MUSCULAR ENDURANCE[a]	
WAnT mean power; isokinetic cycle ergometry power	Decreased
Muscle efficiency	Decreased
MUSCULAR STRENGTH[b]	
Respiratory muscle strength (PImax, PEmax)	Decreased
Peripheral muscle strength	Decreased
BODY COMPOSITION[c]	
Body weight	Decreased
Lean muscle mass	Decreased
Body mass index	Decreased
Percent body fat (BIA, skinfold assessment)	Decreased
FLEXIBILITY[d]	
Peripheral muscle flexibility (hamstrings, quadriceps)	Decreased
Posture—extent of kyphosis	Increased

[a]Muscle endurance decreases as lung disease progresses and may reflect impaired nutritional status and intrinsic cellular deficiencies.
[b]Decreases as disease progresses and may reflect nutritional status and loss of mechanical efficiency.
[c]Decreases in body composition reflect increased caloric expenditure with advanced lung disease, poor oral intake, and release of cachectic mediators.
[d]Reflects deconditioning associated with decreased activity or advanced disease.
WAnT = Wingate anaerobic test; PImax = peak inspiratory pressure; PEmax = peak expiratory pressure; BIA = bioelectric impedance analysis.

Flexibility

Table 21.5 illustrates the typical changes in flexibility observed in patients with CF. For most people with CF, flexibility is not a major limiting factor in exercise performance. As lung disease advances, thoracic kyphosis ensues, and associated mechanical inefficiencies are seen with exercise (49). Some of these postural changes are associated with tight hamstrings, leading to potential exercise limitations and injury (49). Early identification of these abnormalities through routine assessment of large muscle group range of motion, as part of an exercise assessment, can lead to establishment of stretching programs and stabilization of abnormal posture.

Anticipated Responses

Individuals with CF can have impaired exercise tolerance as demonstrated by reduced maximal oxygen consumption and peak work capacity compared with healthy persons (31, 62, 83). The ratio of minute ventilation to the **maximal voluntary ventilation (MVV)**, a marker of ventilatory limitation, may exceed 100% (normal 70%-80%) and worsen as CF lung disease progresses (16, 21). People with CF demonstrate expiratory airflow limitation as evidenced by tidal loop analysis during exercise (4). End-tidal carbon dioxide, another marker of ventilatory limitation, increases with exercise and is related to the severity of lung disease (18, 21,

64). Alveolar ventilation appears normal in patients with mild lung disease with a compensatory increase in the tidal volume (18). But as disease severity increases, alveolar hypoventilation becomes evident as the tidal volume approaches and is limited by the vital capacity (60). As this occurs, breathing frequency increases as a compensatory factor but does not provide the minute ventilation necessary for increased exercise intensity. Gas exchange can also be compromised as evidenced by the lack of increase in the diffusion capacity of the lungs following exercise (108).

In adult patients with moderate to severe disease, dead space ventilation increases. This appears to be secondary to reduced tidal volume in conjunction with increased respiratory rates, resulting in less gas exchange with each breath (102). Finally, the phase II component of oxygen kinetics (increased ventilation secondary to a return of deoxygenated blood from muscles) appears to be slowed in patients with CF, resulting in peripheral adaptations (41). Risk of oxyhemoglobin desaturation was seen to increase significantly in patients with FEV_1 below 50% of predicted or in those demonstrating a reduced baseline saturation, below 90% (32, 46, 59).

The cardiovascular system is generally able to keep up with the oxygen demands of the exercising muscles and becomes compromised only with advanced disease. Heart rate and blood pressure responses to exercise appear normal (36, 93), although a lower peak heart rate is seen as the disease progresses (31). Cardiac hemodynamics have been shown to be reduced in patients with CF, with a decreased oxygen uptake and cardiac index as well as a larger but not significant stroke volume index measured through acetylene rebreathing (104). Measurement of central hemodynamics in CF during exercise through right-heart catheterization suggests an increase in pulmonary pressures with exercise in patients with severe CF, pretransplant (40).

Muscular efficiency in children with CF can be reduced by up to 25%. The reduced efficiency may reflect altered aerobic pathways at the mitochondrial level (27). Studies of muscular strength demonstrate that CF patients have reduced muscle strength when compared with healthy controls (33, 47, 76, 94). Results regarding the trainability of the muscle systems in patients with CF are mixed. Some studies have reported strength gains associated with training programs (42, 47, 55, 76, 95), while others suggest these responses are seen only in compromised patients with the most to gain. One investigation found that patients with CF who are in good general status do not show improvement in muscle testing regardless of type of training (90).

Studies using the WAnT have demonstrated decreased anaerobic performance in individuals with CF that is related to muscle mass quantity (12). Overall oxygen cost of work appears elevated during exercise for people with CF (26). Additionally, it appears that energy metabolism during exercise is abnormal in children who have CF (10).

Basic sensory responses have been assessed to examine why patients with CF discontinue exercise (83). Similar to controls, the greatest reason to terminate activity in patients with mild CF is leg discomfort and not complaints of dyspnea, despite having lower measures of exercise tolerance and ventilation (83). These findings further suggest that nonrespiratory factors may limit exercise capacity and warrant further discussion. This is not entirely surprising in light of nutritional deficits and decreased muscle mass associated with CF. However, it is of interest that these factors may already be contributing to decreased function in patients with only mild disease.

Treatment

Current treatment for CF is complex. Specialized CF centers that offer the multiple-specialty care necessary for these individuals are the standard of care. There are more than 120 Cystic Fibrosis Foundation–accredited centers across the United States that provide the multidisciplinary care required. Both preventive and acute management are required to optimize health for people in this group. Respiratory and gastrointestinal support are the mainstay of therapy for children with CF. The complexity of care increases in adult patients and in those with severe lung disease. Treatment of the pulmonary component can be best viewed in terms of addressing the vicious cycle of infection and inflammation (figure 21.1).

Current strategies are designed to intervene in this process at multiple levels, thus minimizing the progressive loss of lung tissue. A combination of mucolytic agents and daily mucus clearance techniques can help maintain good pulmonary hygiene. Because

bacterial colonization leads to a brisk inflammatory response (e.g., cough, sputum production, increased work of breathing), use of antibiotics becomes necessary. The choice of oral, nebulized, or intravenous antibiotics is determined by the specific organisms and the severity of the acute exacerbation. New therapies targeting the abnormal function of the CFTR protein, based on genotype, have recently become a reality for many, but not all, patients with CF. The first drug on the market, ivacaftor (Kalydeco) increases the amount of time that activated CFTR protein channels are open at the epithelial surface. Patients with CFTR gating mutations who receive ivacaftor have demonstrated improved pulmonary function, improved weight gain, improved glycemic control, decreased pulmonary exacerbations, and decreased sweat chloride concentrations (25, 69, 85). A second drug, lumacaftor, improves the processing of the CFTR protein through the cell. In patients who are homozygous for the delF508 mutation, the combination of lumacaftor and ivacaftor (Orkambi) results in decreased frequency of pulmonary exacerbations and some improvements in pulmonary function and weight gain (105). Efforts continue to develop additional CFTR therapies, particularly for those individuals who have mutations not treatable by the current pharmaceuticals.

Exercise as part of the therapeutic medical regimen is standard care in most specialized CF centers. Approximately 6% of patients with CF use exercise as their only form of airway clearance (ACT), but a much larger percentage use exercise as an adjunct to other techniques such as high-frequency chest wall compressions (HFCWC: the Vest), which is the most common form of ACT in patients with CF (63% usage) (22). The Vest is an easy and portable airway clearance device for use in children and adults with CF. It is similar to wearing a life jacket. An air pulse generator rapidly fills and deflates, gently compressing and releasing the chest wall and assisting in dislodging mucus from the airway. Although not the only technique, the Vest allows for greater independence and eliminates the need for special positioning associated with standard chest percussion. As progressive lung destruction and deterioration occur, the final choice of therapy is lung transplantation. According to recent registry information, about 200 patients with CF undergo lung transplantation annually. Posttransplant survival rates are around 80% at 1 yr and over 50% at 5 yr (22).

Besides considering the pulmonary aspects of CF, the exercise clinician must give attention to the nutritional aspects. The energy needs of patients with CF vary, ranging from normal to 150% of their healthy peers depending on age, CF genotype, and severity of lung disease as well as gastrointestinal impairments. For the 85% of individuals with CF who are pancreatic insufficient, the use of pancreatic enzymes is necessary for the proper digestion and absorption of nutrients. Fat-soluble vitamins are also given as supplements. A high-fat, high-calorie diet to ensure the consumption of sufficient calories is standard care for most patients with CF. Many patients also take nutritional supplements either by mouth or via an enteral feeding tube to optimize their nutritional status. The incidence of cystic fibrosis–related diabetes (CFRD) increases with age. Careful monitoring of blood sugars during times of increased physical exertion should be a common practice in patients with known CFRD. Furthermore, recognition of the signs and symptoms of hyper- and hypoglycemia is essential. Fluid management during exercise poses unique challenges.

People with CF tend to lose more salt in their sweat per surface area than do their non-CF counterparts. Thus, the sodium and chloride levels in the bloodstream often decrease after exercise, whereas these levels are maintained in those without CF (75). In addition, individuals with CF tend to underestimate their fluid needs during exercise. In one study, patients with CF lost twice as much body weight as healthy subjects did when they drank fluid only when thirsty (8). Although children with CF have a tendency to lose salt while exercising, especially in extremely hot, humid weather, most children consume sufficient salt. Ready access to a salt shaker or salty snacks (e.g., pretzels, potato chips), along with liberal fluid intake, usually suffices.

Anthropometric data including height, weight, body mass index (BMI), and percent body fat are vital markers of nutritional status of the patient with CF. Nutritional growth is highly correlated to prognosis. Desirable levels for BMI in both males and females have been established to assist in the nutritional maintenance of these patients. Exercise is routinely recommended for all people with CF, regardless of pulmonary status. In conjunction with regular chest physiotherapy, exercise can enhance clearance of mucus from the bronchial tree, in part by increasing shearing forces and altering fluid transport into the airway of the lumen (107). Exercise alone, however, does not appear to be as effective as standard chest physiotherapy. Nevertheless, exercise therapy with either unsupervised or supervised pulmonary rehabilitation should be recommended because its positive physiologic outcomes have been proven even in patients severely affected with CF. Furthermore, with 21% of adult patients and 2% of children suffering from depression, the psychological effects of exercise also appear to be an important benefit of regular exercise (52).

EXERCISE PRESCRIPTION

Unlike many years ago when CF was predominantly a childhood disease, the life expectancy of a patient with CF has increased considerably. This presents a challenge for the exercise clinician to understand programming across the life cycle. Additionally, as important as formalized exercise training programs are, daily participation in physical activity is paramount to any clinical population, and CF is no different. From the very young (≤6 yr) to the child (7-12 yr), the teen (13-19 yr), and the adult (≥20 yr), each set of patients presents with unique needs that range from encouraging play to establishing habits and ensuring continued adherence. Furthermore, education about the well-established therapeutic benefits of exercise is important in this patient population. One study estimated that the burden of treatment on individuals with CF can exceed 75 min/d, including airway clearance and administration of medications (86). The addition of exercise, especially in the teen and adult setting, may increase the burden, and this should not be discounted. Working to ensure that exercise is both therapeutic and enjoyable may not reduce the time burden but may ease the psychological burden of treatment.

As lung disease progresses, lung function may become significantly compromised. Specific precautions may be needed before initiation of an exercise prescription. Use of supplemental oxygen during exercise for those who are prone to oxyhemoglobin desaturation has been beneficial in allowing successful exercise training and recovery (20, 65, 96). Participation in a formal exercise program or pulmonary rehabilitation can be established after the level of dyspnea and need for supplemental oxygen have been determined.

Special Exercise Considerations

Conditions associated with CF such as CF-related diabetes, exercise-induced asthma, and liver disease may require special consideration. Case-specific guidelines from the CF physician should be developed for these conditions before the beginning of an exercise program. Hypoglycemia and acute bronchospasm can be relatively easy to prevent. Patients who have a gastrostomy tube for additional nutritional support may require special exercise precautions for a brief time after placement so as to prevent damage or infection; subsequently they can resume normal activities provided that general care guidelines are followed. Although severe CF-related liver disease is uncommon, its presence can result in a bleeding tendency. People who have either enlarged visceral organs (e.g., liver, spleen) or a bleeding tendency should avoid contact sports. Prescribing exercise in CF patients with severe lung disease can be challenging. Specific application issues in individuals with advanced lung disease are discussed in practical application 21.1.

Patients with CF often develop secondary conditions that can alter the exercise prescription or affect the training program that is being managed by the exercise clinician. Aside from the obvious pulmonary issues caused by CF, depression, CF-related diabetes, and malnutrition secondary to malabsorption of nutrients are the most common secondary conditions that pose possible exercise concerns.

Approximately one in four adults with CF suffers from anxiety, depression, or both. More specifically, 8% to 29% of children and adolescents and 13% to 33% of adults are affected (82). It is believed that exercise and physical conditioning positively influence individuals with depression. In patients with CF, especially those with advanced disease, exercise may reinforce limitations secondary to their disease, leading to compliance issues and decreased feelings of well-being and a greater sense of their own mortality. The exercise clinician should recognize this possibility and create exercise sessions aimed at achieving desired improvements without magnifying limitations. Alternatives to classic exercises such as low-intensity yoga or Pilates may be desirable.

Patients with CF-related diabetes should be handled like any other patient with known diabetes. Recognizing blood glucose levels preexercise, in addition to the symptoms associated with hypo- and hyperglycemia, is important in working with those with diabetes. It may be necessary to have snacks readily available for patients with low blood glucose. Finally, if blood sugars cannot be normalized, cancellation of the exercise session may be warranted.

Maintaining good nutritional status is a major concern for patients with CF. Malabsorption of nutrients is common in the presence of exocrine pancreatic insufficiency and results in decreased body weight. It is well recognized that patients who maintain an appropriate BMI have improved pulmonary outcomes. A common concern among patients with CF, as well as their families, is that exercise will result in greater loss of calories and thus undesired weight loss. When prescribing exercise, it is important for the exercise clinician to address this concern with the patient and their families. Consultation with the CF team's nutritionist or registered dietitian about supplementing calories through food or a gastrostomy tube can help ease these concerns. It is important to explain that the benefits of exercise in the patient with CF far outweigh the risk of expending calories.

Additional concerns exist for individuals who will be exercising at high altitude. Assessment may be warranted at sea level to determine the risk of desaturation at high altitude (6). Because individuals with CF are susceptible to dehydration, especially when exercising in warm weather, fluid intake should be carefully monitored and encouraged. Fluid intake every 20 to 30 min should suffice.

Exercise Recommendations

The goals of an exercise program for individuals with CF should include enhancing physical fitness, reducing the severity or recurrence of disease, and ensuring safe and enjoyable participation. To maximize compliance with any exercise program, activity should be carefully selected to enhance cardiopulmonary fitness and other exercise goals as previously determined. The age spectrum of patients with CF varies greatly, and each patient requires special considerations based solely on age. Priorities in the very young will be significantly different from adult patients, as will the perceived burden of treatment on the individual and their families.

Young children, especially those under the age of 7, generally do not respond well to formal structured exercise programs, with the primary emphasis of this age set being on physical activity and play. Children do well when the program matches their muscular development, strength, and coordination with age-appropriate activities. Additionally, a gradual progression in the level of physical activity should ensure enjoyment while minimizing the risk of injury and nonadherence. Progression will depend on the fitness parameter being addressed as well as disease severity. Encouraging play consisting of games and diversionary tactics along with reducing nonschool sedentary time by incorporating outside activities or tasks can be helpful. For children aged 7 to 12 yr, the focus should still be on increasing physical activity, but programs can be more structured in terms of mode, intensity, duration, and frequency. There may be considerable differences in the abilities of children with CF. When prescribing exercise in these age groups, including the parent in the development of a program is vital for success because parents often falsely perceive negative consequences of exercise for their children (e.g., weight loss), especially in those with more progressive disease (24).

In teens and adults with CF, many of the same compliance issues seen in the general population exist. Time management issues related to career and daily life responsibilities are compounded by daily treatment regimens and therapies designed to maintain disease stability. Exercise progression should not only follow the guidelines of the American College of Sports Medicine but also emphasize a feeling of well-being and quality of life. It is at this age that a greater discrepancy in exercise goals becomes apparent. For some there may be an emphasis on performance because they may be engaged in sports or organized recreational

---

Practical Application 21.1

EXERCISE RECOMMENDATIONS FOR INDIVIDUALS WITH ADVANCED LUNG DISEASE

Traditionally, patients with severe obstructive pulmonary disease have not received the attention of exercise clinicians because of an extremely conservative approach to their exercise participation. Fears of exercise-induced hypoxia leading to pulmonary hypertensive episodes, cardiac ischemia, and dyspnea, as well as the perception of limited beneficial effects of training, have all been deterrents to regular physical activity. Although concerns about hypoxia exist for individuals with advanced CF, appropriate exercise prescriptions can be safely administered. A consideration for a person with advanced lung disease who wishes to participate in regular physical activity is to determine whether exercise will induce oxyhemoglobin desaturation. A measure of blood oxygenation, using the partial pressure of arterial oxygenation (PaO_2) or percent saturation ($\%SaO_2$), should be obtained during initial testing. Oximetry is also recommended during initial exercise sessions to evaluate exercise-induced desaturation.

activities. For others, a more sedentary lifestyle has become attractive, and this group will need counseling to ensure that physical activity goals are met. Finally, research suggests that individuals with CF who are new to an exercise training program are similar to any other first-time participants in that they stand to see the greatest gains from involvement in physical activity, especially early in the program (36). Capitalizing on this should be a goal of the exercise clinician.

Recommendations for exercise prescription are presented in table 21.6. This chapter does not address specifics of traditional exercise prescription; see chapter 6 for a discussion of classic training principles. Exercise compliance in the child or adolescent with CF depends largely on motivation and encouragement as discussed in practical application 21.2.

Cardiorespiratory Exercise

The main objective of cardiorespiratory training is to improve aerobic capacity. Higher levels of aerobic fitness have been associated with better quality of life and improved survival rates. This section reviews the components of the exercise prescription to optimize the client's cardiorespiratory exercise training program. For the young child (≤6 yr), there is little need for formalized aerobic programming. The general recommendation for this age is 60 min of physical activities that are appropriate for the current stage of development.

Mode

No specific activity has been identified as optimal for patients with CF. Choice of modality depends on the patient's personal preference and age and need not be costly. Cardiorespiratory benefits have been seen with a multitude of activities, some of which require little or no equipment (e.g., walking, jogging). Treadmills, bicycles, or alternative aerobic modalities (e.g., elliptical trainers) can all be incorporated into a successful exercise program. For patients who experience oxyhemoglobin desaturation, the need for supplemental oxygen may limit some choices but should not prohibit participation in an exercise program. For younger patients, activities that involve the family may be preferred. In adults, specialized classes offered through community centers, health clubs, or YMCA programs may lend a social aspect to the conditioning program while still providing structure. In children, participation in sports and activities, as permitted by the CF team, is recommended.

Frequency

In general, daily physical activity should be encouraged in all patients with CF. For formal exercise programs to ensure improvements in cardiorespiratory conditioning, 3 to 5 d of exercise per week appears optimal. Intense exercise beyond five times per week may lead to increased risk of injury. If more than five times per

Table 21.6 Typical Muscle Endurance and Strength, Body Composition, and Flexibility Levels for Individuals With Cystic Fibrosis

Parameter	Change
MUSCULAR ENDURANCE[a]	
WAnT mean power; isokinetic cycle ergometry power	Decreased
Muscle efficiency	Decreased
MUSCULAR STRENGTH[b]	
Respiratory muscle strength (PImax, PEmax)	Decreased
Peripheral muscle strength	Decreased
BODY COMPOSITION[c]	
Body weight	Decreased
Lean muscle mass	Decreased
Body mass index	Decreased
Percent body fat (BIA, skinfold assessment)	Decreased
FLEXIBILITY[d]	
Peripheral muscle flexibility (hamstrings, quadriceps)	Decreased
Posture—extent of kyphosis	Increased

[a]Muscle endurance decreases as lung disease progresses and may reflect impaired nutritional status and intrinsic cellular deficiencies.

[b]Decreases as disease progresses and may reflect nutritional status and loss of mechanical efficiency.

[c]Decreases in body composition reflect increased caloric expenditure with advanced lung disease, poor oral intake, and release of cachectic mediators.

[d]Reflects deconditioning associated with decreased activity or advanced disease.

WAnT = Wingate anaerobic test; PImax = peak inspiratory pressure; PEmax = peak expiratory pressure; BIA = bioelectric impedance analysis.

Practical Application 21.2

CLIENT–CLINICIAN INTERACTION

Motivation may be the most powerful factor in determining whether an exercise program will succeed. Motivation is unique for each person. Because nearly 50% of patients with CF are children or teenagers, motivational issues are especially important given adolescent issues of self-image, fitting in with peer groups, and establishing physical abilities. Adult patients with CF require motivation to maintain appropriate levels of fitness despite increasing time constraints associated with the burden of treatment in addition to both career and daily life responsibilities. Appropriate client–clinician interaction is important in ensuring successful exercise testing and satisfaction with a program that addresses both the physical and the emotional needs of the client.

Situational Motivational Tips for Clinical Testing

- Make the testing experience fun for younger children by creating a gamelike scenario, with cheering and enthusiasm throughout.

- Explain all procedures in detail. Forewarn children about what they will experience during testing (e.g., breathlessness, muscle fatigue, cough).

- Listen to the child's questions and concerns.

- Select apparatus (treadmill, bike) that the client believes will allow them the greatest success.

- Introduce equipment to children in a way that is fun and easily understood. The pulse oximeter might be described as the "tip of a wand" that lights up or as a secret spy decoder that identifies the patient using a fingerprint.

- Use positive motivational phases such as "You can do it!" "You're almost there!" "We are so proud of you!" or "Only one more minute!" Try to avoid using phrases that influence decisions, such as "Do you need to stop?" or "Do you have to quit?"

- It is important to motivate adult patients as well, by ensuring their understanding that the exercise evaluation is a key component in the evaluation and treatment process. For many adult patients, performing exercise tests or assessments of lung function can be

very stressful and emotional, because they understand that poor results are a sign of increasing disease and may see them as a reminder of their mortality. Proper explanation of testing results by the qualified CF team is warranted.

Tips for Exercise Adherence

- Allow children to play an active role in planning. Establish a partnership, and develop the exercise program with the child.

- Address the child's or parents' concerns (e.g., increased weight loss, not being able to keep up with friends, exaggerated cough with exercise, poor body image, presence of a gastrointestinal feeding tube, infection control) associated with exercise programs and facilities.

- Have the individual (adult or child) assist in setting exercise goals.

- In adults, acknowledge the concerns and barriers to adherence to the program and work to find alternative strategies. Goals of improving quality of life along with an improved sense of well-being may provide additional piece of mind in the adult patient.

- Understand that what is successful will be unique for each individual with CF (e.g., completing a marathon, being able to enjoy activities with family, being physically prepared for lung transplantation).

- In both children and adults, communicate frequently and address concerns before they lead to noncompliance, understanding that some days will be better than others.

Patients with cystic fibrosis appear to benefit significantly from exercise programming. Pulmonary function tends to either improve or deteriorate more slowly after training. Muscle strength and endurance improvements are well documented. Additionally, body weight can be successfully maintained or increased during exercise intervention, and patient psychological well-being typically improves.

---

week is desired, cross-training (e.g., strength training, stretching) is advised to allow adequate recovery. Maintenance of body weight should be a priority. Any reduction as a result of training should prompt increased nutritional support or modification of exercise frequency. For adults with CF, 5 d of exercise per week is optimal for enhanced fitness.

Intensity

Exercise intensity should range from 50% to 80% of the measured peak heart rate on a cardiorespiratory test, but lower intensities should be used for beginners. An intensity at or above 70% should induce an adequate stimulus to induce a cough or huff and promote appropriate airway clearance (101). If a maximum heart rate has not been obtained through cardiorespiratory testing, the general rule of using a maximum heart rate of 200 b · min^{-1} for intense running and 195 b · min^{-1} for cycling can be applied for children and adolescents. After a child has completed puberty, the formula of 220 minus age can be applied as it would be for adults. The use of a steady-state protocol (e.g., treadmill or cycle ergometer) can ensure establishment of the appropriate workload but may not be feasible or necessary. In patients with more progressive CF disease, predictions of maximal heart rate may not apply. A maximal exercise test to assess true maximum may be helpful. For the patient with severe CF, standard perceived exertion scales are an alternative way to establish intensities and may be useful in younger patients.

Duration

In all nonadult patients with CF, 60 min of physical activity should be encouraged every day. In adult patients the goal should be to obtain 150 min/wk. Exercise sessions can last anywhere from 20 to 60 min. Two abbreviated sessions are an alternative and can provide similar benefits. Attention span may play a role in a child's ability to perform an activity for longer than 10 min. For the younger patient, varying the exercise sessions by interspersing different activities may minimize boredom and enhance compliance (e.g., 5 min of bike riding, followed by 5 min of jumping rope, followed by 5 min on the treadmill at various speeds and grades).

Progression

In teens and adults, no more than a 10% increase in activity duration should occur after any 2 wk period. Frequency can be gradually increased from three times a week for the beginner to the preferred five times per week over a 3 mo period. With each progression, recognize that the patient may have increased cough and sputum production. This should be encouraged, as it will enhance airway clearance. Finally, the progression of an individual's program should be based on the desired goals and the individual's particular needs.

High-Intensity Interval Training

A growing body of evidence supports the safety and effectiveness of high-intensity interval training (HIIT) for individuals with CF (34, 50). HIIT consists of short bouts of high-intensity exercise followed by periods of lower-intensity exercise, thus allowing for a partial recovery phase (to minimize breathlessness and fatigue) and maximize performance during the higher-intensity exercise periods. Several studies have shown HIIT to be well tolerated and provide improvements in exercise capacity similar to continuous training, with less shortness of breath and muscle fatigue, for individuals with COPD (2, 63). Efforts are underway to perform a randomized controlled trial to determine the effectiveness of HIIT on exercise capacity, health-related quality of life, and feelings of anxiety, depression, and enjoyment for individuals with CF (91).

Muscular Strength and Endurance Training

Although the magnitude of potential gains in strength has been disputed in the literature, there is no question that people with CF may benefit from resistance training through both a generalized increase in muscle strength and a decrease in residual air trapped in pulmonary dead space. Furthermore, strength training in older adolescents and adult patients may slow the development of osteoporosis and osteopenia, enhance body image and promote self-esteem. Researchers have attempted to quantify the added benefits of resistance training in individuals with CF, and special considerations are discussed in this section. For the young child (≤6 yr), there is little need for a formalized resistance training programming. The general recommendation for this age involves body weight activities established in developmental milestones.

Mode

Strength training programs using both supervised and home-based activities have been explored. Free weights, weight machines, and resistance against body weight can all be used to enhance peripheral muscle strength and endurance, provided that proper instruction is given. In children, anaerobic activities that mimic the way children play, such as plyometrics, sprinting, cycling, and other modalities that use the patient's body weight and require high-intensity, short-burst duration, can also develop muscle strength and endurance. In teens and adults, formal resistance training can be performed focusing on large muscle groups in addition to trunk (respiratory) muscles. More recent investigations have looked at specific skill sets such as balance, sports accuracy, flexibility, and walking time in response to an anaerobic conditioning program as a way to quantify gains that may not be represented by true increases in muscular strength. Respiratory muscle training should take place in consultation with a clinician trained in respiratory disorders.

Frequency

A frequency of two or three times a week is usually appropriate. Care should be taken to allow for adequate muscle recovery. Alternating major muscle groups during training minimizes muscle injury while maximizing training effects. Subtle signs of overuse injuries (e.g., muscle soreness, joint pain) should prompt a reduction in frequency or intensity.

Intensity

Muscle strength and endurance can be optimized through high-repetition, low-intensity resistance training or through other modalities described earlier. The American Academy of Pediatrics Committee on Sports Medicine does not recommend high-intensity resistance training for children because of the potential of musculoskeletal injury, epiphyseal fractures, ruptured intervertebral discs, and growth plate injury before a child reaches full maturation (1). But strength training programs that use lower-intensity weights and modalities can be permitted if the planned program is appropriate for the child's stage of maturation (1). Whether adults with CF can safely participate in weightlifting is controversial. Lifting heavy weights can be associated with a Valsalva maneuver that results in increased thoracic pressures. In the susceptible patient with CF (e.g., history of **pneumothorax**, advanced lung disease), this increased thoracic pressure may result in a spontaneous pneumothorax. Consultation with a CF clinician before the start of a weightlifting program is strongly encouraged. After an individual is cleared for participation, a resistance of 60% to 80% of one-repetition maximum is generally used for novices, with the goal of completing two to four sets of 8 to 12 repetitions.

In addition, inspiratory muscle training can significantly improve lung function in many CF patients. Suggested training intensities that produce significant improvements may be at approximately 80% of maximal inspiratory effort.

Duration

In children and adolescents with CF, the duration of each session depends on the number of muscle groups exercised. Generally, 10 to 30 min of properly performed activities can increase muscular strength and endurance. In adults, 30 min of strength training should be sufficient when combined with inspiratory muscle training in addition to routine CF treatments.

Progression

In individuals with CF, the progression of a strength program should be slow. Repetitions or resistance should be increased only when the muscle has adapted to the current workload. The progression for increased resistance should occur when the individual is able to perform 8 to 12 repetitions without fatigue to the muscle. For nonweightlifting activities, activity should progress by no more than 10% during each 2 wk period.

Flexibility Training

Generalized flexibility exercises are a vital component of any exercise program and can benefit patients with CF in ways similar to those for the general population in addition to slowing the development of chest wall rigidity, which can limit chest wall expansion and respiratory function as the disease progresses. Adequate range of motion is essential for minimizing risk of skeletal injury and ensuring healthy aging. Flexibility and alignment exercises are also helpful to prevent and ameliorate postural deformities (66). The increasing popularity of alternative exercise options such as yoga and Pilates has resulted in enhanced flexibility while increasing aerobic and anaerobic fitness; these may be attractive alternatives or adjuncts to a regular fitness routine. At low intensity, these activities may also serve as centering or calming exercise options that allow the patient with anxiety or depression to relax or enhance mood. Finally, activities that enhance flexibility can be completed across the life span, making them a relatively easy activity to incorporate into the routine of the patient with CF.

Mode

Because stretching can be performed with little or no equipment, from a logistics standpoint it is one of the easiest aspects of fitness to address. Stretching may provide a protective mechanism against injury and can be a source of tension release and relaxation. Stretching exercises should focus on large muscle groups and should be included before and after activity as part of an effective warm-up and cool-down. Chest, thoracic, scapular, and shoulder range of motion activities may also assist in respiratory mechanics in older or more severe cases of advancing lung disease.

Frequency

Stretching exercises should be considered a routine component of any exercise program. Stretching can occur before, during, and after an exercise session depending on the activity chosen. A stretching program of 2 or more days per week can yield positive results such as decreased tightening of the hamstrings and quadriceps and more efficient use of respiratory muscles during exercise. Stretching for relaxation or tension relief can be performed daily. Individuals with specific flexibility issues (e.g., posture abnormalities, tight hamstrings) may need a more directed stretching program. Finally, stretching can and should be used before and after aerobic and anaerobic activities as part of a normal warm-up and cool-down.

Intensity

Proper technique will help ensure that an appropriate intensity is used for stretching. A proper stretch often feels like a gentle pull in the muscle. Stretching should not be forced. Proper breathing, including exhaling before the stretch and inhaling afterward, helps minimize injury. Avoid the Valsalva maneuver. Finally, slowly releasing a stretch back into a neutral position allows the muscle to recover.

Duration

A stretch should last between 10 and 30 s. The use of progressive stretching that includes a 10 to 30 s stretch followed by an additional 10 to 30 s has been proposed as well.

Progression

The flexibility of a person gradually improves as the muscle adapts to an increased stretch. Slow progression of a stretching program should occur over a 5 wk period. The time a stretch is held can be gradually progressed from an initial 10 to 30 s to 40 to 50 s by the end of 5 wk.

EXERCISE TRAINING

Despite the pathophysiological manifestations, a number of people with CF accomplish many of the athletic endeavors pursued by their non-CF counterparts. The short-term benefits of exercise include the therapeutic aspects of enhanced mucus clearance, with 6% of CF patients using exercise as their main airway clearance technique. Other effects of exercise include improved cardiorespiratory fitness, maintenance of bone mineral density, and psychological well-being. More recently, fitness levels have been associated with lower risk of hospitalizations (79) as well as a decrease in chronic inflammation and infections (103). The incidence of depression continues to rise in both adolescents and adults with CF, and the documented psychosocial benefits associated with regular physical activity are important. Multiple health benefits in conjunction with an exercise training program have been demonstrated in patients with CF, but because of the variations in duration, intensity, and modality, establishing causal relationships between training and disease progression is difficult. The long-term benefits are more difficult to define because of the complex and multifactorial nature of CF. Despite these frustrations, exercise tolerance and long-term survival have been correlated with one another even though no causal relationship has been established (73).

Cardiorespiratory Exercise

The benefits of exercise training on **static pulmonary function** have produced mixed results. Increases in FEV_1 were seen in response to exercise in studies that used both aerobic and resistance training techniques in both inpatient and outpatient settings (45, 90, 95). Additionally, improvements were observed in forced vital capacity (FVC), FEV_1, FEF_{25-75}, and peak expiratory flow following an intensive 17 d exercise camp (108). Despite these findings, most exercise training programs have not shown significant increases in spirometric indices. Studies taking place in settings ranging from outpatient programs to formalized inpatient rehabilitation clinics have failed to demonstrate significant improvements in lung function but rather have shown slower rates of deterioration compared with a nonexercising cohort (55, 68, 74, 93).

Improvements in dynamic lung function following exercise training are well documented, as evidenced by a lower resting heart rate, increased maximal heart rate, improved maximal oxygen consumption, increased physical work capacity, enhanced ventilatory threshold, and improved maximal minute ventilation. Exercise programs during a hospitalization for an infectious exacerbation can serve as an adjunct to traditional modalities, including chest physiotherapy and bronchial drainage, and have been associated with improvements in peak oxygen consumption and peak work capacity (17, 42, 84). Formal supervised training programs including running, cycling, and swimming, as well as structured camps, have helped maximize adherence to exercise.

Length of participation and intensity of training vary in these studies, and greater training effects were seen in the more intense programs or in programs that included increases in intensity at points throughout (36, 38, 74). These findings in combination with updated consensus reports suggest that the intensity of the prescribed exercise directly relates to the potential therapeutic impact, with higher intensities producing not only exercise gains but also greater impetus for the patient to induce a cough or huff, resulting in increased mucus clearance (101). The greatest improvements

in dynamic function are seen in patients who, by report, were the most sedentary at the beginning of the exercise intervention, suggesting that much like untrained healthy individuals, they have the most to gain from a formal program (36).

Inefficiencies in the mechanism by which individuals with CF accomplish exercise are apparent. It is suggested that much of the ventilatory compromise, especially in patients more severely affected with CF, may be attributed to low tidal volume and resultant hyperventilation (86, 102). Additionally, the resultant increase in dead space results in CO_2 retention and hyperinflation, further magnifying the need to increase breathing rate during exercise (86). This marked increase in the respiratory rate decreases the air actively participating in gas exchange. Furthermore, subtle slowing in the phase II oxygen kinetics has been observed in individuals with CF as deoxygenated blood returns from the periphery (41). Improvement in tidal volume through exercise training, in both an aerobic and an anaerobic fashion, may contribute to increased cardiorespiratory function and improve delayed oxygen kinetics.

Muscular Strength and Endurance

There has been greater attention on the benefits of anaerobic exercise or resistance training in recent years. The benefits of exercise training on anaerobic function have been mixed. A program that included upper body strength training demonstrated increased strength and physical work capacity as well as good compliance throughout, as more than 85% of the subjects completed the program (76). Weight training, home cycling for 6 mo, and interventions including sport participation have increased muscle strength in individuals with CF (38, 42, 100). Increased $\dot{V}O_2$ has been reported along with increased leg strength (55), in addition to peak and mean anaerobic power (42, 55, 95). Studies that have attempted to combine aerobic and anaerobic training principles, ranging from 6 wk to 6 mo, have resulted in gains in FEV_1 and in some measures of sport-related skills, such as balance, flexibility, and vertical jump, without significant increases in pure strength (35, 36, 90).

Muscle limitations exist in CF. Decreased muscle size and metabolic abnormalities may lead to the decreased strength seen in these patients (86). Patients performing resistance training may therefore require a longer break between activities. In the patient with severe disease, resistance training may offer an attractive exercise modality as the ventilatory demands are far less than with traditional dynamic aerobic exercise. Finally, in patients post–lung transplant with potential muscle wasting secondary to routine use of corticosteroids or anti-rejection medications, resistance training becomes an important part of the exercise prescription (101).

Exercise programs focused on respiratory muscle training have also shown training adaptations as demonstrated by increased peak inspiratory pressures (3, 92). The intensity of inspiratory muscle training has become better defined. High-intensity training (80% of maximum) was shown to improve muscle function significantly more than lower-intensity training did (29).

Body Composition and Nutrition

Exercise limitations exist secondary to nutritional concerns and include increased energy expenditure, potential sodium chloride imbalance through increased sweat rates, CF-related diabetes, and a lower overall muscle mass (86). Children with CF who undergo regular exercise training (e.g., swimming, biking, running, weightlifting) are capable of increasing body mass despite increased caloric needs (5, 17, 44). Nutritional supplementation has been associated with improved aerobic exercise tolerance and respiratory muscle strength in some small case reports (19, 98). Randomized controlled studies are mixed in their findings; some showed no difference in anthropometric data (% body fat, BMI) in either a resistance program or a combined program using both aerobic and anaerobic methods (55, 67), while others demonstrated increases in body weight, fat free mass, and BMI (36, 95). Regardless, the exercise clinician needs to be prepared to recognize potential issues that arise because of the hydration or nutritional status of patients with CF.

Psychological Well-Being

The long-term psychosocial benefits of exercise have been fairly well established in children with CF. The Quality of Well-Being Scale, designed to measure daily functioning, has been shown to correlate with exercise capacity in individuals with CF (51, 77). The current rate of depression among children is estimated at between 8% and 29% and among adults is estimated at between 13% and 33%. Recent

Research Focus

Exercise as Part of a Cystic Fibrosis Therapeutic Routine

Expert Review of Respiratory Medicine provides a guide to contributing factors to exercise limitations in patients with CF. The authors provide a multisystem breakdown that can serve as a reference for exercise clinicians as they consider exercise programming for the person with CF. The issues and a brief description of how each contributes to exercise limitations follow.

- *Lung function:* Increases in dead space that lead to changes in ventilation; carbon dioxide retention (>5 mmHg from baseline); static hyperinflation (residual volume / total lung capacity >30%).

- *Nutrition:* Resting energy expenditure that is estimated at 20% to 50% higher than controls due to increased work of breathing, chronic inflammation, and genetics; increased sweat rate and increased sodium and chloride concentrations that reduce osmolarity and alter thirst mechanisms; cystic fibrosis–delated diabetes.

- *Muscular dysfunction:* Decreased maximum strength and power; abnormal bioenergetic pathways both at rest and with exercise; decreased phosphocreatine recovery that may suggest impaired oxidative metabolism; slowed oxygen kinetics.

- *Genotype:* No clearly defined relationship between genotype and oxygen consumption; the CFTR protein may be expressed at the sarcoplasmic reticulum of the skeletal muscle, which may lead to disruptions in muscle function, thereby affecting capacity.

- *Habitual activity:* CF patients do not typically meet activity guidelines despite an understanding of the benefits of exercise on growth and general well-being.

- *Gender:* In general, there is a lower expectation of children of both genders to participate in physical activity or sports; when comparing between sexes, there is greater participation in males than in females.

- *Psychosocial:* Parental perceptions of risk lead to concerns and decreased participation in physical activity; depression and anxiety rates that exceed 25%; burden of treatment issues that exceed 80 min of daily therapies (medications and airway clearance) are compounded with addition of exercise.

The authors provide a detailed discussion of the therapeutic effects of exercise as related to the potential limitations listed here. Exercise will assist in the movement of mucus through body movements and associated shear forces generated in the lungs with the increased work of breathing. Exercise can improve fluid balance and enhance muscle mass and function. Furthermore, evidence suggests enhanced quality of life and well-being associated with increased physical activity. The authors discuss issues associated with modern culture, including increased sedentary time associated with computer games and cellular devices, in addition to the need for increased education and family support that will advocate for increased physical activity within the CF population.

The authors conclude that "exercise is now seen as an important part of the management of individuals with CF," and "it is reasonable to suggest that exercise has the potential to have a positive effect on physiological parameters and outcomes." The authors make compelling arguments for the continuation of long-term controlled intervention studies to further increase understanding; the need for formal exercise testing with measures of maximal oxygen consumption to assess for potential limitations and document declines in physical capacity; and a further commitment to advocate for physical activity and exercise. These are all perfect job responsibilities of the exercise clinician and firmly fit into the mission of this textbook.

Based on Rand and Prasad (86).

CF registry data categorized 1.6% of deaths from suicide. As the prevalence of depression continues to grow, research in the area of quality of life and positive patient perceptions has flourished, and multiple studies cite improvements in overall health-related quality of life (HRQOL) and within the specific domain of physical functioning (55, 93, 95). In 2015, a consensus statement for the screening and treatment of depression and anxiety was published. Universal depression and anxiety screening is recommended for all adults and adolescents with CF, for at least one caregiver of children and adolescents with CF, and in children aged 7 to 11 yr with CF if their caregivers have elevated screening scores. Referrals should be made to appropriate providers based on the depression and anxiety screening scores. Many procedures can be anxiety provoking for patients with CF, including exercise testing. Behavioral approaches are recommended to address this anxiety, which can be present in anticipation of medical procedures and interventions (82).

Exercise Influence on Cystic Fibrosis

Exercise is a vital factor in the care of patients with CF. Evidence suggests that exercise plays an important role in developing strength and aerobic capacity, preventing the deterioration of lung function over time, and decreasing the prevalence of hospitalizations and susceptibility to infection; it is also a significant predictor of long-term mortality. Despite the positive benefits associated with exercise, on average, oxygen consumption fell by 2.1 mL · kg^{-1} · min^{-1} per year, suggesting that disease progression, although delayed, cannot be eliminated. This rate of decline is multifactorial and not just limited to lung pathology but also includes muscle deficits, nutritional compromise, gender, and psychosocial factors. Additionally, new research has begun to focus on the effects of the specific genotype on CF outcomes. It has been estimated that those with an oxygen consumption less than 32 mL · kg^{-1} · min^{-1} had an 8 yr mortality of almost 60%, and those above 45 mL · kg^{-1} · min^{-1} showed no mortality at all, again suggesting the positive and important benefits of physical activity and formalized exercise training in this population. Finally, the potential of enhanced immune modulation is an attractive but speculative theory. The ability of exercise to alter the immune system is now well recognized. Although some studies have shown a beneficial effect of chronic exercise on the immune system (9, 11), this relationship has not been established in people with CF. Table 21.7 provides a review of exercise training guidelines for patients with CF.

Clinical Exercise Bottom Line

- A significant benefit of exercise in this population is that it enhances airway clearance, induces coughing and sputum production, and is often used as an adjunct to more traditional therapies.

- Maintaining exercise performance has been linked to improved long-term prognosis in the CF patient.

- Maintaining an appropriate nutritional status (BMI >23 in males, >22 in females, ≥50th percentile in children, ≥50th percentile weight for length in infants) has resulted in a significant improvement in pulmonary outcomes. Body weight can be maintained or significantly increased through exercise.

- Oxyhemoglobin desaturation is more common in patients with FEV$_1$ <50% predicted and in patients with more severe disease as represented by lower resting O$_2$ saturation.

- Improved survival has resulted in a greater percentage of adult patients with CF. This has led to an increased incidence of depression, bone diseases such as osteoporosis, and CF-related diabetes. Clinicians should consider all of these when creating or maintaining an exercise prescription.

- The median age of survival in patients with CF is approximately 39 yr and has increased significantly in the past two decades.

- Over this time period, exercise has increasingly been shown to significantly retard disease progression.

- Although no absolute contraindications exist, clinicians should take care when prescribing exercise for a CF patient with a history of pulmonary hypertension, acute hemoptysis, pneumothorax, oxygen dependence, bleeding disorders secondary to liver disease, and severe malnutrition. Care should also be taken when exercise is to be performed at altitude.

- In patients with severe disease, maximal exercise testing may result in decreased cardiac responses (lower peak HR and BP) as the patient exhibits increasing ventilator limitations. It is important to take care when creating exercise prescriptions using intensities based on percentage of predicted peak heart rate, because these may exceed patient abilities.

- Assessment of body composition in the CF population is typically via skinfold measures or bioelectrical impedance. Techniques that correct for lung parameters (underwater weighing) tend to be inaccurate because of hyperinflation of the lungs.

- In children and adolescents with mild to moderate CF disease, traditional principles of exercise prescription can be followed (see chapter 6).

Table 21.7   Exercise Training Review in Patients With CF

Parameter	Training specifics
Cardiorespiratory endurance	Static (rest): Spirometry values (FEV$_1$, FEF$_{25-75}$, tidal volume, DL$_{CO}$) are reduced with increased dead space and increased residual volume and RV/TLC. This lung impairment affects endurance activities through impaired ventilation. Lower resting O$_2$ saturation is also common. Dynamic (exercise): Emphasis on increased habitual physical activity in all ages. Aerobic capacity is decreased with an increased breathing response ($\dot{V}_E$, $\dot{V}_E$/MVV, RR, PetCO$_2$). Peak work capacity and maximal oxygen consumption are reduced. HR with exercise may be blunted, making prescriptions based on predicted maximal HRs difficult to use. Moderate to severe patients may desaturate with exercise, resulting in the need for supplemental O$_2$. This may limit exercise choices. Endurance training is critical. The benefit of exercise as an adjunct to traditional airway clearance techniques has been well documented.
Skeletal muscle strength	Peripheral and respiratory muscular strength is reduced as demonstrated by lower values in both 1RM at the periphery and decreased inspiratory and expiratory muscle strength (PImax, PEmax) at the lung. Strength training has been shown to enhance body image and self-confidence. Standard strength training programs have been used. Caution is warranted in young children (1-6 yr) and children (7-12 yr) still undergoing physical development that may be delayed in patients with CF. The use of body weight–appropriate activities may be desirable in younger patients. In advanced disease, avoid the Valsalva maneuver because this increases thoracic pressures.
Skeletal muscle endurance	Muscular endurance is reduced, as is muscular efficiency. Reduced lean body mass significantly affects both muscular strength and endurance. Strength training programs should follow the same guidelines as for individuals without CF with gradual progression. Respiratory muscle training should be incorporated into the program. Severe patients should avoid the Valsalva maneuver, which can increase thoracic pressures. Patients with a history of pneumothorax should not perform resistance training.
Flexibility	Peripheral muscle strength has been shown to be reduced (hamstrings, quadriceps). Poor posture, barrel chest, and kyphosis are common findings. Increases in flexibility are generally seen in individuals with CF using standard stretching techniques. Focusing on upper body range of motion and chest wall mobility may enhance breathing mechanics with exercise.
Body composition	Body weight, lean body mass, BMI, and % body fat are all decreased in individuals with CF. Maintaining proper nutrition is essential because it is strongly correlated with long-term survival. Although it is well documented that individuals with CF can show positive weight gain with exercise training, caution is warranted to ensure proper nutrition. Minimal BMI standards are >23 for adult males, >22 for adult females, ≥50th percentile in children, and ≥50th percentile weight for length in infants. Working with the CF team's clinical nutritionist during exercise training is vital.

FEV$_1$ = forced expiratory volume in 1 s; FEF$_{25-75}$ = forced expiratory flow between 25% and 75%; DL$_{CO}$ = diffusion capacity as measured by carbon monoxide technique; RV/TLC = ratio of residual volume to total lung capacity; $\dot{V}_E$ = minute ventilation; $\dot{V}_E$/MVV = ratio of minute ventilation to maximal voluntary ventilation; RR = respiratory rate; PetCO$_2$ = end-tidal carbon dioxide; HR = heart rate; 1RM = one-repetition maximum; PImax = maximal inspiratory pressure; PEmax = maximal expiratory pressure; BMI = body mass index.

CONCLUSION

There is little doubt that people with CF appear to benefit from exercise training programs. Both static and dynamic pulmonary function either improve or have a significantly slower rate of deterioration following training programs. Reports on muscle strength and endurance following exercise programs have presented varying results regarding their effectiveness but overall appear beneficial. Data are sufficient to support the use of programs that target improvements in muscular strength and endurance in patients with CF. Because nutrition is a major concern in patients with CF, it is critical to understand that body weight can be successfully maintained or increased during an exercise intervention. The psychosocial implications of exercise are considerable, with rates of depression in children at 8% to 29% and adults at 13% to 33%. Improvements in quality of life and sense of well-being have been well documented and may outweigh the physiologic benefits of exercise in some cases. Finally, the potential effect of exercise on improving patient prognosis makes exercise an attractive therapeutic modality.

Online Materials

Visit HK*Propel* to access a link to the references, the case studies with discussion questions, and a quiz to help you review key concepts and test your understanding of the material covered in this chapter.

PART V

Diseases of the Immune System and Oncology

Early-onset fatigue and reductions in exercise tolerance are hallmarks of both cancer and human immunodeficiency virus (HIV). These comorbid characteristics may be related to worsening of the disease or be caused by the medications or other therapies (e.g., radiation) used to treat the disorder. Part V highlights the important role that exercise training has in helping to improve exercise tolerance in patients with cancer or HIV. With continued advancements in therapies, survival rates are increasing for many cancers and for HIV. These diseases provide opportunity for clinical exercise physiologists who are interested in providing exercise treatment to improve fatigue and clinical outcomes.

Chapter 22 focuses on cancer. Today, more than ever, we understand the beneficial role that regular exercise training can play in the care of patients with cancer. Exercise helps to reverse exercise intolerance and can improve mood and quality of life. Still, many questions remain regarding safety, dose, rate of progression, clinical outcomes, and the optimal timing to interject exercise into a patient's treatment plan. This chapter is updated with what we know today about exercise in the prevention and treatment of cancer. If positive results from exercise continue, exercise therapy

for those with cancer may become a guideline treatment, which might lead to reimbursement from health insurers for cancer survivors who undergo exercise treatment.

Chapter 23 reviews the myriad interactions between HIV and exercise. Since 1996 the life expectancy of HIV patients has increased for both men and women. These gains are primarily the result of medications used to attenuate viral load and maintain CD4+ count. However, these treatments often lead to loss of skeletal muscle size, strength, and endurance; increased risk for cardiovascular disease; and body composition changes that require independent treatments. Exercise training is recognized as an important aspect of the treatment process for those who are HIV positive because it can help attenuate these changes. These patients may also enroll in a cardiac rehabilitation program because of their increased risk of coronary artery disease and heart attack. Therefore, the clinical exercise physiologist working with patients who are HIV positive must understand the principles of exercise testing and training as well as the pathophysiology, evaluation, and therapy strategies unique to HIV infection.

CHAPTER 22

Cancer

Dennis J. Kerrigan, PhD

Karen Wonders, PhD

Each of you has been touched by cancer. Either someone close to you or you yourself have heard the frightening words *You have cancer.* For many confronted with this new diagnosis, the mind races ahead to thoughts of chemotherapy, illness, fatigue, and feelings of one's own mortality. With nearly 1.8 million new cases of cancer diagnosed in 2020 in the United States (1), scientists constantly strive to find new ways to prevent, diagnose, and treat it, as well as to lessen the often burdensome side effects that commonly accompany conventional care. However, although cancer is the second leading cause of death in the United States (2, 103), the story is not completely bleak. With continued improvement in diagnosis and treatments, the overall death rates attributable to cancer have declined by 27% over the past decade, leading to an increase in the number of cancer survivors in the country to over 15.5 million (86). To this end, a growing amount of clinical research has focused on the role of regular exercise to counteract the adverse effects of cancer and its treatment and to determine whether exercise can reduce recurrence and death across various cancer populations.

DEFINITION

The words *cancer* and *malignancy* are commonly used for the medical term **neoplasm**, meaning an abnormal growth of tissue. In neoplasm, tissue grows through unregulated cellular proliferation, shows partial or complete lack of structural organization, lacks functional coordination with the normal tissue, and usually forms a distinct mass that may be either benign or malignant (3). Malignant neoplasms are generally fast growing, have the ability to invade host tissue, are associated with large areas of necrosis because the tumor outgrows its blood supply, and can **metastasize** to other parts of the body. They are eventually fatal if untreated.

Benign neoplasms, on the other hand, are generally slower growing, have well organized and well differentiated cells, and usually are not fatal. Cancer is a unique disease because it can originate in any organ system, can spread to other organ systems, and has multiple etiologies. Cancer affects all nationalities, races, and ages, as well as both men and women. The treatment varies with the cancer type and location and includes surgery, chemotherapy, radiation therapy, biotherapy, and targeted therapy, individually or in combination.

The therapeutic role of exercise for patients with cancer is multifaceted. Physical inactivity is a known risk factor for 13 different cancer types, hence exercise may play a role in primary prevention (64). With respect to the adverse effects of anticancer treatments as well as the tumor burden itself, exercise training is now increasingly used as an adjunctive treatment to counteract commonly reported symptoms such as fatigue and muscle weakness. Additionally, preliminary data show an inverse association between regular exercise following diagnosis and cancer recurrence, cancer-specific mortality, and all-cause mortality, suggesting a secondary prevention role for exercise. In this chapter, we focus primarily on the four most common cancers: lung, breast, prostate, and colorectal.

SCOPE

Cancer is found throughout the world, with about a threefold difference between countries with the highest frequency of cancer and those with the least. Geographic variation can be as much as 100-fold for specific cancers. For example, the death rate from upper gastrointestinal cancer is extremely high in South Africa, China, Japan, and Iran. This type of cancer is much less frequent in the United States, except in areas that have a high incidence of alcoholism. The most common cancers in Western countries are

Acknowledgment: Much of the writing in this chapter was adapted from the previous editions of *Clinical Exercise Physiology.* Thus, we wish to gratefully acknowledge the previous efforts of and thank John R. Schairer, DO, Lee W. Jones, PhD, and Kerry S. Courneya, PhD.

lung, large bowel, and breast, whereas in southeast China, nasopharyngeal cancer is the most common malignancy.

Haenszel (41) and Muir and colleagues (66) demonstrated that migrating populations tend to acquire the cancer incidence profile of their new country of residence, suggesting that genetics is less important in the genesis of cancer than environmental influences. Results from epidemiological studies show that more than two-thirds of cancer deaths might be prevented through lifestyle modification. One-third of cancer deaths are due to cigarette smoking, and another one-third are attributed to alcohol use, specific sexual practices, pollution, dietary factors, physical inactivity, and obesity.

In the United States, approximately 596,000 Americans die of cancer annually, ranking it as the second leading cause of death, behind cardiovascular disease. Forty percent of Americans will develop cancer during their lifetimes (about 42% of men and 38% of women), and approximately 1.7 million new cases occur each year (2). There are differences between men and women in the type and frequency of cancer and the likelihood of dying of cancer (1). For nearly all cancers, the incidence rates are higher in men than in women, with the exceptions of thyroid, gallbladder, and, of course, breast and gynecologic cancers. Among men, the most common cancers in order of prevalence are prostate, lung, and colon or rectum. With women, the most common cancers are breast, colon or rectum, and lung. If death from cancer is considered, the order of the cancers changes. More men die of lung cancer, followed by colon or rectum cancer, and finally prostate cancer. Among women, lung cancer is now responsible for more deaths than breast cancer and colon or rectum cancer.

Race affects cancer incidence and cancer death rates. Overall, African Americans have higher mortality than whites for most cancer types. For example, African American females have a lower 5 yr relative breast cancer survival rate than Caucasians (81% vs. 92%). The reasons for this racial disparity include later stage of diagnosis and a difference in receiving high-quality care (1). Asian and Pacific Islanders, American Indians, and Hispanics have lower incidences and death rates than does the white population.

Age also affects the distribution and frequency of cancer. Among patients under the age of 15, the most common cancers are leukemia, brain, and endocrine. For patients over age 75, the most common cancers are lung, colon, and breast. Cancer incidence increases with age. Between birth and age 39 yr, 1 in 58 men and 1 in 52 women will develop cancer. The incidence of cancer in men and women between ages 60 and 79 yr is 1 in 3 and 1 in 4, respectively.

PATHOPHYSIOLOGY

The final common pathway in virtually every instance is a cellular genetic mutation that converts a well-behaved cellular citizen of the body into a destructive renegade that is unresponsive to the ordinary checks and balances of the normal community of cells (16).

The normal growth and proliferation of cells within the body is under genetic control. The stem cell theory is the model developed to explain the orderly proliferation of cells, specialization to perform discrete functions, and cell death within an organ (figure 22.1). The stem cell is pluripotent, which means it is an uncommitted cell with various developmental options still open. The process by which the **pluripotent stem cell** is able to develop special functions and structures within an organ system is called differentiation. Thus, some stem cells are triggered to differentiate and become hair cells, and some cells become cardiac myocytes. The pluripotent stem cell also has the capacity for self-renewal. But after a stem cell becomes committed to a cell line such as a hair cell or cardiomyocyte, it no longer has pluripotent and self-renewal properties and is destined to develop along its specialized pathway of differentiation. The best example is a pluripotent hematopoietic stem cell, with its capacity to form both red blood cells and white blood cells (e.g., neutrophils, basophils). After it commits to a specific cell line, it can no longer differentiate into other cell types or divide to form new cells.

In cancer, the cellular DNA of normal stem cells is damaged, leading to disordered cell growth and specialization. The stem cell model for cancer proposes that tumors arise from carcinogenic-causing events occurring within the normal stem cells of a particular tissue. A cancer-causing insult is believed to produce a defect in the control of normal stem cell function, resulting in abnormalities in self-renewal, differentiation, and proliferation. In other words, the normal quality and quantity control for cell function and growth is lost.

The carcinogenic event for many cancers is unknown, but five broad categories have been identified.

1. Environment
2. Heredity
3. Oncogenes
4. Hormones
5. Impaired immune system function

Cancers most likely arise when a factor from one or more of these categories is present (e.g., radiation exposure from the environment), causing cellular mutations that go unrecognized because of injury to the immune system as well.

Environmental factors are implicated in more than 60% of cancers, partly because of the known association between certain agents and the development of cancer. Common environmental factors that contribute to cancer death include tobacco (25%-30%), obesity and a sedentary lifestyle (25%), infections (15%-20%), radiation (up to 10%), stress, and pollutants (10) (figure 22.2). For example, cigarette smoking, a lifestyle choice, increases the risk for lung cancer because tobacco smoke contains carcinogens. Other lifestyle choices such as a diet high in fat, obesity, and a sedentary lifestyle are associated with an increased likelihood of developing breast and colon cancers. Excessive exposure to sunlight is responsible for a higher incidence of skin cancer in farmers. Lung cancer is more prevalent among miners and those who work with asbestos, chromate, or uranium. Exposure to certain solvents is associated with leukemia.

Some hormones are thought to possess carcinogenic potential. For example, the ovary, breast, and uterus are hormonally sensitive; therefore, estrogen may play a role in the cancers of these organs in women.

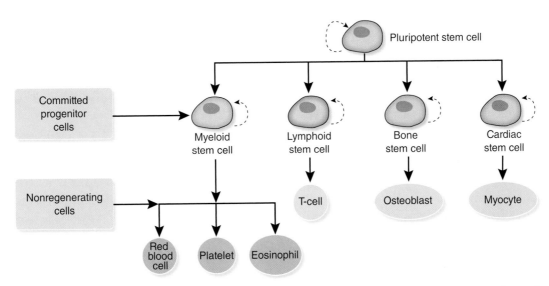

Figure 22.1   Stem cell sequence theory. This schematic presentation of the stem cell theory depicts how the uncommitted pluripotent stem cell develops or differentiates into several committed, nonregenerating cells with specialized functions.

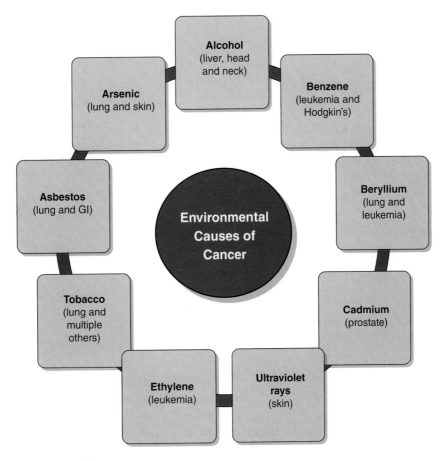

Figure 22.2   Environmental causes of cancer.

Although genetics alone accounts for a small number of cancers, neoplastic pathology often involves mutations from environmental carcinogens on specific genes responsible for cell growth, apoptosis, and regulation (32, 42). Two of the most well-known types of genes associated with cancer are **proto-oncogenes** and **tumor suppressor genes**. Proto-oncogenes act at the cellular level to promote normal cell growth and development. When attacked by specific carcinogens (e.g., radiation, chemicals, viruses), proto-oncogenes are mutated into **oncogenes**, which can lead to uncontrolled proliferation resulting in tumorigenesis (42). However, tumor suppressor genes oppose such unregulated growth through the activation of anticancer pathways, which eventually leads to cellular apoptosis or inhibition of tumor growth (42). Interestingly, mutations to the tumor suppressor gene that regulates the p53 protein, a well-known tumor suppressor, have been found in many cancers (42).

The immune system is responsible for mediating the interaction between an individual's internal and external environment. The immune system is divided into two major categories: innate and acquired responses (table 22.1). The innate immune system response is nonspecific and immediate, beginning within minutes of an insult. The response occurs without memory for the eliciting stimulus. This process is called inflammation. The innate immune system represents our first line of defense against cancer.

The acquired, or adaptive, immune system is characterized by an antigen-specific response to a foreign antigen or pathogen and generally takes several days or longer to activate. A key feature of acquired immunity is memory for the antigen, such that subsequent exposure leads to a more rapid and often more vigorous response.

The overall function of the immune system is to rid the body of foreign agents such as bacteria, viruses, and malignant cells. The immune system recognizes infectious agents and malignant cells because those agents and cells contain abnormal antigens in their cell membranes. The immune system can also inhibit subsequent formation of a tumor by countering factors responsible for its growth. The acute and longer-term effects of exercise training on immune system function are shown in table 22.2.

Table 22.1   Components of the Immune System

Immune system component	Description
Innate immune system Monocytes Macrophages Neutrophils Natural killer cells	Nonspecific response Nonspecific killing response of tumor cells by phagocytosis and cytolysis
Acquired immune system Cytotoxic T lymphocytes	Antigen-specific response Requires tumor antigens in association with class I major histocompatibility antigens

Table 22.2   Effects of Exercise on the Immune System

Component	Effect of acute exercise	Effect of chronic exercise
INNATE IMMUNE SYSTEM		
NK cells	Immediate increase in cell count and cytolytic activity. Depressed for 2-24 h after strenuous exercise.	NK cell count and activity increase, both in blood and in spleen.
Macrophages	Immediate increase in monocyte and macrophage count. Adherence is unchanged. Increased phagocytosis with moderate activity.	Response is unclear. Resting monocyte count is unchanged. May cause adaptations that alter exercise response.
Neutrophils	Most PMN functions decrease significantly after strenuous exercise.	Increase with moderate exercise.
ACQUIRED IMMUNE SYSTEM		
T lymphocytes	Moderate activity enhances cell proliferation, with depressed levels 30 min after exercise. Vigorous activity causes a transient decrease in proliferation.	Regular, moderate exercise enhances cell proliferation.

NK = natural killer; PMN = polymorphonuclear leukocytes.

Cancer therapies such as chemotherapy, radiation therapy, and surgery are known to be immune suppressive. The role this suppression of the immune system plays in the cure and recurrence rates during and after treatment of cancer is unknown. The important role exercise plays in modulating the immune system remains an area of active investigation. Practical application 22.1 discusses the beneficial effects of acute and chronic exercise on immune function.

CLINICAL CONSIDERATIONS

There are four types of cancer: carcinoma, sarcoma, leukemia, and lymphoma. Carcinomas are cancers of epithelial tissues and include cancers of the skin, digestive tract, genitourinary tract, pulmonary system, and so on. Cancer of the breast, colon and rectum, lung, and prostate are examples of carcinoma. Carcinomas represent 90% of all cancers. Sarcomas are tumors of the connective tissue and include cancer of the bone, muscles, and cartilage. Leukemias are cancers of the white blood cells. Lymphomas are cancers of the lymphatic system.

Signs and Symptoms

Early on, the symptoms of cancer are usually nonspecific, such as weight loss, fatigue, nausea, and malaise (65). Only an astute clinician can make the diagnosis of cancer at this time, yet early detection is key to maximizing the patient's chance for survival.

Later, the patient develops symptoms specific to the involved organ, such as shortness of breath in lung cancer or jaundice attributable to biliary obstruction in pancreatic carcinoma. By this time, however, prognosis is much poorer. Social history is also important, revealing occupational exposure to carcinogens or habits such as smoking or ethanol ingestion. The family history may reveal familial predisposition to a cancer, one that requires closer surveillance in the future. The review of systems may reveal symptoms indicating that the cancer has already metastasized.

History and Physical Examination

Because of the diversity of organ system involvement, no single part or segment of the history and physical examination focuses only on cancer. As a result, the physician or physician extender must perform a complete and accurate history and physical examination. Any such examination is usually oriented toward looking for enlargement of an organ, such as an enlarged lymph node in the case of lymphoma or a testicle in testicular cancer. Because most organs (e.g., lung, pancreas, kidney) are deep within the body, the yield is low. Although examining for masses is important, a mass is often a later sign of cancer that occurs after the cancer has metastasized. Drawing upon the historical and physical examination information gathered by others, the clinical exercise physiologist needs to understand the important aspects of each specific cancer in order to assess progress and identify any exercise-related concerns.

Diagnostic Testing

Because prevention of cancer is not always possible, the earliest detection of the disease is the next best strategy to reduce cancer mortality rates. To help accomplish this, the American Cancer Society recommends a series of screening procedures and evaluations depending on age, gender, and risk for specific cancers. When cancer is suspected, the first diagnostic principle is that adequate tissue must be obtained to establish the diagnosis. Because the therapy used for each type and subtype of cancer is often particular to that type or subtype, every effort must be made to obtain appropriate tissue samples even if treatment is delayed for a short time. The process of obtaining a sample of tissue is called a **biopsy**.

The second diagnostic principle is to determine the extent of spread of the cancer, also known as **staging**. In leukemia, staging can be accomplished through routine history and physical examination, laboratory tests, chest X-ray, and bone marrow biopsy. With solid tumors, computed tomography (CT) and magnetic resonance imaging (MRI) in conjunction with a biopsy are often needed to determine the size of the tumor and the extent of its spread. The degree to which the cancer has spread is reflected in its stage, which guides the type of treatment most appropriate for the patient. An example of a simplified staging system is shown in table 22.3. Each cancer has a staging system unique to itself, one that takes into consideration pathogenic features, the modes of spread, and the curability of the disease.

Treatment

There are many treatment options for cancer. The selection of which depends on the type, location, and stage of the cancer. There are five broad treatment categories:

- Surgery
- Radiation therapy
- Chemotherapy
- Biotherapy
- Targeted therapy

Surgery is the oldest and most definitive treatment for cancer. The two types of surgery are curative and palliative. **Curative surgery** involves removing all of the neoplasm and sometimes nearby lymph nodes; it is the primary treatment for cancers that are localized. Radiation or chemotherapy may be given before surgery (i.e., neoadjuvant) to reduce the tumor size and increase the chance for a cure.

In **palliative surgery**, a large tumor mass is removed to make the patient more comfortable, to relieve obstruction of vital organs, and to reduce the tumor burden. For example, a colon cancer may be removed to prevent bowel obstruction, or an ovarian cancer may be removed to prevent obstruction of a ureter. Decreasing tumor burden may also make the tumor more susceptible to radiation or

Practical Application 22.1

EXERCISE AND PRIMARY CANCER PREVENTION

Data show that higher levels of physical activity are associated with lower overall cancer mortality (64, 72, 73). Both physical activity and dietary interventions were identified as strategies to successfully reduce the overall incidence, morbidity, and mortality from certain kinds of cancer (60). In fact, for the nonsmoker, dietary and physical activity interventions are the most important modifiable determinants of cancer risk.

The role of physical activity, either leisure-time activity or occupational physical activity, in reducing overall cancer risk and site-specific cancer risk is gradually being defined (35, 97). A large analysis using pooled data of 1.44 million participants showed a significant inverse and favorable relationship between leisure-time physical activity and risk for the following types of cancer (64):

- Esophageal
- Liver
- Lung
- Kidney
- Gastric cardia
- Endometrial
- Myeloid leukemia
- Myeloma
- Colon
- Head and neck
- Rectal
- Bladder
- Breast

Although some of these associations (i.e., liver, gastric cardia, kidney, and endometrium) may be confounded because of the strong obesity–cancer relationship, for most of these, the exercise–cancer associations are BMI independent. Out of these cancer types, the relationship between colorectal cancer and physical activity has been studied the most. Forty-eight studies with 40,000 patients (97) demonstrated a 10% to 70% reduction in risk for colon cancer in physically active individuals. Decreased bowel transit time caused by physical activity may explain the observation that colon cancer frequency is reduced in physically active individuals, whereas no change occurs in the frequency of rectal cancer.

Forty-one studies including 108,321 women (97) evaluated breast cancer and possible risk reduction with physical activity. Twenty-six studies demonstrated that both occupational and leisure-time activity reduce postmenopausal breast cancer risk by about 30%. In general, the results for breast cancer are less conclusive than those for colon cancer. Of the 12 studies in the literature regarding physical activity and the risk for developing endometrial cancer, 8 demonstrated a 20% to 80% reduction in risk. In the studies that evaluated the possible protective effect of exercise in preventing

ovarian, prostate, and testicular cancer, the data are generally favorable but inconsistent. The association of physical activity with lung cancer has been reported in 11 studies. Six of these support a protective effect of physical activity on lung cancer.

How much physical activity is needed to reduce the risk of cancer? A definitive answer is not currently available. Data from Blair and colleagues (12) indicate that the reduction in cancer risk occurred primarily between the very low fitness group and the moderately fit group, with no further decrease in risk among the more fit subjects. Paffenbarger and colleagues (74) reported a decrease in all-cause mortality for alumni who expended 1,500 or more kcal/wk, but the authors did not break out cancer deaths in this group. Data show that 1,000 kcal/wk (4 h of moderate activity per week or 3 h of vigorous activity) has a protective effect for colon cancer and breast cancer (90, 99). Lee and colleagues (59) also reported that exercising at moderate intensity, greater than 4 to 5 METs for 4 h/wk, decreased the incidence of lung cancer.

Determining a relationship between physical activity and cancer risk is complex because the mechanism through which exercise acts to lower cancer risk is not well understood. It is not known whether risk reduction occurs through lifelong exercise, exercise at certain stages in the life of the patient, moderate activity, or vigorous activity. Exercise may help block initiators of cancer, in which case exercise done consistently at a relatively young age may be most beneficial. This is the rationale behind studies that investigate whether participation in high school and college athletics reduces the risk of cancer during adulthood. Alternatively, exercise may counter the promoters of cancer cell replication, so that exercise during a later phase of the neoplastic process may be preferred to decrease the development of clinically significant disease (89, 94). Therefore, the time point at which exercise occurs during a person's life span may be an important factor relative to its effect on cancer development.

One good example is breast cancer. Estrogen plays a role during four phases of a woman's life: menarche, first pregnancy, menopause, and postmenopause (19). Several studies indicate that breast cancer risk is directly related to the cumulative number of ovulatory menstrual cycles (44, 58, 75). Intense exercise delays menarche, which can be thought of as favorable, because the risk of breast cancer increases twofold in women who experience menarche before age 13 versus at age 13 or older. Epidemiological data indicate that for every year menarche is delayed, breast cancer risk decreases by 5% to 15% (34, 48). Moderate levels of activity have been shown to increase the likelihood for anovulatory cycles threefold. Delayed menarche and anovulatory cycles decrease the woman's lifetime exposure to estrogen and progesterone (11, 36). The first full pregnancy induces differentiation of the breast and may change the sensitivity of the breast to both endogenous and exogenous risk factors (18, 78). And women who experience natural or artificially induced menopause before the age of 45 have a markedly reduced risk of breast cancer compared with women whose menopause occurs after the age of 55 (95). Among postmenopausal women, an increased body mass index is also a significant independent risk factor for breast cancer (45, 74), because fat tissue is the primary source of estrogen. Therefore, excess adiposity increases the woman's exposure to estrogen. Currently, it is not certain at which point or points in a woman's life cycle exercise exerts its greatest anticancer effect.

Another possible explanation for the role exercise plays in primary prevention of cancer is the potential effect on the immune system. Several excellent review articles discuss the effect of exercise on immune system function (46, 69, 88, 89, 101). The working hypothesis in the field of exercise immunology is the inverted-J hypothesis (see figure), which suggests that enhanced immune system function occurs with a chronic, moderate exercise training program, but that immune system function is depressed (even below that of sedentary individuals) after chronic exhaustive exercise. Clinically, moderate to high levels of physical activity seem to be associated with decreased incidence or mortality rates

for some cancers (71, 88, 94), and overtraining or intense competition may lead to immunosuppression as evidenced by an increased incidence of upper respiratory infections in people who train intensely (70, 76). To this point, there is no evidence that exercise directly increases the risk of any cancer. Indirectly, there is a higher incidence in skin cancer among individuals who exercise, although the cause of this association appears to be related to increased UV light exposure from active individuals spending more time outdoors (64). Table 22.2 describes the effects of acute and chronic bouts of exercise on the immune system. Moderate exercise seems to enhance the body's immune system, but the significance of this finding in preventing and treating cancer is unknown.

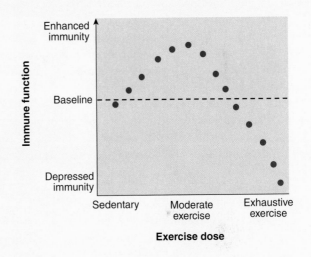

Table 22.3 Tissue Nodal Metastasis (TNM) Classification System for Breast Cancer

Tumor size (T)	Nodal involvement (N)	Metastasis (M)
Ts = in situ	NO = no nodal metastasis	MO = no distant metastasis
T1 = <2 cm	N1 = movable axillary nodes	M1 = distant metastasis
T2 = 2-5 cm	N2 = fixed axillary nodes	
T3 = >5 cm	N3 = internal mammary nodes	

Note: These TNM categories are combined to give the stage (e.g., stage 1 = T1 NO MO).

chemotherapy. Palliative surgery does not usually change overall chances for survival.

Radiation therapy is thought to stop the growth of malignant cells by damaging the DNA within the cell. Radiation can be applied either from implanted internal sources (brachytherapy) or from external machines. Most radiation therapy is applied in small fractions, usually between 180 and 250 rad/d. Doses above 4,000 rad over 4 wk increase the likelihood of developing radiation pericarditis. Radiation therapy can be used alone or in conjunction with surgery or chemotherapy. Radiation is good for a localized tumor that has not metastasized or for tumors that are difficult to reach with surgery (e.g., within the brain).

Cells that grow the fastest are best treated by chemical agents that interfere with cell replication. Known as **chemotherapy**, these agents frequently result in a cure. Some cancer cells, however, may become resistant to chemotherapeutic agents. Using several drugs at the same time (i.e., combination chemotherapy) is one way to minimize the resistance.

Biotherapy uses living organisms (found naturally or made synthetically) to treat the cancer. Cancer cells possess distinct surface protein antigens that are targets for antibody-directed or cell-mediated immunity. Immune therapy is a type of biotherapy that stimulates the immune response of the body to these protein antigens. Biotherapy also involves the production of antibodies outside the body, which are then administered to the cancer patient in an attempt to destroy the tumor. Biotherapy includes bone marrow transplantation and the use of cytokines such as interferon alpha and interleukin-5.

Targeted therapies, also known as precision medicine, take into consideration the variability of individual responses to cancer treatment and use genetic and molecular information about the tumor to formulate an individualized approach (35). An example of targeted therapy is the human epidermal growth factor receptor type 2 (HER2), which is found in approximately 20% to 25% of newly diagnosed breast cancers (13). Individuals who are found to be positive for the HER2 receptor are given the drug Herceptin, which binds to the HER2 receptor and leads to cell cycle arrest.

Hormone therapy is a common type of targeted therapy that deprives cancer cells of the hormones it needs to grow. Many types of breast and prostate cancers use hormone therapy to block either estrogen or testosterone, respectively. In the case of breast cancer, the drug tamoxifen is used in premenopausal women found to have tumors containing the estrogen receptor (ER). Tamoxifen exerts its effects by binding to ER receptors, thus reducing estrogen production. Similarly, Arimidex reduces estrogen levels, but does so by inhibiting aromatase, the enzyme that catalyzes the final step in estrogen production. Common side effects from hormone therapies include vasomotor symptoms (i.e., hot flashes), weight gain, bone and muscle loss, and increased CV risk.

Exercise Testing

Assessment of cardiorespiratory fitness, and thus use of exercise testing, is not routinely performed in the clinical management of cancer patients other than to determine the preoperative physiologic status or risk (i.e., operability) of patients with pulmonary malignancies. Formalized exercise testing guidelines have been issued by several national (e.g., American Thoracic Society/American College of Chest Physicians [ATS/ACCP]) and international (e.g., European Respiratory Society [ERS]) organizations, and the reader is referred to the ATS/ACCP recommendations for a comprehensive overview of exercise testing methodology for clinical populations (4). See also table 22.4.

Theoretically, a cancer diagnosis and the use of conventional and novel therapies may increase the risk of exercise testing–related complications, although the results of a recent systematic review suggest that maximal and submaximal exercise testing are relatively safe procedures in cancer survivors. The ATS and ACCP report that the risk of death and life-threatening complications during exercise testing is 2 to 5 per 100,000 tests (4).

It is not possible to screen or provide cancer-specific recommendations for all known cancer types; however, for certain forms of cancer, knowing which organ system or systems are involved may be immediately informative. For example, patients with lung cancer may be at higher risk for an adverse event, given that both the pathophysiology of the disease and concomitant comorbid conditions are often associated with a history of smoking (52, 53). These clients are considered higher risk, and referral to a physician or other allied health professional is required for electrocardiogram, exercise testing, and other tests, as appropriate. If such testing is unremarkable, clients are cleared for physical activity. If testing is remarkable, depending on the result, clients may be cleared for supervised exercise training in a rehabilitation program with experienced staff.

Preexercise Evaluation

An abundance of evidence shows exercise is a safe and effective intervention for patients with cancer (67). Therefore, outside the usual preparticipation screenings for exercise testing and training for patients free of cancer, the majority of cancer survivors do not need to seek extra medical clearance beyond current ACSM guidelines before participating in a moderate-intensity exercise program. However, for patients with metastatic disease or those currently undergoing cancer treatments, some additional considerations may require further medical clearance.

As part of the preexercise evaluation, the following cancer-specific history should be recorded:

- Cancer type
- TNM stage
- Surgery date(s) and type
- Treatment dates and type
- Cancer-specific medications (e.g., estrogen inhibitor, androgen inhibitor)
- Any new signs or symptoms
- Most recent complete blood count

Table 22.4   Summary of Exercise Testing for Patients With Cancer

Test type	Mode	Protocol specifics	Clinical measures	Clinical implications	Special considerations
Maximal cardiorespiratory	Treadmill Cycle ergometer Arm ergometer	Individualized protocols, based on physical activity history and comorbidities	Peak oxygen consumption METs Peak workload Heart rate Respiratory exchange ratio Treadmill time	Provides basis for determining starting point for exercise training. Used to stratify presurgical risk. Used to evaluate response to training program.	Cancer treatment may result in cardiomyopathy, pulmonary fibrosis, or neuropathy.
Submaximal cardiorespiratory	Cycle ergometer Walking path	6 min walk Constant workload	Distance Heart rate Perceived exertion	Used to evaluate response to training program.	Sensitivity to detect changes may be reduced in more fit individuals.
Muscle strength and endurance	Machine weights	One-repetition maximum (1RM) Multiple-repetition maximum	Kilograms	Provides basis for determining starting point for exercise training. Used to evaluate response to training program.	Modify or avoid 1RM tests with lymphedema or recent surgery.
Flexibility	Goniometry	Active stretching	Degrees	Used to evaluate response to training program.	Avoid pain; assess upper extremity range of motion postmastectomy.

The clinical exercise physiologist should attempt to identify any acute signs or symptoms that could be related to treatment toxicity and thus would require further medical consultation before participation in exercise. For example, fractures related to metastatic bone disease are a particular concern with exercise. Therefore, any complaints of bone tenderness—especially in the pelvis, back, or legs—should be conveyed to the oncologist for suspicion of metastatic lesions to bone.

During radiation therapy (RT), the areas proximal to the radiation treatment field (e.g., skin, lungs, bones) need special attention because the localized nature of RT may cause patients to experience tenderness or reduced flexibility. The presence of these symptoms, which can be due to radiation scarring, may necessitate a plan of light stretching and gentle movement. However, if bone was directly exposed to the radiation field, there may be an increased risk of fracture, so avoiding that area until a medical evaluation is completed would be warranted.

After cancer surgery, some exercises may need to be modified or excluded. An example would be restricting upper body exercises in women who have recently undergone breast surgery. However, even for this example, there can be a large variance (i.e., 1-8 wk) for when to begin or resume upper body exercises. This reinforces the need for good communication with the oncology team since factors such as surgery type (i.e., breast-sparing surgery vs. mastectomy

vs. mastectomy with reconstruction) and patient symptoms (e.g., pain, redness, swelling) will determine when you can introduce or progress certain types of exercise.

Contraindications to Exercise for Patients With Cancer

- Hemoglobin <10.0 g · dL^{-1}
- White blood cells <3,000 · mL^{-1}
- Neutrophil count <0.5 · 10^9 · mL^{-1}
- Platelet count <50 · 10^9 · mL^{-1}
- Fever >38 °C (100.4 °F)
- Unsteady gait (ataxia)
- Cachexia or loss of >35% of premorbid weight
- Limiting dyspnea with exertion
- Bone pain
- Severe nausea
- Extensive skeletal metastases

Other important findings that necessitate communication with the oncologist, or other medical providers, include febrile neutropenia, anemia, the presence of pain and swelling in an arm or leg that could be due to lymphedema, numbness in the extremities associated with neuropathy, the presence of an ostomy bag, and unusual bleeding, which can occur as a result of bone marrow suppression or surgery. Finally, nonspecific complaints of fatigue and weakness may require slow progression of exercise and further investigation if these symptoms progress worsen. All these factors must be considered before a patient begins a structured exercise program and continually reevaluated throughout the program. See the sidebar Contraindications to Exercise for Patients With Cancer.

EXERCISE PRESCRIPTION

In 2010 the American College of Sports Medicine published the first exercise guidelines for cancer survivors (81), emphasizing the importance of avoiding inactivity and developing a regular routine of exercise involving aerobic, resistance, and flexibility training (81). However, the evidence for specific exercise recommendations beyond those for healthy adult populations was limited at that time (81). In 2019 ACSM updated the exercise cancer guidelines (14) with some specific recommendations. Interestingly, these recom-

mendations are based mainly on common symptoms as opposed to specific cancer types. The following health outcomes are given a special exercise prescription:

- Anxiety
- Depressive symptoms
- Fatigue
- Health-related quality of life
- Lymphedema

In general, patients with cancer who are undergoing active treatment should try to achieve 30 min of moderately-intense (60% of heart rate reserve) aerobic exercise on 3 d/wk and 2 d/wk of resistance training (table 22.5). However, based on symptoms the patient presents with, modifications may be needed for this prescription. For example, a patient complaining of fatigue may need to start at lower intensities such as 40% to 50% of heart rate reserve. Conversely, patients who display symptoms of depression may benefit from a higher volume of exercise (i.e., ≥180 min/wk of aerobic exercise). Patients who have completed treatments or who do not have any accompanying symptoms may be able to start at higher intensities, such as 70% to 80% of heart rate reserve. In an excellent review, Maryl Winningham (100) discussed how to get a

Practical Application 22.2

CLIENT–CLINICIAN INTERACTION

Introducing exercise to patients during cancer treatment requires that you, as the clinical exercise professional, accomplish three things. First, become familiar with the type of therapy or therapies your patient is receiving. You may have to do some extra reading when you encounter a therapy you are unfamiliar with. Also, talk with oncologists and surgeons about the agents or interventions used to treat cancer. Such discussions should include the clinical presentation of expected side effects and the natural history of the disease. Ultimately, you will improve your ability to interact with patients in a learned fashion.

Second, develop your ability to ask questions about how well your patient is tolerating therapy and the disease. Understand that patients will exhibit various degrees of side effects to a particular treatment, ranging from severe to none. Further, if a patient tolerates the first few weeks of exercise during treatment, that will not always translate into future symptom-free exercise sessions. Patients often show nonlinear improvements in exercise during treatment, especially chemotherapy regimens. This will require you to titrate the frequency, intensity, time, or modality based on clinical signs and symptoms. The clinical features you should pay special attention to include sudden loss of exercise tolerance, increased shortness of breath with exertion,

bone pain, unusual change in vitals, and a sudden drop in nutritional intake. Obviously, you can take any concerns to the patient's attending oncologist or primary care physician. You play an important role in patient surveillance.

Third, your role as a clinical exercise professional is to help build exercise self-efficacy in your patient with cancer. This is done through a combination of choosing appropriate exercises and building a strong rapport. To accomplish the latter, it is important to connect with your patient through empathy (e.g., "I'm sorry to hear how much of a struggle this has been"), partnership building (e.g., "We are here to help support you and get you stronger"), and presenting a structured plan (e.g., "Based on your treatment and fitness level, here are some steps you can take to improve your fitness and health"). Your choice of exercise will be based on multiple factors such as age, current activity status, fitness level, comorbidities, cancer type, and type of treatment. A good general rule, especially for low-functioning patients, is to start low and go slow. Doing this can help build confidence in your patients while reducing the risk of injury or muscle soreness. In addition to these steps, having the social support of other cancer survivors in your exercise program can be an invaluable source of inspiration and motivation for clients who are just getting started.

patient with cancer started in an exercise program. When in doubt about where to start, the exercise professional should use the 50% rule: Ask the patient how far they can walk before becoming too tired, and start at half that distance or time.

The exercise program for patients with cancer does not typically require electrocardiographic monitoring, although some supervision and instruction about heart rate monitoring, proper exercise techniques, and cancer-specific exercises should be included. Initially, the exercise prescription should be reviewed with the patient, and the patient should be instructed about proper intensity and recognizing common adverse symptoms to exercise. See table 22.5 for a summary of the exercise prescription recommendations for patients with cancer.

EXERCISE TRAINING

Although the goal for all cancer patients is to progress gradually to 150 min/wk (or more) of moderate-intensity aerobic training, one must appreciate that the path to this goal is not always linear. As mentioned previously, there are many factors to consider when one is exercise training a cancer survivor. The exercise professional needs to anticipate setbacks such as side effects due to chemo-

therapy, radiation therapy, and additional surgical procedures. If setbacks occur, exercise goals need to be modified accordingly. At some point it may even be necessary to suspend exercise if the risk–benefit relationship no longer justifies exercise as a component of the treatment. Nevertheless, research has shown that larger volumes of aerobic exercise as well as combined aerobic and strength training exercise programs are feasible during chemotherapy and may result in additional benefits without interfering with chemotherapy completion (22).

Circumstances in which an exercise prescription may need modification include **neutropenia**, **thrombocytopenia**, anemia, neuropathy, lymphedema, and dehydration. Ideally, when working with patients undergoing chemotherapy, bloodwork including platelet and red and white blood counts should be checked on a regular basis to ensure patient safety (see the sidebar Contraindications to Exercise for Patients With Cancer). However, when lab reports are not accessible, it is important to monitor for clinical signs of the toxicities. For example, patients with chemotherapy-induced anemia can experience early fatigue, early-onset exercise dyspnea, and tachycardia resulting in exercise intolerance, as well as a higher heart rate response when exercising at a previously well-tolerated workload.

Table 22.5 Summary of Exercise Prescription for Patients With Cancer During Treatment

Training method	Frequency	Intensity	Time (Duration)	Type (Mode)	Progression	Important considerations
Cardiorespiratory	3 d/wk	60%-80% of heart rate reserve or oxygen uptake reserve, or RPE of 11-14	30 min	Walking, stationary bike, or other exercises that use large muscle groups	As tolerated, although linear progression is not always possible during treatment.	Adjust for the following symptoms: • Anxiety: Consider more vigorous intensity. • Depression: Consider longer duration (>30 min). • Fatigue: Reduce intensity (40%-60% HRR).
Resistance	2 d/wk	60%-75% 1RM	2-3 sets 8-15 reps	Body weight, free weights, weight machines, resistance bands		Patients with metastatic bone disease should avoid excessive weight-bearing exercises and seek medical approval because of increased fracture risk. Consider lower intensity for patients reporting fatigue.
Range of motion	Insufficient evidence beyond recommendations for healthy adults			Yoga has shown some benefit		With approval from surgeon, pay special attention to shoulder mobility stretches in breast cancer survivors.

Note: Guidelines for cancer survivors after treatment are the same as for healthy adults.

The Cost of Cancer Care

The economic impact of cancer is tremendous. In fact, the Agency for Healthcare Research and Quality (AHRQ) estimates that the direct medical cost of cancer in the United States in 2014 (the most recent data available) was $87.8 billion (104). Of this, 58% was for outpatient hospital visits, 27% was for inpatient hospital stays, and 12.4% was for prescription medications (1). To put this into perspective, the median monthly costs of cancer drugs have risen from less than $100 in 1965-69 to more than $10,000 in 2020, with some as high as $30,000 per month (6).

The rising cost of cancer care has a negative impact on many stakeholders involved in the health care system, including insurance providers, hospitals, and patients, making cancer a financial burden for a number of Americans. The term *financial toxicity* has been used to describe this growing concern, as medical costs are the leading cause of personal bankruptcy (33). In fact, people living with cancer are three times more likely to file for bankruptcy than those without cancer (105). Patients are sometimes forced to choose between cancer treatment and paying for food, shelter, and other necessities. In one study of 164 patients, 45% reported cost-related medication nonadherence (7). One-quarter of patients with insurance reported they had used up all or most of their savings to deal with cancer (9).

The escalating cost of health care also affects health insurance companies, who have responded by decreasing the utilization of services and increasing patient financial responsibility through larger co-pays and high deductibles. Finally, the rising cost of treatment influences care providers, whose desire is to provide treatment of the greatest benefit, without regard for cost.

In light of this growing demand to reduce health care costs, high-quality, evidence-based cancer care is vital. To deliver the highest value, care must be patient centered, integrated, and coordinated, achieving the most meaningful outcomes at a sustainable cost (92). Individualized exercise prescription falls in line with an integrated system of medical care, and the present data indicate it would increase cost savings for patients, payers, and providers alike. However, nationally, only 2% of patients are ever referred to oncology rehabilitation services (61). Thus, addressing this gap in care is paramount.

Research has found that exercise during cancer treatment has a significant impact on medical costs. Data revealed a 6% decrease in inpatient hospital stays, a 19% decrease in length of hospital stays (more than one full day shorter), a 27% decrease in emergency room visits, and a 47% decrease in 30-day readmissions. On average, patients saw a reduction of $2,800 in medical expenses in their first 6 mo of participation in a supervised, individualized exercise program. When these benefits are considered in addition to the physiological benefits of exercise on a cancer diagnosis, it becomes clear that exercise should be an integral part of the standard of care (102).

Just as important as monitoring for toxicities related to cancer therapies is anticipating when they are likely to occur. For example, if a patient is on a chemotherapy regimen that causes neutropenia, instructing them to exercise at home—staying away from fitness centers and "gym germs" for at least a week following chemotherapy treatment—may help reduce their chance of contracting an opportunistic infection.

Some patients have cancers that involve the bones, in which case exercise of the affected extremity may cause pain or result in injury. Patients with metastatic disease to the pelvis or legs should avoid high-impact exercises and may benefit from an exercise regimen that allows them to sit down, such as stationary cycling or chair exercises. If a patient reports new-onset bone pain, this could be a sign of metastatic disease, and exercise should be stopped immediately until physician approval is given.

Since most cancer patients receive surgery, special considerations are also needed for the postoperative period (typically 4-6 wk). Gentle stretching and stationary cycle activities are frequently used for patients who are recovering from breast or thoracic surgery so they can maintain joint-specific range of motion and lower body conditioning while avoiding upper body exercise.

Another common treatment, especially in breast and prostate cancer patients, is hormone-directed therapy (e.g., aromatase inhibitors inhibit estrogen production in postmenopausal breast cancer patients who overexpress the estrogen receptor [ER]). Although these treatments have been shown to improve survival, they are also associated with bone and muscle loss as well as central adiposity. Because of these side effects, an exercise prescription that addresses weight management, bone health, and muscular strength is warranted.

Resistance training to maintain or enhance muscle strength is important not only for patients undergoing hormonal therapy but for most cancer patients as well (37). Individuals who are undergoing treatment or are unfamiliar with resistance training should start with resistance bands, light hand weights, or machines (or more than one of these) to avoid any potential for bruising or bone fractures. The recommendations are similar to those for healthy adults—one or two sets of 8 to 12 repetitions, 2 or 3 d/wk. In addition, special precautions are needed for patients with (or at risk for) lymphedema. The current recommendations are to begin a supervised program with very light resistance and progress resistance gradually after two consecutive pain-free sessions. As long as the patient tolerates

TREATMENT-RELATED SIDE EFFECTS: FATIGUE AND LYMPHEDEMA

Cancer-related fatigue (CRF) is the most common side effect in patients with cancer (8). It is defined as sustained physical, emotional, or cognitive exhaustion experienced during and after adjuvant cancer treatments that is not proportional to recent activities and interferes with usual functioning (8). Because CRF is frequently debilitating, it is also associated with reduced compliance to medical treatments, which may increase the risk of recurrence (77). Cancer-related fatigue is experienced in >70% of patients undergoing chemotherapy and persists in 30% of individuals 5 yr or more from the end of cancer treatments (30).

Regular exercise counteracts many of the potential causes of fatigue and is recommended to mitigate this common condition both during and after treatment (15, 21, 23, 29,30, 31, 43, 63, 83, 93, 96). Improvements in fitness may partially explain the benefits of exercise (26), although the multifactorial nature of fatigue makes it difficult to separate (see figure). Additionally, not all exercise studies have found a beneficial effect on fatigue in this population (25, 28). This discordance may be related to the method used to develop the exercise prescription or the timing of when a fatigue questionnaire was administered. For example, under the backdrop of dramatic changes that can occur with some treatments (e.g., hair loss, anemia, anxiety), the impact of exercise on fatigue may at times be masked during treatment compared with after treatment.

Another aspect of the elusiveness in fatigue-related research is the fact that fatigue is mainly a self-reported condition with no clear biomarkers. Research has examined markers of chronic inflammation as either biochemical signals of fatigue or the underlying cause of the fatigue itself (20, 82, 85, 91). Patients with cancer exhibit higher than average inflammatory markers due to local immune responses to the neoplasm, tissue injury from adjuvant treatments (i.e., surgery, chemotherapy, radiotherapy), and the neoplasm itself (27). Much attention has focused on the cytokines released during the inflammation response as chemical mediators of fatigue. Interleukin-6 and tumor necrosis factor-α, in particular, are ubiquitous proinflammatory cytokines that are elevated in patients with cancer.

Interestingly, both prospective and cross-sectional studies have found inverse relationships between regular exercise and proinflammatory cytokines in other chronic disease populations (e.g., people with heart failure, those who are elderly or obese) (38, 40, 57, 68). Although this has yet to be shown in CRF, it does provide a theoretical explanation as to how exercise may alleviate this condition.

Another side effect shown to be improved by regular exercise is lymphedema (79, 80). Lymphedema is a condition in which the lymphatic system is blocked or damaged, resulting in fluid buildup and edema in the affected extremity. Historically, patients who had multiple lymph nodes removed as part of a treatment for their cancer were discouraged from exercising the affected limb for fear that exercise (specifically weight training) might induce or exacerbate lymphedema. However, across many exercise studies involving different modalities (i.e., boat racing, yoga, and weight training), exercise appears safe in individuals who are at risk for lymphedema (20, 79,). In fact, a large, randomized trial involving a cohort of breast cancer survivors with stable lymphedema found that those who performed resistance training twice per week had a 50% reduction in

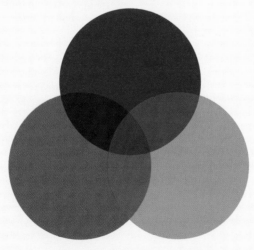

Tumor-related fatigue
Treatment
Medications
Proinflammatory cytokines

Disease-related fatigue
Anemia
Old age
Metabolic abnormalities
Heart disease
Pulmonary disease
Depression

Conditional-related fatigue
Decreased functional capacity
Nutritional deficiency
Dehydration
Sleep deprivation
Infection
Psychological

(continued)

Practical Application 22.3 (continued)

the incidence of lymphedema exacerbations (80). Additionally, a follow-up study showed that resistance training in breast cancer patients who were free of lymphedema after having five or more lymph nodes removed resulted in a 15% lower risk of developing lymphedema in the future (79).

Chemotherapy-induced cardiomyopathy is a serious side effect for which exercise may potentially be protective. Doxorubicin (Adriamycin) is a well-documented chemotherapy agent known to have dose-dependent cardiotoxic effects (84). Although this has not been well studied in humans, animal models have provided preliminary evidence showing that exercise can attenuate doxorubicin-induced myocyte damage (see the sidebar Research Focus). Likewise, the drug trastuzumab (Herceptin) is a very important chemotherapy drug shown to have cardiotoxic effects. In fact, a study involving more than 8,400 patients with breast cancer found the risk of heart failure increases to 20% for patients on both doxorubicin and trastuzumab (13). See Scott and colleagues for a comprehensive review on exercise and cardiotoxicity (84).

---

the strength training program, there is no upper limit for resistance.

As your patient's clinical exercise physiologist, you should be sure to keep in mind the following:

- Many patients with cancer are inactive and experience mood disturbances such as anxiety and depression that are themselves associated with little interest in exercise. Be sure to increase exercise dose progressively in an individualized manner for every patient.

- For the first week or so of exercise training, patients may be a bit tired later in the day. To compensate for this, ask them to limit the amount of other, home-based activities for the first few days.

- Pool activities are not advised for patients with intravenous catheters or those who are receiving radiation therapy.

- A variety of treatment-related obligations and complications often interrupt exercise therapy for patients with cancer. Before patients even start an exercise regimen, inform them that setbacks and interruptions are not uncommon. Instead of not exercising at all or stopping the exercise program, patients should plan around interruptions and continue to adhere to their programs whenever possible.

- Learn about the different types of treatments your patients are receiving.

- Consider developing an exercise buddy system, which matches up cancer survivors who are already exercising with patients who are just starting out. Such group support from patients with similar medical problems may help improve short-term adherence.

- Ask patients daily how they are feeling and what barriers to exercise they may be experiencing.

- When in doubt regarding the safety of a specific exercise, consult with the patient's supervising physician.

Chemotherapy-Induced Peripheral Neuropathy

Chemotherapy-induced peripheral neuropathy (CIPN) is a common side effect that can increase patient anxiety levels as well as negatively affect exercise adherence. Often associated with taxane, vinca alkaloids, or platinum drugs, CIPN causes common symptoms that include pain or numbness in hands and feet, progressing to radiating pain in the limbs (71). Although little evidence is available regarding the use of exercise to prevent or improve CIPN, a study by Kleckner and colleagues showed that exercise did reduce the severity of patient-reported CIPN symptoms in cancer patients receiving taxane-, platinum-, or vinca alkaloid–based chemotherapy (55). The exercise intervention used in this study was a combination of home walking and resistance band exercises modeled after ACSM's Exercise for Cancer Patients (EXCAP) program.

As with diabetic neuropathy, having CIPN does not preclude participation in exercise. The use of non-weight-bearing exercises (e.g., cycle or recumbent stepper) can help with patient tolerance as well as reducing the risk of falling (due to sensory loss). We have also found that patients enjoy performing supervised balance exercises using various stability balls and coordination drills (e.g., catching a ball while standing on one leg). Although it is not always easy to quantify subtle improvements in balance, the use of these exercises as a distraction may also provide a therapeutic benefit. Finally, in our experience, the symptoms of CIPN improve in most patients the further removed they are from chemotherapy. Assuring patients that this condition often resolves over time can also help reduce anxiety.

Practical Application 22.4

EXERCISE AND CANCER RECURRENCE

Of all the studies involving exercise and cancer, observational studies linking increased regular exercise to reduction of cancer recurrence and improved survival have perhaps yielded some of the most intriguing results (35, 47, 49, 50, 51, 62). To date, preliminary data have demonstrated a potential mortality benefit from exercise in patients with existing breast, colon, prostate, ovarian, brain, and lung cancers; the majority of these studies are on breast cancer (35). Specifically, across 26 studies, cancer patients who exercised had a 37% lower risk of dying from cancer compared with those patients who did not exercise (HR = 0.63; 95% CI: 0.54-0.73). Even more intriguing, there is preliminary evidence that the association between exercise and cancer mortality may vary by specific molecular or genetic markers (i.e., perhaps a so-called precision medicine approach to exercise oncology).

In two early landmark studies, Holmes and colleagues and Meyerhardt and colleagues reported reductions in cancer recurrence rates, as well as mortality, for individuals with breast and colon cancer who were the most active following diagnosis. Holmes and colleagues reported, in patients after breast cancer treatment (stage I-III) who averaged over 9 MET-h per week (the equivalent of walking briskly for 30 min, 5 d/wk), an average 50% reduction in recurrence compared with those who were sedentary (47). The incidence of recurrence was similar in the most active patients in Meyerhardt's colon cancer study; however, the volume threshold for these results was double, at 18 MET-h per week (62). Both studies controlled for initial activity level, allowing the investigators to compare the increases in physical activity after time of prognosis.

These results suggest that exercise might prevent recurrence and early death in patients with breast and colon cancer; but because the studies were observational, more research is needed before a definitive answer can be obtained. In the first randomized controlled trial to address this question, Courneya and colleagues (24) reported on an exploratory follow-up of cancer outcomes from the Supervised Trial of Aerobic versus Resistance Training (START), which randomized 242 breast cancer patients to usual care ($n = 82$), supervised aerobic exercise ($n = 78$), or resistance exercise ($n = 82$) during chemotherapy. After a median follow-up of almost 8 yr, the researchers found that disease-free survival (DFS) was 83% for the exercise groups compared with 76% for the control group. Slightly stronger effects were observed for overall survival. Subgroup analyses suggested potentially stronger exercise effects on DFS for women who were overweight or obese, had stage II or III cancer, had ER-positive tumors, had HER2-positive tumors, received taxane-based chemotherapies, and completed ≥85% of their planned chemotherapy. Although preliminary, START provided the first randomized data to suggest that adding exercise to standard chemotherapy may improve long-term breast cancer outcomes.

For health professionals and researchers to incorporate regular exercise as a treatment to reduce cancer recurrence, it is important to identify the mechanism through which exercise acts to exert its effect. There are many plausible theories, including favorable changes in body weight and sex hormones (e.g., estrogen) and effects on various growth factors (e.g., insulin-like growth factor-I), all of which if not mitigated are otherwise associated with increased risk for mortality (49). Additionally, exercise may influence the tumor itself through biological pathways not yet identified. Another potential mechanism by which exercise may mediate cancer-free survivorship is through improved tolerance to treatment. The more a patient is able to tolerate a particular treatment, the greater the log kill (i.e., cancer cell destruction). Courneya and colleagues reported that patients undergoing treatment for breast cancer who performed resistance training were able to tolerate higher doses of chemotherapy (24). Similarly, a Dutch trial (98) showed an improved chemotherapy completion rate with exercise training, suggesting it may be one of the most important clinical benefits of exercise for breast cancer patients receiving chemotherapy.

Research Focus

Might Exercise Someday Be Used to Limit or Prevent the Harmful Effect of Certain Anticancer Agents on the Heart?

Many cancer survivors who complete treatment unfortunately succumb to cardiovascular disease. When compared with cancer-free individuals, cancer survivors have a 1.7 to 18.5 times likelihood of having CV risk factors and a 1.3 to 3.6 times greater likelihood of dying of cardiovascular disease (39). The multiple hit theory attributes the increased risk of CV incidence and mortality to three factors: baseline risk from advanced age and other common risk factors shared between cancer and heart disease (e.g., diabetes), direct injury to myocardial cells or coronary vessels due to the treatment (e.g., direct radiation over the heart), and indirect injury because adverse effects of the treatment led to weight gain, reduced physical activity, or both (39). Exercise training may play an increasing role in reducing these factors, especially as cancer-specific mortality continues to improve in conjunction with a growing, aging cancer population.

Although evidence that exercise attenuates the indirect causes of increased CV risk (e.g., fatigue and depression) continues to grow, there is limited research examining the effects of exercise on indirect causes.

One promising area is the effect of exercise to counteract cardiotoxic drugs, particularly trastuzumab and anthracycline-containing agents (e.g., doxorubicin). Some basic animal research has yielded positive results. Chicco and colleagues examined the cardioprotective effects of a low-intensity exercise training program (20 min/d, 5 d/wk) in Sprague-Dawley rats exposed to doxorubicin (DOX) (17). Compared with a nonex-ercise DOX group, the exercise DOX group displayed preserved left ventricular function. Other similar animal studies have also found exercise training to attenuate the cardiotoxic effects of doxorubicin. At Henry Ford Hospital, a randomized pilot study looked at the effects of exercise in patients on doxorubicin or trastuzumab. Although the exercise group did show improvements in peak $\dot{V}O_2$ and some markers of cardiotoxicity (i.e., left ventricular strain pattern and troponin), these changes (with the exception of $\dot{V}O_2$) were not statistically different from the control group (54).

Although not a primary end point, the PACES trial found individuals randomized to the supervised exercise group had a lower rate of delay or discontinuation of trastuzumab because of reduced left ventricular ejection fraction than individuals in the nonexercise control group (6% versus 28%), suggesting a potential benefit from exercise (98). Scott and colleague developed a construct outlining the possible mechanisms for how exercise could counteract cardiotoxic drugs. (84). These include reduced oxidative stress, protection from apoptosis, improved protein synthesis, favorable myocyte calcium handling, and increased adenosine monophosphate (AMP)-activated protein kinase pathway activity. Identifying the potential mechanisms and performing translational studies in humans are the next steps in determining the applicability of exercise training as a therapy against cardiotoxic chemotherapy. If this is established, it may result in patients' receiving a greater therapeutic dose of these anticancer drugs without greater risk of developing heart failure.

Clinical Exercise Bottom Line

- Cancer, or neoplasm, is an abnormal or unregulated growth of tissue (due almost always to cellular genetic mutation) that shows partial or complete lack of structural organization, lacks coordination with normal tissue, and usually forms a mass that is benign or malignant.

- With cancer, the cellular DNA of normal stem cells is damaged, usually caused by environmental factors, heredity, oncogenes, hormones, or impaired immune function.

- The four types of cancer are carcinomas (e.g., skin, digestive tract), sarcomas (connective tissue such as bones and muscles), leukemias, and lymphomas.

- Early on, symptoms for cancer are quite nonspecific and may include weight loss, fatigue, nausea, and malaise.

- If cancer in a tissue is suspected, the first diagnostic step is to obtain a sample (i.e., biopsy). The second step is to determine the extent of spread of the cancer, known as staging.

- The five broad treatment types for cancer include surgery (curative or palliative), radiation therapy, chemotherapy, biotherapy, and targeted therapy.

- Patients with cancer may experience fatigue, mood disorders (anxiety, depression), and a sense of loss of control. Cancer-related fatigue may be due to the cancer itself or to the radiation or chemotherapy used to treat the disorder.

- Regular aerobic-type exercise and resistance training exercise can attenuate and partially reverse cancer-related fatigue, improve muscular strength and endurance, and lessen anxiety.

- Treatment-related setbacks (e.g., neutropenia, anemia, neuropathy, lymphedema) are common in patients with cancer, and the clinical exercise physiologist needs to be able modify or discontinue the exercise regimen accordingly.

CONCLUSION

Cancer is a constellation of diseases. It can begin in any organ and spread to other organ systems. Treatment includes surgery, radiation therapy, chemotherapy, and, more recently, receptor-targeted drug therapy. Both the disease and the treatment bring emotional and physical challenges to patients with cancer. Exercise benefits these patients primarily through improving (or maintaining) function, reducing fatigue, and countering some of the side effects of cancer-specific therapy. The end result of many of these benefits is an improved quality of life. Additionally, good preliminary evidence is beginning to mount regarding the impact of activity and exercise on survival after being diagnosed with cancer.

Clinical exercise physiologists are often underused relative to assisting patients with cancer; this is analogous to efforts to integrate cardiac rehabilitation programs into the care of heart patients some 30 yr ago. Just as important as educating the prospective patient about exercise is reaching out to oncologists and primary care physicians to communicate the evidence-based benefits of exercise.

Online Materials

Visit HK*Propel* to access a link to the references, the case studies with discussion questions, and a quiz to help you review key concepts and test your understanding of the material covered in this chapter.

CHAPTER 23

Human Immunodeficiency Virus

Vitor H.F. De Oliveira, PhD

Christine Horvat Davey, PhD, RN

Allison R. Webel, PhD, RN

Human immunodeficiency virus (**HIV**), the virus that causes acquired immunodeficiency syndrome (AIDS), was first recognized by the U.S. Centers for Disease Control and Prevention (CDC) in the early 1980s (10). Since that time, HIV has become a leading cause of morbidity and mortality worldwide (27, 128). The global impact of HIV-related infections on social and economic conditions has been substantial (128). Patients living with HIV have higher levels of morbidity and mortality than the general population (22, 128).

Without treatment, in most persons, HIV infection causes a progressive depletion of immune cells and decline in immune function. As untreated HIV disease progresses, people can become increasingly debilitated and prone to opportunistic infections (e.g., pneumocystis pneumonia, cryptosporidiosis), wasting, and other HIV-associated complications. Today, much of the global population living with HIV is on effective HIV medications (**antiretroviral therapy [ART]**). Although there has been a resulting decline in AIDS, people living with HIV (PLWH) are increasingly susceptible to noncommunicable chronic diseases (e.g., cardiovascular disease, cancer, chronic kidney disease). Over time, physical activity becomes increasingly difficult, resulting in a significant loss of physical functioning. Exercise training attenuates the progression of HIV disease as well as complications arising from the current pharmacological treatments. This chapter provides a review of the disease and summarizes the evidence for the use of exercise as a complementary modality in the treatment of HIV infection.

DEFINITION

In 1982, the CDC reported on a cluster of Kaposi's sarcoma and *Pneumocystis carinii* pneumonia among young, previously healthy men in Los Angeles and Orange County, California. These were the earliest cases of what was to become the AIDS pandemic in the United States (10). The cause of this disease was found to be a novel retrovirus, the HIV. That same year, the first case definition was published by the CDC. Since then the case definition of AIDS has evolved, reflecting an enhanced understanding of the disease progression as well as the clinical manifestations of HIV infection.

HIV infection results from the transfer of bodily fluids (e.g., blood, semen, vaginal secretions, breast milk) from an infected individual to an uninfected person. Once chronic infection is established, a complex array of pathogenic mechanisms leads to an increase in viral load and a concomitant decrement in CD4 lymphocytes below the normal range of 500 to 1,400 cells $\cdot$ mm^{-3}. With the scale-up of ART worldwide, survival has improved significantly. Today, people with well-controlled HIV live approximately the same life span as HIV-uninfected individuals (62, 63).

SCOPE

In 2020 the Joint United Nations Programme on HIV/Aids (UNAIDS) estimated that over 38 million people worldwide are living with HIV. The majority are in Africa, with the remainder scattered throughout the world (113, 114, 128). Approximately 1.1

Acknowledgment: Much of the writing in this chapter was adapted from the previous editions of *Clinical Exercise Physiology.* Thus, we wish to gratefully acknowledge the previous efforts of and thank Mansueto Neto, PhD, Edward Archer, PhD, MS, Helmut Albrecht, MD, and Gregory A. Hand, PhD, MPH.

million reside in the United States, more than 2 million in Europe, and about 27,000 in Australia (11a). The UNAIDS estimates that in 2019 about 17 million people were newly infected with HIV, and there were about 690,000 HIV-related deaths. The CDC estimates that in the United States approximately 13% of all PLWH are unaware of their infection (38a).

The demographic characteristics of those newly infected with HIV have changed since the early period of the epidemic. Initially mostly an epidemic among white men who have sex with men, HIV infection now disproportionately affects individuals of low socioeconomic status as well as Black/African American and Hispanic/Latino populations. A detailed description of the demographic characteristics of the HIV/AIDS epidemic is available online at the CDC website, www.cdc.gov. For information on the global epidemic, access the UNAIDS website at www.unaids.org.

PATHOPHYSIOLOGY

HIV infection results from the transfer of infected bodily fluids (e.g., blood, semen, vaginal secretions, breast milk) from an infected individual to an uninfected person. The three primary routes of infection are unprotected penetrative sexual contact, intravenous drug use, and to a lesser degree perinatal exposure (i.e., in utero, during labor, or while breastfeeding).

The time course of HIV infection varies, but it often begins with a primary infection syndrome that usually presents as a moderate to severe case of influenza with associated signs and symptoms such as fever, sore throat, fatigue, lymphadenopathy, rash, myalgia, malaise, and oral or esophageal sores or both. This seroconversion illness (i.e., the time of unchecked viremia before the development of HIV antibodies) may last from a few days to several weeks. During this phase, rapid viral replication leads to a significant increase in HIV viral load concentration. This increase in viral load is often accompanied by a decline in CD4+ T-cell counts, resulting in a decline of cell-mediated immunity. This early CD4+ T-cell decrease is usually reversible with effective ART. It is important to remember that not all individuals with HIV experience primary infection, and some can live many years without symptoms or knowing their HIV status.

Body Composition Alterations

Early in the epidemic (i.e., the 1980s), wasting was a common life-threatening complication of HIV infection (50). Wasting is an unintentional decrease of >5% of body weight that induces a loss of function and weakness via a decrement in protein stores (e.g., lean body mass) and has a significant impact on mortality. As a result of ART, wasting is less of a concern today and will likely be observed only in untreated PLWH (104). Another common complication of ART in PLWH was lipodystrophy, a metabolic effect of some antiretroviral medications (e.g., protease inhibitors) and to a lesser degree of HIV itself, resulting in a loss of subcutaneous fat depots in the arms, legs, and face, with a concomitant increase

in visceral fat. Lipodystrophy is associated with increased risk of cardiometabolic complications, and patients with lipodystrophy demonstrated poorer aerobic capacity compared with PLWH without lipodystrophy (31). Thanks to the lower toxicity of new ART agents, lipodystrophy is less common, but it can still be observed in treated patients who have been living with HIV for many years.

Despite the lower prevalence of wasting and lipodystrophy, body composition changes are still common in PLWH. Globally, an increasing number of PLWH are obese and experiencing the increased burden of metabolic diseases (3). Sarcopenia, a progressive and generalized skeletal muscle disorder, is defined as a loss of both muscle mass and function, and it is more prevalent in PLWH than in people without HIV (76). Sarcopenia is associated with a greater probability of adverse outcomes (i.e., falls, fractures, physical disability, and mortality) and it is prevalent in PLWH, related to both HIV (e.g., longer exposure to specific ART agents) and non-HIV (i.e., aging) risk factors.

Metabolic Complications

Metabolic complications are among the most common problems associated with HIV infection and ART. These metabolic changes confer an increased risk of cardiovascular diseases similar to that of metabolic syndrome and include alterations in glucose and lipid metabolism, insulin resistance, diabetes mellitus, dyslipidemia, lipoatrophy, and lipodystrophy syndrome (129). Certain ART classes (e.g., integrase inhibitors, protease inhibitors) favor the occurrence of multiple abnormalities, including dyslipidemia, insulin resistance, visceral fat accumulation, and metabolic syndrome, which are associated with an increased risk of premature atherosclerosis and myocardial infarction (129).

Cardiovascular Disease

HIV infection is associated with a higher prevalence of cardiovascular disease (CVD), including atherosclerosis, heart failure, stroke, pulmonary arterial hypertension, and peripheral arterial disease (25, 81). Aging, ART use, chronic inflammation, and several other factors contribute to the increased CVD risk noted in PLWH (2). Although adherence to ART and the resulting HIV viral suppression in the bloodstream lead to reduced cardiovascular events in PLWH, there is still residual inflammation, which increases CVD risk (112). This residual inflammation in PLWH results from HIV persistence in cellular and anatomical reservoirs during ART; although HIV is suppressed in the bloodstream, the patient presents with ongoing viral replication and proliferation of latently infected cells, leading to a persistent low level of immune activation (64). HIV infection reduces HDL cholesterol and raises triglycerides, total cholesterol, and vascular inflammation. The prevalence of HIV and the growing prevalence of these CVD risk factors will contribute to an increasing incidence of HIV-associated CVD.

The increase in CVD is hypothesized to be the result of an increased prevalence of traditional CVD risk factors (e.g., smoking, physical inactivity, obesity) in PLWH as well as chronic inflammation, direct viral effects, and factors associated with ART (2). Increased physical activity and exercise training improve these effects (77, 86, 125). Given that CVD is an important contributor to morbidity and mortality in PLWH, its early detection is essential for effective treatment (69). Lifestyle changes (i.e., healthy diet, smoking cessation, and daily physical exercise) reduce the probability of a coronary event by up to 80% in the general population. Thus, dietary modifications and exercise are established interventions to reduce metabolic changes and the relevant CVD risk (74).

CLINICAL CONSIDERATIONS

An increasing number of health care professionals are recommending exercise as a complementary modality to prevent disease and mitigate the signs and symptoms associated with chronic disease states (68, 105). Nonpharmacologic interventions such as exercise training and nutritional behavioral counseling have the potential to improve patient outcomes via the enhancement of health and the augmentation of more traditional medical treatments (123). Complementary modalities such as exercise are low-cost and efficacious additions to the management of chronic diseases. There is evidence that participating in exercise programs might reduce cardiometabolic risk factors and improve quality of life in PLWH (30, 74).

The long-term goals of exercise interventions for PLWH are to decrease morbidity and mortality and to improve quality of life (35, 36, 73, 131). These aims are accomplished by exercise interventions tailored to address the patient's signs and symptoms. The more proximal goals of exercise training are increases in cardiorespiratory fitness (CRF), decreases in CVD risk factors, improvements in body composition and metabolic functioning, and the enhancement of musculoskeletal function. These objectives are most effectively achieved via exercise prescriptions that include both progressive resistance and aerobic training. These modalities improve people's capacity for activities of daily living as well as enhance their ability to remain as physically and mentally active as possible. A systematic review of the effects of different types of exercise on health in PLWH concluded that resistance training, aerobic exercise, and combined aerobic and resistance training are associated with improvements in body composition, muscle strength, and cardiorespiratory fitness (30).

Starting exercise training has been shown to benefit the physiological and psychological functioning of PLWH. Exercise improves mood, quality of life, aerobic fitness, and immune indices and decreases the incidence of comorbid chronic disease states in this population (16, 17, 35, 53, 59, 73, 76, 91, 110). Additionally, exercise significantly increases strength, functionality, and endurance; improves body composition and lipid profiles; decreases resting heart rate; and reverses the nervous system disorders comorbid with HIV infection (24, 26, 73, 111). Thus, HIV infection should be an impetus for the initiation of an exercise program and not a deterrent.

Signs and Symptoms

The clinical exercise professional must be aware that although PLWH will often present as completely asymptomatic, they can also be critically ill. Patterns of physiological, behavioral, environmental, and psychological changes are **idiosyncratic**. Therefore, all patients must be evaluated on an individual basis to assess the effects and development of the disease.

Many patients will develop symptoms of viral illness (e.g., fever, pharyngitis, malaise, myalgias, rash) within 1 mo of HIV seroconversion. The duration of these symptoms varies, with an average of 2 to 4 wk (47). This initial immune response reduces the number of viral particles and marks the advent of the latency period during which infected individuals may experience few symptoms. Latency may last for a few weeks to more than two decades.

Most PLWH seen in a health care setting are on effective ART and will have a suppressed HIV viral load. Thus, the most common signs and symptoms are lipodystrophy, impaired glucose and lipid metabolism, impaired muscle function, and increased CVD risk factors (e.g., obesity). CRF levels of PLWH are among the lowest in comparison with other vulnerable populations. In addition, PLWH in general (124), but in particular those over 50 yr of age and with a higher body mass index, should be considered a high-risk group for low CRF, an important predictor of CVD and premature mortality (119). Fortunately, these manifestations of pharmacological treatment are amenable to exercise training and other changes in lifestyle (e.g., nutrition).

History and Physical Examination

A detailed history combined with general and musculoskeletal examinations will identify individuals with significant changes in body weight or other symptoms of metabolic disorders (e.g., sarcopenia, hypertension), motor abnormalities (e.g., hyperreflexia, loss of equilibrioception), and indicators of CVD (e.g., arrhythmias, edema). Chapter 4 provides general medical history and examination information. The presence of motor disturbances requires vigilance with all subsequent exercise testing. The primary signs to be aware of include immune status (HIV viral load, CD4+ T-cell counts) and functional capacity and fitness. Patients may present with alterations of body composition (e.g., lipodystrophy), changes in body weight (e.g., obesity), loss of skeletal muscle mass, visceral adiposity, and low muscular strength as well as low CRF (87). Strength in the legs and arms should be quantitatively examined, as well as functional strength (e.g., the ability to rise from a chair).

Peripheral neuropathy is a common complication of both HIV and ART regimens. Therefore, both sensory and functional motor testing should be performed. Numerous behavioral and psychological symptoms may also be present, such as fatigue, malaise, depression, anxiety, social isolation, and a decreased quality of life (46, 61, 125). Of particular concern are the reduced levels of physical activity and exercise in PLWH, together with high levels of sedentary behavior (117, 118, 119). High levels of sedentary living combined with low levels of CRF increase the risk of CVD and exacerbate the metabolic complications associated with pharmacological treatments. Thus, the improvement of CRF is a feasible and effective strategy for reducing cardiometabolic risk among PLWH (119).

Diagnostic Testing

The purpose of HIV testing is to ascertain the presence of HIV infection via the presence of antibodies, viral RNA, or both. Modern HIV testing is highly accurate (13). HIV testing has evolved through four generations (classified by the antigen used); the first commercially available immunoassay became available in 1985. Since that time, each generation has improved the specificity (i.e., the percentage of individuals correctly identified as HIV+) or the sensitivity (the percentage of individuals correctly classified as not HIV+) of HIV testing or both. Although some individuals with acute primary HIV infection are asymptomatic, many patients develop symptoms of viral illness (e.g., fever, sore throat, malaise, myalgia, rash). Because they are nonspecific, these symptoms may indicate a wide range of conditions. Therefore, the diagnosis of acute HIV infection requires a thorough assessment of HIV exposure risk and appropriate HIV-related laboratory tests. Individuals with acute infection may have elevated viral loads and be unaware of their infection, and they are therefore more likely to transmit the virus to their sexual partners (94, 122). The ability to adequately and effectively discuss HIV-related issues with patients is important. Practical application 23.1 provides information regarding client and clinician interactions.

Exercise Testing

Exercise training is safe for most PLWH and should be recommended (48). There is consistent and strong evidence that exercise can lead to improvements in CVD risk factors, body composition, muscle function, psychological symptoms, and other adverse outcomes associated with HIV infection (71, 74). Nevertheless, PLWH should undergo a medical evaluation before beginning an exercise program. Patients participating in an exercise intervention should be monitored by a qualified professional health care provider for potential changes in their health status, especially those in more advanced stages of immunosuppression and those unaccustomed to physical exertion, to prevent any potential adverse events of exercise (72).

The detection of conditions that contraindicate exercise is of paramount importance in individuals who have been living with HIV for many years. HIV infection is associated with a condition known as premature or accelerated aging, and PLWH present with major comorbidities up to 16 yr earlier than people without HIV (62, 63, 106). As a result, a number of orthopedic and rheumatic complications and pathologies associated with HIV and ART may preclude exercise training (28, 41, 66, 109, 120). Although screening of asymptomatic patients is not recommended, patients over the age of 50 yr and those experiencing chronic and debilitating joint pain should be evaluated.

Conditions such as extreme fatigue, fever, or chills may indicate an active infection. Those symptoms may be observed in patients during seroconversion or in immunocompromised PLWH not on consistent ART, and exercise should not be performed until the infection is treated. Nausea related to ART may make specific exercises difficult, decrease compliance with exercise prescriptions, or both. As with many populations, PLWH suffer from poor exercise prescription adherence and compliance (77, 92, 117, 118).

Loss of skeletal muscle mass (e.g., wasting or sarcopenia) by itself is not a contraindication to exercise. Indeed, progressive resistance training is recommended to offset the catabolic effects of HIV infection and the metabolic complications associated with ART. Additionally, nutrition counseling may be indicated for individuals beginning exercise training who may be experiencing wasting, sarcopenia, or lipodystrophy (5, 57).

There is a strong association between HIV infection and CVD. Major CVD events include heart failure, peripheral artery disease, stroke, coronary atherosclerotic disease, and myocardial infarction (6, 108). Arterial hypertension and lipid metabolism impairment are also predominant comorbidities in PLWH, and many patients might present themselves on pharmacologic treatment for these problems. Because PLWH have high CVD risk, careful cardiovascular screening is warranted for all patients. Health professionals should periodically assess their patients for risk factors and should closely monitor patients receiving ART, especially those with additional CVD risk factors (9, 12).

Assessment of Functional Status

HIV-related disability has been associated with a decrease in exercise capacity and in a patient's daily activities. Baseline assessments of functional status and fitness are necessary to complement the medical history and physical examination. Baseline measures provide a reference point for evaluating response to the exercise program as well as comparisons in subsequent evaluations. The exercise program should be modified according to the individual's physical function, health status, exercise response, and stated goals. Disease status can be categorized into one of four categories for functional status and fitness testing:

1. Asymptomatic, medically stable, and physically active
2. Asymptomatic, medically stable, and physically inactive
3. Recovering from a medical or disease-related event
4. Symptomatic and suffering from acute illness

Depending on the patient's status, various testing protocols have been proposed to ascertain CRF, body composition, flexibility, strength, and physical functioning. Patients who are symptomatic or are suffering an acute illness should refrain from testing until their condition stabilizes.

CLIENT–CLINICIAN INTERACTION

HIV is not easily transmitted between individuals in normal day-to-day interactions. There is no risk of HIV infection from shaking hands, coming into contact with sweat, regular body contact, or sharing athletic equipment (as long as those accessories do not have blood, semen, or vaginal secretions from a PLWH). The application of general guidelines for disease prevention will provide an adequate level of safety for the clinical exercise physiologist (CEP) in all exercise testing and prescriptive contexts. As with all patients, caution must be taken when collecting blood from a PLWH, and universal precautions should be followed.

The initial contact between a CEP and PLWH is most often for the treatment of established CVD or body composition changes, which occur because of the HIV infection itself, ART exposure, unhealthy daily habits, or a combination of these factors. In the clinical settings that test and treat CVD or musculoskeletal diseases, the vast majority of patients are not living with HIV. Thus, PLWH may be concerned about the disclosure of their HIV status. In non–health care settings, a stigma is often attached to HIV infection. Yet a PLWH on ART usually requires similar health care as that delivered to patients with other chronic conditions. Therefore, in the clinical setting it is helpful to reassure patients that their medical information is confidential and that only those staff members who need to be aware of their HIV status will have that information.

Precautions

- Hands should be washed before and after examining or testing each patient.
- CEPs should routinely use appropriate barrier precautions to prevent skin and mucous membrane exposure when there is the potential for contact with blood or other body fluids. Gloves, goggles, or face shields and gowns provide an appropriate level of protection for the majority of exercise testing contexts (e.g., removing mouthpieces, skin prep for electrocardiogram electrode placement).
- Face shields or goggles and surgical masks should be used for blood sampling, depending on agency protocols. Saliva has not been implicated in the transmission of HIV. Given the potential for emergency mouth-to-mouth resuscitation, resuscitation bags, mouthpieces, or other ventilation devices should be available when the need for resuscitation is anticipated.
- If, after taking all precautions, a CEP is exposed to blood or other body fluids that might contain HIV, there are postexposure prophylaxis protocols and medication regimens available that are effective in preventing HIV seroconversion when taken in a timely manner (i.e., no longer that 72 h after exposure). The CEP can consult the updated U.S. Public Health Service guidelines for the management of occupational exposures to HIV (11b) for complete information on this topic.

Young asymptomatic patients may not differ with respect to age-adjusted norms on graded exercise tests (96). However, they may exhibit a reduction in exercise capacity due to a sedentary lifestyle. Older asymptomatic patients may present with reduced values compared with those expected from age-matched adults without HIV (85). They may exhibit significantly reduced time on the treadmill or bike, lower peak oxygen consumption ($\dot{V}O_2$peak) and ventilatory threshold, and increased heart rate at submaximal work rates. Untreated HIV infection in the presence of AIDS-defining illnesses is associated with dramatically reduced time on the treadmill or bike, much lower peak oxygen consumption ($\dot{V}O_2$peak), and possible failure to reach ventilatory threshold. There is also an increased probability of abnormal neuroendocrine responses at moderate- and high-intensity test stages.

The sidebar Exercise Testing: Important Considerations provides a brief summary of recommendations for exercise testing and the assessment of physical fitness and function in PLWH. This summary is not exhaustive, and specific protocols are described elsewhere (general exercise testing procedures are presented in chapter 5).

As with most populations, the general exercise testing recommendations apply (see chapter 5). Exercise professionals who administer exercise tests or supervise training are at minimal risk of infection. Nevertheless, universal precautions for the transmission of blood pathogens should be followed. See also practical application 23.1.

Cardiorespiratory

Maximal tests carried out on both treadmill and cycle ergometers have been conducted in PLWH and are safe for most patients. Additionally, most PLWH exhibit normal cardiorespiratory values (e.g., blood pressure, ventilation) in response to a graded exercise

Exercise Testing: Important Considerations

- *Cardiorespiratory:* Maximal testing is appropriate for individuals with stable HIV disease.[a] Alternatively, walk tests (i.e., 6 min and 400 m walk tests) are valuable measures in patients with low CRF. Postpone testing in anyone with an active infection. Patients commonly have a significantly lower peak $\dot{V}O_2$ versus age- and sex-matched healthy populations and have an increased risk of CVD and impairment.

- *Strength:* Grip strength and isokinetic dynamometry with comparison against established norms, or a 1RM or multiple RM (2RM-10RM) strength test can be used. Physical function measures (i.e., chair stand and the Short Physical Performance Battery) should also be considered because they are more meaningful for this population than maximal strength.[b]

- *Range of motion:* Sit-and-reach is a good general test. Patients should be able to perform any range of motion assessment.

[a]American College of Sports Medicine (1).

[b]National Institute of Health. www.nia.nih.gov/research/labs/leps/short-physical-performance-battery-sppb

test. Some PLWH, however, suffer from peripheral neuropathy and are at increased risk for autonomic neuropathy and abnormal responses to testing (e.g., attenuated heart rate). Walk tests (i.e., 6 min and 400 m walk tests) are a feasible alternative to maximal tests and are valuable predictive measures in patients with low CRF (78, 84).

Many PLWH are sedentary, and CRF levels in this population are among the lowest in comparison with other vulnerable populations. Low CRF levels are evidenced by low average $\dot{V}O_2$peak and $\dot{V}O_2$max values, with a mean pooled value of 26.4 mL · kg · min^{-1} (obtained from 1,010 subjects with a mean age of 41 yr) (119). As observed in the general population, treadmill tests in PLWH predict higher CRF levels than do cycle ergometer tests. There are few data to compare CRF levels between PLWH and well-matched healthy controls, but PLWH demonstrated $\dot{V}O_2$peak values 44% below age-predicted norms (124). The measurements of $\dot{V}O_2$peak and $\dot{V}O_2$max have been used to assess CRF as well as for the exercise prescription.

Individuals on ART may have a diminished ability to extract and use oxygen in the working skeletal musculature and exhibit a decrement in peak a-$\bar{v}O_2$ difference values. This appears to be a result of the mitochondrial effects of ART rather than true HIV pathogenesis (8). In fact, skeletal muscle mitochondria of men with HIV on ART (including older, more mitochondrial-toxic therapy such as zidovudine) had 17% lower ATP content and reduced oxidative enzyme activity compared with people without HIV (82). Structural and inflammatory muscle abnormalities that may impair the muscle's ability to extract or utilize oxygen during exercise can also be associated with the low CRF (31). The peak $\dot{V}O_2$ that PLWH can attain partially depends on the muscle mass activated by the CRF test. Perfusion of the active tissues can be restricted if the muscles are weakened secondary to disease progression, and lactate begins to accumulate at a low power output, causing local fatigue and limiting the peak effort that can be achieved (97).

Musculoskeletal

Assessing muscular strength and flexibility in PLWH should not differ appreciably from the testing of people without HIV. This population is often untrained and therefore unfamiliar with resistance exercise. Thus, the use of a multiple RM (i.e., 2RM-10RM) strength test protocol may be more appropriate than 1RM for assessing muscular strength and preventing injury or delayed-onset muscle soreness. Isokinetic dynamometry–based tests (i.e., peak torque and maximum voluntary isometric contraction) of the lower limbs are safe in PLWH (79, 80). Alternatively, grip strength has been used as a clinical measure of strength (100).

When assessing musculoskeletal function of PLWH, it is of greatest importance that physical performance be considered. Data suggest that physical performance measures (e.g., chair stand and the Short Physical Performance Battery) may be more relevant for this population than strength measures; such measures of physical performance are more strongly associated with hazard of mortality in this population than grip strength, a common measure of muscle strength (88). Also, contrary to what is usually observed in the general population, changes in physical performance (e.g., slow gait) in PLWH may occur before significant impairment in muscle strength (20). This decline in physical performance may be due to factors intrinsic to HIV infection, such as mitochondrial dysfunction, cognitive impairment, and neuropathy.

Because of the potential for an accelerated aging phenotype, PLWH usually experience accelerated loss of muscle mass and function, presenting lower strength and physical performance than people without HIV. Given that most PLWH are also deconditioned and lead sedentary lives, assessments should be performed on multiple occasions over the course of training. Most individuals may demonstrate substantial improvements during the early stages of exercise training, and reassessment is necessary to ensure the sufficiency of the training stimulus, the progression, or both.

Treatment

The treatment and prevention of HIV infections have been transformed by the use of ART, with current HIV drug regimens consisting of three drugs, generally with once daily dosing. HIV preexposure prophylaxis (PrEP) consists of a two-drug regimen, generally with once daily dosing. There are seven HIV drug classes, categorized by function: non-nucleoside reverse transcriptase inhibitors (NNRTIs), nucleoside reverse transcriptase inhibitors (NRTIs), protease inhibitors (PIs), fusion inhibitors, CCR5 antagonists, integrase strand transfer inhibitors (INSTIs), and post-attachment inhibitors (table 23.1). HIV treatment regimens typically combine drugs from multiple categories.

The life expectancy of PLWH is approaching that of individuals without HIV thanks to pharmacological advancements. PLWH have a high risk for developing chronic comorbidities. Drug–drug interactions between various HIV medications and between HIV medications and treatment for comorbid conditions are common. As such, continuous monitoring for adverse drug interactions should occur.

PLWH have an increased prevalence of glycemic dysregulation, dyslipidemia, and lipodystrophy due to side effects of HIV treatment (86). Regular physical activity and exercise, as well as a balanced diet, is essential for PLWH, especially in the presence of comorbid conditions such as diabetes, CVD, hyperglycemia, lipodystrophy, and hyperlipidemia. It is important to monitor for drug interactions when prescribing an exercise or diet regimen. As with other clinical populations, exercise and nutritional interventions are often employed together to reduce chronic comorbidities. Adequate physical activity and exercise is associated with lower **advanced glycation end products (AGEs)**, lower triglycerides, less lipodystrophy, and lower waist circumference (99, 116). Adequate nutrition is essential for optimal metabolic and immune function in PLWH (33). Various physical activity and nutritional strategies exist for HIV and its comorbidities (18).

Table 23.1 Pharmacology

Medication class	Medication examples	Primary effects	Exercise effects	Important considerations
Protease inhibitors (PIs)	Aptivus, Crixivan, Evotaz, Invirase, Norvir, Prezista, Viracept	Block the enzyme protease, which allows the virus to process itself and be released to infect other cells	Regular exercise and physical activity are indicated for PLWH. Exercise can mitigate many symptoms associated with HIV and ART. There are no specific effects of ART on exercise physiologic responses.	Note that these considerations are general to all HIV medications: potential liver or kidney damage, diarrhea, nausea, GI issues, glycemic dysregulation, dyslipidemia, and lipodystrophy. PIs are associated with multiple metabolic disorders. NRTIs are associated with mitochondrial DNA toxicity and proximal renal tube dysfunction. NNRTIs are associated with hypersensitivity reactions.
Nucleoside reverse transcriptase inhibitors (NRTIs)	Cimduo, Combivir, Descovy, Retrovir, Trizivir, Truvada	Block the enzyme reverse transcriptase, preventing the HIV virus from making copies of itself		
Non-nucleoside reverse transcriptase inhibitors (NNRTIs)	Edurant, Intelence, Rescriptor, Sustiva, Viramune	Block the enzyme reverse transcriptase, preventing the HIV virus from making copies of itself		
Fusion inhibitors (FIs)	Fuzeon	Inhibit the HIV virus from attaching to targeted cells and hinder the initial infection		
Integrase strand inhibitors (INSTIs)	Isentress, Tivicay Vitekta	Block the enzyme integrase, which prevents the integration of HIV genetic material into the infected cell's DNA		
CCR5 antagonists	Selzentry	Block CCR5 coreceptors on the surface of some immune cells where HIV enters		
Post-attachment inhibitors	Ibalizumab	Bind to the CD4 receptor on a host CD4 cell and block the virus from attaching to the CCR5 and CXCR4 coreceptors and entering the cell		

Complications of ART Regimens

Approximately 50% of PLWH are aged 50 yr or older, many of whom experience polypharmacy. Simplification of ART regimens for PLWH is a key priority because it improves patient adherence and quality of life and decreases the incidence of drug interactions (37). Complications of ART regimens are drug-class specific (table 23.1). Non-nucleoside reverse transcriptase inhibitors (NNRTIs) are associated with lipid disorders, rash, and in some first-generation NNRTIs (e.g., Viramune, Sustiva) severe hepatotoxicity. Nucleoside reverse transcriptase inhibitors (NRTIs) are associated with lipodystrophy and lactic acidosis. Lactic acidosis causes symptoms of increased liver function markers due to hepatic stenosis, abdominal pain, nausea, vomiting, diarrhea, and fatigue.

Protease inhibitors (PIs) are associated with lower HDLs; higher triglycerides, total cholesterol, and LDLs; and high **atherogenic index of plasma (AIP)** (65). Fusion inhibitors are associated with neutropenia and increased risk for pneumonia. Integrase strand transfer inhibitors (INSTIs), primarily Isentress, are associated with myopathy and rhabdomyolysis, while Sustiva (an NNRTI) is associated with neuropsychiatric and central nervous system issues (54). CCR5 antagonists, specifically Selzentry, are associated with hepatotoxicity, heart attack, skin reactions, and allergic reactions. Post-attachment inhibitors are associated with diarrhea, dizziness, nausea, and rash. Common drugs that interact with ART are anticonvulsants (e.g., carbamazepine, phenobarbital, phenytoin), cardiac medications (e.g., calcium channel blockers), contraceptives, gastrointestinal medications (e.g., histamine H_2), herbal treatments (e.g., St. John's wort), hormone therapy, phosphodiesterase type 5 inhibitors (e.g., sildenafil, tadalafil), and psychiatric medications (e.g., lorazepam, oxazepam, temazepam). The primary side effects that may arise when a person is performing exercise include gastrointestinal discomfort, dizziness, and physical indisposition.

Lipodystrophy

Lipodystrophy is the abnormal distribution of fat in the body and is associated with insulin resistance and hyperlipidemia. Clinicians can diagnose lipodystrophy by observing body composition changes. Lipodystrophy is multifactorial, and the mechanisms that cause it remain unclear. Drugs in the NRTIs and PI classes are associated with metabolic changes and development of body fat, but through different mechanisms. Risk factors for the development of lipodystrophy include >40 yr of age, advanced HIV disease, and type and duration of NRTI therapy (Stavudine more likely) and PI therapy (Norvir more likely). No treatment exists for body composition changes in PLWH. Change in ART therapy has provided variable results regarding lipodystrophy. Therefore, clinicians should inform patients of the risk of lipodystrophy with use of NRTIs and PIs. Strategies to mitigate lipodystrophy include switching ART therapy and prescribing diet and exercise interventions.

Weight Gain

Evidence suggests that integrase strand transfer inhibitors (INSTIs) (Tivicay and Isentress) have a direct effect on adipose tissue, leading to weight gain and clinical obesity (39, 89). Tivicay exhibits greater levels of weight gain compared with Vitekta and Isentress among treatment-naive individuals, as well as PLWH switching to INSTIs (89). Additionally, Tivicay-containing groups when combined with tenofovir alafenamide demonstrate greater weight gain (89).

Increases in body mass index for PLWH taking integrase inhibitors seem greatest among women. However, a study found inconsistencies regarding weight gain and the specific effects of weight gain with INSTIs (7). Therefore, more research is needed to examine the relationship between INSTIs and weight gain and the ramifications on cardiovascular and metabolic health. Future research should also address the impact of INSTIs on physical activity, on individuals with the highest risk factors for weight gain, and on occurrence of comorbidities associated with weight gain.

Impaired Glucose Metabolism

PLWH on long-term ART have a fivefold greater chance of developing a glucose metabolism disorder (60). Consequently, it is essential for PLWH on ART to be cognizant of their increased risk for impaired glucose metabolism and diabetes. Particularly, NRTIs and PIs are linked to insulin resistance. As many as 80% of PLWH on PIs develop insulin resistance, which often leads to diabetes (34). In order to mitigate impaired glucose metabolism and consequently diabetes, clinicians and PLWH should monitor their blood glucose levels and integrate diet and exercise into their treatment plans.

Insulin Resistance

Insulin resistance remains prevalent in PLWH, even in those receiving modern ART, because of the metabolic complications associated with this treatment. Generalized and abdominal obesity, as well as ectopic fat (103) and pericardial fat, is linked with insulin resistance (56). The mechanism of action for insulin resistance in PLWH is related to inflammation and immune activation even with use of ART (42, 129). Treatment strategies to mitigate altered metabolic function, such as insulin resistance in PLWH, should include pharmacological interventions (e.g., Metformin) and lifestyle modifications (e.g., exercise) (129).

Hyperlipidemia

Hyperlipidemia is associated with HIV infection and ART (especially PIs), physical inactivity, high-fat diet, and smoking. Identification and clinical management of hyperlipidemia in PLWH is essential for cardiovascular health. Hyperlipidemia in PLWH is different from that in the noninfected population, because HIV and ART induce hyperlipidemia but may also interact with lipid-lowering medications. Statin use, except for simvastatin and lovastatin, is effective and safe in treatment of hyperlipidemia (43). Additional treatment strategies for hyperlipidemia include lifestyle modification interventions (e.g., physical activity, healthy diet, smoking cessation). The management of hyperlipidemia and associated CVD risk is complex and may require the involvement of clinical specialists.

Osteonecrosis (Avascular Necrosis)

Osteonecrosis, also known as avascular necrosis, results from death of bone cells and involves bones near a joint. ART, particularly PIs, is linked with osteonecrosis because ART affects bone density and turnover. Osteonecrosis is associated with progressive pain over time and can affect daily living, including physical activity. The treatment goals for osteonecrosis are to prevent further bone loss and attenuate impact on daily living.

Diarrhea, Nausea, and Other Gastrointestinal Tract Disorders

Common side effects of ART include diarrhea and nausea. PIs have the greatest associated cause. Diarrhea is experienced by more than 50% of PLWH and leads to decreased quality of life and lower adherence to ART. HIV infection–associated diarrhea is multifactorial and includes ART-associated factors and gastrointestinal damage related to HIV itself (e.g., HIV enteropathy) (15). HIV-related gastrointestinal tract issues that can lead to diarrhea include HIV enteropathy, small bowel bacterial growth, and intestinal infections such as *Mycobacterium avium* **complex** and *Cryptosporidium*. Persistent diarrhea should be addressed pharmacologically. Crofelemer is FDA approved to treat diarrhea caused by ART. Nausea is a side effect of nearly all ART medications. Pharmaceutical interventions to mitigate nausea should include antiemetics. Clinicians should be cognizant of the ramifications of symptoms such as diarrhea, nausea, and dehydration, which can lead to decreased physical activity.

Increased Risk of Cardiovascular Disease and Type 2 Diabetes

HIV infection and ART are associated with glucose and lipid metabolism disorders, insulin resistance, adipose tissue changes, and lipodystrophy, all of which can contribute to increased risk of CVD and type 2 diabetes mellitus (DM). Studies indicate that DM incidence is decreasing compared with HIV-uninfected populations, possibly because newer ART drugs have fewer metabolic effects (70). However, DM and its complications remain an important factor in the health and well-being of PLWH. Coronary artery disease can be found in PLWH of all ages. Because of various factors including the interaction of viral infection, systemic inflammation and ART, and traditional risk factors, the underlying pathophysiology of coronary artery disease is complex and remains partially unclear (4). Primary prevention should be balanced with ART to suppress the side effects of ART and promote healthy lifestyle choices such as exercise and balanced diet. Exercise and physical activity should be acknowledged as a possible means of mitigating the impact of ART, CVD, and DM in PLWH.

EXERCISE PRESCRIPTION

Participating in regular exercise and maintaining a physically active lifestyle are linked with lower prevalence of CVD risk factors, improved metabolic health, reduced symptoms, and decreased morbidity and mortality in PLWH (74, 86). All PLWH should obtain medical clearance before participating in an exercise program. Exercise prescriptions should be individually tailored for safety and effectiveness.

Physical activity recommendations should be emphasized to PLWH because these individuals typically do not meet sufficient physical activity levels (115, 116). Determining the appropriate exercise mode depends on safety, effectiveness, and patient preference. PLWH may require a different exercise regimen compared with their HIV-uninfected peers related to HIV and exposure to older ART therapy. Limited data exist regarding the optimal dose of exercise needed for maximum benefit in PLWH (93).

The U.S. Department of Health and Human services (HHS) emphasizes that increasing physical activity and decreasing sedentary behavior benefits almost all individuals, with the most sedentary individuals benefiting greatest from minor increases in physical activity (93). Modifications may need to be considered when recommending physical activities for PLWH with specific physical limitations. The goals of exercise training are to improve people's capacity for activities of daily living as well as enhance their ability to remain healthy and active. As such, exercise prescriptions should be targeted toward improving any existing comorbidities. Aerobic, resistance, and low-intensity physical activity (e.g., yoga) are associated with improvements in physiological and psychological functioning in PLWH (68). These include increases in CRF, decreases in CVD risk factors (e.g., blood lipids, obesity), improvements in body composition (e.g., decreased visceral fat depots), augmented metabolic functioning (e.g., insulin resistance), and enhancement of musculoskeletal function (e.g., strength and mobility).

Adherence is an essential factor of successful engagement in physical activity in PLWH, and readiness to engage in physical activity must be addressed. To promote adherence, it is recommended that PLWH adopt behavior change techniques including self-monitoring, goal setting and action planning, social support, and prompts to engage in physical activity (68). Self-monitoring includes tracking one's physical activity through use of technology (e.g., pedometers) or paper diaries. Goal setting involves setting physical activity targets, while action planning refers to a plan executed to accomplish the goal. There is a strong association between physical activity engagement and social support, with health care providers playing an important role in incorporating physical activity into routine HIV care (126). Prompts to engage in physical activity, such as prescheduled physical activities, improve physical activity in PLWH.

Physical activity is one of the most effective interventions to increase medication adherence in this patient population. Barriers that influence adherence to and compliance with exercise recommendations in PLWH include intrapersonal, interpersonal, and environmental. Intrapersonal barriers associated with decreased physical activity in PWLH include older age, fewer years of education, ART use, lower CD4+ count, presence of opportunistic infection, and lipodystrophy, as well as symptoms such as pain and depression (116). Interpersonal barriers include decreased social support and social factors (e.g., stigma, HIV disclosure) (55, 98), and environmental barriers include accessibility (e.g., physical and financial) and stigma related to body image (67). Health care providers can recommend physical activity that does not require a gym, such as walking outdoors or short bouts of activity throughout the day. Other factors influencing adherence to and compliance with exercise recommendations and medications are substance abuse, inadequate transportation, poverty, infection, and the side effects related to treatments (32, 68).

Symptom burden and symptoms (e.g., depression, anxiety, fatigue, neuropathy) lead to decreased compliance with many forms of treatment including pharmacologic and behavioral. Aerobic and resistance exercise, as well as increased cardiorespiratory function, are linked with reduced symptoms including HIV-related fatigue (40, 127). Aerobic exercise at moderate and vigorous intensities is associated with a decrease in depression, including those with diagnosis of a major depressive disorder (101). Given that the prevalence of symptom burden and symptoms (e.g., depression, anxiety, fatigue, neuropathy) is high in PLWH, issues with both pharmacologic and exercise compliance should be continually addressed.

The management of HIV and its related comorbidities is multidimensional and complex because of numerous factors including direct viral effects, lifestyle (e.g., decreased physical activity, diet), and medication regimen. Treatment plans should account for these factors in order to determine effective treatment strategies. Viral load and immune status are predictive markers for the risk of disease progression and development of opportunistic infections. CD4+ T-cell counts and HIV-RNA are used for progression and disease stage assessment (52). The influence of exercise on HIV-specific biomarkers (CD4+ T-cell counts and HIV-RNA) remains inconclusive (58).

Patients with HIV infection, in addition to undergoing ART, may also be subject to psychotropic drugs, chemotherapeutic agents for cancer, or the prophylactic use of antimicrobial agents and vaccines (or more than one of these) to prevent or treat comorbid conditions such as opportunistic infections (e.g., cytomegalovirus, pneumocystis pneumonia). Clinical health professionals must be cognizant of the physiological effects of ART regimens as well as other pharmacologic treatments and allow this perception and understanding to inform exercise prescription.

Cardiorespiratory Exercise

Some CEPs may be tempted to deliver a lower-intensity exercise to PLWH because of their disease status, but evidence has consistently shown that PLWH can follow standard prescription recommendations for the general public (76). However, given the deconditioning due to the sedentary lifestyle of many PLWH, an adequate low-volume and low-intensity familiarization period is essential. Moreover, PLWH can usually progress at the same rate as people without HIV and can also be exposed to higher-intensity exercises, even older PLWH (21, 83). Although a standard relative intensity (e.g., % of $\dot{V}O_2$) can be prescribed, the absolute workload can be lower than usual because of lower baseline fitness values. Precautions must be taken for PLWH who present with CVD risk factors or CVD disease; the CEP should follow the standard recommendations for individuals presenting with these conditions.

Resistance Exercise

PLWH can tolerate generally prescribed relative intensities and can also quickly progress during regular resistance training (21, 76, 90). Thus, the development of a progressive resistance training program should adhere to the ACSM's guidelines while taking into account any complications induced via medication use or disease progression. PLWH usually present with reduced muscle mass and strength compared with people without HIV, so an adequate familiarization period is essential to prevent injuries or excessive delayed-onset muscle soreness. Particular attention must be paid to individuals suffering osteo-related complications or peripheral neuropathies. With osteo-related complications, the joint capsule may be compromised, and resistance training is contraindicated. Peripheral neuropathy may significantly affect coordination, and caution is advised with all weight-bearing exercise. It is advisable to allow participants to experience a varied selection of exercises that allow a pain-free range of motion and appropriate progression.

Range of Motion Exercise

As with any sedentary population, flexibility and range of motion in PLWH may be compromised. Therefore, the initial prescription should be mild, and the progression should be slow but consistent. Refer to chapter 6 for general information on exercise prescription. Table 23.2 is a summary of exercise prescription for PLWH.

EXERCISE TRAINING

Exercise is one of the main nonpharmacologic strategies recommended to help mitigate both the adverse effects of ART and chronic HIV infection. There are numerous benefits of exercise for PLWH, including increased muscle mass, strength, and CRF and enhanced body composition, metabolic profile, psychological profile, and overall health. Interventions using cardiorespiratory exercise or progressive resistance training alone have been conducted over the years, but studies have demonstrated that combined exercise (i.e., cardiorespiratory exercise and resistance training performed together) is the best approach for PLWH because it can provide significant cardiorespiratory and neuromuscular benefits (23). For optimal benefits, all exercise training must be supervised, systematic, and progressive.

Table 23.2 Exercise Prescription Review

Training method	Frequency	Intensity	Time (Duration)	Type (Mode)	Progression	Important considerations
Cardiorespiratory	3-5 d/wk	30%-39% of $\dot{V}O_2R$ or HRR, or 37%-45% of $\dot{V}O_2$ max	10-20 min per session	Varies according to health status and individual interest (e.g., walking, cycling, jogging, swimming)	Slow incremental progression to 40%-80% of $\dot{V}O_2R$, HRR, or $\dot{V}O_2$ max, and 30-60 min per session	Consider the higher cardiovascular risk experienced by these individuals.
Resistance	2 or 3 d/wk	40%-50% of 1RM	1-2 sets per exercise, 10-15 reps per set, 6-8 exercises	Machine weights are safer; free weights and resistance bands can also be used	Slow incremental progression to 2-4 sets at 60%-80% of 1RM, 8-12 reps per set	For individuals living with HIV for many years, suspect osteo-related complications with chronic, unexplained musculoskeletal pain. Also, peripheral neuropathies may be present.
Range of motion	2 or 3 d/wk	Relaxed, comfortable stretch (tightness or slight discomfort); no stretching to the point of muscle spindle activation	Each stretch held for 10-30 s, with 2-4 repetitions of each	Yoga, static stretching, or PNF stretching of the major muscle groups	Stretch should always be relaxed and without significant discomfort	Mild progression is advised because of general deconditioning.

$\dot{V}O_2R$ = oxygen uptake reserve; HRR = heart rate reserve; RM = repetition maximum; PNF = proprioceptive neuromuscular facilitation.

Despite the increasing number of studies testing the effects of an exercise intervention in PLWH, reviews have consistently confirmed the need for further research in this area (30, 44, 48, 58). One reason is that the studies were conducted years ago and their results may not represent the contemporary PLWH population because of a shift in patient demographics and treatment. Prior studies on exercise and HIV were mainly focused on dealing with wasting, lipodystrophy, and immune dysfunction. More recent studies address the increased CVD risk, reduced muscle mass and strength, and impaired psychological profile. For these reasons, we have focused on research (i.e., systematic reviews and original studies) published in 2010 and thereafter to describe the exercise training adaptations in PLWH.

Practical application 23.2 provides a synthesis of the main adaptations associated with exercise training in PLWH. Although individual studies observed improvement in a wide range of variables, the information presented in this summary was derived from meta-analyses and systematic reviews conducted on this topic.

Cardiorespiratory Exercise

Aerobic training results in significant benefits for PLWH, including improved CRF, body composition, and mood; decreased visceral fat depots and waist-to-hip ratio; and reduced risk of CVD (29). O'Brien and colleagues reviewed the literature on aerobic

exercise for PLWH and observed a significant improvement in change of $\dot{V}O_2$max of 2.63 mL · kg · min^{-1} for individuals who performed aerobic exercise when compared with individuals in the nonexercising control group (74). They also observed a significant trend toward a greater improvement in $\dot{V}O_2$max of 4.30 mL · kg · min^{-1} for individuals in the heavy-intensity exercise group compared with the moderate-intensity exercise group. For HRmax, there was an observed nonsignificant trend toward a decrease of 9.81 b · min^{-1} for individuals who performed aerobic exercise. The interventions included in this review were carried out for 6 to 24 wk, and participants performed a variety of exercise types, including the treadmill (walking or jogging), stationary bike, ski machine, cross-country machine, and stair stepper.

Contemporary research on aerobic training for PLWH yields interesting results. Table 23.3 presents full details of selected studies. The long-term benefits of chronic aerobic exercise (i.e., over 24 wk of intervention) have not been fully investigated in this population, but Schlabe and colleagues (102) prospectively monitored 21 PLWH preparing for a marathon over a 1 yr period. The high training volume was tolerated without major complications in the marathon-finisher group. Although the marathoners did not complete a cardiorespiratory parameters assessment, they did report improved metabolic and immunologic parameters after the training period.

Practical Application 23.2

EXERCISE TRAINING ADAPTATIONS SUMMARY

- *Cardiorespiratory endurance:* A significant improvement in $\dot{V}O_2$max of 2.63 mL · kg · min^{-1} after aerobic exercise intervention compared with nonexercisers has been observed. Even greater improvements in $\dot{V}O_2$max were observed after combined aerobic and resistance exercise training compared with nonexercisers (3.71 mL · kg · min^{-1} change), and after heavy-intensity exercise compared with moderate-intensity exercise (4.30 mL · kg · min^{-1} change).[a]

- *Skeletal muscle strength:* After exercise intervention, an overall change in upper body strength from baseline of 18 kg and in lower body strength of 16.8 kg were observed.[b]

- *Body composition:* All types of exercise (i.e., aerobic, resistance, and combination) improve particular body composition outcomes. Aerobic exercise reduces abdominal girth, body mass index (–1.31 kg · m^{-2}), fat mass (–1.12%), triceps skinfold thickness of subcutaneous fat (–1.83 mm), waist circumference, and waist-to-hip ratio. Meanwhile, resistance exercise can increase body weight (+5.02 kg), lean body mass, muscle mass, and peripheral girth. The combination of aerobic and resistance exercise increases leg muscle area (+4.79 cm^2), muscle mass, and mean arm and thigh girth (+7.91 cm).[c]

[a]O'Brien (74)
[b]Pérez Chaparro et al. (90)
[c]Kamitani (48)

Table 23.3 Effects of Exercise Training on Cardiorespiratory Health in PLWH

Authors	Participants	Intervention characteristics	Outcomes
Kocher et al., 2017	From a total of 24 sedentary PLWH, 7 males were compliant with the protocol (>70% of exercise sessions) and had valid data; age range of 36-58 yr	Aerobic exercise (walking or jogging) 3 times per week for 12 wk; duration initially set at 20 min to a maximum of 40 min per session (in 2 min/wk increments); target intensity of 50%-80% of HRmax	Increase in $\dot{V}O_2$max (median of 14% increase); 2.45-fold increase in peripheral blood mononuclear cells mitochondrial respiratory capacity; 5.65-fold increase in spare respiratory capacity; 3.15-fold increase in nonmitochondrial respiration
Schlabe et al., 2017	From 21 PLWH enrolled, 13 completed the intervention period and were analyzed (12 males and 1 female); median age of 42 (27-50) yr	3 or 4 sessions per week during 1 yr while training for a marathon run; first period of training consisted of 3-4 h/wk at 60%-70% of HRmax; second period consisted of 6 h/wk at 70%-80% of HR max; third period consisted of 7-10 h/wk of moderate endurance runs together with extensive sprints; 2 wk before the marathon participants trained at 60%-70% of HRmax	Increase in absolute CD4+ T-cells (20% increase); decrease in systolic blood pressure and cholesterol
Oursler et al., 2018	Total of 22 PLWH ≥50 yr of age randomized to moderate- or high-intensity exercise; 7 participants in the moderate- and 9 in the high-intensity group were analyzed; mean age of 57.4 yr	Aerobic exercise (treadmill or field walk) 3 times per week for 16 wk; high intensity: initially 20-30 min at 50%-60% of HRR, which increased to 40-45 min at 75%-90% of HRR; moderate intensity: self-paced 1 mile walk, initially during 20-30 min, which increased to 45 min	Increase in $\dot{V}O_2$peak in the high-intensity group only (3.6 mL · kg · min^{-1} increase); increase in exercise endurance and 6 min walk test distance in both groups

PLWH = people living with HIV; HRmax = maximum heart rate; HRR = heart rate reserve.

Adapted by permission from G.A. Hand et al., "Impact of Aerobic and Resistance Exercise on the Health of HIV-Infected Persons," *American Journal of Lifestyle Medicine* 3, (2009): 489-499.

Another study assessed alterations in cellular bioenergetics in peripheral blood mononuclear cells after 12 wk of aerobic exercise and reported improvements in respiration at the cellular level, where exercise significantly increased not only $\dot{V}O_2$max but also the number of peripheral blood mononuclear cells (49). Considering the high prevalence of mitochondrial dysfunction among PLWH, studies analyzing cellular bioenergetics at the cellular level can help us understand the underlying mechanisms of alterations following aerobic exercise in this population.

Finally, a pilot study to determine safety and efficacy of high-intensity aerobic training in older PLWH reported that moderate- to high-intensity exercise increased endurance and ambulatory function (83). However, increased CRF was observed only with high-intensity aerobic training despite substantial baseline impairment. Exercise training for older PLWH is a promising research area, but CEPs should be aware of preexisting bone- and joint-related problems. Authors in this study reported a 27% overall dropout rate related to joint pain. This suggests that an adequate and progressive adaptation period is necessary before prescribing high-intensity exercise for this population.

In summary, given that CVD is a significant contributor to morbidity and mortality in PLWH, reductions in CVD risk factors may be the strongest argument for the inclusion of aerobic exercise in the treatment plan. Importantly, aerobic exercise appears to be safe for PLWH who are medically stable, based on the absence of reports of adverse events among the majority of exercisers and the stability of CD4+ count and viral load (74).

Resistance Exercise

Most research on the effects of exercise on PLWH has concentrated on aerobic training or on a combination of aerobic training with resistance training. The few contemporary studies that have examined resistance training as a stand-alone intervention have focused mainly on changes in strength and muscle mass rather than overall fitness or immune function.

Resistance exercise training in PLWH improves outcomes related to body composition, with increases in lean body mass, midthigh cross-sectional muscle area, and bone mineral density, in addition to a reduction in body weight. Greater improvements in weight and body composition were found with resistance training compared with aerobic training. O'Brien and colleagues reviewed the literature on resistance training for PLWH and observed a significant increase in body weight of 4.24 kg among participants in the resistance exercise group compared with the aerobic exercise group (71). Additionally, there was a demonstrated increase in arm and thigh girth of 7.91 cm among participants in the resistance training group compared with the aerobic exercise group.

Resistance exercise training also generated gains in muscle strength and function in PLWH. Poton and colleagues demonstrated a 36% change in muscular strength from pre- to postintervention after reviewing 20 resistance exercise studies (95). Similarly, Pérez Chaparro and colleagues reviewed the literature

and found a change in upper body strength of 17.5 kg and a change in lifted lower body weight of 29.4 kg (90). These results are important because PLWH have lower values of muscle strength when compared with people without HIV (30, 76). Moreover, muscle strength is recognized as a primary outcome when diagnosing **sarcopenia**, a prevalent condition in PLWH (14, 75). Considering that reduced muscle strength and sarcopenia are related to adverse outcomes, such as falls, disability, and mortality, increasing strength in PLWH seems a potentially beneficial approach during treatment.

Overall, results suggest that resistance exercise is safe for medically stable PLWH based on few reports of adverse events with exercise within the existing studies, as well as the lack of change in CD4+ count and viral load (71). Existing resistance exercise interventions for PLWH were carried out for 6 to 52 wk, so the long-term effect of exercise is unclear. However, there is an indication that longer training periods lead to greater increases in muscular strength (95), which can potentially improve physical performance and reduce mortality risk in PLWH.

Combined Aerobic and Resistance Training

Combined exercise (i.e., aerobic and resistance exercise performed together) has been demonstrated as the best approach for PLWH because incorporating both modalities into exercise programs may be more effective for optimizing functional status than programs involving only one component (51, 130). Usually, PLWH are deconditioned and exhibit low CRF, low muscle strength and function, and altered body composition. Thus, improvements in CRF from aerobic exercise and increased strength and muscle mass from resistance exercise have beneficial effects on both the physiological and psychological capacity and functioning of PLWH. Additionally, the improvements in body composition (e.g., reduced visceral fat depots, increased lean tissue mass), metabolic function, blood lipids, and insulin resistance from aerobic and resistance exercise are associated with reductions in CVD risk and all-cause mortality.

Most of the contemporary literature on exercise interventions for PLWH has been conducted using combined exercises. The results of such interventions have been summarized in systematic reviews. Voight and colleagues reviewed seven studies and reported an improvement in all strength, cardiorespiratory, and flexibility outcome measures associated with combined exercise (121). Gomes Neto and colleagues found that combined exercise increased peak $\dot{V}O_2$ (4.48 mL · kg · min^{-1}) as well as muscle strength of the knee extensors (25.06 kg) and elbow flexors (4.44 kg) compared with the nonexercise group (30). Moreover, Lopez and colleagues reported that combined exercise provided the most benefits in fitness and mental health outcomes when compared with other interventions (58).

Other variables (e.g., inflammatory biomarkers, oxidative stress, hormones, bone mineral density, and others) that were examined in individual studies improved after combined exercise but did not attain sufficient power to detect treatment effects in meta-analyses. More research is needed to determine all the benefits associated with combined training in PLWH.

As with aerobic training, recent research has focused on high-intensity combined exercises for older PLWH (19, 20, 45). Authors submitted sedentary adults (50-75 yr) with or without HIV to 24 wk of combined exercise and compared results based on HIV status and exercise intensity. Participants attended exercise sessions three times per week, and after 12 wk of progressive increases in exercise volume and intensity, participants were randomized to either continue moderate-intensity exercise (40%-50% of $\dot{V}O_2$max during 50 min and 60%-70% of 1RM for four exercises) or advance to high-intensity exercise (60%-70% of $\dot{V}O_2$max during 50 min and >80% 1RM for four exercises) for the remaining 12 wk. Subjects had significant improvements in all functional measures, but PLWH randomized to high-intensity exercise gained significantly more strength in the bench and leg press. The rationale for delivering high-intensity exercise for PLWH is based on the fact that this population has very low physical and cardiorespiratory function and may need a higher intensity to return to normal values observed in controls of the same age and sex. In this specific study, PLWH had poorer physical function across nearly all baseline measures than controls, and despite achieving greater relative gains (% of change), postintervention values were still lower than controls.

Considering all the benefits and safety of combined exercise training for PLWH, the inclusion of this modality should be standard practice in HIV treatment unless contraindicated.

Range of Motion Exercise

Flexibility and range of motion in sedentary populations are often compromised, and enhanced joint mobility via flexibility training improves functional outcomes and patient quality of life. Although flexibility training has been incorporated in a few exercise interventions delivered to PLWH, changes in this variable have not been evaluated. However, there is no reason to believe that PLWH would respond differently to flexibility training than people without HIV.

Exercise Training and Immune Function

Early in the HIV epidemic, health care professionals were concerned that exercise might further compromise the already impaired immune function of PLWH. However, exercise has been repeatedly demonstrated to be safe for this population, and there is a consensus that exercise training does not negatively affect immune function or disease progression. Although individual studies demonstrate improvements in the immune function of PLWH after an exercise intervention (48), evaluations using meta-analyses have not observed a significant effect (71, 74). Despite this, exercise training potentially has several positive effects on immune status in that it attenuates psychological stress and reduces depressive symptoms, perceived stress, and symptoms of anxiety, as well as improving quality of life (107).

Research Focus

Effectiveness of Progressive Resistance Training in PLWH

Decreased fitness, including muscular strength, is a hallmark of HIV infection. It is well accepted that aerobic exercise training in stable PLWH provides many benefits. However, less is known about the effects of resistance training. Maintaining skeletal muscle health, including strength and endurance, may be beneficial in combating the negative effects of ART, including an increased risk of CVD and changes in body composition (loss of lean mass and gain of fat mass). Aerobic exercise has been shown to improve skeletal muscle strength and body composition in those with HIV infection. A Cochrane meta-analysis evaluated the effectiveness of progressive resistance training in PLWH.

Cochrane analyses are systematic reviews of the literature focusing on any published manuscript that broadly meets the inclusion criteria. In this case, O'Brien and colleagues searched for papers that compared resistance exercise with either no exercise or another intervention. They analyzed 20 studies with 764 participants. Of these, 12 used resistance exercise combined with aerobic exercise and 8 used resistance exercise alone. The duration of training ranged from as short as 6 wk to as long as 1 yr. The authors reported that resistance exercise improved CRF as measured by oxygen consumption and total exercise time, strength (chest press and knee flexion), and body weight and composition. Additionally, they reported no adverse effects on CD4+ count or viral load.

O'Brien and colleagues concluded that performing resistance exercise alone or in combination with aerobic training at least three times per week for at least 6 wk appeared to be safe and beneficial. Because of the increasing length of life for those living with HIV, the authors recommended that future research be directed at older individuals and those living with comorbid conditions including heart, liver, kidney, and bone and joint disease. They also stated that exercise training in PLWH might be ready to move from the clinical setting to a nonsupervised or community-based setting.

Based on O'Brien et al. (71).

Clinical Exercise Bottom Line

- Metabolic complications are among the most common problems associated with HIV infection and antiretroviral therapy, which confers an increased risk of cardiovascular diseases.

- Exercise testing and training are safe in most people living and aging with HIV and do not negatively affect immune function or HIV disease progression.

- Exercise professionals who administer exercise tests or supervise training are at minimal risk of HIV infection.

- People living with HIV can follow standard exercise prescription recommendations for the general public and do not need to exercise at lower intensity only because of their HIV status.

- Considering the HIV and antiretroviral therapy side effects, combined exercise appears to be the best approach for improving the health and well-being of people living with HIV.

CONCLUSION

Exercise training is safe and effective for PLWH. The scientific literature provides compelling evidence that exercise training confers numerous benefits for PLWH. With the scale-up of effective ART regimens, PLWH are living longer, are more productive, and lead higher-quality lives. Unfortunately, they also experience a large number of complications from the progression of the disease as well as the effects associated with ART (e.g., increased risk of CVD). Research has demonstrated that both aerobic and resistance exercise can attenuate, and in some cases reverse, these detrimental effects so that patients experience improvements in functional capacity (e.g., increased strength and CRF), psychological functioning, and quality of life. As a group, PLWH are sedentary and as a result experience an increased risk of CVD, obesity, insulin resistance, and physical and psychological symptoms. This makes exercise training a vitally important component of every PLWH's treatment plan.

Although some health care professionals still work under the erroneous assumption that exercise training may compromise immune function in PLWH, the scientific evidence strongly suggests otherwise. Therefore, HIV infection should not be a deterrent to starting an exercise program but rather should provide the impetus to begin.

Online Materials

Visit *HKPropel* to access a link to the references, the case studies with discussion questions, and a quiz to help you review key concepts and test your understanding of the material covered in this chapter.

PART VI

Disorders of the Bones and Joints

Part VI contains chapters on three orthopedic and musculoskeletal conditions: arthritis, osteoporosis, and low back discomfort. These chapters were included because of the large numbers of people living with these conditions—55 million with arthritis, 44 million with osteoporosis, and 31 million with low back pain in the United States alone. With the increasing number of people entering the age group greater than 65 yr and rising rates of obesity worldwide, it is anticipated that these conditions will continue to increase in prevalence. The practicing clinical exercise physiologist is likely to deal with these conditions on an almost daily basis.

Chapter 24 focuses on the many types of arthritis. Arthritis is the leading cause of disability in the elderly, and medical treatment costs are substantial. Although arthritis predominantly afflicts older individuals, it can also affect those who are middle aged. This chapter provides a comprehensive review of arthritis and the use of exercise for the treatment of this condition.

Chapter 25 presents osteoporosis, a condition that can affect both men and women. This chapter reviews the importance of exercise training in the prevention and treatment of osteoporosis. This includes information regarding the development and implementation of exercise-training programs for individuals with established osteoporosis, with an emphasis on safety and progression.

Low back pain is the most common reason for work disability in those under the age of 45. Chapter 26 will help the clinical exercise physiologist better understand how aerobic exercise plays a role in both the prevention and the long-term treatment of low back pain. Issues such as improved cardiovascular conditioning and abdominal strength and their relationship to restoration of function are presented.

CHAPTER 24

Arthritis

Melissa Nayak, MD

Andrew K. Cunningham, MD

Contrary to traditional beliefs that individuals with arthritis should avoid vigorous physical activity for fear of worsening joint damage, research findings and clinical experience overwhelmingly demonstrate that regular appropriate exercise reduces disability and improves joint function in these individuals (16, 19, 27, 42, 99, 100, 187). This can be achieved without aggravating symptoms, while improving general health and quality of life (QOL).

While broadly considering the various forms of arthritis, this chapter focuses primarily on osteoarthritis (OA), rheumatoid arthritis (RA), and ankylosing spondylitis (AS) and the effects and benefits of exercise for these conditions.

DEFINITION

Arthritis is a generic term for conditions that involve inflammation of one or more joints. There are more than 100 different forms of arthritis, each characterized by varying degrees of joint damage, restriction of movement, functional limitation, and pain.

SCOPE

In the United States, chronic arthritis affects 22.7% of adults (54.4 million people) (18) and is the leading cause of disability. Annual medical expenditures and earnings losses cost the economy over $300 billion (245). Globally, the economic burden of arthritis is estimated to account for the equivalent of 1% to 2.5% of the gross national product of Western nations. Prevalence is higher for women (26%) than men (19%) and increases with age. For those over age 65, arthritis is the most prevalent medical condition in both women and men (18), affecting 50% of this age group (18).

Because of population aging and the increasing incidence of obesity, arthritis is expected to increase in incidence worldwide; by 2040, 78 million adult Americans are predicted to have these conditions (95). Exercise is identified as one of the principal methods of reducing the incidence of arthritis and attenuating the associated disability and economic burden.

With regard to the specific forms of arthritis, OA occurs most frequently, affecting 13.4% of adult Americans (30.8 million) (37, 89). Osteoarthritis-related surgeries are common, which contributes to OA being the fourth leading cause of hospitalization in the United States (164). Rheumatoid arthritis occurs in approximately 0.7% of Americans (1.5 million) (119). Other common forms of arthritis include gout (4.1%, 8.3 million) (248) and **spondyloarthropathies** (0.6-2.4 million) such as AS, psoriatic arthritis (PsA), reactive arthritis, and enteropathic arthritis. Arthritis is also associated with connective tissue diseases such as systemic lupus erythematosus (SLE), dermatomyositis, and systemic sclerosis (SSc; scleroderma). Although arthritis is most prevalent in the aging population, an estimated 294,000 American children under 18 yr (i.e., 1 in 227) have some form of the condition (198).

Arthritis adversely affects both physical and psychosocial functioning and is the leading cause of disability in later life (230). For example, work disability is reported in 31% of U.S. adults with doctor-diagnosed arthritis (230); the effects of arthritis on physical function are described in detail later in this chapter. Often ignored, however, are the consequences of arthritis on social functioning, which can be dramatic: 25% of people with arthritis either never leave their home or only do so with help, 18% never participate in social activities (15), and individuals with arthritis report a significantly worse QOL than those who do not have arthritis (71,

Acknowledgment: Much of the writing in this chapter was adapted from the previous editions of *Clinical Exercise Physiology*. Thus, we wish to gratefully acknowledge the previous efforts of and thank Andrew B. Lemmey, PhD, and Virginia Kraus, MD.

157, 201). Additionally, arthritis is associated with major depression (attributable risk of 18.1%) (54), and 6.6% of individuals with arthritis report severe psychological distress (94). These factors, combined with pain, fatigue, and the elevated energy cost of performing activities of daily living (ADLs) with increasing impairment (78, 175, 176), contribute to physical inactivity. In turn, this extreme sedentarism negatively affects health by increasing the risks of cardiovascular disease, dyslipidemia, hypertension, diabetes, obesity, and osteoporosis (161, 179).

PATHOPHYSIOLOGY

Although the etiologies, presentation, and clinical manifestations of the various forms of arthritis are generally distinct, they share common features that impair the following:

- Exercise tolerance
- Muscle strength
- Muscular endurance
- Aerobic capacity
- Range of motion (ROM)
- Biomechanical efficiency
- Proprioception

Consequently, all forms of arthritis can potentially be characterized by functional limitation and disability.

Osteoarthritis

Osteoarthritis (OA) involves degradation of joints, which usually develops gradually and particularly affects **articular** cartilage and subchondral bone. Initially, the cartilage becomes pitted, rough, and brittle. In response to this, and to reduce the load on the cartilage, the underlying bone thickens. As a consequence, the synovial membrane swells and increases production of synovial fluid, and the joint capsule and surrounding ligaments thicken and contract (29, 131, 184). These adaptations lead to a narrowing of the joint space, which in advanced OA can result in loss of cartilage, bone rubbing on bone, periarticular muscle loss, and ligaments becoming strained and weakened (131). Although commonly referred to as wear-and-tear arthritis, this term is a misnomer because the pathology of OA involves a process of continuous, abnormal remodeling of joint tissues driven by inflammatory mediators (131). The resultant symptoms are persistent joint pain, stiffness, and deformity, with functional impairment. Patients can present with **crepitus**, joint locking, **effusion**, and bone spurs (**osteophytes**).

The joints most commonly affected by OA are those of the hands, feet, and spine, as well as the large weight-bearing joints (hips and knees). As with the other forms of arthritis, the exact etiology of OA is not known, although, there is a clear genetic contribution (9, 62, 149, 205). The condition is more prevalent in women (occurrence ratio 2-3:1, female to male), and its incidence for both sexes increases with age (OA rarely occurs in persons younger than 40 yr but is evident by radiography [X-rays] in >80% of those over 55 yr). Additionally, obesity and joint injury or trauma are known to predispose an individual to OA (9, 29, 62, 195).

Because joints depend on dynamic loading for maintenance of joint function and nutrition (88, 102), chronic insufficient loading is deleterious, as is chronic excessive loading, as evidenced by the strong association between obesity and OA of the weight-bearing joints (9, 29, 62). As a result of reduced movement secondary to joint pain and stiffness, local muscles atrophy and ligaments become lax. Thus, muscle weakness and joint instability are well-recognized consequences of OA and, in turn, are involved in the development, progression, and severity of the condition (246). This loss of strength is a major contributor to the disability associated with OA (113) and is the best-established correlate of lower limb functional limitation in those affected by knee OA (177). Osteoarthritis is also characterized by loss of joint ROM (104, 177), which exacerbates the reduced function and disability produced by pain and muscle weakness (220).

Rheumatoid Arthritis

Rheumatoid arthritis (RA) is a chronic autoimmune disorder, characterized by systemic inflammation and symmetrical polyarthritis. It affects various tissues and organs but principally targets synovial joints. RA is more prevalent in females than males (occurrence ratio approximately 3:1), with a typical age of onset between 40 and 50 yr. The resultant inflammatory response in joints (**synovitis**) is a consequence of synovial cell hyperplasia, excessive production of synovial fluid, and the development of **pannus**. In time, synovitis results in erosion of articular cartilage and marginal bone, with subsequent joint destruction and **ankylosis**. The joints most affected by RA are the small joints of the hands and feet, followed in order of prevalence by the larger joints of the wrists, elbows, shoulders, and knees, although any joint with a synovial lining is susceptible. In addition to joint pain, stiffness, and damage, RA also has extra-articular effects. Some are specific to RA, such as rheumatoid nodules, while nonspecific effects include muscle loss, increased adiposity (particularly truncal adiposity), fatigue, and increased risk of cardiovascular disease (CVD), metabolic syndrome, type 2 diabetes, and osteoporosis (13, 36, 47, 75, 107, 112, 121, 156, 173, 197, 222).

Ankylosing Spondylitis

Like RA, ankylosing spondylitis (AS) is a chronic inflammatory arthritis and an autoimmune disease. In contrast to RA and OA, AS occurs more commonly in males (3:1 occurrence ratio), and age of onset is usually earlier (20-40 yr) than for either OA or RA. The condition primarily affects the spine and sacroiliac joint. Initially, the ligaments of the lower spine become inflamed at the **entheses**. This process stimulates bone to grow within the ligaments. Gradually these bony growths form bony bridges between adjacent vertebrae, which eventually can cause the vertebrae to fuse, with resultant low back pain and immobility. Additionally, most AS patients suffer synovitis in the larger peripheral joints (predominantly the hips and knees), and common extra-articular features include fatigue, eye inflammation (uveitis or iritis), CVD, and inflammatory bowel disease (IBD) (226). Unlike individuals with other forms of arthritis, AS patients have for decades been encouraged to be physically active, because back pain, which is often severe at rest, usually improves with physical activity, and it was apparent from earlier treatment regimens that bed rest and immobilization accelerated spinal fusion.

Traditionally, arthritis disease stages have been classified at three levels:

1. *Acute:* reversible signs and symptoms in the joint related to synovitis
2. *Chronic:* stable but irreversible structural damage brought on by the disease process
3. *Chronic with acute exacerbation of joint symptoms:* increased pain and decreased ROM and physical function

Each of these stages has disease-specific presentations, treatment considerations, and goals. Table 24.1 provides specifics about the stages of OA, RA, and AS.

CLINICAL CONSIDERATIONS

The various arthritides can be differentiated based on whether symptoms arise from the joint or from a periarticular location, the number of joints involved, their location, whether the distribution is symmetric or asymmetric, the **chronicity** of disease, and extra-articular features (20, 98, 130). Pharmacological treatment of OA, RA, AS, and other forms of arthritis varies according to the condition, between individuals with the same condition, and even for the same individual over time. Despite these treatment nuances, exercise should be included in the routine management of these conditions.

Signs and Symptoms

In the evaluation of individuals with musculoskeletal complaints, the history and physical examination are the most informative elements. Restricted movement of a joint and tenderness to palpation along the axis of joint movement are indicative of arthritis. This contrasts with tenderness around the joint, which is more indicative of periarticular soft tissue involvement. The signs and symptoms of arthritis are as follows:

- Pain
- Stiffness
- Joint effusion
- Synovitis
- Deformity
- Crepitus

Joint pain can arise from pathological changes in the joint capsule and periarticular ligaments, **intraosseous** hypertension, muscle weakness, subchondral microfractures, **enthesopathy**, and **bursitis**, and it may be exacerbated by psychosocial factors including depression (25, 109). Pain, it should be noted, does not emanate from articular cartilage directly, because cartilage is **aneural**. In individuals with inflammatory arthropathies, such as RA and AS, joint pain and stiffness fluctuate directly with disease activity (i.e., the degree of inflammation). The prognosis of recent-onset arthritis is aided by a determination of whether the duration of symptoms has exceeded 4 to 6 wk (98).

History and Physical Examination

The medical history is essential for determining the duration, location, extent, and severity of musculoskeletal symptoms (98). Additionally, because of the genetic component of OA, RA, and especially AS, identifying the presence of these conditions in the family history greatly assists diagnosis. Obtaining information on

Table 24.1 Types of Arthritis, Stages, and Related Impairments

Type of arthritis	Disease stage	Related impairments
Osteoarthritis	Acute joint pain	Often insidious
Chronic radiographic joint disease	Chronic with exacerbation	Increased joint pain and swelling, muscle weakness, and progressively declining functional impairment
Rheumatoid arthritis	Acute disease in multiple joints with pain, limited range of motion, and worsened functional impairment; often symmetrical joint involvement	Joint stiffness, adverse body composition changes (rheumatoid cachexia; muscle loss and fat gain), muscle weakness, fatigue, and increased cardiovascular disease risk
Ankylosing spondylitis	Acute spinal pain and stiffness without significant decrease in mobility	Muscle loss, muscle weakness, and fatigue
Chronic spinal ankylosis predominant with decreased spinal and thoracic mobility	Chronic with exacerbation	Increased pain and stiffness of the back or peripheral joints

the current level of functioning and any previous or ongoing efforts at an exercise intervention, including any barriers or facilitators to exercise, is essential when designing appropriate exercise interventions. The physical examination provides the majority of the information required for establishing an appropriate diagnosis and for recording specific information about any abnormalities of joint ROM, alignment, or function. Additionally, the physical examination may identify the presence of extra-articular features (e.g., rheumatoid nodules in RA, eye disease in AS, and skin disease in psoriatic arthritis), which facilitates formulating the correct diagnosis. The physical examination is also used to assess joints for the four cardinal signs of inflammation: redness, swelling, pain, and heat.

Diagnostic Testing

The American College of Rheumatology (ACR) has developed diagnostic criteria for the classification of hip, knee, and hand OA and for RA (2, 3, 4, 5). Criteria for AS have been developed by the Assessment of SpondyloArthritis international Society (ASAS) (209). Table 24.2 summarizes these criteria.

Currently, although no definitive diagnostic tests or markers of arthritis exist, several specific serum and synovial fluid tests are available that greatly assist in differentiating the arthritic conditions. In combination with joint imaging, these tests contribute to specifying the arthritis diagnosis. For RA, tests for the presence of anti-citrullinated protein antibodies (ACPA; also called anti-cyclic citrullinated protein antibodies [anti-CCP]) and rheumatoid factor (RF) have become routine features of the initial diagnosis. The ACPA test has high specificity (95%) and sensitivity (68%) for RA and for the future development of RA (14). Although RF positivity occurs in about 80% of RA patients, as with ACPA/anti-CCP, a negative test for serum RF should not exclude RA as a possible diagnosis.

For AS, since the HLA-B27 genotype is present in approximately 90% of patients, a positive test for this genotype assists the diagnosis. However, because only 5% of individuals with the HLA-B27 gene develop AS, a positive test for this marker alone is not sufficient to justify a diagnosis of AS. Assessment of disease activity for arthritic conditions is aided (and often defined) by nonspecific measures of systemic inflammation, such as erythrocyte sedimentation rate (ESR) and serum levels of C-reactive protein (CRP). These inflammatory markers are high in active RA and are typically normal or only marginally elevated when the disease is controlled. In contrast, although mild elevations in ESR and CRP are often apparent in active AS and severe OA (178, 180, 213), these markers can also be normal in these conditions despite significant inflammation.

When synovial fluid is aspirated (i.e., removed) to relieve joint swelling, synovial fluid analysis is also often helpful for determining the type of arthritis. For example, the leukocyte count typically increases in synovial fluid from a normal value of 500 cells · mm^{-3} to 2,000 cells · mm^{-3} in OA and to 5,000 to 15,000 cells · mm^{-3} in RA and AS. Additionally, macromolecules originating from joint structures and measured in blood, synovial fluid, or urine reflect arthritic processes taking place locally in the joint (204).

Joint imaging, including radiographs (X-rays), ultrasound (US), and magnetic resonance imaging (MRI), is routinely used to confirm a particular arthritis diagnosis; for example, involvement of the small joints of the hands and feet usually indicates OA or RA, whereas certain skeletal changes in the lower back are suggestive of AS (see figure 24.1). The plain radiograph, which detects bony changes, is the traditional imaging modality. However, since radiological changes are usually evident only in established or advanced disease, MRI and US are the preferred imaging formats for suspected early-onset RA and AS. Radiologic features of OA include **subchondral sclerosis** osteophyte formation, bone cysts, and joint space narrowing. In RA, in addition to joint space narrowing, radiographs may reveal marginal erosions and more pronounced deformity. For early AS, there is a characteristic squaring of the corners of the vertebrae. Later in the disease process there is evidence of bone formation, ossification, or thin vertically oriented outgrowths that bridge the disc space and limit spinal motion (26).

In summary, the information gained from the patient's history, physical examination, serum and synovial fluid tests, and joint imaging allows the clinician to discern patterns that aid in the diagnosis of a particular condition (98).

Exercise Testing

This section reviews special exercise testing considerations for the arthritic population. Modes and protocols may need adjusting depending on the level of disability or the functional limitations of an individual's arthritis.

Individuals with RA have an increased risk of CVD, which largely accounts for higher rates of mortality (13, 223). This heightened CVD risk, which is largely attributed to the inflammatory burden of RA (112), is approximately double that of the general population (181), making RA comparable to diabetes as a CVD risk factor (182, 214, 239). In view of this increased risk, the European League Against Rheumatism (EULAR) task force has advocated that risk scores determined by CVD calculators such as the Framingham and the Systemic Coronary Risk Evaluation (SCORE) methods should be multiplied by 1.5 when RA patients fulfill two out of three of the following criteria: RA disease duration ≥10 yr, presence of extra-articular features, and RF or ACPA/anti-CCP positivity (181). Because AS is also associated with increased CVD occurrence (142), EULAR recommends that this condition also be regarded as a CVD risk factor (181). Moreover, since people with arthritis tend to be even more deconditioned than individuals without arthritis, the risk of CVD is further increased (156). In those with a high CVD risk profile who wish to commence exercise training, a symptom-limited exercise test should be considered, both to screen for the presence of CVD and to assist in developing the exercise prescription (7). An exercise test is also commonly used for assessing cardiovascular status and risk stratification when considering joint replacement surgery.

Table 24.2 Distinguishing Characteristics and ACR and ASAS Diagnostic Criteria for Arthritis

Classification	Measures
OSTEOARTHRITIS	
Distinguishing characteristics	Joint pain Crepitus Gel phenomenon
Presentation	Affects hands, hips, knees, and lumbar and cervical spine Pain worsens throughout the day Affects any traumatized joint
ACR criteria	*Knee clinical* Knee pain and three of the following: • Age >50 yr • Morning stiffness <30 min • Crepitus • Bony tenderness • Bony enlargement • No warmth
	Knee clinical and radiographic Knee pain and one of the following: • Clinical criterion a, b, or c (see above) • Osteophytes on knee X-ray
	Knee clinical and laboratory Knee pain and five of the following: • Clinical criteria a through f (see above) • ESR <44 mm · h^{-1} • RF <1:40 • Synovial fluid compatible with OA
	Hip combined clinical, laboratory, and radiographic Hip pain and one of the following: • ESR <20 mm · h^{-1} • Osteophytes on hip X-ray • Joint space narrowing on hip X-ray
	Hand clinical Hand pain or stiffness and three of the following: • Bony enlargement of two or more DIPs • Bony enlargement of two or more of 2nd and 3rd DIPs, 2nd and 3rd PIPs, 1st CMC • Fewer than three swollen MCPs • Deformity of at least one of 2nd and 3rd DIPs, 2nd and 3rd PIPs, 1st CMC
RHEUMATOID ARTHRITIS	
Distinguishing characteristics	Hand pain Swelling Fatigue Prolonged morning stiffness
Presentation	Affects wrists, MCPs, and PIPs Symmetric

ACR criteria	A score of >6 out of 10 based on the following:
	• Joint involvement[a]
	    Two to 10 large joints[b]: 1
	    One to three small joints[c] (with or without involvement of large joints): 2
	    Four to 10 small joints (with or without involvement of large joints): 3
	    More than 10 joints (at least one small joint): 5
	• Serology
	    Low-positive RF or low-positive ACPA: 2
	    High-positive RF or high-positive ACPA: 3
	Acute-phase reactants, abnormal[d] CRP or ESR: 1
	Duration of symptoms ≥6 wk: 1

(continued)

Table 24.2   *(continued)*

Classification	Measures
ANKYLOSING SPONDYLITIS	
Distinguishing characteristics	Low back pain Low back stiffness
Presentation	Early: chronic low back pain (≥3 mo duration) before age 45 Late: vertebral fusion and sacroiliac joint fusion (via bone formation, ossification)
ASAS criteria	*Clinical* ≥3 mo back pain before age 45 and either of the following: • Sacroiliitis on imaging (X-ray or MRI) + one or more clinical feature • HLA-B27-positive (lab test) plus two or more clinical features *Clinical features* • Inflammatory back pain • Arthritis • Dactylitis • Enthesitis • Good response to NSAIDs • Psoriasis • Inflammatory bowel disease • Uveitis • Positive family history • Elevated CRP

Note: ACR criteria are based on references 2-5. ASAS criteria are based on reference 209.

[a]Joint involvement refers to any swollen or tender joint on examination.

[b]Shoulders, elbows, hips, knees, and ankles.

[c]MCPs, PIPs, 2nd through 5th metatarsophalangeal joints, thumb interphalangeal joints, and wrists.

[d]As determined by local laboratory standards.

ESR = erythrocyte sedimentation rate; RF = rheumatoid factor; DIPs = distal interphalangeal joints; PIPs = proximal interphalangeal joints; CMC = carpometacarpal joint; MCPs = metacarpophalangeal joints; ACPA = anti-citrullinated protein antibody; CRP = C-reactive protein; ESR = erythrocyte sedimentation rate.

Practical Application 24.1

CLIENT–CLINICIAN INTERACTION

As part of the exercise evaluation, individuals with arthritis should be screened for factors that will affect the exercise prescription. Suggested questions for patients include the following topics:

- Individual's age, level of function, medications, personal goals, and lifestyle
- Names and contact information of health care providers including primary care physician, rheumatologist, orthopedist, clinical exercise physiologist, and physical therapist
- Musculoskeletal disease diagnosis
- Pattern of joint involvement: (a) **symmetric** or **asymmetric**, (b) upper or lower extremity involvement, (c) joints affected

- Severity of disease activity (acute, chronic, or chronic with acute exacerbation)
- Comorbidities: other medical conditions, including cardiovascular risk factors (obesity, hypertension, dyslipidemia, insulin resistance or diabetes, systemic inflammation), pulmonary disease, fibromyalgia, **Raynaud's phenomenon**, **Sjogren's syndrome**, osteoporosis
- Surgical history, including joint replacements
- Previous treatment (whether successful or not)
- Presence of fatigue
- Adequacy of footwear

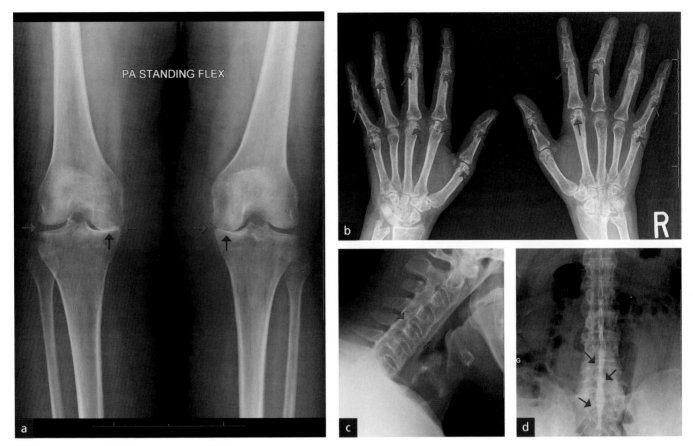

Figure 24.1   X-ray images of *(a)* osteoarthritis of the knee, seen in bilateral knee radiographs demonstrating osteophyte formation (red arrows) and joint space loss in the medial compartments (blue arrows); *(b)* rheumatoid arthritis of the hand, seen in bilateral hand radiographs showing erosions at the metacarpophalangeal (MCP), proximal interphalangeal (PIP), and thumb interphalangeal (IP) joints (red arrows), with ulnar deviation of the fingers noted at the 4th and 5th finger MCP joints of bilateral hands and 3rd finger PIP joint of the right hand; *(c)* ankylosing spondylitis of the cervical spine, seen in a radiograph showing typical appearance of bamboo spine due to ossification of the margins of disc space (red arrows); and *(d)* ankylosing spondylitis of the lumbar spine demonstrating the dagger sign, seen as a single radiodense line due to ossification of the supraspinous and interspinous ligaments (blue arrows).

Photos courtesy of Henry Ford Cardiac Rehabilitation Program.

Although arthritis and musculoskeletal conditions are often listed as relative contraindications to graded exercise testing, 95% of those with severe end-stage hip or knee arthritis attributable to either OA or RA were found to be capable of performing a symptom-limited exercise test using cycle ergometry methods (183). The majority of these patients also achieved a respiratory exchange ratio $(RER = \dot{V}CO_2/\dot{V}O_2)$ greater than 1.0, indicating a metabolically maximal test. Approximately two-thirds of these subjects were capable of completing tests by pedaling with their legs, and the remainder pedaled with their arms. However, during an exercise test, staff must be aware that joint symptoms and fatigue may adversely affect performance and prohibit maximal testing (21, 45). Should directly measuring maximal aerobic capacity prove difficult, assessment of peak $\dot{V}O_2$ or estimating $\dot{V}O_2$max by submaximal tests provides a viable alternative for developing appropriate exercise prescriptions (7, 23, 163; also see chapter 6).

For all testing procedures, guidelines on standard contraindications for exercise testing should be followed (see chapter 5). Exercise testing procedures for individuals with arthritis are similar to protocols recommended for the elderly and deconditioned (7). These tests should have small incremental changes in workload (e.g., increments of 10 to 15 W · min^{-1} on the cycle ergometer or the modified Naughton protocol with use of a treadmill) (23). Cycle ergometry is generally the preferred mode for testing because of the high frequency of lower extremity impairment in arthritis patients. Most individuals with arthritis are able to achieve maximal cardiorespiratory effort during cycle ergometry, which is not weight bearing and less reliant on balance (183). However, equations for estimating $\dot{V}O_2$ by cycle ergometry have not been validated for arthritis patients. In the instance of severe lower extremity joint pain and limitation, and in those with significant deformities that contraindicate cycling, testing by arm ergometry may be necessary. Treadmill testing is usually appropriate for patients with minimal or no functional disability.

The following treadmill-specific equation was developed to predict $\dot{V}O_2$peak in seniors with knee OA or CVD: $\dot{V}O_2$ (mL $\cdot$ kg^{-1} $\cdot$ min^{-1}) = 0.0698 × speed (m $\cdot$ min^{-1}) + 0.8147 × grade (%) × speed (m $\cdot$ min^{-1}) + 7.533 mL $\cdot$ kg^{-1} $\cdot$ min^{-1}. This equation, validated in both men and women, requires that participants use front handrails for support during the exercise test (23). This is the advised method of testing an individual with lower extremity disability for whom standard non–hand support methods of treadmill testing may be hazardous. Recommendations for cardiorespiratory exercise testing are outlined in table 24.3. Recommendations for musculoskeletal and flexibility testing are outlined in table 24.4.

Treatment

A comprehensive treatment strategy for arthritis should strive to counteract physical inactivity, restore a healthier body composition (i.e., increase muscle mass and reduce fat mass), improve physical function (i.e., attenuate disability), and reduce comorbidity-related symptoms or risk. This approach should include appropriate medication to control disease activity and minimize symptoms, as well as exercise.

In the past, the traditional standard of treatment for a symptomatic arthritic joint was rest (30). Current guidelines emphasize increasing physical activity, including exercise training (92, 242), because of conclusive evidence of the beneficial effects and safety of exercise for individuals with arthritis (1, 7, 8, 16, 17, 22, 40, 42, 84, 92, 99, 100, 117, 120, 157, 174, 187, 216, 226, 246). As depicted in figure 24.2, exercise interventions can interact at each stage of arthritis pathology and can help mitigate the effects of the disease process on physical function and disability, as well as negative changes in body composition (141, 216).

The WHO ICIDH-2 update (1999), elaborating on the previous classification of disease effects and their implications, demonstrates a multiperspective approach, including the dynamic nature of various elements (e.g., environmental and personal factors) on activity and participation, among other things.

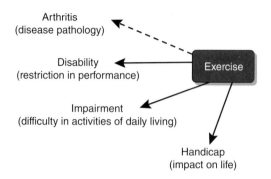

Figure 24.2 The World Health Organization (WHO) classification of impairments, disabilities, and handicaps (ICIDH) demonstrates the potential benefit of exercise as an interactive and mitigating factor in the arthritis disease process, and strong evidence showing benefits of exercise on physical disability and impairment. Although much remains to be learned about the effects of exercise on disease pathology, the overwhelming majority of studies show no worsening of arthritis with exercise, and a substantial proportion demonstrate beneficial effects on clinical aspects of the disease (e.g., reduced pain, joint stiffness, inflammation, bone loss).

Adapted from R.J. Shephard and P.N. Shek, "Autoimmune Disorders, Physical Activity, and Training, with Particular Reference to Rheumatoid Arthritis," *Exercise Immunology Review* 3 (1997): 53-67.

Table 24.3 Cardiorespiratory Exercise Testing in Patients With Arthritis

Mode	Protocol specifics	Clinical measures	Clinical implications	Special considerations
Use a treadmill for those with minimal to mild joint impairment.	Use protocols with small increment increases (i.e., modified Naughton or a ramp protocol) unless disease activity and severity are minimal.	Assess type of arthritis and degree of activity and impairment. Assess comorbidities and past surgical and medical history.	Standard peak $\dot{V}O_2$ prediction equations may overestimate functional capacity because they were developed on healthy (nonarthritic) populations.	With patient using handrails for support, use equation* to predict $\dot{V}O_2$max.
Use cycle ergometry for those with mild to moderate lower extremity impairment.	Use protocols with small increment increases (i.e., 10-15 W $\cdot$ min^{-1}) or ramping protocols.	Assess type of arthritis and degree of activity and impairment. Assess comorbidities and past surgical and medical history.		Additional investigations are needed to improve prediction of peak $\dot{V}O_2$.
Use arm ergometry for those with severe lower extremity impairment.	Use arm ergometry–specific protocols with small increment increases or ramping protocols.	Assess type of arthritis and degree of activity and impairment. Assess comorbidities and past surgical and medical history.		Additional investigations are needed to improve prediction of peak $\dot{V}O_2$. Consider submaximal testing in those with severe impairment.

*$\dot{V}O_2$ (mL $\cdot$ min^{-1} $\cdot$ kg^{-1}) = 0.0698 × speed (m $\cdot$ min^{-1}) + 0.8147 × grade (%) × speed (m $\cdot$ min^{-1}) + 7.533 mL $\cdot$ min^{-1} $\cdot$ kg^{-1} (23).

Table 24.4   Strength, Range of Motion, and Balance Testing in Patients With Arthritis

Test type	Mode	Protocol specifics	Clinical measures	Clinical implications
Lower extremity	Dynamometer	All testing in supine position except knee flexion and extension (while seated)	Reference ranges for 50-79 yr olds (11)	Often (up to 50%) decreased in persons with arthritis
	30 s chair sit-to-stand test	Number of stands completed in 30 s, without using arms, from a chair with a seat height of 17 in.	Reference ranges (189)	
	8RM (8-repetition maximum)	The maximum resistance that can be moved through the full range in a controlled manner for 8 reps (8RM; e.g., leg press, knee extension)	Reference ranges (7)	
Upper extremity and grip	Hydraulic dynamometer (e.g., Jamar)	In seated position with unsupported arm flexed 90° at elbow	Reference ranges for position 2 (143, 169)	Often (up to 50%) decreased in persons with arthritis
	Electronic dynamometer (e.g., Grippit)	Peak grip force Average sustained force	Reference ranges (143, 169, 142)	Usually (up to 90%) decreased in persons with hand arthritis
	30 s arm curl test	Total number of arm curls in 30 s with 5 lb dumbbell for women and 8 lb dumbbell for men	Reference ranges (191)	
	8RM	The maximum resistance that can be moved through the full range in a controlled manner for 8 reps (8RM; e.g., bench press)	Reference ranges (7)	Often (up to 50%) decreased in persons with arthritis
Range of motion	Goniometer	Align device fulcrum with joint fulcrum	Reference ranges	Usually (up to 90%) decreased in persons with arthritis
Balance	Figure-eight walking*	Useful in those with limited or mild impairments* • Track width = 150 mm • Inner diameter = 1.5 m • Outer diameter = 1.8 m		
	Berg balance scale*	Useful in those with moderate to severe impairments* Includes 14 single tasks beginning with sitting unsupported and progressing to standing on one leg	More than two oversteps during two circuits suggestive of decreased balance; median ranges for RA functional classes (189)	Often (up to 50%) decreased in persons with arthritis

*Based on investigations in persons with rheumatoid arthritis (RA) (59, 189).

The implementation of exercise as part of a comprehensive therapeutic management strategy can be challenging. In a study by Law and colleagues looking at RA patients' perceptions of the effects of exercise on joint health, it was noted that patients were aware of the advantages of exercise for their condition but felt that "health professionals lacked certainty and clarity regarding specific exercise recommendations and the occurrence of joint damage" (118, p. 2444). The study also found that physicians and allied health professionals rarely recommend exercise to their arthritis patients, and when they do they usually provide little detailed instruction (118). Additionally, those who do advise their arthritis patients to exercise mostly advocate low-intensity (e.g., ROM) rather than moderate- to high-intensity exercise, despite the proven efficacy and safety of moderate- to high-intensity exercise and the relative ineffectiveness of low-intensity exercise (163). This general failing by clinicians to encourage and provide informed instruction on how best to exercise is a contributor to the physical inactivity that characterizes a patient with arthritis (113). Patients with symptomatic arthritis may not be motivated to perform physical activity on their own because joint pain and fatigue are barriers to exercise. All these factors contribute to adults and children with arthritis being generally less active than healthy individuals (96, 208, 211, 231). This demonstrates the importance of the role of health care professionals in providing specific instructions on exercise and addressing patients' concerns about joint health and joint pain management.

For example, in 2008, 69% of RA patients in the United States reported doing no regular physical activity, and only 17% reported achieving the recommended three or more bouts of physical activity per week (211). In 2009, in the United States, a median of 32% of adults with any form of arthritis reported performing no leisure-time physical activity compared with a median of 21% of adults without arthritis (96). The Osteoarthritis Initiative, which objectively assessed physical activity using accelerometry, found that half of adults with knee OA were inactive, with only 10% sufficiently active to satisfy low or moderate aerobic physical activity recommendations (55). Additionally, a meta-analysis (240) determined that only 13% of individuals with knee OA achieved the current weekly recommendation of 150 min of physical activity. Despite the known benefits of community-based arthritis exercise programs such as the Arthritis Foundation YMCA Aquatics Program (AFYAP) and People with Arthritis Can Exercise (PACE), less than 1% of eligible persons with arthritis access these programs. Thus, individuals with arthritis require education and encouragement to increase and maintain physical activity as well as instruction on how to perform appropriate exercise.

In addition to exercise, multiple nonpharmacological interventions are used in the routine management and rehabilitation of patients with chronic arthritis (44):

- Education
- Physical and occupational therapy
- Braces and bandages
- Canes and other walking aids
- Shoe modification and orthotics
- Ice and heat modalities
- Weight reduction
- Avoidance of repetitive-motion occupations
- Joint surgery (in select circumstances)

Joint replacement surgery is an option for patients with end-stage symptomatic OA who have progressively worsening symptoms, leading to functional limitations that affect ADLs and QOL. Although joint replacement surgery (arthroplasty) is usually performed only when conservative measures have failed, these procedures are increasingly common; by 2010, over 7 million individuals in the United States were living with a total knee (4.7 million) or total hip (2.5 million) replacement, representing 1.5% and 0.8% of the total U.S. population, respectively (35, 137). The prevalence of arthroplasty surgery increases with age, and 10% and 5% of Americans aged 80 yr or more have had total knee or total hip replacements, respectively. Although joint replacement surgery is very effective in resolving pain, physical function of some patients may remain suboptimal 12 mo after surgery. Incomplete rehabilitation is generally attributed to postsurgical rehabilitative exercise that was inadequate (in terms of both volume and intensity) in restoring muscle mass, strength, and aerobic capacity (123, 185, 233).

Pharmacological therapies for arthritis vary according to the form of arthritis and type of symptoms exhibited. Unfortunately, OA is an irreversible process, and the mainstays of treatment remain lifestyle modification (exercise and weight loss) and medications to reduce pain (92). Frequently used and recommended oral agents include nonopioid analgesics such as acetaminophen and aspirin, tramadol (opioid-like), and nonsteroidal anti-inflammatory drugs (NSAIDs) such as ibuprofen and naproxen. In the most recent ACR recommendations for management of knee, hip, and hand OA (92), oral supplementation with glucosamine and chondroitin sulfate (hand), the serotonin–norepinephrine reuptake inhibitor duloxetine (hand, knee, and hip), and topical capsaicin (knee) were all conditionally recommended for use in varying types of OA. Intra-articular corticosteroid injections are commonly used and strongly recommended for pain control in hip and knee OA and can be considered in hand OA. Although commonly used in clinical practice, and FDA approved in the United States for mild to moderate knee OA, intra-articular hyaluronic acid derivatives (viscosupplementation) are not recommended for use (92, 114).

For persons aged 75 yr or older, topical rather than oral NSAIDs are recommended. Dosing of an oral NSAID 1 h before exercise may increase exercise compliance by reducing pain and stiffness during activity. Intra-articular injection of corticosteroids is effective for treating OA flares. After injection, exercise should be avoided for 24 h, as should weight bearing when the injection is into weight-bearing joints; it is generally advised that a patient receive no more than three intra-articular injections into the same joint during any 12 mo period (not given less than 3 mo apart). In persons with OA, being overweight is associated with increased pain and disability (10, 46, 63, 64) and may worsen disease activity (58). Thus the inclusion of dietary intervention in combination with exercise is optimal when weight loss is a treatment goal (154, 207).

In addition to the analgesics and anti-inflammatories mentioned, a variety of **disease-modifying antirheumatic drugs (DMARDs)** are available for treating the autoimmune inflammatory arthropathies, RA and AS. With DMARDs, the clinician aims to control disease activity by suppressing immune function and subsequently inhibiting the inflammatory process. The favored DMARDs for RA include methotrexate (MTX), leflunomide (LEF), sulfasalazine (SSZ), hydroxychloroquine (HCQ), the janus kinase (JAK) inhibitor tofacitinib, and **biologics** such as TNF-α inhibitors (etanercept, infliximab, adalimumab, certolizumab pegol, and golimumab), anti-T cell (anti-CD28) therapy (abatacept), anti-B cell therapy (rituximab), and anti-interleukin-6 receptor (anti-IL-6R) agents (tocilizumab) (210). These agents are administered either singularly (DMARD monotherapy, usually MTX but may also be SSZ, HCQ, or LEF) or in combination: biologic + MTX; double DMARD therapy (e.g., MTX + SSZ, MTX + HCQ); or triple DMARD therapy (MTX + SSZ + HCQ), with the aim of achieving low disease activity, preferably remission. This goal-oriented treatment strategy, which involves regular monitoring of inflammation, with treatment adaptations if the goal is not achieved, is variously called treat to target (T2T) or tight control of disease activity. Additionally, disease flares are often treated by high-dose intravenous or intra-articular corticosteroid injections.

For AS, commonly used medications are locally administered parenteral corticosteroids, DMARDs, and anti-TNF biologics—etanercept, infliximab, adalimumab, certolizumab pegol, and golimumab (242). Key disease-specific clinical outcomes for AS can be assessed by using the Bath Ankylosing Spondylitis Disease Activity Index (BASDAI), the Bath Ankylosing Spondylitis Function Index (BASFI), and the Bath Ankylosing Spondylitis Metrology Index (BASMI; an objective measure of axial mobility).

Because of its numerous benefits, exercise is recommended in the U.S. (6, 7, 92, 242) and European (41, 105, 247, 249, 250) treatment guidelines for OA, RA, and AS. These benefits include increased aerobic capacity, muscle strength, and flexibility; improved physical function, weight control, self-efficacy, and mood; enhanced QOL; reduced pain; and decreased CVD risk (16, 18, 22, 41, 50, 68, 70, 90, 99, 101, 106, 117, 128, 132, 157, 174, 193, 194, 199, 229, 241, 246). Contrary to popular belief, exercise does not increase risk for OA, nor does it exacerbate joint damage in RA patients. These points are discussed in detail later in the chapter. Table 24.5 provides a review of common arthritis medications.

EXERCISE PRESCRIPTION

The following are goals of exercise specific for the treatment of arthritis:

- Maintain or improve physical function by maintaining or improving muscle strength, cardiorespiratory fitness, and ROM

- Improve body composition (i.e., restore muscle mass and reduce fat mass) and, when appropriate, reduce body weight
- Reduce the risk of comorbidities such as CVD, type 2 diabetes, metabolic syndrome, and osteoporosis
- Reduce inflammation and pain
- Prevent contractures and deformities

Immobilization and inactivity amplify the negative systemic and psychological manifestations that accompany arthritis (109, 223). The effects of inactivity include rapid reductions in strength (approximately 3%-8% per week) and aerobic capacity, muscle loss, reduced bone mass, and loss of cartilage matrix components (81, 125, 193). Because cartilage is **avascular**, it depends on normal repetitive loading of the joint for its nutrition and normal physiological function (93). Moreover, joints with effusions may develop synovial ischemia attributable to elevated intra-articular pressure. Walking and cycling increase synovial blood flow in inflamed knees (102). Additionally, in both healthy and OA joints, the intra-articular oxygen partial pressure increases during joint movement, although the increase is diminished in arthritic joints (159).

A person in an exercise program, whether supervised or unsupervised, requires education, skill acquisition, and reinforcement (7, 141). Patients may benefit from occasional monitoring or, in some cases, direct supervision by a health care professional. Supervised exercise training for individuals with arthritis most often occurs in a group setting (vs. personal training), and this is well supported as a treatment modality (38, 58, 141, 163, 170, 235). A supervised group setting may be beneficial because it provides peer social support and facilitates training compliance. Access to a clinical exercise physiologist who has previously evaluated the individual may help decrease anxiety (240). Additionally, the clinical exercise physiologist may play a key role by encouraging regular attendance at sessions early in the treatment regimen, because session attendance in the initiation phase is the strongest predictor of attendance in the later stages of the intervention (237). These are key considerations because exercise training compliance is associated with training response and improved physical function (125, 141, 238). Regular monitoring by clinical exercise professionals also helps ensure safety and appropriate progression of the exercise.

Supplementing supervised classes with home-based exercise may boost improvements in pain and function (42, 51, 148, 191). In addition, participants who perform some home-based exercise during the maintenance stage of the intervention are more likely to adhere to the exercise program than peers who exercise only at a facility (237). Thus, the introduction of unsupervised and independent exercise appears to be important in modifying an individual's behavior so that regular exercise becomes a part of daily routine.

Many exercise options are available for unsupervised programs. Walking is an excellent cardiorespiratory exercise, although performing a variety of cardiorespiratory activities (e.g., cycling, swimming, rowing) may help maintain interest and compliance

Table 24.5   Pharmacology

Medication class or name	Primary effects	Exercise effects	Special considerations
OSTEOARTHRITIS			
Analgesics (e.g., nonopioid analgesics such as acetaminophen, aspirin; tramadol)	Reduce pain Reduce fever (acetaminophen)	None reported	Aspirin may cause GI-related symptoms. Hypersensitivity to aspirin is reported in 1%-2% of the population (rash or hives, respiratory symptoms). Acetaminophen overdose can lead to liver damage. Tramadol is an opioid-like controlled substance with addictive potential; should be used with caution.
Nonsteroidal anti-inflammatory drugs (NSAIDs) (e.g., diclofenac, ibuprofen, naproxen, and selective COX-2 inhibitors such as celecoxib)	Reduce inflammation by inhibiting the enzyme cyclooxygenase, which in turn reduces prostaglandin synthesis Reduce pain	None reported	NSAIDs are associated with GI irritation, nausea, diarrhea, and occasionally ulceration. Avoid use in patients with peptic ulcer disease and history of bariatric surgery. NSAIDs should be used with caution in the elderly, and coincident use of proton pump inhibitors is advised. COX-2 inhibitors have lower risk of GI complications. NSAIDs and COX-2 inhibitors may be associated with increased risk of cardiovascular thrombotic events (MI, stroke).
Intra-articular corticosteroid injections	Reduce inflammation and consequently improve pain and mobility	Avoid exercise involving the affected joint for 24 h after injection	Recommendation is no more than 3 steroid injections per joint per year (i.e., given no less than 12 wk apart).
Glucosamine	Was thought to provide pain relief and stimulate cartilage growth, but actual mechanism of action is unknown Not recommended for knee, hip, or hand OA (92)	None reported	GI side effects are occasionally reported. Glucosamine may interact with warfarin (anticoagulant medication).
Topical capsaicin	Derived from chili peppers; thought to attenuate cutaneous hypersensitivity by defunctionalization of nociceptor fibers Conditionally recommended for knee OA (92)	None reported	Use is associated with warmth, burning, or stinging sensation at application site. Not recommended for hand OA because of lack of direct evidence and risk of potential eye contamination with use (92)
RHEUMATOID ARTHRITIS AND ANKYLOSING SPONDYLITIS			
NSAIDs	See above	See above	See above
Disease-modifying antirheumatic drugs (DMARDs) (e.g., methotrexate, azathioprine, leflunomide, sulfasalazine, hydroxychloroquine) and the biologics (e.g., etanercept, infliximab, adalimumab, golimumab, certolizumab pegol, abatacept, rituximab, and tocilizumab)	DMARDs are immunosuppressants and thus reduce inflammation; the biologics are specific cytokine or T cell or B cell modulators	None reported	Immunosuppressants reduce the body's response to infection. Each DMARD has recognized side effects and toxicities; careful monitoring is essential.
Systemic corticosteroids	Immunosuppression	None reported	In addition to reduced response to infection, chronic corticosteroid use is associated with osteoporosis and, in high doses, muscle loss.
Intra-articular corticosteroid injections	See above	See above	See above

and reduce the likelihood of overuse injuries. An interactive tool featuring exercise programs suited to patients with arthritis is available through the Arthritis Foundation (www.arthritis.org). These exercise options include land-based cardiorespiratory and strength training programs as well as water-based programs (e.g., water walking) and ROM exercises. The American College of Sports Medicine has also produced informational handouts encouraging home exercise regimens in patients with OA and RA through their Exercise Is Medicine campaign (www.exerciseismedicine.org).

Following the Arthritis Foundation's YMCA Aquatic Program (AFYAP), available nationally in YMCAs and in private facilities, has been shown to increase hip ROM, isometric strength, and flexibility when performed two or three times a week over 6 to 8 wk (224, 225). In a meta-analysis that included data from 1,092 individuals with lower limb OA, Waller and colleagues (241) concluded that aquatic exercise achieved small but statistically significant improvements in pain and self-reported function comparable to those gained by land-based exercise or use of NSAIDs. This reported benefit concurs with the findings of the Cochrane review by Bartels and colleagues (19). Similarly, Dunbar and colleagues (53) observed that aquatic exercise performed 5 d/wk for 4 wk significantly attenuated pain and improved QOL in AS patients. However, like numerous other researchers, Bartels and colleagues (19) did not find a benefit regarding muscle strength. This lack of efficacy of aquatic exercise is generally explained by the failure of this form of training to achieve sufficient intensity to increase strength (134), which in turn has led to the recommendation that aquatic exercise be supplemented by more intense strength training performed on land.

Nonfacility-based activities and exercises that can be performed throughout the day include chin tucks, corner pectoral stretches in a doorway, abdominal tightening, checking posture throughout the day in the mirror, and extending walking time by taking the stairs or parking farther from a destination.

Specific Exercise Prescription Considerations

In establishing an exercise prescription, the exercise professional should consider the success of various treatment modalities for particular joint impairments, as well as the individual's affected joints, level of fitness, surgical history, comorbidities, medications, age, personal goals, and lifestyle. Inflammation and joint degeneration associated with the disease process cause a cycle of decreasing function and increasing impairment. It is not yet clear whether therapeutic exercise alters the pathological process of arthritides, although it is clear that exercise (including long-term high-intensity training) does not exacerbate disease activity, and reports of reductions in pain, inflammation, and even joint damage are abundant in the literature.

Exercise can induce muscle and joint pain, especially during initiation of training; this is the primary complaint and disabling factor related to arthritis. If pain occurs, it can become difficult to motivate a person to maintain adherence to an exercise program. In an arthritic population that is generally older, is typically more sedentary than the general population, and may be using systemic corticosteroids, avoidance of injury and pain is an important factor in facilitating exercise compliance. The clinical exercise physiologist's role is to make recommendations that minimize exacerbation of these symptoms. For example, deconditioned patients should begin training programs at a low intensity and progress gradually to reduce the likelihood of muscle soreness (120). Additionally, the clinical exercise physiologist should inquire which joints are affected because this information will allow for a more comprehensive assessment, bringing awareness to decreased ROM, instability, reduced muscle strength and flexibility, poor joint proprioception, and increased pain. This record may be refined with information from the individual's primary care physician, rheumatologist, orthopedist, or physical therapist. Some site-specific exercise recommendations are listed in table 24.6.

Impaired balance and increased fatigue are additional factors that must be considered when developing an exercise program. To increase or maintain patient motivation to start or continue exercising, it is essential that patients be aware that reduction in arthritis pain and fatigue are anticipated benefits of regular exercise (16, 41, 90, 99, 101, 107, 194), and that improved strength and aerobic capacity will reduce the difficulty and resultant fatigue associated with performing ADLs.

Compliance with an exercise program can be challenging for patients with arthritis (61). Efforts on the part of the clinical exercise professional to prevent musculoskeletal injury related to exercise and to appreciate the specific concerns of the individual with arthritis will facilitate overall enjoyment and compliance with the exercise program.

Disease Staging

A consideration in developing an appropriate exercise prescription for persons with arthritis is the disease stage. The focus of exercise therapy for chronic stages of arthritis should be to maintain or improve function while minimizing or preventing symptom exacerbation. Most patients referred for exercise therapy are experiencing arthritis in a chronic stage. Table 24.7 lists the arthritis stages and their associated exercise-related considerations. Table 24.8 provides specific recommendations for the clinical exercise professional to follow when developing an exercise prescription for patients with arthritis who have disease-specific skeletal conditions.

Table 24.6 Site-Specific Recommendations for Exercise Programs in Arthritis Patients

Site	Condition	Presentation	Recommendation
LOWER EXTREMITY			
Foot	Hallux valgus	Lateral deviation of the great toe of the foot, leading to bunion deformity and bursitis at the first metatarsophalangeal (MTP) joint Associated with 1st MTP joint OA	Using a bunion pad between the first and second toe, a gel bunion sleeve, orthotics, or footwear with a wide toe box can provide relief. If moderate to severe, avoid walking or high-impact activities for exercise.
Foot	Plantar fasciitis or heel spur	Sharp heel pain, primarily medially, on weight bearing; worse with first steps upon wakening and with walking	Avoid weight-bearing activities that exacerbate symptoms. Wear shoes with supportive arch; may need soft or semirigid foot orthoses or heel cups. Perform heel cord stretching exercises. Massage ice along plantar fascia. Wear night splints to maintain stretch in dorsiflexion.
Leg	Medial tibial stress syndrome (shin splints)	Commonly due to musculotendinous inflammation of the anterior or posterior tibialis muscle, or medial tibial periostitis Aching pain along distal 2/3 of posteromedial tibial border of lower leg, worse with activity and often improved upon rest Related to overuse	Avoid activities that may exacerbate symptoms (running, prolonged walking, stair climbing). Improved footwear or orthotics may be needed. Evaluate training routine for changes that preceded onset (e.g., increased mileage or frequency of workouts; changes in footwear or running surfaces). Perform targeted stretching exercises. May supplement with low-impact exercise as tolerated. Physical therapy can also be beneficial. If symptoms do not improve, further workup may be indicated to rule out stress fracture or compartment syndrome.
Hip or knee	Hip OA, knee OA	Gait deviation Pain with weight bearing or stair climbing	Patellofemoral OA, hip or groin pain without gait deviation or balance problems: do rearward walking. Patellofemoral OA: perform quadriceps and hamstring strengthening, manual therapy, taping. Hip or groin pain: do hip bridging, free-speed walking, stationary cycling. If gait deviation is caused by pain or decreased joint range of motion (ROM), use of cane or rolling walker in the hand opposite the affected limb may be necessary.
Hip	Bursitis	Lateral hip pain, may extend to lateral knee Lying or single-leg stance on affected side may be uncomfortable or not tolerated	Control inflammation; apply ice or cold pack to lateral hip. Perform targeted stretching exercises. Physical therapy can be beneficial. Patient should see physician for evaluation of bursitis and to rule out other conditions that may produce similar or referred symptoms.
Knee	Valgus or varus deformity	Valgus deformity commonly called knock-knee; varus deformity commonly referred to as bow-legged	Perform exercises that are not limited by deformity and may be done comfortably without exacerbating symptoms.
UPPER EXTREMITY			
Shoulder	Shoulder pathology (bursitis or tendinitis, rotator cuff tear, adhesive capsulitis, OA)	Pain with overhead activities, reaching, or end-range motion Can lead to sleep disturbances	Perform targeted shoulder exercises to help improve strength and ROM. May perform shoulder ROM in pain-free range in pool with upper extremity submerged. If pain disturbs function or sleep, patient should be evaluated by a physician. Physical therapy is commonly prescribed to improve pain, ROM, strength, and function.
Hand	OA of carpometacarpal joint of the thumb	Pain in hand proximal to thumb	Avoid gripping activity during exercise. Enlarge grips.
Hand	Ulnar deviation in RA	Deviation of body of hand and fingers to the small-digit side of the hand	Avoid gripping activity during exercise. Use large muscles and joints for functional activities.

Site	Condition	Presentation	Recommendation
		AXIAL SKELETON	
Lumbar spine	Spinal stenosis	Flexed low back during walking, standing, and sitting Symptoms increase with extension (standing, looking overhead) Often presents with claudication-type symptoms with walking	Perform flexion exercises and seated cardiorespiratory exercise (recumbent bicycle or stair stepper). Perform aquatic exercise. Use rolling walker for household or community ambulation.
Cervical spine	Atlantoaxial subluxation in patients with RA	Facial sensory loss Vertigo, ear pain, headache Numbness or tingling of hands or feet Difficulty walking Loss of control of bowel or bladder Transient loss of consciousness with extension of cervical spine May be asymptomatic	Cervical spine symptoms in RA should be promptly diagnosed and treated because of potential neurological and lethal complications. Avoid any passive or heavy resistive neck ROM. Surgery is indicated if there is cord compression or progression of neurological symptoms.
Cervical or lumbar spine	Nerve compression secondary to OA	Gradual, recurrent pain or pain after activity Numbness, tingling, or pain in the extremities, sometimes only with certain movements	Avoid heavy lifting and activity that results in exacerbation of symptoms (pain, numbness or tingling in extremities). Physiotherapy can strengthen the core and improve function and biomechanics.

Table 24.7   Arthritis Stages, General Signs and Symptoms, and Exercise Prescription Considerations

Stage	Signs and symptoms	Exercise considerations
Acute	Fatigue Joint pain Reduced joint tissue tensile strength attributable to inflammation Reduced joint nutrition	Avoid activities that exacerbate joint pain.
Chronic	Permanent joint damage Pain at end of normal ROM Stiffness after rest Poor posture and ROM Joint deformities Pain with weight bearing Abnormal gait Weakness Contractures or adhesions Reduced aerobic endurance and muscle strength	Perform aerobic, strengthening, and ROM exercises. Perform exercises and intensities during resistance training that do not cause joint pain but are still sufficient to ensure gains in strength and muscle mass. Initiate walking and perform in water if necessary to reduce pain. Perform lower back flexion and abdominal strengthening exercise. Avoid trunk extension (especially with spinal stenosis). Maintain neutral spine position. To reduce risk of osteoporosis and ligament laxity, avoid long-term oral corticosteroids.
Chronic with acute exacerbation	Inflammation and joint size greater than normal Joint tenderness, warmth, swelling Joint pain at rest and with motion Stiffness Functional limitations Hips and spine affected	Normalize gait. Same recommendations as for acute phase.

ROM = range of motion.

Table 24.8   Exercise Prescription Review

Training method	Frequency	Intensity	Time (Duration)	Type (Mode)	Progression	Special considerations
Cardiorespiratory	Begin at 2 or 3 d/wk and increase to 3-5 d/wk, more frequently for moderate-intensity activities; daily exercise is encouraged as tolerated	40%-59% HRR or $\dot{V}O_2$ reserve RPE 12-16 (Borg 6-20 scale) or 3-6 (Borg 1-10 scale); in capable individuals consider vigorous intensity ($\geq$60% HRR)	20-60 min, equaling at least 150 min moderate intensity or 75 min vigorous; can combine intensities	Activities using large muscle groups with repetitive motion: walking, cycling, dancing, swimming or aquatics	Gradually increase duration and frequency of exercise, then increase intensity; if exercise naive or extremely deconditioned, begin at 40% HRR and progress up to 85% HRR.	See text regarding the following: • Injury • Fatigue • Joint replacement • Time of day • Weather effects • Aquatic therapy • Footwear • Cardiovascular and pulmonary manifestations • Ankylosing spondylitis • Corticosteroids • Body composition
Resistance	2 or 3 d/wk, with at least 24 h between each session	Begin at 50%-60% 1RM and progress to 60%-80% 1RM	8-10 exercises, 8-12 reps, 2-4 sets	Isotonic, preferably performed on machines, but free weights can be used in select patients with proper safety precautions	Gradually increase the volume of training, then increase the intensity.	Perform in pain-free range. Whenever possible, use functional movement patterns. Include all major muscle groups, even with unilaterally distributed joint dysfunction. Body weight exercises can also be incorporated.
Range of motion	Daily	To no more than a mild tightness or discomfort of the stretched muscle group	Static and active stretches for 10-30 s for 3-5 reps; dynamic stretches for 10 reps	Static, active, and dynamic stretching* (see chapter 6); functional activities (e.g., sit-to-stand, stairs)	Gradually increase ROM of stretching exercise; gradually increase volume and intensity of functional activities (e.g., sit-to-stand: reduce then eliminate use of arms).	Target shortened muscle groups. Perform active ROM within the normal range for all joints, even with unilaterally distributed joint dysfunction. If pain occurs, reduce range of motion or stop activity.

*Passive ROM involves no muscle work by the individual while an outside force (another person or a passive motion machine) moves the body part through a range of motion. Active range of motion is movement of a body part by the individual performing the exercise without outside forces. Active assisted ROM involves partial assistance with motion by an outside force, whereby a portion of the motion of the limb may be provided by a mechanical device, another limb, or another person. Dynamic stretching is a type of active ROM activity involving controlled movement of a body part with a gradual increase in the range and speed of the movement.

HRR = heart rate reserve; RPE = rating of perceived exertion; RM = repetition maximum; ROM = range of motion.

Preventing Musculoskeletal Injury Secondary to Exercise

Because cardiorespiratory exercise involves high repetition of joint motion, there is a risk of overuse injuries. Fortunately, injuries attributable to supervised exercise are infrequent. It is estimated that 2.2 minor injuries occur per 1,000 h of exercise, and that major injuries (i.e., those necessitating a reduction or discontinuation of exercise for at least 1 wk) occur at a rate of only 0.48 per 1,000 h of exercise (38). Patients with arthritis may minimize overuse by performing interval or cross-training during endurance exercise. Examples include the following:

1. Alternating cycle ergometry between 25% and 75% of the maximum work rate performed during a graded exercise test
2. Alternating between water walking and joint ROM exercises in a pool setting
3. Walking and weight training
4. Walking and higher-intensity cardiorespiratory exercise such as recumbent stair stepping or cycling

The development of strong knee extensors with quadriceps-strengthening exercise decreases impulse loading of the lower limb during walking by slowing the deceleration phase that occurs before heel strike. Therefore, adequate quadriceps strength may help prevent knee injury and slow the progression of knee OA.

Another consideration is that some patients may have laxity in the structures that support a joint due to the rheumatic process or the use of corticosteroids. Under these circumstances, the joint should be protected during exercise or normal activities. For example, patients should stretch cautiously to avoid extending beyond the functional ROM; vigorous stretching or manipulative techniques are contraindicated (111). In many cases, providing external support to a joint may be necessary. To protect smaller joints during activities and exercises, larger muscles and joints should be used.

Fatigue

Fatigue is common with inflammatory rheumatic disease and can profoundly affect QOL. Morning stiffness lasting for 3 or 4 h and fatigue beginning in the afternoon and lasting until evening are common symptoms of inflammatory arthropathies, leaving only a few hours during midday when stiffness or fatigue is not a problem. Fatigue is a complex phenomenon related to exertion, deconditioning, depression, sleep patterns, or a combination of these factors (107, 194). A person who is fatigued is less able to exercise for a long duration, tends to be less motivated, and may become frustrated. Although relatively few studies have investigated the effects of exercise training on fatigue levels in arthritis patients, there is evidence to support a benefit (41, 107, 194). This is consistent with expectations that increased aerobic capacity and strength enable performance of ADLs at a lower percentage of functional capacity, with reduced subsequent fatigue.

Previous Joint Replacement

Total joint replacement surgery is common in patients with advanced arthritis (35, 137), particularly of weight-bearing joints (knee, hip). Individuals who undergo total knee or hip arthroplasty are often deconditioned and overweight; therefore, exercise training is essential to improve physical function and reduce the risk of comorbid conditions (185, 233). Beginning appropriate exercise rehabilitation programs soon after surgery has been shown to improve strength and function, elicit muscle hypertrophy, reduce pain and stiffness, and enhance QOL (109, 123, 221). Postoperative rehabilitation protocols should be followed under the guidance of the patient's orthopedic surgeon.

Time of Day and Weather Effects

Morning stiffness is a symptom for many individuals with arthritis. The clinical exercise professional should have an understanding of the daily variability of symptoms, in particular the difficulties related to early morning activity. For those with inflammatory arthritis characterized by prolonged morning stiffness, exercise should be prescribed for the late morning or early afternoon. Moreover, a change in ability to perform exercise during periods of inclement weather is frequently reported by individuals with rheumatologic conditions. For instance, individuals with arthritis will often report increases in pain and stiffness with variations in temperature, humidity, and barometric pressure.

Aquatic Therapy

Rheumatoid arthritis is commonly associated with Raynaud's phenomenon and Sjogren's syndrome. Raynaud's phenomenon is a **vasospastic** condition presenting as blanching or cyanosis of the hands and feet, particularly when exposed to cold weather or emotional stress. Although the phenomenon typically has a benign course, extreme cases can result in pitting scars and even gangrene. Individuals with Raynaud's phenomenon should be mindful that cold water can exacerbate symptoms. The choice of exercise modality should be tailored to the symptoms attributable to arthritis and Raynaud's phenomenon. Sjogren's syndrome is an autoimmune condition characterized by dry mouth and eyes, caused by lymphocyte infiltration of salivary and lacrimal glands. Patients with this condition may find chlorinated water and the air surrounding pools especially irritating to the eyes and should always wear goggles when in a pool.

Footwear

Use of appropriate footwear can reduce the risk of injury related to poor lower extremity mechanics and repeated impact. Lightweight commercial running shoes that include hindfoot control, a supportive midsole of shock-absorbing materials, a continuous sole, and forefoot flexibility can aid shock attenuation and biomechanics. Individuals with OA, RA, and AS with biomechanical malalignment or deformity of the lower extremity may benefit from prefabricated or custom-fit orthotics. Additionally, with lower extremity arthritis, it is advisable to wear pool shoes to assist mobility and protect feet from injury during aquatic exercise.

Cardiovascular and Pulmonary Manifestations of Rheumatic Disease

Cardiovascular disease risk is increased in RA and AS (112, 181, 223, 226) and is estimated to be 1.5- to 2-fold relative to that of the general population, making inflammatory arthritis comparable to diabetes as a CVD risk factor (173). Pulmonary disease can also be associated with arthritic conditions. For example, RA is associated with interstitial lung disease (10), and those with AS often have restrictive lung disease because of impaired chest expansion and bilateral upper lobe pulmonary fibrosis (66). In most cases, cardiovascular and pulmonary manifestations of rheumatologic disease are not contraindications to exercise, although a vital capacity of 1 L or less is considered a relative contraindication to participating in pool therapy because of the restrictive effects of hydrostatic pressure on the chest wall. Chapters 14 and 15 in part III of this book review exercise for patients with CVD, and chapters 19 and 20 in part IV review exercise for those with pulmonary limitations.

Ankylosing Spondylitis

The bony fusion that occurs in the spine and sacroiliac joint with AS cannot be prevented, but active rehabilitative strategies can improve spinal mobility and physical function (42, 158, 174, 242). Since individuals with AS tend to be younger and more physically active at diagnosis, exercise can generally be started at a higher relative intensity than for RA and OA (66). When peripheral joints are involved, disease pathology is similar to that for RA, and thus exercise recommendations specific for RA should apply. A phenomenon that can occur is the last-joint syndrome, in which bridging ossification between vertebral bodies occurs at every level except one (87). This sole mobile segment is exposed to considerable stresses during exercise and can present with localized pain and **discitis**. In this circumstance, bracing or surgical fusion may be necessary.

Corticosteroids

Systemic corticosteroids are a common treatment for RA. Although clinicians try to limit the use of corticosteroids, chronic use may be unavoidable. Chronic corticosteroid use is associated with a myriad of side effects, including muscle atrophy, myopathy, decreased bone density, increased adiposity and obesity, and development of type 2 diabetes. These effects can put rheumatologic patients at higher risk for fractures in addition to reduced muscle strength.

Body Composition

Obesity is a modifiable factor that is common in most forms of arthritis. Obesity is also a strong risk factor for OA **incidence**, progression, and disability (152, 154) because of the increase in mechanical stress on weight-bearing joints (152); it is estimated that a 5 kg weight loss decreases the risk of developing knee OA within 10 yr by 50% (65). The Arthritis, Diet, and Activity Promotion Trial (ADAPT) (153) found significant improvements in physical function and pain for overweight or obese knee OA patients when exercise and diet restriction were combined. Both diet and exercise improved pain, but the effect was heightened when used in combination, leading authors to conclude that dietary weight loss without exercise was ineffective for improving function, mobility, and pain.

Rheumatoid arthritis is characterized by rheumatoid cachexia, a condition that features reduced muscle mass and elevated adiposity (222). Significant muscle loss is evident in approximately two-thirds of RA patients (120, 121, 197). Obesity among RA patients is even more widespread with a prevalence of approximately 80%, with fat preferentially deposited around the trunk (i.e., central obesity, or truncal adiposity). Both muscle loss and increased adiposity are present in patients with well-controlled disease, including those in remission (120, 121, 217). Because of the coincident loss of muscle and increase in fat mass, body weight often remains stable, rendering body mass index (BMI) a misleading indicator of obesity in this population (215). In studies featuring high-intensity progressive resistance training (PRT), reversal of rheumatoid cachexia (i.e., increased muscle mass and decreased fat mass) has been achieved without exacerbating disease activity (82, 140, 216). Additionally, a study of obese subjects with RA found that a program of moderate physical training, reduced dietary energy intake, and a high-protein, low-energy supplement was successful in achieving significant weight loss without loss of lean tissue (56). Cachexia has also been shown to be a feature of AS; as expected, this muscle loss is associated with reduced strength and physical function (138). Thus, interventions aimed at inducing muscle hypertrophy, such as PRT, may be beneficial for individuals with this condition.

It should be noted that, to date, none of the DMARDs or biologics used to treat the inflammatory arthritides have been shown to be anabolic (i.e., to increase muscle mass), and some medications appear to increase adiposity (high-dose corticosteroids and the anti-TNFs; 57, 91, 139). Hence, it is not surprising that even successful pharmaceutical management of disease activity fails to restore a healthier body composition phenotype to either RA or AS patients. Consequently, an intervention that can improve body composition is required if function is to be improved and comorbidity risk attenuated; exercise training, particularly high-intensity exercise, is considered the best available intervention for achieving these goals (114, 120, 140, 153, 203, 216).

General Exercise Prescription Recommendations

The sequencing of exercises for individuals with arthritis is similar to that for the general population, beginning with a warm-up and ending with a cool-down. A warm-up should be performed to increase tissue temperature, and as a consequence tissue compliance, throughout the body. As decreased stiffness and greater ROM ensue, patients should be taught to judge whether increasing the range through which they are exercising is safe (206). Strengthening and cardiorespiratory conditioning exercises should be performed after a warm-up and should be followed by a cool-down period. Flexibility training is best performed during the cool-down, because muscles and connective tissue are more pliable when body temperature is elevated. Laliberte and colleagues (116) provide extensive and useful examples and illustrations of various sequences of exercise appropriate for people with arthritis.

Exercise intensity, duration, or frequency (or a combination of these variables) should be progressed when the exercise is not as challenging as it was previously (i.e., owing to the training response) and when symptoms do not increase for two or three consecutive sessions. Conservative increments are recommended. In general, after 1 wk of consistent exercise without an exacerbation of symptoms, the training demands should be increased toward the maximum recommended. To ensure that progression in aerobic, resistance, or combined training programs is safe, initially increase the volume (i.e., total duration of exercise, the number of exercises performed, or the time spent exercising with the use of shorter rest periods) and then gradually increase the intensity (e.g., increasing the percent heart rate reserve for aerobic exercise, or the percent one-repetition maximum for PRT). An increase in symptoms may require lowering the intensity or volume of exercise, especially for the affected joint. Once the patient is accustomed to exercising (i.e., muscle soreness is not simply due to unaccustomed activity), the 2 h pain rule is a helpful maxim for regulating exercise intensity. A localized increase in pain that lasts more than 2 h after an activity suggests the need to decrease the exercise intensity or volume for the next training session.

Collecting outcomes data serially to objectively monitor response to therapy is useful. Table 24.4 lists a variety of physical function measures that can be used to evaluate strength, aerobic capacity, physical function, ROM, balance, and body composition in arthritis patients. In addition, a number of standardized and validated instruments (questionnaires) are available for assessing arthritis pain, stiffness, and subjective function, as well as response to therapy, for OA, RA, and AS (21, 126, 151, 182, 200, 202).

Cardiorespiratory Exercise

A key to cardiorespiratory (aerobic) exercise therapy for arthritic conditions is to manipulate the intensity and duration. Cardiorespiratory exercise intensity should be guided by heart rate (i.e., percent heart rate reserve, %HRR) or rating of perceived exertion (RPE) (68, 147); RPE becomes paramount as an indicator of exercise intensity when maximal HR is pharmaceutically restricted by β-blocker or calcium channel blocker treatment. Since most people with arthritis can achieve normal training heart rate ranges (150), standard HRR guidelines can usually be applied (see chapter 6). A cardiorespiratory exercise intensity of 12 to 16 on the 6- to

20-point Borg RPE scale or 3 to 6 on the 10-point Borg RPE scale is recommended.

Aerobic training is ideally achieved using a mode that minimizes the magnitude and rate of joint loading (228). The best types include walking, cycling, and pool exercise (19, 58, 165, 174, 194, 241). Free-speed walking produces less hip joint contractile pressure than isometric or standing dynamic hip exercises (228). Nordic walking (which uses walking poles) offers some advantages over normal walking: It burns more calories, helps keep the body upright, and reduces impact when walking downhill (165, 243). If walking is uncomfortable, if pain lasts more than 2 h after walking, or if the individual has altered biomechanics of the lower extremity, an alternative cardiorespiratory exercise such as cycle ergometry, recumbent stair stepping, upper body ergometry, or water walking should be used (38, 160, 216). Cycle ergometry should be conducted with the seat height and crank length adjusted to limit knee flexion and minimize pedal load, which in turn decrease knee joint stress (135).

Water-based, or aquatic, exercise moderately improves physical function and ROM and reduces pain (19, 99, 241), but it appears to have no effect on strength in OA or RA patients (99, 241) or on balance and fall risk in individuals with knee OA (144). Increased QOL has been widely reported in OA patients after aquatic exercise training (19, 241). Similarly, improvements in pain and QOL have been observed with aquatic exercise in AS patients (53). The buoyant quality of water helps patients perform passive and active joint ROM exercises. Strengthening exercise can be performed in the water, because water offers resistance to motion (i.e., the faster a limb or body moves in water, the greater the resistance and consequently the workload). However, as previously mentioned, the failure to achieve sufficient intensity may explain the lack of effect on strength generally observed after aquatic training.

Many people with arthritis are able to tolerate longer workouts in the water than on land, and this may account for the high levels of training adherence associated with this form of exercise (241). With regard to the pool conditions, compliance with aquatic exercise decreases with water temperatures colder than 84 °F (28.9 °C), and cardiovascular stress increases with temperatures above 98 °F (36.7 °C) (31). Contraindications to hydrotherapy include a history of uncontrolled seizures, bowel or bladder incontinence, pressure sores or contagious skin rashes, and cognitive impairments that would jeopardize the patient's safety. If a great deal of assistance is needed with dressing, or if changing clothes causes fatigue or joint pain, then pool therapy should not be used. In summary, the recommendation of Bartels and colleagues in their Cochrane review is pertinent: "One may consider using aquatic exercise as the first part of a longer exercise programme" for arthritis patients (19).

Resistance Exercise

Isotonic exercise is preferred over isometric exercise for dynamic strength training (175) because it more closely corresponds to everyday activities and therefore promotes improved daily function. Low-intensity isometric exercise, however, may be appropriate for muscle strengthening during the acute arthritic stage because it produces low articular pressures. In this instance, instructions should be to perform a submaximal isometric contraction for 6 s while exhaling. Isometric exercise should target one muscle group at a time.

For resistance training, intensity is determined by the percentage of the one-repetition maximum (1RM) that a load (weight) corresponds to. Although increases in strength and muscle mass can be achieved in previously untrained individuals with loads of 50% 1RM (28, 246), greater effects are seen with relatively heavier loads (127), with 80% 1RM appearing to be optimal (115). At the start of resistance training, the intensity should be approximately 60% 1RM (~15 reps max), with progression to 80% 1RM (~8 reps max) occurring gradually over 6 wk. Low muscle mass and high fat mass are both strongly associated with poor function (74) and increased comorbidity and mortality risk (222, 223). Therefore, resistance training should emphasize improvements in both strength and body composition, because increased strength has been shown to underlie the beneficial effects of training on pain and physical function (113).

Range of Motion Exercise

Regular ROM exercise greatly facilitates maintenance of the degree of joint motion necessary for easy performance of ADLs (7). Thus, 5 to 10 min of active (i.e., executed without assistance) ROM exercise, performed on most or all days of the week, is recommended (7). However, these exercises in isolation typically lack the intensity required to elicit functional improvements in most patients with arthritis (48, 90, 84, 92). Therefore ROM exercises should supplement aerobic and strength training.

Considerable improvements in ROM are unlikely in those with severe joint destruction (e.g., limited or no joint space) and very restricted joint mobility, and therefore forceful stretching is not recommended.

EXERCISE TRAINING

Exercise training for cardiorespiratory, strength, and ROM improvement is very important for patients with arthritis because research and clinical experience demonstrate substantial and clinically significant improvements with each of these types of exercise training. This section reviews the exercise training literature that applies to OA, RA, and AS.

Osteoarthritis

Systematic reviews and meta-analyses of exercise (aerobic or strength training or both) for OA conclude that training is safe and has positive effects on pain and disability (19, 28, 50, 90, 101, 106, 127, 193, 199, 229, 241). In view of this consistent and overwhelming evidence, it is not surprising that in its latest recommendations for hand, hip, and knee OA, the ACR affirmed their strong recommendation for aerobic, resistance, and aquatic exercise training (114).

Cochrane reviews of randomized controlled trials (RCTs) on land-based exercise report improvements in pain and physical function for knee and hip OA. Similarly, a Cochrane review on aquatic exercise for knee and hip OA (19) concludes that this form of training has small to moderate benefit for function and QOL and a minor beneficial effect on pain. A large RCT of exercise for knee OA compared education against exercise interventions (home-based aerobic and resistance exercise with limited supervision) for 15 mo after an initial 3 mo period of supervised exercise (58). In this trial, participation in an aerobic or resistance exercise program showed "improvements in measures of disability, physical performance and pain" (58, p. 25). Additionally, another review of RCTs showed that aerobic walking and quadriceps-strengthening programs provide comparable positive effects on pain and disability in OA patients (192).

Besides reducing pain and disability, therapeutic programs to strengthen knee extensors and hip and ankle flexors in patients with OA have resulted in decreased stiffness, increased QOL and independence, and improved function, including strength, ROM, balance, and gait (28, 127, 229). In a recent RCT, Knoop and colleagues (113) attributed these beneficial effects to improvements in leg strength. This conclusion was also made by Runhaar and colleagues (199), who conducted a systematic review of the mechanisms underlying improved pain and function after exercise training in OA. These findings emphasize the importance of including improved strength as an outcome goal of exercise training. With respect to aerobic exercise, the finding that the degree of severity of knee OA is negatively correlated with the level of cardiorespiratory fitness supports the concept that regular aerobic exercise should be performed by people with OA (190). For aerobic training, relatively low-level exercise intensity (at 40% of heart rate reserve) may be as effective as high-intensity cycling (at 70% of heart rate reserve) in improving function, gait, and aerobic capacity and decreasing pain in OA patients (136).

Weight loss has been shown to reduce knee joint forces in OA (150), and an RCT suggested that a combination of exercise and diet produced superior improvements in pain, function, and mobility compared with either exercise or dietary interventions alone (154). Although exercise-induced weight loss is normally associated with aerobic training, resistance training may be more efficacious. As with aerobic training, resistance training augments daily energy expenditure and is an effective adjunct to restricted energy intake for weight loss in young adults; its efficacy in middle-aged and elderly individuals is questioned. This is because sedentary elderly individuals, particularly those with chronic disease, are usually so deconditioned that they are unable to perform exercise of sufficient duration and intensity to significantly augment energy expenditure (60). In contrast, after 12 wk of PRT, resting metabolic rate (RMR) was increased 15% in elderly men and women as a result of increased lean body mass (34). Given that RMR accounts for 60% to 75% of daily energy expenditure, an increase of 15% is highly relevant to weight loss.

Messier and colleagues investigated the relationship between change in body weight and change in knee-joint loads during ambulation in overweight and obese sedentary older adults with symptomatic knee OA after an 18 mo clinical trial of diet and exercise (152). Participants were assigned to one of four groups (healthy lifestyle [control]; exercise only; dietary weight loss only; and dietary weight loss and exercise). The results demonstrated that "each pound of weight lost will result in a 4-fold reduction in the load exerted on the knee per step during daily activities" (p. 2026).

The Cochrane review on land-based exercise for knee OA concluded that supervised individual exercise programs elicit greater improvements in function and pain than either group classes or home-based programs (69). The MOVE consensus (193) concluded that since group exercise and home-based exercise were equally beneficial, patient preference should be considered in the design of exercise programs. This group also concluded that exercise adherence was an important predictor of exercise benefit, and that adherence is aided by maintenance of an exercise diary, telephone and mail contact, support from family and friends, and working with personal trainers. Alternative modes of exercise such as yoga and tai chi also appear to improve pain, flexibility, and function in patients with OA (186, 212).

The benefits of exercise training for arthritis are listed in table 24.9.

Rheumatoid Arthritis

With RA, combating disability, body composition changes, and risks of CVD and osteoporosis should be primary considerations in the design of appropriate exercise therapies. Both resistance training and aerobic conditioning are recommended in an exercise prescription for patients with RA and have been shown to improve muscle strength, aerobic capacity, functional capacity, and QOL. Additionally, they reduce pain, fatigue, and CVD risk without having detrimental effects on disease activity or joint damage (in some cases, decreasing disease activity) (16, 41, 80, 99, 110, 120, 155, 194, 203, 216).

Exercise is safe for RA patients, even long-term high-intensity (≥2 yr) interventions (48, 49, 84, 85, 167, 168). Historically, a commonly held belief was that high-intensity exercise would accelerate joint destruction in patients with arthritis. Initially, reports from the Rheumatoid Arthritis Patients in Training (RAPIT) trial, which featured 150 patients performing high-intensity aerobic and strength training twice weekly for 2 yr, raised concerns that this exercise program was accelerating joint damage progression in large joints with extensive preexisting damage (48, 163). However, when additional data were collected, the authors concluded that long-term intense weight-bearing exercise was safe for all joints, including large joints that were already extensively damaged. This verdict is in agreement with the findings of other studies (79, 84, 85, 168) and with the conclusion of a meta-analysis on RCTs of aerobic exercise interventions in RA (16).

Whole-body dynamic exercise is preferable to static or isometric exercise because the former elicits greater improvements in body composition, strength, aerobic capacity, and function and is more relevant to the performance of ADLs (79, 82). Additionally, higher-intensity exercise consistently produces significantly better gains than low-intensity ROM exercises (48, 83, 84, 92). An example comes from the RCT conducted by Lemmey and colleagues (122) in which RA patients were randomized to perform 24 wk of either high-intensity progressive resistance training (PRT; two sessions per week, eight exercises, three sets of eight reps at 80% 1RM) or low-intensity ROM exercises. Patients in the PRT group showed significant mean gains in muscle mass (1.5 kg [3.3 lb]), substantial reductions in total fat mass (2.3 kg [5.1 lb]) and trunk fat mass (2.5 kg [5.5 lb]), and improvements in strength (119%) and objective tests of physical function designed to reflect the ability to perform ADLs (17%-30%). In contrast, the patients randomized to ROM exercise demonstrated no changes in body composition or physical function despite good compliance to training. Similarly, in the RAPIT trial, combined high-intensity aerobic and resistance training produced significant improvements in functional ability, aerobic fitness, muscle strength, and emotional health and a decline in bone mineral density loss compared with ROM exercises (47, 48). This high-intensity program had a 78% adherence and satisfaction rate after 2 yr (163).

Another finding from the intervention reported by Lemmey et al. (122) was that after 24 wk of high-intensity PRT, RA patients who had previously been moderately disabled achieved levels of physical function that were equivalent to or better than those of age- and sex-matched healthy controls. A similar restoration of normal physical function was observed by the same group in an earlier PRT intervention study (140).

In another study, 82 RA patients treated exclusively by the current T2T approach, half of whom had achieved clinical remission, were found to have mean reductions of 25% for strength and 28% to 34% for performance of objective functional tests relative to 85 age- and sex-matched healthy sedentary individuals (124). To put this magnitude of functional impairment into context, it is equivalent to that observed with 25 yr of advanced aging (i.e., RA patients of either gender aged 60 yr typically have the physical function of same-sex healthy individuals aged 85 yr). It is now apparent that in addition to effective drug treatment, anabolic exercise is required to normalize physical function in RA patients (124).

Nontraditional modalities such as dance and tai chi may have beneficial effects on depression, anxiety, fatigue, tension, and lower extremity ROM in RA (86, 171, 236). Although earlier investigations of home-based exercise programs failed to demonstrate improvements in physical function (43, 218), Hakkinen and colleagues (83, 84) showed significant improvements in strength, subjectively assessed physical function, and disease activity; additionally noted were trends toward decreased pain and increased bone mineral density (BMD) in RA patients performing high-

Table 24.9 Exercise Training: Important Considerations

Cardiorespiratory endurance	Skeletal muscle strength	Skeletal muscle endurance	Flexibility	Body composition
OSTEOARTHRITIS				
Aerobic exercise improves cardiorespiratory endurance, pain, depression, fatigue, function, health status, and gait and helps reduce fat mass.	High-intensity resistance training improves strength, muscular endurance, function, health status, pain, and stiffness and helps reduce fat and increase muscle mass.	Low-intensity resistance training improves strength, muscular endurance, function, health status, pain, and stiffness and helps reduce fat and maintain muscle mass.	Dynamic exercise improves joint mobility, pain, and function. Aquatic exercise improves knee and hip range of motion, pain, and function. AFYAP and PACE programs improve flexibility and isometric strength (224).	Combined diet and resistance and aerobic training produces weight loss and improves function, mobility, and pain to a greater extent than diet or exercise alone.
RHEUMATOID ARTHRITIS				
Aerobic exercise improves cardiorespiratory endurance, pain, function, and, with resistance training, mood; aerobic training improves fitness without worsening disease activity.	High-intensity resistance training improves strength, muscular endurance, function, and mobility; increases muscle mass; and reduces fat mass. Hand strengthening may improve dexterity and grip strength.	Low-intensity resistance training improves strength, muscular endurance, function, and mobility; maintains muscle mass; and reduces fat mass. Hand strengthening may improve dexterity and grip strength.	Joint mobility improves with dynamic exercise training. AFYAP and PACE programs improve flexibility and isometric strength (224).	Combination of aerobic and resistance training (RT) or RT alone increases muscle mass and decreases fat mass without significant weight loss; also slows bone mineral density loss.
ANKYLOSING SPONDYLITIS				
Supervised physical therapy improves fitness, mobility, and function; improvements in fitness and stiffness may mediate improvements in global health.	Few investigations are relevant, but resistance training aimed at restoring muscle mass and improving function is warranted.	Few investigations are relevant, but resistance training aimed at restoring muscle mass and improving function is warranted.	Home or supervised exercise can improve spinal mobility, physical function, and (supervised only) patient global assessment. Some evidence suggests improvements in pain and depression. Relative to conventional physical therapy, flexibility and strengthening exercises that target specific muscle groups involved in AS improve axial mobility and function.	Relevant studies have not been performed but are warranted, because AS is characterized by muscle loss and increased osteoporotic fracture risk.

AFYAP = Arthritis Foundation YMCA Aquatics Program; PACE = People with Arthritis Can Exercise (land-based community program).

intensity home-based strength training compared with patients performing home-based ROM. These findings demonstrate that if the exercise intervention is appropriately designed, exercising at home can be effective and beneficial for RA patients.

Most investigations evaluating the efficacy of exercise interventions for RA patients have used intensities of 50% to 85% of maximum heart rate for aerobic exercise training; for resistance (strength) training, intensities began at 50% to 60% of 1RM and progressed to 80%. The demonstrated success and safety of these interventions provide the basis for the recommendation of moderate- to high-intensity physical activity for exercise prescription in RA (7, 219). However, even very light and light intensities are shown to confer favorable effects on disability and CVD risk in those with RA (110). Thus, patients who are reluctant to undertake moderately high-intensity exercise can be assured that benefits can still be gained from very light to light physical activity, as is the case for the general population.

As with any patient population, training effects are maintained only as long as exercise continues, and these benefits wane rapidly after termination of exercise (81, 125). Independent training after a supervised program sustains these beneficial effects (84).

In terms of training responses for those with RA, changes in body composition, strength, and aerobic capacity are similar in magnitude to those generally reported for healthy middle-aged or elderly individuals (82). This similarity in response to training is consistent with reports that muscle quality is unaffected by RA, even when pronounced muscle wasting has followed (145, 146).

Ankylosing Spondylitis

Exercise is strongly recommended by the ACR in the latest recommendations for treatment of AS (242), having previously been among the 10 key recommendations for the management of AS developed by the ASsessment in AS (ASAS) International Working Group and the European League Against Rheumatism (EULAR) (249). The literature review from which the ASAS/EULAR recommendations arose suggested that different types of exercise-based intervention could affect disease outcomes in AS (250). Consistent with these guidelines are recommendations made in the evidence-based consensus statement by Millner and colleagues (158) for individual exercise prescriptions for AS patients, which feature stretching, strengthening, cardiopulmonary, and functional fitness components.

A Cochrane review (42) concluded that exercise training improves spinal mobility and physical function in those with AS, albeit with the caveat that this conclusion was based on low-quality research evidence. However, since numerous studies investigating exercise effects on AS have emerged, subsequent systematic reviews and meta-analyses provide convincing evidence that exercise training significantly improves disease activity, physical function, and spinal mobility as well as chest expansion, cardiorespiratory function, pain, QOL, and depression in those with AS

(128, 142, 174). These reviews also indicate that unsupervised home-based exercise interventions can be prescribed to achieve these benefits (128), although better outcomes are generally observed in supervised group exercise programs (174).

Walking for 50 min, three times per week for 12 wk, significantly improves aerobic fitness as well as clinical outcomes (103); Nordic walking at moderate to high intensity (up to 85% maximum HR) for 30 min, twice a week for 12 wk, combined with ROM exercise achieves greater improvements in aerobic capacity and peripheral pain than ROM exercise alone (165); moderate- to high-intensity (50%-80% $\dot{V}O_2$peak and 60%-80% 1RM) home-based combined aerobic, strengthening, and ROM training performed twice weekly for 3 mo elicits superior gains in aerobic capacity and function (BASFI) than ROM training alone (97); and combined aquatic exercise five times a week for 5 wk confers greater benefits in pain management and general health than home-based ROM (53).

Notably, high-intensity (up to 95% maximal HR) aerobic interval training combined with strength training, performed three times per week for 12 wk, is both safe and effective for reducing disease activity and CVD risk (including arterial stiffness). Additionally, improvements are noted in aerobic fitness, physical function, and body composition (increased lean mass, decreased fat mass) in AS patients with moderately to highly active disease (226). These same benefits were also observed by Stavropoulos-Kalinoglou and colleagues (216) in individuals with RA in a similar combined aerobic and strength training program. A further benefit of exercise training for those with AS is that when it is concurrent with anti-TNF therapy, there appears to be a synergistic effect, with enhanced efficacy in improving disease activity, function, spinal mobility, pain, and fatigue compared with biologic therapy alone (133).

Given the goals of maintaining posture and functional ability in this disease, where disability is mainly related to spinal manifestations, most interventions use flexibility and strengthening programs. Achieving functional ROM of the hip joints should be emphasized, because a lack of such capability can be extremely disabling (33). Additionally, muscle strengthening in this population can enable maintenance of proper posture during spinal extension (42).

Daily exercise is considered vital to maintenance of spinal mobility; however, long-term effects have not been studied (42). Rheumatoid cachexia also occurs in AS, and the resultant muscle loss is associated with impaired strength and physical function. Therefore, the high-intensity PRT advocated for RA patients to improve muscle mass and function is also appropriate for individuals with AS (138).

The mode and sequencing of exercise are important considerations for patients with AS. The exercise program should start as soon as possible after diagnosis of the condition, beginning with exercises to improve spinal and peripheral joint motion before moving to a combined aerobic and strengthening program. High-impact activities should be avoided because they are stressful to the spinal and sacroiliac joints. Swimmers with limited spinal and neck motion can use a mask, snorkel, and fins to reduce trunk and neck rotation. Activities that encourage extension are preferred over activities that require flexion (42), and contact sports are contraindicated for those with cervical spine or peripheral joint involvement.

Practical Application 24.2

FIBROMYALGIA

Fibromyalgia is a complex chronic pain syndrome that features widespread pain and is often associated with psychological distress. It is estimated to affect between 2% and 8% of the U.S. population (39). Its prevalence is higher in women than men (ratio 7:1) and increases with age, with initial diagnosis usually made in middle age (76, 119). Although its etiology is unknown, the condition is associated with stressful or traumatic events, repetitive injuries, illness (e.g., viral infection), certain diseases (RA, AS, SLE, chronic fatigue syndrome), genetic predisposition, and obesity (12, 162, 196). It may also involve central **neuromodulatory dysregulation** (150). Fibromyalgia is not a form of arthritis but may easily be confused with arthritis because of its associated widespread musculoskeletal symptoms as well as trigger and tender points, which usually have a **periarticular** location.

Diagnosing fibromyalgia is complicated by its high coexistence (25%-65%) with other rheumatic conditions such as RA, AS, and SLE (117). Currently, fibromyalgia is diagnosed using the 2010 ACR criteria (244), which specifies satisfying the following three conditions: widespread pain, symptoms for at least 3 mo, and no other disorder that would otherwise explain the pain. For pain, tenderness in 19 specific sites over the previous week is considered; symptoms include fatigue, waking unrefreshed, problems with thinking or remembering (sometimes referred to as "fibro fog"), insomnia, and depression. Laboratory tests and joint radiography are typically unremarkable (i.e., muscle and joint inflammation or damage is not associated with fibromyalgia), but a lower pain threshold and altered pain-processing pathways exist, which provides clues to the underlying pathogenesis of this complex syndrome (73). Fibromyalgia is frequently associated with depression (180).

Treatment of fibromyalgia requires a multidisciplinary approach employing exercise, education, and both pharmacologic and behavioral therapies for depression, relaxation, and sleep (76, 232). Unfortunately, despite convincing evidence that regular exercise is safe and beneficial for individuals with fibromyalgia, slightly less than half of newly diagnosed patients are given recommendations and instructions on initiating an exercise program as part of their treatment plan (243). Evidence shows that aerobic exercise improves aerobic capacity, physical function, global well-being, pain and tenderness, and possibly sleep, fatigue, depression, and cognition in persons with fibromyalgia (31,

77, 108). Overall, research suggests that moderately intense aerobic exercise (55%-75% HRmax) appears to be beneficial for fibromyalgia patients (243).

In an RCT conducted by Richards and Scott (188), fibromyalgia patients were allocated to either twice weekly graded aerobic exercise training (walking and cycling at moderate intensity for progressively longer periods) or relaxation and flexibility classes for 12 wk (control treatment). Significant improvements were noted for both groups in tender joint counts, pain scores, fatigue scores, and self-reported disability scores at 3 mo. Improvements were greater in the aerobic exercise group relative to the relaxation group. The fall in tender point counts continued into 12 mo, when fewer of the participants in both groups still met the diagnostic criteria of fibromyalgia (significantly fewer were in the exercise group). The benefits of aerobic training were even more apparent at 1 yr when participants of the exercise group exhibited a decrease in tender point counts and improved scores on the fibromyalgia impact questionnaires.

Similarly, a review of the literature (32) concluded that moderate- to high-intensity strength training is safe and improves muscle strength, physical function, pain, tenderness, and global well-being in women with fibromyalgia. Gavi and colleagues (72) have additionally found that low-intensity strength training (45% 1RM; 12 different exercises, three sets of 12 repetitions, twice weekly for 16 wk) improves strength, aerobic capacity, subjectively assessed physical function, pain, global health, depression, anxiety, and QOL in female fibromyalgia patients; the improvements in strength, aerobic fitness, and pain gained by strength training were superior to those achieved by ROM training.

Aquatic exercise is a commonly recommended training mode for persons with fibromyalgia (243). As with aerobic and strength training, aquatic exercise training is safe and effective in improving function, aerobic capacity, strength, pain, and stiffness in fibromyalgia patients (24, 129). With respect to the relative efficacies of aquatic versus land-based exercise, Bidonde and colleagues (24) found no differences in the gains for function, aerobic capacity, pain, or stiffness but noted that land-based strength training elicited greater strength gains.

In regard to exercise testing, the procedures outlined in chapter 5 can be followed for fibromyalgia. One consideration is that, compared with age- and sex-matched controls, individuals with fibromyalgia report higher levels of perceived exertion for the same relative workload during exercise testing (166) and may not achieve maximal effort (172, 234). In these cases, ventilatory threshold rather than peak $\dot{V}O_2$ is recommended as an indication of fitness levels (234). As they do for persons with arthritis, American College of Sports Medicine (ACSM) prediction equations appear to overestimate peak $\dot{V}O_2$ in persons with fibromyalgia (52). Although slight underestimations and overestimations were noted with the FAST (table 24.3) and FOSTER equations (67), respectively, these equations provide clinically acceptable estimations of peak $\dot{V}O_2$ using the Duke–Wake Forest testing protocol (52).

As with other deconditioned individuals, a prudent approach should be adopted for training initiation and progression with fibromyalgia patients. This includes beginning at a low intensity and progressing slowly and gradually to a moderate intensity (20%-55% of HRR) for 20 to 30 min (31,

77). Some people with fibromyalgia may eventually progress to high-intensity exercise. Although exercise training is usually effective in reducing self-reported pain in fibromyalgia, the clinical exercise physiologist should provide education on transient increases in pain with exercise bouts, methods of adjustment of exercise protocols, and reassurance that appropriate exercise will not worsen fatigue. Such instruction will enhance adherence and long-term benefits of exercise training and should be an essential component of the exercise prescription for patients with fibromyalgia (136).

Recommendations for minimizing pain from strength training sessions include limiting eccentric exercises, performing upper and lower extremity training on alternate days, and resting between repetitions (77). As of yet, it is not known whether combining different types of exercise training, or combining exercise with education, biofeedback, or medication, is optimal for treating fibromyalgia.

Research Focus

Exercise Therapy and Total Hip Replacement

This section highlights the first known study assessing whether exercise therapy affects the need for total hip replacement (THR) surgery in patients with hip OA. The aim of this investigation was to determine the effectiveness of a supervised 12 wk high-intensity exercise program in delaying or preventing the need to undergo total hip replacement (THR) surgery over the following 3.6 to 6.1 (median = 4.8) yr, and to predict 6 yr hip survival (i.e., the likelihood of not requiring THR in the next 6 yr). One hundred and nine patients with mild to moderate symptomatic and radiographic hip OA, not requiring THR, were recruited and randomly assigned to either a gym-based exercise therapy plus patient hip education group ($n = 55$) or education only group (control; $n = 54$).

The exercise group was instructed to train two or three times per week for 12 wk, with at least one of these weekly sessions supervised by a physical therapist. Each session involved a warm-up and strengthening, functional, and flexibility exercises. Specifically, the warm-up consisted of 5 to 10 min of either treadmill walking or cycling on a stationary bike at an intensity rated as 12 to 13 on the Borg RPE scale (i.e., light to somewhat hard). The strengthening exercises were three sets of 8 repetitions at 70% to 80% 1RM of hip extension, hip abduction, leg curl, leg extension, heel raise, and crunch exercises, with progression involving incremental loading. Functional exercises were three sets of 10 repetitions of squats, single-leg stands, forward and sideways lunges, and bench stepping, using both legs, with progression involving using balance pads, graduating from double-leg to single-leg exercises, increasing bench height, and using barbells. Flexibility exercises involved hip ROM exercises.

Statistical analyses were performed using the intention-to-treat principle (i.e., data from all subjects in the exercise group were included irrespective of number of exercise sessions attended). This form of analysis, by including all who are allocated to an intervention and not just those who comply, provides a more realistic real-world perspective of the benefit of an intervention, although it will usually underestimate the benefit of that intervention. (*Note:* Only 53% of the exercisers achieved satisfactory compliance, which was defined as ≥20 sessions.)

During the follow-up period, 22 patients in the exercise therapy and patient education group (40%) and 31 patients in the patient education group (57%) had undergone THR surgery, with a 1.8 times higher probability in the control group. It was estimated that by 6 yr, 59% of the exercise group would require THR compared with 75% of the control group ($p = <0.05$). In addition, the estimated median time to THR was 5.4 yr in the exercise group and 3.5 yr in the control group. Involvement in the exercise class also resulted in sustained (at 29 mo) improvements in self-reported physical function ($p = 0.004$), pain ($p = 0.083$), and stiffness ($p = 0.112$) relative to the controls.

The investigators concluded that an exercise therapy program in addition to patient education "can reduce the need for THR or postpone surgery in patients with hip OA" (p. 169). Given the large and steadily increasing number of individuals undergoing THR surgery, the benefit of relatively short-term exercise interventions in reducing or delaying these operations has potential implications for both health care costs and for patients wishing to avoid surgery and postsurgical complications.

Based on Svege et al. (2015).

Clinical Exercise Bottom Line

- 2019 ACR/Arthritis Foundation guidelines strongly recommend that patients with OA of the hand, hip, and knee participate in a regular ongoing exercise program.

- Exercise programs can be tailored based on the specific arthritic condition and associated functional impairments, other medical conditions, and the patient's preference.

- Exercise programs can be formulated under the guidance of a clinical exercise physiologist and in conjunction with an exercise physiologist to ensure the quality and safety of the interventions.

- Comprehensive approaches to treatment of arthritic conditions include medical management (medications, self-management, weight loss, other modalities) and implementation of exercise programs to improve pain management, functional status, and quality of life.

CONCLUSION

The available data indicate that properly performed exercise is safe and effective for individuals with OA, RA, and AS. In a relatively short period, appropriate exercise can be expected to increase strength, aerobic capacity, and ROM; improve body composition; enhance physical function, thus decreasing, or sometimes even removing, disability; attenuate pain and stiffness; and improve mood and mental health. Precautions must be taken, however, to ensure that the prescribed exercise is appropriate (i.e., safe and tolerable), taking into consideration any underlying arthritic conditions.

Online Materials

Visit HK*Propel* to access a link to the references, the case studies with discussion questions, and a quiz to help you review key concepts and test your understanding of the material covered in this chapter.

CHAPTER 25

Osteoporosis

Lora M. Giangregorio, PhD

Osteoporosis is a disease characterized by reduced bone strength and an increased risk of fractures (50). Osteoporotic fractures, or fragility fractures, can occur as a result of minimal or no trauma. Examples include sustaining a wrist or hip fracture during a fall from standing height or sustaining a vertebral fracture while bending forward to tie one's shoes. Osteoporotic fractures, particularly hip fractures, increase the risk of death and disability and can result in substantial health care costs.

Medical costs for osteoporosis in the United States are estimated at $19 billion per year (9); the estimated overall cost of a single hip fracture is approximately $45,000 (71). The mortality associated with fractures is substantial, although the mortality rate depends greatly on the fracture site (35) as well as sex—men have double the mortality rate of women after a hip fracture (24). A meta-analysis estimated a fivefold increase in all-cause mortality in the 3 mo after a hip fracture in older adults (24). Individuals at moderate or high risk of fractures may be offered medications to slow bone loss or prevent fractures. Exercise is an integral part of fracture prevention in individuals with osteoporosis. It may not be possible to regain all the bone that has been lost. Therefore, prevention of future bone loss and prevention of falls are key therapeutic goals.

DEFINITION

The National Institutes of Health in the United States defines **osteoporosis** as "a skeletal disorder characterized by compromised bone strength predisposing to an increased risk of fracture" (50, p. 768). **Bone mineral density (BMD)**, measured using **dual-energy X-ray absorptiometry (DXA)** at the lumbar spine and hip, is often used to estimate bone strength in clinical settings. However, bone strength is also influenced by bone geometry, microarchitecture, material properties, and other factors. In 1994, the World Health Organization (WHO) defined osteoporosis as a BMD of 2.5 standard deviations or more below the average value for young healthy women, or in other words, a BMD T-score of −2.5 or less (32). Having a BMD T-score between −1.0 and −2.5 was defined as **osteopenia**. However, fracture risk assessment has moved away from defining risk based on BMD alone.

Fracture risk assessment tools, such as FRAX, go beyond BMD to look at other risk factors (e.g., age, gender, use of oral glucocorticoids, history of fracture). In FRAX, the user inputs information on an individual client's risk factors, with or without BMD, and a 10 yr probability of fracture is calculated, based on algorithms derived from large population-based cohort studies (www.shef.ac.uk/FRAX). Assessing and describing fracture risk in terms of risk factors and 10 yr probability is emphasized over using the terms *osteoporosis* and *osteopenia* based on BMD alone. In fact, the term *osteopenia* implies lower risk, where in fact many people with osteopenia will sustain a fracture (67). FRAX has not been validated for use in men and women under the age of 40. For individuals under 40 who are suspected to be at risk of fracture because of medications or medical conditions, position statements from the International Society for Clinical Densitometry (5, 37), the American College of Sports Medicine (46), and others (31) have presented criteria for identifying risk of fracture or low BMD for age; individuals under 40 yr who are identified as at risk may benefit from consultation with a specialist (52). Ultimately, it is not in the scope of practice of the clinical exercise specialist to assess risk of fracture or interpret BMD scans, but they can note risk factors or whether 10-year fracture risk or osteoporosis has been diagnosed and use that information to inform decisions on fitness assessment and exercise prescription.

Acknowledgment: Much of the writing in this chapter was adapted from the previous editions of *Clinical Exercise Physiology.* Thus, we wish to gratefully acknowledge the previous efforts of and thank David L. Nichols, PhD, and Andjelka Pavlovic, MS.

SCOPE

Osteoporosis most often occurs as a result of age-related bone loss, postmenopausal bone loss, or bone loss secondary to other factors (e.g., immobility, medications where bone loss is a side effect, diseases or conditions associated with bone loss), or a combination. At what point age-related bone loss begins has been debated, but in general, after the age of 40, small amounts of bone mass are lost each year (approximately 0.5%-1.0% per year). In women, the rate of bone loss is even greater during a 3 to 5 yr period after menopause. Examples of the many causes of secondary osteoporosis or increased fracture risk include rheumatoid arthritis, malabsorption syndromes (e.g., Crohn's disease, ulcerative colitis), sex hormone deficiency, hyperparathyroidism, chronic kidney or liver disease, diabetes, chronic obstructive pulmonary disease, oral glucocorticoid use, drugs used to treat breast or prostate cancer, proton pump inhibitors, Depo-Provera, and certain mood-altering or anti-seizure medications. Although osteoporosis affects almost one of every two women at some point in life, it is mistakenly thought of as a women's disease—the prevalence of osteoporosis in men can be as high as 15% (10, 31).

PATHOPHYSIOLOGY

The majority of adult bone mass is accrued during childhood and adolescence (59, 72). **Peak bone mass** , or the highest amount of bone mass attained during life, is attained by approximately the end of the second decade. It is influenced to a certain extent by genetics but also by physical activity, diet, and hormonal balance. Whether osteoporosis develops depends on the amount of peak bone mass attained and the extent to which bone loss occurs throughout life (12, 29, 70). Therefore, attempts to maximize peak bone mass during childhood and adolescence may be a first step to preventing osteoporosis.

Bone tissue undergoes **bone remodeling**, where old bone is broken down and new bone is produced in its place, to maintain the strength of the bone and repair fatigue damage. Bone tissue can also be broken down to release calcium if stores are deficient. Bone remodeling, in simple terms, involves **bone resorption** (breakdown of bone tissue by osteoclasts), followed by **bone formation** (production and mineralization of bone tissue by osteoblasts) (56, 72). A bone remodeling cycle takes several months. When bone resorption and formation are balanced, bone mass is maintained. When bone resorption exceeds formation, bone loss occurs. Bone remodeling is distinct from **bone modeling**, where bone formation occurs without prior bone resorption (figure 25.1). Bone modeling and remodeling both occur during growth to increase the size and length of the bone. In adulthood, bone remodeling is predominant. To see an animation of how long bones or flat bones grow, or of the bone remodeling process, explore this site: http://depts.washington.edu/bonebio/ASBMRed/growth.html.

Before puberty, changes in bone mass are influenced primarily by estrogen, growth hormone, and insulin-like growth factor. Bone growth in length occurs at the growth plate and is regulated by factors such as Indian hedgehog, parathyroid hormone-related protein, estrogen, and thyroid hormone (56). During and after puberty and into adulthood, if estrogen deficiency occurs, it results in reduced bone accrual or bone loss. Estrogen deficiency occurs after menopause, which is why osteoporosis becomes increasingly more common in women after menopause. Young amenorrheic women, after having been amenorrheic for an extended period, may not regain sufficient bone mass to return to normal, even with the resumption of regular menstrual periods (46). Further, women who experience estrogen deficiency as a result of surgical removal of one or both ovaries, or women or men who take medications that affect estrogen (e.g., aromatase inhibitors), can experience bone loss. Sex hormone deficiency in men can also result in bone

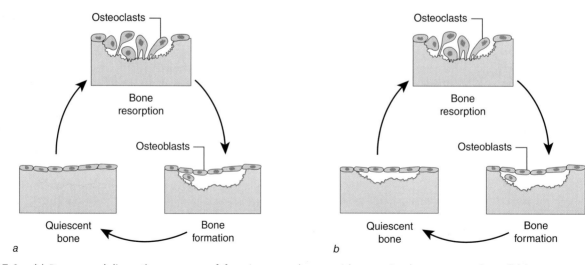

Figure 25.1 (a) Bone modeling: the process of forming new bone without prior bone resorption; (b) bone remodeling: the process of bone formation after bone resorption.

loss. Relative energy deficiency in sport can cause bone loss and fragility fractures in men and women of all ages and contribute to osteoporosis; a comprehensive review of this topic can be found elsewhere (44).

Calcium or vitamin D deficiencies, or inadequate calorie intake, can result in suboptimal bone mass accrual or bone loss. If caloric or calcium intake is insufficient during childhood, when bone modeling is still occurring, the body will sacrifice increases in bone length to maintain bone strength (17, 26). Nutritional deficits can also decrease peak bone mass and strength or contribute to bone loss after late adolescence when bone growth has ceased. Vitamin D deficiency can lead to rickets in children and osteomalacia in adults, which can increase the risk of fracture. Osteomalacia may be mistaken for, or be comorbid with, osteoporosis, particularly among older adults.

Several lines of evidence suggest that mechanical loading can influence bone strength. For example, some research suggests that impact loading of high frequency or magnitude, or that results in high strain rates, can alter the bone remodeling cycle such that increases in bone mass and strength will occur (57, 59). On the other hand, if the habitual loading on bone is decreased, as with immobilization, spaceflight, or neurologic injury, a loss in bone mass and strength will result (18).

CLINICAL CONSIDERATIONS

The most common sites where osteoporotic fractures occur are the wrist, the spine, the humerus, and the hip. Fractures of the rib and pelvis are also common. Although the hallmark of osteoporosis is low BMD, fracture risk is also influenced by the loads applied to bone, so individuals at high risk of falls or who have low body weight (and therefore less soft tissue "padding"), in combination with low BMD, may be at greater risk of fracture. Further, fractures can occur in the spine in the absence of a fall. Spine fractures can occur as a result of compressive or torsional loads. The loads required to cause a fracture decrease as bone strength decreases. Therefore, individuals at high risk of fracture may need to modify the way they move during daily activities or may need to avoid certain movements or activities that increase the risk of falls or increase the loads applied to the spine. The risk of having a subsequent spine fracture is high in the year after a spine fracture, so initiating fracture prevention strategies is critical (38a).

Signs and Symptoms

There are no symptoms during periods of accelerated bone loss or unbalanced bone remodeling; therefore, osteoporosis can be present and go undetected. A fracture is often the first sign that a person has osteoporosis; a fracture of the wrist, spine, hip, humerus, pelvis, or rib from an event that would not normally cause a fracture (e.g., a fall from standing height) may be a red flag. Vertebral fractures are often asymptomatic and come to clinical attention only when there is noticeable height loss or pain. Vertebral fractures can cause height loss, a protruding abdomen, or hyperkyphosis (an exaggerated thoracic kyphosis) or other postural changes (e.g., flattened lumbar lordosis); however, these may not be noticeable until multiple vertebral fractures have occurred. Further, other conditions, such as degenerative disc disease, scoliosis, or ankylosing spondylitis, can cause height loss or postural changes.

History and Physical Examination

When setting physical activity goals and prescribing exercise for a person with osteoporosis, or who is at moderate to high risk of fracture, the clinical exercise physiologist should assess the following in a history and physical examination:

- Medical history and medications, comorbid conditions, and contraindications to exercise
- Fracture risk (see section on diagnostic testing)
- Fall risk
- Physical performance
- Standing posture
- Barriers to and facilitators of physical activity

When assessing fall risk, a fall is often defined as "an event which results in a person coming to rest inadvertently on the ground or floor or other lower level" (75). A person is considered at risk of falls if they present with an acute fall, have had two or more falls in the past 12 mo, or present with gait and balance difficulties (American Geriatrics Society/British Geriatrics Society guidelines) (34). A client who is at risk of falls may need multifactorial assessment and intervention (e.g., home hazards, medication review). Performance-based tests of muscular power and strength, balance, and mobility may provide a more comprehensive assessment and help identify areas for improvement or whether clients can move safely; examples include the Short Physical Performance Battery, the Berg Balance Scale, and the Mini-Balance Evaluation Systems Test (64). However, for community-dwelling individuals, many existing tests of mobility or dynamic balance will exhibit ceiling effects, and more challenging tests, like the Four Step Square Test, could be used (www.rehabmeasures.org). Assessments of muscular strength or power in upper and lower extremities should be considered (e.g., 30 s chair stand test), because capacity to react to a destabilizing event may depend on speed and strength during reactive strategies, such as using upper extremities to break the fall or taking steps to reestablish balance.

When assessing physical performance, consider observation as well as objective performance-based tests, and note pain, weakness, or impairments (e.g., using arms for support during sit-to-stand, body mechanics during movement, slow gait speed). Gait speed is often used as a marker of mobility or sarcopenia. The cut points that have been used to define slow gait speed as it pertains to predicting adverse health outcomes or mortality vary from <0.8 to <1.0 m · s^{-1} (14). Other domains that may be relevant to assess include muscular endurance, neuromuscular or functional performance (e.g., coordination, balance, mobility), and flexibility (e.g., ankle dorsiflexion, hip flexion and extension, shoulder flexion, thoracic extension).

Visual inspection can be used to look for forward head posture, hyper- or hypokyphosis, and hyper- or hypolordosis, as well as protracted scapulae and internal rotation of the shoulders. Occiput to wall distance (OWD), a crude measure of thoracic hyperkyphosis and forward head posture, is determined by having a client stand with their back to the wall—heels, buttocks, and shoulders (if possible) against the wall—and measuring the OWD. Ensure that the client is not hyperextending their neck and that their nose and ears are at the same horizontal level. In the absence of forward head posture or thoracic hyperkyphosis, the head will touch the wall and the OWD will be 0 cm. An OWD of >4 cm has a sensitivity of 41% (95% CI, 31%-52%) and specificity of 92% (95% CI, 87%-95%) for detection of vertebral fractures (66). However, OWD can also be >0 cm in the presence of ankylosing spondylitis and other conditions or impairments.

Rib–pelvis distance measured in fingerbreadths during standing can be used as a crude screening tool for lumbar fractures, where two fingerbreadths or less has been estimated to have good sensitivity (87%) and moderate specificity (47%), but this should be limited to identifying the need for further examination rather than used as a diagnostic tool (65). A flexible ruler could also be used to assess thoracic kyphosis and lumbar lordosis. The Safe Functional Motion test can be used to assess a client's alignment during performance of functional activities (40).

Consider objective measurement or self-report of current physical activity levels as well as questionnaires to assess exercise self-efficacy, available time, whether pain is present at rest or during activity (e.g., with a 0-10 visual analog scale), how comorbid conditions may affect exercise participation, access to exercise resources, and client preferences and confidence in making a change.

Diagnostic Testing

Dual-energy X-ray absorptiometry (DXA) is the most commonly used technology for measuring BMD. DXA uses low-dose X-ray to emit photons at two different energy levels. BMD is calculated based on the amount of energy attenuated by the body, or in other words, the amount of X-rays absorbed. Because of an assumed linear association between attenuation and density, we can obtain an estimate of the amount of **bone mineral content** per unit area; thus BMD measured by DXA is reported in g · cm^{-2}. Scans are usually performed at the lumbar spine and the hip. The hip scans are used to estimate femoral neck BMD, which is used in the diagnosis of osteoporosis or determination of 10 yr fracture risk. Examining lumbar spine BMD T-scores can also inform a physician's discussions with clients on osteoporosis management. DXA is capable of differentiating between bone and **soft tissue**. Trabecular bone score—a measure of bone texture thought to be associated with bone architecture—can be derived from lumbar spine scans to

improve fracture risk assessment. Trabecular bone score is associated with fracture risk even after controlling for BMD and other risk factors. BMD from whole-body scans should not be used to estimate fracture risk in adults, although they are used in research settings to estimate body composition (42).

DXA has a precision of 0.5% to 2.0%, requires short exam times (5-10 min), and provides low radiation exposure. The accuracy of predicting BMD and fracture risk at one site based on measurement of BMD at another site is low (≤50%) (3, 27, 30). A DXA image of a femoral neck scan, along with its accompanying printout, is presented in figure 25.2. The subject's BMD in g · cm^{-2} is provided, as well as T-scores or z-scores calculated using reference standards. When looking at change in BMD, the clinical exercise physiologist should look at the change of the BMD in g · cm^{-2} rather than at changes in T- or Z-scores.

Other methods for estimating BMD and other indices of bone strength include quantitative computed tomography (QCT), peripheral QCT (pQCT), and high-resolution peripheral quantitative computed tomography (HR-pQCT). Because QCT, pQCT, and HR-QCT generate a three-dimensional image, postprocessing of the images allows the user to separate **trabecular bone** and **cortical bone** to get trabecular BMD and cortical BMD, as well as cortical bone geometry and other indices of bone strength. HR-QCT has sufficient resolution to provide estimates of trabecular microarchitectural indices, cortical porosity, and finite element analyses of bone strength. Researchers use QCT devices, but because of cost, radiation exposure, availability, and other factors, they are rarely used in clinical practice. A review of advances in evaluating bone health can be found elsewhere (36).

The clinical exercise physiologist should advise people over the age of 50 with risk factors for osteoporosis to see their doctor for a BMD test. There are guidelines for determining whether someone under the age of 50 should be assessed (52). Diagnostic testing and interpretation by a physician is necessary for confirmation of osteoporosis or 10 yr fracture risk. Fracture risk is determined by a physician using a validated risk calculator (e.g., FRAX) or

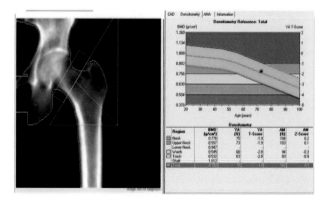

Figure 25.2 Femoral neck scan. Image of a proximal femur scan from a dual-energy X-ray absorptiometer.

Courtesy of Andrew B. Lemmey.

assessment of risk factors (e.g., fracture history at osteoporotic sites after age of 40, age, sex, glucocorticoid use, and BMD T-score at femoral neck if available). Numerous risk factors, both modifiable and nonmodifiable, exist for osteoporosis, and the clinical exercise physiologist should ask about them during a medical history (see the sidebar Risk Factors for Osteoporosis). Oral (not inhaled) glucocorticoid use for greater than 3 mo increases the risk of fracture, even in individuals who do not have osteoporosis based on BMD (8). Individuals with diabetes are also at an increased risk of fracture independent of BMD (19).

In addition to measuring height, it might be informative to record whether there has been historic height loss (i.e., whether current height does not represent peak height) or weight loss greater than 10% of body weight at age 25. A client should be referred back to their physician for X-rays, and a fracture risk assessment should be done in the presence of 6 cm of historic height loss or 2 cm measured height loss, if a fracture risk assessment has not already been performed by a family physician or specialist (52). Notably, individuals with a history of fragility fracture at the spine or hip are automatically considered high risk for future fracture.

A clinical exercise physiologist should understand how to interpret DXA T-scores or FRAX risk assessments and use them in combination with an assessment of the client's other health conditions or impairments to tailor fitness assessment and exercise prescription. If a person has a diagnosis of osteoporosis or osteopenia, moderate or high fracture risk, or history of suspected osteoporotic fracture, the clinical exercise physiologist must consider whether it is necessary to modify fitness assessment or exercise prescription or to identify therapeutic goals related to fall or fracture risk. For example, the clinician might avoid or modify scenarios that could increase the risk of falls or apply compressive or torsional loads to the spine in individuals who are at moderate or high risk of fracture during exercise testing, exercise prescription, or activities of daily living or leisure.

Exercise Testing

Typically, the primary purposes of exercise testing are to detect coronary artery disease or to assess aerobic capacity and determine an appropriate exercise prescription. The clinician must ensure the benefits of doing an exercise testing protocol outweigh the downfalls. For example, the protocol being used must not increase the risk of fracture in a person who is already at risk. Some of the risk factors for osteoporosis may also increase the risk for coronary artery disease (e.g., age, postmenopausal), so appropriate screening and risk management is necessary.

Contraindications

The American College of Sports Medicine does not state that osteoporosis is an absolute or relative contraindication to exercise testing (2). Protocols or exercises that require high-impact loading, such as running or jumping, should be avoided in individuals with osteoporosis or moderate to high risk of fracture, at least initially. Assessments and exercises that contain rapid, repetitive,

---

Risk Factors for Osteoporosis

Nonmodifiable Factors

- Age
- Female sex
- Parental hip fracture
- Postmenopause, hypogonadism, prolonged amenorrhea, or premature menopause (before the age of 45)
- Rheumatoid arthritis
- History of fragility fracture
- Prolonged use of oral glucocorticoids or other high-risk medications[a]
- Other disorders associated with rapid bone loss or fracture[b]

Potentially Modifiable Factors

- Low body weight (<60 kg) or major weight loss (>10% of body weight at age 25 yr)
- Current smoking
- High alcohol intake (three or more units or drinks a day, 8-10 g alcohol per unit)
- Caffeine intake greater than four cups of coffee per day

[a]≥3 mo therapy with oral glucocorticoid at prednisone-equivalent dose of ≥7.5 mg daily; other high-risk medications include aromatase inhibitors or androgen deprivation therapy (19, 52).

[b]Examples include type 1 diabetes, celiac disease, gastric bypass surgery, COPD (chronic obstructive pulmonary disease), chronic liver disease, primary hyperthyroidism, and spinal cord injury.

sustained, end-range, or weighted flexion or twisting of the spine should also be avoided or modified. For example, the assessment of spinal range of motion should be avoided in people with an acute vertebral fracture or multiple fractures. If the fracture has healed, consider weighing the need for assessment with the potential risk, and whether functional mobility can be assessed via observation during functional tasks instead. If it is necessary to assess spinal range of motion, consider a modified version, or cue the movement so it is slow and controlled. Do not continue if the movement is painful. Examples of exercises that might be part of assessments and might be considered risky for individuals at high risk of fracture, including individuals with one or more vertebral fractures are deadlifts, overhead presses, sit-ups, clean and jerk, deep squats, spinal flexion movements, anything involving sudden, end-range, or resisted spinal flexion, sudden or end-range spinal twisting. Some exercises, like deep squats, overhead presses, and modified deadlifts may be acceptable if the patient can perform them with good alignment, or if they could be modified to be safer, ideally supervised by an exercise professional. Progression of an individual client to exercises or assessments that include running, jumping, or other risky movements requires supervision and good clinical judgment in balancing the potential benefits with potential harms. Factors that influence decision making include 10 yr fracture risk, general health and fitness, history of safe performance of the activities being considered, the client's desire to engage in them, and whether the client is taking medications to reduce fracture risk. Being on osteoporosis medication does not eliminate risk.

Sudden onset or acute exacerbation of pre-existing back or radicular pain, decreased mobility due to pain, increase or sudden worsening of thoracic kyphosis, loss of height or shortness of breath might indicate a new fracture or the progression of an existing fracture. Exercise or assessments should cease and the client should be referred back to their physician.

Recommendations and Anticipated Responses

During exercise testing for cardiorespiratory fitness, a protocol using a bicycle causes the least impact on the bones. A walking-only treadmill protocol may also be acceptable. During a bicycle protocol, the client should maintain a neutral spine at all times because sustained spinal flexion or rotation is not advisable in people with osteoporosis. Treadmill protocols are not advisable for individuals with balance impairments or where fall risk or high risk of fracture is a concern. If a treadmill protocol is used, measures should be in place to ensure the client does not trip or fall.

Pain at rest or during activity that is related to prevalent spine fractures may cause a client to want to stop a symptom-limited graded exercise test early. Individuals with hyperkyphosis may have limitations in vital capacity. In instances where a symptom-limited graded exercise test is not likely to be accurate or where the risks outweigh the benefits, pharmacological tests may be a better alternative to aid in diagnosis of coronary artery disease

without the use of exercise. If a client has clearance to exercise from their physician and exercise testing is solely being done to inform exercise prescription, a graded submaximal exercise test could be performed to assess heart rate and blood pressure responses and rating of perceived exertion (RPE). A summary of exercise testing guidelines is presented in table 25.1.

Treatment

Exercise alone is not sufficient to prevent all bone loss. Even young athletes with amenorrhea will continue to lose bone mass, despite their exercise training, as long as their menstrual cycles are absent (46). Individuals at risk of fracture should discuss the risks and benefits of available osteoporosis therapies with their physician, including the following:

- Bisphosphonates (e.g., alendronate, risedronate, zoledronic acid, and ibandronate)
- Denosumab
- Parathyroid hormone or parathyroid hormone–related protein analogs (teriparatide or abaloparatide)
- Romosozumab
- Selective estrogen receptor modulators (SERMs)
- Calcitonin

Calcium is crucial for bone health in the growing years, but data on the usefulness of calcium as therapeutic intervention for either prevention or treatment of osteoporosis are equivocal, and some have argued there is potential for harm, in that supplementation may increase the risk of cardiovascular events (74). Many experts, though, still recommend calcium supplementation or increased calcium intake for individuals at risk of fracture (13, 52). Increasing intake of calcium through diet has not been associated with increased risk for cardiovascular disease. Thus, for clients with low calcium intake, achieving higher calcium intake through the diet is the best option. Clients can be encouraged to use a tool like the calcium calculator (https://osteoporosis.ca/calcium-calculator/) to estimate their daily intake, and if needed, increase intake of calcium, primarily from food sources (49a). Clinical exercise physiologists should defer decisions about supplementation to the client's physician or dietitian, but they can encourage a healthy diet that includes calcium-containing foods.

Vitamin D supplementation is often recommended for individuals at risk of fractures (13). The rationale is twofold: potential maintenance of BMD and fracture risk reduction (74) and potential reduction of falls independent of BMD (43). However, as with calcium supplementation, there is controversy around the efficacy of vitamin D supplementation for fracture prevention (7). Whether a client needs supplemental vitamin D is complicated by the fact that one of our main sources of vitamin D is sunlight exposure, which is difficult to quantify. Therefore, clients with osteoporosis should consult with their physician on whether they need vitamin D supplements.

Table 25.1 Exercise Testing Review

Test type	Mode	Protocol specifics	Clinical measures	Special considerations
Cardiorespiratory	Bicycle or treadmill (walking only)	Standard test (YMCA bike test, standard Balke)	$\dot{V}O_2$, METs, RPE, HR and BP response	Avoid high-impact modes (e.g., running), especially for high-risk patients. Hyperkyphosis can affect gait and balance, and decrease vital capacity. Bicycle preferable to treadmill test for individuals with gait and balance impairments.
Strength	10RM (repetition maximum); 1RM may not be appropriate if high risk of fracture; consider assessment of functional muscle power	Dynamometer (e.g., isokinetic, handheld), 10RM for relevant exercises or movements		May be used to determine intensity of strength training program; strength testing or heavy external resistance (e.g., weights) may be contraindicated in some individuals at high risk of fracture.
Physical function, mobility, and balance	Performance-based tests	30 s chair stand, 40 m fast-paced walk, Short Physical Performance Battery, Berg Balance Scale, Mini-Balance Evaluation Systems Test, Four Step Square Test (for moderate fracture risk with lower fall risk only)	Performance (e.g., time, score) relative to normative data or cutoffs	Ceiling effects will be observed on some gait and mobility tests in community-dwelling adults, but high fall or fracture risk may make more challenging tests risky, so test battery may need to be tailored based on age, fall and fracture risk, and comorbidities.
Flexibility	Goniometer, Thomas test, weight-bearing lunge test, observation during movement (e.g., squat, hinge, back-to-wall shoulder flexion)	Priority: joints where ROM is often limited (e.g., shoulder flexion, hip flexion, ankle dorsiflexion)	ROM, joint angle	Tests that involve end-range forward flexion or trunk rotation may need to be avoided in individuals at high risk of fracture.
Posture	Visual inspection or objective measure of hyperkyphosis or hypo- or hyperlordosis	OWD protocol, rib–pelvis distance, inclinometer, flexible ruler	Subjective evaluation of hyper- or hypolordosis and hyper- or hypokyphosis, distance between occiput and wall or fingerbreadths between ribs and pelvis, degrees of kyphosis	Hyperkyphosis and hypolordosis are common, possibly because of fractures or muscle weakness or imbalance.

METs = metabolic equivalents of task; RPE = rating of perceived exertion; HR = heart rate; BP = blood pressure; ROM = range of motion; OWD = occiput to wall distance.

Many of the current drugs with Food and Drug Administration (FDA) approval for osteoporosis are considered antiresorptive therapy (table 25.2). They halt the loss of bone or even increase bone mass by inhibiting bone resorption or overall bone turnover. The only currently FDA-approved drugs that increase bone formation are romosozumab, teriparatide, and abaloparatide (6, 16). Available drugs and their mechanism of action are discussed in chapter 3. For the premenopausal woman or young male with low bone mass, treatment should always focus on the underlying cause. For example, a premenopausal client who has been diagnosed with an estrogen deficiency could be counseled on ways to regain her menstrual cycles by increasing energy intake. Clients with persistent amenorrhea, or males or females with relative energy deficiency in sport, should engage in bone health management with a physician (44).

Table 25.2 Pharmacology

Medication class and examples	Primary effects	Exercise effects	Special considerations
Bisphosphonates • Alendronate • Ibandronate • Risedronate • Zoledronic acid	Decreases osteoclast activity, thus decreasing bone resorption	Initial use may increase atrial fibrillation, particularly intravenous forms, but no known exercise effects; data are limited	Can cause gastrointestinal distress. Rare side effects include osteonecrosis of the jaw and atypical femoral fractures.
Anti-RANKL agent Denosumab	Inhibits RANK-L, decreasing formation of new osteoclasts	None known, but data are lacking	Rare side effects include osteonecrosis of the jaw and atypical femoral fractures.
Parathyroid hormone or parathyroid hormone–related protein analogs • Teriparatide • Abaloparatide	Increases bone turnover, with increased bone formation relative to resorption	None known, but data are lacking	Treatment duration is up to 2 yr, followed by treatment with antiresorptive medication.
Romosozumab	Targets sclerostin Increases bone formation and decreases bone resorption	None known, but data are lacking	Should not be used in people at high risk of stroke or myocardial infarction. Recommended duration is 12 mo, followed by treatment with antiresorptive medication.
Hormone replacement • Estrogen • Estrogen + progesterone	Decreases osteoclast activity, thus decreasing bone resorption	Acute vasodilator but not known to alter exercise responses	FDA recommends nonestrogen treatments be considered first if use is targeted only at fracture prevention.
SERMs Raloxifene	Decreases osteoclast activity	May increase vasodilation, but no known exercise effects	
Hormone Calcitonin	Decreases osteoclast activity	None known, but data are lacking	

RANK-L = receptor activator of nuclear factor-kappaB ligand; SERMs = selective estrogen receptor modulators.

EXERCISE PRESCRIPTION

The exercise prescription and therapeutic goals for someone with osteoporosis will not be substantially different from national physical activity guidelines for older adults, with a few exceptions. Therapeutic goals specific to osteoporosis are fracture prevention (via prevention of falls and bone loss), and the promotion of spine-sparing strategies. One key difference might be that functional, strength, and balance training should be prioritized over other types of physical activity if the goal is fall or fracture prevention, and the program should be performed at least twice weekly (21, 22). When the goal is improving posture, exercises for back extensor muscles, scapular stabilizers, and abdominal muscles (isometric) are recommended (4). Spine-sparing strategies focus on postural alignment and proper body mechanics to protect the spine from harmful loads during exercise and physical activities of daily living and leisure. This includes prescribing exercises to increase muscular endurance in the back extensor muscles and to mobilize muscles that restrict mobility or optimal alignment (e.g., pectorals, hip flexors). For individuals with a history of vertebral fracture, form and alignment should be the initial focus rather than intensity of aerobic or resistance training, followed by careful attention to safe progression of intensity. Further, the risk of high-intensity aerobic physical activity that involves impact or fall risk may outweigh the benefits for many individuals who have a spine fracture, so moderate intensity should be encouraged.

Studies in animal models suggest the following with respect to mechanical loading regimens aimed at stimulating bone growth (58):

- The exercise program should include load-bearing activities at high magnitude with few repetitions.
- It should be long term and progressive.
- The program should create variable strain distributions throughout the bone structure (i.e., load the bone in directions to which it is unaccustomed).
- Bone responds to loading in a site-specific manner, so exercise should load joints that are at greatest risk for fracture, such as hip, wrist, lower back.

- Added benefit may result from dispersing loading activities throughout the day rather than completing the exercise all at one time.

Although these principles have been shown in animal models and some human studies to increase bone mass, fewer studies have examined the efficacy of different types of exercise for maintaining or increasing BMD in people with established osteoporosis or vertebral fractures (28a). Further, the challenge in individuals with very low bone mass is that the loading characteristics shown to be osteogenic in animal studies, such as high-magnitude loading, have the potential to increase the risk of fractures. Two studies have reported that high-intensity resistance training and high-impact training (e.g., drop jumps) in men and women with low bone mass can increase BMD at the spine and prevent bone loss at the femoral neck, which is very promising (25, 73). The sample reflects relatively healthy older adults without musculoskeletal injuries or conditions or uncontrolled cardiovascular disease, so high-intensity resistance or interval training may not be generalizable to everyone (73). Also, most individuals will need several weeks at a lower intensity to ensure good form and alignment before higher-intensity training is introduced. There is some, albeit limited, evidence that other types of resistance or impact exercise may prevent bone loss in people at risk of fracture (54, 60, 61).

Goal setting in individuals with moderate to high risk of fracture should reinforce the idea that certain types of exercise may help maintain bone mass and prevent falls, rather than making assumptions that all types of exercise will increase BMD. Clinical exercise physiologists considering a progression from moderate- to high-intensity resistance training or progressive increases in impact magnitude in clients with osteoporosis should do so with careful consideration of the risks versus potential benefits for an individual client. For individuals with pain due to spine fractures, encouraging spine-sparing strategies, exercise, or education targeting pain control is recommended. Examples include sitting in erect alignment with appropriate lumbar spine support, spending time (e.g., 15-20 min or more) in supine (to encourage spinal extension and stretching of the pectoral and front shoulder muscles), and spending time prone (to encourage spinal extension and flexibility of the hip flexors). Clients with spine fractures should be educated that the fracture can take up to 12 wk to heal, and that they can gradually resume physical activities of daily life and leisure (including exercise) as pain diminishes. Exercise or activities that involve heavy physical exertion should be avoided for the first 12 wk. Clients with pain that persists beyond 12 wk may need to be referred back to their physician or a pain specialist. Individuals with vertebral fractures may experience early satiety, and thus, experience weight loss. Weight loss may also occur with aging due to loss of appetite. Weight loss is also a common reason to start or persist with an exercise program. However, when you lose weight, you don't just lose fat. Weight loss programs in older adults result in loss of fat, muscle, and bone. If a client with osteoporosis is losing weight, or has a goal to lose weight, communicate that bone loss can occur during weight loss and discuss strategies to minimize bone loss (e.g., include resistance training as part of the exercise program and ensure intake of calcium, vitamin D, and protein meets recommended daily intakes even with reduced energy intake). Individuals who experience unintentional weight loss due to early satiety or for other reasons should be referred to a dietitian.

OSTEOSTRONG® is a for-profit company with centers in the United States and Canada, and has positioned itself as something that "....works for people at all ages and levels of activity to promote skeletal strength which impacts the entire body in many ways using a process known as Osteogenic Loading" (49b). On their website is information that implies their exercise program increases BMD. There are no published randomized controlled trials comparing OSTEOSTRONG® to control with BMD as an outcome. There is one observational study of 55 participants, with high loss to follow-up and missing data, such that they only report BMD data in nine people (28b). The inventor of OSTEOSTRONG® devices was a co-author. The low quality evidence does not support recommendations for using OSTEOSTRONG® to improve bone strength. Despite claims on the website, there are no data to suggest OSTEOSTRONG® is more effective than resistance training or balance and functional training when it comes to bone strength or falls. The LIFTMOR-M trial was a randomized trial in men with low BMD that compared high intensity strength and impact training to isometric axial compression using devices like the ones used at OSTEOSTRONG® centers, as well as comparing it to a nonrandomized control group. The trial suggests that there were either no between group differences in BMD, or that high intensity strength and impact training was superior, depending on the site measured (25). There were 5 new spine fractures in the isometric axial compression group and none in the exercise group or control group (25).

Aerobic Physical Activity

Participation in regular physical activity, particularly moderate or vigorous physical activity, has been associated with reduced disability and mortality (51, 53) and should be encouraged for all adults. Recent guidelines from the American College of Sports Medicine recommend approximately 30 to 60 min of moderate physical activity each day such that the cumulative total for the week is 150 to 300 min (1). Impact exercise (e.g., jumping, running) may increase BMD and improve physical function in people with low bone mass, but the evidence in this population has low to very low certainty because there are few trials (61). Further, the effect of walking alone on BMD is modest, and it is possible that multicomponent exercise programs are more likely to have an effect on hip BMD (60). Therefore, aerobic physical activity should not be prioritized over resistance training and balance training in individuals with osteoporosis; multicomponent programs that include aerobic physical activity with some impact, progressive resistance training, and balance challenges should be encouraged.

Research Focus

Whole-Body Vibration and Bone Health

Whole-body vibration (WBV) has been proposed as an alternative method of exercise. During WBV training, the participant either stands on or performs various exercises on a vibrating platform. It is hypothesized that the sinusoidal vibrations stimulate mechanical loading and reflexive muscular contractions (68). However, many studies focus on postmenopausal women, not all of whom have osteoporosis. Meta-analyses suggest that some studies demonstrate a small but significant effect of WBV on BMD but others do not. A high-quality trial suggested no benefit of WBV for bone in postmenopausal women (69). It is possible that the heterogeneity may vary with WBV frequency, intensity or cumulative dose, body position (e.g., standing versus semiflexed knee), type of vibration, participant age, or methodological quality of the study (47). Some studies suggest WBV may also improve balance and fall risk (39, 68).

A decision to add WBV as a fracture prevention strategy requires careful consideration of the cost (e.g., financial, time, potential risks of WBV) to the client versus the potential for benefits. The ideal WBV protocol is not entirely clear, and there is not enough evidence to strongly recommend it for individuals with osteoporosis.

Mode of aerobic activity needs to be informed by the person's risk of falls, fractures and adverse events, presence of pain, and ability to maintain proper spinal alignment. Weight-bearing activity (e.g., walking, dancing, and other activities where full body weight is supported by limbs) should be encouraged as the mode of aerobic physical activity where possible. Individuals who express a preference or therapeutic need (e.g., comorbid knee osteoarthritis) for non-weight-bearing activities should be encouraged to participate in them if that is the only way they will achieve the aerobic physical activity guidelines. Those who prefer activities like cycling or swimming can be encouraged to add weight-bearing activities in small bouts or on alternate days or as part of a progressive resistance training program (e.g., step-ups, walking lunges). For some people, this may be progressed to include moderate- to high-impact activities (heel drops, drop jumps) if they have sufficient conditioning and it is considered safe for them to do so.

Duration of physical activity should be consistent with general guidelines (i.e., accumulating ≥150 min/wk). If a client is new to exercise, has low endurance, or is at increased risk of cardiovascular disease or adverse events, shorter, more frequent bouts of moderate-intensity aerobic physical activity (e.g., 10 min at a time, three times per day) can be encouraged. Moderate-intensity physical activity (e.g., 5-6 on a 0-10 RPE scale) is a good place to start. Sound clinical judgment is required as to whether it is safe to progress clients to more vigorous physical activities or higher impact.

Progressive Resistance Training (PRT)

In older adults, the effects of PRT alone on fall-related injuries or the number of people who fall are unclear, and the evidence is of very low certainty (62). However, there is moderate certainty evidence that combined interventions (e.g., resistance training combined with functional and balance training) reduces the rate of falls (RaR 0.69, 95% CI: 0.48 to 0.97; 1,084 participants, 8 studies) or risk of falling (RR 0.76, 95% CI: 0.61 to 0.95; 1,375 participants, 13 studies) among older adults (62). Therefore, to reduce falls it may be necessary to combine resistance training with balance and functional exercises.

The effects of PRT on fractures is uncertain because most studies do not have enough participants to measure this effectively. A meta-analysis of three trials that examined the effects of PRT on femoral neck BMD in people with low bone mass suggests small effects, with some heterogeneity (54). It is hard to determine the independent outcomes of PRT because trials often used different intensities or combined interventions (e.g., resistance training combined with walking, aerobic exercise, or impact exercise). Findings were similar using total hip BMD. Small positive effects on femoral neck BMD were reported in a meta-analysis of moderate- to high-intensity resistance training in postmenopausal women (MD +1.03% [95% CI: 0.24 to 1.82]) (28a). However, using low weights had no effect on hip BMD in postmenopausal women. PRT may improve physical functioning, as reported in a systematic review of the effects of PRT on Timed Up and Go test performance in people with low bone mass (MD −0.89 s, 95% CI: −1.01 to −0.78, 12 studies, 886 participants, very low certainty evidence) (54). The findings are consistent with a 2009 Cochrane review reporting that strength training (often at high intensity) can improve measures of physical ability (SMD 0.14, 95% CI: 0.05 to 0.22, 33 trials, 2,172 participants) (38b).

Similar effects on mobility are observed when examining studies in high-risk groups, such as individuals with vertebral fractures (22). Intensity may need to be moderate or high, and longer training durations or time under tension of 6 s may be important training variables related to improving muscle strength (55). PRT may also improve quality of life (54). Minor adverse events may occur with PRT, such as muscle strain, musculoskeletal issues, or pain. One fall-related injury (forearm fracture) attributable to exercise was reported across four studies (414 participants) (54). Thus, when considering PRT for people with osteoporosis, combine it with balance and functional training.

Practical Application 25.1

CLIENT–CLINICIAN INTERACTION

Individuals with osteoporosis are often instructed to avoid bending, twisting, or lifting, but restrictions create fear and are a disincentive to physical activity participation. Clients will have questions about whether it is safe for them to do things like golf, yoga, Pilates, skiing, or other activities. The clinical exercise physiologist should consider the person's risk of fracture, as well as their comorbid conditions, goals, preferences, history of fracture, and history of activity, and tailor activities to minimize risk and maximize benefit. If a client with osteoporosis but no history of spine fractures has a history of doing a certain activity that may be considered risky, and they have a strong preference to continue *and* the activity can be modified so they can do it safely, they should be encouraged to do so. However, if the client is at high risk of fracture (e.g., prevalent vertebral fracture), has never done the activity before or is sedentary, or it cannot be modified to do safely, they should be encouraged to avoid it. With all clients with osteoporosis, the clinical exercise physiologist should outline the types of activities that increase the risk of fracture and discuss risks and benefits.

Here are some recommendations for individuals with osteoporosis and no history of vertebral or hip fracture:

- Very high-impact sports (e.g., high-impact aerobics or plyometrics) may need to be avoided or modified (e.g., a lower step height for step aerobics, brisk walking rather than running).

- Any symptoms consistent with fracture (e.g., acute back pain) that arise after physical activity warrant follow-up with a health care provider. See the Contraindications section.

- Many activities involve rapid, repetitive, weighted, sustained, or end-range twisting or flexion of the spine (e.g., golf swing, bending to retrieve golf balls, yoga) and need to be modified (e.g., using partial swing,

modified golfer's reach, assistive device to pick up balls; choosing alternative yoga poses; avoiding yoga and Pilates postures that involve forward flexion or end-range rotation).

- Modify or avoid activities that have a high fall risk or involve contact (e.g., skiing, racket sports, martial arts, skating). Modifications should reduce potential for injurious contact (e.g., moving at a slower pace, skiing blue or green runs instead of black, wearing hip protectors and wrist guards, wearing shoes with good traction, being careful on slippery surfaces like a pool deck or icy sidewalks, walking on indoor tracks or in malls in icy conditions).

When it comes to individuals at high risk of fracture (e.g., with a spine fracture, or high risk because of age and very low BMD), many sports, exercise classes, and some household activities present risks that may outweigh the benefits, especially if hyperkyphosis, pain, or gait and balance difficulties are present. Even group classes may contain exercises that are not appropriate for individuals at high risk of fracture. Individuals with osteoporosis should speak with the instructor about their condition or take classes that have been adapted for people with osteoporosis.

Clinical exercise physiologists who wish to learn more about advising clients on safe movement during daily activities and leisure are encouraged to read the Too Fit to Fracture consensus papers (21). Osteoporosis Canada has some resources that demonstrate exercises and movement strategies that can be used to inspire individuals with osteoporosis (48). Working with a client to develop a plan that outlines the when, where, how, why, and what of their exercise program and linking the program to their goals may be a way to increase self-efficacy and adherence.

PRT recommendations for individuals with osteoporosis are in line with ACSM recommendations for older adults: exercises for major muscle groups ≥2 d/wk at an intensity of 8RM to 12RM (11). In clients new to PRT, previously sedentary clients, and those at high risk of fractures or other adverse health outcomes, PRT intensity should initially be lower (~8-10 repetitions at 70%-75% of estimated 1RM) and progressively increased over time. Body weight, elastic bands, free weights, and other forms of external resistance can be used. In individuals at high risk of fracture, alignment should be prioritized over intensity, and the clinician might consider how to innovatively increase intensity without requir-

ing the use of large external loads, particularly if the individual will be exercising unsupervised. A PRT program should involve performing the prescribed number of repetitions with one or two repetitions in reserve and progressively increasing the load or difficulty over time. Periodization of intensity and volume should be considered for long-term programming. A supervised approach to progression of PRT or impact exercise to higher intensity or high velocity may be appropriate for some clients at low or moderate risk of fracture, with a goal of building strength or power, but this may be risky for individuals at high risk of fracture; progression requires sound clinical judgment.

For spinal extensors, rotators, and flexors (core muscles), PRT should focus on muscular endurance rather than strength (i.e., lower intensity, longer duration), and daily exercise can be encouraged. PRT for both major muscle groups and core muscles should be performed in positions where the spine is least loaded whenever feasible. Loads on the spine are lowest in supine, followed by prone, standing, and then seated, and highest when seated with trunk flexion (45). Isometric exercises for trunk flexors and rotators (e.g., abdominal bracing, plank or modified plank, and side plank) may be preferable to flexion and rotation against resistance in individuals with osteoporosis. For back extensor muscles, supine shoulder press with or without leg press (where *press* refers to a gentle press into the floor, or making the shoulders feel heavy into the floor), supine arm or leg lengtheners, and bird dog are appropriate (4). Some exercises can be modified to be performed at a wall or counter for individuals who have difficulty getting on or off the floor. To increase the challenge, a client could perform supine back extensor exercises using elastic tubing to add resistance (e.g., thoracic extension while pulling on tubing) or while sitting in a wall squat or lying supine on a foam roller (4). Exercises involving rapid, repetitive, sustained, weighted, or end-range trunk flexion or rotation are contraindicated. The SHEAF trial evaluated an exercise program designed to target hyperkyphosis and reported positive changes in Cobb angle of kyphosis in favor of exercise. The exercises were performed at an intensity described as Borg scale intensity of 4 to 5, 70% to 80% of perceived exertion, and included quadruped arm and leg lift, side lying thoracic rotation/extension, exercises on roller (e.g., unilateral overhead reaching, bilateral pulldown, transversus abdominis activation, chest stretch), wall push-ups, shoulder flexion/extension, single leg stance, daily postural correction practice, diaphragmatic breathing, and stretching for gluteal, quadriceps, chest muscles (4).

When performing back extensor muscle or core exercises as isometric holds, clients can start with one or two sets of 5 to 10 s holds with slow, controlled movements; to increase the difficulty, increase the number of sets, with 3 to 5 s of rest in between each hold (e.g., perform hold for 5 s, rest 3 s, repeat two to four times). To progress the difficulty further, three sets of repeated isometric holds can be performed (e.g., three to five holds for 10 s each with 5 s of rest in between, rest 1 min, repeat two times).

Ensuring safety and optimal alignment during transitions is just as important as during exercise. There are reports of fractures occurring during exercise while turning from supine to prone, falling during exercise, or dropping weights on a foot (15). Resistance training machines often require forward bending or twisting to adjust the machine or to get in and out; clients should be taught how to do this safely, or other exercises that do not require machines should be selected.

Balance Training

Strong evidence from multiple meta-analyses indicates that exercise programs that challenge balance can reduce falls, even in those at risk of falls (23, 63). There is also some evidence that exercise can reduce injurious falls. However, the effect of exercise on falls may vary with the mode of exercise; balance and functional training and tai chi reduce the rate of falls and the risk of being a person who falls, whereas effects of walking, resistance training, or dance are uncertain (62). Emphasizing walking as a fall prevention strategy to the exclusion of balance training is not recommended, but walking training can be included as part of a multicomponent exercise program that includes balance training (63). Walking tends to be a default exercise prescription, but brisk walking may actually increase the risk of falls in individuals already at high risk or with balance impairments, which is why balance and functional training is prioritized. Therefore, all individuals with osteoporosis should be encouraged to challenge their balance daily, while ensuring they are taking precautions to prevent falls.

Static balance refers to the ability to keep the line of gravity (i.e., the line through the body's center of mass) within the body's base of support, having minimal sway. A person's base of support is the area beneath all points of contact that are supporting the body—for someone who is standing, it is the area between the feet; for someone who is standing with a walker, it is the area beneath the feet and the walker. If a person is moving around, the line of gravity, the base of support, or both are moving. The ability to maintain the center of mass over the base of support while the body is moving around, and to maintain stability even when the center of mass moves outside of the base of support, is called dynamic balance.

To challenge balance, clients can do one or a combination of the following:

- Reduce base of support (e.g., placing feet together, standing on one foot)
- Shift body weight within the limits of stability or in three dimensions (e.g., leaning forward and back, shifting weight heel to toe)
- Introduce challenges that require anticipatory adjustments, like standing on one foot and moving an object from one spot to another or catching a ball
- Reduce contact with support objects (e.g., if holding on to an assistive device while standing or doing exercises, reducing that contact gradually)
- Change sensory input (e.g., closing eyes)

Clients can do combinations of these challenges in static standing or can apply the challenges during dynamic activities. For example, tai chi involves three-dimensional movement and weight shifting. Clients can also add balance challenges to daily walks (e.g., tandem walking, walking on one's toes or heels, walking lunges) or to PRT (e.g., choosing resistance training exercises that also challenge balance). Clock Yourself, an app designed by a physical therapist, has the user envision a clock face on the floor and prompts reactive stepping to randomly called numbers. It progresses in intensity by increasing the speed or adding cognitive or physical challenges.

Perturbation training is a novel way to challenge balance (41). It involves requiring the client to respond to a perturbation that induces movement of the center of mass relative to the base of support (e.g., client is pushed gently and taught to react with a step in the direction of the perturbation). The difficulty level of balance challenges should be progressed over time (e.g., prescribing a more difficult exercise, removing vision or contact with support object, or dual-tasking).

Flexibility

Osteoporosis-specific strategies with respect to flexibility include prioritizing the mobilization of muscles that are restricting optimal spinal alignment or ankle mobility (e.g., muscles that internally rotate the shoulder, hip flexors, plantar flexors). Flexibility challenges that involve sustained spinal flexion or rotation, or pushing into end-range spinal flexion or rotation should be avoided. If there is a need to improve spinal mobility, controlled twisting in supine or side-lying is acceptable, as is midrange but not end-range spinal flexion and extension with body weight supported by the upper extremities while in a neutral spine position (e.g., on hands and knees). Yoga and Pilates are activities that challenge balance, strength, and flexibility, and clients may ask whether they are safe or effective. In people with osteoporosis, there is limited to no evidence that yoga or Pilates can increase bone mineral density or reduce falls, but they may improve physical functioning (62). Yoga poses that may be risky include forward fold, spinal rocking, ragdoll, saw, plow, pigeon, and anything that challenges balance that may cause a fall. Clients with a strong preference for yoga or Pilates should look for an instructor who knows how to adapt the moves for osteoporosis, and tell their instructor they have osteoporosis. They should focus on control (rather than pushing intensity or range of motion), have a support object nearby during balance poses, and modify poses to be consistent with spine-sparing principles (described next). For individuals at high risk of fracture (e.g., history of hip or vertebral fracture, very low BMD for age), the risks of some yoga or Pilates classes may outweigh the benefits.

Spine-Sparing Strategies

Avoiding undesirable spinal loading (i.e., application of rapid, repetitive, weighted, sustained, or end-range flexion or twisting torque to the spine, or combined loading) and maintaining good alignment during any type of exercise are important. Spine sparing should also extend to activities of daily living and leisure. When activities require bending, clients can be taught to use a hip hinge and to bend at the knees and ankles, while maintaining the head over the base of support. A hip hinge involves flexing at the hips while bringing the hips posterior to the base of support (like the top portion of a deadlift). Individuals with hyperkyphosis may be able to improve posture to some degree with posture training and exercises for back extensor endurance (5, 33); however, some hyperkyphosis is fixed and not amenable to intervention.

Individuals at high risk of fracture may need to get help with household activities that apply flexion or twisting torque to the spine (e.g., heavy lifting, cleaning gutters, shoveling, or changing light bulbs in ceiling lights). Lifting heavy objects, lifting overhead, lifting combined with twisting or bending, and precarious balancing are other risky activities. Individuals at high risk of fracture may want to consult a physical or occupational therapist on safe movement during daily activities.

People with spine fractures may find sitting or standing for long periods of time uncomfortable (e.g., sitting in a car) and should get up and move around. For those with chronic pain due to fractures, daily practice of ≥20 min periods (may be performed multiple times per day) lying in supine on a firm surface (not a couch) with palms up may reduce the loads on the spine and promote spinal extension and stretching of the pectorals and front shoulder muscles. Individuals with hyperkyphotic posture may require a pillow under the head while lying in supine to prevent hyperextension of the cervical spine. After a spine fracture, finding a comfortable sleeping position can be difficult. One option is to lie flat with knees bent and one or more pillows under the knees. Spinal stenosis, knee pain, wrist or hand pain, and impaired balance may limit the capacity to assume a supine, prone, or quadruped position for pain relief or for back extensor exercises. For example, some older adults may find it easier to get into supine or prone on a bed than getting onto the floor or into a quadruped position. Quadruped arm and leg lifts require sufficient trunk and balance control to prevent rotation of the spine and may be difficult for adults with knee, hand, or wrist osteoarthritis. The following are exercises to avoid or modify for individuals with osteoporosis:

- Exercises that involve hyperextension of the lumbar or cervical spine
- High-impact, high-fall-risk, or contact sports
- Rapid, repetitive, weighted, sustained, or end-range flexion or torsional loading of the spine during physical activity or exercise (e.g., sit-ups, spinal twists in yoga, kettlebell swings), or during transitions (e.g., rolling supine to prone, bending to adjust exercise equipment)
- Exercises or activities that increase the chance of falling, such as using trampolines or exercising on slippery floors or in icy conditions

Table 25.3 provides a summary of exercise prescription guidelines.

EXERCISE TRAINING

Table 25.4 provides a brief review of considerations for exercise training. Spine-sparing strategies extend to daily activities. Individuals with osteoporosis, especially those at high risk, should avoid lifting from or bending to heights below the knees (e.g., floor, low gardens). If bending to lift from a height below the knees is necessary, using a hip hinge while bending at the knees and

Table 25.3   Exercise Prescription Review

Training method	Frequency	Intensity	Time (Duration)	Type (Mode)	Progression	Important considerations
Cardiorespiratory	≥3-5 d/wk	Moderate to vigorous intensity (40%-85% of HRR); only moderate if high risk or not used to moderate intensity	Accumulation of ≥150 min/wk	Ideally involves impact or weight bearing, but consider risk level and client preference	<5% per week	See activity considerations in text. Previously sedentary people or those with comorbid conditions that increase risk of adverse events should start at low intensity and duration.
Strength	≥2 d/wk	8RM-12RM	Exercises for all major muscle groups; start with 1 or 2 sets	Resistance training (e.g., body weight exercises or exercises using elastic tubing, dumbbells, kettle bells)	Progress to 3 sets, then increase intensity over time (<5% per week), split routine, and add exercises	See considerations for progression of intensity and for back extensor and abdominal muscle training in text. Form and alignment are very important.
Balance challenges	Twice weekly minimum but can be performed daily	Difficulty should be sufficient to challenge balance, but precautions must be taken to prevent falls	Variable*	Static or dynamic, tai chi	Progressions include: reduce reliance on upper limb support; reduce base of support; move center of mass or shifting to limits of stability; add perturbation; reduce vision; add cognitive task	Clients should have support objects nearby if needed, wear shoes with good traction, and take precautions to prevent falls.
Range of motion	5-7 d/wk	Self-limiting; should not be painful	Hold position or move through ROM, 2 min per movement	Mobility exercises	Progressively increase ROM during mobility exercise	Avoid stretching exercises involving end-range spinal flexion. Prioritize muscles restricting mobility or alignment (e.g., hip flexors, pectorals, plantar flexors).
Back extensor muscles, scapular stabilizers, abdominal muscles	Minimum 2-3 times weekly	5-10 repetitions of 3-5 s isometric holds	1-3 sets	Supine, prone, body weight exercises, or elastic tubing	Progress duration of holds, then repetitions to 10, then increase challenge (e.g., add tubing or more challenging exercise)	A pillow should be used during supine lying and exercise for people with hyperkyphosis only. Advise isometric only for trunk flexors and rotators. Avoid hyperextension of the spine.

*Balance training can be done as separate exercises or a class (e.g., tai chi) or combined with other domains (e.g., balance challenges as part of aerobic exercise [e.g., walking, aerobics] or resistance training [e.g., by selecting exercises that also challenge balance, such as lunges, step-ups, one-leg squats]).

HRR = heart rate reserve; RM = repetition maximum; ROM = range of motion.

ankles, and keeping the spine neutral and stable, is recommended. Clients should balance loads on either side of the body rather than carrying on one side or out in front. For example, groceries and bags should be carried with weight balanced on either side of the body. Instead of twisting the spine, the client should take a step to turn—that is, they lift one foot and step to the direction they wish to face, such that the toes and front of the torso move to face the same direction in one movement.

Table 25.4 Exercise Training Review

Cardiorespiratory endurance	Skeletal muscle strength	Skeletal muscle endurance	Flexibility	Body composition
Weight-bearing or impact activities are ideal, but consider client preference. Exercises that challenge balance should be incorporated into an aerobic physical activity or resistance exercise program. Endurance training can decrease cardiovascular disease risk factors.	Resistance training will increase muscular strength and power and may help maintain bone mineral density. Prioritize functional and balance training to reduce fall risk.	Endurance will improve with resistance training, and back extensor muscle endurance is an important target to achieve good posture.	Prioritizing muscles that restrict optimal alignment may facilitate better mobility and posture.	Benefits seen in healthy populations are similar to those expected in the client with osteoporosis. If weight loss is a goal and client is reducing energy expenditure, resistance training may, in part, prevent loss of bone and muscle.

Clinical Exercise Bottom Line

- Activities that involve rapid, repetitive, weighted, sustained or end-range flexion or twisting of the spine, whether during assessment, exercise, or physical activities of daily life, may need to be modified or avoided in people with osteoporosis.

- Patient-specific factors that may inform modifications to assessment or exercise prescription include current health and physical functioning, whether they are on medication, whether they are at high risk of fracture or have had a prior fracture, presence of hyperkyphosis or balance impairment, and their desire to do the activity.

- Balance, functional, and strength training at least twice weekly is recommended for all individuals with osteoporosis.

- Patients should be encouraged to perform activities they enjoy for fun or fitness if they can be done safely. The clinical exercise physiologist must be careful not to use language that creates fear and unwarranted avoidance of physical activity. Individuals who are at high risk of fracture, or with a prior fracture or recent fracture, should consider assessment and physical activity advice from a physical therapist or exercise physiologist with training on osteoporosis.

CONCLUSION

Osteoporotic fractures result in a considerable human and economic burden. Multiple lines of evidence highlight the benefits of multicomponent exercise programs for individuals with osteoporosis. An exercise program for someone with osteoporosis should prioritize balance, functional, and strength training. Exercises for the back extensor muscles, scapular stabilizers, and abdominal muscles should be included if better posture or core stability is the client's goal. The clinical exercise physiologist can develop an understanding of a client's fall and fracture risk, medical history, physical function, goals, and preferences and use this information to prioritize where to start as they design a program in which the benefits outweigh the risk of fractures. Modification or avoidance of activities that involve rapid, repetitive, sustained, weighted, or end-range flexion or torsional loads on the spine is important. Consideration of spine-sparing strategies should extend to transitions between exercises and to activities of daily living and leisure.

Online Materials

Visit HK*Propel* to access a link to the references, the case studies with discussion questions, and a quiz to help you review key concepts and test your understanding of the material covered in this chapter.

CHAPTER 26

Nonspecific
Low Back Pain

Peter Ronai, MS

Most people in modern society experience nonspecific low back pain (NSLBP). This complicated and poorly understood phenomenon involves the interaction of a wide range of physical, social, and psychological factors. Clinical exercise physiologists (CEPs) provide exercise testing, prescription, counseling, and education for patients with cardiovascular and other chronic diseases who participate in cardiac rehabilitation (CR) and adult fitness (AF) programs. Nonspecific low back pain is a common comorbidity in these patients and can reduce their physical activity tolerance and function, their performance during exercise testing and workout sessions, and their health-related quality of life (HRQOL). For an individual with comorbid NSLBP, the effect may be devastating; for society, the costs are enormous. A variety of exercise interventions have been shown to help manage symptoms and improve physical function in persons with NSLBP, yet expert consensus on the most effective type or dose of exercise is lacking (40, 46, 77, 84, 87, 88, 101, 104, 116, 125, 126, 135). Furthermore, exercise prescription is complicated by differences between acute, subacute, and chronic NSLBP and by substantial variability in recommended management strategies (17, 45, 76, 84, 87, 95, 105, 135).

DEFINITION

Low back pain (LBP) is considered a symptom rather than a disease and reflects many heterogeneous disorders and causes (37, 38, 45, 54, 55, 60, 76, 84, 85, 87, 95, 135). In the majority of cases, a specific pathoanatomical cause of LBP cannot be determined, and thus the majority of LBP is considered to be nonspecific low back pain (NSLBP) (2, 45, 54, 55, 60, 64, 67, 76, 84, 87, 95, 104, 135). Nonspecific low back pain is described as discomfort and pain localized below the costal margin and above the inferior gluteal folds, with or without leg pain, caused by a number of musculoskeletal dysfunctions but not attributed to recognizable, known, specific pathology (e.g., infection, tumor, osteoporosis, ankylosing spondylitis, fracture, inflammatory process, radicular syndrome, or cauda equina syndrome) (2, 16, 37, 45, 54, 55, 60, 64, 76, 84, 87, 104, 135). Traditionally, LBP is categorized by its etiology (causes), location, and duration of symptoms (2, 45, 55, 60, 64, 66, 67, 76, 84, 87, 95, 104, 135).

As people age, LBP is accompanied by numerous activity limitations (54). Low back pain is often attributed to either nociceptive (sensitization of pain receptors in spinal and mechanical structures and fascial tissues), neuropathic (radicular or nerve-related pain), or central (sensitization within the brain) sources (45, 76, 85).

This chapter focuses primarily on the development and management of exercise strategies for adults engaging in cardiac rehabilitation (CR) and adult fitness (AF) programs who have comorbid recurrent or chronic NSLBP (CNSLBP). Management of the acute phase of an episode of NSLBP mainly involves medical screening to exclude serious conditions, reassurance as to the benign nature of the problem, avoiding prolonged bed rest and encouragement to remain as physically active (without aggravating the condition) as possible, and simple symptom-based treatment (16, 17, 22, 35, 43-45, 66, 67, 76, 84, 87, 95, 119, 134, 135). Therapeutic exercise has not been shown to be any better than other treatments for promoting the resolution of an acute episode of NSLBP, but exercise in general can be helpful, may possibly prevent recurrences, and is suggested most strongly for CNSLBP (17, 45, 66, 67, 76, 77, 84, 87, 95, 135). However, more research is needed because there is insufficient evidence on which forms of

Acknowledgment: I want to acknowledge both Jan Perkins, PT, PhD, and J. Tim Zipple, PT, DSc, for their past work on the first four editions of this chapter. Their strong contributions set the framework for updating this chapter, and I thank them for all their hard work. Portions of this chapter have been reprinted by permission from P. Ronai, "Exercise Recommendations for Cardiac Patients with Chronic Nonspecific Low Back Pain," *Journal of Clinical Exercise Physiology* 8, no. 4. (2019): 144-156.

exercise should be performed (40, 46, 66, 77, 84, 87, 88, 95, 101, 104, 116, 125, 126, 135).

SCOPE

Clinical exercise physiologists provide exercise testing, prescription, counseling, and education for patients with cardiovascular and other chronic diseases who participate in CR and AF programs. Musculoskeletal comorbidities (MSKCs) are common in many of these patients (14, 39, 75, 79, 80, 111). Although coronary artery disease (CAD) is the leading cause of death globally, MSKCs result in the most disability (37, 38, 45, 54, 55, 76, 79, 80, 84, 95). More than 50% of new outpatient CR patients experience some musculoskeletal pain, with comorbid low back pain (LBP) reported by up to 38% of CR patients (37, 75, 79).

Low back pain is the leading cause of disability, and the years lived with disability make it one of the most expensive musculoskeletal conditions around the world (37, 45, 54, 55, 76, 84, 87, 88, 120, 135). More than 540 million people have been affected by activity-limiting LBP (37, 45, 76, 103). LBP affects between 49% and 70% of persons living in westernized countries and 70% to 85% of persons living in the United States in their lifetimes and is one of the most common reasons for physician visits (2, 5, 25, 28, 38, 64, 66, 67, 120). Specific causes of LBP are often unknown, and in approximately 80% to 90% of afflicted patients a specific pain source and cause cannot be identified (2, 37, 45, 55, 60, 64, 67, 76, 84, 87, 88, 104, 135). One-quarter of U.S. adults reported having LBP lasting at least 1 d in the past 3 mo (25). In any given year, between 12% and 14% of the U.S. adult population will visit their primary care physician with a complaint of back pain (120). In the United States, LBP accounts for more lost workdays than any other musculoskeletal condition (120). Low back pain is the leading chronic health problem forcing older Australian workers to retire prematurely (103). Although LBP is the most common cause in Europe of medically certified sick leave and early retirement, occurrence rates vary substantially among European countries (8).

Many adults describe a first episode of NSLBP in high school, frequently precipitated by athletic competition, heavy lifting, or trauma. As people age, LBP is accompanied by numerous activity limitations (54, 55). The burden from LBP has doubled in the last 25 yr, and the **prevalence** of the condition is expected to continue to increase with an aging and increasingly obese population (45). The cost of managing the minority of NSLBP patients whose acute episode becomes chronic is enormous. The direct cost of chronic back pain in the United States is approximately $100 billion each year (26).

Investigators have concluded that persons with CNSLBP experience physical activity (PA) intolerance; lower levels of strength, physical fitness, and function (27, 29-32, 113); increased disability (27, 89); PA avoidance (due to fear of increased pain with activity) (42, 43, 45, 76, 84, 86, 91, 118, 132); lower PA participation levels (18, 123, 128); and reduced health-related quality of life (HRQOL) (89). An association exists between CNSLBP and abdominal obesity (11, 45, 50, 56, 106, 108, 137), smoking (45, 107), and sedentary lifestyle (18, 45). A cross-sectional analysis of data from the National Health and Nutrition Examination Survey (NHANES) found that women participating twice weekly in musculoskeletal exercise training activities had significantly reduced odds of self-reported LBP (3). Most people, in all societal groups, will experience at least one episode of NSLBP during their lifetimes. Many people experience distress and some degree of temporary or mild disability, but only a small percentage become seriously disabled or go on to experience chronic back pain (67, 134).

PATHOPHYSIOLOGY

Low back pain is often attributed to either nociceptive (sensitization of pain receptors in spinal and mechanical structures and fascial tissues), neuropathic (radicular or nerve-related pain), or central (sensitization within the brain) sources (45, 84). An impressive list of structures with pain receptors could cause NSLBP, including the anterior and posterior ligaments, interspinous ligament, yellow ligament, posterior annular fibers of the disc, intervertebral joint capsules, vertebral fascia, blood vessel walls, and paravertebral muscles (133). Traditionally, LBP is classified and categorized by its etiology (causes), location, and duration of symptoms (45, 67, 76, 84, 95).

Most people who see physicians for back pain do not have serious pathology. In 80% to 90% of cases, pain cannot be attributed to recognizable, known, specific pathoanatomical cause, and thus the majority of LBP is considered to be nonspecific low back pain (NSLBP) (2, 45, 54, 55, 60, 64, 67, 76, 104). Classifications of LBP are noted in table 26.1. Additionally, the more common nontraumatic low back injuries are thought to result from a series of cumulative events (82).

CLINICAL CONSIDERATIONS

During medical examinations of patients with LBP, physicians and qualified health care providers conduct screening procedures to rule in or rule out more severe pathology, often referred to as **red flags**. Although worldwide agreement on a uniform list is lacking, some generally recognized red flags exist (8, 13, 14, 16, 17, 45, 48, 60, 66, 67, 76, 84, 87, 95, 122, 124, 125, 134, 135). Their presence or absence can determine whether physicians and health care providers clear patients with CNSLBP for, or exclude them from, exercise participation. Commonly accepted red flags include the following:

- Onset at <20 yr or >55 yr of age
- Nonmechanical pain (unrelated to time or activity)
- Previous history of carcinomas, steroids, or human immunodeficiency virus (HIV)
- Feeling unwell
- Unexplained weight loss

Table 26.1 Classifications of LBP

Classification	Criteria
Etiology	Specific: pain caused by unique or unusual pathophysiologic mechanisms (disc herniation, infection, tumor, ankylosing spondylitis, fracture, osteoporosis, arthritis, diseases, trauma, inflammatory process, radicular symptoms, cauda equina syndrome, or spinal pathology) Nonspecific[a]: pain not caused by a specific disease or spine pathology
Timeline or duration of symptoms	Acute: pain lasting less than 6 wk Subacute: pain lasting 6-12 wk Chronic: pain lasting longer than 12 wk[b] Chronic nonspecific low back pain (CNSLBP): pain lasting for 3 mo or longer[c]

[a]Nonspecific low back pain is described as discomfort and pain localized below the costal margin and above the inferior gluteal folds, with or without leg pain.
[b]25, 45, 66, 67, 76, 84, 85, 95, 104.
[c]2, 28, 45, 54, 60, 64, 66, 67, 76, 84, 95, 104.

- Widespread neurological symptoms (including saddle area numbness)
- Structural spinal deformity
- Spontaneous or persistent pain at night or pain while lying supine
- Indications for nerve root problems
- Unilateral leg pain that is greater than LBP
- Radiating pain to foot or toes
- Numbness and paresthesia in the same sensory distribution
- Straight leg raise test induces increasing leg pain
- Localized neurology (pain or symptoms limited to one nerve)

Some investigators have found that persons with CNSLBP experience delayed activation in the transversus abdominis and multifidus muscles and deficits in motor control (41, 52, 100). Changes in size, composition, and fiber typing of the multifidus, erector spinae, and other paraspinal muscles are considered potential factors in the etiology or recurrence of pain symptoms and related deficits in muscle strength and endurance (19, 34, 51, 58, 78, 100, 104, 110). In addition, gluteus medius weakness and gluteal muscle tenderness are common findings in some patients with CNSLBP (19). These symptoms can cause muscle fatigue, poor movement control, and reduced HRQOL (89). Therefore, an individualized approach to targeting these muscles when designing and progressing the exercise program progression is prudent in these circumstances (74).

Several risk factors are associated with NSLBP, but the strongest predictors of recurring NSLBP are the length of time between episodes (12, 45, 61, 76, 112) and a history of back pain (45, 49, 66, 67).

Signs and Symptoms

The person presenting with NSLBP may complain of localized or generalized lumbosacral region pain of variable intensity, dura-tion, and frequency. Radiating pain with a specific distribution of sensory changes, numbness, or lower extremity weakness can be associated with more serious pathology and can indicate specific tissue involvement. The client may have NSLBP with weight-bearing activities and complain of increasing pain with certain lumbar motions and postures. NSLBP may cause nocturnal discomfort that awakens the client when changing positions in bed, leading to sleep deprivation and impaired mental acuity. Symptoms are usually decreased with rest and anti-inflammatory medication. Any client presenting with any of the previously mentioned red flags or with new undiagnosed symptoms, chest pain, heart palpitations, shortness of breath, hernia, or unremitting spinal pain that is not relieved by rest should consult with a physician or appropriate HCP before beginning an exercise program.

Physical activity (PA) tolerance and exercise responses may be negatively affected by pain severity, location, physical fitness, strength, and body positions required during exercise testing and exercise training. Some individuals with CNSLBP are intolerant of motions such as trunk flexion or extension. In this case, positions such as prolonged standing, sitting, leaning, and reaching forward can cause discomfort and prevent CR patients and AF program participants from achieving their best exercise or testing efforts and results (4, 44, 57, 69, 70, 82, 83, 90, 109, 117).

Patients who are flexion intolerant generally experience pain, discomfort, and fatigue when doing exercises requiring frequent bending at the waist, leaning forward, or prolonged sitting. Practical application 26.1 contains a list of movements that may cause or worsen pain, discomfort, and fatigue in flexion intolerant–related LBP.

Patients and AF participants who are extension intolerant generally experience pain, discomfort, and fatigue when performing exercises requiring frequent standing, backward bending, overhead reaching or lifting, and spinal hyperextension (13). Practical application 26.2 contains a list of movements that may cause or worsen pain, discomfort, and fatigue in extension intolerant–related LBP (44, 57, 69, 70, 83, 90).

Practical Application 26.1

MOVEMENTS THAT MAY CAUSE OR WORSEN PAIN, DISCOMFORT, AND FATIGUE IN FLEXION INTOLERANT–RELATED LOW BACK PAIN

- Seated upper and lower body ergometry
- Inclined treadmill walking (secondary to compensatory forward leaning)
- Recumbent cycle or step ergometry
- Rowing ergometry
- Strength training exercises from a sitting or a bent position (rows, leg presses, deadlifts, knee extensions, hamstring curls)

- Strength training or trunk or core conditioning exercises from the supine position (curl-ups, crunches, full sit-ups)
- Flexibility and range of motion exercises requiring bending and twisting (seated hamstring stretches, toe touches, windmills)

Practical Application 26.2

MOVEMENTS THAT MAY CAUSE OR WORSEN PAIN, DISCOMFORT, AND FATIGUE IN EXTENSION INTOLERANT–RELATED LOW BACK PAIN

- Treadmill walking
- Stair climbing
- Elliptical step ergometry
- Standing strength training exercises and overhead lifting (shoulder presses, squats, rows, biceps curls, triceps extensions, dumbbell shoulder raises)

- Strength training or trunk or core conditioning exercises in the prone position (supermans, swimmers, back hyperextensions)
- Flexibility and range of motion exercises requiring spinal extension or hyperextension (cobra, back bends, overhead reaches)

Clinicians should also be alert to signs and symptoms that may indicate an inflammatory arthritic disorder such as **ankylosing spondylitis**, or another form of spondyloarthropathy, because a disorder of that type will require further medical evaluation and treatment. Chapter 24 provides a thorough description of these disorders and their management.

A number of established risk factors pertain to NSLBP and disability. Prevalence increases with age into midlife. NSLBP is lower in those with greater endurance in back extensor muscles, and people who have had previous episodes of NSLBP are more likely to experience additional back pain. This risk increases as the interval since the last episode shortens. People with recurrent or persistent back pain have decreased flexibility of hamstring and back extensor muscles and lower trunk muscle strength. The best predictor of an episode of back pain (and the only one sufficiently discriminative to be valuable in job selection) is a history of previous episodes, with risk of recurrence being higher the more recent the previous episode (12, 45, 61, 76). Other risk factors for back pain include obesity (45, 106), smoking (11, 45, 50, 56, 107, 108, 137), physical inactivity (18), heavy lifting, and lifting with twisting regardless of the weight (18, 61).

Psychosocial factors are important to consider in management of subacute and CNSLBP, and most authorities suggest they be addressed as part of a conservative plan of action (35, 66). Most current guidelines recommend general aerobic, or cardiorespiratory, exercise (swimming, walking, cycling), stretching, and strengthening exercises as part of a multidisciplinary management plan that also addresses psychosocial factors (10, 35, 45, 46, 66, 67, 76, 78, 84, 87, 101, 104, 115, 125, 127, 135).

History

As mentioned in chapter 4, medical evaluation should have cleared the individual referred with NSLBP for major pathologies, but a clinical exercise physiologist should be alert to the possibility that serious pathologies have been missed.

The history of an individual with NSLBP should first focus on screening for possible serious pathology. Questions should cover all the areas in the Clinical Considerations section of this chapter in an attempt to uncover red flags that indicate when medical evaluation is required. After this, the interview should focus on the mechanisms of injury, both initial and recurrences, because this information may guide the practitioner in selecting management strategies (45, 76, 87, 135). For example, recurrences may be associated with specific situations such as spinal flexion with lifting, prolonged postures as with driving, or prolonged standing and overhead reaching or lifting during occupational, recreational, exercise, or sporting activities (7, 13, 23, 43, 44, 57, 69, 82, 117). Careful questioning and analysis of such patterns can guide physical examinations; suggest postures, movements, and activities to avoid or use during rehabilitation and exercise sessions; and indicate what education is needed during the rehabilitation process (45, 76, 135). For example, a patient with a history of pain recurrences after long periods of driving with the spine in flexion or sitting in a slouched position at a computer may benefit from exercises that avoid this posture, from modifying motor vehicle and workstation seating, and from education on incorporating breaks for stretching and brief walking into any long drives or seated work periods in the future. A history of usual vocational and recreational activities can offer valuable hints for education on self-management strategies after resolution of the current episode. Asking about the benefit of treatments tried for previous episodes is also important in directing treatments for the current episode of NSLBP.

Physical Examination

In addition to the tests and assessments discussed in chapter 4, the physical examination in NSLBP is primarily one of exclusion. An examiner should do a quick scan of posture and general range of motion (45, 76, 135). A neurologic screening should check for any abnormalities of sensation, motor function, or reflexes that may indicate serious pathologies (45, 76, 135). Assuming these screens are clear, the physical examination should include range of motion and flexibility testing that specifically looks at common deficits seen in NSLBP (45, 76, 135). Hamstring and hip flexor tightness is common in people with NSLBP, and spinal flexion and particularly extension are frequently limited. At a minimum, these should be tested. Other recommended assessments include but are not limited to the straight leg raise test, manual muscle testing, sensory dermatome testing, and observation of gait and of heel and toe walking (45, 65, 135). Palpation of paraspinal muscles may reveal increased muscle tone and tenderness. Intervention

strategies for soft tissue abnormalities is outside the scope of this chapter, and consultation is suggested.

Ideally, the examination for CR patients and AF program participants with comorbid NSLBP would incorporate cardiorespiratory testing, muscular strength and endurance testing, and spinal and abdominal muscle strength and endurance testing (4, 82, 90). Back extensor, abdominal, and gluteal muscle weakness is common in individuals with NSLBP. Those who have a long history of recurrences may be considerably deconditioned even when compared against sedentary peers without back pain, and some tests might be uncomfortable for them to do (19, 34, 41, 51, 58, 74, 78, 110). Unfortunately, if the client is being seen soon after a flare-up of pain, strength and cardiorespiratory testing may be difficult and the results may be invalid because of pain limitations (4, 90, 109). In this case, formal testing of these functional capacities may have to be deferred. The examination should instead clear the patient for general safety for exercise using American College of Sports Medicine (ACSM) guidelines (4).

Diagnostic Testing

Numerous medical diagnostic tests can be used to assess NSLBP; however, current treatment and management guidelines recommend reserving their use for situations in which serious spinal pathology or other previously mentioned medical conditions are suspected (45, 66, 67, 76, 84, 87, 93, 95, 122, 135). Following are some common diagnostic tests:

- *Plain radiograph (X-ray):* identifies fractures, arthritis, and changes to skeletal structures
- *Magnetic resonance imaging (MRI):* a computer-generated image that can help identify infection, tumors, inflammation, disc herniation or rupture, or pressure on a nerve
- *Computed tomography (CT) scan:* shows soft tissue structures that cannot be seen on conventional X-rays, such as disc rupture, spinal stenosis, or tumors
- *Nerve conduction study (NCS):* uses two sets of electrodes to record the nerve's electrical signals to detect any nerve damage

Once serious pathology has been ruled out, noninvasive treatment rather than surgery is strongly suggested in most cases.

Exercise Testing

Nonspecific low back pain does not in itself indicate a need for a graded exercise test (GXT). However, it is a common comorbidity in patients who attend CR and AF exercise programs (39, 75, 79, 80). Therefore, a maximal GXT may be warranted before formulation of a comprehensive exercise prescription.

The clinical exercise physiologist should conduct a thorough interview and preactivity risk screening before conducting exercise tests or developing exercise prescriptions for CR patients or AF program participants with comorbid NSLBP (4). Besides measuring and recording vital signs, the clinical exercise physiologist should perform baseline postural and functional movement screening assessments and note basic gait deviations related to velocity, cadence, and base of support. For a more thorough neuromusculoskeletal examination, a physician, physical therapist, or other suitable HCP may be the appropriate referral. For more information on conducting a comprehensive neuromusculoskeletal examination, readers are directed to the University of California San Diego Practical Guide to Clinical Medicine at https://meded.ucsd.edu/clinicalmed/introduction.html. Chapter 4 provides a general overview of patient interview and preactivity screening procedures, while chapter 5 outlines fitness and functional assessments. Table 26.2 gives an overview of exercise testing recommendations for patients with CNSLBP.

Specific Exercise Considerations

The clinical exercise physiologist should perform exercise and functional tests that CR patients and AF program participants with CNSLBP can complete comfortably (4, 90, 109). The chosen assessments should minimize the chances that maximal effort will be preempted by pain, peripheral muscle fatigue, discomfort, or feelings of unsteadiness (4, 90, 109). In all cases, patients should be asked whether a particular movement is known to trigger pain, and assessments should be modified accordingly to accommodate each patient's movement directional preferences (if present) and ensure their comfort.

For some patients, the seated position could be most painful and walking may be preferred, whereas for others a stationary or recumbent bicycle could provoke less pain than treadmill testing does (4, 9, 43, 44, 57, 82, 83, 90, 109). Regardless, early termination of a GXT because of pain exacerbation can cause insufficient cardiorespiratory stress and therefore mask symptoms suggestive of exercise intolerance and potential cardiovascular disease (CVD). In that case, if a maximal GXT is indicated, it is prudent that a physician order a pharmacologic (nonexercise) GXT (4).

Similarly, testing for flexibility can be performed according to usual protocols. In all cases, patients should be asked whether a particular movement is known to trigger pain, and assessments should be modified accordingly. The chosen exercise and functional tests should be ones that CR patients and AF program participants with CNSLBP can complete comfortably (4, 90, 109). Testing should be modified to accommodate each patient's movement directional preferences (if present) and ensure their comfort.

Cardiorespiratory Exercise

When a maximal GXT is not necessary, the clinical exercise physiologist should select a submaximal test that maximizes comfort and minimizes the chances of the desired effort being preempted by pain, peripheral muscle fatigue, discomfort, or feelings of unsteadiness (4, 90, 109).

Treadmill tests using low-level protocols such as the modified Naughton or ramp or discontinuous protocols, along with the field-based 6 min walk test, reduce compensatory trunk flexion thanks to less reliance on elevation. Each of these tests is appropriate for patients with flexion intolerance. Lower body cycle ergometer or recumbent cycle or step ergometer protocols are appropriate for CR patients and AF participants with extension intolerance (4, 90, 109).

Resistance Exercise

Submaximal muscle strength testing using a multiple-repetition maximum is well tolerated and an acceptable alternative to one-repetition maximum (1RM) testing. Submaximal testing is an effective tool for measuring current strength levels, determining training loads, and measuring postprogram strength increases in clients with CNSLBP (59, 62, 63).

The standard Borg rating of perceived exertion (RPE) or the OMNI perceived exertion scale for resistance exercise (OMNI-RES) can approximate intensity of patient effort during muscular strength testing and training (4, 72, 98). The multiple-repetition sit-to-stand test, 30 s arm curl, and handgrip strength test are viable instruments for measuring muscular strength and physical function in patients with CNSLBP who are ≥60 yr old (4, 21, 94).

Range of Motion and Neuromotor Exercise

Patients might experience some discomfort during flexibility assessments if movements are unfamiliar. The spine or any extremity should not be forced into an uncomfortable range. Body positions and postures required for specific field-based flexibility assessments like the sit-and-reach, standing trunk flexion, and prone active and passive trunk extension might cause discomfort for some CR patients and AF program participants with CNSLBP. A comprehensive assessment of range of motion (ROM) should include the trunk and hip flexors, trunk extensors, hamstrings, iliotibial band (tensor fascia latae), quadriceps, latissimus dorsi, and pectoralis major and minor. The patient should perform ROM assessments in a position that ensures their comfort, and CEP should consider measuring ROM with a goniometer. Spine and extremity motions should be compared against normative ranges for age- and sex-matched apparently healthy individuals.

As previously mentioned, CNSLBP is a common comorbidity in CR and AF patients and program participants. After a cardiac event, patients are at risk for deficits in mobility and function due to extended periods of inactivity (e.g., bed rest) (94). Gait speed, the five times sit to stand (STS), and handgrip strength tests are reliable and responsive measures for patients in CR and AF programs and typically well tolerated by those with CNSLBP (94).

Other acceptable tests that assess neuromotor function and balance include the Timed Up and Go (TUG), 6 min walk, and stair climb tests (4, 5, 109). A number of patients in CR and AF programs are ≥65 yr old, and CNSLBP and higher fall risk are common (39). The clinical exercise physiologist should ensure

Table 26.2 Exercise Testing–Specific Considerations

Test type	Mode	Protocol specifics	Clinical measures	Clinical implications	Special considerations
Cardiorespiratory	Treadmill, cycle ergometer, stepping, NuStep recumbent arm and leg ergometer, overground walking	Ramp, or incremental, 6 min walk, NuStep branching protocol	BP, HR, others as indicated by comorbidities or other ACSM guidelines, subjective ratings of intensity and pain	Patients may be deconditioned below healthy sedentary level. Pain or muscular fatigue may cause premature test termination, underestimate fitness levels, and mask symptoms diagnostic of CAD and CVD.	Select recumbent cycling or level walking for difficulties with prolonged flexion or sitting instead of upright cycling. Testing soon after rising is not recommended. Emphasize neutral spine while testing.
Resistance	Free weights or machines and trunk endurance testing with body weight (trunk flexor endurance, lateral flexor endurance, and extensor endurance)	Submaximal testing; use multiple RM tests and anchor effort with OMNI-RES	BP, HR, others as indicated by comorbidities or other ACSM guidelines, subjective ratings of intensity and pain	Many patients experience discomfort, especially early in testing and training.	Ensure that spinal support and good posture are used at all times. Select testing postures to ensure patient comfort.
Range of motion	Standard flexibility testing of multiple joints with a goniometer	Testing of trunk flexors and extensors, hamstrings, quadriceps, and iliotibial band flexibility	Spinal and extremity joint motion within normative ranges for specific populations	Many patients experience discomfort, especially early in testing and training. Avoid forcing movements into uncomfortable range.	Emphasize neutral spine while testing extremity flexibility. Avoid spinal flexion in first 1-2 h after rising.

BP = blood pressure; HR = heart rate; CAD = coronary artery disease; CVD = cardiovascular disease; RM = repetition maximum; OMNI-RES = OMNI perceived exertion scale for resistance exercise.

patient safety and provide appropriate spotting and supervision during these tests.

Treatment

Current emphasis in primary care management of NSLBP is early return to activity, avoidance of needless surgery or unnecessary diagnostic tests, and ultimately cost-effective medical and self-management of back pain (45, 76, 84, 87, 135).

Surgery is considered a last resort when all forms of conservative treatments have failed to manage pain, improve physical function, or prevent worsening nerve damage, serious musculoskeletal injuries, or nerve compression (45, 76, 84, 87, 95, 135). It can take months after surgery before the patient is fully healed, and there may be permanent loss of flexibility. Further, surgery is not always successful. A full explanation of the numerous surgical procedures for managing low back pain is beyond the scope of this chapter. Common surgical procedures the clinical exercise physiologist might encounter include the following:

- In a **laminectomy** (decompression), the bony walls of a vertebra are removed, along with any bone spurs, to relieve pressure on the nerves.
- A **spinal discectomy** involves removing a herniated disc through a small incision to relieve pressure on a nerve root (often done with a laminectomy).
- In **spinal fusion**, the disc between two or more vertebrae is removed and the vertebrae are fused by bone grafts or metal devices and secured by screws. This procedure requires a long recovery period and has been associated with an acceleration of disc degeneration at adjacent levels of the spine.
- Artificial disc replacement has also been associated with an acceleration of disc degeneration at adjacent levels of the spine.

A good source for additional information is the National Institute of Neurological Disorders and Stroke (www.ninds.nih.gov).

Many patients with NSLBP need pharmacologic intervention only briefly; over-the-counter medications are the most often used. Health care providers are encouraged to recommend the lowest effective dosage of nonsteroidal anti-inflammatories (NSAIDS) as first-line treatment for pain (45, 66, 84, 87, 95, 135). NSAIDs, including aspirin and ibuprofen, tend to cause gastrointestinal irritation and ulcers. They may also cause renal pathology, blood thinning, or an allergic reaction (45, 66, 76, 85, 87, 95, 135).

Although not routinely offered, in certain circumstances, some health care providers still prescribe acetaminophen (alone or combined with other medications), selective serotonin reuptake inhibitors (SRIs), tricyclic antidepressants, serotonin–norepinephrine reuptake inhibitors (SNRIs), and anticonvulsants and muscle relaxants for managing CNSLBP (45, 66, 71, 76, 84, 87, 95, 135). They may reduce pain, but a number of side effects limit their use. Many carry the adverse effects of drowsiness, impaired motor function and balance, and increased reaction times, creating potential risks for falling during exercise and when operating motor vehicles, power tools, or machinery (45, 66, 76, 84, 87, 95, 135). Although some physicians still prescribe narcotic analgesics or opioids for relief of musculoskeletal pain, these appear to have no greater benefit than safer analgesics and are generally considered a poor choice for CNSLBP (45, 66, 76, 84, 87, 95, 135). The clinical exercise physiologist should be familiar with the precautions, side effects, and potential safety concerns of these medications.

Commonly prescribed medications for managing acute, subacute, and chronic NSLBP appear in table 26.3.

Prevention and Management

Once serious pathology is ruled out, health care providers should treat patients with CNSLBP-related symptoms with noninvasive, nonpharmacological methods, promote self-management of pain, and encourage patients to become and remain physically active and avoid bed rest (35, 43, 45, 66, 76, 84, 87, 95, 135). Exercise, regardless of type, is recommended for reducing pain and improving function in patients with CNSLBP (40, 46, 101, 104, 125).

CR patients and AF program participants with CNSLBP should be encouraged to engage in **secondary prevention** strategies, which help them avoid recurrences through exercise and patient education. In addition, prevention of first-time LBP (**primary prevention**) for individuals whose occupational, household, and recreational pursuits place them at a higher risk is prudent (66, 84, 87, 90). Some prevention strategies the clinical exercise physiologist can encourage and reinforce with their patients include the following (44, 57, 74, 82, 83, 90):

- Use safety equipment in work and leisure activities.
- Address risk factors of smoking, poor general fitness, obesity, stress, sleep hygiene, and prolonged poor seating postures.
- Perform balanced exercise programs that include both spinal flexion and extension.

Table 26.3 Pharmacology

Drug categories and common generic names	Trade names	Side effects	Contraindications
Nonnarcotic analgesics • Aspirin • Acetaminophen • Celecoxib • Ibuprofen • Naproxen	Bufferin, Empirin Tylenol Celebrex Motrin, Rufen, Nuprin Naprosyn, Anaprox, Advil, Aleve	Gastrointestinal (GI) distress, GI ulcerations, allergic reactions, renal dysfunction, bleeding and bruising	Patients with history of allergic reactions or GI distress with these medications
Narcotic (opiate) analgesics • Meperidine • Hydromorphone • Methadone • Codeine • Morphine • Oxycodone	Demerol Dilaudid Dolophine Codeine Avinza, Roxicodone Oxycontin, OxyIR, Tylox, Percodan, Percocet	Poor tolerance, gastrointestinal disturbances, sleep disturbances, drowsiness, increased reaction time, clouded judgment, misuse or abuse and dependence issues, potentially higher risk of falls	Patients with history of poor tolerance or allergic reactions, dependent personality types
Muscle relaxants • Cyclobenzaprine • Carisoprodol • Metaxalone	Flexeril Soma Skelaxin	Central nervous system depressant, drowsiness, tachycardia, hives, mental depression, shortness of breath, skin rash, itching, potentially higher risk of falls	Allergies, blood disease caused by an allergy or reaction to any other medicine, drug abuse or dependence, kidney or liver disease, porphyria, epilepsy
Antidepressants (SNRIs) Duloxetine Venlafaxine	Cymbalta Effector XR	Drowsiness, insomnia, nausea, dry mouth, dizziness, constipation, excessive sweating, elevated blood pressure, may worsen heart problems	Diabetes, hypertension, heart problems, smoking, liver disease, use of aspirin, nonsteroidal anti-inflammatories, and other blood thinners like warfarin (Coumadin)

SNRIs = serotonin and norepinephrine reuptake inhibitors.

- Avoid long periods in one position—take breaks to move the spine out of fixed positions, and balance postures of flexion and extension.
- Avoid lifting with twisting, and preserve the neutral lordotic spine curve during lifting.
- Avoid lifting activities immediately after rising or prolonged spinal flexion positions without breaks to move into extension.
- Be physically active and keep fit.
- Vary postures during prolonged sitting.
- Balance activity in flexion with activity in extension.
- Perfect fundamental movement competencies like hip hinging, squatting, and lunging with a neutral spine.

Numerous authorities urge physicians and other health care providers to help patients change their perception of NSLBP as a serious disorder that results in permanent disability. They suggest shifting the emphasis back onto the patient with a convincing argument that self-care produces effects comparable to those with traditional medical care (45, 66, 76, 84, 87, 95, 135). Only 5% of people who experience an acute episode of NSLBP go on to chronic pain and disability (67).

Exercise Benefits

A variety of exercise interventions including but not limited to yoga, Pilates, aerobic training, resistance training, and flexibility training have been shown to reduce pain and improve physical function in persons with CNSLBP, yet there is no consensus on the most effective form of exercise (46, 68, 77, 84, 88, 101, 102, 104, 116, 125, 126, 136). Commonly reported benefits of cardiorespiratory and resistance exercise include the following (40, 46, 101, 104, 125):

- Increased physical activity tolerance
- Increased pain tolerance
- Increased range of motion
- Increased physical fitness and strength
- Increased health-related quality of life
- Increased functional capacity
- Increased overall physical activity participation levels

A systematic review and meta-analysis assessed the effects of resistance, endurance, and flexibility exercise in population-based interventions to prevent LBP and associated disability (105). They reported that exercise reduced the risk of LBP by 33% and that severity and disability from LBP were lower in exercise training versus control groups. It was concluded that a combination of strengthening exercise with either stretching or cardiorespiratory exercise performed two or three times per week can reasonably be recommended for the prevention of LBP in the general population (105). Other well-tolerated and effective programs include periodized resistance training, which improves strength

and physical activity participation levels and reduces disability in both sedentary and athletic populations with CNSLBP (20, 24, 47, 59, 62, 63, 81).

EXERCISE PRESCRIPTION AND TRAINING

Most programs of exercise for people with back pain include a combination of several forms of exercise and educational advice regarding lifestyle factors and general back care. Exercise intervention after back pain has occurred is aimed at reducing risk factors and minimizing recurrences. With chronic and recurrent back pain, the evidence for more aggressive intervention is stronger (45, 66, 84, 87, 95, 135). Increasing function and decreasing the severity and frequency of back pain episodes should be the goal (45, 66, 84, 87, 95, 135).

Please refer to chapter 6 for a comprehensive description of general exercise prescription principles. The recommendations for persons in CR exercise and AF programs with comorbid CNSLBP are consistent with those from the American College of Sports Medicine (ACSM) (4) and United States Department of Health and Human Services (USDHHS) (121) for patients participating in outpatient CR exercise and AF programs who are healthy and do not have CNSLBP (4, 36, 90, 109,114, 121).

Common exercise program goals for persons with CNSLBP and patients in outpatient CR are similar and emphasize the importance of the following (4, 36, 121):

- Improving health and well-being
- Improving exercise tolerance
- Improving functional capacity
- Improving HRQOL
- Resuming vocational and recreational pursuits

Aside from performing structured exercise, persons with CNSLBP should adopt an active lifestyle and add routine PA breaks when periods of sitting cannot be avoided (4, 36, 90, 114, 121). An individualized approach to exercise program development that addresses all health-related fitness variables is appropriate when working with patients who have CNSLBP (see table 26.4). The CR or AF program should help enhance functional performance and HRQOL (4, 36, 40, 46, 73, 90, 99, 101, 105, 109, 114, 121).

Components of a comprehensive exercise program for CR patients with CNSLBP should include cardiorespiratory, resistance, flexibility, and neuromotor training (4, 24, 36, 40, 46, 47, 59, 62, 63, 73, 82, 84, 99, 101, 104, 105). The compendium of physical activities can serve as a resource for assisting CR patients and AF participants with CNSLBP find additional recreational and leisure-time activities they can perform comfortably (1). Because of potential physical deconditioning and pain in some patients who have CNSLBP, a slower rate of exercise program progression, volume, and intensity may be warranted (4, 90). The appropriate training adaptations can be summarized in table 26.5.

Table 26.4   Exercise Prescription Review

Training method	Frequency	Intensity	Time (Duration)	Type (Mode)	Progression	Important considerations
Cardiorespiratory	≥5 d/wk of moderate or ≥3 d/wk of vigorous, or a combination of moderate and vigorous exercise on >3-5 d/wk	Moderate, 40%-<60% peak $\dot{V}O_2$, to high, ≥60% peak $\dot{V}O_2$ start at lower range, particularly if deconditioned RPE of 12-16 or 50%-60% of HRR	Build up to ≥30 min; may do multiple ≤10 min bouts throughout day; begin with several 2-5 min bouts if very deconditioned Anticipate 4-6 wk program for substantial improvement	Brisk walking with arm movement ideal, backward walking, cycling (preferably recumbent), swimming, elliptical training as tolerated and consistent with movement directional preferences if present	Increase bouts (if used) to ≥10 min until goal is reached. Use ACSM recommended progression for sedentary low-risk individuals unless comorbidities or symptoms increase restriction.	Low impact is best initially. Exercise in body positions best tolerated. Gradually increase exercise volume and use a "start low and go slow" approach. Add high- intensity interval training (HIIT) as tolerated.
Resistance	8-10 exercises for each major muscle group, 2 or 3 nonconsecutive days a week, with 2-3 min rest intervals between sets and >48 h between training of same muscle groups	60%-70% 1RM intensity for most participants; 40%-50% 1RM intensity for novice, deconditioned, or elderly participants; RPE of 12-13 out of 20 or OMNI-RES of 3-5 out of 10, respectively	Sessions of less than 1 h	Multijoint and single-joint resistance exercises with free weights, machines, resistance tubing, or body weight, addressing all major muscle groups	Follow ACSM guidelines for healthy sedentary individuals. Increase intensity or sets or both as tolerated. Intensities of ≥80% 1RM are appropriate for asymptomatic, highly trained persons.	Exercise in body positions best tolerated. Avoid unstable surfaces (e.g., exercise balls) early in training. Later, unstable surfaces may enhance training. Emphasize good form. Limit spine loading through full range.
Range of motion	2-7 d/wk; higher number preferable with this population; 4 or more stretches per key muscle group	To the point of slight discomfort or tightness but without pain	Each static stretch held 10-60 s per muscle group, session total of at least 10 min	Static stretching exercises or proprioceptive neuromuscular facilitation (PNF) technique stretches	Dictated by intensity.	Balance flexion and extension stretches of the spine. Ensure hamstrings and hip flexors are stretched. Select other stretches based on patient deficits. Avoid flexion stretches soon after rising. Avoid standing toe touches.

RPE = rating of perceived exertion; HRR = heart rate reserve; 1RM = one-repetition maximum; OMNI-RES = OMNI perceived exertion scale for resistance exercise.

Table 26.5   Exercise Training Adaptations Summary

Cardiorespiratory endurance	Skeletal muscle strength	Skeletal muscle endurance	Flexibility	Body composition
Improves performance of functional activities. Enhances the effect of back exercise program. Enhances management of comorbid cardiometabolic diseases and risk factors. Benefits general health. Reduces fear avoidance of physical activities.	Increases tolerance of activities of daily living (ADLs), recreational activities, and occupational performance. Reduces fear avoidance of physical activities.	Back extensor and latissimus dorsi improvements may protect against recurrences. Reduces fear avoidance of physical activities.	Return to normal values (with associated strength gains) may decrease risk of recurrence.	Improves muscle mass and lowers BMI, reducing compressive loads on joints and risk of cardiovascular disease.

Special Exercise Considerations

Regardless of whether a CR patient or AF program participant has a movement directional preference, the clinical exercise physiologist should select warm-up, cardiorespiratory training, resistance training, and flexibility exercise variations that patients can comfortably perform during their CR, AF, or general workout sessions (7, 23, 43, 44, 57, 69, 70, 82, 99, 117). All exercises, regardless of type (resistance, cardiorespiratory, or range of motion), should be performed with the spine in a neutral, pain-free position (82, 83). Examples of exercise modifications for CR patients or AF program participants with CNSLBP who have movement directional preferences are provided in table 26.6.

After a person has been cleared by a medical professional, few contraindications to an exercise program exist. But many patients with recurrent or chronic pain may be so deconditioned and fearful of exercise or movement that progression should be slow and initial exercise levels low.

Supervision is valuable, particularly in the early stages, to ensure use of correct form and technique and encourage exercise adherence (24, 40, 47, 81). Care should be taken to avoid sacrificing proper body alignment and movement quality as well

Table 26.6 Exercise Modifications for Persons With Movement Directional Preferences

Exercise type	Flexion intolerance and extension movement bias, basic	Flexion intolerance and extension movement bias, advanced	Extension intolerance and flexion movement bias, basic	Extension intolerance and flexion movement bias, advanced
Cardiorespiratory	Treadmill or overground walking (level), UBE (standing), walking in deep water, treading water (shallow), backward treadmill walking	Elliptical trainer, stair climber, jogging (level), deep water running, swimming with a mask, backward treadmill walking, HIIT protocols	NuStep,* UBE (seated), recumbent bike, water walking	Swimming on back, walking (incline), upright stationary bike, treading water (deep), deep water running, rowing ergometry (without extension), HIIT protocols
Resistance (lower body)	Chair/potty squat, wall squat with a stability ball, mini lunge	Bench step-up (low height), body weight squat, horizontal machine leg press, squat wearing a light weight vest	Supine gluteal bridges, seated leg press, knee extension, knee flexion	Machine horizontal or 45° leg press, chair/potty squats, wall squats with a stability ball
Resistance (upper body)	Standing or prone: chest press with external support, row, lat pull-down, triceps push-down, reverse shoulder (rear) fly, biceps curl with adjustable cable column or tubing	Assisted, body weight, or TRX/suspension: push-ups, pull-ups, inverted rows; free weights: chest press, bench row (prone), reverse shoulder fly, dumbbell biceps curl with stability ball support (against wall); Smith machine shoulder press	Seated machine, cable, or supine: row, chest press, lateral shoulder raise, reverse shoulder (rear) fly, lat pull-down, triceps push-down, biceps curl	Seated machine or cable exercises, free-weight exercises
Trunk and core conditioning	Hip hinge, standing wall hip extension, standing wall shoulder flexion, quadruped cat and camel, standing wall abdominal plank	Bird dog, stability ball back extension without trunk flexion, standing trunk de-rotation, standing back extensions, hip hinge, Paloff presses (multiple positions and angles), single-leg bridges, "stirring the pot"/sawing with stability ball	Abdominal bracing, dead bug progression, press, gluteal bridge	Abdominal curl-up, bird dog
Cautions: limit or avoid	Bending at the waist, sitting, leaning forward and twisting	Bending at the waist, sitting, leaning forward and twisting	Arching or hyperextending the spine, prolonged standing, overhead reaching and lifting, exercising in the prone position	Arching or hyperextending the spine, prolonged standing, overhead reaching and lifting, exercising in the prone position

*Upper and lower body recumbent step ergometer.

UBE = upper body ergometry.

as to prevent overtraining. Emphasizing form and evaluating postexercise responses and (if present) symptoms are important for determining whether the volume and type of exercises were appropriate. All exercises should provide adequate stimulus to elicit a positive training effect without stressing any tissue beyond tolerable levels or causing abnormal compensatory movement patterns. As mentioned previously, patients with CNSLBP who may have been extremely inactive for a considerable time will benefit from a program of lower intensity and volume to start, with a slower progression than for healthy sedentary persons without CNSLBP (4, 90).

Exercise should be generally tolerable during the activity and should result only in mild musculoskeletal discomfort associated with postexercise delayed-onset muscle soreness (DOMS). Pain with general movements or severe enough to stop exercise, or pain that results in referred or radiating pain into the posterior thigh or down into the lower extremity, indicates that the exercise is too aggressive or is irritating sensitive neuromuscular structures in the lumbar spine. Radiating pain into the lower extremity, in itself, is cause for concern and usually indicates irritation of lumbar nerve roots. Because a common etiology of nerve root irritation is disc herniation that compresses the nervous tissue, or loss of intervertebral disc space that results in narrowing of the space where the nerve root must pass, recurrent **radicular** pain should be referred to the appropriate health care practitioner (90). Other signs and symptoms warranting exercise termination and communication with a physician or other health care practitioner include pain lasting ≥3 h and pain resulting in several days of disability or sleep disturbances (90).

Maintaining an independent exercise program after completing CR or an AF exercise program is a common challenge and may result in recurrence of symptoms in some patients. Support, encouragement, and easy availability of follow-up programs in the community may be especially important until an exercise habit is well established and whenever a program has to be restarted after a lapse for any reason (90).

Cardiorespiratory Exercise

Aerobic endurance training is generally well tolerated by patients with CNSLBP (47). An appropriate goal for CR patients with comorbid CNSLBP is to accumulate ≥30 min of moderate-intensity aerobic activity on most (≥5) days of the week (2-6). Initial bouts of ≤10 min repeated two or three times per day might be best tolerated in deconditioned CR and AF participants. Total daily time during a single bout of aerobic exercise can be increased gradually to meet or exceed the 30 min goal (4, 90, 121).

It is important to increase exercise duration gradually so that longer continuous aerobic exercise bouts can be tolerated (90). An initial intensity equivalent to an RPE of 12 to 13 and progressing to 16 out of 20 over time is appropriate (4, 36, 90, 114). If heart rate is used as a guide, maintaining exercise within the target heart rate range is appropriate. Start at a lower (50%-60%) level and progress to higher percentages (60%-80%). High-intensity interval training (HIIT) has been well tolerated in some people who have CNSLBP (129, 131).

Walking, upright and recumbent cycling, step ergometry, elliptical, seated and recumbent stepping, rowing, swimming, and aquatic exercises are all acceptable forms of aerobic activity. For patients in CR and AF programs, exercise selections should be dictated by their comfort and, if present, movement directional preferences. High-impact exercise such as running should be avoided or introduced gradually with caution (4, 90). Backward walking is a favorable aerobic activity for persons with CNSLBP; it causes greater recruitment of lumbar paraspinal and extensor muscles than does forward walking (6). Figures 26.1 through 26.3 depict recumbent combined arm and leg ergometry, treadmill walking, and upper body ergometry, three types of aerobic exercise commonly prescribed during CR and AF programs.

Resistance Training

Cardiac rehabilitation patients and AF participants with comorbid CNSLBP are encouraged to follow, as tolerated, resistance training guidelines for apparently healthy sedentary adults without CNSLBP (4, 109). Workouts should consist of 8 to 10 exercises that emphasize all major muscle groups on two or three nonconsecutive days of the week. A variety of modalities are appropriate, including free weights, machines, elastic resistance tubing, and body weight exercises (4, 36, 90, 92, 97, 99). Blood pressure monitoring in patients with hypertension is prudent, and terminating sets after 8 to 10 repetitions can limit set duration and reduce the magnitude of blood pressure elevation during a set (73).

Patient comfort and, if present, movement directional preferences should dictate the equipment and body position selected during all resistance training exercises. Standing exercises may be more comfortable for patients who experience pain with trunk flexion, while seated exercises may be more comfortable for patients who experience pain during prolonged standing or with trunk extension (4, 44, 57, 69, 70, 82, 90, 109).

The clinical exercise physiologist may need to emphasize muscular endurance over strength in the early training, adjusting repetitions and resistance to reflect this emphasis (4, 82, 83, 90). Exercises to condition muscles around the trunk and torso should be performed in a neutral, pain-free position. They can be adapted to ensure patient comfort and can be performed while supine, prone, seated, or standing. The clinical exercise physiologist should help patients develop competence in performing fundamental movements like hip hinging and squatting while maintaining a neutral spine (44, 82).

An initial training intensity equivalent to an RPE of 12 to 13 (out of 20) or 4 to 6 (out of 10) on the OMNI-RES scale is appropriate and, if tolerated, may be progressed over time (4, 90, 114). Although single-set resistance programs have produced significant

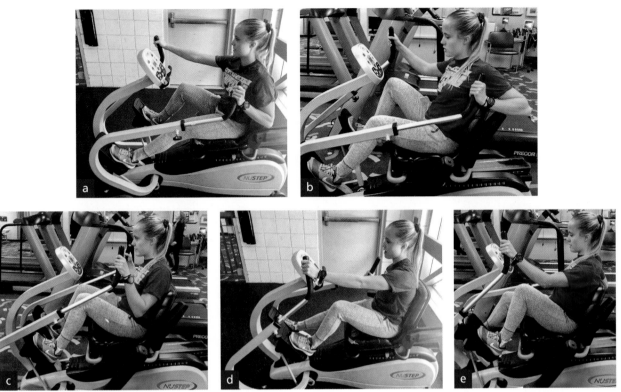

Figure 26.1   Recumbent arm and leg ergometry. *(a)* Improper seat and arm rail adjustments causing overreaching with knee and elbow hyperextension and hyperflexion *(b* and *c)* from improper seat adjustment of the recumbent arm and leg ergometer can exacerbate low back pain. Proper positioning is demonstrated: *(d)* starting and finishing position; *(e)* pulling motion.

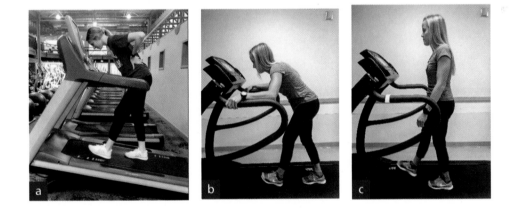

Figure 26.2   Treadmill walking. *(a)* Pain can occur during inclined walking from prolonged compensatory trunk flexion in a person with flexion intolerance. *(b)* Pain during prolonged walking and compensatory trunk flexion in a person with extension intolerance. *(c)* Walking forward and backward on a level surface may be best tolerated by persons with flexion intolerance. Those with extension intolerance may do better with some incline or, if necessary, by substituting seated or recumbent exercises for treadmill walking.

Figure 26.3 Upper body ergometry. Upper body ergometry can be performed *(a)* in a seated position for patients with extension intolerance or *(b)* in a standing position for patients with flexion intolerance.

strength increases in untrained sedentary individuals (4, 36, 97), progressing to a protocol consisting of two to four exercise sets per muscle group is recommended as tolerated (4, 20, 36, 97). Periodized, progressive multi-set RT programs have produced significant increases in strength in persons with CNSLBP, comparable to those in apparently healthy individuals without CNSLBP (20, 59, 62, 63). Intensity and loading of resistance exercises should follow the two for two rule (increase when the patient can properly complete two or more repetitions above the number prescribed in the last set of an exercise for two consecutive workouts) (97). An individualized approach for selecting back, trunk, and core muscle conditioning exercises should be taken. Patients should maintain a neutral and pain-free spine position.

Additional resources for back, trunk, and core muscle conditioning exercise prescription and progression can be found elsewhere (44, 74, 82, 83). Table 26.7 describes sample whole-body resistance exercises for flexion- and extension-intolerant patients with CNSLBP.

Figures 26.4 through 26.7 depict the leg press, row, lat pulldown, and chest press exercises that are commonly performed during CR and AF programs.

Range of Motion Exercises

The focus of flexibility, or range of motion, exercises should be on achieving optimal mobility in order to be able to perform occupational, recreational, exercise, and self-care activities (4, 36, 121). A program of dynamic, static, and proprioceptive neuromuscular facilitation stretching of major muscles is recommended (4, 36, 121).

Activities to enhance whole-body mobility should be emphasized rather than stretches to make the spine more flexible (4, 82, 83). Flexibility exercises for the calf, hamstring, hip flexor, quadriceps, pectoral, and latissimus dorsi muscles are appropriate. Patients should follow the same flexibility and range of motion training recommendations as those for healthy individuals without CNSLBP (4, 82). Stretches should be performed slowly and con-

Table 26.7 Sample Whole-Body Resistance Exercises for Flexion- and Extension-Intolerant Patients

Muscles and muscle groups	Exercises for flexion intolerance	Exercises for extension intolerance
Gluteus maximus, quadriceps, hamstrings	Chair squat or wall squat with a stability ball	Seated or supine leg press (machine)
Quadriceps	Stair step-up	Seated leg extension (machine)
Hamstrings	Standing leg curl (machine)	Seated leg curl (machine)
Latissimus dorsi, teres major	Standing straight-arm cable pull-down or assisted pull-up (machine)	Seated lat pull-down (machine)
Pectorals, deltoids, triceps	Standing cable or resistance tubing chest press	Seated chest press (machine)
Rhomboids, middle and lower trapezius	Standing cable or resistance tubing scapular row	Seated scapular row (machine with a chest support pad)
Deltoids	Standing dumbbell shoulder press (alternating)	Seated lateral shoulder raise (machine)
Biceps	Standing biceps curl (dumbbell)	Seated biceps curl (machine)
Triceps	Standing triceps push-down	Seated triceps push-down or seated dip (machine)

Adapted by permission from G.A. Hand et al., "Impact of Aerobic and Resistance Exercise on the Health of HIV-Infected Persons," *American Journal of Lifestyle Medicine* 3, (2009): 489-499.

Figure 26.4   The selectorized machine seated leg press is a well-tolerated multijoint lower body strength exercise for persons with extension intolerance. Improper seat adjustments can lead to incorrect body positioning and can worsen back pain. *(a)* Overextension of the knees and overreaching during the end of the pushing phase of the leg press can exacerbate low back pain in persons with extension intolerance. *(b)* The knee to chest position is a form of spinal flexion that might need to be modified for persons with flexion intolerance. *(c)* Sitting fully against the back pad will ensure vertical alignment of the head, neck, back, and hips and support a more neutral spine. An alternative can be either a supine leg press (not shown) or *(d)* a standing wall squat, with or without a stability ball behind the back (proper alignment at the bottom position). The wall squat performed with a stability ball is an alternative to the leg press for persons with flexion intolerance.

Figure 26.5   Rowing exercises can be modified to accommodate persons with specific back movement limitations. Selectorized machine scapular rows with extension intolerance: *(a)* starting position and *(b)* pulling motion. Standing scapular rows with an adjustable height cable pulley for persons with flexion intolerance: *(c)* starting position and *(d)* finishing position.

Figure 26.6 Lat pull-down exercise variations. The lat pull-down can be performed from a standing position by those with flexion intolerance: *(a)* start and *(b)* finish. It can be performed seated *(c)* for persons with extension intolerance. *(d)* Excessive trunk flexion and *(e)* excessive trunk extension. A straight-arm lat pull-down is depicted with a slightly wider than shoulder-width grip.

Figure 26.7 The chest press exercise can be modified to accommodate specific back movement limitations. A version of a seated machine chest press for persons with extension intolerance: *(a)* starting position and *(b)* pushing motion. An adjustable cable column version of a standing chest press exercise for persons with flexion intolerance: *(c)* starting position and *(d)* pushing motion.

trolled, held in a comfortable position so that tension but not pain is experienced in the intended muscles. Proper body alignment, patient comfort, and, if present, movement directional preferences should dictate the body position (lying, sitting, standing) selected during all flexibility training. Chapter 6 addresses general flexibility and range of motion exercise training and recommendations.

Many CR patients and AF program participants with CNSLBP are ≥65 yr old. These individuals can benefit from exercise training to improve their balance so they can safely perform exercises like treadmill and overground walking, stair step ergometry, and elliptical step ergometry as well as other activities of daily living (4, 39).

Exercises like single-leg standing, standing side leg raises, weight shifting, tandem standing, tandem (heel–toe) walking, and tai chi are some acceptable activities for improving balance (4). Balance can be challenged by changing body position (lying to sitting to standing), reducing base of support (wide stance to narrow stance to single-leg stance), reducing external support (holding on to something with two hands to one hand to no hands), and challenging the body's balance subsystems (changing standing surfaces, moving the head, and closing the eyes). Frequency and duration of training should be individualized. Chapter 25 provides additional information on balance exercise recommendations.

Research Focus

Exercise Training Adaptations

Although exercise training for the management of CNSLBP is generally recommended in international clinical guidelines (84, 87, 95), effect sizes reported in scientific literature evaluating the role of exercise therapy often remain low. Verbrugghe and colleagues noted that high-intensity training has not been rigorously studied in persons with CNSLBP. They also reported that greater improvements in general health and disease-specific outcomes are experienced in persons with other chronic medical conditions after performing high-intensity exercise than in matched cohorts performing low-intensity exercise training (129, 131).

Therefore, the objective of their study was to compare the effectiveness of a high-intensity multimodal exercise training protocol (HIT, combining cardiorespiratory, general whole-body resistance, and core muscle training) with the same protocol performed at moderate intensity (MIT) on disability (primary outcome), pain, function, exercise capacity, and abdominal and back muscle strength (secondary outcomes) in persons with CNSLBP. During this twice weekly, 24-session intervention, 38 participants (12 males and 26 females) with CNSLBP were randomized into either a HIT (19 participants) or MIT (19 participants) exercise training program.

Participants completed assessments of back pain–related physical disability, pain, function, muscular strength, and cardiorespiratory fitness before and after completing the 12 wk training intervention. Back pain–related disability, pain, and function were assessed with the 10-item (5-point scale) Modified Oswestry Disability Index (MODI) (33), the 10-point numeric pain rating scale (NPRS) (15), and the Patient-Specific Functioning Scale (PSFS) (53). Trunk muscle strength (flexors and extensors), cardiorespiratory fitness, and general muscle strength were assessed by maximal isometric muscle strength testing with an isokinetic dynamometer, a ramping maximal cardiorespiratory exercise test on a cycle ergometer, and one-repetition maximum (1RM) on six (three upper body and three lower body) selectorized resistance training machines, respectively.

Both the HIT and MIT groups performed cardiorespiratory, whole-body resistance, and core muscle training. The HIT group performed five high-intensity cycle ergometry work intervals (HIIT) (110 rpm at 100% $\dot{V}O_2$max workload) followed by 1 min active recovery periods (75 rpm at 50% $\dot{V}O_2$max workload). Cycling bouts increased every two sessions by 10 s, up to 1 min 50 s after 12 sessions. Recovery time (1 min) between bouts remained stable. This protocol was repeated from session 13 to 24 with an updated workload, extracted from a complementary cardiorespiratory exercise test. For the resistance training, participants completed one set of 8 to 12 repetitions (at 80% of their predetermined 1RM) of three upper body exercises (lat pull-down, chest press, arm curl) and three lower body exercises (leg curl, leg press, leg extension) on selectorized machines. Intensity load was progressively increased by 5% when participants were able to perform more than 10 repetitions on two consecutive training sessions (97).

Core muscle training consisted of six static core exercises (glute bridge, resistance band glute clam, lying diagonal back extension, adapted knee plank, adapted knee side plank, elastic band shoulder retraction with hip hinge [standing scapular row]). Participants performed one set of 10 repetitions of a 10 s static hold. They were encouraged to hold the last repetition as long as possible. Exercise difficulty was increased when the participant could execute the exercise with a stable posture for the indicated time on two consecutive training sessions. This was done by increasing the time of the static hold up to 12 s and progressing to more demanding postures through the use of increased body weight bearing, use of external resistance, or both.

Participants in the MIT group initially performed 14 min of constant-load cycling (90 rpm at 60% $\dot{V}O_2$max workload). Duration increased every two sessions from 1 min 40 s up to 22 min 40 s. This protocol was repeated from sessions 13 to 24 with an updated workload, extracted from a complementary cardiorespiratory exercise test. General resistance training was identical to the protocol described in HIT, with the exception of

(continued)

Research Focus (continued)

the exercise intensity. From session 3, one set of 15 repetitions was performed at 60% of 1RM. Core training was identical to the protocol described in HIT, with the exception of the exercise intensity. Participants performed one set of 10 repetitions of a 10 s static hold. Exercises were made more difficult when they were executed with a stable core posture for the indicated time by increasing the time of the static hold every six sessions.

At completion of the intervention, both groups had improved physical performance (cardiorespiratory fitness and trunk muscle strength) as well as low back pain–related outcomes (pain intensity and functional disability). However, the HIT group experienced a significantly greater improvement in functional disability and cardiorespiratory fitness compared with the MIT

group (14.6%, or a 64% relative difference, vs. 6.2%, or a 33% relative difference in disability, and 4.9 mL $\cdot$ kg^{-1} $\cdot$ min^{-1} vs. 1.8 mL $\cdot$ kg^{-1} $\cdot$ min^{-1}). These group differences had a large effect size and exceeded clinically relevant cutoff values. In addition, no adverse events were reported regarding use of the HIT protocol, corroborating the safety and feasibility of this therapy. Participants in this study were highly motivated to train and had moderate pain and disability scores; their results may not be generalizable to all persons with CNSLBP.

These results support the important role of exercise intensity in rehabilitation programs for persons with CNSLBP. Additionally, this study provides evidence for the value of setting up a HIT program for both cardiorespiratory and strength training (131).

Based on Verbrugghe et al. (131); Verbrugghe et al. (130); Fairbank and Pynsen (33); Childs, Piva, and Fritz (15); Horn et al. (53).

Comorbidities

Chronic nonspecific low back pain is a common comorbidity experienced by patients in cardiac rehabilitation programs and participants in adult fitness programs (39, 75, 79, 80). Diabetes, hyperlipidemia, metabolic syndrome, coronary artery disease, arthritis, and osteoporosis are common comorbidities in persons with CNSLBP (18, 96). A chapter has been devoted to each of these conditions (please refer to chapters 8, 11, 12, 14, 24, and 25, respectively). As for CNSLBP, a comprehensive individualized exercise program helps improve strength, endurance, and physical function; lower cardiometabolic disease risk; and prevent disease progression in patients with each of these conditions.

It is prudent for the clinical exercise physiologist to use a more global approach to exercise testing and prescription with patients who have CNSLBP and comorbidities. Exercise testing and exercise prescription should address the overriding goals as well as safety concerns and precautions posed by each condition the patient has. The final exercise program should be effective and safe for each condition. Although exercises to improve trunk motion control, stability, and muscle endurance are important, aerobic endurance, whole-body resistance, range of motion, and

neuromotor training are equally important for reducing or managing weight, reducing cardiometabolic disease risk, and preventing frailty, falls, and fractures.

The exercise program and prescription should address the major consequences of having CNSLBP (e.g., physical inactivity, deconditioning), not just back or leg pain, and of having a comorbid disorder like diabetes (glucose regulation and monitoring, weight loss and management, risk factor reduction, hydration, foot and eye care, and autonomic dysfunction) during every exercise session. Similarly, exercise testing and prescription should address the overriding needs and safety concerns of a patient with CNSLBP who also has other comorbidities such as arthritis, osteoporosis, dyslipidemia, CVD, or metabolic syndrome. As an example, improving muscle strength, endurance, and range of motion can improve joint stability, motion, and mobility, which contribute to better balance, function, and independence. Similarly, exercises to improve cardiorespiratory fitness can promote weight loss, enhance glucose and fat metabolism, reduce cardiometabolic disease risk factors, and enhance physical function. In all cases however, the exercise prescription should be well tolerated and not exacerbate any of the patient's conditions.

Clinical Exercise Bottom Line

- Physical activity tolerance and exercise responses may be negatively affected by pain severity and location, physical fitness levels, and body positions required during exercise testing and training.

- Exercise testing and training of patients with NSLBP in their position of directional preference (if they have one) can maximize their comfort, compliance, and results.

- Clinical exercise physiologists should monitor patients for new or worsening symptoms and excessive fatigue. If present, this warrants immediate exercise cessation and communication with a physician or health care provider.

- Data from recent investigations indicate there are moderate to high levels of comorbidities among NSLBP patients. Diabetes, hyperlipidemia, metabolic syndrome, coronary artery disease, arthritis, and osteoporosis are within the top 10 reported (96).

- Exercise program recommendations for patients with CNSLBP participating in cardiac rehabilitation and adult fitness programs are consistent with those for persons without NSLBP.

- An individualized approach to exercise program development and progression that addresses all health-related fitness variables is appropriate when working with patients who have CNSLBP.

CONCLUSION

Chronic nonspecific low back pain is a common musculoskeletal comorbidity in cardiac rehabilitation patients and adult fitness program participants. However, patients with comorbid CNSLBP can obtain similar improvements in physical activity tolerance, physical function, and health-related quality of life as persons without CNSLBP. Comprehensive exercise program goals and recommendations for persons with comorbid CNSLBP are, with some modifications, consistent with those for improving general fitness in persons without CNSLBP. An individualized approach to exercise program development and progression addressing all health-related fitness variables, comorbidities, and CNSLBP symptoms can reduce disability and maximize exercise program benefits. Clinical exercise physiologists should provide patients with reassurance and education and also encourage them to practice self-management skills learned from their health care providers.

Online Materials

Visit HK*Propel* to access a link to the references, the case studies with discussion questions, and a quiz to help you review key concepts and test your understanding of the material covered in this chapter.

PART VII

Disorders of the Neuromuscular System

The five chapters in part VII focus on the neuromuscular conditions most likely to be encountered by a clinical exercise physiologist. Some of these conditions (spinal cord injury and stroke) require specialized care or rehabilitation following the initial insult, but many persons who are minimally afflicted or who have completed the recovery process may seek exercise training advice or participation in a supervised program. Two of the chapters present congenital conditions (multiple sclerosis and cerebral palsy), and the other a disease most often noted in the aged (Parkinson's). The clinical exercise physiologist should understand these conditions and be able to design, implement, and modify an exercise program specific to each patient's needs.

Chapter 27 addresses spinal cord injury. In addition to the special care needed immediately after the spinal cord injury (not the focus of this chapter), these patients will face a lifetime of physical limitations and involuntary inactivity. In general, those who have significant recovery and a degree of functionality may be able to perform regular exercise training and thus may seek services at facilities staffed by clinical exercise physiologists. This chapter presents the exercise limitations and related effects of spinal injury for these patients and discusses the specifics of designing and implementing regular exercise training.

Multiple sclerosis (MS) is a disease that typically presents in adulthood and has a wide range of presentation and progression. Thus, the effects on physiologic function will vary among individuals. MS patients who are functioning at a higher level may not need to alter a regular exercise training routine much, whereas individuals with greater disability may need significant adaptations to their exercise programs. Chapter 28 focuses on the exercise limitations of individuals with MS and how a clinical exercise physiologist might design and implement an exercise training program.

Cerebral palsy (CP) is typically diagnosed immediately after birth or in very early childhood. As with all neurological diseases, there are degrees of limitation among individuals with CP. As an individual ages, these limitations can worsen because of prolonged inactivity often associated with CP. Clinical exercise physiologists can provide the exercise programming needed by these individuals to maintain functionality and reduce the rate of functional decline. Chapter 29 provides a summary of these functional limitations, with a focus on the design and implementation of an exercise training program.

Textbooks presenting information about strokes often place it within the cardiovascular section, primarily because the initial insult is initiated by the vasculature (often as a result of long-term effects of cardiovascular risk factors). However, the long-term negative effects of a stroke are most often related to a loss of neuromuscular function. A stroke can affect the functioning of any or all limbs and may also affect speech, mentation, and balance. Chapter 30 discusses important considerations the clinical exercise physiologist may encounter with these patients during the implementation of exercise programming.

Chapter 31 reviews Parkinson's disease, which typically affects individuals of older age. The most noticeable effect is an uncontrolled tremor of the head, arms, and hands. This is often preceded by stiffening muscles with a loss of range of motion, slowing movements, and issues with balance and walking ability. Parkinson's disease is a progressive disorder, and the clinical exercise physiologist must continually respond with modifications in exercise programming. This chapter reviews these issues and how to accommodate them when developing and implementing an exercise training program.

CHAPTER 27

Spinal Cord Injury

Sean M. Tweedy, PhD

Emma M. Beckman, PhD

Mark J. Connick, PhD

Anne L. Hart, PT, PhD

Kati Karinharju, PhD

Kelly M. Clanchy, PhD

Timothy Geraghty, FAFRM (RACP)

Spinal cord injury (SCI) profoundly affects functioning at the level of the body systems (e.g., neuromusculoskeletal and cardiovascular functioning), person (e.g., walking, grasping, lifting, carrying), and society (e.g., employment, sports participation, social connectedness) (118). Evidence indicates that people with SCI are profoundly inactive (23, 178), and this inactivity is causally linked to increased risk of a range of preventable diseases that compound the primary effects of SCI (111, 150). Exercise interventions are an effective means of enhancing physical fitness and reducing preventable disease risk in people with SCI. Additionally, specific types of exercise have been shown to positively influence a range of comorbidities commonly associated with SCI. This chapter reviews the most common issues encountered by the clinical exercise physiologist during exercise planning, prescription, and training for persons with SCI and provides clinical exercise physiology practitioners with evidence-based recommendations for prescription of safe, effective exercise interventions for this population.

DEFINITION

The spinal cord serves as the major conduit for the transmission of motor, sensory, and autonomic neural information between the brain and the body. An SCI affects conduction of neural signals across the site of the injury, or lesion. In broad terms, the effects of spinal cord injury can be divided into **tetraplegia (TP)**—the term preferred over *quadriplegia*—and **paraplegia (PP)**. TP refers to injury to the spinal cord that results in impaired arm, trunk, and leg function, while PP refers to injury resulting in impaired function of the trunk and legs, but with arm function fully spared.

Clinically, SCI is classified based on the interaction of two factors: the level (or height) of the injury and the completeness of the injury. The level of injury typically refers to the lowest segment of the spinal cord that is uninjured (i.e., with normal sensory and motor function on both sides of the body). So the sixth thoracic segment will be the lowest intact segment for a client referred with a T6 injury (i.e., T7 will be injured).

The completeness of an injury is described in terms of motor and sensory function. Table 27.1 presents the five-level system used by the American Spinal Injury Association (ASIA) to classify completeness of injury to the spinal cord (103).

The tool used to classify the level and completeness of an SCI is the ASIA Standard Neurological Classification of Spinal Cord Injury, presented in figure 27.1. Motor function is assessed by examining a key muscle function within each of 10 **myotomes** in the upper and lower limbs on each side of the body, with each muscle function assigned a score from 0 to 5 (as described on the tool). Sensory function (light touch and pinprick) is assessed by examining key sensory points within each of 28 **dermatomes**

Acknowledgment: Much of the writing in this chapter built on previous editions of *Clinical Exercise Physiology*. Thus, we wish to gratefully acknowledge and thank David R. Gater Jr., MD, PhD, Stephen F. Figoni, PhD, RKT, Mary P. Galea, PhD, L. Eduardo Cofré Lizama, PhD, and Andisheh Bastani, PhD.

Table 27.1   American Spinal Injury Association Impairment Scale

Classification of cord injury	Description of classification
A	Complete: No motor or sensory function is preserved in the sacral segments S4-S5.
B	Incomplete: Sensory but not motor function is preserved below the neurological level and includes the sacral segments S4-S5.
C	Incomplete: Motor function is preserved below the neurological level, and more than half of key muscles below the neurological level have a muscle grade
D	Incomplete: Motor function is preserved below the neurological level, and at least half of key muscles below the neurological level have a muscle grade of ≥3.
E	Normal: Motor and sensory function are normal.

American Spinal Injury Association: International Standards for Neurological Classification of Spinal Cord Injury, revised 2019; Richmond, VA.

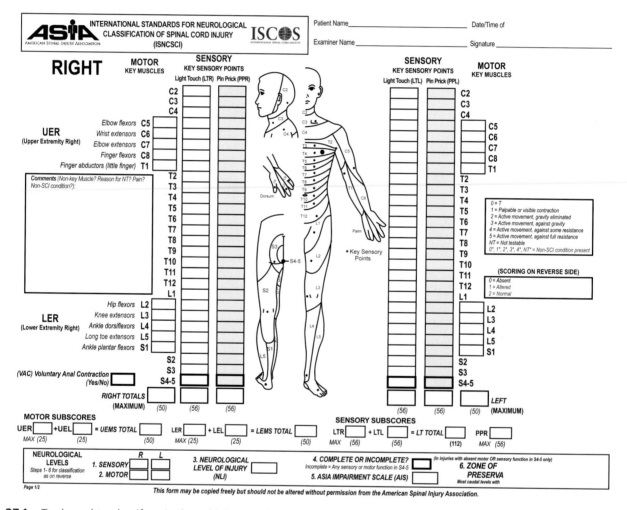

Figure 27.1   Tool used to classify spinal cord injury and describe the resulting motor and sensory deficits.

American Spinal Injury Association: International Standards for Neurological Classification of Spinal Cord Injury, revised 2019; Richmond, VA.

on each side of the body, with each dermatome assigned a score from 0 to 2 (as described on the tool). The injury is classified as complete if no motor or sensory function is spared at the S4 to S5 level; any preservation of function at these sacral levels denotes an incomplete SCI.

Note that the American Spinal Injury Association Impairment Scale (AIS) does not include assessment of autonomic completeness, which must be assessed separately (180). The autonomic nervous system coordinates automatic life-sustaining processes and organizes visceral responses to motor and sensory functioning. It is composed of sympathetic and parasympathetic divisions that regulate the action of smooth muscle and glands. Essential functions of the autonomic nervous system during exercise include modulating heart rate, stroke volume, blood pressure, blood flow, ventilation, thermoregulation, and metabolism (145).

SCOPE

Worldwide, more than 2 million people live with SCI (118). The global incidence rate of SCI is estimated at 23 cases per million (179,312 cases per annum), although there is considerable regional variation, from North America (40 per million) to Australia (15 per million) (118). Almost 40% die before reaching a hospital. Hence, approximately 17,000 Americans will survive a new SCI each year, and more than 90% will return to a private residence after rehabilitation. Men are affected four times as frequently as women, and the average age of injury has increased from 29 yr during the 1970s to 42 yr currently (133). Injury to the spinal cord may be traumatic or nontraumatic. Traumatic injury results primarily from motor vehicle collisions but also from falls (mostly in the elderly), acts of violence (e.g., gunshot wounds, stabbing), and sports and recreation activities. Nontraumatic causes of SCI include spinal stenosis, arthritis, transverse myelitis, spinal tumor, and spina bifida. This chapter focuses primarily on exercise for people with traumatic SCI.

Recent advances in acute SCI care have significantly reduced the number of people with complete SCIs, particularly cervical injuries. Since 2010, 45% of SCIs have resulted in **incomplete tetraplegia** and 13% in complete tetraplegia. In the same period, 21% of SCIs resulted in **incomplete paraplegia**, whereas 20% were **complete paraplegia**. The costs associated with SCI are large. Average lifetime direct health care costs and living expenses for those with complete tetraplegia (levels C5-C8) acquired at age 25 exceed $3.5 million (USD) per person (27), and average lifetime foregone earnings and fringe benefits per case may exceed $71,961 (USD) per year (144). As SCI care and technology have improved, so has life expectancy. A 20 yr old man with newly acquired C5-C8 tetraplegia now has an additional life expectancy of 40 yr or more, whereas the 20 yr old with paraplegia may expect to live an additional 45 yr or more (133).

PATHOPHYSIOLOGY

The spinal cord, a portion of the central nervous system, links the brain with the peripheral and **autonomic nervous systems**. It extends through and is protected by the spinal column, a flexible segment of interdigitating vertebrae and intervertebral discs arranged to maximize mobility and reduce risk of injury. There are 33 vertebrae (7 cervical, 12 thoracic, 5 lumbar, 5 sacral, and 3 to 5 coccygeal). A pair of spinal nerves arises from each vertebral segment. Figure 27.2 presents these segments and nerves. The spinal cord is about 25% shorter than the spinal column; thus, spinal nerves below the first lumbar vertebral level exit from the

cord as a bundle called the **cauda equina**. The vascular supply to the cord comes from one anterior and two posterior spinal arteries at each vertebral level, supplying the anterior two-thirds and posterior one-third of the cord, respectively.

Injury to the cord occurs when a force is applied to the spinal column that is of sufficient magnitude to compromise the integrity of the spinal canal and damage the cord. Common mechanisms include motor vehicle accidents, high falls (from trees or buildings), low falls (particularly in the elderly), and sports and recreation. The initial insult—or primary injury—can damage neural tracts and cell bodies within the cord, as well as vascular structures that supply the cord. Secondary injury occurs because of hemorrhage and local edema within the cord, which can compress the cord and compromise vascular supply, resulting in local ischemia. Infarction of the gray matter happens within 4 to 8 h after injury if blood flow cessation remains. Inevitably, necrosis, or cell death, occurs and can spread over the one or two vertebral levels above and below the area of trauma. **Astrocytic gliosis** and syringomyelia may form during the next several months. Formation of fibrous and glial scarring is the final phase of the injury process.

Direct Effects of Spinal Cord Injury

For the purposes of this chapter, the neurological effects that result directly from injury to the spinal cord are referred to as the *direct effects*. Direct effects can be divided into three categories—motor effects, sensory effects, and autonomic effects. Table 27.2 presents an overview of the motor, sensory, and autonomic function typical of a person with an AIS A injury that is motor, sensory, and autonomically complete.

Note that table 27.2 should be used as a guide only—it is not a substitute for individual clinical assessment. The table illustrates that a complete SCI (i.e., AIS A) results in a functional profile that is generally predictable (lower injury level associated with increased functional level) but highly variable, from ventilator-dependent power wheelchair users with significant cardiorespiratory compromise when the injury is high, to independent walkers with normal cardiorespiratory function. However, when injuries are incomplete (i.e., AIS classification B, C, or D), functioning cannot be predicted, even when segmental level is identical—an injury at segmental level C5 but that is AIS C may include people who do not have sufficient upper limb function to propel a manual wheelchair through to people who are able to stand or walk with aids.

The muscle actions preserved for a given level of AIS A injury as presented in the top row of the table and the next four rows indicate typical motor capabilities of a person with the indicated muscle function in relation to personal mobility, driving, pressure relief, and respiratory function. To obtain an accurate indication of the motor function of an individual client, the clinical exercise physiologist should conduct a physical examination. The methods for doing so are presented in the History section and the Physical Examination section.

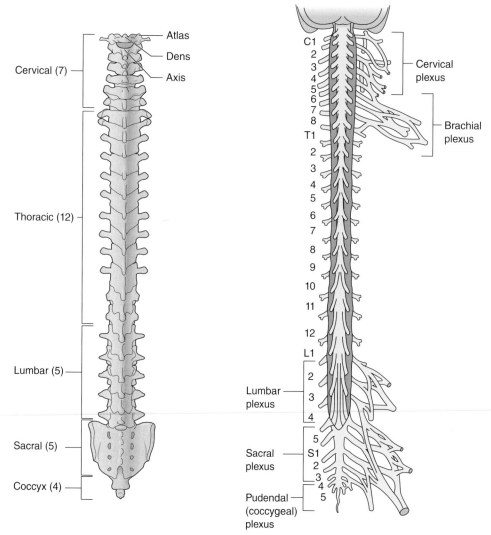

Figure 27.2   Anatomy of the spinal cord.

The main types of sensory function are light touch, proprioception, vibration, pain, and temperature. A general rule of thumb for clinicians is that, for AIS A injuries, sensory function will be preserved above the level of the injury and absent below. Incomplete injuries are much less predictable. To obtain an accurate indication of the sensory function of an individual client, the clinical exercise physiologist should take an accurate history. Important questions in relation to exercise prescription are presented in the History section and the Physical Examination section.

The autonomic effects of SCI have a critical impact on exercise prescription and programming and can influence the exercise capacity of a person with SCI in hot or cold conditions, as well as heart rate and hemodynamic response to exercise. Recommended clinical processes for assessing these effects are presented in the History section and the Physical Examination section as well as in the Exercise Testing section.

Systemic Adaptation to SCI

The direct effects of SCI are neurological, affecting motor, sensory, and autonomic function. However, injury to the spinal cord has widespread effects on all the major body systems—SCI is a truly multisystem condition. Like all effects of SCI, these systemic effects vary significantly from individual to individual, determined largely by the level and completeness of injury. They also vary in their relative importance depending on the length of time postinjury. For example, one of the cardiovascular effects of SCI is orthostatic hypotension (covered in more detail in the History section), which affects most people in the weeks immediately after injury, with many requiring rehabilitation to build tolerance for upright sitting or standing (or both). However, by the time of discharge from hospital, the proportion of people whose daily function is affected by orthostatic hypotension is

Table 27.2 Neurological Effects of SCI

Lowest intact segmental level (not the segmental level of the injury)		C1–C3	C4	C5	C6	C7	C8	T1–T5	T6–T12	L1–L2	L3	L4	L5	S1–S2
Motor function (preserved voluntary muscle actions and key voluntary motor functions)	Muscle actions preserved[a]	Chin	Shoulder shrug and head turn	Shoulder movement and elbow flexion	Wrist extension	Elbow, finger extension	Hand intrinsics	Paraspinal muscles	Anterior trunk stabilizers	Hip flexion	Knee extension	Ankle dorsiflexion	Hip abduction	Plantar flexion, hip extension, knee flexion
	Personal mobility	Power wheelchair (mouth, chin, or voice controlled)		Power wheelchair (arm/hand controlled) or hand-rim propelled wheelchair		Hand-rim propelled wheelchair						Hand-rim propelled wheelchair or walking[b] (limited)		Walking[b] or hand-rim propelled wheelchair (long distance)
	Driving	No		Yes										
	Pressure relief	Dependent			Independent									
	Respiratory function	Ventilator dependent	Unassisted breathing (diaphragm)					Unassisted (diaphragm plus some intercostal)	Normal or near normal	Normal				
Sensory function	Includes light touch, proprioception, vibration, pain, and temperature	Present in head and neck	Present in head, neck, and parts of the arms					Present in head, arms, and parts of the trunk		Present in head, arms, trunk and, parts of the legs				
Autonomic function	Sudomotor and piloerector function	Normal above the level of the injury and absent below the level of the injury												
	Sympathetic vasomotor function	Reduced in arms and absent in splanchnic bed and legs						Normal in arms, absent in splanchnic bed and legs	Normal in arms, reduced in splanchnic bed, and absent in legs	Normal in arms and splanchnic bed, reduced in legs				
	Sympathetic cardiac innervation	None						Reduced	Normal					
	Blood pressure responses	Impaired: susceptible to OH, EH, and AD						Possible OH, EH, or AD	Normal					

[a] The motor profile represents that typically attained by people who are young and otherwise healthy.

[b] Requires lower limb orthoses.

OH = orthostatic hypotension; EH = exercise-induced hypotension; AD = autonomic dysreflexia.
Adapted by permission from American Spinal Injury Association, *International Standards for Neurological Classification of Spinal Cord Injury*, revised 2006 (Chicago, IL: American Spinal Injury Association, 2006).

greatly reduced. Neurogenic osteoporosis follows the opposite time course, becoming of increasing clinical importance as time postinjury increases. Following is an overview of systemic effects of and adaptations to SCI.

- *Cardiovascular:* In individuals without SCI, lower limb muscle contractions propel blood back to the heart (skeletal muscle pump), reducing venous stasis (pooling of blood in the veins) in the legs (53). However, this skeletal muscle pump is impaired in people with SCI who have lower limb muscle paralysis and, in TP and high PP, reduced sympathetic influence on peripheral and splanchnic vascular beds. Together these factors lead to **circulatory hypokinesis**—reduced venous blood return to the central circulation, thus lowering stroke volume, cardiac output, and eventually blood pressure (53). Venous stasis, deep venous thrombosis (DVT), and subsequent pulmonary embolus may occur as a result of circulatory hypokinesis. The risk of orthostatic hypotension and exertional hypotension is also increased, as is maladaptive myocardial atrophy, which can further diminish cardiac output and cardiac reserve.

- *Pulmonary:* Ventilation is impaired in most people with TP or a high thoracic injury because of paralysis of rib cage and abdominal musculature, reduced pulmonary compliance, and reduced diaphragmatic excursion. In some cases, the impairment is so severe that breathing must be ventilator assisted. Tetraplegia below C5 typically spares voluntary control of the diaphragm, although expiration remains impaired. Respiratory function worsens as the level of disability increases (18, 36), although aggressive spirometry and exercise training may result in improvement (38).

- *Bowel and bladder:* Cervical, thoracic, and lumbar spinal cord lesions may increase the risk of gastric and duodenal ulcers, increase bowel motility and **hyperreflexia**, and eliminate voluntary control of defecation (169). Hyperreflexia of the bladder wall and sphincter muscles results in greater risk of vesicoureteral reflux, hydronephrosis, and acute renal failure if not appropriately managed. Bladder spasms and loss of voluntary sphincter control may lead to urinary incontinence. Urinary tract infections, as well as renal or bladder stones, are more frequent in people with SCI than in the non-SCI population.

- *Endocrine:* Impaired motor, sensory, and autonomic function in people with SCI may also affect metabolic and hormonal function, including the sympathomedullary response and the adrenocortical–pituitary axis, resulting in flattened circadian rhythms and poorly regulated corticosteroid responses (174). Glucose intolerance often occurs and is frequently accompanied by **hyperinsulinemia** (3). Although thyroid function may be acutely altered in SCI, thyroid function tests are generally normal in healthy SCI adults (14). Conversely, testosterone and free testosterone levels in men with SCI are often reduced (14, 167), whereas growth hormone release is blunted and chronically depressed (13).

- *Musculoskeletal:* Inability to voluntarily contract muscles affected by the SCI leads to marked muscular **atrophy**. In spastic paralysis, the regular involuntary contractions mitigate atrophy,

but in people with flaccid paralysis (e.g., cauda equina injuries), atrophy can be extremely marked in the lower limbs. The reduction in soft tissue volume increases the exposure of bony prominences (e.g., greater trochanter, ischial tuberosities, and sacrum), increasing the risk of pressure ulcers. Neurogenic osteoporosis after SCI results from the withdrawal of stress and strain of muscle activation on bone below the level of injury (147). Exponential decreases in key bone parameters at the epiphyses and diaphyses of the tibia and femur occur with time postinjury (51). These are typical sites of fragility fractures in chronic SCI. Steady states for bone loss are reached after 3 to 8 yr depending on the parameter, with approximately 50% bone mass loss in the epiphyses and 30% in the diaphyses (51).

CLINICAL CONSIDERATIONS

Of the clinical conditions covered in this textbook, SCI is unique in that it is the only one in which trauma is a major cause. From a clinical exercise physiology perspective this is important because trauma happens at a single, defined point in time, and management proceeds from emergency care and stabilization to acute care and rehabilitation, and finally through a transition process from hospital to community and ongoing chronic management. Over the course of this recovery process, exercise testing and prescription services offered by clinical exercise physiologists follow a trajectory of increasing relevance, from little or no relevance during emergency care through acute in-patient rehabilitation where there is a focus on increasing activity levels, exercise tolerance, and fitness, to the point where evidence indicates that adoption and maintenance of regular exercise training is a highly effective means of achieving optimal long-term health and well-being for people with SCI (122, 123, 168). Furthermore, because SCI is permanent and life expectancy continues to increase, the majority of people with SCI are not in hospital settings but living and working in their communities. For these reasons, coverage in this chapter of the acute and early management of SCI (i.e., <12 mo) is restricted to the sidebar Management Issues During Acute SCI Hospitalization and Rehabilitation.

Another clinically important aspect of the traumatic etiology is that, unlike other neurological conditions such as multiple sclerosis or Parkinson's disease, an SCI results in a static lesion—the neurological damage to the cord does not spread, and there is no ongoing disease process. Many people with SCI are young, healthy, and active at the time of injury. Once they have recovered from the acute effects of their injury and are discharged from hospital, they want professional services that will help them return to active, productive lives. From a clinical perspective, this means the role of the clinical exercise physiologist is not to provide a patient with SCI with a treatment plan for their condition—to date there is no effective remedy for injury to the spinal cord. Rather, the goal should be to help the client optimize their health, fitness, and functioning in the context of spinal cord injury, using exercise interventions that are safe, effective, and sustainable.

Management Issues During Acute SCI Hospitalization and Rehabilitation

Acute Hospitalization

- Spine management: imaging of cervical, thoracic, lumbar, and sacral spine
- Surgical or orthotic stabilization of unstable spinal column injuries and spinal cord decompression
- Range of motion limitations to allow complete bony and soft tissue healing of spinal elements

Respiratory

- Assisted ventilation often required for high cervical injuries
- Secretion management essential because of impaired cough and increased parasympathetic nervous system influence on pulmonary secretions
- Assisted cough required for SCI above T6 attributable to intercostal and abdominal muscle paralysis

Cardiovascular

- Relative bradycardia attributable to impaired sympathetic nervous system in SCI above T6; occasionally requires pacemaker placement
- Hypotension attributable to systemic vasodilation resulting from impaired sympathetic drive
- Venous stasis can result in deep venous thrombosis or pulmonary embolism

Functional Mobility

- Upper extremity range of motion, strengthening, and endurance within limitations of orthotics and medical management
- Bed mobility, including side to side, supine to prone to supine, supine to sit
- Wheelchair mobility, including forward and backward propulsion, turning, uneven terrain, curbs, ramps, hills
- Transfers, including bed to wheelchair to bed, wheelchair to toilet to wheelchair, wheelchair to bath to wheelchair, wheelchair to floor to wheelchair, wheelchair to car to wheelchair
- Activities of daily living (ADLs), including feeding, grooming, dressing, bathing, toileting
- Bladder management training, typically with intermittent catheterization or alternative
- Bowel management training, typically with suppositories and digital stimulation or alternative
- Skin management training with monitoring and pressure relief techniques
- Equipment evaluation for personal care, mobility, and public accessibility
- Home and vehicle evaluation for accessibility
- Psychological and social adjustment to SCI
- Introduction to vocational and recreational opportunities for persons with SCI

The International Classification of Functioning, Disability and Health (ICF) divides the consequences of any given health condition into five domains. These are presented schematically in figure 27.3. SCI profoundly affects all five domains, as illustrated by the text below each domain.

The topmost box in figure 27.3 presents the health condition covered in this chapter—SCI. *Health condition* is an umbrella term for disease, disorder, injury, or trauma. Other health conditions that affect the spinal cord include spina bifida, spinal tumors, and poliomyelitis. If the ICF were applied to those health conditions, there would be some disease-specific differences, but there would be many similarities. In ICF terminology, problems with body structures and body functions (box 1) are referred to as *impairments*. In SCI, the impaired body structure is the spinal cord. An overview of the body functions that may be impaired in a person with SCI has already been presented (table 27.2) and includes motor, sensory, cardiovascular, and respiratory functions.

Box 2 relates to activity/ies, which in ICF terminology is defined as the execution of a task or action, and activity limitations are difficulties an individual with SCI may have (see table 27.2). Box 3 relates to participation, defined as involvement in a life situation; participation restrictions refer to problems a person experiences with participating in life situations (e.g., employment, playing sports, staying fit, access to entertainment venues). Participation restrictions for people with SCI often include access to public venues and high unemployment (20). Box 4 describes the environmental factors that make up the physical, social, and attitudinal environment in which a given individual lives. The presence of paved footpaths and curb cuts for access can vary from country to country, state to state, and street to street and can play an instrumental role in physical activity participation for people with SCI. Box 5 lists personal factors, including age, gender, social status, and life experiences. In the population of people with SCI, young males with a history of risk-taking behavior are overrepresented

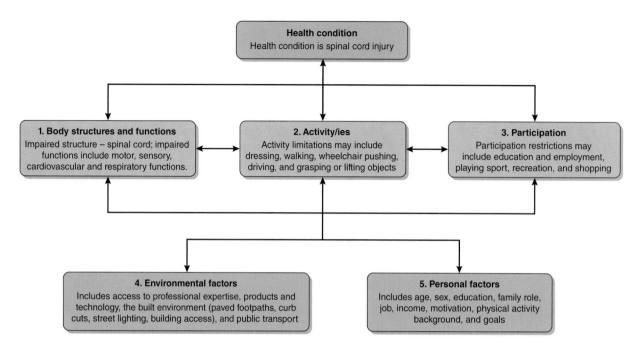

Figure 27.3 Schematic representation of the ICF domains and their interrelationships, as applied in the context of spinal cord injury.

Reproduced from *International Classification of Functioning, Disability and Health: ICF,* World Health Organization (WHO), pg. 7, Copyright 2013.

(118), and the informed clinical exercise physiologist will take such factors into account when developing an exercise intervention for clients from this group.

Figure 27.3 makes it clear that although injury to the spinal cord can result in motor, sensory, cardiovascular, and respiratory impairments that will have a direct impact on the body's responses to exercise, in order to work effectively with people with SCI, the clinical exercise physiologist must be able to physically assess their clients' exercise-specific motor function and understand how personal and environmental factors—such as the ability to afford and drive a car to and from fitness facilities—will be instrumental in developing safe, effective, sustainable exercise interventions.

History

Myriad factors may affect exercise prescription for people with SCI, and an accurate history is one of the most effective ways of determining exactly which factors affect the individual and to what extent. This should be complemented by a systematic physical evaluation (presented in the next section) to provide an indication of a client's individual motor profile. Together, outcomes from the history and physical examination will permit the clinical exercise physiologist to determine whether their client should be referred for a medical assessment before starting an exercise program. This information will also indicate what adjustments may be required to exercise testing and exercise prescription in order to accommodate that client's individual motor profile.

Spinal cord injury has lifelong consequences, and clinical practice guidelines—such as those developed by the Consortium for Spinal Cord Medicine and published by the Paralyzed Veterans of America (www.pva.org)—describe expected outcomes for SCI and provide general guidance on the management of those consequences. However, the most significant issues for a given individual vary considerably, and therefore it's important that the clinical exercise physiologist begin the initial assessment with an accurate and detailed history. In relation to the ICF, an accurate history provides the clinical exercise physiologist with a comprehensive overview of the impairments of body structures and functions that affect their client (see box 1, figure 27.3).

Opening questions should include a description of any medical treatment a client is receiving and who they would consider their primary health care provider—it is this person who would be the point of referral if that client's history indicates they should have further medical assessment before starting a physical activity program. The history should also include medications being taken. Commonly prescribed medications used in the management of SCI are presented in table 27.3, together with their indications and common side effects. These medications are usually taken on a stable, long-term basis, and so most community-dwelling people with SCI will have adapted to their side effects. Some medications used by people with SCI can potentially enhance exercise performance (e.g., midodrine) and are therefore on the World Anti-Doping Agency (WADA) banned list. People using such medications for therapeutic reasons may take part in competitive sport to the highest levels provided they apply for a therapeutic use exemption (TUE).

Table 27.3 Pharmacology

Medication class and name	Primary effects	Exercise effects	Side effects
α-antagonist Phenoxybenzamine / Dibenyline, Dibenzyline	Facilitate bladder emptying Decrease autonomic dysreflexia	Dizziness, hypotension	Nasal congestion, postural hypotension, fatigue, drowsiness
α-blocker • Prazosin / Minipress • Terazosin / Hytrin	Facilitate bladder emptying Decrease autonomic dysreflexia	Dizziness, hypotension	Orthostatic hypotension, syncope, drowsiness, dizziness, fatigue, weakness Weakness, tiredness, stuffy or runny nose, back pain
Anticholinergic Oxybutynin / Ditropan	Decrease bladder hyperreflexia and spasms	Dizziness, hypotension	Drowsiness, dry mouth, constipation, dizziness, blurred vision
Anticoagulant • Enoxaparin / Lovenox, Clexane • Warfarin / Coumadin, Marevan	Anticoagulation, prevent DVT and PE	Easier bleeding, bruising with injury	Bleeding, bruising
Anticonvulsant • Carbamazepine / Tegretol • Gabapentin / Neurontin • Pregabalin / Lyrica	Control seizures Decrease neuropathic pain Reduce neuropathic pain	Dizziness, hypotension	Drowsiness, aplastic anemia, agranulocytosis dizziness, drowsiness, fatigue Weight gain, dizziness, drowsiness, fatigue, dry mouth Dizziness, drowsiness, blurred vision, fatigue, weight gain
Antimuscarinic Tolterodine / Detrol, Detrusitol	Decrease bladder spasm	Dizziness, hypotension	Dry mouth, GI upset, headache, drowsiness
Skeletal muscle relaxant • Dantrolene / Dantrium • Baclofen / Lioresal, Clofen • Diazepam / Valium • Tizanidine / Zanaflex	Relieve anxiety Decrease spasticity	Some or all side effects may occur, depending on drug used Dizziness, hypotension, muscle weakness, spasticity	Drowsiness, dizziness, weakness, liver function disturbance Drowsiness, hallucination, seizure, muscle weakness or pain Impaired concentration, drowsiness, fatigue, muscle weakness Chest pain or discomfort, fever or chills, nausea or vomiting, drowsiness, dry mouth
Sympathomimetic, α-1 agonist Midodrine	Treat hypotension	Increase blood pressure	Blurred vision, headache, stomach pain, urinary problems, chills, dry mouth
Tricyclic antidepressant • Amitriptyline / Elavil, Tryptanol • Imipramine / Tofranil • Nortriptyline / Pamelor	Decrease neuropathic pain Decrease bladder incontinence, depression, and anxiety Decrease neuropathic pain	Dysrhythmia, hypotension	Dry mouth, constipation, dizziness, headache, drowsiness Anxiety, confusion, restlessness, dry mouth, constipation Agitation, insomnia, confusion, dry mouth, constipation

Note: Medications may have brand (trade) names that vary from country to country, or there may be multiple brand names for the same medication within a single country. The medication brand names in this table are presented as examples only.

DVT = deep venous thrombosis; PE = pulmonary embolism.

Screening for cardiorespiratory and metabolic risk factors—as would be done for the general population—should be conducted, bearing in mind that spinal cord injury is associated with increased risk of coronary artery disease, obesity, type 2 diabetes mellitus, and other diseases related to physical activity. A thorough history includes questions specific to the SCI population, which will provide information of importance for exercise testing and training. The most important of these are described next and include etiology, time since injury, autonomic dysreflexia, hypotension (orthostatic or exertional), impaired thermoregulation, pressure ulcers and other sensation-related injury risk, pain, spasticity, syringomyelia, and neurogenic bowel and bladder dysfunction.

Etiology: Traumatic SCI (TSCI) is frequently associated with limb fractures and brain trauma, as well as thoracic contusions and fractures (and associated respiratory complications), **plexopathies**, and peripheral neuropathies, all of which can have significant exercise implications (137). Asking the client about other injuries received at the time of the accident will help in the identification and management of any of these comorbidities should they exist. Similarly, nontraumatic SCI results from a range of health conditions, many of which have specific physiological or biomechanical consequences that may affect exercise (e.g., spina bifida, spinal stenosis, vascular malformations or tumors) (137). It is beyond the scope of this chapter to consider these implications in detail; however, practitioners should be aware that SCI may not be the only (or even the major) impairment affecting exercise for a given client.

Time since injury: Time since injury is associated with a significant decrease in physical functioning and increase in musculoskeletal, neurological, metabolic, and genitourinary comorbidities (29). People with long-standing (>20 yr) SCI require greater physical assistance as they age. The percentage of maximal heart rate during performance of ADLs and community mobility tasks is higher in people with TP than in those with PP, and an inverse relationship exists between physical capacity and physical strain (40, 96). In one study, only 29% of SCI subjects with $\dot{V}O_2$peak less than $15 \text{ mL} \cdot \text{kg}^{-1} \cdot \text{min}^{-1}$ were able to perform independent ADLs (84). Exercise can help prevent these declines. People with long-standing injuries who have been chronically inactive will often have complex and highly individual presentations that can increase the risk associated with exercise (29). Men ≥40 yr and women ≥50 yr who have SCI should be considered at moderate risk for untoward events during exercise, independent of traditional coronary artery disease risk. Practitioners should work cooperatively with such clients to individually tailor exercise interventions and be conservative with exercise progression.

Autonomic dysreflexia: **Autonomic dysreflexia (AD)** affects individuals with SCI above the splanchnic sympathetic outflow (i.e., T6 or above) (107). AD results from a noxious afferent stimulus below the level of the lesion (e.g., overdistended bladder, urinary tract infections, ingrown toenails, cuts, bruises, and pressure sores). The noxious stimulus leads to widespread vasoconstriction below the level of the lesion and vasodilation above (106). Symptoms may be very mild but can be extreme (e.g., blood pressure of 300/220 mm Hg, pounding headache, and blurred vision). Untreated AD can lead to stroke, seizures, and death. Emptying the bladder before exercise will minimize risk (145). Community-dwelling adults with SCI are usually well versed in the recognition and initial self-management of AD and will usually be able to advise what assistance they require. In Australia, people at risk will often carry an autonomic dysreflexia medical emergency card with information on treatment and emergency contact details. If AD occurs during a supervised exercise program, the clinical exercise physiologist must be able to identify that it is occurring and contact emergency services or other assistance as required. If possible, they should try to identify and remove the noxious stimulus (e.g., occluded bladder catheter), loosen tight stockings or bindings, and sit the client upright to minimize intracranial blood pressure (78, 107). Acute AD is an absolute contraindication for exercise training (57). Intentional inducement of AD is used to enhance sport performance. Although it is illegal, evidence indicates that athletes still do this, and practitioners have an important role in educating clients about the previously mentioned dangers of this practice (i.e., stroke, seizures, and death) (19).

Hypotension: **Orthostatic hypotension (OH)** is a decrease of >20 mm Hg in systolic blood pressure (BP) or 10 mm Hg in diastolic BP upon sitting up from a supine position (32). It is caused by blood pooling in the abdominal viscera and legs during lying, compounded by lack of muscular contraction in the paralyzed lower limbs when postural change is initiated (32). Exercise-induced or exertional hypotension (EH) is a fall of ≥10 mm Hg in systolic BP during exercise (120). It is due to a combination of vasodilation in exercising musculature and inadequate vasoconstriction in the splanchnic bed and nonexercising musculature (120). Both EH and OH result in similar symptoms—light-headedness, dizziness, and syncope—and both are best treated with decumbency, leg elevation, and fluid ingestion (145). Risk management includes slow, controlled positional change; a graduated introduction to exercise, particularly for the previously sedentary; and monitoring blood pressure during exercise testing. Regular exercise may improve tolerance of OH (138). Symptomatic hypotension (e.g., dizziness, nausea, pallor) is an absolute contraindication for exercise training (57).

Impaired thermoregulation: Thermoregulatory function may be impaired such that, in hot conditions, inability to sweat and dilate superficial vasculature below the lesion impairs heat loss during exercise. This is particularly significant in TP (106), and consequently exercising core body temperatures tend to be higher in these individuals at a given relative intensity (57). Normal precautions against overheating can be complemented by hand cooling (73), foot cooling (74), ice vests (177), and spray bottles (142). In cold conditions, impaired superficial vasoconstriction, shiver response, and piloerection accelerates heat loss, increasing the risk of exposure (106).

Pressure ulcers and other sensation-related injury risk: Pressure ulcers occur in wheelchair users as the result of shearing forces, friction, and unrelieved pressure, usually over a bony prominence, which damages overlying soft tissue. This problem occurs in persons with SCI because sensory impairment means they are unable to feel the sensations of tingling, discomfort, and pain that, in a neurologically intact person, would typically be the stimulus to relieve the pressure. Sensory impairment also increases the risk of falls (through decreased proprioception) and delays detection and diagnosis of serious injury (e.g., bony fractures) after collisions or falls. Sensation-related risk is increased with new activities, particularly those done out of the client's regular chair (e.g., on new exercise equipment or a new sports wheelchair); when playing contact sports; and when atrophy is severe. Risk management strategies include clear identification of insensate areas during preparticipation screening via client self-report, clients using their own cushions on new equipment, thorough skin checks for red areas after new activities, regular pressure relief during exercise (every 15-30 min) (78), and appropriate protective sports equipment (78). Some evidence indicates that a healthy lifestyle, including exercise, is protective against recurrent pressure ulcers (108).

Pain: The incidence of upper extremity pain is high (50%-70%) for both TP and PP, and the shoulder is the most common site, followed by the wrist (39, 55, 156, 161). This has been attributed to manual wheelchair use and daily transfers, a hypothesis supported by the fact that it is these activities that most commonly exacerbate upper limb pain (161). However, upper extremity pain is similarly prevalent among motorized wheelchair users and those using canes and crutches (41, 93). Evidence also indicates that, compared with athletes, nonathletes are twice as likely to be affected by shoulder pain (61), higher muscular strength is associated with less UE pain in both tetra- and paraplegia (48), and appropriate strength and conditioning training can reduce shoulder pain (128, 132). Management of upper extremity pain should be informed by appropriate preparticipation screening (i.e., does UE pain exist and which activities cause pain) and should involve advice for minimizing upper extremity articular stress during transfers, pressure lifts, and wheelchair propulsion (128), together with an exercise program that avoids painful activity while enhancing strength. Muscle and joint discomfort is a relative contraindication for exercise testing, while severe or uncontrolled pain is an absolute contraindication (57).

Spasticity: **Spasticity** is a velocity-dependent increase in muscle tone with exaggerated tendon jerks. It has been shown that 53% to 78% of persons with chronic SCI experience symptoms of spasticity (1, 173). Spasticity can affect exercise ability and causes significant functional impairments such as a restricted ability to carry out ADLs, pain, fatigue, inhibition of functional ambulation, increased risk of developing contractures and pressure ulcers, and difficulties with self-hygiene (1). Treatment includes the removal of any stimuli-inducing increased tone, daily prolonged stretching of affected muscle groups, physical therapy, and pharmacological or surgical management (101). The extent to which an individual with SCI is affected by spasticity varies widely, and systematic physical examination (described in the next section) will help identify whether, and to what extent, spasticity affects the person's ability to get into and maintain a stable exercising position.

Syringomyelia: Post-traumatic **syringomyelia** (or syrinx) refers to the formation of an abnormal CSF-filled cyst in the spinal cord months or years after the original SCI (22). It affects up to 28% of people with SCI, but only 1% to 9% are symptomatic (22). Initial presentation is segmental sensory loss together with pain above the level of injury that is exacerbated by coughing, sneezing, or straining. Progressive asymmetrical weakness ensues (22). Practitioners can ensure they are appraised of clients with this condition by taking an accurate history. Clients who develop new pain, sensory loss, or weakness should be referred to an appropriate medical specialist (22).

Neurogenic bladder and bowel dysfunction: This condition is common after SCI and may manifest differently in individual clients, but many may be at increased risk of urinary or fecal incontinence, and this risk will be increased further when exercising. People who have sufficient hand function will be taught to self-catheterize (i.e., using a straight catheter) and therefore do not need a permanent urinary catheter. In general terms, all people should be advised to empty the bladder and bowel by their usual method before exercise testing or training. Individuals who require assistance can ask for help. The risk of autonomic dysreflexia may also be increased by having a full bladder or bowel.

Physical Examination

Understanding the level and completeness of SCI and associated medical problems can help the clinical exercise physiologist appreciate the physical limitations of each person with SCI and the degree to which their motor, sensory, and autonomic impairments may affect exercise responses. However, medical records provide relatively limited and often outdated information about the functional mobility of a given client. Therefore, before undertaking any exercise testing or prescription, the clinical exercise physiologist should perform a physical examination of the client to determine how the individual's specific motor impairment profile will affect motor functions commonly required in the exercise context (e.g., grasp, transferring, and shoulder range of movement). Tasks to evaluate the exercise-specific motor function of a client include the following:

- Assessing functional grip strength (participant squeezes practitioner's fingers)
- Assessing shoulder abduction (overhead reaching), unilateral and bilateral
- Writing (own name and address)
- Picking up an object from the ground
- Transferring from chair to ground
- Transferring from ground back to chair
- Transferring to different chairs or benches and back again

MANAGING PRESSURE ULCERS AND HEMODYNAMIC IMPAIRMENT

A pressure ulcer (also called a pressure sore, decubitus ulcer, bedsore, ischemic sore, or skin breakdown) is an area of the skin or underlying tissue (muscle or bone) that is damaged by the occlusion of blood flow to the region. If blood flow is occluded, chronically hypoxic and then necrotic tissue changes can occur. People with SCI are at particular risk of pressure ulcers because of the loss of sensation, particularly afferent input from pressure receptors, and the inability to relieve pressure from the affected body parts because of motor paralysis. This means pressure can be applied to soft tissue for long periods without accompanying discomfort, leading to the aforementioned occlusion of blood and the potential for hypoxic and necrotic soft tissue damage.

The consequences of pressure ulcers can be severe. Infections can develop and spread to the blood and multiple areas of the body. Some severe cases are managed with reconstructive surgery, sometimes by limb amputation, and can be life threatening. Even less severe cases can result in prolonged bed rest (from weeks to 6 mo), keeping people away from work, school, and social activities as well as leading to severe physical deconditioning and loss of fitness. Up to 80% of individuals with SCI will have a pressure sore during their lifetimes, and 30% will have more than one.

Starting a fitness training program will often involve sitting on unfamiliar surfaces; transferring on or off unfamiliar equipment, which increases the risk of getting a bump; and possibly repetitive cyclical movements that can lead to friction injuries. At particular risk is the skin over bony prominences such as the heel, knee, greater trochanter, ischial tuberosity, sacrum, and scapula.

One of the first signs of a pressure ulcer is reddened or pink skin that may feel warm to the touch. All warm, red areas should be taken seriously. When pressure is applied to the red skin for 3 to 4 s, it should turn white; and when the pressure is removed, it should quickly turn red again, indicating prompt return of blood flow to the region. If the area stays white for longer than usual (i.e., compared with other areas of skin), then blood flow is likely to have been impaired and hypoxic damage may have begun. A client with red skin that responds in this unusual way should do their best to stay off the area and seek medical advice as soon as possible.

Clinical exercise physiologists can reduce the risk of pressure ulcers by being aware of the causes and early signs, treating even relatively light bumps and falls seriously, and ensuring the client seeks medical attention if they find a red area. Clients should also be encouraged to conduct regular checks for red or pink skin, particularly in the early stages of a program, and perform regular pressure lifts while sitting and position changes while lying to avoid prolonged pressure in one spot.

People with SCI—particularly those with tetraplegia—can be affected by hypotension (either exertional or orthostatic) or hypertension resulting from autonomic dysreflexia. When these conditions occur, they are managed in radically different ways. If the clinical exercise physiologist is not able to differentiate hypotensive symptoms from hypertensive symptoms and implements the wrong management strategies, the consequences can be fatal. Hypotensive episodes are triggered by either physical exertion (exertional hypotension) or positional change (orthostatic hypotension). Symptoms include the following:

- Light-headedness
- Dizziness
- Fainting
- Blurry vision
- Confusion

To manage hypotension, the clinical exercise physiologist should

- instruct the client to stop exercising,
- help the client lie down if possible, and
- help the client elevate the limbs.

Autonomic dysreflexia is triggered by a noxious stimulus below the level of the lesion (e.g., a blocked urinary catheter, pressure sores, sitting on top of something sharp or hard for an extended period). Symptoms include the following:

- Throbbing, pounding headache
- Sudden rise in blood pressure
- Blurred vision
- Bradycardia
- Dry, pale skin below the level of the lesion
- Sweating, flushing, and goosebumps above the level of the lesion

To manage autonomic dysreflexia, the clinical exercise physiologist should

- instruct the client to stay as upright as possible (client should not lie down),
- call emergency medical services, and
- eliminate the noxious stimulus if possible.

If the client has had a previous episode of either hypotension or autonomic dysreflexia, they can likely describe their particular symptoms and what to look for. They will also be able to indicate how serious the episodes are and how they are best managed. They may carry a treatment card for autonomic dysreflexia or another form of medical alert that gives the clinical exercise physiologist and first responders information about the condition and its treatment.

- Lying on front
- Lying on back
- Moving from sitting to lying and from lying down to sitting up
- Standing to weight bear, if possible, to assess the following:
  - Whether activities can be executed in standing
  - Half squats
  - Sit to stand
  - Stand to sit
  - Walking in parallel bars

The following questions can guide administration of the above tasks:

- Are they willing to try?
- Can they do it safely?
- Can they do it effectively without assistance?
- Is assistance required (personal or equipment)?
- Does the activity elicit pain, discomfort, or spasticity?
- What is the effort or length of time required to complete the task?

In relation to the ICF, the clinical exercise physiologist will already have taken an accurate history to gain information about the body function and structural impairments that affect the client. At the completion of the physical examination, the clinical exercise physiologist will also have a sound grasp of the client's activity/ies limitations (see box 2, figure 27.3).

Referral for Medical Preparticipation Screening

Having obtained an accurate description of the client's current medical management and a thorough client history, the clinical exercise physiologist is in a position to determine whether the client should be referred for medical preparticipation screening. The cardiorespiratory and metabolic risk stratification procedures that apply to the general population also apply to people with SCI criteria, including when referral to an appropriately qualified medical practitioner for a preparticipation medical screen is indicated. Additionally, people with SCI should be referred to their primary medical care provider for individualized medical advice about beginning exercise if they have a history of autonomic dysreflexia, clinically significant orthostatic or exertional hypotension, or pain or spasticity that is exacerbated by exercise or physical exertion. Clinical exercise physiologists should also consider requesting medical advice for clients with SCI who have severe motor impairment, are elderly, have a long history of physical inactivity, or have multiple comorbidities or a combination of these.

Additional investigations that may be warranted pending consultation with the person's medical practitioner include pulmonary function testing, dual-energy X-ray absorptiometry (DXA) to evaluate bone mineral density, X-rays of paralyzed extremities to exclude asymptomatic fractures, blood lipid profiles, and HbA1c to determine whether the person has glucose intolerance or diabetes.

For people undergoing testing to rule out ischemic heart disease, postexercise echocardiography (112) or nuclear imaging studies (13) may improve the sensitivity of the exercise stress test. In a study using standard exercise testing, only 5 of 13 subjects with known myocardial ischemia had ST-segment changes indicative of ischemia (12).

Exercise Testing

Exercise testing to evaluate cardiorespiratory, musculoskeletal, and flexibility fitness can help define areas of need with respect to exercise training. Table 27.4 provides a review of exercise testing specifics related to cardiorespiratory, skeletal muscle, and flexibility fitness.

Cardiorespiratory

Most people with SCI are unable to engage in lower limb exercise and must use upper limb modes such as wheelchair pushing or upper limb ergometers (e.g., arm crank or wheelchair). This constraint is important because, compared with lower limb exercise, upper limb exercise elicits a considerably reduced cardiorespiratory response (166). For example, when people without SCI perform arm cycling, maximal power output (POmax) and $\dot{V}O_2$peak are reduced by approximately 40% and 25%, respectively, compared with leg cycling (166). Efficiency is also reduced—oxygen uptake (and therefore physical strain) for any given power output is greater for arm cycling than for leg cranking.

When exercise capacity is assessed using arm crank ergometry, there is no significant difference between $\dot{V}O_2$peak for people with or without paraplegia (PP) (166). Parity is achieved at least partly because habitual wheelchair use leads to physiological adaptations in the upper limb musculature in people with PP, reducing glycogenolysis and increasing the rate of lipid utilization (166). Even with the limitations inherent to upper limb exercise, SCI athletes can achieve a high $\dot{V}O_2$peak (e.g., a mean of 40 mL · kg^{-1} · min^{-1} for a Paralympic wheelchair basketball team [56]).

This evidence notwithstanding, SCI adversely affects exercise capacity for a number of reasons. First, as lesion level increases, voluntarily activated muscle mass decreases (see table 27.2), reducing oxidative capacity and therefore maximal oxygen uptake and maximum caloric expenditure from exercise. This effect is particularly pronounced in TP because the muscles required for arm exercise are partially paralyzed in addition to paralysis of the trunk and legs (166). Second, as lesion level increases, sympathetic vasoconstrictive capacity gradually diminishes, reducing venous return, and in accordance with the Frank-Starling mechanism, exercising stroke volume does not rise to the same extent as in individuals without SCI. At submaximal exercise levels, cardiac output can be maintained by a compensatory increase in heart

Table 27.4 Exercise Testing Review

Test type	Mode	Protocol specifics	Clinical measures	Clinical implications	Special considerations
Cardiorespiratory	Arm or leg ergometry or both, including wheelchair ergometry	Warm-up; start at 5-25 W; increase by 5-10 W every 2-3 min; pause for BP	HR, BP, RPE, symptoms; 12-lead ECG if heart disease is suspected or documented; monitor through 5-10 min recovery	Signs and symptoms reveal exercise tolerance and source of limitation.	Empty bowel and bladder. Consider handgrip and postural stability in all patients. Use skin protection where needed. Adjust loads according to degree of training and paralysis and other impairments.
Resistance	Combined or isolated joint motions with free weights or machines	Warm-up; 3 trials of manual muscle testing for weaker muscles with grades ≥3; 3 trials of isometric or isotonic testing for stronger muscles; 1 trial of upper body anaerobic power testing for athletes	Manual muscle testing for weakest muscles, if expecting neurological recovery; maximal isometric force production with handheld dynamometer or 1RM- 10RM; peak or average anaerobic power during 30 s maximal effort	Sufficient strength in proximal muscles allows independent mobility. Balancing strength around joints helps prevent contractures, rotator cuff injury, and joint degeneration.	Same as above. Higher joint angular velocities may be impeded by spasticity.
Range of motion	Active and passive assisted stretching	Warm-up; evaluation of major joints; focus on joints with spasticity or contracture	Range of motion measured using a goniometer or other device	Emphasize joints with spasticity or contracture. Support midshaft of long bones with osteopenia.	Care should be taken if spasticity, contracture, or osteopenia are present.

BP = blood pressure; HR = heart rate; RPE = rating of perceived exertion; ECG = electrocardiogram; RM = repetition maximum.

rate, but this is not possible during maximal exercise, reducing maximal exercise capacity.

For TP, two additional factors negatively affect exercise capacity. The first is that cardiac sympathetic innervation is absent, and heart rate increases can be achieved only by parasympathetic withdrawal and circulating catecholamines (166), limiting maximal heart rate to approximately 130 b · min^{-1} (166). However, people with AIS A tetraplegia may retain enough autonomic function to improve cardiovascular responses to exercise (180). The second negative effect is that although respiratory function is normal or near normal in most PP (see table 27.2), it is approximately 60% of normal values in TP (166). The combination of reduced active muscle mass, impaired venous return, neurologically limited maximal heart rate, and decreased respiratory function significantly reduces exercise capacity in TP compared with PP.

Exercise capacity norms have been previously published: For TP, $\dot{V}O_2$peak of <7.6 mL · kg^{-1} · min^{-1} is poor and >16.95 mL · kg^{-1} · min^{-1} is excellent (94); for PP, $\dot{V}O_2$peak of <16.5 mL · kg^{-1} · min^{-1} is poor and >34.4 mL · kg^{-1} · min^{-1} is excellent.

Graded exercise testing can be used to assess aerobic fitness or training effects in asymptomatic or athletic populations, screen individuals at risk for heart disease, determine progress in rehabilitation, demonstrate maximal strength and power capacities, and assist with the exercise prescription (153). Additional benefits to people with SCI include the opportunity to establish a relationship between fitness and post-traumatic return to gainful employment and to determine how the fitness level of a person with SCI changes over time (42).

Field testing is the easiest, least expensive, and most mobility-specific method of evaluation in selected wheelchair users (60). Recent reports, however, have failed to show significant correlations of field testing with actual $\dot{V}O_2$peak, possibly because of variability in terrain, wind speed, temperature, and humidity (172). Arm crank ergometry is the most often used test mode with SCI (37). Wheelchair ergometry is mobility specific for most people with SCI. Several systems have been developed and tested, including wheelchairs mounted on a motorized treadmill (172), low-friction rollers (121), and specialized devices to simulate overground propulsion (35, 69, 113). When compared with arm crank ergometry, wheelchair ergometry results in similar or greater $\dot{V}O_2$peak responses with lower peak power output (70), indicating reduced mechanical efficiency.

Several devices are available to assess all-extremity oxygen consumption (117), which may be appropriate for people with incomplete SCI and for monitoring aerobic fitness of those using combined upper extremity and lower extremity **functional electrical stimulation (FES)**. Inclusion of the lower body muscle pump increases venous return, stroke volume, and cardiac output (129, 140).

Protocols should employ graded increments in resistance or power output requirement with periodic discontinuance for blood pressure, heart rate, and electrocardiogram determination (68, 104). A typical arm ergometry protocol employs an initial resistance of 5 to 25 W, with 5 to 10 W increases every 2 to 3 min to symptom-limited fatigue. People with SCI commonly achieve only 40 to 100 W at peak exercise. Population-specific prediction equations based on heart rate should be used for estimating $\dot{V}O_2$peak (87). Standard test termination rationale should be applied (see chapter 5).

The arm crank or wheelchair ergometer should be adjusted appropriately to allow optimal efficiency and reduce musculoskeletal injuries at the shoulder, elbow, and wrist. Straps applied to the torso improve trunk stability. Wheelchair gloves or flexion mitts with Velcro straps can prevent blisters, lacerations, and abrasions, especially for those with tetraplegia whose hands and fingers are **insensate** or unable to grasp sufficiently. Abdominal binders and leg wraps may improve pulmonary dynamics and venous return, which reduces the risk of hypotension. Conducting testing in a climate-controlled environment will reduce the likelihood of participants overheating because of impaired thermoregulation associated with SCI.

The previously described impairment of the sympathetic innervation of the heart (see table 27.2) has clear implications for exercise testing. Specifically, peak heart rate rarely exceeds 120 $b \cdot min^{-1}$ in those with complete tetraplegia and T1 to T3 paraplegia. Although variable responses occur in T4 to T6 paraplegia, most persons with SCI below T7 are able to reach their age-adjusted peak heart rate. Similar trends are reported for blood pressure responses. In general, $\dot{V}O_2$peak and peak power output are significantly diminished in people with SCI (9, 18, 90, 95, 152, 153). But the lower the injury, the less the impairment. $\dot{V}O_2$peak values range from 12 mL $\cdot$ kg^{-1} $\cdot$ min^{-1} for individuals with tetraplegia to more than 30 mL $\cdot$ kg^{-1} $\cdot$ min^{-1} in persons with low-level paraplegia. In these same groups, peak power output ranges from less than 30 W to more than 100 W, respectively.

People with SCI and a resting systolic blood pressure below 100 mmHg should be closely monitored during exercise because the risk of a hypotensive response increases as peak exercise is approached, despite the use of leg wraps and abdominal binders. Should symptomatic hypotension occur, exercise testing should be halted and the person should be tilted back in the wheelchair to elevate the lower extremities above the level of the heart, promoting venous return.

Standard contraindications for exercise testing noted in chapter 5 should be applied to the SCI population. The following are common contraindications in people with SCI that should be considered absolute reasons not to perform an exercise test:

- Autonomic dysreflexia or the presence of any active condition producing autonomic dysreflexia in a person with an SCI at T6 or above
- Symptomatic hypotension (e.g., dizziness, nausea)
- Unresolved deep vein thrombosis or pulmonary embolism
- Unresolved pressure ulcer

Common relative contraindications include the following:

- Active tendinitis (e.g., rotator cuff, elbow flexors, wrist flexors and extensors)
- Peripheral neuropathy
- Spasticity

Musculoskeletal

In general, testing for muscular strength and endurance for normally innervated muscles in people with SCI is very similar to that in people without SCI. These normal muscles, unaffected by the SCI, can be tested with a 1RM to 10RM (repetition maximum) on resistance machines or free weights, provided the trunk is stabilized, balance is ensured, and excessive axial loading of the spine or weight-bearing skin is avoided. Wingate-type anaerobic power testing with arm ergometry can be performed with younger fit individuals and athletes.

With muscle groups affected by paralysis, manual muscle testing may be necessary to evaluate strength relative to the ability to move the joint through the range of motion with or without gravity-neutral positioning.

Flexibility

In general, testing for joint flexibility in people with SCI is very similar to that in persons without SCI. Joint range of motion can be measured with goniometry by a trained assistant. Joints should be moved passively, with assistance, or actively, but always slowly, gently, and painlessly, especially if spasticity, tightness, or contracture is present. Supporting the long bones will prevent overstress to bone midshafts that are at risk of osteoporosis and osteopenia.

EXERCISE PRESCRIPTION

Overall, exercise prescription for a person with SCI should be based on the principles outlined in chapter 6. Compared with the general population, the quantity of research evaluating the effects of exercise training on health, fitness, and functioning in people with SCI is relatively small, but the findings between the two groups are consistent. In relation to setting exercise guidelines

Practical Application 27.2

CLIENT–CLINICIAN INTERACTION

People with SCI may often need to overcome tremendous physical, emotional, spiritual, and intellectual obstacles to succeed in life. For example, they may need to wake up very early to bathe, groom, perform bowel and bladder care, and dress—all of which may take significantly greater time and effort than for a person without a disability. Seated pressure relief may need to be performed every 15 to 20 min and bodily functions managed according to a timed regimen. Obstacles are everyday occurrences. As a clinical exercise physiologist, you will often find that the person with SCI who consults you is one of the most disciplined, motivated, and enthusiastic clients you will have. You should understand the goals and obstacles of the person with SCI. To do this, you must have direct interaction with the patient. Whenever possible, speak to the person at eye level. Gain an understanding of the individual's daily routine, environmental and transportation barriers that must be overcome, and

concerns they may have about an exercise routine. Provide empathetic listening. Be prepared to discuss the application of exercise benefits to the person's functional abilities. Become familiar with accessible facilities in the community.

You might consider spending a 24 h day without standing. During this time, perform community mobility and seated ADLs (including car, tub, and toilet transfers) from a wheelchair to gain greater appreciation for your SCI client's perspective and needs. This experience can be eye-opening, allowing you a glimpse into the world of someone with SCI, which can be helpful when you are considering how your clients with SCI might incorporate an exercise or physical activity regimen into their daily lives. Remember, most people with SCI are experts in understanding and managing their own bodies—they will usually be able to tell you if there is a problem and how you can help them manage it.

Practical Application 27.3

ESTIMATING OVERWEIGHT AND OBESITY

Resting metabolism in people with SCI is diminished in association with reductions in fat-free body mass (15, 23, 25). Subsequently, energy expenditure during many daily activities is also reduced and, when combined with the generally lower levels of physical activity in the SCI population, there is risk of a net positive energy balance in many clients with SCI (25, 162). A compendium of physical activity energy expenditures for persons with SCI has recently been published, offering additional insights into the weight management issues faced by most individuals in this population (33). Investigations have demonstrated a significant discrepancy between BMI and percent body fat in persons with SCI (23, 95), suggesting that the incidence of overweight and obesity is greatly underestimated (58, 62, 115, 116). Estimates of obesity based on BMI measurement have been shown to be invalid in SCI because height is difficult to measure and obesity tends to be underestimated. However, the **anthropometric index** appears to be a valid proxy measure of obesity in males with SCI, and waist circumference shows

promise, although further research is needed (50).

An investigation is under way to develop a quick clinical tool to assess body composition in people with SCI based on the criterion **four-compartment body composition model**. Since adiposity is now considered one of the major causal factors of the metabolic syndrome (obesity, glucose intolerance or insulin resistance, hypertension, dyslipidemia), a relationship between adiposity and the metabolic syndrome is being sought in the SCI population (63, 64). Several cross-sectional SCI studies have reported that the incidence of metabolic syndrome is at least as high as that in the non-SCI population (63). Most previous exercise research in SCI has focused on physiological improvements in fitness levels, endurance capacity, and strength. The role of energy expenditure in the exercise prescription needs immediate consideration because it may affect overall body composition and health parameters more than changes in aerobic fitness or strength.

for people with SCI, two areas of consistency between people with SCI and the general population are important.

First, the dose–response relationship between exercise and health outcomes is similar. That is, compared with people with SCI who habitually engage in relatively large volumes of exercise, people with SCI who do not exercise are at increased risk of diseases of inactivity and will achieve greater health improvements for a given increase in exercise volume (or dose). To illustrate, if a person with SCI who is completing 20 min of light-intensity aerobic exercise each week increases that by 10 min of light-intensity exercise, the health and fitness benefits they will accrue will be larger than those gained by a person with SCI who is completing 150 min of vigorous-intensity aerobic exercise each week who adds the same amount (i.e., 10 min of light-intensity aerobic exercise). Second, people with SCI who exercise regularly accrue similar health and fitness benefits as people in the general population. Specifically, exercise training positively influences cardiorespiratory fitness, cardiometabolic risk profile, strength, mental health, bone health, and physical functioning in people with SCI.

These consistencies indicate that, in the absence of compelling evidence to the contrary, exercise guidelines for people with SCI should be consistent with those for the general population. This is the approach adopted by reputable international health authorities (26, 170) for the development of exercise recommendations for people with disabilities, as well as two other recently published exercise guidelines for people with SCI (111, 168). Accordingly, the aerobic, strength, and flexibility exercise recommendations presented in table 27.5 are generally consistent with those for the general population but, where available, incorporate SCI-specific evidence.

To apply the guidelines presented in table 27.5 in a clinically appropriate way, it is recommended that practitioners stratify clients with SCI into three broad categories, as follows:

1. *Beginning clients:* People with SCI who are not currently completing the recommended exercise volume for good cardiometabolic health and, based on a conventional and safe rate of exercise progression (145), will be unlikely to reach that exercise volume in the next 3 mo. A disproportionately high number of people with SCI fall into this group (24), and many will have TP and multiple comorbidities. It is well established that the dose–response relationship between exercise and disease risk is curvilinear, such that even modest increases in exercise volume (e.g., 5 min of moderate-intensity activity per day) can meaningfully improve the disease risk profile of beginning clients (141).

2. *Intermediate clients:* People with SCI who are not currently completing the recommended exercise volume for good cardiometabolic health but, based on a conventional and safe rate of exercise progression (145), will be likely to reach that exercise volume in the next 3 mo. Although the exercise recommendations in table 27.5 are somewhat arbitrary, strong scientific evidence substantiates that this volume of exercise will confer a broad spectrum of significant health benefits (111, 141, 168).

3. *Advanced clients:* People with SCI who are currently meeting or exceeding the recommended exercise volume for good health. Although the number of clients in this group is small, it is likely that clinical exercise physiologists will have contact with these clients because they are more likely to seek out expertise. Evidence from the general population indicates that when exercise volumes reach extremely high levels (e.g., Ironman competitors), additional increases in exercise volume can increase health risk, including the risk of musculoskeletal injury (141). Given the high incidence of overuse-related upper limb pain and dysfunction in people with SCI, practitioners working with clients in this stratum must be aware of clinical practice guidelines for preservation and optimization of their upper limb function (34).

Cardiorespiratory Exercise Training

The most appropriate method for monitoring and prescribing aerobic exercise intensity in people with SCI is complicated by the fact that heart rate responses in TP and high-level PP (above T6) are lower than in people without SCI because of reduced active muscle mass and impaired cardiac sympathetic response drive (166). Although variable, 30% to 80% of heart rate reserve appears to correspond to 50% to 85% of $\dot{V}O_2$peak in those with high-level paraplegia and tetraplegia (95, 146). An alternative is to calculate aerobic training intensity as a percentage of peak power output (146). Practitioners who use this method should be aware of the need for regular reevaluation of peak power output because it increases with training.

Rating of perceived exertion (RPE) may also be used to guide exercise training intensity (72). Evidence indicates that RPE can provide the clinical exercise physiologist with a reasonably valid indication of moderate- and vigorous-intensity exercise (72). RPE corresponding to moderate and vigorous intensity is presented in table 27.5, although it should be noted that like many studies in the SCI population, the evidence is preliminary. Another possible method is the talk test. Nondisabled people able to speak during exercise without feeling short of breath are typically performing moderate-intensity exercise; however, this method has not been validated in the SCI population or other groups with sympathetic nervous system dysfunction (4).

As indicated in table 27.5, the frequency and duration of aerobic exercise sessions are inversely related to the intensity of the exercise performed. Specifically, if exercise intensity is moderate, then people with SCI who are categorized as intermediate clients should aim to complete a minimum of five 30 min sessions (150 min) per week; but if exercise intensity is vigorous, then 3 × 25 min (75 min) is an appropriate minimum target (26, 170). People with SCI who are categorized as beginning clients may have $\dot{V}O_2$peak values less than 10.5 mL · kg^{-1} · min^{-1} (<3 METs) and should simply aim to progressively increase moderate-intensity exercise volume. They can start with as little as 5 min of continuous moderate-intensity activity per day, progressing until they are able to tolerate 20 to 30 min sessions.

Table 27.5 Exercise Prescription Guidelines

Training method	Frequency	Intensity	Time (Duration)	Type (Mode)	Progression	Important considerations
Cardiorespiratory	≥5 d/wk if intensity is moderate *or* ≥3 d/wk if intensity is vigorous *or* 3-5 d/wk if doing combined moderate and vigorous	Moderate intensity (3-5.9 METs; RPE 11-13; 40%-59% HR reserve) *or* Vigorous intensity (6-8.9 METs; RPE 14-15; 60%-89% HR reserve	≥30 min if intensity is moderate *or* ≥25 min if intensity is vigorous	Exercise modes involving rhythmic contraction and relaxation of the largest available muscle groups (e.g., arm crank or wheelchair ergometry, hand cycling, community wheeling, seated aerobics, aquatics, wheelchair sports, FES, arm or leg cycling ergometry)	Slow (<5% per week)	Initially monitor for signs of exertional hypotension, autonomic dysreflexia, and thermal stress. Beginner level may initially require short durations (e.g., 5 min). Include warm-up and cool-down.
Resistance	≥2 d/wk	3 sets of 8-12 reps with 2-3 min recovery between each set; moderate intensity (60%-70% 1RM or RPE 12-13)	30-60 min	Free weights, wheelchair-accessible machines (pin weighted or hydraulic resistance), elastic bands or tubing, wrist and ankle weights, body weight; major muscle groups and 4 or 5 upper limb exercises	Increased resistance when 12 reps are achieved	Innervation of scapula stabilizers and posterior shoulder girdle is normal in people with PP and progressively decreases with higher lesion level in TP. Each exercise should be pain free. Where possible, ensure agonists and antagonists are in balance. Avoid internal shoulder rotation when in ≥90° abduction to limit impingement. Avoid Valsalva maneuver.
Range of motion	≥2 d/wk for each of the major muscle groups	Static stretches held for 10-30 s, with 60 s of total stretching time for each flexibility exercise (e.g., 2 × 30 s or 4 × 15 s); stretch to the point of feeling tightness or slight discomfort (active or passive)	15-20 min	Static stretching of the major muscle groups (neck, upper limbs, trunk, and lower limbs); may be assisted or unassisted; standing in standing frame if medically cleared	As tolerated	Stretching should not elicit joint pain. Assisted stretching (with a partner) is superior but less independent than unassisted stretching. Assisted passive stretching of insensate muscles should be undertaken with caution and with real-time verbal feedback from the client. For manual wheelchair users, focus areas for stretching should be internal shoulder rotators, chest, and anterior shoulders. To limit impingement, avoid internal shoulder rotation when completing an overhead range of motion.

Note: These volumes of aerobic, strength, and flexibility exercise are required in order for people with SCI to achieve good cardiometabolic health, physical fitness, and functioning.

METs = metabolic equivalents of task; RPE = rating of perceived exertion (6-20 scale); HR = heart rate; FES = functional electrical stimulation; RM = repetition maximum.

Based on Tweedy et al. (168); Bull et al. (26); Kressler et al. (111); Consortium of Spinal Cord Medicine (34).

Research Focus

Strength Responses in Very Weak Muscles

SCI results in complete or partial disruption of descending motor pathways, resulting in, respectively, paralysis or weakness in muscles innovated below the level of the injury. Progressive resistance training appears to be effective for improving strength in stronger muscles—those muscles able to actively move through full range of movement against gravity (muscle grade 3-5). However, progressive resistance training can be difficult when muscles are very weak. A common alternative is to do a large number of repeated contractions, either isometrically or through the available angular range but without application of external resistance. This study aimed to evaluate whether repeated contractions without external resistance improve the strength of very weak muscles (i.e., muscle grade 1 and 2).

This multicenter study (seven hospitals in Asia and Australia) employed a single-blind randomized controlled trial design. One hundred and twenty people with recent SCI undergoing inpatient rehabilitation were randomized to either a treatment or control group. One major muscle group from an upper or lower limb was selected if the muscle had grade 1 or grade 2 strength on a standard 6-point manual muscle test. Participants in the treatment group performed 10,000 isolated contractions of the selected muscle group, as well as usual care, in 48 sessions over 8 wk. Participants in the control group received usual care alone. Participants were assessed at baseline and at 8 wk by a blinded assessor. The primary outcome was voluntary muscle strength on a 13-point manual muscle test. There were three secondary outcomes capturing therapists' and participants' perceptions of strength and function.

Both the treatment and control groups got stronger. However, the mean between-group difference in voluntary strength at 8 wk was 0.4/13 points (95% confidence interval (CI): −0.5 to 1.4) in favor of the treatment group. There were no notable between-group differences on any secondary outcomes.

Ten thousand isolated contractions of very weak muscles in people with SCI over 8 wk has either no effect on voluntary strength or a very small one. Implications for clinical exercise physiologists are as follows:

- This evidence does not mean that clinical exercise physiologists should not try to strengthen very weak muscles, but the fact that this intervention yielded either no effect or a very small effect indicates that very weak muscles may respond differently to strength training than stronger muscles.

- The strength training employed was repeated contractions—it was not progressive and it did not involve resistance, both of which are common features of strength training for stronger muscles.

- In this study, very weak muscles were classified as grade 1 and 2, but there is an important difference between these grades—maximum contraction of a grade 1 muscle does not lead to movement of a limb segment (only a muscle flicker), whereas maximum contraction of a grade 2 muscle does lead to voluntary movement. Therefore, practitioners have the option of using progressive resistance training for grade 2 muscles but not grade 1.

- Clinical exercise physiologists should be aware of the challenges around trying to strengthen very weak muscles and inform their clients of the relevant implications.

Based on Chen et al. (31).

Resistance Exercise

Exercises to strengthen the scapula stabilizers (trapezius and rhomboids), rotator cuff muscles, pectoralis major, and latissimus dorsi are recommended for all people with SCI capable of voluntary control of these muscles (34). Initial intervention should include two sets of 10 repetitions, with 6 s isometric contractions for shoulder protractors, retractors, elevators, and depressors, as well as for internal and external shoulder rotators. Internal rotation of the shoulder when exercising above shoulder level should be avoided in order to limit impingement (34). As the patient tolerates, progression to dynamic exercise should occur. Resistance band exercises are useful initially, but a plateau in gains can be expected because of the limitation in resistance of this device. Although dumbbells and free weights may be used under close supervision, paralyzed lower extremities and truncal musculature significantly reduce a person's ability to balance even small objects when lying supine or when seated without significant truncal support. When the person is using free weights or isotonic or isokinetic machines, wheelchair brakes should be set before lifting, and care should be taken not to exceed the weight and stress limitations of the wheelchair as provided by the manufacturer.

Standard recommendations for intensity and progression should be followed as discussed in chapter 6. SCI Action Canada recommends strength training exercises two times a week (e.g., free weights, elastic resistance bands, cable pulleys, weight machines, and FES exercise), consisting of three sets of 8 to 10 repetitions of each exercise for each major muscle group (66, 154). The ESSA guidelines recommend that strengthening exercises be performed at a moderate intensity (60%-70% 1RM or 12-13 RPE) (168). Velcro straps and cuffed weights are commonly used to modify resistance training equipment.

Range of Motion Exercise

Range of motion exercise should be performed daily and should focus on all major joints, especially those with contracture and spasticity. Of particular importance is maintaining shoulder range of motion, especially external rotation of the humerus and retraction and upward rotation of the scapula, because loss of range may increase the risk of an upper limb injury and lead to functional limitations (6). Both active and passive assisted methods of static stretching can be used. During passive stretching, carefully working through the range of motion in joints lacking sensation is important because the individual cannot determine when the maximal range of the joint has been reached. Supporting the midshaft of long bones in the patient who has osteopenia may be important to reduce the chance of fracture. If the person is medically cleared for full weight bearing, a standing frame can be used to maintain or increase range of motion of the spinal extensors, hip and knee flexors, and ankle plantar flexors.

In the absence of voluntary grasp, some people with TP may use a tenodesis grip, which is initiated by active wrist extension. This induces passive flexion of the fingers and thumb to produce a functional grip (figure 27.4) (78). The strength and effectiveness of the grip are mainly determined by the passive length–tension characteristics of the extrinsic finger and thumb flexors. Before exercise participation, practitioners should determine whether clients with TP use a tenodesis grip because its effectiveness will be reduced if the finger and thumb muscles are regularly stretched (4, 78).

Special Exercise Considerations

The clinical exercise physiologist is well placed to lead the exercise testing and prescription for community-dwelling people with SCI. However, as indicated earlier, SCI is associated with a range of comorbidities, and consequently a person with SCI might be under the care of a range of other health care professionals. Under these circumstances, consultation with the client's primary health care provider has already been recommended (see the History section), but depending on the issues being managed, the exercise physiologist may need to adjust exercise prescription and programming to accommodate concurrent treatment received from other medical specialists, physiotherapists, occupational therapists, psychologists, social workers, or nurses.

Impairments of Body Systems and Structures

At this point in the assessment, the clinical exercise physiologist will have taken an accurate history and identified the primary clinical concerns for their client. People with SCI are at increased risk of coronary artery disease, obesity, type 2 diabetes mellitus, and other diseases related to insufficient physical activity. A client with SCI who is being medically managed for any of these conditions should be medically screened before starting an exercise program. Other impairments that might affect people with SCI and that are most relevant to the clinical exercise physiologist are presented in the Clinical Considerations section, which also presents an overview of the implications for exercise prescription. These include autonomic dysreflexia, hypotension (orthostatic or exertional), impaired thermoregulation, pressure ulcers and other sensation-related injury risk, pain, spasticity, syringomyelia, osteoporosis, and neurogenic bowel and bladder dysfunction.

Adapted or adaptable equipment, some of which is outlined in the section on exercise testing, is required for safe, effective exercise. Other recommendations include the use of upper armbands or Coban tape to prevent abrasions at the medial upper arm with wheelchair propulsion. Abdominal binders and leg wraps may be used to facilitate improved pulmonary dynamics and greater venous return. Velcro straps and cuffed weights are commonly used for resistance training to improve or create a grip.

Figure 27.4 Tenodesis grip: *(a)* open and *(b)* closed.

Photo courtesy of Sean Tweedy.

Participation Restrictions, Personal Factors, and Environmental Factors

SCI has a profound effect on all the domains of the ICF (see figure 27.3). This section provides the clinical exercise physiologist with a process for systematically assessing the participation restrictions that affect their client, together with those aspects of the personal and environmental domains that will influence exercise adoption and maintenance.

The process is based around the Personal and Environmental Exercise Assessment Tool, found in this chapter's online supplement, which includes components for assessment of the client's community mobility, community environment, home environment, transportation, and living arrangements, including the availability of personal support workers. Each element is critical for prescribing sustainable self-managed exercise for people with SCI. The process uses two modes for gathering information:

- *Tasks to complete:* activities the participant completes and the practitioner observes

- *Questions to ask:* questions the practitioner asks the participant (practitioner records responses)

The practitioner will reflect on their observations and on the participant's responses in order to make judgments and develop an intervention. The assessment is best conducted on location—at the client's home, if possible. If not, a remote assessment of the person's home or community exercise environment can be conducted with the client using a smart phone with a video conferencing app to transmit real-time footage.

Completion of the Personal and Environmental Exercise Assessment Tool will give the exercise physiologist the information required to develop a holistic exercise intervention that is most likely to the safe, effective, and sustainable. A study evaluated behavioral interventions aimed at increasing leisure-time physical activity by community-dwelling adults with SCI at community fitness centers (44). Implementation of this program involved three stages: training exercise professionals on safety and adapted physical activity for people with SCI through a Train the Trainers Spinal Cord Injury (T3-SCI) course; delivering customized physical activity intervention for people with SCI based on their goals, physical strengths, and personality attributes; and performing baseline and three follow-up assessments over 9 mo. Significant improvements in leisure-time physical activity participation and health-related outcomes were observed, especially among those who were the most inactive at baseline. Thus, this program is a valid model for delivering effective physical activity interventions for people with spinal cord injury.

In the general population, devices that provide easily accessible measures of physically active behavior—time and distance of movement, steps taken—can not only promote self-managed exercise but also give exercise physiologists a ready means of evaluating whether their clients are adhering to the self-managed components of an exercise intervention (2). Unfortunately, relatively few options are available for wheelchair users (135). Currently, cycling computers (e.g., the Cateye) are the only commercially available wheelchair-mounted devices configured by the manufacturer to provide measures of wheelchair movement distance and speed (99). A study demonstrated that the Cateye provided a valid estimate of speed and distance for activities typical of wheelchair-based aerobic exercise sessions (e.g., going for a push on a bike path). However, distance and speed estimates were not valid for other activities more typical of everyday, nonexercise wheelchair use, such as when speeds were ≤3 kph or activities comprised maneuvering, stopping, and starting (99). To quantify free-living, nonexercise wheelchair use, the Apple Watch has been shown to provide an acceptable estimate of push counts and could be used in the same way a pedometer is used for the general population (98).

EXERCISE TRAINING

It has already been highlighted that the SCI population is inherently heterogeneous as a result of variations in lesion level and lesion completeness and the presence or absence of comorbidities. This heterogeneity, together with low prevalence in society (relative to the general population), acts as a barrier to recruitment of the sufficiently large, homogeneous samples required for rigorous scientific research and generation of high-quality evidence on the benefits of exercise for people with SCI (158). There are two exercise training adaptations for which there is strong, consistent evidence in the SCI population—aerobic exercise leads to improved cardiorespiratory fitness, and resistance training leads to improved strength in unimpaired muscle groups (66). However, there is also a range of exercise benefits for which there is emerging evidence, which is a broad term referring to evidence that has been published in peer-reviewed literature and is either supported by a plausible mechanism of action or is consistent with findings from exercise training studies in the general population, but that may not have been replicated or may not have used optimal research design (e.g., heterogeneous samples, cross-sectional studies, or single group designs). From a clinical perspective, benefits supported by emerging evidence are likely, but further research is required. Table 27.6 presents the general adaptations that can be expected from exercise training in those with SCI.

Cardiorespiratory Fitness

The general pattern of acute cardiovascular responses in people with SCI during upright exercise is hypokinetic circulation, or circulatory hypokinesis (42, 54, 84, 90), characterized by a lower increase in cardiac output per unit increase in $\dot{V}O_2$ during exercise. This is achieved by a relatively high heart rate and relatively limited increases in cardiac contractility, stroke volume, and arterial blood pressure and decreases in peripheral vascular resistance.

Evidence indicates that cardiorespiratory fitness of people with both TP and PP improves in response to upper limb aerobic training and circuit training (57, 83, 92, 110). Resistance training

Table 27.6 Exercise Training Review

Cardiorespiratory fitness	Skeletal muscle strength	Skeletal muscle endurance	Flexibility	Body composition
Increased exercise tolerance; increased $\dot{V}O_2$peak; increased peak work rate; increased cardiac output; increased stroke volume; decreased submaximal exercise heart rate; decreased total peripheral vascular resistance; potential for increased daily energy expenditure	Increased strength; increased peak and mean anaerobic power output; increased functional mobility	Increased fatigue resistance; increased functional mobility	Maintenance of joint flexibility, in absence of contracture; temporary decrease in spasticity	Decreased % body fat; increased % lean body mass; increased energy expenditure; potential for treatment of obesity, especially with diet modification

also improves cardiorespiratory fitness in PP (91). In a review of 14 exercise training studies, $\dot{V}O_2$peak improved by a mean of 17.6 ± 11.2% and PO by 26.1 ± 15.6%, with a trend for greater gains in PP than TP (57).

Critical review of upper extremity aerobic conditioning studies demonstrates variability in exercise prescription and results, partially attributable to the level of SCI (115). For instance, Gass and colleagues (62) trained seven subjects (four with C5-C6 lesions, three with T1-T4 lesions) five times weekly to exhaustion on a graded exercise test protocol using a wheelchair on a treadmill ergometer. After 7 wk of training with this ergometry system, the mean $\dot{V}O_2$peak had increased from 9.5 to 12.7 mL · kg⁻¹ · min⁻¹ and endurance time had increased by 4.4 min, suggesting considerable change in functional capacity (63). Knutsson and colleagues (104) evaluated 20 SCI inpatients with complete and incomplete SCI between C5 and L1. For the 10 persons assigned to the training group, a 40% increase in peak work rate (40-57 W) was reported, although no significant change was noted in the three subjects with SCI above T6.

Taylor and colleagues (164) assessed the effects of arm crank ergometry (ACE) performed 30 min/d, 5 d/wk for 8 wk, in 10 individuals with paraplegia. The trained group significantly improved $\dot{V}O_2$peak from 22.8 to 26.3 mL · kg⁻¹ · min⁻¹ without significant changes in maximal heart rate, postexercise lactate, or body fat. Davis and Shephard (43) assessed four patterns of ACE training (50% or 70% $\dot{V}O_2$peak, 20 or 40 min per session, three sessions per week, 24 wk) in inactive subjects with paraplegia. Cardiac stroke volume during submaximal exercise and $\dot{V}O_2$peak increased significantly during ACE tests, except in the nonexercise control group and the group with the lowest training intensity and duration.

McLean and Skinner (124) matched 14 subjects with tetraplegia, by peak power output, to either a supine or seated exercise training regimen to assess for the effect of changes in postural position on stroke volume, cardiac output, and exercise capacity. Their subjects performed arm crank ergometry exercise in either a sitting or a supine position at 60% of their $\dot{V}O_2$peak three times

a week for 10 wk, with progressive increments in either duration or resistance; no significant differences were found in stroke volume or cardiac output, although absolute $\dot{V}O_2$peak increased from 720 to 780 mL · min⁻¹, suggesting peripheral adaptations.

Few studies have explored central cardiovascular adaptations to training, such as increased peak stroke volume, cardiac contractility, or cardiac output, in people with TP. Such observed adaptations may well result from at least two mechanisms: increased active muscle mass that exhausts a larger fraction of the cardiac reserve, hence producing higher peak cardiovascular responses, and external lower body compression or supine posture that reduces the lower body venous pool and improves venous return and (via baroreflex) reduces heart rate and increases stroke volume and cardiac output. Overall, training-induced increases in $\dot{V}O_2$peak may not be caused by central cardiovascular limitations but by increased active muscle, venous return, and peripheral O_2 extraction in trained muscles (54).

Muscular Strength and Endurance: Nonparalyzed Muscle Groups

Clinically significant changes in the strength of nonparalyzed muscle groups—e.g., upper limb muscle groups in PP—can be achieved through resistance training and circuit training (57, 83, 91, 110). Because of the large number of shoulder and upper extremity musculoskeletal problems encountered by persons with SCI, a prophylactic, structured, and progressively resistive strengthening program that focuses on scapular, rotator cuff, and pectoral muscles is likely to increase strength and reduce the risk for overuse injury (55). Such a program will probably improve the ability of these people to perform functional tasks in the community.

Increasing information is available on the effects of resistance training on strength, power, muscle mass, and functional abilities in persons with SCI. Clinical studies have recommended resistance training for many years. Nilsson and colleagues (136) reported increases in dynamic strength (16%) and endurance (80%) when comparing bench press before and after a 7 wk combined arm crank ergometry and resistance training program in adults with paraplegia. Chawla and colleagues (30) reported that a resistance training program including bench press, incline press, lateral raises, incline curls, lat pulls, and triceps stretch improved ADL function in 10 people with SCI. Unfortunately, specific ADLs and quantitative measures of strength were not reported. From these limited studies, it appears promising that specific strength training will benefit people with SCI. Increasing information from randomized controlled trials (RCTs) is available on the effects of resistance training on strength, power, muscle mass, and muscle morphology.

Five research groups have used RCTs to test the efficacy of resistance training to improve muscular strength and endurance (71, 77, 82, 88, 134). Needham-Shropshire and colleagues (134) found more significant increases in upper extremity (e.g., triceps) muscle strength (by manual muscle testing) after 8 wk of FES-assisted arm ergometry than after either voluntary arm ergometry or FES-assisted arm ergometry plus voluntary arm ergometry. Hicks and colleagues (82) trained 11 participants with SCI twice weekly for 90 to 120 min per session over 9 mo using progressive resistance exercise and arm ergometry. Participants significantly improved 1RM upper body muscle strength by 19% to 34% and submaximal arm ergometry power output by 81%. Jacobs (88) trained two groups of nine clients with paraplegia over 12 wk, one with arm ergometry and the other with resistance circuit training (three sessions per week, six exercises, three sets of 10 reps per session, 60%-70% of 1RM intensity). Muscular strength significantly increased for all exercises in the resistance training group, with no changes in the arm ergometry group. Mean arm crank anaerobic power increased in both groups by 5% to 8%. Peak arm crank anaerobic power increased significantly by 16% in the resistance training group and 3% in the arm ergometry group.

Muscular Strength and Endurance: Partially Paralyzed Muscles

In relation to strength exercise, current best practice for increasing strength in partially paralyzed muscles adapts established strength training principles, typically starting with unresisted gravity-eliminated movements, progressing to gravity-opposed movements, and finally to gravity-opposed movements with resistance (78). Evidence has demonstrated that circuit training elicits strength gains in people with TP (110). However emerging evidence indicates that nonparalyzed and partially paralyzed muscles (resulting from, for example, incomplete SCI or muscle groups with partial segmental innervation) may respond differently to resistance training.

Chen and colleagues conducted an RCT in which they evaluated the effects of 10,000 voluntary contractions over 8 wk on the strength of very weak muscles in people with spinal cord injury (30). In this study, 120 people with recent SCI undergoing inpatient rehabilitation were randomized to either a control group (usual care) or a treatment group (usual care plus 10,000 isolated contractions of a muscle group with a muscle grade of 1 or 2 on a standard 6-point manual muscle test). The mean between-group difference in voluntary strength at 8 wk was 0.4/13 points, and the authors concluded that 10,000 isolated contractions of very weak muscles in people with SCI over 8 wk has either no effect on voluntary strength or a very small one.

Glinsky and colleagues (71) used progressive resistance exercise to train 32 participants with C5 or C6 tetraplegia during three sessions a week for 8 wk. They showed no clinically significant increases in strength and endurance (8% and 11%, respectively) in paretic wrist extensor and flexor muscle groups. Hartkopp and colleagues (77) used high- and low-resistance FES exercise to train paretic wrist extensors (30 min per session, five sessions per week, 12 wk) of 12 participants with C5 or C6 tetraplegia. Compared with the untrained arm, the trained arm increased strength significantly with the high-resistance protocol but not the low-resistance protocol. Muscular endurance increased with both protocols by 41% and 42%, respectively. Although paradigms varied greatly across trials, most studies report increases in muscular strength and endurance compared with values in control groups. A review, however, found inconclusive evidence of the effectiveness of strength training for partially paralyzed muscles (79).

Evidence is not strong enough to discontinue current best practice, but practitioners should be aware of its possible limitations, communicate with their clients, and stay abreast of research developments.

Range of Motion

Contractures are a common complication of SCI, characterized by marked limitations in passive ROM of affected joints (100). Contracture has a significant impact on ADLs and positioning for exercise, and it is associated with pressure ulcers, sleep disturbances, and pain (100). Stretching and passive movements are widely used for the treatment and prevention of contractures; however, the results from clinical trials have raised doubt about the effectiveness of these interventions, especially when administered for only short periods (100).

Katalinic and colleagues (100) completed a meta-analysis of 35 studies involving 1,391 participants on the efficacy of stretching for the purpose of treating or preventing contractures. In people with neurological conditions, moderate- to high-quality evidence indicated that stretching does not have clinically important immediate (mean difference 3°; 95% CI: 0 to 7), short-term (mean difference 1°; 95% CI: 0 to 3), or long-term (mean difference 0°; 95% CI: −2 to 2) effects on joint mobility. No study performed stretching for more than 7 mo. The results were similar for people with non-neurological conditions. The authors concluded that for all conditions, there is little or no effect of stretching on pain, spasticity, activity limitation, participation restriction, or quality of life. A review concluded that stretching to improve range of motion is ineffective and the evidence for passive movements inconclusive. However, since the level of evidence for stretching is only moderate, further studies should explore the dosage of these interventions (79). The value of stretching may be limited to maintenance of joint range of motion and prevention of contractures.

Other interventions to improve range of motion such as splinting, continuous passive motion, standing in a standing frame, and neuromuscular facilitation techniques have not been as rigorously evaluated. Many fitness and clinically based stretching protocols are available online. Given the severe consequences of contractures, exercise interventions should continue to incorporate stretching and passive movement as described in the Exercise Prescription section, but practitioners should be aware that there is considerable uncertainty about the effectiveness of these exercises, communicate with their clients, and stay abreast of developments in this area.

Other Adaptations

An RCT in people with SCI showed that voluntary upper body aerobic exercise performed at a moderate to vigorous intensity for 20 to 30 min/d, 3 d/wk for at least 8 wk, positively affects many risk factors related to cardiovascular disease, including improved exercise tolerance, $\dot{V}O_2$peak, and power output; reduced LDL, total cholesterol, and serum triglycerides; and enhanced insulin sensitivity and glucose tolerance. The level of research evidence was also strong for improved cardiac output and stroke volume, reduced peripheral vascular resistance, and increased HDL cholesterol. Submaximal and maximal heart rate, as well as cellular adaptations for oxidative metabolism, also likely improve, although the evidence is not as strong. Another RCT exploring the effects of a 6 wk (60 min, three times per week) hand bike exercise program on health and fitness levels found that participants with SCI improved body composition, fasting insulin, and insulin resistance (102).

Lending support to the veracity of these findings from RCTs, cross-sectional evidence indicates that the cardiometabolic risk profile of people with SCI who regularly perform aerobic exercise is significantly better than in those who do not: BMI and percentage fat mass are lower, total daily energy expenditure is higher, and both lipid profile and glucose homeostasis are better (105, 163, 176). Aerobic exercise interventions confer clinically important improvements in lipid lipoprotein and glucose homoeostasis (110). Exercise interventions can augment weight reduction induced by caloric restriction (110), but to date, exercise interventions alone have not been shown to improve body composition or reduce weight in people with SCI (49, 83). Similarly, there is a lack of evidence on the benefits of upper limb resistance training for body composition (i.e., decreasing fat mass) (59).

In people with TP, reduced respiratory function limits exercise capacity and increases the risk of pulmonary infection. Specific training techniques for inspiratory (127, 181) and expiratory muscles (151) can improve respiratory function and may improve peak exercise response (181). Some evidence shows that sufficiently intense aerobic exercise training (>70% maximal heart rate) can improve elements of respiratory function (149).

There is evidence that exercise interventions are an effective means of enhancing functional independence and ADLs in people with SCI (49, 157). Furthermore, higher cardiorespiratory fitness is strongly correlated with increased participation in ADLs such as cleaning and wheeling (80).

The rate of depression among people with SCI is four times that of the general population (109), and health-related quality of life (HRQOL) is significantly lower (119). Exercise interventions can decrease depression (67, 82) and increase quality of life (67, 82) in people with SCI. Moreover, evidence indicates that people with SCI who exercise regularly will have greater life satisfaction than those who do not (160).

SCI is associated with sublesional osteopenia and osteoporosis, leading to bone fragility and increased fracture risk (65). Cross-sectional studies have demonstrated that physically active individuals with TP (28) and PP (97, 125) have significantly better sublesional and general bone health than their inactive counterparts, suggesting that regular exercise may maintain bone health.

Upper limb pain of musculoskeletal origin affects 50% to 70% of people with SCI and has a significant adverse effect on psychological and physical functioning (165). Some believe the pain is primarily due to constant use of the arms for transfers and wheelchair propulsion, although this theory does not explain why nonathletes are twice as likely to develop shoulder pain than athletes (61). Furthermore, higher muscular strength is associated with less upper limb pain in both TP and PP (49), and appropriate strength and flexibility training (e.g., strengthening scapula stabilizers and rotator cuff) has been shown to reduce shoulder pain (132, 165, 171) in TP and PP.

In an RCT, Mulroy and colleagues (128) compared the efficacy of an exercise movement optimization program with an attention control intervention for decreasing shoulder pain in people with paraplegia from SCI. The 12 wk home-based intervention consisted of shoulder strengthening and stretching exercises, along with recommendations on how to optimize the movement technique of transfers, raises, and wheelchair propulsion. Shoulder pain, as measured with the Wheelchair User's Shoulder Pain Index, significantly decreased to one-third of baseline levels after the intervention in the exercise movement optimization group but remained unchanged in the attention control group. Peak torques in scapular elevation and in shoulder adduction and internal and external rotation improved in the exercise movement optimization group (18%-32%) but not in the attention control group. Improvements were maintained at the 4 wk follow-up assessment. Self-reported physical activity did not change in either group.

In addition to musculoskeletal pain, 65% to 85% of people with SCI are affected by neuropathic pain, which can arise from peripheral, spinal, or cerebral mechanisms (155). An RCT of a combined aerobic and resistance training exercise program (60-90 min per session, two sessions per week for 9 mo) found that, compared with a control group, the intervention group significantly reduced pain and depression (46, 67).

Locomotor Training

There are three main modalities for locomotor training: overground, **partial body-weight-supported treadmill training (PBWSTT)**, and robotic-assisted training. Although most studies show these modalities can greatly improve walking ability after SCI (7, 8, 16, 17, 45), it is not clear which is superior. When therapists provide body weight support and manual assistance to SCI subjects during treadmill stepping, task-specific motor learning occurs and gait control and mechanics improve. An observational study of people with chronic incomplete SCI at one center indicated that approximately 80% of wheelchair-reliant individuals became independent walkers after locomotor training (179). Dobkin and colleagues (47) conducted a multicenter clinical trial for subjects with acute incomplete SCI to compare the effect on walking-related outcomes of PBWSTT and overground walking training. Although improvements were noted, PBWSTT was not significantly better than overground walking training in the acute setting, likely because of neural recovery often seen in the acute phase. It has been suggested that future studies could utilize comprehensive gait assessment methods to differentiate changes in gait patterns due to recovery from those due to changes in speed (5).

Despite its possible benefits, only a handful of clinics in the United States offer manually assisted locomotor training with partial body weight support for gait rehabilitation after SCI, presumably because of the high cost of therapist labor and equipment. Reports on overall health parameters including orthostatic instability, aerobic fitness, lipid profiles, body composition changes, insulin sensitivity and glucose tolerance, and lower extremity bone mineral density are promising but sparse to date (76, 81, 131, 139, 175).

Functional Electrical Stimulation

In 1987, medical guidelines were developed for patient participation in FES rehabilitation, including medical criteria for inclusion and exclusion. Computerized FES is a neuromuscular aid used to restore function in upper extremity, lower extremity, and truncal muscles paralyzed by upper motor neuron lesions as well as to restore bladder and respiratory function and prevent pressure ulcers (68, 85). Briefly, FES of the lower extremities can be used to do the following:

- Stimulate skeletal muscle strength (52, 68, 124, 143, 148) and endurance (62, 68, 87, 124)
- Increase energy expenditure and stroke volume (70, 87, 89)
- Increase total body peak power, $\dot{V}O_2$peak, and ventilatory rate (10, 70, 87, 124)
- Reverse myocardial disuse atrophy (130)
- Increase high-density lipoprotein levels and improve body composition (11)
- Improve self-perception (159)
- Increase lower extremity bone mineral density (21, 75, 126)

Despite these encouraging findings, training must be maintained to retain the benefits. In a study by Hooker and colleagues (86), the 45% improvement in $\dot{V}O_2$peak attained by 12 wk of FES plus a leg ergometry training program was reduced by approximately 50% after 8 wk of inactivity, whereas power output returned to pretraining levels. Submaximal heart rate, peak heart rate, and peak ventilatory volume did not change at any time point. It is not yet possible to establish recommendations for FES dosage because the protocols vary greatly across studies. However, it seems clear that long-term interventions yield longer-term benefits for overall health (81, 126).

A logical and intuitive progression in the development of FES lower extremity exercise training has been the combined use of concurrent arm crank ergometry and FES leg cycle ergometry (LCE), termed hybrid exercise. As expected from the combination of upper and lower extremity exercise, peak power, $\dot{V}O_2$peak, stroke volume, and cardiac output significantly increase during hybrid exercise bouts with SCI subjects (56, 129) and during combined upper extremity rowing plus lower extremity FES (114).

Clinical Exercise Bottom Line

- The degree of impairment associated with a spinal cord injury is dependent on the level at which the spinal cord is affected and the completeness of the injury.
- Exercise training can improve all aspects of physical fitness in patients with a spinal cord injury, with a goal of maintaining independence and reducing chronic disease risk.
- Since the legs are always affected, there is a strong focus on upper body aerobic and resistance training.
- Range of motion exercise is necessary for all parts of the body, including those with partial or complete loss of control and feeling, to avoid contracture and reduce spasticity.
- The autonomic nervous system may be disrupted, which can lead to adverse effects on heart rate, blood pressure, temperature control, and breathing.

CONCLUSION

SCI profoundly affects human functioning in all major domains of the ICF—body systems and structures, activity, and participation. The primary effects of SCI and the associated comorbidities are amplified by extremely low levels of physical activity in the SCI population. Exercise interventions have been shown to facilitate a range of positive adaptations—primarily improved cardiorespiratory fitness and improved muscular strength, but other SCI-specific benefits as well—and as time since injury increases, exercise interventions become an increasingly important and effective means of achieving optimal long-term health and well-being for people with SCI. Best practice clinical exercise physiology requires a solid grounding in the principles of exercise testing and prescription, together with systematic assessment of the impairments, activity limitations, and participation restrictions experienced by the presenting client.

Online Materials

Visit HK*Propel* to access a link to the references, the case studies with discussion questions, and a quiz to help you review key concepts and test your understanding of the material covered in this chapter.

CHAPTER 28

Multiple Sclerosis

Ulrik Dalgas, PhD

Lars G. Hvid, PhD

Individuals with multiple sclerosis (MS) have varying degrees of disability. The amount of disability may remain stable for years or even for the rest of one's life. But in some cases, disability may gradually or suddenly worsen and profoundly influence daily living. Exercise is an important aspect of treatment for those with MS. This chapter provides information about using exercise as part of the treatment in people with this disease.

DEFINITION

Multiple sclerosis (MS) is an inflammatory autoimmune disease of the central nervous system. The etiology is unknown, but MS is characterized by nerve **demyelination**. This process may result in multiple plaques, or scleroses (singular: **sclerosis**), in the white matter of the brain and spinal cord (170). These plaques can develop into permanent scars (162) that impair nerve transmission (73, 80, 181) and lead to an array of symptoms, such as muscle weakness, fatigue, and motor function difficulties (28).

SCOPE

According to a 2020 survey conducted by the Multiple Sclerosis International Federation, an estimated 2.8 million people live with MS worldwide. Based on the Atlas of MS 2020 survey, the prevalence of MS is higher in North America, Europe, and Australia; medium in Asia; and lower in Central and South America and Africa. However, there are variations in prevalence within regions (109). Diagnosis of MS generally occurs between the ages of 20 and 50, although young children and older adults also may receive this diagnosis (112). Women appear to be affected at nearly twice the rate as men and make up more than 60% of those diagnosed (40, 75). Of note, 80% of all persons with MS live with the disease for more than 35 yr, underlining the importance of

interventions—such as exercise—that can postpone disability (15). Furthermore, approximately one-third of all people with MS are >60 yr, which should be taken into account when considering exercise (100, 101, 180).

Susceptibility to MS appears to be complex (81), possibly resulting from an interaction of genetic, infectious, and environmental factors. Several genes (HLA [human leukocyte antigen] class II, ApoE [apolipoprotein E], IL [interleukin]-1ra, IL-1β, and TGF [transforming growth factor]-β1) are reported to be associated with the course and severity of MS (56, 81, 153). However, no single identifiable gene is associated with all aspects of MS. A genetic component of MS is supported by studies of twins, which report concordance rates of about 26% in identical twins in contrast with only about 2% in other siblings (149). Additionally, approximately 20% of people with MS have a close relative with the disease (150). Some scientists believe that MS may be triggered by infectious agents such as herpes simplex virus (46) or the Epstein-Barr virus (6). However, no virus is proven to be involved in the initiation of MS. Susceptibility to MS may be due to environmental factors, as the prevalence of the disease is strongly associated with latitude (164). Higher prevalence of MS is observed in northern Europe, the northern United States, Canada, and southern Australia and New Zealand (149). Notably, MS is most common in Caucasians of northern European descent compared with other ethnicities (e.g., Inuit, Norwegian Lapps, Australian Aborigines, New Zealand Maoris) living in the same latitude (112). Geographic clusters (or outbreaks) of MS are observed, but the cause or significance of these clusters is not yet known (69).

MS can be personally devastating, as demonstrated by the fact that within 10 yr of diagnosis more than 50% of individuals with MS will become unemployed (74), primarily because of low functional capacity (70). Lifetime cost of MS health care can exceed $1,000,000 (Canadian dollars) per afflicted person (17), where

Acknowledgment: Much of the writing in this chapter was adapted from the previous editions of *Clinical Exercise Physiology*. As a result, we wish to gratefully acknowledge the previous efforts of and thank Chad C. Carroll, PhD, Charles P. Lambert, PhD, Jane Kent-Braun, PhD, and Linda H. Chung, PhD.

annual direct medical expenditure can be 5.1 times higher for persons with MS (19). In addition, health care costs increase markedly as the severity of neurological dysfunction increases (90, 115).

PATHOPHYSIOLOGY

The pathophysiologic hallmark of MS is the demyelination of neurons in the central nervous system, due to an inflammatory autoimmune response (see figure 28.1). Autoreactive T cells are thought to initiate an immune response against myelin (62, 120, 141). These activated T cells cross the blood–brain barrier, proliferate, and secrete lymphokines or cytokines, which recruit microglia, macrophages, and other immune cells to participate in oligodendrocyte death and myelin destruction (28, 42, 141). At this time, it is not clear whether macrophages or T cells are the primary cells that mediate demyelination (120, 145). As the myelin sheath deteriorates, plaques form, and the end result may be axonal destruction (161). The impairment of normal nerve conduction can lead to an array of symptoms that affect people living with MS, including their ability to carry out typical activities of daily living. Some symptoms may limit exercise performance or be temporarily worsened by exercise. However, this symptom exacerbation is normalized in almost all patients within 30 min of exercise cessation (167).

Multiple sclerosis can follow at least four courses of clinical progression (table 28.1) (98). A fifth course, *relapsing-progressive*, has been defined; but it overlaps other defined clinical courses, and some recommend that this term not be used (98). The most prevalent course of MS is *relapsing-remitting* (RRMS), affecting ~80% of the population with MS (120). The remaining ~20% of the population with MS is diagnosed initially with the *primary progressive* course. More than 50% of individuals initially diagnosed with RRMS develop a steady, progressive form of MS within 10 yr (i.e., convert to secondary progressive) (112). A number of people with MS have occasional attacks or exacerbations, but they frequently recover and may never incur any permanent disability (110). In other cases, individuals may have frequent attacks; and although they do not completely recover, they may retain sufficient function for typical activities of daily living.

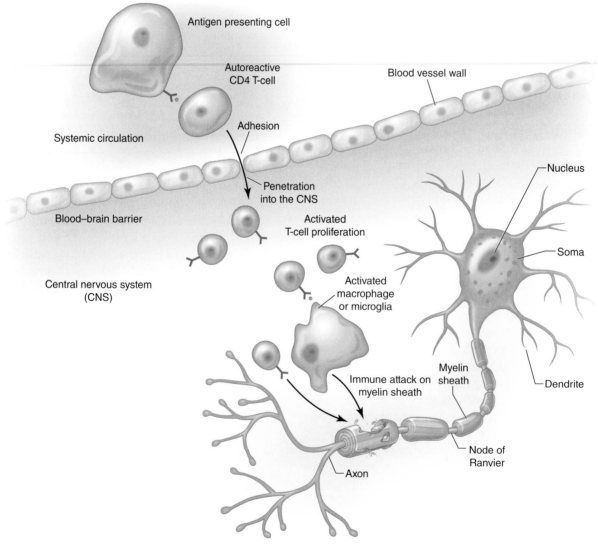

Figure 28.1 Demyelination of neurons in the central nervous system as a result of MS.

Table 28.1 Clinical Courses of Multiple Sclerosis

Type	Characteristic
Relapsing-remitting	Characterized by disease relapses with either a full recovery or a deficit after recovery; no progression of disease symptoms in recovery stage
Primary progressive	Disease progression from onset with infrequent plateaus and only temporary, small improvements; clinical status continuously worsens with no distinctive remissions
Secondary progressive	Begins as relapsing-remitting but progresses either with or without infrequent relapses, plateaus, and remissions
Progressive-relapsing	Progressive from onset with short, definite relapses with or without full recovery

Data from Lublin and Reingold (98).

CLINICAL CONSIDERATIONS

This section reviews the medical and clinical issues that persons with MS encounter and the general methods used to evaluate these individuals.

Signs and Symptoms

Signs and symptoms vary from person to person with MS and are associated with the areas of the central nervous system that are demyelinated to the greatest degree (see the sidebar Common Signs and Symptoms of Multiple Sclerosis). Symptoms of MS can negatively affect daily activities, physical and mental function, and quality of life (188). Common early indicators of MS include visual impairments (e.g., **optic neuritis**, **nystagmus**), motor function difficulties, and **paresthesia** (110, 177). For many people with MS, disease progression is slow, and deterioration occurs over 10 to 25 yr. Ambulation can become increasingly difficult (23, 72), although proper treatment can help delay or diminish disability.

Muscle weakness (4, 20, 26, 77, 93, 116, 118, 131, 155, 159, 176) is a common problem in persons with MS and generally affects the lower extremity muscles more than upper extremity muscles (77, 155). Muscle spasticity may reduce range of motion about a joint and may limit voluntary movement during high-speed flexion or extension (or both) by coactivating antagonist muscles (30). Symptomatic fatigue, defined as an overwhelming sense of tiredness (91), is common and often rated by people with MS as one of the worst symptoms (7, 33, 49, 57, 66). Furthermore, individuals with MS commonly report that fatigue prevents sustained physical functioning, worsens with heat, interferes with responsibilities, and can come on easily (91). Of note, cooling has been reported to improve symptomatic fatigue (91) and other exercise-induced symptom exacerbations (184). Cooling interventions should therefore be considered during exercise for heat-sensitive patients.

Walking, balance, and coordination difficulties are prevalent in MS and are associated with the level of physical disability (23, 72). Walking impairment is another of the worst symptoms, as rated by people with the disease (33, 57, 66). The overall prevalence of falls in persons with MS is ~34%; in those who walk with bilateral support (e.g., two sticks or crutches, frame), it is ~53% (29). Approximately 45% of those with MS use assistive devices

Common Signs and Symptoms of Multiple Sclerosis

Signs
- Optic neuritis
- Nystagmus
- Paresthesia
- Spasticity

Symptoms
- Muscle weakness
- Symptomatic fatigue
- Numbness
- Visual disturbances
- Walking, balance, and coordination problems
- Bladder dysfunction
- Bowel dysfunction
- Cognitive dysfunction
- Dizziness and vertigo
- Depression
- Emotional changes
- Sexual dysfunction
- Pain

(e.g., canes, walkers, wheelchairs) for mobility (48). Although compromised sensory or motor systems (or both) are likely explanations for slowed walking and poor balance control in MS, muscle weakness, spasticity, and lower physical activity level may also contribute to declines in physical function.

Impairments in cognitive function appear in 43% to 65% of persons with MS, including alterations in attention, information-processing speed, working memory, verbal and visuo-spatial memory, and executive functions (68). The decline of cognitive ability and memory loss can occur early after disease onset (133,

177), and weak to moderate associations exist between the level of physical disability and cognitive impairment (99). Depression is also common in MS (110) and can lead to reduced participation in work, social activities, and other endeavors (133).

History and Physical Examination

Many of the signs and symptoms associated with MS are difficult for untrained individuals to recognize. Therefore, the clinical exercise physiologist should consult with a patient's physician before exercise testing and training to discuss the clinical course, current symptoms, and medications. If any signs or symptoms of worsening or intolerance occur during exercise testing or training (e.g., excessive cardiovascular response, heat sensitivity, or fatigue), the clinical exercise physiologist should report this to the patient's physician.

Some people with MS exhibit problems with vision, and issues related to sight during exercise should be addressed. Those with cognitive deficits may have impaired ability to provide informed consent, maintain focus during an exercise test, or follow an exercise prescription. In such cases, a caregiver or support person should accompany the patient. Loss of muscle strength, numbness, symptomatic fatigue, poor flexibility, poor balance and coordination, and gait abnormalities can affect the patient's ability to perform certain types of exercise. These conditions must be considered on an individual basis before any exercise bout.

A review of the medical history and medications an individual is using should be conducted. For example, medications to treat muscle spasticity often result in fatigue and balance perturbations, which may limit maximal exercise capacity. Skeletal muscle fatigue may occur before cardiorespiratory fatigue, and persons with MS who have autonomic dysfunction may not display the typical heart rate or blood pressure responses to graded exercise, rendering those measures less useful for monitoring test progression. For a graded cardiorespiratory exercise test, the result is therefore often a test of $\dot{V}O_2peak$ rather than $\dot{V}O_2max$ in persons with MS (96).

The level of MS-related disability based on the medical evaluation by a neurologist (or trained clinician) can be classified with use of the Kurtzke Functional Systems (table 28.2) and the Kurtzke Expanded Disability Status Scale (table 28.3) (92). In addition, simple functional assessments before and during an exercise training intervention can provide useful insight as to the patient's functional level over time (table 28.4). With this information, the clinical exercise physiologist can objectively rate an individual's ability to perform certain types of exercise.

Diagnostic Testing

The International Panel on the Diagnosis of Multiple Sclerosis has developed criteria for establishing a diagnosis of MS (175). The aim of the criteria is to provide physicians with clearer definitions of *definite*, *possible*, and *not* MS for patients who demonstrate a clinically isolated syndrome or one that is progressive. Diagnostic tools to detect MS include a review of medical history, neurological exam, and various tests (magnetic resonance imaging, evoked potential response, cerebrospinal fluid analysis, blood tests).

Diagnosis initially focuses on the medical history and neurological examination. A medical history provides information about past and present symptoms that may become relevant upon diagnosis. Neurological examination provides insight into the health of the nervous system. Evoked potential tests measure the electrical response to sensory stimulation (e.g., visual, auditory, somatosensory); a lower magnitude and longer response time may indicate the presence of lesions along the nerve. **Magnetic resonance imaging (MRI)** has become an important diagnostic tool for MS, particularly in demonstrating the important criterion of spatial and temporal dissemination of lesions (figure 28.2). Cerebrospinal fluid, obtained by spinal tap, is analyzed to detect levels of immunoglobulins and determine the presence of oligoclonal bands, which can support an elevated immune response. Although there are no specific blood tests for MS, blood tests are useful to rule out other diseases with similar signs and symptoms.

Exercise Testing

Exercise testing is preferably carried out before starting exercise interventions. It is useful in persons with MS to determine their current state of fitness and individual responses to a bout of exercise. It is also helpful for evaluating the effectiveness of (and subsequent adjustments to) exercise training. Each person should have an evaluation before testing, including a review of the medical history and list of medications, as well as a functional assessment combined with a review of the person's training history. Information gathered during an exercise test can be useful in developing an exercise prescription. Chapter 5 contains general information about performing an exercise test.

Clinical exercise physiologists may find the standardized functional tests for MS listed in table 28.4 useful; more details are available at the National Multiple Sclerosis Society website (112). Other tests may also be of use when determining the functional limitations of a person with MS, such as the Berg Balance Scale (10), the Mini-Balance Evaluation Systems Test (53), the Six Spot

Table 28.2 Kurtzke Functional Systems

Category	Rating scale range
Pyramidal	0-6
Bowel and bladder	0-6
Cerebellar	0-5
Visual (optic)	0-6
Brain stem	0-5
Cerebral (mental)	0-5
Sensory	0-6
Other*	0-3

Note: Ratings range from normal function (0) to signs only without disability (1) to increasing levels of disability (2-6).

*Any other neurological findings attributed to multiple sclerosis (e.g., fatigue).

Table 28.3   Kurtzke Expanded Disability Status Scale

Rating	Disability	Functional limitations
0.0	Normal neurological exam	None
1.0	None	Minimal signs in one FS
1.5	None	Minimal signs in more than one FS
2.0	Minimal	Minimal disability in one FS
2.5	Minimal	Minimal disability in two FS
3.0	Moderate, fully ambulatory	Moderate disability in one or minimal in three or four FS
3.5	Moderate, fully ambulatory	Moderate disability in two to four FS
4.0	Mildly severe disability: fully ambulatory without aid; self-sufficient up to 12 h/d	Relatively severe disability affecting one or more FS
4.5	Moderately severe disability: same as 4.0 with some limitation of ADLs or need for minimal assistance	Relatively severe disability affecting one or more FS
5.0	Severe: walks only 200 m without rest or aid; impaired ADLs	Severe disability affecting one or more FS
5.5	Severe: walks only 100 m without rest or aid; impaired ADLs	Severe disability affecting one or more FS
6.0	Severe: needs intermittent or unilateral aid to walk 100 m	Severe disability affecting two or more FS
6.5	Very severe: needs constant aid (cane, crutches, braces) to ambulate 20 m	Severe disability affecting two or more FS
7.0	Extremely severe: unable to walk 5 m; needs aid 12 h/d or more; can transfer self from chair	Severe disability affecting more than one FS
7.5	Extremely severe: takes only a few steps; cannot wheel self; full day in wheelchair	Severe disability affecting more than one FS
8.0	No ambulation; restricted to bed or chair; retains self-care and can use arms	Severe disability affecting many FS
8.5	Bedridden; minimal arm use; some self-care	Severe disability affecting many FS
9.0	Bedridden; no self-care; can talk and eat	Severe disability affecting all FS
9.5	Bedridden; cannot communicate, eat, or swallow	Severe disability affecting all FS
10.0	Death attributable to multiple sclerosis	

FS = functional systems; ADLs = activities of daily living.

Table 28.4   Functional Assessments

	Assessments	Evaluations
MS Functional Composite (MSFC)	Timed 25 ft walk Nine-hole peg test (9-HPT) Paced Auditory Serial Addition Test (PASAT)	Leg function and ambulation Arm and hand function Cognitive function
Modified Fatigue Impact Scale (MFIS)	Self-report questionnaire	Physical fatigue Cognitive fatigue Psychosocial fatigue
Health Status Questionnaire (SF-36)	Self-report questionnaire	Physical component Mental component

Note: These functional assessments can be obtained from the following link: www.nationalmssociety.org.

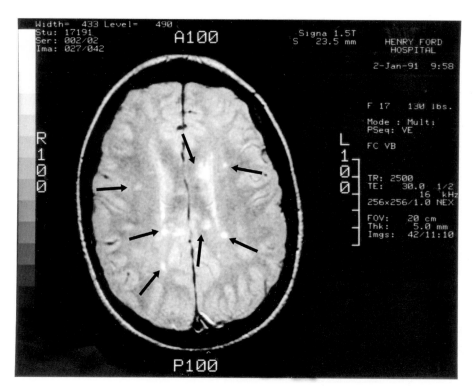

Figure 28.2 Magnetic resonance image depicting multiple areas of demyelination (arrows) within the white matter of the brain in a person with MS.

Reprinted by permission from C.A. Brawner and J.R. Schairer, "Multiple Sclerosis; Case Report From Henry Ford Hospital," *Clinical Exercise Physiology* 2, no. 1 (2000): 17.

Step Test (SSST) (18), and the timed Up and Go (TUG) (79). To help monitor the responses of persons with MS during these functional tests, the Borg rating of perceived exertion (RPE) (16, 27) and the visual analogue fatigue scales (156) can be useful.

Cardiorespiratory

To assess cardiorespiratory fitness, the clinical exercise physiologist should carefully consider the mode of exercise because each mode has specific advantages. A leg cycle ergometer is most common because it requires little balance and can be used by persons with ambulatory impairment (e.g., foot drop, need for assistive devices for mobility, **ataxia**). Additionally, the resistance of a cycle ergometer can be adjusted to a level that is within most patients' capabilities. An excellent option for exercise testing is an ergometer with arm poles (e.g., a Schwinn Airdyne or a NuStep recumbent stepper), which allows the use of all four limbs, thereby providing the opportunity for individuals with weakness in one limb to continue exercising by compensating with the other limbs (130). People with no or little ambulatory impairment can use a treadmill, whereas nonambulatory persons with adequate upper body function can use an arm cycle ergometer. The risk of falling should also be considered. The clinical exercise physiologist can initially estimate the appropriate equipment by observing the

patient's gait when walking from the waiting room, or they can administer balance tests. These types of functional assessments are useful for understanding any limitations a person may have before beginning an exercise intervention. In addition to baseline measures, these assessments are useful for monitoring changes in function over time.

Persons with MS should be closely monitored for any signs of **paresis**, fatigue, or overheating as exercise intensity increases because these are likely to occur in some persons with MS. The use of toe straps with the leg ergometer is recommended for persons with sensory deficits, and an ankle–foot orthotic may be needed for patients in whom foot drop may develop during exercise.

A low-level protocol beginning with a 2 to 5 min warm-up period is recommended. Workload increases should occur at 10 to 25 W for leg ergometry (142) and 8 to 12 W for arm ergometry and should be no more than 2 metabolic equivalents of task (METs) on the treadmill. Each stage should last approximately 2 to 3 min or until a steady state is achieved. Some patients with little disability may tolerate a higher-level protocol. Table 28.5 reviews the physiological responses to exercise in persons with MS. The sidebar Exercise Testing Review provides a summary of exercise testing specifics. Standard graded exercise testing contraindications reviewed in chapter 5 should be followed (137);

no specific contraindications are common among persons with MS. Because of the highly variable nature of the disease, each person may respond somewhat differently to testing, so close monitoring is required.

Individuals with MS may have an attenuated systolic blood pressure response during exercise (127, 158) that may be related to autonomic dysfunction (59, 158, 171), a reduced skeletal muscle metabolic response that affects the skeletal muscle chemoreflex (116), or both. For this reason, the RPE score can be used to help maintain desired intensity during graded exercise testing (13, 27). This approach gives the person with MS a means of adjusting exercise intensity to match current clinical status. Some individuals with MS may also experience abnormal temperature regulation and sweat response (36), increasing the risk of hyperthermia during exercise. Consequently, use of electric fans to improve evaporative and convective cooling, use of cold neck packs, attention to fluid replacement, and control of climate (i.e., room temperature 72-76 °F [22-24 °C], low humidity) are recommended to reduce the risk of heat-related symptoms as well as symptomatic fatigue. Skeletal muscle fatigue may occur before patients reach peak cardiorespiratory levels. This circumstance may be caused by an inability to recruit additional skeletal muscle or by impaired skeletal muscle metabolism (86, 87). A higher energy cost of walking and cycling in persons with MS (25, 44, 52, 124) may also affect their ability to reach peak cardiorespiratory levels.

Musculoskeletal

Standard testing procedures can be used to determine muscle strength and endurance in persons with MS. Because of the heterogeneity of lesion location and impact, some muscle groups may be preferentially weak while others exhibit normal strength. A one-repetition maximum (1RM) test is typically used to measure muscle strength. However, if a person with MS is weak, 3RM to 5RM testing may be more appropriate to estimate the 1RM to minimize the risk of injury (113). Muscle endurance can also be assessed by determining the number of repetitions a person can perform before volitional fatigue while using a light resistance weight (e.g., corresponding to a 20RM or 30RM load).

Exercise Testing Review

Cardiorespiratory

- Cycle ergometry is the most applied modality.
- A common protocol is 2 to 5 min of light warm-up, 2 to 3 min stages, 10 to 25 W workload increase per stage.
- Typical measures include HR, BP, METs or $\dot{V}O_2$, and RPE.
- Persons with MS may have increased CV risk, respiratory dysfunction, very low fitness levels, low physical activity levels, or some combination of these.
- Special considerations include possible attenuated BP and HR responses; impaired thermoregulation; muscle weakness and fatigue.

Strength

- Use machine weights; manual muscle testing may be used in more disabled patients.
- Perform 1RM (or 3RM-5RM) testing of upper and lower body muscle groups by measuring the amount of weight lifted.
- Muscle weakness may contribute to reduced mobility and functional impairment.
- If a person has poor flexibility, spasticity, or muscle fatigue, they should be well supervised during resistance training for safety purposes.

Range of Motion

- Measure ankle, knee, hip, shoulder, and elbow flexibility in degrees using a goniometer, if possible.
- Poor flexibility may contribute to poor ambulation and performance of ADLs.
- Watch for contractures and spasticity during testing.

HR = heart rate; BP = blood pressure; METs = metabolic equivalents of task; $\dot{V}O_2$ = oxygen consumption; RPE = rating of perceived exertion; CV = cardiovascular; RM = repetition maximum; ADLs = activities of daily living.

Table 28.5   Physiological Responses During Exercise

Physiological response	Response of MS relative to non-MS individuals
Submaximal oxygen consumption during treadmill walking	Increased (124)
Submaximal and maximal arterial blood pressure	Same or decreased (127)
Temperature	Same or increased (36)
Skeletal muscle fatigue	Earlier (87, 159)

Flexibility

Range of motion can be assessed with a goniometer. Because flexibility can become reduced in persons with MS, especially in muscles affected by spasticity (139), stretches should be slow and gentle (without bouncing). Patients should perform stretches while seated or lying down to minimize the risk of falling.

Important issues to discuss with patients who have MS are presented in practical application 28.1.

Treatment

There is no cure for MS, but several disease-modifying drugs (table 28.6) are approved by the Food and Drug Administration to treat it. These medications minimize the frequency of exacerbations, reduce the number of lesions in the central nervous system (103, 105, 125, 163, 183), and slow the progression of disability (76, 97, 146). Updated information is available at the National Multiple Sclerosis Society website (111). Three disease-modifying drugs have been approved for secondary progressive MS and one for the primary progressive course. An exacerbation is the result of an increased inflammatory response that promotes demyelination and causes slowing or blocking of neural signaling. Exacerbations are characterized by a mild to severe worsening of symptoms that interferes with a person's functioning and lasts for more than 24 h. Corticosteroid treatment is typically used to ameliorate the inflammation and reduce the recovery time from an exacerbation (28). Side effects of corticosteroid treatments are bone softening, high blood pressure, and transient weight gain. The use and administration of disease-modifying drugs may differ across countries (e.g., due to agency approvals and political treatment priorities).

Persons with MS experience an array of symptoms, as previously discussed. Although many symptoms can be partly managed through medication (table 28.6), nonpharmacological interventions are essential in symptom management, exercise in particular.

Practical Application 28.1

CLIENT–CLINICIAN INTERACTION

The symptomatic fatigue noted by individuals with MS varies from day to day. The clinical exercise physiologist must anticipate the need to reduce exercise training volume (i.e., frequency, intensity, duration) if a person is fatigued. Doing this requires daily assessment for indicators of fatigue. Verbal communication is the best means of assessment. Appropriate questions include "How do you feel today?" and "Are you tired today?" Use of a visual analogue fatigue scale (91, 156) can also be helpful in ascertaining symptomatic fatigue before exercise. Intensity of exercise can be adjusted, and duration monitored, by use of the RPE score. As the RPE approaches the "hard" and "very hard" zones and the individual begins to experience fatigue despite reductions in exercise pace, the exercise session should be ended. Exercise above a moderate intensity, particularly on days when significant symptomatic fatigue is reported, should be carefully considered and perhaps avoided. Although exercise may acutely worsen or lead to increased levels of fatigue later in the day or even in the days soon after the training session, this transient phenomenon (167) normally lessens as the patient becomes more fit. Furthermore, exercise normally has a beneficial effect on the perceived level of chronic fatigue in persons with MS when rated after 2 to 4 wk (67).

Clinical depression or depressive symptoms are somewhat common in persons living with MS. Indicators of stress may be related to depression. These signs include poor sleep habits, noncompliance with lifestyle change, and elevated scores on standardized questionnaires such as the SF-36 and Beck Depression Inventory. The exercise clinician should share information about the potential positive effects of exercise on various psychological variables (e.g., mood, depression, anxiety). People who exhibit signs of depression or are concerned about it should be referred to the mental health professional on their MS management team.

The clinical exercise physiologist should counsel individuals with MS about the risk factors associated with a sedentary lifestyle, including coronary artery and cardiovascular disease, obesity, and type 2 diabetes. Any steps the person can take to include or increase daily activity levels will help with many health outcomes.

The clinical exercise physiologist should help motivate individuals with MS to exercise regularly. Because some people with MS have cognitive problems, the exercise clinician may need to repeat instructions, clarify explanations, or present the training plan in an easy-to-follow format. Reported reasons for nonadherence to exercise training include fatigue, disability, and lack of time (5). The clinical exercise physiologist must communicate the importance of regulating exercise training intensity each day to avoid excessive fatigue; the best way to do this is to use the RPE scale in combination with objective intensity measures (e.g., HR). Additionally, it can be stressed that regular resistance exercise may strengthen the skeletal muscles and allow people to perform and maintain activities of daily living with less overall symptomatic fatigue. This type of information can be helpful for maintaining motivation.

Table 28.6 Common Medications Used for Persons With MS

Medication name [generic (brand)]	Primary effects	Potential side effects relevant to exercise	Other potential side effects
Interferon β-1a (Avonex, Rebif)	Disease modifying	Dyspnea, fatigue, arrhythmias	Flu-like symptoms, headache, nausea
Interferon β-1b (Betaseron, Extavia)			
Glatiramer acetate (Copaxone)			
Teriflunomide (Aubagio)			
Monomethyl fumarate (Bafiertam)			
Dimethyl fumarate (Tecfidera)			
Diroximel fumarate (Vumerity)			
Fingolimod (Gilenya)			
Siponimod (Mayzent)			
Cladribine (Mavenclad)			
Ozanimod (Zeposia)			
Ocrelizumab (Ocrevus)			
Alemtuzumab (Lemtrada)			
Mitoxantrone (Novantrone)			
Natalizumab (Tysabri)			
Ofatumumab (Kesimpta)			
Onabotulinumtoxin A (Botox)	Spasticity	Dizziness, hypotension, fatigue, muscle weakness, hypotonia, ataxia, blurred vision	Nausea, headache, diplopia, dry mouth
Dantrolene sodium (Dantrium)			
Baclofen-intrathecal (Lioresal)			
Diazepam (Valium)			
Clonazepam (Klonopin)			
Tizanidine hydrochloride (Zanaflex)			
Amantadine hydrochloride (Symmetrel)	Symptomatic fatigue	Dizziness	Nausea, headache
Modafinil (Provigil)			
Fluoxetine hydrochloride (Prozac)			
Dalfampridine (Ampyra)	Walking problems	None	None
Clonazepam (Klonopin)	Tremor		Peripheral neuropathy
Isoniazid (Laniazid, Nydrazid)			
Desmopressin (DDAVP nasal spray or tablets)	Bladder dysfunction	Dizziness	Dry mouth, headache
Tolterodine (Detrol)			
Oxybutynin chloride (Ditropan XL or LA, Oxytrol)			
Darifenacin (Enablex)			
Tamsulosin hydrochloride (Flomax)			
Terazosin hydrochloride (Hytrin)			
Prazosin hydrochloride (Minipress)			
Propantheline bromide (Pro-Banthine)			
Trospium chloride (Sanctura)			
Imipramine hydrochloride (Tofranil)			
Solifenacin succinate (VESIcare)			

(continued)

Table 28.6 *(continued)*

Medication name [generic (brand)] and class	Primary effects	Potential side effects relevant to exercise	Other potential side effects
Duloxetine hydrochloride (Cymbalta)	Depression	Dizziness, asthenia, sweating	Nausea, dry mouth, headache
Venlafaxine hydrochloride (Effexor)			
Paroxetine hydrochloride (Paxil)			
Fluoxetine hydrochloride (Prozac)			
Bupropion hydrochloride (Wellbutrin)			
Sertraline hydrochloride (Zoloft)			
Duloxetine hydrochloride (Cymbalta)	Pain	Nystagmus, ataxia, dizziness	Nausea
Phenytoin sodium (Dilantin)			
Amitriptyline hydrochloride (Elavil)			
Clonazepam (Klonopin)			
Gabapentin (Neurontin)			
Nortriptyline hydrochloride (Pamelor)			
Carbamazepine (Tegretol)			
Imipramine hydrochloride (Tofranil)			

Note: Most disease-modifying drugs are used to treat relapsing forms of MS with the exception of mitoxantrone, cladribine, siponimod, and ocrelizumab, which treat progressive or worsening forms of MS. $\dot{V}O_2$ or peak heart rate (or both), to determine a person's target exercise intensity. Moderate to vigorous intensity is the ideal target intensity when symptoms are not present (95, 185).

Rehabilitation is beneficial for maintaining daily function and quality of life. Physical therapy can be important for improving mobility. Occupational therapy is useful in improving motor function and reasoning abilities and for learning to compensate for disabilities (e.g., mitigating fatigue). Speech pathologists and cognitive remediation specialists can help persons with MS improve problems with verbal communication, memory, and cognition. Complementary treatments (e.g., yoga, Pilates, Chinese medicine, naturopathy, relaxation techniques) may provide additional benefits through stress management strategies, nutritional recommendations, exercise, and lifestyle changes, although evidence for efficacy of these treatments is lacking.

Exercise in persons with MS may also have disease-modifying effects, although evidence is preliminary. Specifically, studies indicate that relapses and disability progression can be reduced with exercise (33, 95). Although the underlying mechanisms are still not known, they may include exercise induced anti-inflammatory effects, improved brain perfusion, and an increase in neuroprotective factors (33, 95). As for symptom management, exercise in persons with MS improves impaired bladder and bowel function (128), positively affects psychological health and quality of life (33, 54, 127, 128, 154), improves **muscle weakness** (37, 77, 174, 186), and potentially reduces symptomatic fatigue (3, 130). Flexibility does not appear to improve with exercise, but the existing evidence is sparse (178), and outcome measures of range of motion are not generally included in exercise training studies in persons with MS. The use of assistive devices (e.g., cane, Canadian crutch, ankle–foot orthosis) is an effective strategy to maintain balance control and allow mobility-challenged persons with MS to continue their regular activities of daily living.

EXERCISE PRESCRIPTION

The ultimate goal of exercise in persons with MS is to maintain, reachieve, or even improve general physical and mental health, thus positively affecting the ability to do habitual tasks of daily living, from household chores and walking (without risk of falling) to participation in recreational activities. Overall, lower physical activity levels are observed in persons with MS compared with those without MS (21). In particular, persons with MS with mobility disability tend to participate in more sedentary behaviors than those who are ambulatory (43). Inactivity is known to be associated with muscle weakness (144), lower cardiorespiratory fitness (94), fatigue (67), and potentially lower bone density. Exercise or increased physical activity in general benefits persons with MS by reducing symptoms (3) and improving overall quality of life (54, 128, 154). Increased activity may also lower risk factors for cardiovascular and metabolic disease in this generally sedentary population. The existing evidence is nevertheless of low quality (31).

A number of specific exercise prescription standards (i.e., guidelines) for persons with MS exist (table 28.7). Overall, exercise is safe and does not exacerbate symptoms, and persons with MS can perform standardized (or slightly modified) aerobic and muscle-strengthening activities. Moreover, exercise appears feasible across the disability spectrum (41, 78). When deciding about appropriateness of exercise and level of supervision, the clinical exercise physiologist can use the functional disability systems presented in tables 28.2 and 28.3. In addition, because of the varying nature of some symptoms of MS, the clinical exercise physiologist should have up-to-date knowledge of symptoms before any exercise bout and adjust the exercise prescription accordingly. For example, when someone with MS reports increased symptomatic fatigue, strenuous activities should be reduced or even avoided and fatigue levels monitored regularly using a 10 cm visual analogue fatigue scale during the training session. Those with known impairment of heart rate or blood pressure response during exercise may require monitoring. Persons with MS, most commonly in those with relapsing-remitting MS, should avoid strenuous exercise during an exacerbation to prevent further worsening of symptoms. Moreover, very little is known about optimal exercise levels during and immediately after a relapse (143).

Because people with MS can have problems with thermoregulation and because symptomatic fatigue can be brought on by increased body core temperature, cooling with an electric fan and controlling room temperature are important. Good hydration and fluid replacement during and after exercise at the rate at which fluid is lost from the body are recommended. Preexercise cooling by body immersion in cool water (184) or exercising in cooler waters (8) is beneficial for improving exercise tolerance. Cold immersion can also be helpful following exercise in warm conditions. Cold neck packs may be useful during and after exercise to minimize the effects of symptomatic fatigue. Because of the potential for heat-related illness or temporary worsening of symptoms, persons with MS should be educated about the need to take steps to avoid temperature problems during exercise training. Persons with MS experiencing an exacerbation should avoid exercise training until symptoms improve or stabilize.

A training program that involves intermittent exercise may be necessary for a person with MS in order to avoid excessive buildup of fatigue and heat stress (185). Alternating exercise with short rest periods is recommended, especially for people who have low cardiorespiratory fitness. In addition, symptom worsening is less pronounced in heat-sensitive persons with MS during resistance training than during aerobic training because of a smaller increase in body temperature (165). Clinical exercise physiologists should monitor RPE and fatigue levels during each exercise bout, along with the percentage of peak $\dot{V}O_2$ or peak heart rate (or both), to determine a person's target exercise intensity. Moderate to vigorous intensity is the ideal target intensity when symptoms are not present (95, 185).

Table 28.7 Exercise Prescription Guidelines

Guideline	Source	Content
National Multiple Sclerosis Society (NMSS) (United States)	www.nationalmssociety.org/Living-Well-With-MS-Diet-Exercise-Healthy-Behaviors/Exercise	General information on the benefits of physical activity and exercise in persons with MS
MS Australia	www.msaustralia.org.au/wellbeing-ms/exercise-activity	General information on the benefits of physical activity and exercise in persons with MS
MS Society of Canada (MSSC)	https://mssociety.ca/support-services/programs-and-services/recreation-and-social-programs/physical-activity/the-guidelines	General information on the benefits of physical activity and exercise in persons with MS
Latimer-Cheung et al. Development of evidence-informed physical activity guidelines for adults with multiple sclerosis. *Archives of Physical Medicine and Rehabilitation* (2013)	https://pubmed.ncbi.nlm.nih.gov/23770262/	Physical activity guidelines for adults with multiple sclerosis
Kalb et al. Exercise and lifestyle physical activity recommendations for people with multiple sclerosis throughout the disease course. *Multiple Sclerosis Journal* (2020)	https://pubmed.ncbi.nlm.nih.gov/32323606/?from_single_result=kalb+dalgas	Exercise and lifestyle physical activity recommendations for people with multiple sclerosis throughout the disease course
Motl et al. Exercise in patients with multiple sclerosis. *The Lancet Neurology* (2017)	https://pubmed.ncbi.nlm.nih.gov/28920890/?from_single_result=motl+dalgas+lancet+neurology	Effects on common symptoms; current limitations with existing evidence and future directions
Geidl et al. A systematic critical review of physical activity aspects in clinical guidelines for multiple sclerosis. *Multiple Sclerosis and Related Disorders* (2018)	https://pubmed.ncbi.nlm.nih.gov/30103172/?from_term=streber+r+pfeifer&from_pos=2	Current limitations with existing evidence and future directions

In general, the best approach may be to alternate cardiorespiratory training days and resistance training days to prevent excessive fatigue. However, persons with MS can perform cardiorespiratory and resistance exercise within the same session if necessary or desired, although the effect from each modality may be slightly lower than if performed on separate days.

Cardiorespiratory Exercise

Persons with MS can attain the same types of improvement (e.g., increased peak $\dot{V}O_2$, improved psychological variables, decreased muscle fatigue, and increased skeletal muscle metabolism) with cardiorespiratory training as individuals without MS, and at a similar rate (11, 37, 54, 60, 123, 128, 140, 142, 154, 172). The risk of disease exacerbation is small (127), and symptomatic improvements are observed after training (3, 37, 54, 128, 154, 174). Table 28.8 provides a review of the exercise prescription for cardiorespiratory, resistance, and range of motion training.

To improve cardiorespiratory function, aerobic activity should be performed at a moderate to vigorous intensity (185) for 30 to 60 min on 2 to 5 d/wk (94). If needed, the duration of aerobic exercise can be accumulated in three 10 min bouts interspersed with rest. A goal is to progress initially to 30 min continuous, with an ultimate goal of 60 min continuous. Intensities should range between 40% and 70% of $\dot{V}O_2$ reserve. Persons with MS who have heightened symptoms, autonomic nervous system dysfunction, or poor fitness may benefit from a lower exercise intensity at the outset. Use of Borg's RPE scale and the visual analogue fatigue scale is highly recommended. These scales can easily be applied throughout an exercise session and should be considered in conjunction with the percentage of peak $\dot{V}O_2$ or peak HR for determining a person's exercise intensity.

Stationary cycling with legs or arms (or both), walking, swimming, and aquatic exercise are common methods used to improve aerobic fitness in persons with MS and may be the modality of

Table 28.8 Exercise Prescription Review

Training method	Frequency	Intensity	Time (Duration)	Type (Mode)	Progression	Important considerations
Cardiorespiratory	2-5 times per week	Moderate to vigorous, or 40%-90% of HRmax or $\dot{V}O_2$ reserve,* RPE 12-15, VAFS < 6	30 min continuous or three 10 min bouts, progressing to 60 min as tolerated	Cycle ergometry, walking, swimming, treadmill, elliptical trainer	Careful progression based on the individual's responses to the exercise; avoid excessive fatigue	Balance may be an issue on some equipment; consider seated modes or handrail use on treadmill. Ensure temperature regulation. Monitor perceived exertion and fatigue levels and watch for attenuated HR or BP response. Increase time initially before intensity.
Resistance	2 or more times per week	60%-85% of 1RM, 6-15 reps for 1 or preferably 2 sets	About 30 min total; if fatigued, increase between-set rest time to 2 to 5 min	Machine weights, elastic bands, calisthenics, stair climbing, aquatic training	Careful progression based on the individual's abilities; otherwise follow standard PRT principles	Consider unilateral weakness, balance, muscle spasticity, and excessive fatigue. Work to correct imbalances. Rest at least 1 d between sessions. Avoid free weights unless spotter is used. Use handrail for balance control during stair climbing. Can perform both multi- and single-joint exercises.
Range of motion	Most days of the week; before and after each training session; can perform twice per day if desired	To the point of tightness or mild discomfort	Each stretch held for 30-60 s; 2-4 reps with short rest between	Slow, gentle stretching; may be assisted by a partner or by using a device such as a rope; static technique is preferred	Gradual stretch without bouncing to the end of comfort range	Balance should be considered when doing stretching exercises;. Avoid excessive joint range and muscle spasticity aggravation. Perform while seated or on the floor, after either warm-up or cool-down.

*Largely dependent on patient's ability; assess on an individual basis.

RPE = rating of perceived exertion; VAFS = visual analogue fatigue scale; HR = heart rate; BP = blood pressure; 1RM = one-repetition maximum; PRT = progressive resistance training; ADLs = activities of daily living; ROM = range of motion

choice for those with ambulatory limitations. High-functioning persons with MS may opt to use a treadmill or elliptical trainer instead. Progression over time should be individualized and based on the percentage of peak exercise test heart rate or peak $\dot{V}O_2$, as well as the ability to tolerate the exercise (i.e., perceived exertion and symptomatic fatigue levels).

Resistance Exercise

Muscle strength is critically important for mobility and independence. Resistance training should focus on the major muscle groups and use resistive movements (e.g., weightlifting, stair climbing, calisthenics, and elastic bands) 2 d/wk (67). Resistance should be set at a weight that allows for two sets of 6 to 15 repetitions, resulting in volitional fatigue, for each exercise (64, 102). If a person with MS is weak, lower absolute resistance should be used initially. Seated weightlifting is recommended to prevent the potential risk of falls in people who have problems with balance and coordination. Modifications, such as training muscle groups unilaterally because of differences in strength between the limbs or because of range of motion limitation, may be required. These limitations need to be considered and tailored on an individual basis.

Balance Exercise

Simple upright balance exercises may also be used to improve strength and coordination. The National Institutes of Health recommends five exercises for individuals who are at risk for falls. Balance exercises include standing on one foot, walking heel to toe, balance walking, leg raises to the back, and leg raises to the side. These exercises increase lower body strength and enhance coordination and are simple enough to perform at home. Individuals with good balance control can perform the exercises at an increased level of difficulty, with supervision, by doing them while standing on rubber foam, on uneven surfaces, or with one or more of their sensory systems perturbed (e.g., having eyes closed, leaning). Furthermore, new effective balance training concepts include descriptions of progression models and training intensity (18, 51).

Range of Motion Exercise

Because persons with MS may have reduced range of motion in some joints, a general flexibility program is recommended. Although the evidence is sparse and inconclusive (178), flexibility exercise has been argued to partly counteract spasticity, increase muscle length, and improve posture and balance (185). Stretches should be gentle, slow, and prolonged (30-60 s), for up to four repetitions each, repeated two to four times each. They should not be painful and should focus on all major upper and lower body muscle groups (185). Stretching should be performed before or after a strength or cardiorespiratory training session (or both), as well as during brief periods throughout the day (185). These exercises should be done daily (minimally five times per week).

Some persons with MS may need assistance with passive stretching. The use of a rope or the participation of a trained clinician may be necessary for persons with spasticity. Light massages, yoga, and other relaxation techniques may also complement stretching (185). Chapter 6 provides general information about range of motion training.

Other Exercise Modalities

Aquatic exercise is another training option that has been investigated for persons with MS who are weak, use a wheelchair, or struggle with severe heat sensitivity (2). Aquatic exercise minimizes gravitational effects, provides a greater sense of balance and walking ability, and helps dissipate body heat (in water temperatures at 27-29 °C or 80-84 °F) (185). Limb devices can be used to increase water resistance to movement and strengthen muscles (185).

EXERCISE TRAINING

Exercise training and daily physical activity are important in persons with MS in order to maintain, reachieve, or even improve cardiorespiratory fitness and muscle strength, thus retaining the capacity to perform activities of daily living and minimizing some common symptoms such as symptomatic fatigue (33, 138, 148, 168). Studies indicate that relapses and disability progression can be reduced with exercise (figure 28.3, the sidebar Exercise Training Review) (33, 95, 143). Although the underlying mechanisms are still not known, they may include exercise induced anti-inflammatory effects, improved brain perfusion, and increased neuroprotective factors (33).

Cardiorespiratory Exercise

Peak oxygen consumption (94, 107, 142) and respiratory muscle function (24, 50, 152) are lower in persons with MS versus those without MS. These declines may be partly explained by lower levels of physical activity in this population (117). Lower cardiorespiratory fitness and physical activity levels may place persons with MS (107) at greater risk for diseases related to a sedentary lifestyle (12). Thus, increasing physical activity may help minimize cardiovascular health problems.

As found in the landmark study by Petajan and colleagues (128), endurance exercise can have a significant positive impact on a range of health-related variables in people with MS. Increased leisure-time physical activity can result in a smaller waist circumference, lower blood triglycerides, and lower blood glucose values in this group (166). Also, increased $\dot{V}O_2$peak (45, 128, 132, 140), decreased serum triglyceride levels (128), and increased fatigue tolerance (123, 172) have been shown after aerobic exercise training in persons with MS. Improved health perception and quality of life (88, 128, 154) and increased walking speed and endurance (11, 132, 154) have all been reported in response to aerobic training in

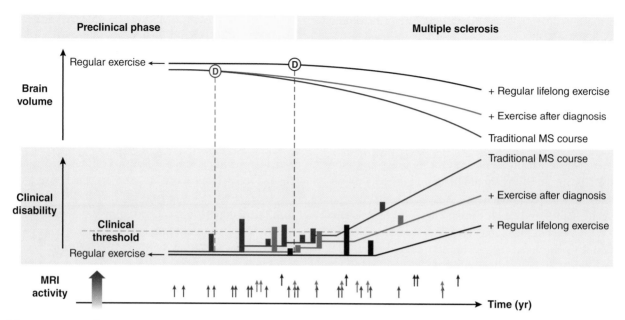

Figure 28.3   Impact of exercise (and physical activity) on brain volume and clinical disability.

Reprinted by permission from Springer Nature: Dalgas et al., "Exercise as Medicine in Multiple Sclerosis—Time for a Paradigm Shift: Preventive, Symptomatic, and Disease-Modifying Aspects and Perspectives," *Current Neurology and Neuroscience Reports* 19, no. 88 (2019): 88.

this population. An increase in perceptual exercise tolerance (i.e., RPE) was observed after 6 mo of combined progressive endurance and resistance training, although no improvements in cardiac autonomic control were shown (63). The sidebar Exercise Training Review provides an overview of the beneficial health effects of cardiorespiratory exercise.

Resistance Exercise

Isometric muscle strength (116, 118, 119, 155, 159), dynamic muscle strength (4, 20, 93, 131, 176), and muscle power (26) are generally lower in persons with MS (77). A number of potential physiological changes in MS may explain reduced strength and power. Compared against those without MS, persons with MS may have decreased central activation (118, 136, 157, 159) and lower motor unit discharge rates (39, 136). Although some studies have shown alterations in contractile function (118, 159) and smaller muscle size (85, 118, 182) in persons with MS, others reported no change in these variables (61, 147). Given the heterogeneity of this disease, variability in strength and power loss is to be expected. Spasticity, characterized by a velocity-dependent increase in tonic stretch reflexes, may also contribute to lower dynamic power. Coactivation of antagonist muscles, possibly due to spasticity, may work against an agonist muscle contraction. An important notion is that these physiological changes are also influenced by decreased physical activity in patients with multiple sclerosis (108, 117) and are independent of disease severity (144).

Resistance training improves strength and power via neural adaptation and muscle hypertrophy (38, 89, 106). Several studies have evaluated the effects of resistance training in persons with MS (37, 60, 65, 89, 142, 173, 174, 186, 187). Increases in muscle strength (14, 34, 35, 47, 169, 174, 186) and power (37, 169) are observed after a progressive resistance training program. Interestingly, White and colleagues (186) observed increases in isometric strength but not muscle size following 8 wk of progressive resistance training (two times a week for 30 min), suggesting that strength gains may be explained by neural adaptations. However, increases in muscle size were noted after a 24 wk progressive resistance training program (two times a week) (89). These studies indicate that perhaps a certain volume (i.e., combination of intensity, frequency, duration) of overall training is necessary to provoke muscle hypertrophy as seen in other resistance training studies in MS (34).

Systematic reviews of progressive resistance training in persons with MS have provided strong evidence for its effectiveness in improving muscle strength, as well as its beneficial impact on functional capacity, balance, symptomatic fatigue, mood, and quality of life in those with MS (33, 89). Specifically, functional gains, such as a decreased Timed Up and Go test (37, 169, 174), faster walking speed (142), an increased number of steps over 3 min (186), greater muscular endurance (173, 174), improved balance (65), and improved gait kinematics (60), are observed after resistance training. Resistance training also reduces symptomatic fatigue (60, 186).

Other types of physical training can safely induce effects resembling those achieved through resistance training in persons with

MS. These include aquatic exercise, whole-body vibration training, and supported treadmill training. A 10 wk aquatic exercise program resulted in improvements in power, muscular endurance, and total work in both the upper and lower extremities (55). In addition, persons with MS significantly improved their level of symptomatic fatigue as well as their health-related quality of life score after 4 and 8 wk of aquatic training (82). A systematic review shows potential benefits of whole-body vibration training on muscle strength and endurance, functional capacity, balance, and coordination (22). Studies using supported treadmill training in persons with MS with high levels of disability show increases in walking speed (129) and a reduction in the percentage of body weight support needed (129, 130). Such training modalities should be considered in combination with resistance training or separately.

Balance Exercise

In persons with MS, walking speed is often slower (20, 26, 104, 124, 155, 176) and demands a high amount of energy expenditure (25, 52, 124). Slower walking coincides with shorter stride lengths, decreased cadence, and a prolonged double-support phase during the gait cycle (9, 104). In addition, longer dual-support times, shorter swing times, and wider stride widths have been observed at fixed walking speeds (135). Gait initiation is also altered in people with MS; smaller posterior shifts in the center of pressure and minimal displacement of the center of mass have been observed (134).

Balance control is also compromised in MS. Greater postural sway is observed when one or more sensory systems (visual, somatosensory, vestibular, proprioception) are perturbed (114). Postural sway is faster compared with that in others when persons with MS stand quietly on stable and foam rubber support surfaces (58). Center-of-pressure displacements are lower during leaning (83, 179) and reaching tasks (179). Higher center-of-pressure variability and asymmetric limb loading during quiet stance have been observed in persons with MS compared with individuals who don't have the disease (26). When vision was limited, subjects with MS showed increased postural sway while leaning backward and greater limb-loading asymmetry during a quiet stance (179). Increasing levels of fatigue also had an effect on postural control during balance tasks (179). Differences in power across limbs are observed and are associated with center-of-pressure variability during quiet stance (26). Thus, a functional adaptation occurs to maintain balance control. The strategies used by persons with MS appear to keep the body within the individual's stability boundaries in an effort to minimize the risk of falling.

Although resistance training in people with MS improves gait kinematics (60) and performance on the Berg Balance Scale (65), few studies have used exercise interventions focusing on multiple sensorimotor processes to improve balance (84). A pilot study using functional balance exercises shows potential benefits on postural sway on an unstable surface and on trunk control during gait (84). A randomized controlled trial investigating the effects of a new training method showed that balance training improved balance. In addition, the study showed that balance training, but not progressive resistance training, was superior to the nontraining control group in improving gait performance, while both interventions improved fatigue (18). However, more well-controlled studies with objective measures are needed to advance our understanding of the benefits of balance-specific training on gait kinematics and postural control in MS.

Range of Motion Exercise

Few studies have examined the effect of exercise on range of motion (flexibility) in persons with MS. Some degree of spasticity is prevalent in MS and can manifest itself as muscle stiffness, muscle spasms, or muscle cramping, which reduces range of motion (71). An active or passive stretching program can minimize spasticity and preserve range of motion (71). Weight-bearing stretches for lower extremities (122) and prolonged static stretches (121) have shown greater effects in reducing muscle tone and increasing range of motion in patients with spasticity. A study utilizing progressive concurrent aerobic and strength training for 6 mo (three times per week; 60 min sessions) in persons with MS shows a trend for improvement in the dynamic range of motion during walking at the hip, knee, and ankle (126). More nonpharmacological studies are needed in persons with MS to understand the benefits of exercise on range of motion and reducing spasticity. Although individual studies may suggest that range of motion exercise can improve flexibility in people with MS, the summary of studies is inconclusive (178).

Other Exercise Modalities

A number of studies have investigated the effects of Pilates and yoga in persons with MS, mostly on fatigue and functional capacity (e.g., walking) (32, 151, 160). Although these training modalities may have some minor beneficial health effects in people with MS, the majority of studies are of low quality, revealing the overall evidence to be weak.

Exercise Training Review

The following is a summary of the effects of exercise modalities on disabling symptoms, physiological impairments, and disease activity and progression. Only findings from reviews and meta-analyses are shown.

a. Campbell E, Coulter EH, Paul L. High intensity interval training for people with multiple sclerosis: A systematic review. *Mult Scler Relat Disord*. 2018;24:55-63.

b. Charron S, McKay KA, Tremlett H. Physical activity and disability outcomes in multiple sclerosis: A systematic review (2011-2016). *Mult Scler Relat Disord*. 2018;20:169-177.

c. Dalgas U, Stenager E, Sloth M, Stenager E. The effect of exercise on depressive symptoms in multiple sclerosis based on a meta-analysis and critical review of the literature. *Eur J Neurol*. 2015;22(3):E443-E434.

d. Demaneuf T, Aitken Z, Karahalios A, et al. Effectiveness of exercise interventions for pain reduction in people with multiple sclerosis: A systematic review and meta-analysis of randomized controlled trials. *Arch Phys Med Rehabil*. 2019;100(1):128-139.

e. Ensari I, Motl RW, Pilutti LA. Exercise training improves depressive symptoms in people with multiple sclerosis: Results of a meta-analysis. *J Psychosom Res*. 2014;76(6):465-471.

f. Gharakhanlou R, Wesselmann L, Rademacher A, et al. Exercise training and cognitive performance in persons with multiple sclerosis: A systematic review and multilevel meta-analysis of clinical trials. *Mult Scler*. 2020:1352458520917935.

g. Gunn H, Markevics S, Haas B, Marsden J, Freeman J. Systematic review: The effectiveness of interventions to reduce falls and improve balance in adults with multiple sclerosis. *Arch Phys Med Rehabil*. 2015;96(10):1898-1912.

h. Hempel S, Graham GD, Fu N, et al. A systematic review of the effects of modifiable risk factor interventions on the progression of multiple sclerosis. *Mult Scler*. 2017;23(4):513-524.

i. Kalron A, Zeilig G. Efficacy of exercise intervention programs on cognition in people suffering from multiple sclerosis, stroke and Parkinson's disease: A systematic review and meta-analysis of current evidence. *NeuroRehabilitation*. 2015;37(2):273-289.

j. Motl RW, Gosney JL. Effect of exercise training on quality of life in multiple sclerosis: A meta-analysis. *Mult Scler*. 2008;14(1):129-135.

k. Paltamaa J, Sjogren T, Peurala SH, Heinonen A. Effects of physiotherapy interventions on balance in multiple sclerosis: A systematic review and meta-analysis of randomized controlled trials. *J Rehabil Med*. 2012;44(10):811-823.

l. Pearson M, Dieberg G, Smart N. Exercise as a therapy for improvement of walking ability in adults with multiple sclerosis: A meta-analysis. *Arch Phys Med Rehabil*. 2015;96(7):1339-1348.

m. Pilutti L, Greenlee T, Motl RW, Nickrent M, Petruzzello SJ. Effects of exercise training on fatigue in multiple sclerosis: A meta-analysis. *Psychosom Med*. 2013;75(6):575-580.

n. Pilutti LA, Platta ME, Motl RW, Latimer-Cheung AE. The safety of exercise training in multiple sclerosis: A systematic review. *J Neurol Sci*. 2014;343(1-2):3-7.

o. Platta ME, Ensari I, Motl RW, Pilutti LA. Effect of exercise training on fitness in multiple sclerosis: A meta-analysis. *Arch Phys Med Rehabil*. 2016;97(9):1564-1572.

p. Sandroff BM, Motl RW, Scudder MR, DeLuca J. Systematic, evidence-based review of exercise, physical activity, and physical fitness effects on cognition in persons with multiple sclerosis. *Neuropsychol Rev*. 2016;26(3):271-294.

q. Taul-Madsen L, Connolly L, Dennett R, Freeman J, Dalgas U, Hvid LG. Is aerobic or resistance training the most effective exercise modality for improving lower extremity physical function and perceived fatigue in people with multiple sclerosis? A systematic review and meta-analysis. *Arch Phys Med Rehabil*. 2021;102(10):2032-2048.

	General exercise[b]	Aerobic training	Resistance training	Yoga, Pilates, and flexibility training
SYMPTOMS				
Fatigue[a]	↓ [3, 67, m]	↓ [67, q]	↓ [q] / → [67]	↓ [32, 160] / → [151]
Pain[a]	↓ [d]	→ [d]		↓ [151]
Depressive symptoms	↓ [c, e]	↓ [c]	↓ [c]	↓ [32] / → [151]
Functional capacity (walking)[a]	↑ [l, 168, b, l]	↑ [95, b, l, q]	↑ [89, 102, l, q] / → [b]	→ [32, 151]
Balance/Falls	↓ [b, g, k]	→ [g]	↓ [89] / → [g]	↓ [151]
Cognition	→ [f, i, p]			↑ [160] / → [32]
Flexibility (range of motion)				→ [178]

	General exercise[b]	Aerobic training	Resistance training	Yoga, Pilates, and flexibility training
PHYSIOLOGICAL IMPAIRMENTS				
Muscle strength	↑ [o]	→ [95, o] / ↑ [95, a, o]	↑ [89,102, b, o, q]	↑ [151]
Aerobic capacity ($\dot{V}O_2max$)	↑ [o]	↑ [95, a, o, q]	→ [89, o]	
DISEASE ACTIVITY/PROGRESSION				
Relapse rate	↓ [143, n]			
Disease progression (EDSS)	→ [h]		→ [89]	
HEALTH-RELATED QUALITY OF LIFE (HRQOL)				
QOL	↑ [j]	↑ [j]	→ [89]	→ [32, 151, 160]

Note: Each arrow has a number and/or letter associated with it. The numbers are from the chapter reference list. The letters are the references in this sidebar.

[a]Symptoms rated among the most important bodily functions by persons with MS (57, 66).

[b]General exercise covers studies that pool findings from different exercise modalities or that use exercise interventions combining different modalities.

↑ = enhances effect of exercise on the listed parameter; → = no effect of exercise on the listed parameter; ↓ = reduces effect of exercise on listed parameter; empty field = no evidence could be located.

Research Focus

Pilates and Functional Changes

Since MS causes continuous neural decline resulting in decreased function, including walking ability and strength, methods to attenuate this decline are an important part of treatment. Exercise and physical activity training thus help individuals with MS maintain their health. Pilates is a common part of a healthy exercise training plan for many individuals but has not been studied much in those with MS. This randomized controlled trial (RCT) enrolled 42 patients with MS, 21 each to a group performing Pilates weekly for 8 wk plus a home exercise program (focused on strength and stability) and to a group performing only the home exercise program (1). A total of 33 participants completed the final assessments.

Although no differences were found between the groups for walking speed, perception of walking ability, and fear of falling, there was a significant difference for walking endurance, postural and core stability, respiratory ability, and cognitive function. Given the increased emphasis on home-based exercise programs (e.g., cardiac and pulmonary rehabilitation) during the 2020-2021 pandemic, this type of program may allow a larger number of individuals with MS to participate. Home exercise removes or reduces barriers to performing exercise, including transportation and time. And Pilates is a low-risk option that can be easily taught or followed on a video. Most important, this type of programming has the potential to reduce progressive declines in motor, cognitive, and respiratory function, which can improve mobility and allow those with MS to maintain their independence and functional ability.

Clinical Exercise Bottom Line

- Exercise is feasible as well as safe in patients with multiple sclerosis and can effectively reduce common symptoms.
- Aerobic and resistance exercise are the most examined modalities and are highly effective in eliciting physiological improvements (e.g., cardiorespiratory fitness and muscle strength) as well as functional performance (e.g., walking).
- A combination of exercise modalities (preferably aerobic and resistance exercise) performed two or three times per week with moderate to vigorous intensity is recommended.

CONCLUSION

There is ample evidence that exercise, when performed properly, can limit the degree of physical disability and ameliorate symptoms in people with MS. Furthermore, exercise may even have disease-modifying effects in MS, suggesting a potential role for exercise as an adjunct to established medical treatments. It is best to begin an exercise routine and increase or maintain daily physical activity levels as early in the disease process as possible. Exercise training recommendations published for the MS population generally suggest that persons with MS may gain cardiorespiratory, strength, and balance benefits using a modified exercise prescription similar to that for adults without MS. It is prudent, however, to adjust the exercise prescription when symptoms are heightened by considering cooling interventions and by lowering exercise intensity, frequency, or duration, or some combination of these. The ultimate goal of regular exercise training in persons with MS is to maintain physical and psychological health, functional capacity, mobility, and quality of life.

Online Materials

Visit HK *Propel* to access a link to the references, the case studies with discussion questions, and a quiz to help you review key concepts and test your understanding of the material covered in this chapter.

CHAPTER 29

Cerebral Palsy

Désirée B. Maltais, PT, PhD

Cerebral palsy is a permanent childhood-onset condition that affects movement and posture and often results in reduced physical activity. Regular exercise training can be beneficial. This chapter reviews the use of exercise to help adults with cerebral palsy live healthier and more active lives.

DEFINITION

Cerebral palsy is defined as "a group of permanent disorders of the development of movement and posture, causing activity limitation, that are attributed to nonprogressive disturbances that occurred in the developing fetal or infant brain. The motor disorders of cerebral palsy are often accompanied by disturbances of sensation, perception, cognition, communication, and behaviour, by epilepsy, and by secondary musculoskeletal problems" (59). Cerebral palsy is therefore best understood as not one disorder but several linked by the timing of the disturbances to the brain (i.e., during fetal development or in infancy) and the resulting permanent motor disorder. Other associated impairments and conditions vary depending on the extent and location of the brain disturbances.

SCOPE

Based on a systematic review and meta-analysis of data from 49 studies (mostly from Europe, North America, and Australia, but with one study each from China, Turkey, and Kenya), the prevalence of cerebral palsy is in the range of 2 to 5 per 1,000 live births (51). The prevalence varies by gestational age at birth and by birth weight. Data from this same systematic review showed the highest prevalence (82.3 per 1,000 live births) is in infants with gestational age at birth of <28 wk, with the lowest prevalence in infants born at >36 wk (0.35 per 1,000 live births). For birth weight, the highest prevalence per 1,000 live births is for infants with a birth weight of 1,000 to 1,499 g (59.2 per 1,000 live births), and the lowest is for those with a birth weight >2,500 g (1.3 per 1,000 live births). Of note, the overall prevalence of cerebral palsy has not changed markedly over the last several years, even though more preterm infants are surviving (51).

Prenatal or perinatal risk factors account for approximately 75% of all cases of cerebral palsy (56). In addition to prematurity (likely the most important risk factor), common risk factors are hyperbilirubinemia, infection of the fetus or mother, trauma during birth, and intracranial hemorrhage (26, 46). Lower socioeconomic status may be an indirect risk factor as there is an increased prevalence for lower birth weight as well as for cerebral palsy in infants born into families of this status (50). Postnatal events that occur before the age of 2 yr (e.g., viral or bacterial meningitis; traumatic head injury from vehicular accidents or abuse; anoxia as a result of near drowning, cerebral vascular accidents, tumors, or surgery; and toxins that cause heavy metal encephalopathy) are also risk factors for cerebral palsy (15). Most individuals (~70%) with cerebral palsy will develop the ability to walk (11). There is, however, an increased economic cost for these individuals. Based on data for American children enrolled in Medicaid, medical costs for children with cerebral palsy are estimated to be 10 times higher than for children without cerebral palsy (33). According to the Centers for Disease Control and Prevention, the lifetime cost of persons born in the year 2000 with cerebral palsy is estimated to reach $11.5 billion (13).

PATHOPHYSIOLOGY

Noxious events lead to damage to brain tissue that was healthy in 90% of cases with cerebral palsy, with the other 10% resulting from abnormal brain development (10). The type and location of the neurological abnormalities and lesions leading to cerebral palsy will differ depending on timing of the noxious events. Both the cerebrum and **cerebellum** can be affected.

Acknowledgment: Much of the writing in this chapter was adapted from the previous editions of *Clinical Exercise Physiology.* As a result, we wish to gratefully acknowledge the previous efforts of and thank Amy E. Rauworth, MS, and James H. Rimmer, PhD.

Brain malformations are often related to noxious events (e.g., cytomegalovirus) occurring before 20 wk gestation (10), because this is the time for brain structure formation. Most of these pregnancies (71%) will go to term (20), however. Common malformations are **microcephalus** and **hydrocephalus** (20). The type and extent of the motor control issues will depend on the specific areas affected and the extent of the abnormalities.

White matter damage, present in about 40% of cases with cerebral palsy (10), typically results when the noxious event occurs from about 26 to about 32 wk gestation, when there is a high rate of white matter development (myelination). Periventricular–intraventricular hemorrhage and periventricular leukomalacia are common white matter lesions that can lead to cerebral palsy (10). White matter damage is often associated with prematurity, hypoxic-ischemic events, and fetal or maternal infection (21). At the end of the third trimester and in the neonate, the gray matter is highly metabolically active, and noxious events such as hypoxia-ischemia at this time can lead to gray matter lesions, such as multifocal or focal cortical lesions (~7% of cases with cerebral palsy) and lesions to the **basal ganglia** or **thalamus** (~13% of cases with cerebral palsy) (10). Lesions to the corticospinal tracts (thus white and gray matter) are associated with spasticity (68). There is a topographic distribution of the corticospinal fibers. Moving from proximal to distal to the lateral ventricles are the fibers involved in control of the lower limb, the upper limb, and the face, respectively (1). Thus, there is a topographic distribution in the clinical presentation of spasticity. These patterns are discussed in the Signs and Symptoms section. Lesions to the basal ganglia, thalamus, and related **extrapyramidal** pathways are associated with **dyskinesia**.

Both the cerebrum and cerebellum can show malformations as well as white and gray matter lesions. Malformations and white and gray matter lesions in the cerebellum (~5% of cases with cerebral palsy) are associated with ataxia (68). Individuals with cerebral palsy can present with a combination of lesions and malformations related to one or more noxious events. Thus they can present with motor disturbances that are a combination of one or more of spasticity, dyskinesia, and ataxia. These motor disturbances also have musculoskeletal and related consequences relevant to the clinical exercise physiologist. Some of the more relevant consequences are described in the next section.

Musculoskeletal and Related Consequences

Individuals with cerebral palsy demonstrate muscle weakness, in part due to motor control issues (decreased central activation, decreased ability to modulate the frequency of motor unit activation, and deficits in reciprocal inhibition) and in part due to cellular changes (smaller muscles overall, smaller muscle fiber diameter, decreased elasticity, and fewer sarcomeres, with those present being overlong with less than optimal actin–myosin filament overlap) (43). Muscle weakness begins in childhood (43) and

---

Definitions of Muscle Tone Impairments Seen in Cerebral Palsy

- Spasticity—A velocity-dependent increase in resistance to passive movement.
- Dyskinesia—Involuntary contractions resulting in atypical postures. There are three types.
  - **Athetosis**: slow, continuous, repetitive torsions; stable postures can be challenging to maintain
  - **Chorea**: sudden, irregular movements of short duration
  - **Dystonia**: sustained or intermittent contractions following a specific pattern
- Ataxia—Loss of full control of bodily movements.

Stanley et al. (68)

---

carries over into adulthood (27, 47). The muscles affected depend on the location and extent of the brain malformations or lesions. Muscles in cerebral palsy also appear to have a blunted response to long bone growth (55). Thus **contractures** can develop and worsen as the child grows (21). Contractures are also seen in adults with cerebral palsy (5, 21). The motor impairments in cerebral palsy are associated with bony torsions, hip subluxation or dislocation, and spinal deformities such as kyphosis and **scoliosis**. These typically appear in childhood but can worsen over time (21).

The increased neck motion seen in adults with athetosis may contribute to premature cervical disc degeneration and instability that progresses more rapidly and is seen in more disc levels than in adults without health problems (22). This degeneration is complicated by the narrow spinal canal common to this group (22). Although this problem is mostly seen in individuals with athetosis, any neurological changes and deterioration in people with cerebral palsy should not be viewed as a typical part of the aging process but as a problem that requires rapid investigation. Bone mineral density is also reduced in both children (23, 24, 25, 29) and adults (41) with cerebral palsy. Requiring assistance for transferring and having a low body mass index may be risk factors for low bone mineral density in adults with the condition (41). See table 29.1 for a list of common musculoskeletal problems and relevant exercise guidelines.

The musculoskeletal impairments seen in cerebral palsy may have consequences that increase over time. Ando and Ueda and others (6, 19, 35, 45) have reported that 35% of adults with cerebral palsy have reduced ability to perform activities of daily living over time. The time sequence for this decline could be as short as

Table 29.1 Conditions Associated With Cerebral Palsy and Relevant Exercise Guidelines

Associated conditions	Exercise guidelines
Incontinence Neurogenic bladder	Bladder should be emptied before exercise session.
Cognitive and behavioral problems	Ask questions that require brief answers. Allow time for accommodation to a new task; repetition is important. Use precise language and simple words. Limit the number of instructions given at one time.
Musculoskeletal problems • Fractures • Hip dislocation • Hypertonia • Leg length discrepancies • Pelvic obliquity • Bone loss	Incorporate weight-bearing and other exercise only when appropriate. Assess balance to choose appropriate exercise modality to prevent falls. Be cautious with exercises that move the hip joint if history of dislocations is present.
Oral motor dysfunction or hypersensitivity • Dysphagia • Drooling • Dysarthria	Provide a towel for possible drooling. Make adaptations to the mouthpiece during exercise testing, or use a face mask. Ask simple questions if speech is difficult to understand.
Asthma	Follow physician's recommendations for use of asthma medication.
Seizures or epilepsy	Consider the effects of medication on exercise and the patient's endurance level.
Sensory and perception impairments	Break down instructions and cues.
Visual impairment	Speak to someone in the appropriate field of vision. Provide an obstacle-free environment to prevent falls.

Data from Gajdosik and Cicirello (19); Krigger (37).

5 yr. Strauss and Shavelle (69) identified a decline in ambulatory function after the age of 60 as being associated with a subsequent increase in mortality. **Secondary conditions** such as poor joint alignment and overuse syndromes may also contribute to the loss of independent walking (45). Fatigue as a limiting factor for ambulation has been reported in several studies (32, 76). Murphy and colleagues noted that approximately 75% of their subjects discontinued ambulation by the age of 25 because of fatigue and inefficient movement (45). This cohort also reported pain in weight-bearing joints that decreased ambulation around age 45 in those who continued to ambulate throughout life.

Pain is commonly reported among people with cerebral palsy, both in childhood and in adulthood. Seventy-nine percent of parents of children with cerebral palsy reported that their child had severe to moderate pain (54). Of those who reported higher levels of pain, poorer health as identified by difficulty in feeding, seizures, and severe motor or intellectual impairments increased the risk of familiar stress. Turk and colleagues reported that 84% of adult women with cerebral palsy identified pain as a deterrent to participation in certain activities of daily living (73). As adults with cerebral palsy age, inactivity levels appear to be directly related to pain. Health care utilization studies suggest that adults with cerebral palsy seek preventive and rehabilitative services less often than others and do not consult physicians regarding pain

(80). The pain level for an adult with cerebral palsy is typically in the moderate to severe range (65). The occurrence of pain is most commonly identified in the lower extremities, back, shoulders, and neck (45, 57, 73). In relation to pain in these areas, limited range of motion is commonly reported in the ankle, hip, and shoulder joints (63). Sandstrom and colleagues (63) reported a high incidence of pain in the lower extremities of persons who could walk with or without assistance.

It is perhaps not surprising that adults with cerebral palsy also have a reduced level of aerobic capacity (27). Also perhaps not surprising is that adults with cerebral palsy show reduced physical activity and increased time spent in sedentary behavior, which may be associated with a higher risk for cardiometabolic diseases (61). The life span of people with cerebral palsy who do not have significant comorbidities, however, approaches that of the general population (69).

Associated Conditions

In addition to motor impairment and associated musculoskeletal impairments with related consequences, the brain lesions and malformations that lead to cerebral palsy may also be linked to several other **associated conditions**, depending on the site and extent of the lesion or malformation. Thus individuals with cerebral palsy may present with intellectual disability, sensory deficits, seizure disorders, feeding problems, behavioral dysfunction, and emotional problems (16). (See table 29.1 for a list of common associated conditions and relevant exercise guidelines.) In addition to varying with the type and severity of cerebral palsy, intellectual disability also increases with the presence of **epilepsy** (50). Seizures occur in approximately half of all children with cerebral palsy. When seizures occur without a direct trigger, the condition is labeled epilepsy.

Visual disturbances frequently occur in cerebral palsy. The most common type is **strabismus**, a condition in which a discrepancy is present between the right and left eye muscles. Individuals with hemiparesis often have an impairment in or loss of sight in the normal field of vision of one eye, referred to as **hemianopia**. This condition can also occur in both eyes and affect the same area of the visual field. Approximately 90% of people with hemiplegic cerebral palsy (i.e., with a unilateral lesion along the corticospinal tract, thus involving white matter or gray matter or both) display significant bilateral sensory deficits (50). Speech impairments such as **dysarthria** and **aphasia** occur most commonly in individuals with **dyskinetic (pertaining to cerebral palsy)** and spastic cerebral palsy affecting all four limbs (17). Dental problems such as malocclusion are prevalent in children with cerebral palsy, resulting in a significant increased overjet (an acute angle of the mandible) (18). For detailed information on the prevalence of associated conditions, see Odding and colleagues (50).

CLINICAL CONSIDERATIONS

As noted already, individuals with cerebral palsy present with a variety of health conditions. The clinical exercise physiologist should have a general familiarity with these conditions to ensure that exercise testing and training are done safely and effectively.

Signs and Symptoms

The information presented here focuses on motor impairments and mobility. However, the individual with cerebral palsy may also present with the previously described musculoskeletal problems, related consequences, and associated conditions.

Depending on the location of the lesion or malformation, individuals with cerebral palsy may present with one or more of the following motor impairments: spasticity, dyskinesia (athetosis or dystonia), or ataxia. For those with spasticity, there will be a topographic distribution that follows each lesion in the corticospinal tract (21). Thus spasticity may be evident on one side of the body (unilateral) or on both sides (bilateral). Unilateral involvement may be evident in one lower limb (**monoplegia**) or in the upper and lower limb on one side (**hemiplegia**). With bilateral spasticity, three limbs or all four limbs may be involved. Spasticity may thus be more evident in the lower limbs compared with the upper limbs (**diplegia**), be present in the lower limbs and one upper limb (**triplegia**), or manifest to a similar degree in all four limbs

(**complete tetraplegia**).

The level of functional mobility of a person with cerebral palsy depends on the extent of the lesions and malformations affecting motor systems and possibly other contextual factors (e.g., the level of physical activity), the latter perhaps becoming increasingly relevant over time. The Gross Motor Function Classification System (GMFCS) is a widely used method to classify and describe the level of mobility in cerebral palsy (52). Although developed for children and adolescents, it is also used with adults (31, 42). There are five levels, with each level showing mobility limitations greater than the previous one, as described in table 29.2.

Adults with cerebral palsy classified in level I present with no walking limitations but show difficulties with more complex tasks, such as tasks that involve high levels of speed, changing direction quickly, or dual-tasking (performing two tasks at once). Those classified in level II can walk without support, but they show limitations nonetheless (e.g., reduced speed) and may have an atypical posture and walking pattern. They may also need to use a handrail when walking up and down stairs, and they may require support (e.g., a cane, crutches, a walker) when walking long distances in the community. Adults classified in level III have even greater limitations than those in level II in that they always need support to walk. Although they may self-propel a manual wheelchair, they often require a powered chair for community mobility. Those classified in level IV may walk for exercise, and if so, they require a walker with extensive support of the trunk and possibly supervision. They may also self-propel a manual wheelchair, but they typically use a powered wheelchair (likely with trunk support) for most mobility needs. Individuals classified in level V have very limited voluntary movement, require personal assistance for all functional mobility, and are typically pushed in a manual wheelchair that has trunk and head support.

Commonly observed postures in the upper limb include shoulder internal rotation, elbow flexion, forearm pronation, wrist flexion, finger flexion, and thumb in palm (56). Common lower limb postures include hip flexion, hip abduction, knee flexion, ankle equinus, **hindfoot valgus**, and toe flexion (56). Common gait pattern deviations include toe walking, crouched gait, jump gait, and scissoring. These postures and patterns are adaptations that allow the individual to perform the mobility task at hand, based on their specific motor impairments and the relevant musculoskeletal and related consequences. When the adaptations have long-term negative consequences on health or function and are modifiable, they are addressed by the appropriate members of the health care team (see the Treatment section).

History and Physical Examination

The clinical exercise physiologist should review the client's health history to have a clear understanding of their predominant motor impairments, musculoskeletal problems and related consequences, and associated conditions as well as their level of mobility and, if applicable, predominant walking pattern.

Table 29.2 Gross Motor Function Classification System

Level	Characteristics of adults with cerebral palsy
I	No major walking limitations Difficulties with motor tasks involving high levels of speed, changing direction quickly, and dual-tasking (performing two tasks at once)
II	Able to walk without support but show limitations to walking such as reduced speed May demonstrate atypical postures and walking patterns May need to use a handrail when walking up and down stairs May need to use support (e.g., a cane, crutches, a walker) when walking long distances in the community
III	Always require support to walk (e.g., a cane, crutches, a walker) May self-propel a manual wheelchair Often require a powered chair for community mobility
IV	May walk for exercise, and if so, require a walker with extensive support of the trunk and possibly supervision May self-propel a manual wheelchair but typically use a powered wheelchair (likely with trunk support) for most mobility needs
V	Limited voluntary movement Require personal assistance for all functional mobility Typically pushed in a manual wheelchair that has trunk and head support

Data from Palisano et al. (52).

The physical examination should screen for contractures (see table 29.3), deformities (see table 29.1), extreme weakness, postural control (balance) issues, pain, and fatigue because these may need to be accommodated for during exercise testing and training (see table 29.1). Any individual with cerebral palsy who has not exercised or who has not exercised in the recent past should undergo a medical examination to identify cardiorespiratory or other conditions that could preclude exercise or limit exercise tolerance.

Diagnostic Testing

According to Samson and colleagues, an accurate diagnosis can be made at 6 mo of age except in the mildest forms of cerebral palsy (62). It has been suggested that because of the progression of signs and symptoms in the early stages of an infant's life, the extent of the condition may not be identified until 2 or 3 yr of age or later (14). The diagnostic process for cerebral palsy begins with a medical history and takes into consideration risk factors and the clinical presentation of the infant or young child (i.e., whether spasticity or other motor control abnormalities are present). Tests such as the Movement Assessment of Infants (MAI), Bayley Motor Scale, and the Test of Infant Motor Performance (TIMP) can be used to screen for motor development abnormalities (12). The American Academy of Neurology and the Child Neurology Society (CNS) have developed evidence-based guidelines for the diagnostic assessment of children with cerebral palsy (7). According to these guidelines, magnetic resonance imaging is preferred to computed tomography for identifying lesions or malformations in the brain in children with possible cerebral palsy. Metabolic and genetic testing is recommended only if the neuroimaging findings identify brain malformations or if the neuroimaging findings, history, or clinical presentation are atypical for cerebral palsy or if they demonstrate factors additional to those typically seen with cerebral palsy. Testing for coagulation disorders is also recommended. An EEG should be performed only if epilepsy or a related condition is suspected. Screening and additional diagnostic testing, if necessary, should be done for common associated conditions such as an intellectual disability, visual or hearing impairment, speech and language disorders, and oral-motor impairment.

Skeletal problems can occur at any age. These are typically identified using radiographic techniques. Clinical gait analysis may also be used in ambulatory individuals to determine the effect of the motor impairment and musculoskeletal abnormalities on various biomechanical and electrophysiological factors. Other musculoskeletal problems and their consequences and associated conditions may be identified using imaging or **electrodiagnostic** techniques or specific test batteries or protocols, depending on the issue at hand.

Practical application 29.1 provides a patient's perspective on ways to work with people who have cerebral palsy.

Exercise Testing

The tests described here are performed in a clinical exercise testing facility, as opposed to being field tests. Clinical exercise testing of patients with cerebral palsy may be required if they have cardiovascular disease risk factors or for exercise prescription purposes or monitoring progress. The choice of modality and the needed adaptations, if relevant, will depend on the purpose of the test, the individual's clinical presentation, and the principle of specificity of training (testing related to what will be trained). Certain exercise tests may not be feasible for some people with

Table 29.3 Common Contractures

Joint	Contracture
Hip	Flexion with adduction and internal rotation
	Flexion with abduction and external rotation
Knee	Flexion
Ankle	Plantar flexion with pronation and eversion
	Plantar flexion with supination and inversion
Shoulder	Flexion
	Adduction
	Internal rotation
Elbow	Flexion
	Pronation
Wrist, MCP, IP, and CMC (thumb) joints	Flexion

MCP = metacarpophalangeal; IP = interphalangeal; CMC = carpometacarpal.

cerebral palsy. Details specific to exercise testing for adults with cerebral palsy follow. For general information regarding exercise testing, see chapter 5.

Cardiorespiratory

The choice of testing modality will depend on the individual's clinical presentation and the purpose of the test. Individuals classified in GMFCS level V will likely not have sufficient motor control to perform cardiorespiratory exercise testing. This may also be the case for some individuals in GMFCS levels II, III, and IV for whom the limiting factors are motor control, contractures, pain, or a marked intellectual disability (or a combination of these). For these individuals, testing will be very individualized and pragmatic and thus based on the goals for progress and what the individual can do.

It is possible that an individual with cerebral palsy can perform a graded exercise testing protocol but cannot tolerate the mask or mouthpiece for expired gas analysis (i.e., cardiopulmonary exercise testing). Problems with tolerance can occur even midtest, so it is important to habituate the individual to the expired gas collection equipment. When expired gas collection is not possible, the outcome measures will be limited to end-test time, stage, and heart rate. A trial test is suggested if an individual with cerebral palsy has a learning or intellectual disability or is somewhat anxious, or if the clinical exercise physiologist is unsure of the best testing protocol.

A pretest may help the clinical exercise physiologist answer the following questions when choosing a protocol: Does the individual have enough motor control and range of motion to perform the test, especially at the higher power outputs or speed and slope combinations? Can deformities be accommodated for, and is there enough clearance between the equipment and the limbs being tested? Should the test involve the upper or lower limbs? What should be the appropriate starting power output or treadmill speed and slope,

and what should be the incremental increases at each stage? Can the individual tolerate and correctly use a mouthpiece or mask?

Protocols are often more individualized with this group than with the general adult population. Graded exercise tests will likely be symptom limited, and thus the main outcome measure may be a peak rather than a maximal value. The sidebar Important Considerations for Exercise Testing reviews some of the specific details for cardiorespiratory exercise testing.

Wheelchair and Upper Body Ergometry In a wheelchair protocol, the test is performed using the individual's own manual wheelchair and a wheelchair roller, which locks the chair in place and provides a stationary means of propelling the wheels. This test may be feasible for individuals who self-propel a manual wheelchair that has specialized trunk support (some individuals in GMFCS levels III and IV). Calculating or controlling workload using wheelchair ergometry with a mechanism to measure power is, however, often difficult.

Arm ergometry is an alternative for individuals who self-propel a manual wheelchair (some individuals in GMFCS level III and those in level II for whom the possibility of aerobic training using the upper limbs is under investigation). An arm ergometer allows for more accurate control of the workload and still provides an accurate measure of oxygen consumption, assuming the test protocol can be performed correctly. However, it is important that the trunk and lower limbs be appropriately stabilized to ensure that any movement the individual makes is directed toward moving the crank handles. Head support may also be needed. The arm ergometer should also be secured in a fixed position. The position of the individual with respect to the ergometer should be such that the correct upper limb movements can be performed during the test. If the individual cannot hold the crank handle because of a contracture or muscle weakness in the hands, straps or an adaptive

Practical Application 29.1

CLIENT–CLINICIAN INTERACTION

June Kailes, author of *Health, Wellness and Aging With Disability*, is a well-known national disability policy consultant. June has ataxic cerebral palsy and uses a scooter to ambulate and to manage musculoskeletal pain. Her job requires long hours on the computer, on the phone, on airplanes, and in meetings. She makes physical activity a priority because her job and lifestyle are sedentary. She works out regularly and can provide insight to the clinical exercise physiologist from both a personal and a professional standpoint.

June's primary exercise goals include pain reduction, prevention of secondary conditions, maintenance of function, and weight control. June works with a personal trainer once a week on average and exercises independently two to four times per week. Her exercise program includes walking on the treadmill for 40 min followed by a comprehensive set of flexibility exercises. June needs a treadmill that can operate at 1.3 mph (2.1 kph). She uses a balance ball to perform strength training exercises that also challenge her balance. Using the treadmill for support, she performs standing exercises to improve her balance. In the evenings, she does floor exercises that focus on strength and flexibility for her back and core muscles.

When asked what motivates her most to continue her exercise program, June suggested that concerns about losing function and preventing increased pain keep her exercise regimen a high priority in her lifestyle. June said, "In the past we didn't age, we just died! We are the first generation to live this long, so the question is not, 'Will we live?' but 'How well will we live?'" She also stated, "Many of us will probably live longer than we think, so we have to think about what we can influence or change in terms of the quality of our years as we age." June advises the clinical exercise physiologist not to minimize the symptoms she is experiencing but to simply relate them to the aging process. "What's coming into sharp focus for many of us is that the changes brought about by regular aging (yet to be clearly defined) can play havoc with a person's ability to function." Like many of her fellow baby boomers, she states that she expects more, not less, with age. When asked what she enjoys the most about exercise, June said, "When it's over . . . a sense of accomplishment!"

June also encourages the clinical exercise physiologist to be specific about the type and amount of exercise recommended. She believes that the more she knows about why she is doing an exercise and what benefits it will bring, the more likely she will be to include it in her activities. She advises clinical exercise physiologists to speak in nonclinical terms, to be creative in exercise design, to ask for the patient's suggestions or thoughts, and to challenge people with cerebral palsy. She stresses the importance of understanding the secondary and associated conditions of cerebral palsy and the importance of tailoring the program to the patient's wants and needs. June often travels and sometimes encounters barriers such as a lack of accessible fitness centers in hotels or equipment that is not accommodating. When fitness centers lack accessible options, June walks the hotel halls and does flexibility and strength exercises in her room. June expects that exercise will always be part of her life. She sees the benefits and believes the investment of time is more important for her than for people without disabilities.

glove may be needed to secure the hand to the crank handles (4, 53). A suggested protocol for arm ergometry begins with a starting power output of 0 to 15 W at 30 to 50 rpm, increasing 5 to 10 W every 1 to 2 min until volitional fatigue is reached or the test is stopped for safety reasons (4). As noted, determination of the exact starting power output and increases may require a pretest.

Leg Cycle Ergometry   Leg ergometry may be feasible for those with sufficient lower limb motor control and range of motion and lack of deformities to allow for safe pedaling at the desired power outputs (most individuals in GMFCS level I, some in GMFCS levels II and III, and perhaps some in GMFCS level IV who used to walk). A suggested protocol is a starting power output between 25 and 50 W at 50 to 60 rpm, with an increase of 15 to 25 W for each 2 min increment until volitional fatigue or until the test is stopped for safety reasons (2, 4). The feet may need to be fixed to the pedals.

Treadmill Testing   The treadmill may be a useful testing device for individuals who can walk on the treadmill without holding the handrails (individuals functioning in GMFCS level I and some in GMFCS level II). At least two testers are needed because one must directly supervise (i.e., "spot") the individual on the treadmill to ensure the test is performed correctly and safely. Even if the individual can correctly perform the test, they should be allowed to practice the testing methods to allow accommodation and therefore produce a more accurate test (34). One accommodation for children and adolescents with cerebral palsy, which may also be used with adults, is a three-stage protocol of 5 min stages, with a few minutes' rest between stages. The first stage

Important Considerations for Exercise Testing

Allow practice time to ensure the individual can safely perform the test and to let them get used to the modality.

Cardiorespiratory

- Ensure enough rest between the practice and the test for valid results.
- Make sure the individual is positioned correctly, especially if an arm or cycle ergometer is used (straps may also be necessary to secure feet and hands; for an arm ergometer, securing the trunk to a back rest will help ensure stability).
- Use a mask if the person drools excessively or has poor oral-motor control.
- Practice with the mouthpiece or mask may be necessary.

Strength

- Ensure the individual has enough range of motion and motor control to perform the test.
- Machines are generally safer than free weights and may allow for greater ease of movement.
- Wide benches, low seats, and trunk and pelvic strapping should be used when necessary to ensure only the desired muscles are contracting.
- Use nonslip handgrips and gloves for safety.

Range of Motion

- Ensure the individual is comfortably positioned and well supported during testing.
- Consider that speed may increase **muscle tone** and attenuate passive range of motion results.
- It may be helpful to measure passive range of motion in two conditions: at the point where resistance is first encountered and at maximal passive range of motion.

is at a comfortable walking speed, holding the handrails if necessary. The second and third stages are of increasing difficulty, with no support and an increase in speed and slope (28).

Other Ergometers   Other ergometers such as recumbent steppers (which use the upper and lower limbs) and the Schwinn Airdyne can be used as testing modalities for individuals with cerebral palsy, depending on their functional ability. An advantage of the Nu-Step is the greater recruitment of musculature as compared with cycling or arm cranking protocols and the ability

to increase workload without increasing cadence. Studies in the nondisabled population show that peak $\dot{V}O_2$ values achieved during treadmill walking correlate strongly with all-extremity exercise testing values and do not differ significantly from those with nonwalking modalities (39).

Musculoskeletal

Individuals classified in GMFCS level V will likely not have enough motor control to perform any repetition maximum, handheld dynamometry, or isokinetic dynamometry testing. This may also be the case for certain muscle groups for certain individuals in GMFCS levels II to IV for whom the limiting factors are motor control, contractures, pain, or a marked intellectual disability (or a combination of these). For those who cannot perform the standardized tests, testing will be very individualized and practical and thus based on what the individual can do as well as the goals for progression of the chosen exercise training program.

Repetition Maximum   According to the American College of Sports Medicine (ACSM), the optimal protocol for determining dynamic muscle strength (involving movement of the body or an external load) is the one-repetition maximum (1RM) (3). Depending on the motor and associated impairments and perhaps the learning or intellectual disability, a 1RM test may not be feasible or safe. In these situations a 6RM or a 10RM may be more appropriate. Extrapolating 6RM or 10RM data and estimating a person's true 1RM may be problematic, however, because of the marked variation in the number of repetitions that can be performed at a fixed percentage of a 1RM for different muscle groups (e.g., leg press vs. bench press). The Holton curve takes an individual's 10RM and adjusts the score to estimate the 1RM. According to the ACSM, an 8RM or 25RM can also be used to determine dynamic muscle strength and endurance of individuals with cerebral palsy (2, 4).

If the clinical exercise physiologist is not sure of test feasibility, a pretest with a very light load is recommended to determine if the individual understands the test and can perform the basic movements. A practical alternative method that indirectly assesses dynamic muscle strength is to record the number of repetitions that can be completed in 1 min of a given load (perhaps a load relevant to the training protocol). Performance level is graded by individual progression from one testing period to the next. Again, a pretest is recommended if feasibility is unknown or uncertain.

Handheld Dynamometry and Isokinetic Dynamometry   Handheld dynamometers can be used to measure isometric muscle strength, and isokinetic muscle strength can be measured using dynamometers such as the Cybex II (8, 71). A pretest is again recommended if feasibility is unknown or uncertain.

Functional Tests of Locomotion   The clinical exercise physiologist may wish to determine whether an exercise training program affected locomotor skills for those classified in GMFCS levels I to III and for those in GMFCS level IV who have some walking ability. Depending on the goals for the exercise program, the 6 min or 10 m walk tests, timed stair climbing, and the Timed Up and Go tests can be used (40).

When testing persons with cerebral palsy, the clinical exercise physiologist should remember several specific issues that can affect the safety and effectiveness of the testing (2, 4):

1. Cocontraction may offset strength in tested muscle groups (agonists).

2. Measure range of motion (ROM) in tested muscle groups to determine the safety of exercise testing and specific exercise prescription.

3. Test muscle groups unilaterally (there may be more spasticity on one side or a significant strength difference).

4. Focus on stability, coordination, ROM, and timing.

5. Adaptations include wide benches, low seats, and trunk and pelvic strapping.

6. Machines are safer than free weights and provide greater fluidity to the movement.

7. Use a metronome to ensure appropriate fluidity. Be sure to use a slow cadence to decrease spasticity.

8. Use nonslip handgrips and gloves if necessary.

9. Always provide adequate practice before testing.

Flexibility

It is important to measure flexibility in individuals with cerebral palsy to ensure that any exercise training program does not have an adverse effect on this factor. Common devices to measure flexibility include goniometers, electrogoniometers, and tape measures. The range of each joint can be measured with a goniometer. This is most commonly done through passive motion but can be done through active motion as well. The measurement is taken throughout one plane of motion at a time. Placing the joint in the zero position is important. The goniometer must be aligned with the bony prominence to measure the range of motion accurately. Other forms of flexibility tests include the sit-and-reach test (which measures low back and hip joint flexibility) and Apley's scratch test. Some of these tests may not be appropriate for all individuals with cerebral palsy.

Treatment

Given that cerebral palsy is a lifelong condition that can affect many different organ systems and have many different effects on function, its management is lifelong and multidisciplinary. Priorities are usually set based on the impact of the issue in question on long-term health and functioning, and the goals of the individual and those of the family or primary caregiver, when appropriate. Exercise interventions play an important role and are addressed in the exercise prescription section.

Childhood is the most intensive period for interventions. The following list of recommended interventions for children with cerebral palsy is based on a review of the evidence (48). Not all interventions will be relevant for all children with cerebral palsy. Some of these interventions may be carried over into adulthood if they are still relevant for the given individual.

- *Seizures:* Anticonvulsants such as phenobarbital, phenytoin, valproic acid, gabapentin, and topiramate are recommended for seizure management (see table 29.4).

- *Spasticity:* Botulinum toxin injections to the affected muscles, diazepam medication, and selective dorsal rhizotomy surgery are highly recommended. Also recommended, but with somewhat less supporting evidence, are the following medications: tizanidine, oral and intrathecal baclofen, and dantrolene. Intramuscular injections of alcohol or phenol are not recommended to address spasticity issues (see table 29.4).

- *Contractures:* Lower limb casting is highly recommended. Also recommended, but with less evidence, are upper limb casting, orthoses for the hand and the ankle or foot, and orthopedic surgery. Manual stretching and neurodevelopmental treatment (NDT) techniques are unlikely to improve existing contractures and thus are not recommended if improvement is the goal (48).

- *Muscle strength:* Exercise is addressed in the exercise prescription section. Other interventions recommended for treating muscle weakness, but with somewhat limited evidence, are electrical stimulation and whole-body vibration. Vojta techniques are not recommended for muscle weakness.

- *Motor activities:* Highly recommended interventions to improve upper limb motor activities are constraint-induced manual therapy, bimanual therapy, and occupational therapy after botulinum toxin injections. Goal-directed and context-focused therapy and similar home programs are also highly recommended for addressing motor activity limitations. Also recommended, but with less evidence, are biofeedback, hippotherapy, hydrotherapy, electrical stimulation, and single-event multilevel surgery with physical therapy. Conductive education, NDT, TheraSuits, hyperbaric oxygen, sensory integration techniques, and Vojta techniques are not recommended for addressing motor activity limitations.

- *Overall function and self-care:* Goal-directed training and similar home programs are highly recommended. Also recommended, but with less evidence, are specialized seating, treadmill training, assistive technologies, and hand orthotics. Certain treatments for spasticity may also influence overall function and self-care, namely botulinum toxin injections, intrathecal baclofen, and selective dorsal rhizotomy. Sensory techniques, NDT, and massage are not recommended for overall function and self-care issues.

- *Associated conditions:* Readers interested in rehabilitation and medical interventions related to associated conditions can consult the review by Novak and colleagues (48).

Table 29.4 Pharmacology Review

Medication class and name	Primary effects	Exercise effects	Important considerations
Antispasmodics and muscle relaxants Valium, clonidine, baclofen, dantrolene, tizanidine, alcohol, phenol	Inhibits reflexes contributing to increased muscle tone	Sedation, nausea, hypotonicity, dizziness, loss of seizure control	These medications act systemically; therefore, if the trunk muscles are already hypotonic, one may see reduced trunk control.
Anticonvulsants Phenobarbital, phenytoin, valproic acid, gabapentin, topiramate	Work through many pathways to decrease transmission of neuronal activity in the brain	Fatigue, ataxia, dizziness, decreased cognitive function; some contribute to liver damage	When beginning the drug, careful BP (phenytoin) and ECG monitoring (phenobarbital, phenytoin, valproic acid) may be required. Monitor for other side effects such as dyspnea (phenobarbital, valproic acid, gabapentin) and apnea (valproic acid). Maintain proper hydration to prevent kidney stones (topiramate).
Antiseizure Phenobarbital, phenytoin, carbamazepine	Suppression of rapid and excessive neuron firing in the brain during seizures	May have a depressant effect on the central nervous system, thus possibly blunting the physiologic responses to exercise	Monitor for side effects such as mental confusion or irritability, dizziness, nausea, weight loss, and sensitivity to sunburn. In some instances, the paradoxical effect of hyperactivity may be a result of these medications. Carbamazepine is considered to have fewer side effects.
Anticholinergics/dopaminergics Benztropine mesylate, carbidopa-levodopa, glycopyrrolate, trihexyphenidyl	Blocks cholinergic neurotransmission; often used to reduce dystonia and drooling	Dry mouth, constipation, agitation, dysuria, syncope; exercise may increase the side effects	When beginning the medication or when exercise testing or beginning exercise training, monitor individuals carefully and have them sit or lie down during initial signs of syncope. Maintain proper hydration.

BP = blood pressure; ECG = electrocardiogram.

EXERCISE PRESCRIPTION

The frequency, intensity, and duration of physical activity used in an exercise prescription for the nondisabled population are well known; refer to chapter 6 for details. Those guidelines serve as the basis for exercise training of individuals with CP who can perform the exercise test associated with the training modality in question. Aerobic and strength training and physical activity guidelines exist for children with cerebral palsy (78), which may be a starting point for many individuals. The clinical exercise physiologist may need to make certain modifications to the training program depending on the individual's clinical presentation and associated conditions. Assuming no cardiorespiratory or other health-related contraindications to exercise, most individuals functioning in GMFCS levels I to III and some functioning in level IV will be able to engage in some form of traditional exercise training. Those who cannot require a practical approach based on their motor functioning. Some guidelines for the modality of exercise and some other adaptations follow. See table 29.5.

Cardiorespiratory Exercise

If running or walking is feasible and desired, such exercise should not be performed on a hard surface, and footwear (and any orthotics) should be chosen carefully to meet both support and cushioning needs. This type of exercise may be appropriate for those functioning in GMFCS level I and for some in levels II and III who do not have (or who are not at risk for) lower limb joint or related soft tissue pain. The individual should be closely monitored (at least at the beginning) for any adverse effects. For individuals at risk for joint and related soft tissue pain, non-weight-bearing activities are recommended. These activities reduce pressure on joints while enabling the individual to engage in exercise. Examples include hand cycling, recumbent stepping, and dual-action cycling or regular cycling (two wheels or three wheels).

To avoid pressure sores, precautions must be taken during prolonged sitting periods during exercise. Proper positioning is important as well. Adaptive straps and supports may be needed to secure parts of the body not engaged in the exercise so an

Table 29.5 Exercise Prescription Review

Training method	Frequency	Intensity	Time (Duration)	Type (Mode)	Progression	Special considerations
Cardiorespiratory	2-5 d/wk	Moderate intensity 40%-50% $\dot{V}O_2$ reserve or HRR for severe CP 50%-80% $\dot{V}O_2$ reserve or HRR as tolerated	Consider interval training as tolerated. May need to incorporate rest periods (e.g., 2 or 3 10 min sessions separated by rest as needed).	Will depend on the ability of the individual. Options include walking, running, cycling, propelling a wheelchair, swimming, hand cycling, stationary cycling, and recumbent stepping.	Slow progression in duration and intensity.	Ensure proper support and positioning, and check skin if there is a chance of pressure points or rubbing. Screen for bone health. If a high fracture risk with poor motor control, consider supported exercises such as swimming.
Resistance	2 or 3 d/wk (avoid consecutive days)	Load should be individualized Each set of repetitions should be done to fatigue. Start with low load, higher number of repetitions (e.g., 15) May use high load and lower number of repetitions with or without velocity training if sufficient motor control and depending on goals	3 sets of 8-15 reps, depending on the load. May need to individualize rest periods (1-3 min, as needed).	Machines or elastic resistance bands may be preferable if limited motor control. Free weights can be used if sufficient motor control.	Increase load as needed to exercise to fatigue.	Ensure sufficient joint stability and range of motion to perform the exercises. Ensure exercises are done correctly. Check skin if there is a chance of rubbing. Ensure bones are healthy enough for exercises to be safely performed.
Range of motion	Used as a warm-up or cool-down for other exercises, so frequency will vary accordingly.	Maintain stretch or ROM to point just before discomfort	30-60 s per stretch with 5-10 reps depending on the number of muscles being stretched and tolerance.	Passive and active stretching, PNF, active and active-assisted ROM activities	For passive stretching, if targeting one specific muscle group (e.g., the hamstrings), and a comfortable position can be found, may progress to fewer repetitions and a longer duration (e.g., 1 repetition of 20 min); otherwise, progress from passive stretching to active stretching.	Ensure joints are stable enough to support range of motion exercises.

HRR = heart rate reserve; PNF = proprioceptive neuromuscular facilitation; ROM = range of motion.

appropriate posture can be maintained. If possible, training on an arm ergometer should include directional changes that work both agonistic and antagonistic upper body muscles (e.g., 5 min forward on the arm ergometer followed by 5 min backward on the arm ergometer). The person should maintain a cadence that does not increase spasticity such that the activity cannot be performed. At least some of these arm cranking and cycling activities will be feasible for most individuals functioning in GMFCS levels I and II, for some functioning in GMFCS level III, and possibly for some in GMFCS level IV.

Aquatic training is an exercise modality that can be used for most individuals with cerebral palsy who do not have medical contraindications or who are not greatly afraid of being immersed in water. It has several advantages (72). Pools with warmer temperatures may help the individual relax and allow for greater movement, and this is especially relevant for those functioning in GMFCS levels III to V. The buoyancy of water also allows for easier movement by reducing the effects of gravity if the individual is supine. This is especially pertinent for those functioning in GMFCS levels IV and V. Since individuals are partially weight supported when standing in a pool, this may allow some who are functioning in GMFCS levels III and IV to walk or walk more easily (with or without support) compared with what they are able to do on dry land. For individuals with cerebral palsy functioning in GMFCS level V and some in GMFCS level IV, using flotation devices that allow for a supine position with the body well supported may allow the most body movement (and thus the most exercise). In these latter cases, one-on-one supervision is necessary.

For some individuals functioning in GMFCS levels IV and V, lifts may be the only method of transferring them in and out of the water. For others, a ramp with bars and a chair may be needed for entry and exit from the pool. If aquatic training is used, maintaining proper skin care is important to ensure that skin does not deteriorate, which could lead to increased risk of pressure sores.

If aquatic training is not practical, then those functioning in GMFCS level IV who cannot use adapted cycling equipment, and those in level V, may be able to exercise on a mat. Lifts may be necessary to transfer the individual on and off the mat. Positioning devices (wedges, firm cushions) can compensate for contractures and ensure the person is comfortable. Trial and error may be required to determine what the individual can do in terms of voluntary movement and what they enjoy doing. This same approach can be used for upper limb exercise with the person comfortably sitting in a wheelchair (for those in GMFCS levels III and IV).

Interval training may be useful (e.g., slow 5 min, fast 3 min, slow 5 min, moderate 7 min, and so on). Intervals may need to be shorter if the person fatigues easily or has very limited voluntary movement. Rest periods, especially when first beginning exercise training, will prevent excessive fatigue and injury. More traditional interval training (exercising for intense bouts of short duration, alternated with active relief periods) may also be appropriate, perhaps with modifications. For example, the level of intensity for the short-duration bouts may not be so high that the entire workout session must be reduced because of secondary conditions (reduced endurance and joint and muscle pain). If reduced cardiorespiratory and muscular endurance is an issue, the duration of exercise may need to be broken into intermittent sessions rather than performed in one continuous session. Instead of one 30 min session of cardiorespiratory exercise, shorter bouts of exercise (3 to 10 min) with frequent rest intervals may be necessary. These exercise sessions can be performed throughout the day and should progress slowly in duration.

Resistance Exercise

People with cerebral palsy can benefit from resistance training using traditional devices, for those able to use them. Other devices such as elastic bands and balance balls, as well as the resistance of water (hydrotherapy), may be useful depending on the ability and interest of the individual. Elastic bands and balance balls provide a variety of resistance exercise options throughout all multijoint movements (e.g., biceps curls, knee extensions, trunk rotations) without the added risk of injury from using free weights. Elastic bands offer many levels of resistance, and the individual can progress as with any other training program (least to most resistant: yellow, red, green, blue, black, and silver). If a lot of involuntary or uncontrolled movement is present (such as seen with athetosis or severe ataxia), elastic bands may be inappropriate because of the involuntary movement and decrease in stability of the exercise band. The most appropriate exercises for people with very poor motor control who are still able to perform the activity may be cuffed weights or exercise machines that can provide fluidity of movement. Active assistive exercise may be needed for smooth performance of the motion. These types of exercises are appropriate for individuals in GMFCS levels I to IV who have sufficient motor control to perform the activity for a specific muscle group.

For individuals who do not have this control for at least some muscle groups (GMFCS levels II to IV) and for those in GMFCS level V, aquatic therapy may be the most appropriate means of engaging in resistance training via the resistance of the water as the limb moves through it. The muscles to target for training will depend on what is feasible for the individual, the goals of exercise, and what muscles are particularly weak. Care should be taken to strengthen both agonist and antagonist muscles to avoid reducing flexibility and range of motion. Persons should be closely supervised, especially when beginning an exercise program, to ensure they are stable and properly performing the exercise.

In general, a high-volume program is more feasible. This means using lower resistance and a higher number of repetitions. The clinical exercise physiologist should be alert for rapid fatigue development related to reduced muscular endurance. Large muscle group work should be completed before small muscle group work (e.g., quadriceps before calf muscles). The use of both multi- and single-joint exercises is not contraindicated (e.g., squat vs. knee extension), provided it is feasible for the individual to perform and does not cause pain. Exercises that emphasize increased speed of muscle contraction may have more functional benefit than those that emphasize a slow, controlled contraction (36, 44), assuming the movement is performed correctly and safely. If the exercise cannot be performed in this manner, then a slow, controlled movement is preferable. To ensure safety, the individual should start slowly with the resistance training program and progress at a comfortable pace. Resistance training programs should also focus on maintaining range of motion relevant to the movements used in the exercise.

Range of Motion Exercise

Although manual stretching is not recommended for improving contractures, it is important that an exercise program for individuals with cerebral palsy not result in a reduced range of motion. Thus active, active-assisted, or passive range of motion activities should be incorporated into the exercise program, perhaps as part of a warm-up or cool-down. When possible, exercises that move joints through their full available range of motion should be performed. Chapter 6 provides general details on range of motion exercise, including how to perform PNF. The clinical exercise physiologist should consider any deformities, subluxations, or dislocations that would contraindicate range of motion exercise or require an accommodation. Tai chi and yoga may be appropriate exercise choices to help maintain flexibility in individuals with cerebral palsy, depending on their motor control impairment and their interest.

EXERCISE TRAINING

Most of the scientific literature on the effects of cardiorespiratory and resistance training for those with cerebral palsy pertains to children. When compared against children with cerebral palsy who did not undergo aerobic exercise training, those who did showed an increase in aerobic capacity (49, 74, 75, 77). In general, the interventions used in these studies focused on functional activities such as walking and running two to four times per week for ≥30 min per session, at 60% to 75% peak heart rate (74, 75, 77) or 50% peak $\dot{V}O_2$ (49). This was done alone or with lower limb resistance training (74) or anaerobic training (75, 77). Both long (8-9 mo) (75, 77) and short (8 wk; 3 mo) (49, 74) programs showed benefits. Improvement in aerobic capacity in the training group, compared with the control group, ranged from 15% to 40%.

When comparing these studies (49, 74, 75, 77), greater improvements were seen with longer interventions (when impairment was similar between the studies) and in those whose subjects had less motor impairment (when duration was similar between the studies). Longer-duration aerobic and anaerobic lower limb exercise (combined) also showed improvements in participation in physical activities and health-related quality of life (77), with a trend (9%, p = 0.07) toward an improved level of habitual physical activity (75).

Isolated lower limb strength training appears to have very little effect on functional mobility in children with CP, despite its positive effect on muscle strength (64). However, lower limb strength training that incorporates increasing speed of the contractions may result in improved walking speed compared with strength training at a fixed contraction speed (44). Exercise training has not been shown to result in lasting changes (77), meaning physical activity

behavior does not appear to change over the long term. It should be noted that these studies were restricted for the most part to children who functioned in GMFCS levels I to III (i.e., those who could walk). As noted previously, passive stretching is not recommended as treatment for contractures because of the lack of evidence.

The limited data for exercise training in adults with cerebral palsy show a fairly similar pattern to that for children. A 6 mo lifestyle intervention (twice weekly aerobic and strength training exercise, with one session per week supervised for the first 3 mo, along with six counseling sessions for physical activity and two to four sessions for sports) for ambulatory older adolescents and young adults with cerebral palsy (20 ± 3 yr, GMFCS levels I-III) showed a significant immediate postintervention increase in peak $\dot{V}O_2$ (8.7%) in the training group compared with a decrease of 5.4% in the control group (67). This between-group difference was no longer present at the 6 mo follow-up.

Objectively measured physical activity was not affected by the intervention (66), nor were cardiovascular risk factors related to body composition and blood cholesterol levels (67). However, there were training-related differences in physical activity and health perceptions. Compared with the control group, there was a significant immediate postintervention improvement in perceptions of physical activity levels, fatigue, participation, and involvement of social support (66). The improvements in perceived physical activity and fatigue were not present at the 6 mo follow-up, although the improvements in participation and involvement of social support were still present. Other changes in perception were present only at the long-term follow-up, such as improvements to quality of life related to bodily pain and mental health. To what extent the perception of an improvement in physical activity and health contribute to the lack of an objective behavior change over the long term warrants investigation.

When lower limb strength training is the focus, ambulatory young adults along with older adolescents with cerebral palsy (GMFCS levels II-III) who underwent 12 wk of strength training following best practice guidelines showed significant (between group) and immediate postintervention improvements in strength (27%) in targeted muscles (preselected via gait analysis) (70). This effect was not present 12 wk later, however. The intervention, which did not focus on speed of muscle contraction, also did not result in a change in mobility (70) or in the level of physical activity (9). However, the training group showed a significant between-group improvement in their perception of their mobility immediately after the intervention (70). This was not maintained at the 12 wk follow-up. Again, the extent to which the difference in perception versus objective measures affects behavior change necessitates further research. Also of interest, similar controlled strength training studies of adults with cerebral palsy that were shorter in duration (8-10 wk) and that did not individually choose the muscle groups to be trained did not show a between-group difference in muscle strength postintervention (60). Further work is perhaps needed to better understand optimal training protocols for this group.

Research Focus

Exercise Training in Young Adults With Cerebral Palsy

Most research on the effects of exercise training in those with cerebral palsy is performed with adolescents. This investigation studied the effects of 12 wk of combined functional anaerobic and strength training on functional capacity on young adults. The 17 subjects (21 ± 4 yr; age range 15-30 yr; nine males, eight females) had the spastic form of cerebral palsy and were randomized to 3 d/wk of high-intensity anaerobic and progressive resistance training or to a no exercise group. Training consisted of progressive resistance training (PRT) performed initially at each training session, followed by a series of functional anaerobic exercises.

- *PRT exercises:* Seated bent-knee calf raise, leg press, seated straight-knee calf press, seated tibialis anterior raise, standing calf raise. Performed 6 to 12 repetitions with assessment for progression every 4 wk.

- *Anaerobic training:* Two or three functional exercises per session performed at maximal intensity. Included stair climbing, bending, changing direction, and stepping over obstacles. The work:rest ratio, number of exercises, and repetitions were progressed per individual every 4 wk.

Compared with the control group, the training group had improvements in muscle volume, range of motion, strength, power, functional ability, and 6 min walk distance. The authors suggested that including functional anaerobic exercise training with PRT is important for those with cerebral palsy because this training is likely to improve daily functional ability. Additionally, they stress that this type of training is inexpensive and readily available, likely within the community.

Based on J.G. Gillett et al., "Functional Anaerobic and Strength Training in Young Adults with Cerebral Palsy," *Medicine and Science in Sports and Exercise* 50, no. 8 (2018):1549-1557.

Clinical Exercise Bottom Line

- Cerebral palsy reduces the ability to move, control movement, and carry out ADLs.

- Although the cause of cerebral palsy is nonprogressive, the direct and secondary effects, combined with aging, act to continually reduce the functional abilities of an individual.

- Traditional exercise testing is not required for those with cerebral palsy to participate in exercise training, but it may be useful if there is a suspicion of cardiovascular disease. If performed, the mode will typically require modification to compensate for physical limitations.

- Maintaining as high a functional capacity as possible, including aerobic, anaerobic, strength, power, and range of motion capacity, will help individuals with cerebral palsy reduce the rate of functional loss.

CONCLUSION

Ambulatory adults with cerebral palsy clearly gain health benefits from exercise training. Thus, the clinical exercise physiologist has an important role to play in the management of these individuals' health needs. Even though information is lacking on the effects of exercise training in nonambulatory adults with cerebral palsy, the clinical exercise physiologist should work with the health care team to provide feasible and safe physical activity programs for this group as well as for those who walk. To better guide clinical exercise physiologists in the future, further research is required to increase our understanding of both optimal exercise training protocols for individuals with cerebral palsy across the severity spectrum and how to change physical activity behavior long term in this group.

Online Materials

Visit HK*Propel* to access a link to the references, the case studies with discussion questions, and a quiz to help you review key concepts and test your understanding of the material covered in this chapter.

CHAPTER 30

Stroke

Christopher J. Womack, PhD

Stroke is a leading cause of both death and disability, making stroke patients a key population for the clinical exercise professional. Paradoxically, stroke patients are often referred only to physical therapy, without consultation about long-term exercise prescription from either their physical therapist or a clinical exercise physiologist. Unfortunately, standard physical therapy does not typically improve functional capacity in this population, and incorporation of a regular exercise program can be a determinant of whether the patient can live independently or perform standard daily tasks without assistance. This chapter provides background on the pathophysiology and diagnosis of stroke, illuminates the importance of exercise for this population, and presents considerations for exercise testing and prescription.

DEFINITION

A stroke is the loss of blood flow to a region of the brain. This loss can occur because of a manifestation of cardiovascular disease, characterized by the buildup of atherosclerotic plaque in cerebrovascular arteries, ultimately resulting in an ischemic stroke. In most ischemic strokes, a blood clot ultimately seals off the narrowing artery. Strokes can also occur because of excessive bleeding in a cerebral artery, also known as a hemorrhagic stroke. The excessive bleeding and swelling in the brain prevent blood from flowing to brain cells downstream of the hemorrhage. Most strokes are ischemic, accounting for approximately 87% of all strokes (50). When a stroke occurs, neurons in the brain die, and the accompanying brain damage is the main cause of subsequent disability in stroke survivors. The brain damage can impair voluntary muscle movement, speech, vision, and judgment. Strokes can also occur in the form of silent cerebral infarctions. These strokes are not coupled with clinically apparent acute symptoms.

SCOPE

Cardiovascular disease is the number one cause of mortality in U.S. adults (50). Of the causes of death from cardiovascular disease, stroke is the second leading cause, encompassing 17% of all deaths attributable to cardiovascular disease (50). This figure amounts to 37.5 deaths per 100,000 people, making stroke the fifth leading cause of death in the United States (50). The incidence of stroke is such that one occurs every 40 s in the United States; thus, about 7 million American adults have had a stroke (50). Each year, almost 800,000 Americans will experience a stroke, and the vast majority of these will be new stroke events (50). Clearly, there are many survivors of stroke. In fact, the number of stroke survivors is expected to increase because of increases in average life expectancy (36). Although this is a positive trend in that stroke is taking fewer lives, stroke is a leading cause of long-term disability (50). In a study evaluating patients after onset of stroke, 26% of individuals were institutionalized, and 30% were unable to walk without some assistance 6 mo after stroke onset (24). In total, over 1 million American adults report functional limitations because of stroke (5). Because of the high rate of mortality and disability associated with this condition, direct medical costs related to stroke are projected to nearly triple from $73.7 billion in 2012 to over $184 billion in 2030 (37).

Gender

Overall, women have a higher lifetime risk of stroke than men. Although stroke incidence in young and middle-aged men is lower than in women, these differences become smaller with increasing age, to the point where incidence is the same or higher in women in the highest age groups (50). The larger number for women is mainly due to the longer life span in women, but it also points out that stroke is an important health concern for women as well as men. The increased risk of stroke is likely due to the increased cardiovascular disease risk associated with menopause.

Although menopause is the primary state that places females at increased risk for stroke, pregnancy also temporarily increases risk. Pregnant women are at nearly a ninefold increased risk for stroke during the 6 wk postpartum period (26). This risk is greater in African American women (19). A study on postpartum stroke showed that 4.1% of women with postpartum stroke died in the hospital, whereas 22% of survivors died at home after discharge (19).

Ethnicity

Table 30.1 summarizes the racial influences on the incidence and mortality rate from stroke. As evidenced by recent data, prevalence is highest among African Americans, followed by Caucasians, Hispanics, and Asian Americans (50). The death rate from stroke is correspondingly higher in African American males and females compared with Caucasians (57.9 deaths per 100,000 for Black males, 48.3 for Black females, 36.0 for white males, and 36.0 for white females) (50). Furthermore, African Americans experience a higher degree of functional limitation after stroke than Caucasians do (4), although the exact cause of these differences is unknown. Residents of southeastern states have a higher prevalence of stroke (4, 50), which likely reflects cultural differences regarding diet and physical activity. The additive effects of these geographical and racial influences combine to make African American males in southeastern states the population at greatest risk for stroke in the United States (4).

PATHOPHYSIOLOGY

The atherosclerotic process that causes cerebrovascular disease, and ultimately an ischemic stroke, proceeds in the same fashion as plaque progression in coronary artery disease (CAD). For a more detailed explanation of this process, see chapter 14. For

Other Conditions

Because hypertension is associated with both hemorrhagic and ischemic stroke, this risk factor is prevalent in stroke patients. As such, the exercise professional would use the same exercise training considerations as for hypertensive patients. An additional consideration is that many of these patients will be on antihypertensive medications, including β-blockers, which would affect the use of heart rate to prescribe exercise. For these patients, a set intensity combined with rating of perceived exertion (RPE) may be the preferred tools for prescribing exercise intensity.

Table 30.1   Stroke Incidence and Death Rate for Selected Ethnic Groups

		Prevalence (% of population)	Death rate
CAUCASIAN			
	Males	2.4	36.0
	Females	2.5	36.0
AFRICAN AMERICAN			
	Males	3.1	57.9
	Females	3.8	48.3
HISPANIC			
	Males	2.0	34.0
	Females	2.2	29.6
AMERICAN INDIAN AND ALASKAN NATIVE			
	Males	NA	33.3
	Females	NA	34.4
ASIAN AMERICAN			
	Males	1.1	32.1
	Females	1.6	28.7

NA = Data not available.
Based on Virani et al. (50).

this reason, the same traditional and nontraditional risk factors that are related to the development and progression of CAD and peripheral artery disease (PAD) are associated with the development of ischemic cerebrovascular disease. **Ischemic strokes** can be further categorized as thrombotic, embolic, and hemorrhagic. In the case of **thrombotic strokes**, in which an occlusive thrombus develops in or outside an ulcerated plaque, hypercoagulable states due to increased coagulation potential or decreased fibrinolytic potential are particularly important risk factors. Emboli that cause **embolic strokes** are typically from the carotid or other arteries. In these cases, the thrombus is not large enough to occlude the large vessel, but the embolus that breaks off ultimately lodges in a smaller cerebral artery or arteriole. Often, major strokes are preceded by transient ischemic attacks (TIAs), which are considered a major predictor of impending stroke (21). In a study of patients who reported to an emergency room with a TIA, approximately 10% experienced a stroke within 90 d. Perhaps even more compelling is the fact that 5% experienced a stroke within 2 d (21). All older populations should be screened for possible prior TIA because about 50% of patients who experience a TIA do not report it to a clinician (21).

Hypertension is the major risk factor for hemorrhagic stroke, which makes up approximately 10% of strokes (47). **Hemorrhagic strokes** can also be caused by aneurysm, drug use, brain tumor, congenital arteriovenous malformations, and anticoagulant medication. Hemorrhagic strokes are classified as either intracerebral, which refers to bleeding inside the brain, or subarachnoid, which refers to bleeding in and around the spaces that surround the brain (47). Unfortunately, there is usually little warning for a hemorrhagic stroke. Acute signs or symptoms include altered consciousness, headache, vomiting, and large elevations in blood pressure. Additionally, patients with subarachnoid hemorrhage may develop neck stiffness (47).

CLINICAL CONSIDERATIONS

Medical and clinical considerations for stroke include signs and symptoms specific to whether damage occurs on the right or left side of the brain. The clinician needs to be aware of the acute signs of the stroke and the clinical manifestation that occur after the event. Finally, the method of determining the definitive diagnosis of stroke is another important clinical decision.

Signs and Symptoms

Memory loss and paralysis are two of the more consistent symptoms of stroke. In the case of paralysis, the brain damage causes paralysis on the opposite side of the body (i.e., right-brain damage causes left-side paralysis). Furthermore, right-brain damage can result in vision problems and awkward or inappropriate behavior, whereas left-brain damage causes speech and language problems and slow or cautious behavior. A patient suffering from acute stroke can have one of the following symptoms: numbness or weakness of the face, arm, or leg; confusion, speech problems, and cognitive defects; impaired bilateral or unilateral vision; impaired coordination and walking; and headache.

History and Physical Exam

One of the main changes for the clinical exercise physiologist to be aware of is the hemiplegic gait of stroke patients. Concomitant risk factors for cardiovascular disease such as hypertension and diabetes are frequently present. Furthermore, underlying CAD is often present (34), so resting electrocardiogram (ECG) changes and symptoms for ischemia should be evaluated. Most stroke patients develop mental depression during the poststroke period (52), so a psychological referral may be necessary.

Diagnostic Testing

Ultrasound, magnetic resonance imaging (MRI), and angiography are the main diagnostic tests used to assess impending occlusions that could cause an ischemic stroke. The major diagnostic tool for determining hemorrhagic stroke is noncontrast computerized tomography (CT) (45). CT can also be used to diagnose ischemic stroke, although the sensitivity of this technique varies across research studies (45). Diffusion-weighted MRI has been shown to be more effective than CT for diagnosing ischemic stroke (45), making this an emerging diagnostic tool.

Exercise Testing

Contraindications for exercise testing are the same as those for all patient populations. Because hypertension is the major risk factor for hemorrhagic stroke, particular attention should be directed toward ensuring that preexercise resting blood pressure is below the contraindicative values, systolic pressure of 200 mmHg and diastolic pressure of 110 mmHg. Furthermore, because ischemic stroke is highly associated with CAD, screening should ensure that symptoms of CAD such as unstable angina are not present.

Considerations for exercise testing in stroke patients are summarized in table 30.2. Not all stroke patients are capable of traditional treadmill graded exercise tests to determine functional capacity. Some researchers have even suggested that submaximal measurements such as lactate or ventilatory threshold or oxygen pulse ($\dot{V}O_2/HR$) should be the criterion measures of cardiorespiratory fitness in stroke patients because these markers are easier to obtain and put the patients at a lower risk (29). Research has shown, however, that if patients achieve a self-selected walking speed of at least 0.5 mph (0.8 kph) during 30 ft (9 m) of floor walking, they are capable of performing a graded treadmill exercise test with handrail support as needed (34). Out of 30 patients with hemiparetic stroke, 29 were able to meet this criterion and achieve an average of 84% of age-predicted maximal heart rate. This intensity was sufficient to allow detection of asymptomatic myocardial ischemia in 29% of patients without previously determined CAD. The treadmill test was at a self-selected walking velocity with grade increases of 2% every 2 min (34).

Table 30.2 Exercise Testing Considerations for Stroke Patients

	Special considerations
Cardiorespiratory	Treadmill testing can be used for all patients able to achieve at least 0.5 mph during floor walking. In an established treadmill protocol, the patient walks at a self-selected speed with increases of 2% grade every 2 min. If testing using a cycle ergometer, begin at 20 W and increase 10 W per minute.
Strength	Testing can involve multiple repetitions to fatigue such as 10RM testing. Handgrip testing is an easy testing mode to implement.
Range of motion	A goniometer or sit-and-reach test can be used. When using goniometers, prioritize testing on the joints affected by paresis.

RM = repetition maximum.

If a patient is unable to perform treadmill or cycle ergometry work, protocols are available that do not require a large amount of leg muscle mass. Tsuji and colleagues (49) developed an incremental protocol that uses an increasing rate of bridging, which involves elevating the pelvis to a point of maximal hip extension. The protocol uses 4 min stages and increases the rate of bridges per minute from a starting point of three to six, with six bridge per minute increases thereafter up to 24. About 89% of patients were able to complete this protocol, and the test–retest intraclass coefficients were over 0.90 for HR, over 0.70 for $\dot{V}O_2$, and 0.98 for the oxygen pulse. However, the peak oxygen consumption does not appear to be higher than 2 metabolic equivalents (METs) for this protocol (49), suggesting that it should be used only for extremely deconditioned patients. Although 6 min walk distance is commonly used to assess cardiorespiratory fitness in several clinical populations, it has been observed that $\dot{V}O_2$peak obtained during cycle ergometry and 6 min walk distance are not associated in stroke patients. Rather, 6 min walk performance is primarily dictated by balance, knee extensor strength, and degree of muscle spasticity (38). Therefore, the 6 min walk may be a good addition to the battery of tests in that it is a functional outcome of these impairments, but it should not be viewed as a measure of cardiorespiratory fitness per se.

Treatment

For advanced atherosclerosis (70%-99% occlusion) in the carotid arteries, carotid endarterectomy is a common surgery (47). In this procedure, most of the plaque is physically removed from the artery wall by the surgeon. However, it has been observed that stenting with an embolic protection device is as effective in both the short and long term (16).

Pharmacological treatment depends on the type of stroke. As shown in table 30.3, this may include anticoagulants, antiplatelet medications, and antihypertensives; the latter are especially important for patients with hemorrhagic stroke. Antihypertensives such as labetalol, enalapril, and clonidine are the most common drugs for subarachnoid hemorrhage (47). Patients with ischemic stroke are often on medication, which could potentially include statins and diabetes medications, to control associated risk factors. Nitroglycerin or other vasodilators may be prescribed following an ischemic hemodynamic stroke to lessen the chance of vasospasm. Patients with subarachnoid hemorrhage are typically given nimodipine to reduce vasospasm every 4 h until symptoms subside (47).

Standard rehabilitation for stroke patients usually includes physical therapy. The primary aim of this treatment is to restore balance, movement, and coordination. Basic strengthening exercises, passive movements to increase range of motion, assisted and unassisted walking, and functional movements such as chair stands and transferring from bed to chair are commonly prescribed. Because of the hemiparesis, patients must often relearn daily activities such as dressing and bathing, particularly if they need to perform these tasks with one hand Therapy will also include instruction and practice using assistive walking devices such as walkers and canes. Speech therapy is commonly necessary because of the effects of the stroke on speech control. Nutritional consultations with a registered dietitian may be necessary if the patient is overweight or obese. Additionally, because depression is common in patients who have had a stroke, psychological referrals may be employed.

Practical application 30.1 provides guidelines for interacting with stroke victims.

EXERCISE PRESCRIPTION

The average functional capacity of a stroke patient is approximately 14.4 mL · kg^{-1} · min^{-1} (33), which is of great concern because 20 mL · kg^{-1} · min^{-1} has been suggested as the minimum necessary for independent living (10). Therefore, a major goal of stroke rehabilitation should be to increase functional capacity. Standard therapy, however, does not provide enough of an aerobic stimulus to achieve an increase in cardiorespiratory fitness (32). In support of this, only 12% of patients receiving physical therapy and 2% of patients receiving occupational therapy received therapy directed toward improving this parameter (10). Therefore, supplemental aerobic exercise should be prescribed for stroke patients. The clinical exercise physiologist should be aggressive about implement-

Table 30.3 Common Medications for Stroke Patients and Their Respective Effects

Medication class and name	Primary effects	Exercise effects	Important considerations
Warfarin	Anticoagulant	None	Patients should avoid activities with high risk for trauma or injury.
Ticlopidine, clopidogrel, aspirin	Antiplatelet	None	
ACE inhibitors (ramipril, enalapril)	Decrease BP	Decrease BP	
Calcium channel blockers (nimodipine)	Decrease BP	Increase exercise capacity in patients with angina	
Diuretics (hydrochlorothiazide)	Decrease BP	Decrease BP	

BP = blood pressure.

Practical Application 30.1

CLIENT–CLINICIAN INTERACTION

Dealing with stroke patients presents some unique challenges that can be best addressed by emphasizing good communication with the patient and associated family members or caregivers. Previously mentioned consequences include depression, inappropriate behavior, or slow or cautious behavior. The clinician should be sensitive to these considerations, establish appropriate boundaries for interaction with the patient, and, if necessary, make appropriate referrals to mental health professionals. As it happens, the exercise itself can help relieve symptoms of depression, although this effect appears to go away once regular exercise is stopped (12), highlighting the need for consistent and long-term participation. Communication with family members or caregivers can help determine whether depression or other psychological conditions may be a concern. Furthermore, because many of the desired outcomes of rehabilitation are improved activities of daily living, some of the outcome data may be qualitative information received from family members or caregivers about the level of independence the patient exhibits when completing daily tasks.

ing this therapy in the patient's rehabilitation regimen. Patient education is important, because many patients may believe their standard physical therapy provides a sufficient amount of exercise.

Besides the decrease in functional capacity, profound decreases occur in muscular strength and endurance, and these should be addressed. In particular, reduced muscle mass is highly correlated with functional capacity (43) in stroke patients. In further support of this, knee extensor strength in the nonparetic limb was shown to be associated with functional capacity in stroke patients with hemiparesis (51). Furthermore, muscular strength is associated with gait and balance (14, 25) in this population. A particular challenge in terms of exercise prescription for a stroke patient who may have led a sedentary lifestyle is the perceived volume of work per week necessary to engender appropriate changes. One suggested approach is that of Rimmer and colleagues (42), who observed significant improvements in body fat percentage, 10RM (repetition maximum) strength, $\dot{V}O_2peak$, and hamstring and lower back flexibility in predominantly African American patients poststroke. The program consisted of 3 d/wk for 60 min each session (30 min of cardiorespiratory training using a variety of modalities, 20 min

of strength training on commercial strength training machines, and 10 min of flexibility training). Therefore, impressive gains can be realized with a total of 3 h/wk of training, a time commitment that should not place an excessive burden on patients. Information on all aspects of exercise prescription is presented in table 30.4.

Cardiorespiratory Exercise

Most studies that have shown improvements in patients with stroke employed a frequency, intensity, and duration similar to those included in the American College of Sports Medicine (ACSM) guidelines for healthy populations. Modes of training are listed in table 30.4 and can include walking (on ground or on a treadmill), water exercise, and cycle ergometry. A mixture of these modalities is warranted because some have specific supplementary benefits. Employing some weight-bearing exercise is highly recommended, because these activities have been shown to maintain bone mineral density in patients poststroke (39). Treadmill exercise not only increases cardiorespiratory endurance but also may improve balance, coordination, and gait abnormalities (27, 40). It also results in

Table 30.4   Exercise Prescription Review

Training method	Frequency	Intensity	Time (Duration)	Type (Mode)	Progression	Important considerations
Cardiorespiratory	3-5 d/wk	40%-80% HRR	15-30 min	Floor and treadmill walking, cycle ergometry, Nu-Step, water exercise	Progress from low to high intensity and to longer durations; because of biomechanical limitations, heart rate recommendations should be superseded by perceived effort	
Resistance	3-5 d/wk	As tolerated, up to 80% of 1RM	30-45 min	Elastic resistance bands, body weight exercises, sandbags, active motion, water exercise, commercial strength training equipment	As tolerated	
Range of motion	3-5 d/wk	Below point of discomfort	10-20 min	Passive movement, PNF	As tolerated	Emphasize stretching muscles on the paretic side, particularly in muscle groups experiencing spasticity.

HRR = heart rate reserve; 1RM = one-repetition maximum; PNF = proprioceptive neuromuscular facilitation.

larger improvements in 6 min walk time than does cycle ergometer training (31). Water-based exercise training has been shown to increase $\dot{V}O_2$peak by approximately 22% and to cause concomitant increases in isokinetic strength (6). Water exercises can be a relatively safe way for these patients to train because risk of falls is greatly reduced. Cycle ergometry training results in improved stair-climbing ability (23), suggesting that this form of training can lead to nonspecific improvements in performance related to ambulation. If either the patient or the clinician is concerned about falls on the treadmill, supported treadmill exercise, as discussed in chapter 25, can be employed (11). By using supported treadmill exercise, patients can train at the highest velocity possible without stumbling and with minimal risk.

Furthering this concept, research suggests that progression of exercise in these patients should focus on intensity. Ivey and colleagues (18) observed that patients who progressively increased intensity over the course of a 6 mo treadmill training intervention increased $\dot{V}O_2$peak to a greater extent than patients who trained at a lower intensity but consistently increased duration (for more details, see the sidebar Research Focus). Other studies have similarly observed that higher-intensity training or even high-intensity intervals cause greater improvements in functional capacity in this population (3, 15). In particular, a meta-analysis confirmed that larger gains can be expected in $\dot{V}O_2$max and 6 min walk time with higher-intensity exercise programs (31). It has also been observed that stroke patients who perform 10 s intervals at higher speeds

experience changes in 10 m walking speed, stride length, and functional ambulation that are superior to those produced with standard treadmill training (40).

Resistance Training and Flexibility

Because of the neuromuscular compromise caused by stroke, resistance training is an important part of the rehabilitative exercise prescription. ACSM recommendations for strength training could be easily followed with possible modification in the mode of training. Because of lowered muscular strength and endurance in these patients, strength training exercise can consist of low-resistance modalities such as elastic bands, body weight exercises, and sandbags. For patients who are extremely weak, exercises against gravity (e.g., shoulder and leg abductions), either with or without assistance, may be necessary until they are able to use any of the other recommended modalities. Rimmer and colleagues (42) successfully employed commercial resistance training equipment in poststroke patients, suggesting that these devices can be used provided the resistance can be decreased enough to accommodate the specific demands in this population. Whatever mode is selected, both concentric and eccentric movements should be included, because concentric-only training does not increase isokinetic strength in stroke patients (13). Additionally, although it would appear self-evident that the focus of strength movements should be on the paretic limbs, bilateral movements engender better improvements on the Berg balance and functional reach tests (20).

A portion of the exercise prescription should focus on exercises designed to improve activities of daily living. Examples include chair stands, stair climbing, ball kicking, balance beam step-ups, and walking through obstacles. These activities require combinations of strength, endurance, and balance. These patients can perform traditional flexibility exercises as tolerated. A raised platform with a stretching mat can be useful because it greatly assists with lying down and returning to a standing position after the exercises. The focus of the flexibility program should be on the paretic limbs, particularly muscle groups in which the patient is experiencing a large degree of muscle spasticity.

EXERCISE TRAINING

Functional capacity can be dramatically improved in stroke patients. The improved $\dot{V}O_2$peak, combined with the lowered oxygen cost of ambulation because of improved economy, results in a profound decrease in the relative intensity of normal ambulation. As an example, Macko and coworkers (35) observed that 6 mo of treadmill aerobic exercise resulted in a 20% reduction in the metabolic cost of ambulating at a constant submaximal intensity because of both increased $\dot{V}O_2$peak and decreased $\dot{V}O_2$ during submaximal walking. This study also suggested that the improvements in exercise economy are maximized within a 3 mo period, while $\dot{V}O_2$peak continues to improve throughout the course of training (35). More recent data further suggest that these benefits are accompanied by increased brain activity in the cerebellum and midbrain that correlates with improvements in walking velocity. In addition, and possibly related to this adaptation, 6 mo of treadmill

training has been shown to improve cerebral vasomotor reactivity, which could have a positive effect on cerebral blood flow (17). Exercise rehabilitation also improves functional ambulation, as evidenced by consistent increases in 6 min walk distances found throughout the literature (1, 8, 11, 44). As stated earlier, 6 min walk performance is largely influenced by balance, strength, and spasticity of the lower paretic limb in stroke patients (38, 39). Therefore, the degree of improvement in the 6 min walk is likely due to improvement in these factors either in place of or in conjunction with improved cardiorespiratory endurance. As evidence of this, a strength training intervention was found to improve gait speed during the 6 min walk (22).

Research using animal models suggests that moderate exercise improves markers of brain repair (2). Furthermore, treadmill training has been shown to increase brain activation in the posterior cerebellum and midbrain during movement of the paretic leg. The increase in activation of the posterior cerebellum was correlated ($R = 0.54$) with the increase in maximal walking speed, suggesting that ambulatory improvements in stroke patients may be at least partially mediated by improvements in brain activation (30). Anticipated physiologic adaptations to exercise training in these patients include the following:

Cardiorespiratory Endurance

- Increased functional capacity
- Improved walking economy
- Increased oxygen pulse
- Decreased submaximal blood pressure
- Improved 6 min walk time
- Increased voluntary physical activity

Research Focus

Importance of Exercise Intensity

Because of the severe functional impairment in stroke patients, including balance issues, treadmill exercise is often prescribed at a lower intensity. Ivey and colleagues (18) observed not only that patients can progress in their training to a higher intensity (80% of heart rate reserve) but also that a higher-intensity program is much more favorable for improvements in functional capacity. Hemiparetic stroke patients were randomized to a low-intensity or high-intensity training group for a 6 mo training intervention. Although the initial intensity (40%-50% of heart rate reserve) and duration (15 min) were similar, the high-intensity group increased their intensity to where they were training at 80% to 85% of heart rate reserve for 30 min by the sixth week of training. In contrast, the low-intensity group progressed duration to the point where they were exercising for 50 min at an intensity less than 50% of heart rate reserve. Despite the fact that the

training protocols were matched for energy expenditure, the high-intensity group experienced a 34% increase in functional capacity, as contrasted with a nonsignificant 5% increase in the low-intensity group. The high-intensity group also experienced significant increases in 6 min walking distance and a trend for improved 48 h step count as measured by a pedometer. The low-intensity group did not significantly improve either of these variables. It is also worth noting that the retention rate for the high-intensity group was even higher than that observed for the low-intensity group (75% vs. 59%, respectively), suggesting that exercising at a higher intensity did not negatively affect participation. These data suggest that stroke patients are capable of training regimens in the high-intensity domain and further suggest that a high-intensity regimen results in larger gains in functional capacity.

Skeletal Muscle Strength

- Increased isokinetic strength
- Improved upper extremity function

Skeletal Muscle Endurance

- Improved Timed Up and Go
- Increased number of chair stands
- Improved upper extremity function
- Increased number of steps in 15 s

Range of Motion

- Improved range of motion
- Decreased spasticity in muscles on the paretic side

Body Composition

- Minimal decrease in body fat percentage
- Large decreases in overweight or obese patients who receive nutritional counseling in addition to exercise

Because of the aforementioned relationship between functional capacity and the ability to live independently (7), these adaptations are extremely meaningful to the quality of life of stroke patients. A direct relationship also exists between improvements in sensorimotor function because of exercise training and the magnitude of improvement in $\dot{V}O_2$peak, which again points to the importance of improving functional capacity. Because blood pressure is an important risk factor for both ischemic and hemorrhagic strokes, another important clinical adaptation is a decrease in blood pressure in response to acute exercise (41). Additionally, exercise training increases HDL and decreases fasting insulin and glucose (9).

Because most studies provide a holistic intervention that includes aerobic and strength training, discerning the exact contribution that resistance training makes to the outcome variables is difficult. Nonetheless, rehabilitation programs that incorporate resistance training typically yield impressive changes in both laboratory and functional evaluations. Isokinetic quadriceps and ankle plantar flexion strength can increase in the paretic limb without increases in muscle spasticity (48). Similar to the aerobic adaptations, strength improvements appear to increase performance in functional activities. Short-distance (7, 10, and 30 m) walking speed (8, 43, 48), Timed Up and Go (8, 46), Berg balance scores (46), number of steps performed in 15 s (8), and force production in the paretic limb when going from a seated to standing position (8) all increase with programs that incorporate resistance exercises, especially those designed to increase functional performance. Combining aerobic and resistance training can drastically improve quality of life and activities of daily living (48). The latter is an especially important finding given that physical activity is associated with risk for stroke (28).

Clinical Exercise Bottom Line

- Monitor progress by testing for $\dot{V}O_2$peak every few months, with the goal of increasing to above $20 \ mL \cdot kg^{-1} \cdot min^{-1}$.
- Include bilateral strengthening exercises because these will contribute to improved functional capacity.
- Progress the program in such a way that some of the weekly exercise sessions are devoted to higher-intensity treadmill training.
- Consider aquatic exercise because this mode has been shown to improve functional capacity and strength in this population.

CONCLUSION

Stroke remains a significant cause of death and disability. Exercise testing and training are important in both the prevention and treatment of patients after stroke. The clinical exercise physiologist will likely become an increasingly demanded allied health professional in the care of patients at risk for stroke or after stroke. This is especially true for poststroke patients with only minimal related loss of motor function. Physical therapy alone is not sufficient to increase functional capacity in stroke patients. The majority of hemiparetic stroke patients can participate in treadmill exercise, and stroke patients can gain larger improvements in functional capacity with higher-intensity treadmill training. The inclusion of strength training, aquatic exercises, or both can be important in improving functional capacity.

Online Materials

Visit *HKPropel* to access a link to the references, the case studies with discussion questions, and a quiz to help you review key concepts and test your understanding of the material covered in this chapter.

CHAPTER 31

Parkinson's Disease

Angela L. Ridgel, PhD

Brandon S. Pollock, PhD

Individuals with Parkinson's disease (PD) have variable types and degrees of symptoms, but in all cases symptoms progress over time. Although there is extensive scientific evidence that exercise improves PD symptoms and quality of life, there is still a lack of consensus for exercise prescription in Parkinson's disease. Furthermore because of the diversity of symptom type and severity among individuals, a personalized exercise program will be most effective for promoting and maintaining exercise-induced benefits (45). This chapter provides evidence and guidance for the use and benefits of exercise therapies for individuals with Parkinson's disease.

DEFINITION

Parkinson's disease (PD) is a progressive neurological disorder caused by loss of dopaminergic neurons in the substantia nigra of the midbrain. The etiology of nongenetic forms of PD is unknown but is characterized by several key features: **bradykinesia** (i.e., slowness of movement), muscular rigidity, 4 to 6 Hz resting tremor, and postural instability (UK Parkinson's Disease Society Brain Bank Criteria) (81).

SCOPE

Approximately 1.04 million people ≥45 yr in the United States were living with diagnosed Parkinson's disease in 2017 (186). It is estimated that by 2037, 1.6 million people in the United States (103) and 12 million people worldwide will have the disease by 2050 (40, 141). There is no known cure for this degenerative disease that results in progressive deterioration of motor skills as well as nonmotor deficits such as cognitive dysfunction and depression. The combined motor and nonmotor symptoms often lead to decreased independence and quality of life as well as increased reliance on the health care system. The economic impact of PD, including medical costs, social security payments, caregiver burden, and loss of work, exceeded $51 billion in 2017 in the United States (94, 184). Current treatments for PD symptoms are medication (levodopa, dopamine agonists), surgical intervention (deep brain stimulation), exercise, and physiotherapy (54).

PATHOPHYSIOLOGY

Parkinson's disease symptoms usually manifest after striatal dopamine is reduced by 70% to 80%. The presymptomatic period is estimated to be 5 to 6 yr (50, 58, 114). The average age of diagnosis of PD is 56 yr, but there is an early-onset form of the disease (108, 177). Most cases of PD (90%) are defined as idiopathic, or due to unknown causes. However, it has been suggested that environmental toxins (26, 63) may increase the risk of developing PD. Several known environmental toxins cause Parkinson's-like symptoms in animal models, such as 1-methyl-4-phenyl-1,2,3,6-tetrahydropyridine (MPTP) (164), paraquat (14), and potentially rotenone (170). Several genes have been shown to increase the risk of developing PD, including autosomal dominant SNCA and LRRK2 and autosomal recessive PARK7, PRKN, and PINK1 (8, 67, 98), but these are rare. Parkinson's disease is progressive, and the average disease duration (time from onset of symptoms to death), when diagnosis is later in life, is 10 to 13 yr (47, 179). Three key milestones in the ongoing progression of PD are the development of postural instability, dementia, and death (179).

CLINICAL CONSIDERATIONS

This section reviews the clinical symptoms that individuals with PD experience and the criteria used to diagnose and evaluate those symptoms. Understanding of these symptoms plays an important role in determining the optimal exercise prescription for this population.

Signs and Symptoms

Parkinson's disease is diagnosed using the UK Brain Bank Criteria (81). The inclusion criteria are as follows:

- Bradykinesia (slowness of voluntary movement with reduction in speed and amplitude of repetitive actions)

and at least one of the following:

- Muscle rigidity
- 4 to 6 Hz resting tremor
- Postural instability not caused by visual, vestibular, cerebellar, or proprioceptive dysfunction

The supportive criteria of three or more of the following symptoms lead to a definitive diagnosis:

- Unilateral onset of symptoms
- Rest tremor present
- Progressive change in symptoms
- Persistent asymmetry of symptoms, with onset side having greater severity
- Response (decrease in symptoms) to levodopa
- Presence of levodopa-induced dyskinesia (involuntary movements)
- Levodopa response ≥ 5 yr
- Clinical course ≥10 yr

Although there is some variation in the types and severity of symptoms, the disease usually manifests as motor difficulties and nonmotor symptoms (see the sidebar Secondary Symptoms of PD), such as loss of sense of smell, sleep difficulties, mood disorders, and orthostatic hypotension (84). Early in the disease process, motor symptoms are detected unilaterally, with later progression to bilateral signs. Heterogeneity in severity and types of symptoms is a common feature of PD. Mehanna and colleagues (107) reported that in individuals diagnosed between 50 and 69 yr of age, 65.9% cited tremor, 23.2% cited bradykinesia, 8.3% cited rigidity, and 13.2% cited gait and posture deficits and falls as the predominant initial symptom of PD (figure 31.1).

The most apparent motor symptom is a 4 to 6 Hz resting tremor present in the hand or the foot (less often in the face or neck) (72, 107, 178). This tremor is greatest while resting and will often decrease during an action (84). Bradykinesia, a cardinal symptom of PD, results in slow movement and slow reaction times (12). This term is also used in conjunction with **akinesia** (poverty of movement) and hypokinesia (slow and small movement). Bradykinesia can manifest as difficulties with fine motor tasks (e.g., buttoning a shirt, grasping, writing), slow walking with lack of arm swing, and lack of facial expression. Muscle rigidity is present as resistance to passive movement of the joint and is most common in the elbow, wrist, knee, ankle, and neck. Postural instability often develops in later stages of the disease and can contribute to the increased incidence of falls and fall-related injuries (3, 6, 80). Furthermore, gait abnormalities such as freezing of gait and shuffle step increase the risk of falls and disability (16, 60).

Several nonmotor symptoms of PD that could limit physical activity and exercise include cognitive dysfunction, depression, anxiety, sleep disorders, autonomic dysfunction, and sensory deficits. Cognitive deficits are present in up to 80% of people with PD (71, 100), with prominent executive dysfunction due to altered dopaminergic processing in the **frontostriatal** cortex (32, 122, 147). Executive dysfunction is prevalent in both early and late

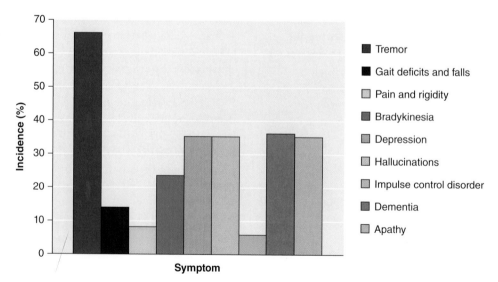

Figure 31.1 Self-reported incidence of motor and selected nonmotor symptoms in individuals diagnosed with Parkinson's disease (50-69 yr).

Data from Mehanna et al. (108).

Secondary Symptoms of PD

Motor Symptoms

- Freezing of gait
- Micrographia (small handwriting)
- Mask-like facial expression
- Stooped posture
- Decreased arm swing
- Softness of voice, slurred speech
- Swallowing problems
- Cramping
- Drooling
- Sexual dysfunction
- Restlessness
- Involuntary muscle contractions (dystonia)

Nonmotor Symptoms

- Constipation
- Bladder issues
- Fatigue
- Depression
- Anxiety
- Cognitive dysfunction
- Impulse problems (related to medication)
- Loss of olfaction (anosmia)
- Autonomic dysfunction
- Proprioceptive deficits

stages of PD, and dopamine replacement therapy has differential effects on cognition depending on the stage of the disease (9, 29, 158). In addition, it has been estimated that 75% of individuals who survive more than 10 yr after diagnosis will develop dementia (23, 100).

Depression occurs in 36% to 50% of individuals with PD (162) and is a significant barrier to therapies aimed at improving quality of life or symptoms, such as exercise (182). A longitudinal study showed that symptomatic depression predicted greater disability and decreased quality of life in PD (130). Antidepressant medications modestly improve health outcomes (185) but are recommended for short-term use only (155). Although standardized diagnostic criteria for anxiety have not been used in literature, a recent review suggested that prevalence of anxiety disorders in PD is 31% (20). Sleep fragmentation and excessive sleepiness is common (16%-74%) and can be exacerbated by levodopa medication (78, 160, 187). Furthermore, sleep fragmentation could be due to autonomic system dysfunction (187). See the Exercise Prescription section for more information regarding autonomic dysfunction in PD.

Lastly, several studies have identified proprioceptive impairment in PD, specifically in muscle spindle responses, load sensitivity, and kinesthesia (2, 12, 31, 87, 93, 104, 189). In addition, dysfunctional sensorimotor integration has been implicated in the etiology of bradykinesia and atypical scaling of movement in PD (12). Levodopa does not appear to improve kinesthetic deficits (86, 181) and has been associated with suppression of sensitivity to joint position (110, 121). This suggests that deficits in peripheral afferent input or sensorimotor integration likely contribute to abnormal motor output in individuals with PD.

History and Physical Exam

There is no definitive cause of PD, so the history should focus around an assessment of motor and nonmotor symptoms and the severity of these signs (46). The level of Parkinson's-related symptoms and severity can be determined with several assessment tests including the Hoehn and Yahr scale (see the sidebar Modified Hoehn and Yahr Staging Scale of Parkinson's Disease) and the Unified Parkinson's Disease Rating Scale (UPDRS). H&Y is a staging system, ranging from 1 to 5, that examines the severity of motor symptoms (79). Although this system is widely used in the scientific literature to determine the degree of disease progression (mild, moderate, severe), it does not take into account nonmotor symptoms or the impact of those symptoms on activities of daily living (ADLs).

The original UPDRS assessment has four parts, each of which can be used independently—Part 1: Mentation, Behavior and Mood; Part 2: Activities of Daily Living; Part 3: Motor Examination; Part 4: Complications of Therapy (115). Each part of the test asks the rater to give a 0 to 4 score (0 = no symptoms and 4 = worse symptoms). This scale assesses multiple aspects of PD, if done in its entirety, and can be completed in 15 to 20 min. Recently, the Movement Disorder Society updated this assessment to include additional PD-related problems (i.e., MDS-UPDRS; 61, 62). This updated version has guided instruction for administration and shows high consistency and correlation with the original version of the UPDRS (62).

Modified Hoehn and Yahr Staging Scale of Parkinson's Disease

0: No signs or symptoms of disease

1: Unilateral symptoms

2: Bilateral symptoms, without impairment in balance

2.5: Mild bilateral disease, with recovery on pull test

3: Mild to moderate bilateral disease, some postural instability, physically independent

4: Severe disability; still able to walk or stand unassisted

5: Wheelchair bound or bedridden unless assisted

Diagnostic Testing

There are no diagnostic tests for Parkinson's disease (PD). However, several brain imaging tests can confirm the diagnosis (73). In 2011 the FDA approved use of DaTscan, which uses dopamine transporter (DaT) single-photon emission computerized tomography (SPEC), in individuals with signs and symptoms of dopaminergic degeneration (7, 70). In addition, ^{18}F-DOPA PET imaging quantifies activity of an enzyme that converts levodopa to dopamine, aromatic amino acid decarboxylase (165). ^{18}F-DOPA uptake by basal ganglia areas (i.e., putamen) shows an annual decline mimicking disease progression (126). Furthermore, ^{18}F-DOPA uptake in the putamen shows an inverse correlation with UPDRS Motor III scores and, more specifically, bradykinesia scores (21). However, these tests are expensive and may not be covered by insurance.

Exercise Testing

Although measurement of peak ($\dot{V}O_2$peak) or maximal ($\dot{V}O_2$max) aerobic capacity during a graded exercise test (GXT) is recommended for assessment of cardiorespiratory fitness, it may be challenging for individuals with Parkinson's disease to complete these tests because of disability, fatigue, and deconditioning (161). Furthermore, autonomic impairment may alter heart rate and blood pressure responses during testing (5).

However, Katzel and colleagues (90) showed that measurement of $\dot{V}O_2$peak, using a progressive GXT treadmill protocol to voluntary fatigue, was reliable and repeatable in individuals with mild to moderate Parkinson's disease (Hoehn and Yahr stages 1-3). In this protocol, the first stage was 2 min at 0% grade, the second stage was 2 min at 4% grade, and then the treadmill grade was advanced by 2% every minute until voluntary exhaustion. When a grade of 10% was reached, the speed of the treadmill was advanced (by 0.2 mph) with the grade, if tolerated by the subject. Oxygen consumption, carbon dioxide production, and minute ventilation were recorded breath by breath using a metabolic analyzer and were averaged over 20 s intervals. The $\dot{V}O_2$peak was recorded as the mean of the last two 20 s intervals during the final stage.

If individuals with PD show deficits in balance or gait, a cycle ergometer can also be used to measure cardiorespiratory fitness. Individuals with mild to moderate PD can tolerate a maximal cycling ergometer test using a metabolic analyzer, and disease severity does not limit aerobic capacity (25). This protocol should begin with a power output of 20 W and progress every minute by 10, 20, or 30 W depending on the individual; pedaling cadence should be set between 50 and 60 revolutions per minute (rpm). The goal is to reach maximum capacity and volitional fatigue between 6 and 12 min.

Several submaximal cycling tests such as the YMCA submaximal test (25, 132, 134, 136) and the Astrand-Rhyming protocol (131, 157, 166) have been used in this population to estimate cardiorespiratory fitness. These can be helpful if individuals have contraindications to maximal workload tests. However, autonomic dysfunction can result in insufficient heart rate responses to increased workload (89), which limits the accuracy of these tests. Nevertheless, they can be useful for measuring changes before and after an exercise prescription as long as assessments are completed at the same time of day and with the same workloads.

The 6 min walk (6MWT) has been used as an indirect measure of cardiorespiratory fitness or aerobic capacity in individuals with PD. The 6MWT assesses distance covered as fast as possible without running or jogging. Individuals can use assistive devices (i.e., canes, walkers) during assessment, but test–retest should be consistent with use of these devices. The 6MWT shows excellent test–retest reliability in PD (24, 168), although use of a PD reference equation (48, 49) is recommended when attempting to compare results to predicted values (based on healthy older adults).

Muscle strength and endurance have been measured using a variety of methods, including one-repetition maximum (1RM) with free weights or machines (39, 69, 77, 124, 156), dynamometers (57, 140), and functional measures (42, 69, 123). Functional measures of strength frequently used in this population include the chair-rise test (69, 123) and the five times sit-to-stand (FTSTS) test (42). However, performance on these tasks will also be affected by balance and fall risk in individuals with PD, so it is important to assess balance before using these tests.

The muscle rigidity often associated with PD limits the flexibility of the joints and the trunk (149, 152). However, range of motion is most often measured with a goniometer using passive or active techniques (1). In addition, sit-up-and-reach and the back scratch test can be used to assess flexibility in the lower back and hamstrings and the shoulder, respectively (139).

Treatment

Parkinson's disease treatments only decrease the symptoms and do not slow the rate of progression (4). The primary treatments are pharmacological (table 31.1). The first line of therapy is usually dopamine precursors or dopamine agonists. However, as the disease progresses, the high dosages of these medications may cause side effects such as dyskinesia (uncontrolled movements) (82) or hallucinations (188). In addition, individuals on high doses of levodopa medications can experience wearing-off, or end-of-dose deterioration (167). This phenomenon results in the reappearance of motor symptoms before the next scheduled dose of medication. In severe cases of Parkinson's disease, surgical treatment (deep brain stimulation, or DBS) is also an option (43, 127). This treatment uses a surgically implanted electrode to provide chronic electrical stimulation to the subthalamic nucleus (STN) or the globus pallidus internus (GPi) (120). However, DBS is reserved for individuals with severe tremor, wearing-off issues, and medication-induced dyskinesias because of the risk of bleeding (51), infection (15), and cognitive dysfunction (105).

EXERCISE PRESCRIPTION

Exercise is a vital adjunct treatment (in conjunction with medication and DBS) for Parkinson's disease. According to the American College of Sports Medicine (ACSM), the primary goals of exer-

Table 31.1 Pharmacology

Medication class and name	Primary effects	Exercise effects	Important considerations
DOPA decarboxylase inhibitors/ Dopamine precursors • Carbidopa/levodopa (Duopa, Parcopa, Rytary, Sinemet, Sinemet CR) • Carbidopa/levodopa/ entacapone (Stalevo)	Improve bradykinesia and rigidity Reduce tremor	Can reduce slowness of movement and rigidity, allowing for greater range of motion and increased intensity during exercise	These medications have limited or no effects on freezing of gait and postural deficits. Possible side effects include nausea, hallucinations, sleepiness, and orthostatic hypotension. Individuals may experience motor and nonmotor fluctuations. Dyskinesia may occur at high dosages.
Dopamine agonists • Apomorphine (Apokyn) • Pramipexole (Mirapex) • Ropinirole (Requip) • Rotigotine (Neupro)	Improve bradykinesia and rigidity Reduce tremor	Can reduce slowness of movement and rigidity, allowing for greater range of motion and increased intensity during exercise	Dopamine agonists are not as effective at relieving symptoms as L-DOPA; used as initial therapy or a complement to L-DOPA. Possible side effects include dizziness, nausea, sedation, hallucinations, ankle swelling, and impulse control disorder.
COMT inhibitors • Entacapone (Comtan) • Tolcapone (Tasmar) • Opicapone (Ongentys)	Inhibit breakdown of L-dopa Used in conjunction with L-DOPA to assist with wearing-off symptoms	Can reduce slowness of movement and rigidity, allowing for greater range of motion and increased intensity during exercise	Possible side effects include dizziness, nausea, headache, dyspepsia, back pain, postural hypotension, and insomnia. If taken with L-DOPA, they can cause dyskinesia and hallucinations.
MAO-B inhibitors • Rasagiline (Azilect) • Selegiline or deprenyl (Eldepryl) • Selegiline hydrochloride (Zelapar)	Improve bradykinesia and rigidity Reduce tremor	Can reduce slowness of movement and rigidity, allowing for greater range of motion and increased intensity during exercise	MAO-B inhibitors are used as an alternative to L-DOPA in early PD. Possible side effects include dizziness, nausea, headache, dyspepsia, back pain, postural hypotension, and insomnia. If taken with L-DOPA, they can cause dyskinesia and hallucinations.
Anticholinergics • Benztropine mesylate (Cogentin) • Trihexyphenidyl (Artane)	Reduce tremor Ease dystonia that could arise during wearing-off period	Can reduce slowness of movement and rigidity, allowing for greater range of motion and increased intensity during exercise	Anticholinergics are used most often in early PD. Possible side effects include dry mouth, constipation, urinary retention, blurred vision, and memory issues.
NMDA antagonist Amantadine (Symmetrel)	Reduce tremor	Reduced tremor would make it easier to hold weights and use machines	Possible side effects include dry mouth, constipation, urinary retention, ankle swelling, skin rash, and hallucinations.

L-DOPA = levodopa; COMT = catechol-O-methyltransferase; MAO-B = monoamine oxidase B.

cise in people with PD are to delay disability, prevent secondary complications, and improve overall quality of life (4). Individuals with PD often lead a lifestyle characterized by excessive sedentary behavior and insufficient physical activity (11), which can create further complications such as coronary heart disease, hypertension, diabetes, and premature death (95). Regular physical activity can decrease or delay secondary symptoms affecting musculoskeletal and cardiovascular systems that occur as a result of decreased physical activity. Furthermore, exercise has many other benefits for those with PD, such as improved cognition, gait performance, and quality of life (13, 75, 116, 128).

Table 31.2 provides a review of the exercise prescription for individuals with PD. Aerobic exercise, resistance exercises, and range of motion exercises all yield significant benefits (150, 156). However, not all types of exercise produce the same benefits. Although limited research has systematically compared multiple modes of exercise within the same study, several different types of exercise have been proven to provide benefits to individuals with Parkinson's disease (156). The exercise prescription for those with PD is similar to that for healthy older adults (see chapter 6 for basics of exercise training), but exercise programming should be tailored to the individual and based around limitations imposed by the disease process (4). In addition, because of the high variability of PD symptoms, the clinical exercise physiologist should have current knowledge of symptoms before beginning exercise. The exercise prescription for individuals with PD should include balance training, flexibility, cardiorespiratory fitness, muscle strength, and functional training to promote improved posture, gait, transfers, mobility, muscle power, and functional capacity (38, 91, 102).

There are several special considerations when prescribing exercise to people with Parkinson's disease. Many individuals with PD suffer from some form of autonomic nervous system dysfunction, and some medications used to treat PD can further impair autonomic responses (64, 143, 153, 154, 190). Orthostatic hypotension (OH), also called postural hypotension, is an abnormal decrease in blood pressure that occurs with standing. OH is one of the most prevalent and disabling nonmotor symptoms associated with Parkinson's disease (146). Senard and colleagues reported that 20% to 50% of PD patients complain about symptoms of OH such as giddiness, dizziness, nausea, and pain, mostly while standing (154). Therefore, sweating, heart rate, and blood pressure should be monitored regularly during exercise (101). Considerable attention should also be paid to the development and management of fatigue (59). Since balance impairment and falls are common problems in individuals with PD, fall prevention education should be incorporated into the exercise program (113). Additionally, individuals with PD often have difficulty with exercise modalities that require multiple tasks (e.g., walking while talking) because of cognitive deficits. Therefore, exercise modalities that require dual-tasking should be introduced gradually to novice exercisers

and individuals who have balance or gait deficits (17, 144, 119).

Cardiorespiratory Exercise

Cardiorespiratory exercise for individuals with PD should be contingent on the individual's clinical presentation of symptoms and symptom severity. To improve cardiorespiratory function, individuals with PD should start an exercise program by performing 30 min of aerobic activity, 3 to 5 d/wk, at an intensity of 60% to 80% of their maximum oxygen uptake ($\dot{V}O_2$) reserve or heart rate (HR) reserve, or until they report a rating of perceived exertion (RPE) of 11 to 13 on a scale of 6 to 20 (4, 18, 111). This activity can be a continuous 30 min or made up of three separate 10 min bouts throughout the day. Progression over time for cardiorespiratory exercise should be individualized, and the exercise prescription should be reviewed and revised as the disease progresses.

Although progression in this population should be slow to minimize excessive fatigue, evidence has shown that moderate to vigorous exercise is safe, tolerated, and effective for reducing motor symptoms and improving quality of life in individuals with PD (85). Several studies have reported greater improvement in motor symptoms of PD (tremor, rigidity, bradykinesia, gait) after high-intensity treadmill exercise (111, 151; see the sidebar Research Focus) and after high-cadence cycling (i.e., forced exercise) (106, 136, 138; for review, see 109) when compared against low- to moderate-intensity treadmill exercise or low-cadence cycling.

Examples of aerobic activities for individuals with PD include treadmill or overground walking (53, 153), cycling, swimming, and dancing (65). Stationary upright cycling, recumbent cycling, and arm ergometer exercise are safer activities for individuals with more advanced PD. For more functional patients with PD, turning-based treadmill exercises are useful for improving turning performance (28). High-intensity treadmill exercise, tandem cycling, and high-cadence stationary cycling have also been shown to improve motor symptoms, gait, and dexterity (106, 136, 145).

Resistance Exercise

Strength improvements are similar in individuals with PD and healthy controls following a period of resistance training (36). Therefore, the recommendations for resistance exercise for healthy older adults can be applied to those with PD (see chapter 6 for basics of exercise training and chapter 33 for older adults) (148). Individuals with PD who are beginning resistance training should perform one set of 10 to 15 repetitions at 40% to 50% of their one-repetition maximum (1RM), 2 or 3 d/wk. This volume and modality may optimally increase muscle strength without excessive fatigue (145). Progression should be carefully based on the individual, and more advanced exercisers can perform more than one set of 8 to 12 repetitions at 60% to 70% of their 1RM, 2 or 3 d/wk.

Table 31.2 Exercise Prescription Review

Training method	Frequency	Intensity	Time (Duration)	Type (Mode)	Progression	Important considerations
Cardiorespiratory	3 d/wk	60%-≤80% $\dot{V}O_2$ reserve or HR reserve, or RPE of 11-13	30 min continuous or accumulated	Stationary cycling, walking, running, dancing, swimming, boxing	Gradual, individualized	Improve walking economy, gait, transfers, balance, and motor symptoms. Selection of type is based on PD severity.
Resistance	2 or 3 d/wk	40%-50% of 1RM for beginners; 60%-70% for advanced	1 set or more of 8-12 reps; 10-15 reps for beginners	Machine weights, resistance bands, hand weights	Begin with 1 set of 10-15 reps at 40%-50% of 1RM; progress to 2 or more sets of 8-12 reps at 60%-70% of 1RM	Improve muscular strength and endurance. Decrease muscle fatigue. Improve performance on ADLs. Emphasize muscles of the trunk and hip (to prevent faulty posture) and major muscles of the lower extremities (to maintain mobility).
Range of motion	1-7 d/wk	Full stretch to the point of slight discomfort	10-30 s	Slow static stretches of all major muscle groups	Begin with static stretching; more functional can progress to yoga, tai chi	Improve flexibility, balance, and range of motion. Emphasize spinal mobility, axial rotation, neck flexibility, and balance.

HR = heart rate; RPE = rating of perceived exertion; 1RM = one-repetition maximum; ADLs = activities of daily living.

Research Focus

SPARX Trial

Research by Schenkman and colleagues (151) suggests that high-intensity exercise is feasible and safe in individuals with early stage (de novo) Parkinson's disease. This phase 2 randomized clinical trial placed individuals with Parkinson's disease into either high-intensity treadmill exercise (80%-85% maximum heart rate), moderate-intensity treadmill exercise (60%-65% maximum heart rate), and a wait-list control group. Over the 6 mo intervention, the high-intensity group was able to reach the target HR zones and to complete the prescribed number of exercise sessions. These data suggest that higher HR goals may be safe and appropriate for some individuals with PD. A phase III multisite clinical trial (Study in Parkinson Disease of Exercise: SPARX3) is investigating the benefits of high-intensity treadmill exercise on MDS-UPDRS Part III score, mobility, walking activity, cognition, quality of life, cardiorespiratory fitness, blood-derived biomarkers of inflammation, and neurotrophic factors. Results from this study will provide additional insight into the role of exercise intensity in improving function and quality of life in individuals with Parkinson's disease.

Resistance training for individuals with PD should focus on major muscle groups and use resistive movements such as body weight exercise, elastic bands, weight machines, or free-weight training (137, 159). Resistance exercise with weight machines or resistance bands should emphasize extensor muscles of the trunk and hip to improve posture and the major muscles of the lower extremities to increase mobility and gait (4, 68, 137). Under the supervision of a physician, trainer, or therapist, resistance training and instability training can be used in combination as long as individuals do not have significant postural or balance issues (159). Individuals who have postural or balance issues should focus on seated exercises or machines that provide postural support (137).

Range of Motion Exercise

For individuals with PD, a general flexibility program is recommended similar to that of a healthy, age-matched individual (see chapter 6 for basics of exercise training). Static stretches should be performed for all major muscle–tendon units on 1 to 7 d/wk. Stretches should be slow, held for 10 to 30 s, and fully extended, flexed, or rotated to the point of slight discomfort. Improvement of neck flexibility should be emphasized to reduce neck rigidity and to improve posture, gait, balance, and mobility (55). Range of motion training programs can include a variety of functional activities, including stepping in all directions, stepping up and down, reaching forward and sideways, turning around, and standing and sitting down (112). For more advanced exercisers with PD, yoga (118) and tai chi (33) can be incorporated into the exercise program to further improve range of motion and balance.

EXERCISE TRAINING

Regular exercise is very important for those with PD in order to reduce disease severity and to maintain and improve cardiorespiratory health, gait performance, and overall quality of life (13, 75, 116). There is also emerging evidence that exercise may have a neuroprotective role in PD (34, 128). This section provides a brief review of the physiological and functional adaptations observed in people with PD as a result of exercise training.

Cardiorespiratory Exercise

Aerobic exercise is imperative for maintaining cardiorespiratory health and minimizing secondary illnesses related to inactivity. Cardiorespiratory training (figure 31.2) appears to have direct benefits for PD patients in a dose-dependent fashion (53), some of which are evident nearly immediately (129, 134). It improves aerobic fitness, motor function, resting HR, walking distance, and lower limb muscle strength and can lead to improvements in body composition in people with PD (35, 53, 135). More complex aerobic exercise such as dance can improve balance, endurance, balance confidence, and quality of life (65). In individuals with mild to moderate PD, aerobic exercise on a treadmill can improve gait performance, including velocity, stride length, stability, stride, rhythmicity, and joint excursion (75, 76).

Although treadmill training predominantly engages the lower extremities, improvements in repetitive upper body motor tasks, as measured by the Unified Parkinson's Disease Rating Scale (UPDRS) motor score, also occur after treadmill and cycle training

---

Practical Application 31.1

IMPORTANT CONSIDERATIONS FOR EXERCISE TRAINING

- *Cardiorespiratory endurance:* Improved aerobic capacity, resting heart rate, motor symptoms, cognitive function, gait, balance, lower limb strength, body composition, and quality of life.

- *Skeletal muscle strength:* Increased muscle hypertrophy, neuromuscular function, muscle strength, and mobility and reduced instability and gait disturbances in people with PD.

- *Skeletal muscle endurance:* Increased muscle endurance of the upper and lower extremities and increased functional mobility.

- *Flexibility:* Improved balance, balance confidence, and range of motion.

- *Body composition:* Improved waist circumference and body composition.

- *Cardiorespiratory:* Exercise type should be individualized and based on symptom severity. Symptoms of autonomic dysfunction should be closely monitored.

- *Strength:* Emphasize muscles of the trunk and hip to prevent faulty posture and major muscles of the lower extremities to maintain mobility.

- *Range of motion:* Neck rigidity is correlated with posture, gait, balance, and functional mobility; therefore, neck flexibility exercises should be emphasized. Balance impairment and falls are major problems in PD, and balance, flexibility, and range of motion training are crucial.

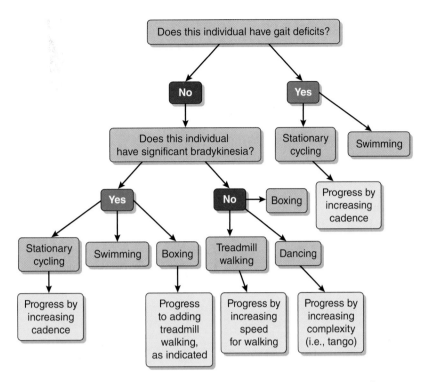

Figure 31.2   Flow chart for choice of cardiorespiratory exercises for PD based on severity of motor symptoms.

(56, 133, 135). One possible explanation for these improvements may be related to the exercise effects on neuroprotection of dopaminergic neurons, or more global effects on the repair or strengthening of overlapping motor circuits (neuroplasticity) (128, 169). Neuroplasticity is a process in which existing neural networks are altered by the addition or modification of synapses in response to changes in behavior or environment, such as exercise (92, 176). Emerging evidence from human studies suggests that exercise modalities with high intensity, high velocity, or complex movements promote neuroplasticity in brain circuits and potentially the recovery of dopaminergic neurons (109, 151, 175).

Evidence has shown that individuals with PD can acquire some degree of automaticity (i.e., the ability to dual-task) after motor skill practice (128, 183). Aerobic exercise using both intense and challenging goal-based practice can restore neuroplasticity in the motor circuitry responsible for automaticity. For example, body-weight-supported treadmill training at walking speeds faster than self-selected while performing problem-solving operations can decrease corticomotor excitability in PD as measured through transcranial magnetic stimulation (TMS) (53). This type of exercise can also increase dopamine D2-receptor binding potential (52). High-cadence cycling combines aspects of cognitive engagement with aerobic training, which leads to improvements in motor symptoms and increased connectivity between brain regions (10, 135, 136). High-cadence cycling is now being implemented in community-based indoor tandem cycling programs, which improve physical performance and are feasible for individuals with mild to moderate PD (106).

There is also growing evidence supporting aerobic exercise as a means to improve cognitive function in PD (116). Various types of exercise, including aerobic, resistance, and dance, can improve cognitive function, although the optimal type, amount, mechanisms, and duration of exercise are unclear (37, 41, 116, 132). Furthermore, the mechanisms behind the cognitive benefits of exercise are also not yet completely understood, but several theories have been proposed. The cerebrovascular reserve hypothesis (174) proposes that chronic exercise maintains integrity of the cerebrovascular system through increased oxygen transport, and that higher levels of physical fitness are associated with better cognitive functioning, and this may be mediated in part by improvements in cerebrovascular reserve. In contrast, the neurotrophin hypothesis suggests that exercise increases levels of two factors known to promote synaptic plasticity and neuron branching in the brain: brain-derived neurotrophic factor (BDNF) and insulin-like growth factor (IGF) (180).

Resistance Exercise

Muscle weakness and motor dysfunction are primary symptoms of PD (27, 88) that can lead to further complications, such as postural instability and problems with gait (83, 117). Although the complete mechanism for this weakness is not fully understood, it likely involves decreased activation of motor neurons due to inadequate basal ganglia stimulation (22). Resistance training (figure 31.3) is effective in improving muscle strength in individuals with PD (142) and therefore should be emphasized as a treatment.

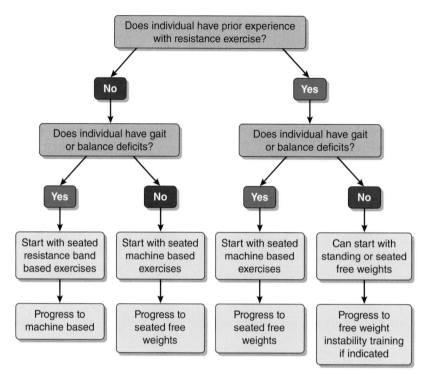

Figure 31.3   Flow chart for choice of resistance exercises for PD based on severity of motor symptoms.

Patients with mild to moderate PD can obtain increases in muscle strength similar to that of normal healthy adults of the same age (68). Resistance exercise using both eccentric and concentric muscle contractions improves muscle hypertrophy, muscle strength, and mobility in people with PD (77). In addition, resistance exercise increases neuromuscular function (69), fat free mass, and muscle endurance, and it improves mobility and performance in functional tasks (19). Regular progressive resistance exercise has also been shown to have a clinically worthwhile effect on walking capacity in people with PD (99), and evidence suggests that resistance training is effective in improving some measures of physical function and mobility, such as knee extension and flexion (30, 142).

Range of Motion Exercise

Parkinson's disease affects neurophysiological function and movement abilities (38, 125). Balance impairments and falls are major problems in those with PD; therefore, balance and range of motion training is crucial (4). Regular physical activity and exercise will result in improvements in postural instability and balance task performance (38). However, compared with general physical exercise, balance training has been shown to elicit superior improvements in balance, postural transfer, and fall prevention (163, 172). Combining resistance training with range of motion or balance training such as stretching, yoga, or tai chi can further improve gait and ambulation in PD, potentially through either altering the ability to control the motor system or helping override faulty proprioceptive feedback (145, 171). Furthermore, balance training is safe and effective for individuals with PD and can yield significant improvements in physical function, balance (66, 96), maximum excursion, directional control, gait, and strength (97). Simple flexibility exercises such as squatting and bending forward may also be useful for improving symptoms of orthostatic hypotension (173).

Overall, data suggest that in mild to moderate PD, balance training may be more effective than stretching or resistance training for improving postural stability (145). However, to significantly address falling, it is recommended that balance training be supplemented with resistance training (77, 145). The long-term benefits of balance training in PD are unclear; however, it is likely that to maintain efficacy, the balance training would need to be ongoing (145).

Clinical Exercise Bottom Line

- Cardiorespiratory, resistance, and range of motion exercises improve motor function and balance in individuals with Parkinson's disease.

- Exercise modalities with high intensity, high velocity, or complex movements appear to promote neuroplasticity in brain circuits and potentially the recovery of dopaminergic neurons.

- Because of the variability in symptom type and severity, exercise prescriptions need to be individualized.

- Autonomic dysfunction may increase the risk of syncope and orthostatic hypotension.

- Balance impairments and falls are frequent in individuals with PD. Appropriate safety protocols should be followed to mitigate these risks.

CONCLUSION

Starting exercise as early in the disease process as possible can limit the degree of disability and alleviate secondary symptoms imposed by the disease process. People with PD may gain cardiorespiratory, strength, and flexibility benefits using a modified exercise prescription similar to that for healthy adults. It is very important to tailor the exercise prescription to the individual because symptoms of PD are highly variable from patient to patient. Individuals with PD can suffer from autonomic nervous system dysfunction and therefore run greater risk of syncope; symptoms of orthostatic hypotension should be watched for during exercise. Heart rate should also be carefully monitored because autonomic dysfunction associated with PD can influence the heart rate response to exercise.

Balance impairment and falls are major problems in PD. The clinical exercise physiologist should take steps to ensure the individual's safety during exercise. The most beneficial exercise regimen for PD likely combines cardiorespiratory training, resistance training, range of motion training, and balance training. Key health outcomes of an exercise program for individuals with PD are improved balance, gait, transfers, joint mobility, muscle power, and functional capacity. The ultimate goals are to delay disability, prevent secondary complications, and improve quality of life as the disease progresses.

Online Materials

Visit HK*Propel* to access a link to the references, the case studies with discussion questions, and a quiz to help you review key concepts and test your understanding of the material covered in this chapter.

PART VIII

Special Populations

Part VIII discusses various populations that have been found to benefit greatly from exercise therapy in terms of preventing health problems and improving quality of life. Often these individuals present with multiple health and functional issues. The clinical exercise physiologist needs to be knowledgeable about the needs of these various special populations in order to maximize exercise-related benefits and minimize potential risks.

Chapter 32 focuses on children. With the continual rise in obesity and sedentary habits observed in children, concern remains about the long-term ramifications of children's health behaviors. These include the risk of developing high blood pressure and diabetes at a young age, along with cardiovascular disease and cancer later in life. To have a favorable influence on children's health, intervention programs need to begin at an early age. By establishing healthy lifestyle habits (such as exercise) at an early age, children have a greater opportunity to avoid chronic diseases as adults. The clinical exercise physiologist should be knowledgeable about developing safe, fun, and effective lifestyle modification programs that address the major factors that negatively influence the health of children.

Addressing the other end of the human life span, chapter 33 deals with the aging process, with a particular focus on those over the age of 65. Since the publication of the second edition of this textbook, approximately half of the large population in the United States that makes up the Baby Boomer generation (those born between 1946 and 1964) has reached age 65. Additionally, worldwide there are now more people living past the age of 90 (known as "the oldest old") than ever before. To meet the medical and social needs of the aging society, we must address chronic health concerns. This can be achieved in large part by encouraging the elderly to remain physically active as a means of maintaining activities of daily living and living an independent lifestyle for as long as possible. When an elderly person plans to begin an exercise program, physician clearance may be important because many older people have risk factors that require a modification of their exercise regimen. The clinical exercise physiologist should be aware of health issues that affect the elderly and the process for developing, modifying, and maintaining an exercise training program for this population.

Chapter 34 provides an in-depth review of exercise among patients with depression. Depression is the most prevalent psychiatric disease or disorder in the world. In addition to chronic depression among the general public, depression is common among patients who have recently experienced a major medical event. Having the knowledge to recognize depression-related symptoms is important for the clinical exercise physiologist, who interacts with these patients more often than other health care providers. For patients who experience depression, regular exercise training can be an effective adjunctive or alternative treatment to medication.

Chapter 35 addresses intellectual disabilities that affect functioning and adaptive behavior. This category includes disorders such as Down syndrome and autism, in which coexisting health problems are common. Individuals with intellectual disabilities are generally inactive and more likely to experience obesity, which increases the risk for diabetes and vascular diseases. This chapter addresses the scope and pathophysiology of intellectual disabilities, as well as considerations for exercise testing and exercise training.

CHAPTER 32

Children

Timothy J. Michael, PhD

Carol Weideman, PhD

Clinical exercise physiologists and other clinical practitioners who evaluate, diagnose, and treat individuals with disease or disorders will more than likely see children in their practices. As seen in adult populations, exercise can be used as a diagnostic tool, as an evaluative tool, and as a treatment for clinical conditions. However, we do need to remind ourselves that children are different from adults in a variety of ways—physically, psychologically, and emotionally. Children's responses to exercise may be different from those of adults, and their adaptations to exercise are predicated, to a certain degree, on growth and maturation. Their maturity also influences their ability to take instruction and to give full efforts when needed. Physical size and fitness may influence test modality and protocol choices (i.e., one size does not fit all).

DEFINITION

Pediatrics is the branch of medicine concerned with children and their diseases. **Children** are the population between infancy and adolescence.

SCOPE

Although children do not, in and of themselves, constitute a clinical population, it is important for clinicians to know about the exercise responses of children and how they may influence exercise testing, exercise training, and exercise prescription. Children can present with many different clinically relevant signs and symptoms, diseases, and disorders. In recent years the number of children who are overweight or obese has increased dramatically. These children are now being diagnosed with chronic diseases that we often see in adult clinical populations, such as type 2 diabetes, hypertension, dyslipidemia, and metabolic syndrome. It is not unusual to see the many different clinical conditions in children

that we would in adults, ranging from cardiovascular to pulmonary, musculoskeletal, and neurological conditions. It is therefore appropriate for clinicians to use exercise with children similarly to the way they do with adult clinical populations, with an understanding of the underlying differences between these two groups.

A review of the literature in regard to children and adolescents has shown an increasing interest in exercise and physical activity in relation to primary and secondary prevention of a variety of clinical conditions. A number of meta-analyses and systematic reviews have been conducted on physical activity and type 1 diabetes (38), exercise and insulin resistance in youth (17), weight-bearing activities and bone health in children (7, 12, 26, 52), and physical activity and cardiovascular and cardiometabolic risk factors in children (10, 16). However, most of the meta-analyses and systematic reviews tend to be in the area of childhood obesity. These reviews have documented issues such as the effect of childhood obesity prevention programs on blood lipids (9), exercise and vascular function in obese children (15), and the effect of exercise in obese children on resting blood pressure (19). Many reviews evaluated exercise interventions on improvement of fitness (strength and endurance), body composition (% body fat), and weight status (BMI z-score) (4, 27, 28, 46, 50, 59). Regardless of the clinical condition, in most cases increased physical activity and participation in regular exercise seem to help in their prevention and treatment. Additionally, there are only a few instances where exercise would be contraindicated for children with chronic health problems.

This chapter does not discuss all clinical conditions that afflict children; readers can refer to the other chapters in this text for information on many of these. The aim of the chapter is to bring to light the differences that children exhibit in relation to adults when exercise is used for evaluation, diagnosis, and treatment. The chapter also addresses the adjustments that may be needed to accommodate children in the clinical exercise laboratory.

Acknowledgment: We would like to acknowledge Dr. William A. Saltarelli's contributions to the previous editions of this chapter.

CLINICAL CONSIDERATIONS

Children may develop a variety of conditions that warrant an exercise test or an exercise program intervention. This section deals with some of the considerations that clinicians need to take into account when conducting exercise testing and prescribing an exercise program for a child. A clinical report (31) detailed the need to conduct physical activity vital sign (PAVS) assessment and counseling as part of the normal clinician interactions with their patients. Additionally, the report advocates for physical activity assessment, testing, and prescription becoming part of the standard education for medical students and residency programs.

Signs and Symptoms

Children may present with a variety of signs and symptoms that would warrant the use of exercise for evaluation, as an aid to diagnosis, or as part of a treatment plan. The signs and symptoms may not be very different from what we see in adults: shortness of breath; chest pain; faintness, dizziness, syncope; unusual fatigue; and exercise intolerance. Paridon and colleagues for the American Heart Association (36) have listed the common reasons why a child should have a stress test:

1. To evaluate specific signs that are induced or aggravated by exercise
2. To assess or identify abnormal responses to exercise in children with cardiac, pulmonary, or other organ disorders, including the presence of myocardial ischemia and arrhythmias
3. To assess efficacy of specific medical or surgical treatments
4. To assess functional capacity for recreational, athletic, and vocational activities
5. To evaluate prognosis, including both baseline and serial testing measurements
6. To establish baseline data for institution of cardiac, pulmonary, or musculoskeletal rehabilitation

History and Physical Examination

The first step, as with adults, is to obtain a complete medical history and physical exam, along with any pertinent laboratory findings (2). The information gleaned from this will help the clinician evaluate the child for risk of exercise by stratifying level of risk and will determine contraindications to exercise testing and exercise participation. The goal is to make sure that, in the case of exercise testing, the benefit of the testing outweighs information to be gained by having the child attempt to complete the exercise test (table 32.1) (36).

Children have many conditions that may elicit a referral for exercise testing, including asthma, cystic fibrosis, diabetes, obesity, and various forms of heart disease (2, 25, 35, 41, 53).

Pharmacology

Drugs are given for specific clinical conditions; refer to the relevant chapters in this textbook for an overview of medications used. Many drugs prescribed for children were not first developed for or tested on children. Children and adults differ in physiologic parameters, and thus pharmacokinetics and pharmacodynamics differ as well (18).

The determination of the correct dose for a child needs to take into account the differing pharmacodynamics and pharmacokinetics of children along with the route of administration and the specific characteristics of the drug being prescribed. In some cases, the amount may need to be more than, less than, or the same as what would be used for an adult. Clinical exercise physiologists need to be cognizant of these differences so as to be aware of possible adverse drug reactions and interactions. Further information on pharmacology can be found in chapter 3, and readers should refer to other chapters to review specific medications associated with various diseases and disorders.

Exercise Testing

Helge Hebestreit (24) lists three reasons that clinical exercise testing of children may be difficult:

- Children have small body size in relation to testing equipment.
- They show poor peak performance.
- They have a short attention span and usually poor motivation during exercise testing; this is especially true with longer exercise protocols.

Furthermore, Hebestreit says, very young children and those with chronic conditions are an even greater challenge when attempting to have them complete a clinical exercise test.

The American College of Sports Medicine (ACSM) (2) offers the following guidelines with regard to exercise testing for children:

- Exercise testing for children is not indicated unless there is a health concern.
- The testing protocol should be based on the reason the test was requested and the capabilities of the child.
- Children should always be familiarized with the test modality and protocol before the actual test; this will help reduce test anxiety as well as increase the chances for success of the exercise test.
- Both a treadmill and a cycle ergometer should be available for exercise testing.
- Children are psychologically and emotionally immature and may require more motivation and support than older people to complete the exercise test.

Table 32.1 Relative Risks for Stress Testing

Lower risk	Higher risk
Symptoms during exercise in an otherwise healthy child who has a normal CVS exam and ECG	Patients with pulmonary hypertension
Exercise-induced bronchospasm in the absence of severe resting airway obstruction	Patients with documented long-QTc syndrome
Asymptomatic patients undergoing evaluation for possible long-QTc syndrome	Patients with dilated or restrictive cardiomyopathy with CHF or arrhythmia
Asymptomatic ventricular ectopy in patients with structurally normal hearts	Patients with a history of hemodynamically unstable arrhythmia
Patients with unrepaired or residual congenital cardiac lesions who are asymptomatic at rest, including these: • Left to right shunts (ASD, VSD, PDA, PAPVR) • Obstructive right heart lesions without severe resting obstruction (TS, PS, ToF) • Obstructive left heart lesions without severe resting obstruction (cor triatriatum, MS, AS, CoA) • Regurgitant lesions regardless of severity	Patients with hypertrophic cardiomyopathy with the following: • Symptoms • Greater than mild LVOTO • Documented arrhythmia
Routine follow-up of asymptomatic patients at risk for myocardial ischemia, including those with the following: • Kawasaki disease without giant aneurysms or known coronary stenosis • After repair of anomalous LCA • After arterial switch procedure	Patients with greater than moderate airway obstruction on baseline pulmonary function tests
Routine monitoring in cardiac transplant patients not currently experiencing rejection	Patients with Marfan syndrome and activity-related chest pain in whom a noncardiac cause of chest pain is suspected
Patients with palliated cardiac lesions without uncompensated CHF, arrhythmia, or extreme cyanosis	Patients suspected to have myocardial ischemia with exertion
Patients with a history of hemodynamically stable SVT	Routine testing of patients with Marfan syndrome
Patients with stable dilated cardiomyopathy without uncompensated CHF or documented arrhythmia	Unexpected syncope with exercise

CVS = cardiovascular system; ECG = electrocardiogram; CHF = congestive heart failure; ASD = atrial septal defect; VSD = ventricular septal defect; PDA = patent ductus arteriosus; PAPVR = partial anomalous pulmonary venous return; TS = tricuspid stenosis; PS = pulmonary stenosis; ToF = tetralogy of Fallot; MS = mitral stenosis; AS = aortic stenosis; CoA = coarctation of aorta; LVOTO = left ventricular outflow tract obstruction; LCA = left coronary artery; SVT = supraventricular tachycardia.

Reprinted by permission from S.M. Paridon et al., "Clinical Stress Testing in the Pediatric Age Group: A Statement From the American Heart Association Council on Cardiovascular Disease in the Young, Committee on Atherosclerosis, Hypertension, and Obesity in Youth," *Circulation* 113, (2006): 1905-1920.

General Considerations for Conducting Exercise Tests With Children

This section considers issues relevant to exercise testing specifically in children. Adjustments often need to be made to fit the equipment to the given child as well as to accommodate the child's condition, maturity (physical and emotional), and fitness.

Researchers (54) have shown (figure 32.1) that cardiopulmonary exercise testing can be used to assess children with medical conditions that decrease functional capacity. Further, improving functional capacity through exercise training can improve the medical condition of the child.

Modalities

The mode of exercise testing for children in a clinical setting is most often either a treadmill or a cycle ergometer. These modes are used to assess functional capacity and to determine exercise prescriptions for aerobic-based exercise plans. Treadmill testing tends to elicit greater oxygen consumption than does cycle ergometer testing. It is particularly important to have an ergometer that can be adjusted appropriately to the child's size (2, 25, 35, 41, 53).

Although the treadmill and the cycle ergometer are the most commonly used modalities for clinical exercise testing of children, other laboratory-based tests can be used to assess factors other than aerobic capacity. Laboratory tests can assess muscular strength and

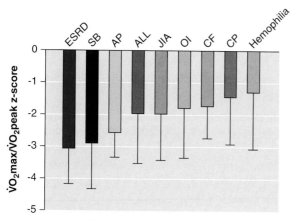

Figure 32.1 Aerobic fitness of children with a chronic condition.

Reprinted by permission from T. Takken et al., "Cardiopulmonary Exercise Testing in Pediatrics," *Annals of the American Thoracic Society* 14, Supplement 1 (2017): S123–S28.

anaerobic power and capacity, as well as flexibility. Again, whatever modality is used must be appropriately adapted to the child.

In 2006, Chang and coworkers reviewed the use of the treadmill versus the cycle ergometer in clinical pediatric exercise testing. Approximately 24% of pediatric cardiology and pulmonology centers use cycle ergometers and 76% prefer to use the treadmills (11).

Protocols

Protocols for adults can usually be used for children. Occasionally adjustments will be needed to accommodate the earlier onset of fatigue in children (i.e., stage time may be shortened), or the large increases in metabolic equivalent (MET) values from stage to stage may be lessened to accommodate children in cases in which large jumps in METs could be discouraging and affect motivation to complete the test. Stephens and Paridon reported that the protocol should be determined based on the kind of information desired by the person requesting the test (2, 25, 35, 41, 53).

The treadmill (TM) and cycle ergometer (CE) are the preferred modes for testing aerobic (cardiorespiratory) function, with the Bruce (TM), Balke (TM), James (CE), and Godfrey (CE) protocols recommended. Variables measured include HR, RPE, BP, ECG, SaO_2, $\dot{V}O_2$, and anaerobic threshold. Adverse signs and symptoms should be documented. Special consideration should be given to arrhythmia and ischemic evaluation as well as echocardiography. To test anaerobic function, the Wingate cycle ergometry protocol is recommended. Key variables are peak power output, mean power output, and the fatigue index, with special consideration for exercise-induced bronchospasm. To test strength, the practitioner can choose between the handgrip dynamometer and isokinetic dynamometer to measure force (kg) and torque (kg). For flexibility, modes include the inclinometer, flexometer, and goniometer to measure degrees (ROM).

Chang and colleagues, as part of the study mentioned previously, also reviewed the use of stress echocardiography in children in the United States (11). They reported that pediatric cardiology and pulmonology centers did not perform stress

The 6 Min Walk Test: Is It Appropriate for Children With Chronic Conditions?

A systematic review (6) of the 6 min walk test in pediatric populations with chronic conditions suggests that results from this test be interpreted with caution. Some of the chronic conditions of patients who performed this test and were included in this review include cystic fibrosis, cerebral palsy, muscular dystrophy, spina bifida, obesity, congenital heart disease, pulmonary hypertension, juvenile idiopathic arthritis, and end-stage renal disease.

- Different patient populations showed different strengths and weaknesses on the 6 min walk test.
- Clinicians should review the reliability and validity associated with this test for a given condition before making any determination of diagnosis or prognosis.
- After reviewing measurement statistics associated with this test and a given patient population, it may be prudent to select a different outcome measure (e.g., GXT).
- Because of incomplete data regarding the measurement statistics of this test across a variety of clinical pediatric conditions, caution should be used when interpreting results.
- Test results may need to be used to measure individual change rather than compared against normative data, at least until more studies show reliability and validity of this test for the population in question.

echocardiography 37% of the time. When ECGs were performed, measurements were taken after dismount from the cycle ergometer (32%) or treadmill (5%), while on the cycle ergometer (3%), during dobutamine infusion (4%), or in some combination of those situations (19%). Kimball (29) encourages the use of stress echocardiography as a means of extending physical examinations for children.

If general health or fitness assessment is all that is needed, use of the Cooper Institute's FitnessGram is suggested. Most of the assessments in this battery are field-based tests but should be used from only a functional, not a diagnostic, point of view. In addition, participation in these field-type tests should be limited to children who are free of contraindications to exercise participation or testing as listed in *ACSM's Guidelines for Exercise Testing and Prescription, Tenth Edition* (2).

Contraindications to Exercise Testing

The safety of the child is always the first priority. As with adults, strict compliance with the contraindications to exercise testing in *ACSM's Guidelines for Exercise Testing and Prescription, Tenth*

CLIENT–CLINICIAN INTERACTION

When referring to responses and adaptations to exercise, the pediatric exercise physiologist Oded Bar-Or (5) often stated in his presentations that "children are not small adults." This is important to keep in mind when evaluating the exercise performance of children. Following is a brief review of special considerations to take into account when testing children.

Laboratory Environment

- The lab must be safe and well illuminated.
- Staff should be trained and have warm, friendly personalities that will help them establish a positive relationship with children as soon as they enter the facility.

Particular Safety Issues

- Two testers are essential to ensure constant visual and verbal contact with the child.
- During treadmill testing, two spotters should be used—one at the subject's side and one behind.

Pretest Protocols

- Establish a relationship with children as they enter the facility.
- Completely explain the test to both parents and children. Be sure that both know exactly what will take place and what the child will be asked to do.
- After the explanation of the procedure, obtain parental consent and complete the child's assent documents. A child's assent document (for children age 6 and older) should be written in age-appropriate language and be short and to the point.
- Although offering an incentive to the child is controversial, rewards can be extremely helpful in eliciting a maximal effort.

Laboratory Equipment

- When appropriate, testing equipment should be modified for the size and maturity of the subject.
- Van Brussel and colleagues (56) have suggested a systematic method for interpreting cardiopulmonary exercise tests in pediatric populations. Figure 32.2 illustrates this methodology.

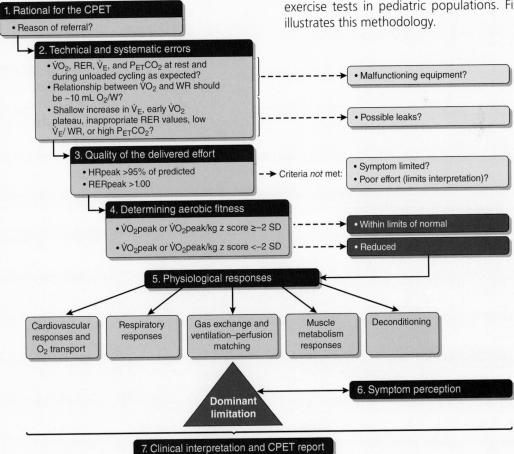

Figure 32.2 The seven-step interpretive strategy for pediatric cardiopulmonary exercise testing.

Reprinted by permission from M. Van Brussel et al., "A Systematic Approach to Interpreting the Cardiopulmonary Exercise Test in Pediatrics," *Pediatric Exercise Science* 31, no. 2 (2019): 194–203.

Research Focus

Safety of Exercise Tests in Children

A study (21) published in the journal *Pediatric Cardiology* presented the outcome of a retrospective review of patients who were evaluated using an exercise stress test. The study reviewed tests conducted over a 4 yr period. The researchers evaluated the data for the prevalence of arrhythmias, predictors for arrhythmias, and the safety of exercise stress testing in children.

The authors reviewed 1,037 exercise tests of children, some with preexisting congenital heart disease. During testing, 28% of the children experienced some sort of arrhythmia; 3% experienced clinically significant arrhythmias, including ventricular

tachycardia, atrial fibrillation, supraventricular tachycardia, second-degree atrioventricular block, and increasing ventricular ectopy. Of the children with clinically important arrhythmias, those with severe left ventricular dysfunction, a pacemaker or implantable cardioverter-defibrillator, a history of arrhythmia or arrhythmia disorder, and a nonsinus baseline rhythm were most likely to develop an arrhythmia during their exercise stress test. It was noted that none of the children required cardioversion, defibrillation, or acute anti-arrhythmic therapy, suggesting it is safe to conduct exercise stress tests on children.

Edition, is warranted (2, 25, 35, 41, 53).

In addition, the criteria for testing termination specified in the ACSM guidelines should be strictly adhered to. Paridon and colleagues for the American Heart Association (36) list three general reasons for terminating a clinical exercise test in children:

- When diagnostic findings have been established (continuing the test would not produce any further information)
- When the monitoring equipment fails
- When signs or symptoms indicate that continuation of the test would put the patient at potential risk for an adverse event

EXERCISE PRESCRIPTION

This section reviews the exercise prescription guidelines relevant to children, specifically those for aerobic exercise, resistance exercise, and flexibility training, and concludes with a review of the response to exercise training. Special attention is given to the differences between children's and adults' responses to exercise.

Aerobic

A review of the literature revealed very little in the way of specific details for aerobic exercise prescriptions. General guidelines for aerobic training are found in a variety of publications and are summarized in the following list and in table 32.2. Note that the guidelines in the table are divided into three objectives: basic health-related fitness, intermediate health-related fitness, and athletic performance. This approach allows for differences in abilities, interests, and fitness objectives of children (2, 34, 48).

- The type, intensity, and duration of exercise activities need to be based on the maturity of the child, medical status, and previous experiences with exercise.
- Regardless of age, the exercise intensity should start low and progress gradually.

- Because of the difficulty in monitoring heart rates with children, modified Borg or OMNI scales are practical methods of monitoring exercise intensity in children.
- Children are involved in a variety of activities throughout the day. A specific time should be dedicated to sustained aerobic activities.
- Adolescents should be physically active daily, or nearly every day, as part of play, games, sports, work, transportation, recreation, physical education, or planned exercise, in the context of family, school, and community activities.
- The intensity or duration of the activities is probably less important than the fact that energy is expended and a habit of daily activity is established.
- Adolescents should engage in three or more sessions per week of activities that last 20 min or more and that require moderate to vigorous levels of exertion. Moderate to vigorous activities are those that require at least as much effort as brisk or fast walking.
- Children and adolescents should do 60 min or more of physical activity each day.

An important omission in the ACSM guidelines is setting proper intensity. Rowland (42, 43) and others have suggested the use of adult guidelines to increase aerobic fitness, and studies have shown higher intensity to be effective.

Note that children perceive and experience exercise differently than do adults. Adult rating of perceived exertion (RPE) scales, such as the original Borg 15-point scale, can pose problems for children. Because of the differences between adults and children with regard to RPE, a new scale, the OMNI RPE scale, was developed (39). Figures 32.3 and 32.4 show specific OMNI RPE scales for running and cycling. Clinicians are advised to use the OMNI RPE scales for children during exercise testing and as a means to prescribe exercise intensities.

Table 32.2 Aerobic Training Prescription for Children

Goals	Frequency	Intensity	Time (Duration)	Type (Mode)	Progression	Important considerations
Basic health-related fitness	3 times per week	HR 50%-60% HRmax Moderate (beginning to sweat) Borg RPE 4-5 or OMNI RPE 2-5	30 min	Walking/jogging Dancing Recreational biking and swimming Active games	Get children moving. It is not necessary to overload. Use minimal progression.	Make activity fun and part of an active lifestyle.
Intermediate health-related fitness	3-5 times per week	HR 60%-75% HRmax Vigorous (breathing hard, sweating) Borg RPE 6-9 or OMNI RPE 6-9	40-60 min	Jogging, running, biking Fitness-based games Intramural and local league sports	Introduce training principles and small-progression overload by increasing tempo or decreasing rest periods.	Be aware of heat stress and overtraining.
Athletic performance	5 or 6 times per week	HR 65%-90% HRmax Vigorous (breathing hard, sweating) Borg RPE 6-9 or OMNI RPE 6-9	60-120 min	Training programs similar to those for adults (running, cycling, aerobics, school and community sports)	Use structured training programs stressing variable intensities and durations.	Be aware of heat stress and overtraining.
Special considerations related to maturation	*Preschool (3-5 yr)*: Aim for at least 180 min/d, 60 min of moderate to vigorous. Activities include unorganized play that focuses on gross motor skills such as walking, running, swimming, and tumbling. *Elementary school (5-10 yr)*: Enjoyment and fun should be the priority, with a goal of 60 min/d. Continue to develop fundamental skills and provide opportunities for play; include new activities such as jump rope and dance. *Middle school (11-14 yr)*: The minimum goal is 60 min/d of activities that encourage socialization and enjoyment. Involvement in a variety of activities should be encouraged rather than specialization. *High school (15-18 yr)*: Build upon the enjoyment and socialization of daily activity, with a goal of 60 min/d. Some individuals will be interested in competition. The guidelines for athletic performance would be appropriate.					

HR = heart rate; RPE = rating of perceived exertion; OMNI RPE = a picture system for perceived exertion.

Vigorous physical activities are rhythmic, repetitive activities that require the use of large muscle groups to elicit a heart response of 60% or more of a subject's maximum heart rate adjusted for age. An exercise heart rate of 60% of maximum heart rate for age is sufficient for cardiorespiratory conditioning. Alternative methods of calculating optimal heart rate in children are available. Most methods of calculating optimal aerobic training zones for children require an estimate of maximum heart rate. Adult formulas such as 220 minus age are usually used for children. As stated by Rowland, however, maximal heart rate determined by treadmill and cycle studies remains constant across the pediatric years (42). Therefore, adult formulas strongly dependent on age may not be appropriate. Direct maximal heart rate is best for determining exercise prescriptions. In a systematic review and meta-analysis of age-based prediction equations of maximal heart rate (MHR) in children, the 220 – age equation overestimated

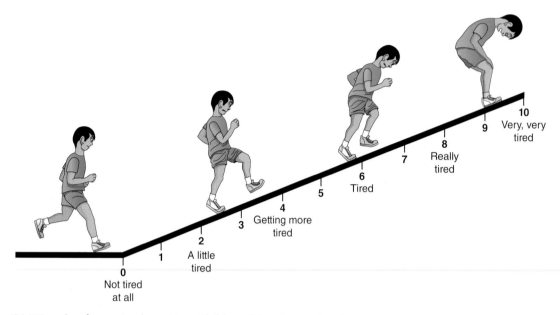

Figure 32.3   OMNI scale of perceived exertion: Child, walking to running format.

Reprinted by permission from R.J. Robertson, *Perceived Exertion for Practitioners: Rating Effort With the Omni Picture System* (Champaign, IL: Human Kinetics, 2004), 146.

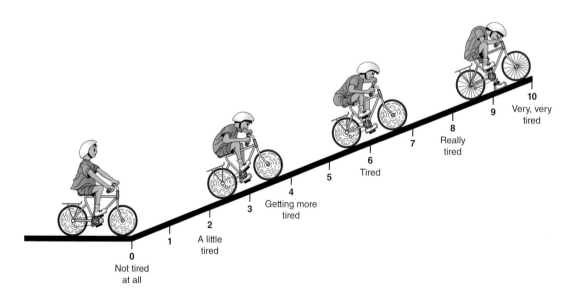

Figure 32.4   OMNI scale of perceived exertion: Child, cycle format.

Reprinted by permission from R.J. Robertson, *Perceived Exertion for Practitioners: Rating Effort With the Omni Picture System* (Champaign, IL: Human Kinetics, 2004), 146.

MHR by 12.4 b · min^{-1} (ES = 0.95, $p < 0.5$) (13). An alternative formula suggested by Tanaka and colleagues (55) is max HR = 208 − (.7 × age) in years. This formula underestimated MHR by 2.7 b · min^{-1} (ES = −0.34, $p < 0.05$) and explained the greater variability in MHR associated with age. The mode of testing and BMI did not explain the differences in MHR; the authors suggest more work is needed to understand the sympathetic response to exercise in children as well as hormonal changes associated with puberty. Adolescents have a slightly lower catecholamine response compared with adults during $\dot{V}O_2$max tests lasting 10 to 15 min but have a similar overall response (44). The Tanaka equation was developed from adult data and requires further study in children (13, 42).

Resistance Training

As recently as the late 1970s, pediatric exercise experts and medical doctors believed that prepubescent children could not benefit from strength or resistance training because this developmental group lacked the prerequisite circulatory hormones (58). Furthermore, many believed that the stress imposed by this type of training was not safe and could injure bones, especially at the growth plates. Since that time, numerous controlled studies have provided compelling evidence that strength or resistance training produces strength gains in both prepubescent girls and boys (37, 51, 60).

As with adults, the effectiveness of training appears to depend on intensity, volume, and duration. But these specific factors have not been established with certainty in children. An excellent reference is a document from a conference sponsored by the American Orthopedic Society for Sports Medicine (8), which concludes that strength training for the prepubescent

- improves muscular strength and endurance,
- improves motor skills,
- protects against injury (sports),
- has positive psychological benefits, and
- provides a forum for the introduction of safe and proper training.

The increase in strength in response to resistance training in children is primarily due to neurologic changes rather than hypertrophy. Factors including increased rate of firing of motor units, increased motor unit recruitment, and increased conduction velocity are similar in girls and boys. However, at puberty, growth factors and anabolic hormones play a role in the change in muscular development in boys (44). Training prescriptions for children show great variability but seem generally to follow adult prescriptions with the exception of lower resistance and higher repetitions. Supervision to ensure proper technique is most important when one is prescribing resistance training to children (47). Practical application 32.2 and table 32.3 review training principles for children with respect to improving muscular strength and endurance for basic health and athletic performance.

Practical Application 32.2

INTEGRATIVE NEUROMUSCULAR TRAINING

The benefits of integrative neuromuscular training (INT) for youth are many. The individual's needs and psychological development and the physiological demands of the sport or activity are integral components to consider when designing the training. When supervised by a qualified professional, INT has been shown to decrease youth sport injuries, provide a new passion for fitness, and improve physical literacy (greater technical ability, competence, and confidence). INT exercises can be introduced as young as 6 yr; however, at this age the focus should be on movement patterns such as skipping, hopping, and static balance. As children progress through puberty, difficulty and intensity can be added, and INT can be incorporated into the warm-up or conditioning portion of the exercise session. When INT exercises are added, the total dose of exercise needs to be considered because it may contribute to chronic repetitive stress on developing skeletomuscular systems. The following are sample exercises based on category, recognizing that the development of the child is first and foremost (61).

- Resistance exercises that use resistance bands, free weights, or kettle bells
- Dynamic stability exercises that target the trunk to improve posture—unilateral balance, Y excursion balance, dynamic lunges
- Core training exercises—standing wood chopper, plank variations, medicine ball toss to rebounder or partner
- Plyometrics—standing broad jump, hopping, skipping, low box jump, squat jump
- Speed and agility exercises—shuttle, diagonal running, karaoke, changing directions

Table 32.3   Resistance Training Prescription for Children

Goals	Frequency	Intensity	Time (Duration)	Type (Mode)	Progression	Important considerations
Basic health-related fitness	2 or 3 times per week	Very light, <40% of projected maximal effort	1 or 2 sets of 6-12 reps	Body weight for major muscle groups (sit-ups, push-ups) Stretch bands	Overload and progression are not essential unless child has interest.	Main objective is introducing resistance training and correct movement patterns.
Intermediate health-related fitness	3 or 4 times per week; alternating upper and lower body segments allows for consecutive days	Light to moderate, 50% of projected maximal effort	1-3 sets of 6-15 reps	Resistance exercise with machines such as leg press, curls, shoulder press Pull-ups	Introduce one or two components of overload 1 or 2 times per week.	Be sure child wants to be involved. Avoid maximal lifts.
Athletic performance	4 or 5 times per week for training activities	Match specific load requirements with sport Max should be <70% projected maximal effort	3-5 sets of 5-20 reps specific to sport	Advanced sport-specific multijoint lifts (clean pull, power press, Olympic-style lifts)	Program design should stress variable intensity and duration to cause overload 2 or 3 times per week.	Be sure child wants to be involved. Avoid maximal lifts.
Special considerations related to maturation:	*Elementary (5-10 yr):* Muscle- and bone-strengthening activities should be included at least 3 d/wk, but focus should be on gross motor skills such as hopping and jumping. *Middle school (11-14 yr):* Muscle- and bone-strengthening activities should be included at least 3 d/wk. Involvement in a variety of activities should be encouraged rather than specialization. *High school (15-18 yr):* Build upon the enjoyment and socialization of daily activity. Some individuals will be interested in competition. The guidelines for athletic performance would be appropriate.					

Flexibility and Range of Motion

Flexibility is defined as the ability to move joints through a full range of motion. A widely accepted concept is that children are extremely flexible and that therefore flexibility should not be a priority in activities or training. Exceptions seem to be children in sports such as gymnastics and dancing, in which flexibility is required and its importance is appreciated. Nevertheless, flexibility training is recommended at all ages to ensure safe activity. Most research studies show a decline in flexibility as children become older. Clark (14) concluded that boys tend to lose flexibility after age 10 and girls after age 12. In fact, Milne and colleagues (33) found that flexibility in both boys and girls declined between kindergarten and 2nd grade. Girls as a group are usually more flexible than boys, which may reflect the activities in which girls participate. In addition, contraindications and specific medical concerns are absent unless the child has preexisting musculoskeletal problems. Table 32.4 summarizes range of motion training guidelines.

EXERCISE TRAINING

Pediatric exercise physiologists are continually challenged to explain basic exercise physiology of children—that is, their responses and adaptations to bouts of exercise. We have learned from Oded Bar-Or that with respect to these responses, children are not small adults. However, despite the desirability of fully understanding children's responses to exercise, exercise research has been hindered by many obstacles that lie in the way of adequately explaining these responses through experiments. For example, ethical issues prevent us from taking multiple muscle biopsies to determine adenosine triphosphate (ATP) use, and the fact that training studies are complicated by developmental changes has limited our knowledge. Often we simply have to rely on what we know about adults and hope it is similar in children. What science has told us about the differences between children and adults with respect to exercise responses and their training implications is summarized in table 32.5. This table also includes exercise prescription implications where appropriate.

Aerobic Training Response in Children

The plasticity of aerobic fitness in children is currently being debated and studied. Plasticity refers to the extent to which normal growth-related changes (improvements) in maximal aerobic capacity can be altered by changes in physical activity. To state this more succinctly, plasticity refers to whether increases in the level of physical activity or physical training improve aerobic fitness ($\dot{V}O_2max$). Plasticity, therefore, can be thought of as trainability. Numerous studies have established that improvement in $\dot{V}O_2max$ after a period of aerobic training is less substantial in children than it is in adults. Rowland (43) reviewed 13 studies that attempted to correlate habitual activity

Table 32.4   Range of Motion Prescription for Children

Goals	Frequency	Intensity	Time (Duration)	Type (Mode)	Progression	Important considerations
Basic health-related fitness	Before and after each activity or exercise session or 3 times per week	To mild tension or slight muscular discomfort	10-15 s, 2 times per stretch	Static stretches for major muscle groups	Overload is not necessary. Start easy into stretch with minimal applied resistance.	Objective is to get children into the habit of stretching. Be sure child is active before beginning stretches.
Intermediate health-related fitness	Before and after each activity or exercise session or 3 times per week	To mild tension or slight muscular discomfort	10-15 s, 3 times per stretch	Static stretches and introduce dynamic stretches	Overload is not necessary. Start with easy multijoint dynamic movements, progressing to more resistive.	Stretch major core muscles first and move to extremities. Introduce dynamic flexibility.
Athletic performance	Before and after each activity or training session	To mild tension or slight muscle discomfort at a level appropriate for sport participation	Depending on static, dynamic, or ballistic (usually conducted by qualified trainer)	Usually dynamic or ballistic; major muscle groups and sport-specific stretches	Overload is not necessary. Start with easy multijoint dynamic movements, progressing to more resistive.	Progression could include moderate static or proprioceptive neuromuscular facilitation stretching.

in children with their level of aerobic capacity ($\dot{V}O_2$max). Only five of the studies concluded that a significant correlation existed between levels of physical activity and aerobic capacity. All these studies measured habitual physical activity of children and were not training studies.

A possible explanation for the lack of support for a significant training effect for habitual physical activity of children can be found in the literature. For example, studies assessing the intensity of physical activity in children have consistently shown that only a small percentage of children meet the guidelines that call for at least 20 min of sustained activity eliciting between 60% and 90% of maximal heart rate. Specifically, Armstrong and colleagues (3) found that only 13% of boys and 6.5% of girls attained heart rates over 160 b · min^{-1} for a 20 min period during a 3 d assessment period. In a systematic review of levels of physical activity during school hours in children and adolescents, 7% to 8% of European and American children reached the recommended 30 min of MVPA during school hours, with boys consistently more likely to be active (4%-8% of total time active for boys, 3%-6% of total time for girls). This difference was maintained in both primary and secondary school; the authors also found a decreasing trend in activity with increasing age (22).

If physical activity is not a major contributor to $\dot{V}O_2$max, then what about formal endurance training? Adult training using ACSM guidelines for intensity, duration, mode, and frequency (large muscle groups, rhythmic activities, 20 to 60 min, 3 to 5 d/wk, equivalent to 65% to 90% maximum heart rate) usually produces an improvement between 5% and 35% in 12 wk (2). Mahon (32) reviewed three controlled training studies on children less than 8 yr of age. In these studies, the experimental group showed a 12.5% increase in $\dot{V}O_2$max, whereas the control group increased only 7.5%. Studies with children 8 to 13 yr old showed an average 13.8% increase in $\dot{V}O_2$max, and the controls increased only 0.7%.

For adolescents 13 yr of age or older, an increase of 6.8% in $\dot{V}O_2$max was found, and the control group exhibited no change in $\dot{V}O_2$max (49). These data show that children and adolescents can adapt to training by increasing $\dot{V}O_2$max but at a much lower rate than adults. When considering $\dot{V}O_2$max changes in children, we must recognize that initial fitness and genetic endowment can also influence responses. In the previously mentioned training studies, the average frequency was 3 d/wk, the average duration was 30 min, and the intensities were greater than 160 b · min^{-1} and as high as 85% maximal heart rate (HRmax). The modes were continuous running, weight training, aerobics, and jumping rope. Clearly, children as young as 8 yr old can increase $\dot{V}O_2$max with training, but their increase will not be as great as that of adults.

Table 32.5　Physiological Responses of Exercising Children Compared With Adults, Their Training Responses, and Prescription Implications

Function	Exercise responses of children compared with adults	Training response and exercise prescription implications
CARDIORESPIRATORY		
$\dot{V}O_2$max (mL · kg^{-1} · min^{-1})	Similar	
$\dot{V}O_2$max (L · min^{-1})	Lower values because of smaller body size and mass	Training response is age dependent, increasing less in younger children. Children improve (1%-16%, mean 9.7%). Training intensity may need to be greater in children to achieve improvements similar to those seen in adults.
Exercise HR	Similar to adult values, lower at all submax levels	As children train, HR intensity should be increased to account for lower response.
Max HR	Similar to adult values (i.e., no change or slight decrease)	
Resting HR	Higher	Training lowers resting HR as in adults. Exercise prescription using Karvonen formula and resting HR should be considered.
Submaximal oxygen demand (economy)	Cycling responses similar; walking and running less efficient, with higher metabolic cost	Running and walking are less economic in children. Consider lower exercise intensity until efficiency improves.
Maximal cardiac output ($\dot{Q}$max) at a given $\dot{V}O_2$	Lower because of smaller cardiac dimensions	Training increases $\dot{Q}$ due to increased SV and possible myocardial hypertrophy.
Maximal stroke volume (SVmax)	Lower because of size and heart volume difference	Training increases SV, reflecting greater blood volume, venous return, or myocardial contractility similar to adults.
Maximal heart rate (HRmax)	Higher because of smaller size	Training may slightly decrease HRmax. HR intensities are therefore higher in children at any % max.
Heart rate at submax work	Higher at a given power output and relative metabolic load	Higher heart rate compensates for lower stroke volume. Training prescription intensity may need to be lowered.
Oxygen-carrying capacity	Blood volume and total hemoglobin lower	Training increases blood volume and total hemoglobin.
a-$\bar{v}O_2$ difference	Somewhat higher	Training does not seem to change a-$\bar{v}O_2$ difference.
Blood flow to active muscle	Higher	Training does not seem to change blood flow.
Systolic and diastolic pressures	Lower maximal and submaximal pressures	Training does not change systolic BP but may increase diastolic BP.
PULMONARY		
Maximal minute ventilation ($\dot{V}_E$max, L · min^{-1})	Lower	Training increases $\dot{V}_E$max.
$\dot{V}_E$max (mL · kg^{-1} · min^{-1})	Same as in adolescents and young adults	Training increases $\dot{V}_E$max.
$\dot{V}_E$, submax	Higher $\dot{V}_E$ at any given $\dot{V}O_2$	Training decreases submax $\dot{V}_E$ as seen in adults.
Respiratory rate, submax	Marked by higher rate (tachypnea) and shallow breathing response	Training decreases respiratory rate, allowing increased intensity for progression.

Function	Exercise responses of children compared with adults	Training response and exercise prescription implications
THERMOREGULATORY AND PERCEIVED EXERTION		
Sweating rate	About 40% lower in children, greater increase in core temperature required to begin sweating	Training does not affect sweating rate. Great risk of heat-related illness on hot, humid days because of reduced capacity to evaporate sweat, lower tolerance time in extreme heat. Exercise prescription should be modified in hot environments.
Acclimatization to heat	Longer and more gradual program of acclimatization required; special attention required during early stages of acclimatization	Training improves heat tolerance somewhat in children. With exercise in hot weather, caution is warranted in prescribing exercise intensity. This is especially true for children who are obese or unfit or who have diabetes.
Body cooling in water	Faster cooling because of higher surface-to-volume ratio and lower thickness of subcutaneous fat	Training does not appear to improve cold tolerance. Caution is needed in prescribing exercise in cold water with children, as potential for hypothermia increases.
Body core heating during dehydration	Greater	For prolonged activity, children must hydrate well before and take in fluid during activity.
Perception (RPE)	Exercising at a given physiological strain perceived to be easier	When exercise intensity is governed by RPE, care must be taken to observe HR and general fatigue to avoid overexertion in children. The OMNI scale (25) is suggested as an alternative to the Borg scale (4).
ANAEROBIC		
Glycogen stores	Lower concentration and rate of utilization of muscle glycogen	Training increases glycogen stores. With interval prescription, lower glycogen stores may initially require smaller number of reps and longer rest intervals.
Phosphofructokinase (PFK)	Glycolysis limited because of lower level of PFK	Training increases PFK. Ability of children to perform intense anaerobic tasks that last 10-90 s is distinctly lower than that of adults. Consider lower number of reps and longer rest intervals.
Phosphagen stores	CP stores lower, ATP stores the same; breakdown of ATP and CRP the same	Training increases high-energy phosphate stores. Consider lower number of reps and longer rest intervals until adaptation.
Oxygen transient	Reach steady state faster and develop smaller oxygen deficit; shorter half-time of oxygen increase	During GXT, stage duration can be shorter (i.e., 2 min) in children. Children also recover faster and therefore are well suited to intermittent aerobic activities.
LAmax	Lower maximal blood lactate levels and lower lactate at given percentage of $\dot{V}O_2$max	Training does not appear to increase LAmax. This may be why children perceive a given workload as easier.
Heart rate at lactate threshold (LT)	Same or higher	Training may lower LT. Because HR at LT is higher in children, exercise HR should be adjusted accordingly.

Data from American College of Sports Medicine (2); Bar-Or and Rowland (5).

Anaerobic Training Response in Children

Little information is available concerning exercise prescription and trainability of anaerobic systems in children using short-burst activities. Rowland (43) indicated that the major reason for this lack of information is that an accurate noninvasive method of assessing anaerobic metabolism similar to $\dot{V}O_2$max in aerobic systems does not exist.

Consequently, short-burst activities, which are common in the habitual activity of children, are poorly understood. The limited available research indicates that children can increase their anaerobic power by training. For example, studies by Grodjinovsky and colleagues (23), Rotstein and colleagues (40), and Sargeant and colleagues (49) have shown that children improve anaerobic performance from 4% to 14% after interval-type training. In the 6 wk training study by Grodjinovsky and coworkers, one group rode a cycle ergometer for five 10 s all-out bouts followed by three

30 s all-out bouts. The other group performed three all-out 40 m runs followed by three all-out 150 m runs during each training session. Both groups trained 3 d/wk. Improvement in anaerobic performance was about 4% in each group (23). Although this study shows that anaerobic performance can be improved by training, it also gives some information on duration, intensity, and frequency of training. Runacres and colleagues (45) compared constant-intensity endurance training (CIET) and high-intensity interval training (HIIT) on aerobic and anaerobic parameters in youth (8-18 yr) over a 3 mo training cycle. Both groups had been training for at least 6 mo and were members in local football and running clubs. The training consisted of 6.5 ± 3.5 h of structured drills or competition each week. After the training, both CIET and HIIT improved the anaerobic parameters, but there were no differences in peak power, mean power, maximum velocity, or mean velocity improvements between training modalities (45). Table 32.5 summarizes the physiological and training responses of children, highlighting the similarities and differences with adults.

Research Focus

Children With Dilated Cardiomyopathy

The use of a cardiopulmonary exercise test to stratify dilated cardiomyopathy patients for heart transplantation has been studied and used for adults. However, very few studies have addressed whether cardiopulmonary exercise testing is useful in children with dilated cardiomyopathy to establish exercise intolerance related to poor short-term prognosis as a means of stratifying pediatric patients for transplantation. A group from Great Ormond Street Hospital for Children in London, United Kingdom (21), reviewed cardiopulmonary exercise test and echocardiography data from 2001 to 2009 for children (N = 82) with dilated cardiomyopathy. The following variables were all related to adverse outcomes: left ventricular shortening fraction, peak heart rate, peak oxygen consumption, peak systolic blood pressure, and ventilator efficiency. From the results of this study, the researchers indicated that cardiopulmonary exercise testing (using a cycle ergometer) in children with dilated cardiomyopathy is possible and that peak oxygen consumption is related to poor outcomes in this clinical population; that is, children with peak $\dot{V}O_2 \leq 62\%$ of predicted had more adverse events in a 24 mo time frame than children with a peak $\dot{V}O_2 >62\%$ of predicted. In addition, peak $\dot{V}O_2$ was the only variable that was associated with the study end point of death without transplantation or listing for urgent transplantation.

Clinical Exercise Bottom Line

- Children's exercise programs should include all components of a fitness program. They should not concentrate on any one component (e.g., resistance exercise).

- When exercise testing children, make sure the equipment fits each child adequately.

- Consider attention span of children when giving instructions, and avoid technical language.

- Be aware that exercise responses of children often differ from that of adults.

CONCLUSION

This chapter brings to light some of the considerations that clinicians need to take into account when working with children. Of particular importance are exercise testing protocols, testing modalities, and exercise responses, all of which affect the exercise prescription for children with various clinical conditions. Children are not small adults, and clinicians must treat them appropriately in order to maximize results from exercise testing and exercise programming.

Online Materials

Visit HK*Propel* to access a link to the references, the case studies with discussion questions, and a quiz to help you review key concepts and test your understanding of the material covered in this chapter.

CHAPTER 33

Older Adults

Jerome L. Fleg, MD

Daniel E. Forman, MD

Over the past century, medical and public health advances have extended the average life expectancy throughout the world, in both wealthy and lower-income countries. As a result, the population of older adults is increasing dramatically. The world population of adults aged 60 yr and older is expected to increase from 900 million in 2015 to two billion in 2050. The subgroup aged 80 yr and older is growing the most rapidly, with an expected increase from 125 million in 2015 to 434 million by 2050 (73). A key manifestation of such advancing age is reduced physical functioning and frailty because of **aging**, deconditioning, and age-associated diseases. These processes often result in an inability to perform **activities of daily living (ADLs)** and a loss of independence. However, regular physical activity and exercise training have the potential to improve functional capacity and quality of life and to reduce dependency and the need for institutionalization.

DEFINITION

Gerontology is the study of the aging process, typically those changes that occur from maturation through old age. Although most physiological processes and structural organ system changes occur gradually over the entire adult age span, these changes are not necessarily linear. Furthermore, the rates of change may vary between organ systems in a given individual and can vary substantially among individuals, depending on genetic factors, lifestyle differences, and superimposed diseases.

Geriatrics is the branch of clinical medicine that encompasses the diagnosis and management of older individuals. This chapter reviews the changes in the cardiovascular (CV) system and other select organs that are commonly observed in older adults and how they alter the approach to exercise testing and training. Particular emphasis is given to the age group 75 yr and older, a period in peoples' lives when the most significant physiological manifestations of advanced aging usually appear.

SCOPE

The societal, public health, and economic impacts of the explosive growth of the **elderly** population are profound. Given that an average 65 yr old has an additional life expectancy of 16 to 19 yr, the ideal is for extended life to be a period of fulfillment and independence, not one of progressive functional decline, **frailty**, and illness. The fact remains, however, that aging predisposes to multiple morbidities that compound one another, often to the detriment of function and well-being. As a result, older adults disproportionately use health care resources. Data from the Medicare Current Beneficiary Survey (MCBS) showed that medical expenses more than double between ages 70 and 90 and that the top 10% of all spenders are responsible for 52% of medical spending in a given year (13). In 2011, beneficiaries aged 80 and older made up 24% of the traditional Medicare population but accounted for 33% of total Medicare spending. In contrast, beneficiaries aged 65 to 69 yr made up 26% of the Medicare population but accounted for just 15% of Medicare spending.

Cardiovascular risk factors increase in both number and exposure duration (e.g., years with hypertension) with advancing age, with cumulative injurious effects that predispose to cardiovascular disease (CVD). For example, hypertension affects ~70% of older adults (17) and is a major risk factor for coronary heart disease, heart failure, atrial fibrillation, stroke, and peripheral artery disease, all of which increase disproportionally with age. Obesity, diabetes, and dyslipidemia are also highly prevalent in older adults, adding to CV risk. At a cellular level, high ambient inflammation is linked to aging itself and to cumulative CV risk factors (11, 18). Furthermore, most CVDs increase the likelihood of a sedentary lifestyle, which itself is a CV risk factor and further reduces mobility because of the effects of deconditioning. A goal would be to limit or end this vicious cycle.

In addition to predisposing to CVD, aging is associated with multiple **comorbidities** that may reduce functional capacity and add to the complexity of management (28). Chronic obstructive lung disease, chronic kidney disease, anemia, and other systemic processes exacerbate functional limitations attributable to CV abnormalities. Arthritis and other musculoskeletal disorders also increase in prevalence among older adults, further reducing mobility and ability to perform daily activities. Loss of skeletal muscle mass and strength, termed sarcopenia, and loss of bone mass accelerate with advanced age, predisposing to increased fatigability, functional decline, falls, and fractures (32, 70). Age-associated sarcopenia involves both the loss of muscle fibers and reduction in their size (12). By age 75, muscle mass typically represents ~15% of body weight compared with 30% in a young adult (47). Sarcopenia is also commonly associated with **dynapenia**, wherein intrinsic muscle strength diminishes (9). The loss of clinical strength and mass results in a much greater effort to perform activities that require lifting, pushing, and pulling (40). Deconditioning exacerbates the impact of age-associated physiologic changes and diseases. Only 20% to 25% of adults aged >65 yr report participation in exercise at least 30 min five times per week (6). The cumulative effects of intrinsic aging changes, comorbidities, and deconditioning often lead to severe functional limitations.

PATHOPHYSIOLOGY

This section focuses on CV pathophysiology because CVD is the leading cause of morbidity and mortality in older adults. Although moderate- and high-intensity physical activities require the integration of multiple organ systems, the CV and musculoskeletal systems are especially critical to such activity. Major age-associated physiological changes in these systems are summarized in table 33.1.

Aging in industrialized societies is typically accompanied by stiffening of the arterial tree, due largely to degeneration of elastic fiber and deposition of nondistensible collagen and calcium in the walls of larger arteries (45). Arterial stiffening is manifest clinically by increased systolic blood pressure (BP) with unchanged or reduced diastolic BP. Isolated systolic hypertension is therefore the dominant form of hypertension in older adults and constitutes a major CVD risk factor in this age group (3).

In response to the higher systolic arterial pressure that occurs with age, as well as to intrinsic cellular aging changes independent of BP, the left ventricle (LV) undergoes a modest concentric wall thickening (33, 35). Accompanying this LV thickening is a reduction of early diastolic LV filling rate with aging despite a well-preserved resting systolic LV function throughout the life span (25, 65). Early LV filling rate declines by approximately 50% between the third and eighth decades, but this is compensated by an increased late LV filling from a more vigorous left atrial contraction, thereby preserving LV end-diastolic volume (15). However, the greater dependence on left atrium–mediated LV filling in older adults means they are more likely than younger adults to develop signs and symptoms of heart failure if they develop atrial fibrillation. These age-associated changes in LV wall thickness and early diastolic function closely resemble those seen with mild hypertension in younger adults. Figure 33.1 is a schematic representation of common age-associated changes in CV structure and resting function (22).

For over a half century, it has been recognized that peak oxygen uptake ($\dot{V}O_2$), the benchmark measure of aerobic capacity, declines approximately 8% to 10% per decade in cross-sectional studies (23, 71). This reduction in peak $\dot{V}O_2$ is due primarily to a reduction in maximal heart rate of ~1 b · min^{-1} per year and a reduced arteriovenous oxygen difference (a-$\bar{v}O_2$ difference). Although earlier cross-sectional studies assumed a linear age-associated decline of peak $\dot{V}O_2$, more recent data suggest that the decline of aerobic capacity accelerates with age, with reductions of 20% or more per decade in persons 70 yr and older (24, 38) (figure 33.2). These accelerated age-associated declines in peak $\dot{V}O_2$ in healthy older adults are magnified by deconditioning and by CV and musculoskeletal diseases, and they have profound implications for performing daily activities and for exercise testing and training in this age group.

Important hemodynamic changes during exercise accompany the substantial reduction in aerobic capacity that occurs with aging. Because of the age-associated increases in large artery stiffness, systolic BP generally rises more briskly in older than in young adults, especially at fixed external work rates. In contrast to the

Normal Changes in Maximal Aerobic Capacity and Its Determinants Between Ages 20 and 80

- Oxygen consumption ↓ 50%
- a-$\bar{v}O_2$ difference ↓ 20%
- Cardiac output ↓ 30%
- Heart rate ↓ 30%
- LV stroke volume: no change
- LV end-diastolic volume ↑ 30%
- LV end-systolic volume ↑ 100%
- LV ejection fraction ↓ 15%
- LV contractility ↓ 60%
- Systemic vascular resistance ↑ 30%
- Plasma catecholamines ↑
- CV β-adrenergic responses ↓

↓ = decrease; ↑ = increase; a-$\bar{v}$ = arteriovenous; LV = left ventricular; CV = cardiovascular.

Based on Fleg et al. (22).

Table 33.1   Physiologic Aging Effects on Organ Systems

Organ system	Effects of aging	Clinical significance
Skeletal muscle	↓ mass of ~1.2 kg/decade from fifth to ninth decade ↓ muscle strength, contractile speed, and power Greater loss of fast-twitch fibers Relative ↑ in slow myosin isoform	↓ ability to perform strenuous activities
Bone Cartilage and connective tissue	Loss of calcium, leading to ↓ bone mass and density, especially in women ↓ thickness, elasticity, and tensile strength; degenerative changes	↑ fracture risk ↑ joint and tendon injury Arthritis
Body composition	↓ lean mass and total body water ↑ % body fat	↓ volume of distribution of water-soluble drugs
Cardiovascular	Arterial stiffening and thickening ↓ vasodilator capacity ↑ left ventricular wall thickness ↓ early left ventricular diastolic filling rate ↓ maximal heart rate and arteriovenous oxygen difference ↓ peak aerobic capacity	Hypertension Diastolic heart failure ↓ ability to perform strenuous activities
Respiratory	↑ chest wall stiffening ↓ vital capacity ↑ residual volume and dead space ↓ maximal voluntary ventilation	Chronic lung disease
Metabolic	↓ resting metabolic rate ↓ insulin sensitivity ↓ glucose tolerance ↓ liver size and blood flow	Obesity Diabetes ↓ metabolism of many drugs
Thermoregulation	↓ thirst sense ↓ skin blood flow ↓ sweat production per sweat gland	↑ risk of dehydration and heatstroke
Renal	Glomerulosclerosis ↓ kidney size ↓ renal blood flow ↓ glomerular filtration rate	↓ renal excretion of drugs Electrolyte disturbances
Central nervous system	↓ β-adrenergic sensitivity ↓ cholinergic sensitivity ↓ brain volume Degenerative brain changes ↓ balance, coordination, hearing, and vision	↓ maximal heart rate ↓ heart rate variability ↓ cognitive function and memory ↑ falls

↓ = decrease; ↑ = increase.

striking reduction in maximal heart rate with age, exercise stroke volume is well maintained in healthy older adults (25). However, the maintenance of exercise stroke volume in the elderly depends more on use of the Frank-Starling mechanism (i.e., increased LV end-diastolic volume) to compensate for a blunted ability to reduce LV end-systolic volume and increase ejection fraction (25). These age-associated changes in exercise hemodynamics occur despite exaggerated increases in plasma catecholamines (27), suggesting they are mediated by reduced β-adrenergic responsiveness, a finding that has been confirmed pharmacologically (26). Prominent age-associated changes in maximal exercise hemodynamics are summarized in the sidebar (22).

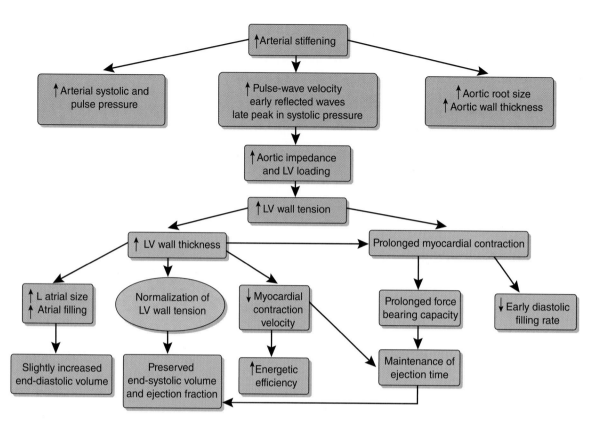

Figure 33.1 Conceptual framework of the major cardiovascular structural and resting functional changes that occur with aging in apparently healthy individuals.

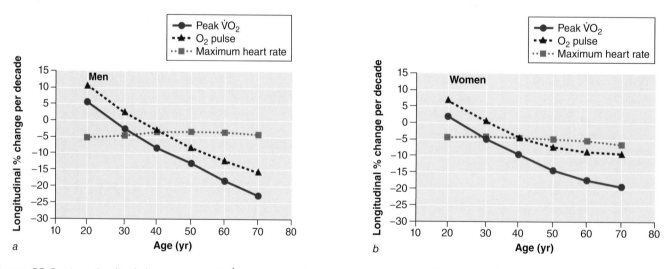

Figure 33.2 Longitudinal changes in peak V̇O₂, maximal heart rate, and oxygen (O₂) pulse in healthy volunteers during treadmill exercise. Although the decline in maximal heart rate across age remains relatively constant at ~5% per decade, an accelerated decline in oxygen pulse with age parallels the decline in peak V̇O₂.

Based on Fleg et al. (22).

CLINICAL CONSIDERATIONS

As described earlier, essentially all common major acquired CV disorders disproportionally affect older adults. These include hypertension, coronary heart disease, peripheral artery disease, cerebrovascular disease, heart failure, and atrial fibrillation; many of these topics are discussed in detail in separate chapters of this text. Chronic pulmonary disease, obesity, arthritis, and other musculoskeletal disorders are also common causes of exercise limitation in the elderly; they are also reviewed in separate chapters. Prior stroke and other neurological and cognitive disorders may pose additional barriers. A screening questionnaire (e.g., the 2014 PAR-Q+) given during the history and physical examination can be useful for identifying significant disorders that warrant careful evaluation and possibly a pretraining exercise test (60). Additionally, the preparticipation screening algorithm of the American College of Sports Medicine (ACSM) may be useful to guide when medical clearance is necessary before embarking on or continuing with an exercise routine (60).

Signs and Symptoms

Symptoms of aging depend primarily on specific physical conditions and chronic diseases rather than on age per se. Older adults often have joint pain, typically related to arthritis, and commonly complain of exertional dyspnea, which may be related to deconditioning as well as to cardiac or lung disease. Since coronary heart disease is common in older adults, they may experience angina during exertion, which often manifests as dyspnea. It is important in this population to follow up on specific symptoms with an interval medical history and exam when indicated.

History and Physical Examination

Although a careful medical history and physical examination are important for all individuals before exercise testing and training, they are especially critical in older adults because of their much higher likelihood of significant CVD and other comorbidities that may affect their evaluation and exercise prescription. The history and physical exam should be performed by an experienced provider, with special attention to the CV and musculoskeletal systems. The patient's typical physical activity pattern and exercise-induced symptoms such as chest pain, marked dyspnea, palpitations, light-headedness, or claudication should be recorded. A history of such symptoms or of prior myocardial infarction, coronary or peripheral revascularization, and other cardiac disorders or interventions places the patient in a higher risk category. A complete list of medications should be obtained, including use of heart rate–slowing drugs such as β-blockers and calcium channel blockers.

In general, the physical examination should focus on body habitus, heart rate and rhythm, BP, cardiac and lung auscultation for evidence of heart failure or significant valvular disease, and evaluation of peripheral pulses. Significant gait disorders may be elicited by a brief corridor walk. Severe cognitive deficits may preclude formal exercise testing or require special precautions to ensure understanding and safety. See chapter 4 for more on the process of conducting a history and physical examination.

Diagnostic Testing

A resting 12-lead electrocardiogram (ECG) should be done on all older individuals. This is often performed immediately before an exercise test. In such cases, it should be reviewed before the test for acute ST-segment changes or significant arrhythmias that contraindicate exercise testing (see chapter 5). Additional CV diagnostic testing to determine cardiac structure and function can be performed as indicated either before or after exercise testing, depending on the clinical situation. A detailed pretraining evaluation such as the one outlined in table 33.2 is useful in assessing gait, balance, range of motion, and coordination in frail older adults.

Exercise Testing

Older adults with no recent regular exercise history and with a well-controlled heart rate and BP and no clinical cardiovascular, metabolic, or renal disease can generally begin a light- to moderate-intensity exercise program such as walking (2-6 metabolic equivalents of task [METs] or 40%-60% of estimated aerobic capacity) without the need for medical clearance but with instruction to seek guidance from a clinician if abnormal symptoms arise (60). For these same older individuals planning to begin a vigorous-intensity training program, or if they have known disease or symptoms, medical clearance is generally recommended. In these cases, this may include a cardiorespiratory exercise test (60). For persons planning to incorporate resistance training, strength and range of motion testing should also be performed (see the sidebar Important Considerations for Exercise Testing). Assessment of balance may be particularly useful in older adults with a history of falling. This can help assess safety during exercise and the need to add balance training as a component of an exercise training program.

Cardiorespiratory Testing

For most older adults, an exercise stress test provides greater value for functional assessment than pharmacological stress testing, which provides no information on aerobic capacity or clinical stability during exercise. However, particular attention to the mode of exercise testing and the protocol is imperative in older individuals.

Treadmill exercise testing is best tolerated among older adults with protocols that use modest increments (e.g., ~1 MET) between successive stages, such as the modified Naughton or Balke protocols (see chapter 5 for specifics). These modest increments help instill confidence in the participant and reduce the likelihood of falling or premature test termination. Individuals with gait disturbances or instability are generally better suited for cycle ergometry than treadmill exercise (55). A modest downside of cycle exercise is the lower peak $\dot{V}O_2$ values (~10% lower) and peak heart rate

Table 33.2 Pretraining Evaluation for Frail Older Adults

Test	Measurement	Outcome	Indicators of risks with testing
Chair stand	Stand from a chair of standard height, unaided and without using arms	Ability Time required	Inability >2.0 s
Step-ups	Step-ups onto a single 23 cm step in 10 s	Ability Number of times	Inability Less than three in 10 s
Walking speed	4 or 5 m walk, starting from rest	Time Gait abnormalities such as asymmetry	$<0.8 \text{ m} \cdot \text{s}^{-1}$
Tandem walk	Walking along a 2 m line, 5 cm wide	Number of errors (off line, touching examiner or another object)	More than eight errors
One-leg stand	Stand on one leg	Ability Time	<2 s
Functional reach	Maximal distance a subject can reach forward beyond arm's length while maintaining a fixed base of support in the standing position	Inches	<6 in. (15.2 cm)
Sit to stand	Maximum number of times a subject can stand up and sit down (without using hands to push up) on a regular chair over 30 s	Number of sit-to-stand maneuvers	<8
Five times sit to stand	Time taken to perform 5 sit-to-stand maneuvers	Time	>13.7 s
Timed Up and Go	Stand up from standard chair, walk distance of 3 m, turn, walk back to chair, and sit down again	Time	>13.5 s
Range of motion	Goniometer used to assess shoulder abduction (SA), flexion (SF), extension (SE); elbow flexion (EF), extension (EE); hip flexion (HF), extension (HE); knee flexion (KF), extension (KE); ankle dorsiflexion (DF), plantar flexion (PF)	Degrees	<90° (SA), <150° (SF), <20° (SE), <140° (EF), <20° (EE), <90° (HF), within 10° (HE), <90° (KF), not within <10° full KE, inability to perform DF and PF

Adapted from E.F. Binder et al., "Peak Aerobic Power is an Important Component of Physical Performance in Older Women," *Journal of Gerontology* 54A, (1999): M353-356; W.J. Chodzko-Zajko and K.A. Moore, "Physical Fitness and Cognitive Function in Aging," *Exercise in Sport and Science Review* 22, (1994): 195-220; R.J. Kuczmarski et al., "Increasing Prevalence of Overweight Among U.S. Adults," *Journal of the American Medical Association* 272, (1994): 205-211; M.C. Nevitt et al., "Risk Factors for Recurrent Non-Syncopal Falls," *JAMA* 261, (1989): 2663-2668; M.L. Pollock et al., "Resistance Exercise in Individuals With and Without Cardiovascular Disease: Benefits, Rationale, Safety, and Prescription. An Advisory From the Committee on Exercise, Rehabilitation, and Prevention, Council on Clinical Cardiology, and the American Heart Association," *Circulation* 101, (2000): 828-833.

generally achieved relative to treadmill exercise, increasing the likelihood of submaximal CV responses (51). A ramp protocol employing frequent, very small increments in work rate is a viable alternative on either a treadmill or a cycle ergometer.

Indications for cardiopulmonary exercise (CPX) testing are generally similar to those in younger adults. By quantifying aerobic capacity (peak $\dot{V}O_2$), respiratory efficiency (minute ventilation divided by exhaled carbon dioxide), and level of effort (respiratory exchange ratio), CPX testing is especially useful in determining the severity of exercise intolerance and whether it is due to cardiac, pulmonary, or motivational factors or simply to deconditioning.

Regardless of the exercise test mode, protocol, or type (standard ECG exercise testing or CPX), each patient's general well-being,

heart rate, BP, and ST-segment response should be closely monitored throughout the test because of the increased risk of an abnormal response in elderly populations. Because exercise-induced arrhythmias are more common and complex in the elderly (7, 49), cardiac rhythm should be continuously monitored. Irrespective of testing type, patients should be encouraged to perform a maximal effort to optimize the detection of ischemia, arrhythmias, unstable hemodynamics, and other abnormalities. However, even a test deemed submaximal by standard criteria (e.g., by heart rate criteria with standard exercise testing or a low respiratory exchange ratio with CPX) provides useful information about an individual's conditioning status and ability to exercise safely to the level achieved during the test. In general, most elderly individuals can perform a maximal effort test.

Important Considerations for Exercise Testing

Cardiorespiratory

- Use: treadmill or ergometer (may be helpful when person has poor balance)
- Perform: low initial intensity with small increments in work rate (i.e., low-level protocol; see chapter 5)
- Measure: estimated METs or peak $\dot{V}O_2$, heart rate, blood pressure, ECG
- Greater vigilance for age-related limitations
- High incidence of undiagnosed CV disease
- Gait deficits may affect safety and ability to test using a treadmill

Strength

- Use: weight machines
- Perform: modified 1RM with a focus on muscles used for ADLs (primarily arms and legs)
- Use low initial load with small increases (≤10% for upper body and ≤20% for lower body)
- Measure: weight lifted, repetitions
- Avoid injury by close supervision
- Agility, balance, and coordination may affect safety and ability to test; setup of testing equipment is important
- Consider muscular endurance assessment

Range of Motion

- Use: goniometer or sit-and-reach box
- Assess: hip, ankle, knee, shoulder, elbow, lower back, and hamstrings
- Measure: degrees, or change in distance, of a motion
- Limited ROM in these joints related to limitations in performing ADLs
- Mental and intellectual impairment may affect ability to perform testing

Balance

- Perform: one-leg stand, tandem walk or others in table 33.2
- Measure: ability, duration, number of errors
- Increased risk of falls while performing
- Useful for assessing risk of falls during ADLs

METs = metabolic equivalents of task; ECG = electrocardiogram; CV = cardiovascular; RM = repetition maximum; ADLs = activities of daily living; ROM = range of motion.

For older adults in whom an assessment of aerobic exercise capacity is desired without the need for heart rate, BP, or electrocardiogram (ECG) monitoring, a 6-min walk test may be a useful and more convenient alternative to traditional exercise testing (59). The test involves traversing a 20 to 30 m course as many times as possible in 6 min. A variation of this test is a timed 400 m walk (68). These submaximal effort tests generally correlate well with measured peak $\dot{V}O_2$ and can be useful in designing a walking program. Their nonintimidating nature, minimal cost, and ease of administration make them particularly attractive for assessing older adults and tracking their responses to training. Furthermore, they more closely reflect a typical older adult's physical activity pattern than does a multistage maximal test.

Musculoskeletal Testing for Strength

Formal strength testing is not always necessary in healthy older adults planning to initiate a low-resistance strength training program. However, strength assessment can often be useful in identifying and quantifying the typical loss that occurs in older age. Baseline measurements of strength may provide additional motivation for the individual to improve function. In frail individuals or those planning a more intensive strength training regimen, a modified or standard one-repetition maximum (1RM) test can be used. Because of the greater loss of fast-twitch compared to slow-twitch muscle fibers with age, reduction in muscular power—the ability to perform rapid muscular contractions—is typically greater than the loss of strength (69), with distinctive risks of falls and disability. These losses in power can be quantified by machines that measure both strength and speed of contraction (i.e., isokinetic dynamometer), with the results used to guide power training to reduce mobility risks (4).

Muscular endurance—the ability to perform repetitive muscle contractions, usually assessed with timed submaximal weight repetitions—also declines with age. Many endurance tests involve common exercises (push-up, squat, sit-up) performed to failure. However, some older individuals may find these tests difficult to perform. For the elderly, common tests that assess functionality include the sit to stand and five times sit to stand (see table 33.2). Time and repetitions for these assessments should be recorded for future comparisons after an exercise program is implemented or after prolonged periods of inactivity. Because everyday activities involve not only muscle strength but also power and endurance, an assessment of all three components may provide more useful information than a 1RM test alone. Special resistance machines, such as those using pneumatic pressure for resistance, are particularly well suited for older adults and help minimize the risk of injury. In frail elders, skeletal muscle endurance testing should focus on the muscles used in typical activities of daily living, including the hip and shoulder flexors, extensors, and abductors, and the knee and ankle flexors and extensors.

Musculoskeletal Testing for Flexibility

The flexibility of the large joints, including the lower back, hips, shoulders, elbows, and knees, is often reduced in older adults because of arthritis, muscle weakness, and disuse. Many routine activities require reasonable flexibility of these joints, so range of motion (ROM) testing is particularly useful in the elderly. Those joints with especially limited ROM can be targeted in a personalized training program. Simple movement through a joint's range can be measured with a goniometer or other measuring device, such as a ruler, tape measure, or yardstick.

Balance Testing

Because balance and gait stability generally decline with age and constitute major risk factors for falls, assessment of balance is often useful, especially in frail older individuals. Simple tests such as a tandem walk or one-leg stand are useful for identifying balance deficits (table 33.2).

Treatment

As previously discussed, today's expanding older population is particularly susceptible not only to CVD but also to multiple chronic diseases in combination with CVD (28, 29). Typically, these diseases compound one another such that senior patients are inherently more unstable and management decisions more complex. For example, it is not uncommon for an older patient to develop heart failure in the context of kidney disease, diabetes, atrial fibrillation, and anemia—a combination of factors that not only predisposes to heart failure but also makes it more difficult to treat (5). Even diseases that seem distinct from one another may have considerable bearing on a patient's management choices; for example, the common occurrence of chronic obstructive lung disease and mild dementia in an older adult might confound considerations for surgical interventions for coronary artery disease (34).

A related concern is polypharmacy; in the United States 44% of older men and 57% of older women take five or more prescription medications (20). A common scenario is an octogenarian taking three or four antihypertensive drugs, three or four heart failure medications, aspirin or warfarin, a statin, and one or two diabetic drugs. The use of multiple medications for coexisting diseases has major implications in older patients in terms of medication costs and adherence as well as the risks of adverse drug reactions (66). Untoward pharmacodynamic effects of multiple medications are especially likely in the context of the changes in drug metabolism, elimination habits (i.e., bowel function), and body composition that occur in old age (41).

Faced with these age- and disease-related limitations, many older adults constrict their physical activity, fearing situations that may threaten their fragile equilibrium. Ironically, these preferences often aggravate health issues (72). In other words, sedentary lifestyles tend to exacerbate medical instability and accelerate progression into frailty and associated susceptibility to falls, diminished independence, and reduced quality of life (31). The challenge for clinical exercise providers is to increase the level of physical activity of senior patients, particularly those with CVD and multiple morbidities. Vigilance and encouragement are required, with the expectation that exercise may initially contribute to instability (e.g., arthritic pain, glucose fluctuations, and increased fatigue), but in the long term becomes beneficial (58). Proper nutrition and hydration, adequate sleep, and appropriate footwear are fundamental.

Exercise training provides a means to potentially forestall age-related morbidity and prevent CV events in healthy seniors (52, 56). It also plays a key role in secondary prevention, modifying symptoms, increasing function, and moderating pathophysiology (e.g., artery plaque stabilization, enhanced vasodilation, improved autonomic function responsiveness) in those with established disease (10, 21, 43). In addition to secondary prevention benefits for CVD, exercise therapy can improve cognition, pulmonary capacity, bone strength, gastrointestinal motility, sleep, mood, and multiple other factors (10, 21, 38). In a very real sense, exercise is a miracle drug, reducing morbidity and mortality and, perhaps most importantly, improving quality of life. Information about interacting effectively with older adults to increase physical activity is provided in practical application 33.1.

EXERCISE PRESCRIPTION AND TRAINING

In the past, exercise recommendations for older adults were primarily extrapolated from those developed for younger individuals (57). However, in 2007, the American College of Sports Medicine (ACSM) and the American Heart Association (AHA) published physical activity and exercise recommendations specifically geared toward seniors (54) (table 33.3); and a document related to these recommendations with a focus on fall risk was released in 2019 (14). Moreover, in 2009, the ACSM published a position stand that extensively elaborated on these principles oriented to exercise benefits in the elderly (10). These efforts emphasize the importance of regular physical activity, including aerobic, strength, range of motion, and balance activities, as part of healthy aging and fall prevention. Parallel literature from the AHA emphasizes the utility of exercise training as part of secondary prevention of CVD (21).

An active lifestyle preserves and enhances cardiorespiratory fitness, skeletal muscle mass, strength and endurance, and flexibility, which otherwise would diminish physical function and increase chronic disease risk with advancing age. Goals are to develop exercise interventions for older adults that augment aerobic capacity, strength, ROM, and balance. Greater physical activity also helps increase energy expenditure, mitigating weight gain, bone and muscle loss, and CV risk factors, as well as preserving, prolonging, and in some cases restoring functional independence.

Practical Application 33.1

CLIENT–CLINICIAN INTERACTION

Despite its numerous benefits, exercise training may evoke complex feelings in many older adults. For those who have never trained, the idea of incorporating exercise behaviors in advanced age is often seen as daunting and unappealing, especially amid fears about illness, medical costs, and mortality. For elders who were vigorous athletes in their youth, the notion of exercise training in old age often evokes unrealistic expectations, making it a priority to teach them about moderation, self-control, and the efficacy of lower-intensity regimens. Thus, the clinical exercise physiologist must perceive the needs and orientation of the patient. Developing an exercise program for an older adult is implicitly personal, requiring tailoring of the training regimen to the circumstances and capabilities of each individual.

For most older individuals, particularly those who are frail, initiating exercise routines, learning safe exercise techniques, and safely advancing exercise duration and intensity require close supervision and teaching. The majority of frail elders will perceive most exercise as high intensity, since each modality is challenging compared with inactivity. Therefore, the clinical exercise physiologist

should employ as much attentiveness and urgency with a 2 lb weight and slow walking in a frail elder as with a 50 lb weight and jogging in a robust older adult. Motivation, adherence, efficacy, and safety depend critically on these important dynamics.

The clinical exercise physiologist must recognize that patients are often struggling with multiple morbidities that confound their best efforts to exercise. Joint or back pain, CV and pulmonary limitations, concerns about urinary or fecal incontinence, and depression or anxiety are among myriad impediments that are more typical than exceptional in frail elders. It is incumbent on the clinical exercise physiologist to be resourceful in helping patients overcome or circumvent these barriers to achieve a successful exercise training routine. This requires working closely with the patient's caregiving team (physicians, nurses, senior living facility staff, family) to ensure safety and efficacy as exercise proceeds. An effective clinical exercise physiologist can play a vital role in facilitating exercise therapy amid challenging medical complexities that might otherwise discourage such activities.

Exercise Training Caveats Specific to the Elderly

An important concept in geriatric medicine is the striking heterogeneity of older individuals; whereas some are quite robust, others may be very frail (10, 54, 64). Such differences must be factored into exercise prescriptions for older adults. The concept of exercise intensity is relative to each person's capacity, such that lifting a 2 lb weight might be vigorous intensity for someone who is frail. Similarly, exercise goals need to be tailored to a patient's unique capabilities. For frail elders, exercise training priorities often center on building strength (including power and endurance), flexibility, and balance before progressing to aerobic modalities. When frailty or a chronic condition prohibits specific activities, exercise should be tailored to those capacities that are preserved. For some frail adults, exercises may be restricted to arm movements, seated machines, or water activities because disabilities preclude other options.

It is also critical to appreciate that most successful exercise regimens match an individual's specific needs and preferences. Exercise adherence is often highest when it corresponds to an activity the patient enjoys. Regimens involving strength or aerobic performance–enhancing exercises can usually be integrated into activities an individual enjoys. For example, one study showed

that dancing was more effective than traditional aerobic training for older patients with heart failure (42).

Warm-up and cool-down activities are particularly important for older individuals (54, 60). In general, warm-ups prompt gradual increases in heart and breathing rates and limb blood flow before higher-intensity exercises are undertaken. For older adults, warm-ups have the added benefit of augmenting chronotropic responses despite diminished β-adrenergic responsiveness, increasing perfusion to muscles and joints despite diminished vasoreactivity and capillary density, and optimizing joint flexibility despite degenerative arthritis and stiffness. Cool-downs are also particularly important for older adults because they attenuate vascular pooling in a population prone to hypotension as well as reduce the risks of arrhythmias and ischemia, which are more likely in older adults. Some general training points are presented in the sidebar Important Considerations for Exercise Training. Note that general exercise training principles are provided in chapter 6.

Specific Training Recommendations

The ACSM/AHA special report on physical activity for older adults recommends aerobic exercise activity 3 (for 20 min) to 5 (for 30 min) d/wk at vigorous and moderate exercise intensity, respectively, or 5 to 8 on a 10-point Borg scale of perceived exer-

Table 33.3   Exercise Prescription Review

Training method	Frequency	Intensity	Time (Duration)	Type (Mode)	Progression	Important considerations
Cardiorespiratory	Moderate-intensity exercise 5 times per week Vigorous-intensity exercise 3 times per week General physical activity performed daily	Moderate: 50%-70% HRR or 5-6 on the 10-point exertion scale Vigorous: >70% HRR or 7-8 on the 10-point exertion scale General physical activity (ADLs) at comfortable intensity	Moderate intensity: 30-60 min intervals, but may be as short as 10 min Vigorous intensity: 20-30 min	Walking, cycling, seated recumbent, pool activity, seated aerobics	As tolerated, with a goal of the upper end of duration and total minutes per week. Typically increase duration before intensity.	Incorporating aerobic activities into daily routines can help achieve these goals. See relevant chapters for further special considerations.
Resistance	2 or more times per week Never on consecutive days	5-6 (moderate) or 7-8 (vigorous) on a 10-point exertion scale	8-12 reps for each muscle group, usually targeting 8-10 muscle groups	Multistation machine-type equipment (e.g., Keiser pneumatic) provides greatest safety. Elastic bands and free weights are less costly and often more available.	Add small amounts of weight as tolerated while maintaining the appropriate intensity.	Frail or deconditioned adults critically benefit from strength training. See chapters that review heart diseases, osteoporosis, and arthritis for further special considerations.
Range of motion	Minimally 2 times per week, but daily is desired	Moderate Mild stretch discomfort without inducing pain	5-30 min total, with two 30 s static stretches or dynamic movements for each major joint	Nothing required. May use the aid of a towel or band to increase and control stretch.	Progress stretch range based on lack of discomfort experienced.	Involve multiple muscle groups (neck, shoulders, arms, lower back, quadriceps, hamstrings, calves, ankles). See chapters that review osteoporosis and arthritis for further special considerations.

HRR = heart rate reserve; ADLs = activities of daily living.

tion, depending on each individual's baseline of habitual activity (54). Once the upper time limit has been achieved, additional time and intensity can be incrementally added. The report also recommends strength training (8-10 exercises for 10-15 repetitions each) at least 2 d/wk as well as biweekly flexibility training and exercises to improve balance.

Although these goals may seem intimidating to many sedentary older people, even small amounts of exercise have benefit. A study by Sattelmair and colleagues corroborates that lower-intensity exercise provides substantial CV benefit (63). For many seniors, bouts of only 10 min of low-intensity activity may be initially realistic, with goals to slowly increase frequency, duration, and intensity as tolerated. Furthermore, for adults who are frail, exercises to build strength and balance often need to precede aerobic exercise

training. This perspective is critical with regard to guiding older adults, for whom activity at any intensity (even light) should never be disparaged, particularly if it is part of a program of increasing duration and intensity.

Low- to moderate-intensity exercises are generally recommended for patients who are exercising on their own, thereby providing a greater margin of safety than vigorous-intensity regimens, as well as for seniors who are primarily trying to maintain their current capacity. Exercise progression, however, is particularly important for frail older adults. In these cases, higher-intensity training goals with greater supervision may be most beneficial. However, it is also important to appreciate that activities that appear to be low intensity (e.g., lifting one's own body weight as part of so-called warm-up activity) may actually represent high-intensity training

Important Considerations for Exercise Training

Cardiorespiratory Endurance

- For adults who are highly deconditioned, are functionally limited, or have chronic conditions that affect their ability to perform physical tasks, intensity and duration of physical activity should be low at the outset, with progression tailored to each person's tolerance and preferences.
- For those who are very limited at the outset, small amounts of exercise may be equivalent to vigorous intensity and should be accorded corresponding duration, frequency, and monitoring.
- Muscle-strengthening activities or balance training (or both) may need to precede aerobic training among very frail individuals.
- Comorbidities such as arthritis, osteoporosis, and heart disease need to be considered.

Skeletal Muscle Strength

- Strength training regimens depend on increasing resistance as well as endurance and velocity; supervision and monitoring are often helpful.
- Free weights may be difficult for a frail person to balance, so instruction and assistance should always be available.
- Proper breathing is important; avoiding the Valsalva maneuver is a safety priority.

Skeletal Muscle Endurance

- Increase number of repetitions (up to 20) for greater endurance.
- Focus on major muscle groups used to perform ADLs (shoulders, legs).

Flexibility

- Avoid ballistic movements and the Valsalva maneuver during the stretching routine. A brief routine performed before or after aerobic or resistance exercise can focus only on the muscle groups used and last as little as 5 min.
- Consider using alternative methods such as seated movements and use of a towel or elastic band to assist those with difficulties performing standard flexibility training.

Balance

- There are currently no specific recommendations regarding frequency, intensity, or type of balance exercises for older adults. However, since balance training requires skeletal muscle effort, it may be prudent to perform it on opposite days to resistance training.
- The ACSM exercise prescription guidelines recommend using activities that include the following: progressively difficult postures that gradually reduce the base of support (e.g., two-leg stand, semi-tandem stand, tandem stand, one-leg stand); dynamic movements that perturb the center of gravity (e.g., tandem walk, circle turns); stressing postural muscle groups (e.g., heel stand, toe stand); or reducing sensory input (e.g., standing with eyes closed).
- Yoga, tai chi, and BOSU ball exercises may be beneficial.

for the frail senior patient, thus necessitating careful supervision and monitoring.

Cardiorespiratory Exercise

Aerobic activity entails repeated movements of large muscle groups (e.g., large-muscle rhythmic activity such as walking, swimming, and bicycling). In a gym-like setting, this usually entails walking (treadmill or indoor track), bicycle ergometry, rowing, or recumbent stepping. However, many physical activities provide similar exercise benefits, including walking outdoors, golfing, gardening, cycling, and swimming (laps or water aerobics). Those who are unable to perform ambulatory activities may be candidates to perform seated aerobic activities, including chair and water exercises.

Enhanced aerobic capacity, cardiac output, peripheral oxygen utilization, vascular responsiveness and perfusion, neuromuscular responses, muscle mass and strength, proprioception and balance, and glucose and lipid metabolism, along with reduced inflammation, have all been attributed to regular aerobic exercise (2, 8, 10, 16, 38, 64). Although the absolute magnitude of exercise training benefit may be greater among robust older adults, even frail and infirm seniors can derive gains that may translate into important clinical benefits, including enhanced independence, confidence, and quality of life (10, 30, 61, 67).

Low- to moderate-intensity exercise sessions can be performed daily but are recommended at least three times a week. Vigorous-intensity exercise sessions (6-8 on a 10-point exertion scale) are recommended only 3 to 5 d/wk, allowing for increased recovery time between sessions, thereby minimizing risk of musculoskeletal injury.

Older individuals with a low exercise capacity often best tolerate (and enjoy) multiple daily sessions of short duration, whereas more robust seniors may prefer higher-intensity exercise bouts performed once per day (10, 64, 74). Older adults are often advised to exercise at a level that feels "somewhat hard" for aerobic and strength training (i.e., a level that usually corresponds to moderate exercise intensity), resulting in many key physiologic and clinical benefits. These individuals can be counseled to exercise below a threshold at which they become short of breath, with a goal to maximize efficacy as well as safety and comfort.

Vigorous-intensity interval training (short bouts of vigorous exercise separated by periods of rest or active recovery) is one form of exercise that incorporates physiologic principles well-suited to many elders. The conceptual advantage is that the higher intensity maximizes physiologic benefit and, when coupled with a rest interval, maximizes recovery and stabilization (68). Additional advantages include enhanced motivation and a similar caloric expenditure in a shorter period than conventional training. Nevertheless, such training programs may not be suitable for frail or markedly deconditioned older adults and those not willing to exercise at high relative work rates. For adults considering vigorous-intensity training, exercise testing before initiating such demanding physiological stress becomes a more important consideration.

Resistance Exercise

Strength training is particularly important in older individuals since the natural susceptibility to sarcopenia and osteoporosis frequently limits the performance of ADLs such as carrying groceries, making beds, and other household tasks. In addition, frail older adults may require strength training before beginning aerobic training. Resistance training programs increase muscle strength and mass through muscle hypertrophy and neuromuscular adaptation, facilitating improvements in functional capacity, gait, balance, and resistance to falls (1, 10, 14, 54). Strength training also increases bone mineral density and content (48, 70), increases metabolic rate (37, 62), assists with maintenance of body weight by decreasing fat mass and increasing lean mass, and improves insulin sensitivity (37, 62), which may be especially useful for older patients with diabetes.

Muscle-strengthening exercises are generally oriented to the legs, hips, chest, back, abdomen, shoulders, and arms, because these muscle groups are all highly engaged during activities of daily living. Strength training sets usually entail 10 to 15 repetitions of moderate to heavy intensity. Although one set of exercises (a series of flexion–extension repetition movements oriented to each of a targeted group of muscles) twice a week is beneficial, achieving two or three sets per session has greater physiological impact. Just as

with aerobic training, physiological benefits must be counterbalanced by musculoskeletal risks and behavioral inertia.

Strength training sessions generally last 20 to 30 min. The recommended intensity level, using the 10-point exertion scale, should reach 5 or 6 for moderate efforts and 7 or 8 for vigorous efforts. Sessions are usually separated by at least 48 h to facilitate recovery. Progress should be carefully monitored, and progression of intensity considered every 2 or 3 wk, depending on the patient and the training goals.

Strength training protocols also vary in relation to specific training goals. In general, increasing capacity to tolerate greater resistance is the primary training end point (i.e., adaptations in muscle size and force). However, dimensions of muscle power and endurance are related concepts that may be associated with even greater clinical relevance. Power implies a dimension of velocity integrated with force, which can provide a critical advantage in reducing risk of falls. Higher-velocity training protocols have been demonstrated to better diminish falling risks than strength emphasis alone (19). Muscle endurance is the ability of muscle to maintain force and power over an extended period; it often plays a critical role in determining a senior's independence and quality of life. Studies suggest that higher-intensity strength training regimens are more likely to achieve significant improvements in endurance (10).

Whereas strength machines generally provide the advantage of structured movements that minimize musculoskeletal risk, free weights are less expensive and more readily available, particularly for a home environment. Common household items such as milk jugs partially filled with water or even cans filled with sand can be used. However, exercise with free weights or these household examples requires greater coordination and has higher potential for injury in older adults. Clinical exercise physiologists must be particularly proactive in guiding use of free weights (or the equivalents) since proper technique substantially reduces the risk of injury. Potentially dangerous situations include overhead exercises in which a weight can be dropped on the head and exercises that shift the center of gravity and increase the risk of loss of balance. Correct breathing patterns should also be emphasized such that participants avoid the Valsalva maneuver (i.e., breath holding), which can reduce venous return and cause dizziness or fainting.

In elders who are too weak to sustain aerobic activities, strength training may be a critical first step toward increasing strength and balance to the minimum needed for aerobic exercise. In such circumstances, training may begin with a patient's own body weight for resistance, then slowly progress to strength training that uses resistance tubing and bands, ankle weights, and other weights over many weeks or months. As strength increases, the best way to advance resistance training is to increase the number of repetitions before increasing the resistance. Such activity is progressively integrated with daily walking and other aerobic activities as tolerated. Supervision, monitoring, teaching, and encouragement are essential, with the recognition that long adaptation periods may be needed as deconditioned elders begin a more active lifestyle.

Range of Motion Exercise

Both flexibility (ROM) and stretching exercises are recommended in the ACSM/AHA report and play a critical role in mitigating some of the natural bone, joint, and muscle changes that occur with aging (54). The recommendation for these exercises is at least 2 d/wk for at least 10 min each session. Most individuals can tolerate daily stretching exercises. Generally, the exercises consist of static stretches (ballistic or bouncing stretches are not recommended) lasting 10 to 30 s per stretch, with three or four repetitions each. Range of motion exercises usually focus on all the major muscle and tendon groups (i.e., all the major joints of the body including the hip, back, shoulder, knee, trunk, and neck region). Performing them in conjunction with aerobic or resistance training is usually well tolerated. In fact, the participant should usually begin with a warm-up period of light-intensity aerobic-type exercise before performing a ROM routine. The degree of stretch slowly increases as tolerated. Chapter 6 reviews general recommendations for ROM exercises.

Balance Training

Balance training complements ROM training with a critical capacity to reduce or reverse falling risks and forestall frailty. Usually balance training is recommended three times per week, but formal dosing recommendations are variable. Balance training options include various walking activities (backward, sideways, heel to toe) as well as standing activities (one-leg stand, heel stand, toe stand) and dynamic movements (circle turns). Tai chi is often touted for enhancing balance in association with strength and aerobic training effects (46). Integrating a BOSU ball into training activities is also a useful balance training technique. In an innovative study of 302 adults aged 60 to 90 yr at high risk for falls, a 6 wk virtual reality intervention simulating real-life obstacles and distractions during treadmill walking significantly reduced falling risk over the subsequent 6 mo compared with treadmill walking alone (50).

Home- and Community-Based Exercise Training

Home-based exercise takes place in an individual's residence, generally is not supervised, and is often conducted alone. As noted earlier, training intensity is typically reduced to mitigate risks. Although this practice is conducive to safety and adherence, potential physiological gains are also usually smaller than those obtained from structured exercise at a health or fitness facility in a supervised setting (53). Furthermore, home-based exercises present particular challenges to frail elders because their baseline capacities are so limited. Such individuals typically require personal instruction and supervision to best ensure the safety and efficacy of activity, but such guidance may be difficult to accomplish at home. Furthermore, lack of emergency care for home-based exercise participants compounds risk and apprehension among patients as well as their families. However, home health aides and physical therapists may be used to initiate and supervise home exercise as covered by Medicare and private insurance. Additionally, an increasing availability of home-based cardiac rehabilitation programs that utilize technology (real-time video and audio) may give more elderly individuals with CVD who cannot attend facility-based cardiac rehabilitation a chance to exercise under supervision.

Community-based exercise generally takes place in a health or fitness center, is usually supervised, and allows several individuals to exercise together or at the same time. In addition to incorporating components of assessment, monitoring, and safety, community-based programs provide opportunities for social reinforcement and facilitate adherence. Community-based supervised resistance training has consistently achieved greater increases in lean mass and muscle strength than home-based programs (53). Furthermore, community-based programs are generally better suited for frail older adults and those with concomitant medical conditions such as neurologic, CV, orthopedic, or pulmonary diseases.

Research Focus

OPACH Study

The Objective Physical Activity and Cardiovascular Disease Health in Older Women (OPACH) study investigated whether higher levels of light physical activity (PA) from 7 d of accelerometry were associated with reduced risks of coronary heart disease (CHD) or cardiovascular disease (CVD) in 5,861 community-dwelling women aged 63 to 90 (mean 78.5 yr). Over a mean follow-up of 3.5 yr, 14 CHD events and 570 CVD events were observed. The hazard ratios (HRs) for CHD in the highest versus lowest quartiles of light PA were 0.58 (95% CI: 0.34 to 0.99; p = .004) after adjustment for age, ethnicity, education, current smoking, alcohol consumption, physical functioning, comorbidity, and self-rated health. Corresponding HR for CVD in the highest versus lowest quartiles of light PA was 0.78 (95% CI: 0.60 to 1.00; p = .004). The HRs for a 1 h/d increment in light PA after additional adjustment for moderate to vigorous PA were 0.86 (95% CI: 0.73 to 1.00; p for trend = .05) for CHD and 0.92 (95% CI: 0.85 to 0.99; p = .03) for CVD.

These findings support the conclusion that all movement, including low-intensity PA, contributes toward the prevention of CHD and CVD in older women. Large, pragmatic randomized trials are needed to determine whether increasing low-intensity PA among older women reduces cardiovascular risk.

Based on LaCroix et al. (44).

Clinical Exercise Bottom Line

- Careful evaluation of older adults before starting an exercise training program should include a detailed history of their comorbidities and medications.

- Exercise testing before starting an intensive training program is important in older adults, both for safety and appropriate exercise prescription.

- Resistance, flexibility, and balance training are especially important in older adults to reduce falls and improve overall function.

- Training programs in older adults should typically start at low exercise levels and progress more slowly than in younger individuals.

CONCLUSION

Aging is typically accompanied by multiple changes that adversely affect central (heart and large arteries) and peripheral (skeletal muscle, bones, joints, and balance) performance, predisposing to constricted functional capacity and frailty. Exercise training can attenuate the progression of age-related physiologic changes and can reverse some of those that have already occurred, improving aerobic, strength, flexibility, and balance capabilities in older individuals. The success of exercise training in older adults usually entails an important relationship between the patient and clinician. Initial assessments are essential, both for determining an appropriate training program and for assessing CV risk and all the other conditions, medications, and circumstances that can affect exercise goals and safety. Most older adults referred for exercise training struggle with a range of medical issues. Whereas some clinicians may assume that chronic medical issues supersede exercise training priorities, the reverse is often the case.

Although many older adults conceptualize exercise only in terms of aerobic activities, it is often critical to implement strength training as the first step in improving aerobic performance. This is because typical aging entails progression of sarcopenia and muscle deconditioning, which erode performance capacity, but can be partially compensated by strength training. The optimal exercise prescription should be carefully tailored and coordinated relative to each person's circumstances. For most elderly people, particularly those who are frail, the capacity to initiate exercise routines, to learn safe exercise techniques, and to advance exercise duration and intensity safely all require close supervision and teaching. Motivation and adherence depend critically on these important dynamics. Despite the challenges of exercise training in older adults, the large potential benefits of such training on physical function, quality of life, and maintaining independence clearly justify the effort.

Online Materials

Visit HK*Propel* to access a link to the references, the case studies with discussion questions, and a quiz to help you review key concepts and test your understanding of the material covered in this chapter.

CHAPTER 34

Depression

Grace M. McKeon, PhD

Simon Rosenbaum, PhD

Depression is the most prevalent mental health condition in the United States. The term *depression* is often used to describe varying levels of emotional distress, ranging from a dysphoric mood state to the diagnosis of a clinical disorder, such as **major depressive disorder (MDD)**. This chapter focuses on the risk of depression, including the clinical syndrome of MDD as well as subclinical depressive symptoms, and on different treatment modalities (including exercise), with an emphasis on the bidirectional relationship between depression and poor physical health.

Since depression often results from, or coexists with, many of the chronic diseases described in this book, students interested in clinical exercise physiology must appreciate the challenges clinicians face when treating people experiencing this disorder. For example, the clinical exercise physiologist must feel comfortable working with not only patients with depression but also patients with depression and comorbid diseases.

DEFINITION

In the United States, psychiatric disorders are currently diagnosed using criteria outlined in the *Diagnostic and Statistical Manual of Mental Disorders, Fifth Edition* (DSM-5) (6). The DSM-5 diagnostic criteria for MDD are presented in the sidebar. A diagnosis of MDD requires endorsement of at least five symptoms, one of which must be either depressed mood or diminished interest or pleasure, that have been present during the same 2 wk period.

Summary of DSM-5 Diagnostic Criteria for Major Depressive Episode

Five (or more) of the following symptoms are present during the same 2 wk period and represent a change from previous functioning. (*Note:* At least one of the symptoms must be either number 1 or number 2.)

1. Depressed mood most of the day, nearly every day
2. Diminished interest or pleasure in all or most activities
3. Significant change in weight or appetite
4. Insomnia or hypersomnia
5. Psychomotor agitation or retardation
6. Fatigue or loss of energy
7. Feelings of worthlessness or guilt
8. Diminished ability to think or concentrate
9. Recurrent thoughts of death or suicide

The symptoms cause clinically significant distress or impairment, are not caused by the effects of a substance or general medical condition, and are not better accounted for by bereavement or another psychiatric disorder.

Acknowledgment: Much of the writing in this chapter was adapted from the first four editions of *Clinical Exercise Physiology*. As a result, we wish to gratefully acknowledge the previous efforts of and thank Benson M. Hoffman, PhD, Krista A. Barbour, PhD, and James A. Blumenthal, PhD.

SCOPE

The lifetime prevalence of MDD in community-based adults is ~17%, and the 12 mo prevalence estimate is ~7% (51). Kessler and colleagues (52) found that among those individuals meeting criteria for MDD, 80% were classified as having moderate or severe episodes. The number of persons with common mental disorders, including MDD, is rising globally, particularly in lower-income countries (67). Thus, MDD is a prevalent disorder with a high burden of disability. Of those with MDD who receive treatment, only a minority receive treatment at a level that meets minimal standards: 1 in 5 in high-income countries and 1 in 27 in low- to lower-middle-income countries (99). If untreated, most episodes of depression will last for several months before remitting (58). Furthermore, MDD tends to be episodic; many individuals who recover from an episode of MDD will experience a recurrence of the disorder (30). At least 50% of those who recover from a first episode of depression will have one or more recurrent episodes in their lifetime, and approximately 80% of those who have at least two episodes will have another recurrence (6).

Elevated depressive symptoms are consistently found to be more common in women than in men, and this gender difference holds across cultures (108). Prevalence varies by age, peaking in older adulthood (67). Depression does occur in children and adolescents under the age of 15 yr but at a lower level than in older age groups (67). Prevalence of MDD between ethnic minorities and Caucasians have yielded mixed results, but both African Americans and Latinos tend to report less depression compared with Caucasians (52). Subgroups of ethnic minorities may be at increased risk for MDD because of difficulties in meeting basic needs, such as food and shelter (72).

Besides the great personal cost of MDD, the public health burden is enormous. For example, MDD has been linked to greater health care utilization, decreased quality of life, lower productivity, increased absenteeism, and increased rates of attempted suicide (48). When compared with patients with chronic cardiometabolic disease (e.g., diabetes, hypertension), people with depression report similar or worse functioning (110). For the year 2010 the economic burden of depression in the United States (including direct treatment costs, mortality costs of depression-related suicide, and indirect costs associated with depression in the workplace) was estimated at approximately $210.5 billion (40). In 2015, depression was ranked by the WHO as the single largest contributor to global disability (7.5% of all years lived with disability) (67). Clearly, MDD is a prevalent, recurrent disorder associated with significant morbidity and economic costs.

People living with a mental illness, including depression, commonly experience poor physical health and subsequently an overall reduced life expectancy (107). A 2019 Lancet commission summarized advances in our understanding of these physical health disparities (36). The commission reviewed data from nearly 100 individual systematic reviews and concluded that regardless of diagnosis, mental illnesses (including depression) are associated with an increased risk of obesity, diabetes, and cardiovascular disease in the magnitude of 1.4 to 2.0 times higher than the general population (36). For example, approximately 30% of people with MDD have comorbid metabolic syndrome, which is 1.5 times greater than in people without MDD (3, 103). Metabolic syndrome is defined by a combination of central obesity, high blood pressure, low levels of high-density lipoprotein cholesterol, elevated triglycerides, and hyperglycemia (1). In addition, people with MDD face an almost 2 times increased risk of type 2 diabetes mellitus; prevalence is approximately 8% (104).

PATHOPHYSIOLOGY

Depression is widely viewed as an interaction between genetics and the environment. As a result, both neurochemical and psychosocial variables have been implicated in its pathophysiology (14). Specifically, three neurotransmitter systems—norepinephrine, serotonin, and dopamine—have been theorized to play a significant role in the onset and maintenance of depression. These theories are complex but generally suggest neurotransmitter deficiencies. In addition, dysregulation in the hypothalamus–pituitary–adrenal (HPA) axis (a marker of reaction to stressors) has been implicated in the pathophysiology of depression (14, 28). HPA axis dysregulation, in turn, has been associated with several of the risk factors for metabolic syndrome and CVD including central obesity, hypercholesterolemia, hypertriglyceridemia, and hypertension (62, 70). Depression has been associated with other pathophysiologic processes as well, such as autonomic nervous system dysregulation as evidenced by reduced heart rate variability (19), increased inflammation (43, 59, 73), elevated levels of cortisol (39, 69), increased platelet reactivity (17), and endothelial dysfunction (87).

The societal determinants of mental health also contribute to the etiology of depression (66). Common mental disorders including MDD are shaped to a great extent by the social, economic, and physical environments that people live in. Risk factors for many mental health conditions are strongly associated with social inequalities, whereby the poor and disadvantaged suffer disproportionately. Adverse social determinants including low socioeconomic status, low education, and unemployment are associated with higher risk of mental health conditions.

There is significant evidence of a bidirectional relationship between physical activity and depression. People living with depression are less likely to engage in physical activity and more likely to engage in sedentary behavior in comparison to the general population (82). Exercise, even in small doses, can provide significant protection against future depression (44). Physical activity may improve depressive symptomatology through alterations in neurotransmitter functioning or improvements in negative evaluations of self (e.g., enhanced self-efficacy and self-worth).

CLINICAL CONSIDERATIONS

The gold standard for the assessment of depression is a structured diagnostic interview administered by a trained mental health professional, but an extensive assessment of that kind is out of the scope of most exercise or rehabilitation settings. Exercise physiologists should be aware of some of the most common valid instruments used to screen for depressive symptoms. This will help exercise physiologists identify at-risk patients and refer them to the appropriate mental health professional for further assessment when necessary.

Signs and Symptoms

Besides administering paper-and-pencil screening tests, exercise physiologists should be cognizant of the common signs and symptoms associated with depression (e.g., weight gain or loss, sleep disturbance, fatigue, feelings of guilt, loss of interest) as well as important symptoms (self-reported symptoms of hopelessness and suicidal ideations) that indicate intervention by a trained mental health professional is urgent. Patients who report such symptoms should be carefully monitored, and the appropriate health care professional (e.g., the patient's primary care provider) should be consulted. In addition, any concerns about the safety of self or others is a key criterion for immediate referral.

History and Physical Exam

A number of instruments are available to assess the presence and severity of depressive symptoms. These are usually self-report questionnaires consisting of a series of written questions that patients complete on their own. Widely used examples include the Beck Depression Inventory-II (BDI-II) (13), the Center for Epidemiological Studies Depression Scale (CES-D) (13) the Hospital Anxiety and Depression Scale (HADS) (115), and the nine-item Patient Health Questionnaire (PHQ-9) (89). For further discussion about the management of depression, see practical application 34.1.

Diagnostic Testing

Given that comorbid physical health conditions are often present in patients with depression, it is important to assess and monitor their physical health outcomes. Exercise physiologists should identify higher-risk individuals, such as patients with cardiovascular disease or type 2 diabetes. These patients should be medically cleared to exercise. Exercise physiologists should assess and monitor vital and anthropometric parameters including blood pressure, body mass index, and waist circumference. The exercise professional should also work within the multidisciplinary mental health team and request any other relevant test results (e.g., glucose and lipid markers) from the appropriate professionals.

Exercise Testing

Intuitively, several of the key symptoms of depression might be expected to affect the assessment of aerobic capacity. For example, a patient with attenuated psychomotor function or fatigue may demonstrate poorer performance relative to patients free from depression. In addition, people with severe mental illness including MDD have been shown to have significantly lower cardiorespiratory fitness compared with healthy controls (105). There is, however, robust evidence from randomized controlled trials showing that physical activity can improve cardiorespiratory fitness in patients with MDD (93). The protocols and procedures for assessing aerobic capacity and muscle strength and endurance are generally the same as those for healthy people. Submaximal fitness tests can be considered for patients with poor physical health, low motivation, lower levels of fitness, or lower levels of energy (54). Patients with increased trait or state anxiety may fear that maximal aerobic tests will provoke physiological reactions similar to a panic attack (e.g., hyperventilation, sweating, and tachycardia) (91).

Finally, exercise testing may not be appropriate for all clients. It is important, however, to assess physical activity and sedentary behavior. Accelerometers can be used, but more commonly interviews and questionnaires are administered. Minutes of physical activity per week may also be useful for other clinicians (e.g., psychologist or psychiatrist) involved in a client's care. The Simple Physical Activity Questionnaire (SIMPAQ) is a clinical tool that can be used to assess physical activity and sedentary behavior, specifically among people with a mental illness (78). This tool is reliable and valid and can be administered by health professionals including exercise physiologists.

Treatment

Many people with depression do not seek treatment, which contributes to the large burden of this disorder. There is an unmet need for the treatment of depression—it is estimated that approximately 36% to 50% of serious cases in developed countries and 76% to 85% in less developed countries received treatment in the previous 12 mo (27). Even among those who do seek help, many do not receive minimally adequate treatment. A study of community household surveys from 21 countries showed that of those who met criteria for DSM-IV MDD and who sought help, only 41% received minimally adequate treatment (99).

Except in cases of severe depression (for which care is usually provided by a psychiatrist, psychologist, or other mental health professional), many depressed patients who seek treatment are often treated by primary care physicians (113). Patients who present with depression in the primary care setting are more likely to be offered medication versus psychotherapy or a combination of medication and psychotherapy (75).

Exercise professionals are now being recognized as important members of the multidisciplinary mental health team, yet their integration has faced considerable barriers. There is evidence in some countries that with an adequately funded health care system, this integration can occur successfully. For example, in Australia exercise physiologists have been embedded within the mental health service, within both the private and public sectors (77). Referrals to exercise physiologists through primary care in Australia increased significantly from 2009 to 2016 (23), and clinical exercise programs and multidisciplinary staff training within

Practical Application 34.1

CLIENT–CLINICIAN INTERACTION

Exercise professionals who work with patients with depression may find the experience both challenging and rewarding. As noted elsewhere in the chapter, symptoms of depression such as fatigue and loss of interest in people and activities may adversely affect a patient's ability to adhere to an exercise program.

It is important to understand the motivating factors and barriers to exercise for people with depression. Clinicians should be aware that the most common barriers to exercise faced by people living with a mental illness are low mood, stress, and lack of social support (33). Because of these additional barriers, people with depression often require more support from the exercise professional compared with individuals without depression. When an exercise prescription for adults with depression includes a motivational component and is delivered by an exercise professional (e.g., an exercise physiologist), the dropout rate is reduced (92).

The exercise professional's input may be especially important during the initiation phase of an exercise program. For example, many patients with depression feel a lack of self-worth and may not believe in their ability to successfully participate in an exercise program. The fatigue and loss of energy that occur with depression may further contribute to a patient's reluctance to begin exercising. Thus, depressed patients may need more general spoken support, reassurance, and positive reinforcement (even for just showing up to an exercise training session) when they begin the program.

Providing one-on-one professional support to help individuals identify and achieve their goals may help them overcome psychological barriers and maintain motivation. For those who are inactive, it may be more useful initially to focus on open-ended goals (e.g., "do your best to increase your daily step count" rather than "increase your step count by 10%") (97). It is important that patients not be discouraged by not achieving their specific goals in the early stages.

For patients with depression who experience fatigue and low motivation, their exercise tolerance is often reduced. A patient's perceived exertion can be derived from the psychophysiological concept of the Borg scale (16). In addition, the cognitive symptoms of depression such as indecisiveness and difficulty in concentrating may lead to problems in recalling the exercise prescription or recalling set exercise goals. Occasional reminders of such exercise-related information can be helpful (and presenting information in both oral and written forms is good practice).

Depression may be accompanied by social problems such as family conflict or unemployment. The exercise professional should be aware that such major life stressors may present barriers to participation in an exercise program.

When clinicians encounter a patient with depressive symptoms that interfere significantly with exercise participation, the appropriate approach may be to refer the person to a mental health provider. Several treatments for depression have shown success, including antidepressant medication and cognitive–behavioral therapy. Patients with depression should also be informed that exercise training can reduce depressive symptoms in both healthy and clinical populations.

In summary, depressed patients may require greater support to ensure adherence to an exercise prescription compared with nondepressed patients. Because depression is associated with increased risk of dropout from exercise programs, staff should closely monitor depressed clients and refer them for specialized treatment when appropriate.

---

psychiatric treatment facilities also increased (76). There is similar and promising recognition of the role of exercise professionals in Europe (95), with examples of successful integration into health services in the United Kingdom and some European countries (90). At a community level, evidence-based mental health programs have been run by exercise professionals around the world, such as the In SHAPE health promotion program in the United States for patients with severe mental illness (102). Patients were mentored by exercise professionals and provided with free access to community-based exercise facilities, funded through community partnerships and not-for-profit agencies. Clinically significant reductions in body weight and improvements in cardiorespiratory fitness were observed (10). This program demonstrated success by engaging with existing community resources; however, appropriate structural support and funding are still required.

Antidepressant Medication

Antidepressant medications are the most widely used pharmacological treatment for depression in the United States (65); see table 34.1 for a list of widely used antidepressant medications. Results from numerous randomized controlled trials (RCTs) have provided evidence for the efficacy of antidepressant medications, both for the treatment of acute depression and for the prevention of relapse (42, 47, 88), although there is uncertainty about which patients stand to benefit most from which treatments. RCTs generally conclude that less than one in two clinically depressed

Research Focus

Exercise for Treating Depression: A Meta-Analysis Adjusting for Publication Bias

The true effect of exercise on depression has been a topic of contentious debate, with meta-analyses showing differing effect sizes. For example, the latest Cochrane review showed only a small effect size in favor of exercise (21). The methodology used in this review has since been criticized (31), including how the exercise intervention was defined, how depression was defined, how high-quality studies were defined, and the comparison group utilized. The review also failed to adjust for possible publication bias, likely resulting in an underestimation of the true effect size. This mixed messaging and uncertainty around the true antidepressant effect of exercise has been detrimental to the field. A landmark meta-analysis by Schuch and colleagues (83) accounted for these limitations, including adjusting for publication bias. This review aimed to establish the true effects of exercise on depression by comparing exercise versus nonactive controls (i.e., studies that did not compare exercise versus another alternative treatment). It also looked at the effects of different variables, including the duration of the trial, frequency, and intensity.

In this meta-analysis, 25 randomized controlled trials (RCTs) looked at exercise interventions in people with depression, including 9 on MDD. It was found that exercise had a large and significant effect on depression overall (standardized mean difference adjusted for publication bias = 1.11). The authors also conducted a fail-safe assessment showing that more than a 1,057 studies with negative results would be needed to nullify the effects of exercise and depression. In participants with MDD, larger effects were found for moderate-intensity aerobic exercise and interventions supervised by exercise professionals. It was suggested that previous meta-analyses underestimated the benefits because of publication bias. The conclusion was that exercise should be prescribed as evidence-based treatment for depression.

---

adults respond to treatment with one antidepressant (46), and less than two in three depressed adults respond to augmented treatment with more than one antidepressant (79, 101). Furthermore, two meta-analyses determined that the effects of antidepressant medications were superior to placebo for severe depression but comparable to placebo for mild to moderate depression (37, 53). Although second-generation antidepressants including SSRIs are usually better tolerated and safer than earlier drugs, they still have significant side effects (11). In addition, adherence to antidepressant medication regimens is often inadequate relative to treatment guidelines (26), and early discontinuation of treatment is associated with recurrence of MDD (63).

It is important to remember that medication is managed by the patient's primary care physician or psychiatrist. The responsibility of an exercise professional is to be aware of potential drug-related side effects, as well as any impact the drug may have on exercise testing or exercise training responses and on exercise testing and prescription. Exercise and psychotherapy are other forms of treatment that may be prescribed alone or in adjunct to medications.

Psychotherapy

There is considerable empirical support for the efficacy of psychotherapies for the treatment of MDD (20). Two of the most well-studied and widely used psychotherapies for depression are **cognitive–behavioral therapy (CBT)** (see table 34.2) and interpersonal therapy (IPT). CBT emphasizes the important influence that negative thoughts have on emotion and behavior (12). For people with MDD, treatment focuses on modifying maladaptive thoughts as well as addressing deficits in behavior (e.g., unassertiveness, isolating oneself from others) that lead to and maintain depression. For example, a depressed patient undergoing CBT might be taught to identify cognitive distortions associated with their depression, to challenge those distortions, and to replace them with more realistic thoughts. Because patients with depression often lack motivation, another therapeutic exercise might involve planning a daily schedule of activities to engage in, with adherence possibly facilitating decreased boredom and loneliness as well as increased motivation.

IPT is a structured and time-limited psychotherapy that emphasizes the interpersonal context in which depressive symptoms occur (e.g., social isolation, interpersonal conflicts, role transition, or loss of a loved one) (109). The IPT therapist forms a supportive relationship with the depressed patient, helps the patient identify interpersonal problems, and uses strategies to help the patient develop and adapt new interpersonal behaviors (e.g., assertive communication, reaching out for social support). CBT, IPT, and other empirically validated psychotherapies have been associated with large effect sizes both in RCTs and in clinical practice (8), and there is little reason to believe that one empirically validated psychotherapy is necessarily superior to another (24).

EXERCISE PRESCRIPTION AND TRAINING

An exercise prescription for people with depression will likely differ little from the prescription used for healthy individuals. Clinicians should be aware, however, that several symptoms of depression (e.g., loss of interest, fatigue, low self-confidence) may

Table 34.1   Pharmacology

Medication	Brand name	Class	Exercise effect	Other effects
Citalopram	Celexa	SSRI	Dizziness	Nausea, diarrhea, sexual side effects, headache, weight gain, nervousness, dry mouth, insomnia
Escitalopram	Lexapro			
Fluoxetine	Prozac			
Paroxetine	Paxil			
Sertraline	Zoloft			
Desvenlafaxine	Pristiq	SNRI	Dizziness	Nausea, sweating, sexual side effects, fatigue, constipation, insomnia, anxiety, headache, loss of appetite
Duloxetine	Cymbalta			
Levomilnacipran	Fetzima			
Venlafaxine	Effexor			
Bupropion*	Wellbutrin	Misc	Rapid heartbeat, dizziness	Restlessness, dry mouth, insomnia, headache, nausea, vomiting, constipation, tremor, excessive sweating, blurred vision, confusion, rash, irritability, ringing in the ears
Mirtazapine	Remeron	Atypical	Dizziness	Somnolence, increased appetite, weight gain
Vilazodone	Viibryd	Atypical		Diarrhea, nausea, vomiting, insomnia
Vortioxetine	Brintellix	Atypical		Nausea, vomiting, constipation

*Bupropion does not share a chemical structure with other types of antidepressant medications and as a result is considered a miscellaneous, or other, antidepressant medication.

SSRI = selective serotonin reuptake inhibitor; SNRI = serotonin and norepinephrine reuptake inhibitor.

Table 34.2   Cognitive–Behavioral Therapy (CBT) Resources

Resource	Description
Feeling Good: The New Mood Therapy	Self-help book written by David Burns that uses the basic principles of CBT
Association for Behavioral and Cognitive Therapies (www.abct.org)	Website that can be used to locate psychotherapists by geographical location
Cognitive Therapy: Basics and Beyond	Book written by Judith Beck that describes the cognitive model and therapy in detail
International Society of Interpersonal Psychotherapy (www.interpersonalpsychotherapy.org)	Website that provides a concise overview of interpersonal psychotherapy

interfere with participation in exercise, and that comorbidities can further complicate matters.

For individuals whose depression is considered stable enough to begin an exercise program (e.g., they are not suicidal; their depression is not so severe that it prevents active participation), the question arises about how to manage depressive symptoms that may affect the course of participation in exercise. First, recognize that significant comorbidity exists between depression and other chronic diseases; people with depression are unlikely to present with depression alone. Furthermore, depression can affect the course of chronic disease. For example, the presence of depression is associated with more complications and increased mortality in patients with diabetes (25, 50). Evidence also identifies depression as a powerful and independent risk factor for

cardiac outcomes in patients with coronary heart disease (such as recurrent myocardial infarction and mortality) (60). For cancer patients, untreated depression has been linked to poorer treatment adherence, increased hospital stays, and mortality (71).

Besides often being present as a comorbidity in patients with another chronic disease, depression is associated with unhealthy lifestyle behaviors such as tobacco and alcohol use (96), poor diet (34), physical inactivity (106), high levels of sedentary behavior (82), and poor sleep (7).

Exercise as a Treatment for Depression

Many systematic reviews and meta-analyses have summarized the antidepressant effect of exercise. A meta-review (a review of

ANXIETY AND EXERCISE

Comorbid anxiety is common in patients with depression (100). Although regular exercise is an effective intervention to improve anxiety symptoms among people with anxiety and stress-related disorders (94), few trials have considered and accounted for how anxiety influences the effects of exercise on depression (15). Comorbid depression and anxiety present unique symptomologies that may alter the effectiveness of exercise, yet only 25% of existing studies on exercise and depression report the prevalence of anxiety among their sample, and 0% accounted for comorbidity in their trial analyses (15).

Exercise professionals should familiarize themselves with the symptoms of anxiety disorders, including excessive anxiety and worry plus at least three of the following six symptoms: restlessness, easy fatigue, difficulty concentrating, irritability, muscle tension, and sleep disturbance (6). Remember that occasional minimal symptoms of anxiety are a normal part of life. To meet criteria for a disorder, symptoms must result in impaired functioning in an aspect of a person's life. If a patient appears to be experiencing significant anxiety, the exercise professional may wish to discuss this with them and make a referral for treatment as needed. Both CBT and anxiolytic medication have been shown to be effective in treating anxiety disorders.

published meta-analyses and therefore one of the highest levels of evidence) conducted in 2019 concluded across eight individual meta-analyses that exercise reduced depressive symptoms in children, adults, and older adults (5). For example, a meta-analysis conducted in 2016 by Schuch and colleagues (83) summarized the results of RCTs on exercise therapy for depression. In 25 RCTs comparing exercise versus control groups, exercise had a large and significant effect on depression. Larger effect sizes were found for outpatients, in samples without other comorbidities, and when exercise interventions were supervised by a trained exercise professional. The review found that previous meta-analyses may have underestimated the benefits of exercise because of publication bias and concluded that exercise can be considered an evidence-based treatment for depression.

Despite the strong evidence base, supervised exercise is not incorporated into clinical practice as often as it should be. It is likely that physicians are unaware of the research or are reluctant to overcome the dualistic treatment approach that underpinned their medical training. Even clinical practice guidelines, which are intended to be based off research evidence, are subject to bias and the potential for overtreatment (86). One study evaluated international clinical practice guidelines for the treatment of major depressive disorder, specifically looking at the recommendation of physical activity (45). Seventeen guidelines were included, of which only four recommended physical activity as a frontline intervention; two did not mention physical activity, and eleven made some mention. The American Psychiatric Association Clinical Practice Guidelines, for example (38), did not suggest exercise as an evidence-based treatment modality, stating only that if a client wishes to engage in physical activity there is little to argue against it. These guidelines have been criticized because they contained

no systematic review of studies on the effects of exercise (22). Conversely, Canada's 2016 CANMAT guidelines (74) suggest that exercise should be recommended as first- or second-line treatment for mild to moderate depression.

Preliminary evidence from meta-analyses suggests that the magnitude of benefit associated with exercise training is similar to that of antidepressant medications (21, 56). However, caution should be taken interpreting these comparisons because their samples were small (usually less than $N = 300$) and fewer than five trials in such analyses were included. Therefore, in summary, the recommendation is that exercise should be prescribed as an add-on treatment to usual care.

Mechanisms

A number of reviews have explored the potential mechanisms of the antidepressant effect of exercise (49), although the research is equivocal. The main reason for this uncertainty is the complex etiology of depression. No single abnormality or altered physiological system occurs in people with depression. Therefore it is unlikely that one mechanism is solely responsible for the antidepressant effects of exercise. One proposed mechanism is through the hypothalamic–pituitary–adrenal (HPA) axis. The HPA axis produces cortisol in response to stress, and among people with depression it is common to see hyperactivity in this pathway. Exercise is believed to influence depression through normalization of the HPA axis (112). Another potential mechanism is the reversal of atrophy of the hippocampus (35), which is consistently affected in people with depression (81). Exercise may also improve symptoms of depression by increasing self-esteem and self-efficacy (32). Specifically, people with depression have lower levels of self-esteem (68) and the relationship may be cyclical, whereby low self-esteem can

Practical Application 34.3

EXERCISE, DEPRESSION, AND SLEEP

Disturbed sleep is a defining feature of depression and a key risk factor for the exacerbation of symptoms. The relationship is bidirectional between depression and poor sleep, with poor sleep being a risk factor for the development of mental health conditions (2). Additionally, the presence of disturbed sleep among depressed patients predicts poor response to treatment and future depressive episodes (61). Many medications used to treat depression may also lead to sleep disruption (4). When sleep is impaired, acute and chronic physical and mental health conditions can develop. There is now evidence to suggest that exercise, known to be an effective treatment for depression, may also improve disturbed sleep (57).

Epidemiological investigations support a significant link between engaging in exercise and better sleep (114). A meta-analysis of studies in the general population found that regular exercise had significant effects on all subscales

of sleep including total sleep time, sleep efficiency, sleep onset latency, and sleep quality in comparison to control groups (55).

Another review and meta-analysis examined the impact of exercise on sleep quality in people with a mental illness (57). It included eight RCTs and a total of 463 participants with severe mental illness, including MDD. Overall, exercise had a large and statistically significant effect on sleep quality ($p = .01$). Of the included studies, patients with a diagnosis of depression demonstrated significant improvements in sleep quality and mood compared with nonactive control groups. This occurred independent of intervention modality, including resistance exercise, yoga, and tai chi.

In summary, there is encouraging evidence that exercise is an effective treatment for improving sleep quality among depressed adults, which is important for physical and mental health outcomes.

increase depressive symptoms and exacerbate self-esteem deficits. Behavioral activation also occurs with regular exercise, which is an important component of CBT for MDD (80). An example of an exercise prescription for a patient with depression appears in table

34.3. The issue of how much exercise and what type is needed to achieve an antidepressant effect is also an important topic, one that is discussed in practical application 34.4.

Table 34.3 Exercise Prescription Review

Training method	Frequency	Intensity	Time (Duration)	Type (Mode)	Progression	Important considerations
Cardiorespiratory	3-5 d/wk	Initially moderate, then increased to 70%-80% of heart rate reserve; as tolerated, train at higher end of heart rate range	30-45 min (progress if necessary)	Gross motor activities such as walking and biking	Begin at lower intensity for markedly deconditioned patients. Gradually progress as tolerated.	Undertreated and untreated depression can negatively affect adherence to exercise. Work with care physician or psychotherapist to prevent depression symptoms from interfering with exercise participation. Depression is an often present comorbidity for patients with chronic diseases such as heart disease, diabetes, obesity, and cancer. Some nonadherence is to be expected. Avoid being judgmental and instead take time to address barriers to participation.
Resistance	2 or 3 d/wk	10RM-15RM RPE of 11-15	2 sets per major muscle group; 1 min rest in between sets	Machines, free weights, elastic bands, calisthenics	Increase weight as tolerated to maintain 10-15 reps per set, at an RPE between 11 and 15.	
Range of motion	Daily	Within comfort	10-30 s per major joint	Static and proprioceptive or passive stretching	As tolerated	

RM = repetitions maximum; RPE = rating of perceived exertion.

Practical Application 34.4

HOW MUCH EXERCISE?

One question that remains unanswered is the dose of exercise required to obtain an antidepressant effect. Specifically, what frequency, intensity, duration, and type are most beneficial in treating patients with depression? A meta-review of eight meta-analyses of RCTs of exercise interventions for MDD concluded that moderate- to vigorous-intensity exercise, alone or in combination with resistance exercise, should be prescribed (5). The review reported that the current evidence base for MDD suggests physical activity should be prescribed for a minimum of 90 min/wk for at least 12 wk. It is, however, still suggested that people adopt the global recommendations of 150 min of moderate to vigorous physical activity per week to achieve the optimal cardiorespiratory benefits.

In addition, a meta-analysis of prospective cohort studies demonstrated that meeting the guidelines of 150 min/wk was also protective against developing depression, reducing the risk by about 22% (85). Although that study was unable to determine the optimal dose for the prevention of depression, it was concluded that higher levels of physical activity were associated with lower risk of developing depression.

Although achieving the exercise recommendations may be optimal for prevention and treatment, mental health benefits can still be obtained from lower doses. Given that more than a quarter of the world's population do not meet the current physical activity guidelines (41), and that those with poor mental health are even less likely to be active, it is important that the guidelines not act as a deterrent. Evidence from reviews of RCTs has shown that short bouts of physical activity (e.g., 10-15 min) can reduce stress and depressive symptoms and improve self-esteem (9). Although it may not suit everyone, there is some evidence supporting self-selection of the exercise intensity (84). Several RCTs have found a significant reduction in depressive symptoms after a self-selected intensity intervention (18, 29).

Does the domain of physical activity matter? Although physiological mechanisms (e.g., regulation of the HPA axis and reduced inflammation) contribute to the anti-depressive effect of exercise, factors such as enjoyment, autonomous motivation, social interaction, skill mastery, and goal achievement are also likely to influence the relationship. These factors, however, are more likely to be present when physical activity is undertaken in the domain of leisure or transport rather than household or occupational purposes (111).

In 2020 Teychenne and colleagues (98) examined the global physical activity recommendations in relation to mental health and concluded that although the optimal dose is unknown, the guidelines should recommend participation in physical activity that includes some leisure time or active travel, prioritizing activities that people enjoy or choose themselves. Second, people should be advised that some physical activity is better than none.

In summary, an exercise training dose consistent with public health recommendations is suggested for the treatment of MDD; however, it is important to remember that even small doses can improve mood and there is no one-size-fits all prescription.

Adherence and Exercise

Patients with depression may find it more difficult to stay engaged in an exercise program compared with patients who are not depressed, and specific symptoms of depression such as fatigue and a loss of interest in people and activities may interfere with adherence to an exercise regimen (see practical application 34.5.

O'Neal and colleagues (64) have offered recommendations for working with depressed people in a supervised exercise setting. First, they emphasize that nonadherence should be expected. Exercise professionals should avoid judging or blaming the patient for their depression because doing so will likely lead to guilt and a sense of failure that may cause the person to drop out of the exercise program. Instead, when nonadherence occurs, it should be viewed as a learning opportunity. That is, lapses in exercise participation can be used to identify an individual's unique barriers to adherence. The exercise professional can then help the patient find ways around these obstacles.

In addition, it is important to adopt a patient-centered approach. This includes helping individuals take personal responsibility for exercise prescription, exercise delivery, and monitoring of compliance. There is no one-size-fits-all prescription for physical activity, and the exercise professional should focus on improving self-efficacy, autonomy, and intrinsic motivation. Finally, when working in exercise settings, it is important to be familiar with the symptoms of depression and have some knowledge of treatment options. When depression is identified, the exercise professional should express warmth and empathy toward the patient while taking care to maintain an appropriate clinician–client boundary. Specifically, the exercise professional should not attempt to be the patient's psychotherapist but should instead have referral sources available.

STRATEGIES FOR IMPROVING EXERCISE ADHERENCE IN PATIENTS WITH DEPRESSION

- Most importantly, work to establish good rapport with patients. Positive feedback and empathy from exercise staff can go a long way toward promoting adherence. Encourage the patient and celebrate progress, even if it is small.

- At the initiation of an exercise program, review with patients their unique barriers to participation (e.g., work responsibilities, family issues). These barriers should then be discussed with patients to find ways to overcome or minimize them.

- Educate patients about the benefits of exercise for physical health and depression. Elicit from patients other benefits of exercise specific to them and periodically remind them of those benefits.

- Patients are more likely to adhere to exercise training if the experience is enjoyable. Work with patients to increase their satisfaction with the program (e.g., switching equipment used, varying location and the time of day). Their programs should be based on their current preferences and expectations, taking into account their initial fitness levels and ratings of perceived exertion during exercise.

- Help patients develop realistic exercise goals (e.g., gradual increase in number of exercise sessions per week). If they are new to exercise, consider open-ended goals (e.g., to increase their daily step count or to break up sitting time). Discuss problem solving around barriers, and encourage modification of goals as needed. Design a plan for when relapses occur.

- Be prepared to identify changes in mood and modify the session accordingly. If a patient is experiencing significant fatigue, low mood, or side effects from their psychotropic medications, this should be taken into consideration.

- Emphasize the short-term benefits, even after a single session, such as reduced stress, increased energy, and distraction from negative thoughts. Many patients are focused on long-term goals such as weight loss or improved self-esteem, so emphasizing short-term benefits can increase adherence.

- Encourage patients to reward themselves for participation in exercise. Emphasize the importance of positive reinforcement for accomplishments. Even simple rewards (e.g., a new book, a pleasant dinner out) can be powerful motivators.

- Recommend to patients that they talk to family members and friends about their exercise program and goals. Such people are often a valuable resource for offering encouragement in support of the patients' participation in exercise.

- Remember that untreated or undertreated depression is likely to have a negative effect on adherence to exercise. Encourage patients to seek treatment for depression if symptoms appear to interfere with exercise participation.

Clinical Exercise Bottom Line

- Physical and mental health are undeniably linked, and people with MDD commonly experience physical comorbidities (e.g., cardiovascular disease and metabolic syndrome).

- People with depression are less likely to engage in sufficient activity and face significant barriers to exercise including low mood, low motivation, and fatigue.

- High levels of exercise are associated with a decreased risk of current and future depression.

- Strategies to increase adherence include providing supervised exercise sessions and finding activities that are enjoyable.

CONCLUSION

Depression is a prevalent condition that can affect exercise testing as well as level of participation in exercise programs. It may be appropriate for an exercise physiologist to screen for depressive symptoms and to subsequently refer clients to specialized mental health treatment if needed. The clinical exercise physiologist should recognize that the depressed person who exercises is at risk for nonadherence, that physical comorbidities (e.g., cardiovascular disease) are common, and that depressive symptoms (e.g., loss of interest, lack of self-confidence, fatigue) may interfere with the enjoyment of exercise and motivation to fully engage in the rehabilitation process. People with depression may require increased support from exercise professionals to ensure adequate adherence to the exercise prescription.

Fortunately, several treatments have shown success in the treatment of depression, including antidepressant medication, cognitive–behavioral therapy, and exercise training. Patients who exhibit significant depressive symptoms should be approached in an empathetic manner and encouraged to seek treatment to improve quality of life and gain maximal benefit from exercise.

Online Materials

Visit HK*Propel* to access a link to the references, the case studies with discussion questions, and a quiz to help you review key concepts and test your understanding of the material covered in this chapter.

CHAPTER 35

Intellectual Disability

Tracy Baynard, PhD

Bo Fernhall, PhD

Individuals with an intellectual or developmental disability may have one of several disorders, all of which are dynamic over time yet have limitations in mental function and personal skills (e.g., ambulation, communication, self-care, social adjustment) in common. An intellectual disability is not a disease and cannot be communicated. It is also not a type of mental illness, like anxiety or depression. There is no cure for intellectual disabilities; and although most children and adults with an intellectual disability can learn to engage in a variety of skills and activities, mastering them takes more time and effort than it does for others without an intellectual disability.

DEFINITION

Intellectual disability (ID) is classified by three different systems, with small differences between them. The American Association on Intellectual and Developmental Disabilities (AAIDD) (7) defines ID as "a disability characterized by significant limitations both in intellectual functioning and in adaptive behavior, which covers many everyday social and practical skills. This disability originates before the age of 18." In general, a person with ID has reduced intellectual functioning (e.g., lower intelligence), which includes reductions in learning, reasoning, and problem solving. Intellectual functioning (i.e., intelligence quotient, IQ) is usually determined from standardized intelligence tests such as the Wechsler Adult Intelligence Scale or the Stanford-Binet scale. The *Diagnostic and Statistical Manual of Mental Disorders, Fifth Edition* (DSM-5) (9, p. 33), defines ID as "is a disorder with onset during the developmental period that includes both intellectual and adaptive functioning deficits in conceptual, social, and practical domains." The DSM-5 diagnosis requires that three criteria be met, similar to the AAIDD's, with deficits in intellectual *and* adaptive functioning that originate during childhood (9). However, both the AAIDD and the DSM-5 have moved away from using IQ scores

to help define the severity of ID (i.e., mild, moderate, severe, or profound), yet IQ scores generally below 70 are indicative of potential ID and are used as a starting point to assess intellectual functioning. The third system to classify ID is from the Social Security Administration. Section 12.05 of the Social Security Administration defines intellectual disorder as subaverage intellectual functioning, adaptive behavioral deficits, and onset prior to the age of 22 yr; this does not include neurocognitive disorders, autism spectrum disorder, or neurodevelopmental disorders, which have different codes (116).

Individuals with ID usually need support services in one or more of the areas of adaptive behavior; the level of support needed is used to plan services but is not part of the definition or diagnosis of ID, with the exception of severe and profound ID. Intellectual disability is not a static, nonchangeable condition. Instead, it is a fluid condition, and early intervention may help some individuals progress to the point where, in fact, they are no longer classified as having ID (11).

SCOPE

It has long been estimated that the prevalence of ID ranges between 1% and 3% of the total population, with the larger estimate from the number of annual births where ID would develop based on IQ test cutoffs (79, 120). A more recent examination of prevalence suggests the global rate may be under 1% (79). As such, there are approximately 9 to 10 million individuals with ID in the United States; over 90% of all persons with ID are classified with mild ID (10, 38). In 2019, a systematic review examined 14 studies that addressed the prevalence of ID or developmental disability since 2000. The authors report prevalence rates from under 1% to 1.2% but note that age differences and operational definitions make it difficult to accurately determine prevalence of ID (10). This uncertainty can ultimately affect research and public policy.

Most individuals with ID live in the community, either at home or in community-based group homes. As a consequence of the deinstitutionalization movement, very few individuals with ID now live in state-supported institutions. People with more severe forms of ID live in community-based facilities that offer more specialized and intensive support (37). Mortality has traditionally been higher in people with ID than in the general population, with estimates varying from 1.5 to 4 times higher than expected in populations without ID (57, 62, 109). The adult (crude) mortality rate is 15 per 1,000 in the United States (73). Mortality has historically been linked to the low IQ and poorer self-care skills of persons with ID, yet more recent data suggest circulatory diseases, respiratory diseases, and neoplasms are the three most common causes of death in ID (57). These may be a direct or indirect consequence related to intellectual functioning and self-care. However, low levels of physical activity, and especially movement ability (which may be related to self-care and low IQ as well), may also contribute to the higher mortality and morbidity of individuals with ID (45, 62). Oppewal and Hilgenkamp (89) reported several physical and fitness factors that independently predict 5 yr survival among older persons with ID (mean age 62 yr), including manual dexterity, visual reaction time, gait speed (comfortable and fast), grip strength, and aerobic fitness estimated from the 10 m shuttle test. Except in persons with **Down syndrome (DS)**, cardiovascular and pulmonary disorders are the most common medical problems in persons with ID (62, 97). For individuals with DS, leukemia, infections, and the early development of Alzheimer's disease are the most frequent causes of both mortality and morbidity (36, 88, 94).

Life expectancy has increased in individuals with ID who do not have DS (94), but the current life expectancy of ~60 yr of age is still ~20 yr lower than for the general population (88). Not surprisingly, those with profound ID have even lower life expectancy and crude mortality rates versus the general population. Comorbidities reduce life expectancy even further, regardless of the severity of ID (88). Over the last 40 yr, life expectancy has also increased rapidly in individuals with DS—the median is now 53 yr of age, with some people living into their 80s (94, 104). The drastic change in life expectancy is likely due to better medical care, better living conditions, and, in the case of individuals with DS, better survival after corrective cardiac surgery. Considering that almost all individuals with ID live in supported living situations, the cost of ID is likely substantial, but little or no information is available on these costs.

PATHOPHYSIOLOGY

The pathophysiology, or cause, of ID is often difficult to identify because it includes any condition that negatively affects brain development before, during, or after birth until the age of 18 yr. Several hundred causes of ID have been identified or suspected, but in one-third to one-half of individuals with ID, the cause is unknown. The three major causes are **fetal alcohol syndrome**, **fragile X syndrome**, and DS (28). In general, causes of ID fit into one of the following categories: genetic, problems during pregnancy, problems at birth, problems after birth, and poverty and cultural deprivation. These categories are summarized in table 35.1.

Genetic conditions that cause ID typically result from either inherited gene disorders such as fragile X syndrome or gene disorders caused during pregnancy. Down syndrome, or trisomy 21, is the most common genetic cause for ID, and in more than 90% of individuals with DS an extra, or third (instead of just two), chromosome is found on chromosome 21. Other causes of DS include disjunction (24 chromosomes in one haploid cell and 22 in the other) and translocation (two chromosomes grown together as one but containing the genetic material of both) (72). Maternal age is the primary risk factor for DS, and up to 50% of all children born with DS are born with congenital heart disease. Although **congenital heart disease** used to be a major cause of early mortality in patients with DS, advances in early diagnosis and surgical techniques have substantially improved the prognosis of these children (37, 134).

Table 35.1 Summary of Potential Causes of Intellectual Disabilities and Related Factors

Potential causes	Factors associated with the potential causes				
Genetic	Maternal age	Infections	Phenylketonuria		Trisomy 21 (Down syndrome)
Problems during pregnancy	Maternal substance abuse	Rubella	Toxoplasmosis	Syphilis	Maternal malnutrition	Maternal exposure to environmental toxins
Problems at birth	Premature delivery, low birth weight	Fetal oxygen deprivation	Certain injuries during birth		
Problems after birth	Meningitis	Encephalitis	Other causes of brain damage (e.g., near drowning)		Environmental toxins (e.g., lead)
Poverty and cultural deprivation	Malnutrition	Inadequate health care	Low educational and cultural stimulation		

When an ID is related to pregnancy, the most common cause is maternal alcohol or drug use. In fact, excessive alcohol consumption is the leading preventable cause of ID. In contrast to the genetic causes of ID, many of the factors during pregnancy that result in ID are preventable (6, 28).

Premature delivery and low birth weight are the leading causes of ID associated with problems at birth. It is also possible for fetal oxygen deprivation during birth to cause ID, in addition to certain birth injuries (6, 28). Later on, several childhood diseases, such as measles, whooping cough, and chicken pox, can lead to meningitis and encephalitis, which may produce brain damage and cause ID. Poverty increases the risk of malnutrition, inadequate health care, and deprivation of educational and cultural experiences, which can also lead to ID over time. Thus, many of the known causes of ID are preventable, but it is important to reemphasize that in up to half of all cases, the cause is unknown (6, 28).

Several features common in individuals with ID affect the physiologic responses during exercise. For example, obesity is more prevalent in persons with ID than in the general population (63, 108), and persons with ID are more likely to present with more advanced stages of obesity. Obesity is even more prevalent in certain subcategories of ID, such as DS and **Prader-Willi syndrome** (12, 58), with many of these individuals categorized as having severe obesity. Most individuals with ID exhibit low or very low levels of cardiorespiratory fitness (90), often coupled with obesity (see sidebar). It is not unusual for people with ID in their mid-20s to have a peak oxygen uptake ($\dot{V}O_2$peak) of only 25 to 30 mL · kg^{-1} · min^{-1}, which is 25% to 35% below expected (12, 20, 47). Interestingly, if individuals with ID are considered separately (with and without DS), patients with ID who do not have DS may have close to expected $\dot{V}O_2$peak levels (12). Yet this may be questionable because recent data suggest they indeed present with lower levels of cardiorespiratory fitness (20). When considering individuals with DS alone, they typically have very low aerobic fitness (even lower than individuals with ID without DS), with $\dot{V}O_2$peak between 15 and 20 mL · kg^{-1} · min^{-1}, and it is very rare to find an individual with DS who exhibits expected age- and sex-specific $\dot{V}O_2$peak. These low levels of cardiorespiratory fitness are not the result of the high incidence of congenital heart disease in this population, because all published studies have been conducted in individuals free of congenital heart disease. Thus, the influence of congenital heart disease on cardiorespiratory fitness among individuals with DS is currently unknown. Given the high prevalence of obesity among individuals with DS, it is important to note that obesity does not appear to have as strong an influence on aerobic capacity among adolescents or adults with DS compared with a non-DS group (69, 70, 132), suggesting DS exerts more of an adverse influence on aerobic capacity than obesity. However, in children with DS, as well as adults with ID without DS, the association between weight status and aerobic fitness is similar to that of the general population (111, 132).

A primary cause for the lower levels of cardiorespiratory fitness in individuals with DS, and potentially in individuals with ID who

---

Physiologic Responses to Acute Exercise in Individuals With Intellectual Disability

Cardiorespiratory Responses to Aerobic Exercise
- Lower maximal heart rate; in persons with DS, ~30 b · min^{-1} lower than age-predicted using typical adult formulas
- Lower cardiac output
- Lower peak aerobic capacity ($\dot{V}O_2$)
- Attenuated plasma catecholamine responses

Responses to Resistance Exercise
- 30% to 50% reductions in expected strength
- Reduced strength even in very active individuals
- Cardiac autonomic dysfunction

---

do not have DS, is a lower maximal heart rate. Although most individuals with ID exhibit maximal heart rates below expected levels, for persons without DS the reduction is not severe enough to classify their exercise heart rate response as chronotropic incompetence (12, 20, 41, 47, 64). However, it is well documented that almost all individuals with DS exhibit chronotropic incompetence (12, 39, 41, 47, 59, 60, 81). In fact, the average maximal heart rate can be expected to be ~30 b · min^{-1} less than in age-matched individuals without disabilities (12, 41). This reduction in peak heart rate leads to a reduced peak cardiac output and subsequently to reduced $\dot{V}O_2$peak.

It is likely that autonomic dysfunction contributes to the reduced maximal heart rate in people with DS (42) because they exhibit reduced vagal withdrawal to handgrip exercise (43, 55), altered orthostasis responses (1, 40), and reduced peripheral blood flow regulation (65). However, the main reason for the reduced maximal heart rate during maximal treadmill exercise appears to be an inability to increase plasma catecholamines (e.g., norepinephrine) during exercise (39). Since catecholamines are an important contributor to increases in heart rate at higher exercise intensities (above the ventilatory threshold), this partly explains the lower maximal heart rates in DS. In one study, a small subset of individuals with DS exhibited normal increases in catecholamines during maximal exercise, and in this subset, both maximal heart rate and $\dot{V}O_2$peak were similar to those of control participants without DS (74). Using heart rate variability analyses, Bunsawat and Baynard (23) provide some evidence supporting the concept there may indeed be a subset of individuals with DS who demonstrate more normal or typical autonomic responses to handgrip exercise. This has not been evaluated using other sympathoexcitatory tasks to date.

Regardless of how muscle strength has been measured, across all age groups, muscle strength in persons with ID is 30% or more below that of age-matched nondisabled individuals (32, 44, 61, 66, 67, 96, 98, 99, 118). People with DS exhibit even lower levels of strength, often 30% to 40% below their peers with ID and less than 50% of the expected strength levels of their nondisabled peers (30, 61, 98). These low levels of muscle strength are present in childhood (44, 66, 99) and persist into adulthood (32, 67). Interestingly, in individuals with ID who were extremely active and exhibited very high aerobic capacities, muscle strength was still ~25% below the expected values for age- and activity-matched subjects without ID (56); this suggests a possible underlying abnormality in skeletal muscle function. The lower levels of muscle strength are important in persons with ID, because muscle strength has been related to both cardiorespiratory fitness (30, 96) and the ability to perform tasks of daily living (30). Furthermore, many individuals with ID exhibit poor coordination and reduced walking economy and walking stability, both of which are also associated with muscle strength (2, 3, 83).

CLINICAL CONSIDERATIONS

Individuals are diagnosed with ID based on intelligence, mental ability, and adaptive skills. Compared with individuals free of an ID, people with an ID have a reduced ability to think and solve problems and to adapt and function independently. Once a diagnosis of ID has been made, an individual's strengths and weaknesses can be evaluated in order to tailor the amount of support or help needed to function at home, in school, and in the community.

Signs and Symptoms

The major signs and symptoms are usually related to delayed developmental stages in infants and children. This can include delays in sitting up, crawling, and walking. Children may also experience delays in or difficulty with talking. Most developmental milestones are delayed, but the amount of delay depends on the severity and cause of the disability. Later in childhood, a child with ID may have difficulties understanding directions and experience problems with logical thinking and problem solving. It is not unusual for children with ID to be unable to understand social rules or the consequences of their actions (28).

History and Physical Examination

The history and physical examination should follow standard formats. It is likely that a history needs to be acquired from a significant caregiver because individuals with ID may not accurately remember their history. For people with DS, it is important to obtain information on congenital heart disease and any joint problems, such as instability of the atlantoaxial joint (**atlantoaxial instability**) in the upper neck (i.e., where the base of the skull meets the neck).

Diagnostic Testing

Chromosomal microarrays are being increasingly used as a first-tier genetic test among individuals with unexplained IDs (85). G-banded **karyotyping** should be reserved for individuals with obvious chromosomal problems or a family history of genetic abnormalities (85). This is a rapidly developing area in relation to ID, and our knowledge and understanding of genetic links to various types of ID will expand substantially in the coming years.

The level of ID is diagnosed through the use of two types of tests: standardized tests of intelligence (e.g., IQ tests) and adaptive behavior (7, 37). The cutoff score used for both is generally 2 standard deviations below the mean for the particular assessment instrument. Well-known IQ tests include the Wechsler Adult Intelligence Scale, the Wechsler Intelligence Scale for Children, the Stanford-Binet Scale, the Woodcock-Johnson Tests of Cognitive Abilities, Raven's Progressive Matrices, and the Kaufman Assessment Battery for Children. Intellectual disability is diagnosed with an overall score at or below 70 points on most IQ tests. It is important to note that IQ scores are fluid because intelligence (and adaptive behavior too) can change over the course of a life span, in either a positive or a negative direction. Therefore, it is recommended that intelligence and adaptive behavior tests be repeated through the life span.

Adaptive behavioral tests address conceptual, social, and practical skills (7, 37). Conceptual skills have to do with issues of language and literacy, money, time, numerical concepts, and self-direction (7). Skills related to the social being include "interpersonal skills; social responsibility; self-esteem; gullibility; naiveté; social problem solving; and the ability to follow rules, obey laws and avoid being victimized" (7). Lastly, practical skills involve areas related to personal care, travel and transportation, job skills, health care, maintaining a schedule or routine, safety, and using money and the telephone. Tests for evaluating adaptive behavior include the Woodcock-Johnson Scales of Independent Behavior (used for children), the Vineland Adaptive Behavior Scale, and the 2017 version of AAIDD's Diagnostic Adaptive Behavior Scale. The Vineland Scale is used for testing social skills in persons from birth through 19 yr of age and is administered to the individual's caregiver. The AAIDD's Diagnostic Adaptive Behavior Scale is used with individuals 4 to 21 yr old and provides specific diagnostic information around the cutoff at which someone is deemed to have significant limitations (7, 121). The Diagnostic Adaptive and Behavior Scale's newest edition utilizes item response theory, which means all test scoring takes place on their online platform (122).

A diagnosis of ID is important for several reasons. It establishes eligibility for special education services, home- or community-based services, and Social Security benefits, as well as allowing for special treatment within the criminal justice system. Additional diagnostic tests should be considered for individuals with ID, in particular those with DS. These include tests for **hypothyroidism** and congenital heart disease, neck X-ray to determine possible atlantoaxial instability, and hearing and vision tests. Individuals with DS have higher rates of infection and respiratory problems; therefore, regular blood tests should be considered to detect any abnormalities with the immune system (94, 109).

Exercise Testing

Exercise testing in individuals with ID appears to be largely safe, and considerations with respect to cardiovascular complications are likely not different from those for the general population (38, 45, 90). However, it is important to keep in mind that few reports, if any, specifically address whether exercise testing in this population is indeed safe. Certainly, this is an area that needs scientific study. Common to a diagnosis of ID is a concern about the person's ability to follow and understand instructions and cooperate with the testing procedures, which is a concern regarding any type of test in patients with ID. Despite these concerns, standard laboratory exercise tests appear to be valid in persons with ID (38, 45, 90). Individuals with ID can undergo a full complement of exercise-based tests, which should include tests of aerobic capacity, muscle strength and endurance, and body composition. A summary of laboratory-based exercise testing recommendations is provided in table 35.2.

There are several key points regarding valid outcomes for exercise testing (37, 38, 51, 53, 54):

1. Pretest screening procedures should follow American College of Sports Medicine (ACSM) guidelines (8). The exception is for individuals with DS. Because nearly half of all individuals with DS have congenital heart disease, a careful health history and physical examination is necessary in this population, regardless of age. Further, there is a relatively high incidence of atlantoaxial instability, which also warrants a physical examination.

2. Appropriate familiarization with both personnel and testing procedures is necessary before conducting the actual test. The level of ID will determine the amount of familiarization visits needed to ensure test validity. Often two or three sessions are necessary, depending on the person and the test being performed. These sessions should include adequate time for the personnel to demonstrate the task(s) and for the individual with ID to practice them. Regular and positive feedback will enhance results.

3. Provide explicit and simple one-step instructions. Allow the individual to move along with each step before proceeding to the next step (e.g., "Step up on the treadmill"; "Hold on"; "Begin walking").

4. Have extra personnel assist with testing on the treadmill, especially in persons with DS, because balance issues are common in these individuals.

Table 35.2 Summary of Exercise Testing for Individuals With Intellectual Disabilities

Test type	Mode	Protocol specifics	Clinical measures	Clinical implications	Special considerations
Cardiorespiratory	Treadmill Dual-action (arm–leg) cycle 20 m shuttle run	Individualized walking speeds on treadmill Dual-action cycle—use both arms and legs, 25 W stages	METs HR $\dot{V}O_2$	To assess progress in a training program	Familiarization is critical. See text for full list of considerations.
Strength	1RM on weight machines Isokinetic testing Isometric MVC	RM testing: progress slowly Isokinetic testing: standard protocols, three or four maximal contraction trials; select best one or average of the three 2 min of isometric contraction at 30% MVC	Kilograms	Same as above	Watch range of motion. Be careful to fit subject to machine; because of short statures, adjustments will need to be made. With MVC, verbal encouragement is needed.
Range of motion	Sit-and-reach Joint-specific goniometry	Active stretching	Distance Degrees	Same as above	Do not allow hyperextension; be watchful of range of motion.
Body composition	Air displacement plethysmography Skinfolds Waist circumference		% body fat Centimeters	Same as above	Estimate lung volume with air displacement plethysmography.

METs = metabolic equivalents of task; HR = heart rate; 1RM = one-repetition maximum; MVC = maximal voluntary contraction.

5. Individualize testing protocols as needed, but do so in as systematic a way as possible. For example, on a graded exercise treadmill test, start at an individualized comfortable walking speed. Then ramp up speed in similar increments between subjects. Then increase the incline similarly between subjects.

6. Be careful with use of an age-predicted maximal heart rate (HR) formula in individuals with ID, especially those with DS (the research focus sidebar). A population-specific formula has been developed for DS (41) and is useful as a guide during exercise testing: estimated maximal HR = 210 − 0.56(age) − 15.5(x) [x = 1 for non-DS and 2 for DS]. Note that this prediction equation is not intended to be used for individual exercise prescriptions (15).

7. Several tests are not recommended for use in individuals with ID because of poor reliability, validity, or both. For cardiorespiratory testing, these include treadmill running protocols, arm ergometry, and the 1 or 1.5 mi runs. Cycle ergometry protocols are also not recommended, with the exception of a dual-action cycle ergometer. In testing for muscle strength or muscle endurance, the one-repetition maximum (1RM) with free weights, push-ups, and the flexed arm hang are not recommended for this population.

8. When field tests are used to estimate cardiorespiratory fitness, it is necessary to apply population-specific formulas, which are outlined in table 35.3.

9. Cardiorespiratory field tests are reliable, but they are not typically valid for predicting peak aerobic capacity in persons with DS; one exception may be the 16 m PACER test, although this equation has not been cross-validated to date (19).

Similar to the exercise testing of any other individual with a physical disability (e.g., osteoarthritis) or metabolic disorder (e.g., diabetes), the exercise testing of individuals with an ID requires that the personnel performing the testing be familiar with the various medical aspects of the specific disability, common secondary treatments, and any unique physiologic adjustments to exercise. In patients with ID, exercise assessments are often modified to accommodate the skills and abilities of the individual.

Cardiorespiratory

Cardiorespiratory testing largely comprises individualized treadmill walking protocols. Fernhall and colleagues (45, 47, 51, 52, 53) have demonstrated that individualized walking protocols are valid in this population. In general, after proper familiarization, the individual begins the treadmill protocol walking at a speed between 0.89 and 1.56 m · s^{-1} (2-3.5 mph) for 4 min at 0% grade. Holding speed constant, grade is increased by 2.5% every 4 min until 7.5% grade is achieved. After this, grade is increased further every 2 min by 2.5%, until 12.5% grade is reached; at this point, speed is increased ~0.22 m · s^{-1} (0.5 mph) until the person reaches exhaustion. At these higher speeds, people will need to jog if they are able. Participants are allowed to hold on to the handrail as needed to maintain balance, especially at the higher speeds. Considering that many individuals with ID are poorly coordinated, holding on to the handrails is usually recommended. Importantly, this means that work capacity cannot be accurately predicted from speed and grade on the treadmill, and actual measurement of $\dot{V}O_2$ is recommended.

Dual-action cycle ergometers that require individuals to use both their arms and legs have also been demonstrated to be valid in this population (100, 101, 102). Stages increase by 25 W, and it is important to focus on keeping the revolutions per minute in a reasonable range (e.g., 40-60 rpm) in order to obtain an accurate work rate. Field tests are another alternative, and some are valid and reliable in estimating aerobic capacity in persons with ID. Because these tests are submaximal and do not require expensive equipment, and because several people can be tested at the same time, they are an attractive alternative.

Leg strength contributes to endurance run performance in this population (44), which is quite different from what is observed in nondisabled individuals. Leg strength is also an independent

Table 35.3 Formulas for Predicting $\dot{V}O_2$max From Field Test Performance in Individuals With ID

Field test	Formula
16 m PACER for individuals with DS (19)	$\dot{V}O_2$max (mL · kg^{-1} · min^{-1}) = 48.23 + 0.32(16 m PACER) − 0.45(BMI) − 2.88(sex; 1 = male, 2 = female) − 0.13(age)
20 m shuttle run (49)	$\dot{V}O_2$max (mL · kg^{-1} · min^{-1}) = 0.35(no. of 20 m laps) − 0.59(BMI) − 4.5(sex; 1 = male, 2 = female) + 50.8
20 m shuttle run for individuals with DS (4)*	$\dot{V}O_2$max (mL · kg^{-1} · min^{-1}) = 21.68 + 0.62(no. of 20 m laps)
600 yd run/walk (49)	$\dot{V}O_2$max (mL · kg^{-1} · min^{-1}) = −5.24(600 yd run time in minutes) − 0.37(BMI) − 4.61(sex; 1 = male, 2 = female) + 73.64
Rockport 1 mi walk test (124)	$\dot{V}O_2$max (L · min^{-1}) = −0.18(walk time in minutes) + 0.03(body weight in kilograms) + 2.90

*BMI and sex did not achieve significance to be added to this model.

PACER = progressive aerobic cardiovascular endurance test; BMI = body mass index (kg · m^{-2}).

contributor to field test performance in children with ID, which is an important issue for exercise training in this population. Interestingly, test–retest reliability of muscular strength was found to be poor for children and adolescents with moderate to severe ID (133), so caution is warranted in this specific cohort.

Resistance Exercise

Use of 1RM testing with free weights is not advised for individuals with ID. However, weight machines are a perfect alternative, and 1RM testing can be achieved in this population across six to eight major muscle groups. Following standard protocols for 1RM testing is recommended. The clinical exercise physiologist should verbally encourage proper breathing and provide physical spotting to ensure proper form. Overextension or overflexion should be avoided.

Isokinetic testing can be performed for measures of strength, endurance, or both in this population. Although no alterations need to be made to the computer-based protocols on these devices, proper fitting to the device is essential. This may be especially difficult among individuals with DS because of their generally shorter stature. Isometric testing can be used in this population, especially for determining cardiorespiratory responsiveness to the adrenergic stress. Visual feedback is important for this task, allowing the individual to make adjustments to the strength necessary to maintain the target resistance. Generally, three or four trials of isometric maximal voluntary contraction are needed; two or three of the trials should be within 1 or 2 kg of each other. Use either a mean of the three trials (assuming they are close enough to each other) or use the highest value obtained as the maximal voluntary contraction. For handgrip testing, 30% of maximal voluntary contraction is commonly set as the target resistance, and this level is generally held for 2 min in individuals with ID.

Flexibility

Both the sit-and-reach test and joint-specific goniometry are used for testing flexibility in this population. A possible concern is joint laxity in individuals with DS.

Field Tests and Tests of Functional Capacity

Field tests and tests of functional capacity include the 20 m shuttle run, the 600 yd run/walk, and the Rockport 1 mi walk test, which have been validated for use and deemed reliable in both children and adolescents with ID (4, 46, 48, 49, 52, 90, 124), and the 16 m PACER and 6 min walk tests, which have been validated for use in adults with DS (19). Boer and Moss (19) reported that the 16 m PACER test was more strongly related to peak aerobic capacity than the 6 min walk test ($r = .78$ vs. .87). Prediction formulas for estimating $\dot{V}O_2$peak with these tests are displayed in table 35.3. Only these population-specific formulas should be used, because they differ from the formulas used for nondisabled individuals. Furthermore, because of the large variability associated with field tests in persons with ID, individual estimates may be less accurate even though group data are very accurate (46, 48).

Functional tests for adolescents and adults with ID, with and without DS, have been demonstrated to be feasible and reliable (5, 24, 123, 125). The functional tests evaluated in these studies include the Timed Up and Go, trunk flexibility, handgrip, timed stand test, 30 s sit-up test, standing broad jump, 4×10 m shuttle run, and 6 min walk test. Tests of functional capacity have also been found reliable for children and adolescents with moderate to severe ID, with stair walking and the 6 min walk test to have good to very good intraclass correlations (133). Although such functional tests do not predict aerobic capacity, these field tests can be useful because laboratory testing in persons with moderate to severe ID has not been established as valid or reliable. Importantly, Terblanche and Boer (125) suggest possible sex differences for some functional testing. Men were found to perform more reliably. Age may also play a role in reliability, with younger groups performing more consistently across a majority of the tests employed (125). Exercise testing considerations include the following:

Cardiorespiratory: Treadmill walking protocols are best; the 20 or 16 m shuttle run is acceptable for field tests.

Strength: 1RM on resistance exercise machines is recommended. Most field tests are not reliable.

Range of motion: Sit-and-reach and goniometry are recommended.

General: Appropriate familiarization is required for all testing.

Treatment

There is no treatment for ID per se. The AAIDD states that the primary reason for evaluation and classification of individuals with ID is to tailor support specific to each individual, whereby detailed strategies and services are outlined over a sustained time (7). The purpose of tailoring the support system is to maximize people's functioning within "their own culture and environment in order to lead a more successful and satisfying life" (7). This helps increase the individual's sense of self-worth, well-being, pride, and social engagement (7). Interestingly, increased physical activity or formal exercise programming (or both) is now recognized as an important aspect of maintaining health in persons with ID (63). Finally, conditions, symptoms, or behaviors that are either directly or indirectly associated with ID are treated as needed. For instance, many individuals with ID are treated for symptoms of depression with antidepressant medications, and many with DS are treated for thyroid disorders as well.

Pharmacology

Individuals with ID may take various antidepressant medications. Many also take anticonvulsive agents. These medications help control inappropriate behavior and physiological symptoms caused by the ID but do not affect ID per se. In severe cases, hypnotics may be used to control psychotic behavior.

It is not unusual for individuals with DS to have hypothyroid conditions, which are usually diagnosed during adolescence or early adulthood. These patients are on thyroxine replacement therapy. Large weight gains should be evaluated in this population, even in persons who have already been diagnosed with hypothyroidism; further titration of the thyroxine dose may be necessary (109).

Antidepressant medications can have a minor effect on heart rate or blood pressure but generally have little effect on how these variables respond to acute or chronic exercise. These agents may also affect the ability of an individual with ID to understand and follow directions, as well as motivation to perform higher-intensity exercise (8, 37). Thyroid medications may increase both heart rate and blood pressure and are potentially arrhythmogenic. Use of β-adrenergic blocking agents is not uncommon, particularly in persons without DS (8). These medications reduce heart rate and blood pressure both at rest and in response to exercise (50). Table 35.4 provides a summary of the pharmacology issues in persons with ID.

The involvement of individuals with ID in exercise testing or training is increasing; thus clinical exercise physiologists are likely to encounter and work with this population more frequently (see practical application 35.1). Consequently, it is important to understand the unique aspects of conducting exercise testing and training in persons with ID. The major differences apply primarily to exercise testing, as exercise training and exercise prescriptions are for the most part similar to those for populations without disabilities.

EXERCISE PRESCRIPTION

The exercise prescription for people with ID is quite similar to that used for nondisabled adults (95). Because of the generally higher prevalence of overweight and obesity and lower physical activity levels, it is important to develop an exercise prescription that progressively increases both physical activity and muscle strength as part of a healthy lifestyle. The exercise prescription is outlined in table 35.5. The following are important special considerations for persons with ID.

1. More positive encouragement will likely be necessary in individuals with ID than in others. Because motivation can be problematic, and given the likelihood of a short attention span, it is important not to ask leading questions (e.g., "Are you tired?"—people are likely to answer yes to such a question even if they have not begun their workout). Also, in planning exercise sessions, consider other tasks or duties that will impose an attentional demand during the day.

2. Exercise training sessions need to be carefully supervised at all times, particularly in the beginning of an exercise regimen, to help individuals gain familiarization.

3. Individualize the training as much as feasible. Although older studies found group exercise to be less effective for individuals with ID, newer research suggests group exercise may be feasible with certain adjustments. For instance, among

Table 35.4 Pharmacology

Medication class or name	Primary effects	Exercise effects	Important considerations
Antidepressants	Used to alleviate mood disorders.	Have minor to negligible effects on HR and BP during acute or chronic exercise.	May produce false-positive or false-negative stress test results; may affect an individual's understanding of test instructions and motivation.
Antiepileptics and anticonvulsants	Primary molecular targets are voltage-gated sodium or calcium channels and parts of the GABA system, which help reduce hyperirritability in parts of the brain's cortex.	Appear safe to use in conjunction with exercise.	May cause hyperinsulinemia and increased appetite, and possibly hormonal changes as well (e.g., testosterone).
Levothyroxine; thyroid replacement hormone	Used to treat hypothyroidism and to treat or suppress goiter (enlarged thyroid) or both.	Normalizes (increases) HR and BP response to exercise compared with that in the hypothyroid condition. Normalizes (increases) LVEF during exercise compared with that in the hypothyroid condition.	Small risk of provoking an arrhythmia, and because of increased cardiac contractility, some individuals may experience brief episodes of angina.
β-blockers	β-adrenergic receptor antagonists; reduce heart rate and blood pressure at rest.	Reduce heart rate and blood pressure during exercise.	—

HR = heart rate; BP = blood pressure; GABA = gamma-aminobutyric acid; LVEF = left ventricular ejection fraction.

Practical Application 35.1

CLIENT–CLINICIAN INTERACTION

Because of the many types and various degrees of ID, one of the most important qualities in the clinical exercise physiologist is patience. Individuals with ID may have difficulties understanding the context surrounding what the clinical exercise physiologist is trying to accomplish and therefore may not understand directions given. It is also likely that task persistence, attention span, and willingness to try something new (as in exercise testing or training) are reduced. It is not unusual for people with ID to also exhibit various forms of behavioral problems, which can make exercise testing and training even more challenging.

Allocate sufficient time in the schedule for exercise testing or training sessions, and be flexible about the degree of availability that individuals with ID have. For instance, because people in this group live primarily in group- or community-based homes, they often rely on caregivers for transportation and therefore their schedules are not uniquely their own. Creativity is often key in successful planning. Furthermore, it is helpful for the clinical exercise physiologist just entering the field to gain experience with this population by working closely with a more experienced CEP before independently supervising either exercise training or testing sessions. Although this chapter includes some tricks of the trade, nothing can replace hands-on experience and the opportunity to learn from an experienced individual or group. Finally, as with all patients or clients seen by the clinical exercise physiologist for exercise testing or training, it is important to preserve modesty and treat the individual with ID respectfully, as well as to provide a positive experience for not only the individual but also the caregiver. This often requires more attention and time than are usually allocated to patients with other health problems and frequently involves some creative scheduling. It should go without saying that being abrupt with a patient or talking down is not effective. Individuals with ID enjoy and expect being treated as an equal.

adolescents with ID and developmental disabilities, compliance was very good for a 12 wk home-based physical activity intervention using video conferencing on tablets, with 27 min of activity achieved per session, 12 min of which was moderate to vigorous (105). Other research suggests it may be better practice to initiate the exercise sessions in a one-on-one fashion before transitioning to small groups for exercise (i.e., two supervisors per seven individuals) (14).

4. To increase motivation, simple games and music can be easily incorporated into an exercise training program, as can participation in activities akin to the Special Olympics. The advent of more advanced gaming technology may also provide opportunities for additional exercise or physical activity with appropriately selected games. For instance, the physically oriented Wii gaming system was found effective for increasing aerobic endurance, explosive leg power, functional mobility, and flexibility after a 2 mo intervention (115). The activities in this study used the Wii Fit Balance Board to perform games such as free run, snowboard slalom, and hula hoop as well as sports-related or dancing games. As new hardware and software gaming combinations become available and evolve, innovative approaches to engage this population in physical activity may follow.

5. Exercises that may involve hyperflexion or hyperextension of the neck are contraindicated because of a high rate of atlantoaxial instability.

6. Individuals with DS often present with skeletal muscle hypotonia and joint laxity. To avoid hyperextension or hyperflexion, ensure the individual is performing the exercise in a smooth and controlled manner and is not jerking through a given exercise with excessive momentum.

7. **Pulmonary hypoplasia** may also be present in individuals with DS. Modifications may be needed to the exercise prescription depending on the severity of the hypoplasia, especially if exercise is performed at altitude. A lower than normal intensity will likely be needed.

8. Specific physical features may limit exercise performance. These include short stature and short limbs and digits. These characteristics may be coupled with additional characteristics such as malformed feet and toes, a small mouth or small nasal cavities or both, and a large or protruding tongue. Poor vision and balance may also contribute to reduced exercise performance.

Cardiorespiratory Exercise

It is important to find an exercise modality that is truly enjoyable for individuals with ID. This will help ensure their attention and increase the likelihood of their continued participation. Acceptable modes of exercise include walking, with the possibility of some jogging; swimming; combined arm and leg ergometry (e.g., elliptical, dual-action [i.e., combined arm and leg] cycle ergometer); and exercising to music or using games.

Table 35.5   Summary of Exercise Prescription for Individuals With Intellectual Disabilities

Training method	Frequency	Intensity	Time (Duration)	Type (Mode)	Progression	Important considerations
Cardiorespiratory	3-7 d/wk, with 3 or 4 d/wk of moderate to vigorous exercise; emphasis on increased activity on the remaining days	40%-80% HRR; RPE may not work	30-60 min/d; accumulated exercise may be used, particularly in the beginning of an exercise program	Walking primarily, with possible progression to running when used intermittently	Increase speed, duration, or both gradually after the initial 2 or 3 wk.	Do not base exercise prescription on predicted maximal heart rate; always use measured maximal heart rate. RPE will likely not work. See text for additional comments.
Resistance	2 or 3 d/wk	12 reps at 15RM-20RM for 1 or 2 wk	2 or 3 sets, 1-2 min rest in between	Swimming	Progress to 8RM-12RM after second week.	Closely supervise for the first 3 mo. See text for additional comments.
Range of motion	5-7 d/wk	To light discomfort, or just shy of discomfort	10-20 min	Combined arm and leg ergometry	Increase or maintain range of motion.	Be very careful of hyperextension and hyperflexion.

HRR = heart rate reserve; RPE = rating of perceived exertion; RM = repetition maximum.

Because obesity is prevalent in this population, particularly in people with DS, some modes of exercise (e.g., jogging) may be less enjoyable or more difficult because of orthopedic limitations. Proper progression is critical, and suggested exercise intensities for cardiorespiratory-based exercise are between 40% and 80% of heart rate reserve. Because of the lower maximal heart rates in persons with DS, this may make programming challenging and also requires that maximal heart rates be measured (vs. predicted using a standard age-dependent formula). Using a rating of perceived exertion scale is not recommended. However, if the clinical exercise physiologist does use a rating of perceived exertion scale in higher-functioning individuals, it is important to concomitantly use an additional marker as well (e.g., heart rate). Exercise should take place most days of the week and should be between 30 and 60 min/d. Accumulated exercise is highly recommended, especially in the beginning of a cardiorespiratory exercise program. The advice is to slowly increase intensity after the initial 2 or 3 wk. The goal should be for individuals with ID to be able to exercise 45 to 60 min/d at a moderate to vigorous intensity. If feasible, it may be helpful for them to experience a variety of cardiorespiratory exercises to increase enjoyment and participation.

Resistance Exercise

Weight machines are recommended for resistance training in individuals with ID. Two advantages of resistance machines are that they require less attention to proper form, although this is not absent entirely, and that full balance capabilities are not as necessary. A majority of individuals with ID benefit from exercising six to eight major muscle groups and may not be as interested in many of the specialized machines available at fitness facilities. Using a load sufficient for 15RM to 20RM for the first few weeks, it is suggested that people complete 12 repetitions per set, across two or three sets, 2 or 3 d/wk, with 2 min rest in between sets. After the first 1 or 2 wk, they can progress to a load of 8RM to 12RM after they are comfortable with the machines, demonstrate sufficient technique, and any delayed-onset muscle soreness has largely disappeared. Close supervision is necessary to ensure proper breathing and provide spotting. This level of supervision may be necessary for 3 mo or longer.

Range of Motion Exercise

People with ID are encouraged to engage in stretching exercises most days of the week, for 10 to 20 min per session. Yoga or similar types of exercise may be especially helpful for individuals suffering from some level of depression. Group exercises may not be ideal in this population for improving ROM because routines would need to be highly individualized depending on attention span and physical characteristics. Also it is important to avoid hyperextension and hyperflexion in individuals with DS with associated joint laxity.

EXERCISE TRAINING

Although a multitude of exercise training studies aimed at improving cardiorespiratory fitness in individuals with ID have been conducted, very few have employed appropriate methodologies. Most early studies did not use either validated tests or control groups, and very few have used a randomized controlled design. Nevertheless, the information in the literature is remarkably uniform and is supported by systematic reviews (discussed later). For instance, it was commonly found that both children and adults with ID improved field test or submaximal exercise performance after a standard endurance training program (13, 22, 126, 127). However, since none of these studies used validated test protocols, it is difficult to evaluate whether the findings are indicative of the actual training response among individuals with ID.

Cardiorespiratory Training

Several early studies also showed substantial improvements in $\dot{V}O_2$peak after endurance training in individuals with ID. However, most of these studies had a small number of subjects and did not include a control group, although the uniform response suggested that improvements in $\dot{V}O_2$peak were real (71, 103, 110). Pitetti and Tan (102) provided more conclusive evidence when they evaluated individuals with ID before and after 16 wk of endurance training. Improvements in $\dot{V}O_2$peak between 16% and 43% were observed in these studies with the use of standard exercise prescription procedures. Interestingly, Boer (16) reported that 3 mo of detraining reduced most gains accrued by endurance training and interval training either back to baseline or somewhere in between baseline and the immediate post-training results. This appears to be the only detraining study in the literature in individuals with ID with DS. It would be logical that the same would follow for individuals with ID without DS.

Ozmen and colleagues (92) evaluated the effect of a 10 wk school-based cardiorespiratory fitness training program in 30 children and adolescents randomized to training and control (8-15 yr of age). The training group exercised for 1 h at 60% to 80% of peak heart rate three times per week. After the 10 wk program, the training group significantly improved 20 m shuttle run performance; however, $\dot{V}O_2$peak was not measured.

These studies included only persons with ID who did not have DS. Two early studies in persons with ID with DS suggested that $\dot{V}O_2$peak does not improve with standard endurance training in DS, but that work capacity may in fact improve (84, 130). These investigations suggested that $\dot{V}O_2$peak is immutable to change in response to a standard endurance program. However, other studies have obtained different results. Significant improvements in $\dot{V}O_2$peak have indeed been reported by several groups after typical aerobic training programs (80, 81, 129). These findings are supported by a meta-analysis showing large effect sizes for improvements in both aerobic capacity and work capacity in persons with ID and with DS. Changes in $\dot{V}O_2$peak following endurance training in individuals with DS range from 0% to 27%, with most recent studies demonstrating increases similar to those expected in populations without disabilities (81). According to the available evidence, individuals with ID, both with and without DS, appear to improve aerobic and work capacity with appropriate endurance training. However, many studies had very small sample sizes; several had no control groups; and most did not use randomized controlled designs.

Interval Training

Boer and colleagues (17, 18) reported all-out sprint interval training to be effective in improving anthropometrics and aerobic capacity among individuals with ID, with and without DS, compared with a continuous aerobic training group. Individuals with ID without DS were also observed to have greater improvements in low-density lipoprotein cholesterol and fasting insulin, suggesting a positive effect of interval exercise training on metabolism in this population, which has not been well studied. Functional improvements were similar for 6 min walk distance when comparing the interval training group with the continuous aerobic training group (17), suggesting that interval training may not impede functional improvements, a result often observed with traditional aerobic interventions.

Resistance Exercise Training

Early studies of the effect of resistance training in individuals with ID used field tests including sit-ups, push-ups, and pull-ups to measure muscle performance, but not strength. Also, neither the push-up nor flexed arm hang (a version of the pull-up) is reliable in persons with ID. Nevertheless, substantial improvements (up to 58%) in sit-ups and chin-ups after training, in both children and adults with ID, have been observed (54, 68, 86). Resistance training using surgical tubing has also been shown to improve muscle strength in individuals with ID (31). Although handgrip strength is often tested in individuals with ID, it does not appear to improve with exercise training (86). In summary, nontraditional muscle strength and endurance programs can improve muscle strength and endurance (primarily endurance) in persons with ID.

Standard circuit training types of resistance programs varying in length from 5 wk to 3 mo, using two or three training sessions per week, also improved muscle strength in individuals with ID to an extent similar to that in persons without disabilities (34, 77, 107, 117, 118, 119). The improvements in muscle strength varied between 8% and 82% depending on the muscle group tested and the type of program. There does not appear to be a difference in findings between studies with and without control groups, and compliance has been reported to be high (90%) (117, 118). Furthermore, individuals with ID have continued to improve muscle strength through participation in a self-directed 9 mo program after an initial 12 wk supervised program. These findings have important implications because they show that self-motivated individuals with mild ID who have learned how to safely conduct strength exercises can maintain their strength gains on their own without supervision.

Because individuals with DS have very low levels of muscle strength, and because muscle strength is associated with activities of daily living and aerobic capacity in persons with DS, interventions designed to improve muscle function may be especially important (29). Interestingly, muscle strength has been improved using exercise regimens other than resistance training in persons with DS. These have included high-intensity sport training, training using jumping and balance activities, agility training via virtual reality in young individuals, and a treadmill walking program in older participants (age ~63 yr) (26, 76, 93, 128). These programs improved muscle strength 7% to 25%, probably because participants had a low level of muscle strength before undergoing training.

Several trials (with and without control groups) have investigated the effect of progressive resistance training on muscle strength in persons with DS and found substantial increases in muscle strength (reviewed earlier in the section on endurance training) or muscle endurance (or both) (33, 81, 106, 113, 131). Interestingly, two studies by Shields and colleagues (112, 114) report significant improvements in leg strength in young adults with ID and DS versus a DS-matched nonexercise control group after resistance training, yet no differences between groups were observed for performance tasks (e.g., box stacking, grocery shelving). One important component of these studies is the unique peer-led resistance training, which provides an important social aspect that may have contributed to increased physical activity among the intervention group (114). Cowley and colleagues (29) also found significant improvement in leg strength and stair climbing performance in the exercise group, suggesting the resistance training program positively influenced a functional activity of daily living. Interestingly, although leg strength improved significantly, there was no change in $\dot{V}O_2$peak. This suggests that the association between leg strength and $\dot{V}O_2$peak in persons with DS does not imply a causal relationship.

Overall, it appears that appropriately conducted resistance exercise training programs will improve muscle strength in persons with ID, including those with DS. It has been suggested that programs using training frequencies of 3 d/wk are required (82), but the study by Cowley and colleagues (29) showed significant improvement with a 2 d/wk regimen. Thus, it would appear appropriate to follow standard guidelines to elicit desired resistance training responses in persons with ID.

Combination Training

Earlier work by Elmahgoub and colleagues (35) found that 10 wk of combined strength and endurance exercise (three times per week, 50 min per session) did not change $\dot{V}O_2$peak among adolescents with ID, even though 6 min walk performance increased and muscle strength improved in the exercise training group. However, only 20 min per session was devoted to endurance training. One study examined the effects of combined aerobic, resistance, and balance training among adults with ID on outcomes related to fitness, strength, and balance (91). The participants completed 1 h of combined exercise 3 d/wk over the course of 14 wk, with all components of the training being progressively increased during the intervention. The intervention group had a small but statistically significant drop in weight and BMI, coupled with a 10% increase in fitness and marked improvements in grip and leg strength (91). Furthermore, flexibility and nearly all balance outcome measures increased in this somewhat large study cohort (N =

37). These data support the earlier work of Calders and colleagues (25), which showed 5 mo of combined endurance and resistance exercise training resulted in improvements in aerobic capacity and strength (upper and lower body, handgrip, abdominal, muscular fatigue) as well as decreases in cholesterol and blood pressure. These data suggest that individuals with ID can experience substantial fitness-related benefits when following a well-balanced exercise prescription.

Little work has been done on combined exercise intervention training studies among individuals with DS. Rimmer and colleagues (106) conducted a combined aerobic and resistance exercise training study in older individuals with DS (mean age 39 yr) and reported significant improvements in aerobic capacity, limb strength, and weight lifted. Unfortunately, they used maximal cycle ergometry to evaluate $\dot{V}O_2$peak, which has not been validated and is not reliable in persons with ID (78). Mendonca and colleagues (81) completed a combined aerobic and resistance exercise training study among adults with DS and reported improved aerobic capacity, strength, and walking economy in both individuals with DS and persons without disabilities. This was especially important for individuals with DS who initially exhibited poor walking economy and low levels of strength. Despite improvements of similar magnitude in aerobic capacity and strength, exercise training did not normalize muscle strength and walking economy; these variables were lower in the group with DS than in individuals without disabilities, both before and after training.

Systematic Review

Li and colleagues (75) completed a systematic review of exercise intervention studies among individuals with DS. With 10 studies fulfilling their inclusion criteria, they concluded that exercise interventions were effective in eliciting moderate to high effect sizes (Cohen's $d = 0.74$-1.10) for improvements in muscular strength and balance. Effects on fitness and body composition were mixed, likely because of the limited number of intervention studies that met the authors' inclusion criteria. A systematic review by Bouzas and colleagues (21) is of particular note. Their review analyzed 44 studies that examined the effects of exercise interventions among individuals with ID, with and without DS. This review included randomized controlled trials ($n = 17$), nonrandomized controlled trials ($n = 11$), comparative studies ($n = 5$), and noncontrolled studies ($n = 11$), which were all individually rated on experimental design between excellent to poor. The authors concluded that exercise training improves aerobic capacity and muscular fitness in adults with mild to moderate ID, with more work needed regarding exercise training effects on body composition in this population (21).

Following is a summary of exercise training adaptations in people with ID:

- *Cardiorespiratory:* Most studies show an 8% to 20% increase in aerobic capacity.

- *Skeletal muscle strength:* Strength can improve up to ~80%.

- *Skeletal muscle endurance:* Muscular endurance can improve up to 50% or more.

- *Range of motion:* ROM and balance likely improve with focused training.

- *Body composition:* Small or no changes can be expected.

Research Focus

Chronotropic Incompetence and Plasma Epinephrine and Norepinephrine Levels During Exercise

In a provocative study (39), epinephrine and norepinephrine responses were evaluated at rest and immediately after maximal treadmill exercise in individuals with DS compared with a control group of individuals without disabilities. The rationale for the study was that maximal heart rate is much lower in persons with DS than in the general population, but the reason for the low maximal heart rate is unclear. Prior research had suggested reduced vagal withdrawal during isometric exercise in persons with DS; however, this did not appear to be the case with treadmill exercise. Since plasma catecholamine concentrations are partly responsible for further increases in heart rate during higher exercise intensities, it is possible that reduced levels of plasma epinephrine and norepinephrine contribute to the low maximal heart rate.

Twenty participants with DS (age ~24 yr) and 21 participants without disabilities (age ~26 yr) completed a maximal treadmill exercise protocol with heart rate measured and oxygen uptake measured using indirect open-circuit spirometry. Blood samples were collected before and immediately after exercise and ana-lyzed for plasma epinephrine and norepinephrine. As expected, both maximal heart rate and $\dot{V}O_2$peak were significantly lower in the group with DS (HRmax 170 vs. 189 b · min^{-1}; $\dot{V}O_2$peak 27 vs. 41 mL · kg^{-1} · min^{-1}). There was no difference in resting catecholamine concentrations between groups; but plasma epinephrine did not change from rest to maximal exercise in individuals with DS, whereas it increased almost 900% in the control group. Similarly, plasma norepinephrine doubled from rest to exercise in persons with DS but increased almost 10-fold in the control group. The changes in catecholamines were also associated with maximal heart rate and $\dot{V}O_2$peak.

This study demonstrated that individuals with DS exhibit chronotropic incompetence and lower aerobic capacity and that the attenuation is caused, in part, by the lack of an increase in plasma epinephrine and norepinephrine during maximal exercise. These findings highlight the importance of measuring maximal heart rate and oxygen uptake in this population and provide a physiologic explanation for the low maximal heart rate in persons with DS.

Clinical Exercise Bottom Line

- *The Diagnostic and Statistical Manual of Mental Disorders, Fifth Edition* (DSM-5), defines an ID as intellectual difficulties, coupled with trouble in conceptual, social, and practical terms. Most definitions of ID have moved away from using intelligence quotient (IQ) scores to help define the severity of ID (i.e., mild, moderate, severe, or profound), yet IQ scores generally below 70 are indicative of potential ID and are used as a starting point to assess intellectual functioning.

- Several hundred causes of ID have been identified or suspected. The three major causes are fetal alcohol syndrome, fragile X syndrome, and Down syndrome (DS). Categorical causes of ID fit include genetic, problems during pregnancy, problems at birth, problems after birth, and poverty and cultural deprivation.

- There is no treatment for ID per se. The primary reason for evaluation and classification of individuals with ID is to tailor support specific to each individual, whereby detailed strategies and services are outlined over a sustained time. Increased physical activity or formal exercise programming (or both) is now recognized as an important aspect of maintaining health in persons with ID.

- Many individuals with ID are treated for symptoms of depression with antidepressant medications, and many with DS are treated for thyroid disorders as well. Many also take anticonvulsive agents. These medications help control inappropriate behavior and physiological symptoms caused by the ID but do not affect ID per se Antidepressant medications can have a minor effect on heart rate or blood pressure but generally have little effect on the acute or chronic exercise response. Thyroid medications may increase both heart rate and blood pressure. Use of β-adrenergic blocking agents is not uncommon, particularly in persons without DS. These medications reduce heart rate and blood pressure both at rest and in response to exercise.

- Cardiorespiratory fitness and muscular strength are typically reduced among patients with ID compared against predicted values for age and sex. The conduct of exercise testing and the prescription of exercise in these patients require an appreciation for their ability to follow and understand instructions and cooperate with requested procedures. Regular and positive feedback will enhance results.

CONCLUSION

Familiarizing individuals with ID with the testing or training is of the utmost importance; repetition, practice, and clear instructions can help with this. It also important for them to become familiar and comfortable with testing and training staff because this will facilitate their ability to pay attention to instructions. Do not treat individuals with ID like children, and be careful not to condescend. They understand when this happens and often act out in response. Be patient; people with ID may have difficulties understanding directions and often have a short attention span. Behavioral problems are more frequent than in the general population.

For most individuals with ID, expect lower work and aerobic capacities, maximal HR, and muscle strength, together with a higher prevalence of overweight and obesity. Very low aerobic capacities and muscle strength levels are often observed in people with DS. Measured values are approximately 50% lower than expected for age and sex. Further, maximal heart rates are often ~30 b · min^{-1} lower than expected for age. Individuals with DS may present with congenital heart disease, atlantoaxial instability, or both.

As mentioned at the beginning of this chapter, an ID is not a disease and there is no cure. However, just like people with other chronic diseases or disabilities, people with an ID can both participate in and enjoy the benefits derived from a regular exercise program. This special population of individuals are well deserving of the skills and abilities of the practicing clinical exercise physiologist.

Online Materials

Visit HK*Propel* to access a link to the references, the case studies with discussion questions, and a quiz to help you review key concepts and test your understanding of the material covered in this chapter.

Glossary

α_1-antitrypsin (AAT)—Protein produced in the liver and found in the lungs that inhibits neutrophil elastase.

α_1-antitrypsin deficiency—A genetic disorder characterized by abnormally low levels of α_1-antitrypsin, thereby predisposing an individual to emphysema.

abdominal obesity—Condition characterized by excessive fat on the trunk, also known as android obesity. Increased fat in the abdominal region increases the risk for development of hypertension, type 2 diabetes, dyslipidemia, coronary artery disease, and premature death compared with gynoid obesity (increased fat in the hip and thigh area).

absolute contraindications for test termination—Conditions that occur during a GXT necessitating that the test be terminated.

absolute oxygen uptake—Oxygen uptake expressed in liters per minute ($L \cdot min^{-1}$).

accredited exercise physiologist—University-qualified allied health professionals equipped with the knowledge, skills, and competencies to design, deliver, and evaluate safe and effective exercise interventions for people with acute, subacute, or chronic medical conditions, injuries, or disabilities.

acquired immunodeficiency syndrome (AIDS)—A disease caused by HIV. See chapter 23 for the CDC case definition.

activities of daily living (ADLs)—Bathing, dressing, grooming, toileting, feeding, and transferring.

acute MI—The initial stages of an evolving myocardial infarction (MI).

Adaptive Sports USA—A national 501(c)3 organization, consisting of grassroots programs arranged into 14 regional sports organizations (RSOs) to support sport competition for athletes with disabilities in the United States.

adult variant CF—Adult diagnosis of cystic fibrosis secondary to presentation of respiratory symptoms. The population with this diagnosis has resulted in the increasing numbers of adults with CF.

advanced glycation end products (AGEs)—Harmful compounds that are formed when protein or fat combine with sugar in the bloodstream, in a process called glycation. AGEs are associated with many pathogenic disorders such as Alzheimer's disease, pathogenesis of diabetes, atherosclerosis, and endothelial dysfunction leading to cardiovascular events.

afterload—The pressure against which the pumping chamber or ventricle of the heart must work to eject blood during systole.

aging—The process of growing old.

agonist—A drug or agent that stimulates or enacts the biologic response for a given receptor.

airflow limitation—The blockage of the flow of air out of the lung that can occur secondary to narrowing of the airway lumen.

airway hyperresponsiveness—The ability of the airway wall to be sensitive to various inhalants.

akinesia—Muscle rigidity that often begins in the legs and neck.

akinetic—Denoting loss of movement of a left ventricular wall during the normal cardiac cycle.

amenorrhea—Absence of normal menses; for most studies, a woman is considered amenorrheic if she has fewer than three menses per year.

anabolic steroids—Testosterone derivatives or steroid hormones resembling testosterone that stimulate the building up of body tissues.

androgenic—Having masculinizing effects (i.e., stimulation of male sex characteristics and male hair characteristics).

anemia—A decrease in the red blood cells that carry hemoglobin, resulting in reduced oxygen-carrying capacity.

aneural—Characterized by absence of nerve fibers.

aneurysm—Dilation of an artery that is connected with the lumen of the artery or cardiac chamber. Usually occurs because of a congenital or acquired (e.g., myocardial infarction) weakness in the wall of the artery or chamber. Forms of aneurysms include true, dissecting, and false.

angina pectoris—Constricting chest pain or pressure, often radiating to the left shoulder or arm, back, or neck and jaw regions, caused by ischemia of the heart muscle.

angina threshold—Point at which the supply of oxygen is less than the demand, leading to ischemia and producing symptoms of angina pectoris. Generally observed during physical or mental exertion in patients with significant coronary artery disease.

angiography—Medical imaging technique used to assess the arterial anatomy, often to confirm the findings of noninvasive studies as to the site of a blockage. Broadly includes the following modalities: invasive angiography, computed tomography angiography, and magnetic resonance angiography.

angioplasty—Reconstitution or recanalization of a blood vessel; may involve balloon dilation, mechanical stripping of intima, forceful injection of fibrinolytics, or placement of a stent.

angiotensin-converting enzyme inhibitor—Medication that prevents the conversion of angiotensin I to angiotensin II, which ultimately decreases blood vessel vasoconstriction.

ankle–brachial index (ABI)—Noninvasive peripheral artery disease screening test that uses a blood pressure cuff and Doppler device to assess the systolic blood pressures of the right and left brachial, posterior tibial, and dorsalis pedis arteries. An index value is computed by dividing the highest ankle pressure for each lower extremity by the higher of the two brachial pressures; a value ≤0.90 is considered abnormal.

ankylosing spondylitis—A chronic rheumatic disease that causes inflammation, stiffness, and pain in the spine; the sacroiliac joints (often an early indicator); and in some cases, the neck, hips, jaw, and rib cage. This disease may be accompanied by fever, loss of appetite, and heart and lung problems. It may cause spinal deformities and eventually causes the spinal segments to fuse (ankylose), with the result that the back assumes a fixed rigid posture.

ankylosis—Immobility and consolidation of a joint due to disease, injury, or surgery.

anorexia nervosa—Loss of appetite associated with intense fear of becoming obese. Can lead to life-threatening weight loss, disturbed body image, hyperactivity, and amenorrhea.

anovulatory—Not accompanied by the discharge of an ovum.

antagonist—A drug or agent that reduces, blocks, or inhibits the biologic response of a given receptor.

anterior cord syndrome—Symptoms of anterior spinal cord injury (T10-L2), including pain, impairment, and temperature deficit below level of injury. Symptoms are the result of an anterior cord lesion.

anthropometric index—A measure of body composition developed for those with a spinal cord injury that uses an estimation of body density using skinfolds and waist and calf circumferences.

anticoagulation therapy—Pharmacological delaying or preventing of blood coagulation (clotting).

antioxidant—An agent that inhibits oxidation and thus prevents the deterioration of other materials through oxidative processes.

antiresorptive therapy—The group of drug therapies currently available for treatment of osteoporosis. The term originates from the fact that the therapies halt the loss of bone by inhibiting bone resorption.

antiretroviral therapy (ART)—A broad category of pharmacologic agents used in HIV treatment regimens.

anuria—Suppression or arrest of urinary output, resulting from impairment of renal function or from obstruction in the urinary tract.

anxiety—A complex psychophysiological response to an environmental stressor, disaster, or trauma; more often manifested in people who are genetically vulnerable to the disorder.

APGAR—A noninvasive clinical test, designed by Dr. Virginia Apgar (1953), carried out immediately on a newborn. The name is also an acronym for activity (muscle tone), pulse, grimace (reflex irritability), appearance (skin color), and respiration. A score is given for each sign at 1 min and 5 min after birth.

aphasia—Partial or total loss of the ability to articulate ideas or comprehend spoken or written language, resulting from damage to the brain caused by injury or disease.

apolipoproteins—Proteins associated with lipoproteins; these proteins stabilize the lipid portion of the lipoprotein in the circulation. Apolipoproteins serve as ligands for cell receptor binding and as cofactors in enzyme reactions.

areflexia—Absence of neurologic reflexes, usually a sign of peripheral nerve damage affecting muscular, bowel, bladder, and sexual function.

arterial intimal layer—The innermost layer of an artery, composed of endothelial cells.

articular—Relating to a joint.

ASCVD—Atherosclerotic cardiovascular disease; cardiovascular disease resulting from the accumulation of atherosclerotic plaque.

associated conditions—Conditions that accompany a primary disability but are not necessarily preventable. They can, however, be controlled with medication, surgery, or medical devices.

asthma—A continuum of disease processes characterized by inflammation of the airway wall.

astrocyte—Star-shaped neural cell that provides nutrients, support, and insulation for neurons of the central nervous system.

astrocytic gliosis—Proliferation of astrocytes in damaged areas of the central nervous system, forming scar tissue.

asymmetric—Denoting lack of symmetry between two or more like parts.

ataxia—Inability to coordinate voluntary muscle movements; unsteady movements and staggering gait or loss of ability to coordinate muscular movement most often caused by disorders of the cerebellum or the posterior columns of the spinal cord; may involve the limbs, head, or trunk.

atherectomy—Procedure used for revascularization of an obstructed coronary artery; uses a catheter tipped with either a metal burr that grinds a calcified atheroma (rotational atherectomy) or a rotating cup-shaped blade housed in a windowed cylinder that cuts or shaves the atheroma (directional atherectomy).

atherogenic—Having the capacity to initiate, increase, or accelerate the process of atherosclerosis.

atherogenic index of plasma (AIP)—A novel index composed of triglycerides and high-density lipoprotein cholesterol. AIP is an indicator involved in dyslipidemia, and it is associated with cardiovascular diseases.

atheroma—An accumulation of lipid in vascular walls. Atheromas are also called fatty streaks and atherosclerotic lesions.

atherosclerosis—An extremely common form of arteriosclerosis in which deposits of yellowish plaques (atheroma) containing cholesterol, lipid material, and lipophages are formed within the intima and inner media of large and medium-sized arteries.

atherosclerotic—Relating to or characterized by atherosclerosis.

athetosis—A constant succession of slow, writhing, involuntary movements of flexion, extension, pronation, and supination of fingers and hands and sometimes of toes and feet.

atlantoaxial instability—Increased flexibility in the atlantoaxial joint, which is actually a composition of three joints: two lateral and one median atlantoaxial joint, the joint where the base of the skull meets the neck. Because of its proximity to the brain stem and importance in stabilization, fracture or injury at this level can be catastrophic. Because of ligament laxity, instability is not uncommon in Down syndrome.

atrophy—Partial or complete wasting away of a part of the body, as from disuse.

auscultation—Listening to the sounds made by various body structures as a diagnostic method.

autoimmune—Referring to cells or antibodies that arise from and are directed against the person's own tissues, as in autoimmune disease.

automatic internal cardioverter-defibrillator (AICD or ICD)—An implantable battery-powered generator used in patients who are at risk for sudden death due to ventricular fibrillation or tachycardia. The device can detect life-threatening cardiac arrhythmias and can deliver a jolt of electricity designed to stop the arrhythmia.

autonomic dysreflexia (AD)—Sudden, exaggerated reflex increase in blood pressure in persons with a spinal cord injury above T6, sometimes accompanied by bradycardia, in response to a noxious stimulus originating below the level of a spinal cord injury.

autonomic nervous system—Components of the nervous system responsible for coordinating life-sustaining processes and organizing visceral responses to somatic reactions.

avascular—Denoting absence of blood vessels.

AV synchrony—The sequence and timing of the atria and ventricles during systole.

balance—The ability to make adjustments to maintain body equilibrium.

ballistic flexibility (stretching)—Stretching using active muscle movement with a bouncing action.

bariatric surgery—Surgical intervention using one of several methods designed to assist morbidly obese people in losing weight.

basal ganglia—The caudate and lentiform nuclei of the brain and the cell groups; all of the large masses of gray matter at the base of the cerebral hemisphere; the large masses of gray matter at the base of the brain that, if damaged, would impair motor abilities.

behavior therapy—Strategies, based on learning principles, that provide tools for overcoming barriers to compliance.

β_2-agonist—A drug or hormone capable of combining with β-receptors to initiate drug actions.

β-blocker—Medication used to block β-receptors in the myocardium, which decreases myocardial work by decreasing heart rate and myocardial contractility.

β-receptor—A cell receptor that is activated by a β-agonist such as epinephrine, norepinephrine, or dopamine.

bifurcation—Point at which an artery branches to form two arteries.

biologics—A recently developed class of drugs that interfere with biologic substances causing or exacerbating inflammation; cytokine or T cell or B cell modulators used to suppress the rheumatic disease process.

biopsy—A surgical procedure whereby a sample of tissue is obtained.

biotherapy—Stimulation of the body's immune response system to cancer-specific protein antigens.

blood pressure (BP)—The force of circulating blood on the walls of the blood vessels as it circulates throughout the body. Chronic elevated blood pressures (systolic blood pressure $\geq$140 mmHg or diastolic blood pressure $\geq$90 mmHg) are associated with increased risk for cardiovascular disease.

body mass index (BMI)—Relative weight for height; weight in kilograms divided by height in meters squared ($kg \cdot m^{-2}$).

bone formation—Also called bone remodeling; the process by which new bone is formed and deposited within the existing bone matrix. Bone formation is accomplished primarily by bone cells called osteoblasts.

bone geometry—The overall cross-sectional area of the bone; the cross-sectional area of the outer cortex, the number of cross-links in the trabecular bone, and other related factors. With reference to the femoral neck, also includes the angle that the neck of the femur makes with the shaft of the femur.

bone mineral content—A measurement of the total amount of hydroxyapatite (calcium phosphate crystal) of bone, expressed as $g \cdot cm^{-2}$; synonymous with bone mass.

bone mineral density (BMD)—Relative amount of bone mineral per measured bone width. Values are expressed as $g \cdot cm^{-2}$.

bone modeling—Alterations in the shape of the bone such as changes in length.

bone remodeling—A constant state of formation and resorption.

bone resorption—The process of eroding old bone from the existing bone matrix so that new bone can be formed in its place. Resorption is accomplished primarily by bone cells called osteoclasts. Bone resorption is greatly increased in estrogen-depleted women.

botulinum toxin—A neurotoxin that blocks the release of acetylcholine from the motor endplates of the lower motor neuron at the myoneural junction, thereby preventing muscle contraction.

bradycardia—A heart rate less than 60 b $\cdot$ min^{-1}.

bradykinesia—Slow movement often associated with an impaired ability to adjust body position. It is a symptom of a nervous system disorder or can be a side effect of medications.

bradypnea—A respiratory rate less than 8 breaths/min.

bronchi—Large airways of the lungs.

bronchiectasis—Chronic dilation of a bronchus usually associated with secondary bacterial infection.

bronchioles—Small airways of the lungs.

bronchodilator—A drug that relaxes the smooth muscles surrounding the bronchi and bronchioles.

bronchoprovocation—A type of pulmonary function testing in which a particular medication (e.g., methacholine) is aerosolized to induce bronchospasm.

bronchospasm—Spasmodic contraction of the smooth muscle of the bronchi, as occurs in asthma.

Brown-Sequard syndrome—Symptoms of a unilateral spinal cord injury, including ipsilateral proprioceptive and motor deficit, as well as contralateral pain impairment and temperature deficit below the level of injury.

bruits—Acquired sounds of venous or arterial origin caused by turbulent blood flow, heard by auscultation.

bulimia—Disorder that includes recurrent episodes of binge eating, self-induced vomiting and diarrhea, excessive exercise, strict diet, and exaggerated concern about body shape.

bursitis—Inflammation of one of the fluid-filled sacs located at sites of friction surrounding the joint.

calcitonin—Hormone that is responsible for calcium regulation and inhibits bone resorption.

cardiac activation—The biological processes resulting in the formation of electrical impulses that regulate the heartbeat, either normal or abnormal.

cardiac rehabilitation (CR)—The provision of comprehensive long-term services involving medical evaluation, prescriptive exercise, cardiac risk-factor modification, education, counseling and behavioral interventions.

cardiac resynchronization therapy (CRT)—The use of a biventricular pacemaker to restore the coordinated (or synchronized) pumping action of the ventricles when electrical conduction is delayed by bundle branch block, a common feature of chronic heart failure.

cardiac tamponade—Accumulation of fluid in the pericardial sac that may compress the ventricles and markedly reduce cardiac output.

cardiogenic shock—Lack of cardiac and systemic oxygen supply resulting from a decline in cardiac output secondary to serious heart disease; typically follows a myocardial infarction.

cardiometabolic risk factors—Collection of health-related variables that increase one's risk for both cardiovascular and metabolic diseases (hypertension, obesity, high triglycerides, glucose intolerance).

cardiorespiratory fitness—Also known as aerobic capacity; the ability of the body to perform higher-intensity activity for a prolonged period without undue physical stress or fatigue; ability of the body to transport and utilize oxygen.

cardiovascular autonomic neuropathy—Neural damage to the autonomic nerves of the cardiovascular system, which can result in a high resting and low peak exercise heart rate and severe orthostatic hypotension.

cardiovascular disease (CVD)—A group of disease conditions that affect the heart, arteries, or veins of the circulatory system or more than one of these; the term most often refers to the atherosclerotic process occurring in the coronary arteries of the heart.

catecholamines—Chemicals released in the body that are major elements in the response to stress and exercise. Two catecholamines of interest are epinephrine and norepinephrine. Both exert, among other effects, a positive inotropic and chronotropic effect on cardiac function.

cauda equina—Lumbosacral spinal nerve roots forming a cluster at the terminal region of the spinal cord that resembles a horse's tail.

cauda equina syndrome—Severe compression of the cauda equina, resulting in loss of bowel or bladder function, loss of sensation in the buttocks and groin, and weakness in the legs. Cauda equina syndrome requires an emergency surgical intervention to prevent permanent damage.

CD4 cells—A membrane receptor found on T-helper lymphocytes (or T4 cells); the preferred target of HIV.

CD4+ T lymphocytes—Immune cells that express the CD4 glycoprotein on their surface. Without treatment, HIV infection leads to a progressive decline in T cells expressing the CD4 glycoprotein.

central cord syndrome—Symptoms of incomplete spinal cord injury including weakness and sensory deficits in the upper extremities (less than in the lower extremities).

central nervous system (CNS)—The brain and the spinal cord.

cerebellum—A trilobed structure of the brain, lying posterior to the pons and medulla oblongata and inferior to the occipital lobes of the cerebral hemispheres; responsible for the regulation and coordination of complex voluntary muscular movement as well as maintenance of posture and balance.

cerebral cortex—The thin, convoluted surface layer of gray matter of the cerebral hemispheres that consists of the frontal, parietal, temporal, and occipital lobes.

chemotherapy—Use of chemical agents to kill rapidly growing cancer cells.

children—Persons between infancy and adolescence.

cholesterol—A fatlike, waxy substance found throughout the body. High levels of cholesterol in the blood ($>200 \text{ mg} \cdot \text{dL}^{-1}$) increase the risk of developing cardiovascular disease.

cholesterol ester transfer protein (CETP)—Protein in the circulation that transfers nonpolar lipids, cholesterol esters and triglycerides between lipoproteins.

chorea—State of excessive, spontaneous movements, irregularly timed; causes nonrepetitive and abrupt motions and inability to maintain voluntary muscle contraction.

chromosomal microarray—A molecular–cytogenetic method for the analysis of copy number changes (gains and losses) in the content of a given subject's DNA.

chronic bronchitis—Disease characterized by the presence of a productive cough on most days during 3 consecutive months in each of 2 successive years.

chronicity—The state of being chronic, or long in duration.

chronic kidney disease (CKD)—A chronic disease condition that is associated with kidney damage.

chronic MI—The latest phase of a myocardial infarction, during which the heart is stable.

chronic obstructive pulmonary disease (COPD)—Presence of airflow obstruction attributable to either chronic bronchitis or emphysema.

chronotropic assessment exercise protocol—Treadmill protocol used to determine whether heart rate response is appropriate throughout the length of the exercise test.

chronotropic incompetence—Lack of an appropriate increase in heart rate with physical exertion. Considered an abnormal response if peak heart rate does not reach two standard deviations of the person's age-predicted maximum heart rate, assuming the patient was highly motivated and not on medications that blunt heart rate response (i.e., β-blockers, calcium channel blockers).

chylomicron—Relatively large, triglyceride-rich lipoprotein secreted by the intestine after digestion and absorption of food. Chylomicrons originate from intestinal absorption of dietary or exogenous triglyceride.

circulating insulin—Insulin available for use by body tissues.

circulatory hypokinesis—Insufficient vascular tone resulting in hypotension; occurs when increased metabolic demands of upper extremity exertion are not matched by appropriate hemodynamic responses.

claudication—Limping, lameness, and pain that occur in individuals who have an ischemia response in the muscles of the legs, which is brought on with physical activity (e.g., walking). A scale can be used to determine the severity of claudication.

claudication onset time—The time point or distance at which the patient first experiences claudication pain in the calves, thighs, or buttocks during walking. Often used as a secondary outcome measure to evaluate walking performance in peripheral artery disease studies.

clinical exercise physiologist—A certified health professional that utilizes scientific rationale to design, implement and supervise exercise programming for those with chronic diseases, conditions, and physical shortcomings.

clonus—An abnormality in neuromuscular activity characterized by rapidly alternating muscular contraction and relaxation; a form of movement marked by contractions and relaxations of a muscle, occurring in rapid succession, after forcible extension or flexion of a part. Also called clonospasm.

clubbing—Rounding and enlargement of the distal-most parts of the fingers, usually most prominently under the fingernails.

clubfoot—Also known as talipes equinovarus, a congenital deformity affecting one or both feet characterized by the heel pointing downward and the forefoot turning inward. The heel cord (Achilles tendon) is tight, causing the heel to be drawn up toward the leg.

cognitive–behavioral therapy (CBT)—For patients with major depressive disorder, treatment that focuses on modifying maladaptive thoughts and addresses deficits in behavior (e.g., unassertiveness, isolating oneself from others).

community mobility—Locomotion and transportation of individuals through their community.

comorbidity—A concomitant but unrelated pathologic or disease process, usually with reference to the coexistence of two or more disease processes.

complete paraplegia—Motor and sensory dysfunction of the trunk, legs, and pelvic organs resulting from spinal cord injury, without motor or sensory sparing below the level of the injury.

complete tetraplegia—Motor and sensory dysfunction of the arms, trunk, legs, and pelvic organs resulting from spinal cord injury, without motor or sensory function sparing below the level of the injury.

computed axial tomography (CAT or CT scan)—Tomography (moving X-ray tube and film) whereby a pinpoint radiographic beam sweeps transverse planes of tissue and a computerized analysis of the variance in absorption produces a precise image of that area.

conduction block—A disease of the electrical system of the heart that, depending on the severity, impedes or completely blocks the conduction of the electrical impulses that initiate the contraction of heart muscle.

congenital heart disease—A defect in the structure of the heart and great vessels present at birth.

congestive or chronic heart failure (CHF)—The symptom complex associated with shortness of breath, edema, and exercise intolerance attributable to abnormal left ventricular function, resulting in congestion of fluid in other bodily organs.

contracture—An abnormal, often permanent shortening, as of muscle or scar tissue, that results in distortion or deformity, especially of a joint of the body. Shortening of a muscle group and tendon is usually observed in persons with spasticity.

conus medullaris syndrome—Symptoms of upper and lower motor neuron damage to the conus medullaris, including bowel, bladder, and lower extremity areflexia and flaccidity and preserved or facilitated reflexes.

coronary dissection—Separation of tissue within the lining of a coronary artery.

coronary revascularization—Establishment of restored blood flow through a stenosed coronary artery via catheter intervention (angioplasty, stenting, thrombectomy) or coronary artery bypass graft surgery.

cortical bone—One of the two main types of bone tissue; hard, compact bone found mainly in the shafts of long bones. The other type is trabecular.

coxa valga—A hip deformity produced when the angle of the head of the femur with the shaft exceeds 120°. The greater the degree of coxa valga, the longer the resulting limb length.

coxa vara—A hip deformity produced when the angle made by the head of the femur with the shaft is below 120°. In coxa vara, it may be 80° to 90°. Coxa vara occurs in rickets or may result from bone injury. The affected leg appears shortened, resulting in a limp.

C-reactive protein (CRP)—A β-globulin found in the serum of persons with certain inflammatory, degenerative, and neoplastic diseases. CRP levels are often detectable in the blood of individuals with metabolic syndrome, suggesting chronic inflammation.

creatinine—End product of creatine metabolism excreted in the urine at a constant rate; a blood marker of renal function.

creatinine clearance—An index of the glomerular filtration rate, calculated by multiplying the concentration of creatinine in a timed volume of excreted urine by the milliliters of urine produced per minute and dividing the product by the plasma creatinine value.

crepitus—Crackling from the joint palpated on examination.

critical limb ischemia (CLI)—The most severe form of peripheral artery disease; generally affects the tibial and pedal arteries of the lower extremity. Can cause lower extremity rest pain, nonhealing ulcer wounds in the foot, and gangrene. Critical limb ischemia may ultimately lead to amputation of the lower extremities.

cross-training—The concept of training in one mode that allows for the development of physiology that will have a carryover effect to another mode; for example, resistance training is often performed to develop sport-specific strength.

culprit lesion—The primary obstruction responsible for decreased blood flow through a coronary artery.

curative surgery—Surgery aimed at complete removal of a tumor along with a small amount of surrounding normal tissue.

cystic fibrosis–related diabetes—Diabetes associated with scarring of the pancreas that prevents insulin from entering the bloodstream and results in abnormal blood glucose levels.

cystic fibrosis transmembrane conductance regulator (CFTR)—A protein that is altered secondary to CF; the gene for CF that is located on chromosome 7 and results in altered production of CFTR, a protein that functions as a chloride channel regulated by cyclic adenosine triphosphate. This protein causes abnormal sodium chloride and water movement across the cell membrane, resulting in thick, dry mucus.

cysts—Abnormal sacs containing gas, fluid, or a semisolid material, with a membranous lining.

cytokines—Nonantibody proteins, released by one cell population on contact with a specific antigen, that act as intercellular mediators.

cytomegalovirus—One of a group of highly host-specific herpes viruses.

dementia—A progressive decline in mental function, memory, and acquired intellectual skills.

demyelination—The loss of the myelin covering that insulates the nerve tissue.

dermatomes—An area of skin that is primarily innervated by a single spinal nerve.

diabetic ketoacidosis—A type of metabolic acidosis caused by accumulation of ketone bodies in diabetes mellitus.

dialysate—The fluid that is on the opposite side of the dialyzer membrane from the blood during dialysis. It contains the substances that can freely diffuse across the membrane and it is meant to maintain a normal electrolyte balance. Small uremic toxins, excess fluid, and abnormally high electrolytes are removed during the process.

dialysis—A method used to remove excess fluid and uremic toxins from the extracellular fluid when the kidneys are no longer capable of performing this function adequately.

diastolic—Referring to the pressure remaining in the arteries after cardiac contraction.

diastolic dysfunction—Most often with reference to left ventricular function, a stiff or less compliant chamber that is partially unable to relax and expand as blood flows in during diastole.

differential diagnosis—The determination of which of two or more diseases with similar symptoms is the one from which the patient is suffering, through systematic comparison and contrasting of the clinical findings.

digoxin—A cardioactive steroid glycoside used to increase myocardial contractility.

diplegia—Paralysis of corresponding parts on the two sides of the body. Fine motor function in the upper extremities may be affected and the trunk may be slightly affected, but primarily the legs are affected.

diplopia—Double vision caused by a disorder of the nerves that innervate the extraocular muscles or by impaired function of the muscles themselves.

disability—Loss of physical function.

discitis—Inflammation of an intra-articular disc.

disease-modifying antirheumatic drugs (DMARDs)—A category of otherwise unrelated drugs that suppress the rheumatic disease process by diminishing the immune response. Can improve not only the symptoms of inflammatory joint disease but also some of the extra-articular manifestations, such as vasculitis.

disimpaction—Manual removal of fecal material from the rectal vault.

disordered eating—Inappropriate eating behaviors leading to insufficient energy intake.

distal—Away from the origin or center line, as opposed to proximal.

Doppler ultrasonography—Application of the Doppler effect in ultrasound to detect movement of scatterers (usually red blood cells) through analysis of the change in frequency of the returning echoes.

dorsal rhizotomy—A surgical procedure used to treat spasticity, particularly in young children, usually between 2 and 8 yr old, with cerebral palsy; often referred to as selective dorsal rhizotomy. This surgical procedure permanently reduces spasticity by selectively cutting the abnormal sensory nerve rootlets.

Down syndrome (DS)—A chromosomal condition caused by the presence of all or part of an extra 21st chromosome. It is named after a British physician, John Langdon Down, who first described the syndrome in 1866.

drug-eluting stent—Stent that slowly releases a drug (sirolimus), resulting in a reduction of restenosis rates.

dual-chamber pacemaker—Pulse generator that can pace or sense in the atrium or ventricle.

dual-energy X-ray absorptiometry (DXA)—A method for measuring bone mineral density and bone mineral content. It is based on the amount of radiation absorption, or attenuation, in body tissues. When bone mass is measured, the higher the attenuation of radiation by the bone, the greater the mass. Radiation exposure is minimal (<5 mR) compared with that from a chest X-ray (100 mR) or lumbar X-ray (600 mR).

dual-photon absorptiometry—A method similar to DXA for measuring bone density but one that relies on a radionuclide source as opposed to X-ray. The photon intensity is not as great as with DXA, and precision is therefore reduced.

Duke nomogram—Five-step tool to estimate a person's prognosis (5 yr survival or average annual mortality rate) following completion of a maximal GXT.

dynamic endurance—Classification of exercise in which concentric–eccentric shifting occurs until muscular fatigue is induced. An example of dynamic endurance is performing biceps curls until fatigue occurs and the subject is unable to continue full motion against resistance.

dynamic flexibility (stretching)—Slow and constant stretch held for a period of time.

dynamic pulmonary function—Assessment of pulmonary status during exercise. Dynamic pulmonary function includes measures of oxygen consumption, minute ventilation, and ventilator equivalents. These measures help identify ventilator limitations to stress.

dynapenia—The age-associated loss of muscle strength that is not caused by neurologic or muscular disease.

dysarthria—Difficulty speaking because of impairment of the tongue and other muscles essential for speech.

dysesthetic—Referring to an abnormal, unpleasant spontaneous or evoked sensation or pain, caused by lesions of the peripheral or central nervous system; involves sensations such as burning, wetness, itching, electric shock, and "pins and needles."

dyskinesia—Abnormality or impairment of voluntary movement.

dyskinetic (pertaining to cerebral palsy)—Characterized by an abnormal amount and type of involuntary motion with varying amounts of tension, normal reflexes, and asymmetric involvement.

dyskinetic (pertaining to the heart)—Denoting an outward or bulging movement of the myocardium during systole; often associated with aneurysm.

dyslipidemia—Plasma lipid disorders resulting in abnormal lipid profiles.

dyslipoproteinemia—Abnormally elevated or reduced lipoprotein concentrations.

dysmenorrhea—Pain in association with menstruation.

dyspareunia—Pain in the labial, vaginal, or pelvic areas during or after sexual intercourse.

dyspepsia—Stomach discomfort, including symptoms such as heartburn, gas, and acid reflux.

dysphagia—Difficulty in swallowing.

dyspnea—Shortness of breath or labored or difficult breathing that is perceived by an individual at rest or with exertion (also referred to as dyspnea on exercise, DOE). A scale can be used to determine the severity of dyspnea.

dystonia—Sustained muscle contractions that result in twisting and repetitive movements or abnormal posture.

eccentric lesion—A blockage that is equal distance away from the center of the artery—around the lining of the artery.

echocardiogram—An investigation of the heart and great vessels with ultrasound technology as a means to diagnose cardiovascular abnormalities.

echocardiography—Use of ultrasound images to evaluate the heart and great vessels.

economy—The rate of oxygen uptake necessary to perform a given activity.

ectopic adiposity—Accumulation of fat in various nonadipose tissue depots, such as the liver and skeletal muscle.

ectopic pregnancy—Implantation of the fertilized ovum outside of the uterine cavity.

edema—A condition in which body tissue contains an excessive amount of fluid.

effusion—Excess synovial fluid within a joint.

ejection fraction (EF)—Percentage of blood that is ejected from the left ventricle per beat (normal 55%-60%); EF = [(EDV − ESV) / EDV] × 100, where EDV = end-diastolic volume and ESV = end-systolic volume. Decreases are noted with systolic heart failure to values below 35% to 40%.

elastin—Structural protein found in the walls of the alveoli.

elderly—Past middle age; relating to later life.

electrodiagnostic—A process that measures the speed and degree of electrical activity in muscles and nerves to diagnose symptoms such as pain, weakness, or numbness in the back, neck, or hands.

electromechanical dissociation (pulseless electrical activity)—Any heart rhythm observed on the electrocardiogram that does not result in a pulse.

electromyography—A diagnostic neurological test to study the potential (electrically measured activity) of a muscle at rest, the reaction of muscle to contraction, and the response to muscle insertion of a needle. The test is an aid in ascertaining whether a patient's illness is directly affecting the spinal cord, muscles, or peripheral nerves.

ELISA—Enzyme-linked immunosorbent assay; the most commonly used test for the presence of HIV antibodies.

embolic stroke—Emboli that cause embolic strokes are typically from the carotid or other arteries. In these cases, the thrombus is not large enough to occlude the large vessel, but the embolus that breaks off ultimately lodges in a smaller cerebral artery or arteriole.

embolism—Obstruction of a blood vessel by foreign substances or a blood clot.

emphysema—Disease characterized by abnormal permanent enlargement of the respiratory bronchioles and the alveoli.

endocrine—Referring to glands that secrete hormones into the bloodstream.

endothelial cell—One of the squamous cells forming the lining of serous cavities, blood, and lymph vessels and the inner layer of the endocardium.

endothelial-derived relaxing factors—Diffusible substances produced by endothelial cells that cause vascular smooth muscle relaxation; nitric oxide is one such substance.

endothelium—A thin layer of cells that line the inner surface of blood vessels.

endovascular revascularization—Percutaneous, catheter-based interventions used to treat obstructive blood vessel disease. Endovascular treatment options include but are not limited to stenting, angioplasty, and atherectomy procedures.

end-stage renal disease (ESRD)—The final stage of chronic kidney disease in which kidney function has deteriorated so drastically that patients must either be dialyzed or receive a kidney transplant or expire from uremia.

enteral—Referring to a route for administration of a drug that is through the gastrointestinal tract.

entheses—Sites where ligaments, tendons, or joint capsules are attached to bone.

enthesopathy—Inflammation at entheses.

environmental factors—Physical and social factors that can influence participation in physical activity (e.g., vehicular traffic, inclement weather, and unsafe neighborhoods).

epilepsy—The paroxysmal transient disturbances of brain function that may be manifested as episodic impairment or loss of consciousness, abnormal motor phenomena, psychic or sensory disturbances, or perturbation of the autonomic nervous system.

epistaxis—Bleeding from the nose.

erythrocyte sedimentation—The sinking of red blood cells in a volume of drawn blood.

erythropoiesis—Stimulation of red blood cell production.

estrogen replacement therapy—Therapy useful for protecting against bone loss in postmenopausal women.

etiology—Cause.

evidence based—Referring to use of the best available clinical research to guide treatment.

evoked response testing—Test in which brain electrical signals are recorded as they are elicited by specific stimuli of the somatosensory, auditory, and visual pathways.

evolving MI—Period of time after the acute onset of an MI when the myocardial tissue is transforming from ischemic to necrotic tissue.

exacerbation—A period of worsening symptoms.

excess body weight—Condition that results when too few calories are expended and too many consumed for individual metabolic requirements; overweight (>25) and obesity (≥30) as defined by body mass index (kg · m^{-2}).

exercise intolerance—Condition in which the individual is unable to perform physical exercise at the level that would be expected for the person's age, gender, and comorbid conditions. It is not a disease in itself but a symptom, either of worse than expected underlying disease or of deconditioning.

exercise physiologist—Fitness professional with a minimum of a bachelor's degree in exercise science qualified to pursue a career in university, corporate, commercial, hospital, and community settings.

exercise prescription—A specific plan of exercise activities that form the basis of an exercise program. The term *prescription* implies that the plan is based on objective criteria from an exercise test and that the activities are designed to restore health (e.g., after a cardiac event) or prevent disease (e.g., by reducing cardiac risk factors).

exercise testing—Measuring the body's reaction to increases to an exercise challenge. Most commonly, testing the cardiovascular response to exercise in terms of ability to perform work; blood pressure, heart rate, and electrocardiographic responses; and development of symptoms.

exertional ischemia—Myocardial ischemic response produced by exerting oneself physically.

extrapyramidal—Denoting the area of the brain that includes the basal ganglia and the cerebellum.

fatigue—After a period of mental or bodily activity, a state characterized by a lessened capacity for work and reduced efficiency of accomplishment, usually accompanied by a feeling of weariness, sleepiness, or irritability; may also supervene when, from any cause, energy expenditure outstrips restorative processes, and may be confined to a single organ.

fatty liver disease—Also known as steatorrhoeic hepatosis. This condition is characterized by the accumulation of triglycerides in the cells of the liver. Fatty liver disease is most commonly associated with excessive alcohol intake or obesity (nonalcoholic fatty liver disease).

fetal alcohol syndrome—A pattern of mental and physical defects that can develop in a fetus in association with high levels of alcohol consumption during pregnancy. Often leads to intellectual disability and behavioral problems.

fibrinolysis—The process of dissolving a coronary artery thrombosis with either an intrinsic thrombolytic peptide or a thrombolytic medication.

fibrinolytic—Causing fibrinolysis, which is the breakdown of fibrin, a blood-coagulating protein.

fibromyalgia—Condition featuring chronic widespread pain and diffuse tenderness at discrete anatomical sites. Generally associated with mood and sleep disturbances and debilitating fatigue.

flaccidity—Lacking muscle tone (opposite of spasticity).

foam cells—Smooth muscle cells that take up intimal lipid when it accumulates in the cytoplasm and have a bubbly appearance when observed microscopically.

follicle-stimulating hormone—Hormone produced by the anterior pituitary to stimulate the growth of the follicle in the ovary and spermatogenesis in the testes.

forced expiratory volume in 1 s (FEV$_1$)—Marker of airway obstruction; the maximum amount of air that can be exhaled in 1 s; may be expressed as an absolute value, a percentage of the forced vital capacity, or a percentage of a predicted value.

forced vital capacity (FVC)—The maximum amount of air that can be exhaled forcefully after a maximal inspiration.

four-compartment body composition modeling—Method of studying body composition by determining the contents of four body compartments: fat, protein, water, and mineral.

fragile X syndrome—Also called Martin–Bell syndrome or Escalante's syndrome. This is a genetic condition that is the most commonly known single-gene cause of autism and the most common inherited cause of intellectual disability. Results in physical, intellectual, emotional, and behavioral limitations.

frailty—The state of having delicate health.

frontostriatal—Circuits that are neural pathways connecting frontal lobe regions with the basal ganglia (striatum) that mediate motor, cognitive, and behavioral functions within the brain, which are part of the executive functions.

functional aerobic impairment (FAI)—Percentage of an individual's observed functional capacity that is below that expected for the person's sex, age, and conditioning level. %FAI = (predicted $\dot{V}O_2$ – observed $\dot{V}O_2$) / (predicted $\dot{V}O_2$) × 100.

functional capacity—A person's maximum level of oxygen consumption; can be measured at maximal effort with the use of a metabolic cart or predicted based on the maximum workload achieved.

functional electrical stimulation (FES)—Externally applied electrical stimulation of neuromuscular elements to activate paralyzed muscles in a precise sequence and at a precise intensity to restore muscular function.

functional fitness—Ability of an individual to perform daily tasks based on balance, risk for falling, muscular function, and mobility; associated with common tasks such as walking, moving from sitting to standing, and moving oneself up from sitting on the floor to standing.

functional food—A food that offers health benefit beyond basic nutrition; usually minimally processed and often contains added nutrients such as vitamins or fiber.

gangrene—Necrosis of body tissues caused by obstruction, loss, or diminution of blood supply.

gastroepiploic artery—An artery with its origin in the stomach region used for coronary revascularization surgery.

gel phenomenon—The sensation of difficulty moving a joint after a period of joint rest or immobility.

genotype—The resultant expression of specific genes.

genu valgum—More commonly referred to as knock-knee deformity; a condition in which the knees angle in and touch when the legs are straightened.

genu varum—More commonly referred to as bowleg deformity; a condition in which the medial angulation of the leg in relation to the thigh results in an outward bowing of the legs.

geriatrics—A branch of medicine that deals with the problems and diseases associated with elderly people (>65 yr) and the aging process.

gestational diabetes—Carbohydrate intolerance of variable severity with onset or first recognition during pregnancy.

ghrelin—A potent appetite-increasing gut hormone.

gliosis—Excess of astroglia in damaged areas of the central nervous system.

glomerular filtration rate (GFR)—The amount of fluid the kidneys filter each minute.

glomerulonephritis—An acute, subacute, or chronic, usually bilateral, diffuse inflammatory kidney disease that primarily affects the glomeruli.

glucagon—A hormone produced by the pancreas that stimulates the liver to release glucose, causing an increase in blood glucose levels and thus opposing the action of insulin.

glucose intolerance—A transitional state between normoglycemia and diabetes. Diagnosed when fasting blood glucose levels are ≥ 100 mg $\cdot$ dL^{-1} but <126 mg $\cdot$ dL^{-1}, or when glucose levels are between 140 and 199 mg $\cdot$ dL^{-1} 2 h after a 75 g oral glucose tolerance test.

GLUT 4—Insulin-regulated glucose transporter responsible for the removal of glucose from blood and delivery to the inner cell membrane.

glycemic goals—A goal range for blood glucose concentration.

Golgi tendon organ—A proprioceptive sensory nerve ending embedded among the fibers of a tendon, often near the musculotendinous junction; it is compressed and activated by any increase of the tendon's tension, caused either by active contraction or by passive stretch of the corresponding muscle.

gonadotropin-releasing hormone—Hormone produced in the hypothalamus that acts on the pituitary and causes the release of gonadotropic substances, luteinizing hormone, and follicle-stimulating hormone.

graded exercise testing (GXT)—Testing that uses a gradual increase in exercise workload to a predetermined point or until volitional fatigue, unless symptoms occur before this point. Generally completed on a treadmill or bicycle ergometer.

growth factor—A category of hormones responsible for stimulating the process of tissue growth.

half-life (t-1/2)—Pertaining to a drug, the time it takes for one-half of the drug concentration to be eliminated.

hazard ratio—Multiplicative measure of association. Exposure to a certain risk factor or certain characteristic is associated with a fixed instantaneous risk compared with the hazard in the unexposed.

HbA1c—Glycosylated hemoglobin. This form of hemoglobin is primarily used to identify the plasma glucose concentration. A very high HbA1c (i.e., >7%) represents poor glucose control.

health belief model—Theory proposing that only psychological variables influence health behaviors.

heart and lung bypass—Device for maintaining the functions of the heart and lungs while either or both are unable to function adequately.

heart failure (HF)—The pathophysiological state in which an abnormality of cardiac function is responsible for failure of the heart to pump blood at a rate commensurate with the requirements of metabolizing tissues.

heart rate reserve (HRR)—The difference between a person's resting heart rate and maximal or peak heart rate.

hematocrit (Hct)—The percentage by volume of packed red blood cells in a sample of blood.

hematuria—Red blood cells in the urine.

hemianopia—Loss of vision for one-half of the visual field in one or both eyes.

hemiparesis—Paralysis affecting only one side of the body.

hemiparetic—Referring to paralysis affecting only one side of the body.

hemiplegia—Paralysis affecting only one side of the body.

hemodialysis—A method of dialysis in which the patient's blood is pumped through an artificial kidney, external to the body. As the blood passes through the dialyzer, its composition is favorably altered.

hemolysis—Alteration or destruction of red blood cells.

hemophilia—A hereditary hemorrhagic diathesis caused by deficiency of coagulation factor VIII. Characterized by spontaneous or traumatic subcutaneous and intramuscular hemorrhages.

hemoptysis—Expectoration of blood arising from the respiratory system; for people with CF, this occurrence reflects further infection or advancing disease.

hemorrhagic—Of or relating to excessive bleeding.

hemorrhagic stroke—Hypertension is the major risk factor for hemorrhagic stroke. It can also be caused by aneurysm, drug use, brain tumor, congenital arteriovenous malformations, and anticoagulant medication. Hemorrhagic strokes are classified as either intracerebral, which refers to bleeding inside the brain, or subarachnoid, which refers to bleeding in and around the spaces that surround the brain. There is usually little warning for a hemorrhagic stroke.

hepatic lipase (HL)—Lipase produced in the liver. Hepatic lipase activity results in the liver uptake of fatty acids by hydrolyzing triglycerides and phospholipids of VLDL and HDL. Hepatic lipase activity also contributes to cholesterol delivery to the liver.

herniation—Development of an abnormal protrusion or projection of an intervertebral disc.

high-density lipoprotein cholesterol (HDLc)—Binds to cholesterol in the arteries and carries the cholesterol back to the liver. Low levels of HDL cholesterol (<40 mg $\cdot$ dL^{-1}) increase the risk of cardiovascular disease.

high tone—Excess tone in a muscle group, often referred to as spasticity or hypertonicity.

hindfoot valgus—Excessive lateral deviation of the talocalcaneal complex relative to the trochanteric knee ankle (TKA) line.

hindfoot varus—Excessive medial deviation of the talocalcaneal complex relative to the trochanteric knee ankle (TKA) line.

HIV negative—Denoting an individual without antibodies to HIV viral proteins. An individual who has been recently infected with HIV may not have yet developed antibodies to the virus.

HIV positive—Denoting an individual with antibodies to HIV viral proteins; often refers to individuals who are infected but who have not yet developed an AIDS-defining condition or whose CD4 cell count is greater than 200 cells $\cdot$ mm^{-3}.

homocysteine—A homolog of cysteine.

hormone therapy—A treatment, usually a drug, used to block or attenuate the body's production of a hormone or block the effect of certain hormones released by the body; usually used to treat breast or prostate cancer.

human immunodeficiency virus (HIV)—The pathogen that leads to AIDS.

hybrid—Referring to concurrent use of both upper extremity exercise and lower extremity functional electrical stimulation ergometry.

hypercholesterolemia—Elevated cholesterol concentration.

hydrocephalus—A condition in which fluid accumulates in the brain, typically in young children, enlarging the head and sometimes causing brain damage.

hydronephrosis—Kidney condition characterized by dilation of the renal pelvis and collecting system attributable to ureteral obstruction or backflow (reflux) from the bladder.

hyperandrogenemia—Elevated androgen hormone levels in the plasma.

hypercapnia—An increased arterial carbon dioxide content.

hyperemia—Increased amount of blood in a part or organ.

hyperglycemia—An abnormally high concentration of glucose in the circulating blood, seen especially in people with diabetes mellitus.

hyperinflation—Overinflation of the lung, resulting in a greater functional residual capacity and total lung capacity.

hyperinsulinemia—Condition characterized by excess levels of circulating insulin in the blood. Also known as prediabetes, insulin resistance, and syndrome X.

hyperkalemia—Excess concentrations of potassium in the bloodstream.

hyperlipidemia—Elevated blood lipid levels that include elevated cholesterol and triglyceride concentrations.

hyperlipoproteinemia—Elevated lipoprotein concentrations.

hyperparathyroidism—A state produced by increased function of the parathyroid glands; results in dysregulation of calcium.

hyperpnea—More rapid and deeper breathing than normal.

hyperreflexia—A condition in which the deep tendon reflexes are exaggerated and are defined by overactive or overresponsive reflexes, which may include twitching and spastic tendencies.

hypertension—A condition in which blood pressure is chronically elevated. Diagnosed by resting systolic blood pressure ≥ 140 mmHg, resting diastolic blood pressure ≥ 90 mmHg, or both on two separate occasions.

hypertensive crisis—An abrupt and excessive elevation in blood pressure, typically greater than 180 mmHg for systolic blood pressure or greater than 120 mmHg in diastolic blood pressure, that can lead to an increased risk for stroke.

hypertonia—Increased rigidity, tension, and spasticity of the muscles.

hypertriglyceridemia—Elevated blood triglyceride concentrations.

hypertrophy—An increase in cell size.

hyperuricemia—Presence of high levels of uric acid in the blood. Diagnosed when uric acid levels are ≥ 7 mg $\cdot$ dL^{-1} in men and are ≥ 6 mg $\cdot$ dL^{-1} in women.

hypoestrogenic—Referring to decreased plasma estrogen levels.

hypoglycemia—Abnormally small concentration of glucose in the circulating blood; symptoms resulting from low blood glucose (normal glucose range 60-100 mg $\cdot$ dL^{-1}, or 3.3-5.6 mmol $\cdot$ L^{-1}) that is either autonomic or neuroglycopenic.

hypokalemia—Extreme potassium depletion in the circulating blood.

hypokinetic—Diminished or reduced muscle function; relative to cardiac muscle, hypokinetic left ventricular wall movement or motion abnormality can easily be observed on the echocardiogram and is often associated with prior myocardial infarction and reduced ejection fraction.

hypotension—Abnormally low blood pressure; typically associated with symptoms.

hypothyroidism—A condition in which the thyroid gland does not make enough thyroid hormone and the person becomes thyroid deficient.

hypotonic—Having a lesser degree of tension.

hypovolemia—Diminished blood volume.

hypoxemia—Insufficient oxygenation of the blood; assessed by arterial blood gas or pulse oximetry.

hypoxia—A state of oxygen deficiency.

iatrogenic—Denoting decrements in health status because of medical treatment. HAART induces numerous adverse (i.e., iatrogenic) effects, such as lipodystrophy, insulin resistance, osteonecrosis, and an increased risk of cardiovascular disease (CVD).

idiopathic—Denoting a disease of unknown cause.

idiosyncratic—Peculiar or individual.

IMAT—Intermuscular adipose tissue; fat that infiltrates the skeletal muscle.

immune system—System that mediates the body's interaction between internal and external environments. Helps rid the body of infectious agents and malignant cells.

immunosuppression—Suppression of immune responses produced primarily by a variety of immunosuppressive agents.

incidence—The frequency of occurrence of any event or condition over a period of time and in relation to the population in which it occurs.

incomplete paraplegia—Incomplete motor and sensory dysfunction of the trunk, legs, and pelvic organs resulting from spinal cord injury.

incomplete tetraplegia—Incomplete motor and sensory dysfunction of the arms, trunk, legs, and pelvic organs resulting from spinal cord injury.

incontinence—Lack of control of urination or defecation.

infarct-related artery—The completely, or nearly completely, stenosed coronary artery that is responsible for the interruption of blood flow and the subsequent myocardial infarction.

insensate—Lacking sensation.

inspiratory resistive loading—The act of inspiring air against a resistance greater than normal.

inspiratory threshold loading—The act of inspiring after attaining and proceeding at a predetermined inspiratory pressure (threshold point).

insulin resistance—A condition in which normal amounts of insulin secreted by the pancreas are inadequate to produce a normal insulin response in the muscle or liver. As a result, the pancreas secretes additional insulin, thereby elevating insulin levels in the plasma. High levels of insulin in the plasma often lead to the development of type 2 diabetes or metabolic syndrome.

internal mammary artery—An artery with its origin in the chest region used for coronary revascularization surgery.

intima—The inner layer of blood vessels containing endothelial cells.

intradialytic exercise—Exercise performed during a hemodialysis treatment session.

intraosseous—Within bone.

invasive cardiology—The branch of cardiology specializing in catheter-based treatment of structural heart disease, such as coronary angiography, angioplasty, and stenting.

ipsilateral—On the same side of the body.

iritis—Inflammation of the iris of the eye.

ischemia—Deficiency of blood flow, attributable to functional constriction or actual obstruction of a blood vessel.

ischemic—Referring to a sustained deficiency in oxygen delivery.

ischemic heart disease—A pathological condition in which blood flow to the myocardium is reduced below the demand, resulting in a lack of oxygen delivery to cardiac tissue (i.e., coronary atherosclerosis).

ischemic stroke—Associated with the development of ischemic cerebrovascular disease; can be further categorized as thrombotic, embolic, and hemodynamic.

isocapnia—Normal arterial carbon dioxide levels.

isokinetic—Denoting the condition in which muscle fibers shorten at a constant speed in such a manner that the tension developed may be maximal over the full range of joint motion.

isometric—Denoting the condition in which the ends of a contracting muscle are held fixed so that contraction produces increased tension at a constant overall length.

isotonic—Denoting the condition in which muscle fibers shorten with varying tension as the result of a constant load.

joint contractures—Reduced passive range of motion at a joint caused by shortened tendons, typically associated with unbalanced spasticity.

joint effusion—Increased fluid in synovial cavity of a joint.

joule (J)—A unit of energy; the heat generated, or energy expended, by an ampere flowing through an ohm for 1 s; equal to 107 erg and to a newton meter. The joule is an approved multiple of the SI fundamental unit of energy, the erg, and is intended to replace the calorie (4.184 J).

Kaposi's sarcoma—Firm, subcutaneous, brown-black or purple lesions usually observed on the face, chest, genitals, oral mucosa, or viscera.

karyotyping—A technique used to quantify the DNA copy number on a genomic scale.

ketone—A substance with the carbonyl group linking two carbon atoms.

kilocalorie (kcal)—A unit of heat content or energy; the amount of heat necessary to raise 1 g of water from 14.5 °C to 15.5 °C times 1,000.

kyphosis—Excessive angulation of the spine resulting in increased anteroposterior diameter of the chest cavity; humpback; may reflect chronic pulmonary disease.

laminectomy—A surgical procedure in which the posterior arch of a vertebra is removed to relieve pressure on the spinal cord or on the nerve roots that emerge from the spinal canal. The procedure may be used to treat a herniated disc or spinal stenosis.

lecithin–cholesterol acyltransferase (LCAT)—An enzyme that hydrolyzes a fatty acid from lecithin and the subsequent esterification of the fatty acid with cholesterol. The products of this reaction are cholesterol esters and lysolecithin. LCAT is essential for the function of HDL and the maintenance of reverse cholesterol transport.

left ventricular dysfunction (LVD)—Abnormal function of the left ventricle (i.e., poor wall motion).

left ventricular ejection fraction (LVEF)—The percentage of the end-diastolic volume ejected per beat; an index of systolic function.

leptin—A protein messenger from adipose tissue to the satiety center in the hypothalamus involved in regulating appetite.

lifestyle-based physical activity—Home- or community-based participation in forms of activity that include much of a person's daily routine (e.g., transport, home repair and maintenance, yard maintenance).

lipid—A fat or derivative of fat that is insoluble in water and must be bound to protein in order to be transported in blood.

lipoprotein—Macromolecule consisting of proteins, phospholipids, cholesterol, and triglycerides that transport lipids in aqueous mediums found in the blood, interstitium, and lymph.

lipoprotein lipase (LPL)—An enzyme, found in skeletal muscle and adipose tissues, that hydrolyzes triglycerides into fatty acids. The fatty acids are taken up by these tissues for energy use or storage.

low-calorie diet—A hypocaloric diet, 1,200 kcal/d or less.

low-density lipoprotein cholesterol (LDLc)—The principal means by which cholesterol is transported throughout the body. Formed in the circulation from lipid and protein exchanges between VLDL and other lipoproteins and tissues.

low energy availability—Also known as energy deficit, energy drain, or negative energy balance; results from low dietary energy intake and high energy expenditure.

lower rate limit—The rate at which the pulse generator begins pacing in the absence of intrinsic activity.

low tone—Often referred to as flaccidity or hypotonia; decreased amount of tone in a muscle group.

luteinizing hormone—Hormone secreted by the anterior lobe of the pituitary to stimulate the development of the corpus luteum.

LV diastolic dysfunction—Clinically, diagnosis is less exact than for systolic dysfunction. Diagnosis is often made when the clinical syndrome of congestive heart failure (fatigue, dyspnea, and orthopnea) requires hospitalization in the presence of a relatively normal ejection fraction.

LV systolic dysfunction—Ejection fraction reduced below 45% (severe considered <30%) as measured by echocardiogram or another quantitative measure.

lymphedema—Swelling as a result of obstruction of lymphatic vessels, resulting in fluid buildup and edema in the affected extremity.

lymphocytes—Any of the mononuclear, nonphagocytic leukocytes found in the blood, lymph, or lymphoid tissues that are the body's immunologically competent cells.

lymphocytopenia—A reduction in the number of lymphocytes in the circulating blood.

macrovascular disease—Atherosclerosis that affects large blood vessels such as the aorta, femoral artery, and carotid artery.

magnetic resonance imaging (MRI)—The diagnostic test that uses principles of magnetism to generate an electromagnetic field around the body, causing certain atoms in the nucleus of the body cells to line up. Then, by sending and receiving radio signals, which are fed into a computer, the device records the position of those atoms, providing a distinct picture of the tissues being investigated. The patient lies inside a large, tunnel-like tube for 30 to 60 min while the images are formulated by the computer. This diagnostic imaging technique has been found to have certain advantages over radiographs and CT scans in the diagnosis of spinal disorders.

major depressive disorder (MDD)—A diagnosis requiring endorsement of at least five symptoms, one of which must be either depressed mood or diminished interest or pleasure, that have been present during the same 2 wk period.

maladaptation—Adaptation to a progressive stimulus (e.g., exercise) that results in an overload to the system to the degree that performance is reduced and the risk of injury is increased.

maximal bone mass—The highest bone mass a person could possibly achieve.

maximal oxygen uptake or consumption—The maximum amount of oxygen consumed (or used) by the body, usually measured under conditions of maximal exercise.

maximal voluntary ventilation (MVV)—Amount of air maximally breathed in, expressed as liters per minute.

maximum sensor rate—The maximum rate for a rate-responsive pacemaker that can be achieved under sensor control.

maximum tracking rate—The maximum rate at which the pulse generator will respond to atrial events.

medical nutrition therapy (MNT)—The use of nutrition as a treatment for a clinical condition or disease.

medication reconciliation—The formal process that compares a patient's current medications to those in the patient's record or medication orders.

menarche—The beginning of menstrual function.

mentation—The process of reasoning and thinking.

metabolic equivalent of task (MET)—An expression of the rate of energy expenditure during seated rest. 1 MET = 1 $kcal \cdot kg^{-1} \cdot h^{-1}$ = 3.5 $mL \cdot kg^{-1} \cdot min^{-1}$.

metabolic syndrome—A constellation of insulin resistance characterized by central obesity, elevated triglycerides, suppressed HDL cholesterol, hypertension, or prediabetes.

metastasize—To spread from one part of the body to another, as when neoplasms appear in parts of the body remote from the site of the primary tumor.

microalbuminuria—A condition in which the kidneys leak a small amount of albumin into the urine. Diagnosed with 24 h urine collections (≥ 20 $\mu g \cdot min^{-1}$) or when levels are ≥ 30 $\mu g \cdot min^{-1}$ on two separate occasions.

microcephalus—A birth defect where a baby's head is smaller than expected when compared to babies of the same sex and age.

microvascular disease—Atherosclerosis that affects small blood vessels such as those of the kidney, eye, heart, and brain.

moderate-intensity physical activity—Activities that cause small increases in breathing and heart rate; ~50% to 70% $\dot{V}O_2$peak.

mode switching—A programmed feature of dual-chamber pacemakers that prevents tracking or matching every atrial impulse with a ventricular pacing pulse; purpose is to prevent tracking of rapid atrial rates to the ventricle.

monoplegia—Paralysis of a single limb, muscle, or muscle group.

monounsaturated fat—Dietary fatty acid that contains one double bond along the main carbon chain.

morbidity—Manifestations of disease other than death.

morphology—Configuration or shape (e.g., shape of the ST segment: downsloping, upsloping, or horizontal).

mortality—Death.

motion artifact—Incidental activity that is picked up on an ECG during body movement.

multiple sclerosis (MS)—A debilitating disease characterized by multiple areas of scar tissue replacing myelin around axons in the central nervous system.

muscle spindle receptors—A fusiform end organ in skeletal muscle in which afferent and a few efferent nerve fibers terminate; this sensory end organ is particularly sensitive to passive stretch of the muscle in which it is enclosed.

muscle tone—Amount of tension in a muscle or muscle group at rest.

muscle weakness—Condition in which skeletal muscle lacks strength and power-generating capability.

muscular strength and endurance—The ability of skeletal muscles to perform hard or prolonged work.

musculoskeletal flexibility—The range of motion in a joint or sequence of joints.

myalgia—Pain in a muscle or muscles.

Mycobacterium avium **complex**—Complex that consists of two predominant species, *M. avium* and *M. intracellulare*. More than 95% of infections in patients with AIDS are caused by *M. avium,* whereas 40% of infections in immunocompetent patients are caused by *M. intracellulare.*

myelography—Radiographic inspection of the spinal cord and nerve roots by use of a radiopaque contrast medium (a substance that causes the absorbing tissues to appear darker or lighter on a radiograph) injected into the intrathecal space. Air or oil dye may be used as contrasting agents.

myelomeningocele—Congenital open neural tube defect with disruption of skin, bone, and neural elements; usually involves spinal cord dysfunction despite surgical closure.

myocardial infarction—Medical term for heart attack in which blood flow through a coronary artery is completely disrupted, leading to myocardial tissue death and placing individual at risk of death.

myocardial ischemia—Temporary lack of oxygen to the heart or myocardium, due to an imbalance between oxygen supply and demand.

myocardium—The heart muscle.

myoglobinuria—Excretion of myoglobin in the urine resulting from muscle degeneration.

myotomes—A group of muscles that a single spinal nerve innervates.

nasal polyposis—Growths of tissue in the nose that may block air passage through the nostril; not life threatening.

neoplasm—Abnormal tissue that grows by cellular proliferation more rapidly than normal and continues to grow after the stimuli that initiated the growth cease. Structural organization and function of neoplastic tissue are partially or completely different from what is seen in normal tissue.

nephron—The functional unit of the kidney (~1 million per kidney), composed of tubules (e.g., descending and ascending limbs of loop of Henle) and vascular (e.g., afferent and efferent arterioles, glomerular capillaries) components; capable of making urine through the processes of filtration, reabsorption, secretion and excretion.

nephropathy—Damage to the kidney, often caused by microvascular disease.

neurectomy—Partial or total excision or resection of a nerve.

neurogenic—Controlled by the nervous system, as in neurogenic bladder, bowel, or sexual function.

neuroglycopenic—Symptoms of hypoglycemia that include feelings of dizziness, confusion, tiredness, difficulty speaking, headache, and inability to concentrate.

neuromodulatory dysregulation—An abnormality or impairment of a physiologic process by which a neuron uses one or more chemicals to regulate a population of neurons.

neuromuscular—Of, relating to, or affecting both nerves and muscles.

neuropathy—A disease involving the cranial nerves or the peripheral or autonomic nervous system.

neuropeptide Y (NPY)—Central nervous system appetite stimulant.

neutral spine—The position in which the trunk and neck, and therefore the joints of the spine, are in neither flexion nor extension.

neutropenia—The presence of abnormally small numbers of neutrophils in the blood.

neutrophilia—An increase in neutrophilic leukocytes in blood or tissue.

nitroglycerine—Medication used to promote vasodilation in patients with angina pectoris.

nonischemic cardiomyopathy—Disease process involving cardiac muscle that is not related to ischemic heart disease; may be attributable to viral infection or alcohol abuse.

normal-weight obesity—A condition in which an individual is classified in the normal range for body mass or BMI but has elevated adiposity or risk factors for metabolic syndrome, atherosclerotic cardiovascular disease, diabetes, or more than one of these.

nuclear perfusion—Radioactive isotope that has the ability to perfuse through tissue so that select organs can be imaged.

nucleoside reverse transcriptase inhibitor (NRTI)—A specific type of antiretroviral medication.

nystagmus—Rhythmic, involuntary movements of the eyes.

obesity—A condition in which the proportion of body fat is abnormally high; defined as a body mass index greater than 30 kg · m^{-2}.

obstructive sleep apnea—Collapse of the airway during sleep resulting in snoring, poor sleep quality, and intermittent complete lack of breathing (apnea).

occlusion—The complete blockage of an artery, resulting in decreased blood flow through the artery to the peripheral muscles of the legs.

old age—Between 65 and 74 yr of age.

oldest old—Older than 85 yr of age.

oligomenorrhea—Scanty or infrequent menstrual flow.

oliguria—A diminution in the quantity of urine excreted; specifically, less than 400 mL in a 24 h period.

omega-3 fatty acids—Long-chain polyunsaturated fatty acids that contain a double bond in the n-3 position.

oncogene—A mutated proto-oncogene that may foster unregulated or malignant cell growth.

opportunistic infections—Infections most commonly seen in people who are immunocompromised, such as individuals with late or advanced HIV-1 disease, cancer, or other immunocompromising conditions.

optic neuritis—Inflammation of the optic nerve.

orthopnea—Labored or difficult breathing while lying flat or supine.

orthostatic—Relating to upright or erect posture.

orthostatic hypotension (OH)—Decrease of at least 20 mmHg in systolic blood pressure when an individual moves from a supine position to a standing position.

ostectomy—Surgical excision of a bone or a portion of one.

osteoarthritis (OA)—Erosion of articular or joint cartilage that leads to pain and loss of function.

osteogenic—Increasing bone mass.

osteopenia—Reduced bone mineral density, defined as between 1 and 2.5 standard deviations below the young adult mean.

osteophyte—A bony excrescence or outgrowth, usually branched in shape.

osteoporosis—A pathological condition associated with increased susceptibility to fracture and decreased bone mineral density, more than 2.5 standard deviations below the young adult mean.

osteoporotic fracture—Broken bone caused by a reduction in the mass of the bone per unit of volume.

osteotomy—Operation for cutting through a bone to improve alignment or correct deformities.

overweight—A body mass index of 25.0 to 29.9 kg · m^{-2}.

oxygen uptake (consumption) ($\dot{V}O_2$)—A measure of a person's ability to transport and use oxygen.

oxyhemoglobin saturation (SaO$_2$)—Percentage of hemoglobin bound to oxygen; assessed noninvasively by pulse oximeter or invasively by arterial blood gas sampling.

pacemaker—Implantable medical device that uses electrical impulses delivered by electrodes to the heart muscle to regulate the beating of the heart. Modern pacemakers can serve multiple functions, such as maintaining an adequate heart rate or improving the synchronization of the chambers of the heart. Some implantable devices also combine a pacemaker with an automatic internal cardioverter-defibrillator. A temporary pacemaker is not implanted and serves primarily to regulate the heart rate.

pacemaker sensor—Sensor incorporated into the pulse generator that detects a physiological stimulus to control the heart rate to match physiological demands.

pacing system—A medical device that uses electrical impulses to regulate the beating of the heart when the heart's natural pacemaker is diseased. The pacing system consists of the pacemaker, or generator, which is the component of the system containing the circuitry that senses when the pacemaker needs to initiate a response, and the battery that powers the device. The pacemaker lead wires are the components of the pacing system that are attached to the pacemaker and deliver the electrical impulses to the heart muscle.

pack-years—Number of packs of cigarettes smoked per day multiplied by the number of years the person smoked; for example, two packs a day for 20 yr equals 40 pack-years.

palliative surgery—Surgery aimed at removal of a tumor to make the patient more comfortable, relieve organ obstruction, or reduce tumor burden.

pancreas—A gland lying behind the stomach that secretes pancreatic enzymes into the duodenum and secretes insulin, glucagon, and somatostatin into the bloodstream.

pancreatic insufficiency—Inadequate exocrine function of the pancreas resulting in little or no production of pancreatic enzymes needed for digestion (i.e., lipase, amylase, protease); results in nutrient malabsorption.

pannus—Inflammatory exudates overlying the lining layer of synovial cells on the inside of a joint.

pansinusitis—Chronic inflammation and infection involving all sinuses; commonly seen in individuals with CF.

papilledema—Swelling of the optic disc in the eye caused by severe hypertension.

paraplegia (PP)—Motor and sensory dysfunction of the trunk, legs, and pelvic organs resulting from a spinal cord injury.

parasympathetic nervous system—Craniosacral portion of the central nervous system that promotes anabolic activity and energy conservation.

parathyroid hormone—A peptide hormone formed by the parathyroid glands; raises the serum calcium when administered parenterally by causing bone resorption.

parenchyma—The essential or primary tissue of the lungs.

parenchymal infiltrates—Deposition or diffusion in lung tissue of substances not normal to it.

parenteral—Relating to the route of administration of a drug that is through the skin or a mucosal membrane.

paresis—Slight or partial paralysis, or partial weakness in one or more limbs.

paresthesia—A subjective feeling such as numbness, "pins and needles," or tingling.

paroxysmal nocturnal dyspnea—Sudden awakening caused by labored or difficult breathing.

partial body-weight-supported treadmill training (PBWSTT)—Mode of exercise training with the person's body weight partially supported over the treadmill by a mechanical suspension system, a therapist, or both.

patency—Of the arterial lumen, the condition of being open, unobstructed.

peak bone mass—The highest amount of bone mass achieved by a person during their lifetime. It is assumed that peak bone mass is achieved in the second or third decade of life. The age at which peak bone mass is achieved also varies based on which bone site is being measured.

peak expiratory flow rate (PEFR)—The highest flow rate (exhalation of gas from the lung) that a person can generate during a forceful expiration.

peak walking time—Most typically, the time point at which a patient cannot walk any further because of severe claudication pain; the most commonly used outcome measure for assessing walking performance in peripheral artery disease studies.

pediatrics—A branch of medicine that is concerned with children and their diseases.

percussion—A diagnostic method that uses short, sharp tapping of the body surface to produce different reflected sounds from the underlying organs.

periarticular—Situated or occurring around a joint of the body.

pericardial effusion—Increased amounts of fluid within the pericardial sac, usually attributable to inflammation.

perineum—The area between the scrotum and anus in males and between the vulva and anus in females.

periodization—A system of fractioning larger periods of muscle training into smaller phases or cycles. Intensity, frequency, sets, repetitions, and rest periods are altered to reduce the risk of overtraining and minimize uncomfortable responses.

peripheral artery disease (PAD, or peripheral vascular disease)—Disease of the vascular system that can be found in the periphery (i.e., commonly observed in the legs, which leads to claudication with physical exertion).

peripheral nervous system—Sensory and motor components of the nervous system that have extensions outside the brain and spinal cord.

peripheral neuropathy—Damage to the nerves of the legs or arms resulting in a loss of sensation (e.g., touch, temperature).

peritoneal dialysis—Dialysis performed through introduction of fluid into the peritoneal cavity. Dialysis fluid can be cycled through the peritoneal cavity by a machine over a 10 to 12 h period daily (intermittent peritoneal dialysis) or exchanged every 4 h, with the fluid staying in the peritoneal cavity between exchanges (continuous ambulatory peritoneal dialysis). The fluid is introduced through a catheter (tube) placed in the abdomen.

pes equinus—A condition marked by walking without touching the heel to the ground.

pharmacodynamic phase—The science that pertains to the effect of a drug on the body.

pharmacokinetic phase—The science that pertains to the movement of a drug's molecules through the body; encompasses four discrete subphases—absorption, distribution, metabolism, and excretion.

physical activity—Any bodily movement produced by skeletal muscles that results in caloric expenditure.

physiological pacing—Setting a pacemaker to regulate the heartbeat to mimic natural physiological processes, such as timed stimuli between the chambers of the heart, to best meet physiological demands. Modern pacemakers can be programmed to provide multiple functions, depending on why the device is needed.

plaque (pertaining to an artery)—A yellow area or swelling on the intimal surface of an artery, produced by the atherosclerotic process of lipid deposition.

plaques (pertaining to the central nervous system)—Scarring of axons in the central nervous system attributable to demyelination.

plasticity (pertaining to aerobic fitness)—The extent to which normal maturation of maximal aerobic power can be altered by changes in physical activity.

platelet aggregation—The congregation of platelets, which are disc-shaped fragments found in the peripheral blood and involved in the clotting process.

plegia—Greater involvement of one or more limbs than paresis; often associated with paralysis.

plexopathies—Disorders affecting a network of nerves, blood vessels, or lymph vessels. The regions of nerves affected are at the brachial or lumbosacral plexus. Symptoms include pain, loss of motor control, and sensory deficits.

pluripotent stem cell—Uncommitted cell with various developmental options pending.

Pneumocystis carinii **pneumonia (PCP)**—An AIDS-defining condition caused by the parasite *P. carinii.*

pneumothorax—An acute collection of air in the pleural space; results in collapse of the affected lung; common in advancing CF lung disease.

poikilothermic—Having body temperature that varies with the environment.

point of maximal cardiac impulse (PMI)—Point identified by palpation and inspection of the chest wall during physical examination as the most prominent location for cardiac apical impulse.

polycystic kidney disease—Hereditary bilateral cysts distributed throughout the renal parenchyma, resulting in markedly enlarged kidneys and progressive renal failure.

polycystic ovarian syndrome—Endocrine disturbance associated with primary anovulation and polycystic ovaries.

polycythemia—An abnormally elevated level of red cells in the blood.

polydipsia—Excessive thirst that is relatively prolonged.

polymorphonuclear leukocytosis—An elevation in neutrophilic leukocyte (white blood cell) count.

polyunsaturated fats—Dietary fatty acids that contain two or more double bonds along the main carbon chain.

polyuria—Excessive excretion of urine.

postprandial—Referring to the time period 1 to 2 h following a meal.

postprandial lipemia (PPL)—Exaggerated levels of triglycerides in the blood and failure to return to baseline levels within 8 to 10 h after consumption of dietary fat.

Prader-Willi syndrome—A rare genetic disorder in which seven genes, or a subset of seven genes, on chromosome 15 are deleted or unexpressed. Characterized by excessive weight gain and obesity.

prediabetes—A precursor to diabetes, defined by the American Diabetes Association as an HbA1c level between 5.7% and 6.5%.

prevalence—The number of cases of a disease present in a specified population at a given time. This number may be given at one identified time (point prevalence) or during a specified period, such as 2 wk or a year (period prevalence).

prevention—Intervention strategies to limit the effect of potential or established disease in the population.

primary amenorrhea—Delay of menarche beyond age 18.

primary prevention—An intervention geared toward removing or reducing the risk factors of disease.

protease inhibitor—A specific type of antiretroviral medication.

proteinuria—The presence of abnormal amounts of protein in the urine.

prothrombotic—Condition or agent that increases the risk of formation or presence of a thrombus.

proto-oncogene—A gene involved in regulation of normal cell growth or proliferation.

pulmonary exacerbation—An episode of worsening lung disease caused by increased infection and inflammation, resulting in increased patient infections, symptoms, and limitations to activities of daily living.

pulmonary hypoplasia—An incomplete development of the lungs, resulting in an abnormally low number or size of bronchioles or alveoli; primarily a congenital condition.

pulmonary rales—Clicking, rattling, or crackling sounds made by one or both lungs during inhalation due to fluid in the alveoli or other pulmonary disease.

pulmonary rehabilitation (PR)—An evidence-based, multidisciplinary, and comprehensive intervention for patients with chronic respiratory diseases who are symptomatic and often have decreased daily life activities.

pulsatile—Characterized by a rhythmical pulsation.

pyelonephritis—The disease process from the immediate and late effects of bacterial and other infections of the parenchyma and the pelvis of the kidney.

Q angle—Acute angle formed by a line from the anterior superior iliac spine of the pelvis through the center of the patella and a line from the tibial tubercle through the patella.

quality of life—Perception of life satisfaction.

quantitative computed tomography—The only method currently available that provides an actual measurement of volumetric bone density.

quantitative ultrasound—A device that measures structural properties of bone with sound waves. Unlike densitometry devices, it uses no ionizing radiation.

radiation therapy—Therapy meant to stop growth of malignant cells by damaging RNA within the cells.

radicular—Relating to pain caused by compression or injury of the root of a nerve.

radionuclide agent—Isotope (natural or artificial) that exhibits radioactivity. Used in nuclear cardiology medicine to image the myocardium for potential ischemia.

radionuclide imaging—A type of cardiac imaging that can detect ischemia and wall motion; uses an injected radioisotope (i.e., thallium 201 or technetium-99m sestamibi) that is scanned using X-ray.

ramping protocol—Continuous gradual increase in workload (treadmill: speed and grade; cycle ergometer: watts) over a select period of time.

range of motion (ROM)—The total degrees of movement that a joint can move through.

rate–pressure product (RPP)—Indirect indication of how hard the heart is working. RPP = systolic BP × HR.

rate-responsive pacing—Function of a pacemaker that changes the rate by sensing a physiological stimulus. This type of pacemaker is also described as modulated, adaptive rate, or sensor driven.

rating of perceived exertion (RPE)—A person's perception of how hard they are working physically. Currently two scales are commonly used to assess RPE (6-20 and 0-10).

Raynaud's phenomenon—Intermittent, bilateral attacks of ischemia of the toes and fingers (and sometimes ears or nose). The classical features are episodic pallor of digits, followed by cyanosis (i.e., blue color due to deoxygenation) and then redness, pain, and tingling.

red flags—Generally, signs and symptoms indicative of serious pathology that is usually beyond the capabilities of the treating health care practitioner to treat; these indicators of disease usually mean an immediate referral to a physician is warranted. Examples include constitutional signs like unexplained nausea, shortness of breath or diaphoresis, unexplained weight loss or gain, spine pain with eating, nocturnal spine pain, and constant spine pain that is not modified by change in position or activity.

refractory—Resilient or resistant to treatment.

rejection—Immune response to foreign tissue (transplanted organ).

relapse—Reversion to an active disease process (e.g., multiple sclerosis) after a remission.

relapse prevention model—A model used to help new exercisers anticipate problems with adherence. Factors that contribute to relapse include negative emotional or physiological states, limited coping skills, social pressure, interpersonal conflict, limited social support, low motivation, high-risk situations, and stress.

relative contraindications for test termination—Conditions occurring during a GXT that require strong clinical judgment concerning the safety of continuing the exercise test.

relative oxygen uptake—Oxygen uptake expressed in milliliters of oxygen per kilogram of body weight per minute ($mL \cdot kg^{-1} \cdot min^{-1}$).

remission—Recovery period from the active disease process.

renal replacement therapy (RRT)—A type of therapy used to replace the functioning of failing kidneys. Medical technologies that serve as substitutes for renal function include hemodialysis, peritoneal dialysis, and transplantation. Without this therapy, the patient with no renal function would die.

reocclusion—Closing again; reclosure.

reperfusion—The process of reinstituting blood flow to an area of tissue previously deprived of normal blood flow.

repetition maximum (RM)—The number of times a weight can be lifted; 1RM is the maximal amount of weight that can be lifted one time only.

resolving MI—The phase of an MI in which necrotic tissue forms a scar.

respiratory failure—Failure of the respiratory system to keep gas exchange at an acceptable level.

restenosis—The recurrence of a narrowing or restriction.

retinopathy—Damage caused to the retina because of retinal vascular disease or abnormal blood flow.

retroviruses—Viruses containing both RNA-dependent and DNA-synthesizing material.

revascularization—Restoration of blood flow to a body part.

rhabdomyolysis—Acute, potentially fatal disease of the skeletal muscle; entails destruction of the muscle as evidenced by myoglobinuria.

rheology—The study of the deformation and flow of liquids and semisolids.

rheumatoid arthritis (RA)—Inflammation of the joints attributable to autoimmune attack; leads to pain and loss of function.

rheumatoid cachexia—Loss of body cell mass, predominantly skeletal muscle, that is characteristic of inflammatory arthritis. Thought to be cytokine driven, the condition is also characterized by increased fat mass; consequently reduced body weight (or BMI) is uncommon (<10%).

roentgenogram—A photograph made with X-rays.

sarcopenia—Degenerative loss of skeletal muscle mass and strength associated with aging.

sclerosis—Tissue hardening that occurs because scar tissue replaces lost myelin around axons in the central nervous system.

scoliosis—Abnormal lateral curvature (side-bending and rotational components) of the spine that may be congenital or acquired by extremely poor posture, disease, or muscular weakness. Usually the curvature consists of two curves, the original abnormal curve and a second compensatory curve in the opposite direction (also referred to as an S-curve).

secondary amenorrhea—Cessation of menses in a woman who has previously menstruated.

secondary condition—An injury, impairment, functional limitation, or disability that occurs as a result of the primary condition or pathology. Secondary conditions include physical problems, social concerns, and mental health difficulties. Secondary conditions also can develop when the primary disability interferes with the delivery of standard health care for the treatment or prevention of a health condition.

secondary coronary prevention—Treatment with medications and lifestyle changes (e.g., exercise, tobacco avoidance, healthy nutrition) in patients after the diagnosis of coronary heart disease.

secondary prevention—An intervention that promotes early detection and treatment of disease with the goals of preventing recurrences or progression, promoting recovery, and avoiding complications.

selective estrogen receptor modulators (SERMs)—Antiresorptive agents that have fewer side effects than estrogen replacement therapy and may be a good alternative to ERT for the woman with a history of breast cancer.

selective serotonin reuptake inhibitors—A class of antidepressant medications commonly used to treat depression.

self-efficacy—A person's belief in their capability to perform a behavior and the perceived incentive to do so.

sensitivity—The proportion of affected people who show a positive test result for the disease that the test is intended to reveal.

set-point theory—A metabolic theory that postulates stability in weight in both overfeeding and calorie-restriction situations.

sick sinus syndrome—Syndrome in which the sinus node is not functioning at an appropriate rate, leading to sinus bradycardia, pauses, arrest, or exit block. Syncopal episodes can be caused by this abnormality.

significant Q wave—Wave that depicts a prior MI on an ECG. For Q waves to be considered significant, they must be either ≥0.04 s wide and one-third the height of the associated R wave.

single-chamber pacemaker—Pulse generator that can pace or sense in the atrium or ventricle.

single-photon absorptiometry—A method for determining bone mineral content through measurement of the absorption by bone of a monoenergetic photon beam.

Sjogren's syndrome—A chronic autoimmune disease of unknown etiology, usually occurring in middle-aged or older women. Features lymphocyte infiltration of exocrine glands (i.e., tear and salivary glands), with resultant dryness of the eyes and mouth, and the presence of a connective tissue disease—usually rheumatoid arthritis but sometimes systemic lupus erythematosus, scleroderma, or polyarthritis.

social cognitive theory—Theory that behavior change is affected by environmental influences, personal factors, and attributes of the behavior itself.

social support—Support and encouragement a person receives from others to maintain behavior change.

soft tissue—The total amount of tissue in the body minus bone mass as determined by DXA.

somatic nervous system—Neural elements over which a person has conscious awareness and control.

somatosensory evoked potentials—A noninvasive diagnostic test to assess the speed of electrical conduction across the spinal cord. The technique involves applying electrical stimulus at specific nerves in the arms and legs and measuring the impulses generated by the stimulus at various points in the body.

spasticity—Of, relating to, or characterized by spasms, an involuntary increase in muscle tone.

specificity—The proportion of people with negative test results for the disease the test is intended to reveal.

spina bifida—Congenital neural tube defect with varying degrees of skin, bone, and neural element involvement.

spinal cord injury (SCI)—Damage involving the spinal cord.

spinal decompression—Surgical intervention to excise bony or soft tissue structures that exert pressure on neural tissues in the spine.

spinal discectomy—Surgical intervention to excise the portion of a herniated disc that is causing compression on neural tissue. The extent of the tissues removed is based on the extent of the intervertebral disc herniation.

spinal fusion—Surgical intervention to fixate unstable hypermobile vertebral segments by the use of metal plates, screws, wires, and autologous bony transplants.

spinal traction—Use of specialized harness systems and electronic winch or manually applied distractive forces on the spine in a variety of spinal and bodily positions. The purpose is to separate vertebrae and stretch the associated soft tissues, thus decompressing nerve roots and relieving symptoms.

spirometry—Measurement of the breathing capacity of the lungs.

spondyloarthropathy—A type of inflammatory arthritis involving ligament or tendon insertion sites (entheses), leading to spinal and peripheral joint arthritis, usually in human lymphocyte antigen B27–positive individuals.

staging—A system used to classify the extent and spread of cancer.

static endurance—Classification of exercise in which isometric contractions of muscle groups lead to anaerobic exhaustion. An example is a trunk extension position prolonged until fatigue occurs and the subject is unable to hold the position.

static pulmonary function—Assessment of pulmonary status at rest. Static pulmonary function includes measures of pulmonary function including FEV_1, FVC, RV/TLC, and diffusion capacity.

steatosis—Pathologic adipose depots.

stenosis—Constriction or narrowing of a passage or orifice. In spinal stenosis, congenital or degenerative narrowing of the intervertebral or vertebral foramen (opening) is present, leading to compressive forces on the nerve roots that travel through these openings.

stent—A stainless steel bridge, expanded by a balloon-tipped catheter, designed to hold open an area of stenosis within an artery.

sternotomy—The operation of cutting through the sternum.

strabismus—A deviation of the eye that the individual cannot overcome. The visual axes assume a position relative to each other different from that required by the physiological conditions. The various forms of strabismus are termed tropias. Their direction is indicated by the appropriate prefix (e.g., cyclotropia, esotropia, exotropia, hypertropia, and hypotropia). Also called cast, heterotropia, manifest deviation, and squint.

strategies to promote adherence—Techniques commonly used among exercise professionals to improve initiation of and compliance to a structured exercise regimen.

stratify—To separate individuals or samples into subcategories based on variables of interest (e.g., sex, age, number of risk factors, symptoms).

strength—Maximal voluntary contractile force of a given muscle group or groups.

stress echocardiogram—Combination of an exercise test and an echocardiogram. Resting and postexercise echocardiogram images are compared for wall motion abnormalities that can suggest an ischemia response.

subarachnoid—Pertaining to space in the brain under the arachnoid membrane containing cerebrospinal fluid.

subchondral sclerosis—Thickening of the bone beneath the cartilage layer of an arthritic joint.

subendocardial—Referring to the endocardial surface of the heart.

subendocardial ischemia—Myocardial ischemic response beneath the endocardium.

subepithelial fibrosis—The structural changes noted beneath the epithelial layer of the bronchus resulting in scar tissue formation in this area.

supervised exercise therapy (SET)—Specific for peripheral artery disease (PAD), it is a physician-referred exercise and education program for those diagnosed with symptomatic PAD. It is designed to help people walk longer without pain in the hope of improving their overall quality of life.

sweat test—Diagnostic test used for CF; usually involves stimulation of the skin's sweat glands by chemical (i.e., pilocarpine) and electrical (i.e., iontophoresis) means; an elevation of greater than 60 mEq $\cdot$ dL^{-1} is highly suggestive of CF.

symmetric—Referring to quality of or correspondence in the form of parts on the opposite sides of any body.

sympathetic nervous system—Lumbosacral portion of the central nervous system that promotes the classic fight-or-flight response to a given stimulus.

syncope—Loss of consciousness caused by diminished cerebral blood flow.

synovial joint—A joint in which the opposing bony surfaces are covered with a layer of hyaline cartilage or fibrocartilage and that is nourished and lubricated by synovial tissue.

synovitis—Swelling within a joint attributable to inflammation of the synovial lining.

syringomyelia—Chronic syndrome characterized pathologically by cavitation and gliosis of the spinal cord (usually cervical or thoracic), medulla, or both.

systemic—Referring to the arterial system supplying the body.

systolic—Referring to the pressure generated in the arteries by contraction of the heart muscle.

systolic dysfunction—Most often with reference to left ventricular function, an inability or lesser ability of the cardiac myofibrils to shorten or contract against a load.

tachycardia—Heart rate greater than $100 \, b \cdot min^{-1}$.

tachypnea—Respiratory rate greater than 20 breaths/min.

targeted therapies—The use of drugs or other substances to block the growth and spread of cancer by interfering with specific molecules (molecular targets) based on genetic and molecular information; sometimes referred to as precision medicine.

technetium-99m sestamibi—A radioisotope, introduced into the bloodstream by a catheter, that tags red blood cells and when imaged using a gamma camera can provide a measure of ventricular volume, ejection fraction, and regional ventricular wall motion at rest and during exercise. Used to depict myocardial ischemia.

tertiary—Characterizing an intervention designed to reduce the functional effect of an illness or disability.

tetraplegia (TP)—Paralysis of all four limbs; also called quadriplegia.

thalamotomy—Destruction of a selected portion of the thalamus for the relief of pain and involuntary movements.

thalamus—Either of two masses of gray matter lying between the cerebral hemispheres on either side of the third ventricle, relaying sensory information and acting as a center for pain perception.

thallium 201—A white metallic substance with radioactivity, introduced into the bloodstream by a catheter, that is perfused into the myocardium; used in conjunction with stress testing (exercise and pharmacological) to image the myocardium to detect transient ischemia and tissue necrosis.

T-helper lymphocyte—Lymphocytes whose secretions and other activities coordinate the cellular and humoral immune responses.

theory of planned behavior—Theory that adds to the theory of reasoned action with the concept of perceived control over the opportunities, resources, and skills necessary to perform a behavior.

theory of reasoned action—Theory that performance of a given behavior (e.g., exercise) is primarily determined by the person's attitude toward the behavior and the influence of the person's social environment or subjective norm (i.e., beliefs about what other people think the person should do as well as the person's motivation to comply with the opinions of others).

therapeutic index—Indicator of a drug's safety; the ratio between lethal dose$_{50}$ (LD$_{50}$) and effective dose$_{50}$ (ED$_{50}$), where LD$_{50}$ equals the dose of drug that is lethal in 50% of the animals tested and ED$_{50}$ is the dose of the drug needed to be therapeutically effective in 50% of a like population of animals.

thrombocytopenia—A condition in which the amount of platelets in the blood is abnormally small.

thrombolysis—The process of dissolving a coronary artery thrombosis with either an intrinsic thrombolytic peptide or a thrombolytic medication.

thrombolytic—Referring to agents that degrade fibrin clots by activating plasminogen, a naturally occurring modulator of hemostatic and thrombotic processes.

thrombosis—The formation, development, or existence of a clot or thrombus within the vascular system.

thrombotic stroke—A stroke where an occlusive thrombus develops in or outside an ulcerated plaque. Hypercoagulable states due to increased coagulation potential or decreased fibrinolytic potential are particularly important risk factors.

thromboxane—Vasoconstrictor and platelet activation substance.

thrombus—An aggregation of blood factors, primarily platelets and fibrin with entrapment of cellular elements, that frequently causes vascular obstruction at the point of formation.

trabecular bone—One of the two main types of bone tissue, also known as cancellous or spongy bone. Trabecular bone is made up of interlacing plates of bone tissue and is found mainly at the ends of long bones and within the vertebrae.

tracking—The concept that risk factors or other conditions expressed in childhood will persist and be expressed in adulthood.

transcutaneous electrical nerve stimulation (TENS)—Use of small battery-operated or plugged-in devices for delivery of electrical current across the skin to provide patients with pain relief, artificial contraction of muscles, fatigue of spastic muscles, and pulsations to decrease swelling in a joint. The stimulation is given through electrode pads placed directly on the skin over the muscles selected for stimulation or inhibition of nociceptive input, or in areas determined by nerve supply or acupuncture points. The underlying theories are based on the gate theory of pain control, in which sensory stimulation inhibits pain transmission at the spinal cord level, or stimulation of AΔ and C fibers, to cause the release of endogenous opiates.

transmural—Referring to effects on all tissue layers of the heart.

transmural ischemia—Myocardial ischemic response that occurs throughout the myocardial wall.

transtheoretical model of behavior change—Model wherein behavior change is conceptualized as a five-stage process or continuum related to a person's readiness to change: precontemplation, contemplation, preparation, action, and maintenance. People are thought to progress through the stages.

tremor—Repetitive, often regular, oscillatory movements caused by alternate, or synchronous, but irregular contraction of opposing muscle groups.

triglycerides—Chemical storage form of fat in the body. Hypertriglyceridemia ($\geq 150 \, mg \cdot dL^{-1}$) increases the risk of cardiovascular disease.

triplegia—A form of cerebral palsy that affects three limbs. The most common pattern is for both legs and one arm to be affected. Triplegia is sometimes thought of as hemiplegia overlapping with diplegia because the primary motor difficulty is with the legs.

true maximal heart rate—The highest heart rate as measured at, or near, the end of an exercise test, as opposed to peak heart rate estimated by the equation 220 − age.

tumor suppressor genes—Genes encoding proteins that normally restrain cell proliferation.

type 2 diabetes mellitus—The most common form of diabetes, affecting approximately 90% to 95% of all those with diabetes. With type 2 diabetes, abnormal blood glucose regulation is often a result of insulin resistance of the peripheral tissues and defective insulin secretion.

ultrasound—Use of sonic wave energy, created by a vibrating quartz crystal, to deliver heat or medication to healing musculoskeletal structures. A variety of machines deliver the ultrasonic waves through a transducer rubbed directly over the skin using gel or water as the transmitting medium.

upper rate limit—The highest rate at which ventricular pacing will track 1:1 each sensed atrial event.

uremia—The medical state that occurs when kidney function is extremely low. The manifold symptoms include confusion, lethargy leading to coma, loss of appetite, and pruritus (itchy skin). Complications include gastrointestinal bleeding, pericarditis, pulmonary edema, and ultimately death.

valgus—Knock-kneed deformity.

Valsalva maneuver—Forced exhalation with the glottis, nose, and mouth closed, resulting in increased intrathoracic pressure, slowing of the heart rate, and decreased return of blood to the heart.

variant angina—Angina pectoris occurring during rest; not necessarily preceded by exercise or an increase in heart rate.

varus—Bow-legged deformity.

vascular pathologies—Manifestations of disease in blood vessels.

vasospastic—Referring to contraction or spasm of the muscular coats of the blood vessels.

ventilatory-derived lactate threshold (V-LT)—The point where a nonlinear increase in blood lactate occurs during exercise; when determined with ventilatory parameters, it is sometimes referred to as ventilatory threshold.

ventilatory muscle training—Specific exercises that are used to increase respiratory muscular strength.

ventricular hypertrophy—Muscle thickening in a pumping chamber of the heart.

vertigo—A sensation of spinning or whirling motion.

very low-density lipoproteins (VLDL)—Precursors to intermediate- and low-density lipoproteins and the primary transport mechanism for endogenous triglyceride to body tissues.

very old age—Between 75 and 84 yr of age.

vesicoureteral reflux—Backflow of urine from the bladder into the upper urinary tracts.

vigorous-intensity physical activity—Activities that result in large increases in breathing and heart rate; >70% $\dot{V}O_2$peak.

viremia—Viral particles in the blood.

virion—A single, encapsulated piece of viral genetic material.

visceral adiposity or fat—One of three compartments of intra-abdominal fat. Others are retroperitoneal and subcutaneous.

wall motion—Movement of the left ventricular segments of the heart; used to describe normal or abnormal movement during contraction and to calculate ejection fraction during two-dimensional echocardiography or some types of nuclear imaging.

wasting—Involuntary loss of more than 10% of body weight.

white matter—Regions of the brain and spinal cord that are largely or entirely composed of nerve fibers and contain few or no neuronal cell bodies or dendrites.

xanthomas—Usually yellow, soft raised bumps that are caused by cholesterol deposits under the skin.

References

As a benefit to the reader, we've moved the chapter references online for this edition. You'll find nearly a hundred pages of references covering each of the 35 chapters, all readily accessible. Visit http://courses.humankinetics.com/references/ehrman5E/index.html or scan the QR code using your mobile device or tablet.

Index of Common Questions

Question	Answer location	Page
CHAPTER 5		
What are the absolute and relative contraindications for graded exercise testing?	Contraindications	58
What is the incidence of death or major events requiring hospitalization during a GXT?	Personnel	59
What are the key skills needed for test supervision?	Personnel	59
What pretest information should be given to the patient to help them prepare for the test?	Patient Pretest Instruction and Preparation	61
When would it be appropriate to use a bicycle ergometer as a mode of testing?	Graded Exercise Testing Protocols and Testing Modalities	61
CHAPTER 6		
What is the target range of suggested physical activity per week? What percentage of the population meets that goal?	Chapter introduction	73
What are the five health-related components of physical fitness that the exercise prescription should address?	Chapter introduction	73
What are some questions to ask when developing an exercise prescription for someone?	Sidebar: Questions to Ask a Person When Developing an Exercise Prescription	75
How do you determine an appropriate target heart rate?	Practical Application 6.2	80
What is HIIT?	Practical Application 6.3	81
CHAPTER 7		
How effective are cardiac rehabilitation programs?	Program Effectiveness	86
What is a typical phase 2 CR program?	Program Structure and Processes	86
What is the effectiveness of pulmonary rehab?	Program Effectiveness	89
What is a typical exercise session for a PAD patient in supervised exercise therapy?	Program Structure and Processes	90
Why do some patients and doctors not want to start an exercise program during the treatment phase of cancer?	Cancer Rehabilitation	91
CHAPTER 8		
What are the ADA's recommendations on macronutrients?	Treatment	105
What are some factors that can affect blood glucose response to physical activity?	Monitoring of Blood Glucose	106
Is exercise safe for individuals with diabetes-related health complications such as peripheral neuropathy?	Peripheral Neuropathy	110
Is there a blood glucose level at which an individual should not exercise?	Practical Application 8.2	111
How can exercise-related hypoglycemia be avoided when an individual takes insulin?	Practical Application 8.2	111
CHAPTER 9		
What are the three elements of therapeutic lifestyle change for obesity?	Figure 9.2	120
What is the estimated cost of obesity in the United States?	Table 9.2	125
What are the health consequences of obesity?	Sidebar: Health Consequences of Obesity	129
Are there specific protocols for exercise testing in obese people?	Exercise Testing	130
What are some health benefits of a 10% weight loss?	Sidebar: Health Benefits of a 10% Weight Loss	132

(continued)

Question	Answer location	Page
CHAPTER 20		
Do athletes have an increased prevalence of asthma and airway hyperresponsiveness compared to the general population?	Asthma in the High-Level Athlete	326
What is the main mechanism underlying exercise-induced asthma?	Exercise-Induced Asthma	326
What conditions have symptoms that often mimic asthma?	Clinical Considerations	327
What tests can be used to confirm a diagnosis of exercise-induced asthma?	Diagnostic Testing	327
Should asthmatic patients avoid exercise?	Exercise Training	332
CHAPTER 21		
What is cystic fibrosis?	Definition	337
How does CF present?	Pathophysiology	337
For individuals with CF, what needs to be monitored during exercise?	Exercise Testing	341
How intensely can those with CF exercise?	Table 21.6	350
Is it safe for an individual with CF to use the HIIT method?	Muscular Strength and Endurance Training: Mode	352
CHAPTER 22		
Although the carcinogenic event for many cancers is unknown, what five broad categories have been identified?	Pathophysiology	362
What are the early signs and symptoms of cancer?	Clinical Considerations	365
A large analysis using pooled data of 1.44 million participants showed a significant inverse relationship between leisure-time physical activity and risk for what types of cancer?	Practical Application 22.1	366
What is the appropriate FITT exercise prescription for cardiorespiratory training for cancer patients?	Table 22.5	371
Should patients with metastatic bone disease participate in resistance training?	Table 22.5	371
CHAPTER 23		
What are the long-term goals of an exercise intervention for someone with HIV?	Clinical Considerations	381
For purposes of fitness testing, what are the four categories used for disease status?	Assessment of Functional Status	382
What precautions does a CEP need to take when testing someone with HIV?	Practical Application 23.1	383
What are important considerations for exercise testing of someone with HIV?	Sidebar: Exercise Testing: Important Considerations	384
What are the appropriate elements of a cardiorespiratory exercise prescription for someone with HIV?	Table 23.2	389
CHAPTER 24		
What impairments are common to all forms of arthritis?	Pathophysiology	398
What are the signs and symptoms of arthritis?	Signs and Symptoms	399
As part of the exercise evaluation, individuals with arthritis should be screened for factors that will affect the exercise prescription. What are some suggested questions for patients?	Practical Application 24.1	402
In addition to exercise, what are some of the nonpharmacological interventions that are used in the routine management and rehabilitation of patients with chronic arthritis?	Treatment	404
What are the exercise goals specific for the treatment of arthritis?	Exercise Prescription	407

(continued)

Question	Answer location	Page
CHAPTER 30		
What are the two types of strokes, and what causes each of these?	Definition	519
What are the signs that a patient is suffering from an acute stroke?	Signs and Symptoms	521
What are some special considerations for cardiorespiratory exercise testing?	Table 30.2	522
What is an appropriate cardiorespiratory training exercise prescription for a poststroke patient?	Table 30.4	524
Is resistance training an important part of stroke rehabilitative exercise?	Resistance Training and Flexibility	524
CHAPTER 31		
Is Parkinson's disease genetic?	Pathophysiology	527
Do Parkinson's disease medications affect the ability to exercise?	Table 31.1	531
How often should a patient with Parkinson's disease exercise?	Table 31.2	533
Can exercise protect someone from developing Parkinson's disease?	Exercise Training	534
What is the optimal exercise mode for an individual with Parkinson's disease?	Exercise Training	534
CHAPTER 32		
What are the common reasons a child should have a stress test?	Signs and Symptoms	542
What are the ACSM guidelines with regard to exercise testing for children?	Exercise Testing	542
Can adult exercise protocols be used with children?	Protocols	544
Is there an RPE scale that is better to use with children?	Aerobic	546
Are children's physiological responses to exercise the same as those of adults?	Table 32.5	552
CHAPTER 33		
What are some common diseases in older adults that may affect their exercise capacity and response to training?	Treatment	564
Why are warm-up and cool-down periods important when training older adults?	Exercise Training Caveats Specific to the Elderly	565
What is an appropriate cardiorespiratory training program for seniors?	Table 33.3	566
For frail and severely deconditioned older adults, how does the initiation of training differ from that for more robust seniors?	Sidebar: Important Considerations for Exercise Training	567
Why is strength training particularly important for this population?	Resistance Exercise	568
CHAPTER 34		
For people with depression, what are the key barriers to engaging in physical activity?	Practical Application 34.1	574
Does exercise help treat depression?	Exercise as a Treatment for Depression	576
What is an appropriate resistance training exercise prescription for a patient with depression?	Table 34.3	578
What is the optimal physical activity prescription for people with depression?	Practical Application 34.4	579
What motivational tools can exercise physiologists use to increase motivation?	Practical Application 34.5	580

(continued)

Question	Answer location	Page
CHAPTER 35		
What are the five general categories of causes of ID?	Pathophysiology	584
What are the physiologic responses to acute exercise in individuals with ID?	Sidebar: Physiologic Responses to Acute Exercise in Individuals With Intellectual Disability	585
Why is the diagnosis of ID important?	Diagnostic Testing	586
What would be an appropriate resistance exercise test for an individual with ID?	Resistance Exercise	589
What is an appropriate cardiorespiratory exercise prescription for an individual with ID?	Table 35.5	592

Index

Note: Page references followed by an italicized *f* or *t* indicate information contained in figures or tables, respectively.

About the Editors

Jonathan K. Ehrman, PhD, is the associate program director of preventive cardiology at Henry Ford Hospital in Detroit, where he also serves as chair of the institutional review board. He has a 36-year background in clinical exercise physiology and is certified by the American College of Sports Medicine (ACSM) as a clinical exercise physiologist and as a program director. He previously served as the chair of the clinical exercise physiologist credentialing committee for ACSM.

Courtesy of Henry Ford Hospital.

Dr. Ehrman is the author of more than 200 manuscripts and abstracts as well as several textbooks and chapters. He currently serves as editor in chief of the *Journal of Clinical Exercise Physiology* and was an associate editor of the 10th edition of *ACSM's Guidelines for Exercise Testing and Prescription*. He is also the coeditor of the sixth edition of the American Association of Cardiovascular and Pulmonary Rehabilitation's *Guidelines for Cardiac Rehabilitation Programs*. He is a fellow of ACSM and the American Association of Cardiovascular and Pulmonary Rehabilitation and is a member of the American Heart Association and the American College of Cardiology. Dr. Ehrman earned his PhD in clinical exercise physiology from The Ohio State University.

Paul M. Gordon, PhD, MPH, is a professor and head of the department of health, human performance, and recreation at Baylor University. He is certified by the American College of Sports Medicine (ACSM) as a clinical exercise physiologist and has over 20 years of experience teaching clinical exercise physiology curricula and directing cardiopulmonary rehabilitation programs. Gordon's areas of expertise include physical activity and lifestyle-based research related to obesity and its comorbidities across the life span. He has published more than 200 papers and abstracts as well as several chapters, including contributions to *ACSM's Guidelines for Exercise Testing and Prescription*. He has also served as an examiner and coordinator for ACSM certification and credentialing.

Dr. Gordon is a fellow of ACSM, the Obesity Society, and the Centers for Disease Control Physical Activity Research Program. He is an international member of the Royal Society of Medicine. He earned his PhD in exercise physiology and an MPH in epidemiology from the University of Pittsburgh.

Paul S. Visich, PhD, MPH, is a professor and chair of the exercise and sports performance department at the University of New England. He has over 20 years of experience in clinical exercise physiology and previously served as director of the Human Performance Laboratory in the College of Health Professions at Central Michigan University. He worked for 12 years in a clinical setting that included cardiac and pulmonary rehabilitation and primary disease prevention. His research interests involve the assessment of cardiovascular disease risk factors in children, the influence of resistance training in elderly populations, and altitude physiology.

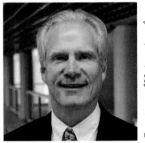

Courtesy of University of New England.

Dr. Visich previously served as a member of the American College of Sport Medicine (ACSM) exercise physiology credentialing committee and as chair of their professional education committee. He is the author of more than 70 published scientific articles and abstracts. He earned a PhD in exercise physiology and an MPH in epidemiology from the University of Pittsburgh.

Steven J. Keteyian, PhD, has more than 40 years of experience working as a clinical exercise physiologist. He is program director of preventive cardiology at the Henry Ford Hospital in Detroit. He is also an adjunct professor in the department of physiology at Wayne State University in Detroit. Over the course of his career, Dr. Keteyian has focused on exercise and physical activity in both healthy individuals and those with chronic diseases. He is the author of more than 250 scientific articles and book chapters, as well as four textbooks, and he previously served as editor in chief for *ACSM's Health & Fitness Journal*.

Courtesy of Henry Ford Hospital.

Dr. Keteyian is a member of the American Association of Cardiovascular and Pulmonary Rehabilitation and the American Heart Association. He earned his PhD from Wayne State University in Detroit.

Contributors

Simon L. Bacon, PhD
Concordia University
Montreal, Québec, Canada

Tracy Baynard, PhD
Department of Kinesiology and Nutrition
College of Applied Health Sciences
University of Illinois at Chicago
Chicago, IL

Anna G. Beaudry, BS
Baylor University
Waco, TX

Emma M. Beckman, PhD
University of Queensland, Australia
Queensland, Australia

Louis-Philippe Boulet, MD
Québec Heart and Lung Institute
Laval University
Montreal, Québec, Canada

Lizbeth R. Brice, MD
Division of Cardiovascular Medicine
Henry Ford Medical Group
Detroit, MI

Ivan P. Casserly, MB, BCh
Cardiac and Vascular Center
University of Colorado Hospital
Aurora, CO

James R. Churilla, PhD, MPH, MS
Brooks College of Health
Jacksonville, FL

Kelly M. Clanchy, PhD
Griffith University
Gold Coast, Australia

Sheri R. Colberg, PhD
Old Dominion University, Emeritus
Norfolk, VA

Mark J. Connick, PhD
University of Queensland, Australia
Queensland, Australia

Andréanne Côté, MD
Laval University
Montreal, Québec, Canada

Kelley Crawford, PT, DPT, MS, CCS
University of New England
Richmond, ME

Stephen F. Crouse, PhD
Department of Health and Kinesiology
Texas A&M University
College Station, TX

Andrew K. Cunningham, MD
Brighton Family Care
Brighton, MI

Ulrik Dalgas, PhD
Department of Public Health, Sport
  Science
Aarhus University
Aarhus, Denmark

Paul G. Davis, PhD
Department of Kinesiology
University of North Carolina–Greensboro
Greensboro, NC

Satvir S. Dhillon, MSc, CCRP
Cardiopulmonary Exercise Physiology
  Laboratory
University of British Columbia
Vancouver, British Columbia, Canada

J. Larry Durstine, PhD
Department of Exercise Science
University of South Carolina
Columbia, SC

Jonathan K. Ehrman, PhD
Division of Cardiovascular Medicine
Henry Ford Medical Group
Detroit, MI

Bo Fernhall, PhD
Department of Kinesiology and Nutrition
College of Applied Health Sciences
University of Illinois at Chicago
Chicago, IL

Jerome L. Fleg, MD
National Heart, Lung, and Blood Institute
National Institutes of Health
Bethesda, MD

Daniel E. Forman, MD
Geriatric Cardiology
University of Pittsburgh
Pittsburgh, PA

Timothy Geraghty, FAFRM (RACP)
Queensland Spinal Cord Injuries Service,
  Princess Alexandra Hospital
Griffith University
Gold Coast, Australia

Michael Germain, MD
Tufts University School of Medicine
Baystate Medical Center
Springfield, MA

Lora M. Giangregorio, PhD
Department of Kinesiology
University of Waterloo
Waterloo, Ontario, Canada

Annie T. Ginty, PhD
Baylor University
Waco, TX

Paul M. Gordon, PhD, MPH
Baylor University
Waco, TX

Peter W. Grandjean, PhD
University of Mississippi
Oxford, MS

Jordan A. Guenette, PhD
Department of Physical Therapy
University of British Columbia
Vancouver, British Columbia, Canada

Anne L. Hart, PT, PhD
IPC Classification Committee
Northern Arizona University
Flagstaff, AZ

Samuel Headley, PhD
Department of Exercise Science and Sport
    Studies
Springfield College
Springfield, MA

Christine Horvat Davey, PhD, RN
Case Western Reserve University
Cleveland, OH

Lars G. Hvid, PhD
Department of Public Health, Sport
    Science
Aarhus University
Aarhus, Denmark

Dennis Jensen, PhD
Department of Kinesiology and Physical
    Education
McGill University
Montreal, Quebec, Canada

Kati Karinharju, PhD
University of Queensland, Australia
Queensland, Australia

Dennis J. Kerrigan, PhD
Division of Cardiovascular Medicine
Henry Ford Medical Group
Detroit, MI

Steven J. Keteyian, PhD
Division of Cardiovascular Medicine
Henry Ford Medical Group
Detroit, MI

Désirée B. Maltais, PT, PhD
Department of Rehabilitation
University of Laval
Quebec City, Quebec, Canada

Ryan J. Mays, PhD, MPH
School of Medicine, Division of
    Cardiology
University of Colorado
Aurora, CO

Grace M. McKeon, PhD
University of New South Wales
Sydney, Australia

Timothy J. Michael, PhD
Department of Exercise Science
Western Michigan University
Kalamazoo, MI

David C. Murdy, MD
Internal Medicine, Dean Medical Center
University of Wisconsin
Janesville, WI

Melissa Nayak, MD
Henry Ford Medical Group
Detroit, MI

Vitor H.F. De Oliveira, PhD
University of Washington
Seattle, WA

Linda S. Pescatello, PhD
Department of Kinesiology
University of Connecticut
Storrs, CT

Brandon S. Pollock, PhD
Exercise Science/Physiology
Kent State University
Kent, OH

Judith G. Regensteiner, PhD
School of Medicine, Division of
    Cardiology
University of Colorado
Aurora, CO

Angela L. Ridgel, PhD
School of Health Sciences
Kent State University
Kent, OH

Peter Ronai, MS
Sacred Heart University
Fairfield, CT

Simon Rosenbaum, PhD
University of New South Wales
Sydney, Australia

Neil A. Smart, PhD
Exercise and Sports Science
University of New England
Armidale, Australia

David D. Spragg, MD
Cardiology
Johns Hopkins School of Medicine
Baltimore, MD

Ray W. Squires, PhD
Department of Cardiovascular Diseases
Mayo Clinic
Rochester, MN

Kerry J. Stewart, EdD
Division of Cardiology
Johns Hopkins School of Medicine
Baltimore, MD

Sean M. Tweedy, PhD
University of Queensland, Australia
Queensland, Australia

Paul S. Visich, PhD, MPH
University of New England

Allison R. Webel, PhD, RN
Case Western Reserve University
Cleveland, OH

Carol Weideman, PhD
Western Michigan University
Kalamazoo, MI

Kenneth Wilund, PhD
University of Illinois at Urbana-
    Champaign
Urbana, IL

Karen Wonders, PhD
Maple Tree Cancer Alliance
Dayton, OH

Christopher J. Womack, PhD
Department of Kinesiology
James Madison University
Harrisonburg, VA

Yin Wu, PhD
Department of Kinesiology
University of Connecticut
Storrs, CT

Danielle A. Young, PsyD
Baylor University
Waco, TX

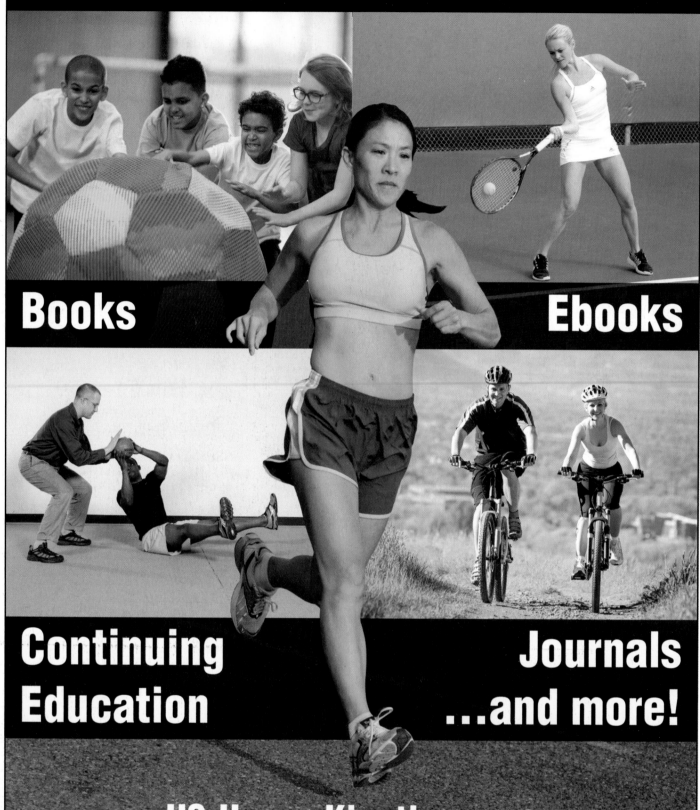